Lecture Notes in Computer Science 7811

Commenced Publication in 1973
Founding and Former Series Editors:
Gerhard Goos, Juris Hartmanis, and Jan van Leeuwen

AF136576

Robin C. Purshouse   Peter J. Fleming
Carlos M. Fonseca   Salvatore Greco
Jane Shaw (Eds.)

Evolutionary Multi-Criterion Optimization

7th International Conference, EMO 2013
Sheffield, UK, March 19-22, 2013
Proceedings

 Springer

Volume Editors

Robin C. Purshouse
Peter J. Fleming
University of Sheffield, Dept. of Automatic Control and Systems Engineering
Mappin Street, Sheffield S1 3JD, UK
E-mail: {r.purshouse; p.fleming}@sheffield.ac.uk

Carlos M. Fonseca
University of Coimbra, Department of Informatics Engineering
Pólo II, Pinhal de Marrocos, 3030-290 Coimbra, Portugal
E-mail: cmfonsec@dei.uc.pt

Salvatore Greco
University of Catania, Department of Economics and Business
Corso Italia 55, 95129 Catania, Italy
E-mail: salgreco@unict.it

Jane Shaw
Unilever Research & Development Port Sunlight
Quarry Road East, Merseyside, CH63 3JW, UK
E-mail: jane.shaw@unilever.com

ISSN 0302-9743   e-ISSN 1611-3349
ISBN 978-3-642-37139-4   e-ISBN 978-3-642-37140-0
DOI 10.1007/978-3-642-37140-0
Springer Heidelberg Dordrecht London New York

Library of Congress Control Number: 2013932870

CR Subject Classification (1998): G.1.6, F.2.1-2, G.1.2, I.2.8, J.1, J.2

LNCS Sublibrary: SL 1 – Theoretical Computer Science and General Issues

Typesetting: Camera-ready by author, data conversion by Scientific Publishing Services, Chennai, India

Printed on acid-free paper

Springer is part of Springer Science+Business Media (www.springer.com)

Preface

EMO is a bi-annual international conference series, dedicated to advances in the theory and practice of evolutionary multi-criterion optimization.

The first EMO was organised in Zürich (Switzerland) in 2001, with later conferences taking place in Faro (Portugal) in 2003, Guanajuato (Mexico) in 2005, Matsushima-Sendai (Japan) in 2007, Nantes (France) in 2009, and Ouro Preto (Brazil) in 2011. Proceedings of every EMO conference have been published as a volume in *Lecture Notes in Computer Science* (LNCS); this history is available in volumes 1993, 2632, 3410, 4403, 5467, and 6576.

The 7th International Conference on Evolutionary Multi-Criterion Optimization took place in Sheffield, UK, during March 19–22, 2013. The event was organized by the Department of Automatic Control and Systems Engineering, University of Sheffield, which was home to some of the earliest work in the EMO field and retains a strong EMO research theme.

In addition to the core content on EMO methods, EMO 2013 aimed to build on the success of the 2011 meeting in Ouro Preto by providing a special track to stimulate cross-fertilization between EMO research and and the wider theory and practice of multiple criteria decision making (MCDM). The conference also promoted a special track on real-world applications (RWA) to provide a new focus on the application of EMO and MCDM research to help real decision-makers solve real problems in government, business, industry and interdisciplinary sciences.

In response to the call for papers, 98 full-length papers were submitted. The papers were subject to a rigorous single-blind peer-review process, with a minimum of three referees per paper. To encourage cross-fertilization, submissions to the MCDM track were reviewed by an International Program Committee member with EMO expertise, and almost all main track and RWA submissions were reviewed by a Committee member with MCDM expertise. Papers where a potential conflict of interest existed with the organizers were handled separately by one of the General Chairs of EMO 2011, Ricardo Takahashi (UFMG, Brazil). Following peer review, 57 papers were accepted for presentation at the conference and publication in this volume of LNCS: 35 main track, six MCDM track, and 16 RWA track. Of the accepted RWA papers, seven included at least one author from an organization outside of the academy.

The conference heard from four outstanding plenary speakers, on themes critical to the success of EMO methods in supporting real decision-makers: Jürgen Branke (University of Warwick, UK) spoke on the incorporation of decision-maker preferences; Kalyanmoy Deb (Indian Institute of Technology Kanpur, India) discussed the discovery of innovative solution principles; Patrick Reed (Pennsylvania State University, USA) spoke on solution visualization methods; and Theodor Stewart (University of Cape Town, South Africa and University of

Manchester, UK) discussed the integration of problem exploration with decision-making processes.

Since its inception, the EMO conference series has led on key developments in the field. EMO 2013 continued methodological advancement of the increasingly popular indicator-based and decomposition-based concepts. Promising novel concepts were introduced that explicitly recognized the dynamics of an optimization problem, in terms of both delays to objective function evaluations during the optimization process and also fundamental issues in the optimization of time-varying systems. The MCDM track witnessed new blendings of EMO with MCDM, including the integration of hypervolume into MCDM methods, and integration of problem exploration into EMO search. The RWA track elicited tangible examples of EMO and MCDM practice, including software tools used to support real decisions, from the design of spacecraft planetary landing systems to maintenance planning in electrical power distribution systems. It is our hope that the new RWA track has both stimulated the uptake of EMO methods by practitioners, and also provoked a renewed consideration of practical decision support as a central tenet of EMO methodology.

We would like to express our appreciation to the plenary speakers for accepting our invitations, all authors who submitted their work to EMO 2013, and to the members of the International Program Committee for their thorough and timely reviews. We would like to acknowledge the support of our sponsors: the University of Sheffield, the International Society on Multiple Criteria Decision Making, the Institute for Operations Research and Management Sciences, the EURO Working Group for Multi-Criteria Decision Aiding, and the Marie Curie International Research Staff Exchange Scheme within the Seventh European Community Framework Programme. We would also like to thank Alfred Hofmann, Frank Holzwarth, and Elke Werner at Springer for their assistance in publishing these proceedings in the *Lecture Notes in Computer Science* series.

March 2013

Robin C. Purshouse
Peter J. Fleming
Carlos M. Fonseca
Salvatore Greco
Jane Shaw

Organization

EMO 2013 was organized by the Department of Automatic Control and Systems Engineering (ACSE) at the University of Sheffield, UK.

General Chairs

Robin C. Purshouse	University of Sheffield, UK
Peter J. Fleming	University of Sheffield, UK
Carlos M. Fonseca	University of Coimbra, Portugal

MCDM Track Chair

Salvatore Greco	University of Catania, Italy

Real-World Applications Track Chair

Jane Shaw	Unilever R&D Port Sunlight, UK

EMO 2013 Advisory Board

Gideon Avigad	ORT Braude College, Israel
John Brazier	University of Sheffield, UK
Carlos A. Coello Coello	CINVESTAV-IPN, Mexico
David Corne	Heriot-Watt University, UK
David Dungate	Tessella Ltd, UK
Val Gillet	University of Sheffield, UK
Salvatore Greco	University of Catania, Italy
Evan Hughes	Cranfield University, UK
Joshua D. Knowles	University of Manchester, UK
John McAlister	PA Consulting Group, UK
Ricardo Takahashi	UFMG, Brazil
Praveen Thokala	University of Sheffield, UK
Elizabeth Wanner	CEFET-MG, Brazil

Local Organizing Committee

Renata Ashton	Maszatul M. Mansor
Peter J. Fleming	Robin C. Purshouse
Ioannis Giagkiozis	Shaul Salomon
Matthew Ham	Rui Wang
Ian Lilley	

Program Committee

Adiel Almeida	Federal University of Pernambuco, Brazil
Maria João Alves	University of Coimbra, Portugal
Silvia Angilella	University of Catania, Italy
Gideon Avigad	ORT Braude College of Engineering, Israel
Sanghamitra Bandyopadhyay	Indian Statistical Institute, India
Helio Barbosa	Federal University of Juiz de Fora, Brazil
Matthieu Basseur	LERIA Lab., France
Juergen Branke	The University of Warwick, UK
Dimo Brockhoff	INRIA Lille - Nord Europe, France
Rafael Caballero	University of Málaga, Spain
Marco Chiarandini	University of Southern Denmark, Denmark
Joao Climaco	INESC-Coimbra, Portugal
Carlos Coello Coello	CINVESTAV-IPN, Mexico
David Corne	Heriot-Watt University, UK
Salvatore Corrente	University of Catania, Italy
Andre de Carvalho	University of São Paulo, Brazil
Alexandre Delbem	University of São Paulo, Brazil
Clarisse Dhaenens	LIFL / INRIA / Université Lille 1, France
Rolf Drechsler	University of Bremen, Germany
Dave Dungate	Tessella Ltd, UK
Matthias Ehrgott	The University of Auckland, New Zealand
Michael Emmerich	Leiden University, The Netherlands
Andries Engelbrecht	University of Pretoria, South Africa
Jonathan Fieldsend	University of Exeter, UK
Jose Rui Figueira	Instituto Superior Tecnico, Portugal
Ioannis Giagkiozis	University of Sheffield, UK
Antonio Gaspar-Cunha	University of Minho, Portugal
Martin Josef Geiger	Helmut Schmidt University, Germany
Christian Grimme	TU Dortmund, Germany
Viviane Grunert da Fonseca	University of Coimbra, Portugal
Frederico Guimarães	Federal University of Minas Gerais, Brazil
Walter Gutjahr	University of Vienna, Austria
Jin-Kao Hao	University of Angers, France
Carlos Henggeler Antunes	University of Coimbra, Portugal
Malcolm Heywood	Dalhousie University, Canada
Evan Hughes	Cranfield University, UK
Masahiro Inuiguchi	Osaka University, Japan
Hisao Ishibuchi	Osaka University, Japan
Alessio Ishizaka	University of Portsmouth, UK
Yaochu Jin	University of Surrey, UK
Dylan Jones	University of Portsmouth, UK
Laetitia Jourdan	University of Lille, France
Naoki Katoh	Kyoto University, Japan
Joshua Knowles	University of Manchester, UK

Mario Koeppen	Kyushu Institute of Technology, Japan
Rajeev Kumar	IIT Kharagpur, India
Mary Kurz	Clemson University, USA
Dario Landa-Silva	University of Nottingham, UK
Per Kristian Lehre	University of Nottingham, UK
Juan Leyva	Universidad de Occidente, Mexico
Arnaud Liefooghe	Université Lille 1, France
José Meriglo Lindahl	University of Manchester, UK
Manuel López-Ibáñez	Université Libre de Bruxelles, Belgium
Jose A. Lozano	University of the Basque Country, Spain
Robert Lygoe	Ford Motor Company, UK
Maszatul Mansor	University of Sheffield, UK
Efrén Mezura-Montes	LANIA, Mexico
Martin Middendorf	University of Leipzig, Germany
Kaisa Miettinen	University of Jyväskylä, Finland
Julian Molina	University of Malaga, Spain
Alec Morton	LSE, UK
Vincent Mousseau	École Centrale Paris, France
Boris Naujoks	Fachhochschule Köln, Germany
Antonio Nebro	University of Málaga, Spain
Guiseppe Nicosia	University of Catania, Italy
Amos Ng	University of Skövde, Sweden
Pedro Oliveira	Universidade do Porto, Portugal
Gisele Pappa	Federal University of Minas Gerais, Brazil
Luis Paquete	University of Coimbra, Portugal
Geoffrey Parks	University of Cambridge, UK
Silvia Poles	University of Padova, Italy
Aurora Pozo	Federal University of Paraná, Brazil
Tapabrata Ray	UNSW, Australia
Peter Rockett	University of Sheffield, UK
Katya Rodriguez-Vazquez	UNAM, Mexico
Günter Rudolph	TU Dortmund, Germany
Francisco Ruiz	University of Málaga, Spain
Sriparna Saha	IIT Patna, India
Javier Sanchis Saez	UPV, Spain
Hartmut Schmeck	Institute AIFB, Germany
Marc Schoenauer	INRIA Saclay Île-de-France, France
Oliver Schuetze	CINVESTAV-IPN, Mexico
Marc Sevaux	University of South Brittany, France
Patrick Siarry	UPEC, France
Johannes Siebert	BWL V, Germany
Karthik Sindhya	University of Jyväskylä, Finland
Ankur Sinha	Aalto University, Finland
Yannis Siskos	University of Piraeus, Greece
Theodor Stewart	University of Cape Town, South Africa

EMO Steering Committee

Table of Contents

III Indicator-Based Methods

IV Aspects of Algorithm Design

V Pareto-Based Methods

VI Hybrid MCDA

VII Decomposition-Based Methods

VIII Classical MCDA

IX Exploratory Problem Analysis

X Product and Process Applications

XI Aerospace and Automotive Applications

XII Further Real-World Applications

XIII Under-Explored Challenges

Many-Objective Visual Analytics: Rethinking the Design of Complex Engineered Systems

(Invited Talk)

Patrick M. Reed

The Pennsylvania State University
212 Sackett Building, University Park, PA, USA 16802
preed@engr.psu.edu

Abstract. Over the past decade our research group has worked to operationlize our "*many-objective visual analytics*" (MOVA) framework for the design and management of complex engineered systems. Successful applications include urban water portfolio planning, satellite constellation design, airline scheduling, and product family design. The MOVA framework has four core components: (1) elicited problem conception and formulation, (2) many-objective search, (3) interactive visualization, and (4) negotiated design selection. Problem conception and formulation is the process of abstracting a practical design problem into a mathematical representation. We build on the emerging work in visual analytics to exploit interactive visualization of both the design space and the objective space in multiple heterogeneous linked views that permit exploration and discovery. Many-objective search produces a Pareto-approximate set of solutions from problem formulations that consider up to ten objectives based on current computational search capabilities. Negotiated design selection uses interactive visualization, reformulation, and optimization to discover desirable designs for implementation. Each of the activities in the framework is subject to feedback, both within the activity itself and from the other activities in the framework. These feedback processes transition formerly marginalized activities of reformulating the problem, refining the conceptual model of the problem, and refining the optimization, to represent the most critical process for innovating real world systems (i.e., learning how to frame the problems themselves). This study demonstrates insights gained by evolving the formulation of a General Aviation Aircraft (GAA) product family design problem. This problem's considerable complexity and difficulty, along with a history encompassing several formulations, make it well-suited to demonstrate the MOVA framework. Our MOVA framework results compare a single objective, a two objective, and a ten objective formulation for optimizing the GAA product family. Highly interactive visual analytics are exploited to demonstrate how decision biases can arise for lower dimensional, highly aggregated problem formulations. As part of our efforts to operationlize the MOVA framework, we have also created rigorous search diagnostics to distinguish the efficiency, controllability, reliability, and effectiveness of multiobjective evolutionary algorithms (MOEAs). These diagnostics have distinguished the auto-adaptive behavior of our recently introduced Borg MOEA relative to a broad sampling of traditional MOEAs when addressing the GAA product family design problem.

R.C. Purshouse et al. (Eds.): EMO 2013, LNCS 7811, p. 1, 2013.
© Springer-Verlag Berlin Heidelberg 2013

Evolutionary Multiobjective Optimization and Uncertainty
(Abstract of Invited Talk)

Jürgen Branke

Warwick Business School, University of Warwick, Coventry, UK
juergen.branke@wbs.ac.uk

Abstract. This talk will look at various aspects of uncertainty and how they can be addressed by evolutionary multiobjective optimization.

If there is uncertainty about the user preferences, evolutionary multi-objective optimization is traditionally used to generate a representative set of Pareto-optimal solutions that caters for all potential user preferences. However, it is also possible to take into account a distribution over possible utility functions to obtain a distribution of Pareto optimal solutions that better reflects the decision maker's likely preferences. And furthermore, it may be possible to elicit and learn the decision maker's preferences by interacting with the decision maker during the optimization process.

If there is a trade-off between a solution's quality and associated risk or reliability, evolutionary multiobjective optimization can simply regard the problem as a two-objective problem and provide a set of alternatives with different quality/risk trade-offs.

If the objective functions of the multi-objective problem are noisy and an accurate evaluation is not possible, for example because the evaluation is done by means of a stochastic simulation, it is no longer possible to decide with certainty whether one solution dominates another. One might calculate the probability of one solution dominating the other, and use this for selection. Still, this is based on noisy observations, and does not allow to make a confident decision about which solutions to keep in an elitist algorithm, because the solution observed as better may only have been lucky in the evaluation process. In order to improve the accuracy of the fitness estimates, it is usually possible to average fitness values over a number of evaluations. However, this is time consuming, and so it raises the question how often each solution should be evaluated such that the algorithm can progress, but at the same time computational effort is minimized. Finally, if the goal is to optimize a quantile or even the worst case, it is not obvious how to even define such a concept in a multi-objective setting.

R.C. Purshouse et al. (Eds.): EMO 2013, LNCS 7811, p. 2, 2013.

Integrating Problem Analysis and Algorithmic Development in MCDA
(Abstract of Invited Talk)

Theodor J. Stewart

Department of Statistical Sciences, University of Cape Town, Rondebosch 7701,
South Africa
Manchester Business School, University of Manchester, Manchester M15 6PB, U.K.
theodor.stewart@uct.ac.za

Abstract. There are two key features to any MCDA intervention,
namely the problem structuring and preference modelling to ensure that
the analysis is directed at solving the right problem (*effectiveness* of
the intervention), and the provision of computationally efficient decision support algorithms (*efficiency* of the intervention). There are, of
course problems where either the computational aspects are unchalleng-
ing so that only effectiveness requires the analyst's attention; or where
the problem is in principle well-defined but computationally complex, so
that efficiency concerns dominate.

Contexts do arise in which structuring and preference modelling (e.g.
identifying criteria, assessing performance in terms of these criteria and
aggregation of preferences across criteria) require careful attention, especially when the numbers of criteria are large, and the resulting models
are computationally complex. In such contexts the two components of
decision support need to work together. On the one hand, the problem
structuring and selection of preference models should balance the need
to represent decision maker preferences faithfully with the need for a
model implementation which is sufficiently responsive and computationally effective to ensure that the decision maker derives useful support.
On the other hand, computational methods and approaches must recognize the cognitive limitations of the decision maker in such complex
settings. For example, unaided or undirected search across the Pareto
set is unlikely to be cognitively meaningful for larger numbers of criteria
even with inventive use of graphics.

This paper will focus primarily on reference point methods both for
problem structuring and representation, and as a guide to computational
identification and exploration of the Pareto frontier. Some comments will,
however, be made on the role of other methods of MCDA in this context.

R.C. Purshouse et al. (Eds.): EMO 2013, LNCS 7811, p. 3, 2013.

Innovization: Discovery of Innovative Solution Principles Using Multi-Objective Optimization

Kalyanmoy Deb

Department of Mechanical Engineering
Indian Institute of Technology Kanpur, PIN 208016, India
deb@iitk.ac.in,
http://www.iitk.ac.in/kangal/

Quest for knowledge has always fascinated man. However, the knowledge associated with scientific problem solving tasks comes in different shades and colors and is largely dependent on the specific problem being solved. Moreover, the methodologies for harvesting knowledge is one of the least standardized domains in the scientific and technological endeavors. An optimization task thrives at finding special solutions (the 'optimal' solutions) in the search space which cannot be bettered by any other solution. Since the optimal solutions are unique in the gamut of all possible feasible solutions for a problem, a two-step standardized knowledge discovery procedure can be derived by using optimization as a vehicle for knowledge discovery:

Task 1: Find a set of Pareto-optimal solutions corresponding to two or more conflicting goals of solving the problem, and

Task 2: Analyze Pareto-optimal solutions to unveil common properties shared by them as valuable knowledge.

Task 1 can be achieved by first identifying two conflicting goals (cost and quality, for example) and solving the resulting problem using a classical generative or an EMO algorithm to find a set of high-performing trade-off solutions ($\mathcal{S}$). Task 2 takes dataset $\mathcal{S}$ consisting of optimized variables $\mathbf{x}$, corresponding objectives $\mathbf{f}$, and constraint values $\mathbf{g}$, as input and performs an intelligent data-mining task to identify hidden relationships in the $\mathbf{x}$-$\mathbf{f}$-$\mathbf{g}$ dataset.

A manual regression analysis for the innovization task has led to the discovery of useful relationships in many problems in the past [2]. Our recent automated innovization procedure [1] based on solving a clustering optimization problem is able to find single and multiple relationships on some of the problems for which the manual innovization procedure was used. Further investigations revealed that basic innovization procedure can be extended to achieve (i) *higher-level* innovization by which common principles between multiple parameterized trade-off fronts can be obtained, (ii) *lower-level* innovization by which relationships local to a part of the trade-off front (and not common to the rest of the front) can be obtained, and (iii) *temporal* innovization by which relative importance of different relationships can be achieved. All these activities elevate the act of optimization to a useful level than simply finding one or more optimal solutions.

R.C. Purshouse et al. (Eds.): EMO 2013, LNCS 7811, pp. 4–5, 2013.
© Springer-Verlag Berlin Heidelberg 2013

References

1. Bandaru, S., Deb, K.: Towards automating the discovery of certain innovative design principles through a clustering based optimization technique. Engineering Optimization 43(9), 911–941 (2011)
2. Deb, K., Srinivasan, A.: Innovization: Innovating design principles through optimization. In: Proceedings of the Genetic and Evolutionary Computation Conference (GECCO-2006), pp. 1629–1636. ACM Press, New York (2006)

'Hang On a Minute': Investigations on the Effects of Delayed Objective Functions in Multiobjective Optimization

Richard Allmendinger[1] and Joshua Knowles[2]

[1] Department of Biochemical Engineering, University College London, UK
r.allmendinger@ucl.ac.uk
[2] School of Computer Science, University of Manchester, UK
j.knowles@manchester.ac.uk

Abstract. We consider a multiobjective optimization scenario in which one or more objective functions may be subject to delays (or longer evaluation durations) relative to the other functions. We motivate this scenario from the viewpoint of experimental optimization problems, and derive several simple strategies for dealing with population and/or archive updates under these conditions. These are embedded in a ranking-based EMO algorithm and tested on the WFG test problems augmented with delayed objective(s). Results indicate that good performance can be achieved when the most recently generated solutions are submitted for evaluation on the delayed objective functions, and missing objective values are approximated using a fitness inheritance-based approach. Also, in general one should wait for all evaluations to complete before resuming search if the delay is short, while a non-waiting strategy should be preferred for longer delays.

1 Introduction

Evolutionary approaches to multiobjective optimization continue to find new applications in a diverse range of scientific and problem-solving contexts, even in areas where existing techniques have considerable history and traction. One area of potential exploitation of EMO methods that is beginning to take off is in applications where the optimization loop involves an experimental (rather than a simulation) step, examples being [13,19,18]. Motivated primarily by this area, this paper considers the problem of optimizing several objectives simultaneously in the case where *at least one of them* requires a relatively longer time to evaluate than the 'cheaper' or 'cheapest' of the objective functions.[1] This kind of problem comes about when one objective, for example, involves a lengthy experimental process such as growth, fermentation or such, or perhaps the involvement of human expert(s) input. We consider here that the objective functions can be evaluated in batches, as is often the case in experimental settings, but that there

---

[1] Ian Stott and Jane Shaw, of Unilever, raised this issue in a recent scientific meeting at University of Manchester [21].

R.C. Purshouse et al. (Eds.): EMO 2013, LNCS 7811, pp. 6–20, 2013.

is not a possibility of speed-up on the expensive objective(s) in terms of the (relative) time taken to evaluate the batch of solutions.

Given this setting, we investigate what strategies one might employ in an EMO algorithm (EMOA) to deal with the delayed objective(s). We simulate the use of surrogate models, which may be appropriate in some contexts, but our main focus is on strategies that do not rely on estimating missing or delayed objective values. In the following section, we note some EA and EMO papers that have looked at similar or perhaps related problems, from which we can and do draw some inspiration. Section 3 defines our problem and reviews some basic properties of dominance relationships in the case of unknown objective values, and this prepares the way for some of the strategies that we go on to detail in Section 4. Experimental results on modified WFG functions are given in Section 5, and Section 6 is a discussion and conclusion.

2   Background and Related Work

All general-purpose approaches to optimization involve a "generate-and-test" loop that must be repeatedly applied in order to discover optimal or high-performing solutions. The iteration of the main loop means that the cost and feasibility of optimization, in any given context, will depend critically on how fast and cheaply solutions can be accurately evaluated. Often, in real applications, it is expensive or time-consuming to evaluate solutions accurately, so that there is much interest in the optimization community in topics around the subject of how to save function evaluations (i.e., designing better optimizers), and also around how to build or use surrogate models of the real evaluation function that are sufficiently accurate to allow optimization to occur but at reduced temporal or financial cost.

Here, we are concerned, in the context of expensive objective functions, with multiobjective optimization: in particular, finding a representative approximation of the entire true Pareto Front. Surrogate modeling is a viable and appropriate approach to tackling many expensive objective functions even in the context of multiple objectives (see [11,8,24]), but is a complicated area with many choices to consider for a proper study. Here, our focus is more on the basic design choices of the multiobjective optimizer.

Given this, the key questions are: how should fitness assignment, and population update occur to account for the fact that one or more objective functions are delayed, i.e. that the fitness estimates of some solutions, at any given time, may only be partial? Should we devise or employ selection and update techniques that can deal with partial fitness information, or should we just use a more standard EMOA, and simply wait for all evaluations to complete, no matter the delay, before going on to the next generation? A look at the literature provides useful clues and ideas to try out.

Some papers have considered the idea of finding 'minimal sets of objective functions' [5], such that the subset does not conflict with the full set, and is not redundant. The context of this work is generally different to ours — mostly the concern has been with reducing the number of objective functions down (online or prior to search) in the context of 'many-objective' optimization [15], to facilitate the optimization process. Nevertheless, clearly the effect of objective function removal is closely related to the effect of objective function delay, where at least some solutions that we wish to rank, or to assign reproductive opportunities, might have a reduced number of objective values (at least temporarily). The difference is that our aim is not to identify which objectives we can neglect, but rather to estimate the effects of neglecting a specific expensive objective in order to determine whether it is worthwhile to use the objective.

Asynchronous evaluation in optimization in the context of grid computing was considered in [12,16]. The problem overlaps but is distinct from ours in that the cloud computing resource is assumed to be heterogeneous and/or unreliable, and the asynchrony happens across the population rather than across objectives. (In contrast, we assume for the moment a rather reliable and homogeneous process for evaluating a whole population en masse, and are concerned only with the fact that some objectives can be evaluated faster than others.) Although the context is a bit different, we think that as Lewis *et al.*[12] found, a strategy based on a moderate amount of waiting for slower evaluations may be competitive in some settings, and we also consider the effect of diversity maintenance might be important (see below).

If we think of using algorithms that have solutions staying in a memory (i.e., an archive or secondary population) to allow that they can be waiting for their expensive objective function values to be computed, and that these solutions are potentially part of the present or future breeding pool, then we may have solutions from some number of generations in the past needing to take part in reproduction. In this context, the use of age-layered populations [9] may be a neat way to handle population update and selection matters. We consider an adaptation of this architecture in our work here.

We have learnt through the development of EMO over the last years that diversity preservation, or methods for ensuring objective space spread are very important to obtain good approximation sets (e.g., if we are considering sensible measures of performance such as hypervolume or epsilon-dominance, at least for external evaluation). But the diversity of solutions in the objective space is going to be difficult to estimate when many solutions are missing one or more objective values; in fact this problem may be more severe than the adaptations necessary to deal with Pareto ranking of solutions. Given this issue, it may be sensible to revert to the use of decision space diversity [17,22] in place of objective space, as we assume that decision space information is quick and cheap to use.

Although we do not wish to cloud our initial studies by complex considerations concerning the use of surrogate modeling techniques, it does seem that one of the most direct approaches to handling delayed objective values is to use estimated objective values in their place (at least until true values become available). In this regard, we consider a few simple methods of estimating objective values, inspired in part by work on fitness inheritance [20,14]. Simple methods of missing value imputation from machine learning might also be used (see [23]).

Finally, we note that the problem of delayed objective function values has some relationship to our recent work on *ephemeral resource constraints* (ERCs) in single-objective optimization [2,3,1,4]. ERCs are temporary limitations in the capacity to evaluate certain otherwise feasible solutions *during the optimization process.* ERCs arise in experimental optimization settings due to external factors such as machine breakdowns, limited availability of certain reagents or chemicals under test, or human experts with limited availabilities, and usually they only affect part of the feasible search region at any given time (often also as a function of previous actions). Delayed objectives, by contrast, prevent (immediate) evaluation of *all* the solutions of a batch, but in only some objectives. A key finding from our work with ERCs is that waiting for objective values is quite often the best thing to do, but it can depend on several other factors about the ERCs, such as how long they will last, how much of the search space is affected, and so on. We expect that also in the case of delayed objectives in EMO, simply applying a standard algorithm and waiting for objective functions to return may be the best thing to do in many cases. But we would like to discover the situations where this is not the case, and what other strategies may be sensible. As we found with ERCs, significant savings may sometimes be possible even with a minimal need for information about the problem [2].

3   Problem Formulation and Pareto Dominance Considerations

We augment the notion of delayed objective functions onto a multiobjective optimization problem as follows:

$$\text{minimize } (f_1(\boldsymbol{x}), ..., f_m(\boldsymbol{x}))^T \tag{1}$$
$$\text{subject to } \boldsymbol{x} \in X,$$

where $\boldsymbol{x} = (x_1, ..., x_l)$ is a *solution vector* and X a *feasible search space*. The static *objective functions* $f_i, i = 1, ..., m$ are to be minimized and each function is associated with some *evaluation delay* of $\Delta t_i \geq 0$ time steps (e.g. hours or days) relative to the objective(s) that is (are) quickest to evaluate. That is, $\Delta t_i = 0$ means that function i is quickest to evaluate and thus has no delay, while $\Delta t_i > 0$ means that function i needs Δt_i time steps longer to be evaluated than the quickest objective function. There is at least one function with delay, i.e. $\exists i \in \{1, ..., m\} : \Delta t_i > 0$.

In this study, each function f_i is evaluated in a batch of $k_i > 0$ solutions, and it takes one time step to evaluate this batch. If not otherwise stated, we assume an optimization scenario with exactly one delayed objective function; this will be always function f_m having a delay of $\Delta t_m > 0$, and we set $\Delta t_i = 0, i = 1, ..., m - 1$ for the other functions. In this setup the following two Pareto dominance relations hold, which we will incorporate later in some of our strategies for dealing with delayed objective functions.

Lemma 1. *Let S be a set of points for which $m - 1$ objective values are all known, and the mth objective values are all unknown. Then if all solutions in S are non-dominated with respect to the $m-1$ objectives, it follows that all solutions in S are non-dominated to each other irrespective of their mth objective values.*

Algorithm 1. Ranking-based EMOA for optimizing subject to delayed objective functions

---

Require: $f_1, ..., f_m, \Delta t_m > 0, (\Delta t_1 = 0, ..., \Delta t_{m-1} = 0), \mu = \lambda, T$ (time limit)
 1: $t = 0$ (time counter), $Pop = \emptyset$
      // Initialize Population:
 2: $Pop = $ random_generate_n_solutions($n = \mu$)
 3: evaluate_pop($Pop, f_1, \ldots f_{m-1}$),   assign_pseudovalues_to_expobjective(Pop, f_m), $t = t + 1$ // evaluation of non-delayed objectives only, and assignment of pseudovalues to the delayed objective f_m
 4: $endtime = $ evaluate_pop_expensive($Pop, f_m, currenttime = t$)   // spawns parallel thread to evaluate Pop on delayed objective; immediately returns the projected end time for spawned process; sets Pop's mth objective value to 'pending'
      // Main Loop:
 5: **while** $t < T$ **do**
 6:      rank(Pop, ranking_method)   // ranking method must account for missing (delayed) objective values of some solutions
 7:      $ParentPop = $ parental_selection(Pop)
 8:      $OffPop = $ crossover_and_mutation($ParentPop$)
 9:      evaluate_pop($OffPop, f_1, \ldots f_{m-1}$), $t = t + 1$ // evaluation of non-delayed objectives only
10:      $Pop = Pop \cup OffPop$
11:      assign_pseudovalues_to_expobjective(Pop, f_m)   // (re)assignment of pseudovalues to delayed objective f_m
12:      **if** ($t = endtime$) **then**
13:          pending objective values are now updated
14:          $EvalPop = $ selection_for_expevaluation(Pop)   // decides which μ solutions from Pop to evaluate on the delayed objective f_m; only selects from solutions that have no value for f_m
15:          $endtime = $ evaluate_pop_expensive($EvalPop, f_m, currenttime = t$)
16: **return** (Pop)

Lemma 2. *Let S be a set of points for which $m - 1$ objective values are all known, and the mth objective values are all unknown. Then (a) the minimum number of different non-dominated sorting (NDS) ranks in S is 1, and (b) the maximum number of different ranks is the number of NDS ranks existing amongst the solutions in S in the $m - 1$ known objectives plus the number of points that are equal with respect to the known objectives.*

4   Strategies for Dealing with Optimizations Problems Featuring Delayed Objective Functions

As the basis for our strategies we use a ranking-based EMOA as shown by Algorithm 1. Unlike standard EMOAs, the size of the population Pop in this EA is not fixed. This way we allow solutions with missing objective function values to influence the search direction and be evaluated at any point in time during the optimization. Assuming that the objective functions are evaluated in a batch of $k_i = \mu, i = 1, ..., m$ solutions, the EA begins the optimization by generating a set of μ solutions at random, evaluating them on the non-delayed objectives only and assigning pseudovalues to the delayed objective (Line 3). At $t = 0$, all solutions are submitted for evaluation on the delayed objective function f_m, and their mth objective values are set to 'pending' (Line 4). The projected end time *endtime* of the delayed objective represents the time step at which the pending objective values are updated (i.e. revealed) (Line 13), and a set of new μ solutions for evaluation on the delayed objective selected (Line 14). Each generation, the population is first ranked (Line 6), and then λ offspring generated by a process of selection, crossover and mutation (Line 7 and 8), and evaluated on the non-delayed objective functions (Line 9). Following this, all offspring are added to *Pop* and pseudovalues are (re)assigned to all solutions in *Pop* that have not been evaluated on f_m (Line 11); reassigning pseudovalues to solutions reduces the risk that these solutions take over the population and potentially misguide the search. Our EMOA ensures that the population *Pop* does not contain duplicate solutions, i.e. offspring are generated until we have a set that has not been evaluated yet.

In the following we describe various modifications to the EMOA we are going to investigate in the presence of delayed objective functions. In particular, we look at different methods for the assignment of pseudovalues to the delayed objective (Line 3 and 11), ranking (Line 6), parental selection (Line 7) and the selection of solutions for evaluation on the delayed objective function (Line 14).

Assignment of Pseudovalues to Delayed Objectives. We investigate three assignment strategies. The first strategy, *random pseudovalue assignment*, assigns to each solution with a missing objective value a pseudovalue drawn at random from the interval $[\min_{Pop} f_m, \max_{Pop} f_m]$, where $\min_{Pop} f_m$ and $\max_{Pop} f_m$ is the minimum and maximum value of objective f_m of all solutions in *Pop* that have actually been evaluated on objective f_m. The second strategy, *noise-based pseudovalue assignment*, draws for each solution with a missing objective value a random solution from *Pop* that has been evaluated on all objectives (including the delayed objective), and adds a small amount of (Gaussian distributed) noise $\mathcal{N}(0, \sigma^2)$ to the value of the delayed objective; the resulting value is used as the pseudovalue for the delayed objective. The third strategy, *fitness inheritance-based pseudovalue assignment*, selects for each solution with a missing objective value a solution from *Pop* that is both closest to it in the decision space (in terms of normalized Euclidean distance) and has been evaluated on all objectives, and then simply takes over the delayed objective value of this solution.

Ranking of Solutions. We investigate two ranking schemes. The first scheme, *performance ranking*, sorts all solutions in *Pop* according to their non-dominated sorting (NDS) ranks only. In contrast, the second ranking scheme, *performance+age ranking*, considers both the NDS rank and the time stamp at which a solution has been generated. More precisely, first the NDS ranks of all solutions in *Pop* are obtained (these ranks are used later as quality criterion in parental selection), and then the population is sorted based on the age of solutions whereby more recently generated solutions are favoured (this sorting affects the truncation selection only).

Determining a Parent Population *ParentPop*. Once the population *Pop* has been ranked we need to decide which solutions should be eligible for parental selection (as our population is not limited in size). In a standard EA setup with a fixed population size, this design choice would correspond to the reproduction scheme (or environmental selection mode). We investigate two schemes: parental selection among (i) the top ranked μ solutions in *Pop* (generational reproduction scheme) (denoted in future by GGA) and (ii) the top ranked $\mu \times 2$ solutions (($\mu + \lambda$)-ES reproduction scheme) (denoted in future by ($\mu + \lambda$)-ES).

Selecting Solutions for Evaluation on the Delayed Objective Function. We investigate two strategies to decide which μ solutions from *Pop* to evaluate on f_m. The first strategy, *sweep selection*, selects always the most recently generated μ solutions for evaluation on f_m. The second strategy, *priority-based selection*, assigns to each solution without a value for f_m, a score representing the solution's priority of being evaluated. We compute this score by first obtaining the NDS ranks, considering the objectives $f_1, ..., f_{m-1}$ only, of all completely evaluated solutions in *Pop*. Then, the priority score of a solution is estimated based on Lemma 1 and 2 (see Section 3), and also on the idea of counting the total amount the ranking of all (completely evaluated) solutions could be changed: If a solution with no value for f_m is dominating all solutions, then potentially it could demote all of these solutions by one rank after revealing the value of objective f_m (i.e. we have a priority score equal to the number of completely evaluated solutions in *Pop*). If a solution with no value for f_m is dominated by all solutions, then it cannot possibly dominate any solution (although it might be non-dominated if the value of f_m is very small) (i.e. we a have a priority score of zero). In all other cases, a solution with no value for f_m can potentially dominate some solutions in *Pop*; here we assign the solution a priority equal to the number of solutions having a lower rank.

The modifications described above enable an EMOA to deal with partial fitness information. We will investigate also an EMOA that waits for all evaluations to complete and thus prevents having to deal with solutions with missing objective values; hence for this algorithm it makes sense to look only at modifications related to the ranking of solutions (Line 6) and the determination of the parent population *ParentPop* (Line 7).

5   Experimental Study

This section describes the test functions and the parameter settings as used in the subsequent experimental analysis, which investigates the performance of the strategies described above when applied to problems with delayed objectives.

5.1   Experimental Setup

Our aim in this study is to understand the effect of delayed objective functions on EA performance. To cover a wide range of problem characteristics, we use the Walking Fish Group (WFG) toolkit [10]. We use the toolkit with 4 distance parameters and 2 position parameters within the standard WFG1-WFG9 test

Table 1. EA parameter settings

Parameter	Setting
Parent population size μ $(= k_i, i = 1, ..., m)$	50
Offspring population size λ	50
Per-variable mutation probability p_m	$1/l$
Crossover probability p_c	0.9
Distribution index (mutation and crossover)	20
Time limit T	40

problems; i.e. we have $l = 4+2 = 6$ continuous decision variables. If not otherwise stated we use the WFG problems with $m = 3$ objectives, with f_3 being the objective function delayed by $\Delta t_3 = 3$ time steps. We set $k_i = \mu, i = 1, ..., m$.

We augment the strategies described in Section 4 on a ranking-based EMOA (see Algorithm 1) that uses binary tournament selection (with replacement) for parental selection, simulated binary crossover (SBX) [6], and polynomial mutation [7]. The parameter settings of the EMOA are given in Table 1.

For the noise-based pseudovalue assignment strategy we use a noise level of $\sigma = \sqrt{(\max_{Pop} f_m - \min_{Pop} f_m) * 0.05}$, where, as in the case of the random pseudovalue assignment method, $\max_{Pop} f_m$ and $\min_{Pop} f_m$ is the maximum and minimum value of objective f_m of all solutions in Pop that have actually been evaluated on objective f_m so far. To reduce the risk of any search bias, the parameters $\max_{Pop} f_m$ and $\min_{Pop} f_m$ are set initially to large positive and negative numbers, here 1000 and -1000, respectively. We set the time limit to $T = 40$ time steps.

Any results shown are average results across 20 independent algorithm runs. We use paired comparison by employing a different seed for the random number generator for each EA run but the same seeds for all strategies described above.

5.2   Experimental Results

Table 2 gives us an initial overview of the performance (hypervolume measurements) of some of the algorithm modifications on all WFG test problems.[2] Results were obtained using an EMOA with random pseudovalue assignment and sweep selection. We can make several observations from the table: (i) optimizing subject to delayed objective function affects the performance negatively on all problems except WFG1; (ii) a generational reproduction scheme without elitism (GGA) tends to perform best in the presence of delayed objectives (when waiting is applied), while there is no clear winner between GGA and $(\mu + \lambda)$-ES in

---

[2] The hypervolume measurements were obtained by normalizing the non-dominated front found by an EA at the end of an optimization run, and then taking the average of the hypervolume measurements across 20 runs. The normalization was done based on the extremal values of the estimated True Front, which is available online at http://jmetal.sourceforge.net/problems.html, and the reference point was set to the minimum and maximum values of the normalized front.

Table 2. Average hypervolume values obtained in an environment with and without (in parenthesis) delayed objective functions for different algorithm setups on the WFG test problems using $m = 3$ objectives. All EAs optimizing subject to delayed objective functions employed random pseudovalue assignment and sweep selection. For each problem instance and optimization environment (with delay vs without delay), we highlighted all algorithm setups in bold face that are not significantly worse than any other setup. A Friedman test revealed a significant difference between the search algorithm setups in general, but differences among the individual setups were tested for in a post-hoc analysis using (paired) Wilcoxon tests (significance level of 5%) with Bonferroni correction.

		GGA		$(\mu + \lambda)$-ES	
		waiting	no waiting	waiting	no waiting
WFG1	Performance	0.1015	**0.1976**	0.0940	0.1682
	ranking	**(0.1939)**		**(0.1916)**	
	Performance+age	0.0885	0.1077	0.0875	0.0992
	ranking	(0.0982)		(0.1016)	
WFG2	Performance	**0.6901**	0.6748	0.6474	0.6655
	ranking	(0.8527)		**(0.8616)**	
	Performance+age	0.6028	0.5603	0.6068	0.5819
	ranking	(0.6822)		(0.6958)	
WFG3	Performance	**0.4292**	0.4030	**0.4214**	0.3915
	ranking	**(0.4639)**		**(0.4624)**	
	Performance+age	0.4129	0.3970	**0.4260**	0.4000
	ranking	(0.4270)		(0.4396)	
WFG4	Performance	**0.3434**	0.2701	0.3362	0.2503
	ranking	**(0.4172)**		**(0.4199)**	
	Performance+age	0.2935	0.2468	0.3015	0.2561
	ranking	(0.3728)		(0.3598)	
WFG5	Performance	**0.3430**	0.2676	0.3353	0.2925
	ranking	**(0.4020)**		**(0.4012)**	
	Performance+age	0.3284	0.2888	0.3290	0.2996
	ranking	(0.3633)		(0.3667)	
WFG6	Performance	**0.3457**	0.2356	0.3240	0.2646
	ranking	**(0.3943)**		**(0.3918)**	
	Performance+age	0.3214	0.2701	0.3252	0.2888
	ranking	(0.3410)		(0.3464)	
WFG7	Performance	**0.3447**	0.2760	**0.3408**	0.3004
	ranking	(0.4096)		**(0.4163)**	
	Performance+age	0.3298	0.2862	0.3346	0.2975
	ranking	(0.3787)		(0.3722)	
WFG8	Performance	**0.3129**	0.2497	0.3021	0.2550
	ranking	**(0.3721)**		**(0.3687)**	
	Performance+age	0.2900	0.2492	0.3047	0.2556
	ranking	(0.3134)		(0.3172)	
WFG9	Performance	**0.3743**	0.3260	0.3596	0.3402
	ranking	**(0.4330)**		**(0.4224)**	
	Performance+age	0.3242	0.2865	0.3332	0.2912
	ranking	(0.3727)		(0.3858)	

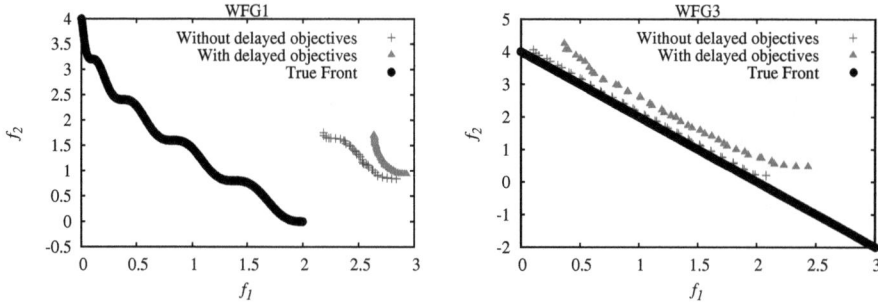

Fig. 1. Plots showing the true Pareto Front, and median attainment surface (across 20 runs) obtained on WFG1 (left) and WFG3 (right) with $m = 2$ objectives in an environment with delayed objective functions (function f_2 was subject to a delay of $\Delta t_2 = 3$ and $k_i = \mu, i = 1, 2$) and without. The EMOA was equipped with a generational reproduction scheme (GGA), sweep selection, random pseudovalue assignment, performance ranking, and waited for all evaluations to complete before resuming search.

an environment without delays; (iii) waiting for all evaluations to complete performs best on all problems except WFG1 and WFG2; (iv) performance ranking generally performs better in an environment with and without delays.

When optimizing subject to delayed objective functions, the convergence speed seems to be reduced and a more diverse population maintained; these properties tend to be amplified as the delay Δt_m becomes larger, ultimately causing the performance to reduce (as will be seen later from Fig. 4). For WFG1, however, the presence of a delayed objective can yield better results than obtainable in an unconstrained environment (observation (i)). In general, we observed that the experimental results obtained on WFG1 are different than on the other WFG problems, which may be due to the structure of this problem (WFG1 is separable, uni-modal and has a dissimilar weight structure [10]). For WFG2-WFG9, assigning pseudovalues to the delayed objective that are too far away from the true objective values can lead to misguidance when performing ranking and parental selection. The risk of misguidance can be reduced when employing an EMOA with a waiting strategy and a generational reproduction scheme (observation (ii) and (iii)). Observation (iv) implies that parental selection should not be limited to a subset (the most recently generated solutions) of the population.

Fig. 1 and 2 show visually the performance impact of delayed objective functions on WFG1 and WFG3 using $m = 2$ and 3 objectives, respectively. Plots are showing the median attainment surface across 20 runs obtained by the best performing EMOA from Table 2. The plots indicate that the impact of a delayed objective function on the performance of an optimizer depends on (i) the characteristics of the fitness landscape of a problem and (ii) the number of objectives to be optimized, an observation we will make again later.

Fig. 3 investigates the effect of different delay lengths Δt_m on the performance of our strategies for WFG1 (top plots) and WFG2 (bottom plots); note that the setting $\Delta t_m = 0$ means there is no delay and thus refers to an unconstrained

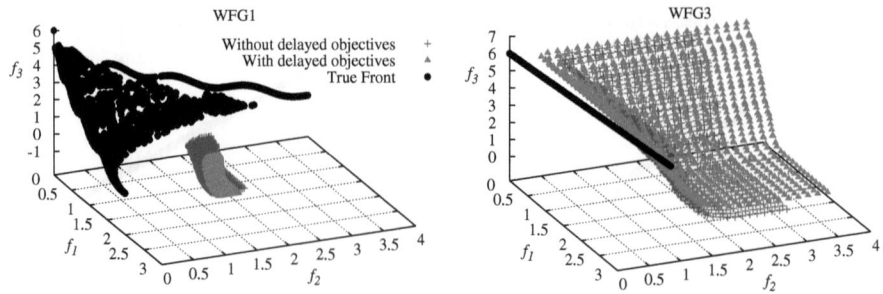

Fig. 2. Plots showing the true Pareto Front, and the median attainment surface (across 20 runs) obtained on WFG1 (left) and WFG3 (right) with $m = 3$ objectives in an environment with and without delayed objective functions. The EMOA was equipped with a generational reproduction scheme (GGA), sweep selection, random pseudovalue assignment, performance ranking, and waited for all evaluations to complete before resuming search.

optimization scenario. The left and right plots show the performance impact for EMOAs employing sweep selection and priority-based selection, respectively. All results were obtained using a generational reproduction scheme (GGA), and performance ranking; this setup yielded best results as shown previously. We make the following observations from the figure.

- Generally, an increase in the delay length Δt_m affects search negatively, and algorithm choice is important. Interestingly, a performance improvement, when compared to an unconstrained optimization scenario, may be obtained for short delays (see range $0 < \Delta t_m < 7$ in the top left plot). It seems to be the case that this is due to an increase in the population diversity, although other factors may be responsible.

- Sweep selection clearly outperforms priority-based selection on both test problems (and the other WFG problems for which the results are not shown here) and for all values of $\Delta t_m > 0$. The reason is that priority-based selection ignores completely the values of the delayed objective f_m when selecting solutions for evaluation on the delayed objective. This may lead to misguidance in the selection and stagnation in the search.

- With respect to the delay length Δt_m, we observe for WFG2 and sweep selection (bottom left plot) that there is a value ($\Delta t_m = 5$) at which one should switch from a waiting strategy to a non-waiting one (when using sweep selection and a random or noise-based pseudovalue assignment; this pattern was apparent for all WFG problems except WFG1). The larger Δt_m the slower is the search progress when employing a waiting strategy. The length of the delay Δt_m at which the switch from a waiting to a non-waiting strategy should be performed depends on the difficulty of the problem at hand, and the optimization time available. For WFG1 and sweep selection (top left plot) a non-waiting strategy should be employed for all values of Δt_m. When using priority-based selection, a waiting strategy should be preferred

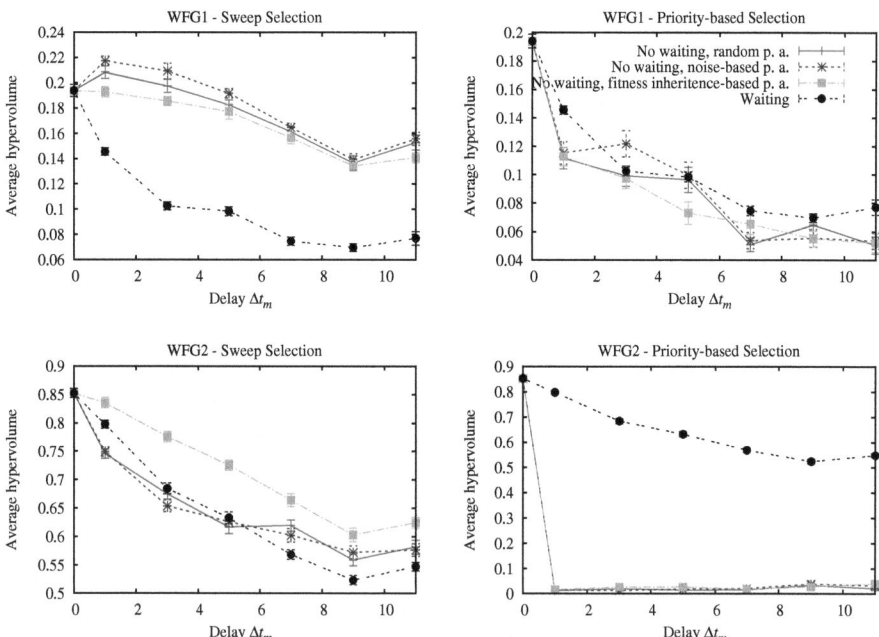

Fig. 3. Plots showing the average hypervolume (and its standard error, indicated by the error bars) obtained on WFG1 (top plots) and WFG2 (bottom plots) with $m = 3$ objectives using EA setups employing sweep selection (left plots) and priority-based selection (right plots). All results were obtained using an EMOA equipped with a generational reproduction scheme (GGA) and performance ranking.

over a non-waiting one because it removes the risk of getting the priority scores wrong and thus submitting non-promising solutions for evaluation on the delayed objective.

- Fitness-inheritance based pseudovalue assignment combined with sweep selection, and no waiting yields best performance on WFG2 (and all the other WFG problems except WFG1). The reason is that the fitness inheritance-based method is able to approximate the value of the delayed objective f_m better than the other two pseudovalue assignment strategies, reducing the risk of misguidance in the search.

Finally, Fig. 4 shows some initial results on how the search performance is affected on WFG2 by the number of objectives m to be optimized (left) and the number of delayed objective functions (right) as a function of the delay length Δt_m. From the left plot we observe that whilst the performance is affected in a similar way for $m = 2$ and 3 objectives, there is a smaller performance gap between different strategies for $m = 2$. This pattern was also apparent for the other WFG problems (results not shown) and may indicate that lower-dimensional problems are easier to deal with in the presence of delayed objectives.

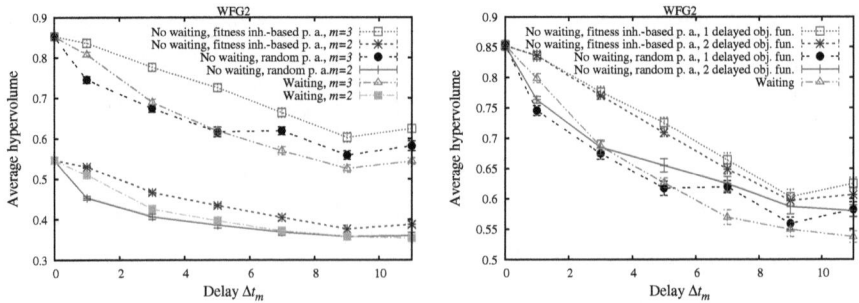

Fig. 4. Plots showing the average hypervolume (and its standard error, indicated by the error bars) obtained by different EA setups on WFG2 with $m = 2$ and 3 objectives (using 1 delayed objective function) (left) and $m = 3$ objectives using 1 and 2 delayed objective functions (in case of 2 delayed objective functions, a delay was on f_2 and f_3 with $k_2 = k_3 = \mu$). All results were obtained using an EMOA equipped with a generational reproduction scheme (GGA), sweep selection, and performance ranking.

From the right plot in Fig. 4 we observe that having two instead of one delayed objective function (i.e. being more uncertain about the quality of a solution) improves the performance of strategies that approximate missing objective values poorly (random pseudovalue assignment) but degrades the performance of otherwise accurate approximation techniques (fitness inheritance-based pseudovalue assignment). This pattern was apparent for all WFG problems except WFG1, and may indicate that there is a trade-off between increasing the risk of misguidance (leading to reduction in the performance when using an accurate approximator) and increasing the probability that truly poor solutions are approximated poorly and thus not considered for evaluation on the delayed objective function (improvement in performance when using a poor approximator).

6   Summary and Conclusion

In this paper we have considered a multiobjective optimization scenario in which at least one objective function may be subject to delays relative to the other functions. In other words, some objective functions take longer to be evaluated than others. This kind of problem can be encountered, for example, when the evaluation of an objective function involves a lengthy experimental process, such as growth or fermentation, or the involvement of human expert(s) input. We have proposed several strategies to deal with this kind of optimization scenario — concerning the pseudovalue assignment to delayed objectives, population ranking, reproduction scheme, and the selection of solutions for evaluation on the delayed objective functions — and assessed them on the (continuous) WFG test problems.

The experimental study revealed that delayed objective functions affect the search performance of an EA — in general, the longer the delay the poorer is the search performance — but that a well-tuned optimizer can damp the performance impact significantly. In particular, when optimizing subject to delays, we can tentatively conclude that one should: (i) employ a fitness inheritance-based pseudovalue assignment (i.e. fill missing objective values of a solution with the objective values of the genetically closest and fully evaluated solution), (ii) select parents for reproduction from a population that is sorted based on the solutions' non-dominated sorting ranks (without accounting for the time stamps at which solutions have been created), (iii) use a generational reproduction scheme (without elitism), and (iv) submit the most recently generated solutions for evaluation on the delayed objectives. Furthermore, we found that, in general, for short delays one should wait for all evaluations to complete before continuing with the next generation. When the delay is long, however, waiting slows down the search and should be avoided. Finally, we have seen that our observations hold for problems with two and three objectives, and that varying the number of delayed objective functions has interesting implications on the performance depending on the algorithm setup employed.

Our study has shown that EA performance crucially depends on the way pseudovalues are assigned and solutions for evaluation on the delayed objective functions selected. We believe there are still some performance improvements to gain by tuning these two aspects, which usually do not need to be considered in the design of an EA. Investigating the effect of delayed objective functions on many-objective optimization problems where several objectives are subject to delays of different durations is another important avenue to be pursued.

Acknowledgments. Thanks to Ian Stott and Jane Shaw of Unilever for inspiring this work on delayed objectives.

References

1. Allmendinger, R.: Tuning Evolutionary Search for Closed-Loop Optimization. PhD thesis, School of Computer Science, The University of Manchester (2012)
2. Allmendinger, R., Knowles, J.: On-Line Purchasing Strategies for an Evolutionary Algorithm Performing Resource-Constrained Optimization. In: Schaefer, R., Cotta, C., Kołodziej, J., Rudolph, G. (eds.) PPSN XI. LNCS, vol. 6239, pp. 161–170. Springer, Heidelberg (2010)
3. Allmendinger, R., Knowles, J.: Policy learning in resource-constrained optimization. In: Proceedings of GECCO, pp. 1971–1978 (2011)
4. Allmendinger, R., Knowles, J.: On handling ephemeral resource constraints in evolutionary search. In: Evolutionary Computation (2013), doi:10.1162/EVCO_a_00097 (posted Online November 19, 2012)
5. Brockhoff, D., Zitzler, E.: Objective reduction in evolutionary multiobjective optimization: theory and applications. Evolutionary Computation 17(2), 135–166 (2009)
6. Deb, K., Agrawal, R.: Simulated binary crossover for continuous search space. Complex Systems 9, 115–148 (1994)
7. Deb, K., Goyal, M.: A combined genetic adaptive search (GeneAS) for engineering design. Computer Science and Informatics 26(4), 30–45 (1996)
8. Emmerich, M., Giannakoglou, K., Naujoks, B.: Single- and multiobjective evolutionary optimization assisted by gaussian random field metamodels. IEEE Transactions on Evolutionary Computation 10(4), 421–439 (2006)

9. Hornby, G.: ALPS: the age-layered population structure for reducing the problem of premature convergence. In: Proceedings of GECCO, pp. 815–822 (2006)
10. Huband, S., Hingston, P., Barone, L., While, L.: A review of multiobjective test problems and a scalable test problem toolkit. IEEE Transactions on Evolutionary Computation 10(5), 477–506 (2006)
11. Knowles, J.: ParEGO: A hybrid algorithm with on-line landscape approximation for expensive multiobjective optimization problems. IEEE Transactions on Evolutionary Computation 10(1), 50–66 (2006)
12. Lewis, A., Mostaghim, S., Scriven, I.: Asynchronous multi-objective optimisation in unreliable distributed environments. Biologically-Inspired Optimisation Methods, 51–78 (2009)
13. O'Hagan, S., Dunn, W., Brown, M., Knowles, J., Kell, D.: Closed-loop, multiobjective optimization of analytical instrumentation: gas chromatography/time-of-flight mass spectrometry of the metabolomes of human serum and of yeast fermentations. Analytical Chemistry 77(1), 290–303 (2005)
14. Runarsson, T.P.: Constrained Evolutionary Optimization by Approximate Ranking and Surrogate Models. In: Yao, X., Burke, E.K., Lozano, J.A., Smith, J., Merelo-Guervós, J.J., Bullinaria, J.A., Rowe, J.E., Tiňo, P., Kabán, A., Schwefel, H.-P. (eds.) PPSN VIII. LNCS, vol. 3242, pp. 401–410. Springer, Heidelberg (2004)
15. Saxena, D.K., Deb, K.: Non-linear Dimensionality Reduction Procedures for Certain Large-Dimensional Multi-objective Optimization Problems: Employing Correntropy and a Novel Maximum Variance Unfolding. In: Obayashi, S., Deb, K., Poloni, C., Hiroyasu, T., Murata, T. (eds.) EMO 2007. LNCS, vol. 4403, pp. 772–787. Springer, Heidelberg (2007)
16. Scriven, I., Ireland, D., Lewis, A., Mostaghim, S., Branke, J.: Asynchronous multiple objective particle swarm optimisation in unreliable distributed environments. In: IEEE Congress on Evolutionary Computation, pp. 2481–2486 (2008)
17. Shir, O.M., Preuss, M., Naujoks, B., Emmerich, M.: Enhancing Decision Space Diversity in Evolutionary Multiobjective Algorithms. In: Ehrgott, M., Fonseca, C.M., Gandibleux, X., Hao, J.-K., Sevaux, M. (eds.) EMO 2009. LNCS, vol. 5467, pp. 95–109. Springer, Heidelberg (2009)
18. Shir, O.M., Roslund, J., Leghtas, Z., Rabitz, H.: Quantum control experiments as a testbed for evolutionary multi-objective algorithms. Genetic Programming and Evolvable Machines 13(4), 445–491 (2012)
19. Small, B., McColl, B., Allmendinger, R., Pahle, J., López-Castejón, G., Rothwell, N., Knowles, J., Mendes, P., Brough, D., Kell, D.: Efficient discovery of anti-inflammatory small molecule combinations using evolutionary computing. Nature Chemical Biology 7(12), 902–908 (2011)
20. Smith, R., Dike, B., Stegmann, S.: Fitness inheritance in genetic algorithms. In: Proceedings of the ACM symposium on Applied computing, pp. 345–350 (1995)
21. Stott, I., Shaw, J.: Presentation for Unilever at the Manchester Institute for Biotechnology. University of Manchester (June 11, 2012)
22. Ulrich, T., Bader, J., Thiele, L.: Defining and Optimizing Indicator-Based Diversity Measures in Multiobjective Search. In: Schaefer, R., Cotta, C., Kołodziej, J., Rudolph, G. (eds.) PPSN XI. LNCS, vol. 6238, pp. 707–717. Springer, Heidelberg (2010)
23. van Buuren, S.: Multiple imputation of discrete and continuous data by fully conditional specification. Statistical methods in medical research 16(3), 219–242 (2007)
24. Voutchkov, I., Keane, A.: Multiobjective optimization using surrogates. In: Adaptive Computing in Design and Manufacture (ACDM 2006), pp. 167–175 (2006)

Optimization of Adaptation - A Multi-objective Approach for Optimizing Changes to Design Parameters

Shaul Salomon[1], Gideon Avigad[2], Peter J. Fleming[1], and Robin C. Purshouse[1]

[1] Department of Automatic Control and Systems Engineering
University of Sheffield
Mappin Street, Sheffield S1 3JD, UK
{s.salomon,p.fleming,r.purshouse}@sheffield.ac.uk
[2] Department of Mechanical Engineering
ORT Braude College of Engineering, Karmiel, Israel
gideona@braude.ac.il

Abstract. Dynamic optimization problems require constant tracking of the optimum. A solution for such a problem has to be adjustable in order to remain optimal as the optimum changes. The manner of changing design parameters to predefined values is dealt with in the field of control. Common control approaches do not consider the optimality of the design, in terms of the objective function, while adjusting to the new solution. This study highlights the issue of the optimality of adaptation, and defines a new optimization problem – "Optimization of Adaptation". It is a multiobjective problem that considers the cost of the adaptation and the optimality while the adaptation takes place. An evolutionary algorithm is proposed in order to solve this problem, and it is demonstrated, first, with an academic example, and then with a real life application of a robotic arm control.

Keywords: dynamic optimization, adaptation, optimal control.

1 Introduction

The competency of living creatures to adapt to a changing environment is a crucial virtue. Adaptation is the evolutionary process that allows species to survive. The properties of a species gradually changes to meet the demands of the changing environment. It is a slow mechanism, and it may take hundreds of generations for a trait to establish among the population [1]. Adaptation is also related to the ability of a specimen to change some of its physical properties during its lifetime. These changes are not genetic changes, and they are not passed on to any offspring. Nevertheless, the ability to adapt is inherent within the species' genotype. Examples of fast adaptation are the changing colours of a chameleon [2], the expansion and contraction of the pupil as light changes [3], and the increasing number of red blood cells as a reaction to low percentage of oxygen in the air in high altitude [4].

R.C. Purshouse et al. (Eds.): EMO 2013, LNCS 7811, pp. 21–35, 2013.

The necessity for adaptation may be also related to engineering (e.g., a product should be adapted to changes in market demands), companies (e.g., change of personnel or facilities, as costs change), politics, and many other fields of interest. In engineering design, adaptation may be ensured by including tunable parameters that can be altered when changes are required.

In the context of *single objective optimization*, an initial design might lose its optimality as time passes due to changes that influence the objective function, or it might become infeasible with the changing of constraints. A tunable design may be able to adapt in order to maintain satisfactory performance, or to remain feasible, when such changes occur. This virtue avoids the need for producing a totally new design whenever an environmental change occurs, and it enables prolonging the lifetime of a product.

These kinds of optimization problems, where the optimal solution changes with time, are known as *dynamic optimization problems* (DOPs). Mathematically, a *dynamic single objective optimization problem* is formulated as follows:

Definition 1. *Dynamic single objective optimization problem:*

$$\max_{\mathbf{x}\in Q} f\left(\mathbf{x},t\right)$$

$$\begin{aligned}
s.t. \quad & g_i\left(\mathbf{x},t\right) \geq 0 , && (i = 1,\ldots,I)\\
& h_j\left(\mathbf{x},t\right) = 0 , && (j = 1,\ldots,J)
\end{aligned}$$

where $Q \subset \mathbb{R}^M$ is the search domain, f is the objective function, and g and h are the inequality and equality constraints, respectively. Since the objective function and the constraints are time dependent, the optimal solution $\tilde{\mathbf{x}}$ also changes with time: $\tilde{\mathbf{x}} \Rightarrow \tilde{\mathbf{x}}(t)$.

Many methods, including evolutionary algorithms, have been developed in order to solve DOPs, and to determine what changes should be performed in order to achieve optimal or satisfactory performance over time [5]. In many cases, the required changes in design over time are continuous and relatively small. In other cases, the optimum might "jump" to a different region of the design space. This can happen when a totally different solution becomes the optimum, or when the current optimal solution becomes infeasible. The former case is illustrated in Figure 1. The dynamic function $f(x,t) = 2sin(x + 2t) + tsin(3x - t) + cos(3xt)$ is presented at several time instants. In Figure 1(a) the optimal solution changes slightly between $t = 0.04$ and $t = 0.12$. On the other hand, in Figure 1(b), a new optimal solution is developed in the region of $x = 7$, and a "jump" of the optimum occurs between the time instants $t = 0.54$ and $t = 0.58$. Since the required changes in design at times of an optimum "jump" are significant, some issues regarding the manner of performing these changes should be considered.

When the required change is a mechanical change (as opposed to tuning an electric signal, for example, which is rapid), the adaptation time might be non-negligible. During that time, the function value of the system changes according to the change in design. Considering the example of Figure 1(b), the function values between the old optimum and the new one are much lower than the optimal value. If the change is continuous (i.e., x goes through all the values in between

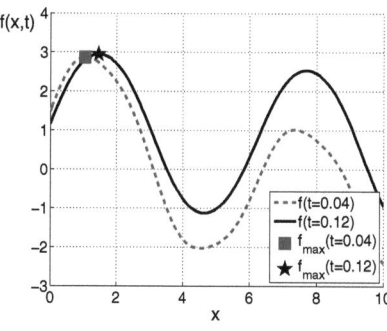

(a) The optimum changes slightly within the same region.

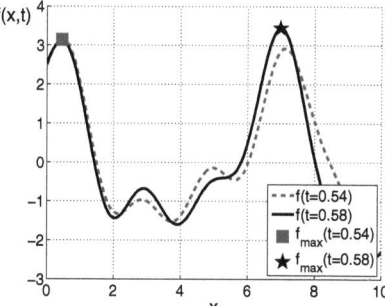

(b) The optimum "jumps" to a different region.

Fig. 1. A dynamic function at four time instants. The optimum changes from the grey square to the black star.

the old and the new solutions), the function values during the adaptation are not optimal.

Although existing research has addressed finding efficient ways to track the moving optimum, no research known to the authors has addressed the optimality during the change itself. This study comes to shed some light on this issue. The aim here is not to suggest a new method for solving DOPs, but to optimize the adaptation of a solution when it requires significant changes. It is assumed here that the DOP is already solved, and the new location of the optimum is known. The aim is at designing a trajectory in phase plane (high level control design), according to its related performances in objective space. This is in contrast to the common approach in control theory, where the optimization is based on the trajectories in phase plane (e.g., minimizing the states' deviation from the new set-point).

Figure 2 depicts two possible trajectories to trace the moving optimum of a single objective DOP with two design variables. Different trajectories in design space are possible in the example, since it consists of two variables. The values of the objective function are illustrated by contour lines. The location of the old optimum x_0 is marked with a square, and the location of the new optimum x_f is marked with a star. In this example, Trajectory 1 (marked with triangles) is along a straight line and passes through a region with very low function values, while Trajectory 2 (marked with diamonds) bypasses this region, and passes through regions with high function values. The trajectories of the objective function over time, for the same example, are shown in Figure 3. Note that the function values along Trajectory 1 are lower than those along Trajectory 2. Therefore, assuming the function is static during the traverse, it can be stated that the optimality of Trajectory 2 is better than the optimality of Trajectory 1 in terms of the function values.

In many cases, changing a system's design has a cost, in terms of money, energy or other resources. The way parameters are changed, especially when more than one parameter is to be adjusted, may affect the cost of the change. The possible conflict between the optimality of the performance during the change, and the cost of the adaptation defines a new optimization problem which is termed here as- *Optimization of Adaptation*. It is a *multi-objective optimization problem* (MOP) by its nature, since the function value during the adaptation process, and the cost of the change are to be optimized at the same time.

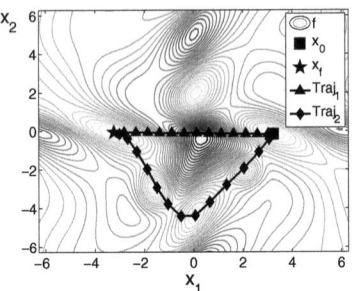

Fig. 2. Two possible trajectories to adapt from x_0 to x_f. Brighter contour lines represent higher function values.

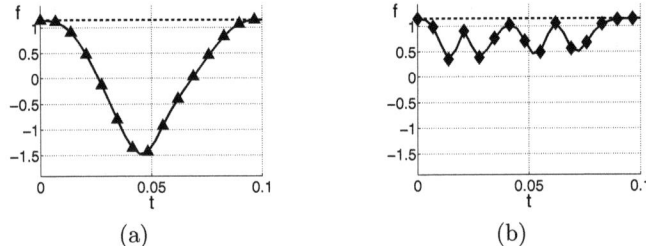

Fig. 3. The trajectories $f(t)$ for the two trajectories of Figure 2. Figure 3(a) refers to trajectory 1, and Figure 3(b) refers to trajectory 2.

The remainder of this paper is organized as follows. In Section 2 the *Optimization of Adaptation Problem* is defined, and a method to assess the objective functions is suggested. In Section 3 an evolutionary algorithm is suggested to solve the problem. In section 4, an academic example and a real-world application for the proposed optimization method are presented. Finally, the paper ends with a discussion in Section 5.

2 Methodology

2.1 Problem Definition

In order to optimize the adaptation of a system at a time instant t_{jump}, when a significant change in design is required, the following assumptions are made:

1. The change takes place over a time interval $[t_0, t_f]$ which is significantly shorter than the time constant of the DOP. Hence, the function value and the constraints can be considered as static for the duration of the change:
   $$f(\mathbf{x}, t_{jump}) = f(\mathbf{x})$$
   $$g_i(\mathbf{x}, t_{jump}) = g_i(\mathbf{x})$$
   $$h_j(\mathbf{x}, t_{jump}) = h_j(\mathbf{x})$$
2. The state of the design parameters prior to the change is $\mathbf{x}(t_0) = \mathbf{x_0}$.
3. The state of the design parameters at the end of the change is the new optimal solution of the DOP: $\mathbf{x}(t_f) = \mathbf{x_f}$, $f(\mathbf{x_f}) = \max f(\mathbf{x})$ (assuming maximization).
4. The system is at a steady state at the beginning and at the end of the adaptation.

Considering the above assumptions, the new optimization problem is defined as follows:

Definition 2. *Optimal adaptation problem (OAP):*

$$\min_{\mathbf{x}(t) \in Q} \{Error(\mathbf{x}(t)), Cost(\mathbf{x}(t))\} , \qquad t \in [t_0, t_f] \qquad (1)$$

$$s.t. \quad \mathbf{x}(t_0) = \mathbf{x_0} \quad , \quad \mathbf{x}(t_f) = \mathbf{x_f} \qquad\qquad\qquad (2)$$

$$\left.\frac{d\mathbf{x}}{dt}\right|_{t_0} = \left.\frac{d\mathbf{x}}{dt}\right|_{t_f} = 0 \qquad\qquad\qquad\qquad (3)$$

$$g_i(\mathbf{x}(t)) \geq 0 , \qquad\qquad (i = 1, \ldots, I) \qquad (4)$$

$$h_j(\mathbf{x}(t)) = 0 , \qquad\qquad (j = 1, \ldots, J) \qquad (5)$$

where $\mathbf{x}(t)$ is a trajectory in the design space for the entire time interval, $Error(\mathbf{x}(t))$ is the difference between the function value at time t and the optimal function value $f(\mathbf{x_f})$: $Error(\mathbf{x}(t)) = f(\mathbf{x_f}) - f(\mathbf{x}(t))$, and $Cost(\mathbf{x}(t))$ represents the invested resources for changing $\mathbf{x}$ from $\mathbf{x_0}$ to $\mathbf{x_f}$. The constraints of the original DOP (Eq. (4) and (5)) must be met at all times.

Note that in spite of the fact that the original DOP is a single objective problem, the new proposed OAP is inherently a MOP. In order to maintain high function values along the trajectory, one might need to invest more energy to the adaptation process. For that reason, the two objectives of the OAP may be conflicting, and the solution of the OAP might be a set of optimal trajectories known as the *Pareto optimal set* (POS).

2.2   Assessment of the Objective Functions

Each solution for the proposed problem is a trajectory, and its resulting performances in the objective space $(Error, Cost)$, as defined previously, are trajectories as well. Some research exists on the optimization of trajectories (e.g. [6]), but the common approach is to represent each trajectory by an auxiliary function (e.g. [7],[8]). In this paper we adopt this approach, and assess the objective functions by using a common concept from optimal control theory:

- The error function $Error\,(\mathbf{x}(t))$ is represented by its *integral absolute* (IAE), which is a well known measure of optimality in control theory:

$$IAE = \int_{t_0}^{t_f} |f(\mathbf{x_f}) - f\,(\mathbf{x}(t))|\ \mathrm{dt}$$

  Recalling the trajectories in Figure 3, the IAE of each trajectory is equal to the area in between the function value and the optimal value marked with the dashed line. It is clear that the IAE of the trajectory in Figure 3(b) is smaller than the IAE of the trajectory in Figure 3(a).
- The cost function $Cost\,(\mathbf{x}(t))$ is represented by the overall cost of the control forces required to follow the trajectory $\mathbf{x}(t)$. It is assumed here that the design variables are controlled by a vector of control forces $\mathbf{u} = [u_1, \ldots, u_p]^T$, where each design variable is controlled by one or more control variables. Every control force has its cost, and the total cost C is calculated as follows:

$$C = \int_{t_0}^{t_f} |\mathbf{w}^T \mathbf{u}(t)|\ \mathrm{dt}$$

  where $\mathbf{w}$ is a cost vector with w_i represents the cost of u_i. Each control force variable u_i has its related saturation values, which define the domain of the control forces.

The OAP may now be reformulated by changing the objectives at Eq. 1 in Definition 2, and by considering the saturation values as additional constraints. Because the objective functions are evaluated based on utility functions, the reformulated problem is presented here as:

Definition 3. *Utility based Optimal Adaptation Problem (UOAP):*

$$\min_{\mathbf{x}(t) \in Q} \{IAE, C\}\ , \qquad\qquad t \in [t_0, t_f] \qquad (6)$$

$$s.t.\quad \mathbf{x}(t_0) = \mathbf{x_0}\ ,\quad \mathbf{x}(t_f) = \mathbf{x_f} \qquad\qquad\qquad (7)$$

$$\left.\frac{\mathrm{dx}}{\mathrm{dt}}\right|_{t_0} = \left.\frac{\mathrm{dx}}{\mathrm{dt}}\right|_{t_f} = 0 \qquad\qquad\qquad (8)$$

$$\mathbf{u}_{min} \le \mathbf{u}(t) \le \mathbf{u}_{max} \qquad\qquad\qquad\qquad (9)$$

$$g_i\,(\mathbf{x}(t)) \ge 0\ , \qquad\qquad (i = 1, \ldots, I) \qquad (10)$$

$$h_j\,(\mathbf{x}(t)) = 0\ , \qquad\qquad (j = 1, \ldots, J) \qquad (11)$$

where $\mathbf{x}(t) = f(\mathbf{u}(t))$.

As is common in control, IAE and C are in conflict with one another. Therefore, the optimal solution for the UOAP, as is the case with the OAP, is expected to be a set rather than a single solution. Note that other reformulations for the OAP are possible.

3 Solving the Problem with an Evolutionary Algorithm

In this section, an evolutionary algorithm (EA) is presented for solving the UOAP, defined in Definition 3.

3.1 The Genotype

For a better visualisation of the problem, the value of the design variables are considered here as their positions. The first and second derivatives in time of the variables are considered as their velocity $\mathbf{v}(t)$ and acceleration $\mathbf{a}(t)$, respectively. Although the solution of the UOAP is the trajectory $\mathbf{x}(t)$, it is coded by $\mathbf{a}(t)$.

The time interval is defined at K time instants- $\mathbf{t} = [t_1, \ldots, t_k, \ldots, t_K]$, where $t_1 = t_0$ and $t_K = t_f$. Each solution is defined by a matrix A of size $M \times K$ of real coded variables. A_{mk} represents the acceleration of the mth design variable at the kth time sample.

Considering the constraints in Eq. (7) and (8), the speed $v_m(t)$ and position $x_m(t)$ can be derived as follows:

$$v_m(t) = \int_{t_0}^{t} a_m(t) \, dt \tag{12}$$

$$x_m(t) = x_{0m} + \int_{t_0}^{t} v_m(t) \, dt \tag{13}$$

Of course, the integrals are evaluated by a discrete approximation method such as the trapezoidal rule, for example.

3.2 Constraint Satisfaction: Repair Method

In order to satisfy the constraints regarding t_f in Eq. (7) and (8), two modifications of the acceleration trajectory are made. The first results in an intermediate acceleration trajectory $\mathbf{a}(t)^{**}$ which satisfies Eq. (8), and the second results in the repaired acceleration trajectory $\mathbf{a}(t)$ which satisfies Eq. (7) as well.

Let a pre-repair acceleration trajectory be $\mathbf{a}^*(t)$. It is clear from Eq. (12) that in order to force the final speed $\mathbf{v}_m(t_f) = 0$, the mean acceleration has to be zero. This can be realized by subtracting from $\mathbf{a}^*(t)$ its mean value $\bar{\mathbf{a}}^*$. The resulting intermediate acceleration trajectory is $\mathbf{a}^{**}(t) = \mathbf{a}^*(t) - \bar{\mathbf{a}}^*$. The intermediate velocity trajectory $\mathbf{v}^{**}(t)$, using Eq. (12), satisfies the constraint of Eq. (8). Then the intermediate final position of the trajectory $\mathbf{x}^{**}(t_f)$ according to Eq. (13) is:

$$\mathbf{x}^{**}(t_f) = \mathbf{x}_0 + \int_{t_0}^{t_f} \mathbf{v}^{**}(t)\, dt$$

At the second stage, in order to satisfy Eq. (7), $\mathbf{a}^{**}(t)$ is scaled by a scaling vector $\mathbf{l}$:

$$\mathbf{l} = [l_1, \ldots, l_m, \ldots, l_M]^T, \qquad\qquad l_m = \frac{x_{fm} - x_{0m}}{x_m^{**}(t_f) - x_{0m}}$$

The repaired final acceleration variable that satisfies both constraints is $\mathbf{a}(t) = \mathbf{a}^{**}(t) \cdot \mathbf{l}$.

Although the suggested gene manipulation may result in violation of the constraint in Eq. (9), it speeds up the evolutionary process since it eliminates the two equality constraints in Eq. (7) and (8).

3.3   The Evolutionary Algorithm

All of the problems of Section 4 are solved using NSGA-II with constraint domination [9]. The genetic operators used are the simulated binary crossover (SBX) operator and polynomial mutation [10], with distribution indexes of $\eta_c = 15$ and $\eta_m = 20$ respectively. A cross-over probability of $p_c = 1$ and a mutation probability of $p_m = 1/MK$ are used. The stopping criterion is a maximal number of generations. A schema of the EA is presented in Algorithm 1.

Algorithm 1. The evolutionary algorithm for solving the UOAP

1: $R_1^* \leftarrow$ generate a random set of solutions of size $2N$
2: $R_1 \leftarrow$ modify R_1^* according to the procedure in Section 3.2
3: $g \leftarrow 1$
4: **while** $g \leq$ number of generations **do**
5:     evaluate R_g
6:     $P_{g+1} \leftarrow$ select N solutions from R_g
7:     $Q_{g+1}^* \leftarrow$ evolve from P_{g+1} (cross-over and mutation)
8:     $Q_{g+1} \leftarrow$ modify Q_{g+1}^* according to the procedure in Section 3.2
9:     $R_{g+1} \leftarrow P_{g+1} \cup Q_{g+1}$
10:     $g \leftarrow g + 1$
11: **end while**

---

4   Test Cases

4.1   Academic Example

Consider the following unconstrained DOP:

$$\max_{\mathbf{x}} f(\mathbf{x}, t) = \frac{\sin(x_1 + x_2 + t/20)}{x_2^2 + 1} - \frac{\cos(x_1 - x_2 + (t/10000)^2 + 4)}{x_2^2 + 1} + \frac{x_1^2}{80}$$

$$x_i \in [-2\pi, 2\pi] \quad, \quad i = \{1, 2\}.$$

When $t = 98$, the optimum "jumps" from $\mathbf{x} = \mathbf{x_0} = [3.24, -0.16]^T$ to $\mathbf{x} = \mathbf{x_f} = [-3.24, -0.12]^T$. The contour of the function at $t = 98$, and the old and new optima were previously depicted in Figure 2. The UOAP for the adaptation in that time instant is solved by the EA described in Algorithm 1. The adaptation's time interval is set to 0.1, and it is defined for $K = 30$ time instants. The population consists of 250 members, and it is evolved over 300 generations.

For this example, the design variables are considered as simple mechanical components that react to the control force according to Newton's 2nd law: $u_m(t) = i_m a_m(t)$, where i_m is a constant number representing the inertia of the mth design variable. Here $i_1 = 20$ and $i_2 = 10$.

The set of non-dominated solutions in the final population is shown in Figure 4. In Figure 4(a), the trajectories of all the non-dominated solutions are illustrated with gray lines. Three different trajectories from this set are marked with diamonds, triangles and circles. The trajectory marked with diamonds does not require investment of much control force, since it is aimed at the new optimum with minimal changes possible. As a result, its cost is the lowest from all solutions. Since it runs through a region with very low function values, its ISE is very high. The trajectory marked with circles passes along the highest function values on the way to the new optimum. As a result, it has a low ISE value. In order to pass through these local optima, a significant force is required to be applied. Therefore, this solution has a high C value. The trajectory marked with diamonds is a compromise between the two objectives. Its path is shorter than the one of the circles, but it passes through only one local optimum. These insights can be depicted from the Pareto front shown in Figure 4(b). All of the non-dominated solutions are marked with gray dots, and the above three solutions are marked with black markers.

The control forces, positions and function values over time of these three solutions are shown in Figure 5. As mentioned in Section 2.2, the area under the control force function indicates the cost of the adaptation, and the area between the function value in time and the optimal function value indicates the ISE. The trade-off between optimality over time and the cost of the adaptation can be seen here. A larger area of the control force is associated with a smaller area of the error, and vice versa.

In order to evaluate the consistency of the algorithm in finding similar approximated sets, the UOAP was simulated for 100 times using the above genetic settings. For three of the 100 simulations the algorithm failed to find the trajectories with the lowest ISE values, such as the triangle associated solution in Figure 4. The lowest ISE value in these cases was 0.04.

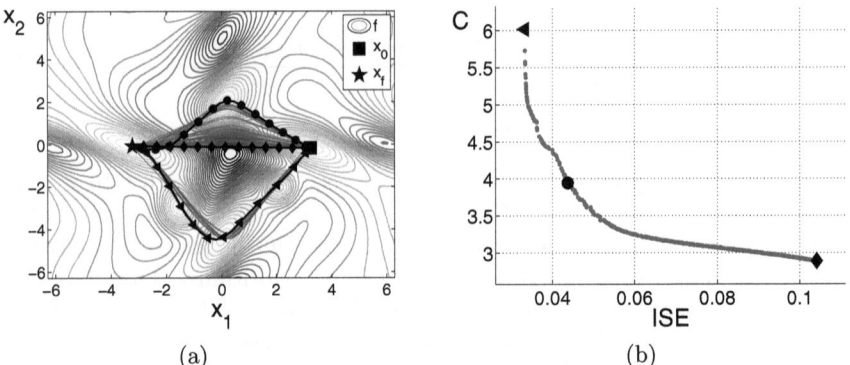

(a)   (b)

Fig. 4. The final approximated set of the UOAP. Three different trajectories are marked in Figure 4(a), and their associated objective values are marked in Figure 4(b).

4.2   Real World Example

To demonstrate the scope of the proposed optimization approach, the following engineering problem is introduced. A planar robot with three rotating links of equal lengths has to carry a load of mass M over a specific route. The angles of the joints are controlled by the user with servo motors that are free to rotate at any angle. The mass of each motor is m. The robot and related parameters are depicted in Figure 6. The desired location of the load, i.e., the end of the robot's third link, is progressing very slowly. Since it has three degrees of freedom, the robot can keep the load at the desired location with an infinite number of configurations. The optimization goal is to follow the path while minimizing the robot's dimensions, i.e., keeping it as folded into itself as possible. This *dynamic optimization problem* can be formulated as follows:

$$\min_{\boldsymbol{\theta}(t) \in \mathbb{R}^3} \phi\left(\boldsymbol{\theta}(t)\right) \tag{14}$$

$$s.t. \quad \mathbf{r_e} = \mathbf{P}(t) \tag{15}$$

where ϕ is the stretch of the robot: $\phi = d_1 + d_2 + d_3$, $\boldsymbol{\theta} = [\theta_1, \theta_2, \theta_3]^T$, $\mathbf{r_e}$ is the location of the manipulator's end, and $\mathbf{P}(t)$ is the desired location of the load at time t. The constraint in Eq. (15) has to be satisfied at all times, other than short periods when the robot has to change a configuration for the reason below.

At some time instants there is a single optimal configuration, while at others, the minimal value of ϕ can be achieved by two different configurations. Figure 7 depicts the solution of the above DOP. The required path of the load

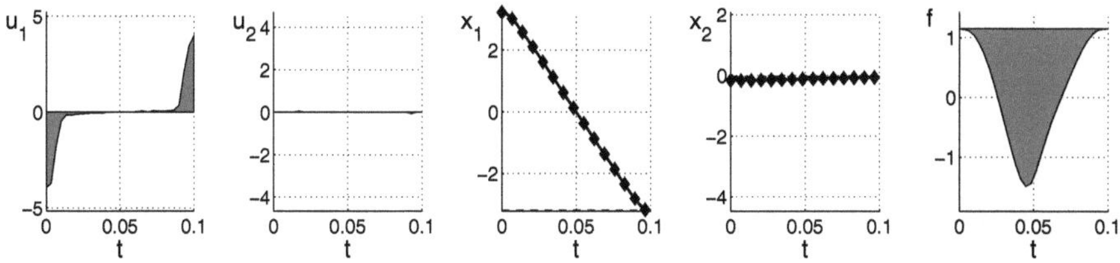

(a) First solution – a high error and low cost.

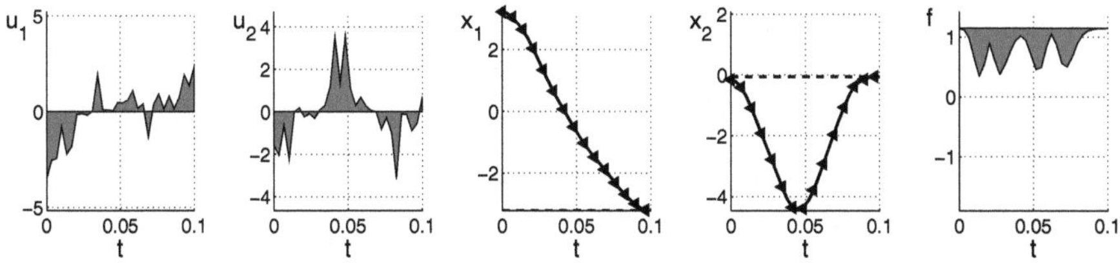

(b) Second solution – a low error and high cost.

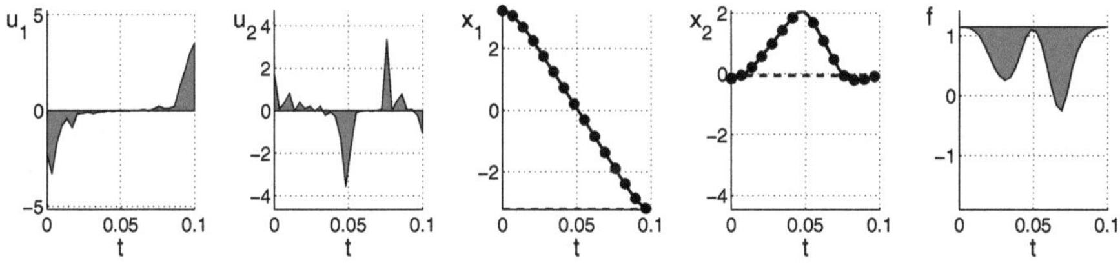

(c) Third solution – a compromise between error and cost.

Fig. 5. The control forces, positions and function values over time for the solutions highlighted in Figure 4

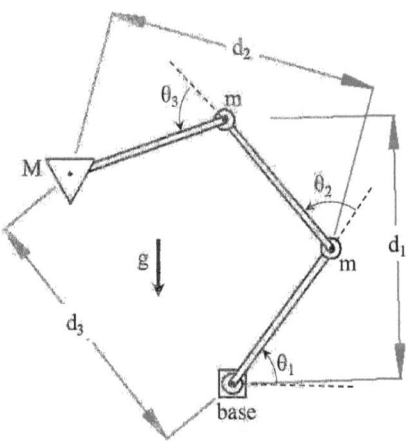

Fig. 6. The robotic manipulator and related parameters

$\mathbf{P}(t)$ is marked with arrows, and the optimal configuration is shown for five time instants. When the robot's end is located far from the base, such as in Figures 7(a) and 7(b), the configuration with minimal dimensions has a "Z" shape. When it is closer to the base, such as in Figures 7(c) and 7(d), the optimal configuration has an "N" shape. For the example given, when $t = 1000s$ the robot has to change its configuration from "Z" to "N". The duration of the change is 4 seconds. The question of how to perform that change is considered as the following *utility optimal adaptation problem*. The UOAP follows the main optimization goal, i.e. minimizing the dimensions of the robot, and it also seeks to minimize the power applied to the motors:

$$\min_{\boldsymbol{\theta}(t) \in \mathbb{R}^3} \{E, C\} , \qquad\qquad t \in [t_0, t_f]$$

$$s.t. \quad \boldsymbol{\theta}(t_0) = \boldsymbol{\theta_0} \quad , \quad \boldsymbol{\theta}(t_f) = \boldsymbol{\theta_f}$$

$$\left.\frac{d\boldsymbol{\theta}}{dt}\right|_{t_0} = \left.\frac{d\boldsymbol{\theta}}{dt}\right|_{t_f} = 0$$

$$-\boldsymbol{\tau}_{sat} \le \boldsymbol{\tau}(t) \le \boldsymbol{\tau}_{sat}$$

where $\boldsymbol{\tau}(t) = [\tau_1(t), \tau_2(t), \tau_3(t)]^T$ is a vector with the required torques in the joints to follow the trajectory $\boldsymbol{\theta}(t)$, $\boldsymbol{\tau}_{sat}$ are the saturation values of the motors, E is the integral of $\phi(t)$, and C the total applied torque over time:

$$E = \int_{t_0}^{t_f} \phi(t)\, dt \quad , \quad C = \int_{t_0}^{t_f} \left(\Sigma_{i=1}^3 |\tau_i(t)| \right) dt.$$

$\boldsymbol{\theta_0}$ and $\boldsymbol{\theta_f}$ are the configurations at Figure 7(b) and 7(c), respectively. $\mathbf{u}(t)$ is calculated by the iterative Newton-Euler dynamic formulation (for more information, see [11]).

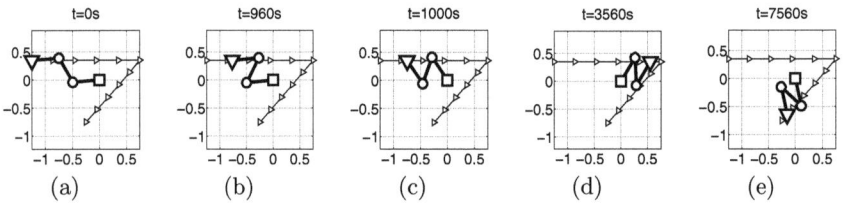

Fig. 7. The configurations with minimal dimensions for five positions of the load. The manipulator's base is marked with a square, and the load is marked with a triangle. The initial configuration of the OAP is the one at Figure 7(b), and the the final configuration is the one at Figure 7(c).

The solutions of the above UOAP are the trajectories of the links and the torques applied by the motors. A controller can be synthesized to follow these trajectories. Note that the above UOAP is different from the common approach arising from using optimal control theory for designing optimal controllers. A controller designed according to optimal control theory considers the final state θ_f as an optimization goal, and tries to minimize the control force and the error from θ_f. In contrast, here, the optimization goal is the original goal of the DOP, i.e., minimizing the dimensions of the robot.

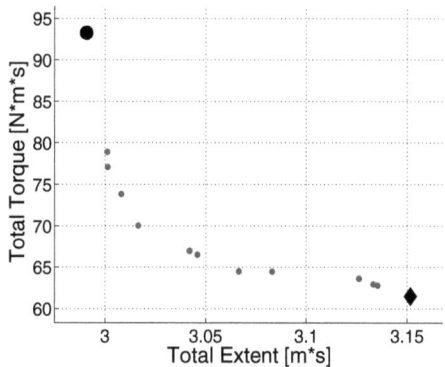

Fig. 8. The approximated set of Pareto optimal solutions for the UOAP, found by the EA.

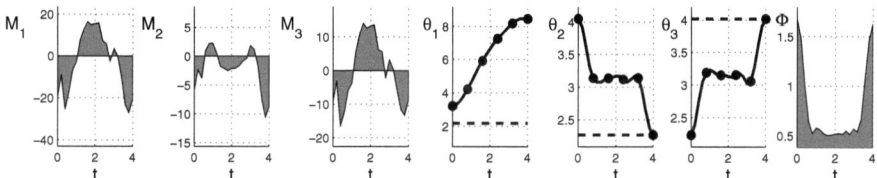

(a) One alternative – small dimensions and high cost. Note that the final θ_1 is 2π larger than the desired final value. That means the manipulator performs a full turn around its base.

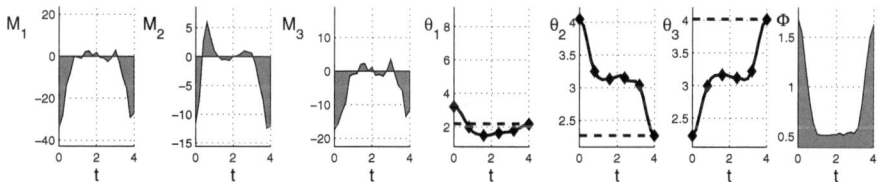

(b) Second alternative – larger dimensions and smaller cost.

Fig. 9. Trajectories of the joint torques and angles and the extent of the robot in time.

The UOAP was solved by the EA described in Algorithm 1, using the same configuration parameters as in the academic example from the previous section. The final approximated set is depicted in Figure 8. The two extreme solutions, marked with a larger circle and diamond, are shown in Figures 9 and 10. The trajectories of the torques, the joint angles and the extent are shown in Figure 9, and the dynamic behavior of the robot is shown in Figure 10. Note that in order to fold more quickly, under the limits of the saturation torque values, the solution marked with a circle rotates around its base. The solution marked with a diamond balances vertically until it needs to stretch again in the other configuration.

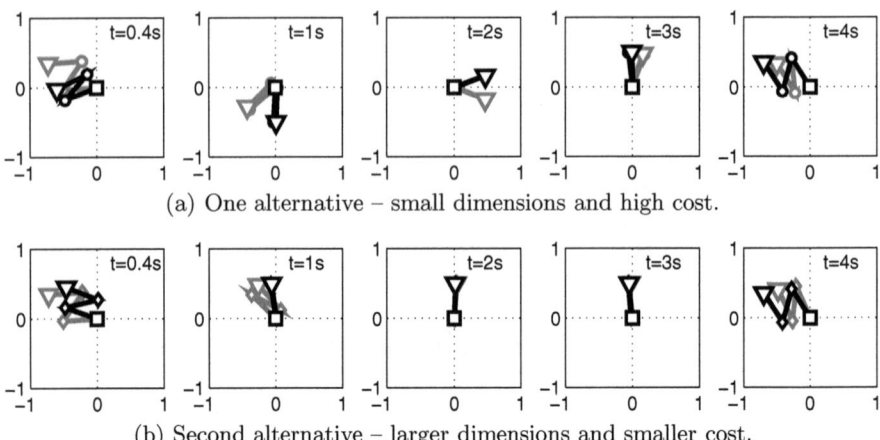

Fig. 10. Positions of two Pareto optimal solutions at different time samples. The lighter configuration is 0.4s prior to the black one.

5   Conclusions and Future Work

In this paper, a new optimization problem was introduced: *Optimization of Adaptation Problem*. The problem deals with situations when a change in design is required in order to remain optimal in a changing environment. The required change is found by solving a DOP prior to the formulation of the OAP. The issues of optimality during the adaptation process and the cost of the adaptation were discussed, and the OAP was defined as a MOP for minimizing the cost and maximizing the function value during the adaptation. A relaxed version of the OAP was defined as a *Utility based Optimization of Adaptation Problem*, and an EA was proposed in order to solve it. The EA was tested on two examples: a theoretic mathematical function and a problem of robotic arm control. For both examples, the EA was able to find a set of trade off solutions that enable the decision maker to choose whether to adapt in a minimum cost manner or to invest more resources in order to maintain high function values along the adaptation.

As future work, the new approach should be integrated and tested on more real life applications. This paper dealt with a single objective DOPs. The issue of optimal adaptation for multiobjective DOPs should be studied as well. The method should be also extended to deal with cases where the future function values and costs are subject to uncertainties.

Acknowledgements. This research was supported by a Marie Curie International Research Staff Exchange Scheme Fellowship within the 7th European Community Framework Programme.

References

1. Rose, M., Lauder, G.: Adaptation. Academic Press, San Diego (1996)
2. Ferguson, G., Murphy, J., Ramanamanjato, J., Raselimanana, A., et al.: The panther chameleon: color variation, natural history, conservation, and captive management. Krieger Publishing Company (2004)
3. Ruseckaite, R., Lamb, T., Pianta, M., Cameron, A.: Human scotopic dark adaptation: Comparison of recoveries of psychophysical threshold and ERG b-wave sensitivity. Journal of Vision 11(8) (2011)
4. Jacobs, R., Lundby, C., Robach, P., Gassmann, M.: Red blood cell volume and the capacity for exercise at moderate to high altitude. Sports Medicine 42(8), 643–663 (2012)
5. Branke, J.: Evolutionary approaches to dynamic optimization problems-updated survey. In: GECCO Workshop on Evolutionary Algorithms for Dynamic Optimization Problems, pp. 27–30 (2001)
6. Avigad, G., Eisenstadt, E., Goldvard, A., Salomon, S.: Transient responses optimization by means of set-based multi-objective evolution. Engineering Optimization 44(4), 407–426 (2011)
7. Matzen, R., Jensen, J.S., Sigmund, O.: Topology optimization for transient response of photonic crystal structures. Optical Society of America. Journal B: Optical Physics 27(10), 2040–2050 (2010)
8. Toscano, R.: A simple robust PI/PID controller design via numerical optimization approach. Journal of Process Control 15(1), 81–88 (2005)
9. Deb, K., Pratap, A., Agarwal, S., Meyarivan, T.: A fast and elitist multiobjective genetic algorithm: NSGA-II. IEEE Transactions on Evolutionary Computation 6(2), 182–197 (2002)
10. Deb, K., Agrawal, R.: Simulated binary crossover for continuous search space. Complex Systems 9(2), 115–148 (1995)
11. Craig, J.: Introduction to Robotics. Pearson Education, Inc. (2005)

Multi-objective AI Planning:
Evaluating DAE$_{\text{YAHSP}}$ on a Tunable Benchmark*

M.R. Khouadjia[1], M. Schoenauer[1], V. Vidal[2], J. Dréo[3], and P. Savéant[3]

[1] TAO Project-team, INRIA Saclay & LRI, Université Paris-Sud, Orsay, France
{mostepha-redouane.khouadjia,marc.schoenauer}@inria.fr
[2] ONERA-DCSD, Toulouse, France
Vincent.Vidal@onera.fr
[3] THALES Research & Technology, Palaiseau, France
{johann.dreo,pierre.saveant}@thalesgroup.com

Abstract. All standard Artifical Intelligence (AI) planners to-date can only handle a single objective, and the only way for them to take into account multiple objectives is by aggregation of the objectives. Furthermore, and in deep contrast with the single objective case, there exists no benchmark problems on which to test the algorithms for multi-objective planning.

Divide-and-Evolve (DAE) is an evolutionary planner that won the (single-objective) deterministic temporal satisficing track in the last International Planning Competition. Even though it uses intensively the classical (and hence single-objective) planner YAHSP (*Yet Another Heuristic Search Planner*), it is possible to turn DAE$_{\text{YAHSP}}$ into a multi-objective evolutionary planner.

A tunable benchmark suite for multi-objective planning is first proposed, and the performances of several variants of multi-objective DAE$_{\text{YAHSP}}$ are compared on different instances of this benchmark, hopefully paving the road to further multi-objective competitions in AI planning.

1 Introduction

An AI Planning problem (see e.g. [1]) is defined by a set of predicates, a set of actions, an initial state and a goal state. A state is a set of non-exclusive instantiated predicates, or (Boolean) atoms. An action is defined by a set of *pre-conditions* and a set of *effects*: the action can be executed only if all pre-conditions are true in the current state, and after an action has been executed, the effects of the action modify the state: the system enters a new state. A plan in AI Planning is a sequence of actions that transforms the initial state into the goal state. The goal of AI Planning is to find a plan that minimizes some quantity related to the actions: number of actions, or sum of action costs in case actions have different costs, or makespan in the case of temporal planning, when actions have a duration and can eventually be executed in parallel. All these problems are P-SPACE.

---

* This work was partially funded by DESCARWIN ANR project (ANR-09-COSI-002).

R.C. Purshouse et al. (Eds.): EMO 2013, LNCS 7811, pp. 36–50, 2013.
© Springer-Verlag Berlin Heidelberg 2013

A simple planning problem in the domain of logistics is given in Figure 1: the problem involves cities, passengers, and planes. Passengers can be transported from one city to another, following the links on the figure. One plane can only carry one passenger at a time from one city to another, and the flight duration (number on the link) is the same whether or not the plane carries a passenger (this defines the *domain* of the problem). In the simplest non-trivial *instance* of such domain, there are 3 passengers and 2 planes. In the initial state, all passengers and planes are in `city` 0, and in the goal state, all passengers must be in `city` 4. The not-so-obvious optimal solution has a total makespan of 8 and is left as a teaser for the reader.

AI Planning is a very active field of research, as witnessed by the success of the ICAPS conferences (`http://icaps-conferences.org`), and its Intenational Planning Comptetition (IPC), where the best planners in the world compete on a set of problems. This competition has lead the researchers to design a common language to describe planning problems, PDDL (Planning Domain Definition Language). Two main categories of planners can be distinguished: *exact planners* are guaranteed to find the optimal solution . . . if given enough time; *satisficing planners* give the best possible solution, but with no optimality guarantee. A complete description of the state-of-the-art planners is far beyond the scope of this paper.

However, to the best of our knowledge, all existing planners are single objective (i.e. optimize one criterion, the number of actions, the cost, or makespan, depending on the type of problem), whereas most real-world problems are in fact multi-objective and involve several contradictory objectives that need to be optimized simultaneously. For instance, in logistics, the decision maker must generally find a trade-off between duration and cost (or/and risk).

An obvious solution is to aggregate the different objectives into a single objective, generally a fixed linear combination of all objectives. Early work in that area used some twist in PDDL 2.0 [2,3,4]. PDDL 3.0, on the other hand, explicitly offered hooks for several objectives x, and a new track of IPC was dedicated to aggregated multiple objectives: the "net-benefit" track took place in 2006 [5] and 2008 [6], . . . but was canceled in 2011 because of the small number of entries. In any case, no truly multi-objective approach to multi-objective planning has been proposed since the very preliminary proof-of-concept in the first *Divide-and-Evolve* paper [7].

One goal of this paper is to build on this preliminary work, and to discuss various issues related to the challenge of solving multi-objective problems with an evolutionary algorithm that is heavily based on a single-objective planner (YAHSP [8]) – and in particular to compare different state-of-the-art multi-objective evolutionary schemes when used within DAE_{YAHSP}. However, experimental comparison requires benchmark problems. Whereas the IPC have validated a large set of benchmark domains, with several instances of increasing complexity in each domain, nothing yet exists for multi-objective planning. The other goal of this paper is to propose a tunable set of benchmark instances, based on a simplified model of the IPC logistics domain ZENO illustrated in Fig. 1. One advantage of this multi-objective benchmark is that the exact Pareto Front is known, at least for its simplest instances.

The paper is organized as follows: Section 2 rapidly introduces *Divide-and-Evolve*, more precisely the representation and variation operators that have been used in the single-objective version of DAE_{YAHSP} that won the temporal deterministic satisficing track at the last IPC in 2011. Section 4 details the proposed benchmark, called MULTIZENO, and gives hints about how to generate instances of different complexities within this framework. Section 3.2 rapidly introduces the 4 variants of multi-objective schemes that will be experimentally compared on some of the simplest instances of the MULTIZENO benchmark and results of different series of experiments are discussed in Section 6. Section 7 concludes the paper, giving hints about further research directions.

2   Divide-and-Evolve

Let $\mathcal{P}_D(I, G)$ denote the planning problem defined on domain D (the predicates, the objects, and the actions), with initial state I and goal state G. In STRIPS representation model [9], a state is a list of Boolean atoms defined using the predicates of the domain, instantiated with the domain objects.

In order to solve $\mathcal{P}_D(I, G)$, the basic idea of DAE_X is to find a sequence of states $S_1, \ldots, S_n$, and to use some embedded planner X to solve the series of planning problems $\mathcal{P}_D(S_k, S_{k+1})$, for $k \in [0, n]$ (with the convention that $S_0 = I$ and $S_{n+1} = G$). The generation and optimization of the sequence of states $(S_i)_{i \in [1,n]}$ is driven by an evolutionary algorithm. After each of the sub-problems $\mathcal{P}_D(S_k, S_{k+1})$ has been solved by the embedded planner, the concatenation of the corresponding plans (possibly compressed to take into account possible parallelism in the case of temporal planning) is a solution of the initial problem. In case one sub-problem cannot be solved by the embedded solver, the individual is said *unfeasible* and its fitness is highly penalized in order to ensure that feasible individuals always have a better fitness than unfeasible ones, and are selected only when there are not enough feasible individual. A thorough description of DAE_X can be found in [10]. The following rest of this section will focus on the evolutionary parts of DAE_X.

2.1   Representation and Initialization

An individual in DAE_X is hence a variable-length list of states of the given domain. However, the size of the space of lists of complete states rapidly becomes untractable when the number of objects increases. Moreover, goals of planning problems need only to be defined as partial states, involving a subset of the objects, and the aim is to find a state such that all atoms of the goal state are true. An individual in DAE_X is thus a variable-length list of partial states, and a partial state is a variable-length list of atoms.

Previous work with DAE_X on different domains of planning problems from the IPC benchmark series have demonstrated the need for a very careful choice of the atoms that are used to build the partial states [11]. The method that is used today to build the partial states is based on a heuristic estimation, for each atom, of the earliest time from which it can become true [12]. These earliest start times are then used in order to restrict the candidate atoms for each partial state: the number of states is uniformly drawn between 1 and the number of estimated start times; For every chosen time, the number of atoms per state is uniformly chosen between 1 and the number of atoms of the corresponding restriction. Atoms are then added one by one: an atom is uniformly drawn in the allowed set of atoms (based on earliest possible start time), and added to the individual if it is not mutually exclusive (in short, *mutex*) with any other atom that is already there. Note that only an approximation of the complete mutex relation between atoms is known from the description of the problem, and the remaining mutexes will simply be gradually eliminated by selection, because they make the resulting individual unfeasible.

To summarize, an individual in DAE_X is represented by a variable-length time-consistent sequence of partial states, and each partial state is a variable-length list of atoms that are not pairwise mutex.

2.2   Variation Operators

Crossover and mutation operators are defined on the DAE_X representation in a straightforward manner - though constrained by the heuristic chronology and the partial mutex relation between atoms.

A simple one-point crossover is used, adapted to variable-length representation: both crossover points are independently chosen, uniformly in both parents. However, only one offspring is kept, the one that respects the approximate chronological constraint on the successive states. The crossover operator is applied with a population-level crossover probability.

Four different mutation operators are included: first, a population-level mutation probability is used; one an individual has been designated for mutation, the choice between the four mutation operators is made according to user-defined relative weights. The four possible mutations operate either at the individual level, by adding (addState) or removing (delState) a state, or at the state level by adding (addAtom) or removing (delAtom) some atoms in a uniformly chose state.

All mutation operators maintain the approximate chronology between the intermediate states (i.e., when adding a state, or an atom in a state), and the local consistency within all states (i.e. avoid pairwise mutexes).

2.3   Hybridization

DAE_X uses an external embedded planner to solve the sequence of sub-problems defined by the ordered list of partial states. Any existing planner can in theory be used. However, there is no need for an optimality guarantee when solving the intermediate problems in order for DAE_X to obtain good quality results [10]. Hence, and because several calls to this embedded planner are necessary for a single fitness evaluation, a sub-optimal but fast planner is used: YAHSP [8] is a lookahead strategy planning system for sub-optimal planning which uses the actions in the relaxed plan to compute reachable states in order to speed up the search process.

For any given k, if the chosen embedded planner succeeds in solving $P_D(S_k, S_{k+1})$, the final complete state is computed by executing the solution plan from S_k, and becomes the initial state of the next problem. If all the sub-problems are solved by the embedded planner, the individual is called *feasible*, and the concatenation of the plans for all sub-problems is a global solution plan for $P_D(S_0 = I, S_{n+1} = G)$. However, this plan can in general be further optimized by rescheduling some of its actions, in a step called compression. The computation of all objective values is done from the compressed plan of the given individual. Finally, because the rationale for DAE_X is that all sub-problems should hopefully be easier than the initial global problem, and for computational performance reason, the search capabilities of the embedded planner YAHSP are limited by setting a maximal number of nodes that it is allowed to expand to solve any of the sub-problems (see again [10] for more details).

3   Multi-objective Divide-and-Evolve

In some sense, the multi-objectivization of DAE_X is straightforward – as it is for most evolutionary algorithms. The "only" parts of the algorithm that require some modification are the selection parts, be it the parental selection, that chooses which individual from the population are allowed to breed, and the environmental selection (aka replacement), that decides which individuals among parents and offspring will survive to the next generation. Several schemes have been proposed in the EMOA literature (see e.g. Section 3.2), and the end of this Section will briefly introduce the ones that have been used in this work. However, a prerequisite is that all objectives are evaluated for all potential solutions, and the challenge here is that the embedded planner YAHSP performs its search based on only one objective.

3.1   Multi-objectivization Strategies

Even though YAHSP (like all known planners to-date) only solves planning problems based on one objective. However, it is possible since PDDL 3.0 to add some other quantities (aka Soft Constraints or Preferences [13]) that are simply computed throughout the execution of the final plan, without interfering with the search.

The very first proof-of-concept of multi-objective DAE_X [7], though using an exact planner in lieu of the satisficing planner YAHSP, implemented the simplest idea with respect to the second objective: ignore it (though computing its value for all individuals) at the level of the embedded planner, and let the evolutionary multi-objective take care of it. However, though YAHSP can only handle one objective at a time, it can handle either one in turn, provided they are both defined in the PDDL domain definition file. Hence a whole bunch of smarter strategies become possible, depending on which objective YAHSP is asked to optimize every time it runs on a sub-problem. Beyond the fixed strategies, in which YAHSP always uses the same objective throughout DAE_{YAHSP} runs, a simple dynamic randomized strategy has been used in this work: Once the planner is called for a given individual, the choice of which strategy to apply is made according to roulette-wheel selection based on user-defined relative weights; In the end, it will return the values of both objectives. It is hoped that the evolutionary algorithm will find a sequential partitioning of the problem that will nevertheless allow the global minimization of both objectives. Section 6.2 will experimentally compare the fixed strategies and the dynamic randomized strategy where the objective that YAHSP uses is chosen with equal probability among both objectives.

Other possible strategies include adaptive strategies, where each individual, or even each intermediate state in every individual, would carry a strategy parameter telling YAHSP which strategy to use – and this strategy parameter would be subject to mutation, too. This is left for further work.

3.2   Evolutionary Multi-objective Schemes

Several Multi-Objective EAs (MOEAs) have been proposed in the recent years, and this work is concerned with comparing some of the most popular ones when used within the multi-objective version of DAE_{YAHSP}. More precisely, the following selection/reproduction schemescan be applied to any representation, and will be experimented with here: NSGA-II [14], SPEA2 [15], and IBEA [16]. They will now be quickly introduced in turn.

The **Non-dominated Sorting Genetic Algorithm** (NSGA-II) has been proposed by Deb et al. [14]. At each generation, the solutions contained in the current population are ranked into successive Pareto fronts in the objective space. Individuals mapping to vectors from the first front all belong to the best efficient set; individuals mapping to vectors from the second front all belong to the second best efficient set; and so on. Two values are then assigned for every solution of the population. The first one corresponds to the rank of the Pareto front the corresponding solution belongs to, and represents the quality of the solution in terms of convergence. The second one, the crowding distance, consists in estimating the density of solutions surrounding a particular point in the objective space, and represents the quality of the solution in terms of diversity. A solution is said to be better than another solution if it has a better rank value, or in case of equality, if it has a larger crowding distance.

The **Strength Pareto Evolutionary Algorithm** (SPEA) [17], introduces an improved fitness assignment strategy. It intrinsically handles an internal fixed-size archive that is used during the selection step to create offspring solutions. At a given iteration of the algorithm, each population and archive member x is assigned a strength value $S(x)$ representing the number of solutions it dominates. Then, the fitness value $F(x)$ of solution x is calculated by summing the

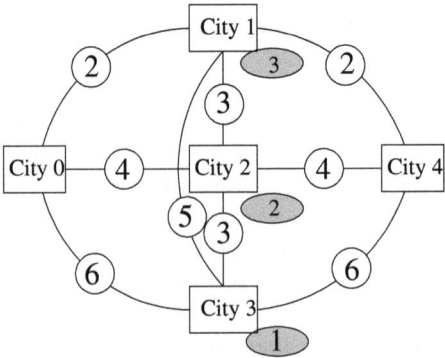

Fig. 1. A schematic view of MULTIZENO, a simple benchmark transportation problem: Durations of available flights are attached to the corresponding edges, costs/risks are attached to landing in the central cities (in grey circles).

strength values of all individuals that x currently dominates. Additionally, a diversity preservation strategy is used, based on a nearest neighbor technique. The selection step consists of a binary tournament with replacement applied on the internal archive only. Last, given that the SPEA2 archive has a fixed size storage capacity, a pruning mechanism based on fitness and diversity information is used when the non-dominated set is too large.

The **Indicator-Based Evolutionary Algorithm** (IBEA) [16] introduces a total order between solutions by means of a binary quality indicator. The fitness assignment scheme of this evolutionary algorithm is based on a pairwise comparison of solutions contained in the current population with respect to a binary quality indicator I. Each individual x is assigned a fitness value $F(x)$ measuring the "loss in quality" that would result from removing x from the current population. Different indicators can be used. The most two popular, that will be used in this work, are the additive ϵ-indicator ($I_{\epsilon+}$) and the hypervolume difference indicator (I_{H^-}) as defined in [16]. Each indicator $I(x, x')$ gives the minimum value by which a solution $x \in X$ can be translated in the objective space to weakly dominate another solution $x' \in X$. An archive stores solutions mapping to potentially non-dominated points in order to prevent their loss during the stochastic search process.

4   A Benchmark Suite for Multi-objective Temporal Planning

This section details the proposed benchmark test suite for multi-objective temporal planning, based on the simple domain that is schematically described in Figure 1. The reader will have by now solved the little puzzle set in the Introduction, and found the solution with makespan 8 (flying 2 passengers to city 1, one plane continues with its passenger to city 4 while the other plane flies back empty to city 0, the plane in city city 4 returns empty to city 1 while the other plane brings the last passenger there, and the goal is reached after both planes bring both remaining passengers to city 4). The rationale for this solution is that no plane ever stays idle.

In order to turn this problem into a not-too-unrealistic logistics multi-objective problem, some costs or some risks are added to all 3 central cities (1 to 3). This leads to two types of problems: In the MULTIZENO$_{Cost}$, the second objective is an additive objective: each plane has to pay the corresponding tax every time it lands in that city; In the MULTIZENO$_{Risk}$, the second objective is similar to a risk, and the maximal value encountered during the complete execution of a plan is to be minimized.

In both cases, there are 3 obvious points that belong to the Pareto Front: the solution with minimal makespan described above, and the similar solutions that use respectively city 2 and city 3 in lieu of city 1. The values of the makespans are respectively 8, 16 and 24, and the values of the costs are, for each solution, 4 times the value of the single landing tax, and exactly the value of the involved risk. For the risk case, there is no other point on the Pareto Front, as a single landing on a high-risk city sets the risk of the whole plan to a high risk. For the cost model however, there are other points on the Pareto Front, as different cities can be used for the different passengers. For instance, in the case of Figure 1, this leads to a Pareto Front made of 5 points, (8,12), (16,8), and (24,4) (going only through city 1, 2 and 3 respectively), plus (12,10) and (20,6). Only the first 3 are the Pareto Front in the risk case.

4.1 Tuning the Complexity

There are several ways to make this first simple instance more or less complex. A first possibility is to add passengers. In this work, only bunches of 3 passengers have been considered, in order to be able to easily derive some obvious Pareto-optimal solutions, using several times the little trick to avoid leaving any plane idle. For instance, it is easy to derive all the Pareto solutions for 6 and 9 passengers – and in the following, the corresponding instances will be termed MULTIZENO3, MULTIZENO6, and MULTIZENO9 respectively (sub-scripted with the type of second objective – cost or risk).

Of course, the number of planes could also be increased, though the number of passengers needs to remain larger than the number of planes to allow for non-trivial Pareto front. However, departing from the 3 passengers to 2 planes ratio would make the Pareto front not easy to identify any more.

Another possibility is to increase the number of central cities: this creates more points on the Pareto front, using either plans in which a single city is used for all passengers, or plans that use several different cities for different passengers (while nevertheless using the same trick to ensure no plane stays idle). In such configuration too the exact Pareto front remains easy to identify: further work will investigate this line of complexification.

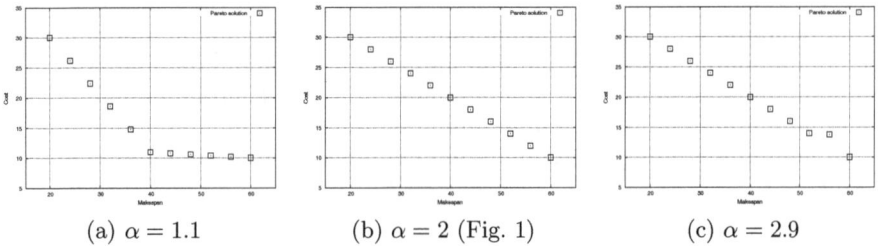

(a) $\alpha = 1.1$ (b) $\alpha = 2$ (Fig. 1)   (c) $\alpha = 2.9$

Fig. 2. The exact Pareto Fronts for the MULTIZENO6 problem for different values of the cost α of `city2` (those of `city1` and `city3` being 3 and 1 respectively)

4.2   Modifying the Shape of the Pareto Front

Another way to change the difficulty of the problem without increasing its complexity is to tune the different values of the flight times and the cost/risk at each city. Such changes does not modify the number of points on the Pareto Front, but does change its shape in the objective space. For instance, simply modifying the cost α of `city2`, the central city in Figure 1, between 1 and 3 (the costs of respectively `city1` and `city3`), the Pareto Front, which is linear for $\alpha = 2$ becomes strictly convex for $\alpha < 2$ and strictly concave for $\alpha > 2$, as can be seen for two extreme cases ($\alpha = 1.1$ and $\alpha = 2.9$) on Figure 2. Further work will address the identification of the correct domain parameters in order to reach a given shape of the Pareto front.

5   Experimental Conditions

Implementation: All proposed multi-objective approaches (see Section 3.2) have been implemented within the PARADISEO-MOEO framework [18]. All experiments were performed on the MULTIZENO3, MULTIZENO6, and MULTIZENO9 instances. The first objective is the makespan, and the second objective either the (additive) cost or the (maximal) risk, as discussed in Section 4. The values of the different flight durations and cost/risks are those given on Figure 1 except otherwise stated.

Parameter Tuning: All user-defined parameters have been tuned using the framework PARAMILS [19]. PARAMILS handles any parameterized algorithm whose parameters can be discretized. Based on Iterated Local Search (ILS), PARAMILS searches through the space of possible parameter configurations, evaluating configurations by running the algorithm to be optimized on a set of benchmark instances, searching for the configuration that yields overall best performance across the benchmark problems. Here, both the parameters of the multi-objective algorithms (including the internal parameters of the variation operators – see [20]) and YAHSP specific parameters (including the relative weights of the possible strategies (see Section 3.1) have been subject to PARAMILS optimization.

For the purpose of this work, parameters were tuned anew for each instance (see [20] for a discussion about the generality of such parameter tuning, that falls beyond the scope of this paper).

Performance Metric: The quality measure used by PARAMILS to optimize DAE$_{\text{YAHSP}}$ is the unary hypervolume I_{H-} [16] of the set of non-dominated points output by the algorithm with respect to the complete true Pareto front (only instances where the true Pareto front is fully known have been experimented with). The lower the better (a value of 0 indicates that the exact Pareto front has been reached).

However, and because the true front is known exactly, and is made of a few scattered points (at most 17 for MULTIZENO9 in this paper), it is also possible to visually monitor when each point of the front is discovered by the algorithm. This allows some deeper comparison between algorithms even when none has found the whole front. Such *attainment plots* will be used in the following, together with more classical plots of hypervolume vs time.

For all experiments, 30 independent runs were performed. Note that all the performance assessment procedures, including the hypervolume calculations, have been achieved using the PISA performance assessment tool suite [21].

Stopping Criterion: Because different fitness evaluations involve different number calls to YAHSP – and because YAHSP runs can have different computational costs too, depending on the difficulty of the sub-problem being solved – the stopping criterion was a fixed amount of CPU time rather than the usual number of fitness evaluation. These absolute limits were set to 300, 600, and 900 seconds respectively for MULTIZENO3, MULTIZENO6, and MULTIZENO9.

6   Experimental Results

6.1   Comparing Multi-objective Schemes

The first series of experiments presented here are concerned with the comparison of the different multi-objective schemes briefly introduced in Section 3.2. Figure 3 displays a summary of experiments of all 4 variants for MULTIZENO instances for both the *Cost* and *Risk* problems.

Some clear conclusions can be drawn from these results, that are confirmed by the statistical analyses presented in Table 1 using Wilcoxon signed rank test with 95% confidence level. First, looking at the minimal values of the hypervolume reached by the different algorithms shows that, as expected, the difficulty of the problems increases with the number of passengers, and for a given complexity, the *Risk* problems are more difficult to solve than the *Cost* ones. Second, from the plots and the statistical tests, it can be seen that NSGA-II is outperformed by all other variants on all problems, SPEA2 by both indicator-based variants on most instances, and $IBEA_{H-}$ is a clear winner over $IBEA_{\varepsilon+}$ except on MULTIZENO6$_{risk}$.

More precisely, Figure 4 show the cumulated final populations of all 30 runs in the objective space together with the true Pareto front for MULTIZENO6-9$_{cost}$ problems: the situation is not as bad as it seemed from Figure 3-(e) for MULTIZENO9$_{cost}$, as most solutions that are returned by $IBEA_{H-}$ are close to the Pareto front (this is even more true on MULTIZENO6$_{cost}$ problem). A dynamic view of the attainment plots is given in Figure 6-(c): two points of the Pareto front are more difficult to reach than the others, namely (48,16) and (56,12).

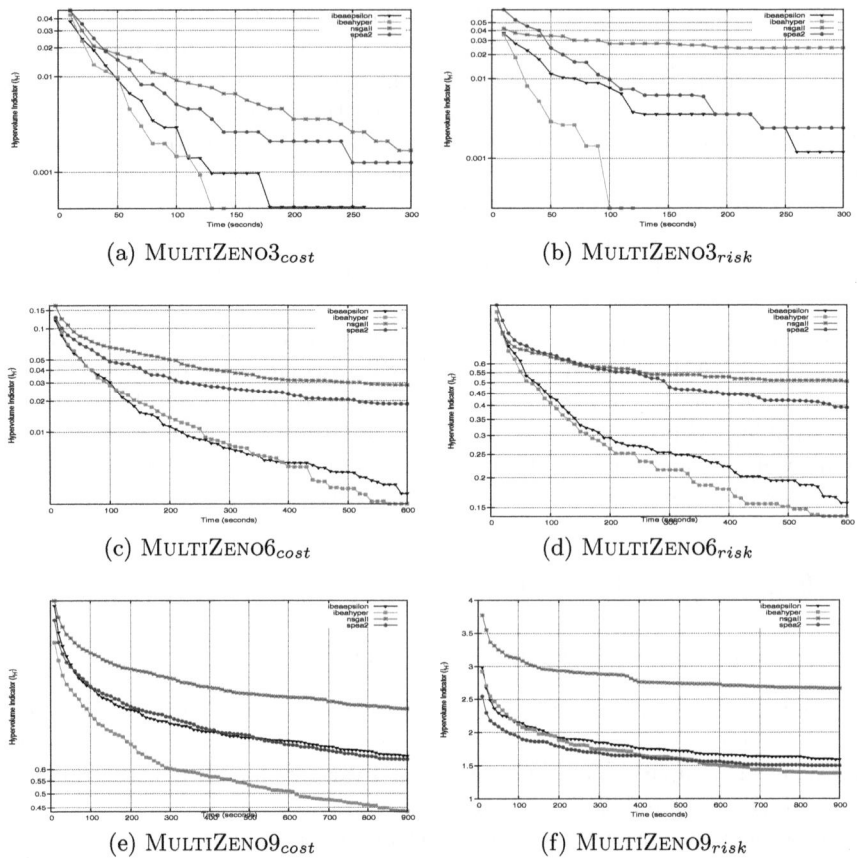

(a) MULTIZENO3$_{cost}$ (b) MULTIZENO3$_{risk}$

(c) MULTIZENO6$_{cost}$ (d) MULTIZENO6$_{risk}$

(e) MULTIZENO9$_{cost}$ (f) MULTIZENO9$_{risk}$

Fig. 3. Evolution of the Hypervolume indicator I_{H-} (averaged over 30 runs) on MULTIZENO instances (see Table 1 for statistical significances)

6.2   Influence of YAHSP Strategy

Next series of experiments aimed at identifying the influence of the chosen strategy for YAHSP (see Section 3.1). Figure 6-(a) (resp. 6-(b)) shows the attainment

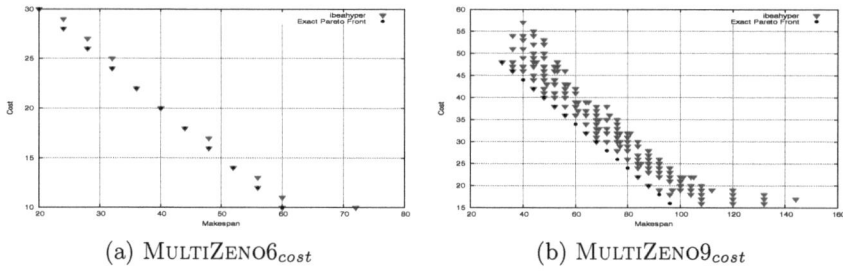

(a) MULTIZENO6$_{cost}$ (b) MULTIZENO9$_{cost}$

Fig. 4. Pareto fronts of IBEA$_{H-}$ on MULTIZENO instances

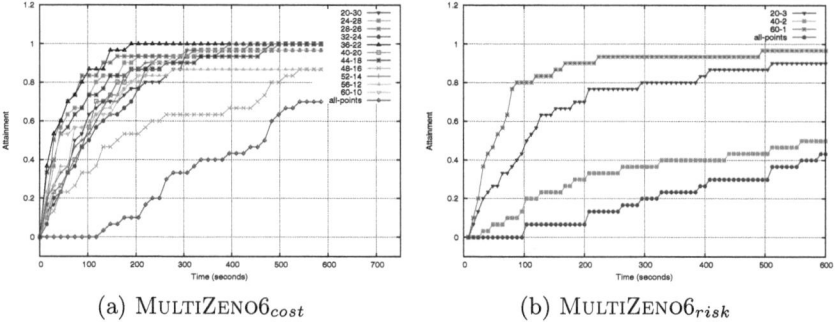

(a) MULTIZENO6$_{cost}$ (b) MULTIZENO6$_{risk}$

Fig. 5. Attainment plots for IBEA$_{H-}$ on MULTIZENO6 instances

plots for the strategy in which YAHSP always optimizes the makespan (resp. the cost) on problem MULTIZENO6$_{cost}$. Both extreme strategies lead to much worse results than the mixed strategy of Figure 5-(a), as no run discovers the whole front (last line, that never leaves the x-axis). Furthermore, and as could be expected, the makespan-only strategy discovers very rapidly the extreme points of the Pareto front that have a small makespan (points (20,30), (24,28) and (28,26)) and hardly discovers the other end of the Pareto front (points with makespan greater than 48), while it is exactly the opposite for the cost-only strategy. This confirms the need for a strategy that incorporates both approaches. best possible choice.

Note that similar conclusion could have been drawn from PARAMILS results on parameter tuning (see Section 5): the choice of YAHSP strategy was one of the parameters tuned by PARAMILS . . . and the tuned values for the weights of both strategies were always more or less equal.

6.3   Shape of the Pareto Front

Figure 7 displays the attainment plots of IBEA$_{H-}$ for both extreme Pareto fronts shown on Figure 2 – while the corresponding plot for the linear case $\alpha = 2$ is that of Figure 5-(a). Whereas the concave front is fully identified in 40% of

Table 1. Wilcoxon signed rank tests at 95% confidence level (I_{H-} metric)

Instances	Algorithms	Algorithms			
		$NSGAII$	$IBEA_{\varepsilon}+$	$IBEA_{H-}$	$SPEA2$
$Zeno3_{cost}$	$NSGAII$	−	≡	≡	≡
	$IBEA_{\varepsilon}+$	≡	−	≡	≡
	$IBEA_{H-}$	≡	≡	−	≡
	$SPEA2$	≡	≡	≡	−
$Zeno3_{risk}$	$NSGAII$	−	≡	≡	≡
	$IBEA_{\varepsilon}+$	≡	−	≡	≻
	$IBEA_{H-}$	≡	≡	−	≻
	$SPEA2$	≡	≺	≺	−
$Zeno6_{cost}$	$NSGAII$	−	≺	≺	≺
	$IBEA_{\varepsilon}+$	≻	−	≡	≡
	$IBEA_{H-}$	≻	≡	−	≡
	$SPEA2$	≻	≡	≡	−
$Zeno6_{risk}$	$NSGAII$	−	≺	≺	≡
	$IBEA_{\varepsilon}+$	≻	−	≺	≻
	$IBEA_{H-}$	≻	≺	−	≻
	$SPEA2$	≡	≺	≺	−
$Zeno9_{cost}$	$NSGAII$	−	≺	≺	≺
	$IBEA_{\varepsilon}+$	≻	−	≺	≡
	$IBEA_{H-}$	≻	≻	−	≡
	$SPEA2$	≻	≡	≡	−
$Zeno9_{risk}$	$NSGAII$	−	≺	≺	≺
	$IBEA_{\varepsilon}+$	≻	−	≺	≺
	$IBEA_{H-}$	≻	≻	−	≡
	$SPEA2$	≻	≡	≡	−

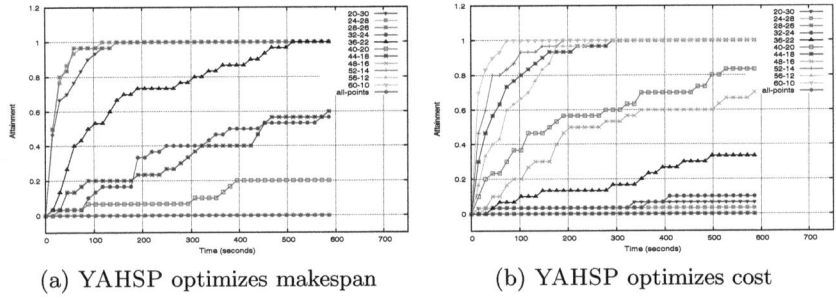

(a) YAHSP optimizes makespan   (b) YAHSP optimizes cost

Fig. 6. Attainment plots for two search strategies on MULTIZENO6$_{cost}$

the runs (right), the complete front for the strictly convex case (left) is never reached: in the latter case, the 4 most extreme points are found by 90% of the runs in less than 200 seconds, while the central points are hardly ever found. We hypothesize that the handling of YAHSP strategy regarding which objective to optimize (see Section 3.1) has a greater influence in the case of this strictly convex front than when the front is linear ($\alpha = 2$) or almost linear, even if strictly concave ($\alpha = 2.9$). In any case, no aggregation technique could ever solve the latter case, whereas it is here solved in 40% of the runs by DAE$_{\text{YAHSP}}$.

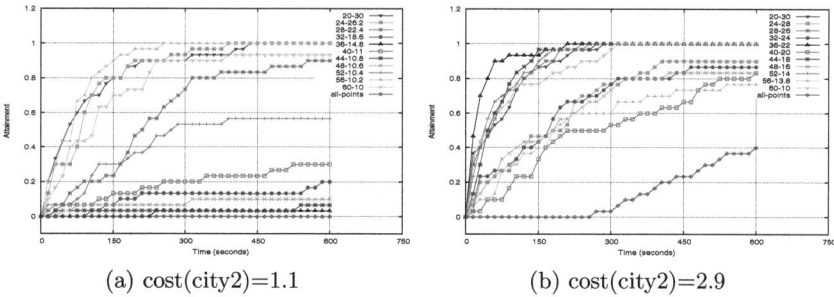

(a) cost(city2)=1.1   (b) cost(city2)=2.9

Fig. 7. Attainment plots for different Pareto fronts for MULTIZENO6$_{cost}$

7   Conclusion and Perspectives

The contributions of this paper are twofold. Firstly, MULTIZENO, an original benchmark test suite for multi-objective temporal planning, has been detailed, and several levers identified that allow to generate more or less complex instances, that have been confirmed experimentally: increasing the number of passengers obviously makes the problem more difficult; modifying the cost of reaching the cities and the duration of the flights is another way to make the problem harder, though deeper work is required to identify the consequences of each modification. Secondly, several multi-objectivization of DAE$_X$, an efficient evolutionary planner in the single-objective case, have been proposed.

However, even though the hypervolume-based IBEA$_{H-}$ clearly emerged as the best choice, the experimental comparison of those variants on the MULTI-ZENO benchmark raises more questions than it brings answers. The sparseness of the Pareto Front has been identified as a possible source for the rather poor performance of all variants for moderately large instances, particularly for the *risk* type of instances. Some smoothening of the objectives could be beneficial to tackle this issue (e.g., counting for the number of times each risk level is hit rather than simply accounting for the maximal value reached). Another direction of research is to combat the non-symmetry of the results, due to the fact that the embedded planner only optimizes one objective. Further work will investigate a self-adaptive approach to the choice of which objective to give YAHSP to optimize. Finally, the validation of the proposed multi-objective DAE$_{YAHSP}$ can only be complete after a thorough comparison with the existing aggregation approaches – though it is clear that aggregation approaches will not be able to identify the whole Pareto front in case it has some concave parts, whereas the results reported here show that DAE$_{YAHSP}$ can reasonably do it.

References

1. Ghallab, M., Nau, D., Traverso, P.: Automated Planning, Theory and Practice. Morgan Kaufmann (2004)
2. Do, M., Kambhampati, S.: SAPA: A Multi-Objective Metric Temporal Planner. J. Artif. Intell. Res. (JAIR) 20, 155–194 (2003)

3. Refanidis, I., Vlahavas, I.: Multiobjective Heuristic State-Space Planning. Artificial Intelligence 145(1), 1–32 (2003)
4. Gerevini, A., Saetti, A., Serina, I.: An Approach to Efficient Planning with Numerical Fluents and Multi-Criteria Plan Quality. Artificial Intelligence 172(8-9), 899–944 (2008)
5. Chen, Y., Wah, B., Hsu, C.: Temporal Planning using Subgoal Partitioning and Resolution in SGPlan. J. of Artificial Intelligence Research 26(1), 323–369 (2006)
6. Edelkamp, S., Kissmann, P.: Optimal Symbolic Planning with Action Costs and Preferences. In: Proc. 21st IJCAI, pp. 1690–1695 (2009)
7. Schoenauer, M., Savéant, P., Vidal, V.: Divide-and-Evolve: A New Memetic Scheme for Domain-Independent Temporal Planning. In: Gottlieb, J., Raidl, G.R. (eds.) EvoCOP 2006. LNCS, vol. 3906, pp. 247–260. Springer, Heidelberg (2006)
8. Vidal, V.: A Lookahead Strategy for Heuristic Search Planning. In: Proceedings of the 14th ICAPS, pp. 150–159. AAAI Press (2004)
9. Fikes, R., Nilsson, N.: STRIPS: A New Approach to the Application of Theorem Proving to Problem Solving. Artificial Intelligence 1, 27–120 (1971)
10. Bibai, J., Savéant, P., Schoenauer, M., Vidal, V.: An Evolutionary Metaheuristic Based on State Decomposition for Domain-Independent Satisficing Planning. In: Brafman, R., et al. (eds.) Proc. 20th ICAPS, pp. 18–25. AAAI Press (2010)
11. Bibai, J., Savéant, P., Schoenauer, M., Vidal, V.: On the Benefit of Sub-optimality within the Divide-and-Evolve Scheme. In: Cowling, P., Merz, P. (eds.) EvoCOP 2010. LNCS, vol. 6022, pp. 23–34. Springer, Heidelberg (2010)
12. Haslum, P., Geffner, H.: Admissible Heuristics for Optimal Planning. In: Proc. AIPS-2000, pp. 70–82 (2000)
13. Gerevini, A., Long, D.: Preferences and Soft Constraints in PDDL3. In: ICAPS Workshop on Planning with Preferences and Soft Constraints, pp. 46–53 (2006)
14. Deb, K., Pratap, A., Agarwal, S., Meyarivan, T.: A fast and elitist multiobjective genetic algorithm: NSGA-II. IEEE Trans. Evol. Comp. 6(2), 182–197 (2002)
15. Zitzler, E., Laumanns, M., Thiele, L.: SPEA2: Improving the Strength Pareto Evolutionary Algorithm for Multiobjective Optimization. In: Evol. Methods Design Optim. Control Applicat. Ind. Prob. (EUROGEN), pp. 95–100 (2002)
16. Zitzler, E., Künzli, S.: Indicator-Based Selection in Multiobjective Search. In: Yao, X., Burke, E.K., Lozano, J.A., Smith, J., Merelo-Guervós, J.J., Bullinaria, J.A., Rowe, J.E., Tiňo, P., Kabán, A., Schwefel, H.-P. (eds.) PPSN 2004. LNCS, vol. 3242, pp. 832–842. Springer, Heidelberg (2004)
17. Zitzler, E., Laumanns, M., Thiele, L.: SPEA2: Improving the Strength Pareto Evolutionary Algorithm. Technical report, ETH Zürich (2001)
18. Liefooghe, A., Basseur, M., Jourdan, L., Talbi, E.-G.: ParadisEO-MOEO: A Framework for Evolutionary Multi-objective Optimization. In: Obayashi, S., Deb, K., Poloni, C., Hiroyasu, T., Murata, T. (eds.) EMO 2007. LNCS, vol. 4403, pp. 386–400. Springer, Heidelberg (2007)
19. Hutter, F., Hoos, H.H., Leyton-Brown, K., Stützle, T.: ParamILS: an automatic algorithm configuration framework. J. Artif. Intell. Res. (JAIR) 36, 267–306 (2009)
20. Bibaï, J., Savéant, P., Schoenauer, M., Vidal, V.: On the Generality of Parameter Tuning in Evolutionary Planning. In: Proc. 12th GECCO, pp. 241–248. ACM (2010)
21. Bleuler, S., Laumanns, M., Thiele, L., Zitzler, E.: PISA — A Platform and Programming Language Independent Interface for Search Algorithms. In: Fonseca, C.M., Fleming, P.J., Zitzler, E., Deb, K., Thiele, L. (eds.) EMO 2003. LNCS, vol. 2632, pp. 494–508. Springer, Heidelberg (2003)

Multi-objective Topic Modeling

Osama Khalifa, David Wolfe Corne, Mike Chantler, and Fraser Halley

Heriot-Watt University, Mathematical and Computer Science,
Riccarton. Edinburgh, United Kingdom
{ok32,d.w.corne,m.j.chantler,f.halley}@hw.ac.uk

Abstract. Topic Modeling (TM) is a rapidly-growing area at the interfaces of text mining, artificial intelligence and statistical modeling, that is being increasingly deployed to address the 'information overload' associated with extensive text repositories. The goal in TM is typically to infer a rich yet intuitive summary model of a large document collection, indicating a specific collection of topics that characterizes the collection – each topic being a probability distribution over words – along with the degrees to which each individual document is concerned with each topic. The model then supports segmentation, clustering, profiling, browsing, and many other tasks. Current approaches to TM, dominated by Latent Dirichlet Allocation (LDA), assume a topic-driven document generation process and find a model that maximizes the likelihood of the data with respect to this process. This is clearly sensitive to any mismatch between the 'true' generating process and statistical model, while it is also clear that the quality of a topic model is multi-faceted and complex. Individual topics should be intuitively meaningful, sensibly distinct, and free of noise. Here we investigate multi-objective approaches to TM, which attempt to infer coherent topic models by navigating the trade-offs between objectives that are oriented towards coherence as well as coverage of the corpus at hand. Comparisons with LDA show that adoption of MOEA approaches enables significantly more coherent topics than LDA, consequently enhancing the use and interpretability of these models in a range of applications, without significant degradation in generalization ability.

Keywords: Multi-objective optimization, Topic Modeling, Latent Dirichlet Allocation, MOEA/D, Pointwise Mutual Information, Perplexity.

1 Introduction

Topic Modeling (TM) is a relatively recent and rapidly-growing area at the interfaces of text mining, artificial intelligence and statistical modeling; it is being increasingly deployed to address the 'information overload' associated with extensive text collections. The growing interest in TM can be associated with the fact that text comprises about 85% of data worldwide [1]. Modern approaches to TM are based on a variety of theoretical frameworks that tend to consider any individual document to be a weighted mixture of *topics*, where each individual topic is a multinomial distribution over words. An inferred *topic model* comprises

R.C. Purshouse et al. (Eds.): EMO 2013, LNCS 7811, pp. 51–65, 2013.

a specific collection of topics, along with an assignment of one of these topics to each word in each document in the corpus at hand. Such a topic model can provide an efficient representation of the corpus and is effective at supporting a wide range of browsing and retrieval strategies (for example, delivering suitable documents in response to queries involving a weighted mixture of topics)[2].

Current TM approaches such as Correlated Topic Models (CTM) [4] and Latent Dirichlet Allocation (LDA), rely on finding a set of topics that maximizes the likelihood that the data were generated by a specific model of document generation. Though commonly returning interpretable results, the inferred models are ultimately aligned to a much-simplified abstraction of the real document generation process, and leave much room for improvement in the intuitive 'real-world coherence' of the resulting models. In current approaches, it is therefore common to evaluate the inferred models via their performance in a specific task such as classification of unseen documents. Such strategies do not represent a fully-rounded evaluation of a topic model, and do not address the question of how more coherent topic models might be inferred in the first place.

A high quality topic model is one that can be expected to score well on a collection of different criteria, concerned with, for example, the coherence of individual topics, the coherence of the collection of topics as a whole, and the extent to which the inferred topics cover the entire collection, as well as the extent to which individual documents are explained by the topics (for example, a poor topic model in the latter respect may leave large portions of many documents unallocated to topics). However, each of these objectives is difficult to evaluate and can only be approximated – meanwhile, the familiar LDA likelihood criterion is a proven successful objective that, similarly, provides an appropriate and alternative approximate measure of quality. Exploiting the multi-criteria nature of topic models, in this article we begin to explore the use of multi-objective evolutionary algorithms (MOEAs) in topic modeling, and we investigate whether MOEA or MOEA/LDA hybrid approaches can be designed that yield better topic models than current approaches, and consequently provide enhanced effectiveness and user experiences in the many applications of TM technologies.

The remainder is organized as follows: in Section 2 we introduce concepts related to the most prominent topic modeling method, LDA, and we describe the evaluation techniques that are generally used. Our MOEA approaches to TM are described in Section 3, and in Section 4 we describe a series of experiments that compare MOEA-TM approaches with LDA on three text corpora. Summary and final reflections are made in Section 5. Meanwhile at `http://is.gd/MOEATM` we provide source code, corpora and associated instructions that are sufficient to replicate our experiments and support further investigations.

2   Topic Modeling

Topic modeling is an approach to analyzing large amounts of unclassified text data [3]. It exploits the statistical regularities that occur in natural language documents in order to match queries to documents in a way that, though entirely statistical, carries strong semantic resonance. Good topic models should connect words with similar meanings (i.e. these words will typically co-occur within topics) and be able to distinguish between multiple meanings of a word depending on context (e.g. the word 'set' will appear with high probability in both a 'tennis' topic, and a 'discrete mathematics' topic).

2.1   Latent Dirichlet Allocation

Latent Dirichlet Allocation (LDA) is among the most prominent of current topic modeling techniques; it considers corpus documents to be underpinned by a mixture of latent topics, where each topic is characterized by a multinomial distribution over words [5]. LDA makes use of Dirichlet distribution which is a continuous multivariate distribution parameterized by a vector of positive reals. In the special case when all this vector's components are the same number, the distribution is called symmetric Dirichlet. A quick summary of LDA using LDA generative process is as follows. Let K be a predefined number of topics, $k \in [1..K]$ a number representing the topic, α a positive K-component vector, η a scalar, $Dir(\alpha)$ a K-dimensional Dirichlet distribution, V the corpus size, $Dir(\eta)$ a V-dimensional symmetric Dirichlet distribution, β_k a topic k distribution over corpus words, θ_d the topics proportion for one document, d a document from the corpus, w a word from the corpus, $w_{d,n}$ the nth word in the document d, and $z_{d,n} \in [1..K]$ the topic assignment for the nth word in document d.

for each topic k
    Choose a distribution over words $\beta_k \sim Dir(\eta)$
for each document d
    Draw a topic proportion $\theta_d \sim Dir(\alpha)$
    for each word w in the document d
        Draw a topic assignment $z_{d,n} \sim Multinomial(\theta_d), z_{d,n} \in 1..K$
        Draw a word $w_{d,n} \sim Multinomial(\beta_{z_{d,n}})$

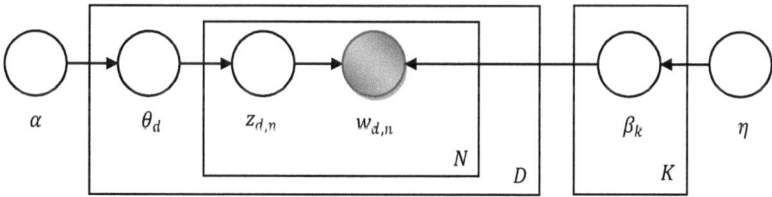

Fig. 1. LDA model graphical representation of generative process

The graphical representation of LDA in Figure. 1 illustrates the relationship between latent and observed variables. The LDA generative process defines a joint probability distribution over these as follows [6]:

$$P\left(\beta,\theta,z,w\right) = \prod_{k=1}^{K} P(\beta_k) \prod_{d=1}^{D} P(\theta_d) \left(\prod_{n=1}^{N} P\left(z_{d,n}|\theta_d\right) P\left(w_{d,n}|\beta, z_{d,n}\right)\right) \qquad (1)$$

where, D is the number of documents, N the number of words inside one document, β is all topics distributions over corpus words, θ is the topics proportions for all documents, and z the topic assignments for all corpus words.

The main computational problem of LDA is to compute the posterior distribution – i.e., the conditional distribution of the latent variables given the observed variables. The posterior is given by the following formula:

$$P\left(\beta,\theta,z|w\right) = \frac{P\left(\beta,\theta,z,w\right)}{P\left(w\right)}. \qquad (2)$$

Unfortunately, the exact posterior calculation is not feasible due to the denominator $P(w)$, calculation of which would involve summing the joint distribution over every possible combination of topic structures. However, there are various methods for approximating this posterior, such as Mean Field Variational Inference, Collapsed Variational Inference, Expectation Propagation and Gibbs Sampling [7]. LDA approaches that use Gibbs sampling are among the most popular methods in the current literature.

2.2 Evaluating Topic Models

The unsupervised nature of topic modeling methods makes choosing one topic model over another a difficult task. Topic model quality tends to be evaluated by performance in a specific application. However, other ways of evaluation are also used in the literature. Topic models can be evaluated based on *perplexity* [8] as a quantitative method; meanwhile, a 'human-evaluation' oriented evaluation method was introduced in [9] by creating a task where humans judge topics in terms of the frequency of apparently irrelevant words.

Perplexity is becoming a standard quality measure for topic models; it measures the topic model's ability to generalize to unseen documents after estimating the model using training documents. Lower perplexity means better generalization ability. Perplexity is calculated for a test corpus D_{test} by calculating the natural exponent of the mean log-likelihood of the corpus words [10] as follows:

$$Perplexity\left(D_{test}|\mathcal{M}\right) = e^{\dfrac{-\sum_{d \in D_{test}} \log P(w_d|\mathcal{M})}{\sum_{d \in D_{test}} N_d}} \qquad (3)$$

where w_d represents the words of test document d, $\mathcal{M}$ is the topic model, N_d is the number of words in document d. Thus, to evaluate two topic models estimated from the same training data, perplexity on test data is calculated, and the model with the lower perplexity value is preferred since it seems to provide a better characterization of the unseen data. However, perplexity does not reflect the topics' semantic coherence [11].

On the other hand, Pointwise Mutual Information (PMI) is an ideal measure of semantic coherence, based on word association in the context of information theory [12,13]. PMI compares the probability of seeing two words together with the probability of observing the words independently. PMI for two words can be given using the following formula:

$$Pmi(w_i, w_j) = \log \frac{P(w_i, w_j)}{P(w_i)P(w_j)}. \tag{4}$$

The joint probability $P(w_i, w_j)$ can be measured by counting the number of observations of words w_i and w_j together in the corpus normalized by the corpus size. PMI-based evaluations correlate very well with human judgment of topic coherence or topic semantics [11,14], especially when Wikipedia is used as a meta-documents to calculate the words co-occurrences within a suitably sized sliding window.

PMI values fall in the range $]-\infty, -\log P(w_i, w_j)]$, hence the higher the PMI value the more coherent the topic it represents. PMI values can be normalized to fall in the range $[-1, 1]$ as shown in [15] using the following formula:

$$nPmi(w_i, w_j) = \begin{cases} -1 & \text{if } P(w_i, w_j) = 0 \\ \frac{\log P(w_i) + \log P(w_j)}{\log P(w_i, w_j)} - 1 & \text{otherwise} \end{cases}. \tag{5}$$

The approach used to evaluate one topic is to calculate the mean of PMI for each possible word pair in the topic T. Consequently, the normalized PMI value for one topic T is given using the following formula:

$$nPmi_T = \frac{\sum_{w_i, w_j \in T} nPmi(w_i, w_j)}{\binom{T_{length}}{2}}. \tag{6}$$

where, T_{length} represents the number of words inside topic T.

3   MOEA Approaches to Topic Modeling

Multi-objective optimization aims to find a set of solutions that represent optimal trade-offs between the objectives. This is the set of *Pareto Optimal* solutions [16]. There are a wide variety of approaches to multi-objective problems, however, many of these may fail when the Pareto front (the geometric structure of the Pareto set in objective space) is concave or disconnected [17]. Multi-objective Evolutionary Algorithms (MOEAs) tend to avoid these drawbacks [17,18], among others, and are currently prominent among state of the art approaches to multi-objective optimization.

Topic models have many applications beyond unstructured text processing and text tagging. They can be used in analyzing genetic data [19], computer vision [20], audio and speech engineering [21], emotion modeling and social affective text mining [22], and financial analysis [23]. Current approaches such as

LDA focus on producing topic models which score well on perplexity as measured over a test set. However, other applications, such as text tagging which is used in digital libraries, require highly coherent topics [11]. Considering the varied requirements of other applications, along with arguments made in Section 1, it is well-worth considering MOEAs in attempt to produce high quality topic models in general, and also in contexts relating to specific applications.

Our first approach ('standalone' MOEA-TM) is to optimize two objectives: PMI and coverage (described in section 3.3). PMI encourages coherent topics, whilst coverage encourages a large proportion of the corpus words to appear in the inferred topics. In 'standalone' MOEA-TM we limit the number of words per topic. This arguably leads to more intuitive topics, and significantly reduces computational load, but means that we cannot use perplexity as an objective, since the perplexity calculation requires all corpus words to be assigned to a topic. Experiments with standalone MOEA-TM are described in section 4.1. In section 4.2, we introduce an alternative approach in which MOEA-TM is used to improve topic models pre-generated by LDA. Here we trade-off the computational load of an unlimited number of words per topic against the optimized starting point, and are able to add perplexity as an additional objective. In each case, MOEA-TM builds on the current prominent 'Multi-objective Evolutionary Algorithm Based on Decomposition (MOEA/D)' [24], and adapts it to this task.

3.1   Encoding and Generation of Initial Population

Each chromosome is a vector of topic variables $T_1, T_2.., T_K$ where K is the number of topics. Each topic variable contains a number of weighted words. Thus, each gene comprises two parts: the word index and a numerical value representing the word's participation in the topic. Chromosome structure is illustrated in Fig. 2. In the standalone case, the population is initialized randomly as each topic variable is initialized on the basis of a randomly chosen document. Topic genes are initialized based on the most frequent words in the chosen document, with random weights. However, when the algorithm is used to enhance an existing model, the population echoes the model itself. Each topic variable is based on its corresponding model's topic, where the genes represent the highest weighted words in that topic.

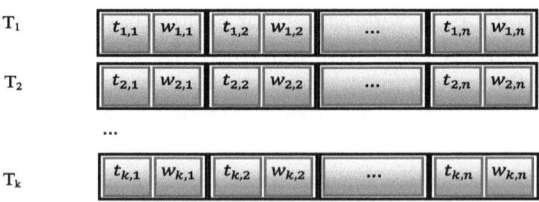

Fig. 2. Chromosome Structure

3.2   Genetic Operators

Crossover in our current approach generates two offspring from two parents. Each child comprises as many topic variables as its parents has, via uniform crossover of the parents' corresponding topic variable genes, ensuring that words and their associated weights are copied together. However, when a word exists in both parents' topic variables, the children have the average word weight. A simple two topics crossover example is illustrated in Fig. 3.

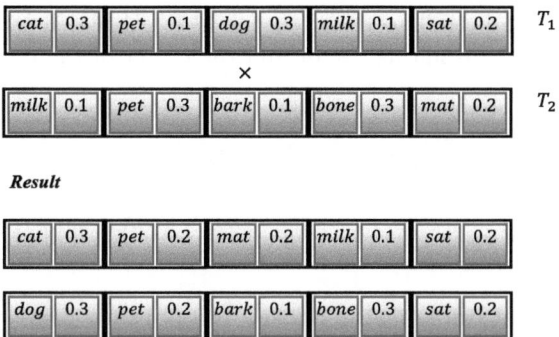

Fig. 3. A simple two topics crossover example

Mutation is applied to a single randomly chosen gene, changing the weight to a new random number, and changing the word to another word from the corpus, ensuing that the newly introduced word occurs together in a document in the corpus with another randomly selected word from the topic variable.

3.3   Objectives

Coverage Score. This objective encourages topic models to represent the whole corpus. For each document, topics are evaluated by calculating the Euclidean distance between the weighted topics and the document itself. This is done by multiplying each topic's word-weight by the document's related topic weight, then calculating the distance between the resulting distribution and the document's word frequencies. Document-related topic weights are calculated using:

$$Prop_d(T) = \frac{\sum_{w \in T} tf_d(w)}{T_{length} - count_{w \in T, d}(w)} \tag{7}$$

where, $tf_d(w)$ gives the frequency of the word w in the document d and $count_{w \in T, d}(w)$ gives the number of words that exist in the topic and document at the same time. Consequently, the coverage score for one document d can be given by:

$$Coverage_d = \sqrt{\sum_{w \in d} \left(tf_d(w) - \sum_{i=1}^{K} T_i(w) Prop_d(T_i) \right)^2} \tag{8}$$

where $T_i(w)$ gives the word's weight if it is present in topic T_i, and zero otherwise. The coverage score can be normalized by its maximum value as follows:

$$nCoverage_d = \sqrt{\frac{\sum_{w \in d} \left(tf_d(w) - \sum_{i=1}^{K} T_i(w) Prop_d(T_i) \right)^2}{\sum_{w \in d} tf_d(w)^2}}. \tag{9}$$

This process is repeated for all corpus documents in order to calculate a coverage score for the corpus. Eventually, there will be a vector of values that need to be minimized. The overall score for corpus D is calculated by measuring the distance between the resulting vector and the center of the representing space using:

$$CovObj = \sqrt{\sum_{d \in D} nCoverage_d^2}. \tag{10}$$

The objective $CovObj$ needs to be minimized in MOEA-TM algorithm.

Pointwise Mutual Information Score. This objective measures the intuitive quality of a topic, in terms of how often words that co-occur in a topic tend to co-occur in general. PMI is calculated for each topic using (6). The higher the PMI value, the more 'coherent' is the topic. For convenience, however, we use $1 - nPmi_T$ as the objective, so that all objectives in MOEA-TM are to be minimized. The overall score for a topic model topics is calculated by measuring the distance between the vector of PMI scores for each topic, and the center of the representing space using:

$$PmiObj = \sqrt{\sum_{i=1}^{K} (1 - nPmi_{T_i})^2}. \tag{11}$$

Perplexity Score. This objective is related to the model's ability to generalize to unseen data. Strictly, the perplexity score requires a topic model which assigns a topic to every word in the entire corpus, so we cannot calculate it for a topics comprising only a subset of corpus words, which is our approach in the 'standalone' MOEA-TM. Consequently, this objective is only investigated when MOEA-TM is used to enhance a pre-calculated topic model, which is the case with our 'LDA-Initialized' MOEA-TM. The Perplexity score is calculated using the following formula:

$$PerpObj = \frac{-\sum_{d \in D_{test}} \log P(w_d | \mathcal{M})}{\sum_{d \in D_{test}} N_d} \tag{12}$$

where, $\mathcal{M}$ is the pre-calculated LDA topic model, D_{test} is a small test corpus, d is a document in the test corpus, w_d is the words of test document d, and N_d number of words in document d. $PerpObj$ objective is calculated using Left to Right method from [8] then normalized dynamically using other calculated values. The minimized negative log-likelihood mean leads to minimized perplexity.

3.4 Best Solution

Our primary aim is to contrast MOEA approaches to topic modeling with the standard single-objective approach, and hence we draw a single solution from each MOEA-TM run. We choose a compromise solution from the (approximated) Pareto front by sorting the Pareto set according to a score representing the Euclidean distance between the objective vector $\overrightarrow{v} = (v_1, v_2 \cdots v_n)$ and the center of the objective space as follows:

$$score(\overrightarrow{v}) = \sqrt{\sum_{i=1}^{n} v_i^2}. \tag{13}$$

4   Experimental Evaluation

A number of experiments were performed to compare MOEA-TM with LDA, arguably the state-of-art in topic modeling. We used the LDA implementation with Gibbs Sampling which is provided by the MALLET package [25]. MOEA implementations utilized the MOEA Framework version 1.11 [26] run by JDK version 1.6 and CentOS release 5.8. Our evaluation uses three corpora: the first is a very small corpus with five documents created from Wikipedia and containing four rather distinct topics (Love, Music, Sport and Government). The second corpus is made from about 15000 documents taken from news articles covering mainly four topics: Music, Economy, Fuel and Brain Surgery. The Third corpus comprises about 800 documents that are summaries of projects in Information and Communication Technology (ICT) funded by the Engineering and Physical Sciences Research Council (EPSRC). Full details of each corpus are available from http://is.gd/MOEATM.

4.1 Standalone MOEA Topic Modeling

Standalone MOEA-TM was run ten times independently on each corpus, using only normalized coverage and normalized PMI objectives. LDA was also run ten times on each corpus. These experiments were done twice, once with number of topics set to 4, and once with number of topics set to 10.

Figure. 4, Figure. 5 and Figure. 6 show all MOEA-TM solutions resulting from ten runs. An averaged MOEA-TM Pareto Front is shown. The 'best' MOEA-TM solution (identified using (13)), is displayed. LDA solutions and their means are also shown. It can be seen that LDA is able to find relatively good solutions with an optimized coverage score; however the PMI (coherence) scores are poor in comparison to those found by MOEA-TM.

Figure. 4 and Figure. 5 show that best MOEA-TM solution optimizes both *PmiObj* and *CovObj* scores for the corpora Wiki and EPSRC respectively. On

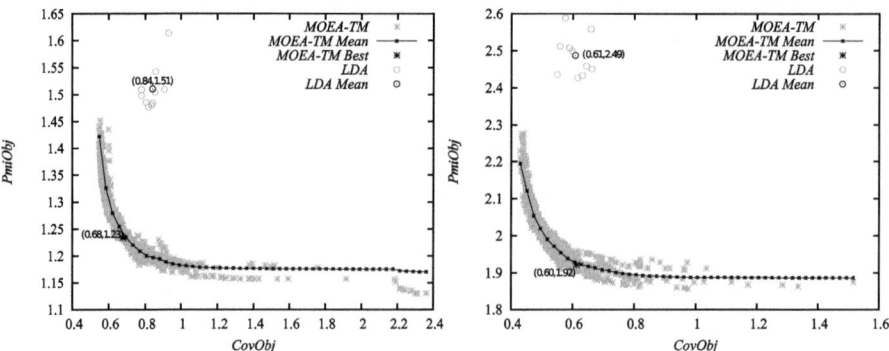

Fig. 4. Wiki Corpus test: MOEA-TM Pareto Front and LDA solutions for ten runs (average is taken), 4 topics left and 10 topics right

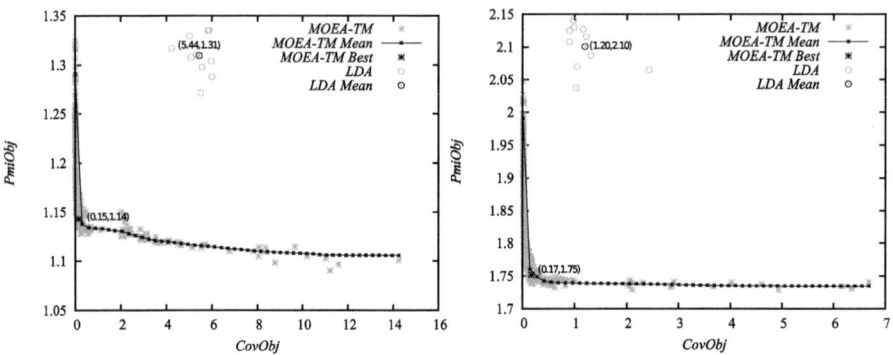

Fig. 5. EPSRC Corpus test: MOEA-TM Pareto Front and LDA solutions for ten runs (average is taken), 4 topics left and 10 topics right

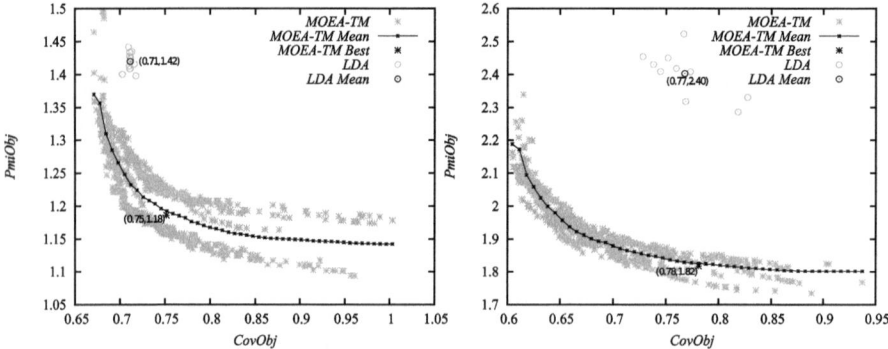

Fig. 6. News Corpus test: MOEA-TM Pareto Front and LDA solutions for ten runs (average is taken), 4 topics left and 10 topics right

the other hand, Figure. 6 shows that for the News corpus MOEA-TM Best Solution was able to optimize the *PmiObj* but not the *CovObj* objective. This means that for this corpus LDA was able to find a higher representing topics but with poor PMI.

Evaluation: Table. 1 and Table. 2 show the mean and sample standard deviations of original PMI metrics from the best MOEA-TM solutions and from LDA for 4 and 10 topic runs respectively. In these tables the higher PMI value is the better as the displayed values are the mean original normalized PMI values for solutions' topics after applying (6) over each topic.

Table 1. PMI for standalone MOEA-TM and LDA, for three corpora / four topics

	MOEA TM		LDA	
	Mean PMI	St. Deviation	Mean PMI	St. Deviation
Wiki Corpus	0.3490	0.0128	0.2460	0.0194
EPSRC Corpus	0.4119	0.0091	0.3457	0.0102
News Corpus	0.3987	0.0178	0.2933	0.0082

Table 2. PMI for standalone MOEA-TM and LDA for, for three corpora / ten topics

	MOEA TM		LDA	
	Mean PMI	St. Deviation	Mean PMI	St. Deviation
Wiki Corpus	0.3483	0.0078	0.2158	0.0163
EPSRC Corpus	0.4264	0.0080	0.3371	0.0106
News Corpus	0.3913	0.0077	0.2448	0.0216

It can be seen that MOEA-TM outperforms LDA in terms of the PMI metric. This means that topic models resulting from MOEA-TM are significantly more coherent than topics resulting from LDA. As suggested by the standard deviations, all MOEA-TM/LDA comparisons are significant with $p < 0.01$. The fact that MOEA-TM outperforms LDA in this respect is of course not very surprising given that LDA does not directly optimize PMI, however it is arguably surprising and interesting that the MOEA-TM approach can show such a marked improvement in topic coherence beyond that which seems achievable by LDA.

4.2 LDA-Initialized MOEA Topic Modeling

In this experiment, similar experiments were run but in this case MOEA-TM is used to enhance a pre-calculated LDA topic model by optimizing three objectives *CovObj*, *PmiObj*, and *PerpObj*. The negative log-likelihood mean of an unseen test corpus words using the updated model is compared with the negative log-likelihood-mean of the same unseen test corpus words using the original LDA model. The model that has lower negative log-likelihood mean (or higher log-likelihood mean) is better as it leads to lower perplexity. LDA-initialized MOEA-TM was run ten times, and compared with (again) the results of ten unenhanced LDA topic models.

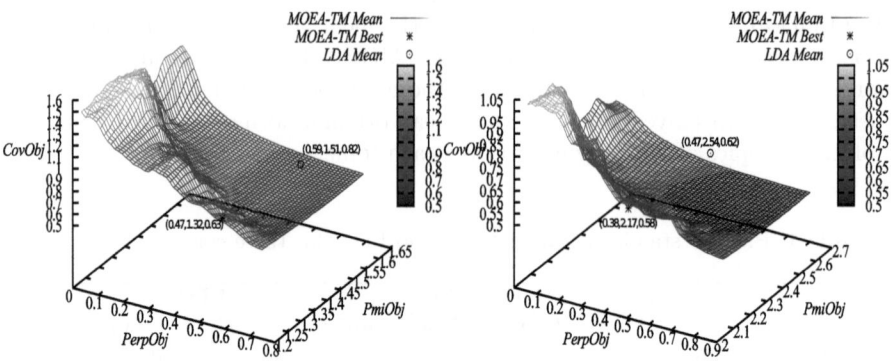

Fig. 7. Wiki Corpus test: LDA-Initialized MOEA-TM Pareto Front and Pure LDA solutions for ten runs (average is taken), 4 topics left and 10 topics right

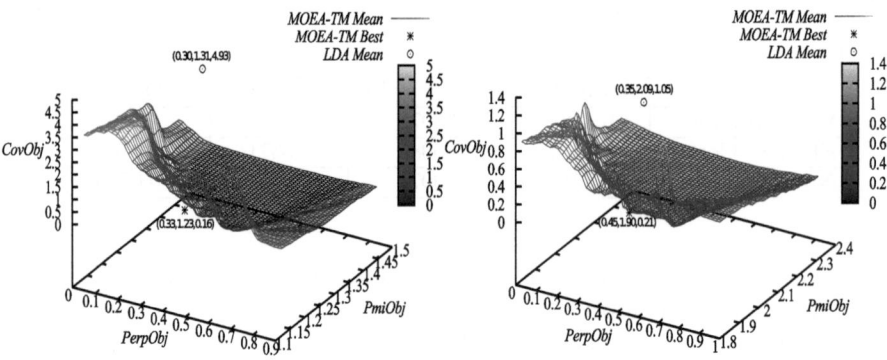

Fig. 8. EPSRC Corpus test: LDA-Initialized MOEA-TM Pareto Front and Pure LDA solutions for ten runs (average is taken), 4 topics left and 10 topics right

Figure. 7, Figure. 8 and Figure. 9 show the average MOEA-TM Pareto Front which is calculated by interpolating each of MOEA-TM Pareto Fronts using Microsphere Projection (the multivariate interpolation method) [27] and then calculating the average surface. The average surface is calculated by substituting each drawing point in each Pareto Front interpolation function then calculating the drawing point average value. Furthermore, best MOEA-TM solution, which is identified using (13), and LDA mean solutions are displayed in the figures. The

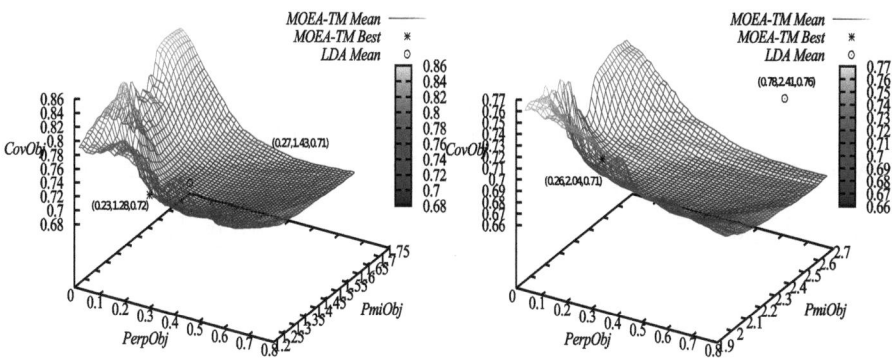

Fig. 9. News Corpus test: LDA-Initialized MOEA-TM Pareto Front and Pure LDA solutions for ten runs (average is taken), 4 topics left and 10 topics right

MOEA-TM solutions and LDA solutions are not displayed for clarity. It can be seen that MOEA-TM was able to find better solution in terms of Coverage ($CovObj$) and PMI ($PmiObj$) for all corpora. In terms of perplexity ($PerpObj$) Figure. 8 shows that LDA was able to find better solutions for EPSRC corpus. Whereas MOEA-TM best solutions have better perplexity for Wiki and News corpora as shown in Figure. 7 and Figure. 9.

Evaluation: Table. 3 and Table. 4 present the original normalized PMI and non-normalized negative Log-Likelihood ($-LL$) metrics for LDA-Initialized MOEA-TM and LDA topic models with four and ten topics, respectively. It can be seen that LDA-Initialized MOEA-TM shows an improvement in terms of PMI values of 39%, 14% and 25% over pure LDA in the corpora Wiki, EPSRC and News, respectively when four topics are learned. When ten topics are learned the PMI improvement is 54%, 14% and 40% in the corpora Wiki, EPSRC and News, respectively. In all cases, a t-test again finds that the MOEA-TM improvement in PMI is significant with $p < 0.01$, while there is in contrast no significance in the difference in log-Likelihood values, suggestion that the improved coherence comes without any significant difference in the perplexity of the enhanced model.

Table 3. PMI scores for LDA-Initialized MOEA TM and Pure LDA for the three corpora with four topics

	MOEA TM				LDA			
	PMI	St. Dev	-LL	st. Dev	PMI	St. Dev	-LL	st. Dev
Wiki Corpus	0.3443	0.1129	8.1417	0.0477	0.2476	0.1932	8.1488	0.0514
EPSRC Corpus	0.3933	0.0107	15.301	0.1128	0.3429	0.0094	15.293	0.1133
News Corpus	0.3653	0.0069	52.680	0.6756	0.2903	0.0142	52.835	0.7976

Table 4. PMI scores for LDA-Initialized MOEA TM and Pure LDA for the three corpora with ten topics

	MOEA TM				LDA			
	PMI	St. Dev	-LL	st. Dev	PMI	St. Dev	-LL	st. Dev
Wiki Corpus	0.3105	0.0135	8.0716	0.0294	0.2013	0.0194	8.0822	0.0262
EPSRC Corpus	0.3889	0.0085	15.034	0.1005	0.3404	0.0101	15.096	0.0960
News Corpus	0.3428	0.0159	51.990	0.5377	0.2445	0.0208	53.261	0.6977

5   Conclusion

To sum up, MOEA-TM shows promising performance in topic modeling. MOEA-TM initialized from LDA models is able to enhance the coherence of the topic models significantly for each of the corpora tested here. A more coherent topic model is one in which the words that tend to appear together in a topic make more sense together to a human being. This can be very useful in many topic modeling applications, such as text tagging in digital libraries, where topic coherence is particularly important [11], while in general we would expect user confidence in inferred topic models, whatever the application, to be boosted when topics are coherent. In general, multi-objective approaches may contribute significantly to topic modeling, providing the ability to specify arbitrary objectives that may be relevant in a given application, and then providing the decision maker with a diverse collection of optimal models from which the most appropriate can be selected.

References

1. Baars, H., Kemper, H.G.: Management Support with Structured and Unstructured Data-An Integrated Business Intelligence Framework. Inf. Sys. Manag. 25(2), 132–148 (2008)
2. Ha-Thuc, V., Srinivasan, P.: Topic Models and a Revisit of Text-related Applications. In: Proceedings of the 2nd PhD Workshop on Information and Knowledge Management (PIKM 2008), New York, pp. 25–32 (2008)
3. Steyvers, M., Griffiths, T.L.: Rational Analysis as a Link Between Human Memory and Information Retrieval. In: Chater, N., Oaksford, M. (eds.) The Probabilistic Mind: Prospects for Bayesian Cognitive Science, pp. 327–347. Oxford University Press (2008)
4. Blei, D.M., Lakerty, J.D.: A correlated topic model of science. Annals of Applied Statistics 1(1), 17–35 (2007)
5. Blei, D.M., Ng, A.Y., Jordan, M.I.: Latent Dirichlet Allocation. J. Mach. Learn. 3, 993–1022 (2003)
6. Blei, D.M.: Probabilistic topic models. Commun. ACM 55(4), 77–84 (2012)
7. Srivastava, A., Sahami, M.: Text Mining: Classification, Clustering, and Applications, 1st edn. Taylor and Francis Group (2009)
8. Wallach, H.M., Murray, I., Mimno, D.: Evaluation Methods for Topic Models. In: Proceedings of the 26th Annual International Conference on Machine Learning, pp. 1105–1112. ACM, Montreal Canada (2009)

9. Chang, J., Boyd-Graber, J., Wang, C., Gerrish, S., Blei, D.M.: Reading Tea Leaves: How Human Interpret Topic Models. In: Advances in Neural Information Processing Systems. NIPS Foundation, Vancouver British Columbia (2009)
10. de Waal, A., Barnard, E.: Evaluating Topic Models with Stability. In: Nineteenth Annual Symposium of the Pattern Recognition Association of South Africa, Cape Town South Africa (2008)
11. Newman, D., Noh, Y., Talley, E., Karimi, S., Baldwin, T.: Evaluating Topic Models for Digital Libraries. In: Proceedings of the 10th Annual Joint Conference on Digital Libraries, pp. 215–224. ACM, Gold Coast (2010)
12. Su, Q., Xiang, K., Wang, H., Sun, B., Yu, S.: Using Pointwise Mutual Information to Identify Implicit Features in Customer Reviews. In: Matsumoto, Y., Sproat, R.W., Wong, K.-F., Zhang, M. (eds.) ICCPOL 2006. LNCS (LNAI), vol. 4285, pp. 22–30. Springer, Heidelberg (2006)
13. Stevens, K., Kegelmeyer, P., Andrzejewski, D., Buttler, D.: Exploring Topic Coherence Over Many Models and Many Topics. In: Proceedings of the 2012 Joint Conference on Empirical Methods in Natural Language Processing and Computational Natural Language Learning, Jeju Island Korea, pp. 952–961 (2012)
14. Newman, D., Lau, J.H., Grieser, K., Baldwin, T.: Automatic Evaluation of Topic Coherence. In: Human Language Technologies: The Annual Conference of the North American Chapter of the Association for Computational Linguistics, Los Angeles California, pp. 100–108 (2010)
15. Bouma, G.: Normalized (Pointwise) Mutual Information in Collocation Extraction. In: Proceedings of The International Conference of the German Society for Computational Linguistics and Language Technology, pp. 31–40 (2009)
16. Pareto, V.: Cours d'Economie politique. Revue Economique 7(3), 426–430 (1896)
17. Coello, C.A.C.: Evolutionary Multi-objective Optimization: a Historical View of the Field. Computational Intelligence Magazine IEEE 1(1), 28–36 (2006)
18. Coello Coello, C.A.: Evolutionary Multi-Objective Optimization: Basic Concepts and Some Applications in Pattern Recognition. In: Martínez-Trinidad, J.F., Carrasco-Ochoa, J.A., Ben-Youssef Brants, C., Hancock, E.R. (eds.) MCPR 2011. LNCS, vol. 6718, pp. 22–33. Springer, Heidelberg (2011)
19. Chen, X., Hu, X., Shen, X., Rosen, G.: Probabilistic Topic Modeling for Genomic Data Interpretation. In: 2010 IEEE International Conference on Bioinformatics and Biomedicine (BIBM), pp. 149–152. IEEE Press, Hong Kong (2010)
20. Malisiewicz, T.J., Huang, J.C., Efros, A.A.: Detecting Objects via Multiple Segmentations and Latent Topic Models. Technical report, CMU Tech (2006)
21. Smaragdis, P., Shashanka, M., Raj, B.: Topic Models for Audio Mixture Analysis. In: Applications for Topic Models: Text and Beyond, Whistler (2009)
22. Shenghua, B., Shengliang, X., Li, Z., Rong, Y., Zhong, S., Dingyi, H., Yong, Y.: Joint Emotion-Topic Modeling for Social Affective Text Mining. In: Proceedings of the Ninth IEEE International Conference on Data Mining, pp. 699–704. IEEE Computer Society, Washington DC (2009)
23. Gabriel, D., Charles, E.: Financial Topic Models. In: Applications for Topic Models: Text and Beyond, Whistler Canada (2009)
24. Zhang, Q., Li, H.: MOEA/D: A Multiobjective Evolutionary Algorithm Based on Decomposition. IEEE Transactions on Evolutionary Comp. 11(6), 712–731 (2007)
25. MALLET: Machine Learning for Language Toolkit, http://mallet.cs.umass.edu
26. MOEA Framework: a Java library for multiobjective evolutionary algorithms, http://www.moeaframework.org
27. Dudziak, W.J.: Multi-Dimensional Interpolation Function for Non-Uniform Data: Microsphere Projection. In: Conf. Computer Graphics and Visualization, Lisbon, pp. 143–147 (2007)

Indicator Based Search in Variable Orderings: Theory and Algorithms

Pradyumn Kumar Shukla and Marlon Alexander Braun

Institute AIFB, Karlsruhe Institute of Technology
Karlsruhe, D-76128, Germany
pradyumn.shukla@kit.edu

Abstract. Various real world problems, especially in financial applications, medical engineering, and game theory, involve solving a multi-objective optimization problem with a variable ordering structure. This means that the ordering relation at a point in the (multi-)objective space depends on the point. This is a striking difference from usual multi-objective optimization problems, where the ordering is induced by the Pareto-cone and remains constant throughout the objective space. In addition to variability, in many applications (like portfolio optimization) the ordering is induced by a non-convex set instead of a cone. The main purpose of this paper is to provide theoretical and algorithmic advances for general set-based variable orderings. A hypervolume based indicator measure is also proposed for the first time for such optimization tasks. Theoretical results are derived and properties of this indicator are studied. Moreover, the theory is also used to develop three indicator based algorithms for approximating the set of optimal solutions. Computational results show the niche of population based algorithms for solving multi-objective problems with variable orderings.

Keywords: variable ordering, hypervolume indicator, approximation, evolutionary algorithms.

1 Introduction

Many complex optimization problems in engineering and mathematical applications involve minimizing a vector-valued objective function $\mathbf{f} := (f_1, \ldots, f_m) : \mathbb{R}^n \to \mathbb{R}^m$. The objective can be wealth, time, cost, safety or performance among others. Such problems are commonly known as multi-objective / vector/ multi-criteria optimization problems. Solving a multi-objective optimization problem (MOP in short) requires specifying an ordering relation in the space $\mathbb{R}^m$ (which lacks a canonical total order). This relation could come from the preferences of a decision maker (DM) interested in solving such a problem. One preference model of a DM characterizes the set of preferred directions in the feasible subset of $\mathbb{R}^m$. Another preference model consist of specifying domination and/ or preference cones and/ or additional axioms (like impartiality, equitability) that need to be satisfied. Pareto-ordering, induced by the nonnegative orthant cone

R.C. Purshouse et al. (Eds.): EMO 2013, LNCS 7811, pp. 66–80, 2013.

$\mathbb{R}^m_+ := \{\mathbf{y} \in \mathbb{R}^m | y_i \geq 0, \forall i = 1, 2, \ldots, m\}$, is a classical way to compare two vector valued objectives. Many times, polyhedral cone orderings are used to include desirable trade-off among the objectives [1].

Yu in his seminal work [2] was the first to propose variable orderings in preference modeling. In this, the ordering relation at a point in the multi-objective space depends on the point itself. In a practical context, this says that preference depends upon the current point in the objective space (so called decisional wealth in [3]). Figure 1 shows an example of a variable ordering in the bi-objective space. We can see that the domination cones at $\mathbf{u}, \mathbf{v}$ and $\mathbf{w}$ are different and $\mathbf{w}$ dominates a portion of the efficient front corresponding to Pareto-ordering. This is a striking difference from usual multi-objective optimization problems, where the ordering is induced by the $\mathbb{R}^m_+$ or a polyhedral cone that remains constant in the entire objective space.

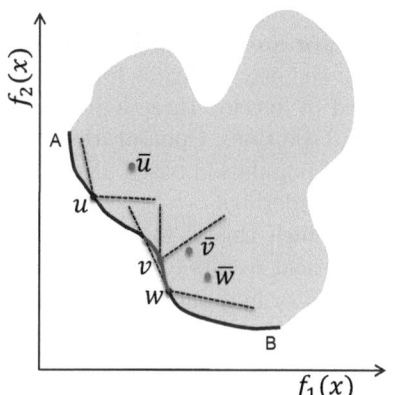 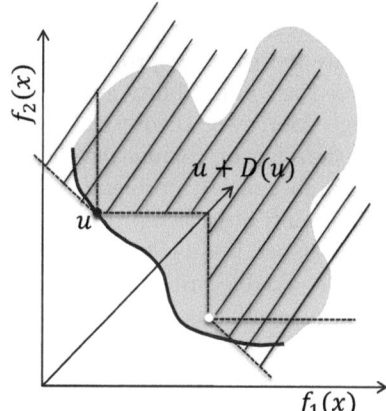

Fig. 1. Schematic of a variable ordering relation. **u** and **w** are optimal points, while **v** is not optimal.

Fig. 2. Schematic of the equitable ordering relation. The shaded area $D(\mathbf{u})$ is a non-convex set and is not a cone.

Variable domination cones are found to be useful in various applications. For example, they have found a recent place in medical engineering where the aim is to merge different medical images obtained by different methods (say computer tomography, ultrasound, positron emission tomography among others). In the problem of medical image registration, one searches for a best transformation map. Variable orderings also arise in a variety of financial applications, general resource allocation models and location theory [4], and in multi-objective n-person cooperative as well as noncooperative games [5]. Equitability [6, 7], for example, is a refinement of Pareto efficiency, where one is interested in the distribution of the outcomes of various objectives rather than their ordering. Figure 2 illustrates equitable variable ordering in the bi-objective space. We can see that the dominated region $D(\mathbf{u})$ is a non-convex set, and is not a cone.

Problems with a variable domination structure give rise to two types of optimality: *minimal* optimality and *nondominated* optimality and these are defined in the next section. These concepts should not be confused with the standard terminology of Pareto-optimal/ nondominated/ efficient (set/ front/ points) [8]. In classical literature [9, 10], we do find some scalarization techniques to find *one* minimal/ nondominated point and some other recent approaches [11, 12] aimed at getting an approximation of minimal/ nondominated solutions. However, all of these assume cone orderings and hence cannot be directly used for set based variable orderings (like equitability). There is a lack of theory for such orderings.

This paper aims to provide theoretical and algorithmic advances for general, non-convex, nonconical, set-based variable orderings. Theoretical results are provided for the most general settings and these results can be used to reduce the pairwise comparisons needed for sorting of a discrete set for optimal points. Assumptions that are needed for the results to hold are also discussed. A hypervolume based indicator measure is also proposed for the first time for such multi-objective optimization with a variable ordering structure. Theoretical results are derived and properties of this indicator are studied. In particular we analyze the compatibility and completeness of the new indicator for variable orderings. Moreover, theoretical results are used to develop three indicator based algorithms for approximating the set of optimal solutions. Computational results on a number of test problems show the niche of population based algorithms for solving multi-objective problems with variable orderings.

This paper is divided into six sections of which this is the first. Section 2 presents variable ordering relations and theoretical results for these. The third section presents a new hypervolume based indicator and analyzes its theoretical properties. Section 4 presents three new indicator based algorithms for variable orderings while Section 5 presents numerical results. Finally, conclusions and extensions of this study are presented at the end of this contribution.

2   Variable Orderings and Their Theoretical Properties

In this section, we first present some formal definitions and concepts and use them later to derive theoretical results.

A nonempty set $C \subseteq \mathbb{R}^m$ is called a *cone* if $c \in C \Rightarrow \lambda c \in C$ for all $\lambda \geq 0$. A cone C is convex if $C + C \subset C$. A cone C is called pointed if it satisfies that $C \cap (-C) = \{0\}$ where 0 is the zero vector in $\mathbb{R}^m$. Hence, the non-negative orthant of $\mathbb{R}^m$, $\mathbb{R}^m_+$ is a closed, convex and pointed cone, and is commonly known as the *Pareto-cone*.

Using the Pareto-cone, we can define a *Pareto-ordering* on the objective space. This allows us to compare vectors in the objective space as follows. We say that a vector $\mathbf{u}$ *Pareto-dominates* a vector $\mathbf{v}$ iff $\mathbf{v} - \mathbf{u} \in \mathbb{R}^m_+ \setminus \{0\}$. If neither $\mathbf{u}$ Pareto-dominates $\mathbf{v}$ nor $\mathbf{v}$ Pareto-dominates $\mathbf{u}$, we call $\mathbf{u}$ and $\mathbf{v}$ *Pareto-nondominated* to each other. Moreover, we call the Pareto-ordering as the standard ordering as it is the most-widely used in multi-objective optimization problems (see [1, 8]).

Yu [13] proposed to use a constant cone to model decision makers' preferences. Using a cone $C \subseteq \mathbb{R}^m$, he defined a different domination structure as follows: the vector $\mathbf{u}$ C-*dominates* the vector $\mathbf{v}$, if $\mathbf{v} - \mathbf{u} \in C \setminus \{\mathbf{0}\}$. The cone C can be non-convex as well and is discussed in [14, 15, 13].

A multi-objective optimization problem involves $m \geq 2$ objectives that need to ne *minimized* over a constraint set. Formally, let $\mathbf{X} \subseteq \mathbb{R}^n$ be a set of feasible alternatives and $\mathbf{f} : \mathbf{X} \to \mathbb{R}^m$ be a vector valued objective function. The spaces $\mathbf{X}$ and $\mathbf{f}(\mathbf{X}) := \{\mathbf{f}(\mathbf{x}) : \mathbf{x} \in \mathbf{X}\}$ are also called the *decision space* and the *objective space*, respectively.

Minimization in $\mathbb{R}^m$ requires a way to compare two points in $\mathbb{R}^m$ and this is given by an ordering relation. An ordering relation, more general than Pareto is based on orderings induced by sets [16] and is defined next. For this, let $\mathrm{Leb}(S)$ denote the Lebesgue measure of a set $S \subseteq \mathbb{R}^m$ and $\mathrm{int}(S)$ be the interior of S [17].

Definition 1. *Let a set $D \subset \mathbb{R}^m$ be an ordering set such that $D \cap (-D) = \{\mathbf{0}\}$ and let $\mathrm{Leb}(D) > 0$. Moreover, let $\mathbf{u}$ and $\mathbf{v}$ be two vectors in $\mathbb{R}^m$. Then,*

1. *$\mathbf{u} \leq_D \mathbf{v}$ ($\mathbf{u}$ weakly D-dominates $\mathbf{v}$) $\iff \mathbf{v} - \mathbf{u} \in D$*
2. *$\mathbf{u} <_D \mathbf{v}$ ($\mathbf{u}$ D-dominates $\mathbf{v}$) $\iff \mathbf{v} - \mathbf{u} \in D \setminus \{\mathbf{0}\}$.*
3. *$\mathbf{u} \ll_D \mathbf{v}$ ($\mathbf{u}$ strictly D-dominates $\mathbf{v}$) $\iff \mathbf{v} - \mathbf{u} \in \mathrm{int}(D)$.*

Note that $\mathrm{Leb}(D) = 0$ is not desired in Definition 1 as then almost all the elements of $\mathbb{R}^m$ are non-dominated.

Definition 2. *A point $\hat{\mathbf{x}} \in \mathbf{X}$ is called D-optimal if no other point in $\mathbf{X}$ D-dominates it. Equivalently, a point $\hat{\mathbf{x}} \in \mathbf{X}$ is D-optimal if and only if*

$$(\{\mathbf{f}(\hat{\mathbf{x}})\} - D) \cap \mathbf{f}(\mathbf{X}) = \{\mathbf{f}(\hat{\mathbf{x}})\}.$$

Let $\mathbf{X}_D$ and $\mathcal{E}_D := \mathbf{f}(\mathbf{X}_D)$ denote the set of D-optimal points and the set of D-efficient points, respectively.

Definition 2 is a solution concept of a multi-objective optimization problem with a *constant* ordering set D. As a special case, any $\mathbb{R}^m_+$-optimal point is called as a Pareto-optimal point. Hence, a point $\hat{\mathbf{x}} \in \mathbf{X}$ is Pareto-optimal if and only if $(\{\mathbf{f}(\hat{\mathbf{x}})\} - \mathbb{R}^m_+) \cap \mathbf{f}(\mathbf{X}) = \{\mathbf{f}(\hat{\mathbf{x}})\}$.

In [3], it is noticed that the importance of objective functions may change during the decision making process depending upon the current objective function value. Some examples of this are also given in [18]. A constant cone is not adequate to compare objectives in such cases. Due to this, several attempts have been made in the classical literature to extend dominance ideas using variable cones and ordering sets and two domination relations are defined as follows.

Definition 3. *Let a (set-valued) variable ordering map $\mathcal{D} : \mathbb{R}^m \rightrightarrows \mathbb{R}^m$ be such that $\mathcal{D}(\mathbf{w}) \cap (-\mathcal{D}(\mathbf{w})) = \{\mathbf{0}\}$ and $\mathrm{Leb}(\mathcal{D}(\mathbf{w})) > 0$ hold for all $\mathbf{w} \in \mathbb{R}^m$. Moreover, let $\mathbf{u}$ and $\mathbf{v}$ be two vectors in $\mathbb{R}^m$. Then,*

1. *$\mathbf{u} \prec^m_{\mathcal{D}} \mathbf{v} \iff \mathbf{v} \in \{\mathbf{u}\} + \mathcal{D}(\mathbf{v})$ (minimality),*
2. *$\mathbf{u} \prec^n_{\mathcal{D}} \mathbf{v} \iff \mathbf{v} \in \{\mathbf{u}\} + \mathcal{D}(\mathbf{u})$ (nondomination).*

The binary relations $\prec_{\mathcal{D}}^m$ and $\prec_{\mathcal{D}}^n$ are reflexive (as $0 \in \mathcal{D}(\mathbf{w})$ for all $\mathbf{w}$) but in general not transitive. These relations (and the ones in Definition 1) can also be extended to approximation sets. For example, $\mathcal{A} \prec_{\mathcal{D}}^m \mathcal{B}$ for two sets $\mathcal{A}, \mathcal{B} \subset \mathbb{R}^m$ if and only if $\forall \mathbf{b} \in \mathcal{B} : \exists \mathbf{a} \in \mathcal{A}$ such that $\mathbf{a} \prec_{\mathcal{D}}^m \mathbf{b}$. From Figure 1, we see that $\mathbf{w} \prec_{\mathcal{D}}^n \mathbf{v}$ but $\mathbf{w} \not\prec_{\mathcal{D}}^m \mathbf{v}$. These variable ordering relations defined in Definition 3 lead to the following two optimality notions.

Definition 4. *A point $\hat{\mathbf{u}} \in \mathbf{f}(\mathbf{X})$ is called a $\mathcal{D}$-minimal point of $\mathbf{f}(\mathbf{X})$ if*

$$(\{\hat{\mathbf{u}}\} - \mathcal{D}(\hat{\mathbf{u}})) \cap \mathbf{f}(\mathbf{X}) = \{\hat{\mathbf{u}}\},$$

or equivalently, if there is no $\mathbf{v} \in \mathbf{f}(\mathbf{X})$ such that $\mathbf{v} \in \{\hat{\mathbf{u}}\} - \mathcal{D}(\hat{\mathbf{u}}) \setminus \{\mathbf{0}\}$. The set of all $\mathcal{D}$-minimal points is called the $\mathcal{D}$-minimal set and is denoted by $\mathcal{E}_{\mathcal{D}}^m$.

Definition 5. *A point $\hat{\mathbf{u}} \in \mathbf{f}(\mathbf{X})$ is called a $\mathcal{D}$-nondominated point of $\mathbf{f}(\mathbf{X})$ if there is no $\mathbf{v} \in (\mathbf{X})$ such that*

$$\hat{\mathbf{u}} \in \{\mathbf{v}\} + \mathcal{D}(\mathbf{v}) \setminus \{\mathbf{0}\}.$$

The set of all $\mathcal{D}$-nondominated elements is called the $\mathcal{D}$-nondominated-set and is denoted by $\mathcal{E}_{\mathcal{D}}^n$.

Let $\mathbf{X}_{\mathcal{D}}^m$ and $\mathbf{X}_{\mathcal{D}}^n$ denote the pre-images (the points in the decision space) of the $\mathcal{D}$-minimal and the $\mathcal{D}$-nondominated sets respectively. This means that $\mathcal{E}_{\mathcal{D}}^m = \mathbf{f}(\mathbf{X}_{\mathcal{D}}^m)$ and $\mathcal{E}_{\mathcal{D}}^n = \mathbf{f}(\mathbf{X}_{\mathcal{D}}^n)$. Note that in Definitions 4 and 5 we could also take an arbitrary set $\mathcal{S} \subset \mathbb{R}^m$ and define $\mathcal{D}$-minimal and $\mathcal{D}$-nondominated sets in a similar way (replacing $\mathbf{f}(\mathbf{X})$ by $\mathcal{S}$). For this scenario, let the set of all $\mathcal{D}$-minimal (nondominated) elements be called as the $\mathcal{D}$-*minimal-(nondominated)set of $\mathcal{S}$* and be denoted by $\mathcal{E}_{\mathcal{D}}^m(\mathcal{S})$ ($\mathcal{E}_{\mathcal{D}}^n(\mathcal{S})$).

The concept of minimal elements is described in [19, 9]. The concept of non-dominated elements is based on [19, 13, 20, 9]. The nondominated-set and the minimal-set are in general different. It is easy to verify that both of these sets are equal if a constant ordering set is used, i.e. if $\mathcal{D}(\mathbf{y}) := D$. Definitions 4 and 5 (and the above concepts of variable orderings) have (mathematical and) real-world applications (see Section 1). For example, recall that the equitable relation, which has many (real-world) applications in allocation problems, is based on a non-convex set (not a cone) based variable ordering. We assume the following throughout this paper.

Blanket Assumption: *The set-valued variable ordering map $\mathcal{D}$ is such that $\bigcap_{\mathbf{w} \in \mathbf{f}(\mathbf{X})} \mathcal{D}(\mathbf{w}) \supseteq \mathcal{K}$, where $\mathcal{K}$ is a closed, convex, nonempty, and pointed cone. Moreover, the set $\mathbf{f}(\mathbf{X})$ is compact.*

The next two lemma relates variable optimality to optimality with respect to a constant cone.

Lemma 1. $\mathbf{X}_{\mathcal{D}}^m \subseteq X_{\mathcal{K}}, \mathbf{X}_{\mathcal{D}}^n \subseteq X_{\mathcal{K}}, \mathcal{E}_{\mathcal{D}}^m \subseteq \mathcal{E}_{\mathcal{K}}, \text{ and } \mathcal{E}_{\mathcal{D}}^n \subseteq \mathcal{E}_{\mathcal{K}}.$

Proof: As we assumed that all the variable sets contain $\mathcal{K}$ (from the Blanket Assumption), the result follows easily from [12, Lemma 3.1]. $\qquad\square$

Lemma 2. $\hat{\mathbf{u}} \in \mathbf{X}_{\mathcal{D}}^m$ *if and only if* $\hat{\mathbf{u}} \in \mathbf{X}_{\mathcal{D}(\hat{\mathbf{u}})}^m$.

Proof: The proof easily follows from the definitions of $\mathbf{X}_{\mathcal{D}}^m$ and $X_{\mathcal{D}(\hat{\mathbf{u}})}^m$. □

Remark 1. *Note that although* $\mathcal{K}$ *in* $\bigcap_{\mathbf{w} \in \mathbf{f}(\mathbf{X})} \mathcal{D}(\mathbf{w}) \supseteq \mathcal{K}$ *is assumed to be a cone in Blanket Assumption, this does not mean that the sets* $\mathcal{D}(\mathbf{w})$ *itself are cones. For example, From Figure 2 (and from [6]) we see that* $\mathbb{R}_+^m$ *can be used as choice for* $\mathcal{K}$. *Hence, Lemma 1 shows that an equitably optimal point is a Pareto-optimal point (another proof of this can be found in [6]).*

The next two theorems are motivated by practical considerations to reduce the computation effort of a population based multi-objective algorithm for finding $\mathcal{D}$-minimal and $\mathcal{D}$-nondominated points.

Theorem 1. *Let* $\mathcal{D}(\mathbf{w}) + \mathcal{K} \subseteq \mathcal{D}(\mathbf{w})$ *be satisfied for all* $\mathbf{w} \in \mathbf{f}(\mathbf{X})$ *and let* $\mathbf{f}(\mathbf{X})$ *be a compact set. Then,* $\hat{\mathbf{u}} \in \mathbf{f}(\mathbf{X})$ *is a* $\mathcal{D}$-*minimal point of* $\mathbf{f}(\mathbf{X})$ *if and only if* $\hat{\mathbf{u}} \in \mathcal{E}_{\mathcal{K}}$ *and* $\hat{\mathbf{u}}$ *is a* $\mathcal{D}$-*minimal point of* $\mathcal{E}_{\mathcal{K}}$, *i.e.,* $\mathcal{E}_{\mathcal{D}}^m = \mathcal{E}_{\mathcal{D}}^m(\mathcal{E}_{\mathcal{K}})$.

Proof: ($\Rightarrow$) Let $\hat{\mathbf{u}} \in \mathcal{E}_{\mathcal{D}}^m$. Hence, $\hat{\mathbf{u}} \in \mathbf{f}(\mathbf{X})$ is a minimal point of $\mathbf{f}(\mathbf{X})$ and, $(\hat{\mathbf{u}} - \mathcal{D}(\hat{\mathbf{u}})) \cap \mathbf{f}(\mathbf{X}) = \{\hat{\mathbf{u}}\}$. $\mathcal{E}_{\mathcal{K}} \subseteq \mathbf{f}(\mathbf{X})$ together with Lemma 1 (as the conditions for Lemma 1 are satisfied) gives that $\hat{\mathbf{u}} \in \mathcal{E}_{\mathcal{K}}$ and moreover

$$(\{\hat{\mathbf{u}}\} - \mathcal{D}(\hat{\mathbf{u}})) \cap \mathcal{E}_{\mathcal{K}} = \{\hat{\mathbf{u}}\}.$$

This means that there is no $\mathbf{v} \in \mathcal{E}_{\mathcal{K}}$ such that $\mathbf{v} \in \{\hat{\mathbf{u}}\} - \mathcal{D}(\hat{\mathbf{u}}) \setminus \{\mathbf{0}\}$ and hence, from Definition 4 (replacing $\mathbf{f}(\mathbf{X})$ by $\mathcal{E}_{\mathcal{K}}$), we obtain that $\hat{\mathbf{u}} \in \mathcal{E}_{\mathcal{K}}$ is a minimal point of $\mathcal{E}_{\mathcal{K}}$. This means that $\hat{\mathbf{u}} \in \mathcal{E}_{\mathcal{D}}^m(\mathcal{E}_{\mathcal{K}})$ and hence, $\mathcal{E}_{\mathcal{D}}^m \subseteq \mathcal{E}_{\mathcal{D}}^m(\mathcal{E}_{\mathcal{K}})$.

($\Leftarrow$) Let $\hat{\mathbf{u}} \in \mathcal{E}_{\mathcal{D}}^m(\mathcal{E}_{\mathcal{K}})$. Hence, $\hat{\mathbf{u}} \in \mathcal{E}_{\mathcal{K}} \subseteq \mathbf{f}(\mathbf{X})$ is a minimal point of $\mathcal{E}_{\mathcal{K}}$. Thus, there is no $\mathbf{v} \in \mathcal{E}_{\mathcal{K}}$ such that $\mathbf{v} \in \{\hat{\mathbf{u}}\} - \mathcal{D}(\hat{\mathbf{u}}) \setminus \{\mathbf{0}\}$. In order to show that $\hat{\mathbf{u}}$ is also a minimal point of $\mathbf{f}(\mathbf{X})$, we take an arbitrary but fixed element $\mathbf{w} \in \mathbf{f}(\mathbf{X}) \setminus \mathcal{E}_{\mathcal{K}}$ and assume that

$$\mathbf{w} \in \{\hat{\mathbf{u}}\} - \mathcal{D}(\hat{\mathbf{u}}) \setminus \{\mathbf{0}\}. \tag{1}$$

Now, as $\mathbf{w} \notin \mathcal{E}_{\mathcal{K}}$ and as the set $\mathbf{f}(\mathbf{X})$ is assumed to be compact, we get the existence of a $\hat{\mathbf{w}} \in \mathcal{E}_{\mathcal{K}}$ which $\mathcal{K}$-dominates it, i.e., $\hat{\mathbf{w}} \in \{\mathbf{w}\} - \mathcal{K} \setminus \{\mathbf{0}\}$. This, together with (1) and that $\mathcal{D}(\hat{\mathbf{u}}) + \mathcal{K} \subseteq \mathcal{D}(\hat{\mathbf{u}})$ (from the assumption), gives that

$$\hat{\mathbf{w}} \in \{\hat{\mathbf{u}}\} - \mathcal{D}(\hat{\mathbf{u}}) \setminus \{\mathbf{0}\} - \mathcal{K} \setminus \{\mathbf{0}\} \subseteq \{\hat{\mathbf{u}}\} - \mathcal{D}(\hat{\mathbf{u}}) \setminus \{\mathbf{0}\}.$$

Using the last inclusion and that $\hat{\mathbf{w}} \in \mathcal{E}_{\mathcal{K}}$, we arrive at a contradiction to the $\mathcal{D}$-minimality of $\hat{\mathbf{u}}$ in $\mathcal{E}_{\mathcal{K}}$. Hence $\mathcal{E}_{\mathcal{D}}^m \supseteq \mathcal{E}_{\mathcal{D}}^m(\mathcal{E}_{\mathcal{K}})$ and the theorem follows. □

Theorem 2. *Let* $\mathcal{D}(\mathbf{w}) + \mathcal{K} \subseteq \mathcal{D}(\mathbf{w})$ *be satisfied for all* $\mathbf{w} \in \mathbf{f}(\mathbf{X})$, *let* $\mathbf{f}(\mathbf{X})$ *be a compact set, and let*

$$\mathbf{v} - \mathbf{u} \in \mathcal{K} \Rightarrow \mathcal{D}(\mathbf{v}) \subseteq \mathcal{D}(\mathbf{u}), \text{ for all } \mathbf{v}, \mathbf{u} \in \mathbf{f}(\mathbf{X}). \tag{2}$$

Then, $\hat{\mathbf{u}} \in \mathbf{f}(\mathbf{X})$ *is a* $\mathcal{D}$-*nondominated point of* $\mathbf{f}(\mathbf{X})$ *if and only if* $\hat{\mathbf{u}} \in \mathcal{E}_{\mathcal{K}}$ *and* $\hat{\mathbf{u}}$ *is a* $\mathcal{D}$-*nondominated point of* $\mathcal{E}_{\mathcal{K}}$, *i.e.,* $\mathcal{E}_{\mathcal{D}}^n = \mathcal{E}_{\mathcal{D}}^n(\mathcal{E}_{\mathcal{K}})$.

Proof: ($\Rightarrow$) Let $\hat{\mathbf{u}} \in \mathcal{E}_{\mathcal{D}}^n$. Hence, $\hat{\mathbf{u}} \in \mathbf{f}(\mathbf{X})$ is a nondominated point of $\mathbf{f}(\mathbf{X})$ and Definition 5 gives that there is *no* $\mathbf{v} \in \mathbf{f}(\mathbf{X})$ such that

$$\hat{\mathbf{u}} \in \{\mathbf{v}\} + \mathcal{D}(\mathbf{v}) \setminus \{\mathbf{0}\}. \tag{3}$$

This obviously means that there is no $\mathbf{v} \in \mathcal{E}_{\mathcal{K}}$ such that $\hat{\mathbf{u}} \in \{\mathbf{v}\} + \mathcal{D}(\mathbf{v}) \setminus \{\mathbf{0}\}$ and hence, from Definition 5 (replacing $\mathbf{f}(\mathbf{X})$ by $\mathcal{E}_{\mathcal{K}}$) we obtain that $\hat{\mathbf{u}} \in \mathcal{E}_{\mathcal{K}}$ is a nondominated point of $\mathcal{E}_{\mathcal{K}}$. Hence, $\mathcal{E}_{\mathcal{D}}^n \subseteq \mathcal{E}_{\mathcal{D}}^n(\mathcal{E}_{\mathcal{K}})$.

($\Leftarrow$) Let $\hat{\mathbf{u}} \in \mathcal{E}_{\mathcal{D}}^n(\mathcal{E}_{\mathcal{K}})$. Hence, $\hat{\mathbf{u}} \in \mathbf{f}(\mathbf{X})$ is a nondominated point of $\mathcal{E}_{\mathcal{K}}$ and there is no $\mathbf{v} \in \mathcal{E}_{\mathcal{K}}$ such that $\hat{\mathbf{u}} \in \{\mathbf{v}\} + \mathcal{D}(\mathbf{v}) \setminus \{\mathbf{0}\}$. This happens if and only if

$$\hat{\mathbf{u}} \notin \{\mathbf{v}\} + \mathcal{D}(\mathbf{v}) \setminus \{\mathbf{0}\} \quad \text{for all } \mathbf{v} \in \mathcal{E}_{\mathcal{K}}. \tag{4}$$

We will show that the set $\mathcal{E}_{\mathcal{K}}$ in (4) can be replaced by $\mathbf{f}(\mathbf{X})$. Taking an arbitrary but fixed element $\mathbf{w} \in \mathbf{f}(\mathbf{X}) \setminus \mathcal{E}_{\mathcal{K}}$ and assuming that $\mathbf{w} \prec_{\mathcal{D}}^n \hat{\mathbf{u}}$ we get the following:

$$\hat{\mathbf{u}} \in \{\mathbf{w}\} + \mathcal{D}(\mathbf{w}) \setminus \{\mathbf{0}\}. \tag{5}$$

Now, as $\mathbf{w} \notin \mathcal{E}_{\mathcal{K}}$ and as $\mathbf{f}(\mathbf{X})$ is assumed to be compact we get the existence of a $\hat{\mathbf{w}} \in \mathcal{E}_{\mathcal{K}}$ which $\mathcal{K}$-dominates $\mathbf{w}$, i.e., $\mathbf{w} \in \{\hat{\mathbf{w}}\} + \mathcal{K} \setminus \{\mathbf{0}\}$. This together with (5) and that $\mathcal{D}(\hat{\mathbf{u}}) + \mathcal{K} \subseteq \mathcal{D}(\hat{\mathbf{u}})$ (from the assumption) gives

$$\hat{\mathbf{u}} \in \{\mathbf{w}\} + \mathcal{D}(\mathbf{w}) \setminus \{\mathbf{0}\} \subseteq \{\hat{\mathbf{w}}\} + \mathcal{K} \setminus \{\mathbf{0}\} + \mathcal{D}(\mathbf{w}) \setminus \{\mathbf{0}\}$$
$$\subseteq \{\hat{\mathbf{w}}\} + \mathcal{D}(\mathbf{w}) \setminus \{\mathbf{0}\}.$$

Now, as $\hat{\mathbf{w}}$ $\mathcal{K}$-dominates $\mathbf{w}$, from (2) we obtain that $\mathcal{D}(\mathbf{w}) \subseteq \mathcal{D}(\hat{\mathbf{w}})$. This shows that $\hat{\mathbf{u}} \in \{\hat{\mathbf{w}}\} + \mathcal{D}(\hat{\mathbf{w}}) \setminus \{\mathbf{0}\}$ for $\hat{\mathbf{w}} \in \mathcal{E}_{\mathcal{K}}$, and we arrive at a contradiction. Hence $\mathcal{E}_{\mathcal{D}}^n \supseteq \mathcal{E}_{\mathcal{D}}^n(\mathcal{E}_{\mathcal{K}})$ and the theorem follows. $\square$

Remark 2. *Theorems 1 and 2 show that in order to check if a point is minimal or nondominated, it is sufficient to check the variable set conditions w.r.t. $\mathcal{K}$-nondominated elements only. For any algorithm, this would drastically reduce the computational effort. Finding a fast algorithm (extensions of efficient divide and conquer based approaches [21, 22]) for sorting based on $\mathcal{D}$ domination structures is a relevant open question.*

Theorems 1 and 2 generalize results from [12, 11] to general variable orderings. It is possible to prove Theorems 1 and 2 under the weaker *external stability* condition [23] rather than compactness of $\mathbf{f}(\mathbf{X})$. For discrete problems, the sets involved are always compact.

The assumptions in the above two theorems seem reasonable. They are shown to be satisfied for equitable cones for example. Moreover, Condition (2) seems reasonable as a larger cone dominates a larger region and if $\mathbf{u}$ Pareto-dominates $\mathbf{v}$, then the dominated region by the point $\mathbf{u}$ should intuitively be larger ($\mathcal{D}(\mathbf{v}) \subseteq \mathcal{D}(\mathbf{u})$). This assumption is in line with the Pareto dominance compliant mechanism (see the discussion in [24]).

Transitivity, antisymmetry, and other properties of the $\prec_{\mathcal{D}}^m$ and $\prec_{\mathcal{D}}^n$ set based binary relations are not yet investigated in detail although there are promising results for cone based variable ordering [12] and constant cone orderings [25].

3   New Indicators and Their Theoretical Properties

In this section, we present new hypervolume based indicators for multi-objective problems with a variable ordering structure. These indicator can be used to compare algorithms and can also be used in the selection mechanism of a population based algorithm (discussed in the next section). For brevity, all the results in this section are presented for $\mathcal{D}$-minimal points but similar results can also be shown for $\mathcal{D}$-nondominated points.

Definition 6. *Let* $\mathcal{K} \supseteq \mathbb{R}^m_+$, *let* $\mathcal{S} \subset \mathbb{R}^m$, *and let* $\mathbf{r} \in \mathbb{R}^m$ *indicate the reference point. The* preferred minimal-hypervolume *is defined by*

$$\mathcal{H}^m(\mathcal{S}, \mathbf{r}) := \text{Leb}\left(\left\{ \mathbf{w} \in \mathbb{R}^m | \exists \mathbf{v} \in \mathcal{E}^m_{\mathcal{D}}(\mathcal{S}) : \mathbf{v} \leq_{\mathbb{R}^m_+} \mathbf{w} \leq_{\mathbb{R}^m_+} \mathbf{r} \right\} \right). \qquad (6)$$

In a similar way, the minimal-hypervolume *w.r.t.* $\mathcal{S}$ *and the* minimal-hypervolume contribution *of a set* $\mathcal{A} \subseteq \mathcal{S}$ *are defined by*

$$\mathcal{H}^m(\mathcal{A}, \mathcal{S}, \mathbf{r}) := \text{Leb}\left(\left\{ \mathbf{w} \in \mathbb{R}^m | \exists \mathbf{v} \in \mathcal{E}^m_{\mathcal{D}}(\mathcal{S}) \cap \mathcal{E}^m_{\mathcal{D}}(\mathcal{A}) : \mathbf{v} \leq_{\mathbb{R}^m_+} \mathbf{w} \leq_{\mathbb{R}^m_+} \mathbf{r} \right\} \right), \text{ and}$$

$$H^m(\mathcal{A}, \mathcal{S}, \mathbf{r}) := \mathcal{H}^m(\mathcal{S}, \mathbf{r}) - \mathcal{H}^m(\mathcal{S} \setminus \mathcal{A}, \mathcal{S}, \mathbf{r}),$$

respectively. Preferred nondominated-hypervolume $(\mathcal{H}^n(\mathcal{S}, \mathbf{r}))$, nondominated hypervolume contribution w.r.t. $\mathcal{S}$ $(\mathcal{H}^n(\mathcal{A}, \mathcal{S}, \mathbf{r}))$, *and* nondominated-hypervolume contribution $(H^n(\mathcal{A}, \mathcal{S}, \mathbf{r}))$ *are defined analogously for* $\mathcal{D}$-nondominated elements. *For empty sets all the above hypervolumes are defined to be zero.*

Definition 6 can be used for finite, countable and uncountable sets, although due to finite size of population based algorithms, we assume that $\mathcal{S}$ is a finite set. The above hypervolume concepts for variable domination structures calculates the Lebesgue measure (the $\mathbb{R}^m_+$ hypervolume- see the excellent reference [26]) using points from the $\mathcal{D}$-minimal set only. In this way, the dominated points w.r.t. the $\prec^m_{\mathcal{D}}$ ordering, contribute zero to $\mathcal{H}^m(\mathcal{S}, \mathbf{r})$. This is an extended hypervolume definition that can be used for any arbitrary binary relation $\precsim$ by computing the usual hypervolume for only points that are not dominated w.r.t. to $\precsim$. The next lemma characterizes points with a positive minimal-hypervolume contribution.

Lemma 3. *Let* $\mathcal{K} \supseteq \mathbb{R}^m_+$ *and let* $\mathcal{S} \subset \mathbb{R}^m$ *be a finite set. Then,* $\forall \mathbf{v} \in \mathcal{S}$ *and* $\forall \mathbf{r} \in \mathbb{R}^m$ *it holds that:*

$$H^m(\{\mathbf{v}\}, \mathcal{S}, \mathbf{r}) > 0 \iff \mathbf{v} \in \mathcal{E}^m_{\mathcal{D}}(\mathcal{S}) \text{ and } \mathbf{v} \ll_{\mathbb{R}^m_+} \mathbf{r} \iff \mathcal{H}^m(\{\mathbf{v}\}, \mathcal{S}, \mathbf{r}) > 0.$$

Proof: Obviously, all the points $\mathbf{w} \notin \mathcal{E}^m_{\mathcal{D}}(\mathcal{S})$ contribute zero to the minimal-hypervolume contribution. Moreover, from Definition 6 we obtain that the contribution of a point is (strictly) positive if and only if $\mathbf{v}$ strictly $\mathbb{R}^m_+$-dominates $\mathbf{v}$ and $\mathbf{v} \in \mathcal{E}^m_{\mathcal{D}}(\mathcal{S})$. The second part follows from the definition of $\mathcal{H}^m(\{\mathbf{v}\}, \mathcal{S}, \mathbf{r})$ (note that $\mathbf{v} \in \mathcal{E}^m_{\mathcal{D}}(\{\mathbf{v}\})$). $\qquad \Box$

The next theorem shows that many properties of the usual hypervolume indicator [26] are preserved in the variable ordering based hypervolume indicator. Let for arbitrary sets $\mathcal{A}, \mathcal{B} \subset \mathbb{R}^m$, $\mathcal{A} \triangleright \mathcal{B}$ mean that $\mathcal{A} \neq \mathcal{B}$ and $\mathcal{A} \prec^m_{\mathcal{D}} \mathcal{B}$.

Theorem 3. *Let $\mathcal{K} \supseteq \mathbb{R}^m_+$ and let $\mathrm{nad}(\mathcal{S})$ denote the nadir point of a set $\mathcal{S} \subset \mathbb{R}^m$ and $\bullet$ be a (fixed) relation out of $\{\leq, \geq, =\}$. Then, for all finite sets $\mathcal{A}, \mathcal{B} \subset \mathbb{R}^m$, and $\mathbf{r} \in \mathbb{R}^m$ holds:*

1. $\forall \mathbf{a}, \mathbf{b} \in \mathbb{R}^m : (\mathrm{nad}(\mathbf{a}, \mathbf{b}) \ll_{\mathbb{R}^m_+} \mathbf{r}) : H^m(\{\mathbf{a}\}, \{\mathbf{a}, \mathbf{b}\}, \mathbf{r}) \bullet H^m(\{\mathbf{b}\}, \{\mathbf{a}, \mathbf{b}\}, \mathbf{r}) \Longleftrightarrow$
   $\mathcal{H}^m(\{\mathbf{a}\}, \{\mathbf{a}, \mathbf{b}\}, \mathbf{r}) \bullet \mathcal{H}^m(\{\mathbf{b}\}, \{\mathbf{a}, \mathbf{b}\}, \mathbf{r})$
2. $\mathcal{A} = \mathcal{B} \;\Rightarrow\; \mathcal{H}^m(\mathcal{A}, \mathbf{r}) = \mathcal{H}^m(\mathcal{B}, \mathbf{r}) = \mathcal{H}^m(\mathcal{A}, \mathcal{A} \cup \mathcal{B}, \mathbf{r}) = \mathcal{H}^m(\mathcal{A}, \mathcal{A} \cup \mathcal{B}, \mathbf{r})$
3. $\mathcal{A} \prec_{\mathcal{D}}^m \mathcal{B} \;\Rightarrow\; \mathcal{H}^m(\mathcal{A}, \mathcal{A} \cup \mathcal{B}, \mathbf{r}) \geq \mathcal{H}^m(\mathcal{B}, \mathcal{A} \cup \mathcal{B}, \mathbf{r})$
4. (▷-*compatibility*) $\mathcal{B} \not\prec_{\mathcal{D}}^m \mathcal{A} \Leftarrow \exists \mathbf{r} \in \mathbb{R}^m : \mathcal{H}^m(\mathcal{A}, \mathcal{A} \cup \mathcal{B}, \mathbf{r}) > \mathcal{H}^m(\mathcal{B}, \mathcal{A} \cup \mathcal{B}, \mathbf{r})$
5. (▷-*completeness*) $\mathcal{A} \prec_{\mathcal{D}}^m \mathcal{B}$ and $\mathcal{B} \not\prec_{\mathcal{D}}^m \mathcal{A} \Rightarrow \forall \mathbf{r} : (\mathrm{nad}(\mathcal{A} \cup \mathcal{B}) \ll_{\mathbb{R}^m_+} \mathbf{r}) :$
   $\mathcal{H}^m(\mathcal{A}, \mathcal{A} \cup \mathcal{B}, \mathbf{r}) > \mathcal{H}^m(\mathcal{B}, \mathcal{A} \cup \mathcal{B}, \mathbf{r})$

Proof: 1.: The statement holds trivially if $\mathbf{a} = \mathbf{b}$ ($H^m(\{\mathbf{a}\}, \{\mathbf{a}, \mathbf{b}\}, \mathbf{r}) = 0$). If $\mathbf{a} \prec_{\mathcal{D}}^m \mathbf{b}$ and $\mathbf{a} \neq \mathbf{b}$, then $\{\bullet\} = \{>\}$, $\mathcal{E}_{\mathcal{D}}^m(\{\mathbf{a}, \mathbf{b}\}) = \{\mathbf{a}\}$, $H^m(\{\mathbf{b}\}, \{\mathbf{a}, \mathbf{b}\}, \mathbf{r}) = \mathcal{H}^m(\{\mathbf{b}\}, \{\mathbf{a}, \mathbf{b}\}, \mathbf{r}) = 0$ and the other terms are strictly positive (analogously if $\mathbf{a}$ and $\mathbf{b}$ are interchanged). If $\mathbf{a} \not\prec_{\mathcal{D}}^m \mathbf{b}$, $\mathbf{b} \not\prec_{\mathcal{D}}^m \mathbf{a}$ and $\mathbf{a} \neq \mathbf{b}$, then both the elements are also incomparable according to $\mathbb{R}^m_+$ cone (as $\bigcap_{\mathbf{w} \in \mathbb{R}^m} \mathcal{D}(\mathbf{w}) \supseteq \mathcal{K} \subseteq \mathbb{R}^m_+$) and the proof follows from [26, Case (iii) Lemma 2.5].

2.: $\mathcal{A} = \mathcal{B}$ means that the sets $\mathcal{E}_{\mathcal{D}}^m(\mathcal{A})$ and $\mathcal{E}_{\mathcal{D}}^m(\mathcal{B})$ are equal and hence 2. follows by Definition 6.

3.: Recall that $\mathcal{A} \prec_{\mathcal{D}}^m \mathcal{B}$ means that $\forall \mathbf{b} \in \mathcal{B} : \exists \mathbf{a} \in \mathcal{A}$ such that $\mathbf{a} \prec_{\mathcal{D}}^m \mathbf{b}$. It is easy to show that would also imply that $\mathcal{E}_{\mathcal{D}}^m(\mathcal{A} \cup \mathcal{B}) \cap \mathcal{E}_{\mathcal{D}}^m(\mathcal{B}) \subseteq \mathcal{E}_{\mathcal{D}}^m(\mathcal{A} \cup \mathcal{B}) \cap \mathcal{E}_{\mathcal{D}}^m(\mathcal{A})$ implying that 3. holds.

4.: If $\mathcal{B} \prec_{\mathcal{D}}^m \mathcal{A}$ then from 2. we obtain $\mathcal{H}^m(\mathcal{B}, \mathcal{A} \cup \mathcal{B}, \mathbf{r}) \geq \mathcal{H}^m(\mathcal{A}, \mathcal{A} \cup \mathcal{B}, \mathbf{r})$ which is the negation of the right hand side of 4. Hence, $\mathcal{H}^m(\mathcal{A}, \mathcal{A} \cup \mathcal{B}, \mathbf{r}) > \mathcal{H}^m(\mathcal{B}, \mathcal{A} \cup \mathcal{B}, \mathbf{r})$ must hold for at least one $\mathbf{r} \in \mathbb{R}^m$.

5.: From 2., $\mathcal{A} \prec_{\mathcal{D}}^m \mathcal{B}$ implies that $\mathcal{H}^m(\mathcal{A}, \mathcal{A} \cup \mathcal{B}, \mathbf{r}) \geq \mathcal{H}^m(\mathcal{B}, \mathcal{A} \cup \mathcal{B}, \mathbf{r})$. Moreover, $\mathcal{A} \prec_{\mathcal{D}}^m \mathcal{B}$ and $\mathcal{B} \not\prec_{\mathcal{D}}^m \mathcal{A}$ means $\mathcal{E}_{\mathcal{D}}^m(\mathcal{A} \cup \mathcal{B}) \cap \mathcal{E}_{\mathcal{D}}^m(\mathcal{B}) \subset \mathcal{E}_{\mathcal{D}}^m(\mathcal{A} \cup \mathcal{B}) \cap \mathcal{E}_{\mathcal{D}}^m(\mathcal{A})$. Hence, there is a $\mathbf{v} \in \mathcal{E}_{\mathcal{D}}^m(\mathcal{A} \cup \mathcal{B}) \cap \mathcal{E}_{\mathcal{D}}^m(\mathcal{A})$ which contributes (strictly) positive to the minimal hypervolume and hence the result follows.   □

There has been lot of work on the compatibility, completeness and other properties of the usual hypervolume indicator [26, 24]. Theorem 3 assumes significance as it shows that these also hold in the variable ordering based hypervolume indicator. This is due to the use of $\mathcal{D}$-minimal sets in Definition 6.

The final result in this section relates the complexity of computing variable ordering based hypervolumes. For this, let $\mathcal{C}(m, n)$ denote the complexity of computing the usual hypervolume (using $\mathcal{D} := \mathbb{R}^m_+$) of a set of n points.

Theorem 4. *Computing the preferred minimal-hypervolume of a set $\mathcal{S} \subset \mathbb{R}^m$ of n points can be done in $\mathcal{O}(\max\{n^2, \mathcal{C}(m, n)\})$ time.*

Proof: From Definition 6 we infer that the computation overhead is to determine the $\mathcal{D}$-minimal set $\mathcal{E}_{\mathcal{D}}^m(\mathcal{S})$. This takes $\mathcal{O}(n^2)$ time as the binary relation might not be transitive and fast divide-and-conquer techniques might not work (see Remark 2 and [12, Lemma 2.1]). Once we obtain $\mathcal{E}_{\mathcal{D}}^m(\mathcal{S})$ the rest is computing the usual hypervolume (as $\leq_{\mathbb{R}^m_+}$ binary relation is used in Definition 6), which can be done in $\mathcal{C}(m, n)$ time (which is $\Theta(n \ln n)$ for $m = 2$ and $m = 3$, $\mathcal{O}(n^2)$ for $m = 4$ [27], and $\mathcal{O}(n^{\frac{m-1}{2}} \ln n)$ in the general case [28]).   □

4   Algorithms

In this section, we introduce three SMS-EMOA [26] based algorithms for finding a set of solutions that maximize the preferred minimal/ nondominated-
hypervolume (from Definition 6). For simplicity, we discuss only the minimal
case (the algorithm for the nondominated case uses $\mathcal{E}_{\mathcal{D}}^n(\mathcal{S})$ instead of $\mathcal{E}_{\mathcal{D}}^m(\mathcal{S})$).

SMS-EMOA is a steady state evolutionary algorithm, which combines
approaches introduced by other multi-objective evolutionary algorithms. It is
particularly designed to maximize the (usual) hypervolume of a population.SMS-
EMOA starts each iteration with generating a single new solution by selecting
parents from the current operation, combining them, and mutating the offspring
solution. The population is then aggregated using non-dominated sorting (used
in NSGA-II). Afterwards, SMS-EMOA eliminates the point of the worst ranked
front contributing least to the hypervolume of this particular front.

We assume a population size of N and a reference point $\mathbf{r} \in \mathbb{R}^m$. At any
generation t, let P_t and q_t denote the parent population and the generated
offspring, respectively. Let R_t be the combined population, i.e., $R_t = P_t \cup \{q_t\}$.
Next, we need to discard one individual from R_t to create P_{t+1}, and this can be
done in three different ways as described next.

Last Front SMS-EMOA (LF-SMS-EMOA)
 L1: Perform a non-dominated sorting to R_t and identify different fronts: $\mathcal{F}_i$,
  $i = 1, 2, \ldots, k$ for some $k \in \mathbb{N}$.
 L2: Let $\mathcal{F}_{k+1} := \mathcal{F}_k \setminus \mathcal{E}_{\mathcal{D}}^m(\mathcal{F}_k)$ and $\mathcal{F}_k := \mathcal{E}_{\mathcal{D}}^m(\mathcal{F}_k)$.
 L3: Let $\mathbf{v} = \operatorname{argmin}_{\mathbf{u} \in \mathcal{F}_{k+1}} H^m(\{\mathbf{u}\}, \mathcal{F}_{k+1}, \mathbf{r})$ and let $= P_{t+1} := R_t \setminus \{\mathbf{v}\}$.

First Front SMS-EMOA (FF-SMS-EMOA)
 F1: Perform a non-dominated sorting to R_t and identify different fronts: $\mathcal{F}_i$,
  $i = 1, 2, \ldots, k$ for some $k \in \mathbb{N}$.
 F2: Let $\mathcal{F}_{k+1} := \mathcal{F}_k, \mathcal{F}_k := \mathcal{F}_{k-1}, \ldots, \mathcal{F}_2 := \mathcal{F}_1, \mathcal{F}_1 := \mathcal{E}_{\mathcal{D}}^m(\mathcal{F}_2) \& \mathcal{F}_2 := \mathcal{F}_2 \setminus \mathcal{F}_1$.
 F3: Let $\mathbf{v} = \operatorname{argmin}_{\mathbf{u} \in \mathcal{F}_{k+1}} H^m(\{\mathbf{u}\}, \mathcal{F}_{k+1}, \mathbf{r})$ and let $= P_{t+1} := R_t \setminus \{\mathbf{v}\}$.

Complete Front SMS-EMOA (CF-SMS-EMOA)
 C1: Let $\mathcal{F}_1 := \mathcal{E}_{\mathcal{D}}^m(R_t)$. Let $\mathcal{F}_i := \mathcal{E}_{\mathcal{D}}^m(R_t \setminus \cup_{j=1}^{i-1} \mathcal{F}_j)$ for all $i = 2, \ldots, k$ until
  $R_t \setminus \cup_{j=1}^{k} \mathcal{F}_j \neq \emptyset$, for some $k \in \mathbb{N}$.
 C2: Let $\mathbf{v} = \operatorname{argmin}_{\mathbf{u} \in \mathcal{F}_k} H^m(\{\mathbf{u}\}, \mathcal{F}_k, \mathbf{r})$ and let $= P_{t+1} := R_t \setminus \{\mathbf{v}\}$.

The above three algorithms are based on Definition 6 which calculates the hyper-
volumes only for the set $\mathcal{E}_{\mathcal{D}}^m(\mathcal{S})$, where $\mathcal{S}$ is the last front (the minimal ordering
is additionally used to split the first front in FF-SMS-EMOA see [29]). Such ap-
proaches require finding $\mathcal{E}_{\mathcal{D}}^m$ for a smaller set ($\mathcal{F}_1$ or $\mathcal{F}_k$, see Theorems 1 and
2) and use Pareto-domination to sort the combined population. The algorithm
(CF-SMS-EMOA) on the other hand, sorts the combined population using minimal ordering, which is more sorting effort (in light of Remark 2) but still has
the advantage of using minimal ordering directly (instead of using it in addition
to Pareto ordering).

The next result shows a desirable property of the algorithms, that the minimal-
hypervolume w.r.t. the combined population cannot decrease.

Lemma 4. *Let* $\mathbf{r}$ *be independent of* t. *Then, for all* $t \geq 1$ $\mathcal{H}^m(P_{t+1}, R_t, \mathbf{r}) \geq \mathcal{H}^m(P_t, R_t, \mathbf{r})$ *holds for all the above three algorithms.*

Proof: The proof follows by noting that only points from $\mathcal{E}_{\mathcal{D}}^m(R_t)$ have a positive contribution to minimal-hypervolume and they are never replaced by a point not belonging to $\mathcal{E}_{\mathcal{D}}^m(R_t)$. In fact, arguing in a similar way it can be shown that

$$\mathcal{H}^m\left(P_{t+1}, \bigcup_{\tau=1}^{t} R_\tau, \mathbf{r}\right) \geq \mathcal{H}^m(P_t, R_t, \mathbf{r})$$

holds, which is a stronger result as $R_t \subseteq \bigcup_{\tau=1}^{t} R_\tau$. $\qquad\square$

5   Experimental Study

Engau [10] proposed a family of so-called *Bishop-Phelps cones* (that have many applications in nonlinear analysis and multi-objective applications) for variable preference modeling and we use these cones in this study. These cones are convex and satisfy several decision-making requirements like monotonicity, local preferences and ideal-symmetry. There are two parameters that fully describe the cone: a scalar γ which controls the angle of the cone (in $\mathbb{R}^m$) and a vector $\mathbf{p} \in \mathbb{R}^m$. Based on these parameters, the variable domination cone $\mathcal{C}(\mathbf{u})$ is defined by

$$\mathcal{C}(\mathbf{u}) := \{\mathbf{d} | \langle \mathbf{d}, \mathbf{u} - \mathbf{p} \rangle \geq \gamma \cdot \|\mathbf{d}\| \cdot [\mathbf{u} - \mathbf{p}]_{\min}\}, \tag{7}$$

where $[\mathbf{u} - \mathbf{p}]_{\min}$ denotes the minimal component of the vector $\mathbf{u} - \mathbf{p}$. The reference point $\mathbf{p}$ is usually taken as the ideal vector, i.e., the vector having the ith component as $\inf\{\mathbf{v}_i | \mathbf{v} \in \mathbf{f}(\mathbf{X})\}$. Although $\mathcal{C}(\mathbf{u})$ is defined for arbitrary $\gamma \in \mathbb{R}$ and any point $\mathbf{p} \in \mathbb{R}^m$ instead of the ideal point, the above conditions guarantee that $\mathcal{C}(\mathbf{u})$ is closed, convex and pointed. The parameter $\gamma \in (0, 1]$ ensures that the Pareto cone $\mathbb{R}_+^m \subseteq \mathcal{C}(\mathbf{u})$. This makes the theory and algorithms that are defined in the preceding sections applicable, as $\mathcal{K} \supseteq \mathbb{R}_+^m$ is assumed in Theorems in Section 3. It is also easy to see that the Blanket Assumption holds for the family of Bishop-Phelps cones.

We tested LF-SMS-EMOA, FF-SMS-EMOA, and CF-SMS-EMOA on 22 test problem instances (11 test problems for both minimal and nondominated domination structures). The test problems chosen are of varying complexity and are include two problems from the CTP suite (bi-objective CTP1, CTP7), one from the DTLZ suite (DTLZ8, 3 objectives), one from the CEC-2007 competition (bi-objective SZDT1), four from the WFG suite (WFG1, WFG2, with both 2 and 3 objectives) and three from the ZDT suite (ZDT3, ZDT4, ZDT6). For all the problems, we use the zero as the $\mathbf{p}$ vector and use $\gamma = 0.5$. For all problems, we compute a well-distributed approximation of $\mathcal{D}$-minimal and $\mathcal{D}$-nondominated set as follows. Corresponding to a problem, we first generate 5,000 well-diverse points on the Pareto-efficient front. From these points, we calculate the minimal and nondominated points (applying Theorems 1 and 2).

In order to evaluate the results, we use the Hölder or power mean based inverted generational distance (IGD_p) [30] (w.r.t. the obtained reference set) and the $\mathcal{H}$ hypervolume metric (minimal and nondominated) metrics. For statistical evaluation, we run each algorithm for 51 times and present summary statistics. For all problems solved, we use a population of size 100 and set the maximum number of function evaluations as 20,000. We use a standard real-parameter SBX and polynomial mutation operator with $\eta_c = 15$ and $\eta_m = 20$, respectively [1]. LF-SMS-EMOA, FF-SMS-EMOA, and CF-SMS-EMOA are written using the jMetal framework and their source codes are available on request. The minimal optimality based algorithms are prefixed with M- while the nondominated ones are with N-. Hence, M-LF-SMS-EMOA means the LF-SMS-EMOA algorithm searching for minimal solutions.

Tables 1 and 2 present the results for the Bishop-Phelps cone based variable ordering using minimal optimality notion and nondominated optimality notion, respectively. IGD_p and $\mathcal{H}_{\mathcal{D}}^m$ (and $\mathcal{H}_{\mathcal{D}}^n$) metrics are able to measure both convergence and diversity, albeit in different ways. IGD_p is a set based, curvature independent measure, in that (power mean based) distances to a reference set are computed. Hypervolumes on the other hand, are curvature dependent measures in that knee points contribute more to the hypervolume.

Based on IGD_p values we see that M-LF-SMS-EMOA and M-FF-SMS-EMOA algorithms perform better than M-CF-SMS-EMOA and the opposite is seen w.r.t. $\mathcal{H}_{\mathcal{D}}^m$ metric. This might be due to a global change in domination that is employed in M-CF-SMS-EMOA that favors more knee points thereby increasing the minimal-hypervolume. However, the advantage of M-LF-SMS-EMOA and M-FF-SMS-EMOA algorithms is that the Pareto-ordering is primarily used for sorting and hence, well-distributed points are not easily replaced by the knee points. Eventually, if we run the algorithms for a long time, then all the three algorithms become the same and achieve the same distribution of points (this is seen from further experiments).

A closer look at the nondominated study brings out additional insights about the two optimality notions. The results from Table 2 reveal a hierarchy among the three algorithms, with N-CF-SMS-EMOA being the best and N-LF-SMS-EMOA the worst. This might be explained as follows. Checking whether a point lies in $\mathcal{E}_{\mathcal{D}}^n$ or not requires comparisons using the binary relation $\prec_{\mathcal{D}}^n$ (see Definitions 3 and 5). This might be more difficult to satisfy as then the variable cones at all the other points need to be tested (or taking into account Theorem 2 variable cones at all the Pareto-nondominated points). Hence, unless we have a good approximation, we could infer falsely that a point is nondominated (in the sense of Definition 5). There might be points in a worse Pareto front, that are still Pareto-nondominated to the point under consideration, and these worse points might be useful in deciding where the nondomination condition in Definition 3 holds or not. In algorithms N-LF-SMS-EMOA and N-FF-SMS-EMOA, Pareto-domination based sorting is used, which might not give a good approximation of $\mathcal{D}$-nondominated points. A similar problem occurs when transitivity does not hold and this is discussed in a recent classical technique [12].

Table 1. Minimal study: Median and interquartile range of Hölder mean based inverted generational distance (IGD_p) and minimal-hypervolume ($\mathcal{H}^m_\mathcal{D}$) metrics

	IGD_p **metric**		
	M-Lf-Sms-Emoa	M-Ff-Sms-Emoa	M-Cf-Sms-Emoa
CTP1	$3.08e - 02_{2.3e-02}$	$3.21e - 02_{2.5e-02}$	$2.66e - 02_{1.8e-02}$
CTP7	$1.34e - 03_{2.9e-05}$	$1.34e - 03_{3.5e-05}$	$1.34e - 03_{2.2e-05}$
DTLZ8	$6.63e - 02_{1.0e-02}$	$7.18e - 02_{1.3e-02}$	$7.30e - 02_{1.4e-02}$
SZDT1	$2.96e - 01_{5.3e-01}$	$2.94e - 01_{5.3e-01}$	$3.01e - 01_{5.4e-01}$
ZDT3	$5.69e - 02_{2.2e-02}$	$4.31e - 02_{1.8e-02}$	$5.31e - 02_{1.5e-02}$
ZDT4	$2.75e - 02_{4.0e-02}$	$3.78e - 02_{7.8e-02}$	$2.89e - 02_{3.3e-02}$
ZDT6	$2.91e - 02_{5.7e-03}$	$3.30e - 02_{6.1e-03}$	$1.35e - 02_{2.0e-03}$
WFG1_2D	$2.84e + 00_{2.3e+00}$	$2.90e + 00_{2.5e+00}$	$3.49e + 00_{2.1e+00}$
WFG1_3D	$2.87e + 00_{2.5e+00}$	$3.01e + 00_{4.1e+00}$	$7.70e + 00_{3.3e+00}$
WFG2_2D	$5.32e - 01_{2.7e-03}$	$5.33e - 01_{2.4e-03}$	$5.33e - 01_{1.8e-03}$
WFG2_3D	$5.98e + 00_{7.7e-02}$	$5.97e + 00_{3.2e-02}$	$5.97e + 00_{4.5e-02}$
	$\mathcal{H}^m$ **metric**		
	M-Lf-Sms-Emoa	M-Ff-Sms-Emoa	M-Cf-Sms-Emoa
CTP1	$5.05e - 01_{7.3e-03}$	$5.06e - 01_{1.3e-02}$	$5.07e - 01_{6.5e-03}$
CTP7	$2.80e - 01_{6.8e-05}$	$2.80e - 01_{9.6e-05}$	$2.80e - 01_{7.4e-05}$
DTLZ8	$4.57e - 01_{3.3e-03}$	$4.57e - 01_{5.4e-03}$	$4.59e - 01_{5.4e-03}$
SZDT1	$6.94e - 01_{3.6e-01}$	$6.89e - 01_{3.7e-01}$	$6.88e - 01_{3.7e-01}$
ZDT3	$5.69e - 01_{1.9e-02}$	$5.55e - 01_{1.4e-02}$	$5.94e - 01_{5.5e-03}$
ZDT4	$4.83e - 01_{3.7e-02}$	$4.65e - 01_{1.1e-01}$	$4.80e - 01_{3.2e-02}$
ZDT6	$4.28e - 01_{8.1e-03}$	$4.22e - 01_{8.9e-03}$	$4.51e - 01_{3.1e-03}$
WFG1_2D	$0.00e + 00_{0.0e+00}$	$0.00e + 00_{0.0e+00}$	$0.00e + 00_{0.0e+00}$
WFG1_3D	$0.00e + 00_{0.0e+00}$	$0.00e + 00_{0.0e+00}$	$0.00e + 00_{0.0e+00}$
WFG2_2D	$6.02e - 01_{4.3e-03}$	$6.03e - 01_{2.9e-03}$	$6.03e - 01_{3.4e-03}$
WFG2_3D	$4.51e - 01_{1.3e-01}$	$4.79e - 01_{5.3e-02}$	$4.75e - 01_{7.8e-02}$

Table 2. Nondominated study: Median and interquartile range of Hölder mean based inverted generational distance (IGD_p) and nondominated-hypervolume ($\mathcal{H}^n_\mathcal{D}$) metrics

	IGD metric		
	N-Lf-Sms-Emoa	N-Ff-Sms-Emoa	N-Cf-Sms-Emoa
CTP1	$3.61e - 02_{1.9e-02}$	$2.88e - 02_{2.3e-02}$	$2.30e - 02_{3.4e-02}$
CTP7	$1.34e - 03_{2.7e-05}$	$1.33e - 03_{1.7e-05}$	$1.34e - 03_{2.7e-05}$
DTLZ8	$1.02e - 01_{2.5e-02}$	$9.97e - 02_{1.4e-02}$	$9.57e - 02_{1.4e-02}$
SZDT1	$1.85e - 01_{3.1e-01}$	$1.85e - 01_{3.1e-01}$	$1.87e - 01_{3.1e-01}$
ZDT3	$8.30e - 01_{1.6e-03}$	$8.31e - 01_{1.8e-03}$	$8.30e - 01_{6.6e-04}$
ZDT4	$4.08e - 02_{3.0e-02}$	$4.46e - 02_{3.6e-02}$	$3.67e - 02_{2.3e-02}$
ZDT6	$8.32e - 02_{6.8e-03}$	$8.21e - 02_{5.5e-03}$	$7.90e - 02_{2.7e-03}$
WFG1_2D	$5.45e - 01_{2.1e-01}$	$5.88e - 01_{2.1e-01}$	$6.49e - 01_{2.9e-01}$
WFG1_3D	$2.54e + 00_{1.8e+00}$	$1.63e + 00_{2.1e+00}$	$3.51e + 00_{2.6e+00}$
WFG2_2D	$5.18e - 01_{2.7e-03}$	$5.18e - 01_{1.7e-03}$	$5.18e - 01_{2.3e-03}$
WFG2_3D	$4.90e + 00_{2.6e-02}$	$4.90e + 00_{3.8e-02}$	$4.91e + 00_{4.6e-02}$
	$\mathcal{H}^n$ **metric**		
	N-Lf-Sms-Emoa	N-Ff-Sms-Emoa	N-Cf-Sms-Emoa
CTP1	$5.03e - 01_{9.8e-03}$	$5.06e - 01_{9.3e-03}$	$5.09e - 01_{1.8e-02}$
CTP7	$2.81e - 01_{1.0e-04}$	$2.81e - 01_{1.4e-05}$	$2.81e - 01_{1.3e-04}$
DTLZ8	$4.50e - 01_{6.9e-03}$	$4.52e - 01_{6.4e-03}$	$4.54e - 01_{5.0e-03}$
SZDT1	$5.31e - 01_{4.0e-02}$	$5.31e - 01_{3.5e-02}$	$5.37e - 01_{3.2e-02}$
ZDT3	$4.01e - 02_{1.2e-03}$	$3.98e - 02_{1.3e-03}$	$4.09e - 02_{4.2e-04}$
ZDT4	$4.68e - 01_{4.7e-02}$	$4.57e - 01_{5.1e-02}$	$4.73e - 01_{3.5e-02}$
ZDT6	$3.92e - 01_{4.3e-03}$	$3.90e - 01_{7.0e-03}$	$4.04e - 01_{2.1e-03}$
WFG1_2D	$1.26e - 01_{1.8e-01}$	$9.12e - 01_{1.8e-01}$	$5.41e - 02_{8.7e-02}$
WFG1_3D	$0.00e + 00_{0.0e+00}$	$0.00e + 00_{0.0e+00}$	$0.00e + 00_{0.0e+00}$
WFG2_2D	$5.82e - 01_{2.9e-03}$	$5.83e - 01_{2.4e-03}$	$5.83e - 01_{3.3e-03}$
WFG2_3D	$4.76e - 01_{4.4e-02}$	$4.78e - 01_{3.6e-02}$	$4.50e - 01_{8.5e-02}$

6 Conclusions

This paper is among the very few works that tackle multi-objective problems defined by a general nonconvex, nonconical, and set-based variable ordering. The classical approach to such problems is to solve these problems using a scalarization approach and get one solution. Even for smooth functions, the variable ordering introduces additional difficulties- like convexity is not preserved, nonsmoothness and discontinuities are additionally introduced (as highlighted in a recent study [12]). The advantages of using population based algorithms are many: getting multiple solutions, tackling non-transitivity of the binary relation, not requiring convexity and smoothness among others.

We studied different optimality notions and presented new theoretical results for problems with a variable ordering structure. Among others, we presented sufficient conditions for a point to be minimal/ nondominated based on pairwise comparisons. Moreover, we proposed new definitions of hypervolume based indicators for such problems. Theoretical results w.r.t. compatibility and completeness of the new indicator were also presented.

In addition to getting a better understanding, the theoretical results on variable orderings and indicators were also of an algorithmic value. We used them and developed three indicator based algorithms for approximating the set of appropriate optimal solutions. Computational results on a number of test problems showed the niche of population based algorithms for solving multi-objective problems with variable orderings. Future works will concentrate on solving real-world financial application problems and other applications where these orderings arise. For such problems, a detailed comparison with a classical [12] and an evolutionary approach [11] will also be carried out.

References

[1] Deb, K.: Multi-objective optimization using evolutionary algorithms. Wiley (2001)
[2] Yu, P.L.: A class of solutions for group decision problems. Management Science 19, 936–946 (1973)
[3] Karasakal, E.K., Michalowski, W.: Incorporating wealth information into a multiple criteria decision making model. European J. Oper. Res. 150, 204–219 (2003)
[4] Ogryczak, W.: Inequality measures and equitable approaches to location problems. European Journal of Operational Research 122, 374–391 (2000)
[5] Bergstresser, K., Yu, P.L.: Domination structures and multicriteria problems in n-person games. Theoy and Decision 8, 5–48 (1977)
[6] Ogryczak, W., Wierzbicki, A.: On multi-criteria approaches to bandwidth allocation. Control Cybernet 33, 427–448 (2004)
[7] Shukla, P.K., Hirsch, C., Schmeck, H.: In Search of Equitable Solutions Using Multi-objective Evolutionary Algorithms. In: Schaefer, R., Cotta, C., Kolodziej, J., Rudolph, G. (eds.) PPSN XI. LNCS, vol. 6238, pp. 687–696. Springer, Heidelberg (2010)
[8] Ehrgott, M.: Multicriteria optimization, 2nd edn. Springer, Berlin (2005)
[9] Eichfelder, G.: Variable Ordering Structures in Vector Optimization. In: Ansari, Q.H., Yao, J.-C., Jahn, J. (eds.) Recent Developments in Vector Optimization. Vector Optimization, vol. 1, pp. 95–126. Springer, Heidelberg (2012)

[10] Engau, A.: Variable preference modeling with ideal-symmetric convex cones. J. Global Optim. 42, 295–311 (2008)

[11] Hirsch, C., Shukla, P.K., Schmeck, H.: Variable Preference Modeling Using Multi-Objective Evolutionary Algorithms. In: Takahashi, R.H.C., Deb, K., Wanner, E.F., Greco, S. (eds.) EMO 2011. LNCS, vol. 6576, pp. 91–105. Springer, Heidelberg (2011)

[12] Eichfelder, G.: Numerical procedures in multiobjective optimization with variable ordering structures. Technical report, Preprint-Series of the Institute of Mathematics, Ilmenau University of Technology, Germany (2012)

[13] Yu, P.L.: Multiple-criteria decision making. Mathematical Concepts and Methods in Science and Engineering, vol. 30. Plenum Press, New York (1985)

[14] Huang, N.J., Rubinov, A.M., Yang, X.Q.: Vector optimization problems with non-convex preferences. J. Global Optim. 40, 765–777 (2008)

[15] Rubinov, A.M., Gasimov, R.N.: Scalarization and nonlinear scalar duality for vector optimization with preferences that are not necessarily a pre-order relation. J. Global Optim. 29, 455–477 (2004)

[16] Bergstresser, K., Charnes, A., Yu, P.L.: Generalization of domination structures and nondominated solutions in multicriteria decision making. Journal of Optimization Theory and Applications 18, 3–13 (1976), doi:10.1007/BF00933790

[17] Rudin, W.: Real and complex analysis. McGraw-Hill, New York (1987)

[18] Wiecek, M.M.: Advances in cone-based preference modeling for decision making with multiple criteria. Decis. Mak. Manuf. Serv. 1, 153–173 (2007)

[19] Huang, N.J., Yang, X.Q., Chan, W.K.: Vector complementarity problems with a variable ordering relation. European J. Oper. Res. 176, 15–26 (2007)

[20] Sawaragi, Y., Nakayama, H., Tanino, T.: Theory of multiobjective optimization. Mathematics in Science and Engineering (1985)

[21] Kung, H.T., Luccio, F., Preparata, F.P.: On finding the maxima of a set of vectors. J. ACM 22, 469–476 (1975)

[22] Jensen, M.: Reducing the run-time complexity of multiobjective eas: The NSGA-II and other algorithms. IEEE Transactions on Evolutionary Computation 7, 503–515 (2003)

[23] Luc, D.T.: Theory of vector optimization. Lecture Notes in Economics and Mathematical Systems, vol. 319. Springer, Berlin (1989)

[24] Zitzler, E., Thiele, L., Laumanns, M., Fonseca, C.M., da Fonseca, V.G.: Performance Assessment of Multiobjective Optimizers: An Analysis and Review. IEEE Transactions on Evolutionary Computation 7, 117–132 (2003)

[25] Noghin, V.D.: Relative importance of criteria: a quantitative approach. Journal of Multi-Criteria Decision Analysis 6, 355–363 (1997)

[26] Beume, N.: Hypervolume based metaheuristics for multiobjective optimization. PhD thesis, Dortmund, Techn. Univ., Diss (2011)

[27] Guerreiro, A.P., Fonseca, C.M., Emmerich, M.T.M.: A fast dimension-sweep algorithm for the hypervolume indicator in four dimensions. In: Proceedings of 24th Canadian Conference on Computational Geometry (2012) (in press)

[28] Yildiz, H., Suri, S.: On klee's measure problem for grounded boxes. In: Proceedings of the 2012 Symposuim on Computational Geometry, SoCG 2012, pp. 111–120. ACM, New York (2012)

[29] Shukla, P.K., Hirsch, C., Schmeck, H.: A Framework for Incorporating Trade-Off Information Using Multi-Objective Evolutionary Algorithms. In: Schaefer, R., Cotta, C., Kołodziej, J., Rudolph, G. (eds.) PPSN XI, Part II. LNCS, vol. 6239, pp. 131–140. Springer, Heidelberg (2010)

[30] Schütze, O., Esquivel, X., Lara, A., Coello, C.: Using the averaged hausdorff distance as a performance measure in evolutionary multiobjective optimization. IEEE Transactions on Evolutionary Computation 16, 504–522 (2012)

Preference Articulation by Means of the $R2$ Indicator

Tobias Wagner[1], Heike Trautmann[2], and Dimo Brockhoff[3]

[1] Institute of Machining Technology (ISF), TU Dortmund University, Germany
wagner@isf.de
[2] Statistics Department, TU Dortmund University, Germany
trautmann@statistik.tu-dortmund.de
[3] DOLPHIN Team, INRIA Lille - Nord Europe, Villeneuve d'Ascq, France
dimo.brockhoff@inria.fr

Abstract. In multi-objective optimization, set-based performance indicators have become the state of the art for assessing the quality of Pareto front approximations. As a consequence, they are also more and more used within the design of multi-objective optimization algorithms. The $R2$ and the Hypervolume (HV) indicator represent two popular examples. In order to understand the behavior and the approximations preferred by these indicators and algorithms, a comprehensive knowledge of the indicator's properties is required. Whereas this knowledge is available for the HV, we presented a first approach in this direction for the $R2$ indicator just recently. In this paper, we build upon this knowledge and enhance the considerations with respect to the integration of preferences into the $R2$ indicator. More specifically, we analyze the effect of the reference point, the domain of the weights, and the distribution of weight vectors on the optimization of μ solutions with respect to the $R2$ indicator. By means of theoretical findings and empirical evidence, we show the potentials of these three possibilities using the optimal distribution of μ solutions for exemplary setups.

1 Introduction

Evolutionary multi-objective algorithms usually approximate the complete Pareto front of a problem in a single run. This is in contrast to classical MCDM approaches, which often apply sequential or hierarchical optimization runs to accomplish this task. In the beginnings of multi-objective optimization, three requirements for the set approximating the true Pareto front were defined: minimization of the distance (convergence), coverage of the extremes (spread), and a good representation of the actual shape (distribution) of the Pareto front [18]. In order to evaluate one or more of these requirements, several performance indicators were introduced [20, 22]. In particular, the Hypervolume (HV) [21] and the $R2$ indicator [11] are two recommended approaches which simultaneously evaluate all these desired aspects. Whereas the HV is a set-based quality indicator by definition, the R indicator family allows Pareto front approximations to be assessed based on a set of utility functions. Thereby, it is possible to find out which of the sets is better for specific preferences encoded in the weight vectors of the utility functions. A unary set-based quality indicator utilizing the mean utility over the weight vectors was proposed just recently [8]. In this paper, the former ideas of assessing regions of specific preferences are thus transferred to the unary variant of the indicator.

R.C. Purshouse et al. (Eds.): EMO 2013, LNCS 7811, pp. 81–95, 2013.

Many popular optimization algorithms are based on set-based quality indicators [5, 14, 17]. In order to learn about the outcome expected from these algorithms, the preferences and bias introduced by the choice of the indicator have to be understood. For the HV indicator, empirical studies [6] and theoretical results [1, 2] do already exist. With respect to the $R2$ indicator, we showed that the indicator tends to place the points more concentrated in the central region of the Pareto front than the HV and that the optimal placement of a point according to the $R2$ indicator only depends on its two neighbors and a subset of the weight vectors in the bi-objective case [8]. Here, we use these insights in order to define and analyze the preferences that can be introduced by a targeted choice of the reference point and the set of weight vectors. For the latter, particularly the influence of the covered domain and the density of the weight vectors on the optimal distributions of μ solutions are assessed. Thereby, we show that it is possible to restrict the search to a subregion and to adjust the focus of the distribution on the center or on the extremes of the Pareto front.

The terms and concepts required for understanding the methodological contributions are provided in the following section 2. Based on these auxiliary means, methods for integrating preferences into the $R2$ indicator are presented in section 3. The conceptual thoughts behind the methods are validated using empirical evidence. This is done in section 4. In the final section 5, the results are summarized and an outlook on further potentials for research on the $R2$ indicator is provided.

2   Foundations

Throughout the paper, we consider, without loss of generality, the simultaneous minimization of k objective functions $f_j : \mathbb{R}^n \rightarrow \mathbb{R}$ ($1 \leq j \leq k$) with respect to the *Pareto-dominance relation*. Since we are interested in optimal distributions of objective vectors, we will further neglect the corresponding decision variables in $\mathbb{R}^n$. Hence, we will use the terms solutions and objective vectors interchangeably in the following. We say a solution $x \in \mathbb{R}^k$ dominates a solution $y \in \mathbb{R}^k$ if $\forall j : x_j \leq y_j$ and $\exists j : x_j < y_j$. The solutions that are non-dominated by any other feasible solution are called Pareto-optimal, and we call the entire set of Pareto-optimal solutions *Pareto front*.

We can formulate the search for a Pareto-optimal solution as a single-objective problem, for example, by means of the achievement scalarizing function (ASF, [16])

$$u_{\mathbf{w}}(\mathbf{y}) = \max_j w_j(y_j - r_j)$$

where $\mathbf{r} = (r_1, \ldots, r_k)$ is a reference point [1] and $\mathbf{w} = (w_1, \ldots, w_k)$ a weight vector. Each Pareto-optimal solution a is associated with a weight vector $\mathbf{w}^a = (\frac{\beta}{a_1 - i_1}, \ldots, \frac{\beta}{a_k - i_k})$ such that the minimization of the ASF with weight $\mathbf{w}^a$ yields the solution a, with $\beta > 0$ being a normalization factor such that $\sum_j w_j = 1$. In case that the reference point is dominating a or dominated by a, the weight vector will be strictly positive. For a proof of this case, see for example [16, Theorem 2.3.4] or [8, Lemma 2].

---

[1] Often, the ideal or a utopian point, is used as reference point, i.e., an objective vector typically better than all feasible solutions.

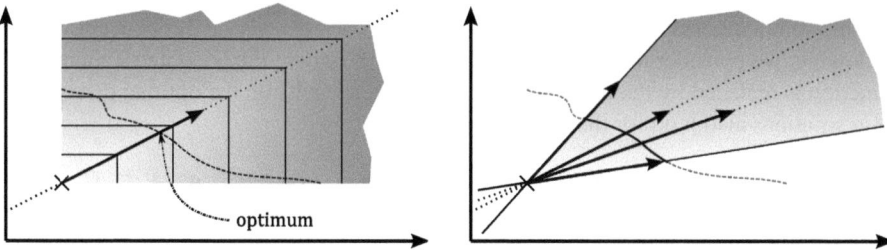

Fig. 1. Left: Illustration of the target direction (dotted line) through the reference point (cross) corresponding to the target vector **t** (black arrow) as well as some lines of equal function value $u_{\mathbf{w^t}}(x) = $ const. Lighter colors in the background indicate a smaller value of the achievement scalarizing function. The optimal solution (tip of dash-dotted arrow) lies at the intersection of the target direction with the Pareto front (dashed line). **Right:** Illustration of a target cone (gray area), given by a set of four weight/target vectors (black arrows). The reference point is denoted again as a cross and the attainable part of the Pareto front is depicted as the solid black part of the gray dashed Pareto front.

This result about the ASF can be interpreted also graphically. Given a so-called *target direction* $\mathbf{r} + m\mathbf{t}$ through the reference point $\mathbf{r}$ in direction of the *target vector* $\mathbf{t} = (t_1, \ldots, t_k)$, the minimization of the ASF $u_{\mathbf{w^t}}(\mathbf{y}) = \max_j w_j^t(y_j - r_j)$ with the weights $\mathbf{w^t} = (\beta/t_1, \ldots, \beta/t_k)$ with $\beta = 1/(\sum_j 1/t_j)$ results in the indifference contours shown in the lefthand plot of Fig. 1 and finally in the optimal solution that is lying at the intersection between the target direction and the Pareto front. In the example of Fig. 1, the target vector is $\mathbf{t} = (2, 1)$ with a *slope* of $t_2/t_1 = 1/2$ while the corresponding normalized weight vector is $\mathbf{w} = \frac{1}{1/t_1 + 1/t_2} \cdot (1/t_1, 1/t_2) = (1/3, 2/3)$.

Definition 1. *Given a target vector* $\mathbf{t} = (t_1, \ldots, t_k)$ *or its associated weight vector* $\mathbf{w} = (\beta/t_1, \ldots, \beta/t_k)$ *with* $\beta = 1/(\sum_j 1/t_j)$, *we call a feasible solution mapped to the minimal value of the ASF* $\max_j w_j(y_j - r_j)$ *an* optimal solution for the target vector $\mathbf{t}$.

Note that we will, in the following, use interchangeably either the weight vector $\mathbf{w} = (w_1, \ldots, w_k)$ or the target direction $\mathbf{t} = (1/w_1, \ldots, 1/w_k)$ to define an ASF problem. In the case of bi-objective problems, we also allow weight vectors of the form $(0, 1)$ and $(1, 0)$ for which the corresponding target directions are $(1, 0)$ and $(0, 1)$ respectively. Furthermore, to make the text more readable, we will also denote any vector pointing in the target direction as *target vector*, cf. again Fig. 1.

Definition 2. *Given a target direction* $\mathbf{r} + m\mathbf{t}$ *through the reference point* $\mathbf{r}$, *we call any vector* $m\mathbf{t}$ *mutually different from the null vector (i.e.* $m > 0$) *a* target vector for the corresponding target direction.

In case we are interested in finding more than one Pareto-optimal solution, we have to either perform several independent optimization runs for different weight or target vectors or we can optimize the achievement scalarizing function for several weight/target vectors simultaneously. The $R2$ indicator [8,11] is based exactly on this concept[2]. In the following, we consider the unary $R2$ indicator of [8] where the weighted Tchebycheff function is replaced by the more general ASF—later results on the optimal distribution of μ solutions also hold for the indicator investigated in [8].

Definition 3. *Given a set of N weight vectors $\mathbf{W} = \{\mathbf{w}^1, \ldots, \mathbf{w}^N\} \subset \mathbb{R}^k$ and a reference point $\mathbf{r} \in \mathbb{R}^k$, the unary $R2$ indicator assigns the following value to a set $S \subset \mathbb{R}^k$ of solutions*

$$R2(S, \mathbf{W}, \mathbf{r}) = \frac{1}{N} \sum_{\mathbf{w} \in \mathbf{W}} \min_{s \in S} \max_j w_j (s_j - r_j) \ . \tag{1}$$

Note that a smaller $R2$ value corresponds to a better solution set and that, following [8], we only consider the unary $R2$ indicator, which does not require a reference set, here. This indicator is popular in multi-objective optimization due to its runtime in $O(Nk|S|)$ which is linear in the number of weights, the objective dimension, and the number of solutions in the set S. In this context, however, it is still unclear how many weight vectors are required for sufficiently covering the weight space. First suggestions for bi-objective problems can be found in [8], but the scaling with number of objectives k is still an open issue. As the volume of the space increases exponentially with k, it would be plausible that the same holds for N, making the indicator as expensive as the hypervolume. In the case of the $R2$ indicator, and more generally, if more than one target direction is involved in the optimization, we generalize the idea of the target direction to *target cones*, see also the righthand illustration in Fig. 1.

Definition 4. *The minimum cone including the target directions $\mathbf{t}^i \in \mathbf{T}$ related to all weight vectors $\mathbf{w}^i \in \mathbf{W}$ is denoted as* target cone.

The target cone therefore defines the region of interest defined by the reference point $\mathbf{r}$ and the set of weight vectors $\mathbf{W}$ (or the corresponding set of target vectors $\mathbf{T}$). For bi-objective problems, we can prove that the solution set which is optimal with respect to the $R2$ indicator lies within the target cone for a suitable set of weight vectors. In order to simplify the proof and to clarify what suitable means, we define what we call a *free target vector* and prove a small technical lemma beforehand.

Definition 5. *Given the $R2$ indicator with reference point $\mathbf{r} \in \mathbb{R}^k$ and weight vectors $\mathbf{W} \subset \mathbb{R}^k$, we call a target vector $\mathbf{t}$ corresponding to a weight vector $\mathbf{w}^t \in \mathbf{W}$* free *with respect to a solution set $S \subseteq \mathbb{R}^k$ iff S does not contain an optimal solution for $\mathbf{t}$.*

Lemma 1. *If a solution $s \in S$ of a solution set S lies outside the target cone and has a positive contribution to the bi-objective $R2$ indicator defined by the reference point $\mathbf{r} \in \mathbb{R}^2$ and a set of weight vectors $\mathbf{W} \subset \mathbb{R}^2$, i.e., if $R2(S, \mathbf{W}, \mathbf{r}) > R2(S \setminus \{s\}, \mathbf{W}, \mathbf{r})$, then s has also a positive contribution to the closest extreme target vector.*

---

[2] Originally, the indicator was introduced as a binary indicator and for no specific utility function, see [11] for details.

Proof. Let us assume without loss of generality that s is lying to the upper left of the target cone and denote by $\mathbf{t}'$ the leftmost target vector, defined by $\mathbf{w^{t'}} \in \mathbf{W}$.

Since s lies on the top left of any target vector $\mathbf{t}$ defined by a weight vector $\mathbf{w^t} = (w_1^t, w_2^t) \in \mathbf{W}$, we have that $w_1^t(s_1 - r_1) < w_2^t(s_2 - r_2)$ (I). Moreover, we know that $w_1^{t'} \geq w_1^t$ and $w_2^{t'} \leq w_2^t$ holds for any weight vector $\mathbf{w^t} = (w_1^t, w_2^t) \in \mathbf{W}$ because $\mathbf{t}'$ is the leftmost target vector (II).

We prove the lemma by contradiction. Assume that s has a positive contribution to a target vector $\mathbf{t}$ associated with $\mathbf{w^t} = (w_1^t, w_2^t) \neq (w_1^{t'}, w_2^{t'})$, i.e., $\max\{w_1^t(s_1 - r_1), w_2^t(s_2 - r_2)\} < \max\{w_1^t(x_1 - r_1), w_2^t(x_2 - r_2)\}$ for any other solution $x \in S$. On the other hand, assume also that s has no positive contribution to the leftmost target vector $\mathbf{t}'$, i.e., that there exists a solution $a \in S \setminus \{s\}$ such that $\max\{w_1^{t'}(a_1 - r_1), w_2^{t'}(a_2 - r_2)\} < \max\{w_1^{t'}(s_1 - r_1), w_2^{t'}(s_2 - r_2)\}$. Then with (I) and (II), we can show a contradiction to the above assumption that s has a positive contribution to the target vector associated with the weight vector $\mathbf{w^t} = (w_1^t, w_2^t)$:

$$\max\{w_1^{t'}(a_1 - r_1), w_2^{t'}(a_2 - r_2)\} < \max\{w_1^{t'}(s_1 - r_1), w_2^{t'}(s_2 - r_2)\}$$

$$\stackrel{(I)}{\Longrightarrow} \max\{w_1^{t'}(a_1 - r_1), w_2^{t'}(a_2 - r_2)\} < w_2^{t'}(s_2 - r_2)$$

$$\Longrightarrow w_1^{t'}(a_1 - r_1) < w_2^{t'}(s_2 - r_2) \text{ and } w_2^{t'}(a_2 - r_2) < w_2^{t'}(s_2 - r_2)$$

$$\stackrel{(II)}{\Longrightarrow} w_1^t(a_1 - r_1) < w_2^t(s_2 - r_2) \text{ and } w_2^t(a_2 - r_2) < w_2^t(s_2 - r_2)$$

$$\Longrightarrow \max\{w_1^t(a_1 - r_1), w_2^t(a_2 - r_2)\} < w_2^t(s_2 - r_2)$$

$$\stackrel{(I)}{\Longrightarrow} \max\{w_1^t(a_1 - r_1), w_2^t(a_2 - r_2)\} \stackrel{!}{<} \max\{w_1^t(s_1 - r_1), w_2^t(s_2 - r_2)\}$$

$\square$

Now, we are finally able to prove that in the bi-objective case, the optimal solution set of size μ that minimizes the $R2$ indicator fully lies within the target cone.

Theorem 1. *Given a bi-objective optimization problem and a set of weight vectors* $\mathbf{W} \subset \mathbb{R}^2$ *for the achievement scalarization functions within the $R2$ indicator with reference point* $\mathbf{r} \in \mathbb{R}^2$. *Then, all objective vectors of a solution set* $S \subseteq \mathbb{R}^k$ *with* $|S| = \mu$ *solutions that minimizes the $R2$ indicator lie within the target cone defined by* $\mathbf{W}$ *if* $\mu \leq |\mathbf{W}|$ *and if at least* μ *target vectors defined by the weight vectors in* $\mathbf{W}$ *intersect with the Pareto front.*

Proof. For the case $\mu = |\mathbf{W}|$, we refer to the proof of Theorem 1 in [8] and prove the case $\mu < |\mathbf{W}|$ by contradiction. To this end, let us assume that the solution set $S \subseteq \mathbb{R}^k$ minimizes the $R2$ indicator with respect to $\mathbf{W}$ and $\mathbf{r}$ and that the solution $s \in S$ is lying outside the target cone. We distinguish two cases: Either s has no contribution to the $R2$ indicator, i.e., $R2(S, \mathbf{W}, \mathbf{r}) = R2(S \setminus \{s\}, \mathbf{W}, \mathbf{r})$ (case 1) or s has a positive contribution to the $R2$ indicator, i.e., $R2(S, \mathbf{W}, \mathbf{r}) > R2(S \setminus \{s\}, \mathbf{W}, \mathbf{r})$ (case 2).

Case 1: If s itself has no contribution to the $R2$ indicator, we can replace s by an optimal solution s^* of a free target vector $\mathbf{t}^*$ which exists due to the pigeonhole principle (we presupposed at least as many intersections of target vectors with the Pareto front as there are solutions in S). Then, the $R2$ indicator for $(S \setminus \{s\}) \cup \{s^*\}$ is larger than for S which is a contradiction to the assumed optimality of S.

Case 2: Let us now assume that solution s, which lies outside of the target cone, has a positive contribution to the $R2$ indicator. Then, we know from Lemma 1 that s has also a positive contribution to the closest extreme target vector $\mathbf{t}'$ of the $R2$ indicator. Hence, $\mathbf{t}'$ must be a free target vector and we can replace s by the optimal solution s' with respect to $\mathbf{t}'$ and improve the overall $R2$ indicator value which contradicts the assumed optimality of S.   □

Note that in case the number of points in the set S is larger than the number of target vectors, the optimal solution sets of size μ can contain solutions outside the target cone due to solutions with no contribution to the $R2$ indicator [8]. As the above theoretical investigations do only show qualitative results, but do not allow concrete solution sets to be proven to correspond to an optimal $R2$ indicator value, we investigate those optimal solution sets of size μ, also called optimal μ-distributions [2], in the following by means of numerical approximations while changing the location of the reference point, the target cone, and the distribution of the target directions within the cone.

3   Integrating Preferences into the $R2$ Indicator

In this section, the concept of the target cone (Definition 4) is further elaborated. We will discuss how the target cone is modified by changing the position of the reference point and by restricting the weight space covered by the weight vectors in $\mathbf{W}$. In addition, the effect of the density of the weight vectors in $\mathbf{W}$ on the distribution of target directions within the target cone is discussed to allow the preferences to be further refined.

3.1   Position of the Reference Point

In the previous section, the special role of the reference vector $\mathbf{r}$ as intersection of all target directions became obvious. By moving the reference point, the target cones are moved accordingly. This is shown in Figure 2. By changing the position of the reference point from $\mathbf{r} = (0,0)^T$ (left) to $\mathbf{r} = (0.2, 0.1)^T$ (right), the focused region on the exemplary Pareto front $f_2 = 0.5 - f_1$ (DTLZ1, [10]) is narrowed significantly.

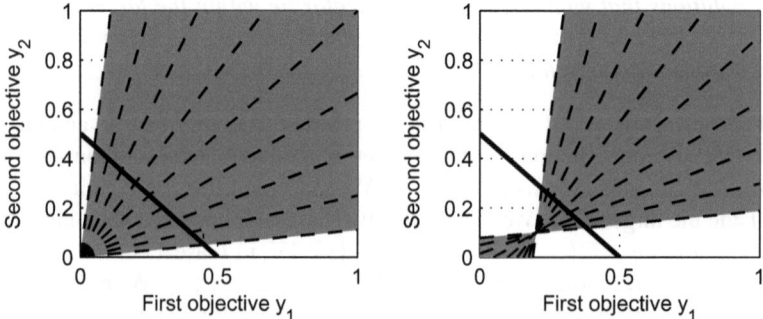

Fig. 2. Moving the target cone (gray area) by changing the position of the reference point from $\mathbf{r} = (0,0)^T$ (left) to $\mathbf{r} = (0.2, 0.1)^T$ (right). The dashed lines correspond to the target directions and the Pareto front of DTLZ1 is indicated by the thick black line.

3.2 Restriction of the Weight Space

The target cone is defined as the envelope of the target directions (cf. Definition 4). As a consequence, it can be narrowed by restricting the components of the normalized weight vectors $\mathbf{w} \in \mathbf{W}$ to subintervals of $[0, 1]$. This is shown in Figure 3.

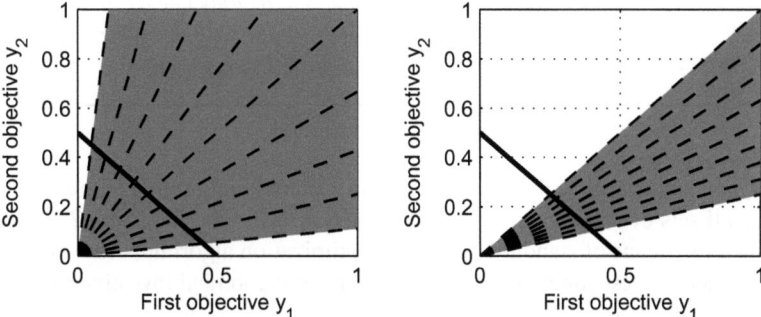

Fig. 3. Narrowing the target cone (gray area) by restricting the first component of the normalized weight vectors from $w_1 \in [0.1, 0.9]$ (left) to $w_1 \in [0.2, 0.5]$ (right). The dashed lines correspond to the target directions and the Pareto front of DTLZ1 is indicated by the thick black line.

3.3 Density of the Weight Vector Distribution

So far, the standard approach is to distribute the weight vectors uniformly within $[0, 1]$, i.e., $w_1 = 0, 0.01, \ldots, 0.99, 1$ and $w_2 = 1 - w_1$ for 101 weight vectors, see e.g. [15]. However, the weight vector distribution influences the optimal distribution of solution sets regarding $R2$. In order to obtain additional flexibility, we propose Algorithm 1 as an exemplary method to generate weight vector distributions which express increased preferences regarding the extremes of the front. These kinds of weight vector distributions might be desired due to the fact that the optimal distributions of μ solutions regarding $R2$ result in more centered point distributions than the HV indicator if the weights are chosen uniformly in the weight space [8]. To accomplish this shifted focus, a power transformation with exponent γ is implemented. With increasing $\gamma > 1$, the initial uniform distribution is more and more skewed. To stick with a symmetric distribution, this skewing is only performed on weights $w \leq 0.5$ which are then mirrored along $w = 0.5$ to obtain the weight components $w > 0.5$. This approach on the one hand increases the effect of the transformation, but on the other hand requires a rescaling for still covering the whole domain of weight vectors. The entire rescaling is provided in Algorithm 1. Fig. 4 shows the weight vector distributions resulting for exemplary values of the skewing factor γ.

4 Results

In the remainder of this paper, we investigate experimentally how the previously described ways to incorporate preferences into the $R2$ indicator change its bias. More

---

Algorithm 1. Generate Weight Vectors

---

function GENERATEWEIGHTVECTORS(γ, n)
$\qquad\qquad\qquad\qquad\qquad\qquad\qquad\quad$ ▷ γ: *skew factor, n: number of vectors, must be odd*
$\qquad$weights.x ← sequence($0, 0.5$,stepsize= $1/(n-1)$) $\qquad\qquad$ ▷ *distribute uniformly*
$\qquad$weights.x ← $0.5^{-\gamma+1}$(weights.x)$^{\gamma}$ $\qquad\qquad\qquad$ ▷ *skew and rescale weights*
$\qquad$diffs ← $1 -$ reverse(weights.x$[1, \ldots, (n-1)/2]$) ▷ *Mirroring differences along w = 0.5*
$\qquad$weights.x ← concatenate(weights.x, diffs) $\qquad\qquad$ ▷ *Build symmetric distribution*
$\qquad$**return** weights.x
end function

---

concretely, we approximate the solution sets of μ points that minimize the $R2$ indicator among all sets of μ points, the so-called optimal μ-distributions, see [2, 8]. Results are obtained by means of standard numerical optimization algorithms for several well-known bi-objective test functions. As we know from theoretical investigations [8], the optimal μ-distributions of the $R2$ indicator lie on the Pareto front if at least μ target vectors of the indicator intersect with the Pareto front. Hence, we are interested in finding the positions of μ points (y_1^i, y_2^i) $(1 \leq i \leq \mu)$ on the Pareto front such that the $R2$ indicator of these points is minimal. It is easy to see that this optimization problem is only of dimension μ due to the fact that the points (y_1^i, y_2^i) have to lie on the Pareto front and are therefore dependent variables [2], for instance $y_2^i = 0.5 - y_1^i$ for DTLZ1.

4.1   Position of the Reference Point

For the validation of the analytical thoughts of subsection 3.1, a simple experiment was performed. Using the experimental setup of a former study [8], the position of the reference point was changed and a corresponding optimal μ-distribution for the $R2$ indicator was empirically determined by optimizing the above mentioned y_1^i values with the CMA-ES [3]. Thereby, the standard setup of the recent MATLAB implementation was used [13]. The results of the best of ten replications on the DTLZ1 test function featuring the linear Pareto front $y_2^i = 0.5 - y_1^i$ are shown in Fig. 5 for solution sets of size $\mu = 10$ and $N = 10$ target directions. As the findings with respect to the movement

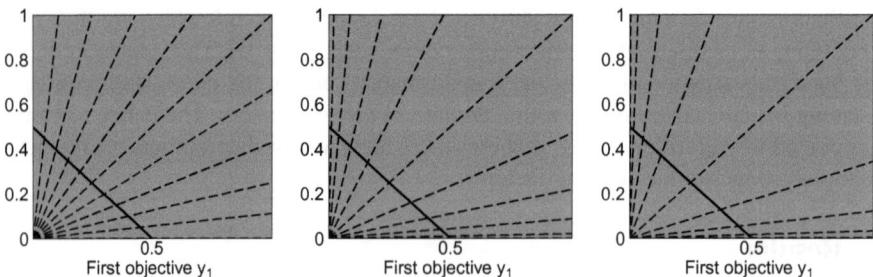

Fig. 4. Different weight vector distributions for eleven weight vectors: Uniform ($\gamma = 1$, left), $\gamma = 2$ (middle) and $\gamma = 3$. The dashed lines correspond to the corresponding target directions. The Pareto front of DTLZ1 is indicated by the solid black line.

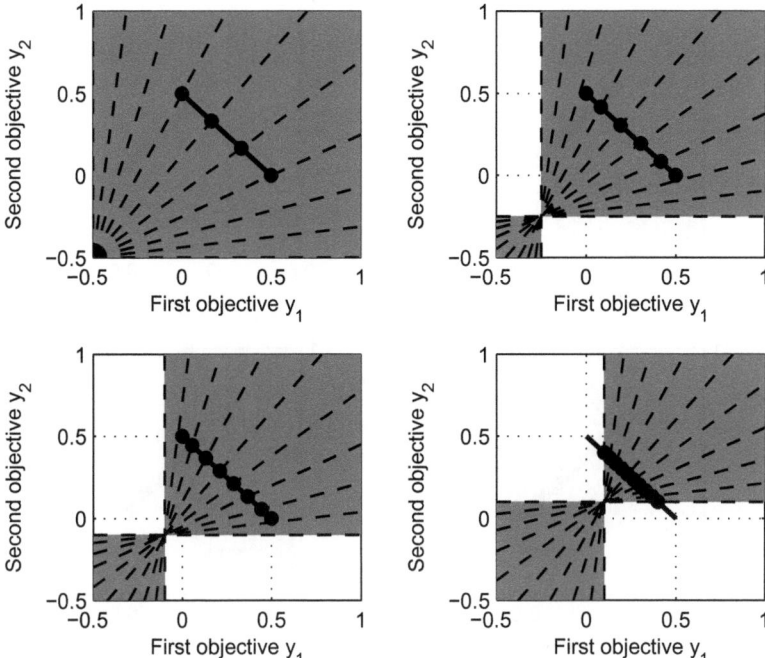

Fig. 5. Experimental results for moving the target cone (gray area) defined by 10 uniformly distributed weight vectors by changing the position of the reference point. Shown are $\mathbf{r} = (-0.5, -0.5)^T$ (upper left), $\mathbf{r} = (-0.25, -0.25)^T$ (upper right), $\mathbf{r} = (-0.1, -0.1)^T$ (bottom left), and $\mathbf{r} = (0.1, 0.1)^T$ (bottom right). The dashed lines correspond to the target directions. The best result of the CMA-ES optimization for $\mu = 10$ solutions is depicted using black dots. The Pareto front of DTLZ1 is indicated by the thick black line.

of the target cone and the distribution of the individuals are general, the choice of the linear front does not represent a restriction. For other fronts with more complex shapes, qualitatively the same results can be expected.

The optimal solutions are located at the intersections between the target directions and the Pareto front. Although the positions of $\mu - N$ solutions are not uniquely determined from a theoretical point of view in the case when less than μ target directions intersect with the Pareto front [8], the used initialization of the CMA-ES result in clusters of solutions at the extremes. As a consequence, one can only distinguish as many different points in the optimal μ-distributions of Fig. 5 as there are intersections between target directions and the Pareto front. Hence, the number of inner mutual solutions decreases with increasing distance to the reference point. If the reference point, however, is moved inside the interval of component values within the front, a focus on the respective region can be realized (bottom right plot of Fig. 5).

The fact that not all μ solutions are uniquely defined in the above examples with $N = \mu = 10$ target vectors, i.e., that they potentially are even dominated in the optimal

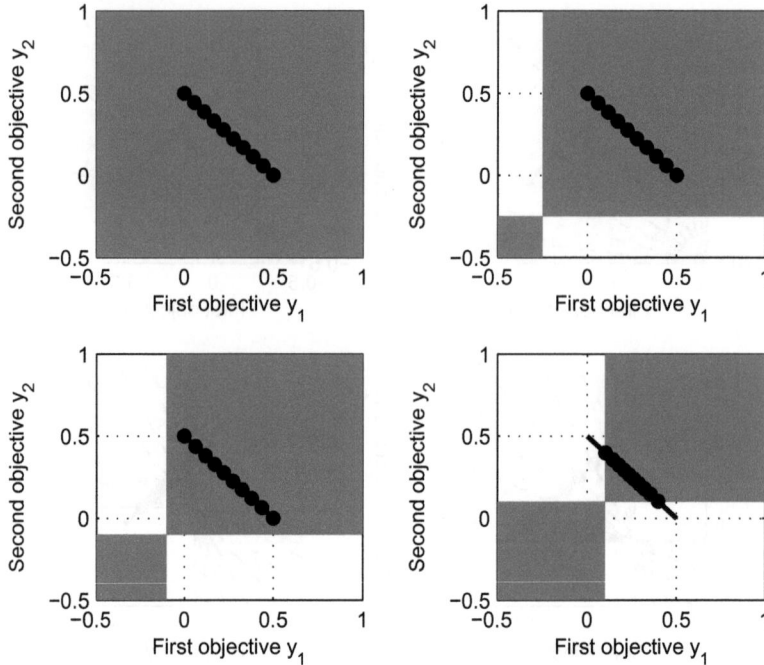

Fig. 6. Experimental results for moving the target cone (gray area) defined by 1000 uniformly distributed weight vectors by changing the position of the reference point. Shown are $\mathbf{r} = (-0.5, -0.5)^T$ (upper left), $\mathbf{r} = (-0.25, -0.25)^T$ (upper right), $\mathbf{r} = (-0.1, -0.1)^T$ (bottom left), and $\mathbf{r} = (0.1, 0.1)^T$ (bottom right). The best result of the CMA-ES optimization for $\mu = 10$ solutions is depicted using black dots. The Pareto front of DTLZ1 is indicated by the thick black line.

μ-distributions, is usually not desired, as in these cases, not all solutions provide additional information about the shape of the Pareto front. In particular, if the true ideal point of the front is not known a priori, it is hard to specify an appropriate reference point allowing the whole front to be covered. A simple solution to this problem is to increase the number of weight vectors. In this case, there are enough intersecting points to locate all individuals on different locations of the Pareto front. This is shown in Figure 6 which only differs from Figure 5 by an increase of the number of weight vectors from $N = 10$ to $N = 1000$. The three cases where the reference point is outside the interval of component values within the front are now resulting in almost the same distribution of individuals.

4.2   Density of the Weight Vector Distribution

Further experiments were conducted in order to experimentally check how the optimal μ-distributions regarding $R2$ are influenced by varying the underlying weight vector distribution. More specifically, it is probable that by shifting the density of the weight vectors away from the center of the Pareto front, the optimal μ-distributions for the $R2$ indicator become more similar to the respective ones for the HV indicator based on former results presented in [8]. Algorithm 1 forms the basis for generating the required weight vector distributions using $\gamma \in \{1, 1.5, 2, \ldots, 4.5, 5\}$.

The size μ of the focused solution set is chosen as 10 to be in line with preceding studies and to ensure a meaningful visualization of the results. The number of weight vectors is set to 501 as former experiments in [8] indicated this as a threshold for generating sufficiently robust optimization results. Three different test functions representing different kinds of Pareto front shapes are addressed, i.e., ZDT1 (convex, [19]), DTLZ1 (linear, [10]) and DTLZ2 (concave, [10]). The definitions of the Pareto fronts are $y_2^i = 1 - \sqrt{y_1^i}$ (ZDT1), $y_2^i = 0.5 - y_1^i$ (DTLZ1), and $y_2^i = \sqrt{1 - (y_2^i)^2}$ (DTLZ2). On each of these test problems, we approximated the optimal μ-distributions using the reference point $(0, 0)^T$. To accomplish this, ten CMA-ES runs with a population size of five and an offspring size of 10 as recommended in [12] enhanced by one L-BFGS-B [9] run were conducted. As starting population, the optimal μ-distributions regarding $R2$ based on uniformly distributed weight vectors [8] were used.

The experimental results are presented in Fig. 7. In order to allow for a visual comparison, the optimal μ-distribution for the hypervolume indicator [1] for $\mu = 10$ solutions is depicted at the top of each figure. It can be clearly seen that the optimal positions of the resulting solutions follow the shifts induced to the weight vectors by increasing γ. For a coarser density of weight vectors at the center of the front, the positions of the optimized solutions indicate movements towards the extremes of the front. Even distributions similar to the HV-optimal ones result which is in line with our expectations. The perturbations in the trends for increasing γ are due to the high problem difficulty for the solvers which sporadically produce slightly suboptimal results, especially for DTLZ2.

4.3 Restriction of the Weight Space

Additionally, experiments were carried out to visualize the effect indicated by Theorem 1 that all solutions of the optimal μ-distributions regarding $R2$ lie within the target cone defined by weight vectors $\mathbf{w}^i \in \mathbf{W}$. By restricting the weight space to predefined intervals (see Sec. 3.2), the solutions y^i are located in between the intersections of the outmost weight vectors with the front. However, as general theoretical results for more than one separate interval in weight space were not yet derived—the optimal μ-distributions depend on the proximity of these intervals—we experimentally analyzed this situation.

The experimental setup coincides with the respective settings of the previous section with respect to the test functions and optimization algorithms considered. However, in the current setup, the 501 weight vectors are uniformly distributed within predefined intervals. Specifically, the situation of two separate intervals is addressed. Thereby, the number of weight vectors is split equally to both intervals. Regardless of this, the initial populations of the CMA-ES are filled by 10 uniformly spaced points on the y_1-axis within the interval defined by the two outmost weight vectors over all considered intervals.

Figure 8 presents the corresponding experimental results. In line with Theorem 1, the optimal μ-distributions regarding $R2$ are located within the intervals defined by

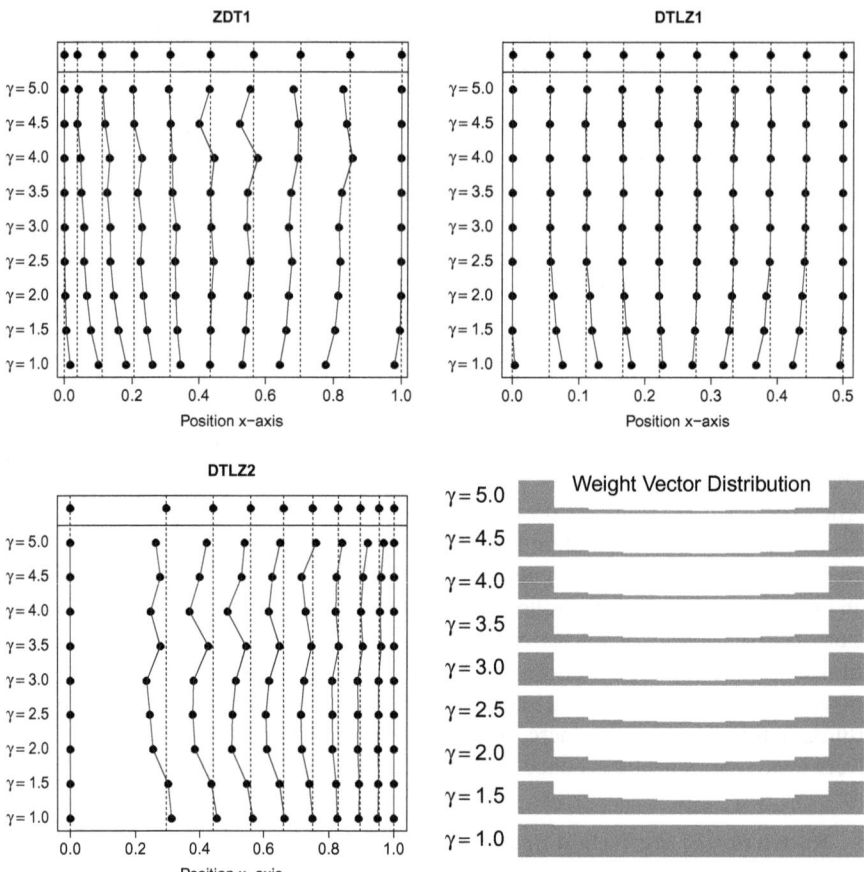

Fig. 7. Experimental results for weight vector distributions with varying density. The best result of the 11 optimization runs is depicted using black dots. The parameter γ of the underlying weight vector distributions is given on the y-axes of the figures. At the top line, the corresponding optimal μ-distribution for the HV is shown for reference. The respective weight vector distributions for different parameter levels γ are visualized using histograms in the lower right plot.

the target cones resulting from the restriction of the weight space. Additionally, as the two separate intervals considered are not close to each other, solutions are concentrated within the two individual target cones in a defined way. Although the points, especially for DTLZ2 and ZDT1, appear to be distributed quite skewed towards the edges of the displayed intervals, the corresponding Euclidean distances in two-dimensional space are much more homogeneous. This is due to the curvature characteristics of the respective fronts with either very small or very high gradients towards the extremes in each dimension and the fact that the plots, as shown here, present only the projections of the solutions onto the y_1-axis of the first objective function.

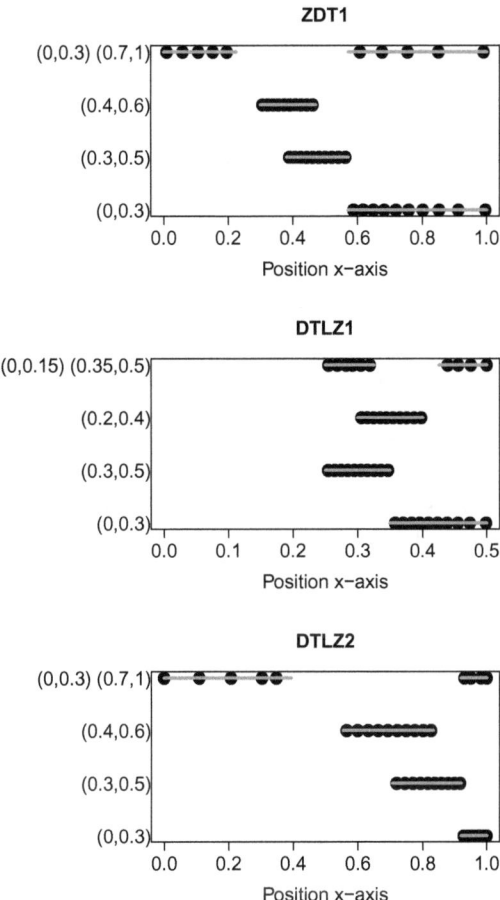

Fig. 8. Experimental results for restricted weight spaces. The best approximation regarding $R2$ out of 11 runs is visualized by black dots. 501 weight vectors are uniformly distributed in the region(s) given on the y-axis of the plot. In case two intervals are considered, the number of weight vectors is split equally to both intervals. The resulting interval(s) on the x-axis bounded by the intersections of the outmost weight vectors with the Pareto front are visualized in orange.

5   Conclusions

In this paper, three variants for integrating preferences into the $R2$ indicator were introduced. These variants exploit that the optimal μ-distributions for the $R2$-indicator are affected by moving the reference point, by restricting the weight space, as well as by skewing the weight vector distribution. In addition, a sound theoretical background for the first two variants was provided by proving a theorem which relates the location of the optimal μ distribution to the target cone of the weight vectors of the $R2$ indicator. Moreover, experiments were conducted to visually illustrate that the choice and the target cone of the weight vector distribution heavily influences the optimal

μ-distribution for the $R2$ indicator. This is of particular interest, as the standard approach is still to uniformly distribute the weight vectors over their complete domain without being aware of the resulting implications. While the results on the one hand demand for a cautious choice of the weight vector distribution, it thereby becomes possible to take into account preferences regarding the distribution of the points on the Pareto front approximation.

The reference point, as a parameter of $R2$, was found to have an influence which is not negligible. Especially, in case it is chosen too close to the Pareto front, i.e., in case it does not dominate the whole Pareto front, the optimal μ-distributions regarding $R2$ will never cover the whole extent of the Pareto front, independent from the number of weight vectors. On the other hand, the number of weight vectors which intersect the Pareto front decreases with increasing distance to the front. Thus, the recommendation is to apply a very conservative reference point (far better than the approximated Pareto front) and to use more weight vectors the higher the uncertainty of the location of the actual ideal point. This increases the probability that a sufficiently high number of weight vectors intersects the complete Pareto front.

Regarding future work, several research directions are worth to be explored. In addition to further quantitative theoretical analyses of concrete optimal μ-distributions for the $R2$ indicator in the bi-objective case, it will be important to investigate the optimal distributions also for problems with more than two objective functions or other related indicator such as the one in [7]. Furthermore, the question arises how the optimal μ-distributions for the (preference-based) $R2$ indicator can be actually obtained in practice. A simple $R2$-indicator-based selection within an evolutionary multi-objective optimization algorithm similar to known HV-indicator-based algorithms, such as [4,5], would be a first step towards this goal.

Acknowledgments. This paper is based on investigations of the project D5 "Synthesis and multi-objective model-based optimization of process chains for manufacturing parts with functionally graded properties" as part of the collaborative research center SFB/TR TRR 30 and the project B4 of the Collaborative Research Center SFB 823, which are kindly supported by the Deutsche Forschungsgemeinschaft (DFG). In addition, the authors acknowledge support by the French national research agency (ANR) within the Modèles Numérique project "NumBBO - Analysis, Improvement and Evaluation of Numerical Blackbox Optimizers".

References

1. Auger, A., Bader, J., Brockhoff, D., Zitzler, E.: Theory of the Hypervolume Indicator: Optimal μ-Distributions and the Choice of the Reference Point. In: Foundations of Genetic Algorithms (FOGA 2009), pp. 87–102. ACM, New York (2009)
2. Auger, A., Bader, J., Brockhoff, D., Zitzler, E.: Hypervolume-based Multiobjective Optimization: Theoretical Foundations and Practical Implications. Theoretical Computer Science 425, 75–103 (2012)
3. Auger, A., Hansen, N.: A Restart CMA Evolution Strategy With Increasing Population Size. In: Congress on Evolutionary Computation (CEC 2005), pp. 1769–1776. IEEE Press (2005)

4. Bader, J., Zitzler, E.: HypE: An Algorithm for Fast Hypervolume-Based Many-Objective Optimization. Evolutionary Computation 19(1), 45–76 (2011)
5. Beume, N., Naujoks, B., Emmerich, M.: SMS-EMOA: Multiobjective selection based on dominated hypervolume. European Journal of Operational Research 181(3), 1653–1669 (2007)
6. Beume, N., Naujoks, B., Preuss, M., Rudolph, G., Wagner, T.: Effects of 1-Greedy S-Metric-Selection on Innumerably Large Pareto Fronts. In: Ehrgott, M., Fonseca, C.M., Gandibleux, X., Hao, J.-K., Sevaux, M., et al. (eds.) EMO 2009. LNCS, vol. 5467, pp. 21–35. Springer, Heidelberg (2009)
7. Bozkurt, B., Fowler, J.W., Gel, E.S., Kim, B., Köksalan, M., Wallenius, J.: Quantitative Comparison of Approximate Solution Sets for Multicriteria Optimization Problems with Weighted Tchebycheff Preference Function. Operations Research 58(3), 650–659 (2010)
8. Brockhoff, D., Wagner, T., Trautmann, H.: On the Properties of the $R2$ Indicator. In: Genetic and Evolutionary Computation Conference (GECCO 2012), pp. 465–472 (2012)
9. Byrd, R.H., Lu, P., Nocedal, J., Zhu, C.: A limited memory algorithm for bound constrained optimization. SIAM Journal on Scientific Computing 16, 1190–1208 (1994)
10. Deb, K., Thiele, L., Laumanns, M., Zitzler, E.: Scalable Multi-Objective Optimization Test Problems. In: Congress on Evolutionary Computation (CEC 2002), pp. 825–830. IEEE Press (2002)
11. Hansen, M.P., Jaszkiewicz, A.: Evaluating The Quality of Approximations of the Non-Dominated Set. Technical report, Institute of Mathematical Modeling, Technical University of Denmark, IMM Technical Report IMM-REP-1998-7 (1998)
12. Hansen, N.: The CMA Evolution Strategy: A Comparing Review. In: Lozano, J., Larrañaga, P., Inza, I., Bengoetxea, E. (eds.) Towards a New Evolutionary Computation. STUDFUZZ, vol. 192, pp. 75–102. Springer, Heidelberg (2006)
13. Hansen, N.: CMA Evolution Strategy Source Code (2012), http://www.lri.fr/~hansen/cmaes_inmatlab.html
14. Igel, C., Hansen, N., Roth, S.: Covariance Matrix Adaptation for Multi-objective Optimization. Evolutionary Computation 15(1), 1–28 (2007)
15. Knowles, J.: ParEGO: A Hybrid Algorithm With On-Line Landscape Approximation for Expensive Multiobjective Optimization Problems. IEEE Transactions on Evolutionary Computation 10(1), 50–66 (2005)
16. Miettinen, K.: Nonlinear Multiobjective Optimization. Kluwer, Boston (1999)
17. Zhang, Q., Li, H.: MOEA/D: A multi-objective evolutionary algorithm based on decomposition. IEEE Transactions on Evolutionary Computation 11(6), 712–731 (2007)
18. Zitzler, E.: Evolutionary Algorithms for Multiobjective Optimization: Methods and Applications. PhD thesis, ETH Zurich, Switzerland (1999)
19. Zitzler, E., Deb, K., Thiele, L.: Comparison of Multiobjective Evolutionary Algorithms: Empirical Results. Evolutionary Computation 8(2), 173–195 (2000)
20. Zitzler, E., Knowles, J., Thiele, L.: Quality Assessment of Pareto Set Approximations. In: Branke, J., Deb, K., Miettinen, K., Słowiński, R. (eds.) Multiobjective Optimization. LNCS, vol. 5252, pp. 373–404. Springer, Heidelberg (2008)
21. Zitzler, E., Thiele, L.: Multiobjective Optimization Using Evolutionary Algorithms - A Comparative Case Study. In: Eiben, A.E., Bäck, T., Schoenauer, M., Schwefel, H.-P. (eds.) PPSN V. LNCS, vol. 1498, pp. 292–301. Springer, Heidelberg (1998)
22. Zitzler, E., Thiele, L., Laumanns, M., Fonseca, C.M., Grunert da Fonseca, V.: Performance Assessment of Multiobjective Optimizers: An Analysis and Review. IEEE Transactions on Evolutionary Computation 7(2), 117–132 (2003)

Do Hypervolume Regressions Hinder EMOA Performance? Surprise and Relief

Leonard Judt[1], Olaf Mersmann[1], and Boris Naujoks[2]

[1] Faculty of Statistics, TU Dortmund University, 44221 Dortmund, Germany
[2] Cologne University of Applied Sciences, 51643 Gummersbach, Germany
{leonard.judt,olaf.mersmann}@tu-dortmund.de,
boris.naujoks@fh-koeln.de

Abstract. Decreases in dominated hypervolume w.r.t a fixed reference point for the $(\mu + 1)$-SMS-EMOA are able to appear. We examine the impact of these decreases and different reference point handling techniques by providing four different algorithmic variants for selection. In addition, we show that yet further decreases can occur due to numerical instabilities that were previously not being expected. Fortunately, our findings do indicate that all detected decreases do not have a negative effect on the overall performance.

Keywords: EMO, hypervolume decreases, reference point handling, numerical instabilities.

1 Introduction

The dominated hypervolume has become the standard performance indicator in evolutionary multi-objective optimisation (EMO) in recent years, cf. Deb [4] or Coello Coello [3] for details on EMO related definitions and vocabulary. The dominated hypervolume was defined by Zitzler and Thiele as the the size of the space covered by a Pareto front with respect to a given reference point [11]. Special properties of this indicator have been proven [12], especially its property of being Pareto compliant, i.e. whenever a Pareto front approximation dominates another approximation, the hypervolume of the former will be greater than the hypervolume of the latter.

Next to becoming the standard performance indicator, the dominated hypervolume has been incorporated for environmental selection in a number of EMO algorithms (EMOA) in the past half decade. Prominent examples are SMS-EMOA [2], MO-CMAES [6], as well as HypE [1]. In all these algorithms as well as publications, the course of the hypervolume was expected to be monotonically decreasing over the number of fitness function evaluations.

It was long thought that decreases in hypervolume could not be experienced over the course of an optimisation process performed for instance with the SMS-EMOA, as individuals would be selected based primarily on their hypervolume contributions, and thus offspring causing a decrease in hypervolume could never be selected.

R.C. Purshouse et al. (Eds.): EMO 2013, LNCS 7811, pp. 96–110, 2013.

Recent results however show that unsuspected decreases in the hypervolume progression during an optimisation run are possible [7]. Even more complex, different reasons for the 2-dimensional and the 3-dimensional case have been identified. We will explain four different strategies for selection and reference point handling to compare the standard selection scheme to more precise ones that were not expected to be influenced by the reasons mentioned above. To this end, we compared SMS-EMOA with decreases and reference point adaptation (the standard case) to the algorithm not accepting decreases. Moreover we test the algorithm with a fixed reference. Thus, we end up with four combinations with respect to decrease acception and reference point usage (cf. Sec. 2.2). From the reasons for hypervolume decreases provided by Judt et al. [7], we do not expect hypervolume decreases within the course of the optimisation runs for a subset of the cases above. Consequently, we expect some algorithmic setting to provide identical results.

Surprisingly, hypervolume decreases could be observed for all algorithmic settings in the 3-dimensional case. Taking a deeper look at generated points, calculated hypervolume values, and the consequences of selection, it was found that numerical issues (highly) influence the selection procedure. We continued our investigation focusing on the effect of such issues.

The goal of this paper is to make researchers a bit more sensible for the numerical issues that may arise in complex calculation like for (higher dimensional) hypervolume values and to examine if omitting hypervolume decreases or using dissimilar reference points for the selection scheme is beneficial or not.

Nevertheless, for the considered cases here, we can give the all-clear. Although different paths through the parameter space are considered, the overall results do not vary significantly. We regard this as a very positive result indicating the robustness of the hypervolume selection approaches.

The following section provides basic elements when talking about hypervolume decreases and summarizes the findings on hypervolume decreases during optimisation runs from prior publications. Moreover, it presents a more detailed description of the four algorithmic settings considered. Sec. 3 provides the experimental setup for our experiments and Sec. 4 summarizes our findings for the 2-dimensional test cases. Moreover, surprising effects are reported here, that are reasoned in a special subsection.

Sec. 5 summarizes all results obtained for the 3-dimensional cases and provides relief that the dominated hypervolume is an adequate selection criterion for indicator based EMOA. The last section summarizes all findings and provides and outlook for further research.

2   Decreases in Hypervolume Progressions

The progression of the hypervolume in a 1-greedy hypervolume selection based EMOA was thought to never decrease in the course of an optimization run. This belief arose from the design of the algorithms, where the individual with the least hypervolume contribution is discarded in every generation. However,

Judt et al. [7] showed that this intuition is wrong for SMS-EMOA featuring an adaptive reference point in the 3-dimensional case and due to a special treatment of boundary solutions in the 2-dimensional case.

2.1   Prior Findings

More precisely, boundary solutions, i.e. solutions with the least fitness function value in one objective, are always kept for the succeeding generation in SMS-EMOA for 2-dimensional objective functions. This was implemented to not lose such solutions with extreme fitness function values. However, these solutions might only supply a minimal hypervolume contribution due to lying very close to another solution on the Pareto front. Since this minimal contribution is kept, a more significant contribution might be lost eliminating an alternative solution in the $(\mu+1)$ selection scheme. This leads to the decrease in hypervolume of the whole Pareto front. For more details we refer to Judt et al. [7].

For 3-dimensional and higher-dimensional objective spaces, comparable decreases are observed, however, the reason for these is different. In higher dimensional objective spaces, boundary solutions are not automatically kept for the succeeding generation since this would fill the whole population within a short period of time. This is also due to employing a different definition for boundary solutions. Here, these solutions are solutions lying next to the boundary of the Pareto front, e.g. the curves where the 1-hypersphere being the Pareto front of DTLZ2, touch the plains spanned by the positive coordinate axes.

An adaptive reference point is used for hypervolume calculations in the 3-dimensional case. This reference point is determined by the solutions contributing the worst (highest) objective function values w.r.t to one objective. These worst values are complemented to a vector and a fixed value of one is added in each direction. This defines the adaptive reference point. As a result, there might be a different reference point within each generation. This is effectively incorporated to calculate the hypervolume contributions of Pareto non-dominated solutions. Thus, this might lead to a decrease in hypervolume of the whole Pareto front observed from the perspective of a reference point fixed for the whole optimisation process.

2.2   Selection Strategies under Investigation

To fairly test the influence of hypervolume decreases on the overall performance of the hypervolume based selection strategies, different strategies have to be considered.

1. The standard implementation, where hypervolume contributions are calculated w.r.t. the adaptive reference point. For our experiments we recalculated all contributions w.r.t. a fixed reference point to identify decreases. However,

the optimisation proceeded accepting the hypervolume regression. For referencing this procedure later on, we abbreviate it `adaptive/with` indication that an adaptive reference point is used and regressions are considered.

2. The repairing implementation, where hypervolume contributions are again calculated w.r.t. an adaptive reference point. However, if the recalculation of the hypervolume yields a hypervolume decrease, the proposed selection step in not accomplished. No selection is performed, the following generation equals the prior one. This procedure is abbreviated `adaptive/without`.

3. The assured implementation, where hypervolume decreases are always calculated w.r.t. the fixed reference point. This way, no decreases in hypervolume are possible for 3-dimensional and higher-dimensional objective spaces. This implementation differs from the repairing one, in such that contributions are not calculated w.r.t different reference points, thus, leading to the selection of different solutions. Just in case and beyond expectation: If hypervolume decreases are discovered, these will not be accomplished. Here, the assured implementation complies with the repairing one. This implementation is abbreviated `fixed/without`.

4. The impossible implementation, where hypervolume decreases calculated to a fixed reference point are accepted during the optimisation process. These decreases are beyond expectation, while this case is called the impossible case, and abbreviated `fixed/with`.

For the third strategy described above, the assured implementation, it is important if a 2-dim test function is considered or a higher-dimensional one. For the latter, using a fixed reference point must not lead not any hypervolume decreases. In such cases, decreases are expected to only show up due to the different reference points considered internally (adaptive reference point) and externally, when comparing hypervolume values from different generations (fixed reference point). This is different for 2-dim. test cases. In this case, decreases are possible even though a fixed reference point is in use. This leads to the question whether decreases should be allowed or not on a different level. For further results we decided to allow decreases in such situations.

The fourth case described above was added to complement all possible combinations. The results received for the third case (`fixed/without`) and the fourth case (`fixed/with`) were expected to be identical for 3- or higher-dimensional test functions. All strategies have been implemented to the SMS-EMOA from the R package `emoa` [9].

3 Experimental Setup

For the paper, a total of 1 600 reproducible runs of the SMS-EMOA were conducted on 2-dimensional, 3-dimensional, and 4-dimensional test cases. More specifically, 800 runs were performed on the four 2-dimensional test functions ZDT1 – ZDT4 (cf. [10]) with 30 decision variables each, 600 runs were conducted with the three 3-dimensional test functions DTLZ1 – DTLZ3 (cf. [5]) with the number of decision variables being reduced to 7 and 12 as suggested by Deb et al. [5] and 200 runs were performed with a 4-dimensional version of DTLZ2 with 13 decision variables.

In addition, as parameters for the Simulated Binary Crossover (SBX, [4]) and the Polynomial Mutation (PM,[4]) operator the combination sbx_n = 10, sbx_p = 0.5, pm_n = 20 and pm_p = 0.1 was considered. All runs consisted of 100 000 fitness function evaluations with a population size μ of 100 individuals.

The fixed global reference points chosen are provided in Tab. 1.

Table 1. Global fixed reference points considered

ZDT1 – ZDT3:	[11, 11]	DTLZ1:	[1000, 1000, 1000]
ZDT4:	[1000, 1000]	DTLZ2:	[11, 11, 11]
4-dim. DTLZ2:	[11, 11, 11, 11]	DTLZ3:	[2000, 2000, 2000]

Such remote choices at times were necessary in order that all generated individuals, including those in the starting populations, would be dominated by these reference points.

Furthermore, the four strategies outlined in Sec. 2 were implemented for these runs. For adaptive/with and adaptive/without, the evaluated hypervolume contributions are based upon an adaptive reference point, whereas the fixed reference point provided above is used for the fixed/with and fixed/without strategies. Moreover, selections that lead to a decrease in hypervolume w.r.t. the fixed reference point are discarded under the adaptive/without and fixed/without strategies, in contrast to being kept under the adaptive/with and fixed/with strategies.

For each combination of strategy, test function, and population size 50 independent runs were conducted. The hypervolume w.r.t. a fixed reference point for each generation was calculated and stored. More details on exact parameterizations and reference points are provided in the supplementary material[1]

4   Results I: 2-dimensional and Surprising 3-dimensional

All results are presented as histograms of the hypervolume distributions. Each entire figure is composed of of several graphics as subfigures that represent histograms of a specific setting. To this end, figure 1 consists of 16 graphics, subdivided into four graphics in each row and column.

The following figures are separated according to the test function considered, the number of fitness function evaluations performed and/or the implemented selection criterion standard, repairing, assured, and impossible (cf. Sec. 2).

Firstly, attention will be paid to the 2-dimensional case, secondly to the 3-dimensional case. For each case, we will first focus on the progression of the hypervolume values during one optimisation run and later on comparing the results received on the different test functions considered.

---

[1] Supplementary Material is available at http://ptr.p-value.net/emo13.

4.1   Progression of the Hypervolume in 2-dimensional Test Cases

Figure 1 presents the histograms described above after 500, 10 000, 30 000, and 100 000 fitness function evaluations for test function ZDT2 from left to right. The different rows present to four different algorithmic variants for selection.

All histograms in each single picture were generated of all optimization runs featuring all mentioned parameterisations, the corresponding test function, algorithm combination, and the addressed way of reference point handling after the given number of fitness function evaluations.

This holds for different points during the progression of the optimization runs as well as for all test cases. ZDT2 was used as a representative example for all other test functions.

Obviously, the results for the schemes `adaptive/with` and `fixed/with` as well as `adaptive/without` and `fixed/without` are identical. As a result, the performance here is solely depending on the handling of the reference point. This is in line with our observations on the reasons for possible hypervolume regressions as described in [7].

With respect to the hypervolume progression, all different variants gain more and more hypervolume with more fitness function evaluations. Even more, the final values after 100 000 fitness function evaluations look more or less identical with a high peak in the left of the graphics and some outliers on the right. However, there is no clear evidence, which is the best variant. A detailed look for different test cases after 30 000 fitness function evaluations is provided in the following section.

4.2   Comparison for Different 2-dimensional Test Functions

The following figure depicts the situation from above after 30 000 fitness function evaluations for all 2-dimensional test function considered. Again, the compliance of the schemes `adaptive/with` and `fixed/with` as well as `adaptive/without` and `fixed/without` can be observed. Since 30 000 fitness function evaluations are the recommended number of fitness function evaluations for these test functions, we decided to judge upon the final performance based on this data.

Comparing different schemes again, no clear winner can be determined. All schemes preform more or less comparably for all different 2-dimensional test cases. Moreover, even differences concerning variances of the distributions are hard to detect either for the different strategies as well as for the different test functions.

As a result, we conclude that no deviating recommendation can be made as an advisable strategy other than the standard implementation `adaptive/with` for calculating the hypervolume contributions in theSMS-EMOA for 2-dimensional functions.

4.3   Progression of the Hypervolume in 3-dimensional Test Cases

In parallel to Figure 1, Figure 3 depicts histograms portraying the obtained hypervolume after 500, 10 000, 30 000, and 100 000 fitness function evaluations

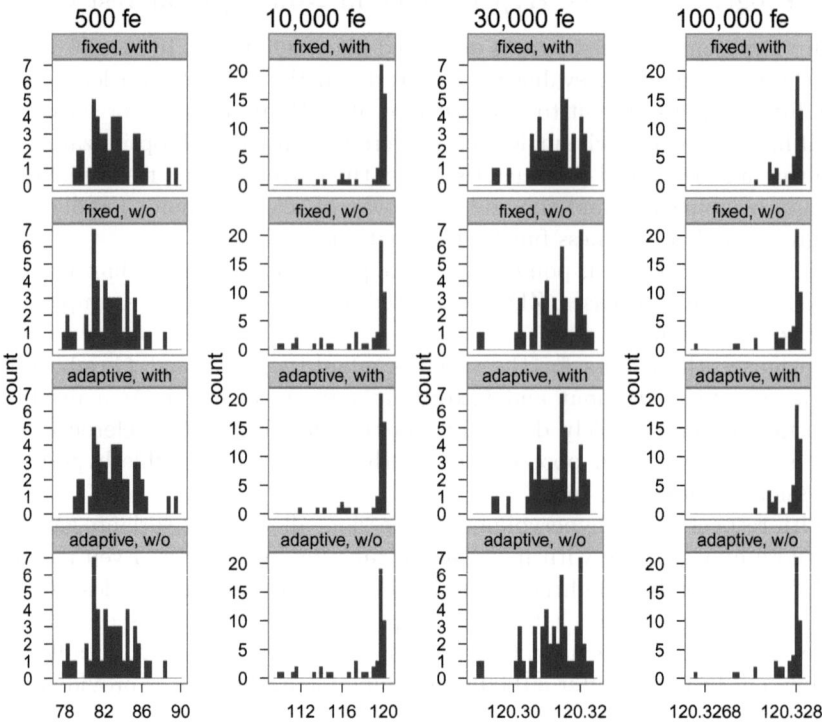

Fig. 1. Histograms of received hypervolume values for ZDT2 within the progression of the optimisation runs, i.e. after 500, 10 000, 30 000, and 100 000 fitness function evaluations (left to right). Within one column, the four different strategies are depicted.

for test function DTLZ2. The four proposed algorithmic variants are aligned in columns again.

DTLZ2 was chosen to represent the 3-dimensional test cases due to the fixed reference point for calculating the hypervolume being comparatively close to the Pareto front. This results in a much lower scale for the achieved hypervolume throughout the optimization process, which in return appears to be more suitable for comparing the different strategies. Nevertheless, similar assessments to the following can be made for the other 3-dimensional test functions, part of which will be shown in Sec. 5.

Most striking, the distribution of the obtained hypervolume for the runs performed under the `fixed/without` selection variant does in fact differ from the distribution achieved with the `fixed/with` variant. This already occurs after 500 fitness function evaluations, hence right at the beginning of the optimization process.

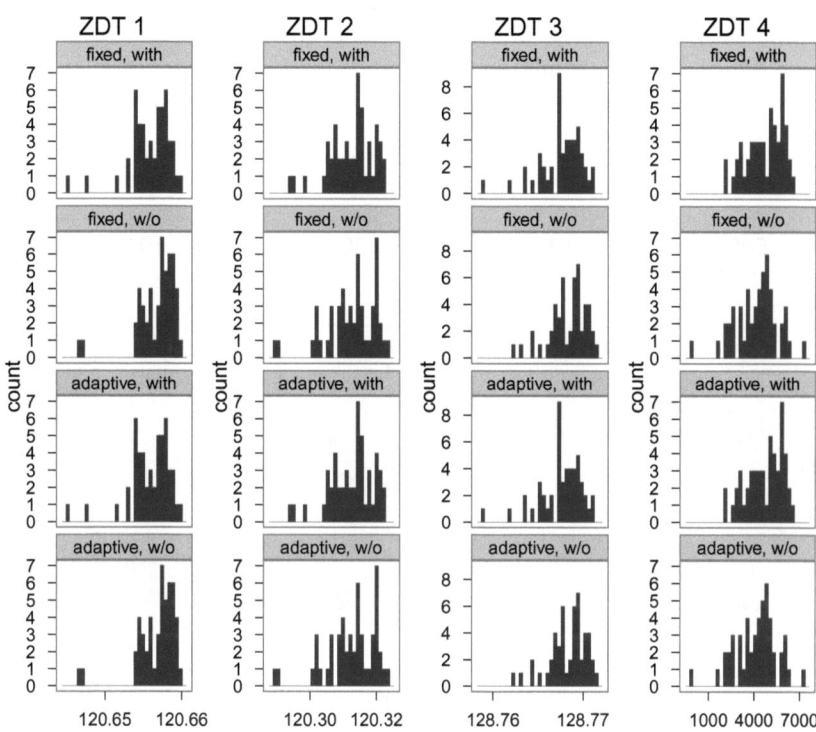

Fig. 2. Histograms of received hypervolume values for the 2-dimensional test functions after 30 000 fitness function evaluations (left to right). Within one column, the four different strategies are depicted. For visualization purposes a hypervolume value of 990 000 is subtracted from the actual hypervolume value for ZDT4 in order to have suitable values for the x-axis.

This is in direct contradiction to our expectations, as 3-dimensional (and higher-dimensional) runs conducted with a selection scheme that uses a fixed reference point should not experience decreases in hypervolume w.r.t. a fixed reference point, thus repairing should not be necessary in practice for fixed/without. It therefore was expected that the fixed/with and fixed/without strategies would lead to identical hypervolume distributions. This issue will be discussed in more detail in the following section.

On a different note, looking at the progression of hypervolume, all four strategies result in a gradual increase of hypervolume with more fitness function evaluations. No significant difference in hypervolume distributions between the different variants in each column can be seen in this figure. While using a version with an adaptive reference point might appear to achieve slightly better results right at the beginning of the runs, they in turn seem to attain a barely inferior hypervolume distribution after 100 000 fitness function evaluations. However, these differences are minimal and therefore no ultimate recommendation can be made on which strategy results in significantly better performances. In Sec. 5, results of runs from other 3-dimensional test functions after 30 000 are used as further reference for comparing the different strategies.

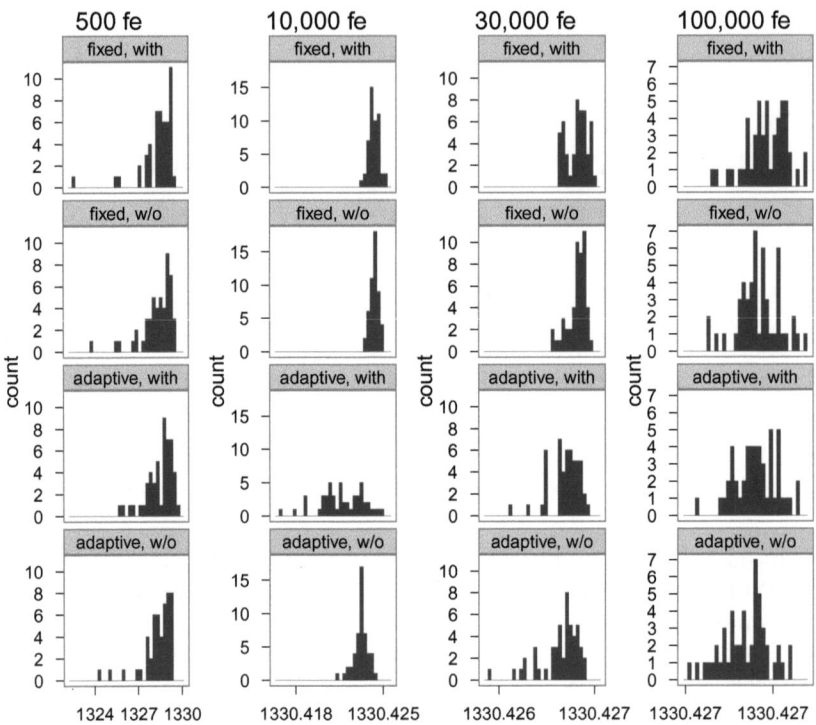

Fig. 3. Histograms of received hypervolume values for 3-dimensional DTLZ2 within the progression of the optimisation runs, i.e. after 500, 10 000, 30 000, and 100 000 fitness function evaluations (left to right). Within one column, the four different strategies are depicted.

4.4   Numerical Issues Involved

During our experiments we observed many numerically identical hypervolume values. In previous work [8] we conjectured that this was likely due to numerical limits of the underlying C code used to calculate hypervolume values. When we started to investigate the current results we noticed another, even stranger, issue. Sometimes the hypervolume of successive Pareto front approximations would sink even though the latter front differed from the former only in a single point and this point dominates another point in the former front. More formally, we found sets of points $\{y_1, \ldots, y_{n+1}\}$ such that

$$\widehat{\mathrm{hyp}}_r(\{y_1, \ldots, y_n\}) < \widehat{\mathrm{hyp}}_r(\{y_1, \ldots, y_{n-1}, y_{n+1}\})$$

even though y_n dominates y_{n+1}. Here $\widehat{\mathrm{hyp}}_r$ denotes the numerical approximation of the hypervolume w.r.t. the reference point r. Clearly, this is in violation of the strong Pareto compliance of the hypervolume.

The observed differences are quite small, usually in the 15th or 16th significant digit. Since all calculations are performed using double precision floating point numbers, these differences are not larger than what one expects. What is surprising is that this *reversal* happens with such high probability (cf. Tab. 2 for details). Even worse, it is likely that there are even more such undetected anomalies because the only reliable way to detect them currently is when the two points y_n and y_{n+1} are ordered. If they are incomparable, we do not have a criterion by which to decide if the numerical approximation of the dominated hypervolume is correct. Given that in the final stages of an optimization run, many changes in hypervolume are of the same order as our observed errors in the numerical approximation. We conjecture that there are also quite a few miniscule decreases in hypervolume that we cannot detect.

The above could have consequences for all indicator based algorithms which use the hypervolume as an indicator - although, as we will see later, there is no strong evidence to suggest that it has a clear detrimental effect. In future, we would like to see further work that studies the numerical properties of the hypervolume indicator and some of the other commonly employed indicators.

Table 2. Mean number of occurred decreases with fixed/with strategy after 500, 10 000, 30 000, and 100 000 fitness function evaluations (fe, left to right). Within one column, the four 3-dimensional and one 4-dimensional test functions are depicted. Note that the numbers shown represent an average of the 50 runs for each of these settings.

	after 500 fe	after 10 000 fe	after 30 000 fe	after 100 000 fe
DTLZ1	7	34	166	1274
DTLZ2	3	4	4	5
DTLZ3	10	44	200	551
4dim. - DTLZ2	3	3	3	3

5   Results II: Relief for the 3-dimensional Cases

5.1   Comparison for Different 3-dimensional Test Functions

After putting the blame for the surprising results received above to numerics, we continue to investigate whether the different algorithmic variant and, thus, also the numerical instabilities really influence the overall performance of the algorithms. To do so in the 3-dimensional case we follow the route already driven for the 2-dimensional case. To this end, we investigate the performance for different 3-dimensional test functions and all four variants after a specific number of fitness function evaluations, i.e. 30 000 again like in the the 2-dimensional case. This number of fitness function evaluations was chosen to be comparable to the 2-dimensional case and because it is the amount of evaluations that is mostly spend on these function in other investigations.

Figure 4, like Fig. 2 before, details these results introduced above. Again, the graphics for one test function are provided in one column while the graphics belonging to one algorithmic variant are presented in one line with `fixed/with` at the top and `adaptive/without` at the bottom.

The difference in performance of the four algorithmic variants can again be observed in the results presented here. This in particular holds for the results for DTLZ2 and DTLZ3, while the results for DTLZ1 look identical for three of the four types. Only the `adaptive/with` variant provides a distribution of results over the presented x-range with some outliers right in the left corner of the graphic. Due to these outliers, the scaling of the x-axis was adapted accordingly, what made, in turn, the other results look like not being distributed at all, i.e. all offering the same results.

A phenomenon a bit similar can be observed for the DTLZ3 test function, where the results for `fixed/without` are much more distributed over the presented range making the other results appear more narrowed around a specific value. However, differences in the histograms for all four algorithmic variants can clearly be observed here.

Note, that the remarkable difference in the presented x-range is due to the choice of different reference points for the test functions under investigation. This as well as the partly different shapes of the Pareto front result in incomparable amounts of hypervolume that can be achieved by the different algorithms/selection variants. Moreover, due to the huge of amount of achievable hypervolume and the desire to have suitable axis labels, amounts of hypervolume were subtracted for DTLZ1 and DTLZ3. These amounts are provided in the caption.

Trying to answer whether a fixed or an adaptive reference point or the acceptance of decreases or not really pays of is not easy. No significant differences can be detected comparing the four variants over all three test functions. Of course, the already mentioned two variants `adaptive/with` and `fixed/without` perform worse on one of the test functions, but `adaptive/without` seems to be a bit below the other performances for DTLZ2 and DTLZ3. As a result, only the `fixed/with` variant seems to perform best in comparison, which is a bit strange, since this exactly is the impossible implementation that differs to the other fixed variant only due to the mentioned numerical instabilities. Therefore, and due to all differences not being really significant, we do not provide a recommendation for one of the variants and will, additionally, take a look at one 4-dimensional instance.

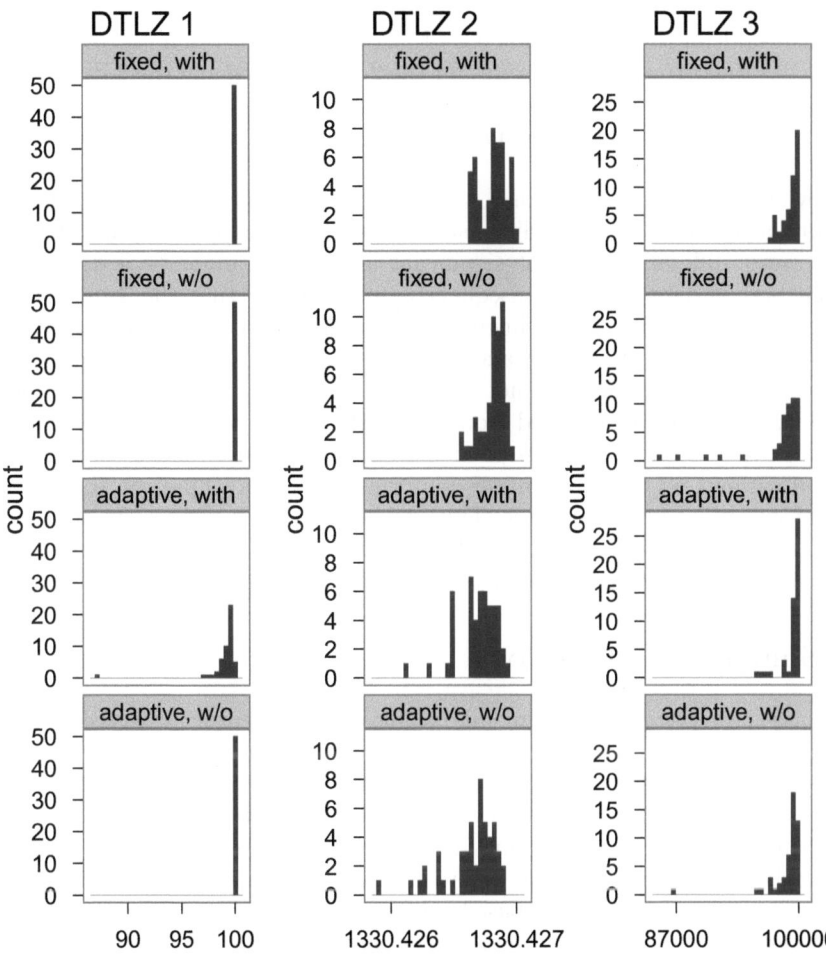

Fig. 4. Histograms of received hypervolume values for the 3-dimensional test functions after 30 000 fitness function evaluations (left to right). Within one column, the four different strategies are depicted. For visualization purposes a hypervolume value of 999 999 900 is subtracted from the actual hypervolume value for DTLZ1 and a hypervolume value of 7 999 900 000 for DTLZ3 in order to have suitable values for the x-axis.

5.2   Comparison for One Higher-Dimensional Test Function

To verity our findings for higher-dimensional objective space we performed experiments corresponding to these in the 2-dimensional and 3-dimensional cases for the 4-dimensional DTLZ2 test function. Of course, only based on these results, we cannot claim portability for other test functions and higher-dimensional spaces. However, these might provide a first hint whether the results are consistent or not.

The results for the 4-dimensional DTLZ2 case are provided in Fig. 5 and this is set up like the corresponding figures for the 2-dimensional and the 3-dimensional cases, i.e. Fig. 2 and 4. Again we have the results after a certain amount of fitness function evaluations in one column and the four different algorithmic variants aligned in one row.

Obviously, and as a main result, there is no consistency w.r.t. the algorithmic variants again. Like in the 3-dimensional case, all variants perform differently. Once more, differences are not significant and there cannot be a decision for a final best variant again. Consequently, we conclude that the 4-dimensional results are in line with the 3-dimensional results achieved and reason that such results hold for other test cases and higher dimensional objective spaces as well.

It can be pointed out, that not accepting decreases in hypervolume is not a solution. This can be reasoned from the performance of the variants yielding an adaptive reference point. In particular for the situation discussed here, the version without accepting decreases performs worse in comparison to its counterpart accepting such decreases. This at least holds for the results at the end of a possible optimisation run, i.e. after 30 000 fitness function evaluations and, more significantly, after 100 000 fitness function evaluations.

6   Summary and Outlook

Decreases in hypervolume during an optimisation run featuring 1-greedy hypervolume selection have not been noticed for a long time. Moreover, they have even been considered impossible. Since they have been detected for the first time, the main question was whether these hinder the overall performance of the algorithm. This was what we aimed to answer by the contribution at hand.

To this end, we proposed four different strategies that have been implemented in a well-known hypervolume based selection EMOA. These variants consider decreases as well as neglect them in combination with using a fixed and an adaptive reference point respectively. All these variants of SMS-EMOA have been tested on 2-dimensional and 3-dimensional mathematical test cases. Moreover, to gain evidence for higher dimensional test cases, we added one 4-dimensional test case to the collection.

As a result, it turned out that decreases in hypervolume do not hinder the overall performance of the algorithm. At least, no significant results have been found that would somehow underpin the claim that they do.

Surprisingly, we stumbled into some other issue: the role of numerical instabilities. We found situations, where apparently better Pareto fronts receive less

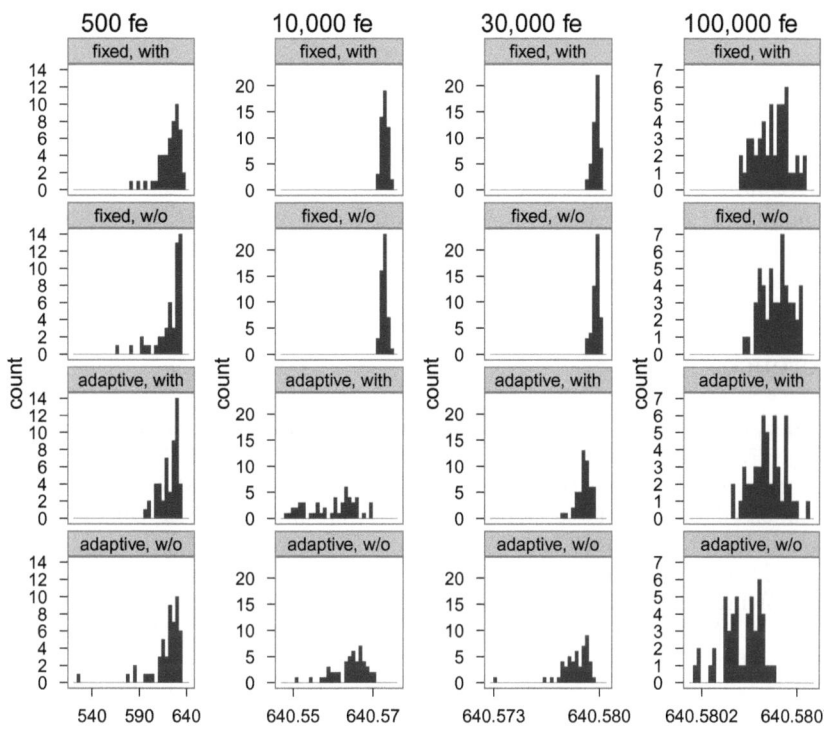

Fig. 5. Histograms of received hypervolume values for 4-dimensional DTLZ2 within the progression of the optimisation runs, i.e. after 500, 10 000, 30 000, and 100 000 fitness function evaluations (left to right). Within one column, the four different strategies are depicted. For visualization purposes a hypervolume value of 14 000 is subtracted from the actual hypervolume in order to have suitable values for the x-axis.

hypervolume values due to issues in precision of the underlying algorithms. The very important role of such numerical issues has been neglected in the EMOA and possibly even in the EA community. It turned out to be a new aim of this contribution to remind researchers of the existence of such issues, the consequences, and possibly to encourage them to address these issues with their research.

A still open question is whether these issues also influence other algorithmic variants like alternative precise hypervolume calculation algorithms and, even more interestingly, the influence on hypervolume approximation techniques, which became famous during the last years. Although these topics are open, we have evidence to believe that even small mistakes in hypervolume values caused by approximation techniques will not worsen the overall performance of the algorithms. The situation is somehow comparable to not having the property of Pareto compliance active for small hypervolume changes like we found in the current research. Hypervolume selection turned out to be a very robust and reliable selection procedure, which is very good and encouraging news.

A possible new research direction, next to investigating the numerical issues in more detail, could be the development of new reference point adaption techniques, which prevent the corresponding hypervolume values to be this large in comparison to the received difference in hypervolume. This could possibly prevent numerical issues. Moreover, we would like to verify our findings on more mathematical and, in particular, on industrial test cases. These test cases seem to be a lot more challenging.

References

1. Bader, J., Zitzler, E.: HypE: An Algorithm for Fast Hypervolume-Based Many-Objective Optimization. Evolutionary Computation 19(1), 45–76 (2011)
2. Beume, N., Naujoks, B., Emmerich, M.: SMS-EMOA: Multiobjective selection based on dominated hypervolume. European Journal of Operational Research 181(3), 1653–1669 (2007)
3. Coello Coello, C.A., Van Veldhuizen, D.A., Lamont, G.B.: Evolutionary Algorithms for Solving Multi-Objective Problems, 2nd edn. Springer, New York (2007)
4. Deb, K.: Multi-Objective Optimization using Evolutionary Algorithms. Wiley, Chichester (2001)
5. Deb, K., Thiele, L., Laumanns, M., Zitzler, E.: Scalable Multi-Objective Optimization Test Problems. In: Congress on Evolutionary Computation (CEC 2002), pp. 825–830. IEEE Press (2002)
6. Igel, C., Hansen, N., Roth, S.: Covariance Matrix Adaptation for Multi-objective Optimization. Evolutionary Computation 15(1), 1–28 (2007)
7. Judt, L., Mersmann, O., Naujoks, B.: Non-monotonicity of Obtained Hypervolume in 1-greedy S-Metric Selection. Journal of Multi-Criteria Decision Analysis (2011), maanvs03.gm.fh-koeln.de/webpub/CIOPReports.d/Judt11a.d/ (accepted for publication, preprint)
8. Judt, L., Mersmann, O., Naujoks, B.: Effect of SMS-EMOA Parameterizations on Hypervolume Decreases. In: Hamadi, Y., Schoenauer, M. (eds.) LION 6. LNCS, vol. 7219, pp. 419–424. Springer, Heidelberg (2012)
9. Mersmann, O.: emoa: Evolutionary Multiobjective Optimization Algorithms (2011), http://CRAN.R-project.org/package=emoa (R package version 0.4-8)
10. Zitzler, E., Deb, K., Thiele, L.: Comparison of Multiobjective Evolutionary Algorithms: Empirical Results. Evolutionary Computation 8(2), 173–195 (2000)
11. Zitzler, E., Thiele, L.: Multiobjective Optimization Using Evolutionary Algorithms — A Comparative Case Study. In: Eiben, A.E., Bäck, T., Schoenauer, M., Schwefel, H.-P. (eds.) PPSN V. LNCS, vol. 1498, pp. 292–301. Springer, Heidelberg (1998)
12. Zitzler, E., Thiele, L., Laumanns, M., Fonseca, C.M., da Fonseca, V.G.: Performance Assessment of Multiobjective Optimizers: An Analysis and Review. IEEE Transactions on Evolutionary Computation 7(2), 117–132 (2003)

Cone-Based Hypervolume Indicators: Construction, Properties, and Efficient Computation

Michael Emmerich[1], André Deutz[1], Johannes Kruisselbrink[1], and Pradyumn Kumar Shukla[2]

[1] LIACS. Leiden University, Niels Bohrweg 1, 2333-CA Leiden, NL
[2] Karlsruhe Institute of Technology, 76128 Karlsruhe, Germany, GE
`emmerich@liacs.nl,deutz@liacs.nl,`
`jkruisse@liacs.nl,`
`Pradyumn.Shukla@aifb.uni-karlsruhe.de`
`http://natcomp.liacs.nl`

Abstract. In this paper we discuss cone-based hypervolume indicators (CHI) that generalize the classical hypervolume indicator (HI) in Pareto optimization. A family of polyhedral cones with scalable opening angle γ is studied. These γ-cones can be efficiently constructed and have a number of favorable properties. It is shown that for γ-cones dominance can be checked efficiently and the CHI computation can be reduced to the computation of the HI in linear time with respect to the number of points μ in an approximation set. Besides, individual contributions to these can be computed using a similar transformation to the case of Pareto dominance cones.

Furthermore, we present first results on theoretical properties of optimal μ-distributions of this indicator. It is shown that in two dimensions and for linear Pareto fronts the optimal μ-distribution has uniform gap. For general Pareto curves and γ approaching zero, it is proven that the optimal μ-distribution becomes equidistant in the Manhattan distance. An important implication of this theoretical result is that by replacing the classical hypervolume indicator by CHI with γ-cones in hypervolume-based algorithms, such as the SMS-EMOA, the distribution can be shifted from a distribution that is focussed more on the knee point region to a distribution that is uniformly distributed. This is illustrated by numerical examples in 2-D. Moreover, in 3-D a similar dependency on γ is observed.

Keywords: Hypervolume Indicator, Cone-based Hypervolume Indicator, Optimal μ-distribution, Complexity, Cone-orders, SMS-EMOA.

1 Introduction

The context in which this work is situated is multiobjective optimization with m objective functions $f_1 : X \to \mathbb{R}$, $f_2 : X \to \mathbb{R}$, ..., $f_m : X \to \mathbb{R}$. Without loss of generality, minimization is assumed to be the goal. A finite approximation to a

R.C. Purshouse et al. (Eds.): EMO 2013, LNCS 7811, pp. 111–127, 2013.
© Springer-Verlag Berlin Heidelberg 2013

continuous Pareto front is searched for, which is assumed to be of dimension $m-1$. To approximate it, a set indicator is maximized over all sets $P \subset \mathbb{R}^m$ of size μ such that $P \subseteq \mathcal{Y} = \mathbf{f}(X)$ (i.e., P is a (finite) subset of the image set $\mathbf{f}(X)$).

The hypervolume indicator (HI), or S-Metric [21], is a common set indicator for measuring the quality of Pareto front approximations and it is often used as a criterion for guiding search algorithms towards Pareto fronts, e.g., [2,7]. For a finite set $P \subset \mathbb{R}^m$ it is defined as

$$\mathrm{HI}(P) = \lambda(\cup_{\mathbf{p} \in P}[\mathbf{p}, \mathbf{r}]), \qquad (1)$$

where λ denotes the Lebesgue measure and $\mathbf{r}$ is a reference point that is usually fixed by the user. Recall that for two vectors $\mathbf{x}, \mathbf{y}$ in some $\mathbb{R}^m$, $\mathbf{x} \preceq \mathbf{y}$ if and only if $\forall i \in \{1, \cdots s\}, x_i \leq y_i$. An equivalent definition is Definition 5 in case the standard cone (i.e., the positive orthant) is taken. This binary relation is known as the weak Pareto order. The reference point should obey $\forall \mathbf{p} \in P : \mathbf{p} \preceq \mathbf{r}$, for all approximation sets P considered in the optimization.

The optimal μ-distribution of the HI is defined as (cf. [1]):

$$P_\mu^* \in \arg \max_{P \subseteq \mathcal{Y}, |P| \leq \mu} (\mathrm{HI}(P)). \qquad (2)$$

For innumerably large Pareto fronts, the optimal μ-distribution is a subset of the Pareto front. For continuous 2-D Pareto curves it is known that in the limit (for large μ) the density of optimal μ-distribution is highest in regions where the slope is close to $-45°$ [1]. Earlier work on cone-based dominance suggests that density of Pareto front approximations can be controlled by using cones of different shapes [3,17]. However, previous work does not yet address the HI for cone-based orders and it seems promising to define the HI also for cones that are different from the Pareto cone. It is hypothesized that optimal μ-distributions with respect to these different cone-based HIs have different densities. It will be interesting to study how this density relates to local trade-offs.

The attention of this paper is mainly focused on a singly parametrized family of symmetrical, polyhedral cones. For these cones we show how to

1. construct the base vectors of a member given its angle-parameter
2. check cone dominance efficiently and compute the cone non-dominated sub-set of a finite set
3. compute the cone based hypervolume efficiently.

Furthermore we study the influence of the cone angle on the optimal μ-distribution when using non-rectangular cone based hypervolume indicators. In particular we point out cases where the optimal μ-distribution is evenly spaced.

In this paper we take the view of studying the (standard) Pareto front by means of cone-based orders, noting that cone-based orders also occur naturally in other contexts, in which the cone-based hypervolume would measure the size of the dominated subspace.

We believe non-rectangular cones to be useful for controlling trade-off sensitivity and spacing (of points in the approximation sets) in search methods guided by hypervolume indicators.

2   Preliminaries

In this section we recall some notions which are used in this paper. We define cones, convex cones, pointed cones, and cone orders according to Ehrgott [6] which in turn builds on the work of Noghin [15]:

Definition 1. *A subset $C \subseteq \mathbb{R}^m$ is called a cone, iff $\alpha \mathbf{p} \in C$ for all $\mathbf{p} \in C$ and for all $\alpha \in \mathbb{R}, \alpha > 0$.*

Definition 2. *A cone C in $\mathbb{R}^m$ is convex, iff $\alpha \mathbf{p}^{(1)} + (1 - \alpha)\mathbf{p}^{(2)} \in C$ for all $\mathbf{p}^{(1)} \in C$ and $\mathbf{p}^{(2)} \in C$ and for all $0 \leq \alpha \leq 1$.*

Definition 3. *A cone C in $\mathbb{R}^m$ is pointed, iff for $\mathbf{p} \in C$, $\mathbf{p} \neq \mathbf{0}$, $-\mathbf{p} \notin C$, i.e., $C \cap -C \subseteq \{0\}$.*

For notational convenience let us also introduce the Minkowski sum:

Definition 4. *Let A and B denote sets in $\mathbb{R}^m$. Then*

$$A \oplus B = \{\mathbf{a} + \mathbf{b} \mid \mathbf{a} \in A \text{ and } \mathbf{b} \in B\}. \tag{3}$$

and

$$A \ominus B = \{\mathbf{a} - \mathbf{b} \mid \mathbf{a} \in A \text{ and } \mathbf{b} \in B\}. \tag{4}$$

Definition 5. *[Cone order]   Let C denote a pointed convex cone, then we define (see also Ehrgott [6])*

$$\mathbf{p} \preceq_C \mathbf{q} \Leftrightarrow \mathbf{q} \in \{\mathbf{p}\} \oplus C \cup \{0\}. \tag{5}$$

This order is a generalization of the Pareto order. The special case of the Pareto order and hypervolume indicator is obtained by choosing the cone

$$C_{\text{Pareto}} = (\mathbb{R}_{\geq 0})^m \setminus \{0\}. \tag{6}$$

By introducing cone orders as in Equation 5, Noghin [15], proves that an order which is irreflexive, transitive, and compatible with addition and scalar multiplication, is cone order derived from cone C if and only if C is a pointed convex cone that does not contain 0. Moreover, the following relationships between the Pareto order and cone orders can be observed:

1. For pointed convex cones C with $C \supset C_{\text{Pareto}}$, minima of the cone order are also Pareto optima, but some Pareto optima may not be minima of the cone order.
2. For pointed convex cones C with $C \subset C_{\text{Pareto}}$, Pareto optima are also minima of the cone order, but some minima of the cone order may not be Pareto optima.

As a consequence, by finding the minima of the cone order, we either compute a subset of the Pareto front (Case 1) or a superset of the Pareto front (Case 2).

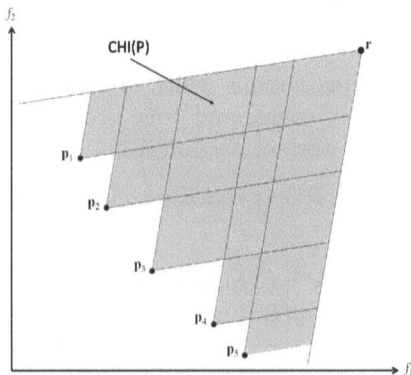

Fig. 1. Cone-based hypervolume example for two objectives

We now propose a definition of the CHI :

Definition 6. *[Cone-based hypervolume] We denote the Lebesgue measure by λ. The cone-based hypervolume for a finite set $P \in \mathbb{R}^m$ and a reference point $\mathbf{r}$ with $\forall \mathbf{p} \in P : \mathbf{p} \preceq_C \mathbf{r}$ is defined as*

$$CHI(P) = \lambda((P \oplus C) \cap (\{\mathbf{r}\} \ominus C)). \tag{7}$$

An equivalent definition is based on the definition of cone orders:

$$CHI(P) = \lambda(\{\mathbf{q} \in \mathbb{R}^m | \exists \mathbf{p} \in P : \mathbf{p} \preceq_C \mathbf{q} \wedge \mathbf{q} \preceq_C \mathbf{r}\}). \tag{8}$$

An example for the cone-based hypervolume indicator for $m = 2$ and a set of 5 points is given in Figure 1.

In the following we restrict our attention to a particular class of polyhedral cones, that will be introduced in Section 3 as γ-cones and its related hypervolume indicator CHI_γ. We will study the following aspects of it:

1. How can γ-cones be defined concisely and constructively (in any dimension) based on the angle γ? (Section 3)
2. How can dominance be checked? How can we obtain non-dominated subsets efficiently? (Section 4)
3. How can the CHI_γ be efficiently computed for these cone-orders and how can individual contributions to the CHI_γ be computed? (Section 5)
4. How does γ influence the optimal μ-distribution?(Section 6)
5. How can the influence of γ be used in hypervolume-based multiobjective optimization algorithm design? (Section 7)

3   Definition and Construction of γ-cones

Informally the γ-cones that we introduce next can be described as symmetrical and polyhedral cones that are spanned by m base vectors, where m denotes the

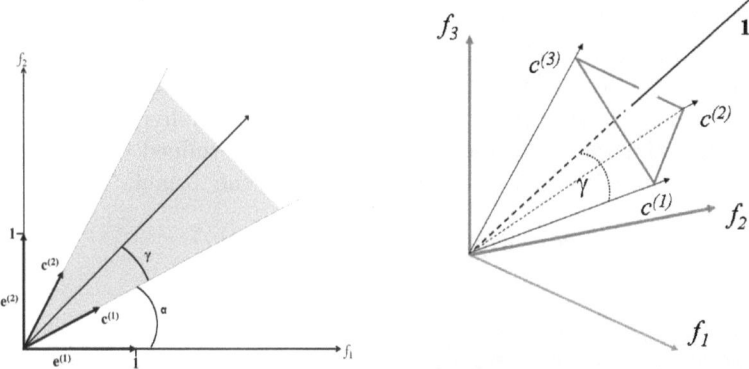

Fig. 2. Left: a two-dimensional γ-cone. Right: a three-dimensional γ-cone.

number of objective functions and the shape of it can be controlled by a single parameter – the opening angle γ. For these types of cones (in addition to the above properties) the hypervolume can be computed as efficiently as the standard hypervolume – a very useful property as for its computation asymptotically efficient algorithms for $m = 2, 3$ are known [12] and also for higher dimensions fast algorithms were proposed [10,13,19,20].

Next, a precise, formal definition of γ-cones is given:

Definition 7 (diagonal line). *The line through the origin of $\mathbb{R}^m$ and the point $\mathbf{1} = (1, \cdots, 1)^T$ is called the* diagonal line *of $\mathbb{R}^m$. For $\mathbb{R}^2$ this is also called the* bisectrix.

Definition 8 (γ-cone). *A cone spanned by m base vectors, $\mathbf{c}^{(1)}, \ldots, \mathbf{c}^{(m)}$ is called a γ-cone, iff*

1. *the angle between the diagonal line and each of the base vectors $\mathbf{c}^{(i)}$ is γ, and*
2. *for all $i \in \{1, \ldots, m\}$ the ith base vector $\mathbf{c}^{(i)}$ is a unit vector in the plane spanned by $\mathbf{1}$ and $\mathbf{e}^{(i)}$, where $\{\mathbf{e}^{(1)}, \mathbf{e}^{(2)}, \ldots, \mathbf{e}^{(m)}\}$ denotes the standard orthonormal basis of $\mathbb{R}^m$. Moreover $0 < \gamma < \pi$ and the angle γ is measured from $\mathbf{c}^{(i)}$ counterclockwise towards $\mathbf{1}$.*

Examples of a two-dimensional and three-dimensional γ-cone are provided in Figure 2.

3.1 Construction of the Base Vectors of γ-cones in Arbitrary Dimension m

For the m-dimensional case, as we show below, the ith base vector $\mathbf{c}^{(i)} = (c_1^{(i)}, \ldots, c_m^{(i)})^T$ has elements

$$c_j^{(i)} = \begin{cases} (1/\sqrt{m-1})\sin(\alpha) & i \neq j \\ \cos(\alpha) & i = j \end{cases} \tag{9}$$

,where α denotes the angle from $\mathbf{e}^{(i)}$ to $\mathbf{c}^{(i)}$ and θ the angle from $\mathbf{e}^{(i)}$ to $\mathbf{a} = \frac{1}{\sqrt{m}}\mathbf{1}$ (thus $\theta = \arccos(1/\sqrt{m})$. With this notation the opening angle is $\gamma = \theta - \alpha$.

To compute the vectors that span the dominance cone for any number of dimensions, geometric algebra offers an elegant coordinate-free calculus.

In geometric algebra a rotation in $\mathbb{R}^m$ $(m \in \mathbb{N})$ determined by a normalized bivector B over an angle α of a vector $\mathbf{x}$ is given by the formula:

$$\mathbf{x}' = \exp(-B\alpha/2)\mathbf{x}\exp(B\alpha/2) \tag{10}$$

where

$$\exp(B\alpha/2) = \cos(\alpha/2) + B\sin(\alpha/2), \text{ and } \mathbf{x}' \text{ is the rotated vector.} \tag{11}$$

We can specialize this to our situation where we want to rotate $\mathbf{e}^{(i)}$ in the plane determined by the normalized bivector $B = (\mathbf{e}^{(i)} \wedge \mathbf{a})/\sin(\theta)$ over an angle α to get $\mathbf{c}^{(i)}$:

$$\mathbf{c}^{(i)} = \exp(-\frac{\alpha}{2}\frac{\mathbf{e}^{(i)} \wedge \mathbf{a}}{\sin(\theta)})\,\mathbf{e}^{(i)}\,\exp(\frac{\alpha}{2}\frac{\mathbf{e}^{(i)} \wedge \mathbf{a}}{\sin(\theta)}). \tag{12}$$

For this particular situation ($\mathbf{e}^{(i)}$ lies in the plane determined by $\mathbf{e}^{(i)}$ and $\mathbf{a}$), we can write for $\mathbf{c}^{(i)}$, $\mathbf{e}^{(i)}\cos(\alpha) + \frac{sin(\alpha)}{sin(\theta)}(\mathbf{a} - \cos(\theta)\mathbf{e}^{(i)})$. So far we did not bother about the coordinates. Next we can compute the coordinates of $\mathbf{c}^{(i)}$:

$$\mathbf{c}^{(i)} = \begin{pmatrix} 0 \\ \vdots \\ 0 \\ \cos(\alpha) \\ 0 \\ \vdots \\ 0 \end{pmatrix} + \frac{\sin(\alpha)}{\sin(\theta)}\begin{pmatrix} \cos(\theta) \\ \vdots \\ \cos(\theta) \\ 0 \\ \cos(\theta) \\ \vdots \\ \cos(\theta) \end{pmatrix} = \begin{pmatrix} \frac{\sin(\alpha)}{\tan(\theta)} \\ \vdots \\ \frac{\sin(\alpha)}{\tan(\theta)} \\ \cos(\alpha) \\ \frac{\sin(\alpha)}{\tan(\theta)} \\ \vdots \\ \frac{\sin(\alpha)}{\tan(\theta)} \end{pmatrix}, \tag{13}$$

or in terms of the dimension m:

$$\mathbf{c}^{(i)} = \begin{pmatrix} 0 \\ \vdots \\ 0 \\ \cos(\alpha) \\ 0 \\ \vdots \\ 0 \end{pmatrix} + \frac{\sin(\alpha)}{\sqrt{\frac{m-1}{m}}}\begin{pmatrix} 1/\sqrt{m} \\ \vdots \\ 1/\sqrt{m} \\ 0 \\ 1/\sqrt{m} \\ \vdots \\ 1/\sqrt{m} \end{pmatrix} = \begin{pmatrix} \frac{\sin(\alpha)}{\sqrt{m-1}} \\ \vdots \\ \frac{\sin(\alpha)}{\sqrt{m-1}} \\ \cos(\alpha) \\ \frac{\sin(\alpha)}{\sqrt{m-1}} \\ \vdots \\ \frac{\sin(\alpha)}{\sqrt{m-1}} \end{pmatrix}. \tag{14}$$

In summary the above stated Equation 9 obtains.

4   Dominance Test and Minimal Subset

Let $\mathbf{q} \in \mathbb{R}^m$ denote a point for which it needs to be determined whether or not it is dominated by a point $\mathbf{p} \in \mathbb{R}^m$ with respect to the cone order defined by a γ-cone C. In other words it needs to be determined, whether or not $\mathbf{q}$ is inside the cone $p \oplus C \cup \{0\}$ with base $\mathbf{C} = (\mathbf{c}^{(1)}, \ldots, \mathbf{c}^{(m)})$. This can be tested by the following well-known lemma:

Lemma 1. *A point* $\mathbf{p}$ *weakly* γ-*cone-dominates a point* $\mathbf{q}$ *(*$\mathbf{p} \preceq_C \mathbf{q}$*) with respect to a* γ-*cone* C *(with base* $\mathbf{C} = (\mathbf{c}^{(1)}, \ldots, \mathbf{c}^{(m)})$*), if and only if* $\exists \lambda_1 \geq 0, \ldots,$ $\lambda_m \geq 0,$ *such that*

$$\mathbf{p} + \lambda_1 \mathbf{c}^{(1)} + \cdots + \lambda_m \mathbf{c}^{(m)} = \mathbf{q}. \tag{15}$$

Lemma 1 can be rephrased into the following lemma.

Lemma 2. *Let* C *be as in Lemma 1. A point* $\mathbf{p}$ *weakly dominates a point* $\mathbf{q}$ *with respect to* C, *if and only if* $\mathbf{0} \leq \mathbf{C}^{-1}(\mathbf{q} - \mathbf{p})$ *(componentwise).*

In order to determine the non-dominated subset of a set of size μ with respect to the cone order based on a γ-cone, it suffices to compute the inverse of $\mathbf{C}$, only once, and for each dominance test to multiply the vector difference with the inverse. By using a coordinate transformation we can reduce the problem of determining the non-dominated subset with respect to C in linear time $\mathbf{O}(m^2\mu)$ (for constant dimension) to the problem of computing the maximal set for the Pareto order. This way, the algorithms by Kung, Luccio, and Preparata [14] can be used to determine non-dominated sets with time complexity in in $\mathcal{O}(m^3 + m^2\mu + \mu\log(\mu))$ for $m = 2, 3$ which is asymptotically optimal in μ and in $\mathcal{O}(m^3 + m^2\mu + \mu(\log(\mu))^{m-2})$, for $m > 3$. The following lemma summarizes the theory for this transformation:

Lemma 3. *Let* $P = \{\mathbf{p}^{(1)}, \ldots, \mathbf{p}^{(\mu)}\}$ *denote a set of* μ *points in* $\mathbf{R}^m$, $\mathbf{C}$ *denote the matrix of base vectors for a* γ-*cone* C, *and*

$$\mathbf{q}^{(i)} = \mathbf{C}^{-1}\mathbf{p}^{(i)}. \tag{16}$$

Then, for subsets of indices $\{i_1, \ldots, i_k\} \subseteq \{1, \ldots, \mu\}$, $\mathbf{p}^{(i_1)}, \ldots, \mathbf{p}^{(i_k)}$ *is a non-dominated subset of* P *in the cone order* $\preceq_C$, *if and only if* $\mathbf{q}^{(i_1)}, \ldots, \mathbf{q}^{(i_k)}$ *is a minimal subset of* Q *with respect to the weak Pareto order* $(\preceq_{Pareto})$.

Proof. The proof follows immediately from Lemma 2. In this lemma the condition $\mathbf{0} \leq \mathbf{C}^{-1}(\mathbf{p}^{(1)} - \mathbf{p}^{(2)})$ can be rewritten as $\mathbf{0} \leq \mathbf{C}^{-1}\mathbf{p}^{(1)} - \mathbf{C}^{-1}\mathbf{p}^{(2)}$ which is equivalent to the condition $\mathbf{C}^{-1}\mathbf{p}^{(1)} \leq \mathbf{C}^{-1}\mathbf{p}^{(2)}$. See also [16] for a similar result.

5   Efficient Computation

Two ways of computing the cone-based hypervolume can be distinguished:

1. Using Cartesian coordinates and computing first the corner points of the m-dimensional trellis and then the volume of the parallelepipeds spanned by these.

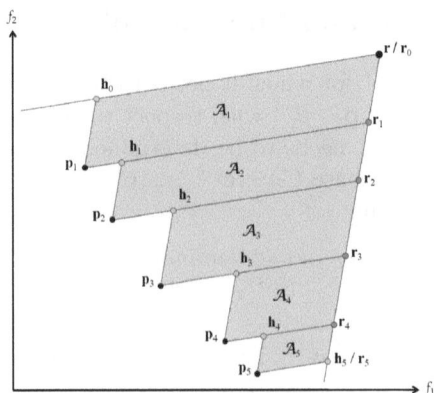

Fig. 3. Computation of $\text{CHI}_\gamma(P)$ for a set of non-dominated points P

2. Transforming the coordinate system in order to obtain an axis-parallel trellis in the transformed coordinate system (for the Pareto cone). Thereafter, compute the hypervolume indicator in this system and correct for the volume change due to the affine transformations.

The computation of the cone-based hypervolume indicator in two dimensions for a non-dominated set P of size μ and reference point $\mathbf{r}$ can be accomplished by processing the slanted staircase trellis. The algorithm is outlined in Figure 4 and illustrated by Figure 3. For the general m-dimensional case, we can transform the coordinate system to an axis-parallel trellis, compute the hypervolume indicator in this system, and correct for the volume change due to the affine transformations. The algorithm shown in Figure 5 describes this procedure. Here we make use of a result on volume change due to affine transformation ([4], Theorem 12.2):

Lemma 4. *We denote the Lebesgue measure by λ. Let $F(\mathbf{x}) = \mathbf{T}\mathbf{x} + \mathbf{x}_0$ denote an non-singular affine transformation, then $\lambda(\mathbf{F}A) = \det(\mathbf{T})\lambda(A)$.*

The computations for the operations required for the computation of the coordinate transformation scales with $\mathcal{O}(\mu m^3)$. Therefore, the overall complexity of the hypervolume computation algorithms is preserved and hence the cone-based hypervolume can be computed in asymptotically optimal time $\Theta(\mu \log \mu)$ for $m = 2$ and $m = 3$ using the algorithm described in Beume et al. [12] and in time $\mathcal{O}(\mu^{(m-1)/2} \log \mu)$ using the algorithm by Yıldız and Suri[20].

5.1 Computation of the Contributions to the Cone-Based Hypervolume

Many algorithms require the computation of hypervolume contributions, e.g., SMS-EMOA [7]. The computation of contributions or more precisely individual contributions to the γ-cone-based hypervolume defined by

---

Algorithm: Two-dimensional cone-based hypervolume computation

Input: Cone base $\mathbf{C}$, non-dominated points $P \subset \mathbb{R}^2$, reference point $\mathbf{r}$

Output: Cone-based hypervolume indicator CHIg(P)

1. Sort P such that $(\mathbf{p}^{(1)}, \ldots, \mathbf{p}^{(\mu)})$, denotes the sequence of points that is sorted in ascending order by the first coordinate of the transformation $\mathbf{C}^{-1}\mathbf{p}^{(i)}$, $i = 1, \ldots, \mu$.
2. For all $i = 1, \ldots, \mu$, let $\mathbf{r}^{(i)}$ denote the intersection point of the ray $\{\mathbf{r} - t_1 \mathbf{c}^{(1)} \mid t_1 \in [0, \infty)\}$ and $\{\mathbf{p}^{(i)} + t_2 \mathbf{c}^{(2)} \mid t_2 \in [0, \infty)\}$.
3. Let $\mathbf{r}^{(0)} = \mathbf{r}$ and let $\mathbf{h}^{(0)}$ denote the intersection point of the ray $\{\mathbf{p}^{(1)} + t_1 \mathbf{c}^{(1)} \mid t_1 \in [0, \infty)\}$ and $\{\mathbf{r} - t_2 \mathbf{c}^{(2)} \mid t_2 \in [0, \infty)\}$.
4. For all $i = 1, \ldots, \mu$, let $\mathbf{h}^{(i)}$ denote the intersection point of the ray $\{\mathbf{h}^{(i-1)} - t_1 \mathbf{c}^{(1)} \mid t_1 \in [0, \infty)\}$ and $\{\mathbf{p}^{(i)} + t_2 \mathbf{c}^{(2)} \mid t_2 \in [0, \infty)\}$.
5. For all $i = 1, \ldots, \mu$, compute the area of the parallelogram with corners $\mathbf{p}^{(i)}$, $\mathbf{r}^{(i)}$, $\mathbf{r}^{(i-1)}$, and $\mathbf{h}^{(i-1)}$ that form the steps or stripes of the slanted staircase, i.e.,

$$\mathcal{A}_i = \det \begin{pmatrix} r_1^{(i)} - p_1^{(i)} & h_1^{(i-1)} - p_1^{(i)} \\ r_2^{(i)} - p_2^{(i)} & h_1^{(i-1)} - p_2^{(i)} \end{pmatrix}.$$

6. Return CHIg(P) $= \sum_{i=1}^{\mu} \mathcal{A}_i$.

Fig. 4. Computing the two-dimensional γ-cone-based hypervolume computation

---

Algorithm: m-dimensional cone-based hypervolume computation

Input: Cone base $\mathbf{C}$, non-dominated points $P \subset \mathbb{R}^m$, reference point $\mathbf{r}$

Output: Cone-based hypervolume indicator CHIg(P)

1. Let $\mathbf{r}' = \mathbf{C}^{-1}\mathbf{r}$.
2. For all $i = 1, \ldots, \mu$, let $Q = \{\mathbf{q}^{(1)}, \ldots, \mathbf{q}^{(\mu)}\}$, with $\mathbf{q}^{(i)} = \mathbf{C}^{-1}\mathbf{p}^{(i)}$.
3. Compute the standard hypervolume HI($Q, \mathbf{r}'$).
4. Return CHIg(P) $= (1/\det \mathbf{C}^{-1}) \cdot$ HI($Q, \mathbf{r}'$).

Fig. 5. Computing the m-dimensional cone-based hypervolume computation

$\Delta\mathrm{CHI}_\gamma(\mathbf{p}, P) := \mathrm{CHI}_\gamma(P) - \mathrm{CHI}_\gamma(P \setminus \{p\})$ with $\mathbf{p} \in P$ requires, given the above, only little modifications. For the two-dimensional case using the Cartesian coordinates approach, we can modify the algorithm of Figure 4 by computing for each point $\mathbf{p}$ the area of the parallelogram with corners $\mathbf{p}^{(i)}$, $\mathbf{h}^{(i)}$, $(h_1^{(i)} + (h_1^{(i-1)} - p_1^{(i)}), h_2^{(i-1)} + (h_2^{(i)} - p_2^{(i)}))^\mathsf{T}$, and $\mathbf{h}^{(i-1)}$ that form the steps or stripes of the slanted staircase, i.e.,

$$\mathcal{A}_i = \det \begin{pmatrix} h_1^{(i+1)} - p_1^{(i)} & h_2^{(i+1)} - p_2^{(i)} \\ h_1^{(i)} - p_1^{(i)} & h_2^{(i)} - p_2^{(i)} \end{pmatrix}.$$

This follows immediately from the relation between determinant and the parallelogram area (e.g. [11], page 168).

Likewise, for general dimension m using the same coordinate transformation as in Figure 5, the individual contributions of points to the γ-cone-based hypervolume, $\Delta CHI_\gamma(\mathbf{p}, P)$, can be computed, yielding asymptotically optimal algorithms with running time $\Theta(\mu \log \mu)$ for $m = 2$ and $m = 3$ for computing all contributions using the algorithm described in [9]. For fast algorithms for $m > 3$ see [5].

5.2 Choosing the Reference Point

As with the standard hypervolume indicator, also for the cone-based hypervolume indicator, the issue of choosing an appropriate reference point remains. Two observations:

- When using acute cones (i.e, $\gamma < \pi/4$), the reference point should be located farther away than for the standard hypervolume indicator in order to assure that the whole Pareto front is contained in the cone.
- When using obtuse cones (i.e, $\gamma > \pi/4$), the reference point can move closer to the Pareto front.

How to construct algorithmically a proper reference point for a given γ is still an open question.

6 Optimal μ-distribution

This section states first theoretical results on the optimal μ-distribution for CHI_γ (see Equation 2). In particular it is studied which point sets maximize CHI_γ subject to the constraint that all points need to be nondominated in the Pareto order.

6.1 Linear Case

The following statement summarizes a result for the 2-D linear case:

Lemma 5. *Let us consider a fixed value of γ with $0 < \gamma < \pi/2$ and a Pareto front that is given by some straight line segment with $\mathbf{a}$ as a lower right and $\mathbf{b}$ as a upper left end point. Moreover assume that all points on the line segment dominate the reference point in the γ-cone order. Let P^* denote the optimal μ-distribution for CHI_γ. Then $P^* \cup \{\mathbf{a}, \mathbf{b}\}$ is evenly spaced.*

Proof. We show the contrapositive is false. Assume one inner point, that is a point in P^*, has not the same distance to its two direct neighbors. Then its individual contribution can be improved by moving it to the middle, while the remaining part of the dominated hypervolume remains unchanged.

To show, that its contribution can be improved we look at the geometrical situation depicted in Figure 6. Let $\mathbf{l}$, $\mathbf{p}$, and $\mathbf{r}$ denote three consecutive points in $P^+ = P \cup \{\mathbf{a}, \mathbf{b}\}$ in ascending x-coordinate. Assume $\mathbf{l}$ and $\mathbf{r}$ are fixed, and we want to determine the position of $\mathbf{p}$ such that the value of $\Delta CHI_\gamma(\mathbf{p}, P^+)$ is maximal.

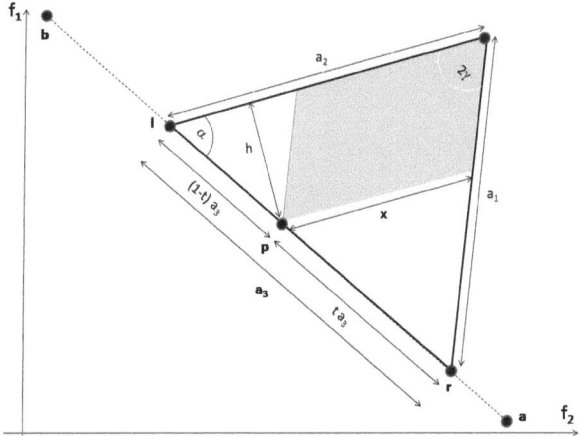

Fig. 6. Geometrical situation for linear Pareto front

This corresponds to finding the maximally sized inscribed parallelogram of an triangle in Figure 6. We use the symbols from Figure 6 and obtain: $1/t = a_2/x$ and $x = ta_2$ (intercept theorem). Moreover $h = \cos(\alpha)(1 - t)a_3$. We need to maximize xh (size of inscribed parallelogram); $xh = Ct(t - 1)$ for the positive constant $C = \cos(\alpha)a_2a_3$, yielding $t^* = 0.5$ as the sole maximum. Hence, by setting $\mathbf{p} = \frac{1}{2}(\mathbf{r} + \mathbf{l})$ its contribution can be improved. See also Auger et al [1], Beume et al [12] for the analogous statement on the standard hypervolume.

6.2   Evenly Spaced Distribution for Small Angles

The lemma discussed in this subsection explains that for small angles of γ, the distribution tends to become a more evenly spaced distribution. In this subsection we assume $m = 2$. To state the theorem in a compact form, some definitions need to be made first:

Definition 9. *Let Below($\mathbf{x}$) denote the set of points which lie below the line through $\mathbf{x}$ in the direction of the symmetry axis of the cone (i. e., for γ cones the direction of the separatrix). Similarly, define Above($\mathbf{x}$) the set of points which lie above this line.*

Definition 10. *Let Incomparable($\mathbf{x}, \gamma$) denote the set of incomparable points with respect to $\mathbf{x}$ and a γ-cone.*

Definition 11. *Let PF denote a Pareto front and $\mathbf{u}, \mathbf{l} \in \mathbb{R}^2$. Then the point $\mathbf{m}(\mathbf{u}, \mathbf{l}, PF)$ is the point on PF that has the same Manhattan distance to $\mathbf{u}$ and $\mathbf{l}$, provided it exists.*

Remark 1. The point $\mathbf{m}(\mathbf{u}, \mathbf{l}, PF)$ is unique and can be constructed as follows: First the midpoint on the line segment between $\mathbf{u}$ and $\mathbf{l}$ is determined. Then a

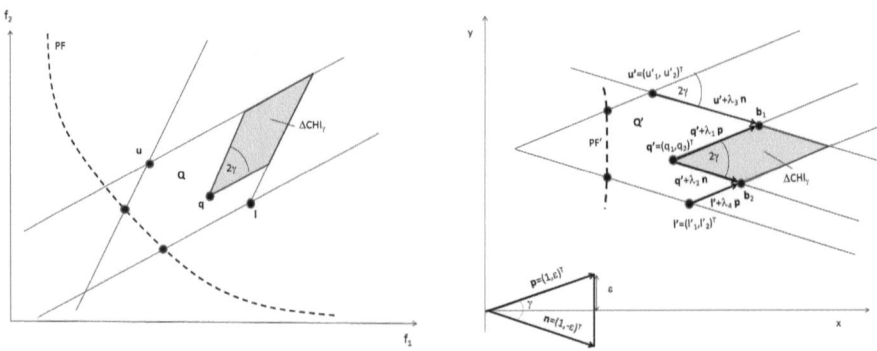

Fig. 7. Determination of $\Delta HI_\gamma(\mathbf{q}, \{\mathbf{u}, \mathbf{l}\})$: Before (left) and after rotation (right)

line that is parallel to the bisectrix is constructed through this point. All points on this line share the same Manhattan distance to $\mathbf{u}$ and $\mathbf{l}$. Where the line intersects with the Pareto front this holds, too. The intersection point, provided it exists, is *unique*, as for any two points on the line, the point with the lower y-coordinate dominates for any $\gamma > 0$) the point with the higher y-coordinate.

Let X be equipped with a weak Pareto order (denoted by $\preceq$) (see also Page 2) and let $A \subseteq X$. Then we say that A weakly Pareto dominates (or simply dominates) an element $x \in X$ if and only if $\exists a \in A$ such that $a \preceq x$

Lemma 6. *Let* Dom *denote the set of points weakly Pareto dominated by a compact and connected Pareto front. Moreover let* $\mathbf{u} = (u_x, u_y)^T \in$ Dom *and* $\mathbf{l} = (l_x, l_y)^T \in Below(\mathbf{u}) \cap$ Dom. *Let*

$$Q = Incomparable(\mathbf{u}) \cap Below(\mathbf{u}) \cap Above(\mathbf{l}) \cap Incomparable(\mathbf{l}).$$

(Q is the parallelogram spanned by $\mathbf{u}$ and $\mathbf{l}$ and the two directions of the cone (see Fig. 7 (left)). It is clear that Q depends on the opening angle γ.) We assume, that for any γ a sufficiently far away reference point, which is dominated by all points in Q, is chosen. Then

$$\lim_{\gamma\downarrow 0} \arg \max_{\mathbf{q}\in Q\cap Dom} \Delta CHI_\gamma(\mathbf{q}, \{\mathbf{u}, \mathbf{l}\}) = \mathbf{m}(\mathbf{u}, \mathbf{l}, PF). \tag{17}$$

Proof. First, an expression for $\Delta CHI_\gamma(\mathbf{q}, \{\mathbf{u}, \mathbf{l}\})$ will be derived. Without loss of generality, we look at the congruent parallelogram after a rotation (see Figure 7). The area of this parallelogram can be determined as

$$\Delta CHI_\gamma(\mathbf{q}', \{\mathbf{u}', \mathbf{l}'\}) = \det(\lambda_2\mathbf{n} \circ \lambda_1\mathbf{p}) \tag{18}$$

with $\mathbf{p} = (1, \epsilon)^T$ and $\mathbf{n} = (1, -\epsilon)^T$ for some $\epsilon > 0$ denoting the vectors that span the cones and $\lambda_1\mathbf{p}$ is the vector from $\mathbf{q}'$ to $\mathbf{b_1}$, where $\mathbf{b_1}$ is the intersection of the straight lines $\ell_1 = \mathbf{q}' + t\mathbf{p}, t \in \mathbb{R}$ and $\ell_2 = \mathbf{u}' + s\mathbf{n}, s \in \mathbb{R}$. In a similar way, we get $\lambda_2\mathbf{n}$: it is the vector from $\mathbf{q}'$ to $\mathbf{b_2}$ (again $\mathbf{b_2}$ is the intersection of the lines

$\ell_3 = \mathbf{q}' + t\mathbf{n}, t \in \mathbb{R}$ and $\ell_4 = \mathbf{l}' + s\mathbf{p}, s \in \mathbb{R}$). Written as two equation systems we get the following.

$$\mathbf{q}' + \lambda_1 \mathbf{p} = \mathbf{u}' + \lambda_3 \mathbf{n} \tag{19}$$

$$\mathbf{q}' + \lambda_2 \mathbf{n} = \mathbf{l}' + \lambda_4 \mathbf{p}. \tag{20}$$

The solutions of these equation systems (19 and 20) give rise to $\lambda_2 \mathbf{n}$ and $\lambda_1 \mathbf{p}$ and in turn these can be used to determine the sought after area of the parallelogram ($= \det(\lambda_2 \mathbf{n} \circ \lambda_1 \mathbf{p}$ (18)). After some manipulation we obtain the formula for the area (times 2ϵ: $2 \cdot \epsilon \cdot \Delta\mathrm{CHI}_\gamma(\mathbf{q}', \{\mathbf{u}', \mathbf{l}'\}) = \epsilon^2(-l_1' q_1' + (q_1')^2 + l_1' u_1' - q_1' u_1') + \epsilon^1(l_2' q_1' - l_1' q_2' - l_2 u_1' + q_2' u_1' + l_1' u_2' - q_1' u_2') + \epsilon^0(l_2' q_2' - (q_2')^2 - l_2' u_2' + q_2' u_2')$ For bounded Pareto fronts we get that the coefficient of ϵ^2 in the above expression is bounded from below. As q_1' is bounded from below and the bound is independent of ϵ and we can assume that q_1' is bounded from above with a bound that does not depend on ϵ, for instance, such a bound would be u_1'. The coefficient of ϵ^1 is also bounded. Both q_1' and q_2' can be assumed bounded from below and above. For q_1' we already gave a reason and $l_2' < q_2' < u_2'$. So for small ϵ the coefficient of ϵ^0 determines the maximum. In view of this we get after some algebraic manipulations:

$$\lim_{\epsilon \downarrow 0} 2\epsilon \Delta\mathrm{CHI}_\gamma(\mathbf{q}', \{\mathbf{u}', \mathbf{l}'\}) = (q_2' - u_2')(q_2' - l_2') \tag{21}$$

and, hence,

$$(q_1^*, q_2^*) = \arg \max_{q' \in Q'} \lim_{\epsilon \downarrow 0} 2\epsilon \Delta\mathrm{CHI}_\gamma(\mathbf{q}', \{\mathbf{u}', \mathbf{l}'\}) = (-, \frac{1}{2}(u_2' + l_2')). \tag{22}$$

This means, that the optimal solution lies on the horizontal line between $\mathbf{u}'$ and $\mathbf{l}'$ which after rotation becomes the line through the midpoint between $\mathbf{u}$ and $\mathbf{l}$ that is parallel to the bisectrix. We have used that $\epsilon = \sin(\gamma)$ and when $\gamma \downarrow 0$ also $\epsilon \downarrow 0$ and vice versa (i.e., $\lim_{\gamma \to 0} \frac{\sin(\gamma)}{\gamma} = 1$).

7 Numerical Studies

The following results are based on the RODEOlib implementation which can be found on the sourceforge net (http://sourceforge.net/projects/rodeolib/) with time stamp 2012-01-19 under the GNU open source license. With Cone-based SMS-EMOA algorithm we denote a generalization of the SMS-EMOA that uses the CHI_γ-Metric as a performance indicator. Apart from using the cone-order and the cone- based hypervolume, it works in the same way as the SMS-EMOA introduced in [7] and uses non-dominated sorting based on the Pareto order. It will be studied for some basic problems with a variety of Pareto front shapes, what kind of distributions are achieved after a large number of iterations (50000) for different dominance cones (represented by different values of γ). For 2-D problems we evolve a population of 30 and for 3-D problems a population of 50 points.

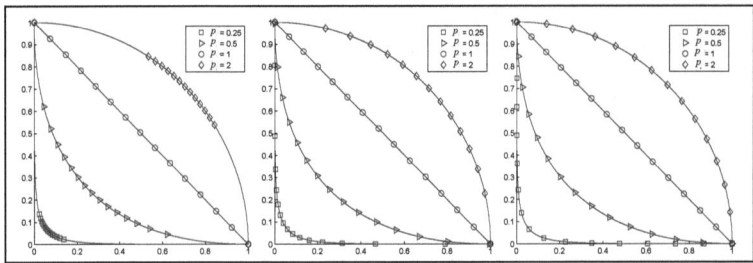

Fig. 8. The approximation sets of the Pareto front for the GSP problem of three instances of the cone-based SMS-EMOA. Left: an obtuse cone, $\gamma = \pi/3$. Center: a Pareto cone, $\gamma = \pi/4$. Right: an acute cone $\gamma = \pi/8$.

The generalized Schaffer test problems (GSP) were introduced in [8] as basic problems for analysis:

$$f_1(\mathbf{x}) = \frac{1}{d^p}(\sum_{i=1}^{d} x_i^2)^p \to \min, \; f_2(\mathbf{x}) = \frac{1}{d^p}(\sum_{i=1}^{d}(x_i - 1)^2)^p \to \min, \; \mathbf{x} \in \mathbb{R}_+^m. \quad (23)$$

Their curvature can be scaled by means of p. For different values of p the shape of the Pareto front equals the arc of a 2-D Lamé supercircle that lies in the positive quadrant. Similarly, in 3-D the following supersphere problems are equipped with a shape parameter p:

$$g = \sqrt{\sum_{i=3}^{n} x_i^2},$$

$$f_1 = ((\cos(x_1))^2)^p(1 + g(\mathbf{x}))$$

$$f_2 = ((\sin(x_1)\cos(x_2))^2)^p(1 + g(\mathbf{x})) \text{ and}$$

$$f_3 = ((\sin(x_1)\sin(x_2))^2)^p(1 + g(\mathbf{x}))$$

and $p = 0.4$ (convex), $p = 0.6$ (convex), $p = 1$ (linear), and $p = 2$ (concave).

Figure 8 and 9 show results of the runs of three instances of the Cone-based SMS-EMOA, using different cones: a closed cone, $\gamma = \pi/8$, a normal cone, $\gamma = \pi/4$, an open cone $\gamma = \pi/3$ and figure 9 shows results on the supersphere problems for different γ-values. As reference point we took $5 \cdot \mathbf{1}$ and the input dimension is 5. The sampling pattern is very regular and points are located on the true Pareto fronts, indicating that the algorithm converged to a near optimal solution. In the concave and convex case distributions on the three fronts differ strongly for the three shapes. *While for obtuse cones (big γ) the distribution is more concentrated on the knee point and on the boundary, for acute cones (small γ) it tends to be more evenly distributed.* In the linear case points are distributed evenly across the true Pareto front, which can be explained by the theoretical result of Section 6.1. Also, evenly spaced distributions are observed for small angle, conforming with theoretical results of Section 6.2.

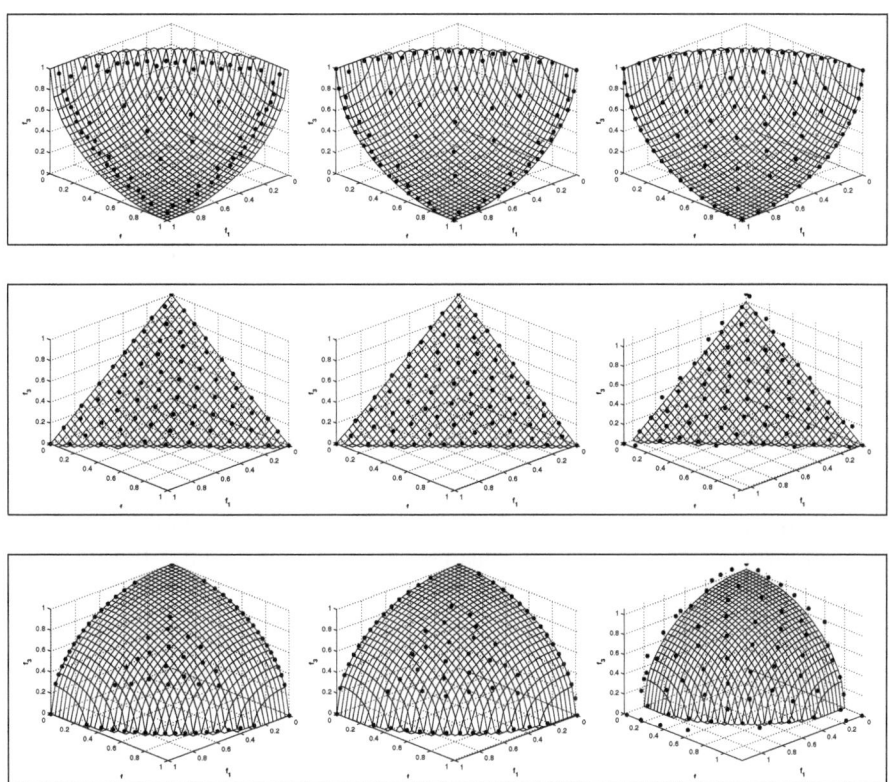

Fig. 9. Approximations of μ-maximal sets (obtained with SMS-EMOA) and Pareto fronts for the supersphere problem of three instances of the cone-based SMS-EMOA. Left: an obtuse cone, $\gamma = \pi/3$. Center: a Pareto cone, $\gamma = \pi/4$. Right: an acute cone $\gamma = \pi/8$. The top row shows results for the convex mirrored supershere problem, the middle row for a linear supersphere problem, and the bottom row for the concave problem.

Another effect that can be observed is that convergence towards the Pareto front seems to slow down in case of acute cones. This might be explained by the positive contribution of points in dominated parts of the boundary.

8   Conclusion and Outlook

The CHI extends the HI to cone orders. With the γ-cones a family of cones spanned by m coordinate vectors that is symmetric to the diagonal line of $\mathbb{R}^m$ was introduced. A construction of the base vectors of these cones was described for m dimensions and an algorithm for the computation of its nondominated set. The CHI$_\gamma$ and its contributions can be efficiently computed by means of a transformation reducing the problem to computations of the standard HI. Theoretical results (in 2-D), and empirical results (in 2-D and 3-D) with a modified

SMS-EMOA indicate that γ is useful for controlling the knee-point focus of the distribution of points on the Pareto front. While small values of γ give rise to more evenly spaced optimal μ-distributions, higher values of γ yield distributions with a higher concentration of points near the knee-point, and for 3-D, also a higher concentration at the boundary. Interesting future work will include a more detailed study of optimal μ-distributions and the influence of the reference point. Moreover, a generalization to more general families of cones, as well as the study of monotonicity properties and indicator-based algorithm designs using cone-based hypervolume indicators will be of interest.

References

1. Auger, A., Bader, J., Brockhoff, D., Zitzler, E.: Theory of the hypervolume indicator: optimal μ-distributions and the choice of the reference point. In: FOGA 2009, pp. 87–102. ACM, NY (2009)
2. Bader, J., Zitzler, E.: HypE: An algorithm for fast hypervolume-based many-objective optimization. Evolutionary Computation 19(1), 45–76 (2011)
3. Batista, L.S., Campelo, F., Guimarães, F.G., Ramírez, J.A.: Pareto cone - dominance: Improving convergence and diversity in multiobjective evolutionary algorithms. In: Takahashi, et al. (eds.) [18], pp. 76–90
4. Billingsley, P.: Probability and Measure, 3rd edn. Wiley (1995)
5. Bringmann, K., Friedrich, T.: Approximating the Least Hypervolume Contributor: NP-Hard in General, But Fast in Practice. In: Ehrgott, M., Fonseca, C.M., Gandibleux, X., Hao, J.-K., Sevaux, M. (eds.) EMO 2009. LNCS, vol. 5467, pp. 6–20. Springer, Heidelberg (2009)
6. Ehrgott, M.: Multicriteria Optimization, 2nd edn. Springer (2005)
7. Emmerich, M., Beume, N., Naujoks, B.: An EMO Algorithm Using the Hypervolume Measure as Selection Criterion. In: Coello Coello, C.A., Hernández Aguirre, A., Zitzler, E. (eds.) EMO 2005. LNCS, vol. 3410, pp. 62–76. Springer, Heidelberg (2005)
8. Emmerich, M.T.M., Deutz, A.H.: Test Problems Based on Lamé Superspheres. In: Obayashi, S., Deb, K., Poloni, C., Hiroyasu, T., Murata, T. (eds.) EMO 2007. LNCS, vol. 4403, pp. 922–936. Springer, Heidelberg (2007)
9. Emmerich, M.T.M., Fonseca, C.M.: Computing hypervolume contributions in low dimensions: Asymptotically optimal algorithm and complexity results. In: Takahashi, et al. (eds.) [18], pp. 121–135.
10. Guerreiro, A.P., Fonseca, C.M., Emmerich, M.T.M.: A Fast Dimension-Sweep Algorithm for the Hypervolume Indicator in Four Dimensions. In: CCCG 2012, pp. 77–82 (2012)
11. Fischer, G.: Lineare Algebra, 11th edn. Vieweg Studium (1997)
12. Beume, N., Fonseca, C.M., López-Ibáñez, M., Paquete, L., Vahrenhold, J.: On the complexity of computing the hypervolume indicator. Transaction IEEE Evolutionary Computation 13(5), 1075–1082 (2009)
13. Beume, N.: S-Metric Calculation by Considering Dominated Hypervolume as Klee's Measure Problem. Evolutionary Computation 17(4), 477–492 (2009)
14. Kung, H.T., Luccio, F., Preparata, F.P.: On finding the maxima of a set of vectors. J. ACM 22(4), 469–476 (1975)
15. Noghin, V.D.: Relative importance of criteria: a quantitative approach. Journal of Multi-Criteria Decision Analysis 6(6), 355–363 (1997)

16. Sawaragi, Y., Nakayama, H., Tanino, T.: Theory of multiobjective optimization. Academic Press Inc. (1985)

17. Shukla, P.K., Hirsch, C., Schmeck, H.: Towards a Deeper Understanding of Trade-offs Using Multi-objective Evolutionary Algorithms. In: Di Chio, C., Agapitos, A., Cagnoni, S., Cotta, C., de Vega, F.F., Di Caro, G.A., Drechsler, R., Ekárt, A., Esparcia-Alcázar, A.I., Farooq, M., Langdon, W.B., Merelo-Guervós, J.J., Preuss, M., Richter, H., Silva, S., Simões, A., Squillero, G., Tarantino, E., Tettamanzi, A.G.B., Togelius, J., Urquhart, N., Uyar, A.Ş., Yannakakis, G.N. (eds.) EvoApplications 2012. LNCS, vol. 7248, pp. 396–405. Springer, Heidelberg (2012)

18. Takahashi, R.H.C., Deb, K., Wanner, E.F., Greco, S. (eds.): EMO 2011. LNCS, vol. 6576. Springer, Heidelberg (2011)

19. While, R.L., Bradstreet, L., Barone, L.: A Fast Way of Calculating Exact Hypervolumes. IEEE Trans. Evolutionary Computation 16(1), 86–95 (2012)

20. Yıldız, H., Suri, S.: On Klee's measure problem on grounded boxes. In: Proceedings of the 28th Annual Symposium on Computational Geometry (SoCG), pp. 111–120 (June 2012)

21. Zitzler, E., Thiele, L.: Multiobjective Optimization Using Evolutionary Algorithms - A Comparative Case Study. In: Eiben, A.E., Bäck, T., Schoenauer, M., Schwefel, H.-P. (eds.) PPSN V. LNCS, vol. 1498, pp. 292–301. Springer, Heidelberg (1998)

Many-Objective Optimization Using Taxi-Cab Surface Evolutionary Algorithm

Hans J.F. Moen[1,2], Nikolai B. Hansen[1,3], Harald Hovland[1], and Jim Tørresen[2]

[1] Information Management Division, Norwegian Defence Research Establishment, Norway
{jonas.moen,harald.hovland}@ffi.no
[2] Department of Informatics, University of Oslo, Norway
jimtoer@ifi.uio.no
[3] Department of Mathematics, University of Oslo, Norway
nikolabh@student.matnat.uio.no

Abstract. Optimization of problems spanning more than three objectives, called many-objective optimization, is often hard to achieve using modern algorithm design and currently available computational resources. In this paper a multi-objective evolutionary algorithm, called the Surface Evolutionary Algorithm, is extended into many-objective optimization by utilizing, for the first time, the taxi-cab metric in the optimizer. The Surface Evolutionary Algorithm offers an alternative to multi-objective optimizers that rely on the principles of domination, hypervolume and so forth. The taxi-cab metric, or Manhattan distance, is introduced as the selection criterion and the basis for calculating attraction points in the Surface Evolutionary Algorithm. This allows for fast and efficient many-objective optimization previously not attainable using this method. The Taxi-Cab Surface Evolutionary Algorithm is evaluated on a set of well-known many-objective benchmark test problems. In problems of up to 20 dimensions, this new algorithm of low complexity is tested against several modern multi-objective evolutionary algorithms. The results reveal the Taxi-Cab Surface Evolutionary Algorithm as a conceptually simple, yet highly efficient many-objective optimizer.

Keywords: Multi-objective optimization, multi-objective evolutionary algorithms, many-objective optimization.

1 Intoduction

Many real-world optimization problems are naturally cast as multi-objective optimization problems [1-4]. In these problems one tries to find the optimal trade-off solutions between different objectives, known as the search for the optimal Pareto front [1]. Evolutionary algorithms like Genetic Algorithms [5], [6], Genetic Programming [7] and Particle Swarm Optimization [8] are especially suited for optimizing multi-objective problems. This is due to their population based approach [1] seeking to spread their solutions uniformly onto the entire optimal Pareto front. An analytical expression of the true optimal Pareto front is often difficult to obtain in

R.C. Purshouse et al. (Eds.): EMO 2013, LNCS 7811, pp. 128–142, 2013.

a multi-objective optimization problem. In a multi-objective evolutionary algorithm one seeks to distribute the population of solutions in objective space as close as possible to the true optimal Pareto front. The Pareto optimal set of solutions is an approximation of the optimal Pareto front. The non-dominated set of the entire search space is the global Pareto optimal set. Ideally, the global Pareto optimal set approximating the true optimal Pareto front should be on the true optimal Pareto front and have a uniform distribution of samples on the entire true optimal Pareto front. Also, the global Pareto optimal set should be found in the shortest time possible. However, optimizing problems of more than three objectives has proven to be a challenging task, even for modern multi-objective evolutionary algorithms [1], [2], [9-12].

Today, there are several approaches to multi-objective evolutionary optimization. One of them is the dominance-based approach, relying on the principle of Pareto optimality. Multi-objective solvers like PAES [13], SPEA2 [14] and NSGA-II [15] are good examples of modern dominance-based optimizers that perform well on many multi-objective problems. In particular, NSGA-II is arguably the most successful and widely applied multi-objective solver in use today. Due to its success, NSGA-II is the de facto benchmark in the field of multi-objective optimization and is thus used in numerous comparative optimization studies. Even though optimizers relying on the principle of dominance perform well on many low-dimensional problems, they suffer from loss of selection pressure when solving many-objective problems. This is mainly due to the fact that as dimensionality increases, the fraction of non-dominated solutions in the population increases. And, by the principle of Pareto optimality, all non-dominated solutions are indistinguishable in terms of fitness selection. Insufficient selection pressure may result in stagnation of the algorithm when optimizing many-objective problems.

The SMS-EMOA [16] and MO-CMA-ES [17] algorithms are good examples of hypervolume based optimizers relying on the calculation of the hypervolume indicator to rank non-dominated solutions in a population. These optimizers, currently attracting a lot of research attention, are highly efficient in solving low-dimensional multi-objective problems but suffer from time-consuming computations when optimizing many-dimensional problems. In most cases, given the current level of computational resources available, the high complexity of the hypervolume metric calculations [18] will render them impractical as general many-objective optimizers.

Other multi-objective evolutionary algorithms rely on more direct search methods to drive the solutions towards the true optimal Pareto front. For instance, MODELS [19] uses local gradient information in combination with evolutionary methods. Furthermore, methods like MSOPS [20] and MSOPS-II [21] rely on performing many parallel searches of multiple conventional target vector based optimizations. These methods can be used to generate the Pareto set and analyze problems with large numbers of objectives. The MSOPS algorithm is regarded as the first true many-objective evolutionary optimizer [21]. The MSOPS-II algorithm provides two important extensions to MSOPS. These extensions allow MSOPS to work as a general-purpose multi-objective evolutionary algorithm. The first extension is the inclusion of an automatic target vector generator, removing the need for designer intervention. The second extension improves the fitness assignment for simplified analysis and better constraint handling.

The Surface Evolutionary Algorithm (SEA) [22], on the other hand, relies on a somewhat different method than MSOPS and MSOPS-II for converging towards the optimal Pareto front. The main idea behind SEA is to calculate a set of evenly distributed attraction points and use these attraction points for driving the solutions in the population towards the true optimal Pareto front. The attraction points are usually calculated in every generation using some kind of suitable interpolation surface located somewhere between the solution set and the optimization target. After the attraction point calculation, each solution in the population is associated with the nearest attraction point. Then, all similarly associated solutions compete locally for a way out to the next generation by using some kind of single value metric evaluation. By selecting the best solutions from each attraction point group the population should converge towards the true optimal Pareto front and diversity should be maintained throughout evolution.

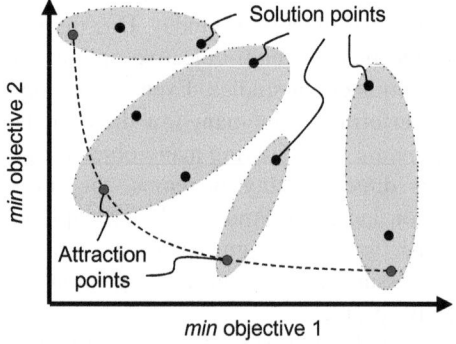

Fig. 1. Illustration of the general SEA principle in a bi-objective minimization problem

In Fig. 1 the general SEA principle is illustrated. For visualization purposes a bi-objective minimization problem is assumed. In this example 8 solution points (black) are associated with 4 attraction points (red) evenly distributed on a suitable bounding interpolation surface (dashed black line) somewhere between the solution set and the optimization target (e.g. the origin). Solution points are grouped into attraction point cohorts represented by the different shaded regions in the figure. Solutions are selected for the next generation based on local competition among the grouped solutions.

The main issue in the SEA is how to proceed in order to obtain a suitable set of attraction points that are computable in a reasonable amount of time, even when optimizing many-dimensional problems. Also, the choice of single value metric for solution selection must be computable in reasonable time when optimizing many-dimensional problems. If done properly, the SEA should be able to guide a multi-objective population towards the true optimal Pareto front, both in terms of convergence and in terms of diversity.

In [22] the SEA principle is used to span the entire true optimal Pareto front of a bi-objective real-world radar problem. In this problem the SEA is applied using the convex hull of the solution set as the interpolation surface and the selection is evaluated using several different single value metrics of varying complexity. Nevertheless, extending this bi-objective SEA procedure directly into many-objective optimization problems could prove difficult. First, computing the convex hull of a many-dimensional solution set is of high computational complexity, rendering many-objective optimization impractical with currently available resources. Secondly, interpolating evenly distributed attraction points on a convex hull is not a trivial task for problems of more than 2 dimensions. The constantly changing convex hull of the population, in terms of surface size and location during evolution, might introduce attraction point instabilities possibly disrupting optimization efforts. Consequently, interpolating attraction points on the convex hull of the solution set seems like a poor strategy for optimizing many-dimensional problems using the SEA method.

In this paper, the Taxi-Cab (TC) metric, or Manhattan distance as it is also called, is introduced in the SEA method as a suitable metric for many-dimensional optimization. The novel optimizer is called the Taxi-Cab Surface Evolutionary Algorithm (TC-SEA). The TC metric is very fast to calculate, being the sum of all absolute values of a vector, i.e. a solution point. The TC metric is introduced both as the basis for generating the interpolation surface, and hence the attraction points in the TC-SEA method, and as the single value metric for selecting solutions for the next generation. A TC interpolation surface, based on the TC values of the solution points, is easy to calculate and construct. It is also easy to produce a good distribution of attraction points on this surface, even for high-dimensional problems. Furthermore, intra cohort competition among solutions, for selection into the next generation, can be efficiently done by evaluating the TC value of the grouped solutions. Thus, it will be demonstrated how the introduction of the simple and efficient TC metric in the TC-SEA method can improve performance when optimizing several well-known many-objective benchmark test functions.

The outline of the paper is as follows. In section 2, the TC-SEA procedure is described. In section 3, the TC-SEA is benchmarked on a set of well-known multi-objective test problems of up to 20 dimensions, proving its efficiency against several established multi-objective methods. Finally, in section 4, conclusions are drawn.

2   Taxi-Cab Surface Evolutionary Algorithm

In this paper, the single value metric and bounding interpolation surface in the SEA is based on the simplest single value metric obtainable; namely the taxi-cab metric.

2.1   Taxi-Cab Metric

The taxi-cab metric can formally be defined as

$$TC = \sum_{d=1}^{D} |P_{i,d}| \qquad (1)$$

where $P_{i,d}$ is the d'th objective value of a vector from origin to solution point $P_i \in N$ population vectors. In short, TC is just the sum of all D objective values of a solution, taken as absolute values. Its calculation is linear in complexity, both in terms of population size and in terms of number of objectives, making it extremely fast to calculate. Furthermore, a value of TC defines a plane bounded by the coordinate planes in objective space. A TC plane is an equilateral triangle making it easy to interpolate evenly distributed attraction points by using the triangular number as the distribution measure. The triangular number is the number of objects that can form an equilateral triangle.

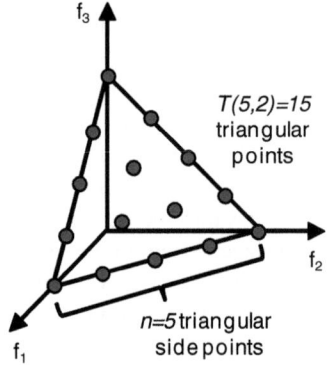

Fig. 2. A set of 15 triangular interpolation points on a TC plane

In Fig. 2 a TC plane with 15 triangular interpolation points having 5 points on each triangular side is depicted. The generalized hypertriangular number is given by the following equation

$$T_D(n, D') = \frac{n(n+1)\cdots(n+D'-1)}{(D')!} \qquad (2)$$

where T is the hypertriangular number, n is the number of points on the triangular sides and $D'=D-1$ is the dimension of the triangular hyperplane.

2.2   The TC-SEA Procedure

The TC-SEA procedure presented here is primarily based on two important concepts: First, the TC metric is used as a suitable single value metric for converging towards the true optimal Pareto front. Secondly, the interpolation surface generating attraction points is based on the fixed TC value of the best solution in the population. Assuming non-negative objective values and minimization of all objectives, this would be the lowest TC metric value in the population. The triangular attraction points are considered to be stable in the sense that they are stationary from the perspective of the origin. This is important for avoiding disruptive fluctuations in the population during evolution.

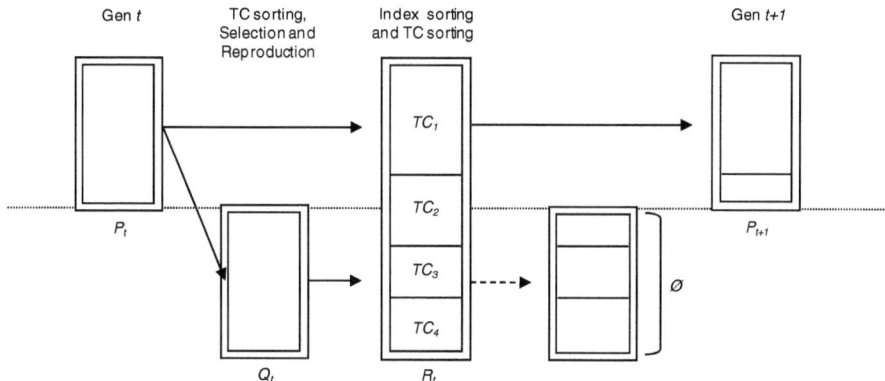

Fig. 3. The TC-SEA generation-to-generation procedure. TC_1 are the best solutions from each index cohort sorted by TC value. TC_2 are the second best solutions from each cohort sorted by TC value and so forth.

In Fig. 3 the TC-SEA generation-to-generation procedure is illustrated. Initially, the TC-SEA creates a random parent population P_t of size N. This population is then sorted according to its TC values and scaled by fitness rank. A stochastic uniform selection operator is used to select a population Q_t from the parent population, also of size N. Q_t is then crossed and mutated upon. The merger of parent and children population $R_t=P_t+Q_t$ ensures elite preservation in the algorithm. This merger resembles the fast and efficient recombination operator found in NSGA-II [15]. The population R_t of size $2N$ is formed as a selection pool for the next generation P_{t+1} of N individuals again. The selection procedure is as follows. First, the TC value of all individuals in R_t is calculated. Then, each solution is indexed according to its nearest attraction point or index point on the interpolated hypertriangular surface using Euclidian distance. These index points I^* are calculated using the triangular number of equation (2) and spread out on the TC plane constructed using the lowest TC value of the R_t population. Additional attraction points on the coordinate planes could be included when optimizing special problems like asymptotic convex functions. Furthermore, all objective values could be scaled if required. Now, by sorting the population R_t according to the best TC value from each index cohort and then the second best TC value from each index cohort and so forth, the N best solutions can be put into the next generation P_{t+1}, finalizing the process of producing a new generation.

The complexity of the TC-SEA is dominated by the computation of the distance to the nearest attraction point, having a complexity of $O(DN^2)$ if $|I^*|=|N|$, where N is the population size and D is the number of objective dimensions. The complexity of sorting R_t according to cohort number and TC value is approximately $O(N\log(N))$.

3   Benchmark Problems

In multi-objective evolutionary algorithm design and evaluation, several multi-objective benchmark test problems with known true optimal Pareto front have been proposed and tested. Deb *et al.*'s [23] DTLZ multi-objective test problems are popular references in a great number of comparison studies [24-26] and are also used in this paper.

3.1   Benchmark Problem Formulations

The DTLZ test suite has been specifically developed for evaluating the ability of different multi-objective evolutionary algorithms to solve optimization problems with three or more objectives. The test functions have been designed with special emphasis on simplicity of construction, scalability of number of decision variables and objectives, and knowledge of exact shape and location of the true optimal Pareto front [23]. Thus, by using suitable multi-objective performance metrics, it should be possible to evaluate the ability of an algorithm to converge to the true optimal Pareto front and to maintain a uniformly distributed set of solutions during evolution. The DTLZ2, DTLZ3 and DTLZ4 benchmark test problems have been selected and used for evaluating the performance of the multi-objective algorithms tested.

The DTLZ2 test problem is used to investigate the ability of an algorithm to scale up its performance in large number of objectives. This problem has a spherical Pareto front of radius 1. The DTLZ3 introduces many locally optimal Pareto fronts which could trap a multi-objective algorithm during evolution. Consequently, this problem is used for evaluating an algorithm's ability to converge to the globally optimal Pareto front. This problem has the same optimal Pareto front as the DTLZ2 but is much harder to solve requiring more optimization time in order to reach the true optimal Pareto front. The DTLZ4 is designed to investigate an algorithm's ability to maintain a good distribution of solutions. The optimal Pareto front of this problem is the same as the DTLZ2 but due to its increased difficulty, the DTZL4 requires more optimization time to reach the true optimal Pareto front.

NSGA-II, MSOPS-II and random search have been selected for benchmarking the TC-SEA optimization method. The NSGA-II has been included since it is the standard method of comparison in multi-objective studies. The MSOPS-II has been included as the state-of-the-art method for many-objective optimization. The MSOPS-II code employed here is based on the code found at [27] but altered to include the same genetic operators as in the other methods tested. The number of target vectors in the MSOPS-II algorithm is always set equal to the population size. Random search is included as the natural baseline optimizer. The TC-SEA has been run using the same number of interpolation points as population size and a rank fitness scaling of $1/(rank)^{1/2}$ is employed. All algorithms have been tested on 3, 6, 10 and 20 dimensional versions of the DTLZ problems using the same real-coded genetic operators. The polynomial mutation operator [28] of mutation rate $p_m=1/chromosome$ *length* and $\eta_m=5$ together with a simulated binary crossover operator [29] of parameters $\eta_c=5$ and varying $p_{ic}=[0.5\ 0.1\ 0.05\ 0.01]$ for 3D, 6D, 10D and 20D objective size respectively, have been employed in accordance with [26] and determined by extensive parameter studies too lengthy to be included here. The NSGA-II tests on DTLZ2 have been conducted using the same optimal parameter values as in [26] η_m=20 and η_c=15.

The algorithms tested have been run using similar population sizes for all dimensions. The appropriate triangular numbers are; $T_{3D}(20,2)$=210, $T_{6D}(6,5)$=252, $T_{10D}(4,9)$=220 and $T_{20D}(3,19)$=210. This ensures that the scalability of the algorithms is properly tested. The DTLZ2 tests are run for 500 generations and the more difficult problems DTLZ3 and DTZL4 are run for 1000 generations. All results reported are the average performances of 35 parallel runs on a Matlab Distributed Computing Server [30] of HP x8600 and z800 multi-core workstation clients.

3.2   Performance Metrics

The different search goals of a multi-objective evolutionary algorithm are difficult to express using one single performance measure [1]. Hence, independent performance metrics for each of the multi-objective optimization criteria identified have to be employed. Moreover, the multi-objective performance metrics should have sufficiently low complexity in order to be able to compute in a reasonable time, especially when dealing with many-objective problems of high dimensionality. In the following, two such independent multi-objective performance metrics are reviewed.

The Convergence Metric

The generational distance [15], [31], [32] is used as the convergence measure in these tests. The generational distance has been used in numerous multi-objective optimization studies [1], [25] including several analyses of many-objective optimization problems [26], [33]. The generational distance characterizes the search efficiency towards the optimal Pareto front when the true optimal Pareto front is *known* in advance. In these tests the generational distance expresses the average distance from the Pareto optimal set of solutions to the true optimal Pareto front, known as the γconvergence measure. It is computed in the following manner

$$\gamma = \frac{1}{|PO|}\sum_{i=1}^{|PO|} l_i \quad \text{and} \quad l_i = norm\left(\bar{l}_{P_i} - \bar{l}_{PF_i}\right) \tag{3}$$

where i is the index of a Pareto optimal solution $P_i \in PO$ in the Pareto set PO of size $|PO|$ and the convergence measure γ is the mean error of the distances l_i, which is the norm of the vector to point P_i minus the vector in the same direction but to the true Pareto front.

The Distribution Metric

It is often difficult to construct a good measure of the distribution quality of the solutions found in a many-dimensional optimization problem [26], [33]. Several metrics [34-36] have been proposed to assess the density distribution of the Pareto optimal set and have been tested on bi- and tri-objective problems with varying degree of success. Most importantly, the distribution metric should be able to discriminate between Pareto optimal solution sets that are grouped into clusters and sets that are sparsely populated over the rest of the Pareto front. Good solution sets should be characterized by being uniformly distributed over the entire accessible state space of the Pareto front.

An entropy based metric [33] is employed here for measuring the distribution quality of the benchmark test results. This metric has its roots in Shannon's entropy metric [37] measuring the flatness of a statistical distribution. This entropy metric relies on the quantification of all accessible states bounded by a projected image of the Pareto front solutions. The image resolution is determined by the indifference region. This is the region where a decision maker is indifferent to the choice between two projected solutions. By using an appropriate indifference function for each solution projected, the density of each image cell can be computed. The entropy H of the entire image is then computed as

$$H = -\sum_{k=1}^{K} p_k \log(p_k)$$ (4)

where p_k is the density of cell k in an image of K cells. Maximal entropy $H_{max}=\log(K)$ is given when $p_k=1/K$ for all k, assuming a normalized image density of $\sum p_k=1$. A solution point located at the centre of a resolution cell is often assumed to have a Gaussian influence function that vanishes at the boundaries of the cell as given by

$$\Omega(r) = \frac{1}{(2\pi\sigma^2)^{D'/2}} e^{\left(-\frac{r^2}{2\sigma^2}\right)}$$ (5)

where r is the Euclidian distance from the solution point P_i projected onto the image, σ is the standard deviation and $D'=D-1$ is the dimension of the projected image. The image coordinates are often normalized in the sense of having image sides of length 1.

The entropy metric is calculated by camera projecting the Pareto set solutions onto an image bounded by the coordinate planes. This is possible since all objective values are non-negative for the DTLZ problems. The image is then a hypertriangle of size $K=N$ image cells distributed according to the appropriate triangle number. This ensures a constant size indifference region throughout evolution. The influence function (5) is assumed to vanish over the indifference region having σ $=0.5\times\tan(30°)/3n$, where n is the number of points on the triangular sides. The entropy image density is thus fast to compute due to the limited number of image cells, even for high-dimensional problems.

3.3   Benchmark Results

In Figs. 4 to 6 the test results from the different multi-objective algorithms are given, consecutively representing DTLZ2 test results to DTLZ4 test results.

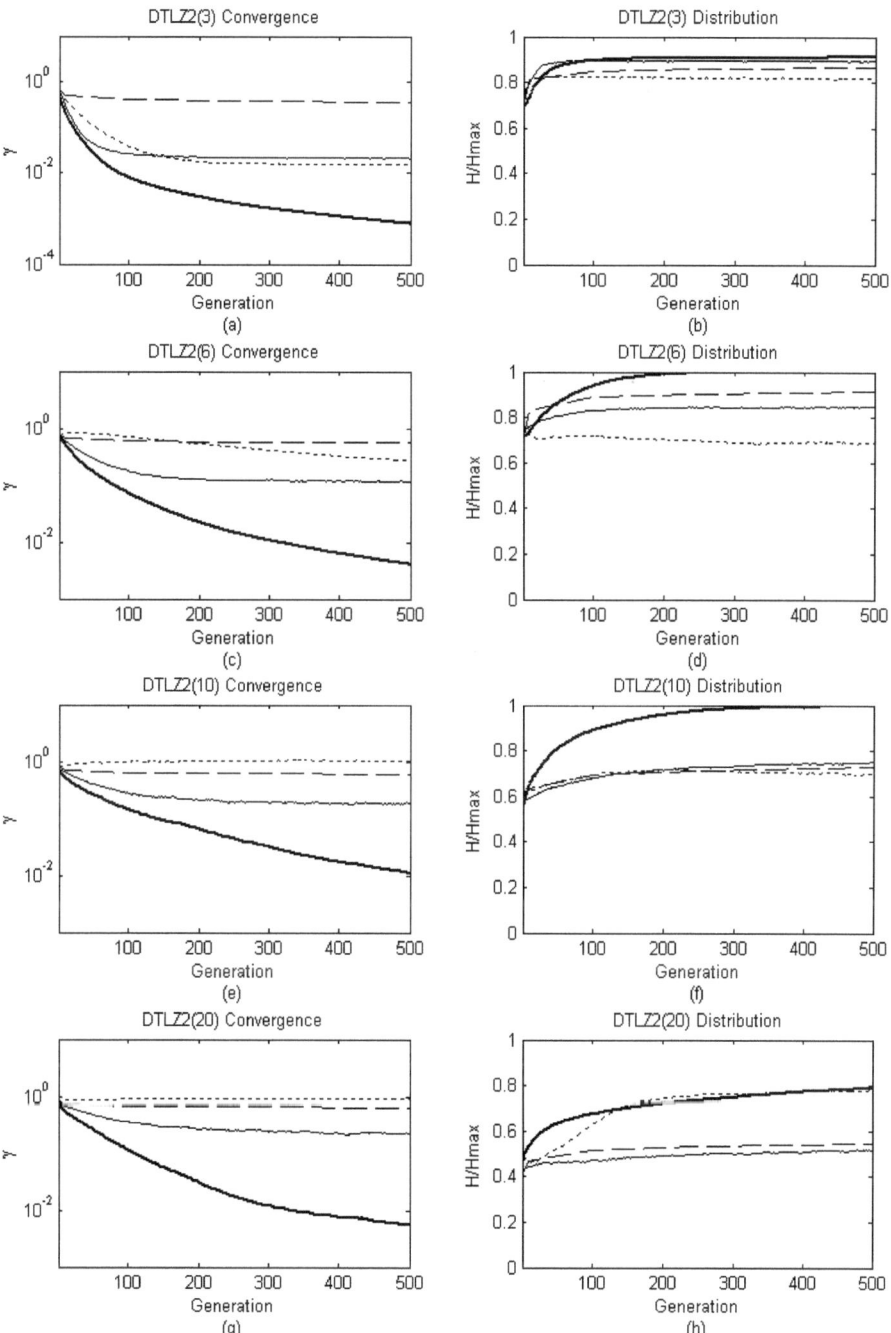

Fig. 4. DTLZ2 test results. Thick solid lines are TC-SEA results, thin dotted lines are NSGA-II, thin solid lines are MSOPS-II and thin dashed lines are random search results.

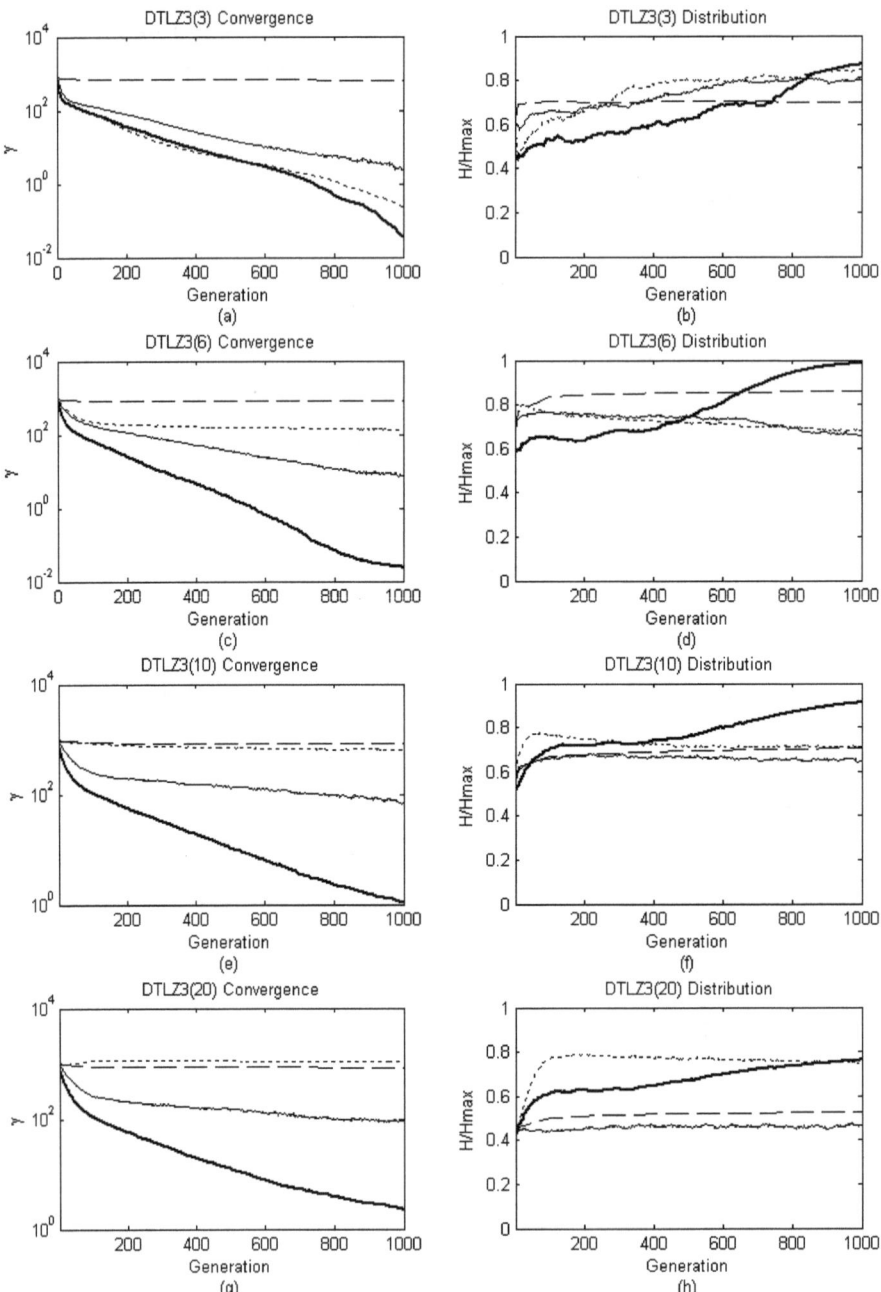

Fig. 5. DTLZ3 test results. Thick solid lines are TC-SEA results, thin dotted lines are NSGA-II, thin solid lines are MSOPS-II and thin dashed lines are random search results.

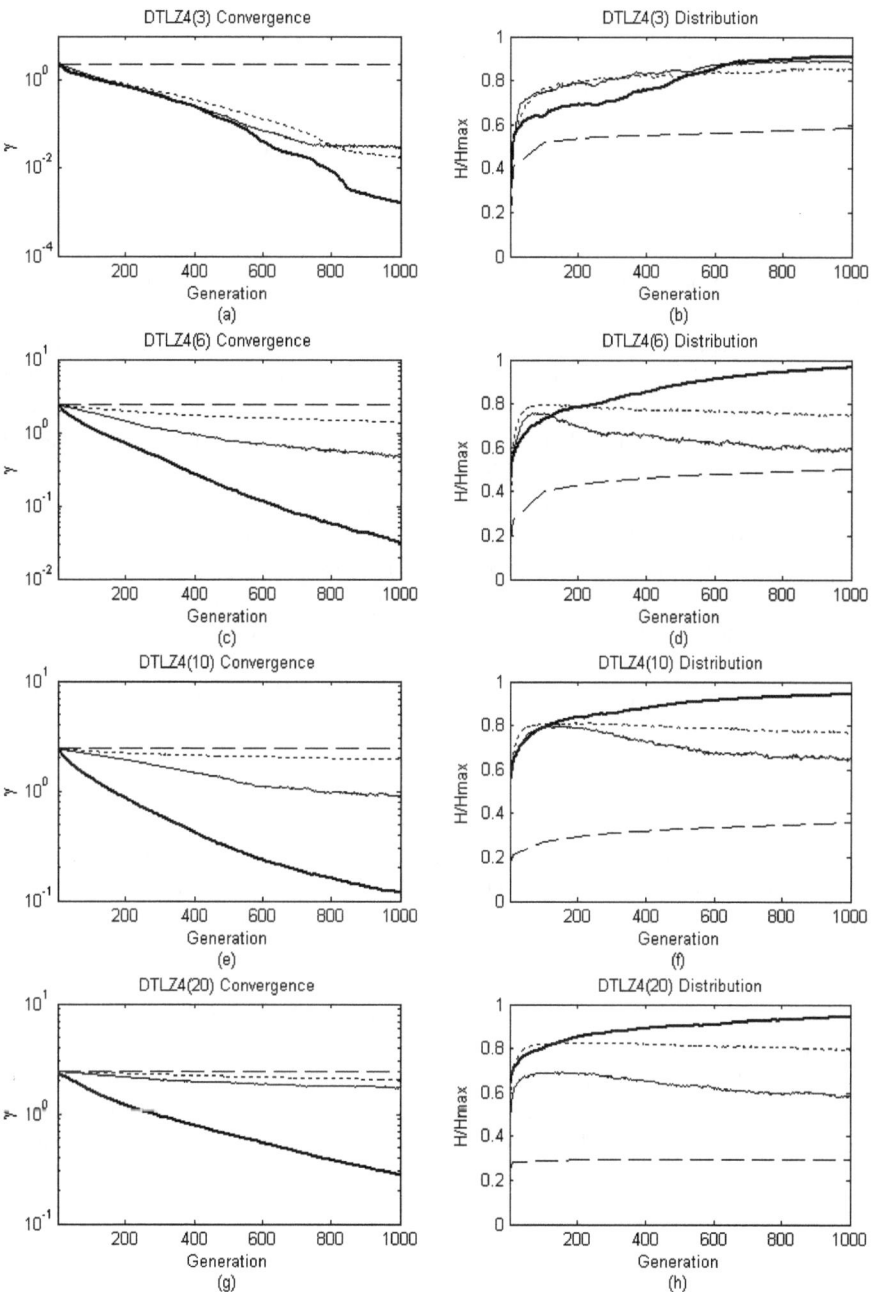

Fig. 6. DTLZ4 test results. Thick solid lines are TC-SEA results, thin dotted lines are NSGA-II, thin solid lines are MSOPS-II and thin dashed lines are random search results.

In all plots the results obtained using the TC-SEA method are drawn using thick solid lines, NSGA-II test results are drawn using thin dotted lines, MSOPS-II test results are drawn using thin solid lines and random search test results are drawn using thin dashed lines. In Figs. 4 to 6 all convergence results are given in the first column and the performance of the normalized entropy distribution metric is given in the second column. The number in parenthesis, next to the problem name, is the dimension of the benchmark problem tested.

As a first remark, it is evident from all the plots that the DTLZ test problems are generally hard to optimize using random search. It is also evident that it is harder to achieve a good distribution of solutions as dimensionality increases. The DTLZ3-4 require more optimization time than DTLZ2 to reach the true optimal Pareto front.

In Fig. 4 the DTLZ2 test results are shown. Only the TC-SEA and the MSOPS-II optimize the problem for all the dimensions tested. The NSGA-II only optimizes well on the 3D version of the problem, but seems to have some convergence pressure left in the 6D problem, as indicated in Fig. 4c. The TC-SEA is able to approximate the true optimal Pareto front on all dimensions tested. The MSOPS-II reaches the same level of convergence on the 3D problem only. The TC-SEA performs well in terms of distribution performance in all dimensions. The MSOPS-II performs on the same level as random search in terms of the distribution characteristics. The NSGA-II keeps an approximately constant level of distribution performance regardless of dimensionality.

DTLZ3 results are shown in Fig. 5. Once again, the TC-SEA and MSOPS-II optimize for all dimensions tested, but only the TC-SEA approximates the true optimal Pareto front for the 3D and 6D problems, requiring more than 1000 generations to do the same for 10D and 20D problems. The NSGA-II almost reaches the true optimal Pareto front on the 3D problem and to some extent optimizes the 6D version of the problem. The NSGA-II performs similarly to a random search on higher dimensional problems. In DTLZ3 all the algorithms take longer time, compared to DTLZ2, to outperform the random search distribution performance. The random search distribution level is only consistently outperformed by the TC-SEA when testing on all dimensions of the problem. The MSOPS-II performs similarly to a random search and, also here, the NSGA-II performs at a fixed level in terms of distribution performance.

The DTLZ4 results are given in Fig. 6. The TC-SEA approximates the true optimal Pareto front in all dimensions tested. The MSOPS-II approximates the true optimal Pareto front in the 3D and 6D cases of the problem. The NSGA-II performs well in 3D, and to some degree in 6D, but behaves like a random search in higher dimensional versions of the problem. The distribution characteristics of all methods tested outperform random search for all dimensions. This indicates that all methods employ good diversity operators for many-objective optimization.

In summary, the DTLZ test results reveal how the TC-SEA method efficiently optimizes and approximates the true optimal Pareto front in almost all the many-objective problems tested. The advantageous convergence results of the TC-SEA are also obtained with very good distribution characteristics. This indicates that the TC-SEA is a well performing many-objective optimizer when compared to the other algorithms tested. The MSOPS-II also optimizes all the many-objective problems but at a slower rate than the TC-SEA. Furthermore, the MSOPS-II is not able to achieve the same level of distribution performance as the TC-SEA. NSGA-II performs well on the 3D problems but loses much of its convergence pressure as dimensionality increases.

4   Conclusions

The introduction of the TC metric as the basis of the SEA method is here demonstrated to transform the optimizer into an efficient many-objective evolutionary algorithm. The excellent scaling capabilities of the new TC-SEA procedure are evident in the good convergence and distribution results obtained when testing on a set of well-known DTLZ multi-object benchmark test problems. The new TC surface employed in the TC-SEA method is found to produce the stable attraction points necessary for efficient many-objective optimization. Furthermore, the low complexity of the TC metric enables fast multi-objective optimization, even for high dimensional problems.

References

1. Deb, K.: Multi-Objective Optimization using Evolutionary Algorithms. John Wiley & Sons Ltd., West Sussex (2001)
2. Knowles, J., Corne, D., Deb, K. (eds.): Multiobjective Problem Solving from Nature. From Concepts to Applications, Natural Computing Series. Springer, Berlin (2008)
3. Rahmat-Samii, Y., Christodoulou, C.: Special Issue, IEEE Trans. Antennas Propag. 55(3) (2007)
4. Hughes, E.J.: Radar Waveform Optimisation as a Many-Objective Application Benchmark. In: Obayashi, S., Deb, K., Poloni, C., Hiroyasu, T., Murata, T. (eds.) EMO 2007. LNCS, vol. 4403, pp. 700–714. Springer, Heidelberg (2007)
5. Holland, J.H.: Adaption in Natural and Artificial Systems. The University of Michigan Press, Ann Arbor (1975)
6. Goldberg, D.E.: Genetic Algorithms in Search, Optimization, and Machine Learning. Addison-Wesley Longman Inc. (1989)
7. Koza, J.R.: Genetic Programming. On the Programming of Computers by Means of Natural Selection. The MIT Press (1992)
8. Kennedy, J., Eberhart, R.: Particle swarm optimization. In: Proc. IEEE International Conference on Neural Networks, vol. 4, pp. 1942–1948 (1995)
9. Purshouse, R.C., Fleming, P.J.: Conflict, Harmony, and Independence: Relationships in Evolutionary Multi-criterion Optimisation. In: Fonseca, C.M., Fleming, P.J., Zitzler, E., Deb, K., Thiele, L. (eds.) EMO 2003. LNCS, vol. 2632, pp. 16–30. Springer, Heidelberg (2003)
10. Purshouse, R.C., Fleming, P.J.: Evolutionary many-objective optimization: An exploratory analysis. In: Proc. of IEEE CEC 2003, vol. 3, pp. 2066–2073 (2003)
11. Hughes, E.J.: Evolutionary many-objective optimization: Many once or one many? In: Proc. of IEEE CEC 2005, vol. 1, pp. 222–227 (2005)
12. Wagner, T., Beume, N., Naujoks, B.: Pareto-, Aggregation-, and Indicator-Based Methods in Many-Objective Optimization. In: Obayashi, S., Deb, K., Poloni, C., Hiroyasu, T., Murata, T. (eds.) EMO 2007. LNCS, vol. 4403, pp. 742–756. Springer, Heidelberg (2007)
13. Knowles, J.D., Corne, D.W.: Approximating the Nondominated Front Using the Pareto Archived Evolution Strategy. Evolutionary Computation 8(2), 149–172 (2000)

14. Zitzler, E., Laumanns, M., Thiele, L.: SPEA2: Improving the Strength Pareto Evolutionary Algorithm. TIK-Report 103, ETH, Zurich, Switzerland (2001)
15. Deb, K., Pratap, A., Agarwal, S., Meyarivan, T.: A Fast and Elitist Multiobjective Genetic Algorithm: NSGA-II. IEEE Trans. Evol. Comput. 6(2), 182–197 (2002)
16. Beume, N., Naujoks, B., Emmerich, M.: SMS- EMOA: Multiobjective selection based on dominated hypervolume. Eur. J. Oper. Res. 181, 1653–1669 (2007)
17. Igel, C., Hansen, N., Roth, S.: Covariance Matrix Adaptation for Multi-objective Optimization. Evol. Comput. 15(1), 1–28 (2007)
18. Beume, N., Fonseca, C.M., Lopez-Ibanez, M., Paquete, L., Vahrenhold, J.: On the Complexity of Computing the Hypervolume Indicator. IEEE Trans. Evol. Comput. 13(5), 1075–1082 (2009)
19. Hughes, E.J.: Many-Objective Directed Evolutionary Line Search. In: Proc. of GECCO 2011, pp. 761–768. ACM (2011)
20. Hughes, E.J.: Multiple Single Objective Pareto Sampling. In: Proc. of IEEE CEC 2003, vol. 4, pp. 2678–2684 (2003)
21. Hughes, E.J.: MSOPS-II: A general-purpose Many-Objective optimizer. In: Proc. of IEEE CEC 2007, pp. 3944–3951 (2007)
22. Moen, H.J.F., Kristoffersen, S.: Spanning the Pareto Front of a Counter Radar Detection Problem. In: Proc. of GECCO 2011, pp. 1835–1842. ACM (2011)
23. Deb, K., Thiele, L., Laumanns, M., Zitzler, E.: Scalable Multi-Objective Optimization Test Problems. In: Proc. of IEEE CEC 2002, pp. 825–830 (2002)
24. Deb, K.: A Robust Evolutionary Framework for Multi-Objective Optimization. In: Proc. of GECCO 2008, pp. 633–640. ACM (2008)
25. Bandyopadhyay, S., Saha, S., Maulik, U., Deb, K.: A Simulated Annealing-Based Multiobjective Optimization Algorithm: AMOSA. IEEE Trans. Evol. Comput. 12(3), 269–283 (2008)
26. Adra, S.F., Fleming, P.J.: Diversity Management in Evolutionary Many-Objective Optimization. IEEE Trans. Evol. Comput. 15(2), 183–195 (2011)
27. Hughes, E.J.: http://code.evanhughes.org
28. Deb, K., Goyal, M.: A Combined Genetic Adaptive Search (GeneAS) for Engineering Design. Computer Science and Informatics 26(4), 30–45 (1996)
29. Deb, K., Agrawal, R.B.: Simulated Binary Crossover for Continuous Search Space. Complex Systems 9, 115–148 (1995)
30. Matlab Distributed Computing Server. The MathWorks Inc. (1994-2011)
31. Van Veldhuizen, D.A.: Multiobjective Evolutionary Algorithm: Classifications, Analyses, and New Innovations. Ph.D. thesis, Air Force Institute of Technology (1999)
32. Zitzler, E.: Evolutionary Algorithms for Multiobjective Optimization: Methods and Applications. Ph.D. thesis, ETH, Zurich, Switzerland (November 1999)
33. Farhang-Mehr, A., Azarm, S.: Diversity Assessment of Pareto Optimal Solutions: An Entropy Approach. In: Proc. of CEC 2002, pp. 723–728 (2002)
34. Zitzler, E., Thiele, L.: Multiobjective Optimization Using Evolutionary Algorithms - A Comparative Case Study. In: Eiben, A.E., Bäck, T., Schoenauer, M., Schwefel, H.-P. (eds.) PPSN V. LNCS, vol. 1498, pp. 292–301. Springer, Heidelberg (1998)
35. Schott, J.R.: Fault Tolerant Design Using Single and Multicriteria Genetic Algorithm Optimization. M.Sc. thesis, Cambridge, Massachusetts (1995)
36. Bandyopadhyay, S., Pal, S.K., Aruna, B.: Multiobjective GAs, Quantitative Indices, and Pattern Classification. IEEE Trans. Syst. Man Cybern. Syst.:B 34(5), 2088–2099 (2004)
37. Shannon, C.E.: A Mathematical Theory of Communication. Bell Systems Technical Journal 27, 379–423, 623–656 (1948)

IPESA-II: Improved Pareto Envelope-Based Selection Algorithm II

Miqing Li[1], Shengxiang Yang[2], Xiaohui Liu[1], and Kang Wang[3]

[1] Department of Information Systems and Computing, Brunel University,
Uxbridge, Middlesex, UB8 3PH, United Kingdom
{miqing.li,xiaohui.liu}@brunel.ac.uk
[2] School of Computer Science and Informatics, De Montfort University
Leicester LE1 9BH, United Kingdom
syang@dmu.ac.uk
[3] Institute of Information Engineering, Xiangtan University,
Xiangtan, 411105, China
kangkangfly@126.com

Abstract. The Pareto envelope-based selection algorithm II (PESA-II) is a classic evolutionary multiobjective optimization (EMO) algorithm that has been widely applied in many fields. One attractive characteristic of PESA-II is its grid-based fitness assignment strategy in environmental selection. In this paper, we propose an improved version of PESA-II, called IPESA-II. By introducing three improvements in environmental selection, the proposed algorithm attempts to enhance PESA-II in three aspects regarding the performance: convergence, uniformity, and extensity. From a series of experiments on two sets of well-known test problems, IPESA-II is found to significantly outperform PESA-II, and also be very competitive against five other representative EMO algorithms.

Keywords: Evolutionary multiobjective optimization, PESA-II algorithm, convergence, uniformity, extensity.

1 Introduction

Many real-world problems involve multiple competing objectives that should be considered simultaneously. Because of the conflicting nature of the objectives, in these multiobjective optimization problems (MOPs) there is usually no single optimal solution but rather a set of Pareto optimal solutions (or called Pareto front in the objective space). Evolutionary algorithms (EAs), which use natural selection as their search engine, have been recognized to be well suitable for MOPs due to their population-based property of achieving an approximation of the Pareto front in a single run. Over the past two decades, a number of state-of-the-art evolutionary multiobjective optimization (EMO) algorithms have been proposed [6], [2]. Generally speaking, these algorithms share three common goals—minimizing the distance from the resulting solutions to the Pareto front

R.C. Purshouse et al. (Eds.): EMO 2013, LNCS 7811, pp. 143–155, 2013.

(i.e., convergence), maintaining the uniform distribution of the solutions (i.e., uniformity), and maximizing the distribution range of the solutions along the Pareto front (i.e., extensity).

The Pareto Envelope-based Selection Algorithm II (PESA-II) [3] and its predecessor PESA [5], proposed by Corne et al. at the turn of the century, are a kind of classic EMO algorithms [2], which have been widely applied in many fields [25], [15]. The most attractive characteristic of them is their grid-based fitness assignment mechanisms that maintain diversity in both environmental selection and mating selection. From some comparative studies, they have been found to be competitive on some artificial test functions [3], [12] and real-world problems [10]; additionally, for problems with a high number of objectives, they have been shown to significantly outperform their contemporary competitors (e.g., nondominated sorting genetic algorithm II (NSGA-II) [9] and strength Pareto evolutionary algorithm 2 (SPEA2) [31]) regarding the ability of searching towards the Pareto front [17].

Despite the obvious usefulness of PESA-II and its predecessor, there is some room for improvement in this kind of algorithms. The first issue is concerned with the distribution uniformity. Since the grid environment of the archive set may need to be adjusted after the entry of each individual into the archive, the uniformity of the final population may be negatively affected to some extent (a detailed explanation will be given in Section 3.1). The second issue is related to the distribution extensity. Unlike some other classic EMO algorithms, such as NSGA-II and SPEA2, which explicitly or implicitly assign boundary solutions better fitness than internal solutions, PESA-II and its predecessor have no boundary solution preservation mechanism, which largely reduces the distribution range of their final solution set [31], [20]. The last issue is about the convergence. Although, compared with their contemporary algorithms, PESA-II and its predecessor appear to be competitive, they are beaten by some recent state-of-the-art algorithms, such as indicator-based evolutionary algorithm (IBEA) [30], ϵ-dominance [19] based multiobjective evolutionary algorithm (ϵ-MOEA) [7], S metric selection evolutionary multiobjective optimization algorithm (SMS-EMOA) [1], decomposition-based multiobjective evolutionary algorithm (MOEA/D) [28], and territory defining evolutionary algorithm (TDEA) [16].

In this paper, an improved PESA-II is presented, called IPESA-II. IPESA-II attempts to address the above issues by introducing three simple but effective improvements. The remainder of this paper is organized as follows. Section 2 reviews the original PESA-II. Section 3 contains the detailed description of our proposed improvements. The experimental results and discussions are given in Section 4. Finally, Section 5 indicates our conclusions and notes for further work.

2 PESA-II

PESA-II follows the standard principles of an EA, maintaining two populations: an internal population of fixed size, and an external population (i.e., archive set) of non-fixed but limited size. The internal population stores the new solutions generated from the archive set by variation operations, and the archive set only contains the nondominated solutions discovered during the search. A grid division of the objective space is introduced to maintain diversity in the algorithm. The number of solutions within a hyperbox is referred to as the density of the hyperbox, and is used to distinguish solutions in two crucial processes of an EMO algorithm: mating selection and environmental selection.

Unlike most EMO algorithms (including its predecessor PESA), the mating selection process of PESA-II is implemented in a region-based manner rather than in an individual-based manner. That is, a hyperbox is first selected and then the resulting individual for genetic operations is randomly chosen from the selected hyperbox—thus highly crowded hyperboxes do not contribute more individuals than less crowded ones.

In the environmental selection process, the candidate individuals in the internal population are inserted into the archive set one by one, thus the grid environment updated step by step. A candidate may enter the archive if it is nondominated within the internal population, and is not dominated by any current member of the archive. Once a candidate has entered the archive, corresponding adjustment of the archive and grid environment will be implemented. Firstly, the members in the archive which the candidate dominates are removed to ensure that only nondominated individuals exist in the archive. Secondly, the grid environment is checked to see whether its boundaries have changed with the addition or removal of individuals in the archive[1]. Finally, if the addition of a candidate renders the archive overfull, an arbitrary individual in the most crowded hyperbox will be removed.

3   IPESA-II

IPESA-II is proposed as an enhanced version of PESA-II that introduces three simple but effective improvements in the algorithm's environmental selection:

- Maintaining the archive after all individuals in the internal population have entered it, instead of doing step by step.
- Extending the distribution range of the solution set by keeping the boundary individuals.
- Improving the convergence of the solution set by removing the worst-performed individual in the most crowded hyperbox.

These three improvements, which focus on the performance of the solution set in terms of uniformity, extensity, and convergence respectively, are explained in the following sections.

---

[1] In PESA-II, the setting of grid environment is adaptive according to the archive set so that it can just envelop all the individuals in the archive set [5].

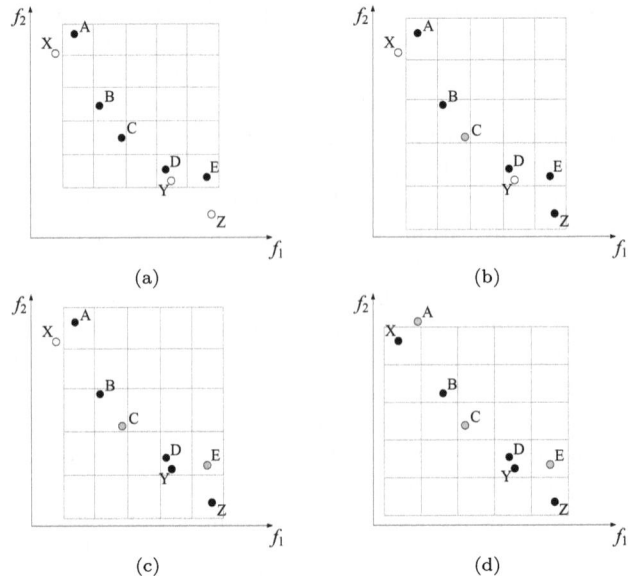

Fig. 1. An example of environmental selection in PESA-II, where individuals **A–E** are the current members in the archive set (the size of the archive set is five), and individuals **X–Z** are the candidates to be archived. (b)–(d) show the archiving process of the candidates in the order of **Z**, **Y**, and **X**. Black points correspond to the current individuals in the archive, hollow points stand for the candidates, and gray points denote the individuals removed from the archive.

3.1   Uniformity Improvement

As stated in the previous section, for the environmental selection process PESA-II adopts an incremental update mode. Once a candidate has entered the archive, the grid environment will need to be checked to see if it has changed. Any excess of the grid boundary or the upper limit of the archive size will lead to the adjustment (or even reconstruction) of the grid environment. This not only causes extra time consumption, but also affects the uniformity of the final archive set since different sequences that the candidates enter the archive result in different distributions. Figure 1 gives an example to illustrate this issue.

Supposing that individuals **A–E** are the current members in the archive set, and individuals **X–Z** are the candidates to be archived (cf. Fig. 1(a)). Figures 1(b)–(d) show the entry of the candidates into the archive in the order of **Z**, **Y**, and **X**. Here the archive size is set to five. Firstly, the candidate **Z** enters the archive and the grid environment is re-constructed. Accordingly, either **B** or **C** will be deleted since they reside in the most crowded hyperbox. Figure 1(b) shows the archiving result, assuming that individual **C** is removed. Then, individuals **E** and **A** will be eliminated in turn since they are dominated by **Y** and **Z** respectively (cf. Figs. 1(c) and (d)). The final archive set is formed by **X**, **B**, **D**, **Y**, and **Z**.

Note that the archive obtained above is not an ideal distribution result. A better archive is that individual **C** is preserved and either **D** or **Y** is removed, which, in fact, is the result of the entry of the candidates into the archive in the order of **X**, **Y**, and **Z**. In addition, if the enter order is **Y**, **Z**, and **X**, both the above results may occur with equal probability.

The above case clearly indicates that different entry orders of individuals into the archive may lead to different distribution results. In fact, it may be unreasonable to adjust the grid environment to maintain diversity whenever one candidate has entered the archive because the locations of the later candidates are unknown. Here, we make a simple improvement by only adjusting the grid environment after all candidates have entered the archive. Specifically, we first put all candidates into the archive, then delete the dominated individuals in the archive and construct the grid environment according to all individuals in the archive, and finally remove the individuals located in the current most crowded hyperbox one by one until the upper limit of the archive size is achieved. Let's revisit the example of Fig. 1 using the improved approach. Firstly, the archive set contains all eight individuals. Then, **A** and **E** are eliminated, and the grid environment is constructed by the rest individuals. Finally, either **D** or **Y** is removed, considering that they are located in the most crowded hyperbox.

3.2   Extensity Improvement

Extreme solutions (or boundary solutions) of a solution set refer to those solutions that have a maximum or minimum value on at least one objective for an MOP. Extreme solutions of a nondominated solution set are very important since they determine its distribution range. Many EMO algorithms preserve extreme solutions by explicitly or implicitly assigning them better fitness than internal solutions. However, in PESA-II, extreme solutions do not get special treatment—they could be eliminated with the same probability as other solutions, which leads the solution set of PESA-II to have a poorer distribution range than that of other algorithms [21].

In this paper, we preserve the extreme solutions of the archive set in the environmental selection. To be specific, extreme solutions will not be removed unless they are the only solution (or solutions) in the selected hyperbox. In other words, if a hyperbox has both extreme and internal solutions, the extreme solutions will always be eliminated after the internal ones.

3.3   Convergence Improvement

In environmental selection of PESA-II, when the archive set is overfull, a nondominated solution will be randomly eliminated in the most crowded hyperbox. Although this elimination strategy seems to be reasonable, it still has room for improvement in the context of convergence. The Pareto dominance relation is a qualitative metric of distinguishing individuals, which fails to give a quantitative difference of objective values among individuals. That is, two individuals are incomparable even if the former is largely superior to the latter in most of the objectives but only slightly inferior to the latter in one or a few objectives.

Inspired by ϵ-MOEA [7], we adopt a distance-based elimination strategy in IPESA-II, instead of the random elimination strategy in PESA-II. For the most crowded hyperbox, the proposed strategy removes the individual that is farthest away from the utopia point of the hyperbox (i.e., the best corner of the hyperbox) among all the individuals in the hyperbox. For example, consider individuals **D** and **Y** in Fig. 1. If one individual needs to be eliminated in the hyperbox where they are located, **D** will be removed since it is farther from the lower left corner of the hyperbox.

It deserves to be pointed out that the calculation of the distance is based on the normalized difference of objective values between individuals and the corresponding utopia point according to the size of a hyperbox, which is consistent with the adaptive grid environment. This means that the distance-based comparison strategy is unaffected by the range difference of different objective functions.

4   Experimental Results

This section validates the performance of IPESA-II by comparing it with PESA-II and five other well-known EMO algorithms: NSGA-II [9], SPEA2 [31], IBEA [30], ϵ-MOEA [7], and TDEA [16]. First, we briefly introduce the performance metrics and test problems used in the experiments. Second, the general experimental setting is assigned for the seven algorithms. Then, a comparative study between IPESA-II and PESA-II is presented, in terms of convergence, uniformity, and extensity. Finally, we further compare the comprehensive performance between IPESA-II and the other five algorithms.

4.1   Performance Metrics and Test Problems

This paper considers four performance metrics to examine the proposed algorithm. They are Generational Distance (GD) [26], Spacing (SP) [24], Maximum Spread (MS) [11], and Hypervolume (HV) [32]. The first three metrics are used to assess the convergence, uniformity, and spread of a solution set, respectively. The last one involves the comprehensive performance of the above three aspects.

The convergence metric GD calculates the average distance of the obtained solutions set away from the Pareto front, and a lower value is better. The uniformity metric SP measures the standard deviation of distance from each solution to its closest neighbor in the obtained set, and also a lower value is preferable. The spread metric MS is an improved version of the original Maximum Spread [32] and considers the distribution range of the Pareto front. The original MS, which measures the length of the diagonal of the hypercube formed by the extreme objective values in the obtained set, may be influenced heavily by convergence

Table 1. *div* setting in PESA-II and IPESA-II, and ϵ and τ settings in ϵ-MOEA and TDEA, respectively

	ZDT1	ZDT2	ZDT3	ZDT4	ZDT6	DTLZ1
div	50	50	70	50	50	9
ϵ	0.0076	0.0076	0.0030	0.0075	0.0065	0.0340
τ	0.0090	0.0090	0.0070	0.0075	0.0060	0.0600
	DTLZ2	DTLZ3	DTLZ4	DTLZ5	DTLZ6	DTLZ7
div	8	8	7	50	9	10
ϵ	0.0630	0.0630	0.0150	0.0050	0.0300	0.0500
τ	0.1050	0.0200	0.0400	0.0110	0.0250	0.0600

of an algorithm. For easing this effect, the new MS introduces the comparison between the the extreme values of the obtained solutions and the boundaries of the Pareto front. It takes the value between zero and one, and a larger value means a broader range of the set. The HV metric is a very popular quality metric due to its good properties, one of which is that it can assess the comprehensive performance on convergence, uniformity, and extensity. HV calculates the volume of the objective space between the obtained set and a given reference point. A larger HV value is preferable. More details of these metrics can be referred to in their original papers.

To benchmark the performance of the seven algorithms, two commonly used test problem suites, the ZDT problem family [29] and DTLZ problem family [8], are invoked. The former, including ZDT1, ZDT2, ZDT3, ZDT4, and ZDT6, is used to test the algorithms on bi-objective problems, and the latter, including DTLZ1 to DTLZ7, is employed to test the algorithms on tri-objective problems. All the test problems have been configured as in the original paper where they are described.

4.2   General Experimental Setting

All compared algorithms are given real-valued decision variables. A crossover probability $p_c = 1.0$ and a mutation probability $p_m = 1/l$ (where l is the number of decision variables) are used. The operators for crossover and mutation are simulated binary crossover (SBX) and polynomial mutation with the both distribution indexes 20 [6].

We independently run each algorithm 30 times for each test problem. The termination criterion of the algorithms is a predefined number of evaluations. For biobjective problems, the evaluation number is set to 25,000, and for problems with three objectives, the evaluation number is 30,000. The population size, for the generational algorithms IPESA-II, PESA-II, NSGA-II, SPEA2, and IBEA, is set to 100, and the archive is also maintained with the same size if existing. For the steady-state algorithms ϵ-MOEA and TDEA, the regular population size is set to 100.

In the calculation of HV, similar to [18], we select the integer point slightly larger than the worst value of each objective on the Pareto front of a problem as its reference point. As a consequence, the reference points for the ZDT and

Table 2. Comparison results between PESA-II and IPESA-II regarding GD, SP, and MS, where the top and bottom values in each cell correspond to PESA-II and IPESA-II, respectively, and the better mean is highlighted in boldface

Problem	GD	SP	MS
ZDT1	$2.048e{-}4_{(1.4e-4)}^{\dagger}$ $\mathbf{4.930e{-}5}_{(3.3e-5)}$	$6.771e{-}3_{(5.7e-4)}^{\dagger}$ $\mathbf{4.468e{-}3}_{(5.2e-4)}$	$9.762e{-}1_{(1.9e-2)}^{\dagger}$ $\mathbf{9.997e{-}1}_{(4.7e-4)}$
ZDT2	$2.040e{-}4_{(8.1e-5)}^{\dagger}$ $\mathbf{4.733e{-}5}_{(2.7e-5)}$	$6.793e{-}3_{(6.2e-4)}^{\dagger}$ $\mathbf{4.268e{-}3}_{(5.3e-4)}$	$9.779e{-}1_{(1.5e-2)}^{\dagger}$ $\mathbf{9.994e{-}1}_{(3.0e-4)}$
ZDT3	$1.216e{-}4_{(4.5e-5)}^{\dagger}$ $\mathbf{5.120e{-}5}_{(1.9e-5)}$	$\mathbf{8.181e{-}3}_{(1.1e-3)}$ $8.415e{-}3_{(7.7e-4)}$	$9.817e{-}1_{(1.0e-2)}^{\dagger}$ $\mathbf{9.995e{-}1}_{(5.7e-4)}$
ZDT4	$4.628e{-}2_{(8.7e-2)}^{\dagger}$ $\mathbf{2.805e{-}4}_{(1.5e-4)}$	$3.669e{-}1_{(7.5e-1)}^{\dagger}$ $\mathbf{6.202e{-}3}_{(7.0e-4)}$	$9.726e{-}1_{(3.6e-2)}^{\dagger}$ $\mathbf{9.964e{-}1}_{(1.2e-3)}$
ZDT6	$1.014e{-}2_{(9.5e-3)}^{\dagger}$ $\mathbf{3.472e{-}4}_{(7.7e-4)}$	$7.236e{-}2_{(8.0e-2)}^{\dagger}$ $\mathbf{5.112e{-}3}_{(7.3e-3)}$	$9.930e{-}1_{(3.8e-3)}^{\dagger}$ $\mathbf{9.978e{-}1}_{(2.9e-4)}$
DTLZ1	$6.746e{-}2_{(1.0e-1)}^{\dagger}$ $\mathbf{2.293e{-}2}_{(3.5e-2)}$	$1.463e{-}1_{(2.0e-1)}^{\dagger}$ $\mathbf{5.997e{-}2}_{(6.4e-2)}$	$9.862e{-}1_{(2.3e-2)}^{\dagger}$ $\mathbf{9.963e{-}1}_{(6.4e-3)}$
DTLZ2	$1.500e{-}3_{(2.0e-4)}^{\dagger}$ $\mathbf{8.186e{-}4}_{(2.3e-4)}$	$3.674e{-}2_{(3.4e-3)}^{\dagger}$ $\mathbf{2.905e{-}2}_{(4.6e-3)}$	$9.901e{-}1_{(5.4e-3)}^{\dagger}$ $\mathbf{9.998e{-}1}_{(4.9e-4)}$
DTLZ3	$1.039e{+}0_{(8.0e-1)}^{\dagger}$ $\mathbf{5.153e{-}1}_{(6.2e-1)}$	$3.065e{+}0_{(5.0e+0)}$ $\mathbf{1.172e{+}0}_{(2.6e+0)}$	$9.999e{-}1_{(1.8e-4)}$ $\mathbf{1.000e{+}0}_{(0.0e+0)}$
DTLZ4	$1.209e{-}3_{(5.7e-4)}^{\dagger}$ $\mathbf{7.622e{-}4}_{(2.7e-4)}$	$\mathbf{3.460e{-}2}_{(1.6e-2)}$ $3.489e{-}2_{(1.6e-2)}$	$8.313e{-}1_{(3.7e-1)}$ $\mathbf{9.031e{-}1}_{(2.0e-1)}$
DTLZ5	$5.703e{-}4_{(4.4e-5)}^{\dagger}$ $\mathbf{4.577e{-}4}_{(3.5e-5)}$	$7.385e{-}3_{(9.3e-4)}^{\dagger}$ $\mathbf{6.232e{-}3}_{(6.2e-4)}$	$9.894e{-}1_{(1.1e-2)}^{\dagger}$ $\mathbf{9.989e{-}1}_{(1.8e-3)}$
DTLZ6	$7.980e{-}2_{(1.9e-2)}^{\dagger}$ $\mathbf{4.961e{-}2}_{(1.1e-2)}$	$8.172e{-}2_{(4.5e-2)}^{\dagger}$ $\mathbf{5.577e{-}2}_{(3.1e-2)}$	$9.889e{-}1_{(3.0e-2)}^{\dagger}$ $\mathbf{1.000e{+}0}_{(1.5e-6)}$
DTLZ7	$3.315e{-}3_{(1.1e-3)}^{\dagger}$ $\mathbf{2.077e{-}3}_{(6.8e-4)}$	$\mathbf{4.206e{-}2}_{(1.4e-2)}$ $4.643e{-}2_{(7.5e-3)}$	$8.688e{-}1_{(1.4e-1)}^{\dagger}$ $\mathbf{9.979e{-}1}_{(2.4e-3)}$

"$\dagger$" indicates that the p-value of 58 degrees of freedom is significant at a 0.05 level of significance by a two-tailed t-test.

DTLZ problems are $(2, 2)$ and $(2, 2, 2)$ respectively, except $(1, 1, 1)$ for DTLZ1 and $(2, 2, 7)$ for DTLZ7.

PESA-II and IPESA-II require the user to set a grid division parameter (call *div* here). Base on some trail-and-error experiments, both algorithms can work well under the settings in Table 1. In addition, the algorithms ϵ-MOEA and TDEA require the user to set the size of a hyperbox in grid (i.e., ϵ and τ). In order to guarantee a fair comparison, we set them so that the archive of the two algorithms is approximately of the same size as that of the other algorithms (also given in Table 1).

For each problem, we have executed 30 independent runs. The values included in the tables of results are mean and standard deviation. By the central limit theorem, we assume that the sample means are normally distributed. Then, we test the following hypothesis at a 95% significance level to check whether there exists a statistically significant difference between the performances of IPESA-II (denoted as I) and its competitors (denoted as C).

$$H_0 : \mu^I = \mu^C \tag{1}$$

$$H_1 : \mu^I \neq \mu^C \tag{2}$$

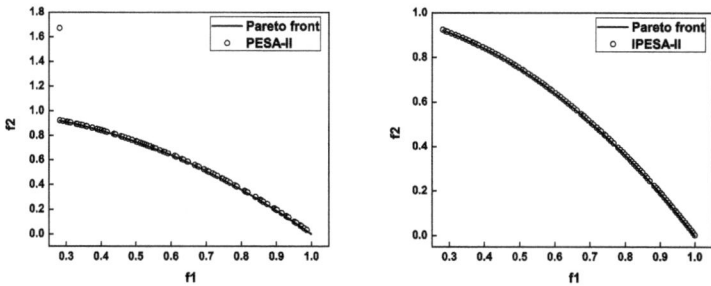

Fig. 2. The final solution set of PESA-II and IPESA-II on ZDT6

In the tables, the symbols "†" indicates that the p value of 58 degrees of freedom is significant at a 0.05 level of significance by a two-tailed t-test.

4.3 Comparison with PESA-II

Table 2 shows the comparison results between IPESA-II and PESA-II in terms of convergence, uniformity, and extensity. The values in the table correspond to mean and standard deviation. It is clear that IPESA-II performs significantly better than PESA-II on all the considered metrics. For the convergence metric GD, IPESA-II always achieves lower value with statistical significance on all the test problems, and especially on the problems ZDT4 and ZDT6, the result is improved by around two orders of magnitude. Figure 2 shows a typical distribution of the final solution set of the two algorithms on ZDT6. Clearly, of the solutions obtained by PESA-II, a boundary solution fails to approximate the Pareto front.

Concerning the uniformity metric SP, IPESA-II performs better for 9 out of the 12 problems. PESA-II has a better SP value on ZDT3, DTLZ4, and DTLZ7. Also, the difference on most of the test problems where IPESA-II performs better than PESA-II has statistical significance (8 out of the 9 problems), whereas for all problems where PESA-II outperforms IPESA-II, the difference has not statistical significance. In fact, for the problems DTLZ4 and DTLZ7, the solution set obtained by IPESA-II has better uniformity than that obtained by PESA-II (the typical distribution of the final solutions of the two algorithms on DTLZ7 is shown in Fig. 3). The misleading results on SP are due to the influence of the extensity of a solution set on this uniformity metric: a solution set with less distribution range may reach a lower SP value than another solution set even if the former performs worse in terms of uniformity [23]. Therefore, the solution set obtained by PESA-II, which fails to cover the whole Pareto front (cf. the MS results of PESA-II on DTLZ4 and DTLZ7 in the table), may result in better SP values.

Regarding the extensity of solution sets, IPESA-II also performs significantly better PESA-II, and with statistical significance for all the problems except one problem (DTLZ4). The MS results of IPESA-II in Table 2 very approximate the

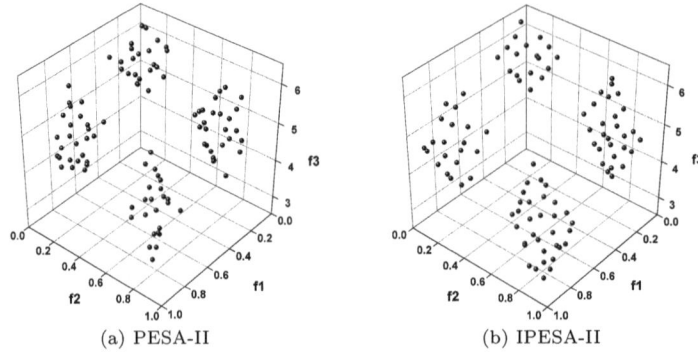

(a) PESA-II   (b) IPESA-II

Fig. 3. The final solution set of PESA-II and IPESA-II on DTLZ7

ideal value (i.e., MS = 1) on most of the tested problems. This means that the proposed algorithm can effectively extend the distribution range of a solution set.

4.4 Comparison with other EMO Algorithms

Table 3 shows the HV comparison results between IPESA-II and the other five algorithms, NSGA-II, SPEA2, IBEA, ϵ-MOEA, and TDEA. Apparently, IPESA-II is a very competitive algorithm, and achieves the best result in 6 out of the 12 problems. TDEA, IBEA, ϵ-MOEA, and SPEA2 perform the best in 2, 2, 1, and 1 out of the 12 problems, respectively. Concerning the statistical significance of results between IPESA-II and the other five algorithms, the number of the problems where the proposed algorithm outperforms NSGA-II, SPEA2, IBEA, ϵ-MOEA, and TDEA with statistical significance is 8, 7, 9, 8, and 7, respectively; the number of the problems where IPESA-II performs worse with statistical significance is 1, 2, 3, 3, and 1, respectively. This means that IPESA-II can provide a good tradeoff among convergence, uniformity, and extensity on most of the problems.

Considering the test problems with different numbers of objectives, IPESA-II performs significantly better than the other algorithms on biobjective problems. Its HV value achieves the best on all the problems except on ZDT6 where IBEA performs the best and IPESA-II takes the second place. For tri-objective problems, the advantage of IPESA-II over the other algorithms is not as clear as that for biobjective problems. The proposed algorithm obtains the best values in 2 out of the 7 problems. On some difficult problems, such as DTLZ3 and DTLZ6, the algorithm IBEA and ϵ-MOEA perform clearly better than IPESA-II. One of the reasons is that in contrast to the Pareto dominance relation, the hypervolume indicator used in IBEA and the ϵ dominance relation used in ϵ-MOEA can provide more selection pressure searching towards the Pareto front.

In addition, it is worth to point out that IPESA-II is capable of dealing with MOPs with many objectives. Although most grid-based EMO algorithms encounter difficulties (e.g., exponential increase of the computational cost) when

Table 3. HV comparison of the six EMO algorithms. The best mean for each problem is highlighted in boldface.

Problem	NSGA-II	SPEA2	IBEA	ϵ-MOEA	TDEA	IPESA-II
ZDT1	$3.659e+0^\dagger$ (4.1e-4)	$3.659e+0^\dagger$ (4.7e-4)	$3.659e+0^\dagger$ (8.0e-4)	$3.648e+0^\dagger$ (1.7e-3)	$3.657e+0^\dagger$ (1.8e-3)	**3.661e+0** (2.5e-4)
ZDT2	$3.325e+0^\dagger$ (5.8e-4)	$3.325e+0^\dagger$ (8.9e-4)	$3.324e+0^\dagger$ (2.8e-4)	$3.323e+0^\dagger$ (1.6e-3)	$3.319e+0^\dagger$ (3.1e-3)	**3.327e+0** (5.6e-4)
ZDT3	$4.812e+0^\dagger$ (4.7e-4)	$4.812e+0^\dagger$ (5.1e-4)	$4.806e+0^\dagger$ (2.1e-4)	$4.809e+0^\dagger$ (1.1e-3)	$4.804e+0^\dagger$ (3.6e-3)	**4.813e+0** (6.6e-4)
ZDT4	3.651e+0 (7.8e-3)	3.650e+0 (8.6e-3)	$2.482e+0^\dagger$ (2.1e-1)	$3.635e+0^\dagger$ (2.1e-2)	$3.631e+0^\dagger$ (4.1e-2)	**3.652e+0** (9.9e-3)
ZDT6	$3.022e+0^\dagger$ (2.7e-3)	$3.023e+0^\dagger$ (2.1e-3)	**3.037e+0†** (5.7e-4)	$3.028e+0^\dagger$ (2.2e-3)	$3.024e+0^\dagger$ (2.7e-3)	3.035e+0 (8.4e-4)
DTLZ1	$9.673e-1^\dagger$ (5.3e-4)	**9.721e-1†** (1.0e-3)	$8.977e-1^\dagger$ (1.0e-2)	$9.561e-1^\dagger$ (1.2e-2)	9.707e-1 (5.1e-4)	9.695e-1 (2.8e-3)
DTLZ2	$7.351e+0^\dagger$ (1.9e-2)	$7.391e+0^\dagger$ (6.9e-3)	$5.702e+0^\dagger$ (2.9e+0)	$7.389e+0^\dagger$ (9.0e-3)	**7.401e+0** (1.2e-2)	7.398e+0 (1.0e-2)
DTLZ3	6.919e-1 (2.3e+0)	6.369e-1 (1.7e+0)	$6.228e+0^\dagger$ (2.1e+0)	**6.326e+0†** (2.0e+0)	9.825e-1 (2.2e+0)	1.668e+00 (2.6e+0)
DTLZ4	6.888e+0 (5.9e-1)	6.912e+0 (5.5e-1)	$6.411e+0^\dagger$ (1.2e+0)	7.013e+0 (4.0e-1)	6.942e+0 (4.5e-1)	**7.128e+0** (6.3e-1)
DTLZ5	$6.099e+0^\dagger$ (6.8e-4)	$6.100e+0^\dagger$ (7.2e-4)	$6.088e+0^\dagger$ (1.1e-3)	$6.100e+0^\dagger$ (2.1e-3)	**6.101e+0†** (1.2e-3)	6.097e+0 (6.4e-3)
DTLZ6	4.098e+0 (2.6e-1)	$4.044e+0^\dagger$ (2.6e-1)	**5.974e+0†** (9.7e-2)	$5.292e+0^\dagger$ (1.2e-1)	$5.255e+0^\dagger$ (1.1e-1)	5.101e+0 (1.3e-1)
DTLZ7	$1.326e+1^\dagger$ (5.7e-2)	$1.333e+1^\dagger$ (2.0e-1)	$1.026e+1^\dagger$ (2.7e+0)	$1.312e+1^\dagger$ (2.0e-2)	$1.328e+1^\dagger$ (4.8e-2)	**1.340e+1** (5.0e-2)

"$\dagger$" indicates that the p-value of 58 degrees of freedom is significant at a 0.05 level of significance by a two-tailed t-test.

the number of objectives scales up [4], some effective measures can be adopted [22], [27]. Interested readers are referred to the three measures in [27] which make grid-based algorithms suitable in many-objective optimization.

5 Conclusion

In this paper, an enhanced version of PESA-II, called IPESA-II, is presented. IPESA-II introduces three operations to improve the performance of PESA-II in terms of convergence, uniformity, and extensity, respectively. Simulation experiments have been carried out by providing a detailed comparison with PESA-II and five well-known EMO algorithms, NSGA-II, SPEA2, IBEA, ϵ-MOEA and TDEA. The results reveal that IPESA-II has a clear advantage over PESA-II in finding a near-optimal, uniformly-distributed, and well-extended solution set. Additionally, IPESA-II is also found to be very competitive with the other five algorithms, considering the fact that it performs better than them on the majority of the tested problems.

Future work includes the investigation of IPESA-II on more test problems, e.g., on some MOPs with a high number of objectives [14], [13]. Moreover, a deeper understanding of the algorithm behavior may be obtained. In this context, the setting of the grid division parameter div and the separate contributions of the three simple improvements will be first investigated.

References

1. Beume, N., Naujoks, B., Emmerich, M.: SMS-EMOA: Multiobjective selection based on dominated hypervolume. European Journal of Operational Research 181(3), 1653–1669 (2007)
2. Coello, C.A.C., Veldhuizen, D.A.V., Lamont, G.B.: Evolutionary Algorithms for Solving Multi-Objective Problems, 2nd edn. Springer, Heidelberg (2007)
3. Corne, D.W., Jerram, N.R., Knowles, J.D., Oates, M.J.: PESA-II: Region-based selection in evolutionary multiobjective optimization. In: Proceedings of the Genetic and Evolutionary Computation Conference (GECCO 2001), pp. 283–290. Morgan Kaufmann, San Francisco (2001)
4. Corne, D.W., Knowles, J.D.: Techniques for highly multiobjective optimisation: some nondominated points are better than others. In: Proceedings of the 9th Annual Conference on Genetic and Evolutionary Computation (GECCO 2007), pp. 773–780 (2007)
5. Corne, D.W., Knowles, J.D., Oates, M.J.: The Pareto Envelope-Based Selection Algorithm for Multiobjective Optimization. In: Deb, K., Rudolph, G., Lutton, E., Merelo, J.J., Schoenauer, M., Schwefel, H.-P., Yao, X. (eds.) PPSN VI. LNCS, vol. 1917, pp. 839–848. Springer, Heidelberg (2000)
6. Deb, K.: Multi-Objective Optimization Using Evolutionary Algorithms. John Wiley, New York (2001)
7. Deb, K., Mohan, M., Mishra, S.: Evaluating the ϵ-dominated based multi-objective evolutionary algorithm for a quick computation of Pareto-optimal solutions. Evolutionary Computation 13(4), 501–525 (2005)
8. Deb, K., Thiele, L., Laumanns, M., Zitzler, E.: Scalable test problems for evolutionary multi-objective optimization. In: Abraham, A., Jain, L., Goldberg, R. (eds.) Evolutionary Multiobjective Optimization, pp. 105–145 (2005); Theoretical Advances and Applications
9. Deb, K., Pratap, A., Agarwal, S., Meyarivan, T.: A Fast and Elitist Multiobjective Genetic Algorithm: NSGA-II. IEEE Trans. Evol. Comput. 6(2), 182–197 (2002)
10. Diosan, L.: A multi-objective evolutionary approach to the portfolio optimization problem. In: Proceedings of the International Conference on Computational Intelligence for Modelling, Control and Automation, pp. 183–187 (2005)
11. Goh, C.K., Tan, K.C.: An Investigation on Noisy Environments in Evolutionary Multiobjective Optimization. IEEE Trans. Evol. Comput. 11(3), 354–381 (2007)
12. Hu, J., Seo, K., Fan, Z., Rosenberg, R.C., Goodman, E.D.: HEMO: A Sustainable Multi-objective Evolutionary Optimization Framework. In: Cantú-Paz, E., Foster, J.A., Deb, K., Davis, L., Roy, R., O'Reilly, U.-M., Beyer, H.-G., Kendall, G., Wilson, S.W., Harman, M., Wegener, J., Dasgupta, D., Potter, M.A., Schultz, A., Dowsland, K.A., Jonoska, N., Miller, J., Standish, R.K. (eds.) GECCO 2003, Part I. LNCS, vol. 2723, pp. 1029–1040. Springer, Heidelberg (2003)
13. Hughes, E.J.: Radar Waveform Optimisation as a Many-Objective Application Benchmark. In: Obayashi, S., Deb, K., Poloni, C., Hiroyasu, T., Murata, T. (eds.) EMO 2007. LNCS, vol. 4403, pp. 700–714. Springer, Heidelberg (2007)
14. Ishibuchi, H., Tsukamoto, N., Nojima, Y.: Evolutionary many-objective optimization: A short review. In: Proc. IEEE Congr. on Evol. Comput., pp. 2419–2426 (2008)
15. Jarvis, R.M., Rowe, W., Yaffe, N.R., O'Connor, R., Knowles, J.D., Blanch, E.W., Goodacre, R.: Multiobjective evolutionary optimisation for surface-enhanced Raman scattering. Analytical and Bioanalytical Chemistry 397(5), 1893–1901 (2010)

16. Karahan, I., Köksalan, M.: A territory defining multiobjective evolutionary algorithm and preference incorporation. IEEE Transactions on Evolutionary Computation 14(4), 636–664 (2010)
17. Khare, V., Yao, X., Deb, K.: Performance scaling of multi-objective evolutionary algorithms. In: Proc. 2nd Int. Conf. Evol. Multi-Criterion Optim., pp. 376–390 (2003)
18. Kukkonen, S., Deb, K.: A Fast and Effective Method for Pruning of Non-dominated Solutions in Many-Objective Problems. In: Runarsson, T.P., Beyer, H.-G., Burke, E.K., Merelo-Guervós, J.J., Whitley, L.D., Yao, X. (eds.) PPSN IX. LNCS, vol. 4193, pp. 553–562. Springer, Heidelberg (2006)
19. Laumanns, M., Thiele, L., Deb, K., Zitzler, E.: Combining convergence and diversity in evolutionary multi-objective optimization. Evolutionary Computation 10(3), 263–282 (2002)
20. Li, M., Yang, S., Zheng, J., Liu, X.: ETEA: A Euclidean minimum spanning tree-based evolutionary algorithm for multiobjective optimization. Evolutionary Computation (2012) (in press)
21. Li, M., Zheng, J.: Spread Assessment for Evolutionary Multi-Objective Optimization. In: Ehrgott, M., Fonseca, C.M., Gandibleux, X., Hao, J.-K., Sevaux, M. (eds.) EMO 2009. LNCS, vol. 5467, pp. 216–230. Springer, Heidelberg (2009)
22. Li, M., Zheng, J., Shen, R., Li, K., Yuan, Q.: A grid-based fitness strategy for evolutionary many-objective optimization. In: Proceedings of the Genetic and Evolutionary Computation Conference (GECCO 2010), pp. 463–470 (2010)
23. Li, M., Zheng, J., Xiao, G.: Uniformity Assessment for Evolutionary Multi-Objective Optimization. In: Proceedings of IEEE Congress on Evolutionary Computation (CEC 2008), Hongkong, pp. 625–632 (2008)
24. Schott, J.R.: Fault tolerant design using single and multicriteria genetic algorithm optimization. Master's thesis, Department of Aeronautics and Astronautics, Massachusetts Institute of Technology (1995)
25. Talaslioglu, T.: Multi-objective Design Optimization of Grillage Systems According to LRFDAISC. Advances in Civil Engineering (2011) (in press)
26. Van Veldhuizen, D.A., Lamont, G.B.: Evolutionary computation and convergence to a Pareto front. In: Koza, J.R. (ed.) Late Breaking Papers at the Genetic Programming Conference, pp. 221–228 (1998)
27. Yang, S., Li, M., Liu, X., Zheng, J.: A grid-based evolutionary algorithm for many-objective optimization. IEEE Transactions on Evolutionary Computation (2012) (in press)
28. Zhang, Q., Li, H.: MOEA/D: A multiobjective evolutionary algorithm based on decomposition. IEEE Transactions on Evolutionary Computation 11(6), 712–731 (2007)
29. Zitzler, E., Deb, K., Thiele, L.: Comparison of multiobjective evolutionary algorithms: Empirical results. Evolutionary Computation 8(2), 173–195 (2000)
30. Zitzler, E., Künzli, S.: Indicator-Based Selection in Multiobjective Search. In: Yao, X., Burke, E.K., Lozano, J.A., Smith, J., Merelo-Guervós, J.J., Bullinaria, J.A., Rowe, J.E., Tiño, P., Kabán, A., Schwefel, H.-P. (eds.) PPSN VIII. LNCS, vol. 3242, pp. 832–842. Springer, Heidelberg (2004)
31. Zitzler, E., Laumanns, M., Thiele, L.: SPEA2: Improving the strength Pareto evolutionary algorithm for multiobjective optimization. In: Evolutionary Methods for Design, Optimisation and Control, pp. 95–100. CIMNE, Barcelona (2002)
32. Zitzler, E., Thiele, L.: Multiobjective evolutionary algorithms: A comparative case study and the strength Pareto approach. IEEE Transactions on Evolutionary Computation 3(4), 257–271 (1999)

Theory and Algorithms for Finding Knees

Pradyumn Kumar Shukla, Marlon Alexander Braun, and Hartmut Schmeck

Institute AIFB, Karlsruhe Institute of Technology
Karlsruhe, D-76128, Germany

Abstract. A multi-objective optimization problem involves multiple and conflicting objectives. These conflicting objectives give rise to a set of Pareto-optimal solutions. However, not all the members of the Pareto-optimal set have equally nice properties. The classical concept of proper Pareto-optimality is a way of characterizing good Pareto-optimal solutions. In this paper, we metrize this concept to induce an ordering on the Pareto-optimal set. The use of this metric allows us to define a *proper knee* region, which contains solutions below a user-specified threshold metric. We theoretically analyze past definitions of knee points, and in particular, reformulate a commonly used nonlinear program, to achieve convergence results. Additionally, mathematical properties of the proper knee region are investigated. We also develop two multi-objective evolutionary algorithms towards finding proper knees and present simulation results on a number of test problems.

Keywords: knee regions, proper Pareto-optimality, ordering relations, evolutionary algorithms.

1 Introduction

Many industrial, engineering, and economic problems involve multiple and conflicting objectives. Due to a lack of a total order in two and higher dimensional real coordinate spaces, there are multiple solution concepts for these problems. The classical Pareto ordering [1] plays a central role in multi-objective optimization. This ordering is used to define a Pareto-optimal set and, the image of this set in the objective space is known as the efficient front. However, not all the members of the Pareto-optimal set (or of the efficient front) have equally nice properties. The trade-off (used to express preferences), for example, is dictated by the curvature of the efficient front and varies over the front. Starting with the classical work of Kuhn and Tucker [2], the concept of proper Pareto-optimality, is a way of characterizing *good* Pareto-optimal solutions. There are many notions of proper Pareto-optimality [3], depending on the nice property desired: trade-offs, geometrical properties, stability of the efficient front are a few of these.

In this paper, we metrize a well known trade-off based concept of proper Pareto-optimality by Geoffrion [4], and use it to induce an ordering relation on the Pareto-optimal set. This is a point-to-set relation instead of the usual point-to-point binary relation induced by the Pareto order. The motivation is to have a procedure to facilitate comparisons *within* the proper Pareto-optimal

R.C. Purshouse et al. (Eds.): EMO 2013, LNCS 7811, pp. 156–170, 2013.
© Springer-Verlag Berlin Heidelberg 2013

set, rather than just answer the question of whether a Pareto-optimal solution is good (proper) or not. The use of the metric also allows us to define a *proper knee region*, which contains solutions below a user-defined threshold metric. Mathematical properties of the proper knee region are investigated and special results for bi-objective problems are derived. We theoretically analyze past definitions of knees and relate them to the proper knee region. In particular, we reformulate a commonly used nonlinear program used in a direction based approach to find knees. This reformulation removes additional (hard) non-convex and nonlinear equality constraints that are inherent in the existing approach, thereby guaranteeing convergence to knees and making the formulation amenable to evolutionary algorithms.

Multi-objective evolutionary algorithms have an inherent population based advantage in obtaining a well diverse set of Pareto-optimal solutions [5]. In recent years, these approaches have also shown their applicability in finding a well-diverse set of user preferred regions. The point-to-set ordering relation is difficult to deal with classical methodologies. However, this can be efficiently integrated into the search mechanism of a population based algorithm. With this idea in mind, we develop two NSGA-II based multi-objective evolutionary algorithms towards finding proper knees. One of them uses an ordering to give the best proper knee point while the other uses a user defined threshold, together with a customized non-dominated sorting, to concentrate on the proper knee region. Simulation results are also presented on a number of test problems.

The paper is structured as follows. The next section presents various existing notions of knee solutions. The concept of proper knee region is described and theoretically evaluated in Section 3. Section 4 presents the two new algorithms and the fifth section presents extensive simulation results. Finally, conclusions as well as extensions which emanated from this study are presented at the end of this contribution.

2   Preliminaries and Existing Knee Notions

Let $\mathbf{f}(\mathbf{x}) : \mathbb{R}^n \to \mathbb{R}^m$ and $X \subseteq \mathbb{R}^n$ be given. The multi-objective optimization problem (MOP) and the definition of Pareto-optimality are as follows:

$$\min_{\mathbf{x}} \mathbf{f}(\mathbf{x}) := (f_1(\mathbf{x}), f_2(\mathbf{x}), \ldots, f_m(\mathbf{x})) \qquad \text{s.t. } \mathbf{x} \in X.$$

Definition 1. *A point* $\mathbf{x}^* \in X$ *is called* Pareto-optimal *if no* $\mathbf{x} \in X$ *exists so that* $f_i(\mathbf{x}) \le f_i(\mathbf{x}^*)$ *for all* $i = 1, \ldots, m$ *with strict inequality for at least one* i.

Usually, there is no point in X that maps to the ideal point $\mathbf{f}^*$ (that simultaneously minimizes all the objectives). Let X_p and $\mathcal{E} := \mathbf{f}(X_p)$ denote the set of Pareto-optimal and efficient solutions, respectively. A criticism of Pareto-optimality is that it allows unbounded trade-offs among the objectives. Hence, starting with the classical works of Kuhn and Tucker [2], and Geoffrion [4], various stronger optimality notions have been defined.

Definition 2. *A point* $\mathbf{x}^* \in X$ *is* Geoffrion proper Pareto-optimal *if there exists a number $M > 0$ such that for each $(\mathbf{x}, i) \in X \times \{1, 2, \ldots, m\}$ satisfying $f_i(\mathbf{x}) < f_i(\mathbf{x}^*)$, there exists an index j with $f_j(\mathbf{x}^*) < f_j(\mathbf{x})$ and $(f_i(\mathbf{x}^*) - f_i(\mathbf{x}))/(f_j(\mathbf{x}) - f_j(\mathbf{x}^*)) \leq M$.*

Proper Pareto-optimality is a way to reduce the set of Pareto-optimal solutions, by removing solutions having an unbounded trade-off between objective values. For a continuous problem, the set of Geoffrion proper Pareto-optimal solutions is known to be *dense* in the set of Pareto-optimal solutions [3]. This means that any Pareto-optimal point is either proper or is the limit of proper Pareto-optimal points. This only removes countably finite points from the Pareto-optimal set, and hence, Definition 2 is more a mathematical construct. One practical definition, using a general bound rather than just the existence of it, is defined in [6]. However we see that an M bound does exist in Definition 2, and we can use this to metrize the concept of proper Pareto-optimality in that smaller M values are better than larger ones. Formal definitions will be provided in the next section while here we look at a related concept of a *knee* solution.

Knees have been investigated for real world decision problems since a long time, although more for bi-objective problems [7,8,9]. Generally, a knee is a preferred solution, as moving away from these offers small gain in an objective but a large sacrifice in another objective. For bi-objective problems, the knee (if present) is visible in the plot of the efficient front (a curve), and various characterizations exist for higher dimensional problems [9,10,11,12,13,14,15].

A linear marginal utility function is used in [12] to search for a *utility knee* solution. For bi-objective problems, the same study also proposed a *reflex angle knee*, where the angle is calculated for a Pareto-optimal point and two of its neighbors. An extended notion of *bend angle knee*, using a left and right Pareto-optimal point, and an *edge-knee* is discussed in [9]. A *trade-off knee*, using a cone [9] is motivated from practical trade-off concepts for bi-objective problems.

All of the above notions except the utility knee have been investigated for bi-objective problems although these could be extended to higher dimensions. A *bulge knee* using the normal boundary intersection technique [16] is proposed by Das [13] to characterize knees in arbitrary dimensions. This uses a normal to the convex hull of individual objective function minimum (CHIM) as the search direction. Starting from an arbitrary point $\mathbf{u}$ on the CHIM hyperplane, a search is performed along the normal direction $\hat{\mathbf{n}}$, to find a boundary point $B(\mathbf{u})$ and is illustrated in Figure 1. The aim is to find a boundary point having a maximal distance to its corresponding point on CHIM. We see that the point $B(\mathbf{u}^*)$ is the bulge-knee point corresponding to the point $\mathbf{u}^*$. The following nonlinear program (assuming $\mathbf{f}$ to be nonlinear) is used for the search of the bulge knee point.

$$\max_{\mathbf{x}, t, \mathbf{u}} t \quad \text{subject to } \mathbf{u} + t\hat{\mathbf{n}} = \mathbf{f}(\mathbf{x}) - \mathbf{f}^*, \, \mathbf{x} \in X \,, \mathbf{u} \text{ lies on CHIM}, t \in \mathbb{R}. \quad (1)$$

The bulge knee depends on the curvature of the efficient front and has been discussed by many researches (see the citations of [13]). Although the nonlinear program (1) can be used to find the bulge knee point, the formulation (1) is neither theoretically sound (as no optimality can be guaranteed) nor a computationally favorable one. The reason is as follows. Let us assume that $\mathbf{f}$ is nonlinear and convex. Then, the equality constraint $\mathbf{u} + t\hat{\mathbf{n}} = \mathbf{f}(\mathbf{x}) - \mathbf{f}^*$ is nonlinear and *nonconvex* (as only *linear* equality constraints are convex). Hence, one has to solve a nonconvex problem to find the knee point, even if the original objectives are convex and convergence cannot be guaranteed using KKT solving based techniques (like Newton methods). The use of a nonlinear equality constraint is also a concern when using evolutionary algorithms. A simple way to get rid of the equality constraint is to use the following reformulation.

$$\max_{\mathbf{x},t,\mathbf{u}} t \quad \text{subject to } \mathbf{u} + t\hat{\mathbf{n}} \geq \mathbf{f}(\mathbf{x}) - \mathbf{f}^*, \, \mathbf{x} \in X, \mathbf{u} \text{ lies on CHIM}, t \in \mathbb{R}. \quad (2)$$

Theorem 1. *If $\mathbf{f}$ is convex, then the nonlinear program (2) is convex. Moreover, any KKT point $(\bar{t}, \bar{\mathbf{x}})$ of (2) is such that $\mathbf{f}(\bar{\mathbf{x}})$ is the bulge knee.*

Proof: The convex feasible region of (2) is illustrated in Figure 2 for $m = 2$, and the proof is based on the results from [13, Sections 4 and 7.2] and [17].   □

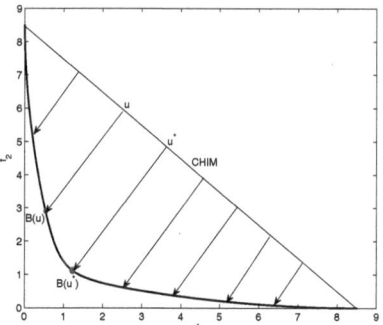 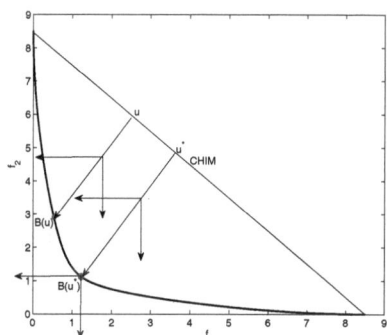

Fig. 1. Schematic showing the bulge-knee finding procedure   **Fig. 2.** Schematic showing the improved bulge-knee finding procedure

3   Proper Knee Region and a Theoretical Investigation

There is a parameter involved in many of the existing knee notions and these can be used to quantify the knee notion. There are some works [9,18] which focus on this. Quantization permits a theoretical investigation and has also algorithmic implications (discussed in the next section).

Following the ideas in [6], we could use the M in Definition 2 to reduce the set of Pareto-optimal points. Smaller M values lead to a greater reduction. For every Pareto-optimal point we could find a threshold value $\bar{M}$, such that the point is not a proper Pareto-optimal solution if $M < \bar{M}$ is chosen. In this spirit, we go on to propose the notion of *proper utility* of a point in a set and the notion of a *proper-knee*. For this, let the index set I, and, for any two vectors $\mathbf{u}, \mathbf{v} \in \mathbb{R}^m$, the index sets $I_<(\mathbf{u}, \mathbf{v})$, and $I_>(\mathbf{u}, \mathbf{v})$ be defined by

$$I := \{1, 2, \ldots, m\},$$
$$I_<(\mathbf{u}, \mathbf{v}) := \{i \in I | u_i < v_i\},$$
$$I_>(\mathbf{u}, \mathbf{v}) := \{i \in I | u_i > v_i\}.$$

Moreover, let $\mathbb{R}^m_+$ denote the non-negative orthant in $\mathbb{R}^m$.

Definition 3. *Let $\mathcal{S} \subseteq \mathbf{f}(X)$ be an arbitrary but fixed set. The* proper-utility $\mu(\mathbf{u}, \mathcal{S})$ *of a point $\mathbf{u} \in \mathcal{S}$ is defined by*

$$\mu(\mathbf{u}, \mathcal{S}) := \sup_{v \in \mathcal{S}} \sup_{i \in I_>(u,v)} \inf_{j \in I_<(u,v)} \frac{u_i - v_i}{v_j - u_j}. \tag{3}$$

A point $\mathbf{w}^* \in \mathcal{E}$ is said to be a proper-knee *if it is the solution of*

$$\inf \mu(\mathbf{u}, \mathcal{E}) \qquad \text{subject to } \mathbf{u} \in \mathcal{E}. \tag{4}$$

The set of all points satisfying $\mu(\mathbf{u}, \mathcal{S}) \leq \epsilon$, corresponding to a threshold $\epsilon > 0$, is termed as the ϵ proper-knee region.

Proper utility defined above can be used to quantify the goodness of a Geoffrion proper Pareto-optimal solution. Points with smaller utility values are preferred as for these the maximal of the trade-off term appearing in Definition 2 is small. Hence, if the classical Geoffrion notion is considered then the proper knee is the best solution. The proper knee region on the other hand consists of all points having an utility value less than a threshold (similar to the M bound in Definition 2). The ϵ threshold here can also be related to approximate minimizers of (4) and it can be used in situations where the knee is not *visible* (no bulge).

As optimization over open sets are involved in Definition 3, supremum and infimum are used. Hence, calculating exact proper utilities can be quite difficult using nonlinear optimization algorithms (both classical or nature-inspired ones, like evolutionary algorithms). Computing the set of approximate minima on the other hand is simpler, and one can use this to approximate the proper knee region. Application of population based algorithmic paradigms, like evolutionary algorithms make more sense for such tasks.

Definition 3 can be used for any type of problems, linear or nonlinear, discrete or continuous. In general, (3) involves solving a fractional program. For continuous problems, the proper utility $\mu(\mathbf{u}, \mathcal{S})$ is *not* defined if $\mathbf{u} \in \text{int}(\mathcal{S})$, where $\text{int}(\mathcal{S})$ is the interior of the set $\mathcal{S}$. This is not a concern as the efficient front lies on the boundary of the feasible subset of $\mathbb{R}^m$. In the discrete case, minimization and maximization are involved. This is also the case if one uses (finite) population based algorithms for solving continuous problems. The next result show that the proper utility of a point in set $\mathcal{S} \subset \mathbb{R}^m$ is the same as its utility in an appropriate subset of $\mathcal{S}$. For this, let $\mathcal{N}(\mathcal{S})$ be the nondominated set of $\mathcal{S}$ (i.e., the efficient front of a problem consisting of $\mathcal{S}$ as the feasible set).

Theorem 2. *Let X^d denote a discrete subset of the set X (which itself can be continuous or discrete) and let $\mathbf{u} \in X^d$. Then, it holds that*

$$\mu(\mathbf{u}, X^d) = \mu\left(\mathbf{u}, \mathcal{N}\left(X^d \setminus \left(\left(\{\mathbf{u}\} - \mathbb{R}_+^m \setminus \{0\}\right) \cup \left(\{\mathbf{u}\} + \mathbb{R}_+^m \setminus \{0\}\right)\right)\right)\right). \quad (5)$$

Proof: A rigorous mathematical proof is beyond the scope of this paper and we provide here a sketch of the proof. The sets $\{\mathbf{u}\} - \mathbb{R}_+^m \setminus \{0\}$ and $\{\mathbf{u}\} + \mathbb{R}_+^m \setminus \{0\}$ are the sets consisting of points that dominate and are dominated by the point $\mathbf{u}$, respectively. Hence, the set $\overline{X}^d := X^d \setminus \left(\left(\{\mathbf{u}\} - \mathbb{R}_+^m \setminus \{0\}\right) \cup \left(\{\mathbf{u}\} + \mathbb{R}_+^m \setminus \{0\}\right)\right)$ consists exactly of those points that are nondominated w.r.t. $\mathbf{u}$. This is the set that appears in problem (3). Note that $\mathbf{u} \in \overline{X}^d$ and hence $\mu(\mathbf{u}, \overline{X}^d)$ is defined. Studying the feasibility and infeasibility of an appropriate system of inequalities (based on [6, Lemma 1] and the results from [19, Chapter 4] and [20]), we can show that the set of minimizers of $\mu(\mathbf{u}, X^d)$ (as X^d is a discrete set) is a subset of the set of nondominated points of the set $\overline{X}^d$. $\quad\square$

Remark 1. *Theorem 2 says that in order to calculate the proper utility of a point $\mathbf{u}$, we need to consider only the nondominated set of the set of solutions that are nondominated with $\mathbf{u}$. This is of algorithmic importance and is used later in the ranking mechanisms of the algorithms .*

Remark 2. *We conjecture that Theorem 2 can also be extended for the case when X^d contains an uncountably infinite number of points. Apart from pure theoretical reasons this might be of importance to investigate convergence behavior of algorithms or archiving schemes that allow the size to grow with generations.*

Theorem 3. *Proper utility is a point-to-set order relation and is not Pareto-compliant. In other words, if $\mathbf{v} - \mathbf{u} \in \mathbb{R}_+^m \setminus \{0\}$ (i.e., if $\mathbf{u}$ Pareto dominates $\mathbf{v}$), then it does not necessarily hold that $\mu(\mathbf{u}, \mathcal{S}) \leq \mu(\mathbf{v}, \mathcal{S})$, for a given set $\mathcal{S}$.*

Proof: We can easily find a contradiction by considering the following situation. Let $\mathcal{S}$ be the set consisting of the following points: $\mathbf{v} = (0,0)$, $\mathbf{u} = (-\frac{1}{2}, -\frac{1}{2})$, $\mathbf{v}^1 = (-\frac{1}{2}, 4)$, $\mathbf{v}^2 = (4, -\frac{1}{2})$, $\mathbf{u}^1 = (-\frac{3}{4}, -\frac{1}{4})$, and $\mathbf{u}^2 = (-\frac{1}{4}, -\frac{3}{4})$. We see that $\mathbf{u}$ Pareto dominates $\mathbf{v}$ but $\mu(\mathbf{u}, \mathcal{S}) = 1 > \mu(\mathbf{v}, \mathcal{S}) = \frac{1}{8}$ $\quad\square$

Remark 3. *Theorem 3 shows an unfavorable property of the proper utility relation in that it is not compliant with the Pareto order. Algorithmically, this means that we have to impose the Pareto order on top of the proper utility order.*

The proper utility of a point $\mathbf{u}$ depends on the set of points that are nondominated wrt. $\mathbf{u}$ and hence can be only used to differentiate between a set of nondominated points rather than all the points.

In general, the notion of a proper knee is not equivalent to any of the existing knee notions. In contrast to utility knee there is no linear marginal utility function that can be used to characterize a proper knee. The bi-objective notions of the reflex, bend angle, and trade-off based knees on the other hand could be related to the proper knee, under some conditions (like bend angle being calculated from all the neighbors instead of just two). Similarly, the bulge knee is equivalent to the proper knee for bi-objective convex problems under appropriate conditions. Some of these characterizations are discussed next.

Theorem 4. *If $m \geq 3$, then the proper knee is not equivalent to the utility knee, any cone angle based extension of the reflex knee, or any polyhedral or convex cone based extension of the bi-objective trade-off knee.*

Proof: The main idea of the proof (a rigorous mathematical proof is beyond the scope) is as follows. Any polyhedral cone uses a fixed trade-off on all or some pairs of the objectives. Assuming any polyhedral cone based extension of the notion of trade-off knee, we can construct counterexamples showing the difference between a proper knee. The essential element comes here from Definition 2, where the existence of *one* j is sufficient to bound the trade-off. In the case of a general convex cone $\mathcal{C}$, we can construct an inner and outer polyhedral cone based approximation of $\mathcal{C}$ and use these to come up with counterexamples.   □

Theorem 5. *Let $\mathbf{f} \in \mathbb{R}^2$ and let $\mathcal{E}$ be considered as an implicit function of f_1 and f_2. Moreover, let u^L, u^R, u be the leftmost, rightmost, and an arbitrary efficient points. If $\mathbf{f}$ is differentiable and convex then*

$$\mu(\mathbf{u}, \mathcal{E}) = \max\left\{ \frac{d\mathcal{E}}{df_1}\bigg|_{\mathbf{u}}, \frac{d\mathcal{E}}{df_2}\bigg|_{\mathbf{u}} \right\} \tag{6}$$

holds. If $\mathbf{f}$ is concave, then $\mu(\mathbf{u}, \mathcal{E}) = \max\left\{ \dfrac{u_1 - u_1^L}{u_2^L - u_2}, \dfrac{u_2 - u_2^R}{u_1^R - u_1} \right\}$

Proof: For a convex bi-objective problem the efficient front $\mathcal{E}$ is a convex curve. Let the convex function g, denote the efficient front in an explicit form, i.e., let $f_2 = g(f_1)$ be the efficient front curve. For convex functions of one variable it is known that the function $r(t_1, t_2) = \dfrac{g(t_1) - g(t_2)}{t_1 - t_2}$ is monotonically non-decreasing in t_1, for a fixed t_2 and vice versa [21]. Using this we can show that, the maximum of the ratio in (3) is the maximum of the slope of the efficient curve at $\mathbf{u}$ and its inverse. Hence, (6) follows. If $\mathbf{f}$ is concave, the results follows by a similar argument (and the maximum occurs at one of the end points).   □

Even if we assume that $\mathbf{f}$ is nonsmooth, we can use subdifferential techniques from [22] and get a result similar to that in Theorem 5.

Corollary 1. *If $\mathbf{f} \in \mathbb{R}^2$ is differentiable, convex, and normalized (so that the nadir point is $(1,1)^\top$), then, the bulge knee point is also the proper knee point.*

Proof: The conditions imply that Theorem 5 holds, and if the functions are additionally normalized, then the slope of the efficient front curve at the proper knee point is -1. The rest of the proof follows by elementary convex analysis techniques and by noting that the bulge knee also has the same slope.   □

For two dimensional problems, Corollary 1 show the equivalence of the bulge knee to the proper knee. However, the relations are not clear in higher dimensions. We note that Theorem 5 and Corollary 1 also hold for a convex efficient curve (instead of $\mathbf{f}$ being convex, which is a stronger condition).

4   Algorithms

Using the theoretical results of the last section we develop two algorithms for finding proper knee regions. The algorithms use the algorithmic framework of the nondominated sorting based genetic algorithm NSGA-II [23]. Non-dominated sorting is an elite ranking scheme, the use of which can result in optimal solutions even when a random tournament selection (we call this as nRandom algorithm) is used. Figure 3 shows the sample run of the nRandom algorithm on ZDT1.

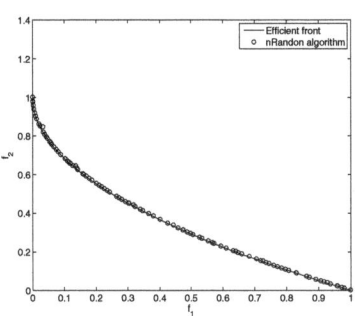
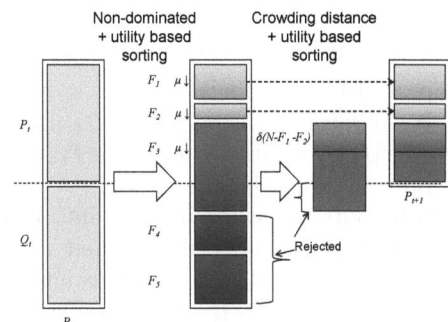

Fig. 3. Simulation results of nRandom algorithm on ZDT1

Fig. 4. Schematic showing creation of new parent population in PKEA

4.1   Proper Knee Based Evolutionary Algorithm (PKEA)

The (PKEA) algorithm uses two parameters to find the proper knee region. The first of them is a lower bound on the utility value, this is analogous to ϵ in Definition 3 and is called as M instead (similar to Definition 2). A second parameter $\delta \in [0, 1]$ is employed to ensure diversity in the population.

We assume a population size of N. At any generation t, let P_t and Q_t denote the parent and the offspring populations, respectively. Let R_t be the combined population, i.e., $R_t = P_t \cup Q_t$. The ranking of the solutions in the set R_t and creation of P_{t+1} is defined next.

PKEA Ranking

Step 1: Perform a non-dominated sorting to R_t and identify different fronts: $\mathcal{F}_i$, $i = 1, 2, \ldots$, etc.

Step 2: For an arbitrary solution $\mathbf{u} \in R_t$, such that $\mathbf{u} \in \mathcal{F}_k$ for some $k \in \mathbb{N}$ ($\mathcal{F}_k$ is the k-th non-dominated front in which $\mathbf{u}$ lies), let $\mu(\mathbf{u}) > 0$ be defined by

$$\mu(\mathbf{u}) = \max \left\{ \max_{\mathbf{v} \in \mathcal{F}_k} \max_{i \in I_>(\mathbf{u},\mathbf{v})} \min_{j \in I_<(\mathbf{u},\mathbf{v})} \frac{u_i - v_i}{v_j - u_j}, M \right\}. \tag{7}$$

The normalized utility $\bar{\mu}(\mathbf{u})$ is defined by

$$\bar{\mu}(\mathbf{u}) := \frac{\mu(\mathbf{u})}{\max_{\mathbf{w} \in \mathcal{F}_k} \mu(\mathbf{w}) + 1}. \tag{8}$$

Step 3: The rank of $\mathbf{u}$ is $k + \bar{\mu}(\mathbf{u})$.

Steps 1-3 are used to rank the entire combined population R_t. The normalized utility (8) is used here to make sure that the ranks are Pareto-compatible. Guided by Theorems 2 and 3, the utility of a point is computed based on the non-dominated front in which it lies.

Creation of New Parent Population Using PKEA

Step a: Set new population $P_{t+1} = \emptyset$. Set a counter i=1. While $|P_{t+1}| + \mathcal{F}_i <$ N, perform $P_{t+1} = P_{t+1} \cup \mathcal{F}_i$ and $i = i + 1$. Now, we need to include $N - |P_{t+1}|$ solutions more into P_{t+1} from $\mathcal{F}_l$ (hence, the l-th front cannot be fully accommodated).

Step b: Sort the solutions of $\mathcal{F}_l$ into increasing order of their ranks (or equivalently, increasing order of their μ values as k is the same). Take the first $\bar{N} := \lceil \delta(N - |P_{t+1}|) \rceil$ solutions into P_{t+1} from the sorted list.

Step c: The $N - |P_{t+1}| - \bar{N}$ remaining solutions are taken considering the crowding distance values (in $\mathcal{F}_l$).

Figure 4 shows the schematic of creating the new parent population. The non-dominated sorting gives five fronts, and each of the fronts are additionally sorted based on (normalized) utility values. The third front cannot be fully accommodated and here, both the utility and the crowding distance values are used. The next offspring population is created from P_{t+1} using a lexicographic (rank, crowding distance) tournament selection, crossover and mutation operators.

4.2   Proper Utility Based NSGA-II (UNSGA-II)

This algorithm changes the domination definition from Pareto-domination to a *utility* based domination in the following way.

Definition 1. *A solution* $\mathbf{u} \in \mathbb{R}^m$ *U-dominates a solution* $\mathbf{v} \in \mathbb{R}^m$ *if either* $\mathbf{u}$ *Pareto-dominates* $\mathbf{v}$*, or if* $\mathbf{u}$ *and* $\mathbf{v}$ *are nondominated and additionally*

$$\max_{i \in I_>(\mathbf{u},\mathbf{v})} \min_{j \in I_<(\mathbf{u},\mathbf{v})} \frac{u_i - v_i}{v_j - u_j} < \max_{i \in I_>(\mathbf{v},\mathbf{u})} \min_{j \in I_<(\mathbf{v},\mathbf{u})} \frac{v_i - u_i}{u_j - v_j} \tag{9}$$

holds.

The U-domination is used in UNSGA-II instead of the usual Pareto-domination and the rest of the algorithm is the same as the usual NSGA-II. This domination has a global effect and can be seen as the utility based definition (3) if the population consists of only two solutions $\mathbf{u}$ and $\mathbf{v}$ (hence inf and sup are replaced by min and max, respectively).

U-domination is a binary relation and its axiomatic properties (like total order, transitivity and antisymmetry) are currently being investigated. Hence, checking of U-domination can be done for every pair of points.

5 Simulation Results

Both the algorithms described in the last section were tested on various artificial test problems. The test problems include problems from various test suits proposed over the last years. Specifically, we included problems from the ZDT suite [24], the DTLZ family (three dimensional) [25], and from the *Knee suite* (DEB2DK and DEB3DK test problems) [12]. For all problems solved, we use a population of size 100 and set the maximum number of function evaluations as 20,000 (i.e., 200 generations). Moreover, we use a standard real-parameter SBX and polynomial mutation operator with $\eta_c = 15$ and $\eta_m = 20$, respectively [24]. Both PKEA and UNSGA-II are written using the jMetal framework [26] and their source codes are available on request. We ran the algorithms on all the test problems for 51 times, (PKEA was run for all M and δ combinations discussed in this paper). The proper utilities of the efficient points are calculated either numerically from the efficient front or analytically.

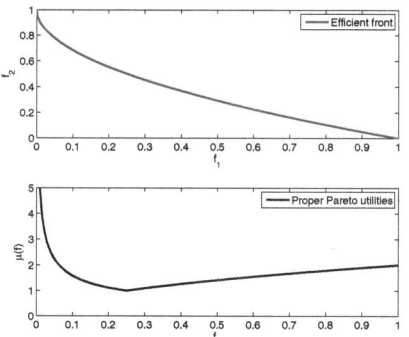

Fig. 5. Convex efficient front of ZDT1 (top) and its proper utility (bottom)

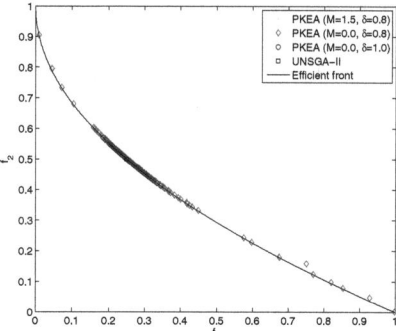

Fig. 6. Sample run of the algorithms on ZDT1

Figure 5 show the efficient front and its proper utilities plotted against f_1, for ZDT1. The efficient front of ZDT1 is given by the curve $f_2 = 1 - \sqrt{f_1}$ and from the proof of Corollary 1, we obtain that proper knee point occurs where the slope $\frac{-1}{2\sqrt{f_1}} = -1$. Hence, the proper knee is the point $(\frac{1}{4}, \frac{1}{2})$. ZDT1 is a well studied problem but it has no apparent bend knees. A bend knee is characterized by a maxima of the second derivative (see [9, Definition 6.1]), which occurs at $f_1 = 0$, in this case. However, this point is the worst point based on proper utility. This is an interesting difference between existing knee notions. The sample run of the algorithms in Figure 6 shows that UNSGA-II is able to find this knee, while PKEA founds the proper knee region corresponding to different thresholds.

It is to be noted that the aims of PKEA and UNSGA-II are not the same and hence a comparison between them is not justified. UNSGA-II searches for the proper knee point while PKEA is designed to find a well-distributed proper knee

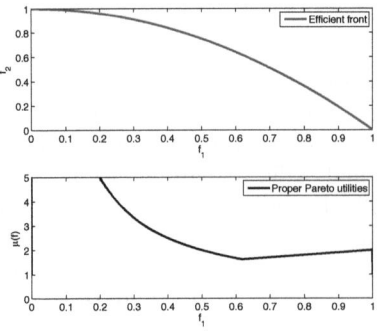
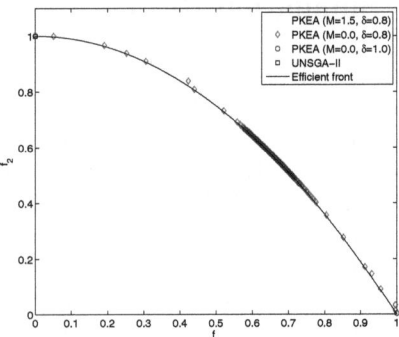

Fig. 7. Concave efficient front of ZDT2 (top) and its proper utility (bottom)

Fig. 8. Sample run of the algorithms on ZDT2

region. A value of $\delta < 1$ in PKEA ensures that solutions lying in a less crowded region are also carried over to the next generation. There is no explicit diversity mechanism in UNSGA-II (although one could think of such a technique) as the idea was to investigate the effect of using the proper utility based domination in a global sense. This lead (successfully), to a faster and accurate convergence to the proper knee point.

Figure 7 show the concave efficient front and its proper utilities plotted against f_1, for ZDT2. Using Theorem 5 we can easily obtain that $\mu(\mathbf{u}) = \max\left(\frac{1}{u_1}, 1 + u_1\right)$ for all $u_1 \in (0, 1)$, and that $\mu(0, 1) = \mu(1, 0) = 1$. Hence, the end points are the proper knees and moreover, the point $\left(\frac{\sqrt{5}-1}{2}, \frac{\sqrt{5}-1}{2}\right)$ is a local minimum of the proper utility. These results are in accordance with the simulation results shown in Figure 8 where PKEA founds a biased distribution of points and UNSGA-II finds the proper knee.

Next, we consider various knee problems from [12]. These problems are tunable in the number of knees. Figure 9 shows the convex efficient front and its proper utilities plotted against f_1, for the DO2DK problem with $k = 1$ and $s = 0$. We see one knee where the utility is minimal (equals 1 from Corollary 1). However, as soon as we increase the number of knees by changing the parameters to $k = 4$ and $s = 1$, we see from Figure 10, that the proper utilities has multiple local minima. These correspond to the (visible) knees in Figure 10 (top) with the global minimum corresponding to the knee lying between $f_1 = 1.5$ and $f_1 = 2.0$.

The concept of proper utilities defines the *strength* of a knee and the global minimum of the proper utility function is the strongest knee. Additional robustness measures on the proper utility function could also be defined, similar to ones in [27]. Doing this focusses the search on *robust* proper knees, as we might expect that the proper knee in Figure 10 is less robust than the local minimum close to $f_1 = 1.0$. From the sample runs of the algorithms in Figures 11 and 12 show that all the algorithms are able to find the proper knee.

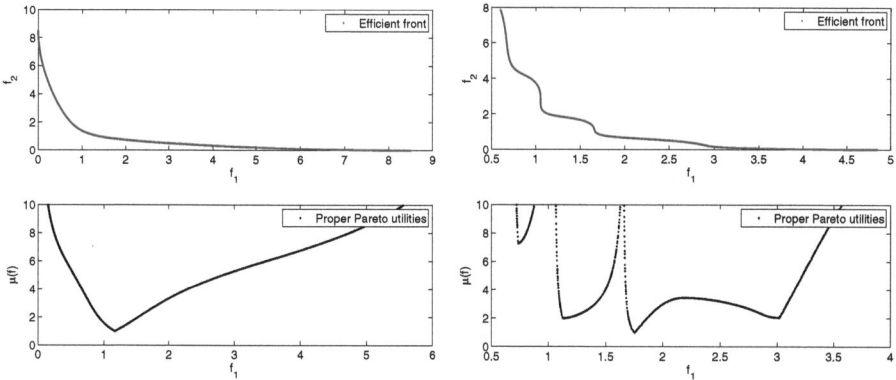

Fig. 9. Efficient front of DO2DK ($k = 1, s = 0$) (top) and its proper utility (bottom)

Fig. 10. Efficient front of DO2DK ($k = 4, s = 1$) (top) and its proper utility (bottom)

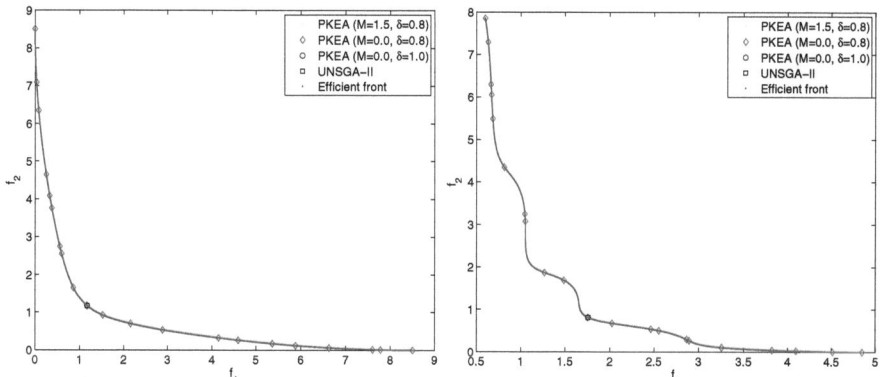

Fig. 11. Sample run of the algorithms on DO2DK ($k = 1, s = 0$)

Fig. 12. Sample run of the algorithms on DO2DK ($k = 4, s = 1$)

Next, we present simulation results on a couple of three dimensional problems. As discussed earlier, for three and higher dimensional problems, we do not know the theoretical relationships with the bulge knee and analytical calculation of the proper knee is cumbersome. Let us first consider the test problem DEB3DK with $k = 2$. The efficient front of this problem looks like a butterfly and there are three *bulges* that can be seen if we plot and rotate in three dimensions. The set of all points having a utility value less than 4 are plotted in Figure 13 and these correspond to the three bulges. Figure 14 shows the simulation run of PKEA (with $M = 4$ and $\delta = 0.8$) and we can see that a diverse set of the three bulge regions are found.

Finally, we consider another three dimensional problem DTLZ7. The efficient front of this problems consist of four disconnected surfaces (see Figure 15). Each

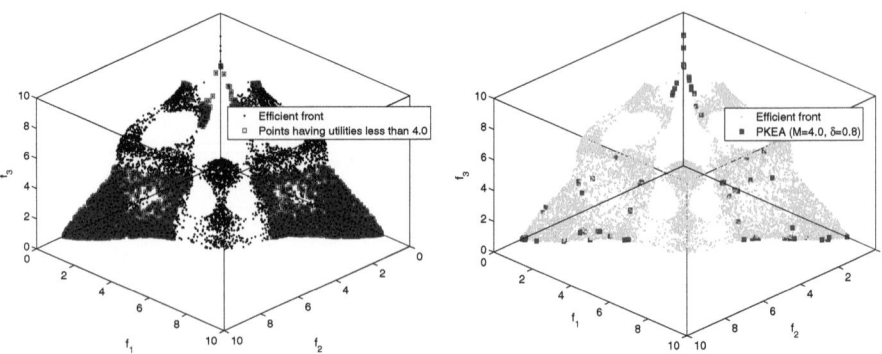

Fig. 13. Efficient front of DEB3DK with $k = 2$. The points in blue are the ones have utilities less than 4.

Fig. 14. Sample run of PKEA on DEB3DK with $k = 2$

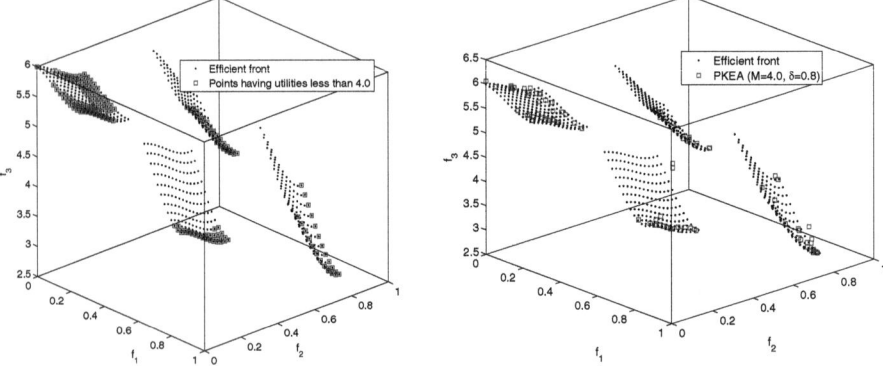

Fig. 15. Efficient front of DTLZ7. The points in blue are the ones have utilities less than 4.

Fig. 16. Sample run of PKEA on DTLZ7

of these have different curvature (a simple look reveals different angles of the four (almost linear) surfaces, with the $f_3 = 0$ plane. Figure 15 also shows the set of all points (in blue) having a utility value less than 4. More blue points can be seen in the patch having the largest angle with $f_3 = 0$ plane as in the other surfaces the gain to loss trade-offs are higher. It might be interesting to relate this to the maximum bulge corresponding to the CHIM plane (used in formulation (2)). Figure 16 shows the simulation run of PKEA (with $M = 4$ and $\delta = 0.8$) and we can see that a diverse set of solutions (with square markers) are found in all the four patches. During out simulations, we also found that a larger δ results in more solutions having a better (smaller) utility value.

6   Conclusions and Future Works

This study presented a way to compare and characterize good Pareto-optimal solutions. For this, we took the classical concept of proper Pareto-optimality and defined an ordering relation. This lead to a point-to-set binary relation and to the notion of proper utility. This notion was used to define a proper knee region. Moreover, we reviewed various knee definitions and reformulated a nonlinear program used to find a bulge-knee. For convex problems, this changed a non-convex equality constraint to a favorable convex equality constraint.

A theoretical result for computing proper utilities was presented. It was also shown that the point-to-set relation is not Pareto compatible. Additional theoretical results related existing knee definitions to that based on proper Pareto-optimality. A characterization of proper knee for bi-objective convex and concave problems were also presented. We used proper utilities to develop two population based algorithms. This first one used a threshold to integrate a diversity measure, which was not inherent in proper utility. The second algorithm used a new domination definition to sort the entire population. This leads to a faster convergence towards the proper knee. Theoretical results were used to justify the ranking schemes of both the algorithms. Towards the end, we presented simulation results on a number of two and three dimensional test problems.

Many new theoretical and algorithmic avenues are opened by this study. Properties of the point-to-set ordering relations could be explored (not only in the context of proper knee). Transitivity, antisymmetry and other axiomatic properties of the utility based domination needs to be studied. Lack of transitivity would prohibit any fast non-domination based sorting scheme, like the ones presented in [24,28]. Relationships between proper and bulge-knees, and also the extended notions from [9]) is worthwhile for higher dimensional problems. It would be also interesting to investigate other proper knee notions e.g., based on stability properties of the efficient front [3].

References

1. Pareto, V.: Cours d'Economie Politique. Droz, Genève (1896)
2. Kuhn, H.W., Tucker, A.W.: Nonlinear programming. In: Proceedings of the Second Berkeley Symposium on Mathematical Statistics and Probability, pp. 481–492. University of California Press, Berkeley (1951)
3. Makarov, E.K., Rachkovski, N.N.: Unified representation of proper efficiency by means of dilating cones. J. Optim. Theory Appl. 101(1), 141–165 (1999)
4. Geoffrion, A.M.: Proper efficiency and the theory of vector maximization. Journal of Mathematical Analysis and Applications 22, 618–630 (1968)
5. Shukla, P.K., Deb, K.: On finding multiple pareto-optimal solutions using classical and evolutionary generating methods. European J. Oper. Res. 181(2), 1630–1652 (2007)
6. Shukla, P.K., Hirsch, C., Schmeck, H.: A Framework for Incorporating Trade-Off Information Using Multi-Objective Evolutionary Algorithms. In: Schaefer, R., Cotta, C., Kołodziej, J., Rudolph, G. (eds.) PPSN XI, Part II. LNCS, vol. 6239, pp. 131–140. Springer, Heidelberg (2010)
7. Merrill, H.: Cogeneration - a strategic evaluation. IEEE Transactions on Power Apparatus and Systems 102(2), 463–471 (1983)

8. Burke, W., Merrill, H., Schweppe, F., Lovell, B., McCoy, M., Monohon, S.: Trade off methods in system planning. IEEE Transactions on Power Systems 3(3), 1284–1290 (1988)
9. Deb, K., Gupta, S.: Understanding knee points in bicriteria problems and their implications as preferred solution principles. Engineering Optimization 43(11), 1175–1204 (2011)
10. Mattson, C.A., Mullur, A.A., Messac, A.: Smart pareto filter: obtaining a minimal representation of multiobjective design space. Engineering Optimization 36(6), 721–740 (2004)
11. Rachmawati, L., Srinivasan, D.: Multiobjective evolutionary algorithm with controllable focus on the knees of the pareto front. IEEE Transactions on Evolutionary Computation 13(4), 810–824 (2009)
12. Branke, J., Deb, K., Dierolf, H., Osswald, M.: Finding Knees in Multi-objective Optimization. In: Yao, X., Burke, E.K., Lozano, J.A., Smith, J., Merelo-Guervós, J.J., Bullinaria, J.A., Rowe, J.E., Tiňo, P., Kabán, A., Schwefel, H.-P. (eds.) PPSN VIII. LNCS, vol. 3242, pp. 722–731. Springer, Heidelberg (2004)
13. Das, I.: On characterizing the "knee" of the Pareto curve based on Normal-Boundary Intersection. Structural Optimization 18(2-3), 107–115 (1999)
14. Miettinen, K.: Nonlinear Multiobjective Optimization. Kluwer, Boston (1999)
15. Bechikh, S., Ben Said, L., Ghédira, K.: Searching for knee regions in multi-objective optimization using mobile reference points. In: Proceedings of the 2010 ACM Symposium on Applied Computing, SAC 2010, pp. 1118–1125. ACM, New York (2010)
16. Das, I., Dennis, J.: Normal-boundary intersection: A new method for generating the Pareto surface in nonlinear multicriteria optimization problems. SIAM Journal of Optimization 8(3), 631–657 (1998)
17. Gembicki, F., Haimes, Y.: Approach to performance and sensitivity multiobjective optimization: The goal attainment method. IEEE Transactions on Automatic Control 20(6), 769–771 (1975)
18. Schütze, O., Laumanns, M., Coello Coello, C.: Approximating the Knee of an MOP with Stochastic Search Algorithms. In: Rudolph, G., Jansen, T., Lucas, S., Poloni, C., Beume, N. (eds.) PPSN X. LNCS, vol. 5199, pp. 795–804. Springer, Heidelberg (2008)
19. Kaliszewski, I.: Quantitative Pareto analysis by cone separation technique. Kluwer Academic Publishers (1994)
20. Kaliszewski, I.: Soft computing for complex multiple criteria decision making. Springer, New York (2006)
21. Conway, J.B.: Functions of one complex variable, 2nd edn. Graduate Texts in Mathematics, vol. 11. Springer (1978)
22. Clarke, F.H.: Optimization and nonsmooth analysis, 2nd edn. Classics in Applied Mathematics, vol. 5. SIAM, Philadelphia (1990)
23. Deb, K., Agrawal, S., Pratap, A., Meyarivan, T.: A fast and elitist multi-objective genetic algorithm: NSGA-II. IEEE Transactions on Evolutionary Computation 6(2), 182–197 (2002)
24. Deb, K.: Multi-objective optimization using evolutionary algorithms. Wiley (2001)
25. Deb, K., Thiele, L., Laumanns, M., Zitzler, E.: Scalable test problems for evolutionary multi-objective optimization. In: Abraham, A., et al. (eds.) Evolutionary Multiobjective Optimization, pp. 105–145. Springer, London (2005)
26. Durillo, J.J., Nebro, A.J.: jmetal: A java framework for multi-objective optimization. Advances in Engineering Software 42(10), 760–771 (2011)
27. Deb, K., Gupta, H.: Introducing robustness in multi-objective optimization. Evol. Comput. 14(4), 463–494 (2006)
28. Kung, H.T., Luccio, F., Preparata, F.P.: On finding the maxima of a set of vectors. Journal of the Association for Computing Machinery 22(4), 469–476 (1975)

Knowledge Transfer Strategies for Vector Evaluated Particle Swarm Optimization

Kyle Robert Harrison[1], Beatrice Ombuki-Berman[1],
and Andries P. Engelbrecht[2]

[1] Department of Computer Science, Brock University
St. Catharines, Canada
{kh08uh,bombuki}@brocku.ca
[2] Department of Computer Science, University of Pretoria
Pretoria, South Africa
engel@cs.up.ac.za

Abstract. Vector evaluated particle swarm optimization (VEPSO) is a multi-swarm variant of the traditional particle swarm optimization (PSO) algorithm applied to multi-objective problems (MOPs). Each sub-objective is allocated a single sub-swarm and knowledge transfer strategies (KTSs) are used to pass information between swarms. The original VEPSO used a ring KTS, and while VEPSO has shown to be successful in solving MOPs, other algorithms have been shown to produce better results. One reason for VEPSO to perform worse than other algorithms may be due to the inefficiency of the KTS used in the original VEPSO. This paper investigates new KTSs for VEPSO in order to improve its performance. The results indicated that a hybrid strategy using parent-centric crossover (PCX) on global best solutions generally lead to a higher hypervolume while using PCX on archive solutions generally lead to a better distributed set of solutions.

Keywords: Vector evaluated particle swarm optimization (VEPSO), multi-swarm particle swarm optimization, multi-objective optimization (MOO), knowledge transfer strategy (KTS), global guide selection.

1 Introduction

Optimization problems occur frequently in everyday real-world situations, many of which require simultaneous optimization of two or more conflicting sub-objectives. Problems of this nature are referred to as multi-objective problems (MOPs) and occur in a wide variety of fields including computer engineering [1], aerodynamic design [2], power system performance [3], and scheduling [4].

Vector evaluated particle swarm optimization (VEPSO) was introduced by Parsopoulos et al. [5,6] as one of the first applications of particle swarm optimization (PSO) [7] for MOPs. VEPSO, being a multi-swarm variant of the original PSO, dedicates a single sub-swarm to each objective and effectively optimizes each sub-objective separately. However, to retain the integrity of the

R.C. Purshouse et al. (Eds.): EMO 2013, LNCS 7811, pp. 171–184, 2013.
© Springer-Verlag Berlin Heidelberg 2013

MOP as a whole, information is shared between sub-swarms. The way in which information is shared is dictated by a knowledge transfer strategy (KTS).

The original VEPSO algorithm made use of a ring KTS, whereby the swarms passed information to their immediate neighbor according to a directed ring topology. Recently [4], a random global best KTS was developed and shown to improve VEPSO performance, as each sub-swarm now conveys information to any other sub-swarm rather than just its immediate neighbor, as seen with the ring KTS. With the ring KTS, information from a sub-objective may take many iterations to propagate to other sub-swarms. However, with the random global best KTS, this information can be shared with any other sub-swarm immediately.

This paper proposes new KTSs for VEPSO, and empirically analyzes performance under these KTSs. Four random-based approaches, inspired by the success of the random global best KTS, as well as hybrid strategies borrowing crossover operators from genetic algorithms are presented. Proposed KTSs are shown to outperform the existing strategies with respect to both the hypervolume [8] and distribution [9] measures on a number of MOPs.

The remainder of the paper is organized as follows: Section 2 provides an overview of multi-objective optimization. The PSO algorithm and the VEPSO variant are discussed in Sect. 3. Section 4 contains a discussion of the proposed KTSs. Section 5 describes the experiments carried out to evaluate the performance of VEPSO with the proposed KTSs. The experimental results and a discussion of the results are given in Sect. 6. Finally, conclusions and areas of future work are presented in Sect. 7.

2   Multi-objective Optimization

This section provides an overview of multi-objective optimization problems and the set of solutions to such problems, known as the Pareto front.

Multi-objective optimization problems are quite intuitively harder than single objective problems as they have a number of conflicting objectives. MOPs can have a potentially infinite number of solutions, representing tradeoffs among the sub-objectives. These tradeoffs represent solutions which cannot improve further in any sub-objective without worsening in another sub-objective. The goal of multi-objective optimization is to find a well distributed set of such tradeoff solutions, referred to as non-dominated or Pareto optimal solutions.

A dominance relation[1], $\prec$, is defined such that for objective vectors f^* and f, $f^* \prec f$ indicates that f^* is no worse in all objectives than f and is strictly better in at least one objective [10]. In this case, f^* is said to (strictly) dominate f. The Pareto front is then defined as the set of all vectors in objective space which are not dominated by any other vector. The corresponding set of vectors in decision space is referred to as the Pareto set.

---

[1] Without loss of generality, minimization is assumed for this section. An analogous definition, $\succ$, exists for maximization problems.

More formally [1], if $\boldsymbol{f}$ has n_o objectives then:

$$\boldsymbol{f}^* \prec \boldsymbol{f} := \forall k : f_k^* \leq f_k \wedge \exists k : f_k^* < f_k \quad \text{where } k \in \{1, 2, ..., n_o\}. \tag{1}$$

The Pareto front, PF^*, is then defined as:

$$PF^* = \{\boldsymbol{f}^* \in S | \nexists \boldsymbol{f} \in S : \boldsymbol{f} \prec \boldsymbol{f}^*\} \tag{2}$$

where S denotes the objective space corresponding to feasible solutions [1].

3   Vector Evaluated Particle Swarm Optimization

Kennedy and Eberhart [7] developed the PSO algorithm in 1995, inspired by a simple bird flocking model. PSO is a stochastic, population based algorithm influenced by social dynamics. The basis of the algorithm is that there is a collection of particles, called a swarm, which each move based on two very simple behaviors: move towards the best particle in a neighborhood and move back towards its own best position. From this emerges a more complex behavior in that particles converge on a single solution.

The original PSO algorithms were developed to solve continuous, real-valued, single-objective optimization problems. However, various multi-objective formulations of PSO exist [11,12]. VEPSO [5,6] is one such multi-objective variant of PSO inspired by the vector evaluated genetic algorithm (VEGA) [13].

In VEPSO, each sub-objective is assigned its own sub-swarm with a primary goal of optimizing this single objective. A mechanism, the KTS, is needed to exchange information regarding the different sub-objectives among the different sub-swarms. This is needed in order to find the tradeoffs among the sub-objectives.

The original VEPSO made use of a ring KTS [5,6] where the sub-swarms are organized in a ring structure such that each sub-swarm has one arbitrarily assigned neighboring sub-swarm. When the velocity of particles is updated, the global best position is selected as the global best of the neighboring sub-swarm. The ring KTS information flow is visualized in Fig. 1. While the ring KTS has shown to solve MOPs [4,5,6,14], it suffers from the following problems:

- Each sub-swarm shares information with only one other sub-swarm. Therefore, particles within a sub-swarm are updated with respect to two sub-objectives.
- If there are more than two sub-objectives, i.e. more than two sub-swarms, information about the different sub-objectives is slowly transferred among all of the sub-swarms, with each sub-swarm's search behavior heavily biased with respect to only two of the sub-objectives.

Recently, Grobler [4] proposed the random global best KTS where a random sub-swarm is selected (which may include the current sub-swarm) and the global guide is taken as the global best position of the selected sub-swarm. The random global best KTS allows each sub-swarm to exchange information with all other sub-swarms, eliminating some of the drawbacks of the ring KTS.

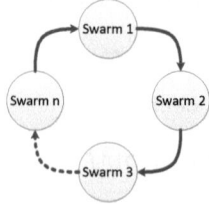

Fig. 1. Ring KTS information flow used by the original VEPSO [5,6]

4   Knowledge Transfer Strategies

This section proposes new strategies to handle knowledge transfer for VEPSO. Sect. 4.1 proposes random and probabilistic approaches to selecting the global guide while Sect. 4.2 presents hybrid approaches to computing a global guide.

4.1   Random and Probabilistic Strategies

This section discusses the four proposed random and probabilistic strategies. Note that for the roulette, tournament, and rank-based strategies, the selection is done with respect to the randomly selected sub-swarm's sub-objective.

1. **Random Personal Best KTS**–The global guide for a sub-swarm is a randomly selected personal best position from a randomly selected sub-swarm. This selection technique still focuses on best solutions, however, it improves exploration.
2. **Roulette Wheel Personal Best KTS**–The global guide for a sub-swarm is a personal best position from a random sub-swarm using roulette wheel selection. The probability of selecting a given individual is given as its proportion of the overall fitness. This allows increased exploration while putting more selection pressure on better solutions.
3. **Tournament Personal Best KTS**–The global guide for a sub-swarm is a personal best solution from a random sub-swarm using tournament selection. The tournament size is set to 10% of the sub-swarm size. This strategy attempts to balance the tradeoffs between exploration and exploitation by allowing increased exploration while still selecting the best solution of the chosen group.
4. **Rank-Based Personal Best KTS**–The global guide for a sub-swarm is a personal best solution from a random sub-swarm using rank-based selection. The selection technique is given as follows: the particles are each assigned a ranking based on their fitness. A uniform random number, $k \in \{1, 2, .., n\}$ where n is the number of particles in the chosen sub-swarm, is selected and the best k particles are chosen. From these k particles, one is selected at random. This strategy prevents selection pressure from becoming overly large, as seen in roulette wheel selection.

4.2 Hybrid Strategies

This section discusses the two proposed hybrid strategies, which make use of a PCX operator [15] to compute a global guide.

1. **PCX Archive KTS**–The global guide for a sub-swarm is computed as the offspring of PCX applied to three randomly selected non-dominated solutions from the archive. This strategy works under the assumption that areas in decision space surrounding non-dominated solutions are also worth exploring.

2. **PCX GBest KTS**–The global guide for a sub-swarm is computed as the offspring of PCX applied to the global best position of three randomly selected sub-swarms. This strategy focuses on the best solutions found so far, but assumes that solutions near the global bests are of high quality.

For both PCX-based approaches, the guide is updated if and only if the offspring had a better fitness with respect to the sub-objective of the current sub-swarm, otherwise the process is repeated up to a maximum of 10 times. If there are less than three valid parent choices available, a random personal best solution from a randomly selected sub-swarm is used. The three parents are guaranteed to be unique. The offspring calculation phase of PCX uses two zero-mean, normally-distributed random variables with standard deviations of σ_1 and σ_2, respectively. The effects of these sigma values are examined later in this study.

5   Experimental Setup

This section describes the experiments which were carried out in order to evaluate the performance of the proposed KTSs in comparison with the existing ring and random global best KTSs. The test functions and performance measures which were used to assess performance are presented and the statistical analysis procedure used to rank strategies is described.

Each experiment consisted of 30 independent runs using the Computational Intelligence library (CIlib)[16]. Each sub-swarm was initialized with 100 particles and ran for 250 iterations. PSO parameters shown to lead to convergence [17] are used, namely $\omega = 0.729844$ and $c_1 = c_2 = 1.496180$. Particles use a clamping boundary strategy preventing them from exiting the feasible region. Personal best positions are updated without using Pareto dominance. That is, a new personal best position is not required to dominate the previous personal best position, it is only required to be better with respect to the current sub-swarm's sub-objective. An archive of size 500 was maintained and where necessary, the solution with the smallest nearest-neighbor distance was replaced in an attempt to increase distribution of solutions along the discovered front.

5.1   Benchmark Functions

This section discusses the benchmark functions used in this paper. Huband et al. [18] proposed the Walking Fish Group (WFG) toolkit which defines functions in terms of a vector of parameters derived from a series of transition vectors. These transition vectors add complexity in various forms, such as multi-modality and non-separability. There are several shape functions, which allow control over the geometry of the Pareto front, and several transformation functions, which allow control over the transitions.

This toolkit consists of nine minimization problems, each of which are used in this paper. For all problems, $M = 3$, $k = 4$, and $l = 20$, where M is the number of objectives and $k + l$ is the number of decision variables, with the first k being position-related and the last l being distance-related. All problems have a decision space with 24 dimensions and an objective space of dimension 3. Properties of the WFG functions are summarized in Table 1. Note that the first $M - 1$ objective functions are unimodal for WFG 2 and the last is multimodal. For WFG 3, a degenerate geometry is one in which the dimension of the Pareto front is less than the dimension of the objective space.

Table 1. Properties of WFG Functions

Function	Separability	Modality	Pareto Front Geometry
WFG1	separable	uni	convex, mixed
WFG2	non-separable	uni/multi	convex, disconnected
WFG3	non-separable	uni	linear, degenerate
WFG4	separable	multi	concave
WFG5	separable	multi	concave
WFG6	non-separable	uni	concave
WFG7	separable	uni	concave
WFG8	non-separable	uni	concave
WFG9	non-separable	multi, deceptive	concave

5.2   Performance Measures

The goal of the performance measures in this paper is to provide a fair assessment of performance without assuming a known Pareto front, as real world scenarios will generally not provide this information. Each of the measures used are based purely on the obtained approximation front with no external information. A good front is defined as one which is close to the true front and has a large number of solutions with good distribution. The measures used in this study address these constraints as described below.

Zitzler and Thiele [19] defined the hypervolume indicator, I_H, as a measure of space covered by a set. Fleischer [20] proved that this metric is maximized if and only if the approximation front contains only maximally diverse Pareto optimal solutions. To understand the importance of this measure, a relation, $\triangleleft$, is defined such that for approximation fronts A and B, $A \triangleleft B$ denotes that every objective vector $\boldsymbol{f}_b \in B$ is weakly dominated by at least one objective vector $\boldsymbol{f}_a \in A$, $A \neq B$ [10]. Furthermore, if $A \triangleleft B$, then $I_H(A) > I_H(B)$. Thus if $I_H(A) < I_H(B)$, it can be said that A cannot be better than B.

The distribution measure [9], D, gives an indication of the distribution of the solutions along the discovered front and is calculated as:

$$D = \frac{1}{n_s} \sqrt{\frac{1}{n_s} \sum_{i=1}^{n_s} (d_i - \overline{d})^2}, \quad \overline{d} = \frac{1}{n_s} \sum_{j=1}^{n_s} d_j \tag{3}$$

where n_s is the number of non-dominated solutions and d_i is the Euclidean distance between non-dominated solution i and its nearest neighbor in objective space. It is noteworthy that equidistant points and a larger number of solutions are both contributing factors to a better distribution score.

5.3   Statistical Analysis

This section describes the statistical analysis which was performed on the obtained results. Pairwise Mann-Whitney U tests [21] were performed between obtained measures for KTSs in order to determine if a significant difference in performance existed. All tests were performed at the 95% confidence level. The null hypothesis for the Mann-Whitney tests was that there existed no significant difference between the performance of the KTSs, with respect to the current metric. The alternative hypothesis was that there was a significant difference in performance, favoring the first KTS. For each comparison, if the Mann-Whitney test indicated a significant difference existed, a win was recorded for the better KTS and a loss for the other. The term "difference" is used to denote the subtractive difference between the number of pairwise wins and losses.

6   Experimental Results and Discussion

This section presents the results of the experiments described in Sect. 5. Experiments were carried out in four phases as discussed in Sects. 6.1–6.4.

6.1   Comparison of Proposed Random-Based Knowledge Transfer Strategies with Existing Strategies

This section discusses the results of the pairwise Mann-Whitney tests between proposed random-based KTSs and the existing ring and random global best KTSs, as summarized in Table 2.

The first observation is that the existing random global best KTS performed consistently well with respect to the hypervolume, with the exception of WFG 9 where it was outperformed by all other compared KTSs for both measures. For WFG 1 to 4, 7, and 8, the random global best KTS had a pairwise difference, as described in Sect. 5.3, of +5 with reference to the hypervolume measure, indicating the random global best KTS significantly outperformed all the other strategies for this measure.

The most interesting results for hypervolume shown in Table 2 were seen in the functions where the global best strategy was not dominant, namely WFG 5, 6, and 9. For these functions, the top performing strategy was the existing ring KTS. It is noted that for all three of these functions, the ring KTS also performed

Table 2. Mann-Whitney Wins and Losses for Proposed Random-Based Knowledge Transfer Strategies vs. Existing Strategies

Knowledge Transfer Strategy	Metric	Result	WFG Function								
			1	2	3	4	5	6	7	8	9
Ring KTS	Hypervolume	Wins	1	0	0	1	4	5	0	0	4
		Losses	1	5	5	2	0	0	5	5	0
		Difference	0	-5	-5	-1	+4	+5	-5	-5	+4
		Rank	3	6	6	3	1	1	6	6	1
	Distribution	Wins	1	1	2	0	1	5	0	3	5
		Losses	0	0	1	0	0	0	0	0	0
		Difference	+1	+1	+1	0	+1	+5	0	+3	+5
		Rank	1	1	2	1	1	1	1	1	1
Random Global Best KTS	Hypervolume	Wins	5	5	5	5	3	3	5	5	0
		Losses	0	0	0	0	0	2	0	0	5
		Difference	+5	+5	+5	+5	+3	+1	+5	+5	-5
		Rank	1	1	1	1	2	3	1	1	6
	Distribution	Wins	0	1	5	0	0	3	0	0	0
		Losses	5	0	0	0	5	2	0	1	5
		Difference	-5	+1	+5	0	-5	+1	0	-1	-5
		Rank	6	1	1	1	6	3	1	4	6
Random Personal Best KTS	Hypervolume	Wins	0	1	2	0	0	0	1	1	1
		Losses	3	2	2	5	3	4	3	2	2
		Difference	-3	-1	0	-5	-3	-4	-2	-1	-1
		Rank	6	4	3	6	4	5	5	3	3
	Distribution	Wins	1	1	0	0	1	0	0	0	3
		Losses	0	0	2	0	0	3	0	1	1
		Difference	+1	+1	-2	0	+1	-3	0	-1	+2
		Rank	1	1	4	1	1	5	1	4	2
Tournament Personal Best KTS	Hypervolume	Wins	2	3	3	4	3	4	4	4	4
		Losses	1	1	1	1	1	1	1	1	0
		Difference	+1	+2	+2	+3	+2	+3	+3	+3	+4
		Rank	2	2	2	2	3	2	2	2	1
	Distribution	Wins	1	1	0	0	1	4	0	0	1
		Losses	0	0	1	0	0	1	0	3	2
		Difference	+1	+1	-1	0	+1	+3	0	-3	-1
		Rank	1	1	3	1	1	2	1	6	4
Roulette Wheel Personal Best KTS	Hypervolume	Wins	0	1	1	1	0	0	1	1	1
		Losses	2	1	3	2	3	4	2	2	2
		Difference	-2	0	-2	-1	-3	-4	-1	-1	-1
		Rank	5	3	5	3	4	5	4	3	3
	Distribution	Wins	1	0	0	0	1	0	0	1	1
		Losses	0	0	1	0	0	4	0	0	1
		Difference	+1	0	-1	0	+1	-4	0	+1	0
		Rank	1	5	3	1	1	6	1	2	3
Rank-Based Personal Best KTS	Hypervolume	Wins	0	1	1	1	0	2	2	1	1
		Losses	1	2	1	2	3	3	2	2	2
		Difference	-1	-1	0	-1	-3	-1	0	-1	-1
		Rank	4	4	3	3	4	4	3	3	3
	Distribution	Wins	1	0	0	0	1	1	0	1	1
		Losses	0	4	2	0	0	3	0	0	2
		Difference	+1	-4	-2	0	+1	-2	0	+1	-1
		Rank	1	6	4	1	1	4	1	2	4

the best in terms of the distribution metric. WFG 5 is a deceptive problem, WFG 6 contains a non-separable reduction, and WFG 9 is both deceptive and contains a non-separable reduction. It was hypothesized that these types of problems respond well to the ring KTS when the hypervolume is being measured as the information propagation is much slower, allowing the deception to have a lesser effect. Also, the ring KTS attained the most distributed approximation fronts in seven of the nine functions, and the second best in the remaining two functions.

Furthermore, in only one pairwise test, on WFG 3, was any KTS found to be significantly better than the ring KTS in terms of distribution.

It was noted that tournament personal best KTS consistently ranked well in terms of hypervolume, being the second best KTS for seven of the nine test functions. It was outranked by only one other KTS on functions WFG 1 to 8, and ranked highest on WFG 9. When looking at the distribution measure the tournament personal best KTS also performed very well, being the top KTS for five of the nine functions. Overall, the proposed random-based KTSs did not outperform existing KTSs.

6.2   Comparison of PCX GBest Knowledge Transfer Strategy Parameters

This section discusses the results of a study on the effects of the sigma parameters for the PCX GBest KTS. The results are summarized in Table 3. The key observation made from Table 3 is that, with respect to the hypervolume, the PCX GBest KTS favored lower sigma values of 0.01 and 0.05, whereas the distribution was generally better with slightly higher sigma values of 0.05 and 0.10.

The first observed behavior in Table 3 is that the lowest tested sigma value, 0.01, lead to the highest hypervolume for six of the nine functions, where in four the sigma values of 0.01 outperformed all other sigma value settings. There were only four recorded instances of a sigma value combination attaining a significantly higher hypervolume value than with the sigma values set at 0.01. Similarly, the highest sigma value of 1.00 lead to the worst hypervolume ranking for eight of the nine functions, being significantly outperformed by all other values for seven of the functions. Note that the rankings for hypervolume degraded linearly with increases in sigma values.

Far less consistent results are observed for the distribution measure than with the hypervolume, but observations are that the middle range of sigma values, 0.10 and 0.20, were the most consistent in terms of ranking and were the only two sigma value settings with nearly all pairwise differences greater than or equal to zero. The distribution ranking peeked when sigma values were set to 0.20, degrading at both higher and lower values. Again, when sigma was set to 1.00, the distribution measure was the worst in general, being significantly outperformed by all other sigma settings in four of the nine functions, and being the worst overall in seven functions.

6.3   Comparison of PCX Archive Knowledge Transfer Strategy Parameters

This section discusses the results of a study on the effects of the sigma parameters for the PCX archive KTS. The observations are summarized in Table 4. With this strategy, far less pronounced trends are observed. Nevertheless, some key observations were still made.

Table 3. Mann-Whitney Wins and Losses for PCX GBest KTS with Various Sigma Values

Sigma Values ($\sigma_1 = \sigma_2$)	Metric	Result	WFG Function								
			1	2	3	4	5	6	7	8	9
0.01	Hypervolume	Wins	5	4	1	5	4	1	5	5	4
		Losses	0	0	2	0	0	1	0	0	1
		Difference	+5	+4	-1	+5	+4	0	+5	+5	+3
		Rank	1	1	3	1	1	2	1	1	2
	Distribution	Wins	0	4	0	1	1	1	1	4	3
		Losses	5	0	3	3	2	0	1	0	0
		Difference	-5	+4	-3	-2	-1	+1	0	+4	+3
		Rank	6	1	4	5	3	1	3	1	1
0.05	Hypervolume	Wins	2	4	4	4	4	1	3	3	1
		Losses	1	0	0	1	0	1	1	1	2
		Difference	+1	+4	+4	+3	+4	0	+2	+2	-1
		Rank	2	1	1	2	1	2	2	2	3
	Distribution	Wins	1	1	3	4	1	1	4	2	3
		Losses	4	1	0	0	2	0	0	1	0
		Difference	-3	0	+3	+4	-1	+1	+4	+1	+3
		Rank	5	3	1	1	3	1	1	3	1
0.10	Hypervolume	Wins	2	3	1	3	3	1	2	3	5
		Losses	1	2	2	2	2	1	1	1	0
		Difference	+1	+1	-1	+1	+1	0	+1	+2	+5
		Rank	2	3	3	3	3	2	3	2	1
	Distribution	Wins	2	1	3	3	4	1	1	2	0
		Losses	1	0	0	0	1	0	0	0	4
		Difference	+1	+1	+3	+3	+3	+1	+1	+2	-4
		Rank	4	2	1	2	2	1	2	2	5
0.20	Hypervolume	Wins	2	1	4	2	2	1	2	2	1
		Losses	1	3	0	3	3	1	2	3	2
		Difference	+1	-2	+4	-1	-1	0	0	-1	-1
		Rank	2	4	1	4	4	2	4	4	4
	Distribution	Wins	2	1	3	2	5	1	1	2	2
		Losses	0	1	0	1	0	0	1	1	0
		Difference	+2	0	+3	+1	+5	+1	0	+1	+2
		Rank	2	3	1	3	1	2	3	3	3
0.50	Hypervolume	Wins	1	1	1	1	1	0	0	1	1
		Losses	4	3	2	4	4	5	4	4	2
		Difference	-3	-2	-1	-3	-4	-5	-3	-3	-1
		Rank	5	4	3	5	5	6	5	5	4
	Distribution	Wins	3	0	0	1	1	0	1	1	2
		Losses	0	1	3	2	2	5	1	4	2
		Difference	+3	-1	-3	-1	-1	-5	0	-3	0
		Rank	1	5	4	4	3	6	3	5	4
1.00	Hypervolume	Wins	0	0	0	0	0	5	0	0	0
		Losses	5	5	5	5	4	0	5	5	5
		Difference	-5	-5	-5	-5	-4	+5	-5	-5	-5
		Rank	6	6	6	6	5	1	6	6	6
	Distribution	Wins	2	0	0	0	0	1	0	0	0
		Losses	0	4	3	5	5	0	5	5	4
		Difference	+2	-4	-3	-5	-5	+1	-5	-5	-4
		Rank	2	6	4	6	6	1	6	6	5

With respect to both measures, setting sigma values to 1.00 lead to overall poor performance. For the distribution measure the pairwise difference was less than or equal to 0 for all functions, and less than or equal to 0 for six functions with respect to the hypervolume. For five of the nine functions, sigma values of 1.00 were recorded as the overall worst sigma value settings for both measures. For the distribution measure, setting sigmas to 1.00 tied for worst ranking on an additional two functions. However, for the hypervolume, setting sigma to 1.00 ranked the highest on three of the functions, namely WFG 1, 5, and 6.

Table 4. Mann-Whitney Wins and Losses for PCX Archive KTS with Various Sigma Values

Sigma Values ($\sigma_1 = \sigma_2$)	Metric	Result	WFG Function								
			1	2	3	4	5	6	7	8	9
0.01	Hypervolume	Wins	0	2	1	2	0	0	1	3	1
		Losses	0	0	0	0	1	1	0	0	1
		Difference	0	+2	+1	+2	-1	-1	+1	+3	0
		Rank	1	1	1	2	5	2	1	1	4
	Distribution	Wins	2	2	1	1	0	0	0	0	2
		Losses	0	1	0	0	0	0	0	0	0
		Difference	+2	+1	+1	+1	0	0	0	0	+2
		Rank	1	3	1	1	2	2	2	2	1
0.05	Hypervolume	Wins	0	2	0	1	0	0	1	1	0
		Losses	0	0	0	1	2	1	0	0	2
		Difference	0	+2	0	0	-2	-1	+1	+1	-2
		Rank	1	1	5	4	6	2	1	2	5
	Distribution	Wins	0	2	1	1	0	0	1	0	2
		Losses	0	0	0	0	1	0	0	0	0
		Difference	0	+2	+1	+1	-1	0	+1	0	+2
		Rank	2	2	1	1	6	2	1	2	1
0.10	Hypervolume	Wins	0	2	1	2	0	0	1	1	0
		Losses	0	0	0	0	0	1	0	0	4
		Difference	0	+2	+1	+2	0	-1	+1	+1	-4
		Rank	1	1	1	2	3	2	1	2	6
	Distribution	Wins	0	4	1	1	0	0	0	0	2
		Losses	1	0	0	0	0	0	0	0	0
		Difference	-1	+4	+1	+1	0	0	0	0	+2
		Rank	5	1	1	1	2	2	2	2	1
0.20	Hypervolume	Wins	0	2	1	3	1	0	1	1	3
		Losses	0	0	0	0	0	1	0	1	0
		Difference	0	+2	+1	+3	+1	-1	+1	0	+3
		Rank	1	1	1	1	2	2	1	4	1
	Distribution	Wins	0	2	1	1	0	1	0	1	2
		Losses	0	1	0	0	1	0	0	0	0
		Difference	0	+1	+1	+1	-1	+1	0	+1	+2
		Rank	2	3	1	1	4	1	2	1	1
0.50	Hypervolume	Wins	0	1	1	1	0	0	1	1	1
		Losses	0	4	0	3	0	1	0	1	0
		Difference	0	-3	+1	-2	0	-1	+1	0	+1
		Rank	1	5	1	5	3	2	1	4	3
	Distribution	Wins	0	0	1	1	3	0	0	0	1
		Losses	0	4	0	0	0	1	0	0	4
		Difference	0	-4	+1	+1	+3	-1	0	0	-3
		Rank	2	5	1	1	1	6	2	2	5
1.00	Hypervolume	Wins	0	0	0	0	2	5	0	0	2
		Losses	0	5	4	5	0	0	5	5	0
		Difference	0	-5	-4	-5	+2	+5	-5	-5	+2
		Rank	1	6	6	6	1	1	6	6	2
	Distribution	Wins	0	0	0	0	0	0	0	0	0
		Losses	1	4	5	5	1	0	1	1	5
		Difference	-1	-4	-5	-5	-1	0	-1	-1	-5
		Rank	5	5	6	6	4	2	6	6	6

Although not very well pronounced, there was a correlation between mid-range sigma settings and higher rankings for both measures. Sigma values of 0.20 were the highest ranked in general and showed the highest ranking in six of the nine functions for the hypervolume, and five of the nine functions for the distribution measure. Note that, for the distribution measure, the lower sigma values of 0.01, 0.05, and 0.10 all recorded only one pairwise loss for the distribution measure, whereas 0.20 recorded two losses, and for the hypervolume, 0.20 recorded only two losses, being the lowest overall.

6.4 Comparison of Proposed Hybrid Knowledge Transfer Strategies with Existing Strategies

This section discusses the results of a comparison between the proposed hybrid KTSs and the existing KTSs. Table 5 provides the results of a pairwise comparison of four strategies, ring KTS, random global best KTS, PCX GBest KTS, and PCX archive KTS. The purpose of this comparison is to determine if the proposed hybrid KTSs outperform the existing KTSs.

For each of the functions, the highest ranked sigma value configurations for the PCX KTSs were selected for the comparison. In the event of ties for one measure, the second measure was used to break the tie. If a tie was observed for the second measure, equivalence was assumed and one was sigma value combination was selected at random to be used for the pairwise comparison.

With respect to the hypervolume, Table 5 shows that the PCX GBest KTS was the highest ranking KTS for seven of the nine functions. An interesting result, which is contrary to previously observed results for dynamic environments [22], was that for the two functions where PCX GBest KTS was not the highest ranking KTS, namely WFG 5 and 6, the ring KTS was the highest ranking KTS.

Table 5. Mann-Whitney Wins and Losses for Proposed Hybrid Knowledge Transfer Strategies vs. Existing Strategies

Knowledge Transfer Strategy	Metric	Result	WFG Function								
			1	2	3	4	5	6	7	8	9
Ring KTS	Hypervolume	Wins	0	0	1	0	1	3	0	0	1
		Losses	3	3	2	3	0	0	3	3	1
		Difference	-3	-3	-1	-3	+1	+3	-3	-3	0
		Rank	4	4	4	4	1	1	4	4	2
	Distribution	Wins	2	0	1	0	3	2	0	2	1
		Losses	1	1	1	2	0	1	2	0	1
		Difference	+1	-1	0	-2	+3	+1	-2	+2	0
		Rank	2	3	2	3	1	2	4	1	2
Random Global Best KTS	Hypervolume	Wins	1	1	2	1	1	0	1	1	0
		Losses	2	2	1	1	0	2	1	1	3
		Difference	-1	-1	+1	0	+1	-2	0	0	-3
		Rank	3	3	2	2	1	4	2	2	4
	Distribution	Wins	1	0	3	0	0	1	0	1	0
		Losses	2	2	0	2	3	2	1	2	3
		Difference	-1	-2	+3	-2	-3	-1	-1	-1	-3
		Rank	3	4	1	3	4	3	3	3	4
PCX GBest KTS	Hypervolume	Wins	3	3	3	3	0	0	3	3	3
		Losses	0	0	0	0	3	1	0	0	0
		Difference	+3	+3	+3	+3	-3	-1	+3	+3	+3
		Rank	1	1	1	1	4	3	1	1	1
	Distribution	Wins	0	1	0	2	1	0	1	0	1
		Losses	3	1	1	0	1	3	0	3	1
		Difference	-3	0	-1	+2	0	-3	+1	-3	0
		Rank	4	2	3	1	2	4	2	4	2
PCX Archive KTS	Hypervolume	Wins	2	2	1	1	1	1	1	1	1
		Losses	1	1	1	1	0	1	1	1	1
		Difference	+1	+1	0	0	+1	0	0	0	0
		Rank	2	2	3	2	1	2	2	2	2
	Distribution	Wins	3	3	0	2	1	3	2	2	3
		Losses	0	0	2	0	1	0	0	0	0
		Difference	+3	+3	-2	+2	0	+3	+2	+2	+3
		Rank	1	1	4	1	2	1	1	1	1

Another observation was that in all instances where the PCX GBest KTS was ranked highest, the sigma values were 0.10 or lower, whereas the two functions on which it was ranked the worst had larger values of 0.20 and 1.00, respectively.

In terms of the distribution measure, we see that PCX archive KTS was the highest ranking strategy for seven of the nine functions. Overall, the ring, random global best, and PCX GBest KTSs were not overly different in terms of ranking when the distribution measure was examined.

7   Conclusion

This paper proposed six new knowledge transfer strategies for vector evaluated particle swarm optimization. Four of these strategies select a global guide from existing particles using random and probabilistic selection mechanisms, while two hybrid strategies compute the global guide based on a parent centric crossover (PCX) operator. The results showed that the hybrid approaches outperformed the existing ring and random global best KTSs using two well-known performance measures.

The results showed that the tournament personal best KTS was the best proposed random-based KTSs, but does not outperform the existing random global best KTS. A comparison of the hybrid strategies with the existing ring and random global best KTSs determined the PCX GBest KTS as the best KTS with respect to the hypervolume. When the distribution measure was compared, the PCX archive KTS was determined as the best performing KTS.

Future work involves determining which function properties were responsible for the observed behavior and comparing VEPSO using the proposed strategies against other well-known MOO algorithms. Lastly, this paper focused on functions with three objectives. An immediate further investigation is to evaluate the performance of the proposed strategies with greater than three objectives.

References

1. Zitzler, E.: Evolutionary Algorithms for Multiobjective Optimization: Methods and Applications. PhD thesis, ETH Zurich, Switzerland (1999)
2. Stewart, T., Bandte, O., Braun, H., Chakraborti, N., Ehrgott, M., Göbelt, M., Jin, Y., Nakayama, H., Poles, S., Di Stefano, D.: Real-World Applications of Multiobjective Optimization. In: Branke, J., Deb, K., Miettinen, K., Słowiński, R. (eds.) Multiobjective Optimization. LNCS, vol. 5252, pp. 285–327. Springer, Heidelberg (2008)
3. Vlachogiannis, J., Lee, K.: Multi-objective based on parallel vector evaluated particle swarm optimization for optimal steady-state performance of power systems. Expert Systems with Applications 36(8), 10802–10808 (2009)
4. Grobler, J.: Particle swarm optimization and differential evolution for multiobjective multiple machine scheduling. Master's thesis, University of Pretoria, Pretoria, South Africa (2008)
5. Parsopoulos, K.E., Vrahatis, M.N.: Particle swarm optimization method in multiobjective problems. In: Proceedings of the 2002 ACM Symposium on Applied Computing, SAC 2002, pp. 603–607. ACM, New York (2002)

6. Parsopoulos, K.E., Tasoulis, D.K., Vrahatis, M.N.: Multiobjective optimization using parallel vector evaluated particle swarm optimization. In: Proceedings of the IASTED International Conference on Artificial Intelligence and Applications. AIA 2004, vol. 2, pp. 823–828. ACTA Press (2004)
7. Kennedy, J., Eberhart, R.C.: Particle swarm optimization. In: IEEE Int'l Conference on Neural Networks, vol. IV, pp. 1942–1948 (1995)
8. Zitzler, E., Deb, K., Thiele, L.: Comparison of multiobjective evolutionary algorithms: Empirical results. Evol. Comput. 8(2), 173–195 (2000)
9. Goh, C.K., Tan, K.C.: An investigation on noisy environments in evolutionary multiobjective optimization. IEEE Trans. on Evolutionary Computation 11(3), 354–381 (2007)
10. Zitzler, E., Thiele, L., Laumanns, M., Fonseca, C., da Fonseca, V.: Performance assessment of multiobjective optimizers: an analysis and review. IEEE Trans. on Evolutionary Computation 7(2), 117–132 (2003)
11. Padhye, N.: Comparison of archiving methods in multi-objective particle swarm optimization (mopso): empirical study. In: Proceedings of the 11th Annual Conference on Genetic and Evolutionary Computation, GECCO 2009, pp. 1755–1756. ACM, New York (2009)
12. Reyes-sierra, M., Coello, C.A.C.: Multi-objective particle swarm optimizers: A survey of the state-of-the-art. International Journal of Computational Intelligence Research 2(3), 287–308 (2006)
13. Schaffer, J.D.: Multiple objective optimization with vector evaluated genetic algorithms. In: Proceedings of the 1st International Conference on Genetic Algorithms, pp. 93–100. Erlbaum Associates Inc., Hillsdale (1985)
14. Greeff, M., Engelbrecht, A.: Solving dynamic multi-objective problems with vector evaluated particle swarm optimisation. In: IEEE Congress on Evolutionary Computation (2008)
15. Deb, K., Anand, A., Joshi, D.: A computationally efficient evolutionary algorithm for real-parameter optimization. Evolutionary Computation 10(4), 371–395 (2002)
16. Pampara, G., Engelbrecht, A.P., Cloete, T.: Cilib: A collaborative framework for computational intelligence algorithms - part i. In: Proc. of IEEE World Congress on Computational Intelligence, Hong Kong, pp. 1750–1757 (2008), https://github.com/cilib/cilib (last accessed July 8, 2012)
17. Van Den Bergh, F.: An analysis of particle swarm optimizers. PhD thesis, University of Pretoria, Pretoria, South Africa (2002)
18. Huband, S., Hingston, P., Barone, L., While, L.: A review of multiobjective test problems and a scalable test problem toolkit. IEEE Trans. on Evolutionary Computation 10(5), 477–506 (2006)
19. Zitzler, E., Thiele, L.: Multiobjective evolutionary algorithms: A comparative case study and the strength pareto approach. IEEE Trans. on Evolutionary Computation 3(4), 257–271 (1999)
20. Fleischer, M.: The Measure of Pareto Optima Applications to Multi-objective Metaheuristics. In: Fonseca, C.M., Fleming, P.J., Zitzler, E., Deb, K., Thiele, L. (eds.) EMO 2003. LNCS, vol. 2632, pp. 519–533. Springer, Heidelberg (2003)
21. Mann, H.B., Whitney, D.R.: On a test of whether one of two random variables is stochastically larger than the other. Annals of Mathematical Statistics 18(1) (March 1947)
22. Greeff, M., Engelbrecht, A.P.: Dynamic Multi-objective Optimisation Using PSO. In: Nedjah, N., dos Santos Coelho, L., de Macedo Mourelle, L. (eds.) Multi-Objective Swarm Intelligent Systems. SCI, vol. 261, pp. 105–123. Springer, Heidelberg (2010)

Hypervolume-Based Multi-Objective Path Relinking Algorithm

Rong-Qiang Zeng[1,2], Matthieu Basseur[1], and Jin-Kao Hao[1]

[1] LERIA, Université d'Angers, 2, Boulevard Lavoisier,
49045 Angers Cedex 01, France
[2] Chengdu Library, Chinese Academy of Sciences, 610041 Chengdu, Sichuan, China
{zeng,basseur,hao}@info.univ-angers.fr, zengrq@clas.ac.cn

Abstract. This paper presents a hypervolume-based multi-objective path relinking algorithm for approximating the Pareto optimal set of multi-objective combinatorial optimization problems. We focus on integrating path relinking techniques within a multi-objective local search as an initialization function. Then, we carry out a range of experiments on bi-objective flow shop problem and bi-objective quadratic assignment problem. Experimental results and a statistical comparison are reported in the paper. In comparison with the other algorithms, one version of our proposed algorithm is very competitive. Some directions for future research are highlighted.

Keywords: multi-objective optimization, hypervolume contribution, path relinking, local search, flow shop problem, quadratic assignment problem.

1 Introduction

Local search is an effective search strategy for both single objective optimization and multi-objective optimization. Particularly, local search requires a method to generate initial solutions. However, how to set the initialization methods still remains an open question in many cases, especially in multi-objective optimization. In this paper, we investigate path relinking [8] as an initialization method for hypervolume-based multi-objective local search (HBMOLS) [3].

The HBMOLS algorithm aims to generate a Pareto approximation set by improving an initial population. In this work, we use path relinking to construct paths and then select from each path a set of solutions to initialize a new population for HBMOLS. In order to evaluate the effectiveness of our proposed method, we show experimental results on the bi-objective flow shop problem and bi-objective quadratic assignment problem, and we compare them with the HBMOLS algorithm which initializes a new population using random mutations or crossover operator.

The remainder of this paper is organized as follows. In Section 2, we present some basic notations and definitions related to multi-objective optimization. Then, in Section 3, we briefly review the literature using the path relinking techniques to solve multi-objective optimization problems. Afterwards, in Section 4, we describe the hypervolume-based multi-objective path relinking algorithm. Section 5 reports the computational results and analyzes the behavior of the proposed algorithm. Finally, the conclusions and perspectives are given in the last section.

R.C. Purshouse et al. (Eds.): EMO 2013, LNCS 7811, pp. 185–199, 2013.

2   Multi-Objective Optimization

In this section, we recall some useful notations and definitions of multi-objective optimization. Let X denote the search space of the optimization problem under consideration and Z the corresponding objective space. Without loss of generality, we assume that $Z = \Re^n$ and all n objectives are to be minimized. Each $x \in X$ is assigned exactly one objective vector $z \in Z$ on the basis of a vector function $f : X \to Z$ with $z = f(x)$. The mapping f defines the evaluation of a solution $x \in X$, and often one is interested in those solutions that are Pareto optimal with respect to f. The relation $x_1 \succ x_2$ means that the solution x_1 is *preferable* to x_2. The dominance relation between two solutions x_1 and x_2 is usually defined as follows:

Definition 1. *A decision vector x_1 is said to dominate another decision vector x_2 (written as $x_1 \succ x_2$), if $f_i(x_1) \le f_i(x_2)$ for all $i \in \{1, \ldots, n\}$ and $f_j(x_1) < f_j(x_2)$ for at least one $j \in \{1, \ldots, n\}$.*

Definition 2. *$x \in S\ (S \subset X)$ is said to be non-dominated if and only if there does not exist another solution $x' \in S$ such that x' dominates x. When $S \equiv X$, x is said to be Pareto optimal.*

Definition 3. *S is said to be a non-dominated set if and only if S is composed of non-dominated solutions. When S is composed of all the Pareto optimal solutions, S is said to be a Pareto optimal set.*

In multi-objective optimization, there usually does not exist one optimal but a set of Pareto optimal solutions, which keeps the best compromise among all the objectives. Nevertheless, in most cases, it is not possible to compute the Pareto optimal set in a reasonable time. Then, we are interested in computing a non-dominated set, which is as close to the Pareto optimal set as possible. Therefore, the goal is often to identify a good Pareto approximation set.

3   Related Works

Path Relinking (PR) was initially proposed by Glover [8] as an effective search strategy, which has proved its efficiency in single objective optimization [8]. Its objective is to explore the search space by creating paths within a given set of high-quality solutions. In the following paragraphs, we focus on the studies dealing with multi-objective optimization problems.

Basseur et al. [2] propose a multi-objective approach to integrate PR techniques into an adaptive genetic algorithm, which is dedicated to obtaining a first well diversified Pareto approximation set. Based on this set, two solutions are randomly selected to generate a path. According to the distance measure defined in [2], there are many intermediate solutions which can be generated at each step of the PR procedure. Then, the authors apply a random aggregation of the objectives to determine which solution is selected from the possible eligible solutions. After linking these two solutions, a Pareto local search is applied in order to improve the quality of the non-dominated set generated by the PR algorithm. Experimental results on bi-objective flow shop problem show that this PR approach is very promising and efficient.

In [13], Pasia et al. present three PR approaches for solving a bi-objective flow shop problem. By using a straightforward implementation of the ant colony system, they first generate two pools of initial solutions, where one pool contains solutions that are good with respect to the makespan and the other one contains solutions that are good with respect to the total tardiness. Based on random insertion, all the solutions in both pools are improved by local search in order to obtain a non-dominated set. Then, the authors randomly select two solutions from this non-dominated set to construct a path. Along the path, some of the solutions are submitted for improvements. The authors propose three different strategies to define the heuristic bounds. Each strategy allows the solutions to undergo local search under the conditions based on the local nadir points. Computational results demonstrate that their proposed approaches are competitive.

In addition, two different versions of iterated Pareto local search (IPLS) algorithms, which are path-guided IPLS (pIPLS) and a combination of IPLS and pIPLS named rIPLS, are presented in [6]. The authors propose a path-guided mutation that generates solutions on the path linking two local optimal individuals. This mutation generates individuals at a certain distance from the initial solution to the guiding solution. Then, Pareto local search is restarted from the individual generated on the path. Experiments on bi-objective quadratic assignment problem show that pIPLS and rIPLS both outperform the multi-restart Pareto local search algorithm.

4   Hypervolume-Based Multi-Objective Path Relinking Algorithm

This section describes the hypervolume-based multi-objective path relinking algorithm, which is a combination of the Hypervolume-Based Multi-Objective Local Search algorithm (HBMOLS) and the Multi-Objective Path Relinking algorithm (MOPR). The outline of the proposed algorithm is illustrated in Algorithm 1 and depicted in Fig. 1.

In this algorithm, all the solutions in an initial population are randomly generated. Then, each solution in the population is optimized by the HBMOLS algorithm [3], which is based on the Hypervolume Contribution Selection illustrated in Algorithm 2. The HBMOLS algorithm achieves the fitness assignment by using the hypervolume contribution indicator $HC(x, P)$ defined in [3]. Afterwards, we randomly choose two solutions (an initial solution and a guiding solution) from the Pareto approximation set generated by HBMOLS, and we define a distance between these two solutions to construct a path. At each step, we generate only one new solution and make sure the distance between the new solution and the guiding solution decreases by 1.

After the path generation, a subset of solutions in the path are selected and used to initialize a new population P for HBMOLS. These solutions are potentially inserted into P, according to their corresponding hypervolume contribution. Actually, we propose four mechanisms to select a set of solutions from the generated path. These mechanisms are illustrated in Fig. 2 and described in detail below.

All: All the solutions in the path are selected to be inserted into the population P (solutions represented both in circle and in square in Fig. 2).

Algorithm 1. Hypervolume-Based Multi-Objective Path Relinking Algorithm

---

Input: N (Population size)

Output: A (Pareto approximation set)

Initialization: $P \leftarrow N$ randomly generated solutions

               $A \leftarrow$ Non dominated solutions of P

while Running time is not reached **do**

  Local Search (HBMOLS):

    1) Fitness Assignment: Calculate a fitness value for each $x \in P$, i.e., $Fit(x) = HC(x, P)$

    2) **For** each $x \in P$ **do**:

      repeat

        a) $x^* \leftarrow$ one randomly chosen unexplored neighbors of x

        b) Progress $\leftarrow$ **Hypervolume Contribution Selection** (P, x^*)

      until all neighbors are explored or Progress $=$ True

    3) $A \leftarrow$ Non dominated solutions of $A \bigcup P$. If A does change, back to step 2

  Path Relinking (MOPR):

    1) $P^{'} \leftarrow N$ randomly generated solutions

    2) randomly choose an initial solution x_i and a guiding solution x_j from A

    3) compute the distance d_{ij} between x_i and x_j

    4) generate a set of solutions: $T = \{t_1, t_2, \cdots, t_{d_{ij}-1}\}$ along a path linking x_i to x_j

    5) select n_{pr} solutions: $T^{'} = \{y_1, y_2, \cdots, y_{n_{pr}}\}$ from the set T

    6) **for** $i \leftarrow 1, \ldots, n_{pr}$ **do**

      Hypervolume Contribution Selection $(P^{'}, y_i)$

    end for

end while

Return A

Best: The solutions in the path are divided into two sets, according to their Pareto dominance relations. The solutions belonging to the non-dominated set are selected. In Fig. 2, the solutions represented in square are selected, since they belong to the non-dominated set.

Middle: The solutions located at the beginning or at the end of the path are similar to the initial solution or the guiding solution. These solutions could not be very useful, since HBMOLS will search the explored areas alike. One way to avoid this problem is to select a single solution, which is located in the middle of the path (solution represented in black circle in Fig. 2). In fact, this mechanism can be seen as a kind of crossover operator.

K-**Middle:** Here, we also aim to avoid the problem of proximity of intermediate solutions to the initial solution and the guiding solution. Then, we propose to select a set of solutions located in the middle of the path. The number N_{KM} of these solutions is defined according to the length of generated path. We define this number by using the formula $N_{KM} = \sqrt{N_{All}}$, where N_{All} being the number of the solutions in the path, and N_{KM} is the greatest integer that is not bigger than $\sqrt{N_{All}}$ (solutions located in the dashed circle in Fig. 2).

Algorithm 2. Hypervolume Contribution Selection

Step:
1) $P \leftarrow P \bigcup x^*$
2) compute x^* fitness: $HC(x^*, P)$, then update all $z \in P$ fitness values:
$Fit(z) = HC(z, P)$
3) $w \leftarrow$ worst individual in P
4) $P \leftarrow P \backslash \{w\}$, then update all $z \in P$ fitness values: $Fit(z) = HC(z, P \backslash \{w\})$
5) **if** $w \neq x^*$, **return** True

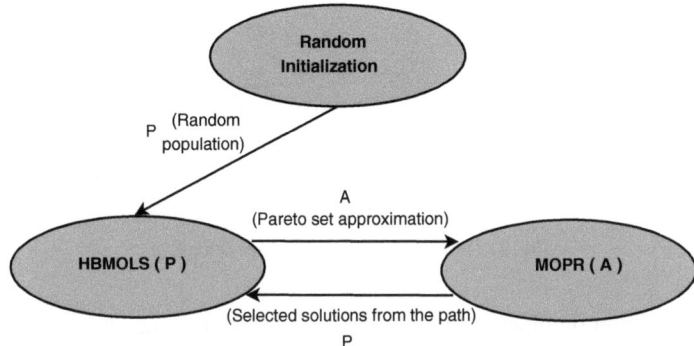

Fig. 1. A random population is initialized and provided as an entry to to HBMOLS, which generates a Pareto approximation set by improving the initial population. Then, MOPR generates a path between two solutions belonging to the Pareto approximation set provided by HBMOLS. A subset of solutions in the path is selected to initiate a new HBMOLS execution.

5   Computational Results

In order to evaluate the efficiency of our proposed algorithms, we carry out experiments on the bi-objective flow shop problem and bi-objective quadratic assignment problem. We compare four versions of hypervolume-based multi-objective path relinking algorithm (named PR_A, PR_B, PR_M and PR_KM) with two versions of HBMOLS (named RM and CO), which use random mutation and crossover operator as the initialization functions [1]. All the algorithms are programmed in C and compiled using Dev-C++ on a PC running Windows XP with Pentium 2.61 GHz CPU and 2 GB RAM.

5.1   Performance Assessment Protocol

We evaluate the effectiveness of multi-objective optimization algorithms by using a test procedure that has been undertaken with the performance assessment package provided by Zitzler et al.[1]

The quality assessment protocol works as follows: we first create a set of 20 runs with different initial populations for each algorithm and each benchmark instance. Afterwards, we calculate the set PO^* in order to determine the quality of k different sets

---
[1] http://www.tik.ee.ethz.ch/pisa/assessment.html

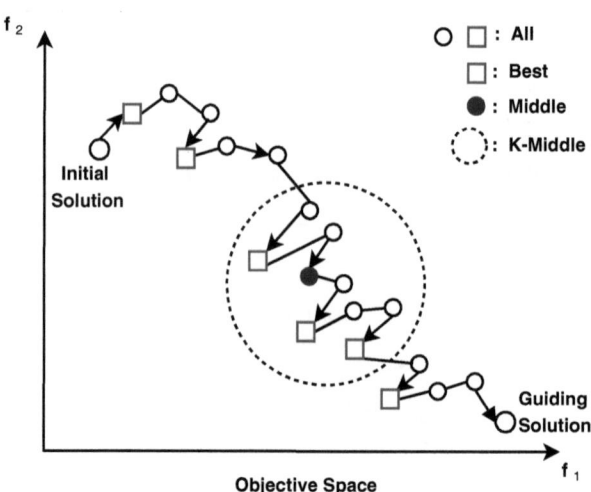

Fig. 2. The mechanisms of subset selection

$A_0 \ldots A_{k-1}$ of non-dominated solutions (The set PO^* is generated by removing the dominated solutions from the union of k different sets, more details can be found in [19]). Furthermore, we define a reference point $z = [w_1, w_2]$, where w_1 and w_2 represent the worst values for each objective function in $A_0 \cup \cdots \cup A_{k-1}$. Then, the evaluation of a set A_i of solutions can be determined by finding the hypervolume difference between A_i and PO^* [19], which has to be as close to zero as possible.

For each algorithm, we compute 20 hypervolume differences corresponding to 20 runs, and perform the Mann-Whitney statistical test on the sets of hypervolume difference. In our experiments, we say that an algorithm $\mathcal{A}$ outperforms an algorithm $\mathcal{B}$ if the Mann-Whitney test provides a confidence level greater than 95%. The computational results are summarized in Tables 2 and 4 respectively. In these two tables, each line contains at least a value **in grey** for each instance, which corresponds to the best average hypervolume difference obtained by the corresponding algorithm. The values both **in italic** and **bold** mean that the corresponding algorithms are **not** statistically outperformed by the algorithm which obtains the best result (with a confidence level greater than 95%).

5.2   Application to Bi-objective Flow Shop Problem

The Flow Shop Problem (FSP) is one of the most thoroughly studied machine scheduling problems, which schedules a set of jobs on a set of machines according to a specific order. In this paper, we focus on optimizing two objectives: total completion time and total tardiness.

5.2.1   Bi-objective Flow Shop Problem

Generally, the FSP deals with n jobs $\{J_1, J_2, ..., J_n\}$ and m machines $\{M_1, M_2, ..., M_m\}$, where each job has to be processed on all the machines in the same machine sequence. Each machine could only process one job at a time, and the machines can not be interrupted once they start processing a job. As soon as the operation is finished, the machines become available.

Specifically, each job J_i is composed of m consecutive tasks $\{t_{i1}, t_{i2}, ..., t_{im}\}$, where t_{ij} represents the jth task of the job J_i requiring the machine m_j. Each task t_{ij} is associated with a processing time p_{ij}, which is scheduled at the time s_{ij} and should be achieved before the due date d_j. Actually, we aim to minimize two objective functions: total completion time C_{max} and total tardiness T, which are formally defined as follows:

$$f_1 = C_{max} = \max_{i \in [1...n]} \{s_{im} + p_{im}\} \tag{1}$$

$$f_2 = T = \sum_{i=1}^{n} [\max(0, s_{im} + p_{im} - d_i)] \tag{2}$$

Both of them have been proven to be NP-hard [9,7]. In addition, all the FSP instances used in this paper are taken from Taillard benchmark instances and extended into bi-objective case [17][2].

5.2.2   Path Generation

A candidate solution to FSP can be encoded as a permutation $\mathcal{P}$ composed of $\{0, ..., n-1\}$ values, such that $\mathcal{P}(i)$ denotes the job to be executed at the ith position. As proved in [15], the insertion operator, which inserts a selected job to a designated position, is more effective than other operators in solving FSP. Moreover, the authors in [4] show the insertion operator is also very efficient in solving multi-objective FSPs.

Therefore, we decide to define our distance measure directly related to the insertion operator. This property allows us to to compute the minimum number of moves, which have to be applied on an initial solution to reach a guiding solution. As suggested in [2], we use the Longest Common Subsequence (LCS) between two solutions as a distance measure for path generation. The LCS can be calculated in $O(n^2)$ by a dynamic programming algorithm, which is similar to the well known Needleman-Wunsch algorithm [5,14]. Then, the distance between two solutions is defined as the length of permutation minus the length of LCS.

After the distance computation, we generate a path in a random way. In this method, we randomly select a candidate job, and insert this job into a randomly selected position. In fact, this method consists of four main steps:

Step 1: We randomly select a candidate job from an initial solution. For example, in Fig. 3, the longest common subsequence between an initial solution and a guiding solution is colored in black, the remaining jobs are candidate jobs. In this example, the candidate job 15 is randomly selected.

---

[2] Benchmarks available at
http://www.lifl.fr/ liefooga/benchmarks/index.html

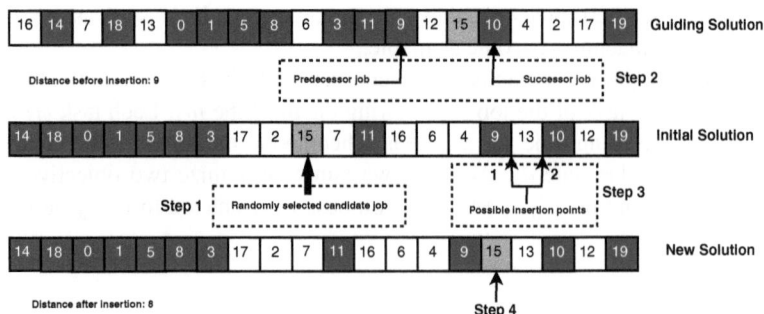

Fig. 3. Path generation for flow shop problem

Step 2: We find the position of the selected candidate job in the LCS of the guiding solution. In Fig. 3, the candidate job 15 is located between two jobs 9 and 10.

Step 3: We find the insertion position for the selected candidate job in the LCS of the initial solution. As shown in Fig. 3, there are two possible insertion positions for the job 15: (9 13) and (13 10).

Step 4: We insert the selected candidate job into a randomly selected insertion position to generate a new solution in the path. As illustrated in Fig. 3, we insert the job 15 into the randomly selected insertion position (9 13) to obtain a new solution. We continue the process in this manner until the distance between the new solution and the guiding solution equals to 0.

5.2.3   Parameters Settings

The proposed algorithms require to set a few parameters, we mainly discuss two important ones: running time and population size.

Running time: The running time T is a key parameter in the experiments. We define the time T for each instance by Equation 3, in which N_{Job} and N_{Mac} represent the number of jobs and the number of machines of one instance, N_{Obj} represents the number of objectives (see Table 1).

$$T = \frac{N_{Job}^2 \times N_{Mac} \times N_{Obj}}{100} \; sec \tag{3}$$

T is defined according to the "difficulty" of instance. Indeed, N_{Job} defines the size of search space, which is $N_{Job}!$. Moreover, the roughness of landscape is strongly related with N_{Mac}. Then, we use this formula to obtain a good balance between the problem difficulty and the time allowed.

Population size: According to the results obtained in [1], the experiments realized previously on the IBMOLS algorithm showed that the best results are achieved with a small population size N. We set this size from 10 to 40 individuals by Equation 4, relative to the size of tested instance (see Table 1).

$$|N| = \begin{cases} 10 & : & 0 < |N_{Job} \times N_{Mac}| < 500 \\ 20 & : & 500 \leq |N_{Job} \times N_{Mac}| < 1000 \\ 30 & : & 1000 \leq |N_{Job} \times N_{Mac}| < 2000 \\ 40 & : & 2000 \leq |N_{Job} \times N_{Mac}| < 3000 \end{cases} \tag{4}$$

Table 1. Parameter values used for bi-objective FSP instances (i_j_k represents the kth bi-objective FSP instance with i jobs and j machines): population size (N) and running time (T)

Instance	Dim	N	T	Instance	Dim	N	T
20_05_01_ta001	20 × 5	10	40"	50_15_01	50 × 15	20	12'30"
20_10_01_ta011	20 × 10	10	80"	50_20_01_ta051	50 × 20	30	16'40"
20_15_01	20 × 15	10	2'	70_05_01	70 × 5	10	8'10"
20_20_01_ta021	20 × 20	10	2'40"	70_10_01	70 × 10	20	16'20"
30_05_01	30 × 5	10	1'30"	70_15_01	70 × 15	30	24'30"
30_10_01	30 × 10	10	3'	70_20_01	70 × 20	30	32'40"
30_15_01	30 × 15	10	4'30"	100_05_01_ta061	100 × 5	20	16'40"
30_20_01	30 × 20	20	6'	100_10_01_ta071	100 × 10	30	33'20"
50_05_01_ta031	50 × 5	10	4'10"	100_15_01	100 × 15	30	50'
50_10_01_ta041	50 × 10	20	8'20"	100_20_01_ta081	100 × 20	40	66'40"

5.2.4 Experimental Results

The computational results are summarized in Table 2. In this table, we observe that RM has a good performance on the first eight instances from 20_5_01 to 30_20_01. It obtains the best average hypervolume differences on these instances. On the other hand, PR_KM outperforms the other algorithms on the remaining instances from 50_5_01 to 100_20_01, where almost all the best results are obtained by this algorithm. Additionally, CO is less effective in comparison with RM and PR_KM.

From Table 2, we can see the path relinking techniques have a limited contribution on the small instances from 20_5_01 to 30_20_01. We suppose that, when the instance size is small, the length of the path is so short that it is difficult to find a set of solutions far enough from the initial and guiding solutions to initialize a new population. In this case, it is more useful to perform random moves in the search space as done in RM. When we consider the instances with more than 30 jobs, the length of the path is longer, which means we have more possibilities to explore new high quality areas in the search space. Therefore, PR_KM has a good performance on the large instances from 50_5_01 to 100_20_01.

Table 2. Comparison of four versions of hypervolume-based multi-objective path relinking algorithm (PR_A, PR_B, PR_M and PR_KM) with two versions of HBMOLS (RM and CO) on 20 bi-objective FSP instances from 20_5_01 to 100_20_01. Each value in the table represents an average hypervolume difference.

Instance	Algorithm					
	PR_A	PR_B	PR_M	PR_KM	RM	CO
20_05_01_ta001	0.050496	0.076627	0.093801	0.067028	**0.000260**	0.005152
20_10_01_ta011	0.023355	0.055498	0.048349	0.034595	**0.000739**	0.027353
20_15_01	0.032433	0.073174	0.070448	0.037654	**0.002330**	0.037131
20_20_01_ta021	0.009737	0.034508	0.024761	0.010079	**0.000077**	0.044826
30_05_01	0.049260	0.081154	0.099705	0.040607	**0.011844**	0.062030
30_10_01	0.100098	0.200979	0.176367	0.088794	**0.041814**	0.116553
30_15_01	0.052479	0.096203	0.105293	0.048227	**0.028186**	0.054050
30_20_01	0.048423	0.064844	0.071167	*0.040580*	**0.035835**	0.051028
50_05_01_ta031	0.031220	0.083466	0.090345	**0.022628**	0.041017	0.056559
50_10_01_ta041	0.103891	0.149919	0.132192	**0.079505**	0.089703	0.116051
50_15_01	0.131563	0.173639	0.156972	**0.091552**	0.114880	0.131505
50_20_01_ta051	0.129671	0.176523	0.146388	**0.093540**	0.117150	0.141695
70_05_01	0.110650	0.191452	0.152058	*0.096111*	**0.084047**	0.146741
70_10_01	0.131195	0.177933	0.157369	**0.119054**	0.146445	0.172327
70_15_01	0.149831	0.174514	0.164179	**0.134607**	0.156965	0.178769
70_20_01	0.139377	0.183869	0.147617	**0.102067**	0.135491	0.137697
100_05_01_ta061	0.199309	0.359023	0.236139	**0.157834**	*0.169815*	*0.175162*
100_10_01_ta071	0.093883	0.121682	0.104086	**0.071063**	*0.080287*	0.086577
100_15_01	0.187296	0.205879	0.175943	**0.128876**	0.163312	0.174849
100_20_01_ta081	0.205930	0.220908	0.187275	**0.131843**	*0.137246*	0.180406

Compared with other versions of hypervolume-based multi-objective path relinking algorithms, the advantages of PR_KM are very clear. As N_{KM} is smaller than N_{All}, in most cases, PR_KM saves a lot of time during the initializing process, then it performs more effectively than PR_A, especially on the large instances. Considering PR_B, we select a set of non-dominated solutions from the path. However, these solutions are often close to the initial solution and the guiding solution. The similar search areas have little contribution in initializing a new population, which decreases the global effectiveness of PR_B. For PR_M, only one intermediate solution is selected from the path at each step, which means this algorithm spends a little time in the initializing process. Then, it is not very helpful to reinforce the population's diversity. For this reason, the effectiveness of PR_M is affected.

5.3   Application to Bi-objective Quadratic Assignment Problem

The quadratic assignment problem (QAP) is a classical combinatorial optimization problem both in theory and in practice. As one of the most difficult problems in the NP-hard class, it models many real-life problems in many areas such as the facility location, parallel and distribute computing, and combinatorial data analysis [11]. In our case, we concentrate on bi-objective quadratic assignment problem.

5.3.1 Bi-objective Quadratic Assignment Problem

The quadratic assignment problem can be described as the problem of assigning a set of facilities to a set of locations with given distances between the locations and given flows between the facilities [12]. Given n facilities and n locations, three $n \times n$ matrices D, F_1 and F_2, where d_{ij} is the distance between location i and j, and f_{rs}^1 and f_{rs}^2 are two flows between two facilities r and s. The goal is to minimize the sum of the product between flows and distances. The objective of the QAP can then be formulated as follows:

$$\min_{\phi \in \Phi} \sum_{i=1}^{n} \sum_{j=1}^{n} d_{ij} f_{\phi_i \phi_j}^k, \ k \in \{1, 2\} \tag{5}$$

where Φ is the set of all permutations of $\{1, \ldots, n\}$, and ϕ_i gives the location of item i in a solution $\phi \in \Phi$.

In this paper, all the tested instances of QAP are provided by R. E. Burkard et al.[3] In our case, a bi-objective QAP instance is generated by keeping the distance matrix of the first instance and using two different flow matrices. Moreover, we denote a bi-objective instance as N_i_ab (N represents the name of instance such as "esc") with a matrix of size i respectively. For example, esc_32_ab denotes a bi-objective instance named "esc", which is generated by two single-objective instances esc_32_a and esc_32_b.

5.3.2 Path Generation

A candidate solution to QAP can be encoded as a permutation $\mathcal{P}$ composed of $\{1, \ldots, n\}$ values, such that $\mathcal{P}(i)$ denotes the facility to be assigned at the ith location. As proved in [16], the swap operator, which exchanges two facilities in a permutation, is very effective for solving QAP. Then, we define the distance between two solutions directly related to the swap operator.

For QAP, we use the permutation distance and the cycle distance [18,14] as the distance measure. Actually, the distance between two solutions is defined as the permutation distance minus the cycle distance. Afterwards, we construct a path by randomly selecting an element from one cycle in a permutation and applying the swap operator to this element to obtain a new solution.

An example of path generation for QAP is illustrated in Fig. 4. In this example, there is one integer element (11) located at the same position in an initial solution and a guiding solution, then the permutation distance is 10. On the other hand, there are three cycles ($\{3, 1, 2, 7, 8\}$, $\{4, 5, 6\}$ and $\{10, 9\}$) between these two permutations, so the cycle distance is 3. Therefore, the distance between the initial solution and the guiding solution is equal to 7.

Furthermore, there are 7 steps starting from the initial solution P_x to the guiding solution P_y, which allows us to generate 6 solutions on the path. For instance, we first randomly select a facility 2 from one cycle $\{3, 1, 2, 7, 8\}$ in P_x, and we can observe the facility 2 is located at the second position in P_y. Then, we apply the swap operator to

---

[3] Benchmarks available at http://www.seas.upenn.edu/qaplib/inst.html

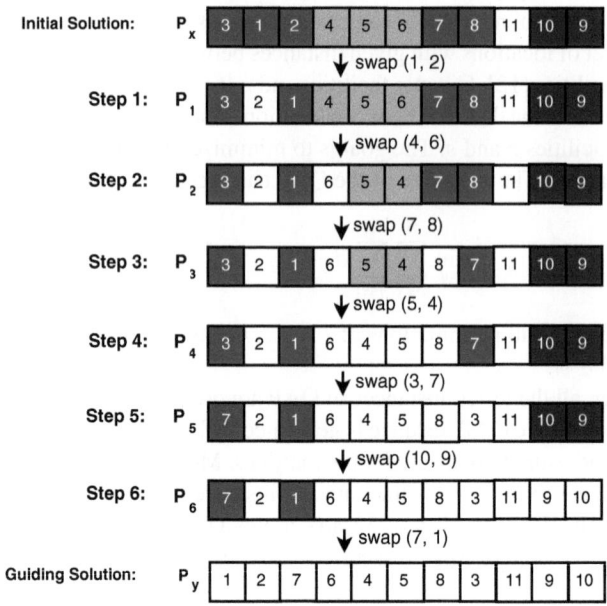

Fig. 4. Path generation for quadratic assignment problem

two facilities 1 and 2 in P_x in order to generate a new solution. We continue this process until the distance between the new solution P_i and the guiding solution P_y is equal to 0.

5.3.3   Parameters Settings

Similar to the parameter settings in FSP, we consider two important parameters: running time and population size.

– **Running time:** We define the running time T for each instance by Equation 6, in which N_{Dis}, N_{Flow} and N_{Obj} represent respectively the size of the distance matrix, the size of the flow matrix and the number of objectives in an instance (see Table 3).

$$T = N_{Dis} \times N_{Flow} \times N_{Obj} \; sec \tag{6}$$

– **Population size:** Here, we set this size from 10 to 30 individuals according to Equation 7, relatively to the size of the tested instance (see table 3).

$$|N| = \begin{cases} 10 & : & 0 < |N_{Dis} \times N_{Flow}| < 500 \\ 20 & : & 500 \leq |N_{Dis} \times N_{Flow}| < 1000 \\ 30 & : & 1000 \leq |N_{Dis} \times N_{Flow}| < 2000 \\ 40 & : & 2000 \leq |N_{Dis} \times N_{Flow}| < 3000 \end{cases} \tag{7}$$

Table 3. The instances of bi-objective quadratic assignment problem (Parameters: population size N, running time T)

Inst 1	Inst 2	Dim	N	T
chr_12_a	chr_12_b	12×12	10	4'48"
chr_15_a	chr_15_b	15×15	10	7'30"
chr_20_a	chr_20_b	20×20	10	13'20"
esc_16_a	esc_16_b	16×16	10	8'32"
esc_32_a	esc_32_b	32×32	30	16'20"
Lipa_30_a	Lipa_30_b	30×30	20	15'
Ste_36_a	Ste_36_b	36×36	30	21'36"
tai_40_a	tai_40_b	40×40	30	26'40"
tai_50_a	tai_50_b	50×50	40	41'40"

5.3.4   Experimental Results

The computational results for the bi-objective QAP are presented in Table 4. From this table, we can see RM has a good performance almost on all the instances. Particularly, it obtains the best average hypervolume differences on five instances. Moreover, PR_KM also obtains very competitive results on all the instances, especially on the large instances, such as Lipa_30_ab, tai_40_ab and tai_50_ab. However, CO is statistically outperformed by RM and PR_KM on most of the instances.

Table 4. Comparison of four versions of hypervolume-based multi-objective path relinking algorithm (PR_A, PR_B, PR_M and PR_KM) with two versions of HBMOLS (RM and CO) on 9 bi-objective QAP instances. Each value in the table represents an average hypervolume difference.

Instance	Algorithm					
	PR_A	PR_B	PR_M	PR_KM	RM	CO
chr_12_ab	**0.000000**	**0.000000**	**0.000000**	**0.000000**	**0.000000**	0.013407
chr_15_ab	0.002988	0.010994	**0.000000**	0.002271	**0.000000**	0.026494
chr_20_ab	0.014042	0.025258	0.004827	0.005560	**0.001899**	0.017890
esc_16_ab	**0.000000**	**0.000000**	**0.000000**	**0.000000**	**0.000000**	**0.000000**
esc_32_ab	0.006312	0.008839	0.002594	0.003003	**0.002433**	0.007948
Lipa_30_ab	0.001956	0.002159	*0.001369*	0.001347	0.001433	0.003047
Ste_36_ab	0.304087	0.356747	0.669592	0.364021	*0.215314*	**0.203776**
tai_40_ab	0.037541	0.041880	0.031969	**0.027092**	0.046076	0.080534
tai_50_ab	0.038647	0.030516	0.040565	**0.027077**	0.048410	0.046201

According to the experimental results in table 4, RM has a better performance than PR_KM on the first four instances. Since these instances are small and relatively easy to solve, the PR_KM and RM algorithms achieve the best results on three instances (chr_12_ab, chr_15_ab and esc_16_ab), where the average hypervolume differences are equal to 0. Furthermore, on these small instances, it is not easy for PR_KM to construct a long path to find enough diversified solutions for initializing a new population. Then it is better to perform random moves in the search space or to select only one solution from the generated path as done in PR_M. When the size of instance becomes larger, we can construct a longer path and select more useful solutions from the path, which means we have more chances to explore high quality areas in the objective space. Therefore, PR_KM obtains the best value on the large instances such as tai_50_ab and a competitive value on the instance esc_32_ab. However, the instance Ste_36_ab is an exception, CO obtains the best value on this instance. In fact, only several non-dominated solutions are found in the population. We suppose that the search procedure is often trapped in some local optimums, then using crossover operator is a better way to be out of these traps.

6   Conclusions and Perspectives

In this paper, we present a hypervolume-based multi-objective path relinking algorithm, which is applied to the bi-objective flow shop problem and bi-objective quadratic assignment problem. This algorithm integrates the path relinking techniques into hypervol-umebased multi-objective local search as an initialization function, in order to find a Pareto approximation set. Actually, we provide a general scheme of path relinking algorithm, which can be used to deal with other multi-objective optimization problems.

Experimental results indicate one version of our proposed algorithms is very competitive in comparison with other algorithms. The performance analysis gives us a few directions for future research. The first possibility is to generate more intermediate solutions at each step, then one can construct several different paths simultaneously. Especially, for each path, it could give birth to another path in reverse direction. Second, it is worth proposing other mechanisms of subset selection. The new mechanisms could have the potential to obtain a better Pareto approximation set.

On the other hand, it should be very interesting to integrate MOPR into other metaheuristics such as tabu search, in order to evaluate its overall effectiveness. The cooperation of MOPR with exact methods can be also a promising search area. For instance, MOPR could be used to link Pareto optimal solutions found by an exact approach. Several approaches between MOPR and exact approaches could be defined, as those described in the taxonomy of Jourdan et al. [10].

Acknowledgment. The work is partially supported by the "Pays de la Loire" Region (France) within the RaDaPop (2009-2013) and LigeRO (2010-2013) projects. We thank the reviewers of the paper for their helpful comments.

References

1. Basseur, M., Liefooghe, A., Le, K., Burke, E.: The efficiency of indicator-based local search for multi-objective combinatorial optimisation problems. Journal of Heuristics 18(2), 263–296 (2012)

2. Basseur, M., Seynhaeve, F., Talbi, E.-G.: Path Relinking in Pareto Multi-objective Genetic Algorithms. In: Coello Coello, C.A., Hernández Aguirre, A., Zitzler, E. (eds.) EMO 2005. LNCS, vol. 3410, pp. 120–134. Springer, Heidelberg (2005)
3. Basseur, M., Zeng, R.Q., Hao, J.K.: Hypervolume-based multi-objective local search. Neural Computing and Applications 21(8), 1917–1929 (2012)
4. Brizuela, C., Aceves, R.: Experimental Genetic Operators Analysis for the Multi-objective Permutation Flowshop. In: Fonseca, C.M., Fleming, P.J., Zitzler, E., Deb, K., Thiele, L. (eds.) EMO 2003. LNCS, vol. 2632, pp. 578–592. Springer, Heidelberg (2003)
5. Cormen, T.H., Leiserson, C.E., Rivest, R.L.: Introduction to Algorithms. The MIT Press, Cambrige (1990)
6. Drugan, M.M., Thierens, D.: Path-Guided Mutation for Stochastic Pareto Local Search Algorithms. In: Schaefer, R., Cotta, C., Kołodziej, J., Rudolph, G. (eds.) PPSN XI. LNCS, vol. 6238, pp. 485–495. Springer, Heidelberg (2010)
7. Du, J., Leung, J.Y.-T.: Minimizing total tardiness on one machine is NP-hard. Mathematics of Operations Research 15, 483–495 (1990)
8. Glover, F., Laguna, M.: Fundamentals of scatter search and path relinking. Control and Cybernetics 29, 653–684 (1999)
9. Graham, R.L., Lawler, E.L., Lenstra, J.K., Rinnooy Kan, A.H.G.: Optimization and approximation in deterministic sequencing and scheduling: A survey. Annals of Discrete Mathematics 5, 287–326 (1979)
10. Jourdan, L., Basseur, M., Talbi, E.: Hybridizing exact methods and metaheuristics: A taxonomy. European Journal of Operational Research 199(3), 620–629 (2009)
11. Loiola, E.M., de Abreu, N.M.M., Boaventura-Netto, P.O., Querido, P., Querido, T.: A survey for the quadratic assignment problem. European Journal of Operational Reasearch 176, 657–690 (2007)
12. Pardalos, P., Rendl, F., Wolkowicz, H.: The quadratic assignment problem: A survey and recent developments. In: Proceedings of the DIMACS Workshop on Quadratic Assignment Problems. DIMACS Series in Discrete Mathematics and Theoretical Computer Science, vol. 16, pp. 1–42 (1994)
13. Pasia, J.M., Gandibleux, X., Doerner, K.F., Hartl, R.F.: Local Search Guided by Path Relinking and Heuristic Bounds. In: Obayashi, S., Deb, K., Poloni, C., Hiroyasu, T., Murata, T. (eds.) EMO 2007. LNCS, vol. 4403, pp. 501–515. Springer, Heidelberg (2007)
14. Schiavinotto, T., Stützle, T.: A review of metrics on permutations for search landscape analysis. Computers and Operations Research 34(10), 3143–3153 (2011)
15. Taillard, E.: Some efficient heuristic methods for flow-shop sequencing. European Journal of Operational Research 47, 65–74 (1990)
16. Taillard, E.: Robust taboo search for the quadratic assignment problem. Parallel Computing 17, 443–455 (1991)
17. Taillard, E.: Benchmarks for basic scheduling problems. European Journal of Operational Research 64, 278–285 (1993)
18. Thierens, D.: Exploration and Exploitation Bias of Crossover and Path Relinking for Permutation Problems. In: Runarsson, T.P., Beyer, H.-G., Burke, E.K., Merelo-Guervós, J.J., Whitley, L.D., Yao, X. (eds.) PPSN IX. LNCS, vol. 4193, pp. 1028–1037. Springer, Heidelberg (2006)
19. Zitzler, E., Thiele, L.: Multiobjective evolutionary algorithms: A comparative case study and the strength pareto approach. Evolutionary Computation 3, 257–271 (1999)

Multiobjective Path Relinking for Biclustering: Application to Microarray Data

Khedidja Seridi, Laetitia Jourdan, and El-Ghazali Talbi

INRIA Lille-Nord Europe/LIFL/CNRS
40 Avenue Halley 59650 Villeneuve d'Ascq France
{Khedidja.seridi,laetitia.jourdan}@inria.fr, talbi@lifl.fr

Abstract. In this work we deal with a multiobjective biclustering problem applied to microarray data. $MOBI_{nsga}$ [21] is one of the multiobjective metaheuristics that have been proposed to solve a new multiobjective formulation of the biclustering problem. Using $MOBI_{nsga}$, biclusters of good quality can be extracted. However, the generated front approximation contains a lot of gaps. Using path relinking strategies, our aim is to improve the generated front's quality by filling the gaps with new solutions. Therefore, we propose a general scheme PR-$MOBI_{nsga}$ of different possible hybridization of $MOBI_{nsga}$ with path relinking strategies. A comparison of different PR-$MOBI_{nsga}$ hybridizations is performed. Experimental results on reel data sets show that PR-$MOBI_{nsga}$ allows to extract new interesting solutions and to improve the Pareto front approximation generated by $MOBI_{nsga}$.

1 Introduction

Biclustering (also called co-clustering, or two-mode clustering) is a well-known data mining task. It has been widely applied in a broad range of domain such as marketing, psychology and bioinformatics. Within the field of bioinformatics, important applications have appeared with regard to the study of microarrays data analysis. Microarray technologies allow studying thousands of genes behavior under several conditions. These studies result in a large amount of data that is usually presented in 2D matrices, where rows represent genes and columns represent experimental conditions. Given a matrix data, biclustering performs simultaneously the selection of rows and columns of a data matrix leading to the discovery of *biclusters*. The first biclustering algorithm applied to the analysis of microarray data was proposed by Cheng and Church[5].

Extracting biclusters from a microarray data can be formulated as a combinatorial optimization problem, where two objectives are to be maximized: the similarity (coherence) between the bicluster's elements and its size. As dissimilarity measure for microarray data, *Mean Squared Residue* (MSR) [5] is widely used.

Furthermore, in microarray data analysis, biologists are usually interested in extracting biclusters that present some fluctuations in the rows (non-flat biclusters). Hence, *mean rows variances* can be considered as a third objective.

R.C. Purshouse et al. (Eds.): EMO 2013, LNCS 7811, pp. 200–214, 2013.

Since these criteria (size, coherence and mean rows variances) are usually conflicting, some multiobjective models have been proposed to formulate the biclustering problem [11,14,15,16,17,20].

For real-life or classical optimization, combining metaheuristics with other methods such as complementary metaheuristics and exact methods, provides very powerful search algorithms [22]. One of the methods that can be combined to metaheuristics is Path Relinking (PR) strategy. Given two solutions s and t, PR consists in creating a path of solutions that links them.

$MOBI_{nsga}$ [21] is a hybrid MOEA (Multi Objective Evolutionary Algorithm) for solving biclustering problem in the case of microarray data. $MOBI_{nsga}$ allows to extract biclusters of good quality. However, the generated front approximation is discontinuous i.e. several gaps exits along the entire front. In this work we aim to improve the front's quality by filling the gaps by new solutions generated using PR strategies. For that, we proposed a general scheme (PR-$MOBI_{nsga}$) of metaheuristics composed of $MOBI_{nsga}$ in conjunction with different PR variants. Moreover, we evaluate the effectiveness of PR-$MOBI_{nsga}$ variants and perform a comparison between them.

This paper is organized as follows: section 2 gives some details about the problem modeling and $MOBI_{nsga}$ algorithm. Then, definitions related to the multiobjective path relinking and its different variants are given. Experimental results and a comparative analysis are discussed in section 3. The last section concludes the paper and gives some perspectives.

2 Multiobjective Path Relinking for Gene Expression Data

In this section, we present our multiobjective model for the biclustering of microarray data, and give some details about $MOBI_{nsga}$. After that, PR strategy is presented and the different ways of integrating it in a multiobjective metaheuristic are detailed. Finally we give the general scheme of PR-$MOBI_{nsga}$ algorithms.

2.1 Multiobjective Biclustering

Let $X - (C, G)$ be a microarray data matrix, where $C = \{C_1, C_2, ...C_N\}$ represents a set of N conditions and $G = \{G_1, ..., G_M\}$ a set of M genes, and $a_{ij} \in A$ $(i \in X, i \in Y)$ represents the expression level of gene i under condition j. A bicluster B is a submatrix of X defined by a subset of conditions $I \subset C$ and a subset of genes $J \subset G$: $(B = (I, J))$. The MSR value of a bicluster B=(I,J) is defined by:

$$MSR(I, J) = \frac{1}{|I| \times |J|} \times \sum_{i \in I, j \in J} (a_{ij} - a_{iJ} - a_{Ij} + a_{IJ})^2$$

where a_{iJ} represents the mean of the i-th row of B, a_{Ij} represents the mean of the j-th column of B and a_{IJ} the mean of all the elements in B.

In order to optimize the three conflicting criteria (size, coherence and mean rows variances) we consider respectively three objective functions f_1, f_2 and f_3, where:

$$f_1(I, J) = \frac{1}{2} \times \frac{|I|}{|X|} + \frac{1}{2} \times \frac{|J|}{|Y|}$$

$$f_2(I, J) = \begin{cases} \frac{MSR(I,J)}{\delta} & \text{if} MSR(I, J) \leqslant \delta \\ 0 & \text{else} \end{cases}$$

$$f_3(I, J) = \frac{1}{Rvar(I,J)+1} \quad (\text{Rvar(I,J): mean of rows variances})$$

The objectives f_1 and f_2 are to be maximized whereas f_3 has to be minimized. δ is user-threshold that represents the maximum dissimilarity allowed within the bicluster. As the size and similarity criterion are conflicting, we allow the function f_2 to be maximized as long as the residue does not exceed the threshold δ, while f_1 (size) is always maximized and f_3 is always minimized. More details about the proposed model are given in [21].

In order to solve this model, we have proposed a multiobjective evolutionary algorithm called $MOBI_{nsga}$ based on NSGA-II and hybridized with a local search inspired from Cheng and Church's heuristic. All details about $MOBI_{nsga}$ are given in [21].

2.2   Solutions Encoding

In our approach, each solutions represents a bicluster. We choose to represent a bicluster as a list compound of four parts: the first part is an ordered rows indexes, the second part is an ordered columns indexes, the third part is the rows number and the fourth part is the columns number. By this representation we aim to reduce time and memory space especially for local search based metaheuristics.

Given the data matrix presented in Figure 1, the string {1 3 2 3 2 2} represents the bicluster compound of the rows (1 and 3) and the columns (2 and 3). The last numbers (2 and 2) indicate that the bicluster contains 2 rows and 2 columns.

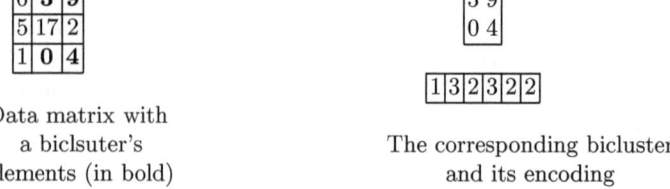

Data matrix with
a biclsuter's
elements (in bold)

The corresponding bicluster
and its encoding

Fig. 1. Example of a solution encoding

In literature, a solution (bicluster) is usually represented by a fixed size binary string, with a bit string for genes appended by another bit string for conditions. A bit is set to one if the corresponding gene/condition is present in the bicluster, and set to zero otherwise. Other representation is considered in [19], where a solution represents a set of biclusters. A solution is compound of two parts: one for clustering the genes, and another for clustering the conditions. The first M positions represent the M gene cluster centers, and the remaining N positions represent the N condition cluster centers.

2.3   Multiobjective Path Relinking

Path Relinking (PR) approach was originally proposed by Glover et al. [9] within the framework of scatter search. It allows exploring paths connecting elite solutions found by scatter search. However, this strategy may be generalized and applied to any population-based metaheuristic generating a pool of "good" solutions such as evolutionary algorithms, greedy adaptive search procedure (GRASP), ant colony, and iterative local search. A general scheme is given in Figure 2.3.

Starting from a solution s, PR approach generates and explores the trajectory in the neighborhood space that leads to a *target* solution t. A sequence of neighboring solutions in the decision space is generated from the *starting* solution to the *target* solution. The best found solution in the sequence is returned. Adding some (good) solutions contained in the trajectory allow integrating intensification and diversifications strategies.

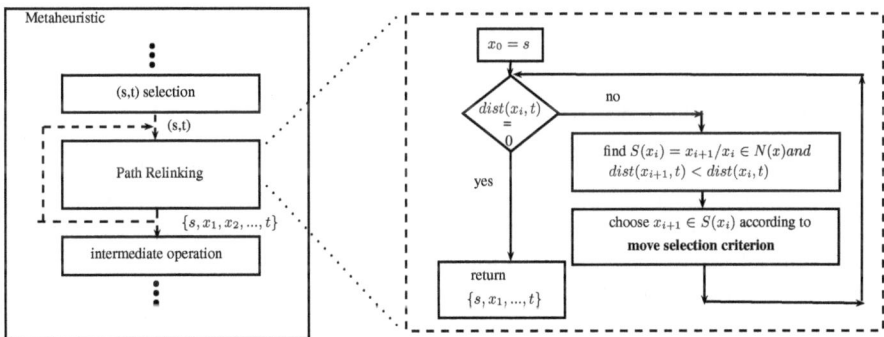

Fig. 2.   Path   Relinking   strategy   within   a   metaheuristic   framework $\{x_1, ..., x_n\}$ is a set of solutions that links s and t

Some studies used PR in the multiobjective context [2,3,18]. In [3], [2] and [18] the PR is used in order to improve the Pareto approximation found by using respectively a Genetic Algorithm, Iterative Local Search and GRASP algorithm. In fact, applying PR strategy on solutions chosen from the Pareto front approximation may improve its diversification and convergence toward the Pareto

Optimal Front. On the other hand, if there is a gap of solutions along the Pareto front, the path exploration feature would search for non-dominated solutions between solutions at either side of the gap.

2.4 PR-$MOBI_{nsga}$ Algorithms

In our work, we integrate PR strategy in order to improve the results given by $MOBI_{nsga}$. In PR-$MOBI_{nsga}$ the PR strategy is applied over some solutions selected from the Pareto front approximation generated by $MOBI_{nsga}$.

Before integrating path relinking, multiple *strategies* in the different PR steps have to be fixed, mainly:

- pre-PR: Before starting the application of PR, two solutions (starting and target solutions) have to be selected. This selection is done according to a criterion that defines the goal of integrating the PR strategy, *i.e* intensification, diversification, etc.
- During PR: in order to generate the path that links *starting* and *target* solutions, a neighborhood operator has to be defined. On the other hand, a distance operator is required to be able to measure the progress while generating the path. Once the neighborhood and distance operators are defined, we need to define a **move selection criterion**.
- Post-PR: Some of the obtained solutions may be considered to undergo an intermediate operation in order to improve their quality. Usually a local search is involved.

In the following we detail the different strategies to be fixed in combining a population-based multiobjective metaheuristic with path relinking.

Selection of Linked Solutions

The choice of the linked solutions is very important in designing the PR as different choices lead most of the times to different results. In mono-objective optimization, generally *good* solutions are chosen to be linked, which corresponds, in the multiobjective case, to non-dominated solutions.

In the case where the distances in the objective and decision space are correlated, favoring distant solutions may favor the exploration of the search space, while favoring adjacent solutions may favor intensification of the search process. However, if the distances are not correlated we cannot predict the position of the new solutions generated by PR relative to the linked solutions in the objective space. As a result, it is useless to favor particular way in selecting linked solutions.

As distance measure, we have considered *Euclidean distance* in the objective space and *Edit distance* in the decision space. The *Edit distance* is defined by the minimal cost to transform one string into the other via a sequence of edit operations (usually insertions, deletions and replacements), where insertions and deletions have equal cost and replacements have twice the cost of an insertion [8].

For our case, studying the correlation between the two distances in the two spaces revealed that they are not correlated. Hence, we deduce that the distance cannot be used as a selection criterion of the starting and the target solutions. Consequently, a random selection of the solutions seems to be the best strategy of selection.

Neighborhood and Distance Operators

The neighborhood operator has a high influence on the quality of the algorithm. In order to perform the operation of 'linking' the initial and the target solutions, it is necessary to be able to generate all (or at least a part of) the neighbors to *get closer* to the target solution. Iterating the *getting closer* mechanism allows the construction of one, several, or all the paths linking the two solutions. Besides the neighborhood operator it is necessary to define a distance operator in order to guarantee that at each step of the path construction, the distance to the target solution decreases.

Given two solutions s and t. Let Ind_{s-t} be the set of rows and columns indexes in s and not present in t and symmetrically, let Ind_{t-s} be the set of rows and columns indexes in t and not present in s. The trajectory between s and t can be defined by adding (and removing) sequentially the elements of: Ind_{t-s} (Ind_{s-t}). Figure 3 gives an example of possible moves.

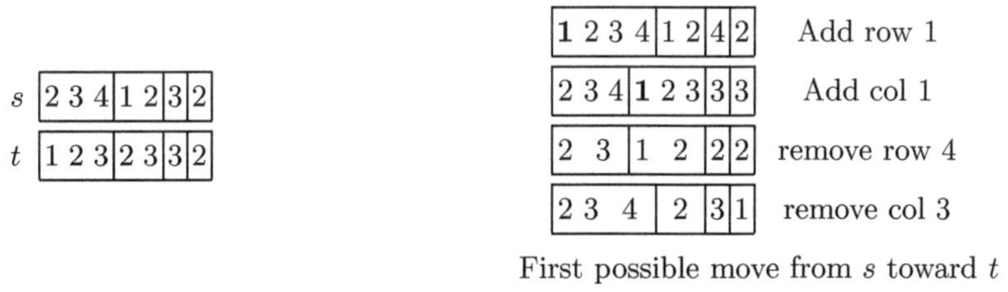

s : 2 3 4 | 1 2 | 3 | 2

t : 1 2 3 | 2 3 | 3 | 2

1 2 3 4	1 2	4	2	Add row 1
2 3 4	1 2 3	3	3	Add col 1
2 3	1 2	2	2	remove row 4
2 3 4	2	3	1	remove col 3

First possible move from s toward t

Fig. 3. Example of the possible moves that can be applied to s to get closer to t

As distance operator between the solutions, we choose the Edit distance. Hence, the number of moves (distance) that have to be applied to join t starting form s is defined by the sum of Ind_{s-t} and Ind_{t-s} cardinalities.

$$dist(s,t) \; = \; editDistance(s,t) \; = \; card(Ind_{s-t}) \; + \; card(Ind_{s-t})$$

Let x_i and x_{i+1} be two successive solutions of the path that links s to t. Thus, x_{i+1} is obtained by adding (or removing) an element j of the Ind_{t-s} (or Ind_{s-t}) set.

$$x_{i+1} = x_i \cup \{j\}, \text{ where } j \in Ind_{t-s} \text{ or } x_{i+1} = x_i - \{j\}, \text{ where } j \in Ind_{s-t}$$

Therefore, $dist(x_{i+1}) = dist(x_i) - 1$, i.e. each move applied allows to reduce the distance to the target solution.

Move Selection

Using the neighborhood operator, we can generate several paths that link solutions s and t. As the number of explored solutions grows exponentially with the distance between the two considered solutions, it is time consuming task to generate all the paths. Thus, at each step in path construction (to obtain x_{i+1} form the current solution x_i), we have to decide which move to apply (which $j \in (Ind_{t-s} \cup Ind_{s-t})$ will be added or removed). Different strategies may be considered, such as:

1. Random selection: where each move in the path construction is randomly chosen from the all the possible moves.
2. Single objective selection: at each move of the path construction, we choose the one producing the *best* solution optimizing one objective function.
3. Multiobjective selection:
   - Pareto based selection: we evaluate all the possibilities for $j \in Ind_{s-t}$ to be removed and $j \in Ind_{t-s}$ to be added and extract the non-dominated set of solutions in the neighborhood space. Once the non-dominated set is found, another strategy is considered to reduce the size of the exploration. Basseur et al. [3] have considered a random aggregation of the objectives to select only one solution in the set of non-dominated solutions.
   - Aggregation selection: the selection of a solution is done be evaluating all the possibilities and choosing the one that optimizes the weighted objective function: $f(x) = \sum_{i \in n} w_i f_i(x)$.
   - Sequential selection: the selection of a solution is done by alternating the objective function used. In this way, if f_k is used to select x_i, then $f_{k+1} \bmod n$ is used to select x_{i+1}, with n is the number of objectives.
4. Hybrid selection: for some problems, it may be important to guide the selection by a combination of several strategies.

Intermediate Operation

After the generation of all the intermediate solutions in the path, we select a set of solutions from this path to integrate the Pareto front approximation. The solutions generated using PR may not be as good as the solutions generated by $MOBI_{nsga}$. In other words, they may be dominated by them. Hence, integrating the new solutions to $MOBI_{nsga}$'s Pareto front approximation may not be possible. However, an intermediate operation can be applied over the selected path solutions in order to improve their convergence toward the Pareto front. Different strategies could be considered in choosing solutions from the path. In [23], Zeng proposed four different criteria: all solutions, non-dominated solutions, middle solution and k-middle solutions. By choosing middle (and k-middle) solutions, the author aim to avoid the problem of proximity of intermediate solutions to initial and guiding solutions. In fact, the solutions located at the beginning or at the end of the path are similar to the initial solution or the guiding solution respectively and they could be not very useful. In our case, we consider all solutions with non-null values in the second objective function, *i.e.* solutions with MSR value below the threshold. In fact, the objective function f_2 of the model is set to 0 when the MSR value of the relevant solution is above a threshold δ, which means that this solution is not interesting.

As an intermediate operation, we use a *Dominance-based Multiobjective Local Search* DMLS [13] algorithm which consists in iteratively improving and updating an archive of non-dominated solutions. At each iteration, one or more non-visited solutions are selected (solutions with non-explored neighborhood) from the *archive*. Then, the neighborhood of the selected solutions is explored looking for new dominating solutions. Several DMLS variants exist depending in the selection and exploration strategies. In our case, we use DMLS(1,*) which corresponds to the well-known PLS-1 algorithm [7]. At each generation, PLS-1 selects *one* non-visited solution and the whole neighborhood.

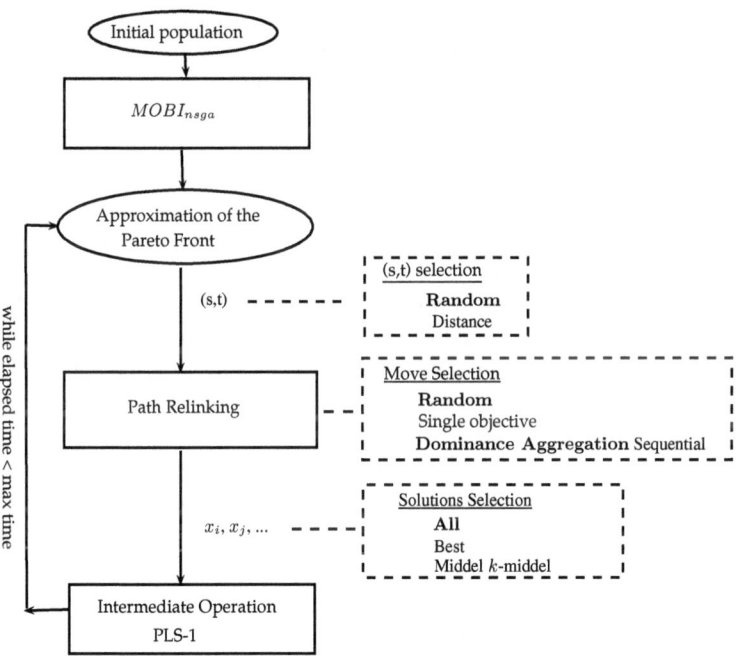

Fig. 4. General scheme of PR-$MOBI_{nsga}$ algorithms. The different options related to the chosen variants RPR, APR and PPR are written in bold.

General Scheme

In our work, we compare different PR strategy variants with different *move selection* strategies. Figure 4 presents the general scheme of the PR-$MOBI_{nsga}$

algorithms. Starting by a random initial population, $MOBI_{nsga}$ generates a first approximation of the Pareto Front. After that, according to a *selection (s,t)* criterion, a starting and a target solution are selected from the Pareto approximation. Then, a PR is applied over the selected solutions based on a *move selection* criterion. Some of the solutions that comprise the path are selected to serve as initials solutions for the PLS-1. In our case, we chose only solutions with MSR value below the threshold δ. The process: (s,t) selection, PR, PLS-1, is repeated while a maximum time is not reached.

3 Experiments and Results

In this study, we will compare the performance of the different PR variants when combined with $MOBI_{nsga}$. We have chosen to study the different variants related to the different multiobjective move selection, mainly: Pareto based selection and aggregation selection. Moreover, we consider the random selection. The different resulting algorithms are PPR-$MOBI_{nsga}$, APR-$MOBI_{nsga}$ and RPR-$MOBI_{nsga}$ respectively. All the implementation have been done thanks ParadisEO framework[1] and in particular with ParadisEO-MOEO [12]. Furthermore, we will determine the biological relevance of some biclusters that have been found through PR for the *Yeast cell-cycle* data.

3.1 Data and Parameters

In our experiments we use the well-known *Yeast Cell Cycle data* [6] (2884 genes and 17 conditions), *Human B-cell expression data* [6] (4026 genes and 96 conditions) and *Colon data* [1] (2000 genes and 62 conditions).

The mechanism of missing data replacement is explained in [21]. For each dataset, we ran 20 times each algorithm and set the population size to 200. The crossover and mutation probabilities are set to: 0.5 and 0.4 respectively. These parameters have been set experimentally. For each data set, the runs are realized with a specific time limit (700 s for Yeast data, 2000 s for Colon data and 3000 s for Human data). For the combined approach, we chose to run $MOBI_{nsga}$ for 10% then 90 % of the total run time. A maximum number of generations is used as a stopping criterion for PLS-1. The number of generations of PLS-1 is set to 5 when the rate of time granted to PR is 10% of total time, and 45 when it is 90%.

3.2 Performance Assessment

In order to evaluate the quality of the non-dominated front approximations obtained for a specific test instance, we follow the protocol given by Knowles *et al.* [10]. For that, we consider the hypervolume difference indicator (I_H^-) and the additive ϵ-indicator ($I_{\epsilon+}^-$) in order to assess the performance of the different algorithms. To this end, for each data set, we compute 20 hypervolume differences and 20 epsilon measures, corresponding to the 20 runs, per algorithm. As

---

[1] http://paradiseo.gforge.inria.fr

suggested by Knowles *et al.*[10] , once all these values are computed, we perform a statistical analysis on pairs of optimization methods for a comparison on a specific test instance. For this purpose, we use the Fisher-Matched statistical test.

Note that all the performance assessment procedures have been achieved using the performance assessment tool suite provided in PISA [4].

3.3   Results and Discussion

Tables 1 and 2 (3 and 4) give a comparison of $MOBI_{nsga}$ and different PR-$MOBI_{nsga}$ algorithms with regard to the hypervolume indicator (I_H^-) and the epsilon indicator ($I_{\epsilon+}$) respectively where the time accorded to $MOBI_{nsga}$ in the combined algorithms represents 90% (10%) of the total run time. According to the metric under consideration, either the results of the algorithm located at a specific row are significantly better than those of the algorithm located at a specific column ($\succ$), either they are worse ($\prec$) or there is no significant difference between both ($\equiv$).

For the three studied data sets, Tables 1 and 2 show that almost all combined metaheurstics PR-$MOBI_{nsga}$ outperform $MOBI_{nsga}$ with regard to the hypervolume and epsilon indicators. We can see that $MOBI_{nsga}$ do not outperform any PR-$MOBI_{nsga}$ algorithm. In the case of Human data set, $MOBI_{nsga}$ is equivalent to APR-$MOBI_{nsga}$ and PPR-$MOBI_{nsga}$ and for colon data set, $MOBI_{nsga}$ is equivalent to PPR-$MOBI_{nsga}$.

Table 1. Comparison of the different metaheuristics for the I_H^- metrics by using Fisher-Matched statistical test with a *p*-value of 10%. For the combined metaheuristics, the time allocated to PR is **10%** and to $MOBI_{nsga}$ is **90%** of the execution time respectively.

		$MOBI_{nsga}$	APR-$MOBI_{nsga}$	PPR-$MOBI_{nsga}$	RPR-$MOBI_{nsga}$
Human	$MOBI_{nsga}$	-	$\equiv$	$\equiv$	$\prec$
	APR-$MOBI_{nsga}$	$\equiv$	-	$\succ$	$\equiv$
	PPR-$MOBI_{nsga}$	$\equiv$	$\prec$	-	$\prec$
	RPR$MOBI_{nsga}$	$\succ$	$\equiv$	$\succ$	-
Yeast	$MOBI_{nsga}$	-	$\prec$	$\prec$	$\prec$
	APR-$MOBI_{nsga}$	$\succ$	-	$\succ$	$\prec$
	PPR-$MOBI_{nsga}$	$\succ$	$\prec$	-	$\prec$
	RPR-$MOBI_{nsga}$	$\succ$	$\succ$	$\succ$	-
Colon	$MOBI_{nsga}$	-	$\prec$	$\equiv$	$\prec$
	APR-$MOBI_{nsga}$	$\succ$	-	$\succ$	$\equiv$
	PPR-$MOBI_{nsga}$	$\equiv$	$\prec$	-	$\prec$
	RPR-$MOBI_{nsga}$	$\succ$	$\equiv$	$\succ$	-

Table 2. Comparison of the different metaheuristics for the I_ϵ metrics by using Fisher-Matched statistical test with a p-value of 10%. For the combined metaheuristics, the time allocated to PR is **10%** and to $MOBI_{nsga}$ is **90%** of the execution time respectively.

		$MOBI_{nsga}$	APR-$MOBI_{nsga}$	PPR-$MOBI_{nsga}$	RPR-$MOBI_{nsga}$
	$MOBI_{nsga}$	-	≡	≡	≺
Human	APR-$MOBI_{nsga}$	≡	-	≻	≡
	PPR-$MOBI_{nsga}$	≡	≺	-	≺
	RPR-$MOBI_{nsga}$	≻	≡	≻	-
	$MOBI_{nsga}$	-	≺	≺	≺
Yeast	APR-$MOBI_{nsga}$	≻	-	≺	≺
	PPR-$MOBI_{nsga}$	≻	≻	-	≻
	RPR-$MOBI_{nsga}$	≻	≻	≺	-
	$MOBI_{nsga}$	-	≺	≡	≺
Colon	APR-$MOBI_{nsga}$	≻	-	≻	≺
	PPR-$MOBI_{nsga}$	≡	≺	-	≡
	RPR-$MOBI_{nsga}$	≻	≡	≻	-

Table 3. Comparison of the different metaheuristics for the I_H^- metrics by using Fisher-Matched statistical test with a p-value of 10%. For the combined metaheuristics, the time allocated to PR is **90%** and to $MOBI_{nsga}$ is**10%** of the execution time.

		$MOBI_{nsga}$	APR-$MOBI_{nsga}$	PPR-$MOBI_{nsga}$	RPR-$MOBI_{nsga}$
	$MOBI_{nsga}$	-	≡	≻	≻
Human	APR-$MOBI_{nsga}$	≡	-	≡	≡
	PPR-$MOBI_{nsga}$	≺	≡	-	≡
	RPR-$MOBI_{nsga}$	≺	≡	≡	-
	$MOBI_{nsga}$	-	≡	≡	≺
Yeast	APR-$MOBI_{nsga}$	≡	-	≻	≡
	PPR-$MOBI_{nsga}$	≡	≺	-	≡
	RPR-$MOBI_{nsga}$	≻	≡	≡	-
	$MOBI_{nsga}$	-	≡	≡	≡
Colon	APR-$MOBI_{nsga}$	≡	-	≡	≡
	PPR-$MOBI_{nsga}$	≡	≡	-	≡
	RPR-$MOBI_{nsga}$	≡	≡	≡	-

Table 1 shows that RPR-$MOBI_{nsga}$ is not outperformed by any algorithm with regard to the hypervolume indicator. In the case of Human and Colon data sets, RPR-$MOBI_{nsga}$ is equivalent to APR-$MOBI_{nsga}$. Concerning epsilon indicator, Table 2 shows that RPR-$MOBI_{nsga}$ is equivalent to PPR-$MOBI_{nsga}$ in the case of Colon data and outperformed by it in the case of Yeast data.

Table 4. Comparison of the different metaheuristics for the I_ϵ metrics by using Fisher-Matched statistical test with a p-value of 10%. For the combined metaheuristics, the time allocated to PR is **90%** and to $MOBI_{nsga}$ is **10%** of the execution time.

		$MOBI_{nsga}$	APR-$MOBI_{nsga}$	PPR-$MOBI_{nsga}$	RPR-$MOBI_{nsga}$
Human	$MOBI_{nsga}$	-	≡	≻	≻
	APR-$MOBI_{nsga}$	≡	-	≡	≡
	PPR-$MOBI_{nsga}$	≺	≡	-	≡
	RPR-$MOBI_{nsga}$	≺	≡	≡	-
Yeast	$MOBI_{nsga}$	-	≡	≡	≺
	APR-$MOBI_{nsga}$	≡	-	≻	≡
	PPR-$MOBI_{nsga}$	≡	≺	-	≡
	RPR-$MOBI_{nsga}$	≻	≡	≡	-
Colon	$MOBI_{nsga}$	-	≡	≡	≡
	APR-$MOBI_{nsga}$	≡	-	≡	≡
	PPR-$MOBI_{nsga}$	≡	≡	-	≡
	RPR-$MOBI_{nsga}$	≡	≡	≡	-

From that, we can conclude that according 10% of time to PR strategy after $MOBI_{nsga}$ improves the quality of the extracted Pareto Front approximation with regard to the hypervolume and epsilon indicators. Furthermore, RPR-$MOBI_{nsga}$ gives the best results compared to the other strategies. This can be explained by the importance of applying several times randomly PR instead of spending time to choose the best path, either using Pareto selection or aggregation selection. Tables 3 and 4 show that providing less time to $MOBI_{nsga}$ (10%) in the combined metaheuristics reduces the quality of the extracted Front approximation.

These results show the interest of combining $MOBI_{nsga}$ with PR, and specifically with the *Random move selection Path Relinking*. In the other hand, giving the role of $MOBI_{nsga}$ in converging toward the optimal Pareto front is of a significant importance.

3.4   Biological Relevance

In this section, we are interested in showing the biological value of biclusters extracted using PR strategy. We determined the biological relevance of relatively small biclusters for the Yeast cell-cycle data, with $\delta = 150$. The idea is to determine whether the set of genes discovered by PR strategy during R-$MOBI_{nsga}$ algorithm show significant enrichment with respect to a specific Gene Ontology (GO) annotation. For that, we use the web-tool Gene Ontology Term Finder [2]. In fact, GO Term Finder searches for significant shared GO terms, or parents of

---

[2] http://db.yeastgenome.org/cgi-bin/GO/goTermFinder

those GO terms, used to describe the genes of a given bicluster to help us discovering what the genes may have in common. Here genes are assigned to three structured, controlled vocabularies (ontologies) that describe gene products in terms of associated biological processes, components and molecular functions in a species-independent manner Table 5 shows the significant shared GO terms (or parent of GO terms) used to describe the set of genes (154 and 228) in the biclusters, for the process, function and component ontologies. The values within parentheses after each GO term in Table 5, such as (46; 6.09e-05) in the first bicluster, indicate that out of 154 genes in the first bicluster 46 belong to this process, and the statistical significance is provided by a p-value of 6.09e-05 (highly significant). Note that the genes in the biclusters share other GO terms also, but with a lower significance (i.e., have higher p-value).

Table 5. Significant shared GO Terms of two selected biclusters using RPR-$MOBI_{nsga}$ for *yeast* data

Bicluster's size	Process	Function	Component
14x154	cellular componen biogenesis (46; 6.09e-05)	DNA-directed RNA polymerase activity (5;0.03119)	macromolecular complex (86,5.89e-11)
	cellular process (132; 0.00061)		intracellular part (136,9.31e-05)
11x228	nucleic acid metabolic process (85; 7.70e-06)	DNA-directed RNA polymerase activity (6; 0.02735)	intracellular part (200; 9.19e-07)
	cellular macromolecule (metabolic process (122,0.00214)		membrane-bounded organelle (159,4.14e-05)

4   Conclusion

In this work, we have studied the hybridization of multiobjective metaheuristics with path relinking techniques. As an application, we choose $MOBI_{nsga}$ algorithm: a multiobjective metaheuristic for biclustering problem applied to the analysis of microarray data. By integrating Path Relinking technique, we aim to improve the quality of the front generated by $MOBI_{nsga}$. Several path relinking variants can be defined depending on the *solution selection* and *move selection* strategies.

In our work, we studied three path relinking variants namely RPR, APR and PPR which correspond to *random* move selection, *aggregation-based* move selection and *Pareto-based* move selection. The results showed that combining PR technique allow improving the front quality in terms of hypervolume and epsilon indicators. In the other hand, the results showed the importance of $MOBI_{nsga}$ in the convergence of the algorithms PR-$MOBI_{nsga}$. In terms of move selection strategy, experiments showed that random selection outperforms the advanced strategies such that: Pareto based and Aggregation based strategies. This can be explained by the low time consuming of random strategy which allows applying PR more times compared to the advanced strategies.

In future works, other hybridization schemes between $MOBI_{nsga}$ and PR strategy will be considered. Especially, integrating PR within $MOBI_{nsga}$ by applying it in each generation over the updated archive.

References

1. Alon, U., Barkai, N., Notterman, D.A., Gish, K., Ybarra, S., Mack, D., Levine, A.J.: Broad patterns of gene expression revealed by clustering analysis of tumor and normal colon tissues probed by oligonucleotide arrays. Proceedings of the National Academy of Sciences of the United States of America 96(12), 6745–6750 (1999)
2. Arroyo, J.E.C., Santos, A.G., dos Santos, P.M., Ribeiro, W.G.: A Bi-objective Iterated Local Search Heuristic with Path-Relinking for the p-Median Problem. In: Takahashi, R.H.C., Deb, K., Wanner, E.F., Greco, S. (eds.) EMO 2011. LNCS, vol. 6576, pp. 492–504. Springer, Heidelberg (2011)
3. Basseur, M., Seynhaeve, F., Talbi, E.-G.: Path Relinking in Pareto Multi-objective Genetic Algorithms. In: Coello Coello, C.A., Hernández Aguirre, A., Zitzler, E. (eds.) EMO 2005. LNCS, vol. 3410, pp. 120–134. Springer, Heidelberg (2005)
4. Bleuler, S., Laumanns, M., Thiele, L., Zitzler, E.: PISA – A Platform and Programming Language Independent Interface for Search Algorithms. In: Fonseca, C.M., Fleming, P.J., Zitzler, E., Deb, K., Thiele, L. (eds.) EMO 2003. LNCS, vol. 2632, pp. 494–508. Springer, Heidelberg (2003)
5. Cheng, Y., Church, G.M.: Biclustering of expression data. In: Proc. of the 8th ISMB, pp. 93–103. AAAI Press (2000)
6. Cheng, Y., Church, G.M.: Biclustering of expression data (supplementary information). Technical report (2006), http://arep.med.harvard.edu/biclustering
7. Chiarandini, M., Stützle, T., Paquete, L.: Pareto local optimum sets in the biobjective traveling salesman problem: An experimental study. Metaheuristics for Multiobjective Optimisation 535 (2004)
8. Esposito, F., Di Mauro, N., Basile, T.M.A., Ferilli, S.: Multi-dimensional relational sequence mining. Fundam. Inf. 89(1), 23–43 (2009)
9. Glover, F., Laguna, M.: Fundamentals of scatter search and path relinking. In: Control and Cybernetics, pp. 653–684 (1999)
10. Knowles, J., Thiele, L., Zitzler, E.: A tutorial on the performance assessment of stochastic multiobjective optimizers. Technical report, Computer Engineering and Networks Laboratory (TIK), ETH Zurich, Switzerland (2006) (revised version)

11. Lashkargir, M., Monadjemi, S.A., Dastjerdi, A.B.: A new biclustering method for gene expersion data based on adaptive multi objective particle swarm optimization. In: Proceedings of the 2009 Second International Conference on Computer and Electrical Engineering, ICCEE 2009, pp. 559–563. IEEE Computer Society, Washington, DC (2009)
12. Liefooghe, A., Jourdan, L., Talbi, E.-G.: A software framework based on a conceptual unified model for evolutionary multiobjective optimization: Paradiseo-moeo. European Journal of Operational Research 209(2), 104–112 (2011)
13. Liefooghe, A., Mesmoudi, S., Humeau, J., Jourdan, L., Talbi, E.-G.: A Study on Dominance-Based Local Search Approaches for Multiobjective Combinatorial Optimization. In: Stützle, T., Birattari, M., Hoos, H.H. (eds.) SLS 2009. LNCS, vol. 5752, pp. 120–124. Springer, Heidelberg (2009)
14. Liu, J., Li, Z., Hu, X., Chen, Y.: Biclustering of microarray data with mospo based on crowding distance. BMC Bioinformatics, 1 (2009)
15. Liu, J., Li, Z., Hu, X., Chen, Y.: Multi-objective ant colony optimization biclustering of microarray data. In: GrC 2009, pp. 424–429 (2009)
16. Liu, J., Li, Z., Liu, F., Chen, Y.: Multi-objective particle swarm optimization biclustering of microarray data. In: IEEE International Conference on Bioinformatics and Biomedicine, pp. 363–366 (2008)
17. Liu, J., Li, Z., Liu, F., Chen, Y.: Multi-objective particle swarm optimization biclustering of microarray data. In: IEEE International Conference on Bioinformatics and Biomedicine, pp. 363–366 (2008)
18. Marti, R., Campos, V., Resende, M.G.C., Duarte, A.: Multi-objective grasp with path-relinking. Computers and Operations Research, 498–508 (2010)
19. Maulik, U., Mukhopadhyay, A., Bandyopadhyay, S.: Finding multiple coherent biclusters in microarray data using variable string length multiobjective genetic algorithm. Trans. Info. Tech. Biomed. 13(6), 969–975 (2009)
20. Mitra, S., Banka, H.: Multi-objective evolutionary biclustering of gene expression data. Pattern Recognition 39(12), 2464–2477 (2006)
21. Seridi, K., Jourdan, L., Talbi, E.-G.: Multi-objective evolutionary algorithm for biclustering in microarrays data. In: IEEE Congress on Evolutionary Computation, New Orleans, USA, pp. 2593–2599 (2011)
22. Talbi, E.-G.: Metaheuristics: From Design to Implementation. Wiley Publishing (2009)
23. Zeng, R.: Multi-Objective Metheuristics based on Neighborhoods for the Approximation of the Pareto Set. PhD thesis, École Doctorale d'Angers, Juillet (2012)

Selection Operators Based on Maximin Fitness Function for Multi-Objective Evolutionary Algorithms

Adriana Menchaca-Mendez and Carlos A. Coello Coello*

CINVESTAV-IPN (Evolutionary Computation Group)
Departamento de Computación
México D.F. 07300, México
adriana.menchacamendez@gmail.com, ccoello@cs.cinvestav.mx

Abstract. We analyze here some properties of the maximin fitness function, which has been used by several researchers, as an alternative to Pareto optimality, for solving multi-objective optimization problems. As part of this analysis, we identify some disadvantages of the maximin fitness function and then propose mechanisms to overcome them. This leads to several selection operators for multi-objective evolutionary algorithms which are further analyzed. We incorporate them into an evolutionary algorithm, giving rise to the so-called Maximin-Clustering Multi-Objective Evolutionary Algorithm (MC-MOEA) approach. Our proposed approach is validated using standard test problems taken from the specialized literature, having from two to eight objectives. Our preliminary results indicate that our proposed approach is a good alternative to solve multi-objective optimization problems having both low dimensionality (two or three) and high dimensionality (more than three) in objective function space.

1 Introduction

The use of evolutionary algorithms for solving multi-objective optimization problems (MOPs) has become very popular in the last few years [7]. When designing multi-objective evolutionary algorithms (MOEAs), there are two main types of approaches that are normally used as selection mechanism: (i) those that incorporate the concept of Pareto optimality, and (ii) those that do not use Pareto dominance to select individuals.

In this work, we are interested in the *maximin fitness function* [2] (belonging to the type (ii)). This technique assigns a fitness to each individual in the population. Such fitness value encompasses Pareto dominance (we can know which individuals are non-dominated), distance to the non-dominated individuals, and clustering between individuals (it penalizes individuals that are too close from each other). This scheme has the advantage of requiring very simple operations to calculate the fitness and is, thus, computationally efficient (its complexity is

---

* The second author acknowledges support from CONACyT project no. 103570.

R.C. Purshouse et al. (Eds.): EMO 2013, LNCS 7811, pp. 215–229, 2013.

linear with respect to the number of objectives). A preliminary study allowed us to design some selection operators based on the maximin fitness function which are incorporated into a MOEA that uses a simulated binary crossover (SBX) and a polynomial mutation operator (PM), giving rise to the main proposal of this paper, which is called: *Maximin-Clustering Multi-Objective Evolutionary Algorithm (MC-MOEA)*. The proposed approach is validated with several standard test problems using the hypervolume and the additive epsilon-indicator. Our proposed MC-MOEA approach is compared with respect to the NSGA-II (which is a very competitive Pareto-based MOEA), with respect to SMS-EMOA (which is a hypervolume-based MOEA), and with respect to a version of SMS-EMOA that uses Monte Carlo simulation to approximate the exact hypervolume (we called it APP-SMS-EMOA) [1]. Our preliminary results indicate that our proposed approach is a viable alternative, particularly when dealing with a high number of objectives, since it produces results that are similar in quality to those obtained with SMS-EMOA (low dimensionality) and APP-SMS-EMOA (high dimensionality), but at a very low computational cost.

The remainder of this paper is organized as follows. The maximin fitness function is described and studied in Section 2. Section 3 presents the proposed mechanisms to improve the maximin fitness function and describes in detail three selection operators based on it. In Section 4, we present a full description of our proposed MC-MOEA approach. Our experiments and the results obtained are shown in Section 5. Finally, we provide our conclusions and future work in Section 6.

2   Maximin Fitness Function

The maximin fitness function was proposed by Richard Balling and Scott Wilson in [2],[4] and, it works as follows. Let's consider a MOP with K objectives and an evolutionary algorithm whose population size is P. Let f_k^i be the normalized value of the kth objective for the ith individual in a particular generation. Assuming minimization problems, we have that the jth individual weakly dominates the ith individual if:

$$min_k(f_k^i - f_k^j) \geq 0 \qquad (1)$$

The ith individual, in a particular generation, will be weakly dominated by another individual, in the generation, if:

$$max_{j \neq i}(min_k(f_k^i - f_k^j)) \geq 0 \qquad (2)$$

Then, the maximin fitness function of individual i is defined as:

$$fitness^i = max_{j \neq i}(min_k(f_k^i - f_k^j)) \qquad (3)$$

where the min is taken over all the objectives from 1 to K, and the max is taken over all the individuals in the population from 1 to P, except for the same individual i. From eq. (3), we can say the following:

---

[1] We approximate the hypervolume using the approach proposed in HyPE [1].

1. Any individual whose maximin fitness is greater than zero is a dominated individual,
2. Any individual whose maximin fitness is less than zero is a non-dominated individual.
3. Finally, any individual whose maximin fitness is equal to zero is a weakly-dominated individual.

2.1   Reviewing the Properties of the Maximin Fitness Function

Let's review the properties of the maximin fitness function as presented in [3]:

1. The maximin fitness function penalizes clustering of non-dominated individuals. In the limit, the maximin fitness of duplicate non-dominated individuals is zero. See Figure 1.
2. The maximin fitness function rewards individuals at the middle of convex non-dominated fronts, see Figure 2. Also, it rewards individuals at the extremes of concave non-dominated fronts, see Figure 3. The maximin fitness function is a continuous function of objective values.
3. The maximin fitness of dominated individuals is a metric of the distance to the non-dominated front. See Figure 4.
4. The *max* function in the maximin fitness of a dominated individual is always controlled by a non-dominated individual and is indifferent to clustering. The *max* function in the maximin fitness of a non-dominated individual may be controlled by a dominated or a non-dominated individual. See Figure 4.

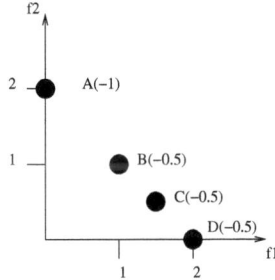

Fig. 1. We can see that the maximin fitness function penalizes individuals B, C and D because they are close from each other. It also rewards individual A, because it is far away from the other individuals.

Analyzing Property 1, we can see that although the maximin fitness function penalizes the clustering between individuals, it has the following disadvantage. In Figure 1, we can observe that individuals B, C and D have the same maximin fitness. Then, if we use the maximin fitness function, we can not know which of the three is the best individual to form part of the next generation.

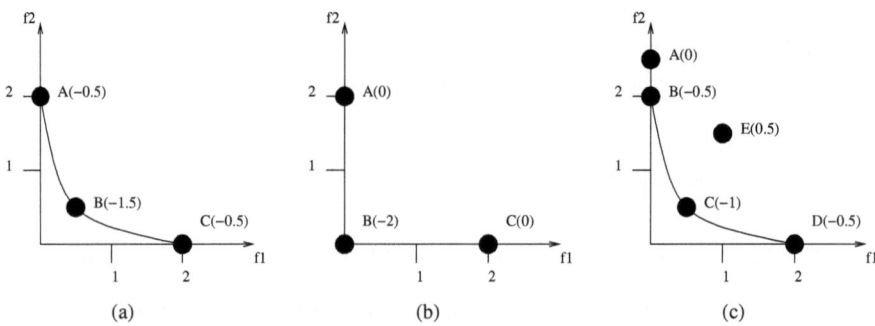

Fig. 2. In all cases, we can see that the maximin fitness function rewards individuals at the middle of convex non-dominated fronts. In (c), individual A has a maximin fitness equal to zero because it is a weakly dominated solution, and individual E has a positive maximin fitness equal to 0.5 because it is a dominated solution.

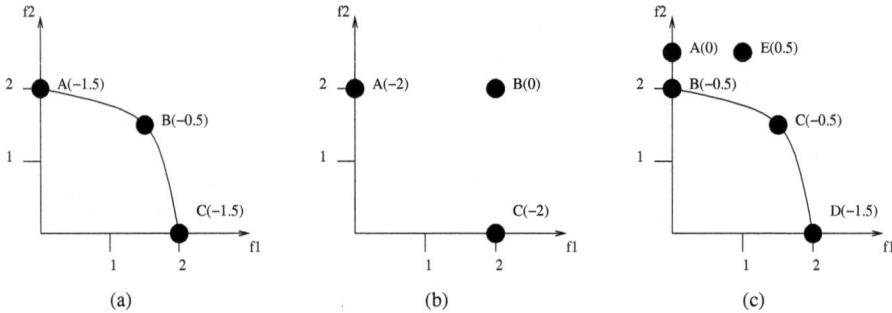

Fig. 3. In all cases, we can see that the maximin fitness function rewards individuals at the extremes of concave non-dominated fronts. In (c), individual A has a maximin fitness equal to zero because it is a weakly dominated solution, and individual E has a positive maximin fitness equal to 0.5 because it is a dominated solution.

To review Property 4, let's see Figure 4. In this case, we can see that the fitness of the non-dominated individual B is affected by the dominated individual D. Then, the maximin fitness function penalizes non-dominated individuals if they are close to another individual (no matter whether or not it is a dominated solution). The author of the maximin fitness function proposed in [4] the following modified maximin fitness function:

$$fitness^i = max_{j \neq i, j \in P}(min_k(f_k^i - f_k^j)) \tag{4}$$

where P is the set of non-dominated individuals. Using eq. (4) to assign the fitness of each individual, we guarantee that the fitness of a non-dominated individual is controlled only by non-dominated individuals and then we only penalize clustering between non-dominated individuals.

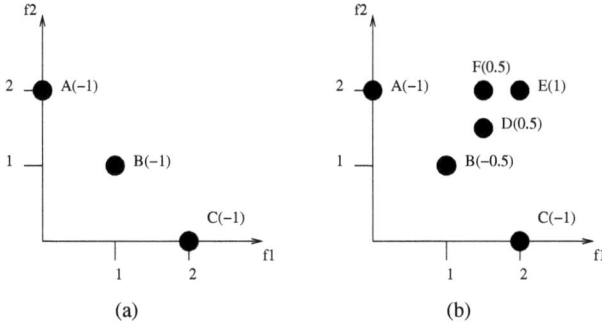

Fig. 4. In (b), we can see that the fitness of individuals D, E and F is controlled by the non-dominated individual B, and the value of their fitness is a metric of the distance to the individual B. Also, we can see that the fitness of B is affected by the dominated individual D.

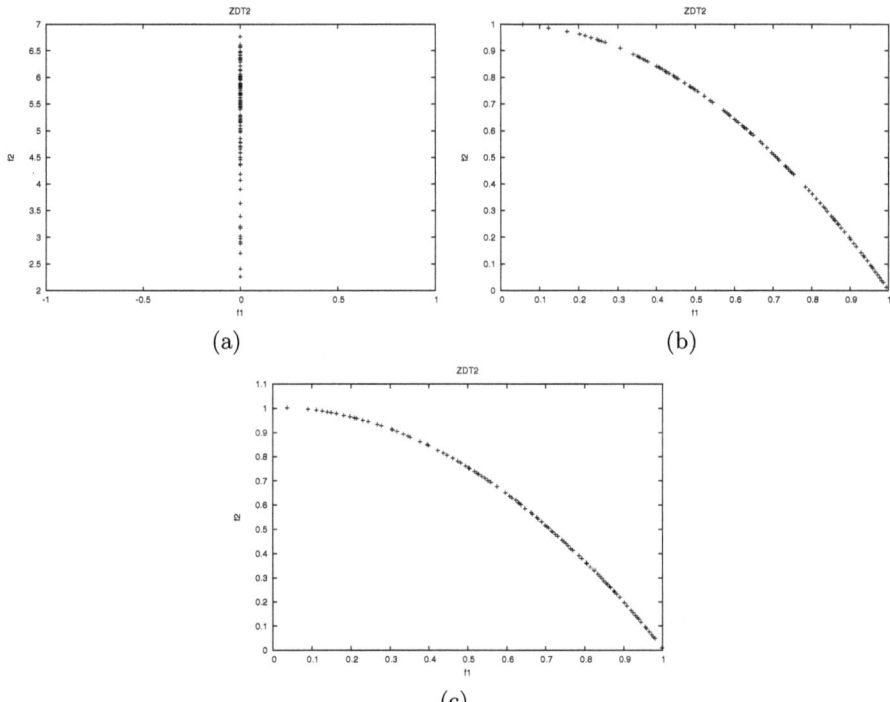

Fig. 5. In all cases, we use an evolutionary algorithm based on Differential Evolution coupled to any of the 3 selection mechanisms proposed here. In (a), we use only the maximin fitness to select individuals. In case (b), we use the maximin fitness and the constraint that prevents us from selecting similar individuals (in objective function space). Finally, in (c) we use the full selection operator proposed in this work.

On the other hand, it is important to analyze if it is better to prefer weakly dominated individuals than dominated individuals. In the study that we will include next, we show that it is not good to prefer weakly dominated individuals or individuals which are close to being weakly dominated (even if they are weakly dominated by any dominated individual). For example, in Figure 2 (c), solution A is a weakly dominated individual and solution E is a dominated individual. To guarantee convergence to the Pareto optimal set, we must choose individual E. Otherwise, it is possible that the evolutionary algorithm converges to a weak Pareto optimal solution. Problem ZDT2 is an example of this:

$$f_1(\boldsymbol{x}) = x_1$$
$$f_2(\boldsymbol{x}) = g(\boldsymbol{x}) \left(1 - (x_1/g(\boldsymbol{x}))^2\right)$$
$$g(\boldsymbol{x}) = 1 + \frac{9}{n-1} \sum_{i=2}^{n} x_i) \tag{5}$$

where $x_i \in [0,1]$, and the problem has 30 decision variables. If we set the fitness of each individual with the maximin fitness function, into an evolutionary algorithm, and after sorting the individuals with respect to their fitness, we perform selection. Then, at the end of the generations, we obtain only weakly Pareto points, see Figure 5 (a). This happens because f_1 is easier to optimize than f_2 and then, we quickly obtain weakly dominated solutions in this extreme of the Pareto front.

3   Selection Operators Based on the Maximin Fitness Function

Considering the properties of the maximin fitness function and its disadvantages, we propose here three possibles selection operators.

3.1   Operator I

In order to deal with the problem of the weakly dominated individuals, we proposed in [10] the following constraint: Any individual that we want to select must not be similar (in objective space) to another (selected) individual. The process to verify similarity between individuals is shown in Algorithm 1 and the full selection process is shown in Algorithm 2. So, we avoid selecting solutions that are weakly dominated by non-dominated solutions (see individual A in Figure 2 (c)) or solutions which are weakly dominated by dominated solutions (see individual F in Figure 4 (b)). In Figure 5 (b), we can observe that by imposing this constraint, we can find the true Pareto front of the ZDT2 function.

In order to deal with the disadvantage of Property 1, we proposed in [10] a technique based on maximin fitness and clustering. Such a technique works as follows. If we want to select S individuals from a population of size P, then, we choose the best S individuals with respect to their maximin fitness, and use

---

Input : min_dif (Minimum difference), x (individual), Y (population), P (population
            size), K (number of objectives).
Output: Returns 1, if the individual x is similar to any individual in the population Y;
            otherwise, returns 0.
for $i \leftarrow 1$ **to** P **do**
    for $k \leftarrow 1$ **to** K **do**
        if $|x.f[k] - Y[i].f[k]| < min\_dif$ **then**
            | return 1;
        end
    end
end
return 0;

Algorithm 1. IsSimilarToAny

them as centers of their clusters. Then, we proceed to place each individual in the nearest cluster. Finally, for each of the resulting clusters, we recompute the center, and choose the individual closest to it. It is important to note that we don't iterate many times to improve the distribution of the centers, we only execute one time the correction. This procedure is shown in Algorithm 3.

Input : X (Population), P (population size), K (number of objectives), S (the number of
            individuals to choose) and min_dif (minimum difference between objectives).
Output: Y (Selected individuals).
$s \leftarrow 1,\ i \leftarrow 1$;
/*Sorting with respect the maximin fitness of each individual   */
$X_{sorted} \leftarrow Sort(X)$;
/*Fill up the new population with the best copies according to maximin fitness,
  verifying that there is not a similar one   */
while $s \leq S$ AND $i \leq P$ **do**
    while $IsSimilarToAny(min\_dif, X_{sorted}[i], Y, s, K) = 1$ AND $i \leq P$ **do**
        | $i \leftarrow i + 1$;
    end
    if $i \leq P$ **then**
        $Y[s] \leftarrow X_{sorted}[i]$;
        $s \leftarrow s + 1$;
    end
end
/*Fill up the new population with the best copies according to maximin fitness   */
$i \leftarrow 1$;
while $s \leq S$ **do**
    if $X_{sorted}[i]$ has not been selected **then**
        $Y[s] \leftarrow X_{sorted}[i]$;
        $s \leftarrow s + 1$;
    end
    $i \leftarrow i + 1$;
end
return Y;

Algorithm 2. Maximin-Selection

With the maximin-clustering technique, if we return to Figure 1 and assume that we want to choose two individuals, we can see that regardless of the individual (B, C or D) that we choose as an initial center of the cluster, we always obtain two clusters: one of them contains individual A, and the other one contains individuals B, C and D. After applying this procedure, we always choose

Input   : X (Population), $NonDom$ (number of non-dominated individuals), K (number of
          objectives) and S (number of individuals to choose).
Output: Y (individuals selected).
/*Choose the best S individuals, according to maximin fitness, as centers of the
  clusters C */
$X_{sorted} \leftarrow Sort(X)$, $C_j = \{X_{sorted}[j]\}$;
/*Do one iteration of clustering   */
for $i \leftarrow S+1$ to $NonDom$ do
    if $X_{sorted}[i]$ is closer to C_j then
        | $C_j \leftarrow C_j \cup X_{sorted}[i]$;
    end
end
/*Obtain the new centers of the clusters   */
for $j \leftarrow 1$ to S do
    $\mu_j \leftarrow \frac{1}{|C_j|} \sum\limits_{X[i] \in C_j} X[i]$;
end
/*Select to individuals who are closest to the centers of the clusters   */
for $j \leftarrow 1$ to S do
    if $X[i] \mid X[i] \in C_j$ is the nearest to the center μ_j then
        | $Y[j] \leftarrow X[i]$;
    end
end
Returns Y;

Algorithm 3. Maximin-Clustering Selection

individuals A and C. It is important to note that the above technique, which
is used to improve the distribution of the selected individuals, is only effective
in cases when all individuals are non-dominated. For example, if we analyze
Figure 4 (b), and we want to select three individuals, our technique selects individuals A, D and C, penalizing individual B. This is clearly not good because
individual B dominates individual D. In Figure 5 (c), we can see that if we use
the maximin-clustering technique, we obtain a better distribution of solutions.
In Algorithm 4, we describe the full selection operator.

Input   : X (Current population), P (population size), K (number of objectives) and S
          (number of individuals to choose).
Output: Y (individuals selected).
MaximinFitnessFunction(X, P, K);
if The number of nondominated individuals is greater to S then
    | $Y \leftarrow$ Maximin-Clustering Selection(X_{sorted}, P, K, S);
else
    | $Y \leftarrow$ MaximinSelection(X_{sorted}, P, K, S);
end
Returns Y;

Algorithm 4. Operator I

3.2   Operator II

The clustering technique, that we propose to address the disadvantage of Property 1, makes the correction of the centers only once. Then, if we choose more
efficiently the initial centers, we hope to obtain a better distribution of solutions

along the Pareto front. Considering Property 4, we believe that it is a good idea to use the maximin fitness function at the beginning of the search and use the modified fitness function when we have many non-dominated individuals [2] because in this part of the search process, we are interested in obtaining a better distribution between non-dominated individuals and, the modified maximin fitness guarantees to penalize clustering only between non-dominated individuals. We show the full selection operator in Algorithm 5.

Input : X (Current population), P (population size), K (number of objectives) and S
        (number of individuals to choose).
Output: Y (individuals selected).
if *The number of nondominated individuals is greater to S* **then**
    | ModifiedMaximinFitnessFunction(X, P, K);
    | $Y \leftarrow$ Maximin-Clustering Selection(X_{sorted}, P, K, S);
else
    | MaximinFitnessFunction(X, P, K);
    | $Y \leftarrow$ MaximinSelection(X_{sorted}, P, K, S);
end
Returns Y;

Algorithm 5. Operator II

3.3   Operator III

For the last operator, we decided to apply the modified maximin fitness function only when we use our clustering technique, because at the beginning of the search process we need to penalize the clustering between individuals regardless of their dominance. This is because we want to explore all the search space and, therefore, we decided to use the original maximin fitness function at the beginning of the evolutionary process. However, in the third operator, we propose to use the modified maximin fitness function since the beginning of the search in order to analyze the behavior of the operator when we use the modified maximin fitness all the time. The third operator is described in Algorithm 6.

4   Maximin-Clustering Multi-Objective Evolutionary Algorithm

In order to compare the three operators based on the maximin fitness function, we designed a multi-objective evolutionary algorithm using a simulated binary crossover (SBX) and a polynomial mutation operator (PM) to create new individuals combined with the previously described selection operators as follows: If the size of the population is P, then we create P new individuals. The parents are selected as follows: We use a binary tournament. At each tournament, two individuals are randomly selected and the one with the higher fitness value is

---

[2] Considering a $(\mu + \mu)$ selection scheme, we say that we have many non-dominated individuasl if more than μ individuals are non-dominated.

chosen. After that, we combine the population of parents and offspring to obtain a population of size $2P$. Then, we use one of the selection operators to choose the P individuals that will take part of the following generation.

Input   : X (Current population), P (population size), K (number of objectives) and S
          (number of individuals to choose).
Output: Y (individuals selected).
ModifiedMaximinFitnessFunction(X, P, K);
if *The number of nondominated individuals is greater to S* then
  | $Y \leftarrow$ Maximin-Clustering Selection(X_{sorted}, P, K, S);
else
  | $Y \leftarrow$ MaximinSelection(X_{sorted}, P, K, S);
end
Returns Y;

Algorithm 6. Operator III

5   Experimental Results

Aiming to validate the selection mechanism of our proposed approach with respect to other types of mechanisms, we chose the following MOEAs: NSGA-II [8] (based on Pareto dominance) and (SMS-EMOA) [5] (based on the hypervolume performance measure [12] combined with the non-dominated sorting procedure adopted in NSGA-II). Due to the high computational cost required to calculate the hypervolume, we decided to use also an approximate calculation of the hypervolume. For this sake, we used the source code of HyPE available in the public domain [1] adopting 10^3 as our number of samples.

5.1   Experiments

For all our experiments, we used the two following sets of problems: The first consists of five bi-objective test problems taken from the Zitzler-Deb-Thiele (ZDT) test suite [13]. The second consisted of seven problems having three or more objectives, taken from the Deb-Thiele-Laumanns-Zitzler (DTLZ) suite [9]. For the DTLZ test problems, we used $k = 5$ for DTLZ1 and DTLZ6 and $k = 10$ for the remaining test problems and, three, four, five, six, seven and eight objective functions (i.e., $M = 3, 4, 5, 6, 7$ and 8). For each test problem, we performed 30 independent runs. For all algorithms, we adopted the parameters suggested by the authors of NSGA-II: crossover probability $p_c = 0.9$, mutation probability $p_m = 1/n$, where n is the number of decision variables. Both for the crossover and mutation operators, we adopted $\eta_c = 15$ and $\eta_m = 20$, respectively. For our proposed selection operators we used $min\_dif = 0.0001$ in all cases. All approaches performed the same number of objective function evaluations. For the ZDT test problems, we performed 20,000 evaluations (we used a population of 100 individuals and we iterated for 200 generations). For the DTLZ test problems we performed 125,000 evaluations (we used a population of 250 individuals and we iterated for 500 generations). In the case of SMS-EMOA, we adopted five hours as our maximum computation time (SMS-EMOA requires more than five hours when dealing with 4 or more objectives).

In order to assess performance, we adopted the *hypervolume indicator* (φ).[3] and the additive ϵ-indicator [4] because these two indicators are Pareto compliant and SMS-EMOA is based in φ and our MC-MOEA approach is based on the maximin fitness function, which can be considered as the binary ϵ-indicator of a solution with respect to a reference set defined by the remaining non-dominated solutions of the population. To compute φ, we used the following reference points: For the ZDT test problems, we used $y_{ref} = [1.1, 1.1]$. For DTLZ1, we used $y_{ref} = [y_1, \ldots, y_M] \mid y_i = 0.7$. For DTLZ(2-6), we used $y_{ref} = [y_1, \ldots, y_M] \mid y_i = 1.1$. For DTLZ7, we used $y_{ref} = [y_1, \ldots, y_M]$ $y_M = 6.1$ and $y_{i \neq M} = 1.1$.

5.2   Results

In Table 1, we present the results with respect to the hypervolume indicator and we can see that our MC-MOEAs obtained competitive results with respect to SMS-EMOA and APP-SMS-EMOA. One important thing is that our proposed MC-MOEAs presented a consistent behavior when we increased the number of objectives unlike NSGA-II. To validate the results in our experiments, we performed a statistical analysis using Wilcoxon's rank sum on our MC-MOEAs with respect to SMS-EMOA and APP-SMS-EMOA and, we obtained that only in the problems DTLZ3 (with 3, 5, 6, and 7 objectives) and DTLZ6 (with 4 and 5 objectives) the null hypothesis ("medians are equal") can be rejected at the 5% level. If we check these problems in Table 1, we can see that our MC-MOEAs obtained better results than SMS-EMOA and APP-SMS-EMOA. This means that in these problems our MC-MOEAs significantly outperformed both SMS-EMOA and APP-SMS-EMOA. Note, however, that we could not include the table with the results of the Wilcoxon's rank sum due to space limitations.

Table 3 shows the results with respect to the additive epsilon indicator. In this case, we only compared our MC-MOEAs with respect to APP-SMS-EMOA because, as noted in Table 1, NSGA-II did not have a consistent behavior when we increased the number of objectives and SMS-EMOA required a very large computational time, making the comparison unfair. The results show that our MC-MOEAs outperformed APP-SMS-EMOA in most cases. For these experiments, we also performed a statistical analysis and we obtained that in the test problems in which APP-SMS-EMOA outperformed our MC-MOEAs, we can not reject the null hypothesis ("medians are equal") and, in many cases, when our MC-MOEAs outperformed APP-SMS-EMOA, the null hypothesis can be rejected at the 5% level. An important advantage of our MC-MOEAs is that

---

[3] The hypervolume was originally proposed by Zitzler and Thiele in [14], and it's defined as the size of the space covered by the Pareto optimal solutions. φ *rewards convergence towards the Pareto front as well as the maximum spread of the solutions obtained.* The disadvantage of this indicator is its high computational cost (the running time for calculating φ is exponential in the number of objectives).

[4] Given two approximate sets, A and B, the ϵ-indicator measures the smallest amount, ϵ, that must be used to translate the set, A, so that every point in B is covered [7].

Table 1. Results obtained with respect to the hypervolume indicator. The value in parentheses in the first column indicates the number of objectives. We show average values over 30 independent runs. The values in parentheses of the other columns correspond to the standard deviations.

	NSGA-II	SMS-EMOA	APP-SMS-EMOA	MC-MOEA(v1)	MC-MOEA(v2)	MC-MOEA(v3)
ZDT1 (2)	0.867920 (0.000489)	**0.871433 (0.000093)**	0.867986 (0.000421)	0.862576 (0.001036)	0.864293 (0.000873)	0.864215 (0.000937)
ZDT2 (2)	0.534002 (0.000553)	**0.537417 (0.001047)**	0.533294 (0.002330)	0.520382 (0.002330)	0.525339 (0.002581)	0.526132 (0.001956)
ZDT3 (2)	**1.325397 (0.000708)**	1.320005 (0.024943)	1.305149 (0.027317)	1.307519 (0.025795)	1.312000 (0.024431)	1.306731 (0.030536)
ZDT4 (2)	**0.857773 (0.010887)**	0.804517 (0.050998)	0.782593 (0.069529)	0.818084 (0.062050)	0.829491 (0.057508)	0.840295 (0.035847)
ZDT6 (2)	0.481464 (0.003094)	**0.490964 (0.002359)**	0.485441 (0.002728)	0.467497 (0.006066)	0.468000 (0.005054)	0.468757 (0.007073)
DTLZ1 (3)	0.315526 (0.000568)	0.319138 (0.000008)	0.306568 (0.006481)	0.315955 (0.000714)	0.316679 (0.000791)	0.316622 (0.000703)
DTLZ2 (3)	0.721960 (0.010450)	**0.776854 (0.000021)**	0.761623 (0.003126)	0.734580 (0.005829)	0.738114 (0.004938)	0.738805 (0.005594)
DTLZ3 (3)	0.059242 (0.179471)	0.000000 (0.000000)	0.000000 (0.000000)	0.715957 (0.011959)	0.716155 (0.013048)	**0.718686 (0.010902)**
DTLZ4 (3)	0.722625 (0.014284)	**0.776861 (0.000026)**	0.763312 (0.002203)	0.736373 (0.005998)	0.742850 (0.004866)	0.744456 (0.004385)
DTLZ5 (3)	0.440252 (0.000208)	**0.441267 (0.000003)**	0.438059 (0.012219)	0.430387 (0.005275)	0.431899 (0.003225)	0.432158 (0.003319)
DTLZ6 (3)	0.406129 (0.024792)	0.414851 (0.019473)	0.350072 (0.101829)	0.390702 (0.030285)	0.404420 (0.019828)	0.392088 (0.018441)
DTLZ7 (3)	2.007096 (0.005727)	**2.037787 (0.089016)**	1.923641 (0.120558)	1.992563 (0.011031)	2.002309 (0.005891)	1.984739 (0.085183)
DTLZ1 (4)	0.199714 (0.072810)	0.227113 (0.019905)	0.225110 (0.004836)	0.232710 (0.000636)	**0.232832 (0.000753)**	0.232626 (0.000641)
DTLZ2 (4)	0.916315 (0.050375)	**1.074969 (0.000534)**	1.048196 (0.005676)	0.965540 (0.011771)	0.972949 (0.009525)	0.972032 (0.008293)
DTLZ3 (4)	0.000000 (0.000000)	0.000000 (0.000000)	0.000000 (0.000000)	0.940267 (0.027559)	**0.945889 (0.021329)**	0.937558 (0.028903)
DTLZ4 (4)	0.919121 (0.053369)	**1.076531 (0.000323)**	1.049698 (0.025284)	0.983616 (0.008471)	0.990356 (0.007794)	0.991357 (0.008939)
DTLZ5 (4)	0.427341 (0.002086)	**0.434905 (0.002811)**	0.411700 (0.012550)	0.246597 (0.021079)	0.252224 (0.025367)	0.262790 (0.019642)
DTLZ6 (4)	0.082789 (0.025430)	0.005002 (0.003624)	0.017953 (0.015585)	0.188079 (0.018994)	0.245121 (0.013218)	**0.245153 (0.013223)**
DTLZ7 (4)	0.673851 (0.019135)	**0.799224 (0.179936)**	0.516516 (0.139265)	0.697724 (0.011902)	0.718840 (0.012895)	0.715425 (0.012490)
DTLZ1 (5)	0.000000 (0.000000)	0.000000 (0.000000)	0.160563 (0.002329)	0.158865 (0.028230)	0.163938 (0.000865)	**0.164163 (0.000575)**
DTLZ2 (5)	0.607031 (0.362156)	0.810554 (0.081746)	1.292202 (0.005203)	1.129832 (0.016540)	1.137390 (0.016104)	1.138317 (0.014105)
DTLZ3 (5)	0.000000 (0.000000)	0.000000 (0.000000)	0.000000 (0.000000)	**1.127122 (0.038275)**	1.125465 (0.030709)	1.108675 (0.048260)
DTLZ4 (5)	0.552563 (0.394310)	0.619604 (0.094095)	1.288836 (0.008553)	1.163695 (0.015934)	1.173994 (0.015513)	1.173374 (0.016415)
DTLZ5 (5)	0.418359 (0.004244)	**0.253338 (0.015809)**	0.419592 (0.012492)	0.190355 (0.025803)	0.193947 (0.022278)	0.201230 (0.020934)
DTLZ6 (5)	0.000000 (0.000000)	0.000000 (0.000000)	0.002493 (0.003237)	0.071664 (0.017627)	0.113595 (0.020441)	0.113595 (0.020441)
DTLZ7 (5)	0.000000 (0.000000)	0.003702 (0.005092)	0.071404 (0.047761)	0.116535 (0.007517)	0.120316 (0.008210)	**0.120386 (0.008311)**
DTLZ1 (6)	0.105390 (0.008146)	—	**0.114030 (0.001116)**	0.096169 (0.041014)	0.103155 (0.032221)	0.102862 (0.033258)
DTLZ2 (6)	0.000000 (0.000000)	—	**1.497165 (0.012818)**	1.240839 (0.031766)	1.243283 (0.032054)	1.251246 (0.031293)
DTLZ3 (6)	0.014439 (0.028545)	—	0.000000 (0.000000)	**1.181907 (0.234392)**	1.118765 (0.380109)	1.009064 (0.459079)
DTLZ4 (6)	0.000000 (0.000000)	—	**1.511018 (0.005841)**	1.272827 (0.036356)	1.290147 (0.035050)	1.290147 (0.035050)
DTLZ5 (6)	0.015083 (0.033135)	—	**0.423127 (0.016780)**	0.171457 (0.020233)	0.172391 (0.022213)	0.171830 (0.018877)
DTLZ6 (6)	0.402419 (0.022651)	—	**0.000017 (0.000060)**	0.000000 (0.000000)	0.000000 (0.000000)	0.000000 (0.000000)
DTLZ7 (6)	0.000000 (0.000000)	—	0.002785 (0.004630)	**0.099829 (0.019956)**	0.010246 (0.002577)	0.010804 (0.002256)
DTLZ1 (7)	0.010739 (0.002179)	—	**0.080705 (0.001011)**	0.050365 (0.035001)	0.057601 (0.031938)	0.052974 (0.031998)
DTLZ2 (7)	0.000000 (0.000000)	—	**1.663548 (0.020716)**	1.266017 (0.083504)	1.307874 (0.073108)	1.313721 (0.050521)
DTLZ3 (7)	0.002535 (0.007792)	—	0.000000 (0.000000)	0.666609 (0.658681)	0.841286 (0.642597)	**0.901860 (0.597147)**
DTLZ4 (7)	0.000000 (0.000000)	—	**1.721242 (0.010973)**	1.290180 (0.092052)	1.322271 (0.085731)	1.322271 (0.085731)
DTLZ5 (7)	0.000049 (0.000228)	—	**0.435360 (0.024210)**	0.141655 (0.023779)	0.157740 (0.025355)	0.173623 (0.290061)
DTLZ6 (7)	0.426098 (0.017081)	—	0.000000 (0.000000)	0.000000 (0.000000)	0.000000 (0.000000)	0.000000 (0.000000)
DTLZ7 (7)	0.000000 (0.000000)	—	0.056576 (0.000741)	0.000416 (0.000230)	0.000372 (0.000232)	0.000360 (0.000220)
DTLZ1 (8)	**0.000894 (0.000231)**	—	0.056576 (0.000741)	0.027953 (0.021725)	0.024212 (0.024790)	0.025604 (0.022569)
DTLZ2 (8)	0.000000 (0.000000)	—	**1.811587 (0.035750)**	1.061160 (0.251401)	1.277504 (0.107501)	1.281810 (0.111406)
DTLZ3 (8)	0.000221 (0.001189)	—	0.000000 (0.000000)	**0.384426 (0.636684)**	0.372337 (0.571584)	0.226089 (0.460711)
DTLZ4 (8)	0.446433 (0.028464)	—	**1.938835 (0.014867)**	0.898659 (0.382592)	1.443849 (0.270982)	1.443849 (0.270982)
DTLZ5 (8)	0.000000 (0.000000)	—	**0.464482 (0.023309)**	0.136816 (0.037378)	0.155077 (0.028200)	0.152089 (0.030343)
DTLZ6 (8)	0.000000 (0.000000)	—	0.000000 (0.000000)	0.000000 (0.000000)	0.000000 (0.000000)	0.000000 (0.000000)
DTLZ7 (8)	**0.000085 (0.000026)**	—	0.000000 (0.000000)	0.000015 (0.000019)	0.000016 (0.000013)	0.000016 (0.000013)

Table 2. Running time required per run, s = seconds. All algorithms were implemented in the C programming language and they were executed on PCs with the same hardware and software platform.

Set of problems	Objectives	NSGA-II	SMS-EMOA	APP-SMS-EMOA	MC-MOEAs
ZDT	2	$\lesssim 1s$	$5s - 10s$	$5s - 10s$	$\lesssim 1s$
DTLZ	3	$2s - 4s$	$4568s - 8468s$	$231s - 307s$	$3s - 9s$
DTLZ	4	$3s - 4s$	$14448s - 14650s$	$378s - 423s$	$5s - 12s$
DTLZ	5	$4s - 5s$	$15423s - 18000s$	$472s - 499s$	$9s - 14s$
DTLZ	6	$5s - 6s$	-	$531s - 584s$	$8s - 16s$
DTLZ	7	$5 - 6s$	-	$536s - 583s$	$9 - 18s$
DTLZ	8	$5s - 7s$	-	$525s - 583$	$9s - 16s$

computing the maximin fitness function is an inexpensive process, since their complexities are linear with respect to the number of objectives. In Table 2, we can see that our MC-MOEAs require much less time than SMS-EMOA and even much less time than APP-SMS-EMOA. Thus, we argue that our MC-MOEAs can be a good alternative for dealing with many objective optimization problems.

6   Conclusions and Future Work

In this work, we have studied the maximin fitness function and its properties with the aim of identifying its advantages and disadvantages. Then, we proposed some mechanisms to improve it. Our study encompassed three selection operators (one of them was proposed in [10] and the other two were proposed here). These operators were incorporated into a MOEA that uses simulated binary crossover (SBX) and parameter-based mutation (PM), giving rise to the main proposal of this paper, which is called: *Maximin-Clustering Multi-Objective Evolutionary Algorithm (MC-MOEA)*. We compared our proposed MC-MOEA with respect to a Pareto-based MOEA (NSGA-II) and with respect to two hypervolume-based MOEAs (SMS-EMOA and APP-SMS-EMOA). Our results showed that our MC-MOEA outperformed NSGA-II in most cases and that it was competitive with respect to SMS-EMOA and APP-SMS-EMOA with respect to the hypervolume indicator, but at a much lower computational cost. Also, it was better than APP-SMS-EMOA in most cases with respect to additive epsilon indicator. Thus, we believe that our proposed selection operators can be a viable alternative for dealing (at an affordable computational cost) with many-objective optimization problems. As part of our future work, we plan to study the behavior of our selection operators if we allow that the clustering technique iterates for a longer time. We also plan to incorporate our selection operator into a different approach (e.g., particle swarm optimization) in order to assess the impact of the search engine in the results. Finally, we plan to compare our approach with respect to AGE, which is based on Maximin fitness [6], and with respect to MOEA/D, which is based on decomposition and is known to be very competitive [11].

Table 3. Results obtained with respect to the additive epsilon indicator. The value in parentheses in the first column indicates the number of objectives. We show average values over 30 independent runs. The values in parentheses of the other columns correspond to the standard deviations.

	MC-MOEA(v1)	APP-SMS-EMOA	MC-MOEA(v2)	APP-SMS-EMOA	MC-MOEA(v3)	APP-SMS-EMOA
ZDT1 (2)	0.002122 (0.001941)	0.266333 (0.027070)	0.001478 (0.000885)	0.316856 (0.025420)	0.001500 (0.001088)	0.327322 (0.028306)
ZDT2 (2)	0.001022 (0.001199)	0.380433 (0.038094)	0.000467 (0.000452)	0.460300 (0.038666)	0.000556 (0.000482)	0.420622 (0.038305)
ZDT3 (2)	0.002122 (0.002642)	0.290033 (0.042676)	0.001022 (0.001229)	0.326000 (0.036908)	0.001467 (0.002229)	0.308822 (0.036630)
ZDT4 (2)	0.319067 (0.157283)	0.310033 (0.165624)	0.347045 (0.154540)	0.291311 (0.182824)	0.308167 (0.145411)	0.321345 (0.220199)
ZDT6 (2)	0.002811 (0.008674)	0.918600 (0.057853)	0.001633 (0.003862)	0.931322 (0.051950)	0.003433 (0.007189)	0.899189 (0.070228)
DTLZ1 (3)	0.000004 (0.000024)	0.003804 (0.000797)	0.000004 (0.000024)	0.006236 (0.000774)	0.000004 (0.000024)	0.006609 (0.000824)
DTLZ2 (3)	0.000000 (0.000000)	0.080467 (0.002483)	0.000000 (0.000000)	0.088733 (0.003025)	0.000000 (0.000000)	0.090404 (0.003393)
DTLZ3 (3)	0.000400 (0.000238)	0.000391 (0.000119)	0.000373 (0.000116)	0.001600 (0.000391)	0.000365 (0.000091)	0.001276 (0.000299)
DTLZ4 (3)	0.000000 (0.000000)	0.079884 (0.003326)	0.000000 (0.000000)	0.096458 (0.003444)	0.000000 (0.000000)	0.090964 (0.002840)
DTLZ5 (3)	0.000000 (0.000000)	0.047689 (0.002008)	0.000000 (0.000000)	0.064009 (0.002219)	0.000000 (0.000000)	0.063933 (0.002449)
DTLZ6 (3)	0.182560 (0.110139)	0.285391 (0.264605)	0.161787 (0.084851)	0.293862 (0.313067)	0.139133 (0.071011)	0.368169 (0.353202)
DTLZ7 (3)	0.000000 (0.000000)	0.052507 (0.003122)	0.000004 (0.000024)	0.064120 (0.043345)	0.000009 (0.000033)	0.060058 (0.003881)
DTLZ1 (4)	0.000000 (0.000000)	0.008422 (0.000779)	0.000000 (0.000000)	0.020658 (0.001978)	0.000004 (0.000024)	0.026813 (0.002857)
DTLZ2 (4)	0.000271 (0.000064)	0.102338 (0.005304)	0.000329 (0.000136)	0.107911 (0.004461)	0.000325 (0.000095)	0.109924 (0.004891)
DTLZ3 (4)	0.000000 (0.000000)	0.000307 (0.000085)	0.000000 (0.000000)	0.001991 (0.000666)	0.000000 (0.000000)	0.002360 (0.000945)
DTLZ4 (4)	0.000124 (0.000059)	0.110169 (0.004690)	0.000124 (0.000059)	0.121022 (0.004768)	0.000124 (0.000048)	0.124916 (0.005537)
DTLZ5 (4)	0.000000 (0.000000)	0.249071 (0.013026)	0.000000 (0.000000)	0.242676 (0.011803)	0.000000 (0.000000)	0.234964 (0.013292)
DTLZ6 (4)	0.000031 (0.000056)	0.135711 (0.018356)	0.000031 (0.000056)	0.157236 (0.019663)	0.000027 (0.000053)	0.155605 (0.019630)
DTLZ7 (4)	0.000053 (0.000065)	0.087022 (0.013190)	0.000027 (0.000063)	0.080956 (0.011480)	0.000027 (0.000072)	0.089262 (0.012833)
DTLZ1 (5)	0.170471 (0.041638)	0.089964 (0.008869)	0.211160 (0.046707)	0.076747 (0.014855)	0.210382 (0.046157)	0.070053 (0.012736)
DTLZ2 (5)	0.000027 (0.000053)	0.134591 (0.006910)	0.000035 (0.000059)	0.155960 (0.008473)	0.000018 (0.000045)	0.141640 (0.007720)
DTLZ3 (5)	0.000000 (0.000000)	0.000533 (0.000138)	0.000000 (0.000000)	0.000462 (0.000082)	0.000000 (0.000000)	0.000458 (0.000149)
DTLZ4 (5)	0.000000 (0.000000)	0.210578 (0.008651)	0.000000 (0.000000)	0.194676 (0.008037)	0.000000 (0.000000)	0.187938 (0.007237)
DTLZ5 (5)	0.000124 (0.000059)	0.164760 (0.007025)	0.000124 (0.000059)	0.176467 (0.076686)	0.000124 (0.000048)	0.182858 (0.008700)
DTLZ6 (5)	0.000000 (0.000000)	0.042333 (0.009989)	0.000000 (0.000000)	0.041249 (0.024001)	0.000000 (0.000000)	0.041249 (0.009875)
DTLZ7 (5)	0.000031 (0.000056)	0.099422 (0.023848)	0.000027 (0.000063)	0.095249 (0.024001)	0.000027 (0.000072)	0.083547 (0.022144)
DTLZ1 (6)	0.000178 (0.000173)	0.254507 (0.012576)	0.142182 (0.037791)	0.239484 (0.018511)	0.142182 (0.037791)	0.233298 (0.018348)
DTLZ2 (6)	0.000000 (0.000000)	0.218396 (0.017481)	0.000240 (0.000240)	0.222662 (0.018911)	0.000004 (0.000024)	0.207076 (0.016790)
DTLZ3 (6)	0.000040 (0.000092)	0.000067 (0.000096)	0.000018 (0.000045)	0.000604 (0.000315)	0.000000 (0.000000)	0.000973 (0.000239)
DTLZ4 (6)	0.000000 (0.000000)	0.354707 (0.010429)	0.000000 (0.000000)	0.337271 (0.008929)	0.000000 (0.000000)	0.337271 (0.008929)
DTLZ5 (6)	0.000000 (0.000000)	0.150840 (0.006913)	0.000000 (0.000000)	0.136347 (0.073308)	0.000000 (0.000000)	0.157738 (0.006627)
DTLZ6 (6)	0.000000 (0.000000)	0.990302 (0.005175)	0.000000 (0.000000)	0.987258 (0.006683)	0.000000 (0.000000)	0.987258 (0.006683)
DTLZ7 (6)	0.000169 (0.000179)	0.084836 (0.027049)	0.000227 (0.000176)	0.077058 (0.025112)	0.000151 (0.000178)	0.078431 (0.024091)
DTLZ1 (7)	0.000009 (0.000033)	0.553269 (0.027195)	0.000035 (0.000059)	0.507302 (0.035147)	0.000000 (0.000000)	0.582867 (0.031450)
DTLZ2 (7)	0.000000 (0.000000)	0.327400 (0.026559)	0.000000 (0.000000)	0.283538 (0.023866)	0.000000 (0.000000)	0.274733 (0.023531)
DTLZ3 (7)	0.000000 (0.000000)	0.004080 (0.003656)	0.000000 (0.000000)	0.003151 (0.001397)	0.000000 (0.000000)	0.003169 (0.001462)
DTLZ4 (7)	0.000000 (0.000000)	0.525089 (0.015005)	0.000000 (0.000000)	0.496098 (0.119074)	0.000253 (0.000177)	0.496098 (0.011994)
DTLZ5 (7)	0.000800 (0.000958)	0.135000 (0.006122)	0.000755 (0.000825)	0.144400 (0.006197)	0.000013 (0.000040)	0.133316 (0.005925)
DTLZ6 (7)	0.000013 (0.000040)	0.935022 (0.019142)	0.000004 (0.000024)	0.932467 (0.190074)	0.000000 (0.000000)	0.932467 (0.019074)
DTLZ7 (7)	0.000000 (0.000000)	0.042213 (0.019122)	0.000000 (0.000000)	0.038800 (0.017292)	0.000711 (0.000816)	0.040049 (0.017645)
DTLZ1 (8)	0.000000 (0.000000)	0.752924 (0.041588)	0.000000 (0.000000)	0.739569 (0.032785)	0.000000 (0.000000)	0.779698 (0.039500)
DTLZ2 (8)	0.000000 (0.000000)	0.480196 (0.024395)	0.000000 (0.000000)	0.370102 (0.019897)	0.000018 (0.000045)	0.368333 (0.019622)
DTLZ3 (8)	0.000000 (0.000000)	0.007791 (0.006625)	0.000000 (0.000000)	0.004436 (0.002156)	0.000000 (0.000000)	0.006440 (0.004454)
DTLZ4 (8)	0.000000 (0.000000)	0.778618 (0.018189)	0.000000 (0.000000)	0.691338 (0.021952)	0.000000 (0.000000)	0.691338 (0.021952)
DTLZ5 (8)	0.000000 (0.000000)	0.144769 (0.005696)	0.000000 (0.000000)	0.141418 (0.004752)	0.000000 (0.000000)	0.144693 (0.004958)
DTLZ6 (8)	0.000000 (0.000000)	0.831440 (0.040860)	0.000000 (0.000000)	0.814929 (0.043509)	0.000000 (0.000000)	0.814929 (0.043509)
DTLZ7 (8)	0.000000 (0.000000)	0.035400 (0.012263)	0.000000 (0.000000)	0.039831 (0.012084)	0.000000 (0.000000)	0.039831 (0.012084)

References

1. Bader, J., Zitzler, E.: HypE: An Algorithm for Fast Hypervolume-Based Many-Objective Optimization. Evolutionary Computation 19(1), 45–76 (2011)
2. Balling, R.: Pareto sets in decision-based design. Journal of Engineering Valuation and Cost Analysis 3, 189–198 (2000)
3. Balling, R.: The Maximin Fitness Function; Multi-objective City and Regional Planning. In: Fonseca, C.M., Fleming, P.J., Zitzler, E., Deb, K., Thiele, L. (eds.) EMO 2003. LNCS, vol. 2632, pp. 1–15. Springer, Heidelberg (2003)
4. Balling, R., Wilson, S.: The Maximin Fitness Function for Multi-objective Evolutionary Computation: Application to City Planning. In: Spector, L., Goodman, E.D., Wu, A., Langdon, W.B., Voigt, H.-M., Gen, M., Sen, S., Dorigo, M., Pezeshk, S., Garzon, M.H., Burke, E. (eds.) Proceedings of the Genetic and Evolutionary Computation Conference (GECCO 2001), pp. 1079–1084. Morgan Kaufmann Publishers, San Francisco (2001)
5. Beume, N., Naujoks, B., Emmerich, M.: SMS-EMOA: Multiobjective selection based on dominated hypervolume. European Journal of Operational Research 181(3), 1653–1669 (2007)
6. Bringmann, K., Friedrich, T., Neumann, F., Wagner, M.: Approximation-guided evolutionary multi-objective optimization. In: Proceedings of the Twenty-Second International Joint Conference on Artificial Intelligence, IJCAI 2011, vol. 2, pp. 1198–1203. AAAI Press (2011)
7. Coello, C.A.C., Lamont, G.B., Veldhuizen, D.A.V.: Evolutionary Algorithms for Solving Multi-Objective Problems, 2nd edn. Springer, New York (2007) ISBN 978-0-387-33254-3
8. Deb, K., Pratap, A., Agarwal, S., Meyarivan, T.: A Fast and Elitist Multiobjective Genetic Algorithm: NSGA-II. IEEE Transactions on Evolutionary Computation 6(2), 182–197 (2002)
9. Deb, K., Thiele, L., Laumanns, M., Zitzler, E.: Scalable Test Problems for Evolutionary Multiobjective Optimization. In: Abraham, A., Jain, L., Goldberg, R. (eds.) Evolutionary Multiobjective Optimization. Theoretical Advances and Applications, pp. 105–145. Springer, USA (2005)
10. Menchaca-Mendez, A., Coello, C.A.C.: Solving Multi-Objective Optimization Problems using Differential Evolution and a Maximin Selection Criterion. In: 2012 IEEE Congress on Evolutionary Computation, CEC 2012, Brisbane, Australia, June 10-15, pp. 3143–3150. IEEE Press (2012)
11. Zhang, Q., Li, H.: MOEA/D: A Multiobjective Evolutionary Algorithm Based on Decomposition. IEEE Transactions on Evolutionary Computation 11(6), 712–731 (2007)
12. Zitzler, E.: Evolutionary Algorithms for Multiobjective Optimization: Methods and Applications. PhD thesis, Swiss Federal Institute of Technology (ETH), Zurich, Switzerland (November 1999)
13. Zitzler, E., Deb, K., Thiele, L.: Comparison of Multiobjective Evolutionary Algorithms: Empirical Results. Evolutionary Computation 8(2), 173–195 (2000)
14. Zitzler, E., Thiele, L.: Multiobjective Optimization Using Evolutionary Algorithms - A Comparative Case Study. In: Eiben, A.E., Bäck, T., Schoenauer, M., Schwefel, H.-P. (eds.) PPSN V. LNCS, vol. 1498, pp. 292–301. Springer, Heidelberg (1998)

Difficulty in Evolutionary Multiobjective Optimization of Discrete Objective Functions with Different Granularities

Hisao Ishibuchi, Masakazu Yamane, and Yusuke Nojima

Department of Computer Science and Intelligent Systems, Graduate School of Engineering, Osaka Prefecture University, 1-1 Gakuen-cho, Naka-ku, Sakai, Osaka 599-8531, Japan
{hisaoi@,masakazu.yamane@ci.,nojima@}cs.osakafu-u.ac.jp

Abstract. Objective functions are discrete in combinatorial optimization. In general, the number of possible values of a discrete objective is totally different from problem to problem. That is, discrete objectives have totally different granularities in different problems (In this paper, "granularity" means the width of discretization intervals). In combinatorial multiobjective optimization, a single problem has multiple discrete objectives with different granularities. Some objectives may have fine granularities with many possible values while others may have very coarse granularities with only a few possible values. Handling of such a combinatorial multiobjective problem has not been actively discussed in the EMO community. In our former study, we showed that discrete objectives with coarse granularities slowed down the search by NSGA-II, SPEA2, MOEA/D and SMS-EMOA on two-objective problems. In this paper, we first discuss why such a discrete objective deteriorates the search ability of those EMO algorithms. Next we propose the use of strong Pareto dominance in NSGA-II to improve its search ability. Then we examine the effect of discrete objectives on the performance of the four EMO algorithms on many-objective problems. An interesting observation is that discrete objectives with coarse granularities improve the search ability of NSGA-II and SPEA2 on many-objective problems whereas they deteriorate their search ability on two-objective problems. The performance of MOEA/D and SMS-EMOA is always deteriorated by discrete objectives with coarse granularities. These observations are discussed from the following two viewpoints: One is the difficulty of many-objective problems for Pareto dominance-based EMO algorithms, and the other is the relation between discrete objectives and the concept of ε-dominance.

Keywords: Evolutionary multiobjective optimization, many-objective problems, discrete objectives, ε-dominance, combinatorial multiobjective optimization.

1 Introduction

Evolutionary multiobjective optimization (EMO) has been a hot research area in the field of evolutionary computation for the last two decades [2], [3], [24]. Whereas a large number of various EMO algorithms were proposed, Pareto dominance-based algorithms such as NSGA-II [5], SPEA [29] and SPEA2 [28] have always been the

R.C. Purshouse et al. (Eds.): EMO 2013, LNCS 7811, pp. 230–245, 2013.

main stream in the EMO community since Goldberg's suggestion [7]. However, the use of scalarizing function-based algorithms (e.g., MOEA/D [27]) and indicator-based algorithms (e.g., SMS-EMOA [1]) have also been actively examined in recent studies, especially for difficult multiobjective problems with complicated Pareto fronts [20] and many-objective problems [25].

In multiobjective optimization, the ranges of values of each objective can be totally different. Those objective values are often normalized in the application of EMO algorithms to multiobjective problems so that the range of values of each objective becomes the same over all objectives. For example, a normalization mechanism was included in the crowding distance calculation of NSGA-II [5]. The importance of the normalization of objective values is widely recognized in the EMO community. This is because almost all elements of EMO algorithms except for Pareto dominance (e.g., crowding mechanisms, hypervolume calculations and scalarizing functions) depend on the magnitude of objective values of each objective.

Objective values are discrete in combinatorial multiobjective optimization due to the combinatorial nature of decision variables. The number of possible values of each objective is totally different. For example, in pattern and feature selection for nearest neighbor classifier design [11], the number of patterns usually has more possible values than the number of features. This is because classification problems usually have more patterns than features (e.g., a magic data set in the UCI Machine Learning Repository has 19,020 patterns with 20 features). In multiobjective genetics-based machine learning [12], the total number of antecedent conditions has more possible values than the number of rules. This is because each rule has a different number of antecedent conditions. In multiobjective flowshop scheduling [16], the maximum flow time has more possible values than the maximum tardiness. This is because a large number of different schedules have the same value of the maximum tardiness even when they have different values of the maximum flow time.

As these examples show, each discrete objective has a different number of possible values (i.e., different granularity). Some discrete objectives have fine granularities with many possible values while others have coarse granularities with only a small number of possible values. We have various examples of multiobjective problems where discrete objectives have totally different granularities. The handling of discrete objectives with different granularities, however, has not been actively studied for EMO algorithms. In our former work [15], we examined the effect of discrete objectives with different granularities on the search behavior of EMO algorithms through computational experiments on two-objective problems. For example, when two objectives had coarse granularities, the search by EMO algorithms was severely slowed down in comparison with the case of two objectives with fine granularities. When two objectives had different granularities, the search was biased towards one objective with a finer granularity. That is, a population was biased towards the edge of the Pareto front with the best value of that objective. The search along the other objective with a coarser granularity was severely slowed down. Whereas we clearly reported those interesting observations in our former work [15], we could not explain why the search by EMO algorithms was affected in such an interesting manner. The main aim of this paper is to explain the reasons for the above-mentioned observations.

This paper is organized as follows. In Section 2, we briefly show the above-mentioned interesting observations in our former work [15]. In Section 3, we clearly explain why those interesting observations were obtained. Based on the explanations in Section 3, we suggest the modification of NSGA-II by the use of strong Pareto dominance in Section 4. It is shown that the suggested modification improves the search ability of NSGA-II on two-objective problems with coarse granularities. In Section 5, we examine the performance of EMO algorithms on many-objective problems with discrete objectives. Experimental results show that discrete objectives with coarse granularities improve the performance of NSGA-II and SPEA2 on many-objective problems while they severely deteriorated the performance on two-objective problems. We also discuss why discrete objectives with coarse granularities have such a positive effect on many-objective optimization from the following two viewpoints: One is the difficulty of many-objective problems for Pareto dominance-based EMO algorithms, and the other is the relation between discrete objectives and the concept of ε-dominance [19]. In Section 6, we conclude this paper.

2   Effect of Discrete Objectives on Two-Objective Optimization

In our former work [15], we examined the effect of discrete objectives with different granularities on the search behavior of NSGA-II [5], SPEA2 [28], MOEA/D [27] and SMS-EMOA [1] on the following four types of two-objective problems:

(i) Two-objective 500-item knapsack problem in Zitzler and Thiele [29],
(ii) 500-bit one-max and zero-max problem,
(iii) Modified 500-bit one-max and zero-max problem with a convex Pareto front,
(iv) Modified 500-bit one-max and zero-max problem with a concave Pareto front.

Similar effects of discrete objectives were observed on the search behavior of the four EMO algorithms on the three types of two-objective problems in our former work. Here we only show experimental results of NSGA-II on the two-objective 500-item knapsack problem in Zitzler and Thiele [29].

The two-objective 500-item knapsack problem with two constraint conditions in Zitzler and Thiele [29] is written as follows:

$$\text{Maximize} \quad f_i(x) = \sum_{j=1}^{n} p_{ij} x_j, \quad i = 1, 2, \tag{1}$$

$$\text{subject to} \quad \sum_{j=1}^{n} w_{ij} x_j \le c_i, \quad i = 1, 2, \tag{2}$$

$$x_j = 0 \text{ or } 1, \quad j = 1, 2, ..., n. \tag{3}$$

In (1)-(3), n is the number of items (i.e., $n = 500$ in this paper), x is a 500-bit binary string, p_{ij} is the profit of item j according to knapsack i, w_{ij} is the weight of item j according to knapsack i, and c_i is the capacity of knapsack i. The value of each profit p_{ij} in (1) was randomly specified as an integer in the interval [10, 100]. As a result, each objective has integer objective values. We use exactly the same two-objective 500-item knapsack problem as in Zitzler and Thiele [29].

In Fig. 1, we show randomly generated 200 solutions of this two-objective 500-item knapsack problem together with its Pareto front. In Fig. 1, we used a greedy repair method based on the maximum profit/weight ratio in Zitzler and Thiele [29] when randomly generated solutions were infeasible. The greedy repair method in [29] was always used in our computational experiments in this paper. As shown in Fig. 1, randomly generated solutions are not close to the Pareto front. Thus a high selection pressure towards the Pareto front is needed to efficiently search for Pareto optimal or near Pareto optimal solutions. At the same time, a strong diversity improvement mechanism is also needed to find a wide variety of solutions along the entire Pareto front. That is, EMO algorithms for the knapsack problem in Fig. 1 need strong convergence and diversification properties. Multiobjective knapsack problems have been frequently used to evaluate the performance of EMO algorithms in the literature (e.g., Jaszkiewicz [17] and Sato et al. [22]).

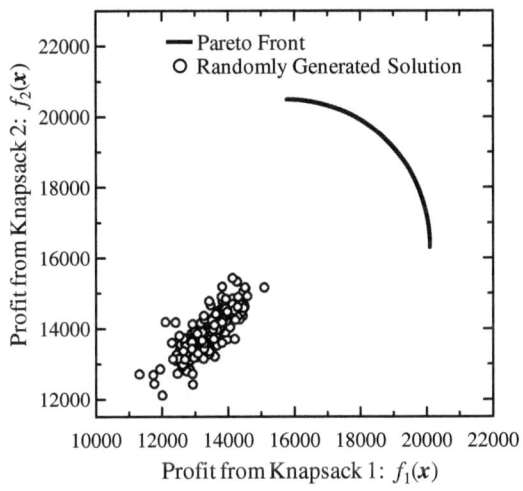

Fig. 1. Pareto front and randomly generated 200 solutions [15]

In our former work [15], NSGA-II, SPEA2, MOEA/D and SMS-EMOA with the following parameter specifications were applied to the knapsack problem in Fig. 1:

Coding: Binary string of length 500 (i.e., 500-bit string),
Population size: 200,
Termination condition: 2000 generations (400000 solution evaluations in MOEA/D),
Parent selection: Random selection from the population (SMS-EMOA),
                   Random selection from the neighborhood (MOEA/D),
                   Binary tournament selection with replacement (NSGA-II and SPEA2),
Crossover: Uniform crossover (Probability: 0.8),
Mutation: Bit-flip mutation (Probability: 1/500),
Number of runs for each test problem: 100 runs.

The origin (0, 0) of the two-dimensional objective space was used as a reference point for hypervolume calculation in SMS-EMOA. In MOEA/D, the weighted Tchebycheff function was used in the same manner as in Zhang and Li [27]. The neighborhood size in MOEA/D was specified as 10.

The four EMO algorithms were also applied to discretized problems with the discretization interval of width 100. For example, objective values in [20001, 20100] and [20101, 20200] were rounded up to 20100 and 20200, respectively. It should be noted that the width of the discretization interval for each objective in the original knapsack problem in Fig. 1 is 1. This is because each profit p_{ij} in the two objective functions was randomly specified as an integer in the interval [10, 100]. We denote discretized problems using the width of the discretization interval for each objective such as G100-G1 and G100-G100. In the G100-G100 problem, both objectives were discretized by the discretization interval of width 100. The original knapsack problem is denoted as G1-G1. Only the first objective of G100-G1 (only the second objective of G1-G100) was discretized by the discretization interval of width 100.

In Fig. 2, we show experimental results by NSGA-II on the four knapsack problems (i.e., G1-G1, G1-G100, G100-G1 and G100-G100). In each plot of Fig. 2, all solutions at the final generation of a single run of NSGA-II are shown together with the 50% attainment surface [6] over its 100 runs.

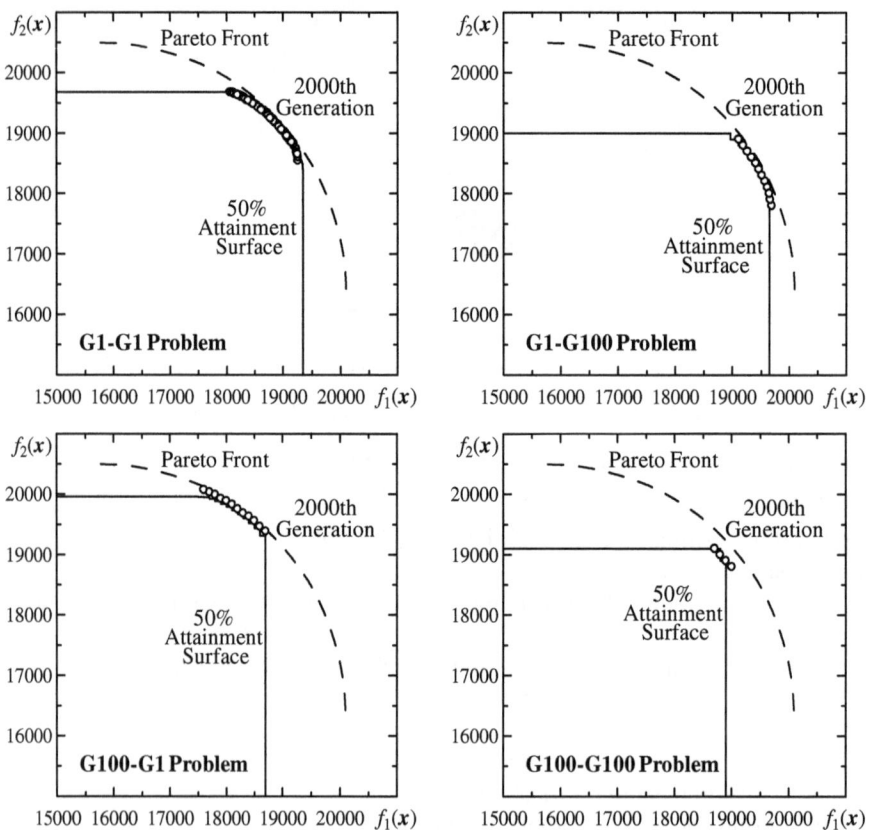

Fig. 2. Experimental results by NSGA-II on four variants of the knapsack problem

3   Discussions on the Search Behavior of EMO Algorithms

From each plot in Fig. 2, we can see that the following results were obtained about the search behavior of NSGA-II on each problem:

(i) G1-G1: NSGA-II found many solutions around the center of the Pareto front.
(ii) G1-G100: The search was biased towards the bottom-right part of the Pareto front.
(iii) G100-G1: The search was biased towards the top-left part of the Pareto front.
(iv) G100-G100: The performance of NSGA-II was severely deteriorated.

Let us discuss why these results were obtained. First we address the search behavior of NSGA-II on the G100-G100 problem. In the bottom-right plot of Fig. 2, only four solutions of the G100-G100 problem were obtained by a single run of NSGA-II. We checked all the 200 solutions in the final population. Then we found that they were overlapping on the four solutions in the discretized objective space. We also found that the number of different strings in the final population was eleven. All of them had different objective vectors in the original objective space (i.e., the objective space of the G1-G1 problem). The four solutions in the discretized objective space and the eleven solutions in the original objective space are shown in an enlarged view in the left plot of Fig. 3.

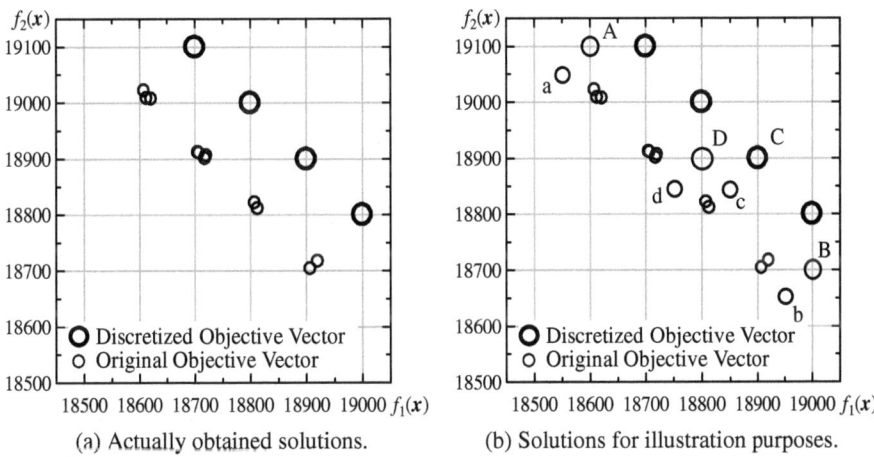

(a) Actually obtained solutions.   (b) Solutions for illustration purposes.

Fig. 3. Obtained Solutions by a single run of NSGA-II on the G100-G100 problem

Using the right plot of Fig. 3, we explain why the search ability of NSGA-II on the G100-G100 problem was severely deteriorated. Let us assume that new solutions "a" and "b" are generated by crossover and mutation. Whereas those solutions increase the diversity, they are not likely to survive because they are dominated solutions in the discretized objective space (i.e., because solutions "A" and "B" are dominated solutions). This explains why the diversity of obtained solutions for the G100-G100 problem was very small in Fig. 2. Let us also assume that a new solution "c" is generated by crossover and mutation. Whereas this solution is better than the two solutions in the same box, all of them are discretized to the same solution "C" in the discretized objective space. Thus the difference between the new solution "c" and the existing two solutions in the same box disappears by the discretization. This explains why the search of NSGA-II towards the Pareto front of the G100-G100 problem was slow in Fig. 2. Further we assume that a new solution "d" in the right plot of Fig. 3 is generated by crossover and mutation. This solution is not likely to survive because its discretized solution "D" is a dominated solution. The situation of the new solution "d" also explains the deterioration in the search ability of NSGA-II.

Next, let us address the search behavior of NSGA-II on the G100-G1 problem. The first objective of this problem has a very coarse granularity (i.e., G100) while its second objective is a fine granularity (i.e., G1). The objective space of the G100-G1 problem is illustrated in Fig. 4. Ten solutions in Fig. 4 are non-dominated with each other in the objective space of the original G1-G1 problem. However, six of them are dominated solutions in the objective space of the G100-G1 problem. For example, let us consider the four solutions "e", "f", "g" and "h" in the objective space of the original G1-G1 problem in Fig. 4. They are discretized to the solutions "E", "F", "G" and "H" in the objective space of the G100-G1 problem, respectively. Whereas "e", "f", "g" and "h" are non-dominated with each other, "F", "G" and "H" are dominated by "E". In the fitness evaluation of NSGA-II, the ranks of these solutions are as follows: E: Rank 1, F: Rank 2, G: Rank 3, H: Rank 4. Thus the solutions G and H are likely to be removed in the generation update phase of NSGA-II. This explains why the multiobjective search of NSGA-II on the G100-G1 problem was biased towards the top-left part of the Pareto front in Fig. 2. In the same manner, we can explain why the multiobjective search of NSGA-II on the G1-G100 problem was biased towards the bottom-right part of the Pareto front in Fig. 2.

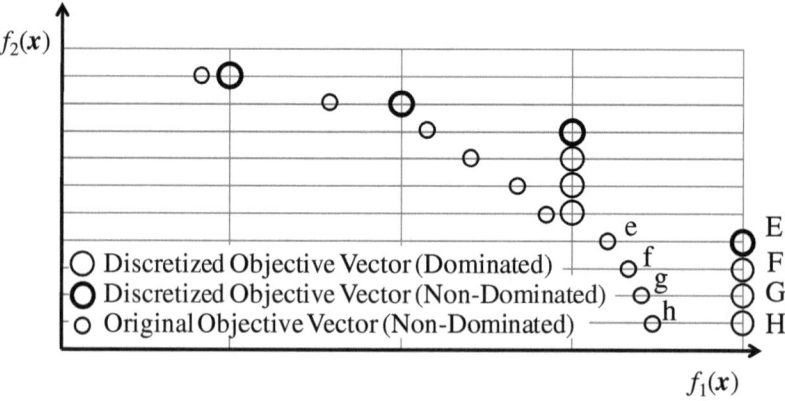

Fig. 4. Explanations of the search behavior by NSGA-II on the G100-G1 problem

4   Use of Strong Pareto Dominance in NSGA-II

As explained in Section 3, many non-dominated solutions of the original G1-G1 problem become dominated solutions by the use of the coarse granularity G100. For example, the non-dominated solutions "f", "g" and "h" in Fig. 4 were discretized to the dominated solutions "F", "G" and "H" by the use of G100 for the first objective. In the right plot in Fig. 3, the non-dominated solutions "a", "b" and "d" were discretized to the dominated solutions "A", "B" and "D". Such a solution status change seems to have a lot of negative effects on the search ability of NSGA-II. In other words, the handling of those dominated solutions as non-dominated ones may prevent the deterioration in the performance of NSGA-II.

Motivated by these discussions, let us examine the modification of NSGA-II by using the following strong Pareto dominance in the fitness evaluation of NSGA-II:

Strong Pareto Dominance

For multiobjective maximization, an objective vector $\boldsymbol{f}(\boldsymbol{x}) = (f_1(\boldsymbol{x}), f_2(\boldsymbol{x}), ..., f_n(\boldsymbol{x}))$ is defined as being strongly dominated by another objective vector $\boldsymbol{f}(\boldsymbol{y}) = (f_1(\boldsymbol{y}), f_2(\boldsymbol{y}), ..., f_n(\boldsymbol{y}))$ if and only if $f_i(\boldsymbol{x}) < f_i(\boldsymbol{y})$ holds for all $i = 1, 2, ..., n$.

When we use this definition, solutions "A", "B", "D", "F", "G" and "H" in Fig. 3 and Fig. 4 are handled as non-dominated solutions. We applied NSGA-II with this definition to the G1-G1, G1-G100, G100-G1 and G100-G100 problems in the same manner as in Section 2. Experimental results are shown in Fig. 5.

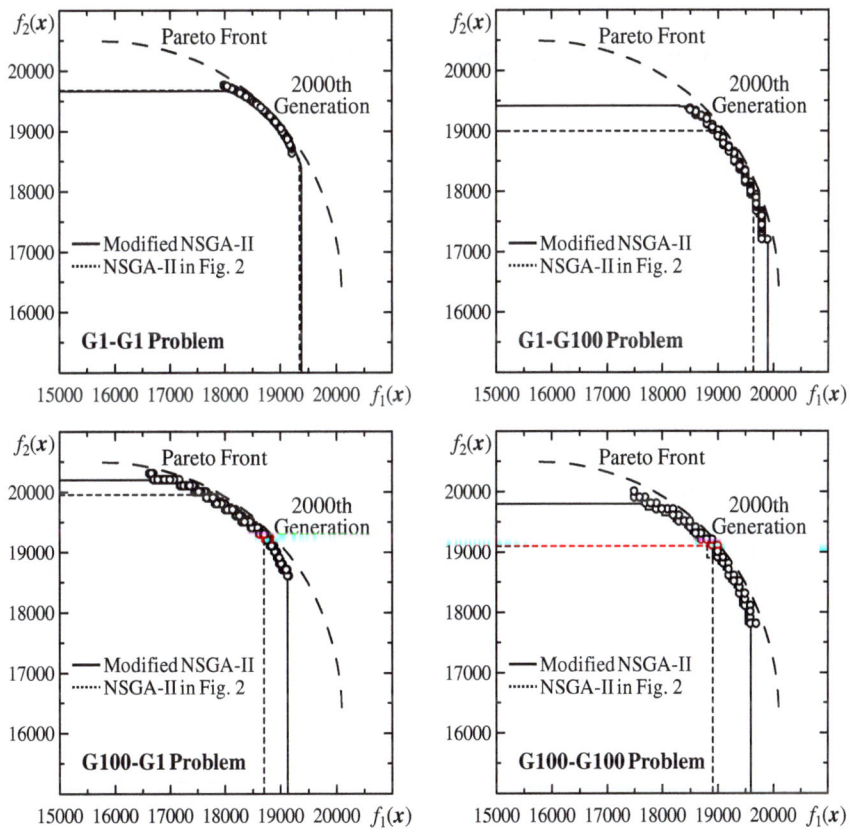

Fig. 5. Experimental results by NSGA-II with strong Pareto dominance

From the comparison between Fig. 2 and Fig. 5, we can obtain the following observations about the search ability of NSGA-II:

(i) The use of strong Pareto dominance had almost no effect on the search ability of NSGA-II on the G1-G1 problem (see the top-left plot of each figure). This is because the difference between the standard and modified NSGA-II algorithms is very small when the granularity is fine (i.e., because $f_i(x) = f_i(y)$ is not likely to hold for many pairs of different solutions x and y when the granularity is fine).

(ii) The search ability of NSGA-II on the G100-G100 problem was clearly improved by the use of strong Pareto dominance with respect to both the convergence and the diversity (see the bottom-right plot of each figure). That is, the intended positive effect of strong Pareto dominance was observed for the G100-G100 problem.

(iii) The search ability of NSGA-II on the G1-G100 and G100-G1 problems was also clearly improved by the use of strong Pareto dominance with respect to the diversity of obtained solutions. The bias of the search of NSGA-II toward a part of the Pareto front was somewhat remedied by the use of strong Pareto dominance. That is, the intended positive effect of strong Pareto dominance was also observed for the G1-G100 and G100-G1 problems.

5   Computational Experiments on Many-Objective Problems

We have already examined the effect of discrete objectives with different granularities on the search behavior of NSGA-II on the four two-objective knapsack problems. We have also demonstrated that the use of strong Pareto dominance improved the performance of NSGA-II on the three two-objective knapsack problems with the coarse granularity. In this section, we examine the performance of NSGA-II, SPEA2, MOEA/D and SMS-EMOA on many-objective knapsack problems with fine and coarse granularities using the hypervolume measure.

Many-objective optimization with four or more objectives is usually very difficult for Pareto dominance-based EMO algorithms [10], [18], [21]. This is because almost all solutions in a population quickly become non-dominated within a small number of generations in evolutionary multiobjective search for many-objective problems. As a result, Pareto dominance-based selection pressure quickly becomes very weak. Various approaches have been proposed to improve the search ability of Pareto dominance-based EMO algorithms on many-objective problems [13], [14]. Recently it has also been shown that many-objective problems are not necessarily difficult for Pareto dominance-based EMO algorithms in the literature [23].

As test problems, we generated multiobjective 500-item knapsack problems with up to eight objectives by adding randomly generated objectives $f_i(x)$, $i = 3, 4, ..., 8$ to the original two-objective 500-item knapsack problems in the previous sections:

$$f_i(x) = \sum_{j=1}^{500} p_{ij} x_j, \quad i = 3, 4, ..., 8, \tag{4}$$

where p_{ij} is a randomly specified integer in the interval [10, 100].

Using the same parameter specifications as in Section 2, we applied NSGA-II, its modified version, SPEA2, MOEA/D and SMS-EMOA to our test problems with two, four, six and eight objectives. Only in MOEA/D, the population size was specified as 220, 252 and 120 for our test problems with four, six and eight objectives, respectively. This is due to the combinatorial nature in the number of weight vectors in MOEA/D (for details, see [27]). The population size was specified as 200 in all the other cases. We used a fast hypervolume calculation method by While et al. [26] in SMS-EMOA. Each algorithm was applied to each test problem 100 times to calculate the average hypervolume except for SMS-EMOA on the eight-objective problem due to its heavy computation load: 30 runs of SMS-EMOA on the eight-objective problem.

Each test problem was discretized using the discretization intervals of width 10 (i.e., granularity: G10) and width 100 (i.e., granularity: G100). The granularity of our test problems before the discretization was G1 since each profit value in the objective functions was randomly specified as an integer in the interval [10, 100]. We used the origin of the objective space as a reference point for hypervolume calculation. The hypervolume calculation was always performed in the original objective space with the granularity G1. That is, discretized objective values with the granularities G10 and G100 were restored to their original values with G1 for fair comparison.

In Tables 1-5, we summarize our experimental results by each EMO algorithm. Each table shows the average hypervolume value and the standard deviation by each EMO algorithm on each test problem. The best result (i.e., the largest average hypervolume) among the three granularity specifications is highlighted by boldface for each test problem in each table.

Now, let us examine experimental results in each table. In Table 1, experimental results by NSGA-II are summarized. As we have already demonstrated in Fig. 2 in Section 2, the use of the coarse granularity G100 deteriorated the performance of NSGA-II on the two-objective problem in Table 1 (i.e., 5.6% decrease in the average hypervolume from 3.800×10^8 to 3.588×10^8). However, it improved the performance of NSGA-II on the eight-objective problem by 3.0% from 1.100×10^{34} to 1.133×10^{34}.

Let us discuss these observations from the viewpoint of ε-dominance which was proposed by Laumanns et al. [19] and used in ε-MOEA [4]. The comparison between boxes in the discretized objective space in ε-MOEA [4] is the same as the Pareto dominance-based comparison between discretized objective vectors in this paper. Horoba and Neumann [8], [9] theoretically explained that the use of ε-dominance deteriorates the search ability of their EMO algorithm when many Pareto-optimal solutions are included in a single box. This may be related to the performance deterioration by the use of G100 for the two-objective problem in Table 1 because a large number of Pareto-optimal solutions are included in a single box with G100.

Table 1. Average hypervolume and standard deviation by NSGA-II

Problem	Granularity: G1		Granularity: G10		Granularity: G100	
	Average	Stand. Dev.	Average	Stand. Dev.	Average	Stand. Dev.
2-Objective	**3.800E+08**	1.703E+06	3.779E+08	1.416E+06	3.588E+08	1.797E+06
4-Objective	1.227E+17	0.925E+15	**1.235E+17**	0.918E+15	1.209E+17	1.306E+15
6-Objective	3.729E+25	3.885E+23	3.751E+25	4.065E+23	**3.813E+25**	4.149E+23
8-Objective	1.100E+34	1.722E+32	1.112E+34	1.452E+32	**1.133E+34**	1.475E+32

Horoba and Neumann [8], [9] also proved that the use of ε-dominance improves the search ability of their EMO algorithm when the number of Pareto-optimal solutions exponentially increases with the problem size. This may be related to the performance improvement by the use of G100 for the six-objective and eight-objective problems in Table 1. The performance improvement can be also explained by the decrease in the number of non-dominated solutions. As we demonstrated in Section 3 using Fig. 3, many non-dominated solutions become dominated after the discretization of objective values. This means that the number of non-dominated solutions is decreased by the discretization. Since the difficulty of many-objective problems is caused by the increase in the number of non-dominated solutions, the discretization of objective values is likely to work as a countermeasure for improving the performance of NSGA-II on many-objective problems.

Whereas we have explained the effect of discrete objectives using the concept of ε-dominance, there is a clear difference between the discretization in this paper and the use of ε-dominance. In this paper, we assume that each objective has discrete values. For example, a discrete objective with the granularity G100 is assumed to be measured in multiples of 100. Thus its objective values are always multiples of 100 such as 500 and 600. However, the objective space discretization by ε-dominance is used only for the comparison between boxes. Objective values are not discretized. Thus EMO algorithms based on ε-dominance can use the standard Pareto dominance between objective vectors as well as ε-dominance between boxes. For example, two objective vectors (525, 550) and (531, 566) can be compared even when the objective space is discretized by ε-dominance using the discretization interval of width 100. This is not the case in this paper because these two objective vectors are always handled as the same objective vector (600, 600) in the case of the granularity G100. That is, they cannot be compared in this paper when the granularity is G100.

Table 2 shows experimental results by NSGA-II with strong Pareto dominance. As we have already demonstrated in Fig. 5, the use of strong Pareto dominance improved the performance of NSGA-II on the two-objective problem with G100 from Table 1 to Table 2 (i.e., 7.2% increase from 3.588×10^8 in Table 1 to 3.843×10^8 in Table 2). However, it deteriorated the performance of NSGA-II on the eight-objective problem with G100 by 4.3% from 1.133×10^{34} in Table 1 to 1.084×10^{34} in Table 2. This is because the use of strong Pareto dominance increases the number of non-dominated solutions, which is the main reason for the difficulty in the handling of many-objective problems by Pareto dominance-based EMO algorithms. That is, the increase in the number of non-dominated solutions by the use of strong Pareto dominance has a positive effect on the two-objective problem and a negative effect on the eight-objective problem.

Table 2. Average hypervolume and standard deviation by the modified NSGA-II

Problem	Granularity: G1		Granularity: G10		Granularity: G100	
	Average	Stand. Dev.	Average	Stand. Dev.	Average	Stand. Dev.
2-Objective	3.806E+08	1.603E+06	3.819E+08	1.419E+06	**3.843E+08**	1.925E+06
4-Objective	1.228E+17	1.016E+15	1.230E+17	0.841E+15	**1.290E+17**	1.032E+15
6-Objective	3.728E+25	4.020E+23	3.726E+25	4.078E+23	**3.795E+25**	4.276E+23
8-Objective	1.093E+34	1.924E+32	**1.097E+34**	1.793E+32	1.084E+34	2.046E+32

Table 3 shows experimental results by SPEA2. We can obtain similar observations from Table 1 (NSGA-II) and Table 3 (SPEA2). That is, the discretization of objective values by the granularity G100 deteriorated the performance of SPEA2 on the two-objective problem and improved its performance on the eight-objective problem in Table 3. This is because fitness evaluation in NSGA-II and SPEA2 is based on Pareto dominance (i.e., because they are Pareto dominance-based EMO algorithms).

Table 3. Average hypervolume and standard deviation by SPEA2

Problem	Granularity: G1		Granularity: G10		Granularity: G100	
	Average	Stand. Dev.	Average	Stand. Dev.	Average	Stand. Dev.
2-Objective	**3.786E+08**	1.162E+06	3.777E+08	1.657E+06	3.623E+08	2.098E+06
4-Objective	**1.218E+17**	0.801E+15	1.218E+17	0.889E+15	1.161E+17	1.148E+15
6-Objective	3.553E+25	4.411E+23	3.558E+25	3.863E+23	**3.619E+25**	3.600E+23
8-Objective	1.029E+34	1.362E+32	1.032E+34	1.396E+32	**1.068E+34**	1.295E+32

Table 4 shows experimental results by MOEA/D. We can observe totally different effects of discrete objectives in Table 4 on MOEA/D from Table 1 on NSGA-II and Table 3 on SPEA2. That is, the use of the coarse granularities G10 and G100 monotonically deteriorated the performance of MOEA/D on all the test problems with 2-8 objectives in Table 4. This is because the discretization of objective values simply makes single-objective optimization of scalarizing functions difficult independent of the number of objectives in MOEA/D.

Table 4. Average hypervolume and standard deviation by MOEA/D

Problem	Granularity: G1		Granularity: G10		Granularity: G100	
	Average	Stand. Dev.	Average	Stand. Dev.	Average	Stand. Dev.
2-Objective	**4.009E+08**	0.931E+06	3.993E+08	1.034E+06	3.610E+08	4.382E+06
4-Objective	**1.430E+17**	0.705E+15	1.421E+17	0.745E+15	1.249E+17	2.114E+15
6-Objective	**4.525E+25**	3.778E+23	4.484E+25	3.912E+23	3.763E+25	9.70E+23
8-Objective	**1.355E+34**	1.335E+32	1.340E+34	1.563E+32	1.022E+34	2.191E+32

Table 5 shows experimental results by SMS-EMOA. Effects of discrete objectives on SMS-EMOA in Table 5 are similar to those on MOEA/D in Table 4 (i.e., the discretization of objective values deteriorated the performance of SMS-EMOA). This may be because their fitness evaluation is not based on Pareto dominance.

Table 5. Average hypervolume and standard deviation by SMS-EMOA

Problem	Granularity: G1		Granularity: G10		Granularity: G100	
	Average	Stand. Dev.	Average	Stand. Dev.	Average	Stand. Dev.
2-Objective	**3.760E+08**	1.734E+06	3.730E+08	2.042E+06	3.571E+08	1.695E+06
4-Objective	**1.285E+17**	8.128E+14	1.277E+17	9.275E+14	1.129E+17	2.356E+15
6-Objective	**4.146E+25**	3.322E+23	4.141E+25	3.611E+23	3.784E+25	3.845E+23
8-Objective	1.305E+34	1.227E+32	**1.309E+34**	1.400E+32	1.205E+34	1.292E+32

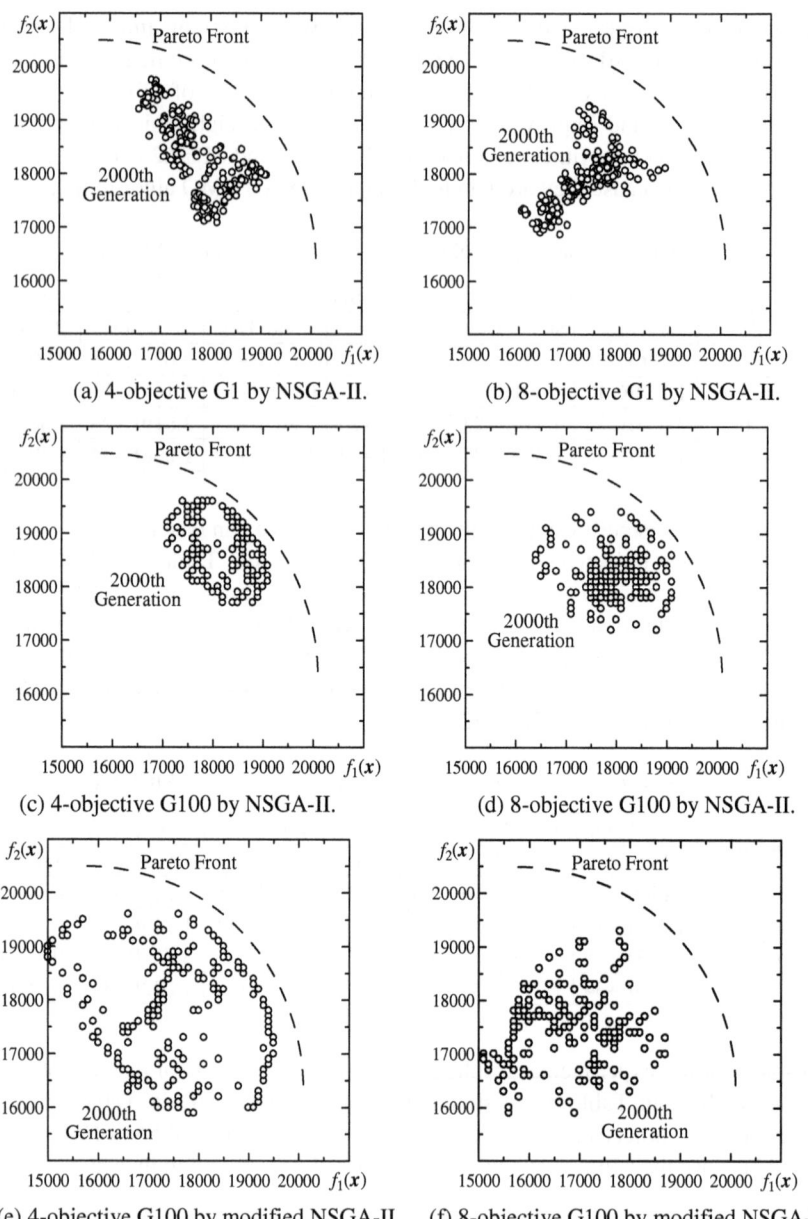

(a) 4-objective G1 by NSGA-II.   (b) 8-objective G1 by NSGA-II.

(c) 4-objective G100 by NSGA-II.   (d) 8-objective G100 by NSGA-II.

(e) 4-objective G100 by modified NSGA-II.   (f) 8-objective G100 by modified NSGA-II.

Fig. 6. Projection of solutions in the final generation of NSGA-II and the modified NSGA-II

Fig. 6 shows the projection of a final population on the two-dimensional subspace with $f_1(x)$ and $f_2(x)$ in a single run of NSGA-II and its modified version on the four-objective and eight-objective problem with G1 and G100. Fig. 6 shows the increase in the diversity of solutions and the deterioration in their convergence by the use of strong Pareto dominance for many-objective problems with G100.

6   Conclusions

In this paper, we first explained why discrete objectives with coarse granularities deteriorated the search ability of EMO algorithms on two-objective problems. Next we proposed the use of strong Pareto dominance, which improved the performance of NSGA-II on discrete two-objective problems with coarse granularities. Then we examined the effect of discrete objectives on the performance of EMO algorithms on many-objective problems. Finally we discussed why the use of coarse granularities improved the performance of NSGA-II and SPEA2 on many-objective problems whereas it deteriorated the performance of MOEA/D and SMS-EMOA. The performance improvement of NSGA-II and SPEA2 on many-objective problems was explained from the difficulty of many-objective problems (i.e., the increase in the number of non-dominated solutions). Since the use of coarse granularities decreases the number of non-dominated solutions, it remedies the difficulty of many-objective problems for Pareto dominance-based EMO algorithms. We also discussed the effect of discrete objectives on the performance of EMO algorithms from the viewpoint of the concept of ε-dominance. Our observations were compared with some theoretical studies [8], [9]. It is an interesting future research topic to examine the performance of ε-dominance EMO algorithms on discrete many-objective problems with different granularities in comparison with Pareto dominance-based EMO algorithms.

References

1. Beume, N., Naujoks, B., Emmerich, M.: SMS-EMOA: Multiobjective Selection based on Dominated Hypervolume. European Journal of Operational Research 181, 1653–1669 (2007)
2. Coello, C.A.C., Lamont, G.B.: Applications of Multi-Objective Evolutionary Algorithms. World Scientific, Singapore (2004)
3. Deb, K.: Multi-Objective Optimization Using Evolutionary Algorithms. John Wiley & Sons, Chichester (2001)
4. Deb, K., Mohan, M., Mishra, S.: Evaluating the ε-Domination Based Multi-Objective Evolutionary Algorithm for a Quick Computation of Pareto-Optimal Solutions. Evolutionary Computation 13, 501–525 (2005)
5. Deb, K., Pratap, A., Agarwal, S., Meyarivan, T.: A Fast and Elitist Multiobjective Genetic Algorithm: NSGA-II. IEEE Trans. on Evolutionary Computation 6, 182–197 (2002)
6. Fonseca, C.M., Fleming, P.J.: On the Performance Assessment and Comparison of Stochastic Multiobjective Optimizers. In: Ebeling, W., Rechenberg, I., Voigt, H.-M., Schwefel, H.-P. (eds.) PPSN IV. LNCS, vol. 1141, pp. 584–593. Springer, Heidelberg (1996)
7. Goldberg, D.E.: Genetic Algorithms in Search, Optimization, and Machine Learning. Addison-Wesley, Reading (1989)
8. Horoba, C., Neumann, F.: Benefits and Drawbacks for the Use of ε-Dominance in Evolutionary Multi-Objective Optimization. In: Proc. of 2008 Genetic and Evolutionary Computation Conference, pp. 641–648 (2008)

9. Horoba, C., Neumann, F.: Additive Approximations of Pareto-Optimal Sets by Evolutionary Multi-Objective Algorithms. In: Proc. of 10th ACM SIGEVO Workshop on Foundations of Genetic Algorithms, pp. 79–86 (2009)
10. Hughes, E.J.: Evolutionary Many-Objective Optimization: Many Once or One Many? In: Proc. of 2005 IEEE Congress on Evolutionary Computation, pp. 222–227 (2005)
11. Ishibuchi, H., Nakashima, T.: Multi-Objective Pattern and Feature Selection by a Genetic Algorithm. In: Proc. of 2000 Genetic and Evolutionary Computation Conference, pp. 1069–1076 (2000)
12. Ishibuchi, H., Nojima, Y.: Analysis of Interpretability-Accuracy Tradeoff of Fuzzy Systems by Multiobjective Fuzzy Genetics-Based Machine Learning. International Journal of Approximate Reasoning 44, 4–31 (2007)
13. Ishibuchi, H., Tsukamoto, N., Hitotsuyanagi, Y., Nojima, Y.: Effectiveness of Scalability Improvement Attempts on the Performance of NSGA-II for Many-Objective Problems. In: Proc. of 2008 Genetic and Evolutionary Computation Conference, pp. 649–656 (2008)
14. Ishibuchi, H., Tsukamoto, N., Nojima, Y.: Evolutionary Many-Objective Optimization: A Short Review. In: Proc. of 2008 IEEE Congress on Evolutionary Computation, pp. 2424–2431 (2008)
15. Ishibuchi, H., Yamane, M., Nojima, Y.: Effects of Discrete Objective Functions with Different Granularities on the Search Behavior of EMO Algorithms. In: Proc. of 2012 Genetic and Evolutionary Computation Conference, pp. 481–488 (2012)
16. Ishibuchi, H., Yoshida, T., Murata, T.: Balance between Genetic Search and Local Search in Memetic Algorithms for Multiobjective Permutation Flowshop Scheduling. IEEE Trans. on Evolutionary Computation 7, 204–223 (2003)
17. Jaszkiewicz, A.: On the Computational Efficiency of Multiple Objective Metaheuristics: The Knapsack Problem Case Study. European Journal of Operational Research 158, 418–433 (2004)
18. Khare, V., Yao, X., Deb, K.: Performance Scaling of Multi-objective Evolutionary Algorithms. In: Fonseca, C.M., Fleming, P.J., Zitzler, E., Deb, K., Thiele, L. (eds.) EMO 2003. LNCS, vol. 2632, pp. 376–390. Springer, Heidelberg (2003)
19. Laumanns, M., Thiele, L., Deb, K., Zitzler, E.: Combining Convergence and Diversity in Evolutionary Multiobjective Optimization. Evolutionary Computation 10, 263–282 (2002)
20. Li, H., Zhang, Q.: Multiobjective Optimization Problems with Complicated Pareto Sets, MOEA/D and NSGA-II. IEEE Trans. on Evolutionary Computation 13, 284–302 (2009)
21. Purshouse, R.C., Fleming, P.J.: On the Evolutionary Optimization of Many Conflicting Objectives. IEEE Trans. on Evolutionary Computation 11, 770–784 (2007)
22. Sato, H., Aguirre, H.E., Tanaka, K.: Controlling Dominance Area of Solutions and Its Impact on the Performance of MOEAs. In: Obayashi, S., Deb, K., Poloni, C., Hiroyasu, T., Murata, T. (eds.) EMO 2007. LNCS, vol. 4403, pp. 5–20. Springer, Heidelberg (2007)
23. Schütze, O., Lara, A., Coello, C.A.C.: On the Influence of the Number of Objectives on the Hardness of a Multiobjective Optimization Problem. IEEE Trans. on Evolutionary Computation 15, 444–455 (2011)
24. Tan, K.C., Khor, E.F., Lee, T.H.: Multiobjective Evolutionary Algorithms and Applications. Springer, Berlin (2005)
25. Wagner, T., Beume, N., Naujoks, B.: Pareto-, Aggregation-, and Indicator-Based Methods in Many-Objective Optimization. In: Obayashi, S., Deb, K., Poloni, C., Hiroyasu, T., Murata, T. (eds.) EMO 2007. LNCS, vol. 4403, pp. 742–756. Springer, Heidelberg (2007)
26. While, L., Bradstreet, L., Barone, L.: A Fast Way of Calculating Exact Hypervolumes. IEEE Trans. on Evolutionary Computation 16, 86–95 (2012)

27. Zhang, Q., Li, H.: MOEA/D: A Multiobjective Evolutionary Algorithm Based on Decomposition. IEEE Trans. on Evolutionary Computation 11, 712–731 (2007)
28. Zitzler, E., Laumanns, M., Thiele, L.: SPEA2: Improving the Strength Pareto Evolutionary Algorithm. TIK-Report 103, Computer Engineering and Networks Laboratory (TIK), Department of Electrical Engineering, ETH, Zurich (2001)
29. Zitzler, E., Thiele, L.: Multiobjective Evolutionary Algorithms: A Comparative Case Study and the Strength Pareto Approach. IEEE Trans. on Evolutionary Computation 3, 257–271 (1999)

Solution of Multi-objective Min-Max and Max-Min Games by Evolution

Gideon Avigad[1], Erella Eisenstadt[1], and Valery Y. Glizer[2]

[1] Mechanical Engineering, Ort Braude College of Engineering
[2] Mathematics Department, Ort Braude College of Engineering
{Gideona,erella,valery48}@braude.ac.il

Abstract. In this paper, a multi-objective optimal interception problem is proposed and solved using a Multi-Objective Evolutionary Algorithm. The traditional setting of an interception engagement between pursuer and evader is targeted either at minimizing a miss distance for a given interception duration or at minimizing an interception time for a given miss distance. Such a setting overlooks an important aspect — the purpose of launching the evader in the first place. Naturally, the evader seeks to evade the pursuer (by keeping away from it), but what about hitting its target? In contrast with the traditional setting, in this paper a multi-objective game is played between a pursuer and an evader. The pursuer aims at keeping a minimum final distance between itself and the evader, which it attempts to keep away from its target. The evader, on the other hand, aims at coming as close as possible to a predefined target while keeping as far away as possible from the pursuer. Both players (pursuer and evader) utilize neural net controllers that evolve during the proposed evolutionary optimization. The game is shown to involve very interesting issues related to the decision-making process while the dilemmas of both opponents are taken into consideration.

Keywords: Differential games, evolutionary algorithms, worst-case evolution.

1   Introduction

Consider a planar engagement between two objects moving at a constant speed: a pursuer and an evader. The geometry of this engagement, shown in Figure 1, defines the variables of the engagement. The following set of nonlinear differential equations describes the dynamics of the engagement for these variables.

$$\dot{R} = V_e \cos(\theta_e - \psi) - V_p \cos(\theta_p - \psi), \tag{1}$$

$$\dot{\psi} = [V_e \sin(\theta_e - \psi) - V_p \sin(\theta_p - \psi)]/R, \tag{2}$$

The engagement (pursuit-evasion) starts at $t = 0$ with the following initial conditions:

$$R(0) = R_0 , \ \psi(0) = \psi_0 . \tag{3}$$

R.C. Purshouse et al. (Eds.): EMO 2013, LNCS 7811, pp. 246–260, 2013.
© Springer-Verlag Berlin Heidelberg 2013

The directions of the objects' velocity vectors are governed by the dynamic controllers:

$$\dot{\theta}_p = \Omega_p u_p, \quad \theta_p(0) = \theta_{p0}, \tag{4}$$

$$\dot{\theta}_e = \Omega_e u_e, \quad \theta_e(0) = \theta_{e0}, \tag{5}$$

$$\text{where} \quad \Omega_p = \frac{V_p}{r_p}, \quad \Omega_e = \frac{V_e}{r_e} \tag{6}$$

are the maximal admissible turning rates of the pursuer and the evader, respectively; r_p and r_e are the smallest turning radii of the objects; and $u_p(t)$ and $u_e(t)$ are non-dimensional controls of the pursuer and the evader, respectively, satisfying the constraints:

$$|u_p(t)| \le 1, \quad t \ge 0, \tag{7}$$

$$|u_e(t)| \le 1, \quad t \ge 0, \tag{8}$$

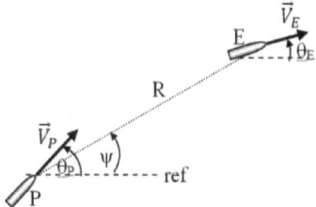

Fig. 1. The pursuer and the evader

The usually considered differential game associated with dynamics (1)-(6) and control constraints (7)-(8) (see e.g. [1] and references therein) is formulated as:

$$\min_{u_p} \max_{u_e}(t_f) \quad \text{subject to a given} \quad R(t_f) = D_f > 0, \tag{9}$$

$$\text{or as} \quad \min_{u_p} \max_{u_e}(R(t_f)) \quad \text{subject to a given} \quad t = t_f > 0. \tag{10}$$

This setting is clearly not a multi-objective setting. The main drawback of this setting is that the evader does nothing but evade the pursuer. Nevertheless, this seems adequate for evaders with low-level control. Yet it only seems reasonable that future evaders will be smart attackers rather than sitting ducks and will be designed to evade their pursuer and hit their target.

In this paper, the pursuit-evasion interception game is reformulated and posed as a multi-objective game. The setting of this new problem is shown in Figure 2.

The new setting bears some resemblance to the classical setting (shown in Figure 1), with one clear difference, and that is the existence of a target. The distance from the evader to the target, L, serves as a base for defining the new problem. The problem solved by the pursuer is:

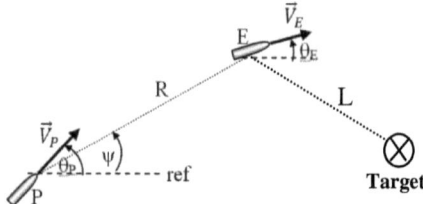

Fig. 2. The new multi-objective game

$$\min_{u_p}\max_{u_e}(R, \frac{1}{L}),$$ (11)

while the evader solves:

$$\max_{u_e}\min_{u_p}(R, \frac{1}{L})$$ (12)

where D_{is} and T are the boundaries of the game.

In the current paper we solve these games and highlight the importance of the results to the control designers both of the pursuer and of the evader.

2   Background

Most existing studies that deal with multi-objective games (MOGs) are associated with differential games and deal with definitions of the equilibrium for these games. The predominant definition of this equilibrium is the Pareto-Nash equilibrium, e.g., [2, 3]. The notion of the Pareto-Nash equilibrium is based upon the concept of cooperative games. According to the Pareto-Nash equilibrium, sub-players in the same coalitions should optimize their vector functions over a set of strategies. On the other hand, this notion also takes into account the concept of non-cooperative games, because coalitions interact on a set of situations and are interested in preserving Nash equilibrium between coalitions. Linear programming is commonly used for solving MOGs, e.g., [4], where a multi-objective zero-sum game is solved. A non-zero-sum version of a MOG was solved in [5]. Detailed algorithms for finding strategies related to the Pareto-Nash equilibrium can be found in [3]. A detailed mathematical step-by-step solution of a MOG using the Kuhn-Tucker equation can be found in [6].

In all of these algorithms, weights are altered in order to search for one equilibrium solution at a time. This means the algorithms must be executed sequentially to reveal more equilibrium points. Evolutionary algorithms have been used to search for optimal strategies in a single objective game setting e.g., [7]. Evolutionary multi objective optimization algorithms have been also utilized for such a purpose e.g., [8] however, not in a MOG setting where the objectives of the opponents are contradicting, as considered here. Artificial intelligence-based approaches have been applied to MOGs within the framework of fuzzy MOGs. In such studies (e.g., [9]), the objectives are aggregated to a surrogate objective in which the weights (describing the players' preferences regarding the objectives) are modeled through fuzzy functions. Avigad et al. [10] used the worst case evolutionary algorithm proposed in [11] to evolve a set of optimal strategies for a discrete MOG. In the worst case analysis, each solution is associated with a set of scenarios. These scenarios are evolved in an embedded MOEA to find the worst scenario, which in a multi-objective problem (MOP) setting may be a set of worst scenarios. These worst scenarios represent the solution in an outer set-based evolutionary algorithm in order to find the optimal solutions.

In contrast to discrete games, in differential games the players' moves are found simultaneously by solving the related model, which is described by differential equations. To the best of our knowledge, the first time a multi-objective differential game was solved using an evolutionary multi-objective algorithm is in our recently introduced study [12], in which a game between a pursuer and an artificial opponent was solved. The pursuer aimed at intercepting an evader having a constant trajectory, while simultaneously minimizing the interception distance and time. The artificial opponent aimed at maximizing these objectives by maximizing the harmful effect of measuring uncertainty. Both opponents utilized neural net controllers, which were tuned by the evolutionary algorithm.

The current study introduces a new multi-objective game between pursuer and evader. The game involves objectives that seem reasonable for real life applications. Moreover, an evolutionary algorithm is proposed for optimizing the controllers of both opponents. The study examines the optimization of the game from the perspectives of the pursuer and of the evader, and highlights unique issues involved in multi-objective games.

3   Methodology

3.1   Problem Formulation

In this study the controllers of the pursuer and the evader are artificial, single layer, neural network controllers having d and e nodes, respectively, such that:
$u_p^{(s)} = [\omega_{1,1}^{p(s)},...,\omega_{i,j}^{p(s)},...,\omega_{d,3}^{p(s)}]^T$, $u_e^{(m)} = [\omega_{1,1}^{e(m)},...,\omega_{i,j}^{e(m)},...,\omega_{e,2}^{e(m)}]^T$. Let U and C be the sets of all possible S and M controllers for the pursuer and the evader, respectively, such that:

$$U_p = \{u_p^{(1)},.....,u_p^{(s)},....,u_p^{(S)}\}, U_e = \{u_e^{(1)},.....,u_e^{(s)},....,u_e^{(M)}\}, \quad \text{where} \quad u_p^{(s)} \in \Omega \subseteq R^S,$$

$u_e^{(m)} \in \Gamma \subseteq R^M$. A game outcome is a result of the opponent's choice of controls $u_p^{(s)}$ and $u_e^{(m)}$. Thus, the game between the s-th pursuer and the m-th evader is reduced to a game $g_{s,m} = (u_p^{(s)}, u_e^{(m)}) \in \Phi = \Omega \times \Gamma$ between two players using the controls $u_p^{(s)}$ and $u_e^{(s)}$. The optimization of Equations 11- 12 may be reformulated as:

For the pursuer: $\min\limits_{u_p^{(s)}} \max\limits_{u_e^{(m)}} F(g_{s,m})$,   (13)

For the evader: $\max\limits_{u_e^{(m)}} \min\limits_{u_p^{(s)}} F(g_{s,m})$ (14)

where $F(g_{s,m}) = (R, \frac{1}{L}) \in \Psi \subseteq R^2$. The value of the function F is in fact the result of solving the game's differential equations that utilize the sets of weights of the two controls.

3.2   Problem Solution

Assessing the R and L associated with the game is based on the players' related trajectories attained during the game. The trajectory of a pursuer is given by $Tr^p(x,t)$, where x is the pursuer's position coordinates (two dimensional) and $0 < t \le t_f$ where t_f is predefined (estimated time of the game, based on availability of propulsion fuel, etc.) In the same manner, $Tr^e(x,t)$ is the trajectory of the evader. The distances R and L for this game are:

$$R = \min\left\|Tr^p(x,t) - Tr^e(x,t)\right\| \text{ for all } 0 < t \le t_f.$$ (15)

$$L = \min\left\|Tr^e(x,t) - X\right\| \text{ for all } 0 < t \le t_f.$$ (16)

where X are the coordinates of the target. When considering the pursuer problem (Equation 13) for the pursuer controller (say the s-th controller), M controls are possible for the evader. Therefore, there may be M games in which M is either finite or infinite. All such possible games for the s-th pursuer form a set of possible games: $G_{u_p^{(s)}} \subseteq \Phi = [g_{s,1} \ldots g_{s,m} \ldots g_{s,M}]^T$. The m-th evader aims at maximizing the distance R and the inverse of the distance L attained in the game with the s-th pursuer, that is:

$$\max\limits_{u_e^m} F(g_{s,m}), \quad m = 1, \ldots M,$$ (17)

In general, these objectives clearly contradict one another. Hence, a set of evader controls exist that may serve as solutions to this problem. These will form a Pareto set of evader controls playing against the s-th pursuer controller:

$$C_{u_p^{(s)}}^* : \{u_e^{(m)} \in U_e \mid \neg \exists u_e^{(m')}, m' \in \{1, \ldots, M\} : F(g_{s,m'}) \succ F(g_{s,m})\}, \quad \forall m = 1, \ldots M$$ (18)

Note that the evaders are aiming at maximization and therefore to designate maximization form minimization, $\succ$ is used in (18) instead of $\prec$. Mapping these optimal evader controllers to the objective space forms a Pareto front $F^{*}_{u^{(s)}_p}$, where

$F^{*}_{u^{(s)}_p} := \{ F(g_{s,m}) \in \Psi \mid u^{(m)}_e \in C^{*}_{u^{(s)}_p} \}$ is associated with the s-th pursuer's controller. Now we can search for the optimal pursuer control by solving:

$$\min_{u^s_p \in U_p} \quad F^{*}_{u^{(s)}_p} \quad , s = 1,..., \ S \tag{19}$$

This means that the best pursuer control is searched for while considering the best (most problematic from the pursuer's perspective) control used by the evader. This optimization results in a set of optimal pursuer controls:

$$U_p^{*} := \{ u^{(s)}_p \in U_p \mid \forall u^{(m)}_e \in U_e \ \neg \exists u^{(s')}_p \in U_p : F(g_{s',m}) \prec F(g_{s,m})\} \tag{20}$$
$$\text{for all } m \text{ and for all } s \neq s'$$

Here we take the high reliability for interception one step further by representing each cluster by its related ideal point (the utopia point related to the extremes of the performances in the objective space). That is, we represent each pursuer control by a point in the objective space such that:

$$\bar{F}(u^{(s)}_p) = [\max R, \max(\tfrac{1}{L})]^T \text{ over all } u^{(s)}_p \in U_p^{*}. \tag{21}$$

Refer to Section 2.3 for a discussion of the pros and cons of this decision.

Now that the worst performances of each $u^{(s)}_p$ are represented by $\bar{F}(u^{(s)}_p)$, the solution to Equation (14) may be reformulated as:

$$U_p^{*} := \{ u^{(s)}_p \in U_p \mid \forall u^{(m)}_e \in U_e \ \neg \exists u^{(s')}_p \in U_p : \bar{F}(u^{(s')}_p) \prec \bar{F}(u^{(s)}_p)\} \tag{22}$$
$$\text{for all } m \text{ and for all } s \neq s'$$

In the same manner, when considering the problem for the evader (Equation 18), the following may be attained. For each of the evader controls (say the m-th controller), S controls are possible for the pursuer. Therefore, S games are possible in which S is either finite or infinite. All such possible games for the m-th evader form a set: $G_{u^{(m)}_e} \subseteq \Phi = [g_{m,1}...g_{m,s},...g_{m,S}]^T$. The s-th pursuer aims at minimizing the distance R and the inverse of the distance L attained in a game with the m-th evader, that is:

$$\min_{u^{(s)}_p} F(g_{s,m}), \quad s = 1, ... \ S. \tag{23}$$

Solving (23) will form a Pareto set of pursuer controls playing against the m-th evader controller:

$$C^*_{u_e^{(m)}} := \{ u_p^{(s)} \in U_p \mid \neg \exists u_p^{(s')} : F(g_{s',m}) \prec F(g_{s,m}) \}, \quad \forall s, s' = 1, \dots, S. \tag{24}$$

Mapping these optimal pursuer controllers to the objective space forms a Pareto front $F^*_{u_e^{(m)}}$, where $F^*_{u_e^{(m)}} : \{ F(g_{s,m}) \in \Psi \mid u_p^{(s)} \in C^*_{u_e^{(m)}} \}$ is associated with the m-th evader controller. Now we can search for the optimal evader control by solving:

$$\max_{u_e^{(m)} \in U_e} F^*_{u_e^m}, m = 1, \dots, M. \tag{25}$$

This means that the best evader control is searched for, while considering the best (most problematic from the evader's perspective) control used by the pursuer. This optimization results in a set of optimal evader controls:

$$U_e^* := \{ u_e^{(m)} \in U_e \mid \forall u_p^{(s)} \in U_p \; \neg \exists u_e^{(m')} \in U : F(g_{s',m}) \succ F(g_{s,m}) \} \tag{26}$$

$$\text{for all } s \text{ and for all } m \neq m'$$

Furthermore

$$\bar{F}(u_e^{(m)}) = [\min R, \min(1/L)]^T \text{ over all } u_e^m \in U_e^*. \tag{27}$$

Now that the worst performance of each $u_e^{(m)}$ is represented by $\bar{F}(u_e^{(m)})$, the solution to Equation (14) may be reformulated as:

$$U_e^* := \{ u_e^{(m)} \in U_e \mid \forall u_p^{(s)} \in U_p \; \neg \exists u_e^{(m')} \in U_e : \bar{F}(u_e^{(s')}) \succ \bar{F}(u_e^{(s)}) \} \tag{28}$$

$$\text{for all } s \text{ and for all } m \neq m'$$

It is assumed that the pursuer's controller and the evader's controller will be chosen from the optimal sets (Equations 22 and 26, respectively). Therefore, the optimal games are an outcome of games played between optimal pursuer controllers and optimal evader controllers. The performances on these games establish a set of optimal game performances within the objective space. This set, termed here as the Optimal Games Set (OGS), is defined as:

$$OGS := \{ F(g_{s,m}) \in \Psi \mid \exists F(g_{s',m'}) \in \bar{F}(u_p^{(m)}) \wedge \exists F(g_{s'',m''}) \in \bar{F}(u_e^{(s)}) : \tag{29}$$

$$F(g,m) \preceq F(g_{s',m'}) \wedge F(g,m) \succeq F(g_{s'',m''}) \}.$$

From a non-mathematical perspective, these optimal games are those whose performances are dominated (with respect to maximization) by optimal pursuer performances and also are dominated (with respect to minimization) by optimal evader performances.

To elucidate these hereby introduced notions, consider Figure 3, which depicts the performances associated with games among four pursuers and three evaders. For the sake of this example we assume that all twelve games are possible, that is S=4 and M=3. The pursuers' controls are depicted by striped, bold, gray and white circles, whereas the evaders' controls are represented by stars, squares and triangles. The performances of the different games within an artificial bi-objective space are represented by combinations of the above representations (e.g., black circle inside a square).

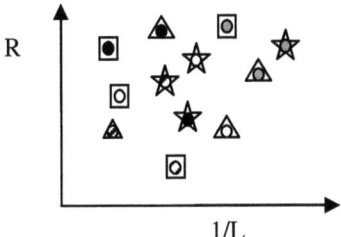

Fig. 3. The performances of the twelve games in the bi-objective space

For the pursuer, the problem is posed as a min-max problem, defined in (13). Starting with (17), Figure 4a shows the performances of the Pareto sets for the four pursuers. For example, maximizing the performances of the evaders with respect to the gray pursuer results in two optimal performances (star and square), while the third vanishes (triangle) because it is dominated when maximization is considered. The figure also depicts the related ideal points (Equation 21) designated by the pursuers' symbols. Now, when minimization is implemented (Equation 23), the resulting optimal pursuer is represented by the circle with stripes (indicated by an arrow) as it dominates all others in the min sense. The left panel of Figure 5 depicts the area dominated by the performances of that game. The implication of this area is that if the pursuer were to choose the controller represented by the circle with stripes, the evader could not choose any controller that could remove the performances from that area (by choosing either a triangle, a square or a star). This can easily be verified by observing that all games associated with the circle with stripes are included in that area.

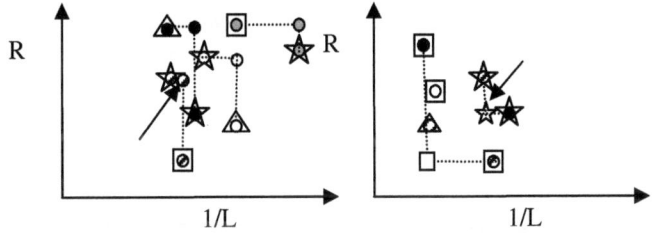

Fig. 4. Left panel - min-max fronts, related ideal points and front; Right panel - max-min fronts, related ideal points and front

When the optimization is considered from the evader's perspective (Equation 14) by using Equations 24-28, the optimal controller is attained for the evader, represented by the star. Figure 4b highlights the domination of the star over the other two controllers (when maximization is considered). By choosing the star controller, the evader ensures that the pursuer cannot attain performances beyond the gray boundary, as shown in the right panel of Figure 5.

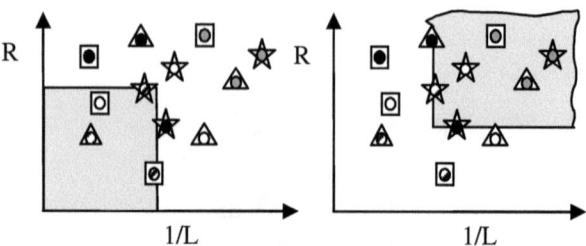

Fig. 5. Optimal pursuer and evader fronts and dominated regions, shown in the left and right panels, respectively

In the example shown in Figure 5, choosing the optimal controllers (pursuer would choose the circle with stripes and evader would choose the star) would result in a single game. This means that solving (17) would yield the same game as solving (18). In other words, solving the min-max problem would yield the same result obtained by solving the max-min problem.

This situation, however, is not the usual case. It is easy to construct an example in which the result is not a single game but rather a set of games. As an example of this case, consider the performances of the 16 games depicted in Figure 6. The games are those shown in Figure 3 with the addition of four games. These additional games are the result of adding the controllers of another evader, designated by diamonds.

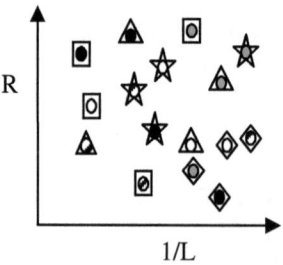

Fig. 6. Performances for sixteen games

Following the same procedure results in three optimal pursuer controllers (black, white and striped), shown in the left panel of Figure 7 with their dominated region (gray area). Clearly, deciding which pursuer controls to use involves multi-objective decision making. For example, the black circle controller would be chosen if the interception were aimed at interception at long distances from the target at the expense of greater interception distance.

Solving the problem for the evader side, results in two optimal evader controllers (star and diamond). The right panel of Figure 7 shows the domination of these controllers and the related domination area.

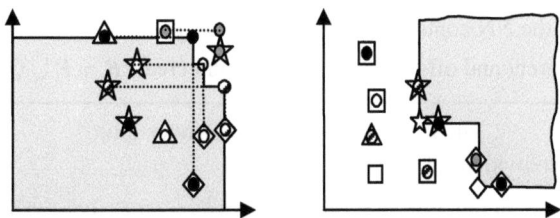

Fig. 7. The min-max fronts, related ideal points and front (left panel). The max-min fronts, related ideal points and front (right panel).

Figure 8 depicts the two regions within the same figure. The overlapping area designates the area in which the games will take place if the opponents choose to play only with optimal controllers. The games resulting from the opponents using the optimal controllers are highlighted with a dotted background.

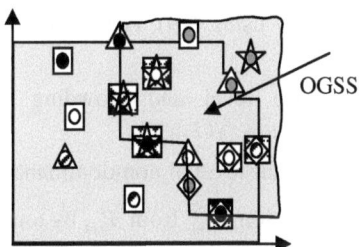

Fig. 8. The two fronts and related OGS

This figure highlights the obtained results, namely that the pursuer should not choose the gray circle controller and the evader should not choose the square. This is because the opponent may drive their performances to unwanted performances. For example, if the pursuer chooses the gray circle it puts itself in danger, which would be realized if, for example, the evader chooses the star. Nevertheless, the white circle is outperformed by the black and striped circles (from the pursuer side). This means that solving the min-max problem will not yield the same results as would solving the max-min problem.

3.3   The Evolutionary Search

In the current paper the optimal controls are searched for by using evolutionary algorithms. The following describes the algorithm for optimization of the pursuer's control as well of the evader's control (designated in italics). The algorithm includes an inner-loop algorithm (designated in gray) in which the representative ideal point for each pursuer/evader is found based on the evolved front. The ideal points are then used within the outer loop to evolve the related optimal controls.

a. Initialize a population P_t of size N= |P_t| of pursuer/*evader* controls u_p/ u_e
    (weights of the NN controllers). Also, set Qt = P_t .

b. Combine parent and offspring populations and create $R_t = P_t \cup Q_t$

The Embedded Algorithm (inner-loop)

c. For each individual of R_t:

c.1 Initialize a population G_t of size $ng = |G_t|$ for the evaders/*pursuers'*
    controllers u_e/ u_p (weights of NN controllers).

c.2 Run NSGA-II on the reversed optimization problem to find, for each u_p/ u_e of R_t, the corresponding $C^*_{u_p^{(s)}}$ / $C^*_{u_e^{(s)}}$ using $F(g_{s,m})$.

c.3 For each u_p/u_e of R_t use $F^*_{u_p^{(s)}}$ / $F^*_{u_e^{(s)}}$ to assign $F(u_p^{(s)})$/ $F(u_e^{(s)})$.

---

d. Initialize a new parent population $P_{t+1} = \varnothing$

e. Compute $F(u_p^{(s)})$/ $F(u_e^{(s)})$ by using (19), (20) on the results of simulations that solve (1)-(8).

f. Assign a non-dominance level and crowding value (NSGA-II) to all individuals of R_t by using $F(u_p^{(s)})$/ $F(u_e^{(s)})$.

g. Fill in P_{t+1} according to their level of non-dominance and crowding measure.

h. Create an offspring population Q_{t+1}^* from P_{t+1} by tournament selection.

i. Perform crossover on Q_{t+1}^* to obtain Q_{t+1}^{***} .

j. Perform mutation to obtain Q_{t+1} .

k. If the last generation has not been reached, go-to 'b'.

3.4   Remarks on Using the Ideal Point as Representative of an Opponent

In the current paper we chose to represent the Pareto set of each pursuer/evader by the related ideal point (see Section 2.2). As mentioned, the initial idea behind this decision was that, to be on the safe side, the worst of both objectives is considered. Such a representation has pros and cons, as discussed in the following. Apart from being on the safe side by considering the worst, using the ideal point reduces computational complexity with respect to the complexity of using entire Pareto sets. If instead of the ideal point, the Pareto set of each pursuer/evader had been used, the complexity would be much greater (see e.g., [11]). Furthermore, considering one point as representative of the set makes both the explanations (with relation to

Figures 3-9) and the decision making much easier. Moreover, utilization of the ideal point confirms with most of the demands of an evolutionary multi-objective algorithm, as elucidated through the panels of Figure 9. The figure shows that based on the ideal points, the pursuer's controller represented by the black circle would be preferred (minimization) over the one represented by the striped circle in a case where the former Pareto front dominates the latter, when it is less spread out and when it is less crowded. Nevertheless, considering the ideal point may lead to not preferring the black controller in all cases in which it should have been preferred, as shown in Figure 10.

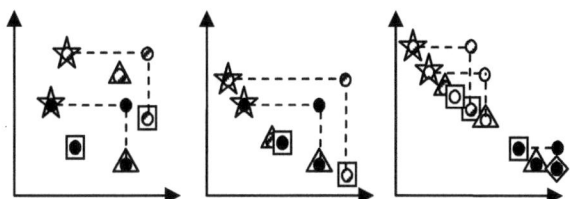

Fig. 9. Without loss of generality for any minimization problem, using the ideal point allows applying pressure towards dominating (left panel), small spread (middle panel) and diverse (right panel) sets

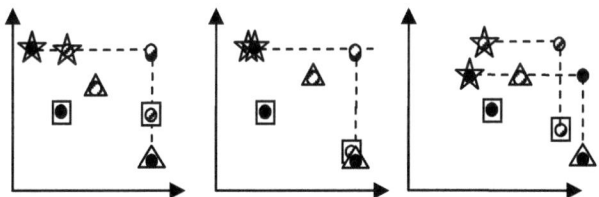

Fig. 10. Three examples in which, without loss of generality, the black circle should be preferred for any minimization problem, though based on the ideal point it is not

4   Test Case

In this test case, the problem is solved using the algorithm in Section 2.2. The velocities of the pursuers and the evaders are set to 2 and 1, respectively. The controllers are neural net controllers, each having two nodes with two *"tansig"* activation functions. The inputs to the controllers of the pursuers and evaders are R, ψ, θ_p, L and R, ψ, θ_e, L respectively. The objective space is set as $\Psi = (R, -L)$. Using $-L$ instead of $1/L$ diffracts the original objective space less. The initial population includes 50 decoded evaders (embedded loop) and pursuers (outer loop) that are run 100 and 200 hundred generations, respectively. The controllers' weights are given by real-valued decision variables. A simulated binary crossover (SBX) operator and polynomial mutation with distribution indexes of $\eta_c = 15$ and $\eta_m = 20$ respectively are used. A crossover probability of $pc = 50\%$ and a mutation probability of $pc = 5\%$ are used.

The left panel of Figure 11 uses gray circles to depict the initial population of evaders associated with one of the pursuer controllers. Clearly the related performances associated with the initial population are not very good and are scattered throughout the objective space. The middle panel of Figure 11 depicts the trajectories of one of these evaders and the pursuer it plays against. As expected, the pursuer does not attempt to intercept the evader, nor does the evader try to evade the pursuer. This is due to the poor weights associated with the initial generation. The left panel of Figure 11 also uses bold circles to depict the pursuer's evolved front (evolved by the algorithm's inner loop), as well as the related ideal point (depicted by a star). A set of the fronts of the first generation pursuers is depicted in the figure's right panel.

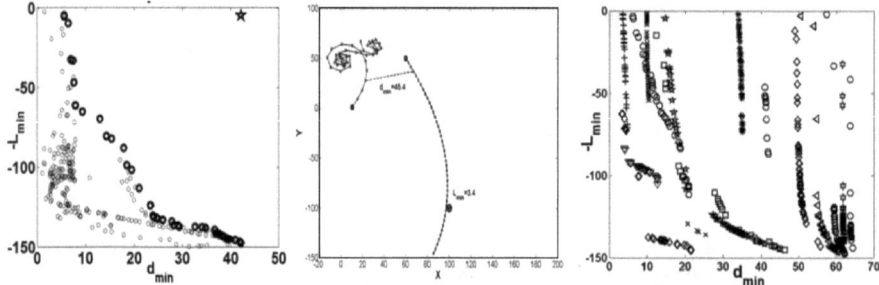

Fig. 11. First generation of one pursuer, related front and ideal point (left panel); example of a bad interception (middle panel) and some of the first generation pursuer fronts (right panel)

Figure 12 depicts the evolved fronts of the outer loops for the pursuer (stars) and for the evader (diamonds). It also depicts the region where optimal games are played. These results are in accordance with the simple visual examples given in Section 2.2.

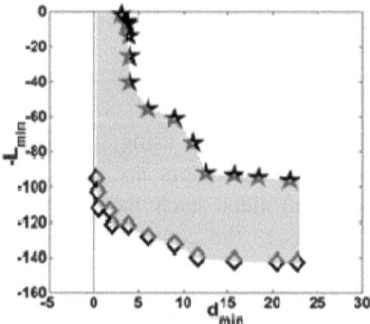

Fig. 12. Pursuer- and evader-related fronts (stars and diamonds, respectively). The gray region designates the region of the OGS.

The panels of Figure 13 depict two optimal games.

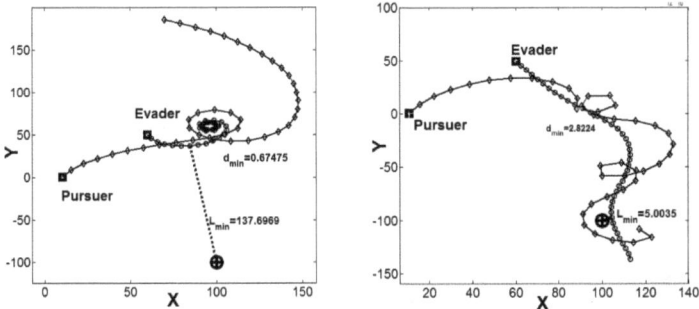

Fig. 13. Two optimal games showing different interception distances. Smaller R with larger L (left panel), larger R with smaller L (right panel).

Please note that the situation in this case, in which the performances of the optimal games do not dominate each other, is not compulsory. However, the non-domination relation holds among the related ideal points. It should also be noted that attaining the two fronts depicted in Figure 12 involved running the algorithm on a Power Edge R710 for almost 150 hours!

5   Conclusions and Future Work

In this study we have proposed a rational multi-objective game between pursuer and evader by adding a trivial but reasonable objective, namely that the evader aims at minimizing its distance to a target. This paper has introduced several novelties, including:

- The min-max versus max-min in multi-objective games is considered. While this issue has been treated within the context of games, treating it in the context of multi-objective games is new.
- A region of optimal games is defined. This is an outcome of considering min-max versus max-min in multi-objective games. It was shown that there is a region of optimal games bounded by the fronts of these two problems.
- A multi-objective differential game is optimized directly, with no modifications. This means that linearized fuzzy representation of the opponent's actions has not been used, nor have other modifications been made in the differential equations.
- An evolutionary algorithm is implemented for solving both problems (min-max and max-min).

The drawbacks of the current paper include the high computational cost and the use of the ideal points rather than the Pareto fronts of the pursuers/evaders. These two issues should be considered before further investigating this type of games. It seems that the new insights presented here should open the way for exploring a wide scope of applications, for example in economics. Although most real-world applications would be related to discrete games rather than to differential games, we envisage that the inherent nature of the results would not change. This means that it is expected that solving discrete multi objective games, would also involve e.g., an OGS.

Acknowledgements. This research was supported by a Marie Curie International Research Staff Exchange Scheme Fellowship within the 7th European Community Framework Programme.

References

1. Glizer, V.Y., Shinar, J.: On the structure of a class of time-optimal trajectories. Optimal Control Applications and Methods 14(4), 271–279 (1993)
2. Lozovanu, D., Solomon, D., Zelikovsky, A.: Multiobjective games and determining pareto-nash equilibria. Buletinul Academiei de Ştiinţe a Republicii Moldova. Matematica 3(49), 115–122 (2005)
3. Somasundaram, K., Baras, J.: Pareto nash replies for multi-objective games. The Institute for Systems Research ISR, TR 2009-4 (2008)
4. Zeleny, M.: Games with multiple payoffs. International Journal of game theory 4(4), 179–191 (1975)
5. Contini, B., Olivetti, I., Milano, C.: A decision model under uncertainty with multiple payoffs. In: Theory of Games; Techniques and Applications, pp. 50–63 (1966)
6. Li, S.: Interactive strategy sets in multiple payoff games. Computers & Industrial Engineering 37(3), 613–630 (1999)
7. Choi, H., Ryu, H., Tahk, M., Bang, H.: A co-evolutionary method for pursuit evasion games with non-zero lethal radii. Engineering Optimization 36(1), 19–36 (2004)
8. Brown, M., An, B., Kiekintveld, C., Ordóñez, F., Tambe, M.: Multi-objective optimization for security games. In: Proceedings of the 11th International Conference on Autonomous Agents and Multiagent Systems, AAMAS 2012, vol. 2, pp. 863–870. International Foundation for Autonomous Agents and Multi-agent Systems, Richland (2012)
9. Sakawa, M.: Genetic Algorithms and Fuzzy Multiobjective Optimization. Kluwer, Boston (2002)
10. Avigad, G., Eisenstadt, E., Cohen, M.: Optimal strategies for multi objective games and their search by evolutionary multi objective optimization. In: 2011 IEEE Conference on Computational Intelligence and Games, CIG, pp. 166–173. IEEE (2011)
11. Avigad, G., Branke, J.: Embedded evolutionary multi-objective optimization for worst case robustness. In: Proceedings of the 10th Annual Conference on Genetic and Evolutionary Computation, pp. 617–624. ACM (2008)
12. Avigad, G., Eisenstadt, E., Glizer, V.Y.: Evolving a Pareto Front for an Optimal Bi-objective Robust Interception Problem with Imperfect Information. In: Schütze, O., Coello Coello, C.A., Tantar, A.-A., Tantar, E., Bouvry, P., Del Moral, P., Legrand, P. (eds.) EVOLVE - A Bridge Between Probability. AISC, vol. 175, pp. 121–136. Springer, Heidelberg (2012)

A Comparative Study on Evolutionary Algorithms for Many-Objective Optimization

Miqing Li[1], Shengxiang Yang[2], Xiaohui Liu[1], and Ruimin Shen[3]

[1] Department of Information Systems and Computing, Brunel University,
Uxbridge, Middlesex, UB8 3PH, United Kingdom
{miqing.li,xiaohui.liu}@brunel.ac.uk
[2] School of Computer Science and Informatics, De Montfort University
Leicester LE1 9BH, United Kingdom
syang@dmu.ac.uk
[3] Institute of Information Engineering, Xiangtan University, Xiangtan, 411105, China
srmxp@126.com

Abstract. Many-objective optimization has been gaining increasing attention in the evolutionary multiobjective optimization community, and various approaches have been developed to solve many-objective problems in recent years. However, the existing empirically comparative studies are often restricted to only a few approaches on a handful of test problems. This paper provides a systematic comparison of eight representative approaches from the six angles to solve many-objective problems. The compared approaches are tested on four groups of well-defined continuous and combinatorial test functions, by three performance metrics as well as a visual observation in the decision space. We conclude that none of the approaches has a clear advantage over the others, although some of them are competitive on most of the problems. In addition, different search abilities of these approaches on the problems with different characteristics suggest a careful choice of approaches for solving a many-objective problem in hand.

1 Introduction

Recently, many-objective optimization, which refers to the simultaneous optimization of four or more objectives, has been gaining increasing attention in the evolutionary multiobjective optimization (EMO) community. An important reason is due to distinctly different behaviors of EMO algorithms in many-objective optimization against in optimization regarding two or three objectives. Many current EMO algorithms, which work well on 2- and 3-objective problems, encounter difficulties in high-dimensional objective spaces. This brings a big challenge for researchers and practitioners in the area.

Since the pioneering studies appeared at the beginning of the century [9], a number of EMO algorithms have been proposed in the literature for addressing many-objective optimization problems [4], [15], [17]. Some of them concentrate on the investigation or improvement of the existing approaches, such as Pareto-based, aggregation-based, and indicator-based approaches [24], [23], [13],

R.C. Purshouse et al. (Eds.): EMO 2013, LNCS 7811, pp. 261–275, 2013.
© Springer-Verlag Berlin Heidelberg 2013

while the others are dedicated to developing new techniques specially for many-objective optimization problems [19], [22], [25].

It is well known that Pareto-based approaches, such as the nondominated sorting genetic algorithm II (NSGA-II) [5], fail to provide sufficient selection pressure in the evolutionary process of many-objective optimization, despite belonging to the most popular class of approaches in the EMO community. However, two other classes of approaches, aggregation-based and indicator-based approaches, have been found to be very promising in many-objective optimization. The former uses the idea of single-objective aggregated optimization and combines the objectives into a scalar function that is used for fitness calculation. In fitness calculation, each solution can be evaluated by a unique weight vector (e.g., in the multiobjective evolutionary algorithm based on decomposition (MOEA/D) [26]) or by a set of weight vectors (e.g., in the multiple single objective Pareto sampling (MSOPS) [12]). The latter adopts a single performance indicator to optimize a desired property of the evolutionary population [28]. The indicator *hypervolume* is an effective tool to balance convergence and diversity in both multi- or many-objective optimization [3], [24]. Despite a large computation requirement regarding the exact calculation of the hypervolume indicator in a high-dimensional space, algorithms based on approximate estimation of the indicator using Monte Carlo sampling have recently been developed, such as the hypervolume estimation algorithm (HypE) [2].

On the other hand, a lot of effort of improving Pareto-based approaches for many-objective problems is being made. According to the angle of dealing with the issue, these efforts can be divided into two classes. The first class is to modify or enhance the Pareto dominance relation to increase the selection pressure towards the Pareto front. Most approaches belong to this class. Among them, ϵ-dominance is a representative example [7], although it is not developed particularly for many-objective optimization. The other class is concerned with improving the diversity maintenance mechanism of Pareto-based EMO algorithms. As a result of the increase of the objective space in size, the conflict between the convergence and diversity requirements is gradually aggravated [23]. Apparently, a viable way for solving this problem is to decrease the diversity requirement in the evolutionary process. An interesting attempt along this direction has been made, called the diversity management operator (DMO) [1].

In addition, approaches using some non-Pareto dominance techniques to rank individuals (e.g., *average ranking* [4]) have also been shown to provide competitive results. In spite of the risk of leading a solution set to converge into a sub-area of the Pareto front [17], [21], they provide a new alternative for evolutionary many-objective optimization.

Overall, the above approaches provide a variety of alternatives to handle many-objective optimization problems. However, a series of questions arise: which approaches are suited to which sort of problems, what are the strengths and drawbacks of each approach, and is there an approach clearly outperforming other approaches in most problems? To date, few comparative studies consider all the above questions. Early studies on many-objective optimization mainly

Table 1. List of some existing comparison studies on many-objective optimization

	Algorithm Class	Test Problem
Khare et al. (2003) [18]	C1	DTLZ
Hughes (2005) [13]	C1, C2	Custom
Purshouse and Fleming (2007) [23]	C1	DTLZ
Corne and Knowles (2007) [4]	C4, C6	TSP
Wagner et al. (2007) [24]	C1, C2, C3, C4	DTLZ
Ishibuchi et al. (2008) [14]	C1, C2, C4, C5, C6	Knapsack
Jaimes and Coello (2009) [17]	C1, C4, C6	DTLZ
Hadka and Reed (2012) [10]	C1, C2, C3, C4	DTLZ, WFG, UF

C1: Pareto-based approaches; C2: Aggregation-based approaches;

C3: Indicator-based approaches; C4: Improved Pareto dominance-based approaches;

C5: Improved diversity maintenance-based approaches; C6: Non-Pareto-based approaches.

focus on the investigation of ineffectiveness of the Pareto dominance relation. Recent studies fail to involve all the six classes of the approaches mentioned previously (i.e., Pareto-based, aggregation-based, indicator-based, improved Pareto dominance-based, improved diversity maintenance-based, and Non-Pareto-based approaches), and also are often restricted on only a few test problems. Table 1 lists some existing comparative studies on many-objective optimization, including the class of tested algorithms and the considered problems.

In this paper, we provide a systematic comparison of eight representative EMO algorithms selected from the six classes of approaches of dealing with many-objective problems. The basis of this empirical study is formed by several groups of well-defined continuous and combinatorial test functions that allow an extensive investigation of algorithms on problems of various kinds, such as having convex, concave, multimodal, disconnected, degenerate, biased, and non-separable (variable linkage) Pareto fronts. Additionally, the Pareto-Box problem [19] is also included in order to investigate the algorithms' ability of maintaining the diversity of solutions in the decision space.

2   Eight Algorithms Investigated

In this section, we briefly describe the eight EMO algorithms selected from the six classes of approaches for many-objective problems. Readers seeking more details on these algorithms should refer to their original works.

- **NSGA-II** [5]. As a representative algorithm of Pareto-based approaches, NSGA-II is considered in our comparative study. The main characteristic of NSGA-II is its nondominated sorting and crowding distance-based density estimation in fitness assignment and environmental selection.
- **MOEA/D** [26]. MOEA/D is one of the most popular EMO algorithms developed recently. Using a predefined set of weight vectors to maintain a diverse set of solutions, MOEA/D converts a multiobjective problem into many single-objective problems and tackles them simultaneously. Two aggregation functions, Tchebycheff and penalty-based boundary intersection (PBI), can be used in the algorithm, and each of them works well on different classes of problems. Here, the PBI function is selected since the algorithm

with PBI has been found to be more competitive when solving problems with a high-dimensional objective space [6].

- **MSOPS** [12]. MSOPS, using the idea of single-objective aggregated optimization to search in parallel, also belongs to the class of aggregation-based approaches. Unlike MOEA/D where an individual corresponds to only one weight vector, MSOPS specifies an individual with a number of weight vectors. MSOPS is a popular algorithm to deal with many-objective optimization problems since it can achieve a good balance between convergence and diversity [13], [24].

- **HypE** [2]. HypE is an indicator-based algorithm for many-objective optimization. HypE adopts Monte Carlo simulation to approximate the exact hypervolume value, significantly reducing the time cost of the HV calculation and enabling hypervolume-based search to be easily applied on many-objective optimization, even when the number of objectives reaches 50 [2].

- **ϵ-dominance Multiobjective Evolutionary Algorithm (ϵ-MOEA)** [7]. ϵ-MOEA is a steady-state algorithm using ϵ-dominance to strengthen selection pressure towards the Pareto front. Dividing the objective space into many hyperboxes, ϵ-MOEA assigns each hyperbox at most a single solution based on ϵ-dominance and the distance from solutions to the utopia point in the hyperbox. Although not specifically designed for many-objective optimization, ϵ-MOEA has been found to perform well on many-objective problems [24], [10].

- **DMO** [1]. DMO is an attempt to manage the use of diversity preservation operators in dealing with many-objective problems. Based on the basic framework of NSGA-II, DMO modifies the diversity maintenance operation by adaptively tuning it according to the requirement of the evolutionary population. If the diversity result is smaller than 1 by the *Maximum Spread* test [27], the diversity promotion mechanism (i.e., crowding distance) is activated; otherwise, deactivated.

- **Average Ranking (AR)** [4]. AR is regarded as a good non-Pareto-based approach to solve many-objective problems. In contrast to Pareto-based approaches, AR compares all solutions in each objective and independently ranks them. The final rank of a solution is obtained by summing its ranks of all objectives. AR is found to be successful in evolving towards the optimal direction in many-objective optimization [4], [17], despite the risk of leading a solution set to converge into a sub-area of the Pareto front due to the lack of a diversity maintenance scheme [17], [21].

- **Average Ranking combined with Grid (AR+Grid)** [21]. AR+Grid is a hybrid method which uses grid to enhance diversity for AR in many-objective optimization. In AR+Grid, the AR strategy is employed to provide the selection pressure towards the Pareto front, and the grid is introduced to prevent solutions from being crowded in the objective space. Since an adaptive punishment of solutions located in neighboring cells is implemented, a balance between convergence and diversity can be obtained even when a large number of objectives are involved in a problem [21].

Table 2. Properties of test problems and parameter setting in ϵ-MOEA and AR+Grid. The settings of ϵ and div correspond to the different numbers of objectives of a problem. M and L denote the number of objectives and decision variables, respectively

Problem	M	L	Properties	ϵ	div
DTLZ2	5, 10	$M+9$	Concave	0.158, 0.275	15, 13
DTLZ3	5, 10	$M+9$	Concave, Multimodal	0.158, 0.8	10, 10
DTLZ5	5, 10	$M+9$	Concave, Degenerate	0.071, 0.14	25, 27
DTLZ7	5, 10	$M+19$	Mixed, Disconnected	0.1595, 0.73	10, 5
WFG1	5, 10	$2\times(M-1)+20$	Mixed, Biased	0.092, 0.105	15, 8
WFG8	5, 10	$2\times(M-1)+20$	Concave, Nonseparable	0.8, 2.1492	13, 10
WFG9	5, 10	$2\times(M-1)+20$	Concave, Multimodal, Nonseparable	0.72, 1.761	25, 14
TSP(-0.2)	5, 10	30	Convex, Negative correlation	1.5, 4.3	19, 14
TSP(0)	5, 10	30	Convex, Zero correlation	1.1, 3.15	16, 14
TSP(0.2)	5, 10	30	Convex, Positive correlation	0.75, 2.26	16, 14
Pareto-Box	10	2	Convex	6.085	40

3   Experimental Design

3.1   Test Problems

Eleven test functions selected from the four problem suites DTLZ [8], WFG [11], multiobjective TSP [4], and Pareto-Box [19] (listed in Table 2) are used in this study. DTLZ and WFG are two very popular continuous problem suites and can be scaled to any number of objectives and decision variables. Due to space limitations, we do not consider all test instances in them; instead, seven representative problems are selected to challenge different abilities of algorithms.

The multiobjective TSP problem is selected since combinatorial problems show different behaviors from continuous ones. In multiobjective TSP, a parameter is used to control the correlation between objectives of the problem, called TSP correlation parameter ($TSPcp \in (-1,1)$) [4]. When $TSPcp < 0$, $TSPcp = 0$, or $TSPcp > 0$, it introduces negative, zero, or positive interobjective correlations, respectively. In our study, $TSPcp$ is assigned to -0.2, 0, and 0.2 to represent different characteristics of the problem.

The Pareto-Box problem, proposed by Köppen and Yoshida [19] and extended by Ishibuchi et al. [16], is a simple and interesting many-objective function. The most important characteristic is that its Pareto optimal set in the decision space is a (or several) two-dimensional closure(s). Moreover, since the crowding in its decision space is closely related to the crowding in its objective space, a visual investigation of the distribution of solutions in the former will facilitate the understanding of behavior of algorithms in the latter.

3.2   Performance Metrics

Three widely-used performance metrics, convergence measure (CM) [18], inverted generational distance (IGD) [26], and hypervolume (HV) [27], are introduced to compare the tested algorithms. CM assesses the convergence of a solution set by calculating the average normalized Euclidean distance from the obtained set to the Pareto front. Both IGD and HV can give a comprehensive assessment in terms of convergence and diversity [20]. The former two require a reference set of representing the Pareto front, and are used to evaluate algorithms on DTLZ since their optimal fronts are known. The latter one is used on WFG and TSP whose Pareto front is unknown or unavailable.

In the calculation of HV, two crucial issues are the scaling of the search space and choice of the reference point. Since the objectives in the WFG problem take different ranges of values, we normalize the objective value of the obtained solutions according to the range of the Pareto front of the problem. Following the recommendation in [16], the reference point is set to be at 1.1 times the upper bound of the Pareto front (i.e., $r = 1.1^M$ for WFG) to emphasize the balance between convergence and diversity of the obtained solution set. As to the TSP problem, since the range of Pareto front is unknown, we regard the point with 22 for each objective (i.e., $r = 22^M$) as the reference point, considering that it is slightly larger than the worst value of the mixed nondominated solution set constructed by all the obtained solution sets. In addition, since the exact calculation of the hypervolume metric is infeasible for a solution set with 10 objectives, we approximately estimate the hypervolume result of a solution set by the Monte Carlo sampling method used in [2]. Here, 10,000,000 sampling points are used to ensure accuracy [2].

3.3 General Experimental Setting

All the results presented in this paper are obtained by executing 30 independent runs of each algorithm on each problem with the termination criterion of 100,000 evaluations. Following the practice in [16], the population size is set to 200 for general EMO algorithms, and the archive is also maintained with the same size if required. Note that the population size, in MOEA/D, is the same as the number of weight vectors. Here, we use the closest number to 200 among the possible values as the population size (i.e., 210 and 220 for 5- and 10-objective problems, respectively). In ϵ-MOEA, the size of the archive set is determined by the ϵ value. In order to guarantee a fair comparison, we set ϵ so that the archive of ϵ-MOEA is approximately of the same size as that of the other algorithms (shown in Table 2).

Parameters need to be set in some tested algorithms. According to their original papers, the neighborhood size and the penalty parameter in MOEA/D are specified as 10% of the population size and 5 respectively, and the number of sampling points in HypE is set to 10,000. Since increasing weight vectors with the number of objectives benefits the performance of the algorithm, 200 weight vectors in MSOPS are selected according to the experimental results in [24]. In AR+Grid, a grid division parameter (div) is required. The setting of div in Table 2 can make the algorithm obtain a good balance between convergence and diversity of solutions on the considered problems.

A crossover probability $p_c = 1.0$ and a mutation probability $p_m = 1/L$ (where L denotes the number of decision variables) are used. For continuous problems, the operators for crossover and mutation are simulated binary crossover (SBX) and polynomial mutation with both distribution indexes 20 [6]. As to the combinatorial problem TSP, the order crossover (OX) and inversion operator are chosen as crossover and mutation operators, respectively.

4   Results and Discussion

4.1   The DTLZ Test Problems

Tables 3 and 4 show the CM and IGD results of the eight EMO algorithms on DTLZ2, DTLZ3, DTLZ5, and DTLZ7. The values in the tables are the mean and standard deviation.

First, consider the DTLZ2 problem with a spherical Pareto front satisfying $f_1^2 + f_2^2 + ... + f_M^2 = 1$ in the range $f_1, f_2, ..., f_M \in [0, 1]$. We observe that even for this relatively easy test instance, not all the algorithms work well. NSGA-II has the difficulty in approximating the Pareto front, and obtains poor CM and IGD results. Although performing the best in terms of convergence, AR struggles to maintain a diverse set of solutions, resulting in a poor IGD value. In fact, only three algorithms, MOEA/D, ϵ-MOEA, and AR+Grid, can achieve a good balance between convergence and diversity for both 5- and 10-objective cases.

Despite the same optimal front as DTLZ2, DTLZ3 is modified by introducing a vast number of local optima. This brings a stiff challenge to search towards the global optimal front, especially when the number of objectives becomes large. For this problem, the final solution set of all the algorithms except MOEA/D fails to approach the Pareto front for both 5- and 10-objective cases. ϵ-MOEA performs slightly better than MOEA/D on the 5-objective problem but struggles on the higher-dimensional instance.

DTLZ5 tests the ability of an algorithm to find a lower-dimensional Pareto front while working with a higher-dimensional objective space. From the convergence results in Tables 3, none of the tested algorithms can reach the Pareto front for both instances. The two aggregation-based algorithms especially MSOPS appear to be competitive on this problem. MSOPS significantly outperforms the other algorithms in terms of the comprehensive performance metric IGD when the number of objectives reaches 10. Note that MSOPS has a large CM value. This is because a small part of the algorithm's solutions are located far away from the Pareto front, thereby resulting in a poor convergence assessment result, although the rest of its solutions are distributed widely over the Pareto front.

With many disconnected Pareto optimal regions, DTLZ7 tests an algorithm's ability to maintain subpopulations in disconnected portions of the objective space. Interestingly, the Pareto-based algorithm NSGA-II on this problem outperforms some algorithms designed specially for many-objective optimization. Performing slightly worse than DMO, NSGA-II can be in the second place for the 5-objective instance. On the other hand, the two aggregation-based algorithms struggle on this problem, and MOEA/D even obtains the worst IGD

Table 3. CM results of the eight algorithms on DTLZ2, DTLZ3, DTLZ5, and DTLZ7

Obj.	Alg.	DTLZ2	DTLZ3	DTLZ5	DTLZ7
5	NSGA-II	5.718E–1 (9.7E–2)	8.287E+2 (6.0E+1)	1.302E+0 (4.3E–2)	1.589E–1 (1.2E–2)
	MOEA/D	3.552E–4 (4.5E–5)	6.607E–2 (2.0E–1)	1.213E–1 (5.5E–3)	3.305E–2 (4.6E–3)
	MSOPS	8.696E–2 (1.1E–2)	1.640E+2 (2.4E+1)	8.918E–1 (6.6E–2)	4.890E+0 (2.7E–1)
	HypE	2.601E–3 (6.3E–4)	1.890E+1 (6.9E+0)	1.749E–1 (2.9E–2)	1.150E–1 (3.8E–2)
	ϵ-MOEA	2.185E–2 (2.5E–3)	6.350E–2 (2.1E–2)	6.720E–1 (7.4E–2)	3.067E–2 (9.6E–4)
	DMO	4.837E–2 (5.5E–3)	1.329E+2 (3.8E+1)	4.729E–1 (7.7E–2)	1.151E–1 (8.1E–3)
	AR	7.548E–5 (7.0E–5)	4.653E+0 (7.4E+0)	4.336E–3 (5.8E–4)	4.513E–2 (4.8E–3)
	AR+Grid	7.328E–3 (1.5E–3)	8.919E+1 (4.4E+1)	7.216E–1 (6.5E–2)	7.224E–2 (8.3E–3)
10	NSGA-II	2.308E+0 (2.6E–2)	1.858E+3 (4.1E+1)	2.090E+0 (4.9E–2)	1.976E+0 (1.2E–1)
	MOEA/D	1.088E–3 (1.4E–4)	2.326E–2 (1.5E–2)	1.510E–3 (4.4E–4)	9.316E–2 (1.8E–2)
	MSOPS	5.498E–1 (1.3E–1)	5.966E+2 (1.0E+2)	1.272E+0 (5.1E–2)	5.756E+0 (8.0E–2)
	HypE	2.227E–2 (7.0E–3)	3.794E+0 (3.5E+0)	3.628E–1 (5.9E–2)	3.631E–1 (5.0E–2)
	ϵ-MOEA	4.766E–2 (1.1E–2)	4.479E+1 (8.5E+1)	8.176E–1 (2.6E–1)	2.817E–1 (1.3E–2)
	DMO	2.549E–1 (3.2E–2)	6.611E+2 (1.0E+2)	5.559E–1 (1.2E–1)	8.374E–1 (8.1E–2)
	AR	1.020E–4 (1.3E–4)	3.793E–1 (7.3E–1)	2.383E+0 (2.5E–1)	8.288E–2 (1.2E–2)
	AR+Grid	2.516E–2 (5.0E–3)	5.914E+1 (2.3E+2)	1.759E+0 (6.8E–2)	5.142E–1 (1.2E–2)

Table 4. IGD results of the eight algorithms on DTLZ2, DTLZ3, DTLZ5, and DTLZ7

Obj.	Alg.	DTLZ2	DTLZ3	DTLZ5	DTLZ7
5	NSGA-II	3.101E–1 (2.7E–2)	1.683E+2 (5.4E+1)	5.165E–2 (1.1E–2)	3.367E–1 (1.5E–2)
	MOEA/D	1.332E–1 (7.0E–5)	1.785E–1 (1.7E–1)	2.009E–2 (2.7E–5)	3.503E+0 (7.0E–1)
	MSOPS	1.960E–1 (8.9E–3)	4.025E+1 (1.3E+1)	1.927E–2 (1.7E–3)	2.828E+0 (8.3E–1)
	HypE	2.366E–1 (4.7E–2)	1.404E+0 (1.1E+0)	1.251E–1 (3.8E–2)	4.253E–1 (4.4E–2)
	ϵ-MOEA	1.625E–1 (4.5E–3)	1.753E–1 (1.2E–2)	8.421E–2 (1.0E–2)	4.958E–1 (1.1E–1)
	DMO	2.310E–1 (3.0E–2)	2.023E+1 (9.1E+0)	4.049E–1 (8.4E–2)	3.250E–1 (1.8E–2)
	AR	6.264E–1 (2.5E–2)	1.522E+0 (1.7E+0)	6.054E–1 (5.1E–2)	1.867E+0 (2.1E–1)
	AR+Grid	1.454E–1 (3.4E–3)	7.805E–1 (2.2E–1)	6.268E–2 (1.2E–2)	1.220E+0 (3.3E–1)
10	NSGA-II	2.112E+0 (1.5E–1)	5.928E+2 (1.9E+2)	1.887E–1 (1.0E–0)	2.288E+0 (6.1E–1)
	MOEA/D	4.921E–1 (6.9E–5)	4.950E–1 (5.0E–3)	6.495E–2 (2.2E–6)	4.193E+0 (1.2E+0)
	MSOPS	6.852E–1 (4.9E–2)	8.303E+1 (1.7E+1)	1.769E–2 (2.0E–3)	2.208E+1 (3.0E+0)
	HypE	6.294E–1 (9.4E–2)	2.242E+0 (1.3E+0)	1.472E–1 (3.2E–2)	1.010E+0 (3.3E–2)
	ϵ-MOEA	4.048E–1 (4.9E–3)	1.273E+1 (2.0E+1)	1.714E–1 (1.7E–2)	1.879E+0 (1.8E–1)
	DMO	5.280E–1 (1.8E–2)	2.368E+2 (8.2E+1)	4.656E–1 (1.1E–1)	6.561E+0 (2.5E+0)
	AR	1.169E+0 (8.1E–3)	1.251E+0 (2.4E–1)	2.582E+0 (4.2E–1)	8.440E+0 (1.5E–1)
	AR+Grid	4.835E–1 (5.9E–3)	1.735E+0 (2.9E+0)	8.854E–1 (1.5E–1)	1.381E+0 (1.6E–1)

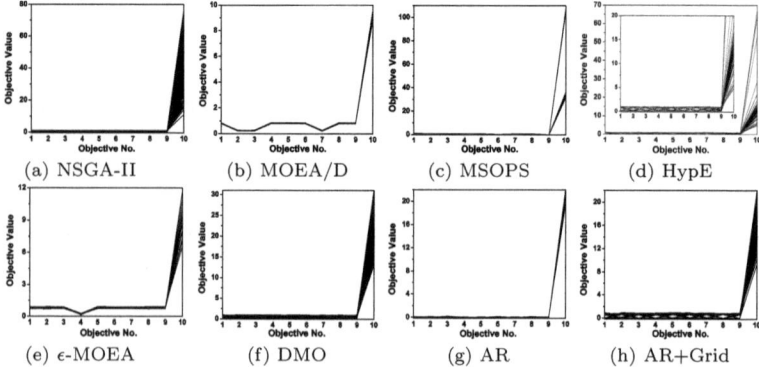

(a) NSGA-II   (b) MOEA/D   (c) MSOPS   (d) HypE

(e) ϵ-MOEA   (f) DMO   (g) AR   (h) AR+Grid

Fig. 1. The final solution set of the eight algorithms on the ten-objective DTLZ7, shown by parallel coordinates

Table 5. HV results of the eight algorithms on WFG1, WFG8, and WFG9

Obj.	Alg.	WFG1	WFG8	WFG9
5	NSGA-II	4.984E−1 (1.3E−2)	8.507E−1 (1.5E−2)	8.665E−1 (1.2E−2)
	MOEA/D	1.157E+0 (1.8E−1)	8.311E−1 (1.3E−1)	1.012E+0 (2.5E−2)
	MSOPS	4.274E−1 (1.2E−2)	8.309E−1 (1.1E−2)	9.310E−1 (9.5E−3)
	HypE	5.481E−1 (1.0E−1)	7.615E−1 (1.6E−2)	8.511E−1 (4.9E−2)
	ϵ-MOEA	4.075E−1 (3.0E−2)	9.404E−1 (7.5E−3)	9.440E−1 (1.6E−2)
	DMO	5.024E−1 (9.9E−3)	6.748E−1 (1.9E−2)	8.718E−1 (1.7E−2)
	AR	3.234E−1 (1.4E−2)	2.380E−1 (2.1E−2)	1.896E−1 (8.8E−2)
	AR+Grid	7.220E−1 (3.9E−2)	1.098E+0 (3.1E−3)	1.052E+0 (2.7E−3)
10	NSGA-II	8.409E−1 (1.3E−2)	2.456E+0 (1.4E−2)	2.371E+0 (1.6E−3)
	MOEA/D	1.031E+0 (2.8E−2)	7.091E−1 (4.0E−1)	3.661E−1 (1.9E−1)
	MSOPS	7.611E−1 (1.8E−2)	2.395E+0 (8.4E−2)	2.373E+0 (6.6E−4)
	HypE	1.174E+0 (1.4E−1)	2.587E+0 (2.1E−3)	2.563E+0 (8.3E−3)
	ϵ-MOEA	7.402E−1 (4.3E−2)	4.497E−1 (8.1E−1)	1.122E+0 (6.2E−1)
	DMO	8.355E−1 (1.5E−2)	2.226E+0 (5.1E−2)	2.321E+0 (5.4E−3)
	AR	7.153E−1 (1.2E−1)	2.525E+0 (4.5E−3)	2.329E+0 (1.7E−4)
	AR+Grid	1.132E+0 (7.4E−2)	2.577E+0 (8.2E−4)	2.373E+0 (8.0E−4)

value for the 5-objective DTLZ7. Figure 1 plots the final solutions of the eight algorithms on the 10-objective DTLZ7. Clearly, the solution set of MOEA/D, ϵ-MOEA, AR, and AR+Grid can approximate the Pareto front (the upper bound of the last objective in the Pareto front of DTLZ7 is equal to $2 \times M$, i.e., $f_{10} \leq 20$ for the 10-objective instance). However, the former three converge into a subarea of the optimal front, thus obtaining a worse IGD value than AR+Grid. It is interesting to note that HypE, with several solutions far away from the Pareto front, obtains the best IGD value. This occurrence can be attributed to the fact that most of its solutions have good convergence and diversity, which can be observed from the scaling plot of the solutions in Figure 1(d) (the lower bound of the solutions in the last objective of the problem reaches around 5).

4.2   The WFG Test Problems

By a series of transition of introducing complexity (such as flat bias, multimodality, and nonseparability), the WFG problem suite poses a big challenge for algorithms to obtain a well-converged and well-distributed solution set. Table 5 gives the HV results of the eight EMO algorithms on WFG1, WFG8, and WFG9 with 5 and 10 objectives. We observe that these algorithms show different behaviors on different test instances.

Unlike the case on the DTLZ suite, where DMO outperforms NSGA-II for most of the instance, on the WFG suite they present similar performance. Especially for the three 10-objective instances, NSGA-II even performs slightly better than DMO. This means that decreasing the diversity requirement in the evolutionary process of many-objective optimization may not always improve the performance of an algorithm.

MOEA/D shows different search abilities on different WFG problems. It appears to be competitive with the other algorithms on the 5-objective WFG1 and WFG9 and 10-objective WFG1, but performs poorly on the 10-objective WFG8 and WFG9. In fact, all the tested algorithms have difficulty on the WFG1 problem which has the characteristic of flat bias. Although clearly superior to the

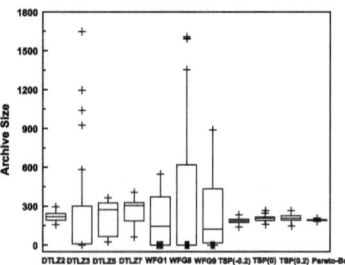

Fig. 2. The box plot of the archive size of ϵ-MOEA on 30 runs for all 10-objective problems

other algorithms on the 5-objective WFG1, MOEA/D fails to cover the whole region of the Pareto front—the average value of the final solution set of MOEA/D is 0.371 by the improved *Maximum Spread* (MS') test [21] (MS' $\in [0, 1]$, and one is the optimal value which indicates the obtained set reaches the all the boundaries of the Pareto front[1]).

Although both are aggregation-based approaches, MSOPS shows different behavior from MOEA/D. The former is significantly inferior to the latter on the 5- and 10-objective WFG1, but clearly superior on the 10-objective WFG8 and WFG9. For the rest of the problems, the difference between them is slight.

An interesting result of the HypE algorithm on the WFG problem suite is observed from the table. HypE performs poorly for the three 5-objective WFG problems, but achieves the best HV value on the 10-objective instances. This means that HypE becomes more competitive as the number of objectives increases. In contrast, ϵ-MOEA works well on the 5-objective WFG8 and WFG9, but struggles on the higher-dimensional WFG problem. Note that large standard deviation of HV is obtained by ϵ-MOEA for the 10-objective cases. This means an unstable search ability of ϵ-MOEA on these problems. An important reason for this occurrence is the instability of the archive size of the algorithm. In fact, when the number of objectives reaches 10, the size of the final archive set for each run may be completely different no matter what setting the parameter ϵ. Figure 2 gives the archive size of ϵ-MOEA on 30 runs for all 10-objective problems by box plot. Clearly, severe instability of the archive size is shown on all the DTLZ and WFG problems except on DTLZ2. For some runs, the size reaches over 1600, yet for some other runs, the archive set has only one solution.

Similar to the case on DTLZ, the AR approach, in general, performs the worst of all on the WFG problems, obtaining the lowest HV value in 4 out of the 6 instances. However, AR+Grid, an improvement of AR by using grid to enhance diversity, is very competitive with the other algorithms. AR+Grid performs the

---

[1] Here, MS' is an improved version of the original MS [27] to ease the influence of convergence of a solution set. It evaluates the spread of a solution set by introducing the comparison between the extreme values of the solutions in the set and the boundaries of the Pareto front.

Table 6. HV results of the eight algorithms on multiobjective TSP with different correlation parameter ($TSPcp$) values

Obj.	Alg.	$TSPcp = -0.2$	$TSPcp = 0$	$TSPcp = 0.2$
	NSGA-II	3.616E+05 (3.3E+04)	4.406E+05 (3.4E+04)	5.245E+05 (4.3E+04)
	MOEA/D	1.226E+06 (2.8E+04)	1.038E+06 (2.4E+04)	9.056E+05 (2.0E+04)
	MSOPS	8.511E+05 (4.0E+04)	7.970E+05 (3.3E+04)	7.456E+05 (2.7E+04)
5	HypE	3.547E+05 (2.5E+04)	3.800E+05 (4.1E+04)	5.070E+05 (8.0E+04)
	ϵ-MOEA	1.092E+06 (3.9E+04)	9.743E+05 (3.8E+04)	8.753E+05 (2.7E+04)
	DMO	3.522E+05 (3.3E+04)	4.345E+05 (4.4E+04)	5.271E+05 (5.1E+04)
	AR	3.924E+05 (8.9E+04)	3.947E+05 (5.9E+04)	3.812E+05 (5.1E+04)
	AR+Grid	1.109E+06 (3.7E+04)	9.558E+05 (2.7E+04)	8.636E+05 (2.8E+04)
	NSGA-II	1.040E+10 (1.8E+09)	1.801E+10 (2.6E+09)	2.545E+10 (4.5E+09)
	MOEA/D	2.572E+11 (1.2E+10)	1.969E+11 (1.3E+10)	1.580E+11 (1.0E+10)
	MSOPS	1.980E+11 (1.7E+10)	1.898E+11 (1.1E+10)	1.662E+11 (9.2E+09)
10	HypE	1.033E+10 (1.2E+09)	1.613E+10 (9.8E+08)	3.536E+10 (7.2E+09)
	ϵ-MOEA	1.337E+11 (1.2E+10)	1.440E+11 (9.3E+09)	1.530E+11 (1.3E+10)
	DMO	8.084E+09 (1.6E+09)	1.691E+10 (3.2E+09)	2.504E+10 (5.2E+09)
	AR	5.306E+10 (1.2E+10)	7.649E+10 (2.1E+10)	4.946E+10 (9.8E+09)
	AR+Grid	3.159E+11 (1.0E+10)	2.496E+11 (7.8E+09)	2.080E+11 (5.9E+09)

best on the 5-objective WFG8 and WFG9, and takes the second place for the rest of the test instances.

4.3   The TSP Test Problem

One important property of the multiobjective TSP problem is that the conflict degree among the objectives can be adjusted according to the parameter $TSPcp \in (-1, 1)$, where a lower value means a greater degree of conflict. Table 6 gives the HV results of the eight EMO algorithms on TSP with different correlation parameter settings and numbers of objectives.

As can be seen from the table, MOEA/D and AR+Grid generally outperform the other algorithms. More specifically, MOEA/D performs best on the three 5-objective instances, and AR+Grid keeps a clear advantage for the 10-objective TSP with different $TSPcp$ values. Among the remaining algorithms, MSOPS and ϵ-MOEA are competitive. For the lower-dimensional problems, ϵ-MOEA obtains a larger HV value, and MSOPS performs better when the higher-dimensional TSP instances are considered. The other four algorithms seem not to work well on the TSP problem suite. But an interesting phenomenon can be observed—AR, which generally performs the worst in terms of the comprehensive assessment metrics on DTLZ and WFG, obtains a significant better HV result than HypE and DMO on the three 10-objective TSP instances.

In addition, note that the tested algorithms show different trends with the change of the correlation parameter $TSPcp$. Most of the "good" algorithms (i.e., MOEA/D, AR+Grid, and MSOPS) have a decreasing HV value with the increase of $TSPcp$ for both 5- and 10-objective cases; conversely, most of the "poor" algorithms (i.e., NSGA-II, HypE, and DMO) obtain an increasing result. This means that the search ability of algorithms on TSP is reflected more clearly when a greater degree of conflict among objectives of the problem is involved.

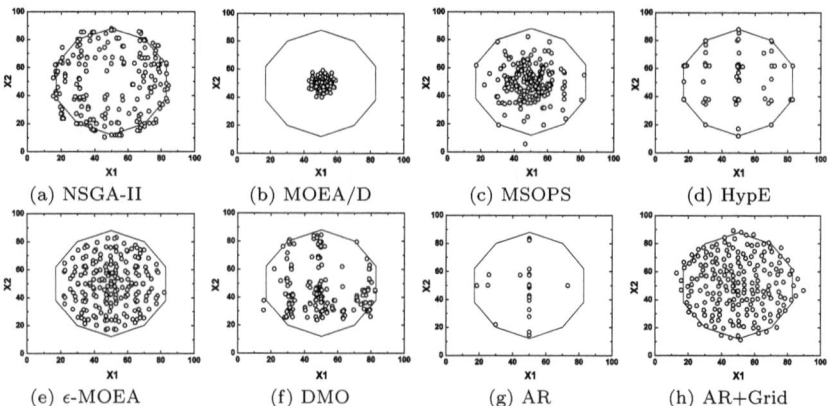

Fig. 3. The final solution set of the eight algorithms in the decision space on the ten-objective Pareto-Box problem

4.4 The Pareto-Box Test Problem

With a high-dimensional objective space and a two-dimensional decision space, the Pareto-Box problem is used to visually investigate the distribution of solutions in the decision space. Due to the corresponding relation of the crowding between the objective space and decision space of the problem, this can also facilitate the understanding of the ability for an algorithm to maintain diversity in the objective space [19]. Figure 3 shows the 10-objective Pareto-Box problem considered here (the Pareto optimal region is the inside of the regular decagon), as well as the final solution set of the eight algorithms on the problem.

It is clear from the figure that none of the algorithms works well in terms of both convergence and diversity, although they show different behavior characteristics. Many solutions of NSGA-II fail to converge into the Pareto optimal region. The solutions of MOEA/D, AR, and HypE concentrate (or even coincide) in one (or several) region(s) of the decagon. Specifically, The solutions obtained by MOEA/D gather around the center of the decagon, and most of the solutions of AR are located in the central axes of the figure. Although the solutions of HypE reach each angle of the decagon, there are a lot of vacant places. MSOPS and DMO perform better than the previous four algorithms, but struggle to maintain uniformity of their solutions.

ϵ-MOEA and AR+Grid are the two best algorithms on this problem. However, they also have their own shortcomings—the former fails to keep the boundary solutions and the latter has difficulty in making all solutions converge into the optimal region.

4.5 Summary

Based on the above examination on four groups of test functions, the summary observation of the eight algorithms can be made:

- NSGA-II does not always perform the worst on all many-objective problems. On some problems with relatively low dimensions, such as 5-objective DTLZ7 and WFG8, NSGA-II outperforms some algorithms designed specially for many-objective optimization.
- The search ability of MOEA/D has sharp contrasts on different problems. It works very well on DTLZ2, DTLZ3, WFG1, and TSP, but encounters great difficulties on the DTLZ7, WFG8, and Pareto-Box problems. From the results on the WFG and TSP suites, MOEA/D appears to be more competitive in relatively low-dimensional problems.
- Similar to MOEA/D, MSOPS struggles on the problem with the disconnected Pareto front (DTLZ7). But MSOPS performs the best on the degenerate problem DTLZ5.
- Although favoring the boundary solutions, HypE shows advantages in a higher-dimensional objective space. This can be obtained from the results of the DTLZ7 and three WFG test instances.
- ϵ-MOEA performs well on most of the 5-objective test instances. However, the instability of the archive size will count against the evolutionary process of the algorithm as the number of objectives further increases.
- By adaptively controlling the diversity maintenance mechanism, DMO has a clear advantage over NSGA-II on the DTLZ suite. However for the WFG and TSP suites, the advantage vanishes, NSGA-II even outperforming DMO on WFG8 and most of the TSP instances.
- Due to the lack of diversity maintenance, AR is the algorithm with poor comprehensive performance on all the test problems, except for the 10-objective TSP, where AR is clearly superior to HypE and DMO.
- Despite being competitive on most of the test instances, AR+Grid has difficulty on the problem with many local optima, such as DTLZ3. This is because the neighbor punishment strategy in AR+Grid may make some "bad" individuals rank higher than their better competitors.

5   Conclusions

This paper has compared eight state-of-the-art EMO algorithms from the six classes of approaches of dealing with many-objective problems. The benchmark has been composed of four groups of continuous and combinatorial test problems. The behavior of the algorithms has been understood by three performance metrics, and also with the help of a visual investigation in the decision space.

Our study has revealed that there is not a clear performance gap between algorithms for all the tested problems. These algorithms have their own strengths on different test instances. This means that a careful choice of algorithms must be made when dealing with a many-objective problem in hand.

Despite various adaptability, several algorithms demonstrate their competitiveness on most of the test instances. AR+Grid, MOEA/D, and ϵ-MOEA are three such algorithms—AR+Grid works well on DTLZ2, WFG, TSP and Pareto-Box, MOEA/D on DTLZ2, DTLZ3, WFG1 and TSP, and ϵ-MOEA on DTLZ,

Pareto-box and the 5-objective WFG8, WFG9 and TSP instances. Among the rest, MSOPS and HypE are two more promising approaches. The former is impressive on DTLZ5 and the 10-objective TSP, and the latter on the 10-objective DTLZ7 and WFG. DMO outperforms NSGA-II for the DTLZ problem suite. Finally, NSGA-II and AR struggle to make solutions approximate the Pareto front and to maintain a diverse set of solutions, respectively.

Another observation of our study is that none of the tested algorithms can produce a well-converged and well-distributed solution set even for some "easy" problems, such as the Pareto-Box problem. This indicates the infancy of evolutionary many-objective optimization and highlights the need for further development in the area.

One area for further investigation is to compare the eight algorithms under varying parameter settings, including the dimensions of decision space, population size, and computational budget. In addition, statistical analysis will be applied to evaluate the confidence level of the obtained results.

References

1. Adra, S.F., Fleming, P.J.: Diversity management in evolutionary many-objective optimization. IEEE Trans. Evol. Comput. 15(2), 183–195 (2011)
2. Bader, J., Zitzler, E.: HypE: An algorithm for fast hypervolume-based many-objective optimization. Evolutionary Computation 19(1), 45–76 (2011)
3. Beume, N., Naujoks, B., Emmerich, M.: SMS-EMOA: Multiobjective selection based on dominated hypervolume. European Journal of Operational Research 181(3), 1653–1669 (2007)
4. Corne, D.W., Knowles, J.D.: Techniques for highly multiobjective optimisation: some nondominated points are better than others. In: Proc. 9th Ann. Conf. Genetic and Evol. Comput., GECCO 2007, pp. 773–780 (2007)
5. Deb, K., Pratap, A., Agarwal, S., Meyarivan, T.: A fast and elitist multiobjective genetic algorithm: NSGA-II. IEEE Trans. Evol. Comput. 6(2), 182–197 (2002)
6. Deb, K., Jain, H.: An improved NSGA-II procedure for many-objective optimization part I: solving problems with box constraints. KanGAL, Indian Institute of Technology, Tech. Rep. 2012009 (2012)
7. Deb, K., Mohan, M., Mishra, S.: Evaluating the ϵ-domination based multi-objective evolutionary algorithm for a quick computation of Pareto-optimal solutions. Evolutionary Computation 13(4), 501–525 (2005)
8. Deb, K., Thiele, L., Laumanns, M., Zitzler, E.: Scalable test problems for evolutionary multiobjective optimization. In: Evolutionary Multiobjective Optimization. Theoretical Advances and Applications, pp. 105–145 (2005)
9. Farina, M., Amato, P.: On the optimal solution definition for many-criteria optimization problems. In: Proc. NAFIPS Fuzzy Information Processing Society 2002 Annual Meeting of the North American, pp. 233–238 (2002)
10. Hadka, D., Reed, P.: Diagnostic assessment of search controls and failure modes in many-objective evolutionary optimization. Evol. Comput. (2012) (in press)
11. Huband, S., Hingston, P., Barone, L., While, L.: A review of multiobjective test problems and a scalable test problem toolkit. IEEE Trans. Evol. Comput. 10(5), 477–506 (2006)
12. Hughes, E.J.: Multiple single objective Pareto sampling. In: Proc. Congress Evolutionary Computation CEC 2003, vol. 4, pp. 2678–2684 (2003)

13. Hughes, E.J.: Evolutionary many-objective optimisation: many once or one many? In: Proc. IEEE Congress Evolutionary Computation, CEC 2005, pp. 222–227 (2005)
14. Ishibuchi, H., Tsukamoto, N., Hitotsuyanagi, Y., Nojima, Y.: Effectiveness of scalability improvement attempts on the performance of NSGA-II for many-objective problems. In: Proc. 10th Annual Conf. Genetic Evol. Comput., GECCO 2008, pp. 649–656 (2008)
15. Ishibuchi, H., Tsukamoto, N., Nojima, Y.: Evolutionary many-objective optimization: A short review. In: Proc. IEEE Congress Evolutionary Computation, CEC 2008, pp. 2419–2426 (2008)
16. Ishibuchi, H., Hitotsuyanagi, Y., Tsukamoto, N., Nojima, Y.: Many-Objective Test Problems to Visually Examine the Behavior of Multiobjective Evolution in a Decision Space. In: Schaefer, R., Cotta, C., Kołodziej, J., Rudolph, G. (eds.) PPSN XI. LNCS, vol. 6239, pp. 91–100. Springer, Heidelberg (2010)
17. Jaimes, A.L., Coello Coello, C.A.: Study of preference relations in many-objective optimization. In: Proc. 11th Annual Conf. Genetic Evol. Comput., GECCO 2009, pp. 611–618 (2009)
18. Khare, V., Yao, X., Deb, K.: Performance Scaling of Multi-objective Evolutionary Algorithms. In: Fonseca, C.M., Fleming, P.J., Zitzler, E., Deb, K., Thiele, L. (eds.) EMO 2003. LNCS, vol. 2632, pp. 376–390. Springer, Heidelberg (2003)
19. Köppen, M., Yoshida, K.: Substitute Distance Assignments in NSGA-II for Handling Many-objective Optimization Problems. In: Obayashi, S., Deb, K., Poloni, C., Hiroyasu, T., Murata, T. (eds.) EMO 2007. LNCS, vol. 4403, pp. 727–741. Springer, Heidelberg (2007)
20. Li, M., Yang, S., Zheng, J., Liu, X.: ETEA: A Euclidean minimum spanning tree-based evolutionary algorithm for multiobjective optimization. Evol. Comput. (2013) (in press)
21. Li, M., Zheng, J., Li, K., Yuan, Q., Shen, R.: Enhancing Diversity for Average Ranking Method in Evolutionary Many-Objective Optimization. In: Schaefer, R., Cotta, C., Kołodziej, J., Rudolph, G. (eds.) PPSN XI. LNCS, vol. 6238, pp. 647–656. Springer, Heidelberg (2010)
22. Li, M., Zheng, J., Shen, R., Li, K., Yuan, Q.: A grid-based fitness strategy for evolutionary many-objective optimization. In: Proc. 12th Annual Conf. Genetic Evol. Comput., GECCO 2010, pp. 463–470 (2010)
23. Purshouse, R.C., Fleming, P.J.: On the evolutionary optimization of many conflicting objectives. IEEE Trans. Evol. Comput. 11(6), 770–784 (2007)
24. Wagner, T., Beume, N., Naujoks, B.: Pareto-, Aggregation-, and Indicator-Based Methods in Many-Objective Optimization. In: Obayashi, S., Deb, K., Poloni, C., Hiroyasu, T., Murata, T. (eds.) EMO 2007. LNCS, vol. 4403, pp. 742–756. Springer, Heidelberg (2007)
25. Yang, S., Li, M., Liu, X., Zheng, J.: A grid-based evolutionary algorithm for many-objective optimization. IEEE Trans. Evol. Comput. (2013) (in press)
26. Zhang, Q., Li, H.: MOEA/D: A multiobjective evolutionary algorithm based on decomposition. IEEE Trans. Evol. Comput. 11(6), 712–731 (2007)
27. Zitzler, E., Thiele, L.: Multiobjective evolutionary algorithms: a comparative case study and the strength Pareto approach. IEEE Trans. Evol. Comput. 3(4), 257–271 (1999)
28. Zitzler, E., Künzli, S.: Indicator-Based Selection in Multiobjective Search. In: Yao, X., Burke, E.K., Lozano, J.A., Smith, J., Merelo-Guervós, J.J., Bullinaria, J.A., Rowe, J.E., Tiño, P., Kabán, A., Schwefel, H.-P. (eds.) PPSN VIII. LNCS, vol. 3242, pp. 832–842. Springer, Heidelberg (2004)

Effect of Dominance Balance in Many-Objective Optimization

Kaname Narukawa

Honda Research Institute Europe GmbH,
Carl-Legien-Strasse 30, 63073 Offenbach am Main, Germany
kaname.narukawa@honda-ri.de
http://www.honda-ri.de/

Abstract. This paper examines the effect of dominance balance in many-objective optimization. The dominance balance can be defined by the ratio of the dominating space to the objective space. Here, CDAS, which is one of the most powerful evolutionary many-objective optimization algorithms, is known to be able to change the ratio of the dominating space by relaxing the definition of Pareto dominance with its user-specified parameter. However, the dominance balance is too difficult to control for the parameter in the higher-dimensional objective space. Therefore, we analyze the performance of CDAS by changing the ratio of the dominating space directly in even steps from the minimum to the maximum according to the number of objectives. The corresponding user-specified parameter in CDAS can be obtained from an equation which we assume in the paper. As benchmark test problems, we use DTLZ1, DTLZ2, DTLZ3, and DTLZ4 with two to ten objectives. From computational experiments, we can conclude that the optimal ratio of the dominating space differs depending on the problem at hand. It can be also said that the performance of CDAS is good especially when the ratio of the dominating space is small enough but not the minimum. Based on these observations, we propose a new version of CDAS called CDAS-D which controls the ratio of the dominating space dynamically during optimization.

Keywords: Pareto dominance, dominance relation, dominance area, many-objective optimization, multi-objective optimization.

1 Introduction

Recently many-objective optimization has attracted much attention in evolutionary multi-objective optimization (EMO) which is one of the most active research areas in evolutionary computation [9]. In the past two decades, a lot of EMO algorithms have been proposed in the hope of working for optimization problems with an arbitrary number of objectives [2,3]. However, it can be found these days that they can hardly handle optimization problems with more than three objectives [9]. For example, NSGA-II, which is one of the most reputable EMO

R.C. Purshouse et al. (Eds.): EMO 2013, LNCS 7811, pp. 276–290, 2013.

algorithms, compares solutions based on their rank values which generate search pressure in the objective space [4]. In the higher-dimensional objective space, however, only the first rank may be assigned to every solution as almost all solutions in the population will become non-dominated. Without a variety of rank values, NSGA-II can not keep the search pressure anymore in many-objective optimization. Thus, we have to adopt totally different approaches.

There are mainly three approaches to tackle this issue in many-objective optimization. One is to use a scalarizing function. The scalarizing function yields a scalar value from multiple objective values with a weight vector. MOEA/D is one of representative algorithms that use the scalarizing function with multiple weight vectors in a single run of optimization [17]. Another approach is to have an indicator for comparing solutions sets. IBEA [20] and SMS-EMOA [1] are well-known indicator-based algorithms. As the indicator, the hypervolume is often used. The third approach is to modify the dominance balance. CDAS [13,14,15], for example, can keep the dominance balance by relaxing the definition of Pareto dominance. Although the performance of CDAS is good for both benchmark test problems and real-world applications [12], its theoretical understanding is not enough in the literature.

In the present paper, we examine the effect of the dominance balance on the performance of CDAS, theoretically and experimentally. The dominance balance is meant by the ratio of the dominating space to the objective space. CDAS can change the ratio of the dominating space by relaxing the definition of Pareto dominance with its parameter S. However, the ratio of the dominating space (i. e., dominance balance) is too hard to handle for the parameter S in the higher-dimensional objective space as the ratio is proportional to a^{d-1}, where $a = \pi/2 - S \cdot \pi$, d is the number of objectives. Therefore, we analyze the performance of CDAS by changing the ratio of the dominating space directly in even steps from the minimum $((1/2)^d)$ to the maximum $(1/2)$. The corresponding parameter S in CDAS can be calculated by an equation which we assume in the paper. As benchmark test problems, we use DTLZ1, DTLZ2, DTLZ3, and DTLZ4 with two to ten objectives. From computational experiments, we see that the optimal ratio of the dominating space differs depending on the problem at hand. It can be also said that the performance of CDAS is improved especially when the ratio of the dominating space is small enough but not the minimum. Based on these observations, we propose a new version of CDAS called CDAS-D which controls the ratio of the dominating space dynamically during optimization. In computational experiments, we examine the performance of CDAS-D with two different settings for a range of the ratio of the dominating space.

This paper is organized as follows. We give a detailed description of the dominance balance with a brief introduction to multi-objective optimization and CDAS in section 2. The relation among the parameter S in CDAS, the ratio of the dominating space, and the number of objectives is also given in the form of an equation and a figure. In section 3, we examine the performance of CDAS on benchmark test problems. Based on our observations in section 3, we propose CDAS-D in section 4. The paper concludes in section 5 with discussions on results and future works.

2   Dominance Balance

2.1   Multi-objective Optimization

In this paper we consider minimization problems. Generally a d-objective minimization problem is formulated as follows:

$$\text{Minimize } \mathbf{f}(\mathbf{x}) = (f_1(\mathbf{x}), f_2(\mathbf{x}), ..., f_d(\mathbf{x})), \text{ subject to } \mathbf{x} \in \mathbf{X}, \tag{1}$$

where $\mathbf{f}(\mathbf{x})$ is the d-dimensional objective vector, $f_i(\mathbf{x})$ is the i-th objective value to be minimized, $\mathbf{x}$ is the decision vector, $\mathbf{X}$ is the feasible region. When two feasible solutions $\mathbf{a}$ and $\mathbf{b}$ of (1) satisfy the following conditions, we can say that $\mathbf{a}$ dominates $\mathbf{b}$.

$$\forall i : f_i(\mathbf{a}) \leq f_i(\mathbf{b}) \text{ and } \exists j : f_j(\mathbf{a}) < f_j(\mathbf{b}). \tag{2}$$

If $\mathbf{a}$ is not dominated by any other feasible solutions, $\mathbf{a}$ is referred to as a Pareto-optimal solution. The entire set of Pareto-optimal solutions forms the Pareto front. The task in multi-objective optimization is to find a set of non-dominated solutions that approximates the whole Pareto front well [4,18].

2.2   Controlling Dominance Area of Solutions (CDAS)

As explained in section 1, CDAS can change the ratio of the dominating space (i. e., dominance balance) by relaxing the definition of Pareto dominance indirectly with its parameter S [13,14,15]. In order to relax the definition implicitly, objective values are transformed in CDAS. The transformation equation for the i-th objective value of $\mathbf{f}(\mathbf{x})$ is given with the parameter S as follows:

$$f_i'(\mathbf{x}) = r\cos(\omega_i) + r\sin(\omega_i) \cdot \frac{\cos(\varphi)}{\sin(\varphi)}$$
$$= \frac{r\sin(\varphi + \omega_i)}{\sin(\varphi)}, \ \varphi = S \cdot \pi \left(\frac{1}{4} \leq S \leq \frac{1}{2}\right), \tag{3}$$

where r is the norm of $\mathbf{f}(\mathbf{x})$, and ω_i is the declination angle between $f_i(\mathbf{x})$ and $\mathbf{f}(\mathbf{x})$. The transformation is also illustrated in Fig. 1 for the two-objective case. Based on transformed objective values by (3), CDAS assigns rank values and crowding distances to solutions in the same manner as NSGA-II. It should be noted that the algorithm of CDAS is exactly the same as that of NSGA-II except that transformed objective values are used to calculate rank values and crowding distances in CDAS. Although CDAS can handle multiple parameters S_i ($i = 1, 2, ..., d$), only one parameter S is considered in this paper for the sake of simplicity.

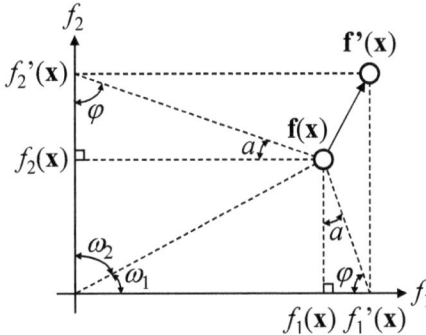

Fig. 1. Transformation of objective values in the two-objective space

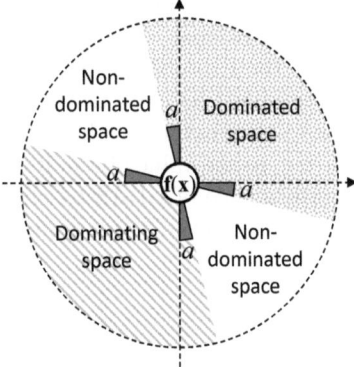

Fig. 2. Dominating, non-dominated, dominated spaces after transforming objective values are shown

2.3 Ratio of Dominating Space

In order to understand the ratio of the dominating space to the objective space, the dominating space that dominates $\mathbf{f}(\mathbf{x})$, the non-dominated space that is non-dominated in comparison with $\mathbf{f}(\mathbf{x})$, and the dominated space that is dominated by $\mathbf{f}(\mathbf{x})$ after the transformation by (3) are illustrated in Fig. 2 with $a = \pi/2 - S \cdot \pi$ ($0 \leq a \leq \pi/4$, see (3) and Fig. 1). Here, we assume a hypersphere to consider these spaces in the higher-dimensional objective space. From Fig. 2, we can see that the ratio of the dominating space to the objective space (i. e., circle) is proportional to the value of a for the case of two objectives. In the three-dimensional objective space, the ratio of the dominating space to the objective space (i. e., sphere) is proportional to the value of a to the second power. Therefore, we can assume that the ratio of the dominating space is proportional to the value of a to the $(d-1)$-th power in the d-dimensional objective space. It can be also assumed that the minimum and the maximum of the ratio of the dominating space is $(1/2)^d$ and $1/2$, respectively. From these assumptions, we obtain the following relational expression of the ratio of the dominating space, the value of a, and the number of objectives d.

$$\text{Ratio}_{\text{Dominating Space}} = \left\{\frac{1}{2} - \left(\frac{1}{2}\right)^d\right\} \cdot \left(\frac{4}{\pi}\right)^{d-1} \cdot a^{d-1} + \left(\frac{1}{2}\right)^d \quad (4)$$

We can also visualize the equation (4) in Fig. 3. From Fig. 3, we can see that the ratio of the dominating space is nearly zero in the objective space with more than ten objectives when the value of a is specified as $a = 0$. In other words, search pressure will be lost in the higher-dimensional objective space as the ratio of the non-dominated space is nearly one, thereby almost all solutions become non-dominated. However, when the value of a is specified as close to the maximum (i. e., $\pi/4$), the ratio of the dominating space is still more than zero even in the higher-dimensional objective space. It should be noted that the ratio of the dominating space (i. e., dominance balance) is difficult to specify for the value of a in the higher-dimensional objective space. In the 30-dimensional objective space, for example, the ratio of the dominating space is nearly zero when the value of a is not specified as close to the maximum (i. e., $\pi/4$). On the other hand, if it is close to the maximum, the ratio of the dominating space changes dramatically according to the value of a in its narrow range. In the next section, we examine the performance of CDAS by changing the value of a and the ratio of the dominating space in even steps, respectively.

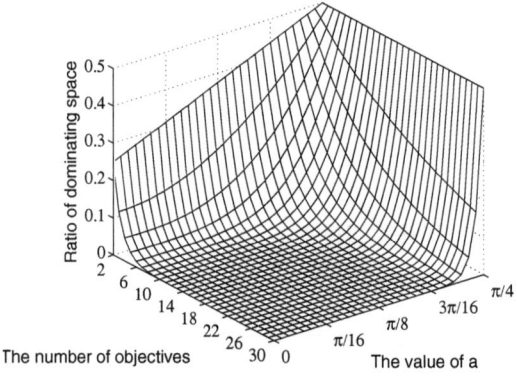

Fig. 3. Ratio of dominating area

3   Effect of Dominance Balance

In the paper, we use DTLZ1, DTLZ2 ,DTLZ3, and DTLZ4 with two to ten objectives as benchmark test problems [5,6]. They are well-known benchmark test problems where the number of objectives can be specified by the user.

We use the relative hypervolume (also called as covered fraction in [7]) to assess the performance of CDAS on benchmark test problems. The relative hypervolume is calculated as follows. First the hypervolume of non-dominated solutions in a population is calculated with a reference point. The reference point for calculating the hypervolume should be specified in advance. In this paper, we use the reference point $\mathbf{p} = 0.7^d$ for DTLZ1, $\mathbf{p} = 1.1^d$ for DTLZ2, DTLZ3, and DTLZ4, where d is the number of objectives. We can also calculate the hypervolume of the Pareto front analytically as the Pareto front of each problem is known. Then the hypervolume of non-dominated solutions divided by the hypervolume of the Pareto front yields the relative hypervolume. When the relative hypervolume is one, it means that the Pareto front is obtained completely. On the other hand, when the relative hypervolume is zero, no solutions which dominate the reference point are found. It should be noted that the hypervolume and its relative value give the same conclusion because the hypervolume is just divided by a constant. Regarding the other settings for our computational experiments like a crossover probability and a mutation probability and so on, we use the same settings as [16] except that the number of generations is set to 1000.

First, we show the relative hypervolume of non-dominated solutions obtained by CDAS for combinations of $a = (0, 0.025\pi, \ldots, 0.25\pi)$ and $d = (2, 3, \ldots, 10)$ in Fig. 4. The average of the relative hypervolume over 20 trials is given throughout the paper. We can see from Fig. 4 that the relative hypervolume is larger for each problem when the number of objectives is smaller and the value of a is $a = 0$. It should be noted that results with $a = 0$ correspond to results of NSGA-II as objective values do not change when the value of a is specified as $a = 0$. When the number of objectives is larger, the performance of CDAS with $a = 0$ (i. e., NSGA-II) easily deteriorates. This is justified by the fact that the performance of NSGA-II deteriorates with an increase in the number of objectives [9]. From Fig. 4(a), the relative hypervolume is nearly one in almost all cases for DTLZ1 when the value of a is set to $a \neq 0$. On the other hand, we obtain different results for DTLZ2, DTLZ3, and DTLZ4 from Fig. 4(b), Fig. 4(c), and Fig. 4(d), respectively. From these results, when the number of objectives is smaller, the best performance is achieved with small values of a. However, the best relative hypervolume is obtained by large values of a excluding the maximum when the number of objectives is larger. As discussed in Section 2.3 for Fig. 3, the ratio of the dominating space is still more than zero even in the higher-dimensional objective space when the value of a is specified as large values in its range. Therefore, we can say that the ratio of the dominating space is related to the performance of CDAS.

Although CDAS can change the ratio of the dominating space, it is not changed in even steps in Fig. 4. Therefore, we measure the performance of CDAS for combinations of $r = (r1, r2, \ldots, r11)$ and $d = (2, 3, \ldots, 10)$, where r is the ratio of the dominating space, d is the number of objectives, respectively. Here, $r1$ corresponds to the minimal ratio of the dominating space (i. e., $(1/2)^d$), $r11$ corresponds to the maximal ratio of the dominating space (i. e., $1/2$). The values of $r = r1, r2, \ldots, r11$ are evenly distributed. It should be noted that the

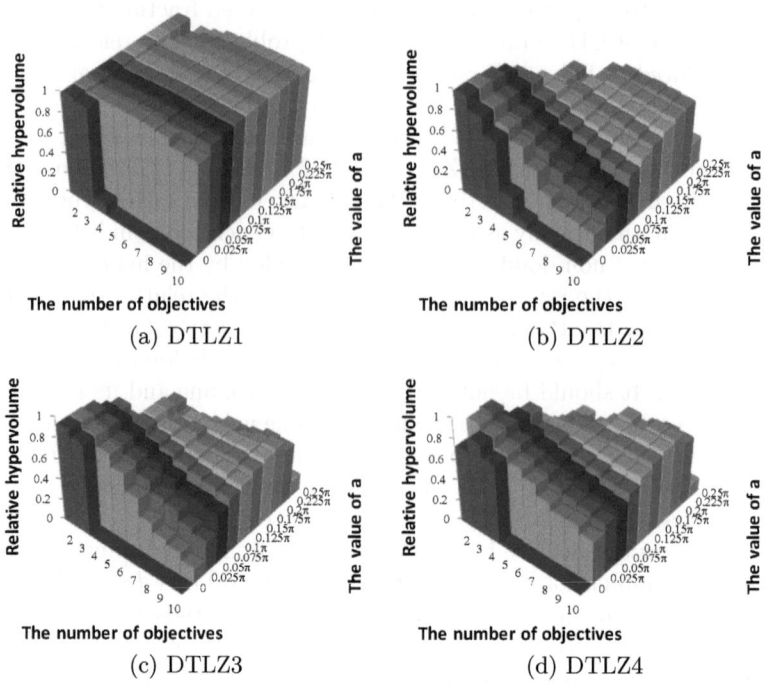

Fig. 4. The relative hypervolume of solutions obtained by CDAS for each benchmark test problem is shown for the value of a and the number of objectives

corresponding parameter S is calculated from (4), which is needed by CDAS to transform objective values (see (3)). We summarize the performance of CDAS on benchmark test problems in Fig. 5. From Fig. 5(a) for DTLZ1, we can say that the relative hypervolume is nearly one for $r = r1, r2, \ldots, r10$ when the number of objectives is two. In the three-dimensional objective space, the relative hypervolume decreases with an increase in the ratio of the dominating space. When we consider the objective space with more than four objectives, CDAS with $r = r1$ (i. e., NSGA-II) does not find any non-dominated solutions which dominate the reference point. We can also see that the relative hypervolume is almost one for $r = r2, r3, \ldots, r11$ when the number of objectives is more than three. On DTLZ2 in Fig. 5(b), when the number of objectives is specified as two to four, $r = r1, r2, r3$ gives good results. In the more than four-dimensional objective space, $r = r1$ yields the worst performance for each objective. On the other hand, $r = r2, r3, \ldots, r10$ gives similar nice results. Here, we should notice that worse results are obtained when the ratio of the dominating space is specified as the maximum. Regarding the performance of CDAS on DTLZ3, the best relative hypervolume is obtained by $r = r1$ in the objective space with two and three objectives in Fig. 5(c). In the more than three-objective space, $r = r1$ gives the worst performance and the optimal ratio of the dominating space is different depending on the number of objectives. With respect to DTLZ4 from Fig. 5(d),

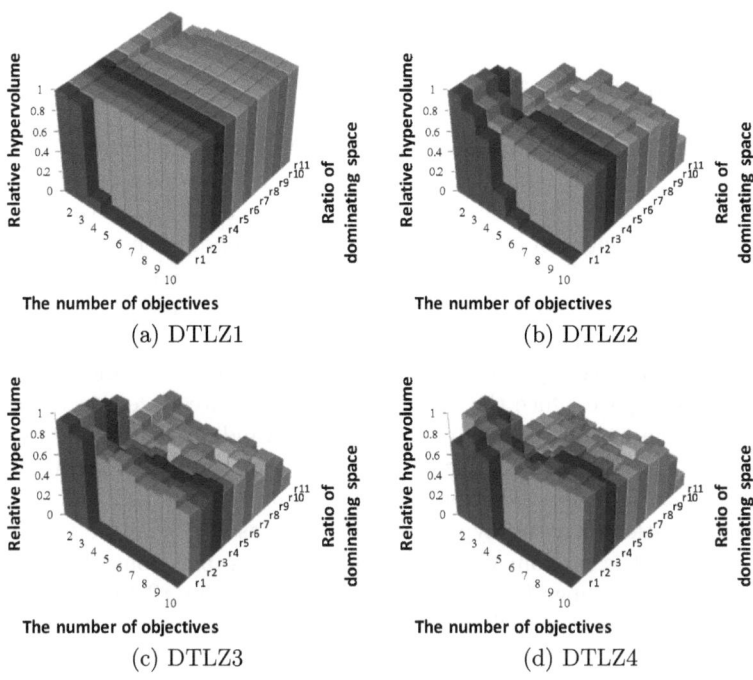

Fig. 5. The relative hypervolume of solutions obtained by CDAS for each benchmark test problem is shown for the ratio of the dominating space and the number of objectives

surprisingly, $r = r3$ gives the best performance when the number of objectives is two. In the three- and four-objective space, $r = r1$ gives the best relative hypervolume. The optimal ratio of the dominating space is different depending on the number of objectives as well as DTLZ3 with more than four objectives. It should be noted that CDAS with $r = r1$ (i. e., NSGA-II) does not work anymore for more than five objectives here too.

From Fig. 4 and Fig. 5, our consideration on the dominance balance is summarized as follows:

- CDAS with $a = 0$ or $r = r1$ (i. e., NSGA-II) performs well for a few objectives whereas it does not work anymore in the higher-dimensional objective space.
- CDAS with $a = \pi/4$ or $r = r11$ (i. e., when the ratio of the dominating space is $1/2$) gives poor results except for DTLZ1.
- In the higher-dimensional objective space, CDAS shows poor results for smaller values of a and nice results for larger values of a excluding the maximum except for DTLZ1.
- In the higher-dimensional objective space, CDAS shows stable and good results for the ratio of the dominating space $r = r2, r3, \ldots, r10$.
- CDAS with $r = r2$ often shows good performance.

4   CDAS-D

4.1   Proposal of CDAS-D

From the previous section, we can see that CDAS with the ratio of the dominating space $r = r2, r3, \ldots, r10$ frequently shows more stable and better performance than that with the value of a. However, we do not know the optimal ratio of the dominating space in advance as it changes depending on the problem and the number of objectives. Therefore we propose a new version of CDAS with the dynamic ratio of the dominating space, which we call CDAS-D in this paper. In CDAS-D, the hypervolume of solutions is calculated at each generation. As long as the hypervolume is improved (i. e., increased), we do not change the ratio of the dominating space. Once the hypervolume deteriorates (i. e., decreases), the ratio of the dominating space is replaced with a random value from a uniform distribution. The idea behind this is that we do not want the ratio to be fixed during optimization. As we do not know the optimal ratio, the random value is tried for the ratio when the hypervolume is decreased. There are two settings for the range of the random value. One is the wide range from $r1$ to $r11$. The other is the narrow range from $r1$ to $r2$. The values of $r1, r2, r11$ are explained in the previous section. The flow chart of CDAS-D is shown in Fig. 6.

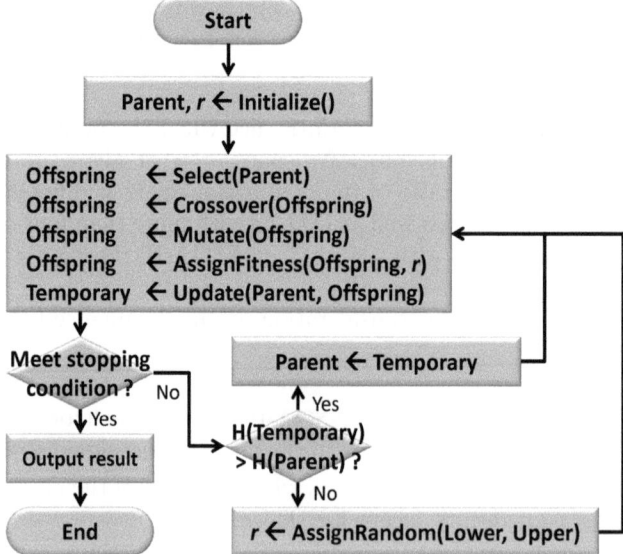

Fig. 6. The flow chart of CDAS-D is illustrated. In the chart, H(S) gives the hypervolume of S, and AssignRandom(Lower, Upper) returns a random value uniformly distributed between Lower and Upper

4.2 Computational Experiments

We apply our CDAS-D to benchmark test problems in the same manner as the previous section. Computational results are summarized in Fig. 7. It should be noted that Fig. 7 is obtained by adding the results of CDAS-D to Fig. 5 for comparison. From Fig. 7 we can see that our CDAS-D is comparable to CDAS with the optimal ratio of the dominating space. It can be also said that CDAS-D with the narrow range shows better performance than that with the wide range. It should be noted that we do not have to specify the ratio of the dominating space in CDAS-D. In order to see a process of optimization, we show the relative hypervolume during optimization in the case of three, six, and nine objectives for CDAS with $r = r1$ (i. e., NSGA-II), $r = r2$, CDAS-D with the wide range (dw), and CDAS-D with the narrow range (dn) on DTLZ1 and DTLZ2 in Fig. 8, on DTLZ3 and DTLZ4 in Fig. 9, respectively. From Figs. 8(a)

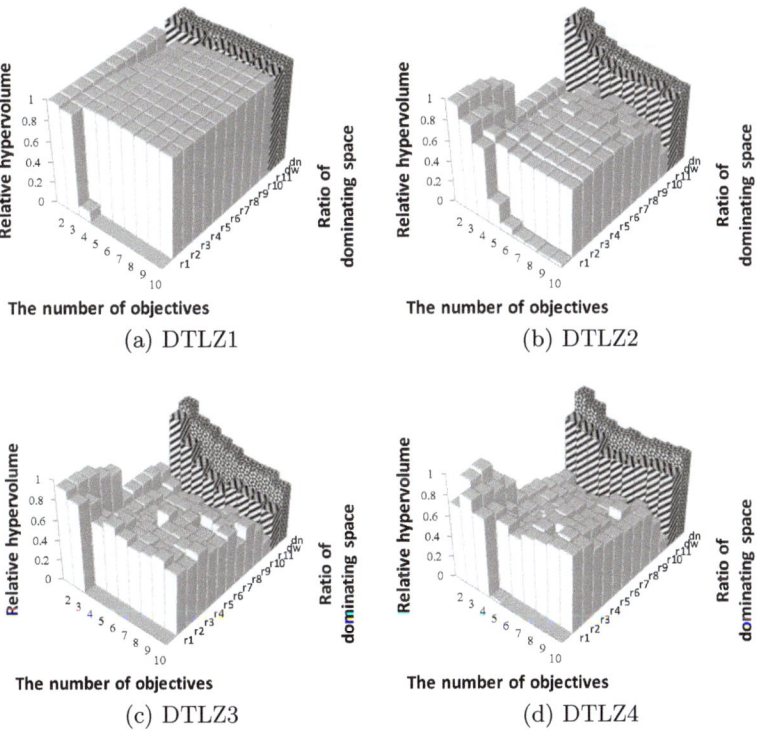

Fig. 7. The relative hypervolume of solutions obtained by CDAS-D for each benchmark test problem is shown for the ratio of the dominating space and the number of objectives. Results of CDAS-D are shown for the wide range (dw) setting by bars with diagonals, for the narrow range (dn) setting by bars with confetti, respectively. The results of CDAS (i. e., Fig. 5) are also included by bars in white.

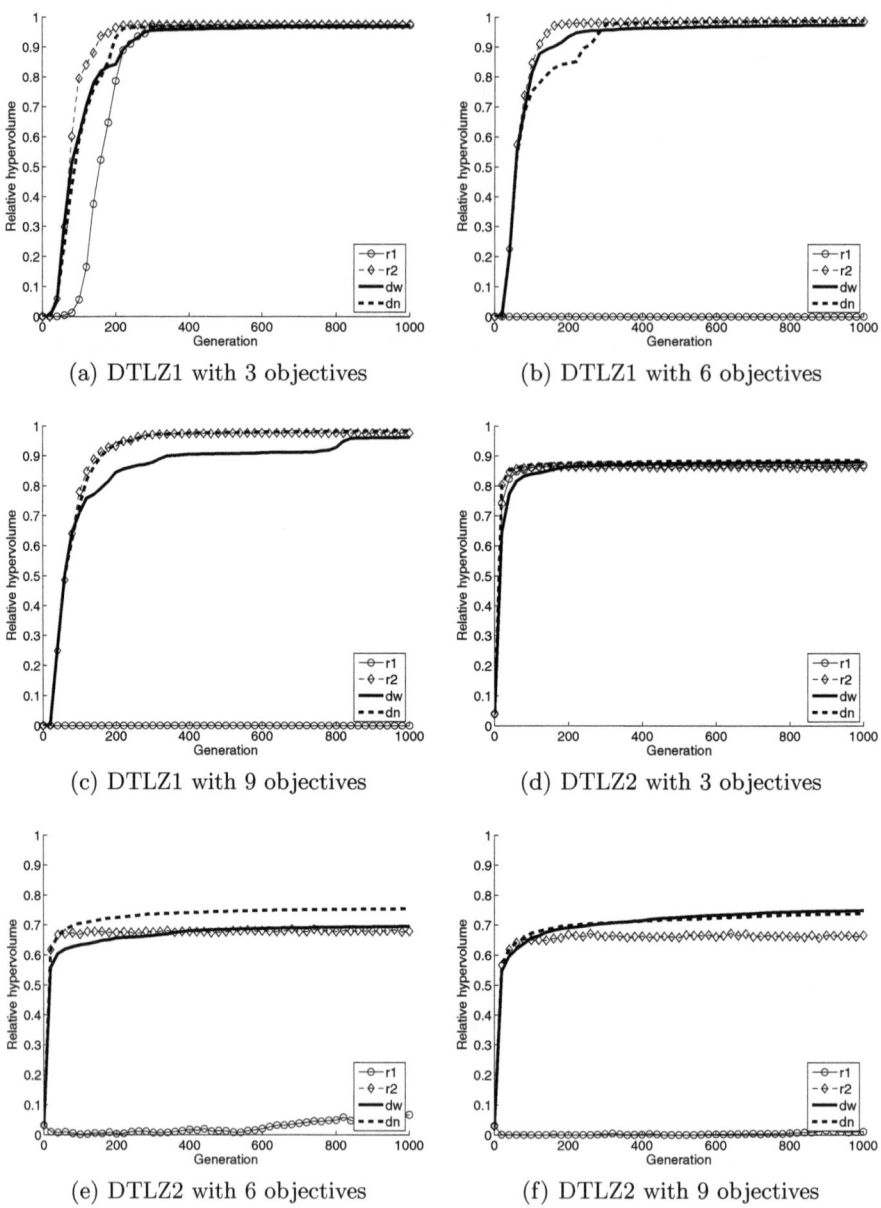

(a) DTLZ1 with 3 objectives   (b) DTLZ1 with 6 objectives

(c) DTLZ1 with 9 objectives   (d) DTLZ2 with 3 objectives

(e) DTLZ2 with 6 objectives   (f) DTLZ2 with 9 objectives

Fig. 8. CDAS-D for DTLZ1 and DTLZ2

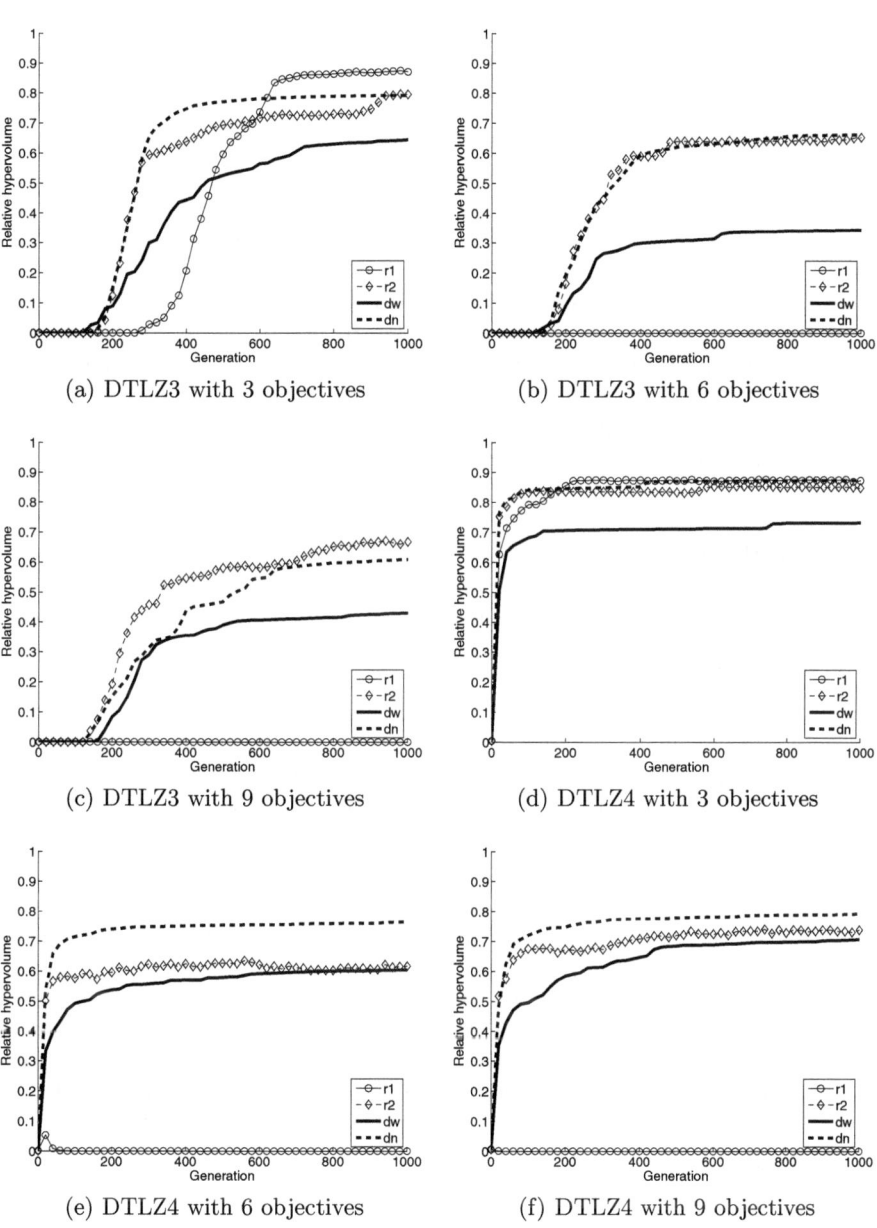

(a) DTLZ3 with 3 objectives

(b) DTLZ3 with 6 objectives

(c) DTLZ3 with 9 objectives

(d) DTLZ4 with 3 objectives

(e) DTLZ4 with 6 objectives

(f) DTLZ4 with 9 objectives

Fig. 9. CDAS-D for DTLZ3 and DTLZ4

to 8(c), we can see that the convergence speed of CDAS with $r = r2$ is very fast for all DTLZ1 problems. On the other hand, CDAS with $r = r1$, CDAS-D with the narrow range, and CDAS-D with the wide range is very slow in the three-, six-, and nine-objective space, respectively. The relative hypervolume at the last generation (i. e., 1000th generation) is almost one for every CDAS and CDAS-D. On DTLZ2 in Figs. 8(d) to 8(f), there is no remarkable difference on the performance among CDAS and CDAS-D in the case of three objectives. However, CDAS-D with the narrow range outperforms CDAS on DTLZ2 with six and nine objectives. Regarding DTLZ3 from Figs. 9(a) to 9(c), CDAS with $r = r2$ performs quite well. We can also say that the performance of CDAS-D with the narrow range is good although CDAS-D with the wide range gives the worst results. From Figs. 9(d) to 9(f) on DTLZ4, CDAS-D with the narrow range performs the best. It is also interesting to point out that CDAS-D with the wide range performs the worst. From Figs. 8 and 9 we can see that CDAS with $r = r1$ (i. e., NSGA-II) does not work anymore in six- and nine-objective space. We can also see that the relative hypervolume of CDAS-D (i. e., dw and dn in Figs. 8 and 9) monotonically increases in all cases.

5 Summary

In this paper we examined the effect of dominance balance on the performance of CDAS for DTLZ1, DTLZ2, DTLZ3, and DTLZ4 benchmark test problems. The dominance balance is defined by the ratio of the dominating space to the objective space. In CDAS, the ratio of the dominating space can be changed by specifying the parameter S which is used to transform objective values. However, it is very difficult for the parameter S to control the dominance balance properly in the higher-dimensional objective space. In order to show this difficulty, we examined the performance of CDAS by changing the value of $a = \pi/2 - S \cdot \pi$ ($0 \leq a \leq \pi/4$) in even steps (see Fig. 4). Then we can see that the optimal value of a that gives the best performance of CDAS increases with an increase in the number of objectives except for DTLZ1. In DTLZ2 with more than five objectives, for example, we can also see that the optimal value of a is large enough but not the maximum in its range (i. e., $a = 0.225\pi$) from Fig. 4(b). This is because the performance of CDAS will be affected by the ratio of the dominating space directly and it is considered to be proportional to the value of a to the $(d-1)$-th power in the d-dimensional objective space. Therefore, next we showed the performance of CDAS by changing the ratio of the dominating space from the minimum to the maximum in even steps directly. The corresponding parameter S is calculated from an equation which we propose in the paper. From our computational experiments in Fig. 4 and Fig. 5, we can see that CDAS with the ratio of the dominating space $r = r2, r3, \ldots, r10$ frequently shows more stable and better performance than that with the value of a in the higher-dimensional objective space. In this paper, we also proposed a new version of CDAS with the dynamic ratio of the dominating space, which we call CDAS-D. In CDAS-D, the hypervolume of solutions is calculated at each generation. As long as the hypervolume is increased, we do not change the ratio of the dominating space. Once the hypervolume is decreased, the ratio of the dominating space is replaced with a random value from a uniform distribution. The idea behind this is that we do not want the ratio to be fixed during optimization. As we do not know the optimal ratio, the random value is tried for the ratio when the hypervolume is decreased. There are two settings for the range of the random value; the wide range and the narrow range. From computational experiments, we can see that our CDAS-D is comparable to CDAS with the optimal ratio of the dominating space. It can be also said that CDAS-D with the narrow range shows better performance than that with the wide range. It should be noted that the ratio of the dominating space in our CDAS-D does not need to be specified by the user in advance and during optimization although in the original CDAS it needs to be fixed properly before optimization. In our CDAS-D, when the hypervolume of solutions is decreased, the ratio of the dominating space is updated with the random value. However, there may be much smarter ways like adaptive approaches to specify the ratio. This issue will be left for one of further works. We should also consider the performance of CDAS-D on real-world applications in the future.

References

1. Beume, N., Naujoks, B., Emmerich, M.: SMS-EMOA: Multiobjective Selection Based on Dominated Hypervolume. European Journal of Operational Research 181, 1653–1669 (2007)
2. Coello, C.A.C., Lamont, G.B.: Applications of Multi-Objective Evolutionary Algorithms. World Scientific, Singapore (2004)
3. Deb, K.: Multi-Objective Optimization Using Evolutionary Algorithms. John Wiley & Sons, Chichester (2001)
4. Deb, K., Pratap, A., Agarwal, S., Meyarivan, T.: A Fast and Elitist Multiobjective Genetic Algorithm: NSGA-II. IEEE Trans. on Evolutionary Computation 6, 182–197 (2002)
5. Deb, K., Thiele, L., Laumanns, M., Zitzler, E.: Scalable Multi-Objective Optimization Test Problems. In: Proc. of 2002 IEEE Congress on Evolutionary Computation, pp. 825–830 (2002)
6. Deb, K., Thiele, L., Laumanns, M., Zitzler, E.: Scalable Test Problems for Evolutionary Multiobjective Optimization. In: Abraham, A., Jain, L.C., Goldberg, R. (eds.) Evolutionary Multiobjective Optimization: Theoretical Advances and Applications, pp. 105–145. Springer, Heidelberg (2005)
7. Grunert da Fonseca, V., Fonseca, C.M.: The Relationship between the Covered Fraction, Completeness and Hypervolume Indicators. In: Hao, J.-K., Legrand, P., Collet, P., Monmarché, N., Lutton, E., Schoenauer, M. (eds.) EA 2011. LNCS, vol. 7401, pp. 25–36. Springer, Heidelberg (2012)
8. Goldberg, D.E.: Genetic Algorithms in Search, Optimization, and Machine Learning. Addison-Wesley, Reading (1989)
9. Ishibuchi, H., Tsukamoto, N., Nojima, Y.: Evolutionary Many-Objective Optimization: A Short Review. In: Proc. of 2008 IEEE Congress on Evolutionary Computation, pp. 2424–2431 (2008)

10. Ishibuchi, H., Tsukamoto, N., Sakane, Y., Nojima, Y.: Indicator-Based Evolutionary Algorithm with Hypervolume Approximation by Achievement Scalarizing Functions. In: Proc. of 2010 Genetic and Evolutionary Computation Conference, pp. 527–534 (2010)
11. Ishibuchi, H., Hitotsuyanagi, Y., Tsukamoto, N., Nojima, Y.: Many-Objective Test Problems to Visually Examine the Behavior of Multiobjective Evolution in a Decision Space. In: Schaefer, R., Cotta, C., Kołodziej, J., Rudolph, G. (eds.) PPSN XI. LNCS, vol. 6239, pp. 91–100. Springer, Heidelberg (2010)
12. Narukawa, K., Rodemann, T.: Examining the Performance of Evolutionary Many-Objective Optimization Algorithms on a Real-World Application. In: Proc. of 2012 International Conference on Genetic and Evolutionary Computing, pp. 316–319 (2012)
13. Sato, H., Aguirre, H.E., Tanaka, K.: Controlling Dominance Area of Solutions and Its Impact on the Performance of MOEAs. In: Obayashi, S., Deb, K., Poloni, C., Hiroyasu, T., Murata, T. (eds.) EMO 2007. LNCS, vol. 4403, pp. 5–20. Springer, Heidelberg (2007)
14. Sato, H., Aguirre, H.E., Tanaka, K.: Pareto Partial Dominance MOEA and Hybrid Archiving Strategy Included CDAS in Many-Objective Optimization. In: Proc. of 2010 IEEE Congress on Evolutionary Computation, pp. 3720–3727 (2010)
15. Sato, H., Aguirre, H.E., Tanaka, K.: Self-Controlling Dominance Area of Solutions in Evolutionary Many-Objective Optimization. In: Deb, K., Bhattacharya, A., Chakraborti, N., Chakroborty, P., Das, S., Dutta, J., Gupta, S.K., Jain, A., Aggarwal, V., Branke, J., Louis, S.J., Tan, K.C. (eds.) SEAL 2010. LNCS, vol. 6457, pp. 455–465. Springer, Heidelberg (2010)
16. Wagner, T., Beume, N., Naujoks, B.: Pareto-, Aggregation-, and Indicator-Based Methods in Many-Objective Optimization. In: Obayashi, S., Deb, K., Poloni, C., Hiroyasu, T., Murata, T. (eds.) EMO 2007. LNCS, vol. 4403, pp. 742–756. Springer, Heidelberg (2007)
17. Zhang, Q., Li, H.: MOEA/D: A Multiobjective Evolutionary Algorithm Based on Decomposition. IEEE Trans. on Evolutionary Computation 11, 712–731 (2007)
18. Zitzler, E., Thiele, L.: Multiobjective Evolutionary Algorithms: A Comparative Case Study and the Strength Pareto Approach. IEEE Trans. on Evolutionary Computation 3, 257–271 (1999)
19. Zitzler, E., Thiele, L., Laumanns, M., Fonseca, C.M., Fonseca, V.G.: Performance Assessment of Multiobjective Optimizers: An Analysis and Review. IEEE Trans. on Evolutionary Computation 7, 117–132 (2003)
20. Zitzler, E., Künzli, S.: Indicator-Based Selection in Multiobjective Search. In: Yao, X., Burke, E.K., Lozano, J.A., Smith, J., Merelo-Guervós, J.J., Bullinaria, J.A., Rowe, J.E., Tiňo, P., Kabán, A., Schwefel, H.-P. (eds.) PPSN VIII. LNCS, vol. 3242, pp. 832–842. Springer, Heidelberg (2004)

An Alternative Preference Relation to Deal with Many-Objective Optimization Problems

Antonio López[2], Carlos A. Coello Coello[1], Akira Oyama[2], and Kozo Fujii[2]

[1] CINVESTAV-IPN* (Evolutionary Computation Group)
Departamento de Computación, México D.F. 07300, México
ccoello@cs.cinvestav.mx

[2] Institute of Space and Astronautical Science, Japan Aerospace Exploration Agency,
Sagamihara 252-5210, Japan
{antonio,oyama,fujii}@flab.isas.jaxa.jp

Abstract. In this paper, we use an alternative preference relation that couples an achievement function and the ϵ-indicator in order to improve the scalability of a Multi-Objective Evolutionary Algorithm (MOEA) in many-objective optimization problems. The resulting algorithm was assessed using the Deb-Thiele-Laumanns-Zitzler (DTLZ) and the Walking-Fish-Group (WFG) test suites. Our experimental results indicate that our proposed approach has a good performance even when using a high number of objectives. Regarding the DTLZ test problems, their main difficulty was found to lie on the presence of dominance resistant solutions. In contrast, the hardness of WFG problems was not found to be significantly increased by adding more objectives.

1 Introduction

Since the first implementation of a Multi-Objective Evolutionary Algorithm (MOEA) in the mid 1980s, a wide variety of new MOEAs have been proposed, gradually improving in both their effectiveness and efficiency to solve Multiobjective Optimization Problems (MOPs) [1]. However, most of these algorithms have been evaluated in problems with only two or three objectives, in spite of the fact that many real-world problems have more than three objectives.

Recently, the Evolutionary Multiobjective Optimization community has devoted important efforts to investigate the performance of MOEAs in problems with a high number of objectives. These MOPs are usually known as Many-objective Optimization Problems (MOPs). One of the first findings in this area [2,3] is that MOEAs based on Pareto optimality scale poorly with respect to the number of objectives. Currently, two main difficulties that make a problem harder when the number of objectives is increased have been suggested:

- Increase of the proportion of nondominated solutions. Since in MOPs almost all solutions are equivalent in terms of Pareto optimality, many researchers

---

*The second author acknowledges support from CONACyT project no. 103570.

R.C. Purshouse et al. (Eds.): EMO 2013, LNCS 7811, pp. 291–306, 2013.

have suggested [4,5,6] that in such problems, the selection of the appropriate individuals for steering the population towards the Pareto optimal set gets more difficult. However, as pointed out by Schütze et al. [7], the increase of the number of nondominated individuals is not a sufficient condition for an increase of the hardness of a problem. They found that in a class of uni-modal problems, the difficulty was marginally increased when more objectives are added.

- Effectiveness of crossover operators. In a combinatorial class of MOPs, Sato et al. [8] observed that solutions in decision variable space become more distant[1] from each other as more objectives are added. As a result, even if two parents close to the Pareto front are recombined, the generated offspring might be far from the Pareto front.

Although not related with the search ability of the MOEA, there are other important difficulties associated with a MOP. For example, the visualization of the Pareto front in high dimensional spaces, or the generation of an accurate sample of the Pareto front, since the required number of points increases exponentially with the number of objectives.

Although the rise of the proportion of incomparable solutions might not significantly determine the difficulty of a MOP *per se*, it seems that the addition of objectives aggravates some particular difficulties observed in the context of 2 or 3 objectives. This is the case of the so called dominance resistant solutions (DRSs) or outliers [9,10,11]. DRSs are non Pareto optimal solutions with a poor value in at least one of the objectives, but with near optimal values in the others. These kinds of solutions represent potential difficulty since the number of DRSs grows as the number of objectives is increased.

In this paper, we propose the use of the recently introduced Chebyshev preference relation [12] in order to improve the scalability of a MOEA in MOPs. That new preference relation divides the objective space in two regions. In the region farther from the ideal point, the solutions are compared using an achievement scalarizing function, whereas in the region near the ideal point, solutions are compared using the usual Pareto dominance. The idea behind this proposal is to increase the selection pressure when the solutions are far from the Pareto front. This way, we have a discriminative criterion to evaluate nondominated solutions.

Additionally, we introduce the idea of coupling the Chebyshev relation with two preference relations based on the ϵ-indicator. These new preference relations show that a straightforward use of the ϵ-indicator produces a good approximation of the Pareto front.

The experiments are concentrated in evaluating the performance of the Chebyshev preference relation and also in the sources of difficulty when the number of objectives is increased. For the experiments we employed 5 problems from the DTLZ test suite, and 2 problems from the WFG test suite.

---

[1] In terms of Hamming distance between binary encoded solutions.

2   Basic Concepts and Notation

This section briefly presents the concepts and notation used throughout the rest of the paper.

2.1   Multiobjective Optimization Problems

Definition 1. *A MOP is defined as:*

$$\text{Minimize} \quad \mathbf{f}(\mathbf{x}) = [f_1(\mathbf{x}), f_2(\mathbf{x}), \dots, f_k(\mathbf{x})]. \tag{1}$$
$$\mathbf{x} \in \mathcal{X}$$

The vector function $\mathbf{f} : \mathcal{X} \rightarrow \mathbb{R}^k$ is composed by $k \geq 2$ *objective functions* $f_i : \mathcal{X} \rightarrow \mathbb{R}$ $(i = 1, \dots, k)$. The image of the *feasible set* $\mathcal{X} \subseteq \mathbb{R}^n$ under the function $\mathbf{f}$ is a subset of the objective function space denoted by $\mathcal{Z} = \mathbf{f}(\mathcal{X})$. The sets $\mathbb{R}^n$ and $\mathbb{R}^k$ are known as *decision variable space* and *objective function space*, respectively.

   In multiobjective optimization, the *Pareto dominance relation* is usually adopted to compare vectors in $\mathbb{R}^k$.

Definition 2. *A vector $\mathbf{z}^1 \in \mathbb{R}^k$ is said to dominate vector $\mathbf{z}^2 \in \mathbb{R}^k$ (denoted $\mathbf{z}^1 \prec_{\text{par}} \mathbf{z}^2$) if and only if: $z_i^1 \leq z_i^2$ $(i = 1, \dots, k)$, and $\mathbf{z}^1 \neq \mathbf{z}^2$.*

Definition 3. *A solution $\mathbf{x}^* \in \mathcal{X}$ is Pareto optimal if there is no solution $\mathbf{x} \in \mathcal{X}$ such that $\mathbf{f}(\mathbf{x}) \prec_{\text{par}} \mathbf{f}(\mathbf{x}^*)$.*

Definition 4. *The Pareto optimal set, P_{opt}, is composed by all the Pareto optimal solutions.*

Definition 5. *The image of P_{opt} under the vector function $\mathbf{f}(\mathbf{x})$ is called the Pareto optimal front and is denoted by PF_{opt}.*

In practice, the goal of a MOEA is finding the best approximation set of the Pareto optimal front. We denote an approximation set by PF_{apx}. Currently, it is well accepted that the quality of an approximation set is determined by the closeness to the Pareto optimal front, and the spread over the entire Pareto optimal front.

   In some cases it is useful to know the lower and upper bounds of the Pareto front. The *ideal point* represents the lower bound and is defined by the point $z_i^\star = \min_{\mathbf{z} \in \mathcal{Z}}(z_i)$ for all $i = 1, \dots, k$. In turn, the upper bound is defined by the *nadir point*, which is given by $z_i^{\text{nad}} = \max_{\mathbf{z} \in PF_{\text{opt}}}(z_i)$ for all $i = 1, \dots, k$.

2.2   Achievement Scalarizing Functions

The preference relation adopted in this paper is based on the achievement scalarizing function approach proposed by Wierzbicki [13].

Definition 6. *An achievement (scalarizing) function is a parameterized function* $s(\mathbf{z}|\mathbf{z}^{\mathrm{ref}}) : \mathbb{R}^k \to \mathbb{R}$, *where* $\mathbf{z}^{\mathrm{ref}} \in \mathbb{R}^k$ *is a reference point representing the desired aspiration levels.*

The augmented Chebyshev achievement function [14] is one of the most common achievement functions.

Definition 7. *The augmented Chebyshev achievement function is defined by*

$$s_\infty(\mathbf{z}|\mathbf{z}^{\mathrm{ref}}) = \max_{i=1,\ldots,k}\{\lambda_i(z_i - z_i^{\mathrm{ref}})\} + \rho\sum_{i=1}^{k}\lambda_i(z_i - z_i^{\mathrm{ref}}), \tag{2}$$

where $\mathbf{z}^{\mathrm{ref}}$ *is a reference point,* $\boldsymbol{\lambda} = [\lambda_1,\ldots,\lambda_k]$ *is a vector of weights such that* $\forall i\ \lambda_i \geq 0$ *and, for at least one* i, $\lambda_i > 0$, *and* $\rho > 0$ *is a sufficiently small augmentation coefficient.*

3   Related Work

In the current literature, some alternative preference relations have been used to deal with MOPs. However, the optimal solution set induced by these preference relations is a subset of PF_{opt}. As a consequence, when one of these preference relations is applied, for example, on the current population of a MOEA, the optimal solutions regarding the alternative preference relation would belong to a portion of PF_{opt}. Thus, some parts of the Pareto front will not be generated.

Among the alternative preference relations that have been proposed we can find the following. The Average Ranking and Maximum Ranking relations [15] which have the drawback of favoring extreme solutions. These preference relations have been used in [16] to deal with MOPs. Drechsler et al. [17] proposed the *favour relation* which also emphasizes extreme solutions.

The Preference Order Relation, developed by di Pierro [18], compares two solutions by discarding objectives until one of them dominates the other. The disadvantage of this approach is its high computational cost.

Sato et al. [19] proposed a preference relation to control the dominance area of a solution. This relation emphasizes solutions in the middle region of the Pareto front.

4   Solving MOPs Using an Alternative Preference Relation

In this section we first present the Chebyshev preference relation introduced in [12] and we describe how to use this relation to approximate the entire Pareto front.

4.1   The Chebyshev Preference Relation

The Chebyshev preference relation combines the Pareto dominance relation and an achievement function to compare solutions in objective function space. First, this relation defines a Region of Interest (RoI) with respect to a given reference point. This region contains all solutions with an achievement value $s_\infty(\mathbf{z}|\mathbf{z}^{\text{ref}}) \leq s^{\text{min}} + \delta$, where $s^{\text{min}} = \min_{\mathbf{z} \in \mathcal{Z}} s_\infty(\mathbf{z}|\mathbf{z}^{\text{ref}})$, and δ is a threshold that determines the size of the RoI. Fig. 1 shows the RoI defined by means of the achievement function. Solutions in this region are compared using the usual Pareto dominance relation, while solutions outside of the RoI are compared using their achievement function value.

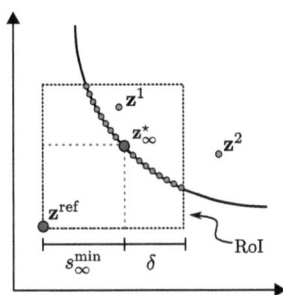

Fig. 1. Nondominated solutions with respect to the Chebyshev relation

The Chebyshev preference relation is formally defined as follows:

Definition 8. *A solution $\mathbf{z}^1$ is preferred to solution $\mathbf{z}^2$ with respect to the Chebyshev relation ($\mathbf{z}^1 \prec_{\text{ch}} \mathbf{z}^2$), if and only if:*

1. $s_\infty(\mathbf{z}^1|\mathbf{z}^{\text{ref}}) < s_\infty(\mathbf{z}^2|\mathbf{z}^{\text{ref}}) \wedge \{\mathbf{z}^1 \notin R(\mathbf{z}^{\text{ref}}, \delta) \vee \mathbf{z}^2 \notin R(\mathbf{z}^{\text{ref}}, \delta)\}$, *or,*

2. $\mathbf{z}^1 \preceq_{\text{par}} \mathbf{z}^2 \wedge \{\mathbf{z}^1, \mathbf{z}^2 \in R(\mathbf{z}^{\text{ref}}, \delta)\}$,

where $R(\mathbf{z}^{\text{ref}}, \delta) = \{\mathbf{z} : s_\infty(\mathbf{z}|\mathbf{z}^{\text{ref}}) \leq s^{\text{min}} + \delta\}$ is the Region of Interest with respect to a given reference point $\mathbf{z}^{\text{ref}}$.

As an illustration of the preference relation, consider solutions $\mathbf{z}^1$ and $\mathbf{z}^2$ presented in Fig. 1. Since $\mathbf{z}^2 \notin R(\mathbf{z}^{\text{ref}}, \delta)$ and $s_\infty(\mathbf{z}^1, \mathbf{z}^{\text{ref}}) < s_\infty(\mathbf{z}^2, \mathbf{z}^{\text{ref}})$, then $\mathbf{z}^1 \prec_{\text{ch}} \mathbf{z}^2$.

Since, in general, the objective ranges of PF_{opt} might be different, the weight vector $\boldsymbol{\lambda}$ (Eq. 2) is used for normalizing each objective function. The weights are set as $\lambda_i = 1/(z_i^{\text{nad}} - z_i^\star)$, for all $i = 1, \ldots, k$. As the ideal and nadir points are not usually known in advance, these values are approximated using the current PF_{apx}. In order to approximate these bounding points, the Chebyshev relation always considers extreme solutions as nondominated in order to keep them in the population. This way, the approximation of the bounding points can be improved during the course of the search. To approximate $\mathbf{z}^\star$, the following set must be updated at each generation: $\Phi = \{\mathbf{z}^1, \ldots, \mathbf{z}^k \mid \mathbf{z}^i = \arg\min_{\mathbf{z} \in PF_{\mathrm{apx}}}(z_i)\}$. That is, the solutions having the best value for each objective. The approximation of the ideal point is then $\dot{\mathbf{z}}^\star = \{z_i^i, \ldots, z_k^k\}$ with $\mathbf{z}^i \in \Phi$. Similarly, to approximate $\mathbf{z}^{\mathrm{nad}}$, the following set is computed: $\Theta = \{\mathbf{z}^1, \ldots, \mathbf{z}^k \mid \mathbf{z}^i = \arg\max_{\mathbf{z} \in PF_{\mathrm{apx}}}(z_i)\}$. Thus, the normalized Chebyshev relation is defined by:

Definition 9. *A solution $\mathbf{z}^1$ is preferred to $\mathbf{z}^2$ with respect to the normalized Chebyshev preference relation ($\mathbf{z}^1 \prec_{\mathrm{n\text{-}ch}} \mathbf{z}^2$) if and only if: $\mathbf{z}^1 \prec_{\mathrm{ch}} \mathbf{z}^2$, and $\mathbf{z}^2 \notin \{\Phi \cup \Theta\}$.*

Additionally, the threshold δ can be normalized using the current range of the achievement function. Thus, the user can provide a normalized $\delta' \in [0, 1]$, and the actual value used for computing the Chebyshev relation is $\delta = \delta' \cdot (s^{\mathrm{max}} - s^{\mathrm{min}})$, where $s^{\mathrm{max}} = \max_{\mathbf{z} \in PF_{\mathrm{apx}}} s_\infty(\mathbf{z}|\mathbf{z}^{\mathrm{ref}})$ and $s^{\mathrm{min}} = \min_{\mathbf{z} \in PF_{\mathrm{apx}}} s_\infty(\mathbf{z}|\mathbf{z}^{\mathrm{ref}})$.

In order to incorporate the (normalized) Chebyshev relation into a MOEA we only have to replace the usual Pareto dominance checking procedure by the procedure that implements the new relation.

4.2 Using the Chebyshev Relation to Approximate the Entire Pareto Front in Many-Objective Problems

Although the Chebyshev relation was proposed to guide the search towards a subset of PF_{opt}, in this section we propose the use of this relation to approximate the entire range of the Pareto front.

As previously mentioned, the Chebyshev relation ranks solutions outside the region of interest using the achievement function. This way it can help to rank solutions considered as incomparable by the Pareto dominance relation. In order to approximate the entire Pareto front we used as reference point the approximation of the ideal point maintained by the Chebyshev relation. In addition, we adopted a threshold $\delta' = 0.9$, comparing this way most of the solutions using Pareto dominance, while solutions far from the current PF_{apx} will be compared using their achievement function value. The basic idea is to use a stringent criterion for solutions far from the Pareto front for guiding the solutions towards the ideal point, and when the solutions are near to the Pareto front, then we use Pareto dominance to cover the entire Pareto front.

Furthermore, since the Chebyshev relation preserves the vectors that generate the approximations of $\mathbf{z}^\star$ and $\mathbf{z}^{\mathrm{nad}}$ in the current population, the extreme solutions of the Pareto front will be found.

Additionally, since the relation used inside RoI is not essential for the mechanism of the Chebyshev relation, a different preference relation can be used as the second criteria. In this paper we investigate the performance of two preference relations derived from the additive ϵ-indicator [20]:

$$I_\epsilon(A, B) = \inf_{\epsilon \in \mathbb{R}} \{\forall \mathbf{z}^2 \in B \; \exists \mathbf{z}^1 \in A : z_i^1 \leq \epsilon + z_i^2 \; \text{ for } i = 1, \ldots, k\},$$

where A and B are two nondominated sets. In other words, $I_\epsilon(A, B)$ is the minimum ϵ value such that added to any vector in B, then $A \preceq B$. As shown in [21], the ϵ-indicator is dominance preserving since if $\mathbf{z}^1 \prec_{par} \mathbf{z}^2$, then $I_\epsilon(\{\mathbf{z}^1\}, \{\mathbf{z}^2\}) < I_\epsilon(\{\mathbf{z}^2\}, \{\mathbf{z}^1\})$.

In order to use the information provided by the ϵ-indicator we need to define a function for measuring the performance of a solution $\mathbf{z}^1 \in P$ with respect to the members in the population P. In this paper we have adopted two functions for this purpose. The first function uses the minimum value of $I_\epsilon(\{\mathbf{z}^2\}, \{\mathbf{z}^1\})$ among every $\mathbf{z}^2$ in the current population. That is, $F_\epsilon^{min}(\mathbf{z}^1) = \min_{\mathbf{z}^2 \in P\backslash\{\mathbf{z}^1\}} I_\epsilon(\{\mathbf{z}^2\}, \{\mathbf{z}^1\})$. This function is also known as *maximin* fitness function[2] [22].

The second fitness function was proposed by Zitzler and Künzli [21] and it is defined by $F_\epsilon^{sum}(\mathbf{z}^1) = \sum_{\mathbf{z}^2 \in P\backslash\{\mathbf{z}^1\}} -\exp(-I_\epsilon(\{\mathbf{z}^2\}, \{\mathbf{z}^1\})/(c\cdot\kappa)$, where c is a normalizing factor given by $c = \max_{\mathbf{z}^1, \mathbf{z}^2 \in P} |I_\epsilon(\{\mathbf{z}^2\}, \{\mathbf{z}^1\})|$, and κ is a scaling factor that regulates the influence of the dominating solutions over dominated ones. In our computations we used $\kappa = 0.05$ since this value yielded good results in [21].

Using these different fitness functions we can define appropriate preference relations in order to integrate them into the Chebyshev preference relation.

Definition 10. *A solution $\mathbf{z}^1$ is preferred to solution $\mathbf{z}^2$ with respect to the I_ϵ^{sum}-relation ($\mathbf{z}^1 \prec_\epsilon^{sum} \mathbf{z}^2$), if and only if: $F_\epsilon^{sum}(\mathbf{z}^1) > F_\epsilon^{sum}(\mathbf{z}^2)$.*

Definition 11. *A solution $\mathbf{z}^1$ is preferred to solution $\mathbf{z}^2$ with respect to the I_ϵ^{min}-relation ($\mathbf{z}^1 \prec_\epsilon^{min} \mathbf{z}^2$), if and only if: $F_\epsilon^{min}(\mathbf{z}^1) > F_\epsilon^{min}(\mathbf{z}^2)$.*

5   Experimental Evaluation and Analysis

In this section, we analyze the Chebyshev relation coupled with each preference relation derived from the ϵ-indicator, i.e., solutions outside the RoI are compared using their achievement value, while solutions inside the RoI are compared employing the relations I_ϵ^{min} or I_ϵ^{sum}, respectively.

5.1   Algorithms and Parameter Settings

The experiments presented in this section were designed with two goals in mind. First, to investigate whether the Chebyshev relation is able to improve the scalability of Nondominated Sorting Genetic Algorithm II (NSGA-II) w.r.t. the number of objectives. Secondly, to analyze the effect of DRSs on the performance of Pareto-based MOEAs.

For the first goal, we compare the performances of NSGA-II using three different preference relations, namely: usual Pareto dominance, Chebyshev relation with I_ϵ^{sum}, and Chebyshev relation with I_ϵ^{min}. We evaluated the cases with 3, 4, 6, 8, 10, 12 and 14 objectives.

---

[2] Since the maximin fitness is to be minimized, the value $-F_\epsilon^{min}(\mathbf{z}^1)$ is used instead.

The Chebyshev relation relies on two key elements: the evaluation of the solutions far from the Pareto front using the achievement function, and the approximation of the ideal point and the nadir point. Therefore, the selection of the test problems was made in order to evaluate whether the pressure selection biased towards solutions near the ideal point might lead to premature convergence in problems with several local Pareto fronts. Besides, we want to test the quality of the approximation of the bounding points in problems with disconnected Pareto fronts and different objective ranges.

We adopted 7 test problems presented in Table 1 taken from the DTLZ [10], and WFG [11] test suites . The variables of these problems are divided in position-related and distance-related parameters.

For the second goal of the experiments we kept the same number of distance-related variables for any number of objectives in order to isolate the effect of the number of objectives, namely, $k - 1$ position-related variables and we fixed the number of distance-related variables to 5 for DTLZ1, and for the other test problems to 20. Similarly, we carried out the same number of function evaluations in every problem in order to observe variations in performance when more objectives are added. In Table 2, we can see the standard parameter values used for NSGA-II. For all the configurations we carried out 30 runs for each MOEA. The results presented were averaged over the total of this number of runs.

Table 1. Adopted MOPs

Problem	Features
DTLZ1, DTLZ3	Multiple local Pareto fronts.
DTLZ4	Nonuniform solution density.
DTLZ7, WFG2	Disconnected PF_{opt}.
WFG6	Nonseparable MOP.

Table 2. NSGA-II parameters

Parameter	Value
Population size	200
Generations	200
Crossover rate	0.9
Mutation rate	$1/n$
Crossover index	20
Mutation index	20

Another reason for our selection of MOPs is that the generational distance (GD) can be computed without the need of having a discrete representation of the Pareto optimal front. For these problems we took advantage of their geometrical shape or their known Pareto optimal set.

For computing GD for DTLZ1 we used $GD = (\|\mathbf{z}\|/|P|) - 0.5$ since its Pareto front is a hyperplane that intersects each axis in 0.5, while for DTLZ2, DTLZ3 and DTLZ4 we used $GD = (\|\mathbf{z}\|_2/|P|) - 1$ since its Pareto front is a sphere of radius 1. In DTLZ7, we used the value of the auxiliary function $g(\mathbf{x}) \geq 1$ (see [10] for details). The Pareto optimal front of DTLZ7 is achieved when $g(\mathbf{x}) = 1$. Thus, we use this function to compute a variant of GD, defined by $GD_g = g(\mathbf{x}) - 1$. Since the optimal solutions of WFG2 and WFG6 are those for which the distance-related variables are equal to 0.35, we adopted another variant of GD, denoted by $GD_{\mathbf{x}}$, which measures distance in decision variable space. For the sake of clarity, in the following discussion we refer to all these variants just as GD.

Additionally, to evaluate distribution we employed the inverted generational distance (IGD). As reference set, we used the nondominated set resulting from the union of all the PF_{apx} sets generated in the experiments for each problem.

In order to directly compare the performance of the MOEAs we used the additive ϵ-indicator previously presented. Roughly speaking, A is better than B if $I_\epsilon(A, B) < I_\epsilon(B, A)$.

5.2   Discussion of the Results

From observing the GD values obtained (Table 3 and Fig. 2) we can confirm that the convergence ability of the original NSGA-II deteriorates as the number of objectives is increased. In contrast, when the Chebyshev relation is employed the performance is degraded by some small degree. In particular, the performance achieved by using I_ϵ^{sum}-relation or I_ϵ^{min}-relation is very similar in most of the test problems. Only on DTLZ1 (Fig. 2) we can see that I_ϵ^{min}-relation achieved a bad GD on some objectives. This suggests that I_ϵ^{min}-relation can lead to get stuck in local Pareto fronts in some runs. The results obtained using the ϵ-indicator confirm that the performance of NSGA-II is greatly improved by introducing the Chebyshev relation (see Fig. 5 for problem DTLZ1). Although not shown here, the results for the other DTLZ problems showed a tendency similar to that of DTLZ1. Specifically, in all the DTLZ problems we observed that $I_\epsilon(\text{NSGA2-}I_\epsilon, \text{NSGA2}) < I_\epsilon(\text{NSGA2}, \text{NSGA2-}I_\epsilon)$.

With respect to the distribution, the results of IGD suggest that the Chebyshev relation was able to cover the full range of the Pareto front in all the test problems considered in this paper. In Table 4 we show a representative selection of the obtained results.

The results obtained in problems WFG2 and WFG6 deserve a more detailed analysis since according to the ϵ-indicator (Fig. 5), the incorporation of the Chebyshev relation yielded a small improvement for NSGA-II. However, by inspecting the GD values of the WFG problems (Table 3 and right plot of Fig. 2), NSGA-II's performance is not as remarkably deteriorated as we observed in the DTLZ problems, especially in problem WFG2.

By analyzing some plots and performance indicator results we hinted that the divergence problems of the Pareto-based MOEAs when the number of objectives increases was due to the so-called DRSs. Fig. 3 shows an example of DRSs generated by NSGA-II in problem DTLZ3. Although the pointed DRSs in the figure have poor values in objective f_3, for example, they are nondominated solutions because they have values close to zero in objectives f_1 and f_2.

As Figs. 3 and 4 suggest, an important source of the scalability issues observed in the DTLZ test problems might be due to the generation of DRSs. Other DTLZ test problems not included in this paper have a similar feasible search space to that of DTLZ2 or DTLZ3. Therefore, we can expect that other DTLZ test problems will also have DRSs.

In order to evaluate in a quantitative manner the effect of DRSs in problems DTLZ and WFG we suggest using the distribution of the *maximum tradeoff* of the solutions of PF_{apx}. We define the maximum tradeoff of solution **z** as

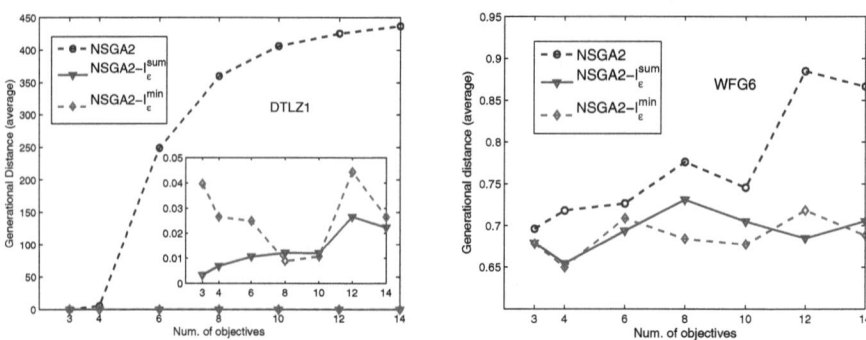

Fig. 2. *GD* values for DTLZ1 and WFG6 varying the number of objectives from 3 to 14

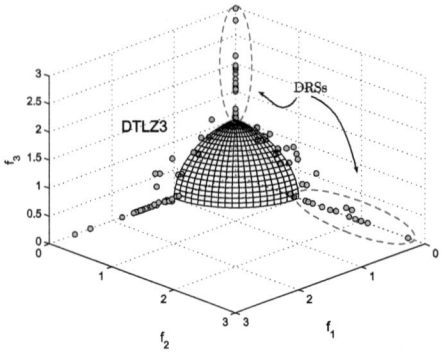

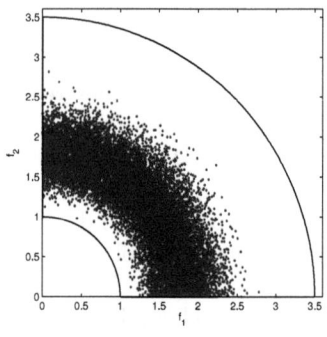

Fig. 3. Illustration of dominance resistant solutions in DTLZ3 using NSGA-II (objs. values are divided by 20 to observe the distribution wrt the Pareto front)

Fig. 4. Feasible objective function space of 2-objective DTLZ2 and 20 000 solutions generated at random

$\Lambda^{\max}(\mathbf{z}) = \max_{i=1,...,k}(z_i)/(\min_{i=1,...,k}(z_i)+1)$. By using this value, DRSs would receive a very large $\Lambda^{\max}$ value since they have in at least one objective a small value and in at least another objective a large value. It is worth noting that solutions far from the Pareto front but located in the middle region of the objective space would not obtain a large $\Lambda^{\max}$, since they have large values in all the objectives.

For computing this measure, we used the exact $\mathbf{z}^\star$ and $\mathbf{z}^{\mathrm{nad}}$ points for normalizing the achieved PF_{apx} by each MOEA. Therefore, for every MOP, we have that $\max_{\mathbf{z} \in PF_{\mathrm{opt}}}\{\Lambda^{\max}(\mathbf{z})\} = 1$. Any solution with $\Lambda^{\max} > 1$ is a potential DRS.

In Figs. 6–8 we show the distribution of $\Lambda^{\max}$ for the solutions generated by NSGA-II and NSGA-II with $I_\epsilon^{\mathrm{sum}}$-relation. For DTLZ1 (Fig. 6) we can clearly see that the proportion of DRSs not removed by NSGA-II is very high when the number of objectives is large. In contrast, by using the Chebyshev relation almost all DRSs are eliminated from the population even for 12 objectives. In the case of DTLZ2 (Fig. 7) the effect of the number of objectives on NSGA-II is more clear since the number of DRSs drastically increases with the number of objectives. WFG6 is an interesting test problem (Fig. 8), since in this case, regardless of the number of objectives, DRSs are not maintained by NSGA-II.

NSGA-II is specially sensitive to DRSs since they are spread in a very large space and, therefore, their crowding distance is larger compared to that of solutions nearby the Pareto front. As a consequence, DRSs are preferred over good solutions to compose the next generation.

On the other hand, when the Chebyshev relation is used, solutions far from the Pareto front are compared using the achievement function value. Thus, although DRSs are equally ranked by the Pareto relation, the Chebyshev relation ranks DRSs worse than other nondominated solutions located nearby the Pareto front. As a result, as it was shown in the experiments using DTLZ test problems, the Chebyshev relation can effectively discard dominance resistant solutions. Finally, the results suggest that WFG2 and WFG6 do not induce the rise of dominance resistant solutions.

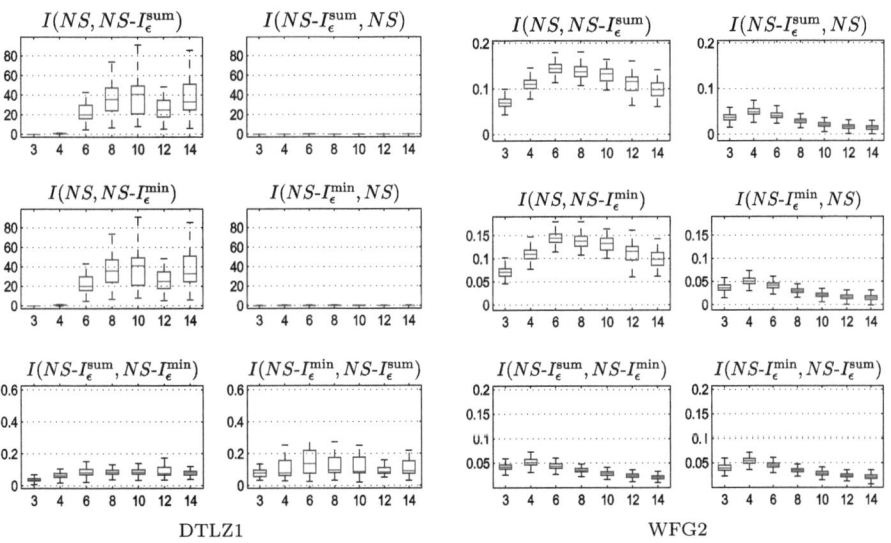

Fig. 5. Results of the ϵ-indicator for DTLZ1 and WFG2. Each subplot presents the values for 3 to 14 objectives (NS is the short for NSGA-II). Hint: A is better than B if $I_\epsilon(A, B) < I_\epsilon(B, A)$.

Table 3. Results of GD for 3 to 14 objectives. The first line for each MOEA is the mean of GD (best values are shown in bold type) and the second line, the standard deviation.

MOP	MOEA	3	4	6	8	10	12	14
DTLZ1	NSGA-II	0.0095	5.2103	248.836	359.990	406.337	425.471	436.947
		0.0058	3.3177	20.6124	17.4954	16.5937	13.9222	12.5937
	NSGA-II-I_ϵ^{sum}	**0.0034**	**0.0069**	**0.0106**	0.0122	0.0120	**0.0266**	**0.0223**
		0.0018	0.0028	0.0090	0.0121	0.0107	0.0925	0.0742
	NSGA-II-I_ϵ^{min}	0.1397	0.0265	0.0249	**0.0088**	**0.0107**	0.0444	0.0264
		0.4940	0.0897	0.0890	0.0079	0.0087	0.1201	0.0898
DTLZ2	NSGA-II	**0.0085**	**0.0290**	0.7404	3.3684	3.9733	4.1150	4.1956
		0.0008	0.0030	0.1456	0.2080	0.0965	0.0832	0.0657
	NSGA-II-I_ϵ^{sum}	0.0104	**0.0275**	0.0500	**0.0575**	**0.0661**	0.0735	0.0778
		0.0011	0.0026	0.0044	0.0063	0.0067	0.0094	0.0078
	NSGA-II-I_ϵ^{min}	0.0105	0.0276	**0.0482**	0.0577	0.0671	**0.0722**	**0.0759**
		0.0010	0.0027	0.0063	0.0064	0.0061	0.0082	0.0097
DTLZ3	NSGA-II	39.41	193.98	1436.66	2557.56	3058.35	3310.95	3409.70
		11.5519	29.1555	117.9495	146.3092	76.1268	68.8680	57.1938
	NSGA-II-I_ϵ^{sum}	**10.9479**	**11.0799**	**14.5876**	**17.8609**	**23.6319**	**28.3081**	29.1092
		3.4523	3.8608	4.8746	6.6429	7.9697	8.2998	12.3618
	NSGA-II-I_ϵ^{min}	55.0847	70.4200	17.1429	19.5645	24.7109	28.4519	**28.5141**
		26.6606	16.3303	4.5782	7.2080	7.9627	8.2112	9.3669
DTLZ4	NSGA-II	0.0063	0.0257	1.4083	3.9525	4.2261	4.3055	4.3426
		0.0031	0.0075	0.3612	0.1322	0.0687	0.0717	0.0562
	NSGA-II-I_ϵ^{sum}	**0.0041**	**0.0120**	0.0305	0.0337	0.0416	0.0440	0.0491
		0.0044	0.0099	0.0086	0.0073	0.0066	0.0058	0.0078
	NSGA-II-I_ϵ^{min}	0.0060	0.0190	**0.0230**	**0.0309**	**0.0377**	**0.0434**	**0.0464**
		0.0046	0.0082	0.0112	0.0081	0.0056	0.0077	0.0091
DTLZ7	NSGA-II	0.0145	0.0534	0.2565	0.8974	1.7709	2.3915	2.7311
		0.0018	0.0042	0.0297	0.1338	0.1491	0.1747	0.1816
	NSGA-II-I_ϵ^{sum}	**0.0094**	**0.0198**	0.0343	**0.0428**	0.0506	0.0694	**0.0718**
		0.0009	0.0017	0.0034	0.0033	0.0050	0.0062	0.0086
	NSGA-II-I_ϵ^{min}	0.0099	**0.0198**	**0.0332**	0.0433	**0.0484**	**0.0680**	0.0748
		0.0011	0.0023	0.0032	0.0042	0.0040	0.0078	0.0079
WFG2	NSGA-II	0.0524	0.0722	0.0999	0.1192	0.1351	0.1353	0.1280
		0.0307	0.0178	0.0157	0.0128	0.0254	0.0322	0.0278
	NSGA-II-I_ϵ^{sum}	0.0374	**0.0564**	**0.0605**	**0.0509**	**0.0490**	**0.0451**	**0.0469**
		0.0065	0.0079	0.0130	0.0124	0.0158	0.0128	0.0135
	NSGA-II-I_ϵ^{min}	**0.0372**	0.0578	0.0634	0.0514	**0.0448**	**0.0431**	0.0471
		0.0066	0.0084	0.0132	0.0101	0.0095	0.0126	0.0118
WFG6	NSGA-II	0.6961	0.7180	0.7265	0.7761	0.7454	0.8848	0.8664
		0.1967	0.1534	0.1686	0.1428	0.1921	0.1478	0.1510
	NSGA-II-I_ϵ^{sum}	0.6793	0.6545	**0.6939**	0.7310	0.7048	**0.6851**	0.7055
		0.1895	0.1457	0.1415	0.1176	0.1094	0.1157	0.1471
	NSGA-II-I_ϵ^{min}	**0.6780**	**0.6498**	0.7085	**0.6842**	**0.6774**	0.7183	**0.6884**
		0.1405	0.1318	0.1229	0.1186	0.1331	0.1218	0.1253

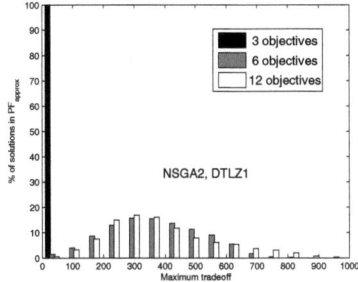 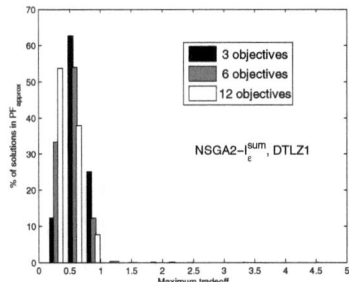

Fig. 6. Maximum tradeoff distribution for NSGA-II and NSGA-II-I_ϵ^{sum} in DTLZ1.

Table 4. Results of *IGD* for 3 to 14 objectives. The first line for each MOEA is the mean of *IGD* (best values are shown in bold type) and the second line, the standard deviation.

MOP	MOEA	3	4	6	8	10	12	14
	NSGA-II	0.0151	0.0163	0.0239	0.0380	0.0469	0.0421	0.0443
		0.0010	0.0010	0.0030	0.0077	0.0113	0.0111	0.0104
DTLZ2	NSGA-II-I_ϵ^{sum}	**0.0079**	0.0105	0.0132	**0.0148**	0.0183	0.0230	0.0331
		0.0025	0.0036	0.0047	0.0043	0.0091	0.0112	0.0185
	NSGA-II-I_ϵ^{min}	0.0087	**0.0104**	**0.0121**	0.0159	**0.0165**	**0.0189**	**0.0220**
		0.0027	0.0039	0.0028	0.0060	0.0070	0.0071	0.0099
	NSGA-II	0.0896	0.0759	0.3300	2.3732	12.7540	23.0939	38.0094
		0.2740	0.0071	0.0830	0.6462	4.8488	7.9811	9.5044
DTLZ7	NSGA-II-I_ϵ^{sum}	**0.0138**	**0.0295**	**0.2226**	1.2982	**4.9511**	**11.3958**	**9.1685**
		0.0029	0.0038	0.0083	0.0323	0.3701	0.1847	0.3093
	NSGA-II-I_ϵ^{min}	0.0150	0.0304	0.2229	**1.2806**	4.9978	11.4871	9.2333
		0.0061	0.0053	0.0073	0.0250	0.2918	0.1453	0.1852
	NSGA-II	0.0075	0.0081	0.0120	0.0146	0.0187	0.0230	0.0271
		0.0034	0.0023	0.0049	0.0047	0.0052	0.0066	0.0071
WFG6	NSGA-II-I_ϵ^{sum}	**0.0073**	**0.0075**	0.0091	0.0119	0.0141	0.0189	0.0232
		0.0024	0.0024	0.0020	0.0051	0.0038	0.0093	0.0240
	NSGA-II-I_ϵ^{min}	0.0075	0.0079	**0.0087**	**0.0114**	**0.0132**	**0.0180**	**0.0209**
		0.0015	0.0029	0.0025	0.0051	0.0050	0.0075	0.0085

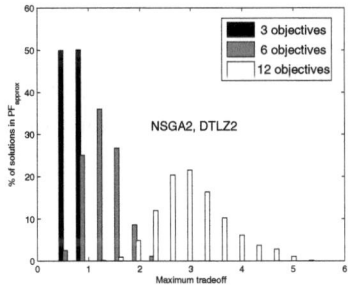 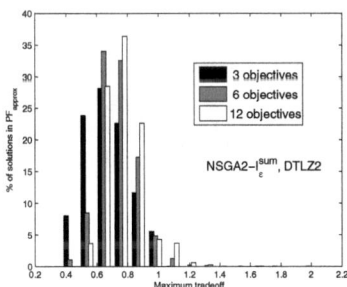

Fig. 7. Maximum tradeoff distribution for NSGA-II and NSGA-II-I_ϵ^{sum} in DTLZ2

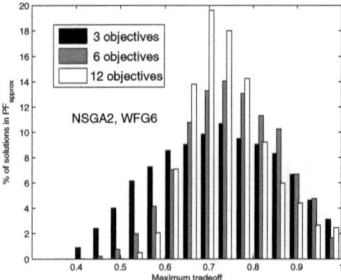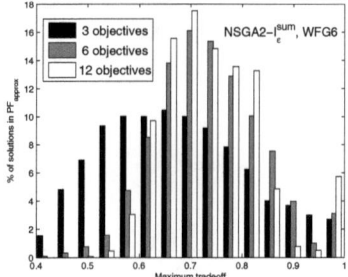

Fig. 8. Maximum tradeoff distribution for NSGA-II and NSGA-II-I_ϵ^{sum} in WFG6

6   Conclusions and Future Work

In this paper we replaced the Pareto dominance by a new preference relation that combines a Chebyshev achievement function and an ϵ-indicator based relation. The resulting relation, called Chebyshev relation, was initially proposed as a technique to incorporate preferences. Nonetheless, in this paper we showed that it is also useful to deal with Many-objective Optimization Problems (MOPs). The Chebyshev relation improves drastically the convergence of NSGA-II without sacrificing distribution. One important finding is that the main source of difficulty of DTLZ problems is the presence of dominance resistant solutions (DRSs) which are equally ranked by the Pareto dominance. However, since the Chebyshev relation compares solutions using either the achievement function or the ϵ-indicator value, it was able to eliminate DRSs, preserving this way, the search ability in MOPs. On the other hand, since WFG problems do not induce DRSs, even the standard NSGA-II was able to maintain a similar level of performance despite the number of objectives. Although these problems are hard for other reasons (e.g., nonseparability, multimodality), it seems that the number of objectives does not significantly affect their difficulty. We are aware that there are other sources of difficulty for MOPs. However, since DRSs might be present in other problems, we suggest that the development of a MOEA integrates mechanisms to overcome these types of solutions.

In the future we plan to apply the Chebyshev preference relation in real-world problems in order to investigate if DRSs are also present. Finally, we will compare the performance of the Chebyshev relation against other optimization techniques that have shown good scalability in MOPs.

References

1. Coello Coello, C.A., Lamont, G.B., Van Veldhuizen, D.A.: Evolutionary Algorithms for Solving Multi-Objective Problems, 2nd edn. Springer, New York (2007) ISBN 978-0-387-33254-3
2. Hughes, E.J.: Evolutionary Many-Objective Optimisation: Many Once or One Many? In: CEC 2005, Edinburgh, Scotland, vol. 1, pp. 222–227 (September 2005)

3. Wagner, T., Beume, N., Naujoks, B.: Pareto-, Aggregation-, and Indicator-Based Methods in Many-Objective Optimization. In: Obayashi, S., Deb, K., Poloni, C., Hiroyasu, T., Murata, T. (eds.) EMO 2007. LNCS, vol. 4403, pp. 742–756. Springer, Heidelberg (2007)
4. Purshouse, R.C., Fleming, P.J.: Evolutionary Multi-Objective Optimisation: An Exploratory Analysis. In: CEC 2003, Canberra, Australia, vol. 3, pp. 2066–2073 (December 2003)
5. Knowles, J., Corne, D.: Quantifying the Effects of Objective Space Dimension in Evolutionary Multiobjective Optimization. In: Obayashi, S., Deb, K., Poloni, C., Hiroyasu, T., Murata, T. (eds.) EMO 2007. LNCS, vol. 4403, pp. 757–771. Springer, Heidelberg (2007)
6. Ishibuchi, H., Tsukamoto, N., Nojima, Y.: Evolutionary many-objective optimization: A short review. In: CEC 2008, Hong Kong, pp. 2424–2431 (June 2008)
7. Schütze, O., Lara, A., Coello Coello, C.A.: On the Influence of the Number of Objectives on the Hardness of a Multiobjective Optimization Problem. IEEE Transactions on Evolutionary Computation 15(4), 444–455 (2011)
8. Sato, H., Aguirre, H.E., Tanaka, K.: Genetic Diversity and Effective Crossover in Evolutionary Many-Objective Optimization. In: Coello, C.A.C. (ed.) LION 5. LNCS, vol. 6683, pp. 91–105. Springer, Heidelberg (2011)
9. Ikeda, K., Kita, H., Kobayashi, S.: Failure of Pareto-based MOEAs: Does Nondominated Really Mean Near to Optimal? In: CEC 2001, Piscataway, New Jersey, vol. 2, pp. 957–962 (May 2001)
10. Deb, K., Thiele, L., Laumanns, M., Zitzler, E.: Scalable Multi-Objective Optimization Test Problems. In: CEC 2002, Piscataway, New Jersey, vol. 1, pp. 825–830 (May 2002)
11. Huband, S., Barone, L., While, L., Hingston, P.: A Scalable Multi-objective Test Problem Toolkit. In: Coello Coello, C.A., Hernández Aguirre, A., Zitzler, E. (eds.) EMO 2005. LNCS, vol. 3410, pp. 280–295. Springer, Heidelberg (2005)
12. López Jaimes, A., Arias-Montaño, A., Coello Coello, C.A.: Preference Incorporation to Solve Many-Objective Airfoil Design Problems. In: CEC 2011, New Orleans, USA (June 2011)
13. Wierzbicki, A.: The use of reference objectives in multiobjective optimisation. In: Fandel, G., Gal, T. (eds.) Multiple Criteria Decision Making Theory and Application. Lecture Notes in Economics and Mathematical Systems, Vol. 177, pp. 468–486. Springer (1980)
14. Ehrgott, M.: Multicriteria Optimization, 2nd edn. Springer, Berlin (2005)
15. Bentley, P.J., Wakefield, J.P.: Finding Acceptable Solutions in the Pareto-Optimal Range using Multiobjective Genetic Algorithms. In: Chawdhry, P.K., Roy, R., Pant, R.K. (eds.) Soft Computing in Engineering Design and Manufacturing, pp. 231–240. Springer, London (1997)
16. Kukkonen, S., Lampinen, J.: Ranking-Dominance and Many-Objective Optimization. In: CEC 2007, Singapore, pp. 3983–3990 (September 2007)
17. Drechsler, N., Drechsler, R., Becker, B.: Multi-Objected Optimization in Evolutionary Algorithms Using Satisfyability Classes. In: Reusch, B. (ed.) International Conference on Computational Intelligence, Theory and Applications, 6th Fuzzy Days, Dortmund, Germany, pp. 108–117 (1999)
18. di Pierro, F., Khu, S.T., Savić, D.A.: An Investigation on Preference Order Ranking Scheme for Multiobjective Evolutionary Optimization. IEEE Transactions on Evolutionary Computation 11(1), 17–45 (2007)

19. Sato, H., Aguirre, H.E., Tanaka, K.: Controlling Dominance Area of Solutions and Its Impact on the Performance of MOEAs. In: Obayashi, S., Deb, K., Poloni, C., Hiroyasu, T., Murata, T. (eds.) EMO 2007. LNCS, vol. 4403, pp. 5–20. Springer, Heidelberg (2007)
20. Zitzler, E., Thiele, L., Laumanns, M., Fonseca, C.M., da Fonseca, V.G.: Performance Assessment of Multiobjective Optimizers: An Analysis and Review. IEEE Transactions on Evolutionary Computation 7(2), 117–132 (2003)
21. Zitzler, E., Künzli, S.: Indicator-Based Selection in Multiobjective Search. In: Yao, X., Burke, E.K., Lozano, J.A., Smith, J., Merelo-Guervós, J.J., Bullinaria, J.A., Rowe, J.E., Tiño, P., Kabán, A., Schwefel, H.-P. (eds.) PPSN VIII. LNCS, vol. 3242, pp. 832–842. Springer, Heidelberg (2004)
22. Balling, R.: The Maximin Fitness Function; Multi-objective City and Regional Planning. In: Fonseca, C.M., Fleming, P.J., Zitzler, E., Deb, K., Thiele, L. (eds.) EMO 2003. LNCS, vol. 2632, pp. 1–15. Springer, Heidelberg (2003)

An Improved Adaptive Approach for Elitist Nondominated Sorting Genetic Algorithm for Many-Objective Optimization

Himanshu Jain and Kalyanmoy Deb

Department of Mechanical Engineering,
Indian Institute of Technology, Kanpur
{hjain,deb}@iitk.ac.in
http://www.iitk.ac.in/kangal/

Abstract. NSGA-II and its contemporary EMO algorithms were found to be vulnerable in solving many-objective optimization problems having four or more objectives. It is not surprising that EMO researchers have been concentrating in developing efficient algorithms for many-objective optimization problems. Recently, authors suggested an extension of NSGA-II (NSGA-III) which is based on the supply of a set of reference points and demonstrated its working in three to 15-objective optimization problems. In this paper, NSGA-III's reference point allocation task is made adaptive so that a better distribution of points can be found. The approach is compared with NSGA-III and a previous adaptive approach on a number of constrained and unconstrained many-objective optimization problems. NSGA-III and its adaptive extension proposed here open up new directions for research and development in the area of solving many-objective optimization problems.

Keywords: Many-objective optimization, NSGA-II, adaptive optimization, evolutionary optimization.

1 Introduction

Over the years, NSGA-II [2] has been applied to various practical problems and was adopted in various commercial softwares. However, NSGA-II, like other evolutionary multi-objective optimization (EMO) algorithms, suffers from its ability to handle more then three objectives adequately. When the so-called 'curse of dimensionality' thwarted the progress of algorithm development in the EMO field, researchers took interests in devising new methodologies for solving many-objective optimization problems, involving four or more objectives [7,11,10,8]. Progressing towards the Pareto-optimal front and simultaneously arriving at a well-distributed set of trade-off solutions in a high-dimensional space were found to be too challenging tasks for any algorithm to be computationally tractable. Earlier in 2012, authors of this paper suggested a new extension of NSGA-II(MO-NSGA-II) specifically for solving many-objective optimization problems. MO-NSGA-II [5] starts with a set of automatically or user-defined reference

R.C. Purshouse et al. (Eds.): EMO 2013, LNCS 7811, pp. 307–321, 2013.

points and then focuses its search to emphasize the EMO population members that are non-dominated in the population and are also "closest" to each of the reference points, thereby finding a well-distributed and well-converged set of solutions. In later studies [3,4] MO-NSGA-II was further modified and extended to solve constrained problems, this new algorithm was named as NSGA-III. The latter study also suggested an adaptive approach (A-NSGA-III) that was capable of identifying reference points that do not correspond to a well-distributed set of Pareto-optimal points.

In this paper, we extend the concept of relocation of reference points and attempt to remove some of the shortcomings of A-NSGA-III algorithm and suggest an efficient adaptive NSGA-III approach (A^2-NSGA-III) for this purpose. In the remainder of this paper, we first provide a brief overview of NSGA-III and A-NSGA-III approaches. Thereafter, we motivate the reasons for improving the adaptive approach and suggest our proposed procedure (A^2-NSGA-III). Simulation results are shown on constrained and unconstrained problems using the proposed procedure and are compared with original NSGA-III and the A-NSGA-III approaches. Conclusions of this study are then made.

2   Many Objective NSGA-II or NSGA-III

The basic framework of the NSGA-III [3] is similar to the original NSGA-II algorithm [2]. First, the parent population P_t (of size N) is randomly initialized in the specified domain, then the binary tournament selection, crossover and mutation operators are applied to create an offspring population Q_t.

Thereafter, both populations are combined and sorted according to their domination level and the best N members are selected from the combined population to be the parent population for the next generation. The fundamental difference between NSGA-II and NSGA-III lies in the way the niche-preservation operation is performed.

Unlike NSGA-II, NSGA-III starts with a set of reference points Z^r. After non-dominated sorting, all acceptable front members and the last front F_l which could not be completely accepted are saved in a set S_t. Members in S_t/F_l are selected right away for the next

---

Algorithm 1. Generation t of NSGA-III procedure

Input: H reference points Z^r, parent population P_t
Output: P_{t+1}
1: $S_t = \emptyset$, $i = 1$
2: $Q_t = $ Recombination+Mutation(P_t)
3: $R_t = P_t \cup Q_t$
4: $(F_1, F_2, \ldots) = $ Non-dominated-sort(R_t)
5: **repeat**
6:     $S_t = S_t \cup F_i$ and $i = i + 1$
7: **until** $|S_t| \geq N$
8: Last front to be included: $F_l = F_i$
9: **if** $|S_t| = N$ **then**
10:     $P_{t+1} = S_t$, break
11: **else**
12:     $P_{t+1} = \cup_{j=1}^{l-1} F_j$
13:     Points to be chosen from F_l: $K = N - |P_{t+1}|$
14:     Normalize objectives
15:     Associate each member **s** of S_t with a reference point: $[\pi(\mathbf{s}), d(\mathbf{s})] =$Associate($S_t, Z^r$)   % $\pi(\mathbf{s})$: closest reference point, d: distance between **s** and $\pi(\mathbf{s})$
16:     Compute niche count of reference point $j \in Z^r$: $\rho_j = \sum_{\mathbf{s} \in S_t/F_l} ((\pi(\mathbf{s}) = j) ? 1 : 0)$
17:     Choose K members one at a time from F_l to construct P_{t+1}: Niching($K, \rho_j, \pi, d, Z^r, F_l, P_{t+1}$)
18: **end if**

generation, however the remaining members are selected from F_l such that a desired diversity is maintained in the population. Original NSGA-II used the crowding distance measure for selecting well-distributed set of points, however in NSGA-III the supplied reference points (Z^r) are used to select these remaining members. To accomplish this, objective values and reference points are first normalized so that they have an identical range. Thereafter, orthogonal distance between a member in S_t and each of the reference lines (joining the ideal point and a reference point) is calculated. The member is then associated with the reference point having the smallest orthogonal distance. Next, the niche count ρ for each reference point, defined as the number of members in S_t/F_l that are associated with the reference point, is computed for further processing. The reference point having the minimum niche count is identified and the member in front last front F_l that is associated with the identified reference point is included in the final population. The niche count of the identified reference point is increased by one and the procedure is repeated to fill up population P_{t+1}. The entire procedure is presented in algorithmic form in 1. In NSGA-III, a different tournament selection operator is used. If both competing parents are feasible, then one of them is chosen at random. However, if one is feasible and the other is infeasible, then the feasible one is selected. Finally, if both are infeasible, then the one having the least constraint violation is selected. After applying this tournament selection operator, usual crossover and mutation operations are carried out to create the offspring population and the above-mentioned niching operation is applied again to the combined population. These steps are continued until a termination criterion is satisfied.

Some interesting features of NSGA-III are as follows: (i) it does not require any additional parameter setting, just like its predecessor NSGA-II, (ii) the population size is almost same as the number of reference points, thereby making an efficient computational effort, (iii) it can be used to find trade-off points in the entire Pareto-optimal front or focused in a preferred Pareto-optimal region, (iv) it is extended easily to solve constrained optimization problems, (v) it can be used with a small population size (such as a population of size 100 for a 10-objective optimization problem) and (vi) it can be used for other multi-objective problem solving tasks, such as in finding the nadir point or other special points.

3   Adaptive NSGA-III

A little thought will reveal the fact that not all reference points may be associated with a well-dispersed Pareto-optimal set and carrying on with a predefined set of reference points from start to finish may be a waste of computational efforts. To clarify, let us consider a three-objective optimization problem where the Pareto-optimal front is shown as the shaded portion in Figure 1 and 91 initial reference points are marked as open circles. Clearly, only 28 reference points have a corresponding Pareto-optimal point, while the rest 63 points are then found to randomly distributed, as shown in small circles in Figure 1. One possible remedy to this problem is to first identify all those reference points that are not associated

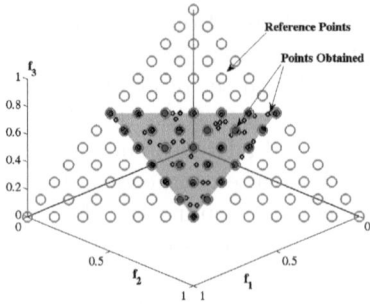

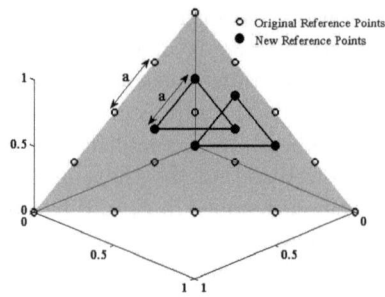

Fig. 1. Only 28 out of 91 reference points find a Pareto-optimal solution

Fig. 2. Addition of reference points

with a population member. Then, instead of eliminating these reference points, they can then be relocated so as to find a better distribution of Pareto-optimal points. The following addition and deletion strategies were proposed in A-NSGA-III.

Algorithm 2. Add Reference Points(Z^r, ρ, p) procedure

Input: Z^r, p, ρ, $Flag(Z^r)$
Output: $Z^r (updated)$
1: nref $= |Z^r|$
2: **for** $j = 1$ **to** nref **do**
3:    **if** $\rho_j > 1$ and $Flag(Z^r_j) = 0$ **then**
4:       $Z^r_{new} = Structured - points(p = 1)$
5:       $Z^r_{new} = Z^r_{new}/p + (Z^r_j - 1/(M * P))$
6:       **for** $i = 1$ **to** M **do**
7:          **if** already-exist$(Z^r_{new,i}) = $ FALSE and $Z^r_{new,i}$ lie in first quadrant **then**
8:             $Z^r = Z^r \cup Z^r_{new,i}$
9:             $Flag(Z^r_{new,i}) = 0$
10:          **end if**
11:       **end for**
12:       $Flag(Z^r_j) = 1$
13:    **end if**
14: **end for**

Note that, in NSGA-III, after the niching operation P_{t+1} population is created and the niche count ρ_j (the number of population members that are associated with j-th reference point) for each reference point is updated. As the number of reference points (H) is kept almost equal to population size (N), every reference point is expected to be associated with one population member. Thus, if $\rho_j \geq 2$ is observed for any reference point, this means that some other reference point has a zero niche count value. Hence a reference point having zero ρ value is relocated close to the j-th reference point. The relocation procedure is shown in Figure 2. Consider the situation in $M = 3$ objective case. For adding extra reference points, a $(M-1)$-dimensional simplex having M points at its vertices is added. The side length of the simplex is equal to the distance between two consecutive reference points (which is controlled by parameter p) on the original specified hyperplane and the centroid of the simplex is kept on the j-th reference point. If there are more than one reference points for which $\rho_j \geq 2$, the above inclusion step is executed for each of these reference points. Before a new added reference point is accepted, two checks are made: (i) if it lies outside the boundary of original simplex, it is not accepted, and (ii) if it already exists in the set of reference points, it is also not accepted. This addition procedure is presented in algorithmic form in 2. Using this procedure it may happen that after some generations too many reference points are added and many of them eventually become *non-useful* again. To avoid this, deletion of non-useful reference points is also carried out simultaneously, as described in the following paragraph:

After the inclusion operation is performed, the niche count of all reference points are updated. Now, if there exist no reference point whose niche count is greater than one (that is, $\rho_j = 0$ or 1 for all reference points), this means that each and every population member is associated with a single unique reference point. In this case, the reference points ($\rho_j = 0$) that are not associated with any population members are simply deleted. In this way the inclusion and deletion operations adaptively relocate reference points based on the niche count values of the respective reference points. The A-NSGA-III worked well on a number of problems in the previous study [4], however, the concept deserves more attention.

4   Limitations of A-NSGA-III

Following limitations of adaptive strategy discussed above are observed:

1. In problems where the entire Pareto-optimal front is concentrated in a small region or in case we start with only few reference points and a sufficiently large population size, the above addition procedure may not be able to introduce enough reference points so that the entire population may be evenly distributed.
2. The above followed addition procedure does not allow introduction of extra reference points around the corner reference points of the hyperplane.
3. Since the addition procedure is carried out right from first generation when the population is far from the actual Pareto-front we may not have given enough time for the algorithm to spread the population evenly in various regions which may lead to premature introduction of extra reference points in unwanted regions.
4. Since the removal procedure is only carried out when the ρ value for all the reference points is less than or equal to one, in some cases (specially large-dimensional problems) it may happen that this condition is never satisfied and the algorithm keeps on adding extra reference points, thus increasing the computational cost.

In order to overcome these limitations, we modify the above approach for addition and deletion of reference points in the following section.

5   Efficiently Adaptive NSGA-III Procedure (A²-NSGA-III)

Let us suppose that the extra reference points are to be added around the j-th reference point (marked as P in Figure 3). As done earlier, here also we use an

$(M-1)$-dimensional simplex (shown as ABC) but having a side length equal to the half of the distance between two consecutive reference points on the original normalized hyperplane. This simplex is called the primary simplex that will be added to the reference point P. However, instead of adding the simplex around the reference point as it was done in A-NSGA-III, it is now added by keeping the j-th reference point as one of the corners of the simplex as shown in Figure 3, thus adding $(M-1)$ new reference points. Since there are M points in the simplex, there are a total of M such ways of adding the simplex. To implement, we randomly select one of the corner points and overlay the simplex with the selected corner point falling on the reference point. For example in Figure 3, if we choose the corner A of simplex and coincide it with the reference point P then we get configuration 1 shown in the figure. On the other hand, if we select corner B then we get configuration 2, and so on. Like before, before accepting a configuration, all new locations of reference points are checked for the following two conditions:(i) if any newly located reference point lies outside the original hyperplane, the configuration is not accepted (for example in Figure 3 if extra points are to be added around reference point Q then configurations 1 and 3 are

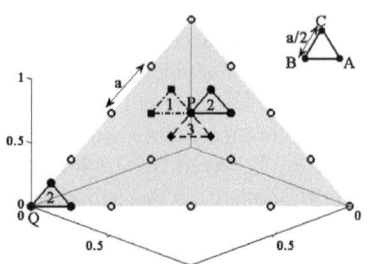

Fig. 3. Proposed approach for adding extra reference points

Algorithm 3. Add Reference Points(Z^r, ρ, p, λ) procedure for A^2-NSGA-III

Input: Z^r, p, ρ, $Config(Z^r)$, λ { %$Config(Z^r)$ *contains the configuration number of simplex to be added to each reference point, if* $Config(Z^r_k) > M$ *it means all configurations are added around* k^{th} *ref point.*}
Output: Z^r *(updated)*
1: nref $= |Z^r|$
2: **for** $j = 1$ to nref **do**
3: Flag $= 0$
4: **if** $\rho_j > 1$ **then**
5:   **while** $Config(Z^r_j) \leq M$ and $Flag == 0$ **do**
6:    $Z^r_{new} = Structured - points(p = 1)$
7:    $Z^r_{new} = Z^r_{new}/(\lambda * p) + (Z^r_j - Z^r_{new,Config(Z^r_j)}/(\lambda * p))$
8:    **for** $i = 1$ to M **do**
9:     **if** $i \neq Config(Z^r_j)$ and already-exist$(Z^r_{new,i}) =$ FALSE and $Z^r_{new,i}$ lie in first quadrant **then**
10:      $Z^r = Z^r \cup Z^r_{new,i}$
11:      $Config(Z^r_{new,i}) = 1$
12:      $Flag = 1$
13:     **end if**
14:    **end for**
15:    $Config(Z^r_j) = Config(Z^r_j) + 1$
16:   **end while**
17: **end if**
18: **end for**
19: **if** $\exists$ j(=1:nref) s.t $\rho_j > 1$ and $\forall$ j s.t $\rho_j > 1$ $Config(Z^r_j) > M$ and no new reference point is added **then**
20: $\lambda = \lambda * 2$
21: **end if**

---

not acceptable), and (ii) if a newly located reference point already exists in the set of reference points, that point is not duplicated. The procedure is continued with all reference points having $\rho_j > 1$. After one simplex is added to such a reference point, if ρ_j is still greater than one, other allowable configurations are continued to be added until all M configurations are added. Thereafter, for further $\rho_j > 1$ occurrences, simplexes half of its current size are introduced one at a time, this is controlled by a scaling factor λ whose initial value is kept to be 2 which denotes simplexes having a side length equal to the half of the distance between two consecutive reference points on the original normalized hyperplane are to be added. Whenever the above stated condition arrives λ value is increased by a factor of 2. This process is repeated till none of the reference point has a niche count greater than one. This improved addition procedure is presented in algorithmic form in 3. The ability to introduce more and more simplexes but with reduced size alleviates the first limitation described above with the previous approach. The above procedure of using simplexes half the size of original simplexes enable more concentrated reference points to be introduced near the vertices of the original hyperplane, thereby alleviating the second limitation mentioned above. To cater the third limitation, we introduce a condition before new reference points can be added. The number of reference points having $\rho > 1$ is monitored and only when the number has settled down to a constant value in the past τ generations, addition of reference points is allowed. This check will ensure that enough time has been spent by the algorithm to evenly spread its population members with the supplied set of reference points before any new reference points are introduced. In this study, we have used $\tau = 10$ generations. Now the last limitation is alleviated by putting a cap over the maximum number of reference points that can ever be handled by the algorithm. Thus, if the total number of reference points shoots up beyond this value the deletion process (described in subsection 3) is carried out, thereby lowering the burden of carrying forward with a large number of reference points. Here, we have used 10 times the number of originally supplied reference points as the cap for maximum number of reference points.

6   Results

We now present the simulation results of both adaptive NSGA-III approaches on a number of three to eight-objective test problems. The population sizes and number of reference points are kept as mentioned in Table 1. Other parameters are kept identical for both approaches: (i) SBX crossover probability of one, (ii) polynomial mutation probability of $1/n$ (where n is the number of variables), (iii) SBX crossover and polynomial mutation indices are kept as 30 and 20, respectively. As a performance metric, we have used the hypervolume indicator as it captures both convergence and distribution ability of an algorithm. In

Table 1. Number of ref. points and population sizes used

No. of obj. (M)	Ref. pts. (H)	Pop. size (N)
3	91	92
5	210	212
8	156	156

each case, 20 runs are carried out and the best, median and worst hypervolume values are reported.

6.1 Unconstrained Test Problems

Inverted DTLZ1 and DTLZ2 Problem: First of all we take two problems: DTLZ1 and DTLZ2 from scalable DTLZ suite[6] and modify them so that there are certain reference points on normalized hyperplane corresponding to whom there is no point on Pareto-optimal front. To accomplish this in both problems the objective functions are calculated using the original formulation, however after calculating the objective function values, following transformations are made: for DTLZ1:

$$f_i(\mathbf{x}) \leftarrow 0.5(1 + g(\mathbf{x})) - f_i(\mathbf{x}), \ \ for \ i = 1, \ldots, M$$

for DTLZ2:

$$f_i(\mathbf{x}) \leftarrow (1 + g(\mathbf{x})) - f_i(\mathbf{x})^4, \ \ for \ i = 1, \ldots, M - 1 \ \ and \ \ f_M(\mathbf{x}) \leftarrow (1 + g(\mathbf{x})) - f_M(\mathbf{x})^2.$$

where $g(\mathbf{x})$ is calculated as in the original DTLZ1 and DTLZ2 formulation respectively [6]. This transformation inverts the original Pareto-optimal front thereby rendering several reference points as non-useful.

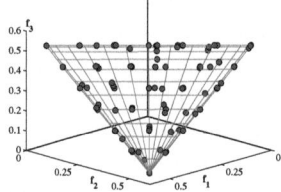
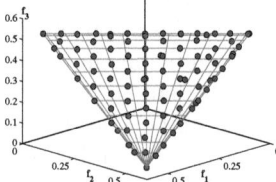
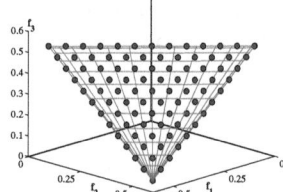

Fig. 4. NSGA-III solutions for Inv-DTLZ1

Fig. 5. A-NSGA-III solutions for Inv-DTLZ1

Fig. 6. A²-NSGA-III solutions for Inv-DTLZ1

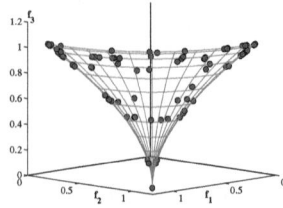
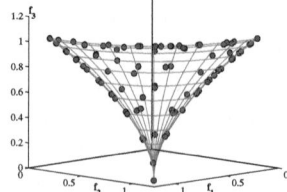
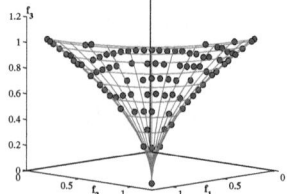

Fig. 7. NSGA-III solutions for Inv-DTLZ2

Fig. 8. A-NSGA-III solutions for Inv-DTLZ2

Fig. 9. A²-NSGA-III solutions for Inv-DTLZ2

All the approaches are tested against three, five, and eight objective versions of both problems. Pareto-optimal fronts obtained in the case of three-objective version are plotted in Figures 4, 5, 6, 7, 8, and 9 (plotted fronts correspond to median hypervolume values as tabulated in Table 2). As evident from Figure 4 for DTLZ1, in the case of A^2-NSGA-III, all 91 points are uniformly distributed, while with NSGA-III only 28 population members are distributed and with A-NSGA-III, the number increases to 81 and rest are randomly dispersed over the entire front. Similar observation is made in case of inverted DTLZ2 problem.

Hypervolume values for three to eight-objective cases are tabulated in Table 2 for both the problems. It is clear that in all cases, the use of proposed adaptive method improves the performance of algorithm as compared to the basic and previously suggested adaptive algorithm. The performance of the proposed method gets better with an increase in the number of objectives.

Table 2. Best, median and worst hypervolume values on M-objective inverted DTLZ1 and DTLZ2 problems. Best values are maked in bold.

Function	M	Gen	NSGA-III	A-NSGA-III	A^2-NSGA-III
Inv-DTLZ1			$6.1115e-02$	$6.0945e-02$	$\mathbf{6.3738e-02}$
	3	400	$6.2294e-02$	$6.5404e-02$	$\mathbf{6.5693e-02}$
			$6.3052e-02$	$6.5967e-02$	$\mathbf{6.6206e-02}$
			$1.5475e-03$	$1.9610e-03$	$\mathbf{2.6257e-03}$
	5	600	$1.9821e-03$	$2.1255e-03$	$\mathbf{2.7117e-03}$
			$2.1409e-03$	$2.2413e-03$	$\mathbf{2.8273e-03}$
			$3.0186e-06$	$3.0186e-06$	$\mathbf{4.2066e-06}$
	8	750	$3.3864e-06$	$3.4220e-06$	$\mathbf{4.8353e-06}$
			$3.8361e-06$	$3.8361e-06$	$\mathbf{5.5693e-06}$
Inv-DTLZ2			$1.0530e-01$	$1.1593e-01$	$\mathbf{1.1790e-01}$
	3	250	$1.0951e-01$	$1.1951e-01$	$\mathbf{1.2342e-01}$
			$1.1197e-01$	$1.2137e-01$	$\mathbf{1.2550e-01}$
			$1.5083e-03$	$1.5365e-03$	$\mathbf{2.2707e-03}$
	5	350	$1.9100e-03$	$2.0241e-03$	$\mathbf{3.0822e-03}$
			$2.0894e-03$	$2.2904e-03$	$\mathbf{3.1760e-03}$
			$1.2168e-06$	$1.2168e-06$	$\mathbf{1.2814e-06}$
	8	500	$1.4818e-06$	$1.4818e-06$	$\mathbf{2.3228e-06}$
			$1.7822e-06$	$1.7822e-06$	$\mathbf{3.7236e-06}$

6.2   Constrained Test Problems

After demonstrating the efficacy of proposed approach on a couple of unconstrained test problems, next we consider two constrained problems. These problems are designed by adding constraints to the original scalable DTLZ1 and DTLZ2 problems so that some portions of the original Pareto-optimal front become infeasible.

C-DTLZ1 and C-DTLZ2 Problem: In case of C-DTLZ1 problem, we add a hyper-cylinder (with its central axis passing through the origin and equally inclined to all the objective axes) as a constraint so that the region inside the hyper-cylinder is feasible. In case of C-DTLZ2 we add a constraint to the original DTLZ2 problem, thereby making the entire region of objective space lying in between $0.1 < f_M < 0.9$ infeasible. Due to these changes in both problems , not all reference points initialized on the normalized hyperplane will have an associated feasible Pareto-optimal point. In such problems, the original NSGA-III may waste its computations in dealing with such non-productive reference points and the previously suggested adaptive A-NSGA-III may not be able to fully relocate all reference points to find a well-distributed set of Pareto-optimal points. Figures 10, 11, and 12 show the obtained fronts using the three approaches, respectively for problem C-DTLZ1. As one can see here the distribution of points obtained using the proposed A^2-NSGA-III is better than that obtained using NSGA-III but relocations of reference points in both A-NSGA-III and A^2-NSGA-III are comparable. The current approach allows a greater density in solutions, but this may not happen uniformly across the entire front, as evident from the Figure 12 which may lead to very less improvement in hypervolume value. Table 3 shows that for three-objective version, the hypervolume is slightly better for A-NSGA-III, while for larger objective cases, A^2-NSGA-III has better hypervolume values.

For C-DTLZ2, as shown in Figures 13, 14, and 15, we get a better distribution of points using A^2-NSGA-III and the same is reflected in the hypervolume values, tabulated in Table 3 for three-objective case. In five-objective version, A^2-NSGA-III performs the best, but in eight-objective version of the problem, A-NSGA-III performs slightly better.

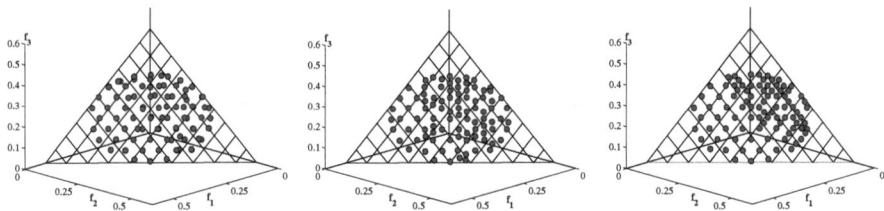

Fig. 10. NSGA-III solutions for DTLZ1   **Fig. 11.** A-NSGA-III solutions for DTLZ1   **Fig. 12.** A^2-NSGA-III solutions for DTLZ1

6.3 A^2-NSGA-III with Large Population Size

NSGA-III and its adaptive version A-NSGA-III were developed with the principle that a population size almost equal to the number of reference points is to be used. This enabled a direct control of the maximum number of obtained trade-off

Table 3. Best, median and worst hypervolume values on M-objective constrained DTLZ1 and DTLZ2 problems

Function	M	Gen	NSGA-III	A-NSGA-III	A²-NSGA-III
C-DTLZ1-hole			$1.7099e-01$	$\mathbf{1.7135e-01}$	$1.7096e-01$
	3	400	$1.7357e-01$	$\mathbf{1.7458e-01}$	$1.7418e-01$
			$1.7501e-01$	$\mathbf{1.7576e-01}$	$1.7535e-01$
			$7.4929e-02$	$\mathbf{7.4934e-02}$	$7.4846e-02$
	5	600	$\mathbf{7.5151e-02}$	$7.5128e-02$	$7.5141e-02$
			$7.5224e-02$	$7.5246e-02$	$\mathbf{7.5310e-02}$
			$1.5683e-02$	$1.5724e-02$	$\mathbf{1.5878e-02}$
	8	750	$1.5916e-02$	$1.5842e-02$	$\mathbf{1.6118e-02}$
			$1.6120e-02$	$1.6022e-02$	$\mathbf{1.6306e-02}$
C-DTLZ2			$6.2968e-01$	$6.2998e-01$	$\mathbf{6.3056e-01}$
	3	250	$6.3091e-01$	$6.3091e-01$	$\mathbf{6.3267e-01}$
			$6.3230e-01$	$6.3187e-01$	$\mathbf{6.3483e-01}$
			$1.2340e+00$	$\mathbf{1.2350e+00}$	$1.2348e+00$
	5	350	$1.2375e+00$	$1.2377e+00$	$\mathbf{1.2400e+00}$
			$1.2398e+00$	$1.2405e+00$	$\mathbf{1.2422e+00}$
			$1.8209e+00$	$1.9114e+00$	$\mathbf{1.9151e+00}$
	8	500	$1.8409e+00$	$\mathbf{1.9245e+00}$	$1.9211e+00$
			$1.8576e+00$	$\mathbf{1.9286e+00}$	$1.9273e+00$

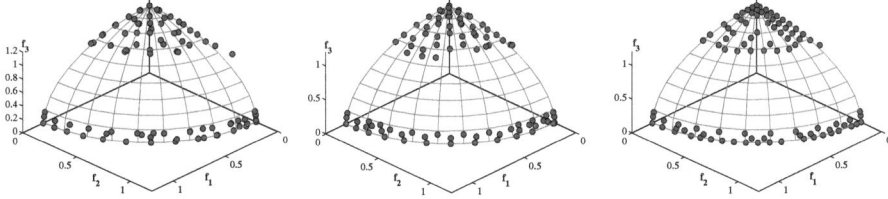

Fig. 13. NSGA-III solutions for DTLZ2

Fig. 14. A-NSGA-III solutions for DTLZ2

Fig. 15. A²-NSGA-III solutions for DTLZ2

points that is expected from the algorithms. If more points are needed, previous algorithms allowed a larger population to be used, but the obtained points may not be distributed well. However, with our proposed A²-NSGA-III modification, reference points now can be considered as seed points and a much larger set of trade-off points can be obtained by simply using a larger population size. To illustrate, we consider two scalable test problems DTLZ1 and DTLZ2 and use population sizes larger than number of reference points as tabulated in Table 4.

We have used three, five, and eight objective versions of both problems. Figures 16, 17, and 18 show the obtained fronts for three-objective DTLZ1 problem using all three

Table 4. Number of ref. points and population sizes used

No. of obj. (M)	Ref. pts. (H)	Pop. size (N)
3	28	92
5	35	212
8	44	156

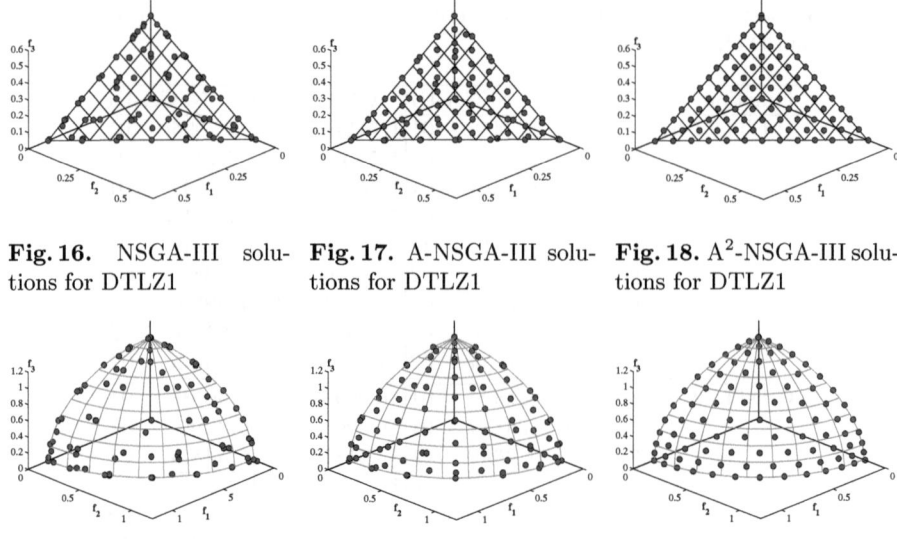

Fig. 16. NSGA-III solutions for DTLZ1

Fig. 17. A-NSGA-III solutions for DTLZ1

Fig. 18. A^2-NSGA-III solutions for DTLZ1

Fig. 19. NSGA-III solutions for DTLZ2

Fig. 20. A-NSGA-III solutions for DTLZ2

Fig. 21. A^2-NSGA-III solutions for DTLZ2

approaches. Clearly, with A^2-NSGA-III we get an excellent distribution having all 92 population members, despite the initialization of only 28 reference points on the normalized hyperplane. With A-NSGA-III, the distribution is better than that obtained using NSGA-III, but due to the limitation of only one configuration per reference point, the approach cannot utilize the available pool of population members adequately.

Similarly, for the three-objective DTLZ2, we get the best distribution using A^2-NSGA-III as shown in Figure 21 while with rest of the approaches (Figures 19 and 20) not all the points are well distributed. Table 5 shows the best, median, and worst hypervolume values for three, five and eight objective versions of both problems obtained using all three approaches. A^2-NSGA-III performs the best in all cases in both problems with an identical number of function evaluations, thereby showing the superiority of the proposed adaptive procedure.

6.4   Engineering Optimization Problems

Next, we apply the proposed algorithm on four engineering problems ranging from three to five objectives.

Crashworthiness Problem: This is a three-objective unconstrained problem aimed at structural optimization of the frontal structure of vehicle for crashworthiness [9]. Thickness of five reinforced members around the frontal structure are chosen as design variables, while the mass of vehicle, deceleration during the full frontal crash (which is proportional to biomechanical injuries caused

Table 5. Best, median and worst hypervolume for DTLZ1 and DTLZ2 problems

Problem	M	Gen	NSGA-III	A-NSGA-III	A²-NSGA-III
DTLZ1			$1.8569e-01$	$\mathbf{1.8810e-01}$	$1.8671e-01$
	3	400	$1.8674e-01$	$1.8871e-01$	$\mathbf{1.8928e-01}$
			$1.8731e-01$	$1.8912e-01$	$\mathbf{1.8962e-01}$
			$\mathbf{7.6252e-02}$	$7.6252e-02$	$7.6235e-02$
	5	600	$7.6361e-02$	$7.6361e-02$	$\mathbf{7.6726e-02}$
			$7.6449e-02$	$7.6449e-02$	$\mathbf{7.6742e-02}$
			$8.9663e-03$	$8.9663e-03$	$\mathbf{1.6761e-02}$
	8	750	$1.6753e-02$	$1.6753e-02$	$\mathbf{1.6766e-02}$
			$1.6760e-02$	$1.6760e-02$	$\mathbf{1.6770e-02}$
DTLZ2			$7.0956e-01$	$7.2851e-01$	$\mathbf{7.4288e-01}$
	3	250	$7.1800e-01$	$7.3040e-01$	$\mathbf{7.4379e-01}$
			$7.2399e-01$	$7.3320e-01$	$\mathbf{7.4414e-01}$
			$1.2583e+00$	$1.2583e+00$	$\mathbf{1.3017e+00}$
	5	350	$1.2660e+00$	$1.2660e+00$	$\mathbf{1.3039e+00}$
			$1.2780e+00$	$1.2780e+00$	$\mathbf{1.3049e+00}$
			$3.8935e-03$	$3.8935e-03$	$\mathbf{6.4729e-01}$
	8	500	$1.6799e-02$	$1.6799e-02$	$\mathbf{2.1428e+00}$
			$2.0983e+00$	$2.0983e+00$	$\mathbf{2.1435e+00}$

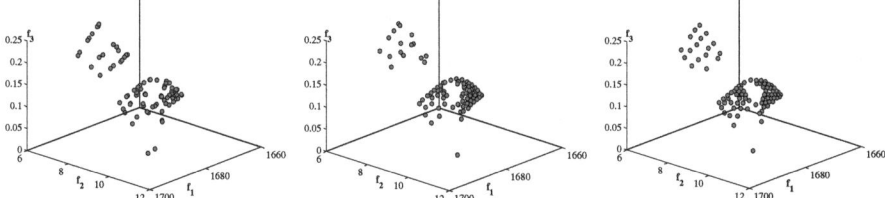

Fig. 22. NSGA-III solutions for Crashworthiness problem   **Fig. 23.** A-NSGA-III solutions for Crashworthiness problem   **Fig. 24.** A²-NSGA-III solutions for Crashworthiness problem

to the occupants) and the toe board intrusion in the offset-frontal crash (which accounts for the structural integrity of the vehicle) are taken as objectives. Mathematical formulation for the problem can be found elsewhere [3]. We solve this problem using all three approaches keeping a population size of 92 and the number of reference points as 91. Rest all parameters are kept the same as before and each algorithm is run for 500 generations. Figures 22, 23, and 24 show the obtained front using NSGA-III, A-NSGA-III, and A²-NSGA-III, respectively. Clearly, the distribution obtained using A²-NSGA-III is the best as compared to others. This practical problem demonstrates that even in a complicated shape of non-dominated front, the proposed A²-NSGA-III can be effective.

Car Side Impact Problem: This is also a three-objective problem but has 10 constraints. The problem aims at minimizing the weight of car, the pubic force experienced by a passenger, and the average velocity of the V-Pillar responsible for withstanding the impact load [3]. We choose 91 reference points and use a population size of 92. We run all the three approaches for 500 generations. The obtained fronts are shown in Figures 25, 26, and 27. As we can see the distributions obtained using A-NSGA-III and A²-NSGA-III are similar, but are considerably better than that obtained using NSGA-III.

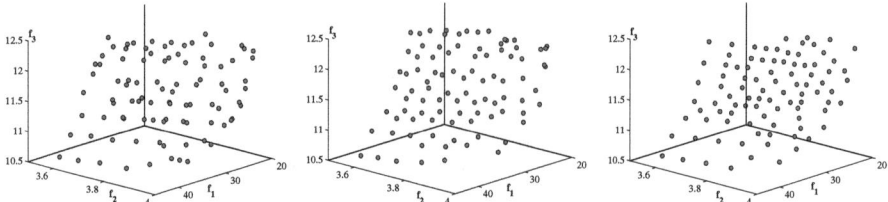

Fig. 25. NSGA-III solutions for car side impact problem   **Fig. 26.** A-NSGA-III solutions for car side impact problem   **Fig. 27.** A²-NSGA-III solutions for car side impact problem

Machining Problem: This a four-objective, three-variable constrained problem aimed at optimizing machining performance subject to four constraints. 165 reference points are initialized uniformly over the normalized hyperplane. The problem is solved using all three algorithms keeping a population size of 168. Each algorithm is run for 750 generations and 20 such runs are made. The best, median, and worst hypervolume values of the obtained points are shown in Table 6. A²-NSGA-III performs much better than the other two approaches, thereby indicating its efficacy.

Table 6. Best, median and worst hypervolume for machining and water problems

Problem	M-NSGA-II	AM-NSGA-II	A²M-NSGA-II
Machn.	2.2339	2.2730	**2.2925**
	2.2688	2.2952	**2.3052**
	2.2856	2.3162	**2.3188**
Water	0.5280	0.5349	**0.5402**
	0.5306	0.5365	**0.5429**
	0.5341	0.5396	**0.5455**

Water Problem: Finally, we consider a five-objective, three-variable, seven-constraint problem taken from the literature [1]. 210 reference points and 212 population are used. 20 different runs are made for all three algorithms and the best, median, and worst hypervolume values are tabulated in Table 6. It is clear from the table that A²-NSGA-III performs the best.

7   Conclusions

In this paper, we have suggested a new and adaptive relocation strategy for reference points in the recently proposed many-objective NSGA-II procedure. The proposed method attempts to alleviate the limitations of a previously proposed adaptive strategy. On a number of unconstrained and constrained three to eight-objective optimization problems, it has been found that the proposed strategy is able to find a better distribution of trade-off points on the entire Pareto-optimal frontier. On a set of four engineering many-objective optimization problems, the proposed A^2-NSGA-III procedure is also found to find a better distribution of points, both visually and in terms of hypervolume measure. The algorithm is also found to produce better and more dense distribution with a supply of more population size. The suggestion of NSGA-III and results with its current adaptive version open up new directions for handling many-objective optimization problems efficiently. Further studies should now be made to make the approaches more computationally efficient and worthy of their use in practice.

References

1. Deb, K.: Multi-objective optimization using evolutionary algorithms. Wiley, Chichester (2001)
2. Deb, K., Agrawal, S., Pratap, A., Meyarivan, T.: A fast and elitist multi-objective genetic algorithm: NSGA-II. IEEE Transactions on Evolutionary Computation 6(2), 182–197 (2002)
3. Deb, K., Jain, H.: An improved NSGA-II procedure for many-objective optimization Part I: Problems with box constraints. Technical Report KanGAL Report Number 2012009, Indian Institute of Technology Kanpur (2012)
4. Deb, K., Jain, H.: An improved NSGA-II procedure for many-objective optimization, Part II: Handling constraints and extending to an adaptive approach. Technical Report KanGAL Report Number 2012010, Indian Institute of Technology Kanpur (2012)
5. Deb, K., Jain, H.: Handling many-objective problems using an improved NSGA-II procedure. In: Proceedings of World Congress on Computational Intelligence, WCCI 2012, Brisbane, Auatralia, June 10-15, pp. 10–15 (2012) (in press)
6. Deb, K., Thiele, L., Laumanns, M., Zitzler, E.: Scalable test problems for evolutionary multi-objective optimization. In: Abraham, A., Jain, L., Goldberg, R. (eds.) Evolutionary Multiobjective Optimization, pp. 105–145. Springer, London (2005)
7. Hughes, E.J.: Evolutionary many-objective optimisation: Many once or one many? In: IEEE Congress on Evolutionary Computation, CEC 2005, pp. 222–227 (2005)
8. Ishibuchi, H., Tsukamoto, N., Nojima, Y.: Evolutionary many-objective optimization: A short review. In: Proceedings of Congress on Evolutionary Computation, CEC 2008, pp. 2424–2431 (2008)
9. Liao, X., Li, Q., Zhang, W., Yang, X.: Multiobjective optimization for crash safety design of vehicle using stepwise regression model. Structural and Multidisciplinary Optimization 35, 561–569 (2008)
10. López, J.A., Coello, C.A.C.: Some techniques to deal with many-objective problems. In: Proceedings of the 11th Annual Conference Companion on Genetic and Evolutionary Computation Conference, pp. 2693–2696. ACM, New York (2009)
11. Zhang, Q., Li, H.: MOEA/D: A multiobjective evolutionary algorithm based on decomposition. IEEE Transactions on Evolutionary Computation 11(6), 712–731 (2007)

Adaptive ε-Sampling and ε-Hood
for Evolutionary Many-Objective Optimization

Hernán Aguirre[1], Akira Oyama[2], and Kiyoshi Tanaka[1]

[1] Faculty of Engineering, Shinshu University
4-17-1 Wakasato, Nagano, 380-8553 Japan
[2] Institute of Space and Astronautical Science, Japan Aerospace Exploration Agency
{ahernan,ktanaka}@shinshu-u.ac.jp, oyama@flab.isas.jaxa.jp

Abstract. Many-objective problems are becoming common in several real-world application domains and there is a growing interest to develop evolutionary many-objective optimizers that can solve them effectively. Studies on selection for many-objective optimization and most recently studies on the characteristics of many-objective landscapes, the effectiveness of operators of variation, and the effects of large populations have proved successful to advance our understanding of evolutionary many-objective optimization. This work proposes an evolutionary many-objective optimization algorithm that uses adaptive ε-dominance principles to select survivors and also to create neighborhoods to bias mating, so that solutions will recombine with other solutions located close by in objective space. We investigate the performance of the proposed algorithm on DTLZ continuous problems, using a short number of generations to evolve the population, varying population size from 100 to 20000 individuals. Results show that the application of adaptive ε-dominance principles for survival selection as well as for mating selection improves considerably the performance of the optimizer.

1 Introduction

Many-objective problems are becoming common in several real-world application domains and there is a growing interest to develop evolutionary many-objective optimizers (EMyOs) that can solve them effectively. One such application domains is multi-objective design exploration for real-world design optimization. Here, a large number of Pareto optimal solutions that give a good representation of the true Pareto front in terms of convergence, spread, and distribution of solutions along the front are essential to extract relevant knowledge about the problem. This knowledge, rather than a particular precise solution, is valuable to establish trade-offs and hotspots regions for objectives and design variables in order to provide useful guidelines to designers during the selection of alternative designs, and to facilitate the implementation of the finally chosen design. In these applications it is also common that the evaluation of solutions is computationally expensive and takes a long time to calculate it, which prohibits running the evolutionary algorithm for a large number of generations. Thus, in real-world

R.C. Purshouse et al. (Eds.): EMO 2013, LNCS 7811, pp. 322–336, 2013.

applications of many-objective optimization we have to face the difficulties imposed by the increased complexities of large-dimensional spaces and are often constrained by time.

It is well known that conventional evolutionary multi-objective optimizers (EMOs) [1, 2] scale up poorly with the number of objectives of the problem, which is often attributed to the large number of non-dominated solutions and the lack of effective selection and diversity estimation operators to discriminate appropriately among them, particularly in dominance-based algorithms. Selection, indeed, is a fundamental part of the algorithm and has been the subject of several studies, leading to improve the performance of conventional EMOs on many-objective problems. However, finding trade-off solutions that satisfy simultaneously the three properties of convergence, spread, and distribution is especially difficult on many-objective problems. In fact, most EMOs with improved selection strategies for many-objective optimization proposed recently compromise one in favor of other [3]. In addition to selection, recent studies on the characteristics of many-objective landscapes, the effectiveness of operators of variation, and the effects of large populations [4–7] have proved successful to advance our understanding of evolutionary many-objective optimization.

In previous work, we have studied the behavior of Adaptive ε-Box with Neighborhood Recombination [6], an algorithm built under the framework of NSGA-II [8]. The algorithm uses ε-box non-dominated sorting with a logarithmic function that maps solutions to a grid [9] and selects for survival just one solution per non-dominated ε-box. It also keeps a list of neighbors for each solution to restrict mating. This algorithm improves considerably the performance of NSGA-II. However, the logarithmic function originally proposed in [9] introduces a too strong bias towards the edges of objective space. In addition, though nearby individuals recombine, in this approach the search effort is not evenly balanced towards all regions of objective space. Moreover, it uses crowding distance as secondary ranking, which does not scale up well in high dimensional spaces.

In this work, we propose an EMyO that uses adaptive ε-dominance principles to select survivors and also to create neighborhoods to bias mating for recombination. The method used for survival selection is based on ε-sampling that selects solutions randomly from the set of non-dominated solutions, eliminating solutions that are ε-dominated by the sampled solutions. The motivation to use ε-sampling is that surviving solutions are spaced following the distribution implicit in the mapping function used for ε-dominance and the search effort could be balanced according to such distribution. In this work we use an additive mapping function that induces a uniform distribution of solutions, aiming to cover all regions of objective space. The method to create neighborhoods is also based on ε-dominance. Here, a randomly sampled solution from the surviving population and its ε-dominated solutions determine the neighborhood, so that recombination can take place between individuals located nearby in objective space. The motivation to restrict mating is to enhance the effectiveness of recombination in many-objective problems, where the difference in variable space between individuals in the population is expected to be larger than in multi-objective problems and therefore more disruptive for recombination. In addition, the method gives more reproductive opportunities to individuals located in under-represented regions to balance the search effort towards all regions.

We investigate the performance of the proposed algorithm on DTLZ continuous problems, varying population size from very small to large -from 100 to 20000 individuals. We assume scenarios in which computational time to calculate fitness could be very large and thus use a very short number of generations to evolve the population. We also assume that all individuals in the population can be evaluated simultaneously in parallel. As a reference for comparison, we include results by Adaptive ε-Box with Neighborhood Recombination showing that the proposed algorithm performs significantly better in terms of convergence and distribution of solutions.

2   Proposed Method

2.1   Concept

In many-objective landscapes the number of solutions in the Pareto optimal set increases almost exponentially [10, 11] with the number of objectives. Keeping fixed the size of the variable space, an increase in the number of objectives also implies that these large number of Pareto optimal solutions become spread over broader regions in variable space [4]. Analysis of many-objective landscapes show that this is the case not only for the global Pareto set containing optimal solutions, but also for local Pareto sets containing suboptimal solutions [10, 11]. These characteristics of many-objective landscapes are reflected on the dynamics of the optimizer and are directly correlated to the effectiveness of the operators of selection and variation. For example, the large number of non-dominated solutions makes dominance-based selection random and their spread on variable space causes recombination to recombine distant solutions making it too disruptive and ineffective. Hence, the characteristics of many-objective landscapes must be carefully considered when we design our algorithms.

The proposed method aims to perform an effective search on many-objective landscapes by using dominance and ε-sampling for survival selection to get a well distributed subset of non-dominated solutions, and ε-hood creation and ε-mating for parent selection to enhance the effectiveness of recombination. In multi-objective optimization, dominance is used for survival selection and to rank solutions in the elite surviving population, so that mating selection can give more reproductive opportunities to dominant individuals. However, this is impractical in many-objective optimization. In the proposed method, dominance acts as a mean to eliminate dominated solutions during survival selection and leave the non-dominated ones for further processing, but it has no role ranking the surviving population. For most part of the evolution, survival selection is achieved by ε-sampling, which samples randomly from the large set of non-dominated solutions and eliminates solutions ε-dominated by the samples. The aim is to get a set of surviving solutions spaced according to the distribution implicit in the mapping function $\boldsymbol{f}(\boldsymbol{x}) \mapsto^{\varepsilon} \boldsymbol{f}'(\boldsymbol{x})$ used for ε-dominance. Only during the few initial generations, where the number of non-dominated solutions is smaller than the size of the surviving population, ε-sampling plays no role during survival.

After survival selection, in the proposed method there is not an explicit ranking that could be used to bias mating. Rather, we use a procedure called ε-hood creation to cluster solutions in objective space. This method is also based on ε-dominance and is adaptive too. Here, a randomly sampled solution from the surviving population and its ε-dominated solutions determine the neighborhood, so that recombination can take place between individuals located nearby in objective space. The motivation to restrict mating is to enhance the effectiveness of recombination in many-objective problems, where the difference in variable space between individuals in the population is expected to be larger than in multi-objective problems and therefore more disruptive for recombination. In addition, to balance the search effort towards all regions, individuals located in under-represented regions are given more reproductive opportunities.

Summarizing, dominance acts as a mean to eliminate inferior solutions, ε-sampling gets a set of well distributed solutions from the large set of non-dominated solutions so that search effort could be uniformly distributed, ε-hood creation clusters elite solutions in objective space, ε-mating pairs nearby solutions to enhance the effectiveness of recombination, and reproduction gives more reproductive opportunities to individuals in under-represented regions.

2.2 Adaptive ε-Sampling and ε-Hood Evolutionary Many-Objective Optimizer ($A\varepsilon_s\varepsilon_h$EMyO)

In this section we explain the general flow of the proposed algorithm $A\varepsilon_s\varepsilon_h$EMyO illustrated in **Procedure 1** and in the next sections we explain in detail its distinctive features. The proposed method uses ε-dominance principles to truncate the population by sampling from the set of non-dominated solutions and also to create neighborhoods to bias mating for recombination, using parameters ε_s and ε_h, respectively. The ε_s parameter for sampling during truncation is dynamically adapted to keep the number of sampled solutions N_S close to the population size P_{size}. Similarly, the ε_h parameter for neighborhood creation is dynamically adapted to keep the number of neighborhoods N_H close to a user specified number N_H^{Ref}. In addition to ε_s and ε_h, their steps of adaptation are also adapted to properly follow the dynamics of the search. Thus, before its main loop, the algorithm sets the reference number of neighborhoods N_H^{Ref}, initial values for ε_s and its step of adaptation Δ_s, and initial values for ε_h and its step of adaptation Δ_h. Next, it creates randomly the initial population. Then it iterates the main evolutionary loop.

The main loop starts by evaluating the offspring population $\mathcal{Q}$. After offspring $\mathcal{Q}$ is evaluated, non-dominated sorting is performed on the population that results from joining the current population $\mathcal{P}$ and its offspring $\mathcal{Q}$. The population of size $2P_{size}$ sorted in non-dominated fronts $\mathcal{F}$ is then truncated to obtain the surviving population $\mathcal{P}$ of size P_{size} using a ε-sampling truncation procedure set with parameter ε_s.

Procedure 1. $A\varepsilon_s\varepsilon_h EMyO$

Require: Population size P_{size}, reference neighborhood size H_{size}^{Ref}
Ensure: $\mathcal{F}_1$, set of Pareto non-dominated solutions

1: $N_H^{Ref} \leftarrow P_{size}/H_{size}^{Ref}$ // set reference number of neighborhoods
2: $\varepsilon_s \leftarrow 0$, $\Delta_s \leftarrow \Delta_0$ // set ε_s-dominance factor and its step of adaptation
3: $\varepsilon_h \leftarrow 0$, $\Delta_h \leftarrow \Delta_0$ // set ε_h-dominance factor and its step of adaptation
4: $\mathcal{P} \leftarrow \emptyset$, $\mathcal{Q} \leftarrow random$ // initial populations $\mathcal{P}$ and $\mathcal{Q}$, $|\mathcal{Q}| = P_{size}$
5: **repeat**
6:    evaluation($\mathcal{Q}$)
7:    $\mathcal{F} \leftarrow$ non-dominated sorting($\mathcal{P} \cup \mathcal{Q}$)   // $\mathcal{F} = \{\mathcal{F}_i\}$, $i = 1, 2, \cdots, N_F$
8:    $\{\mathcal{P}, N_S\} \leftarrow \epsilon$-sampling truncation($\mathcal{F}, \epsilon_s, P_{size}$)   // $|\mathcal{P}| = |\mathcal{Q}| = P_{size}$
9:    $\{\epsilon_s, \Delta_s\} \leftarrow$ adapt ($\epsilon_s, \Delta_s, P_{size}, N_S$)
10:   $\{\mathcal{H}, N_H\} \leftarrow \epsilon$-hood creation ($\mathcal{P}, \epsilon_h$) // $\mathcal{H} = \{\mathcal{H}_j\}$, $j = 1, 2, \cdots, N_H$
11:   $\{\epsilon_h, \Delta_h\} \leftarrow$ adapt ($\epsilon_h, \Delta_h, N_H^{Ref}, N_H$)
12:   $\mathcal{P}' \leftarrow \epsilon$-hood mating($\mathcal{H}, Psize$)
13:   $\mathcal{Q} \leftarrow$ recombination and mutation($\mathcal{P}'$)
14: **until** termination criterion is met
15: **return** $\mathcal{F}_1$

The number of sampled solutions N_S and the population size P_{size} are used as reference to adapt ε_s and its step of adaptation Δ_s. Next, neighborhoods are created from the surviving population using a ε-hood creation procedure set with parameter ε_h. Similar to ε_s, ε_h and its step of adaptation Δ_h are adapted so that the number of created neighborhoods N_H would be close to a user specified reference number N_H^{Ref}. After the neighborhoods have been created, ε-hood mating creates a pool of mates $\mathcal{P}'$ by selecting solutions within the neighborhoods, so that a solution would recombine only with a solution that is close by in objective space. Next, the already defined mates are recombined and mutated to create the offspring population $\mathcal{Q}$ and the algorithm continues with the next generation until a termination criterion has been met.

2.3 ε-Sampling Truncation

Survival selection is implemented by the ε-sampling truncation method, illustrated in **Procedure 2**. This method receives the sets of solutions $\mathcal{F}$ created by non-dominated sorting and selects exactly P_{size} surviving solutions from them. In case the number of non-dominated solutions $|\mathcal{F}_1| > P_{size}$, it calls ε-sampling with parameter ε_s to get from $\mathcal{F}_1$ its extreme solutions $\mathcal{E}$, a subset of randomly sampled solutions $\mathcal{S}$ and their ε_s-dominated solutions $\mathcal{D}^{\varepsilon_s}$, as illustrated in **Procedure 3**.

The surviving population $\mathcal{P}$ always includes extreme solutions $\mathcal{E}$ and it is complemented with solutions from $\mathcal{S}$ and possibly from $\mathcal{D}^{\varepsilon_s}$. If $\mathcal{S}$ overfills $\mathcal{P}$, solutions in $\mathcal{S}$ are randomly eliminated as survivors. Otherwise, if after adding

$\mathcal{S}$ to $\mathcal{P}$ there is still room for some solutions, the required number are randomly chosen from $\mathcal{D}^{\epsilon_s}$. On the other hand, is $|\mathcal{F}_1| < P_{size}$, while there is room in $\mathcal{P}$ the sets of solutions $\mathcal{F}_i$ are copied iteratively to $\mathcal{P}$. The remaining solutions are chosen randomly from the set that did not fit completely in $\mathcal{P}$.

Procedure 2. ϵ-sampling truncation ($\mathcal{F}$, ϵ_s, P_{size})

Require: sets of non-dominated solutions $\mathcal{F} = \{\mathcal{F}_i\}, i = 1, 2, \cdots, N_F$, ϵ-dominance parameter ϵ_s and desired population size after truncation P_{size}
Ensure: Truncated population $\mathcal{P}$ obtained from $\mathcal{F}$ and number of sampled solutions including the extremes N_S

1: $\mathcal{P} \leftarrow \emptyset, N_S \leftarrow 0$
2: **if** $|\mathcal{F}_1| > P_{size}$ **then**
3:      $\{\mathcal{E}, \mathcal{S}, \mathcal{D}^{\epsilon_s}\} \leftarrow \epsilon$-sampling($\mathcal{F}_1, \epsilon_s$)
4:      $N_S \leftarrow |\mathcal{E}| + |\mathcal{S}|$
5:      **if** $N_S > P_{size}$ **then**
6:          $\mathcal{X} \leftarrow \{x_r \in \mathcal{S} \mid r = rand(1, |\mathcal{S}|), |\mathcal{X}| = N_S - P_{size}\}$
7:          $\mathcal{P} \leftarrow \mathcal{E} \cup \mathcal{S} \setminus \mathcal{X}$
8:      **else**
9:          $\mathcal{X} \leftarrow \{x_r \in \mathcal{D}^{\epsilon_s} \mid r = rand(1, |\mathcal{D}^{\epsilon_s}|), |\mathcal{X}| = P_{size} - N_S\}$
10:       $\mathcal{P} \leftarrow \mathcal{E} \cup \mathcal{S} \cup \mathcal{X}$
11:     **end if**
12: **else**
13:      $\mathcal{P} \leftarrow \bigcup_{i=1}^{k} \mathcal{F}_i, \quad \sum |\mathcal{F}_i| < P_{size}$
14:      $\mathcal{X} \leftarrow \{x_r \in \mathcal{F}_{k+1} \mid r = rand(1, |\mathcal{F}_{k+1}|), |\mathcal{X}| = P_{size} - \sum_{i=1}^{k} |\mathcal{F}_i|\}$
15:      $\mathcal{P} \leftarrow \mathcal{P} \cup \mathcal{X}$
16: **end if**
17: **return** $\mathcal{P}$ and N_S

Procedure 3. ϵ-sampling ($\mathcal{F}_1, \epsilon_s$)

Require: Non-dominated solutions $\mathcal{F}_1$, ϵ-dominance parameter ϵ_s
Ensure: $\mathcal{E}$, $\mathcal{S}$ and $\mathcal{D}^{\epsilon_s}$, $\mathcal{E} \cup \mathcal{S} \cup \mathcal{D}^{\epsilon} = \mathcal{F}_1$. $\mathcal{E}$ and $\mathcal{S}$ contain extreme solutions and a randomly chosen sample of solutions from $\mathcal{F}_1$, respectively, whereas $\mathcal{D}^{\epsilon_s}$ contains solutions ϵ_s-dominated by those in $\mathcal{S}$. Maximization in all objectives is assumed

1: $\mathcal{E} \leftarrow \{x \in \mathcal{F}_1 \mid f_m(x) = \max(f_m(\cdot)), m = 1, 2, \cdots, M\}$ // extremes
2: $\mathcal{F}_1 \leftarrow \mathcal{F}_1 \setminus \mathcal{E}$
3: $\mathcal{D}^{\epsilon_s} \leftarrow \emptyset$
4: **while** $\mathcal{F}_1 \neq \emptyset$ **do**
5:      $z \leftarrow x_r \in \mathcal{F}_1 \mid r = rand(1, |\mathcal{F}_1|)$
6:      $\mathcal{S} \leftarrow \mathcal{S} \cup \{z\}$ // add randomly chosen solution z to sample
7:      $\mathcal{Y} \leftarrow \{y \in \mathcal{F}_1 \mid z \succeq^{\epsilon_s} y, z \neq y\}$ // solutions ϵ_s-dominated by z
8:      $\mathcal{D}^{\epsilon_s} \leftarrow \mathcal{D}^{\epsilon_s} \cup \mathcal{Y}$
9:      $\mathcal{F}_1 \leftarrow \mathcal{F}_1 \setminus \{\{z\} \cup \mathcal{Y}\}$
10: **end while**
11: **return** $\mathcal{E}, \mathcal{S}, \mathcal{D}^{\epsilon_s}$

2.4   ε-Hood Creation and ε-Hood Mating

Neighborhoods are created from the surviving population by the ε-hood creation procedure, which is also based on ε-dominance as illustrated in **Procedure 4**. This procedure randomly selects an individual from the surviving population and applies ε-dominance with parameter ε_h. A neighborhood is formed by the sampled solutions and its ε_h-dominated solutions. Neighborhood creation is repeated until all solutions in the surviving population have been assigned to a neighborhood.

Procedure 4. ε-hood creation ($\mathcal{P}$, ε_h)

Require: Population $\mathcal{P}$, ε-dominance parameter ϵ_h for neighborhood creation
Ensure: Neighborhoods $\mathcal{H} = \{\mathcal{H}_i\}$, $i = 1, 2, \cdots, N_H$

1: $\mathcal{H} \leftarrow \emptyset$
2: $i \leftarrow 0$
3: **while** $\mathcal{P} \neq \emptyset$ **do**
4:       $z \leftarrow x_r \in \mathcal{P} \mid r = rand(\ 1,\ |\mathcal{P}|\ )$ // z, a randomly chosen solution
5:       $\mathcal{Y} \leftarrow \{y \in \mathcal{P} \mid z \succeq^{\varepsilon_h} y, z \neq y\}$ // solutions ε_h-dominated by z
6:       $i \leftarrow i + 1$
7:       $\mathcal{H}_i \leftarrow \{\{z\} \cup \mathcal{Y}\}$ // z and its ε_h-dominated solutions form the hood
8:       $\mathcal{H} \leftarrow \mathcal{H} \cup \mathcal{H}_i$
9:       $\mathcal{P} \leftarrow \mathcal{P} \setminus \mathcal{H}_i$
10: **end while**
11: $N_H \leftarrow i$
12: **return** $\mathcal{H}$, N_H

Procedure 5. ε-hood mating ($\mathcal{H}$, P_{size})

Require: Neighborhoods $\mathcal{H} = \{\mathcal{H}_i\}$, $i = 1, 2, \cdots, N_H$, and population size P_{size}
Ensure: Pool of mated parents $\mathcal{P}'$, $|\mathcal{P}'| = 2P_{size}$

1: $\mathcal{P}' \leftarrow \emptyset$
2: $i \leftarrow 1$
3: $j \leftarrow 0$
4: **while** $j < P_{size}$ **do**
5:       $\{y, z\} \leftarrow \{x_{r_1}, x_{r_2} \in \mathcal{H}_i \mid r_1 \wedge r_2 = rand(\ 1,\ |\mathcal{H}_i|\ ), r_1 \neq r_2\}$
6:       $\mathcal{P}' \leftarrow \mathcal{P}' \cup \{y, z\}$
7:       $i \leftarrow 1 + (i \bmod N_H)$
8:       $j \leftarrow j + 1$
9: **end while**
10: **return** $\mathcal{P}'$

Mating for recombination is implemented by the procedure ε-hood mating illustrated in **Procedure 5**. Neighborhoods are considered to be elements of a list. To select two mates, first a neighborhood from the list is specified deterministically in a round-robin schedule. Then, two individuals are select randomly within the specified neighborhood, so that an individual will recombine with other individual that is located close by in objective space. Due to the round-robin schedule, the next two mates will be selected from the next neighborhood in the list. When the end of the neighborhood lists is reached, mating continues with the first neighborhood in the list. Thus, all individuals have the same probability of being selected within a specified neighborhood, but due to the round-robin scheduling individuals belonging to neighborhoods with fewer members have more recombination opportunities that those belonging to neighborhoods with more members. Once the pool of all mates $\mathcal{P}'$ has been established, they are recombined and mutated according to the order they were selected during mating.

2.5 Additive Epsilon Mapping $f(x) \mapsto^\epsilon f'(x)$

In this work we use an evenly spaced Additive mapping function $f(x) \mapsto^\epsilon f'(x)$ [5] for both ε-sampling and ε-hood creation. The Additive function maps $f(x)$ to $f'(x)$ by adding the same value ϵ to all coordinates f_i, independently of the position of $f(x)$ in objective space. This mapping in ε-sampling induces a distribution of solutions evenly spaced by ϵ. The expression for Additive mapping is as follows

$$f'_i(x) = f_i(x) + \epsilon, \quad i = 1, \cdots, m \tag{1}$$

2.6 Adaptation

The number of sampled solutions N_S by ε-sampling depends on the value set to ϵ_s (≥ 0). Larger values of ϵ_s imply that sampled solutions ϵ_s-dominate larger areas, increasing the likelihood of having more ϵ_s-dominated solutions excluded from the sample. The proposed algorithm adapts ϵ_s at each generation so that N_S is close to the population size P_{size}. The closer N_S is to P_{size}, the larger the number of surviving solutions that will be spaced according to the distribution implicit in the mapped function used for ε-dominance.

Similarly, the number of created neighborhoods N_H depends on the value set to ϵ_h (≥ 0). Larger values of ϵ_h imply that sampled solutions ϵ_h-dominate larger areas, increasing the likelihood of having more ϵ_h-dominated solutions that form its neighborhood, and therefore less created neighborhoods. The proposed algorithm adapts ϵ_h at each generation so that N_H is close to a user specified number N_H^{Ref}.

The adaptation rule, similar for both processes, is as follows. If $N > Ref$ it increases the step of adaptation $\Delta \leftarrow \min(\Delta \times 2, \Delta_{max})$ and $\epsilon \leftarrow \epsilon + \Delta$. Otherwise, if $N < Ref$ it decreases $\Delta \leftarrow \max(\Delta \times 0.5, \Delta_{min})$ and $\epsilon \leftarrow \max(\epsilon - \Delta, 0.0)$. In this work we set initial values $\epsilon_0 = 0.0$ and $\Delta_0 = 0.005$. Also, $\Delta_{max} = 0.05$ and $\Delta_{min} = 0.0001$.

In the case of adapting the parameter ϵ_s used for truncation, the above rule is called with $\epsilon = \epsilon_s$, $\Delta = \Delta_s$, $N = N_S$, and $Ref = P_{size}$. On the other hand, in the case of the parameter ϵ_h used for neighborhood creation, the above rule is called with $\epsilon = \epsilon_h$, $\Delta = \Delta_h$, $N = N_H$, and $Ref = N_H^{Ref}$.

3   Test Problems, Performance Indicators, and Experimental Setup

We study the performance of the algorithms in continuous functions DTLZ2, DTLZ3, and DTLZ4 of the DTLZ test functions family [12]. These functions are scalable in the number of objectives and variables and thus allow for a many-objective study. In our experiments, we vary the number of objectives from $m = 4$ to 6 and set the total number of variables to $n = (m-1)+10$. DTLZ2 has a non-convex Pareto-optimal surface that lies inside the first quadrant of the unit hyper-sphere. DTLZ3 and DTLZ4 are variations of DTLZ2. DTLZ3 introduces a large number of local Pareto-optimal fronts in order to test the convergence ability of the algorithm. DTLZ4 introduces biases on the density of solutions to some of the objective-space planes in order to test the ability of the algorithms to maintain a good distribution of solutions. For a detailed description of these problems the reader is referred to [12].

To evaluate the Pareto optimal solutions obtained by the algorithms we use the Generational Distance (GD) [13], which measures the convergence of solutions to the true Pareto front using equation 2, where P denotes the set of Pareto optimal solutions found by the algorithm and x a solution in the set. Smaller values of GD indicate that the set P is closer to the Pareto optimal front. That is, smaller values of GD mean better convergence of solutions.

$$GD = \underset{x \in P}{average} \left\{ \left[\sum_{i=1}^{m} (f_i(x))^2 \right]^{\frac{1}{2}} - 1 \right\} \qquad (2)$$

To visually assess the distribution of solutions in objective space, we plot solutions projected to a two dimensional plane.

We run the algorithms 30 times and present average results, unless stated otherwise. We use a different random seed in each run, but all algorithms use the same seeds. The number of generations is set to 100 generations, and population size varies from to 100 to 20000, $|\mathcal{P}| = |\mathcal{Q}|$. As variation operators, the algorithms use SBX crossover and polynomial mutation, setting their distribution exponents to $\eta_c = 15$ and $\eta_m = 20$, respectively. Crossover rate is $pc = 1.0$, crossover rate per variable $pcv = 0.5$, and mutation rate per variable is $pm = 1/n$.

For $A\varepsilon_s\varepsilon_h$EMyO ($A\varepsilon_s\varepsilon_h$ for short) we set the reference neighborhood size H_{size}^{Ref} to 20 individuals. On the other hand, for Adaptive ε-Box with Neighborhood Recombination (AεBox-NR), we set the size of the neighborhood to 10% of the population size, a value that gave the best results in [6].

4   Simulation Results and Discussion

4.1   Convergence

Fig.1 shows GD over population size by $A\varepsilon_s\varepsilon_h$ and AεBox-NR at generation $T = 100$ on problem DTLZ2 for $m = 5$ and $m = 6$ objectives. It can be seen

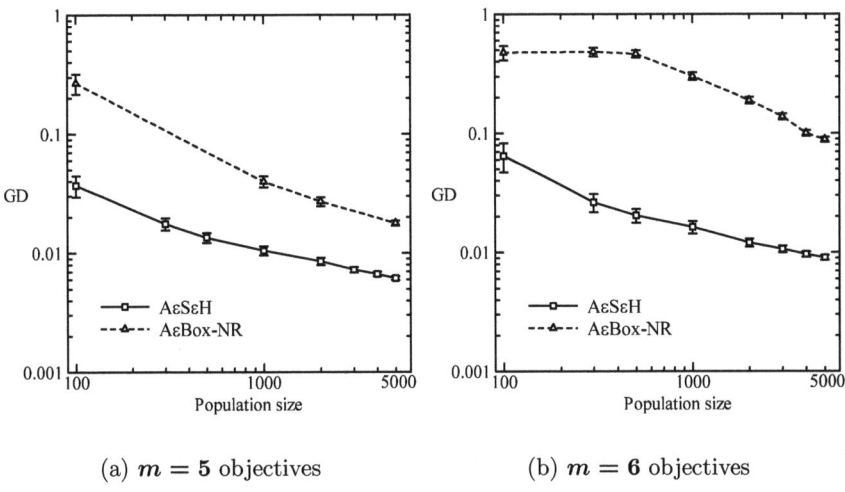

(a) $m = 5$ objectives   (b) $m = 6$ objectives

Fig. 1. GD after 100 generations for various population sizes, DTLZ2

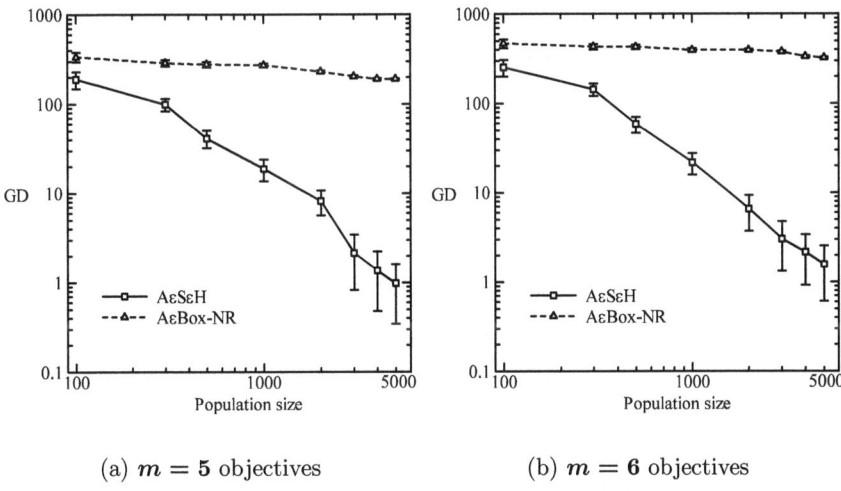

(a) $m = 5$ objectives   (b) $m = 6$ objectives

Fig. 2. GD after 100 generations for various population sizes, DTLZ3

that smaller values of GD are achieved by using larger populations. Note that GD by the proposed $A\varepsilon_s\varepsilon_h$ is substantially better than by AεBox-NR for any population size. A larger performance difference between the two algorithms can be seen by increasing the number of objectives from **5** to **6**.

Similarly, **Fig.2** shows results for DTLZ3 problem. Analogous to problem DTLZ2, the proposed algorithm $A\varepsilon_s\varepsilon_h$ performs better than AεBox-NR for any population size. In fact, in DTLZ3 the improvement in performance by the proposed algorithm is more notorious than in the case of DTLZ2. Note that increasing population size does not help much AεBox-NR, whereas GD by $A\varepsilon_s\varepsilon_h$ decreases in two orders of magnitude when population size is increased from $|\mathcal{P}| = 100$ to $|\mathcal{P}| = 5000$. However, it should be noticed that variance is larger and increases considerably with population size for the DTLZ3 problem. This is because convergence is harder in this problem and a larger number of generations are required to facilitate convergence of most individuals of the population towards the true Pareto front.

Results for DTLZ4 are similar to DTLZ2. Also, results on $m = 4$ objectives for all problems follow a similar trend to those observed on $m = 5$ and $m = 6$ objectives. Due to space limitations, those results are not included here.

4.2   Distribution

Fig.3 shows the f_1-f_2 objective values of the obtained non-dominated solutions by the proposed $A\varepsilon_s\varepsilon_h$ at generation 100. Results are shown for DTLZ2, $m = 6$ objectives, running the algorithm with some representative population sizes. Similarly, **Fig.4** shows solutions by AεBox-NR. In DTLZ2 the sum of the squares of the fitness values of a Pareto optimal solution is one. Thus, on a problem with more than 2 objectives the f_1-f_2 values of Pareto optimal solution fall within the positive quadrant of the circle of radius one. From **Fig.3** note that objective values are close to or within the positive quadrant of the circle of radius one. This is in accordance with the good convergence values observed for GD discussed above. Increasing the population size there is a better coverage of the quadrant, which implies a better distribution of solutions in objective space, and fewer solutions are located outside the quadrant. On the other hand, from **Fig.4** it can be seen that solutions by AεBox-NR tend to focus on extreme regions of objective space, where one or more objective values are close to 0, and many of them are far away from the optimal front. This effect reduces when a large population size is used, such as $|\mathcal{P}| = 5000$, but still there are many extreme solutions away from the optimal front and those located within the positive quadrant are not able to fully cover it.

4.3   GD over the Generations and Larger Population Sizes

Fig.5 shows the transition of GD over the generations by $A\varepsilon_s\varepsilon_h$ on $m = 6$ objectives DTLZ2 and DTLZ3 problems, varying population sizes from $|\mathcal{P}| = 100$ to $|\mathcal{P}| = 20000$ individuals. Note that from early generations the algorithm with a larger population shows better convergence. This is a clear indication that population size is very important to support appropriately the evolutionary search on many-objective problems. On DTLZ2, note that for large populations initially there is a fast convergence, but after 50 generations or so the algorithm slows down significantly. On DTLZ3, convergence at the beginning is slower than on DTLZ2 but after some generations the effect of population size becomes more evident. Note that after **50** generations convergence speeds up significantly for populations $|\mathcal{P}| = 5000$ and $|\mathcal{P}| = 10000$ individuals. For

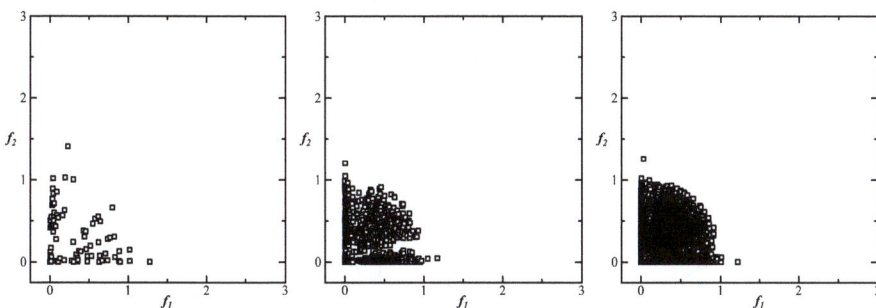

(a) $|\mathcal{P}| = 100$ individuals (b) $|\mathcal{P}| = 1000$ individuals (c) $|\mathcal{P}| = 5000$ individuals

Fig. 3. Obtained non-dominated solutions after 100 generations by $A\varepsilon_s\varepsilon_h$ for various population sizes, DTLZ2. Projection in plane f_1-f_2.

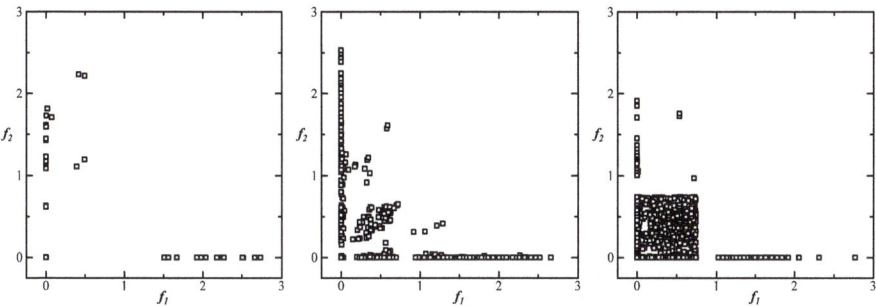

(a) $|\mathcal{P}| = 100$ individuals (b) $|\mathcal{P}| = 1000$ individuals (c) $|\mathcal{P}| = 5000$ individuals

Fig. 4. Obtained non-dominated solutions after 100 generations by $A\varepsilon$Box-NR for various population sizes, DTLZ2. Projection in plane f_1-f_2

$|\mathcal{P}| = 20000$ individuals convergence speeds up earlier, at generation **20**. However, after generation **80** convergence slows down similar to DTLZ2. Although in both problems, DTLZ2 and DTLZ3, there is still room for converging closer to the optimal Pareto front the algorithm in final generations seems to stagnate. This suggests that the operators of variation themselves might need to be improved, particularly for the latest stage of the search when the population is approaching the Pareto optimal front. We would like to look into this in a future work.

4.4   Adaptation

Fig.6 (a) shows the adaptation of ε_s for ε-sampling and **Fig.6 (b)** shows the number of solutions on the first front $\mathcal{F}_1$ after non-dominated sorting together with the number of sampled solutions N_S by the ε-sampling procedure using

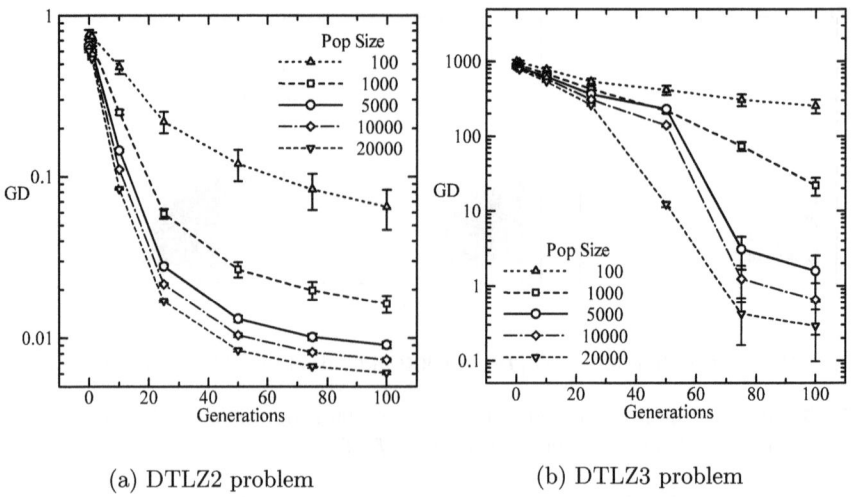

(a) DTLZ2 problem   (b) DTLZ3 problem

Fig. 5. GD over the generations for various population sizes, $m = 6$ objectives

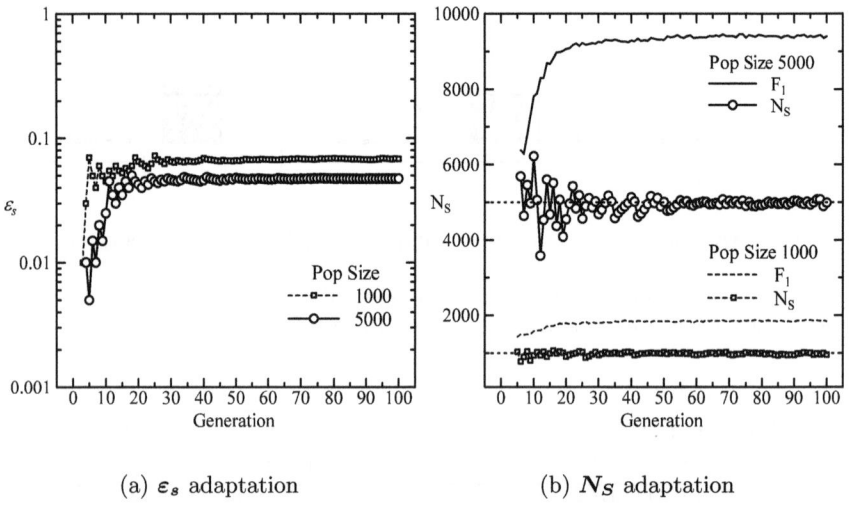

(a) ε_s adaptation   (b) N_S adaptation

Fig. 6. Adaptation for ε-sampling, DTLZ2, $m = 6$ objectives

the adapted ε_s parameter. Results are shown for population sizes $|\mathcal{P}| = 1000$ and $|\mathcal{P}| = 5000$. From **Fig.6 (b)** note that $\mathcal{F}_1$ is larger than $\mathcal{P}$ since early generations and quickly approaches $2|\mathcal{P}|$. The number of solutions N_S obtained after ε-sampling from $\mathcal{F}_1$ is kept around the desired number $|\mathcal{P}|$ thanks to the adaptation of ε_s, as shown in **Fig.6 (a)**. From **Fig.6 (a)** note that ε_s is quickly adapted from its initial value so that N_S approaches the desired value $|\mathcal{P}|$. Also note that the value that ε_s takes depends on the population size.

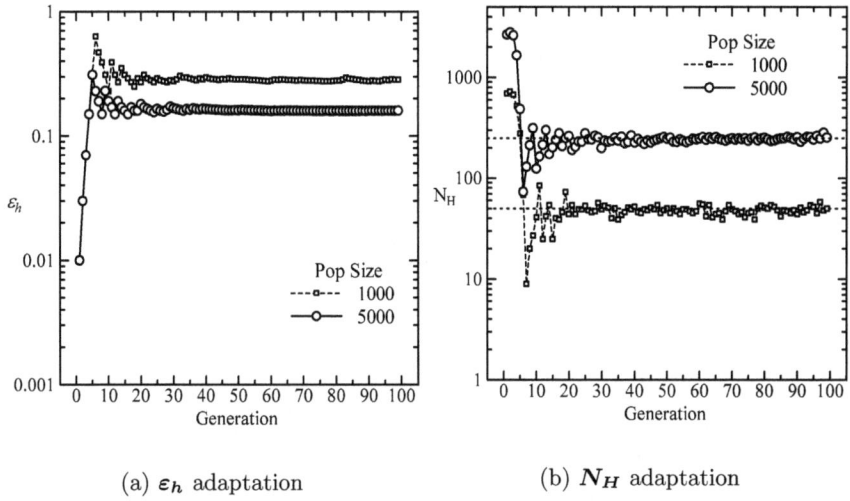

(a) ε_h adaptation   (b) N_H adaptation

Fig. 7. Adaptation for neighborhood creation, DTLZ2, $m = 6$ objectives

Similar to the previous figure, **Fig.7 (a)** shows the adaptation of ε_h for neighborhood creation and **Fig.7 (b)** shows the number created neighborhoods N_H by the ε-hood creation procedure using the adapted ε_h parameter. From **Fig.7 (a)** note that ε_h is quickly adapted from its initial value, it takes values larger than ε_s as should be expected, and depends on population size. From **Fig.7 (b)** note that at the beginning the number of neighborhoods is quite large but thanks to the quick adaptation of ε_h the number of created neighborhoods N_H approaches the specified number N_H^{Ref}, **50** for $|\mathcal{P}| = 1000$ and **250** for $|\mathcal{P}| = 5000$.

The above results show that adaption is working properly for both ε-sampling and ε-hood creation.

5   Conclusions

This work has proposed an evolutionary many-objective optimizer that uses adaptive ε-sampling to select a subset of well distributed solutions for the surviving population. The method also uses adaptive ε-dominance to create neighborhoods of surviving solutions and performs mating between individuals of the same neighborhood to enhance the effectiveness of recombination. We verified the performance of the algorithm using DTLZ problems, observing the effects of increasing the population size on convergence and distribution of solution. We showed that for any population size the proposed method achieves substantially better quality of solutions in terms of convergence and distribution compared to Aε-Box with Neighborhood Recombination. We also showed that the method can successfully adapt the ε parameters used for truncation and neighborhood creation. In the future we would like to look into the operators of variation, aiming to further improve convergence. Also, we should analyze with more detail the impact of on performance of neighborhood size. In addition, we would like to test the proposed method on other classes of problems.

References

1. Deb, K.: Multi-Objective Optimization using Evolutionary Algorithms. John Wiley & Sons, Chichester (2001)
2. Coello, C., Van Veldhuizen, D., Lamont, G.: Evolutionary Algorithms for Solving Multi-Objective Problems. Kluwer Academic Publishers, Boston (2002)
3. Ishibuchi, H., Tsukamoto, N., Nojima, Y.: Evolutionary Many-Objective Optimization: A Short Review. In: Proc. IEEE Congress on Evolutionary Computation (CEC 2008), pp. 2424–2431. IEEE Press (2008)
4. Sato, H., Aguirre, H., Tanaka, K.: Genetic Diversity and Effective Crossover in Evolutionary Many-objective Optimization. In: Coello, C.A.C. (ed.) LION 5. LNCS, vol. 6683, pp. 91–105. Springer, Heidelberg (2011)
5. Aguirre, H., Oyama, A., Tanaka, K.: Adaptive Epsilon-Ranking and Distribution Search on Evolutionary Many-objective Optimization. In: Proc. 2012 Italian Workshop on Artificial Life and Evolutionary Computation: Artificial Life, Evolution and Complexity (WIVACE 2012), CD-ROM, pp. 1–12 (2012)
6. Kowatari, N., Oyama, A., Aguirre, H., Tanaka, K.: A Study on Large Population MOEA Using Adaptive ϵ-Box Dominance and Neighborhood Recombination for Many-Objective Optimization. In: Hamadi, Y., Schoenauer, M. (eds.) LION 6. LNCS, vol. 7219, pp. 86–100. Springer, Heidelberg (2012)
7. Kowatari, N., Oyama, A., Aguirre, H., Tanaka, K.: Analysis on Population Size and Neighborhood Recombination on Many-Objective Optimization. In: Coello, C.A.C., Cutello, V., Deb, K., Forrest, S., Nicosia, G., Pavone, M. (eds.) PPSN XII, Part II. LNCS, vol. 7492, pp. 22–31. Springer, Heidelberg (2012)
8. Deb, K., Agrawal, S., Pratap, A., Meyarivan, T.: A Fast Elitist Non-Dominated Sorting Genetic Algorithm for Multi-Objective Optimization: NSGA-II, KanGAL report 200001 (2000)
9. Laumanns, M., Thiele, L., Deb, K., Zitzler, E.: Combining Convergence and Diversity in Evolutionary Multi-objective Optimization. Evolutionary Computation 10(3), 263–282 (2002)
10. Aguirre, H., Tanaka, K.: Insights on Properties of Multi-objective MNK-Landscapes. In: Proc. 2004 IEEE Congress on Evolutionary Computation, pp. 196–203. IEEE Service Center (2004)
11. Aguirre, H., Tanaka, K.: Working Principles, Behavior, and Performance of MOEAs on MNK-Landscapes. European Journal of Operational Research 181(3), 1670–1690 (2007)
12. Deb, K., Thiele, L., Laumanns, M., Zitzler, E.: Scalable Multi-Objective Optimization Test Problems. In: Proc. 2002 Congress on Evolutionary Computation, pp. 825–830. IEEE Service Center (2002)
13. Zitzler, E.: Evolutionary Algorithms for Multiobjective Optimization: Methods and Applications. PhD thesis, Swiss Federal Institute of Technology, Zurich (1999)

"Whatever Works Best for You"- A New Method for a Priori and Progressive Multi-objective Optimisation

Rui Wang[1,2], Robin C. Purshouse[1], and Peter J. Fleming[1]

[1] Department of Automatic Control & Systems Engineering, University of Sheffield
{cop10rw,r.purshouse,p.fleming}@sheffield.ac.uk
http://www.shef.ac.uk/acse
[2] Department of Systems Engineering, College of Information Systems and Management, National University of Defence Technology, ChangSha, China, 410073

Abstract. Various multi-objective evolutionary algorithms (MOEAs) have been developed to help a decision maker (DM) search for his/her preferred solutions to multi-objective problems. However, none of these approaches has catered simultaneously for the two fundamental ways that DM can specify his/her preferences: weights and aspiration levels. In this paper, we propose an approach named iPICEA-g that allows the DM to specify his preference in either format. iPICEA-g is based on the preference-inspired co-evolutionary algorithm (PICEA-g). Solutions are guided toward regions of interest (ROIs) to the DM by co-evolving sets of goal vectors exclusively generated in the ROIs. Moreover, a friendly decision making technique is developed for interaction with the optimization process: the DM specifies his preferences easily by interactively brushing his preferred regions in the objective space. No direct elicitation of numbers is required, reducing the cognitive burden on DM. The performance of iPICEA-g is tested on a set of benchmark problems and is shown to be good.

Keywords: Preferences, interactive, decision making, co-evolution.

1 Introduction

Multi-objective optimization problems (MOPs) arise in many real-world applications, where multiple conflicting objectives must be simultaneously satisfied. Over the last two decades, multi-objective evolutionary algorithms (MOEAs) have become increasingly popular for solving MOPs since: (1) their population based nature is particularly useful for approximating trade-off surfaces in a single run; and (2) they tend to be robust to underlying cost function characteristics [1].

The fundamental goal of solving MOPs is to help a DM to consider the multiple objectives simultaneously and to identify one final Pareto optimal solution that pleases him/her the most [2]. Most of the proposed MOEAs aim to obtain a good approximation of the whole Pareto optimal front and subsequently let the DM choose a preferred one, i.e. a posteriori decision making [2]. Such a

R.C. Purshouse et al. (Eds.): EMO 2013, LNCS 7811, pp. 337–351, 2013.
© Springer-Verlag Berlin Heidelberg 2013

process is effective on small-scale problems. However, it has difficulties on large-size problems (e.g. MOPs with many objectives), because approximation of the whole Pareto optimal front is computationally difficult and DM is usually only interested in some regions of the Pareto front.

To facilitate the process of decision making, a better choice is to consider DM preferences in an a priori (preferences are specified before the start of the search) or an interactive (preferences are articulated during the search) way [2]. In these cases, the search can be guided by the preferences toward the ROIs of the Pareto front and away from exploring non-interesting solutions. Since the final decision making process is based on a set of preferred solutions the burden to DM can be reduced significantly.

In the multi-criteria decision making (MCDM) community, various ways have been proposed to represent the preferences of a DM e.g. aspiration levels (goals), weights (search directions), local trade-off, utility function, outranking, fuzzy logic, etc. [2]. the The most frequently used ways are weights (or search direction) and aspiration levels (or goals) [3]. By using weights [4,5] it is easy to articulate DM's bias toward some objectives yet difficult to obtain a precise ROI. By using aspiration levels [6] it is easy to obtain a precise ROI yet difficult to incorporate the DM's bias. Moreover, in some cases it is easier for the DM to express the preferences by weights and in some cases by aspiration levels. The most flexible approach would be to develop a unified approach which enables the DM to articulate both types of the preferences.

In this paper, we describe such a unified approach. Three parameters: reference point (R), weight (W) and search range (θ) are introduced. Then a new interactive evolutionary multi-bjective optimization and decision-making algorithm, iPICEA-g, is proposed that incorporates the unified approach within the existing algorithm PICEA-g [7,8]. Similar to PICEA-g, in iPICEA-g candidate solutions are co-evolved with goal vectors and so guided toward the Pareto front. However, in iPICEA-g the co-evolved goal vectors are exclusively generated in the ROIs that are defined by the three parameters. Moreover, a very friendly interactive technique is developed with which the DM need not use any numeric values to specify his preferences; rather he describes his preferences by interactively brushing his preferred regions in objective space. iPICEA-g automatically configures the required parameters according to the brushed regions and therefore guides the solutions toward the ROIs.

The reminder of the paper is organized as follows: in Section 2 a brief review of preference based MOEAs is presented. This is followed, in Section 3, by an elaboration of the proposed approach iPICEA-g. Section 4 introduces the simulation results of using iPICEA-g to solve different problems in an a priori way or an interactive way. Section 5 concludes and discusses the future research.

2   Review of a Selection of Preference Based MOEAs

A variety of MOEAs that have integrated MCDM methods for preference articulation have been proposed in literature. In this section, we briefly review some representative preference based MOEAs. Two comprehensive survey papers can be found in [9,3].

MOGA [10,6] proposed by Fonscea and Fleming includes probably the earliest attempt to incorporate DM preferences. In their studies, preferences are expressed with goals and priorities. The incorporation of the preferences can be in either a priori or interactive manner. Candidate solutions are ranked based on the Pareto dominance relation together with the specified preferences and therefore the search space of interest gradually becomes smaller during the evolution. MOGA has been successfully used in a variety of applications, including the low-pressure spool speed governor of a Pegasus gas turbine engine [11,12]. The main disadvantage of this approach is that it cannot explore multiple ROIs at the same time. However, exploring multiple ROIs simultaneously is useful when the DM cannot decide which particular region to explore to be explored at the beginning, also for group decision making (different DMs can search for their preferred solutions and select the final solution at the end).

Molina et al. [13] suggested a dominance relation called g-dominance. Solutions satisfying all aspiration levels and solutions fulfilling none of the aspiration levels are preferred over solutions satisfying some aspiration levels. In [13] an approach that couples g-dominance and NSGA-II is proposed to search for ROIs. This algorithm works regardless of whether the specified goal vector is feasible or infeasible and also it is also easy to extended in an interactive manner. However, the g-dominance relation does not preserve a Pareto based ordering. Also, the performance of the algorithm is degraded as the number of objectives increases [14].

Branke et al [15] proposed a guided MOEA (G-MOEA). In the algorithm, considering DM preferences are expressed by modifying the definition of dominance using specified trade-offs between objectives: that is, how much improvement in one or more objective(s) is comparable to a unit degradation in another objective. G-MOEA works well for two objectives; however, providing all pair-wise information in a problem with many objectives is cognitively intensive.

In addition to the above Pareto related approaches, a large body of works are based on the use of reference point, reference direction and light beam search [2]. Two representative reference point based MOEAs are R-NSGA-II [16] and PBEA [17]. R-NSGA-II hybridized reference point with NSGA-II. Reference point is not applied in a classical way, i.e. together with an achievement scalarizing function [18], but rather to establish a biased crowding scheme. Specifically, solutions near reference points are emphasized by the selection mechanisms. The extent and the distribution of the solutions is maintained by an additional parameter ϵ. PBEA is hybridization of reference point method and the indicator based evolutionary algorithm (IBEA [19]). The preference is incorporated by a binary quality indicator (the ϵ-indicator) which is also Pareto dominance preserving. However, since the spread range of the obtained solutions are controled by an additional fitness scaling factor, it is difficult to control the range of the obtained solutions.

Deb et al. [4] combined the reference direction method with NSGA-II. Preferences are modelled by the reference direction (weights) encoded by a staring point and a reference point. This approach is able to find Pareto optimal solutions corresponding to reference points along the reference direction. Multiple ROIs can be obtained by using multiple reference directions. Deb et al. [5] also hybridized NSGA-II with the light beam search method, which enables searching part(s) of the Pareto optimal regions illuminated by the light beam emanating from the starting point to the reference point with a span controlled by a threshold.

Researchers from the MCDM community also developed some interactive MCDM approaches based on MOEAs. For example, Kaliszewski et al [20] proposed to incorporate the DM preference (expressed by search directions) with a Chebyshef scalarizing function and to execute the optimization search from both below (lower bounds) and above (upper bounds). The bounds are approximated based on the objective values of the solutions which are of interest to the DM.

All the above approaches have merit and are able to find Pareto optimal solutions in a ROI. However, none of the above approaches can simultaneously deal with preferences in the form of weights or in the form of aspiration levels. Moreover, among these approaches, some cannot explore multiple ROIs, e.g. MOGA; some do not perform well on many-objective problems [21,22,14], e.g. g-dominance based MOEA; some cannot search for a precise ROI, e.g. R-NSGA-II and PBEA.

3   A Unified New Approach for Articulating Decision Maker's Preference

In this section we introduce in detail the iPICEA-g algorithm. Since the iPICEA-g is based on PICEA-g [7,8], we firstly give a short introduction to PICEA-g.

3.1   Preference-Inspired Co-evolutionary Algorithms Using Goal Vectors

Preference-inspired co-evolutionary algorithms (PICEAs) represent a new class of MOEAs that were proposed by Purshouse et al. [7]. In PICEAs, incorporating concepts from Lohn et al [23], a population of candidate solutions are co-evolved with a set of preferences during the optimization process. Note that the co-evolved preference are not the real decision-maker preferences but are used as a means of comparing solutions for the purposes of a posteriori decision making.

Co-evolution of goal vectors (PICEA-g) is one realization of a PICEA [8]. In PICEA-g, a family of goal vectors and a population of candidate solutions are co-evolved as the search progresses. Candidate solutions gain fitness by meeting (weakly dominating [1]) a particular set of goal vectors in objective-space, but the fitness contribution is shared between other solutions that also satisfy those goals. Goal vectors only gain fitness by being satisfied by a candidate solution, but the fitness is reduced the more times the goals are met by other solutions in the population. The overall aim is for the goal vectors to adaptively guide the candidate solutions towards the Pareto optimal front. That is, the candidate solution population and the goal vectors co-evolve towards the Pareto optimal front. For more details readers are referred to [7,8].

3.2 Interactive Preference-Inspired Co-evolutionary Algorithms Using Goal Vectors

As argued earlier, in some cases it is easier for the DM to specify his preferences in the form of weights (reference/search direction) while other times it is more convenient for the DM to specify an aspiration level (goal). To meet the needs of both types of DM, a unified approach is proposed in this section.

The Unified Approach. Three parameters are defined for the unified approach: a reference point in objective space (R), a search direction (W) and a search range (θ). R is to describe the aspiration levels; W is to introduce the DM's bias toward some objectives where $\sum_{i=1}^{M} w_i = 1$, $\forall i, w_i \geq 0$. M is the number of objectives; θ is to control the range of the ROI. An example in the bi-objective case is shown in Figure 1. Note that R can also be unattainable; this will be described later.

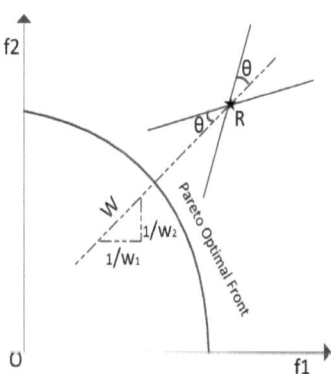

Fig. 1. Illustration of the parameters R, W and θ

Using the three parameters, DM preferences can be expressed either by weights or aspiration levels. If the DM specifies weights then R is set to the *ideal* point (or the coordinate origin, O), W represents specified weights and θ could be any value within the range $[0, \frac{\pi}{2}]$ radians. If the DM specifies aspirations then R is set as the aspiration levels, $w_i = 1/M, i = 1, \cdots, M$, $\theta = \arccos(\frac{\sqrt{M-1}}{\sqrt{M}})$, e.g., when $M = 2$, $w_1 = w_2 = 0.5$ and $\theta = \frac{\pi}{4}$.

The Proposed Algorithm: iPICEA-g. Using the concepts from PICEA-g, it is easy to imagine that if the goal vectors are exclusively generated in a region then candidate solutions inside this region will be encouraged in the evolution. The reason is that these candidate solutions can meet (weakly dominate) more goal vectors and so result in higher fitness, while candidate solutions outside this region can only meet(weakly dominate) few goal vectors and so have a lower fitness. Therefore, over the generations more and more candidate solutions will be guided toward the specified region. For example, in Figure 2, goal vectors are generated in regions $G1$ and $G2$. The objective vector $f(s_1)$ of solution s_1 is inside the region $G1$ while $f(s_1)$ of s_2 is outside the $G1$. Compared to $f(s_2)$, $f(s_1)$ can meet more goal vectors. That is, $f(s_1)$ would obtain a higher fitness than $f(s_2)$, thereby, $f(s_1)$ is more likely to be retained in the search process while $f(s_2)$ is likely to be disregarded.

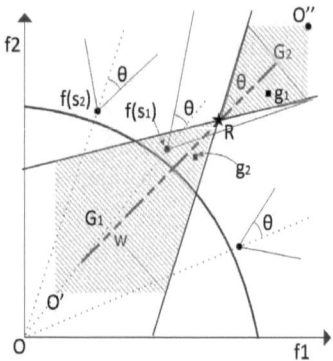

Fig. 2. Illustration of iPICEA-g

Inspired by this thinking, in iPICEA-g goal vectors are not generated in the whole objective space but somewhere which is related to the given ROIs (see the shaded regions in Figure 2). By co-evolving candidate solutions with these specially generated goal vectors, candidate solutions would be guided toward the ROIs. In details, goal vectors are generated in both the shaded regions ($G1$ and $G2$) that are determined by R, W, θ, O' and O''. The region extends both toward and away from the coordinate origin in order to handle the case where the supplied R is unattainable. O' and O'' are the lower and upper bounds of the regions that are to generate goal vectors. O and O' are estimated based on $f(S^*)$ (where S^* represents the current non-dominated solutions) and the specified reference point, R: see equation 1:

$$
\begin{aligned}
O' &= \alpha \times \min\left(R_i \uplus f_i(S^*)\right), i = 1, 2, \cdots, M, 0 < \beta < 1 \\
O'' &= \beta \times \max\left(R_i \uplus f_i(S^*)\right), i = 1, 2, \cdots, M, \beta > 1
\end{aligned}
\tag{1}
$$

where α and β are two scaling parameters, here, we use $\alpha = 0.5$ and $\beta = 1.5$. Note O' and O'' can be set equal to *ideal* and *nadir* point, if they are known.

A modified Pareto dominance relation named Pareto cone-dominance is applied in iPICEA-g. A goal vector, g_i is said to be satisfied (Pareto cone-dominated) by a candidate solution, $f(s_i)$ if and only if the angle between the vector $\overrightarrow{f(s_i)g_i}$ and the vector $\overrightarrow{Of(s_i)}$ is not larger than the specified search range, θ. For example, in Figure 2, g_1 is satisfied (Pareto cone-dominated) by candidate solution $f(s_1)$ while g_2 is not.

Apart from the benefit that iPICEA-g can handle the DM preference either as weights or aspiration levels, another major benefit of iPICEA-g is that multiple ROIs can be explored by simultaneously generating goal vectors for all the ROIs. Besides, we anticipate that iPICEA-g performs well on many-objective problems because PICEA-g has a good performance on MOPs with many objectives [7,8].

4   Experiments

In this section, we illustrate the performance of iPICEA-g on different benchmarks from the ZDT [24] and DTLZ [25] test suites. In all the experiments the population size of candidate solutions and goal vectors of iPICEA-g are set as $N = 100$ and $Ngoal = 100$, respectively. Simulated binary crossover (SBX, $p_c = 1, \eta_c = 15$) and polynomial mutation (PM, $p_m = \frac{1}{nvar}$ per decision variable and $\eta_m = 20$, where $nvar$ is the number of decision variables) [26] are applied as genetic variation operators. Firstly, we show the effects of R, W and θ. Secondly, we show the performance of iPICEA-g on searching for ROIs in an a priori and progressive way. Note that all the results are illustrative rather than statistically robust.

4.1   Demonstrations of the Effects of R, W and θ

The bi-objective 20-variable DTLZ2, which has a concave Pareto optimal front is selected as test problem to study the effect of the three parameters.

The Effect of R. Assuming the DM would like to have solutions around a point then we set R as the specified point. For example, the DM specify (1) one infeasible (0.6,0.6) reference point; (2) one feasible (0.8,0.8) reference point; and (3) two reference points (0.7,0.9) and (0.8,0.3). Figure 3 shows the obtained results after performing iPICEA-g for 200 generations. During the simulation the search direction is set as $W = [0.5, 0.5]$ which means there is no bias for any objective and the search range $\theta = \frac{\pi}{4}$ radians shows a range that is close to 50:50 emphasize. From Figure 3, we observe that in all cases iPICEA-g can find a set of well converged solutions. It illustrates that iPICEA-g is able to handle both the feasible and infeasible aspiration level, moreover, it can explore multiple ROIs simultaneously.

The Effect of W. Assuming that DM would like to specify a preference for one objective over another we use W. For example, the DM specifies that (1)

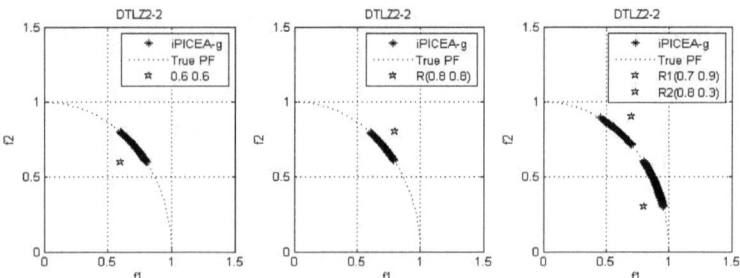

Fig. 3. The solutions obtained by iPICEA-g with different reference points

both the objectives are equally important then $W = [0.5, 0.5]$ or (2) objective f_1 is twice as important as f_2 then $W = [0.67, 0.33]$ or (3) objective f_1 is half as important as f_2 then $W = [0.33, 0.67]$. Figure 4 shows the obtained results after performing iPICEA-g for 200 generations with $W = [0.5, 0.5]$, $W = [0.33, 0.67]$ and $W = [0.67, 0.33]$, respectively. During the simulation, $R = (0.5, 0.5)$ and $\theta = \frac{\pi}{6}$ radians. From the Figure, we observe that the obtained solutions are along the given search direction, W. In other words, the obtained solutions are biased with different W. For example, in the case of $W = [0.67, 0.33]$ f_1 is more optimized.

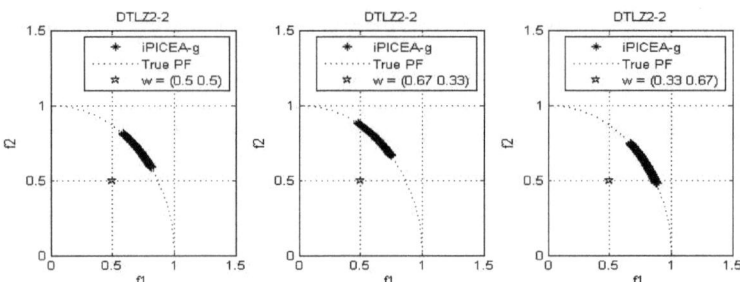

Fig. 4. Solutions obtained by iPICEA-g with different search directions

The Effect of θ. If the DM would like to obtain a large spread range of solutions then θ could be a large value e.g. having $\theta = \frac{\pi}{2}$ radians, the whole Pareto front can be obtained. If the DM would like to obtain some solutions that are exactly along the specified W then θ is set to $\frac{\pi}{180}$ radians. Figure 5 show the obtained results after performing iPICEA-g for 200 generations with $\theta = \frac{\pi}{2}, \frac{\pi}{4}$ and $\frac{\pi}{180}$, radians, respectively. During the simulation, $R = (0.3, 0.3)$ and $W = [0.5, 0.5]$. Clearly, the range of the obtained solutions decreases as θ decreases.

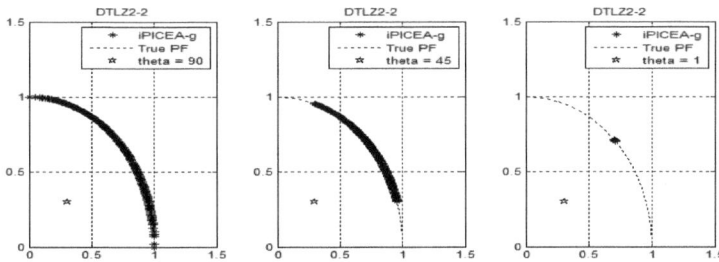

Fig. 5. The distribution of solutions obtained by iPICEA-g with different θ values

4.2   Results for a Priori Preference Expression

Search With Weights. Here we consider the case where the DM states the relative importance of each objective. Bi-objective ZDT2 and 4-objective DTLZ2 are used in the simulation and. iPICEA-g is run for 200 generations on each problem.

For ZDT2 assuming that DM specifies f_1 is twice as important as f_2 then $W = [\frac{2}{3}, \frac{1}{3}]$; correspondingly, θ is set as $\frac{\pi}{4}$ radians in order to obtain a moderate range of solutions; R is set as the O'. From Figure 6.a we can clearly observe the obtained solutions are biased to objective f_1. Also, the solutions are near the true Pareto front.

For DTLZ2 we assume the DM specifies that f_1 is four times as important as f_4, f_2 is three times as important as f_4, f_3 is twice as important as f_4 then $W = [0.4, 0.3, 0.2, 0.1]$; correspondingly, θ is set as $\frac{\pi}{12}$ radians so as to obtain a close range of solutions; R is set to the O'. Observed from Figure 6.b (parallel coordinates plots [12]), a set of solutions are obtained, which are located around the projected point Q shown as $-\star-$. Q is the projection of the coordinate origin to the Pareto optimal front along the direction $[0.4, 0.3, 0.2, 0.1]$. The true Pareto front of DTLZ2 is the surface of hyper-sphere with radius 1 ($\sum_{i=1}^{M} f_i^2 = 1$) in the first quarter [25]. Having computed $\sum_{i=1}^{4} f_i^2$ for all the obtained solutions, we find all values lies within the range $[1.0391, 1.0903]$ which confirm that the obtained solutions have almost converged to the true Pareto front.

Search With Aspiration Levels. Here we consider the case where the DM specifies preferences as aspiration levels. Again, the bi-objective ZDT1 and 4-objective DTLZ2 problems are used in the simulation.

For ZDT1 we assume that DM specifies his aspiration level as $[0.7, 0.7]$ and so $R = (0.7, 0.7)$; correspondingly, W is set as $[0.5, 0.5]$ and θ is set as $\frac{\pi}{4}$ radians. After running iPICEA-g for 200 generations a set of satisfied solutions are obtained shown in Figure 7.a. We can see that visually all the obtained solutions are very close to the true Pareto front.

For DTLZ2, we assume the DM specifies that f_1, f_2, f_3 and f_4 should be better (smaller) than 0.58, 0.7, 0.6 and 0.5, respectively. Therefore, we set

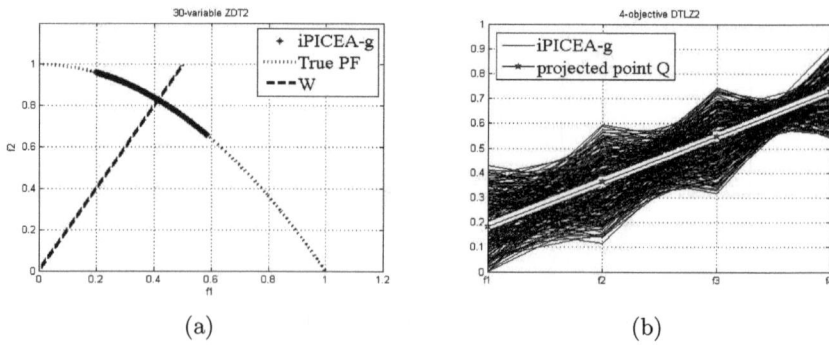

Fig. 6. Illustration of searching with weights

$R = (0.58, 0.7, 0.6, 0.5)$. Correspondingly, W is configured as $[0.25, 0.25, 0.25, 0.25]$ and θ is set as $\arccos(\frac{\sqrt{M-1}}{\sqrt{M}}) = \frac{\pi}{6}$ radians. After running iPICEA-g for 200 generations a set of solutions is found as shown in Figure 7.b. All the solutions have met the aspiration level. After computing $\sum_{i=1}^{4} f_i^2$ for all obtained solutions, the values lie within the range $[1.0141, 1.0528]$, therefore indicating that all solutions have converged close to the true Pareto front.

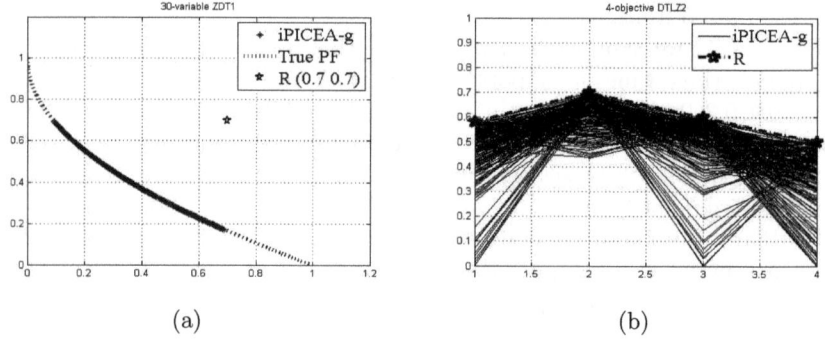

Fig. 7. Illustration of searching with aspirations

4.3   Results for a Progressive Preference Expression

Cognitively, DM may find it easier to specify preferences visually by drawing rather than using numbers. iPICEA-g allows DM to brush existing solutions or regions of the objective space that are of interest. These preferences would be automatically converted into R, W and θ parameters. Consider a 2-objective minimization example, see Figure 8. The brushed region is labelled as A.

Firstly, we find the extreme points of A, i.e. P_{f1} and P_{f2}. P'_{f1} and P'_{f2} are the normalized vector of P_{f1} and P_{f2}, respectively, i.e. $P_{fi} = \frac{P_{fi}}{L_i}, i = 1, 2$, where L_i is the Euclidean distance from O' to P_{fi}. $|O'P'_{f1}| = |O'P'_{f2}| = 1$. Then the search direction W is determined by vector $\overrightarrow{O'P}$, where P is the center of P'_{f1} and P'_{f2}. θ is then calculated by $\arccos(\overrightarrow{O'P} \cdot \overrightarrow{O'P'_{f1}})$. R is set as the O' (which can be obtained by equation 1). The co-evolved goal vectors are then generated in the shaded region closed by points R, P_{f1}, P_{f2} and O''.

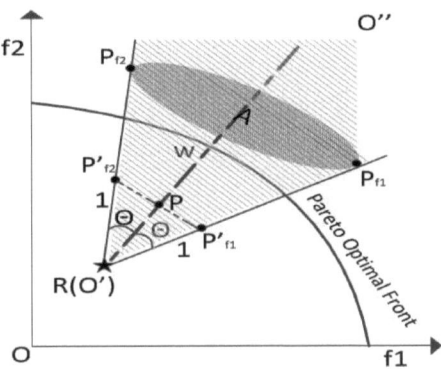

Fig. 8. Illustration of parameter calculation

To describe the working process, we solve the bi-objective ZDT1 and 4-objective DTLZ4 problems by simulating an interactive search process.

Bi-objective ZDT1. Firstly, iPICEA-g is run for 10 generations without incorporating any preferences. The aim is to roughly know the range of the objectives so as to give better preferences. The obtained solutions are shown in Figure 9.a.

Secondly, the DM brushes his preferred regions, i.e. the shaded regions in Figure 9.a. The related parameter settings of iPICEA-g are then calculated based on the brushed region, which are $W = [0.25, 0.75]$, $\theta = \frac{5\pi}{36}$ radians and $W = [0.75, 0.25]$, $\theta = \frac{5\pi}{36}$ radians for region A and B, respectively. After running iPICEA-g for 50 more generations, two sets of improved solutions are found. See Figure 9.b.

Thirdly, we assume that the DM is not satisfied with either of the two sets of solutions. However, he/she is interested in exploring a nearby region, C. The related parameter settings are $W = [0.6, 0.4]$, $\theta = \frac{\pi}{12}$ radians. By running iPICEA-g for another 50 generations, a set of solutions are found in C shown in Figure 9.c.

Fourthly, the DM is still dissatisfied. He/She would like to exploit these solutions. The preferred solutions are then brushed (See Figure 9.c) and iPICEA-g is run for 50 more generations. The related parameters are configured as

$W = [0.5, 0.5]$, $\theta = \frac{\pi}{18}$ radians. A set of better solutions are found. The DM is now happy to choose a single solution from this set. The solution D is selected; see Figure 9.d.

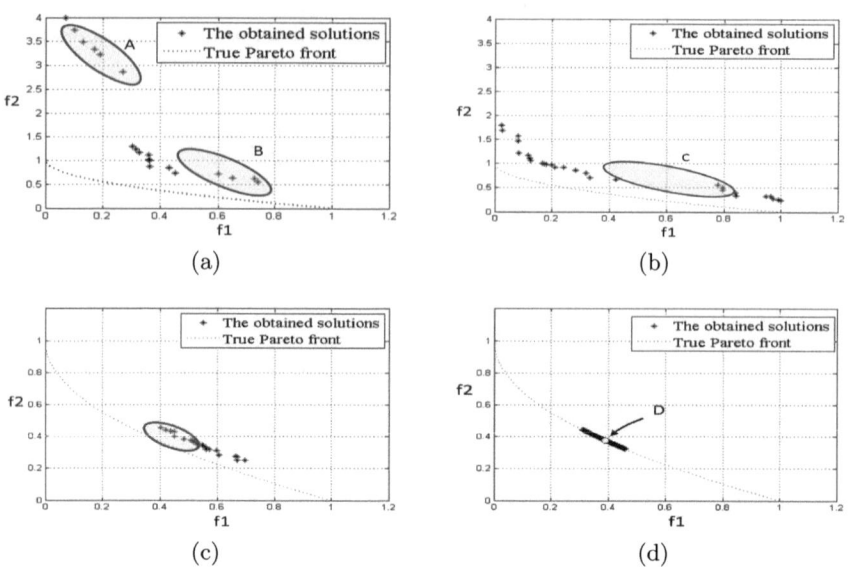

Fig. 9. Interactive scenario on 30-variable ZDT1

4-objective DTLZ2. Similarly, iPICEA-g is run without introducing any preference for 10 generations. A set of solutions are found as shown in Figure 10.a.

Secondly, DM brushes the preferred solutions for each objective (see Figure 10.a). Parameters W and θ are then calculated as $[0.25, 0.25, 0.25, 0.25]$ and $\frac{\pi}{6}$ radians. iPICEA-g is run for 50 more generations. An improved set of solutions are obtained (see Figure 10.b).

Thirdly, assuming DM is dissatisfied with the obtained solutions. He/She brushes some solutions that are of interest. Based on the brushed solutions, two ROIs are identified. The related parameters are configured by $W = [0.3986, 0.3500, 0.1105, 0.1409]$, $\theta = \frac{\pi}{12}$ radians and $W = [0.1124, 0.2249, 0.3498, 0.3128]$, $\theta = \frac{7\pi}{90}$ radians. The brushed solutions are shown in Figure 10.c. After running iPICEA-g for another 50 generations, more solutions are found. See Figure 10.d.

Fourthly, the DM is still not satisfied with the obtained solutions. He/she decides to explore one set of the obtained solutions. Again, he/she brushes his preferred solutions which are shown in Figure 10.e and run iPICEA-g for 50 more generations. W is set as $[0.3691, 0.2773, 0.1383, 0.2153]$, θ is set as $\frac{\pi}{36}$ radians. Seen from Figure 10.f, a set of refined solutions are found in this preferred region. We compute $\sum_{i=1}^{4} f_i^2$ for all the obtained solutions. The value lies within the range of [1.0190,1.041] which means the obtained solutions have well converged to the true Pareto front. The DM is now happy to choose a single solution from this set. The solution shown as the white dash line is selected; see Figure 10.d.

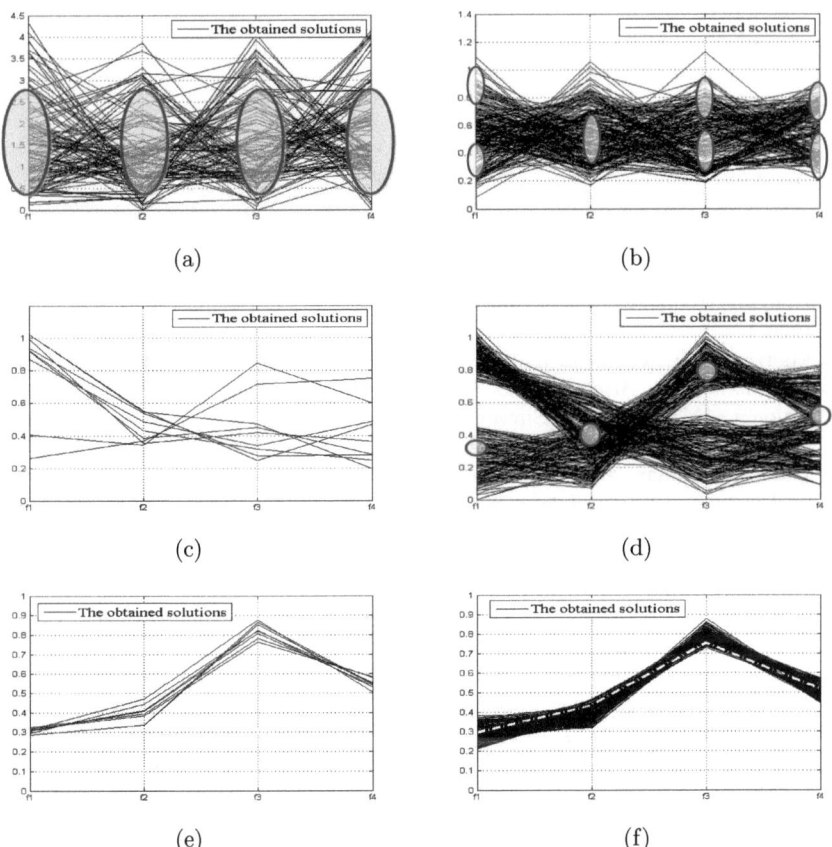

Fig. 10. Interactive scenario on 4-objective DTLZ2

5   Conclusions

Incorporation of DM preference is an important part of a real-world decision support system. However, current methods for preference-based multi-objective optimisation are unable to handle, comprehensively, the range of ways in which a DM likes to articulate his/her preferences. In this paper, we have presented, to the best of our knowledge, the first method that is simultaneously able to handle preferences expressed as weights or as aspirations and that is also able to support multiple regions of interest. We also enhance the DM-friendliness by allowing preferences to be expressed either numerically or by interactively drawing on cartesian coordinate plots or parallel coordinates plots. Simulation results have shown the effectiveness of the method.

There are three core directions for future research, firstly, since decision-making is often a group rather than individual activity, it would be useful to develop the method in order to support group decision making. Secondly, since the DM's preference is often expressed in fuzzy linguistic terms [27], it is important to study how to handle fuzzy preferences. Thirdly, the method should be trialled in a real decision making problem.

Acknowledgments. The work of Rui Wang was supported by the National Science Foundation of China (No. 70971132).

References

1. Deb, K.: Multi-objective optimization using evolutionary algorithms, vol. 16. Wiley (2001)
2. Miettinen, K.: Nonlinear multiobjective optimization, vol. 12. Springer (1999)
3. Rachmawati, L., Srinivasan, D.: Preference Incorporation in Multi-objective Evolutionary Algorithms: A Survey. In: The 2006 IEEE Congress on Evolutionary Computation, pp. 962–968. IEEE (2006)
4. Deb, K., Kumar, A.: Interactive evolutionary multi-objective optimization and decision-making using reference direction method. In: GECCO 2007: Proceedings of the Genetic and Evolutionary Computation Conference, vol. 1. ACM, London (2007)
5. Deb, K., Kumar, A.: Light beam search based multi-objective optimization using evolutionary algorithms. In: The 2007 IEEE Congress on Evolutionary Computation, pp. 2125–2132. IEEE (2007)
6. Fonseca, C., Fleming, P.J.: Multiobjective optimization and multiple constraint handling with evolutionary algorithms. I. A unified formulation. IEEE Transactions on Systems, Man and Cybernetics, Part A: Systems and Humans 28(1), 26–37 (1998)
7. Purshouse, R.C., Jalbă, C., Fleming, P.J.: Preference-Driven Co-evolutionary Algorithms Show Promise for Many-Objective Optimisation. In: Takahashi, R.H.C., Deb, K., Wanner, E.F., Greco, S. (eds.) EMO 2011. LNCS, vol. 6576, pp. 136–150. Springer, Heidelberg (2011)
8. Wang, R., Purshouse, R.C., Fleming, P.J.: Preference-inspired Co-evolutionary Algorithms for Many-objective Optimisation. IEEE Transactions on Evolutionary Computation (to appear) (accepted)
9. Coello, C.: Handling preferences in evolutionary multiobjective optimization: A survey. In: The 2000 IEEE Congress on Evolutionary Computation, vol. 1, pp. 30–37. IEEE (2000)
10. Fonseca, C., Fleming, P.J.: Genetic algorithms for multiobjective optimization: Formulation discussion and generalization. In: Proceedings of the 5th International Conference on Genetic Algorithms, pp. 416–423. Morgan Kaufmann Publishers Inc. (1993)
11. Fonseca, C., Fleming, P.J.: Multiobjective optimization and multiple constraint handling with evolutionary algorithms. II. Application example. IEEE Transactions on Systems, Man and Cybernetics, Part A: Systems and Humans 28(1), 38–47 (1998)

12. Fleming, P.J., Purshouse, R.C., Lygoe, R.J.: Many-Objective Optimization: An Engineering Design Perspective. In: Coello Coello, C.A., Hernández Aguirre, A., Zitzler, E. (eds.) EMO 2005. LNCS, vol. 3410, pp. 14–32. Springer, Heidelberg (2005)

13. Molina, J., Santana, L.V., Hernández-Díaz, A.G., Coello Coello, C.A., Caballero, R.: g-dominance: Reference point based dominance for multiobjective metaheuristics. European Journal of Operational Research 197(2), 685–692 (2009)

14. Said, L.B., Bechikh, S.: The r-dominance: a new dominance relation for interactive evolutionary multicriteria decision making. IEEE Transactions on Evolutionary Computation 14(5), 801–818 (2010)

15. Branke, J.: Guidance in evolutionary multi-objective optimization. Advances in Engineering Software 32(6), 499–507 (2001)

16. Deb, K., Sundar, J.: Reference point based multi-objective optimization using evolutionary algorithms. In: GECCO 2006: Proceedings of the Genetic and Evolutionary Computation Conference, pp. 635–642. ACM, New York (2006)

17. A preference-based evolutionary algorithm for multi-objective optimization. Evolutionary Computation 17(3), 411–436 (2009)

18. Wierzbicki, A.: The use of reference objectives in multiobjective optimization-theoretical implications and practical experiences. In: Proceedings of the Third Conference on Multiple Criteria Decision Making: Theory and Application, vol. 1979, p. 468. Springer, Hagen Konigswinter (1979)

19. Zitzler, E., Künzli, S.: Indicator-Based Selection in Multiobjective Search. In: Yao, X., Burke, E.K., Lozano, J.A., Smith, J., Merelo-Guervós, J.J., Bullinaria, J.A., Rowe, J.E., Tiňo, P., Kabán, A., Schwefel, H.-P. (eds.) PPSN VIII. LNCS, vol. 3242, pp. 832–842. Springer, Heidelberg (2004)

20. Kaliszewski, I., Miroforidis, J., Podkopaev, D.: Interactive multiple criteria decision making based on preference driven evolutionary multiobjective optimization with controllable accuracy. European Journal of Operational Research 216(1), 188–199 (2012)

21. Purshouse, R., Fleming, P.: Evolutionary many-objective optimisation: an exploratory analysis. In: The 2003 IEEE Congress on Evolutionary Computation, vol. 3, pp. 2066–2073. IEEE (2003)

22. Purshouse, R.C., Fleming, P.J.: On the Evolutionary Optimization of Many Conflicting Objectives. IEEE Transactions on Evolutionary Computation 11(6), 770–784 (2007)

23. Lohn, J., Kraus, W., Haith, G.: Comparing a coevolutionary genetic algorithm for multiobjective optimization. In: The 2002 IEEE Congress on Evolutionary Computation, vol. 2, pp. 1157–1162. IEEE (2002)

24. Zitzler, E., Thiele, L.: Multiobjective evolutionary algorithms: A comparative case study and the strength pareto approach. IEEE Transactions on Evolutionary Computation 3(4), 257–271 (1999)

25. Deb, K., Thiele, L., Laumanns, M., Zitzler, E.: Scalable multi-objective optimization test problems. In: The 2002 IEEE Congress on Evolutionary Computation, pp. 825–830. IEEE (2002)

26. Deb, K., Pratap, A., Agarwal, S., Meyarivan, T.: A fast and elitist multiobjective genetic algorithm: NSGA-II. IEEE Transactions on Evolutionary Computation 6(2), 182–197 (2002)

27. Rachmawati, L., Srinivasan, D.: Incorporation of imprecise goal vectors into evolutionary multi-objective optimization. In: The 2010 IEEE Congress on Evolutionary Computation, pp. 1–8. IEEE (2010)

Hypervolume-Based Multi-Objective Reinforcement Learning

Kristof Van Moffaert, Madalina M. Drugan, and Ann Nowé

Computational Modeling Lab, Vrije Univesiteit Brussel,
Pleinlaan 2, 1050 Brussels, Belgium
{kvmoffae,mdrugan,anowe}@vub.ac.be

Abstract. Indicator-based evolutionary algorithms are amongst the best performing methods for solving multi-objective optimization (MOO) problems. In reinforcement learning (RL), introducing a quality indicator in an algorithm's decision logic was not attempted before. In this paper, we propose a novel on-line multi-objective reinforcement learning (MORL) algorithm that uses the hypervolume indicator as an action selection strategy. We call this algorithm the *hypervolume-based MORL* algorithm or *HB-MORL* and conduct an empirical study of the performance of the algorithm using multiple quality assessment metrics from multi-objective optimization. We compare the hypervolume-based learning algorithm on different environments to two multi-objective algorithms that rely on scalarization techniques, such as the linear scalarization and the weighted Chebyshev function. We conclude that HB-MORL significantly outperforms the linear scalarization method and performs similarly to the Chebyshev algorithm without requiring any user-specified emphasis on particular objectives.

Keywords: multi-objective optimization, hypervolume unary indicator, reinforcement learning.

1 Introduction

Multi-objective optimization (MOO) is the process of simultaneously optimizing multiple objectives which can be complementary, conflicting or independent. MOO is omnipresent in real-life and comprises a large part of the current research landscape involving optimization techniques.

Most of the research concerning this domain is being focused on evolutionary algorithms (EAs), such as NSGA-II [1]. A popular approach to solving MOO problems is to transform the multi-objective problem into a single-objective problem by employing *scalarization* functions. These functions provide a single figure indicating the quality over a combination of objectives, which allows a simpler and fast ordering of the candidate solutions. Recently, quality indicators, such as the hypervolume measure that are usually used for performance assessment, are introduced into the decision making process of these EAs. Searching the decision space using quality indicators is a fruitful technique in EAs, but in

R.C. Purshouse et al. (Eds.): EMO 2013, LNCS 7811, pp. 352–366, 2013.
© Springer-Verlag Berlin Heidelberg 2013

reinforcement learning, this approach remained untouched. This paper fills this gap by proposing a novel reinforcement learning algorithm based on Q-learning that uses the hypervolume metric as an action selection strategy.

Contributions. There exist several algorithms that focus on *multi-objective reinforcement learning* (MORL) [2,3,4], but they only take into account the linear scalarization function. We combine ideas from two machine learning techniques (i.e. optimization and reinforcement learning) that have different goals in exploiting the multi-objective environments. We propose a novel multi-objective reinforcement learning algorithm where the hypervolume unary indicator is used to evaluate action selection. We call it *hypervolume MORL* (HB-MORL) and conceptually compare it two two scalarization-based MORL algorithms on environments consisting of two and three objectives. The experimental results show that HB-MORL outperforms the linear scalarization algorithm and performs similar to the Chebyshev-based algorithm.

Outline. In Section 2, we provide an overview of background concepts such as multi-objective optimization and we introduce reinforcement learning in Section 3. Subsequently, in Section 4, we reveal our novel algorithm, HB-MORL and conduct experiments in Section 5. Finally, we draw conclusions in Section 6.

2   Preliminaries

A multi-objective optimization problem optimizes a vector function whose elements represent the objectives. A maximization multi-objective problem is $\max \mathbf{F}(x) = \max\{f^1(x), f^2(x), ..., f^m(x)\}$, where m is the number of objectives, and f^i is the value for the i-th objective. A solution x_1 is said to *dominate* another solution x_2, $\mathbf{F}(x_2) \prec \mathbf{F}(x_1)$, iff for all objectives j, $f^j(x_2) \leq f^j(x_1)$, and there exists a objective i, for which $f^i(x_2) < f^i(x_1)$.

2.1   Scalarization Functions

Scalarization functions transform a multi-objective problem to a single-objective problem. The scalarization functions often take into consideration weighting coefficients which allow the user some control over the chosen policy, by placing more or less emphasis on each objective. In this paper, we consider two instances of scalarization functions:

Linear Scalarization Function. In the linear weighted-sum method a weighted coefficient w_i is associated with each objective function. A weighted-sum is performed over all objectives and their corresponding weights. The value of a solution x is $\sum_{i=1}^{m} w_i f_i(x)$. The benefit of the linear scalarization functions is its simplicity and intuitive representation.

Chebyshev Scalarization Function. Also for this scalarization, we have weights associated to each objective. The Chebyshev metric [5] calculates for each objective the weighted distance between a reference point, $\mathbf{z}^*$ and a point of interest in the multi-objective environment. For maximization problems, it chooses the greatest of these distances. The scalarized value for a solution x is $\max_{i=1,\ldots,m} w_i |f_i(x) - z_i^*|$. The reference point $\mathbf{z}^*$ is a parameter that is constantly being updated with the best value for each objective of solutions in the current Pareto set plus a small constant, i.e. $\mathbf{z}^*$ is $z_i^* = f_i^{best}(x) + \epsilon$, where ϵ is a small number.

2.2 Indicator-Based Evolutionary Algorithms

Indicator-based evolutionary algorithms or IBEA is the class of algorithms that rely on a quality indicators in their selection process. Let $\Psi \subseteq 2^X$ be the set of all possible Pareto set approximations. A unary quality indicator is a function $I : \Psi \to \mathbb{R}$, assigning a Pareto set approximations, A_1, a real value $I(A_1)$. Many quality indicators exist, but the one that is most interesting for our context is the hypervolume indicator. This metric calculates the volume of the area between a reference point and the Pareto set obtained by a specific algorithm.

The hypervolume measure is of particular interest in this context as it is the only single set quality measure known to be strictly increasing with regard to Pareto dominance. The drawback of calculating the exact hypervolume remains its computation time, as it is an NP-hard problem [6]. Over the years, several hypervolume-based EAs for MOO have been proposed, such as MO-CMA-ES [7] and SMS-EMOA [8].

3   Multi-Objective Reinforcement Learning

Evolutionary methods optimize an explicit objective function where reinforcement learning (RL) optimizes an *implicit* objective function. More precisely, RL involves an agent operating in a certain environment and receiving reward or punishment for certain behaviour. The focus of this paper is on multi-objective reinforcement learning (MORL) and how to combine it with the hypervolume unary indicator. In the following sections, we give a brief overview of existing multi-objective reinforcement learnings algorithms that utilize scalarization functions to transform the multi-objective search space of a problem into a single-objective environment.

Markov Decision Process. The principal structure for RL is a Markov Decision Process (MDP). An MDP can be described as follows. Let the set $S = \{s_1, \ldots, s_N\}$ be the state space of a finite Markov chain $\{x_l\}_{l \geq 0}$ and $A = \{a_1, \ldots, a_r\}$ the action set available to the agent. Each combination of starting state s_i, action choice $a_i \in A_i$ and next state s_j has an associated transition probability $T(s_j, s_i, a_i)$ and and immediate reward $R(s_i, a_i)$. The goal is to learn a policy π, which maps each state to an action so that the expected discounted reward is maximized. [9] proposed Q-learning, an algorithm that expresses this goal by using Q-values which explicitly store the expected discounted reward for every state-action pair. Each entry contains the value for $\hat{Q}(s, a)$ which represents the learning agent's current hypothesis about the actual value of $Q(s,a)$. The $\hat{Q}$-values are updated according to the following update rule:

$$\hat{Q}(s,a) \leftarrow (1 - \alpha_t)\hat{Q}(s,a) + \alpha_t[r(s,a) + \gamma \max_{a'} \hat{Q}(s',a')] \tag{1}$$

where α_t is the learning rate at time step t and $r(s,a)$ is the reward received for performing action a in state s.

Multi-objective MDPs. In MOO, MDPs are replaced by multi-objective MDPs or MO-MDPs [10]. These extend MDPs by replacing the single reward signal by a vector of rewards, i.e. $\vec{R}(s_i, a_i) = (R_1(s_i, a_i), \ldots R_m(s_i, a_i))$, where m represents the number of objectives. Since the reward vector consists of multiple components, each representing different objectives, it is very likely that conflicts arise between them. In such case, trade-offs between these objectives have to be learned, resulting in a set of different policies compared to a single optimal policy in single-objective learning. The overall goal of solving MO-MDPs is to find a set of policies that optimize different objectives. The set of optimal policies for each objective or a combination of objectives is referred to as the *Pareto optimal set*.

Multi-objective Reinforcement Learning. There are several MORL frameworks proposed in literature. For instance, [3] suggests a multi-objective algorithm that uses a lexicographic ordering of the objectives. More precisely, by placing minimal thresholds on certain objectives, policies are discovered that take into account these constraints. [4] proposes a batch Convex Hull Value Iteration algorithm that learns all policies in parallel, defining the convex hull of the optimal Pareto set. [2] also proposes a batch MORL approach, based on the linear scalarization function, to identify which actions are favoured in which parts of the objective space. Notwithstanding their results, they all consists of *off-line* algorithms, which involve sweeping over a set of collected data. Therefore, the aspects of these algorithms on using and adapting their policy during the learning process (i.e. *on-line* learning) were not studied.

Scalarization-Based MORL. To the best of our knowledge, all MORL algorithms are currently focusing on the linear scalarization function. Therefore, the most general MORL algorithm that allows a fair comparison in this paper is an on-line multi-objective Q-learning algorithm (MO Q-learning) employed with the linear and the Chebyshev scalarization functions, presented in Section 2. These novel multi-objective reinforcement learning algorithms [11] are an extenstion to the single-objective Q-learning algorithm [9] that can accommodate for any scalarization function. The main change compared to standard Q-learning and the work in [2] is the fact that scalarization functions are applied on Q-values in contrast to reward signals. Thus, the standard Q-table, used to store the expected reward for the combination of state s and action a, is extended to incorporate objectives, i.e. $Q(s, a, o)$. This has the advantage that non-linear functions, such as the Chebyshev function, can be utilized in the same framework.

Algorithm 1. Scalarized ϵ-greedy action selection, $scal$-ϵ-greedy()

1: $SQList \leftarrow \{\}$
2: **for** each action $a_i \in A$ **do**
3:      $\vec{o} \leftarrow \{Q(s, a_i, o_1), \ldots, Q(s, a_i, o_m)\}$
4:      $SQ(s, a) \leftarrow scalarize(\vec{o})$ ▷ Scalarize Q-values
5:      Append $SQ(s, a)$ to $SQList$
6: **end for**
7: **return** ϵ-greedy($SQList$)

In Algorithm 1, we present the scalarized action selection strategy for MO Q-learning. At line 4, the *scalarize* function can be instantiated by any scalarization function to obtain a single indication for the quality of the combination of state s and action a, $SQ(s, a)$ over the Q-values for each objective. Furthermore, the standard ϵ-greedy strategy from RL can be applied after we transform the multi-objective problem to a single-objective problem and decide the appropriate action, based on these individual indications in $SQList$. The new multi-objective Q-learning algorithm is presented in Algorithm 2. At line 1, the Q-values for each triple of states, actions and objectives are initialized. Each episode, the agent starts in state s (line 3) and chooses an action based on the multi-objective action selection strategy of Algorithm 1 at line 5. Upon taking action a, the agent is being transitioned into the new state s' and the environment provides it with the vector of rewards $\vec{r} \in \vec{\mathbb{R}}$. At line 10, the $Q(s, a, o)$ are updated with a multi-objective version of Eq. 1. This process is repeated until the Q-values converge.

Algorithm 2. MO Q-learning algorithm

1: Initialize $Q(s, a, o)$ arbitrarily
2: **for** each episode T **do**
3:      Initialize state s
4:      **repeat**
5:          Choose action a from s using policy derived from Q (e.g. $scal$-ϵ-greedy)
6:          Take action a and observe state $s' \in S$, reward vector $\vec{r} \in \vec{\mathbb{R}}$
7:          $\max_{a'} \leftarrow$ Call Scal. greedy action selection   ▷ Get best scal. action in s'
8:
9:          **for** each objective o **do**   ▷ Update Q-values for each objective
10:             $Q(s, a, o) \leftarrow Q(s, a, o) + \alpha[\vec{r}(s, a, o) + \gamma Q(s', \max_{a'}, o) - Q(s, a, o)]$
11:         **end for**
12:
13:         $s \leftarrow s'$ ▷ Proceed to next state
14:     **until** s is terminal
15: **end for**

Algorithm 3. Greedy Hypervolume-based Action Selection, HBAS(s, l)

1: $volumes \leftarrow \{\}$ ▷ The list collects hv contributions for each action
2: **for** each action $a_i \in A$ of state s **do**
3:     $\vec{o} \leftarrow \{Q(s, a_i, o_1), \ldots, Q(s, a_i, o_m)\}$
4:     $hv \leftarrow calculate\_hv(l + \vec{o})$ ▷ Compute hv contribution of a_i to l
5:     Append hv to $volumes$
6: **end for**
7: **return** argmax$_a$ $volumes$ ▷ Retrieve the action with the maximal contribution

---

4 Hypervolume-Based Multi-Objective RL

In this section, we present our novel *hypervolume-based* MORL algorithm (HB-MORL) that combines the hypervolume unary indicator as a novel action selection mechanism. This action selection mechanism is similar to the selection strategy utilized for the MO Q-learning algorithm (Algorithm 1).

The proposed strategy is presented in Algorithm 3, while the entire HB-MORL algorithm is presented in Algorithm 4. The outline of the HB-MORL algorithm is similar to the MO Q-learning algorithm in Algorithm 2, but has an additional parameter, l. Each episode, the agent maintains a list l of Q-values of already visited states and actions. Initially, this list is empty (Algorithm 4, line 3).

In the action selection strategy, the agent consults this list (Algorithm 3) by employing the hypervolume metric. For each action a_i of state s, the vector of Q-values is retrieved from the table at line 3, whereafter the contribution of each action to the list of visited state-action pairs is calculated (line 4) and stored in the *volumes* list. In the greedy selection case, the action with the largest contribution is retrieved from *volumes* and selected (line 7), while in the ϵ-greedy case a random action is selected with a probability of ϵ (not shown in Algorithm 3). Subsequently, the Q-values of the selected action are appended to the list l (line 8, Algorithm 4) and the learning proceeds.

Differences Between MORL Algorithms. The HB-MORL algorithm, presented in this paper, resembles in quite a few places to the scalarization framework, presented in Algorithm 2. They are both based on Watkins' Q-learning algorithm and its update rule. This offers the advantage that we can rely on the same convergence proof and no exotic or problem-specific algorithm is proposed. On the contrary, their correspondence allows the same generality that Q-learning has been offering for decades. As presented, the main difference to the scalarization framework lies in the action selection strategy. The scalarization framework transforms the vector of Q-values into a single indicator, whereas the hypervolume-based algorithm performs searches directly into the objective space. Furthermore, HB-MORL does not rely on weights, defined a priori to guide the search process, as opposed to the scalarized algorithms. When the policies obtained by different runs of the algorithm are collected, the user can still make her/his decision on which policies or trade-offs are preferred, but the advantage is that emphasis on particular objectives is not required beforehand.

Algorithm 4. Hypervolume-based Q-learning algorithm

1: Initialize $Q(s, a, o)$ arbitrarily
2: **for** each episode T **do**
3:     Initialize s, $l = \{\}$
4:     **repeat**
5:         Choose a from s using policy derived from Q (e.g. ϵ-greedy HBAS(s, l))
6:         Take action a and observe state $s' \in S$, reward vector $\vec{r} \in \vec{\mathbb{R}}$
7:         $\vec{o} \leftarrow \{Q(s, a, o_1), \ldots, Q(s, a, o_m)\}$
8:         Add $\vec{o}$ to l ▷ Add Q-values of selected action a to l
9:         $max_{a'} \leftarrow$ greedy HBAS(s', l)   ▷ Get greedy action in s' based on new l
10:
11:         **for** each objective o **do**   ▷ Update Q-values for each objective
12:             $Q(s, a, o) \leftarrow Q(s, a, o) + \alpha[\vec{r}(s, a, o) + \gamma Q(s', max_{a'}, o) - Q(s, a, o)]$
13:         **end for**
14:
15:         $s \leftarrow s'$ ▷ Proceed to next state
16:     **until** s is terminal
17: **end for**

---

5   Results

In this section, we experimentally evaluate the performance of the HB-MORL algorithm on two benchmark environments for different quality measures. These results are then compared to two instances of MORL algorithms that use scalarization-based action selection strategies, i.e. the linear and the Chebyshev Q-learning algorithm.

5.1   Testing Environments

Recently, [12] proposed empirical evaluation techniques for multi-objective reinforcement learning, together with a few benchmark environments. We build further on this work and perform our experiments on the same worlds, such as the *Deep Sea Treasure* and the *Multi-Objective Mountain Car* environments to compare the two scalarization functions and the HB-MORL algorithm in detail. The optimal Pareto sets of each world were provided by the same researchers.

5.2   Parameter Setting

In the experiments, presented below, we relied on identical configurations for each of the testing environments. We applied an ϵ-greedy exploration strategy with ϵ set to 0.1 and the Q-values were initialized randomly for each objective. The learning rate α was set to 0.1 and the discount factor γ to 0.9. Results are collected and averaged over 50 trials of each 500 runs.

Table 1. The reference points (RP) used to calculate the hypervolume indicator in the learning algorithm and in the quality assessment for each environment

	Deep Sea Treasure	MO-Mountain Car
Learning RP	$(-25.0, -5.0)$	$(-10.0, -10.0, -10.0)$
Quality Assessment RP	$(-25.0, 0.0)$	$(-350.0, -250.0, -500.0)$

Each scalarization-based MORL algorithm was tested using different sets of the weights sets ranging from 0 to 1, with a steps of 0.1 for each objective, i.e. there are 11 and 64 tuples of weights (and experiments) for the worlds with two and three objectives, respectively. Each scalarized MORL algorithm using a particular weight tuple could be regarded as an individual learning agent. As the hypervolume metric does not require any prior knowledge and to ensure a fair comparison in the results, we ran the HB-MORL algorithm for exactly the same number of experiments as the scalarized algorithms. Per iteration, each agent individually tests its learned policy by greedily selecting actions, whereafter each agents' policy is stored in a set. This set is called the Pareto approximation set for a particular iteration number of the learning phase.

We employed the hypervolume metric also as a quality assessment tool in the comparisons of the different algorithms. Table 1 presents the reference points (RP) used for calculating the hypervolume in both the learning phase and the testing phase for each of the environments. These values were determined empirically by examining the bounds on the reward structure of each testing environment in a straightforward way.

5.3   Performance Experiment

In this experiment, we compare the performance of the linear and Chebyshev-based algorithms to our novel hypervolume-based MORL algorithm. In Table 2 and 3, we relied on the Wilcoxon rank test [13] to indicate a significant difference on the mean performance between the indicator-based MORL algorithm and both scalarized methods on each environment. We present the learning curves for each of the environments in Fig. 1(a) and 1(b) by applying the hypervolume indicator for quality assessment purposes. For the Deep Sea Treasure world (Fig. 1(a)), the linear scalarization-based MORL is not capable of improving the hypervolume after 100 runs whereas the Chebyshev-based algorithm slightly improves its performance until 250 runs. The HB-MORL is gradually improving its policy and improves the Chebyshev algorithm after 400 runs. It is interesting to note that the HB-MORL algorithm is increasingly its performance gradually until the end of the learning phase. In other preliminary tests, not included in this paper, we ran the experiments for a longer period of time, but no algorithm was able to further improve its policy after 500 runs.

Table 2. The Wilcoxon rank test denoted a significant difference on the mean performance of the linear scalarized algorithm and HB-MORL on both testing worlds for a threshold p value of 0.05

	Linear scalarization	HB-MORL	p-value	Significant?
Deep Sea Treasure	762	1040.24	7.0338^{-4}	√
MO-Mountain Car	15727946.18	23984880.12	$5.9282e^{-5}$	√

Table 3. Also for the Chebyshev-based algorithm and HB-MORL, a significant difference is noted by the Wilcoxon rank test on the two benchmark instances (threshold p = 5%)

	Chebyshev	HB-MORL	p-value	Significant?
Deep Sea Treasure	938.29	1040.24	$1.5256e^{-17}$	√
MO-Mountain Car	23028392.94	23984880.12	$5.0379e^{-16}$	√

Finally, we observed the largest difference between the two scalarized algorithms in the complex 3D MO-Mounting Car world (Fig. 1(b)). In this benchmark environment, the Chebyshev-based algorithm is stabilized after 100 runs, whereas the linear scalarization-based MORL algorithm slowly increases until the last learning steps. Nevertheless, a considerable difference in quality is kept between the two scalarized MORL methods. The hypervolume-based Q-learning algorithm is stabilising approximately as fast as the Chebyshev algorithm, but the gradual improvement phase is also observed in this complex world and a significant improved performance is achieved.

In Fig. 2(a) and 2(b), we elaborate into more detail the gradual learning phase achieved by each of the learning methods. We restricted the information in the learning curves to capture the performance every 100 runs. For each of the environments, the linear scalarized algorithm is performing the worst and its performance stagnates after only a few iterations. The Chebyshev method is able to escape the local maxima in a much better way and is improving until the end of the learning phase. Finally, the HB-MORL algorithm is able to improve even further and achieves an enhanced performance in its finals runs.

In Fig. 2(c), we show the frequency probability of the 10 Pareto dominating goals (i.e. treasures) reached in the Deep Sea world. This plot provides us with an idea on the spread that each learning method can obtain amonst Pareto dominating solutions. We note that the linear scalarization-based algorithm only finds extreme solution, i.e. solutions that maximize only one of the objectives. More precisely, the two results found are the treasures with value 1 and 124, i.e. the treasures that minimize the *time* objective and maximize the *treasure* objective, respectively, but no real compromising solutions were obtained. The Chebyshev algorithm however, obtains a larger spread in the results compared

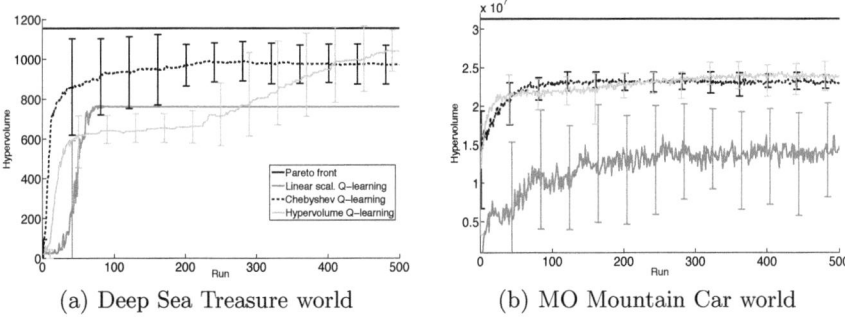

(a) Deep Sea Treasure world   (b) MO Mountain Car world

Fig. 1. Learning curves on the Deep Sea Treasure and the MO Mountain Car worlds, respectively

to the linear case. Out of the 10 possible goals, it found 8 on a regular basis, with increasing probability near the extreme solutions. Without being initialized with any prior preferences, the hypervolume-based MORL performed acceptable and focused on 5 solutions.

5.4 Quality Indicator Comparison

In multi-objective research, quality indicator studies are a popular approach for conducting algorithm comparisons. The performance indicators applied in this experimental study are the (inverted) generational distance , the generalized spread indicator, the cardinality and the hypervolume distance. The former three are minimization metrics, while the latter two are to be maximized. In detail, the generational distance and the inverted generational distance were both proposed by [14]. The former measures how far the elements in the set of non-dominated vectors, generated by a learning algorithm, are from those in the Pareto optimal set. The latter calculates how far each element of the Pareto optimal set is from the non-dominated vectors in the approximation set. The spread indicator [1] on the other hand is a diversity metric that measures the extent of spread achieved among the obtained solutions. The cardinality measure simply counts the number of elements found in the Pareto set.

The results are presented in Table 4. On the Mountain Car (MC) world, the HB-MORL obtained overall the best results out of the three algorithms, except for the fact that the Chebyshev method found one extra solution. The linear scalarized algorithm obtained the best value for the generalized spread, but as this metric only uses the members on the boundaries of the Pareto optimal set (i.e. the extreme solutions) in its calculations, this metric is biased towards the linear method that exclusively finds these solutions (see Fig. 2(c)).

On the Deep Sea (DS) world, the Chebyshev method found 8 out of 10 distinct results and obtained the best value for the inverted generational distance. Closely followed by HB-MORL that without any prior information (i.e. no weights)

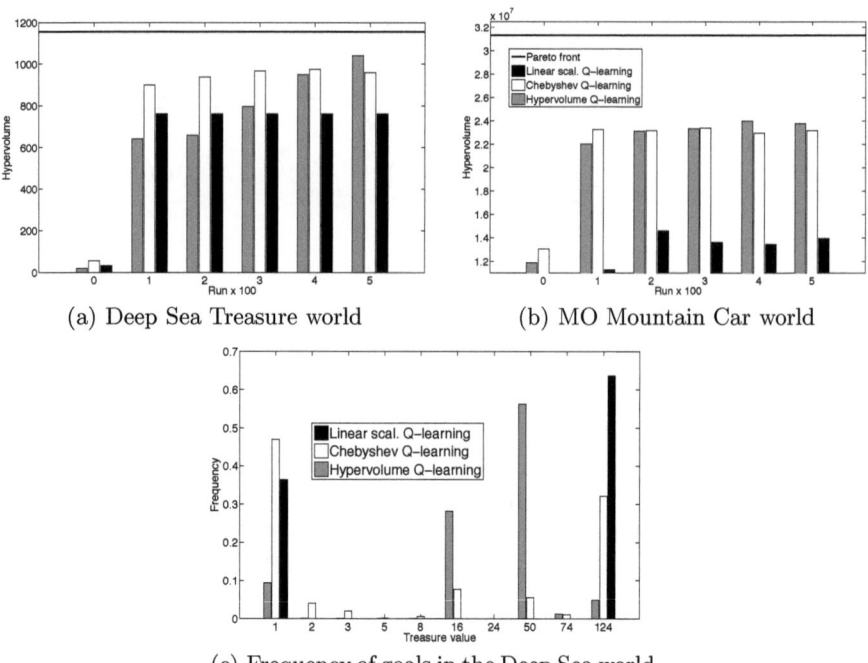

(a) Deep Sea Treasure world   (b) MO Mountain Car world

(c) Frequency of goals in the Deep Sea world

Fig. 2. Fig. 2(a) and 2(b) depict the performance of each learning algorithm each 100 runs. In Fig. 2(c), the frequency probabilities of each of the 10 Pareto dominating goals in the Deep Sea Treasure world are presented. The Chebyshev-based algorithm obtained the best spread, closely being followed by HB-MORL. The linear scalarized algorithm only found two (extreme) solutions.

obtained 5 distinct results, but a larger hypervolume was obtained. This means that HB-MORL was much more consistent in finding good solutions frequently (i.e. the increased hypervolume), but its results were not as spread around in the search space as the Chebyshev-based algorithm (i.e. the cardinality). The linear scalarized algorithm only obtained 2 (extreme) results that are located at the largest possible distance from each other, resulting in the best generalized spread value. Each of the results found by any of the algorithms were an element of the optimal Pareto set, meaning that the generational distance is 0.

To conclude, on each environments, HB-MORL outperformed the linear scalarization algorithm and obtained the best results on the most important quality indicator, i.e. the hypervolume metric. On the other indicators in the Deep Sea Treasure world, the HB-MORL algorithm obtained good results but was not always the best performing algorithm. We can conclude that the HB-MORL algorithm was very consisting in finding solutions that maximize the hypervolume metric, but could be improved by more spread results.

Table 4. Five quality indicator for each of the three algorithms on the two benchmark instances. The first three are to be minimized, while the latter two are maximization indicators. The best values are depicted in bold face.

		Linear	Chebyshev	HB-MORL
Inverted Generational distance	DS	0.128	**0.0342**	0.0371
	MC	0.012	0.010	**0.005**
Generalized spread	DS	**3.14e^{-16}**	0.743	0.226
	MC	**0.683**	0.808	0.701
Generational distance	DS	0	0	0
	MC	0.0427	0.013824	**0.013817**
Hypervolume	DS	762	959.26	**1040.2**
	MC	15727946	23028392	**23984880**
Cardinality	DS	2	**8**	5
	MC	15	**38**	37

Weights Vs. Quality Indicator. In the following test, we investigate into more detail the results of HB-MORL to the results obtained for each weighted tuple for the scalarization-based algorithms (Table 5). It is important to note that the HB-MORL algorithm is not set up with any information on how the objectives should be balanced or weighed. Therefore, in the table, its values remain identical. Note that the generational distance was omitted because every result obtained was an element of the Pareto set. We focus on the differences between the Chebyshev method and the HB-MORL algorithm and notice that there is still a significant portion of the weighted tuples for which the Chebyshev algorithm achieved better performance in term of the inverted generational distance than was presumed in Table 4. Although, there are four cases (weights W6, W8, W9 and W10) where the HB-MORL algorithm obtained improved performance and 2 tuples (weights W3 and W4) that perform similarly.

The hypervolume measure indicated that for some weights, the Chebyshev algorithm obtained a large portion of the Pareto set, but on the larger portion of the experiments the results are less efficient. Especially when assigning a very low weight to the *treasure* objective (e.g. weight W10), a limited hypervolume is achieved. In those cases, the *time* objective is being minimized with the result that treasures with a limited value, located near the starting position to minimize the time and distance, are favored. The generalized spread indicator showed that when focusing on the performance of particular tuples of weights, the values become more clear and the HB-MORL algorithm is performing intrinsically better. Note that the Chebyshev algorithm found the two extreme points in the objective space for weight W9, thus resulting in the best possible value of 0. The same can be concluded for the cardinality indicator as for particular weights, very few solution points are obtained.

To conclude, based on the empirical results, for a large portion of weights, the Chebyshev MORL algorithm is considered a well-performing algorithm by many quality indicator measures, such as spread and cardinality (see Table 4). We have seen that for some weights (Table 5), the Chebyshev algorithm obtained

Table 5. Five quality indicator for each of the 11 weights in the Deep Sea world. The inverted generational distance (IGD) and spread indicator are to be minimized, while the other two are maximization indicators. The best values are depicted in bold face.

	Weight	Linear	Chebyshev	HB-MORL		Weight	Linear	Chebyshev	HB-MORL
	W0	0.3234	**0.0227**	0.0371		W0	744	**1145**	1130
	W1	0.3234	**0.0253**	0.0371		W1	744	**1143**	1130
	W2	0.3234	**0.0296**	0.0371		W2	744	**1136**	1130
	W3	0.3234	**0.0365**	0.0371		W3	744	1094	**1130**
	W4	0.3234	**0.0360**	0.0371		W4	744	**1140**	1130
IGD	W5	0.32344	**0.0260**	0.0371	HV	W5	744	1024	**1130**
	W6	0.32344	0.0451	**0.0371**		W6	744	1082	**1130**
	W7	0.2201	**0.0260**	0.0371		W7	24	**1140**	1130
	W8	0.2201	0.0616	**0.0371**		W8	24	1018	**1130**
	W9	0.2201	0.1279	**0.0371**		W9	24	762	**1130**
	W10	0.2201	0.2201	**0.0371**		W10	24	24	**1130**
	W0	1	0.5481	**0.2257**		W0	1	**8**	5
	W1	1	0.4003	**0.2257**		W1	1	**7**	5
	W2	1	0.3714	**0.2257**		W2	1	**6**	5
	W3	1	0.6234	**0.2257**		W3	1	**6**	5
	W4	1	0.6830	**0.2257**		W4	1	**6**	5
Spread	W5	1	0.4159	**0.2257**	Cardinality	W5	1	**7**	5
	W6	1	**0.1949**	0.2257		W6	1	4	**5**
	W7	1	0.4159	**0.2257**		W7	1	**7**	5
	W8	1	0.7121	**0.2257**		W8	1	4	**5**
	W9	1	**0**	0.2257		W9	1	2	**5**
	W10	1	1	**0.2257**		W10	1	1	**5**

very good results for many quality indicators, while for other weights the method operates ineffectively. Thus, the principal drawback of these scalarization-based algorithm remains the fact that the user has to predefine its preferences by placing greater or lesser emphasis on each objective. This is a task that should not be underestimated. The main benefit of scalarization techniques remains their simplicity, but this does not compensate for the inconvenience of manually guiding the nature of the policy and the fact that these methods are very biased to the actual weights used.

5.5 Discussion

By randomly initializing the Q-values, the HB-MORL method obtains an acceptable notion of spread and found a large portion of Pareto dominating solutions. Thus, approaching the Chebyshev method for the spread indicator. But more importantly, HB-MORL improved the hypervolume indicator on every benchmark, indicating the method's robustness as good results are found frequently. Furthermore, HB-MORL does not require any direct input from the user on its actual preferences and solves this burden by employing the hypervolume quality indicator directly in its search process. This makes the main advantage of the HB-MORL algorithm its simplicity. Also, unlike the Chebyshev method, no specific reference point z^* is to be specified and updated in every run of the algorithm, making it easier for the developer to conduct experiments with HB-MORL instead of scalarization-based methods.

On both benchmark instances, the linear scalarization algorithm failed to achieve decent performance and got stuck in local optima while obtaining only two extreme solution points. These outcomes are in accordance with previous findings [12] on the linear scalarization algorithm, stating that the method is unable of finding solutions near non-convex regions of the optimal Pareto set. Independent of the running time, the algorithm gets stuck in local optima from which it can not escape. The Chebyshev-based algorithm and HB-MORL are able to gradually improve their hypothesis. Especially the latter is able to improve in final stages of its training phase.

Note that the HB-MORL algorithm is more of a slow starter compared to the Chebyshev-based algorithm. We believe that this is caused by the lack of diversity in the Pareto set explored with the hypervolume based indicator. This is also noticed in other approaches that conduct searches using the hypervolume metric and therefore [15] proposes to include a mechanism to increase the diversity of the Pareto sets and to encourage exploration. The reason why the Chebyshev-based method does not have this problem, is because the algorithm is restarted with different weights forcing the exploration of different regions of the multi-objective environment.

6   Conclusions

In this paper, we have successfully built a bridge between two machine learning techniques that rely on different solution approaches given an certain environment. More precisely, we have included a technique from multi-objective optimization, i.e. the hypervolume unary based indicator, into reinforcement learning. We have conceptually and experimentally compared our novel hypervolume-based MORL (HB-MORL) algorithm to two other scalarization-based learning algorithms, which require weights to be defined beforehand. In contrast, the HB-MORL algorithm does not contain preference-based parameters to be specified. For our experiments, we performed performance assessment tests on two benchmark instances with two and three objectives. We have noted that the suggested algorithm significantly improved the linear scalarization-based algorithm and performed similarly to the Chebyshev-based algorithm. Especially on indicators that asses the robustness of an algorithm on finding high-quality solutions frequently, the hypervolume-based algorithm turned out to be the best performing. We believe that HB-MORL is especially useful in cases where it is difficult to define user-preferences beforehand or in cases where it is complex to tune an algorithm specifically for a particular problem instance. In those situations, HB-MORL would allow to obtain a significant amount of high-quality solutions without requiring any weights parameters to be defined.

Acknowledgement. This research is supported by the IWT-SBO project PER-PETUAL (grant nr. 110041).

References

1. Deb, K.D., Pratap, A., Agarwal, S., Meyarivan, T.: A fast and elitist multiobjective genetic algorithm: NSGA-II. IEEE Transactions on Evolutionary Computation 6(2), 182–197 (2002)
2. Lizotte, D.J., Bowling, M., Murphy, S.A.: Efficient reinforcement learning with multiple reward functions for randomized controlled trial analysis. In: Proceedings of the Twenty-Seventh International Conference on Machine Learning (ICML), pp. 695–702 (2010)
3. Gábor, Z., Kalmár, Z., Szepesvári, C.: Multi-criteria reinforcement learning. In: Shavlik, J.W. (ed.) ICML, pp. 197–205. Morgan Kaufmann (1998)
4. Barrett, L., Narayanan, S.: Learning all optimal policies with multiple criteria. In: Proceedings of the 25th International Conference on Machine Learning, ICML 2008, pp. 41–47. ACM, New York (2008)
5. Miettinen, K.: Nonlinear Multiobjective Optimization. International Series in Operations Research and Management Science, vol. 12. Kluwer Academic Publishers, Dordrecht (1999)
6. Bringmann, K., Friedrich, T.: Approximating the volume of unions and intersections of high-dimensional geometric objects. Comput. Geom. Theory Appl. 43(6-7), 601–610 (2010)
7. Igel, C., Hansen, N., Roth, S.: Covariance matrix adaptation for multi-objective optimization. Evol. Comput. 15(1), 1–28 (2007)
8. Beume, N., Naujoks, B., Emmerich, M.: Sms-emoa: Multiobjective selection based on dominated hypervolume. European Journal of Operational Research 181(3), 1653–1669 (2007)
9. Watkins, C.: Learning from Delayed Rewards. PhD thesis, University of Cambridge, England (1989)
10. Wiering, M.A., de Jong, E.D.: Computing Optimal Stationary Policies for Multi-Objective Markov Decision Processes. In: 2007 IEEE International Symposium on Approximate Dynamic Programming and Reinforcement Learning, pp. 158–165. IEEE (April 2007)
11. Van Moffaert, K., M. Drugan, M., Nowé, A.: Multi-objective reinforcement learning using scalarization functions. Technical report, Computational Modeling Lab, Vrije Universiteit Brussel, Brussels, Belgium (2012)
12. Vamplew, P., Dazeley, R., Berry, A., Issabekov, R., Dekker, E.: Empirical evaluation methods for multiobjective reinforcement learning algorithms. Machine Learning 84(1-2), 51–80 (2010)
13. Gibbons, J., Chakraborti, S.: Nonparametric Statistical Inference. Statistics, Textbooks and monographs. Marcel Dekker (2003)
14. Veldhuizen, D.A.V., Lamont, G.B.: Multiobjective evolutionary algorithm research: A history and analysis (1998)
15. Ulrich, T., Bader, J., Zitzler, E.: Integrating decision space diversity into hypervolume-based multiobjective search. In: Proceedings of the 12th Annual Conference On Genetic and Evolutionary Computation, GECCO 2010, pp. 455–462. ACM, New York (2010)

A Theoretical Analysis of Curvature Based Preference Models

Pradyumn Kumar Shukla[1], Michael Emmerich[2], and André Deutz[2]

[1] Karlsruhe Institute of Technology – Institute AIFB
76128 Karlsruhe, Germany
pradyumn.shukla@kit.edu
[2] LIACS, Leiden University
2333-CA Leiden, Netherlands
{emmerich,deutz}@liacs.nl

Abstract. Various notions of preferences exist in multi-objective optimization and the decision making community. On the one hand, preferences appear as domination relations that are stronger than the classical Pareto-domination, while on the other hand, they introduce relative importance on the objective functions. In this way, preferences can appear in both domination relations and objectives. In this paper, we analyze and put together different preference models and classify them into two groups. We theoretically analyze many preference models within these groups. In particular, we are interested in curvature/ slope based models where the preferred set depend upon the curvature of efficient front. This amounts to having a direct control on trade-offs among the objective functions. A related concept of cone-based hypervolume is also theoretically investigated in this paper. Special emphasis is placed on equitable efficiency and its applications. Furthermore, we present two algorithms for finding solutions that are compatible with a given preference model.

Keywords: Preference models, Cone-based hypervolume indicator, Theoretical analysis, Trade-offs.

1 Introduction

Many real-world, mathematical and economical problems are characterized by the presence of several conflicting objective functions (see [17,8]). These problems involve minimization of all the objective functions and are called multi-objective optimization problems (MOPs). In the absence of further information, Pareto-ordering is used to compare two vectors $\mathbf{u}, \mathbf{v} \in \mathbb{R}^m$. $\mathbf{u} := (u_1, u_2, \ldots, u_m)$ is said to Pareto-dominate (and if preferred over) a vector $\mathbf{v} := (v_1, v_2, \ldots, v_m)$ if $u_i \leq v_i$ for all $i = 1, 2, \ldots, m$ and $\mathbf{u} \neq \mathbf{v}$. This way to compare two points can be seen as a preference model in that those points are sought that are not Pareto-dominated by any other points.

Although the Pareto preference model is central to solving MOPs and satisfies many basic properties like reflexivity, transitivity and strict monotonicity, many engineering design approaches use additional requirement like fixing trade-offs.

R.C. Purshouse et al. (Eds.): EMO 2013, LNCS 7811, pp. 367–382, 2013.
© Springer-Verlag Berlin Heidelberg 2013

Trade-offs are a basic tool in decision making and we find various notions in the classical multiple criteria decision making (MCDM) community also (see [17]). Apart from trade-offs, there exists many kinds of preferences that can be used to enrich the Pareto preference model. The decision maker (DM) who is interested in solving the problem might be biased towards certain regions in the objective (or variable) space, for example. Broadly, we could classify preference models into two categories. The first of these are direction or region based models where the user has a basic idea where he want look for solutions. This could be a region (like intervals for objective values) or direction (like biased weights) based. The second class is based on slope or curvature based models that have inherent trade-off information. The solutions based on these models depend upon the shape of the efficient front (defined as the image of non Pareto-dominated points), like knees for example [24].

This paper is concerned with curvature based preference models. We find curvature based models more interesting as the curvature of efficient front amounts to having a direct control on the trade-offs among the objective functions [25]. These models are based on a user specified preference. This preference can either be based on trade-offs or could come by imposing conditions that require the Pareto preference model to additionally satisfy some axioms (like impartiality or principle of transfers [20]). Curvature based preferences appear as domination or ordering relations that are stronger the classical Pareto ordering. Trade-offs can be appended to Pareto-ordering and this induces a polyhedral cone ordering. Curvature based preferences also come when relative importance on the objective functions in introduced (see Noghin [19] for a formal definition). Hence, preferences can appear in both domination and objectives.

In this paper, we analyze and put together different curvature based preference models and classify them into two groups: smooth and nonsmooth. Smooth preference models use a closed, convex, and pointed cone to order the space $\mathbb{R}^m$. In this way, many classical algorithms can be easily applied to find one solution. Many preference models within these groups, like proper Pareto-optimality and equitability, are theoretically analyzed and shown to be nonsmooth. Nonsmoothness can be difficult to tackle using classical algorithms (like Newton's method) and for these preference models, population based algorithms are more suitable.

The concept of cone-based hypervolume [12] is theoretically investigated in this paper as this can be used to find control the density of points based on curvature of the efficient front. Furthermore, we present two algorithms for finding solutions that are compatible with a given preference model.

This paper is divided into five sections of which this is the first. The next section presents a classification of various curvature based preference models. We also theoretically analyze them in the same section. Section 3 presents two indicator based algorithms for finding solutions belonging to the different classes of preference models. Section 4 presents some numerical results for equitably efficient points, Finally, conclusions and outlook are presented in the last section.

2   Classification and Analysis of Preference Models

A multi-objective optimization problem (MOP) deals with minimizing $m > 2$, objective functions $F_1, \ldots, F_m : \mathbb{R}^n \to \mathbb{R}$ restricted to a constraint set $X \subseteq \mathbb{R}^n$, and can be written as follows:

$$\min \mathbf{F}(\mathbf{x}) := (F_1(\mathbf{x}), F_2(\mathbf{x}), \ldots, F_m(\mathbf{x})) \qquad \text{s.t. } \mathbf{x} \in X.$$

As the space $\mathbb{R}^m$ lacks a canonical total ordering among its elements, the notion of Pareto-ordering induced by the nonnegative orthant cone $\mathbb{R}^m_+ := \{\mathbf{y} \in \mathbb{R}^m | y_i \geq 0, \forall i \in \mathcal{I} := \{1, 2, \ldots, m\}\}$, is usually used to compare two points in $\mathbb{R}^m$. Based on this, various optimality notions are defined. A point $\mathbf{x}^* \in X$ is called *Pareto-optimal* if no $\mathbf{x} \in X$ exist so that $F_i(\mathbf{x}) \leq F_i(\mathbf{x}^*)$ for all $i \in \mathcal{I}$ with strict inequality for at least one index i. In a similar way, a point $\mathbf{x}^*$ is called *weakly Pareto-optimal* if no $\mathbf{x} \in X$ exists so that $F_i(\mathbf{x}) < F_i(\mathbf{x}^*)$ for all $i \in \mathcal{I}$. Let $X_p(\mathbf{F}, X, \mathbb{R}^m_+)$ and $X_w(\mathbf{F}, X, \mathbb{R}^m_+)$ denote the set of Pareto-optimal and weakly Pareto-optimal points of the above MOP, respectively.

Although the concept of weak Pareto-optimality is needed for many classical algorithms [17], it is not satisfactory in solving real-world problems as it ignores the possibility of improvement wrt. some of the objectives. A criticism of Pareto-optimality is that it allows unbounded trade-offs. To avoid this, starting with the classical work of Geoffrion [13], various stronger optimality notions, known as proper Pareto-optimality, have been defined.

Definition 1. *A point* $\mathbf{x}^* \in X$ *is* Geoffrion proper Pareto-optimal *if there exists a number* $M > 0$ *such that for each* $(\mathbf{x}, i) \in X \times \mathcal{I}$ *satisfying* $f_i(\mathbf{x}) < f_i(\mathbf{x}^*)$, *there exists an index* j *with* $f_j(\mathbf{x}^*) < f_j(\mathbf{x})$ *and* $(f_i(\mathbf{x}^*) - f_i(\mathbf{x}))/(f_j(\mathbf{x}) - f_j(\mathbf{x}^*)) \leq M$. *Let* $X_{pp}(\mathbf{F}, X, \mathbb{R}^m_+)$ *denote the set of Geoffrion proper Pareto-optimal points.*

The practical idea of properly Pareto-optimal solutions is that to a decision maker solutions with an unbounded trade-off is essentially a weakly Pareto-optimal solution. Different classes of properly Pareto-optimal exists [17] and the notion of trade-off in them is inherent (see [10] for more details).

Pareto optimality is a way to reduce elements from its weak counterpart. Proper Pareto optimality goes further and reduces the set of Pareto optimal solutions (by removing unbounded solutions). If we compare the reduction from weak Pareto to Pareto, multi objective problem instances can be constructed such that $X_w(\mathbf{F}, X, \mathbb{R}^m_+)$ is much larger than $X_p(\mathbf{F}, X, \mathbb{R}^m_+)$. For a continuous problem, if we compare the reduction from Pareto to proper Pareto, it is known that the set of proper Pareto optimal solutions is dense in the set of Pareto optimal solutions [16]. Hence, any Pareto-optimal point is either proper Pareto optimal or is the limit of proper Pareto optimal points. This only removes countably finite points from the Pareto-optimal set, and hence, proper Pareto-optimality is more a mathematical construct. One practical modification is to bound M.

Apart from the above three optimality notions, a closed and nonempty subset D is sometimes used to define an ordering in the space $\mathbb{R}^m$ (see [4,5]). For this, let $\lambda(S)$ denote the Lebesgue measure of a set $S \subseteq \mathbb{R}^m$ and $\text{int}(S)$ be the interior of S [22].

Definition 2. *Let a set $D \subset \mathbb{R}^m$ be an set such that $D \cap (-D) = \{0\}$ and let $\lambda(D) > 0$. Moreover, let $\mathbf{u}$ and $\mathbf{v}$ be two vectors in $\mathbb{R}^m$. Then,*

1. $\mathbf{u} \leq_D \mathbf{v}$ *($\mathbf{u}$ weakly D-dominates $\mathbf{v}$)* $\iff \mathbf{v} - \mathbf{u} \in D$
2. $\mathbf{u} <_D \mathbf{v}$ *($\mathbf{u}$ D-dominates $\mathbf{v}$)* $\iff \mathbf{v} - \mathbf{u} \in D \setminus \{0\}$.
3. $\mathbf{u} \ll_D \mathbf{v}$ *($\mathbf{u}$ strictly D-dominates $\mathbf{v}$)* $\iff \mathbf{v} - \mathbf{u} \in \text{int}(D)$.

Note that $\lambda(D) = 0$ in not desired in Definition 2 as then almost all the elements of $\mathbb{R}^m$ are non-dominated.

Definition 3. *A point $\hat{\mathbf{x}} \in X$ is called D-optimal if no other point in $\mathbf{F}(X)$ D-dominates the point $\mathbf{F}(\hat{\mathbf{x}})$. Equivalently, a point $\hat{\mathbf{x}} \in X$ is D-optimal if and only if*

$$(\{\mathbf{F}(\hat{\mathbf{x}})\} - D) \cap \mathbf{F}(X) = \{\mathbf{F}(\hat{\mathbf{x}})\}.$$

$\mathbf{F}(\hat{\mathbf{x}})$ is known as a D-efficient point. Let $X_p(\mathbf{F}, X, D)$ and $\mathcal{E}(\mathbf{F}(X), D)$ denote the set of D-optimal and the set of D-efficient points, respectively.

It is common to assume that D is a closed, convex, and pointed cone defined as follows.

Definition 4. *A subset $\mathcal{C} \subseteq \mathbb{R}^m$ is called a cone, iff $\alpha\mathbf{p} \in \mathcal{C}$ for all $\mathbf{p} \in \mathcal{C}$ and for all $\alpha \in \mathbb{R}, \alpha > 0$. The cone $\mathcal{C}$ is closed and convex, iff $\mathcal{C}$ is a closed set and moreover $\alpha\mathbf{p} + (1 - \alpha)\mathbf{q} \in \mathcal{C}$ for all $\mathbf{p} \in \mathcal{C}$ and $\mathbf{q} \in \mathcal{C}$ and for all $0 \leq \alpha \leq 1$. The cone $\mathcal{C}$ in $\mathbb{R}^m$ is said to be pointed, iff for $\mathbf{p} \in \mathcal{C}, \mathbf{p} \neq 0, -\mathbf{p} \notin \mathcal{C}$, i.e., $\mathcal{C} \cap -\mathcal{C} \subseteq \{0\}$.*

In many real-world MOPs from engineering design and financial applications, it is common to impose additional restrictions on the Pareto-optimal set, more than what proper Pareto-optimality could provide. This can be done by specifying preferences like trade-offs limits, or additional properties/ axioms (like fairness or equitability). Consolidating these additional requirements with the axioms of Pareto-ordering gives a preference model $\mathcal{P}$, a general definition of which is provided in the following.

Definition 5. *A preferred solution set, denoted by $X_{\mathcal{P}}(\mathbf{F}, X)$, is a proper subset of $X_p(\mathbf{F}, X, \mathbb{R}^m_+)$ such that $X_{\mathcal{P}}(\mathbf{F}, X)$ is not dense in $X_p(\mathbf{F}, X, \mathbb{R}^m_+)$. The set $X_{\mathcal{P}}(\mathbf{F}, X)$ is said to be induced by a preference model $\mathcal{P}$.*

The preference model $\mathcal{P}$ can be used in an algorithm to narrow the search to those Pareto-optimal solutions that satisfy additional requirements coming from the preference model. The preference model can enrich Pareto-optimality, by using a stronger ordering relation for example. This can include pairwise trade-offs among the objectives, marginal rates of substitution among others. These trade-offs might also vary from point to point (so called decisional wealth in [14]). Another preference model comes by using the notion of relative importance of objectives [19]. Although many different preference models have been proposed in literature a categorization of these is useful, more so as this has an algorithmic value (see also the discussion in [27]). One way is to classify these into two different categories and is discussed next.

2.1 A Theoretical Analysis of Smooth Preference Models

Smooth preference models are algorithmically *easy to handle* models (analogous to smooth functions) and are introduced in this section. In additional to this, we theoretically analyze this class of preference models.

Definition 6. *A preference model $\mathcal{P}$ will be called as* smooth *if any one of the following conditions are true:*

1. *there exists a closed, convex, and pointed cone $\mathcal{C} \supset \mathbb{R}_+^m$ such that*

$$X_{\mathcal{P}}(\mathbf{F}, X) = X_p(\mathbf{F}, X, \mathcal{C}), \quad or$$

2. *there exists a (Fréchet) differentiable function $\mathbf{T} : \mathbb{R}^m \to \mathbb{R}^k$, for some $k \geq m$, such that*

$$X_{\mathcal{P}}(\mathbf{F}, X) = X_p(\mathbf{T} \circ \mathbf{F}, X, \mathbb{R}_+^m),$$

where $\mathbf{T} \circ \mathbf{F}$ denotes the composite function ($\mathbf{T}$ of $\mathbf{F}$).

From 1. in Definition 6, we see that $\mathcal{P}$ is called smooth if the preferred solution set $X_{\mathcal{P}}(\mathbf{F}, X)$ can be obtained by changing the ordering cone from $\mathbb{R}_+^m$ to a closed, convex and pointed cone C. A convex cone has many nice properties. For example, from [19], we know that for any binary ordering relation R is irreflexive, transitive and invariant with respect to positive linear transformation if and only if R is induced by a pointed convex cone $\mathcal{C}$ (without the origin, see Definition 2). Hence, many familiar properties (like transitivity, scalar multiplication) of Pareto-ordering are preserved by such a C cone ordering.

2. in Definition 6 requires the existence of a (Fréchet) differentiable function $\mathbf{T} : \mathbb{R}^m \to \mathbb{R}^k$, such that $X_{\mathcal{P}}(\mathbf{F}, X) = X_p(\mathbf{T} \circ \mathbf{F}, X, \mathbb{R}_+^m)$. This means that in order to find the preferred set induced by $\mathcal{P}$, we need to search for Pareto-optimal solutions of a transformed problem. The transformation $\mathbf{T}$ is assumed to be smooth (or Fréchet differentiable), as then the composite function $\mathbf{T} \circ \mathbf{F}$ is smooth for a smooth $\mathbf{F}$ (and no nonsmoothness is additionally introduced). The next two lemma relates the two conditions in Definition 6.

Lemma 1. *If $\mathbf{T} := \mathcal{A}$, where $\mathcal{A}$ is a m by k (real) matrix, then 1. and 2. in Definition 6 are equivalent and, $\mathcal{C}$ is the polyhedral cone $\{d \in \mathbb{R}^m | Ad \geq \mathbf{0}\}$. In general, 1. and 2. do not imply each other.*

Proof: The first statement in the lemma can be easily proved using ideas from [23, Lemma 2.3.4]. Moreover, one could easily come up with counterexamples assuming that 1. and 2. are equivalent. □

Lemma 2. *Let the objective function $\mathbf{F}$ be convex and let X be a closed, convex, and compact set. Then, for any closed, convex and pointed cone $\mathcal{C} \supset \mathbb{R}_+^m$, and any $\epsilon > 0$, there exists differentiable functions $\mathbf{T}_\epsilon^i, \mathbf{T}_\epsilon^o$ satisfying*

$$X_p(\mathbf{T}_\epsilon^i \circ \mathbf{F}, X, \mathbb{R}_+^m) \supset X_{\mathcal{P}}(\mathbf{F}, X) = X_p(\mathbf{F}, X, \mathcal{C}) \supset X_p(\mathbf{T}_\epsilon^o \circ \mathbf{F}, X, \mathbb{R}_+^m), \quad (1)$$

and moreover,

$$d_H(X_p(\mathbf{T}_\epsilon^i \circ \mathbf{F}, X, \mathbb{R}_+^m), X_{\mathcal{P}}(\mathbf{F}, X)) \leq \epsilon, \text{ and}$$
$$d_H(X_p(\mathbf{T}_\epsilon^o \circ \mathbf{F}, X, \mathbb{R}_+^m), X_{\mathcal{P}}(\mathbf{F}, X)) \leq \epsilon,$$

hold, where d_H denotes the Hausdorff distance between two sets.

Proof: We sketch the proof idea as a detailed proof is beyond the scope of this paper. Under the assumptions of this lemma, the set is $X_{\mathcal{P}}(\mathbf{F}, X)$ is convex and connected [10]. The idea is to use inner and outer polyhedral approximations of the convex cone $\mathcal{C}$ to a desired accuracy and employ Lemma 1 to define $\mathbf{T}_\epsilon^i$ and $\mathbf{T}_\epsilon^o$. The sequences can be shown to be convergent using [15]. □

Lemma 2 shows that it possible to use inner and outer approximations of a convex cone to generate the set $X_{\mathcal{P}}(\mathbf{F}, X)$. For an arbitrary convex cone, however, it might happen that k appearing in Lemma 1 is a very large number, especially if a small ϵ is desired.

Many different trade-off notions can be shown to be equivalent to an appropriate smooth preference model. Some of these are discussed next.

Definition 7 (Allowable tradeoff [28]). *An allowable tradeoff between criteria i and j, with $i, j \in \mathcal{I}$ denoted by a_{ij}, is the largest amount of decay in criterion i considered allowable to the decision maker to gain one unit of improvement in criterion j. Also, $a_{ij} \geq 0$ for all i and j, $i \neq j$.*

If $a_{ij} = 0$, then the decision maker's preference model is based on the classical Pareto-cone domination structure [17]. A trade-off between two criteria incurred along a direction $\mathbf{d}$ is called *directional trade-off*.

Definition 8 (Directional trade-off [28]). *A directional trade-off between criteria i and j, $i, j = 1, 2, \ldots, m$, $i \neq j$, denoted by $t_{ij}(\mathbf{d})$, is defined as follows:*

$$t_{ij}(\mathbf{d}) = 0, \quad \text{if } d_i \leq 0 \text{ and } d_j \leq 0$$
$$t_{ij}(\mathbf{d}) = \frac{d_i}{-d_j}, \quad \text{if } d_i > 0 \text{ and } d_j < 0$$
$$t_{ij}(\mathbf{d}) = \infty, \quad \text{if } d_i \geq 0 \text{ and } d_j \geq 0, \mathbf{d} \neq 0$$
$$t_{ij}(\mathbf{d}) \quad \text{is undefined otherwise.}$$

A direction $\mathbf{d} \in \mathbb{R}^m$ is an attractive direction if $t_{ij}(\mathbf{d}) \leq a_{ij}$ for every pair of criteria $i, j = 1, 2, \ldots, m$, $i \neq j$.

Based on the above definition Wiecek et al. [28] construct a model where they assume that the decision maker allows one criterion i to decay only if all the other criteria $j \neq i$ improve. The values a_{ij} come from the decision maker. It may be of interest to repeat the process with more than one selection of criteria and the model includes that. Let P_i, for a given criterion i, be the set of all attractive directions. All the attractive directions are appended with $-\mathbb{R}_{\geq}^m$ so as to obtain a set (which is a cone) of attractive directions given by

$$P := \bigcup_i P_i \cup (-\mathbb{R}_{\geq}^m), \tag{2}$$

where, for all $i \in \mathcal{I}$, the cone P_i is mathematically defined by

$$P_i := \{\mathbf{d} \in \mathbb{R}^m | d_i > 0, \, d_j < 0, \text{ for all } j \in \mathcal{I}, \, j \neq i \text{ and } t_{ij}(\mathbf{d}) \leq a_{ij} \text{ for all } j \neq i\}.$$

The MOP with a domination structure given by the cone P is termed as *preference model 1* in [28] and it has been used in many civil and mechanical engineering applications (like truss design [29]). For an m-dimensional problem, under certain conditions, P can be represented with the help of an $m(m-1)$ by m matrix A defined next [28].

Definition 9. *Let A_1 be an $m(m-1)$ by m matrix described by m blocks of $m-1$ rows and m columns, where A_1^{ij} represents row $j \in \mathcal{I} \setminus \{m\}$ of block $i \in \mathcal{I}$ of A_1, and $(A_1^{ij})_k$ represents the element of A_1^{ij} in column $k \in \mathcal{I}$. The elements of A_1 are defined as follows:*

$$
\begin{aligned}
(A_1^{ij})_i &= 1 && \text{for all } i \in \mathcal{I}, \, j \in \mathcal{I} \setminus \{m\} \\
(A_1^{ij})_j &= a_{ij} && \text{if } j < i, \, i \in \mathcal{I}, \, j \in \mathcal{I} \setminus \{m\} \\
(A_1^{ij})_{j+1} &= a_{i(j+1)} && \text{if } j \geq i, \, i \in \mathcal{I}, \, j \in \mathcal{I} \setminus \{m\} \\
(A_1^{ij})_i &= 0 && \text{otherwise, } i \in \mathcal{I}, \, j \in \mathcal{I} \setminus \{m\}.
\end{aligned}
$$

As an example, for $m = 3$, the matrix A_1 can be represented as

$$A_1 := \begin{pmatrix} 1 & 1 & a_{21} & 0 & a_{31} & 0 \\ a_{12} & 0 & 1 & 1 & 0 & a_{32} \\ 0 & a_{13} & 0 & a_{23} & 1 & 1 \end{pmatrix}^\top, \tag{3}$$

and the decision maker has to supply six a_{ij}'s.

Lemma 3. *If $a_{ij}a_{ji} \leq 1$ for any $i \neq j$ and $i, j \in \mathcal{I}$ and $m \geq 3$ then preference model 1 described above is a smooth preference model.*

Proof: Since P is the union of all P_i's, in general the cone P might be non-convex. From [28, Corollary 3.1], P is non-convex if $a_{ij}a_{ji} > 1$ for any $i \neq j$. The rest of the proof follows from the discussion in [28, Page 161]. □

There are other smooth preference models that we find in literature. The notion of using relative importance of objectives (see Noghin [19]), for example, can also be shown to be smooth. In this case also $\mathcal{A}$ is such that $k > m$.

2.2 A Theoretical Analysis of Nonsmooth Preference Models

A preference model $\mathcal{P}$ will be called as *nonsmooth* if it is not smooth. In other words, there exists neither a convex cone $\mathcal{C}$ nor a smooth differentiable function $\mathbf{T}$ such that 1. and 2. in Definition 6 holds.

Nosmoothness in a preference model can be introduced in different ways. For example, it might happen that there is a *non-convex* cone C, such that $X_{\mathcal{P}}(\mathbf{F}, X) = X_p(\mathbf{F}, X, C)$ or it also might happen that there exists a *nonsmooth* function $T : \mathbb{R}^m \to \mathbb{R}^k$, for some $k \geq m$, such that $X_{\mathcal{P}}(\mathbf{F}, X) = X_p(T \circ \mathbf{F}, X, \mathbb{R}_+^m)$. Both non-convexity and nonsmoothness of T might also occur at the same time.

This poses additional problems, and convergence results are difficult to obtain, if one uses standard nonlinear programming techniques.

The next lemma shows that the classical concept of proper Pareto-optimality is in general nonsmooth.

Lemma 4. *Let M in Definition 1 be fixed beforehand. Then, if $m \geq 3$, then Geoffrion proper Pareto optimality is a nonsmooth preference model. However, if $m = 2$, then it is a smooth preference model.*

Proof: The main idea of the proof (as a rigorous mathematical proof is beyond the scope) is as follows. Let us assume that Geoffrion proper Pareto-optimality is smooth. Lemma 2 shows that any smooth preference model can be approximated to a desired accuracy by inner and outer polyhedral cone approximations of $\mathcal{C}$. Any polyhedral cone uses a fixed trade-off on all or some pairs of the objectives. Assuming a polyhedral ordering cone, we can construct counterexamples showing that the set $X_{pp}(\mathbf{F}, X, \mathbb{R}^m_+)$ cannot be obtained completely. The essential element comes here from Definition 1, where the existence of *one* j is sufficient to bound the trade-off. On other hand, any closed, convex, and pointed, cone in $\mathbb{R}^2$ is a polyhedral cone and hence, the index $j \neq i$ is fixed (as there are just two indices). Hence the lemma follows. $\square$

The notion of equitability [20] is another example of a nonsmooth preference model and this is discussed next. Let the map $\Theta : \mathbb{R}^m \to \mathbb{R}^m$ be so that $\Theta(\mathbf{y}) = (\theta_1(\mathbf{y}), \theta_2(\mathbf{y}), \ldots, \theta_m(\mathbf{y}))$, where $\theta_1(\mathbf{y}) \geq \theta_2(\mathbf{y}) \geq \ldots \geq \theta_m(\mathbf{y})$ and there exists a permutation τ of the set $\mathcal{I}$ such that $\theta_i(\mathbf{y}) = y_{\tau(i)}$ for all $i \in \mathcal{I}$. Moreover, let $\Gamma : \mathbb{R}^m \to \mathbb{R}^m$ and $\mathbf{q} := (q_1, \ldots, q_m)$ be so that $\Gamma(\mathbf{q}) = (\gamma_1(\mathbf{q}), \gamma_2(\mathbf{q}), \ldots, \gamma_m(\mathbf{q}))$, where $\gamma_i(\mathbf{q}) = \sum_{j=1}^{i} q_j$ for all $i \in \mathcal{I}$.

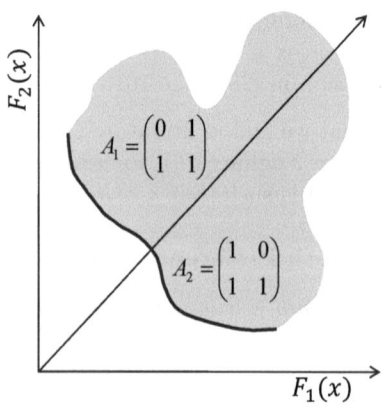
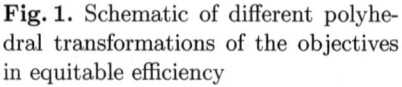
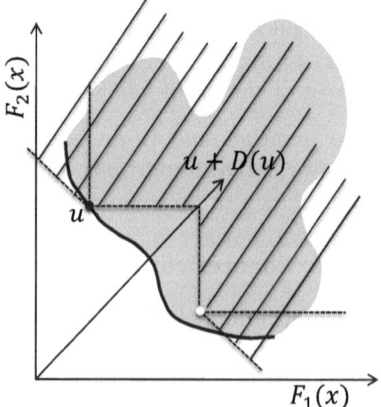

Fig. 1. Schematic of different polyhedral transformations of the objectives in equitable efficiency

Fig. 2. Schematic of equitable ordering relation. The shaded area D is a nonconvex set and is not a cone.

Composition of Θ and Γ gives us the *cumulative ordering map* $\mathbf{T} = (t_1, t_2, \ldots, t_m)$ defined as $\mathbf{T}(\mathbf{y}) = \Gamma(\Theta(\mathbf{y}))$, i.e.,

$$t_i(\mathbf{y}) = \sum_{j=1}^{i} \theta_j(\mathbf{y}) \quad \text{for all } i \in \mathcal{I}. \tag{4}$$

From [20], we obtain that equitable domination is equivalent to Pareto-domination on the modified set of objectives. This is used to transform the goal of finding equitably optimal points of the original problem to that of finding Pareto-optimal points of the following multi-objective problem (see also [26]):

$$\min \mathbf{T}(\mathbf{F}(\mathbf{x})) := \left(\bar{\theta}_1(\mathbf{x}), \bar{\theta}_2(\mathbf{x}), \ldots, \bar{\theta}_m(\mathbf{x}) \right) \quad \text{s.t. } \mathbf{x} \in X. \tag{5}$$

If we look carefully at (5), we see that now all the first $m-1$ objective functions are *non-differentiable*. This non-differentiability is inherent in the Θ mapping.

Figure 1 shows the nonsmooth transformation (2. in Definition 6 is not satisfied) that in inherent in equitable efficiency. $\mathbf{T}$ can be represented by a collection of polyhedral matrices A_i, that depend upon the region (sector) relative to the line $y = x$ in $\mathbb{R}^2$. From Figure 2 we can see that the ordering cone here D is a non-convex set, and is not even a cone (1. in Definition 6 is not satisfied). The next lemma summarized this situation.

Lemma 5. *Equitable efficiency is a nonsmooth preference model. However, if we weaken differentiability to local Lipschitzness in Conditions 2. (in Definition 6), then Condition 2. is satisfied and equitable efficiency is a smooth preference model.*

Proof: The proof easily follows by the definition of $\mathbf{T}$ (Eq. 4) and by noting that $\mathbf{T}$ defined in such a way is locally Lipschitz (piecewise polyhedral) [7]. $\square$

There are many cone based extended notions of equitable efficiency (see Mut and Wiecek [18]) and it can shown that all of these are nonsmooth.

3   Indicator Based Algorithms for Preference Models

The hypervolume indicator (HI), or S-Metric [31], is a common indicator for measuring the quality of Pareto front approximations [32] and it is often used as a criterion for guiding search algorithms towards Pareto fronts [2,1,11]. For a finite set $S \subset \mathbb{R}^m$ it is defined as

$$\text{HI}(S) = \lambda(\cup_{\mathbf{p} \in S}[\mathbf{p}, \mathbf{r}]), \tag{6}$$

where recall that λ denotes the Lebesgue measure, $\mathbf{r}$ is a reference point that is usually fixed by the user, and $[\mathbf{p}, \mathbf{r}]$ denotes the hypercuboid spanned by $\mathbf{p}$ and $\mathbf{r}$. Recently [12], the classical hypervolume indicator has been extended to a cone-based hypervolume indicator (CHI). Moreover, a family of polyhedral cones with scalable opening angle γ were studied to obtain a well-diverse set of solutions. In this section, we will theoretically analyze CHI in the context of smooth and nonsmooth preference models. Before this some formal definitions (of Minkowski sum and CHI) are presented.

Algorithm: Cone-based hypervolume computation using a polyhedral cone.

Input: An m by k matrix $\mathcal{A}$, points $S \subset \mathbb{R}^m$, reference point $\mathbf{r}$

Output: Cone-based hypervolume indicator $\mathrm{CHI}(S)$

1. Let $\mathbf{r}' = \mathcal{A}\mathbf{r}$.
2. For all $i = 1, \ldots, |S|$, let $Q = \{\mathbf{q}^{(1)}, \ldots, \mathbf{q}^{|S|}\}$, with $\mathbf{q}^{(i)} = \mathcal{A}\mathbf{s}^{(i)}$.
3. Compute the standard hypervolume $\mathrm{HI}(Q, \mathbf{r}')$.
4. Return $\mathrm{CHI}(S) = (1/\det(\mathcal{A}^\top \mathcal{A})) \cdot \mathrm{HI}(Q, \mathbf{r}')$.

Fig. 3. Computing CHI for a smooth preference model using a polyhedral cone

Definition 10. *Let A and B denote sets in $\mathbb{R}^m$. Then*

$$A \oplus B = \{\mathbf{a} + \mathbf{b} \mid \mathbf{a} \in A \text{ and } \mathbf{b} \in B\}. \tag{7}$$

and

$$A \ominus B = \{\mathbf{a} - \mathbf{b} \mid \mathbf{a} \in A \text{ and } \mathbf{b} \in B\}. \tag{8}$$

Definition 11 (Cone-based hypervolume [12]). *The cone-based hypervolume for a finite set $S \in \mathbb{R}^m$, a cone $C \subset \mathbb{R}^m$, and a reference point $\mathbf{r}$ with $\forall \mathbf{p} \in S : \mathbf{p} \leq_C \mathbf{r}$ is defined as*

$$CHI(S) = \lambda((S \oplus C) \cap (\{\mathbf{r}\} \ominus C)). \tag{9}$$

or, based on Definition 2, as

$$CHI(S) = \lambda(\{\mathbf{q} \in \mathbb{R}^m | \exists \mathbf{p} \in S : \mathbf{p} \leq_C \mathbf{q} \wedge \mathbf{q} \leq_C \mathbf{r}\}). \tag{10}$$

The quantity $H(\mathbf{u}, S, C, \mathbf{r}) := \mathrm{CHI}(\mathbf{S}) - \mathrm{CHI}(\mathbf{S} \setminus \{\mathbf{u}\})$ will be termed as the cone-based hypervolume contribution of the point $\mathbf{u}$ to the set S corresponding to cone C (and reference point $\mathbf{r}$).

The next theorem is the main result of this section. Recall that $\mathcal{E}(\mathcal{S}, D)$ denotes the D-efficient points of a set $\mathcal{S} \subseteq \mathbb{R}^m$.

Theorem 1. *Let the ordering cone C be the union of a finite number of closed and pointed (possibly non-convex) cones, i.e., $C = \bigcup_{i=1}^{\ell} C_\ell$ such that C is a (possibly non-convex) pointed cone. Then, for a finite set $S \in \mathbb{R}^m$ and a reference point $\mathbf{r}$ with $\forall \mathbf{s} \in S : \mathbf{s} \leq_C \mathbf{r}$*

$$\lambda((P \oplus C) \cap (\{\mathbf{r}\} \ominus C)) = \lambda((\mathcal{E}(P, C) \oplus C) \cap (\{\mathbf{r}\} \ominus C)) \tag{11}$$

and

$$\lambda((P \oplus C) \cap (\{\mathbf{r}\} \ominus C)) = \lambda\left(\left(\bigcap_{i=1}^{\ell} \mathcal{E}(P, C_\ell) \oplus C\right) \cap (\{\mathbf{r}\} \ominus C)\right), \tag{12}$$

hold.

Proof: Let $\mathbf{u} \in (P \oplus C) \cap (\{\mathbf{r}\} \ominus C)$ be an arbitrary but fixed element. The CHI hypervolume contribution of $\mathbf{u}$ to the set P corresponding to cone C is (strictly) positive iff $\mathbf{u} \ll_C \mathbf{r}$ (i.e, the hypercuboid spanned by point $\mathbf{u}$ and $\mathbf{r}$ is full dimensional) and $\nexists \mathbf{w} \in P$ such that $\mathbf{w} \neq \mathbf{u}$ and $\mathbf{w} \leq_C \mathbf{u}$. On the other hand, the cone-based hypervolume contribution is zero if there is a $\mathbf{v} \in S$, such that $\mathbf{v} \leq_C \mathbf{u}$. Hence, exactly points from the set $\mathcal{E}(P, C)$ have a positive contributions and hence (11) holds.

For the second part, let $\mathbf{u} \in \mathcal{E}(P, C)$. Hence,

$$(\{\mathbf{u}\} - C \setminus \{\mathbf{0}\}) \cap S = \emptyset. \tag{13}$$

We claim that $\mathbf{u} \in \mathcal{E}(P, C_i)$ for all $i = 1, 2, \ldots, \ell$. If this is not the case then, there exists an index $k \in \{1, 2, \ldots, \ell\}$ such that $\mathbf{u} \notin \mathcal{E}(P, C_k)$. Hence, $\mathbf{v} \in S$ such that $\mathbf{v} \in \{\mathbf{u}\} - C_k \setminus \{\mathbf{0}\}$ exists. This implies that $\mathbf{u} - \mathbf{v} \in C_k \setminus \{\mathbf{0}\} \subseteq C \setminus \{\mathbf{0}\}$ which is a contradiction to (13). Hence, we obtain that $\mathcal{E}(P, C) \subseteq \bigcap_{i=1}^{\ell} \mathcal{E}(P, C_\ell)$.

Next, we show that $\mathcal{E}(P, C) \supseteq \bigcap_{i=1}^{\ell} \mathcal{E}(P, C_\ell)$. For this, let $\mathbf{u} \in \bigcap_{i=1}^{\ell} \mathcal{E}(P, C_\ell)$. This means that

$$(\{\mathbf{u}\} - C_i \setminus \{\mathbf{0}\}) \cap S = \emptyset \text{ for all } i = 1, 2, \ldots, \ell. \tag{14}$$

In this case, one can easily show that (14) implies (13). Hence $\mathcal{E}(P, C) = \bigcap_{i=1}^{\ell} \mathcal{E}(P, C_\ell)$ and the result of the theorem follows. □

Remark 1. Convexity is not assumed in Theorem 1 and hence it is useful in cases where the nonsmooth preference model can be written as a union of finite number of non-convex cones. Only efficient front contributes to the hypervolume and the above result can be used to find the efficient front as intersections of different efficient front.

Next, we present an algorithm for computing CHI (and CHI contributions) corresponding to a smooth preference model. The aim is to find a subset of at most μ points such that CHI is maximized. We start with $\mathbf{T} := \mathcal{A}$, where $\mathcal{A}$ is a m by k real matrix. In such a case, [23, Lemma 2.3.4] shows that in order to find $X_{\mathcal{P}}(\mathbf{F}, X)$ we could search for Pareto-optimal solutions of the transformed objectives $\mathcal{A}\mathbf{F}$. We compute the hypervolume indicator in $\mathbb{R}^k$ space and correct for the volume change due to the affine transformations (see [3]). The algorithm shown in Figure 3 describes this procedure.

Figure 4 describes an algorithm for computing the individual hypervolume contributions of a point if the underlying transformation map $\mathbf{T} : \mathbb{R}^m \to \mathbb{R}^m$ is smooth. In this case the determinant of the Jacobian matrix $J_{\mathbf{T}}(\mathbf{u})$ gives the local scaling factor [3] and this is used to find an approximate individual hypervolume contribution of a point $\mathbf{u} \in S$.

The above two algorithms use a volume correction term. However, we note that the ordering of points w.r.t. there hypervolume contributions does not depend on this correction term if polyhedral cones are used. In the case of nonlinear transformation, we have to use the Jacobian based correction term since the correction depends on the point. If the preference model uses a nonlinear and convex

---

Algorithm: Approximate hypervolume contribution using a smooth transformation function **T**.

Input: Function **T**, points $S \subset \mathbb{R}^m$, reference point **r** and a point $\mathbf{u} \in S$
Output: Cone-based hypervolume contribution of **u** in S

1. Let $\mathbf{r}' = T(\mathbf{r})$.
2. For all $i = 1, \ldots, |S|$, let $Q = \{\mathbf{q}^{(1)}, \ldots, \mathbf{q}^{|S|}\}$, with $\mathbf{q}^{(i)} = \mathbf{T}(\mathbf{s}^{(i)})$.
3. Compute the standard hypervolume contribution $H(\mathbf{u}, Q, \mathbb{R}^m_+, \mathbf{r}')$.
4. Return $(1/\det(J_{\mathbf{T}}(\mathbf{u})) \cdot H(\mathbf{u}, Q, \mathbb{R}^m_+, \mathbf{r}')$.

Fig. 4. Computing an approximate value for the hypervolume contribution of a point **u** using a smooth transformation function **T**

cone C, then we could still use the algorithm for cone-based hypervolume computation using a polyhedral cone, with inner and outer approximation of C (see Lemma 2). The mean value of such could be an approximate measure of CHI.

Finally, we note that nonsmooth preference models are the hardest to deal with. If we know additional properties like piecewise polyhedral (equitability structure) or Lipschitzness we could resort to objects from the Clarke generalized Jacobian and use the algorithm described in Figure 4. However, computing the Clarke generalized Jacobian is difficult and is beyond the scope of this paper.

4   Experimentation

In this section, we present preliminary numerical results. We used an equitable ordering (see Figure 2) based SMS-EMOA [6] algorithm for finding the set of solutions that maximize CHI. CHI computation is done by the algorithm described in Figure 3. The only difference it now we have piecewise polyhedral cones that depend on the position in the objective space (see Figure 1). The source code of the algorithms is written using the jMetal framework [9] and is available on request. We ran SMS-EMOA using a population size of 30 for 400 generations. Although, we tested the algorithm on many two and three dimensional problems, due to space limitation we only present illustrative result on few test problems.

Figure 5 shows simulation results on the two dimensional ZDT1 test problem. The equitable front for this problem consists of a part of the original convex front given by $F_2 = 1 - \sqrt{F_1}$. In the original objective space, we see a biased distribution of points, more towards the point $(0.38, 0.38)^\top$. This is due to the equitable nature of the ordering (in that more equal distributions are preferred, see [20]). In the (transformed) equitable space however, we see uniformly distributed points (seen in Figure 6). This in interesting, as the use of a cone-based (or set-based in general) hypervolume indicator can be implemented in different spaces depending on where it is desirable. This is also important for inner and outer approximation based approaches, where k could be much large than m, if one requires an accurate approximation of a nonlinear and convex cone.

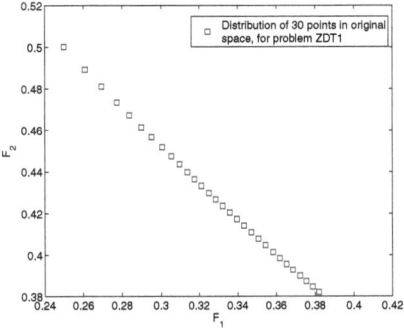

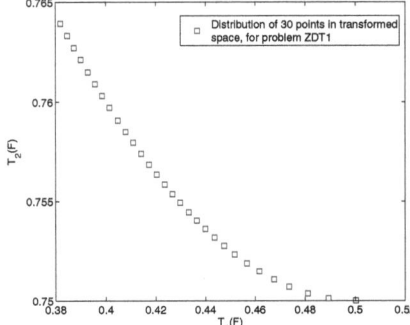

Fig. 5. Distribution of 30 points in original space on test problem ZDT1

Fig. 6. Distribution of 30 points in the equitable space on test problem ZDT1

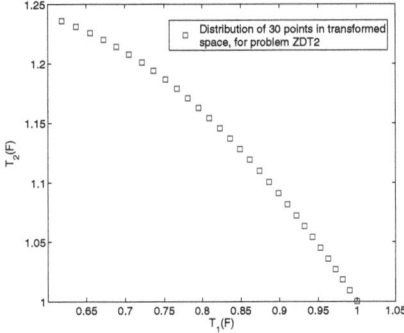

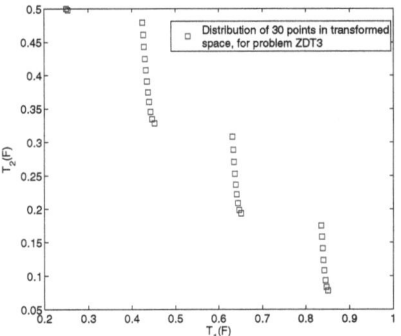

Fig. 7. Distribution of 30 points in the equitable space on test problem ZDT2

Fig. 8. Distribution of 30 points in the equitable space on test problem ZDT3

Controlling k is very important as hypervolume computation algorithms are exponential in k (for n points in $\mathbb{R}^m$, the complexity of computing the hypervolume is $\Theta(n \log n)$ for $m = 2$ and $m = 3$ [11] and $\mathcal{O}(n^{d-1/2}) \ln n$ for $m > 3$ [30]. Also in many smooth polyhedral cone models k can be larger than m (in Section 2, for example, $k = m(m - 1)$). This is one disadvantage of a direct cone based hypervolume computation. In such cases, Monte Carlo based approaches could be used (for example, the technique in [2]).

Figure 7 shows simulation results on the non-convex, two dimensional ZDT2 test problem, in the transformed space. One could see that the front is almost linear and hence the points are equispaced, although one could observe that the points are slightly more dense near the point $(1, 1)^\top$. The reason for this is the same as in the case of ZDT1. Figure 8 shows results on the non-convex, disconnected, two dimensional ZDT3 test problem and here also we can see that the algorithm is able to find a well-diverse set of solutions.

5   Conclusions

This study is towards a better understanding of various preference models that are present in the multi-objective community. There has been a lot of work in preference modeling by the classical MCDM researchers and this has been a subject of active research since the late sixties. The first contribution of this work is a classification of preference models. This has algorithmic implications, especially if one is interested in finding a well-diverse set of preferred solutions. Evolutionary algorithms have an inherent population based advantage (see [21] for more on this interesting aspect) and they can be properly tailored to find the complete preferred area. This is only possible if we know beforehand what kind of preference models are amenable to evolutionary algorithms and what is needed to make them fit. The classification makes this possible, in that it shows how polyhedral cones can be used to model decision makers preferences. This also introduces theoretical challenges, in terms of sorting or hypervolume computation (if $k > m$ or for a nonpolyhedral cone, for example). The cone based hypervolume is an interesting concept and, it can used to control the density of solutions. In equitable spaces, we showed that such an indicator based algorithm can be successfully applied. Future works will emphasize efficient computation schemes more and on finding distribution of points that yield maximal hypervolume, for general cone orders.

References

1. Auger, A., Bader, J., Brockhoff, D.: Theoretically Investigating Optimal μ-Distributions for the Hypervolume Indicator: First Results For Three Objectives. pp. 586–596. Springer (2010)
2. Bader, J., Zitzler, E.: Hype: An algorithm for fast hypervolume-based many-objective optimization. Evolutionary Computation, 1–32 (2010)
3. Ben-Israel, A.: A volume associated with m x n matrices. Linear Algebra and its Applications 167, 87–111 (1992)
4. Bergstresser, K., Charnes, A., Yu, P.L.: Generalization of domination structures and nondominated solutions in multicriteria decision making. Journal of Optimization Theory and Applications 18, 3–13 (1976)
5. Bergstresser, K., Yu, P.L.: Domination structures and multicriteria problems in n-person games. Theoy and Decision 8, 5–48 (1977)
6. Beume, N.: Hypervolume based metaheuristics for multiobjective optimization. Ph.D. thesis, Dortmund, Techn. Univ., Diss (2011)
7. Clarke, F.H.: Optimization and nonsmooth analysis, Classics in Applied Mathematics, 2nd edn., vol. 5. SIAM, Philadelphia (1990)
8. Deb, K.: Multi-objective optimization using evolutionary algorithms. Wiley (2001)
9. Durillo, J.J., Nebro, A.J.: jmetal: A java framework for multi-objective optimization. Advances in Engineering Software 42(10), 760–771 (2011)
10. Ehrgott, M.: Multicriteria optimization, 2nd edn. Springer, Berlin (2005)
11. Emmerich, M., Beume, N., Naujoks, B.: An EMO Algorithm Using the Hypervolume Measure as Selection Criterion. In: Coello Coello, C.A., Hernández Aguirre, A., Zitzler, E. (eds.) EMO 2005. LNCS, vol. 3410, pp. 62–76. Springer, Heidelberg (2005)

12. Emmerich, M., Deutz, A., Kruisselbrink, J., Shukla, P.: Cone-Based Hypervolume Indicators: Construction, Properties, and Efficient Computation. In: Purshouse, R.C., Fleming, P.J., Fonseca, C.M., Greco, S., Shaw, J. (eds.) EMO 2013. LNCS, vol. 7811, pp. 111–127. Springer, Heidelberg (2013)
13. Geoffrion, A.M.: Proper efficiency and the theory of vector maximization. Journal of Mathematical Analysis and Applications 22, 618–630 (1968)
14. Karasakal, E.K., Michalowski, W.: Incorporating wealth information into a multiple criteria decision making model. European J. Oper. Res. 150(1), 204–219 (2003)
15. Lemaire, B.: Approximation in multiobjective optimization. J. Global Optim. 2(2), 117–132 (1992)
16. Makarov, E.K., Rachkovski, N.N.: Unified representation of proper efficiency by means of dilating cones. J. Optim. Theory Appl. 101(1), 141–165 (1999)
17. Miettinen, K.: Nonlinear Multiobjective Optimization. Kluwer, Boston (1999)
18. Mut, M., Wiecek, M.M.: Generalized equitable preference in multiobjective programming. European J. Oper. Res. 212(3), 535–551 (2011)
19. Noghin, V.D.: Relative importance of criteria: a quantitative approach. Journal of Multi-Criteria Decision Analysis 6(6), 355–363 (1997)
20. Ogryczak, W.: Inequality measures and equitable locations. Annals of Operations Research 167, 61–86 (2009)
21. Prügel-Bennett, A.: Benefits of a population: Five mechanisms that advantage population-based algorithms. IEEE Transactions on Evolutionary Computation 14(4), 500–517 (2010)
22. Rudin, W.: Real and complex analysis, 3rd edn. McGraw-Hill Book Co, New York (1987)
23. Sawaragi, Y., Nakayama, H., Tanino, T.: Theory of multiobjective optimization, vol. 176. Academic Press Inc., Orlando (1985)
24. Shukla, P.K., Braun, M.A., Schmeck, H.: Theory and Algorithms for Finding Knees. In: Purshouse, R.C., Fleming, P.J., Fonseca, C.M., Greco, S., Shaw, J. (eds.) EMO 2013. LNCS, vol. 7811, pp. 156–170. Springer, Heidelberg (2013)
25. Shukla, P.K., Hirsch, C., Schmeck, H.: A Framework for Incorporating Trade-Off Information Using Multi-Objective Evolutionary Algorithms. In: Schaefer, R., Cotta, C., Kołodziej, J., Rudolph, G. (eds.) PPSN XI, Part II. LNCS, vol. 6239, pp. 131–140. Springer, Heidelberg (2010)
26. Shukla, P.K., Hirsch, C., Schmeck, H.: In Search of Equitable Solutions Using Multi-objective Evolutionary Algorithms. In: Schaefer, R., Cotta, C., Kołodziej, J., Rudolph, G. (eds.) PPSN XI, Part I. LNCS, vol. 6238, pp. 687–696. Springer, Heidelberg (2010)
27. Shukla, P.K., Hirsch, C., Schmeck, H.: Towards a Deeper Understanding of Trade-offs Using Multi-objective Evolutionary Algorithms. In: Di Chio, C., Agapitos, A., Cagnoni, S., Cotta, C., de Vega, F.F., Di Caro, G.A., Drechsler, R., Ekárt, A., Esparcia-Alcázar, A.I., Farooq, M., Langdon, W.B., Merelo-Guervós, J.J., Preuss, M., Richter, H., Silva, S., Simões, A., Squillero, G., Tarantino, E., Tettamanzi, A.G.B., Togelius, J., Urquhart, N., Uyar, A.Ş., Yannakakis, G.N. (eds.) EvoApplications 2012. LNCS, vol. 7248, pp. 396–405. Springer, Heidelberg (2012)
28. Wiecek, M.M.: Advances in cone-based preference modeling for decision making with multiple criteria. Decis. Mak. Manuf. Serv. 1(1-2), 153–173 (2007)
29. Wiecek, M.M., Blouin, V.Y., Fadel, G.M., Engau, A., Hunt, B.J., Singh, V.: Multi-scenario Multi-objective Optimization with Applications in Engineering Design. In: Barichard, V., Ehrgott, M., Gandibleux, X., T'Kindt, V. (eds.) Multiobjective Programming and Goal Programming. LNEMS, vol. 618, pp. 283–298. Springer, Heidelberg (2009)

30. Yildiz, H., Suri, S.: On klee's measure problem for grounded boxes. In: Proceedings of the 2012 Symposuim on Computational Geometry, SoCG 2012, pp. 111–120. ACM, New York (2012)
31. Zitzler, E., Thiele, L.: Multiobjective Optimization Using Evolutionary Algorithms - A Comparative Case Study. In: Eiben, A.E., Bäck, T., Schoenauer, M., Schwefel, H.-P. (eds.) PPSN V. LNCS, vol. 1498, pp. 292–301. Springer, Heidelberg (1998)
32. Zitzler, E., Thiele, L., Laumanns, M., Foneseca, C.M., Grunert da Fonseca, V.: Performance assessment of multiobjective optimizers: An analysis and review. IEEE Transactions on Evolutionary Computation 7(2), 117–132 (2003)

Force-Based Cooperative Search Directions in Evolutionary Multi-objective Optimization

Bilel Derbel[1,2], Dimo Brockhoff[1], and Arnaud Liefooghe[1,2]

[1] Inria Lille - Nord Europe, DOLPHIN Project-team, 59650 Villeneuve d'Ascq, France
[2] Université Lille 1, LIFL, UMR CNRS 8022, 59655 Villeneuve d'Ascq cedex, France
firstname.lastname@inria.fr

Abstract. In order to approximate the set of Pareto optimal solutions, several evolutionary multi-objective optimization (EMO) algorithms transfer the multi-objective problem into several independent single-objective ones by means of scalarizing functions. The choice of the scalarizing functions' underlying search directions, however, is typically problem-dependent and therefore difficult if no information about the problem characteristics are known before the search process. The goal of this paper is to present new ideas of how these search directions can be computed *adaptively* during the search process in a *cooperative* manner. Based on the idea of Newton's law of universal gravitation, solutions attract and repel each other *in the objective space*. Several force-based EMO algorithms are proposed and compared experimentally on general bi-objective ρMNK landscapes with different objective correlations. It turns out that the new approach is easy to implement, fast, and competitive with respect to a $(\mu + \lambda)$-SMS-EMOA variant, in particular if the objectives show strong positive or negative correlations.

1 Introduction

Besides established *Pareto-based* EMO algorithms, such as NSGA-II, SPEA2, or ε-MOEA, and the recently proposed *indicator-based* algorithms such as IBEA, SMS-EMOA, MO-CMA-ES, or HypE, a third group of *aggregation-based* algorithms, containing e.g. MSOPS [5] and MOEA/D [17], for solving multi-objective optimization problems can be identified [14]. Aggregation-based algorithms reformulate the multi-objective optimization problem as a set of single-objective problems by means of multiple scalarizing functions [10] that are typically solved independently from each other. Standard scalarizing functions such as weighted sum or achievement functions are thereby defining a search direction in the objective space in which the solutions evolve during the search process. As the a priori definition of these search directions is difficult if no further information is known about the problem at hand, this paper proposes a new force-based approach to cooperatively adapt the (single-objective) search directions.

Throughout the paper, we assume the maximization of a vector-valued objective function $\phi : X \to Z$ that maps a solution x from the feasible *search space* X to $\phi(x) = (\phi_1(x), \ldots, \phi_M(x))$ in the *objective space* $Z \subseteq \mathbb{R}^M$. We say, an objective vector $z \in Z$ is *dominated* by objective vector $z' \in Z$, denoted by $z \prec z'$, if for all $i \in \{1, \ldots, M\}$ $z_i \leq z'_i$ and there exists a $j \in \{1, \ldots, M\}$ such that $z_j < z'_j$. Similarly, a solution $x \in X$ is dominated by $x' \in X$, denoted by $x \prec x'$, if $\phi(x) \prec \phi(x')$. An

R.C. Purshouse et al. (Eds.): EMO 2013, LNCS 7811, pp. 383–397, 2013.

objective vector $z \in Z$ is *non-dominated* with respect to a set S if there does not exist any other objective vector $z' \in S$ such that $z \prec z'$. A solution $x \in X$ is *Pareto optimal* if $\phi(x)$ is non-dominated with respect to Z. The set of all Pareto optimal solutions is the *Pareto set* and its mapping in the objective space is the *Pareto front* of which an approximation is sought.

Contribution Overview. Generally speaking, we consider to evolve a set of solutions towards the Pareto front by computing, dynamically at each generation, a force-based direction in the objective space with respect to each solution. Each direction is used to define a single-objective optimization problem to be optimized by each solution independently of the others. Inspired by particle physics and more precisely by Newton laws, we define the direction relative to a solution as an aggregation of forces exerted by other solutions in the objective space. For each pair of solutions, we propose to compute their respective forces according to their dominance relation, while adjusting force magnitudes according to the distance between solutions in the objective space. One specificity of this approach is to evolve a set of solutions in a dynamic and local manner. In fact, search directions are not fixed and evolve throughout generations in an attempt to adaptively fit the search process and better approach the Pareto front. Furthermore, while search directions are computed in a cooperative manner depending on the relative position of solutions at some point of the search, each solution uses its own direction in parallel to other solutions in order to evolve towards a new solution. Thus, maintaining the set of solutions is done in a straightforward manner while avoiding sophisticated data structures and costly operations. Besides being extremely simple to implement, and through extensive experiments on ρMNK landscapes, our approach is also proved to be efficient in dynamically finding good directions leading to a good approximation of the Pareto front.

Related Work. While the force-based approach presented in this paper share similarities with particle swarm optimization, there are few other studies that are even more related and also compute forces among solutions to steer the search and maintain diversity. In [15,16], for example, the authors present a constraint multi-objective artificial physics optimization algorithm for continuous problems extending on previous single-objective techniques based on virtual force computations. Such an algorithm considers to move each individual using a velocity vector driven by the total force from other individuals *in the search space*. Similar ideas are investigated in the so-called gravitational search algorithm [4,11,12]. One can also find other related studies where forces are used in a *problem-specific* manner in order to move solutions in the search space [3,8]. However, the way forces are computed and used by those algorithms differ mainly in the fact that the approach proposed here is *problem-independent*. Indeed, it computes the forces and search directions *in the objective space* and then evolves individuals *adaptively* in a *cooperative manner* on the basis of those computed directions.

Outline. The paper is organized as follows. Section 2 proposes the general template of the force-based EMO algorithm together with different instantiations of its components. In particular, a number of force-based search direction schemes and different selection and replacement strategies are introduced. Section 3 presents the setup of the experimental analysis. Section 4 discusses the dynamics of the algorithm with the aim

---

Algorithm 1. Force-Based EMO generic scheme

1   $P \leftarrow \{x^1, x^2, \ldots, x^\mu\}$: initial population;
2   $Z \leftarrow \phi(P) = \{z^1, z^2, \ldots, z^\mu\}$: initial outcome vector in the objective space;
3   **repeat**
4   **for** $i \in \{1, \ldots, \mu\}$ **do**
5        $\overrightarrow{d}^i \leftarrow$ Force $(z^i, Z \setminus \{z^i\})$;
6   **for** $i \in \{1, \ldots, \mu\}$ **do**
7        $S^i \leftarrow$ Pool $(x^i) \cup \{x^i\}$ /* variation */
8        $x^i \leftarrow$ Select $(S^i, \overrightarrow{d}^i)$ /* replacement selection */
9   **until** STOPPING_CONDITION;

---

of better understanding its general behavior. Section 5 gives a detailed experimental analysis. Finally, Section 6 concludes the paper and discusses further research.

2   Algorithm Description

The basic idea behind our approach is to view the evolution of individuals from one generation to another as a set of particles moving *in the objective space* due to virtual forces exerted by other particles in the population. Algorithm 1 gives a high-level description of such a force-based EMO algorithm. Starting with an initial population, the algorithm proceeds in generations in which the population is evolved by means of three main steps. First, for each individual x^i, we compute a search direction $\overrightarrow{d}^i$ using function Force and the position of other particles in the objective space. Then, for each individual x^i, a sample of candidate solutions S^i is, independently of the computed search directions, generated by means of some variation operator(s), denoted by Pool. Finally, the selection of the new candidate solution from the old individual x^i and its offsprings S^i, denoted by Select, is based on an underlying scalar sub-problem defined by the force-based direction $\overrightarrow{d}^i$ previously computed for x^i. In the following, we give a detailed description of the functions Force and Select. Function Pool is independent of the direction and typically problem-specific.

2.1   Force-Based Search Direction Strategies

The way we define the search directions $\overrightarrow{d}^i$, $i \in \{1, \ldots, \mu\}$, is crucial to control the movement of particles, and to efficiently guide them towards the Pareto front. Compared against a simple strategy where search directions are fixed initially, for instance following a random distribution, we consider to dynamically and adaptively compute search directions following some attraction-repulsion force-based rules. Roughly speaking, the closer particles corresponding to non-comparable individuals are in the objective space, the more particles should move away from each others to increase diversity—meaning that search directions of individuals should, in general, be relatively repulsive. On the other hand, particles should also move towards better objective function values and

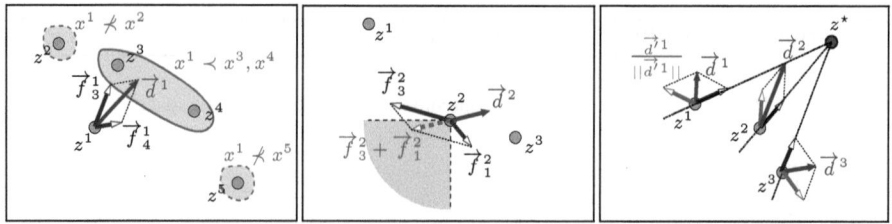

Fig. 1. Illustration of force strategies: D-D (left), NB-D (middle) and BH-D (right)

search directions should be attractive with respect to dominating individuals. In this study, we use different Force functions in order to compute directions which shall induce different dynamics of the particles, and thus provide different results in terms of approximation quality. In the following, we assume that we are given a norm function $\|\cdot\|$ in the objective space and a scaling factor $\alpha \in \mathbb{R}^+$. We denote the force exerted by particle z^j (w.r.t. solution x^j) on particle z^i by $\overrightarrow{f}^i_j$ and define the five following strategies.

Repulsive Force-Based Directions (R-D). Within R-D forces, particles are pairwisely repulsive. More precisely, for every $i, j \in \{1, \ldots, \mu\}$, we set:

$$\overrightarrow{f}^i_j = \frac{z^i - z^j}{\|z^i - z^j\|^\alpha} \ . \qquad (1) \qquad \overrightarrow{d}^i = \sum_{j \in \{1,\ldots,n\}\setminus\{i\}} \overrightarrow{f}^i_j \ . \qquad (2)$$

Repulsive-attractive Directions (RA-D). Here, a particle is attracted by any other particle which is dominating it and repelled otherwise. More precisely, for every $i \in \{1, \ldots, \mu\}$, the direction $\overrightarrow{d}^i$ is given by Eq. 2 and the forces $\overrightarrow{f}^i_j$ are defined as:

$$\overrightarrow{f}^i_j = \begin{cases} \dfrac{z^j - z^i}{\|z^i - z^j\|^\alpha} & \text{if } x^i \prec x^j \\[4mm] \dfrac{z^i - z^j}{\|z^i - z^j\|^\alpha} & \text{otherwise} \end{cases} \ . \qquad (3)$$

Dominance-Based Directions (D-D). In the case of the D-D Force function, a particle, that is dominated by at least one other particle, is attracted only by those particles dominating it and neither attracted nor repelled by the others. Otherwise, if a particle is not dominated by any other particle, then its direction is computed following the previous repulsive-attractive (RA-D) strategy. This strategy is motivated by the intuitive idea that particles should exclusively follow those that are dominating them, in an attempt to intensify the search process (see Fig. 1). More formally, let $\text{Dom}(x^i)$ be true whenever there exists a $k \neq i$ such that $x^i \prec x^k$, and false otherwise. Direction $\overrightarrow{d}^i$ for every particle is then given by Eq. 2 where forces $\overrightarrow{f}^i_j$ are now defined as follows:

$$\vec{f}^i_j = \begin{cases} \dfrac{z^j - z^i}{\|z^i - z^j\|^\alpha} & \text{if } \mathrm{Dom}(x^i) \text{ and } x^i \prec x^j \\[2mm] 0 & \text{if } \mathrm{Dom}(x^i) \text{ and } \neg \left(x^i \prec x^j \right) \\[2mm] \dfrac{z^i - z^j}{\|z^i - z^j\|^\alpha} & \text{otherwise} \end{cases} \qquad (4)$$

Non-backward Directions (NB-D). The repulsive-attractive force-based rule of Eq. 3 can lead to situations where a direction $\vec{d}^i$ computed for a particle z^i, which is *not dominated* by other particles, points away from the Pareto front. In the case of Fig. 1, for example, the direction computed for z^2 may lead to a situation where it is replaced by a dominated particle during the search. To counteract this situation, the NB-D strategy inverses those directions which are going backward and uses the direction $\vec{d}^i = -\vec{1} \cdot \vec{d'}^i$ if $\vec{d'}^i \cdot z^i \prec z^i$ and $\vec{d}^i = \vec{d'}^i$ otherwise, instead of the standard repulsive-attractive directions $\vec{d'}^i$ computed with Eq. 2 and 3.

Black Hole Directions (BH-D). Here, we consider an imaginary fixed particle which acts as a black hole attracting all the others. More precisely, we consider a virtual particle at the position of a utopian point $z^\star$ such that for every i, $z^i \prec z^\star$ [10] which additionally attracts all other particles when computing search directions. With the directions $\vec{d'}^i$ computed via Eq. 3 and Eq. 2, the BH-D strategy then uses the directions:

$$\vec{d}^i = \vec{d'}^i / \|\vec{d'}^i\| + \left(z^\star - z^i \right) / \|z^\star - z^i\| \ . \qquad (5)$$

2.2 Selection and Replacement Strategies

Once the direction of each particle is computed, Algorithm 1 proceeds to the computation of the next generation. Each solution x^i is replaced by a new one from set S^i, using its direction $\vec{d}^i$ and function **Select**. We shall consider several strategies, which are essentially a mix between two ideas. Firstly, we shall use one of two scalarizing (single-objective) functions denoted by $\mathcal{W}$ and $\mathcal{A}$ to evaluate the candidate solutions. These two functions are based on a weighted sum and an achievement scalarizing function [10], where the weighting coefficient vector used for each solution x^i is determined by the corresponding direction $\vec{d}^i = \{d^i_1, \ldots, d^i_M\}$ of particle $z^i = \{z^i_1, \ldots, z^i_M\}$. More formally, given a candidate solution $x \in S^i$ w.r.t. individual x^i, we let:

$$\mathcal{W}(x) = \sum_{m=1}^{M} d^i_m \cdot \phi_m(x) \quad \text{and} \quad \mathcal{A}(x) = \max_{m \in \{1, \ldots, M\}} \left\{ w^i_m \cdot \left(z^i_m - \phi_m(x) \right) \right\} \quad (6)$$

where $w^i_m = 1/d^i_m$ if $d^i_m \neq 0$ and $w^i_m = 0$ otherwise. Notice that with function $\mathcal{W}$, the lines of equal fitness values are orthogonal to the search direction d^i, no matter the sign of d^i components. With $\mathcal{A}$, they are half-lines reaching out from the line through z^i in direction $\vec{d}^i$ (see Fig. 2).

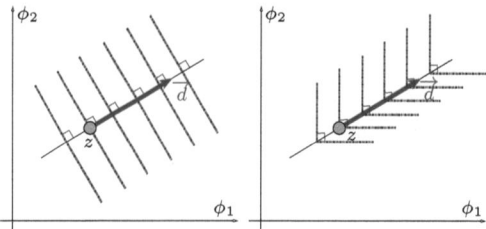
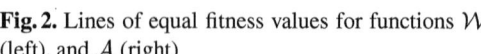
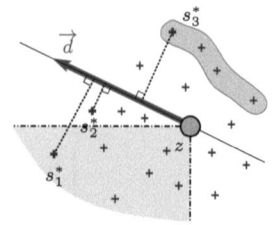

Fig. 2. Lines of equal fitness values for functions $\mathcal{W}$ (left), and $\mathcal{A}$ (right)

Fig. 3. Selection strategies (i), (ii), and (iii) based on a weighted sum scalarizing function with the selected objective vectors s_1^* for strategy (i), s_2^* for (ii), and s_3^* for (iii)

Secondly, we shall choose from a subset $S'^i \subseteq S^i$ of candidate solutions, the solution that optimizes the so-defined scalar problem. We here consider three different possibilities with an increasing inherent focus towards the Pareto front, (see also Fig. 3 for an example): (i) $S'^i = S^i$, (ii) $S'^i = \{x \in S^i \mid x \not\prec x^i\}$ the subset of candidate solutions not dominated by x^i, or (iii) $S'^i = \{x \in S^i \mid \forall x' \in S^i \setminus \{x\}, x \not\prec x'\}$ the set of candidate solutions from S^i which are not dominated. The output of function Select can then be formalized using the simple single-objective problem:

$$\text{Select}\left(S^i, \overrightarrow{d}^i\right) = \text{argopt}_{x \in S'^i} G(x) \tag{7}$$

where $G \in \{\mathcal{W}, \mathcal{A}\}$ and S'^i is one of the previously defined subsets. Notice that overall we have six possible combinations leading to six variants of function Select.

3   Experimental Design

In order to better understand and compare the different force, selection, and replacement strategies, we follow two separate lines of presentation. In Section 4 we show some exemplary runs in detail while Section 5 presents quality performance from 30 independent runs and the results of statistical tests. With this section, we provide details on the experimental design of those comparisons with respect to the used algorithms and test problem instances, the parameter setting, and performance assessment.

Competing Algorithms. Besides the five force variants R-D, RA-D, NB-D, D-D, and BH-D with their two different scalarizing functions (weighted sum, achievement) and the three proposed selection and replacement strategies, we use two baseline algorithms in our comparisons. On the one hand, we have a simple strategy, denoted by I-D, which assigns a fixed search direction to each of the μ population members. The search directions are *initially fixed* to a set of μ direction vectors chosen with equal distances in the weight space. The same functions Pool and Select as for the force-based strategies are used for algorithm I-D, except that the weights of the functions $\mathcal{W}$ and $\mathcal{A}$ are fixed throughout the search and instead of z^i in Eq. 6, a utopian point z^* is used. This corresponds to the classical way of defining multiple independent scalarizing functions and allows us to appreciate the gain we get when adapting search directions.

The second baseline algorithm is a $(\mu + \lambda)$-SMS-EMOA [2] variant with one-shot replacement strategy—denoted by SMS. At each iteration of the algorithm, λ solutions are selected (with replacement) and mutated. The μ solutions forming the population of the next iteration are chosen from the old μ and the new λ solutions by means of their hypervolume contributions after a non-dominated sorting and using the contributing hypervolume as the second-level sorting criterion. All μ solutions are chosen at once without new hypervolume calculations (one-shot scenario). Compared to our *local* approach where each solution of the population is replaced independently of the other ones, SMS uses a *global* strategy to evolve the whole population of individuals—thus allowing us to better appreciate the locality property of our approach.

ρ**MNK-landscapes.** The family of ρMNK-landscapes is a problem-independent model used for constructing multi-objective multi-modal landscapes with objective correlation [13]. It extends single- and multi-objective NK-landscapes [1,7]. Feasible solutions are represented as binary strings of size N, *ie.* the decision space is $X = \{0,1\}^N$. Parameter K refers to the number of variables that influence a particular position from the bit-string (the epistatic interactions). Each objective function $\phi_m : \{0,1\}^N \to [0,1)$ is defined by $\phi_m(x) = \frac{1}{N} \sum_{i=1}^{N} c_i^m(x_i, x_{i_1}, \ldots, x_{i_K})$ where $c_i^m : \{0,1\}^{K+1} \to [0,1)$ defines the multidimensional component function associated with variable $i \in \{1, \ldots, N\}$, and where $K < N$ [13]. By increasing the number of variable interactions K from 0 to $(N-1)$, ρMNK-landscapes can be gradually tuned from smooth to rugged. In this work, we choose the positions of these interactions uniformly at random and the same for all objective functions. Component values are distributed in the range $[0,1)$ and follow a multivariate uniform distribution of dimension M, defined by a correlation coefficient $\rho > \frac{-1}{M-1}$. The positive (resp. negative) data correlation decreases (resp. increases) the degree of conflict between objectives.

Parameter Setting. In this paper, we consider three bi-objective $(M = 2)$ ρMNK-landscapes of size $N = 128$, non-linearity $K = 4$, and correlation $\rho \in \{-0.7, 0.0, +0.7\}$. One instance, generated at random, is considered per parameter setting. For all competing algorithms, we consider an independent bit-flip mutation operator, where each bit is mutated at random with a probability $1/N$. We use the Euclidean distance and $\alpha = 2$ to compute forces. The utopia point $z^\star$ is set to $(1, 1)$. For R-D, RA-D, NB-D, D-D, BH-D, and I-D, the candidate solution set, i.e., Pool, is obtained by mutating each solution N times. For algorithm SMS, λ is set to N. All force-based algorithms are run for a fixed number of generations denoted by $gen \le 128$. In order to compare the algorithms in terms of function evaluations fairly, one generation of Algorithm 1 is equivalent to μ generations of SMS. For each algorithm run, we use an unbounded archive to record the computed Pareto front approximations. We shall use $\mu \in \{8, 16, 32, 64, 128, 256, 512\}$ as population sizes. We use the notation $\mathsf{F} \in \{\mathcal{W}, \mathcal{A}\}$ to refer to weighted sum ($\mathsf{F} = 0$) and achievement scalarizing function ($\mathsf{F} = 1$). Similarly, $\mathsf{S} \in \{0, 1, 2\}$ indicates which selection strategy of Section 2.2 is used ($\mathsf{S} = 0$: all, $\mathsf{S} = 1$: solutions not dominated by the parent, $\mathsf{S} = 2$: solutions not dominated by the parent and the offspring).

Performance Assessment. A set of 30 runs *per* instance is performed for each algorithm using a randomly generated initial population. For each ρMNK-landscape and a given number of function evaluations, we compute a reference set $Z_N^\star$ containing the

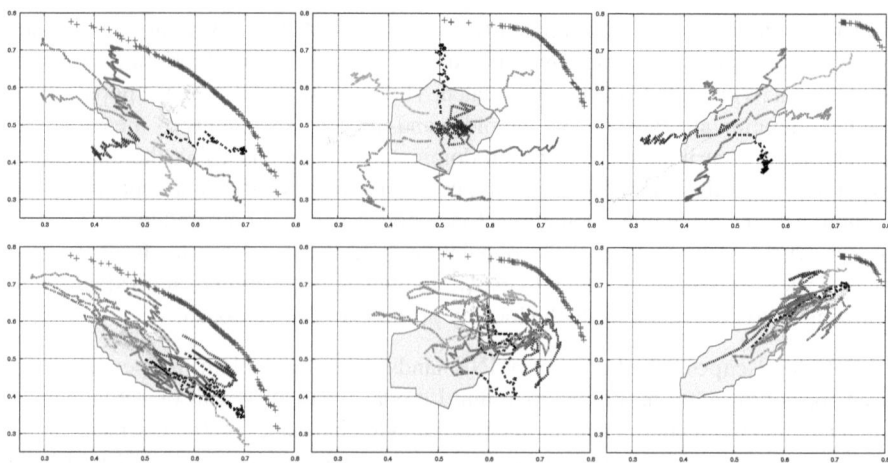

Fig. 4. Exemplary trajectories of R-D (top) and RA-D (bottom) strategies. From left to right: $\rho = -0.7$, $\rho = 0.0$, and $\rho = +0.7$. Polygons are obtained by random sampling for an equivalent number of evaluations. Runs are for $\mu = 8$, $\mathsf{F} = 0$, $\mathsf{S} = 0$, $gen = 64$. Crosses represent the reference set $Z_N^\star$.

non-dominated points among all solutions visited during all experiments. To measure the quality of a Pareto front approximation A in comparison to $Z_N^\star$, we use both the difference hypervolume indicator (I_H^-) and the multiplicative epsilon indicator ($I_\varepsilon^\times$) [18]. The I_H^--indicator gives the portion of the objective space that is dominated by $Z_N^\star$ and not by A. The reference point is set to the origin. The $I_\varepsilon^\times$-indicator gives the minimum multiplicative factor by which an approximation A has to be translated in each dimension of the objective space in order to dominate the reference set $Z_N^\star$ (both I_H^-- and $I_\varepsilon^\times$-values are to be minimized). The experimental results report the average indicator value together with the results of pairwise Mann-Whitney tests with a p-value of 0.05.

To explore the difference between the multiple variants of the competing algorithms, we also consider their empirical attainment functions (EAFs) [9]. The EAF provides the probability, estimated from several runs, that an arbitrary objective vector is dominated by, or equivalent to, a solution obtained by a single run of the algorithm. The difference between the EAFs for two different algorithms enables to identify the regions of the objective space where one algorithm performs better than another. The magnitude of the difference in favor of one algorithm is plotted within a gray-colored graduation.

4   Understanding the Algorithm

In this section, we report some exemplary runs to better understand the idea of our approach and the role of the different selection strategies and search directions.

4.1   Feasible Objective Space vs. Non-dominated Set Approximation

Fig. 4 shows the trajectories of the $\mu = 8$ population members in exemplary runs of the R-D and RA-D strategies on ρMNK landscapes with different correlation ρ and up to 64

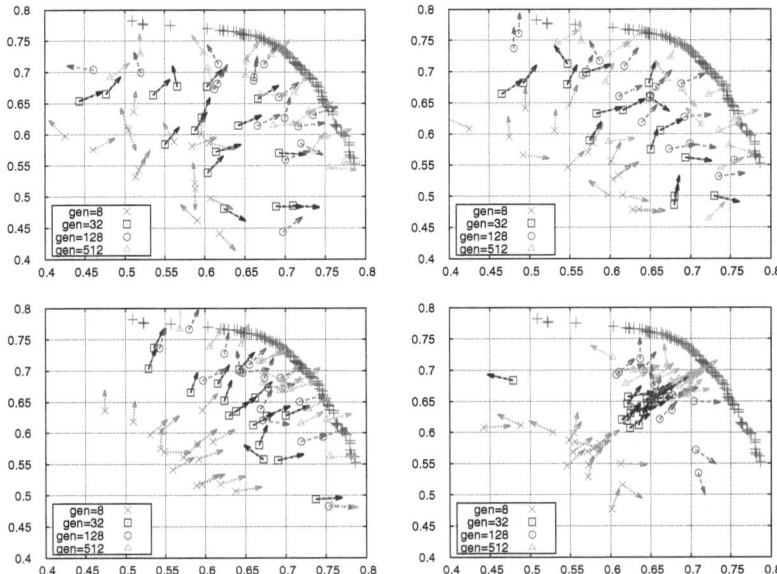

Fig. 5. Exemplary directions (unitary arrows) at different generations: RA-D (top left), NB-D (top right), BH-D (bottom left) and D-D (bottom right). Runs are for $\rho = 0.0$, $\mu = 16$, F $= 0$, S $= 0$.

generations. The polygons in the background show the spread[1] of $gen \cdot N \cdot \mu = 65,536$ random samples which correspond to the same amount of function evaluations as for the force-based strategies after $gen = 64$ generations.

We can make two main observations. First, the attracting forces of the RA-D strategy seem to "boost" the population towards the Pareto front while only using repelling forces in R-D allows to get a good idea about the feasible objective space region which can be considered as of independent interest [6]. Secondly, Fig. 4 shows how much more efficient the force-based algorithms are when compared to naive random sampling. In the exemplary results, the R-D strategy covers an objective space area more than four times larger when compared to random sampling while the RA-D variant roughly halves the distance to the best known Pareto front approximation—independently of the objective correlation ρ. Note that the plots are conceptionally similar when looking at different runs or different settings of μ, F, or S.

4.2 Comparing Force-Based Search Directions

In Fig. 5, we report exemplary runs showing how solutions and their directions evolve through generations. We can observe that the direction distribution of all strategies is rather stable over time—except for D-D where directions get clustered around the diagonal as soon as some dominating solutions are found. However, between the considered

---

[1] The polygon corners are defined by the non-dominated solutions found when maximizing (minimizing) both objectives and by the non-dominated solutions found when maximizing one and minimizing the other.

strategies, differences can be noticed with respect to both population distribution and directions. It appears that the strategies RA-D and NB-D cover a larger area of the objective space when compared to BH-D and D-D. The directions produced by RA-D and NB-D also appear to be more diversified. The D-D strategy has a tendency to cluster solutions in the middle, while solutions near the extremes of the reference set $Z_N^\star$ are rare. This is without surprise since the D-D strategy follows the idea that solutions are attracted by the dominating ones. Another interesting observation is that for BH-D and NB-D, the search directions appear to be well correlated with the solution positions— pointing right towards the closest solutions in the reference set. However, compared to RA-D, solutions are more clustered near the center. As it will be shown later, the relative performance of our different force strategies are directly related to the behaviors observed informally in this set of exemplary runs.

5   Experimental Analysis

In this section, we go into a deeper and more detailed experimental analysis of our approach by considering its performance over multiple runs. We first start with Table 1, providing the values of indicators I_H^- and $I_\varepsilon^\times$ for the instance with correlation $\rho = -0.7$, and for different algorithms under different configurations.

Influence of the Neighborhood Selection Strategy. A very first observation can be made on the minor effect of the neighborhood selection strategy. In fact, only in exceptional cases does the strategy have a strong influence on the indicator values and the algorithm ranking. For instance, this is the case for a low number of generations or for the I-D strategy (fixed independent weights) where the strategy of selecting the non-dominated solutions ($S = 2$) appears to be better than the two others. When examining the results for correlations $\rho = 0.0$ and $\rho = 0.7$, essentially the same observation can be made. Thus, for these correlations we show in Table 2 the results using the most simplest implementation only ($S = 0$, i.e., selection among all candidates).

Weighted Sum Vs. Achievement Scalarizing Function. When comparing the algorithm variants employing weighted sum ($F = 0$) with the ones using achievement function ($F = 1$), we can clearly see that the weighted sum consistently yields better results in terms of hypervolume and ε-indicator differences (see Tables 1 and 2). To understand this outperformance, let us again consider the lines of equal utility in Fig. 2. For a search direction pointing towards the Pareto front, it can be observed that the weighted sum allows a particle to move to incomparable solutions while with the achievement function, a particle can only move to a dominating solution. Thus, assuming that for the ρMNK landscapes, the bit-flip operator is likely to produce more incomparable neighbors than dominating ones, we can reasonably claim that it is more difficult for our algorithms to escape local optima using the achievement scalarizing function.

Comparison Between the Five Scalarizing Strategies. One observation is that the algorithm based on N independently fixed scalarizing functions results in the worst hypervolume and ε-indicator values—except for low and medium numbers of function evaluations if only the non-dominated portion of the offspring is considered for

Table 1. Comparison of the different algorithms with respect to indicators I_H^- and $I_\varepsilon^\times$ (lower is better) for the anti-correlated instance ($\rho = -0.7$), $\mu = 128$ and gen $\in \{8, 16, 32, 64, 128\}$. Column r is the rank of the algorithm under consideration—computed as the number of algorithms that statistically outperform it with the given indicator. Bold style refers to algorithms that are not outperformed by any other. Gray cells refer to algorithms which are analyzed in Fig. 8 and Fig. 9.

	F	S	gen=8 $I_\varepsilon^\times$	r	I_H^-	r	gen=16 $I_\varepsilon^\times$	r	I_H^-	r	gen=32 $I_\varepsilon^\times$	s	I_H^-	r	gen=64 $I_\varepsilon^\times$	r	I_H^-	r	gen=128 $I_\varepsilon^\times$	r	I_H^-	r
RA-D	0	0	1.211	5	0.148	4	1.118	2	0.073	6	1.082	2	**0.050**	0	**1.057**	0	**0.036**	0	**1.043**	0	**0.028**	0
RA-D	0	1	1.207	2	0.146	2	1.118	2	0.072	2	1.083	4	**0.051**	0	1.059	2	**0.038**	0	1.044	0	**0.028**	0
RA-D	0	2	1.207	3	0.145	2	1.119	2	0.072	3	**1.079**	0	**0.051**	0	**1.057**	0	**0.038**	0	1.044	0	**0.029**	0
RA-D	1	0	1.224	13	0.156	14	1.135	13	0.084	13	1.100	12	0.067	13	1.083	13	0.055	12	1.070	12	0.049	12
RA-D	1	1	1.222	11	0.154	13	1.135	13	0.085	13	1.100	12	0.067	14	1.083	13	0.057	15	1.070	12	0.050	13
RA-D	1	2	1.221	11	0.155	13	1.135	12	0.085	13	1.099	12	0.069	17	1.083	13	0.059	18	1.069	11	0.051	13
BH-D	0	0	1.215	6	0.147	4	1.121	2	0.071	2	**1.080**	0	**0.050**	0	1.062	6	0.039	1	1.049	6	0.031	2
BH-D	0	1	1.208	3	0.145	2	1.116	2	0.069	1	**1.077**	0	**0.050**	0	**1.057**	0	**0.037**	0	1.049	3	0.031	2
BH-D	0	2	1.211	4	0.147	4	1.126	9	0.073	6	**1.082**	0	0.052	1	**1.058**	0	0.039	1	1.051	6	0.032	6
BH-D	1	0	1.290	24	0.181	24	1.192	24	0.118	24	1.166	24	0.109	24	1.155	24	0.104	24	1.146	26	0.096	24
BH-D	1	1	1.284	24	0.179	24	1.193	24	0.117	24	1.164	24	0.107	24	1.151	24	0.102	24	1.146	26	0.097	24
BH-D	1	2	1.290	24	0.181	24	1.201	24	0.121	25	1.167	24	0.109	24	1.150	24	0.102	24	1.146	26	0.096	24
D-D	0	0	1.223	11	0.153	13	1.136	12	0.083	13	1.102	12	0.066	13	1.080	11	0.054	11	1.070	11	0.046	10
D-D	0	1	1.229	15	0.155	13	1.133	12	0.084	13	1.095	11	0.065	13	1.074	10	0.052	10	1.059	10	0.043	10
D-D	0	2	1.215	5	0.149	7	1.125	4	0.076	9	1.086	5	0.058	11	1.065	8	0.046	9	1.053	7	0.037	9
D-D	1	0	1.230	16	0.158	17	1.143	20	0.087	18	1.105	18	0.069	14	1.088	19	0.060	20	1.080	19	0.055	20
D-D	1	1	1.235	14	0.159	18	1.149	22	0.090	21	1.103	17	0.070	17	1.090	19	0.061	20	1.079	19	0.055	20
D-D	1	2	1.234	13	0.156	13	1.142	15	0.089	19	1.109	18	0.073	23	1.089	18	0.061	20	1.077	19	0.053	18
NB-D	0	0	1.205	2	0.145	3	1.119	2	0.071	2	1.083	4	0.052	2	**1.057**	0	**0.037**	0	**1.043**	0	**0.027**	0
NB-D	0	1	1.209	3	0.146	3	1.120	2	0.071	2	**1.079**	0	**0.049**	0	**1.057**	0	**0.036**	0	**1.045**	0	0.029	2
NB-D	0	2	1.203	2	0.144	2	1.115	1	0.071	2	**1.079**	0	**0.051**	0	**1.056**	0	**0.038**	0	**1.043**	0	**0.029**	0
NB-D	1	0	1.220	11	0.154	13	1.132	11	0.083	13	1.099	12	0.066	13	1.082	13	0.056	12	1.071	12	0.050	13
NB-D	1	1	1.222	12	0.156	13	1.134	13	0.084	13	1.099	12	0.066	13	1.083	13	0.056	12	1.070	12	0.050	13
NB-D	1	2	1.227	13	0.155	13	1.134	13	0.085	14	1.100	12	0.070	18	1.082	13	0.057	15	1.070	12	0.052	13
I-D	0	0	1.350	27	0.212	27	1.288	28	0.166	27	1.235	27	0.156	27	1.178	27	0.129	27	1.128	23	0.097	24
I-D	0	1	1.350	27	0.211	27	1.281	27	0.165	27	1.240	29	0.155	27	1.180	27	0.129	27	1.126	23	0.097	24
I-D	0	2	1.200	1	0.140	1	1.111	1	0.067	1	1.085	6	0.053	7	1.075	10	0.050	10	1.067	11	0.047	11
I-D	1	0	1.348	27	0.212	27	1.276	27	0.165	27	1.229	27	0.154	27	1.178	27	0.133	29	1.149	26	0.106	29
I-D	1	1	1.349	27	0.212	27	1.275	27	0.165	27	1.229	27	0.154	27	1.176	27	0.133	29	1.150	26	0.106	29
I-D	1	2	1.193	1	0.143	2	1.129	10	0.071	2	1.105	18	0.057	11	1.095	22	0.052	11	1.085	22	0.049	13
SMS			**1.151**	0	**0.070**	0	**1.063**	0	**0.035**	0	1.103	12	**0.048**	0	1.120	23	0.053	11	1.130	23	0.053	18

selection. This indicates that choosing the search directions cooperatively seems to be a good choice. Secondly, we can observe that, with a few exceptions, in particular for $\rho = 0.7$ and larger generations, the D-D variant performs badly when compared to the other three force-based algorithms. In these exceptional cases where the D-D variant produces good results for one indicator, the other indicator often shows a medium performance while typically a high positive correlation between the two indicators can be reported. When looking more carefully at how this algorithm approaches the Pareto front (Fig. 5), we can better understand why often the hypervolume values are quite good but the ε-indicator values are not. In fact, due to the absence of backwards directions and the attraction of dominated by dominating points, the D-D variant loses diversity during the run and approaches the Pareto front via the diagonal of the objective space. Like that, the large number of points in the middle of the objective space results in quite high hypervolume values. However, the absence of extreme solutions in the population yields lower ε-indicator values.

Table 2. Comparison of the different algorithms for instance correlation $\rho \in \{0.0, 0.7\}$. Same settings and notations than in Table 1 are used. The selection strategy is $\mathsf{S} = 0$.

ρ		F	gen=8 $I_e^\times$	r	I_H^-	r	gen=16 $I_e^\times$	r	I_H^-	r	gen=32 $I_e^\times$	r	I_H^-	r	gen=64 $I_e^\times$	r	I_H^-	r	gen=128 $I_e^\times$	r	I_H^-	r
	RA-D	0	1.215	5	0.176	3	1.138	7	0.109	9	1.085	6	0.063	9	1.058	9	0.038	1	1.041	6	0.027	4
	RA-D	1	1.236	17	0.193	16	1.156	15	0.128	14	1.106	16	0.080	14	1.080	14	0.056	14	1.069	13	0.045	12
	BH-D	0	1.207	2	0.174	3	1.123	1	0.105	3	1.072	1	0.057	1	1.047	1	0.037	1	1.041	6	0.030	7
	BH-D	1	1.227	11	0.189	14	1.149	12	0.132	18	1.114	20	0.097	24	1.101	23	0.084	24	1.093	23	0.077	24
	D-D	0	1.225	11	0.185	13	1.149	12	0.128	14	1.095	13	0.086	18	1.071	12	0.064	21	1.058	12	0.053	20
0.0	D-D	1	1.247	22	0.197	21	1.163	21	0.135	19	1.113	20	0.084	18	1.093	20	0.060	18	1.087	20	0.052	19
	NB-D	0	1.215	6	0.177	3	1.136	5	0.109	7	1.086	7	0.063	9	1.057	9	0.039	5	1.043	7	0.028	4
	NB-D	1	1.235	17	0.192	15	1.154	14	0.128	14	1.102	14	0.079	14	1.081	15	0.055	13	1.070	13	0.045	12
	I-D	0	1.339	27	0.252	27	1.289	27	0.221	27	1.238	27	0.191	27	1.186	27	0.156	27	1.136	27	0.117	27
	I-D	1	1.338	27	0.253	27	1.292	27	0.222	27	1.256	29	0.193	27	1.223	29	0.158	27	1.177	29	0.120	27
	SMS		**1.051**	0	**0.043**	0	**1.034**	0	**0.027**	0	**1.031**	0	**0.024**	0	**1.033**	0	**0.027**	0	**1.032**	0	**0.026**	0
	RA-D	0	1.208	3	0.180	3	1.118	3	0.103	3	1.060	4	0.048	1	**1.040**	0	**0.035**	0	**1.032**	0	**0.024**	0
	RA-D	1	1.226	22	0.191	20	1.133	21	0.119	20	1.078	21	0.071	18	1.060	14	0.056	16	1.052	13	0.046	14
	BH-D	0	1.208	3	0.180	3	1.112	2	0.101	3	1.055	1	0.045	1	**1.038**	0	**0.036**	0	1.035	3	0.030	6
	BH-D	1	1.212	6	0.184	4	1.119	5	0.108	10	1.068	14	0.062	15	1.060	14	0.058	18	1.055	17	0.053	21
	D-D	0	1.212	5	0.183	4	1.117	3	0.106	5	1.060	3	0.053	9	**1.041**	0	0.037	4	**1.034**	0	0.031	5
0.7	D-D	1	1.219	19	0.189	19	1.125	16	0.115	17	1.071	14	0.065	15	1.063	17	0.058	18	1.058	20	0.053	21
	NB-D	0	1.209	4	0.180	3	1.117	3	0.106	5	1.062	5	0.051	5	1.041	1	**0.034**	0	**1.032**	0	**0.024**	0
	NB-D	1	1.216	14	0.184	4	1.131	21	0.117	18	1.074	16	0.069	17	1.057	14	0.053	15	1.051	13	0.046	14
	I-D	0	1.341	27	0.262	27	1.288	27	0.226	27	1.234	27	0.194	27	1.179	27	0.161	27	1.128	27	0.119	27
	I-D	1	1.344	27	0.262	27	1.294	27	0.228	27	1.237	28	0.194	27	1.188	29	0.161	27	1.140	29	0.118	27
	SMS		**1.052**	0	**0.048**	0	**1.037**	0	**0.030**	0	**1.034**	0	**0.027**	0	**1.037**	0	**0.032**	0	**1.036**	0	0.031	3

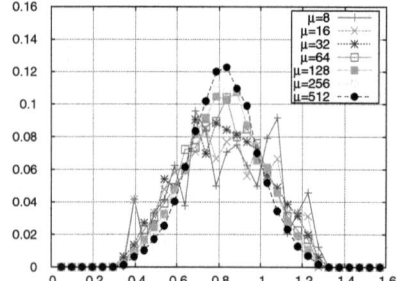 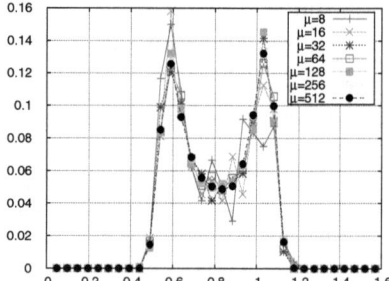

Fig. 6. Distribution of population members in the objective space for the weighted sum version of RA-D (left column), and I-D (right column) after 128 generations in polar coordinates (x-axis: angles from 0 to $\pi/2$), for $\rho = -0.7$, $\mathsf{S} = 0$, $\mu \in \{8, \dots, 512\}$. The y-axis refers to the empirical probability of having a solution in the corresponding angle.

When looking closer at the typical algorithm behavior (Fig. 6 and Fig. 8), one can also observe another obvious difference between the four force-based strategies: The BH-D variant produces consistently more solutions in the middle of the Pareto front while the other strategies, in particular RA-D, are more balanced and outperform the BH-D variant at the extremes (Fig. 8). Overall, we therefore recommend to use BH-D when a focus on the middle of the front is desired and RA-D otherwise because of its simple implementation and the resulting more uniform distribution.

Distribution of the Population Over the Objective Space. Several interesting observations can be made about the dynamics of the population in polar coordinates. In Fig. 6, we can see that, for RA-D, solutions are distributed following a bell-shaped

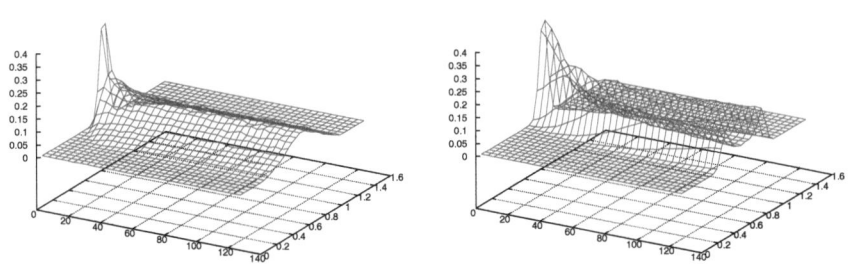

Fig. 7. Empirical distribution of population members over generations for RA-D (left) and I-D (right). Runs are for $\rho = -0.7$, $\mu = 128$, F $= 0$, S $= 0$.

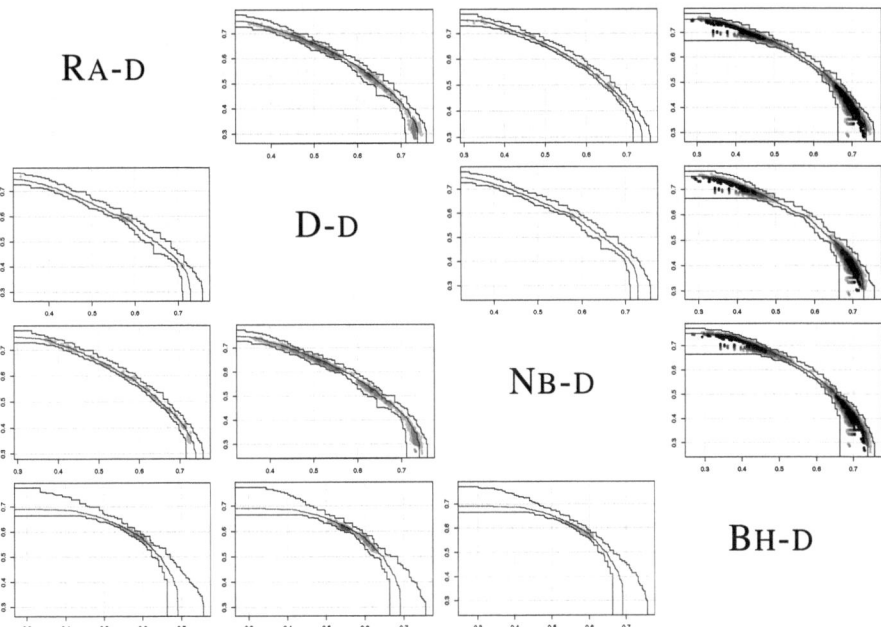

Fig. 8. EAF differences for RA-D, D-D, NB-D, and BH-D, with $\rho = -0.7$, $\mu = 128$, $gen = 128$, F $= 1$ (achievement), S $= 0$. The plot in row i and column j shows the areas where algorithm i improves over algorithm j.

distribution , with the mean being the middle of the Pareto front ($0.8 \simeq \pi/4$). However, For I-D, for which the μ directions are uniformly distributed over $[0, \pi/2]$ and constant, the distribution of individuals is seemingly different. In fact, since the Pareto front lies in a smaller range than $[0, \pi/2]$, searching along more than one fixed search direction results in reaching the extreme points of the Pareto front, especially for anti-correlated instances. All this is not only observed after the specific generations of Fig. 6 but it can also be seen in Fig. 7 that the point distribution converges rather quickly with time and stays roughly the same after about 60 generations.

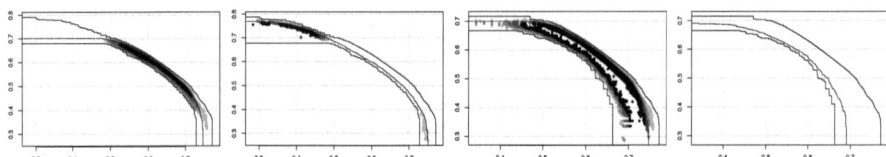

Fig. 9. EAFs comparing (from left to right): SMS with RA-D (F = 0, S = 0), RA-D (F = 0, S = 0) with SMS, SMS with BH-D (F = 1, S = 0), and BH-D (F = 1, S = 0) with SMS. Runs are for $\rho = -0.7$, $gen = 128$.

Comparison with SMS. From Tables 1 and 2, it appears that SMS outperforms the other algorithms for few generations while for a larger number of generations, the force-based approach is competitive with SMS on all instances, and it even outperforms it for the two instances with positive and negative correlation—although the force-based approach uses a simpler selection strategy compared to the global one of SMS. Notice that the differences between SMS and the proposed approach are rather small for both indicators. When examining the EAF differences between SMS and our approaches, as depicted in the examples of Fig. 9, we observe that SMS produces substantially less solutions at one side of the front while performing substantially better at the rest.

In addition, our comparison between SMS and the force-based approach uses the same number of function evaluations while ignoring the actual costs of maintaining the population and any other complexity issues. In the current implementation, the actual runtimes of the two approaches differ by a factor of about two as SMS needs more costly *global* operations to evolve the population such as sorting and hypervolume computations while in the force-based approach the population replacement can be done *locally* at the level of each solution once the forces are computed.

6   Conclusion

In this paper, we proposed a force-based EMO paradigm, and we studied its properties in different configurations. The originality of our approach stems from the fact that each solution computes, dynamically at each generation, a search direction in the objective space cooperatively, and in parallel with other solutions. Some variants of our approach are proved to be efficient in finding those directions that lead to a good covering of the Pareto front. However, defining what would be the optimal directions that solutions should follow at each step of the search process is a difficult open question. Furthermore, our approach is inherently *local*, in the sense that, although the computation of the directions are "synchronized" for all solutions, the selection is performed locally in parallel for the population members. An interesting open question is to design better direction-based localized strategies, and to come up with inherently distributed approaches. Last but not least, while being in general directly applicable to problems with more than two objectives, further investigations on those problems are needed.

Acknowledgements. This work has been supported by the French national research agency (ANR) within the 'Modèles Numériques' project NumBBO.

References

1. Aguirre, H.E., Tanaka, K.: Working principles, behavior, and performance of MOEAs on MNK-landscapes. Eur. J. Oper. Res. 181(3), 1670–1690 (2007)
2. Beume, N., Naujoks, B., Emmerich, M.: SMS-EMOA: Multiobjective selection based on dominated hypervolume. Eur. J. Oper. Res. 181(3), 1653–1669 (2007)
3. Chen, J.H., Kang, C.W.: A force-driven evolutionary approach for multi-objective 3D differentiated sensor network deployment. Int. J.of Ad Hoc and Ubi. Comp. 8(1/2), 85–95 (2011)
4. Hassanzadeh, H.R., Rouhani, M.: A multi-objective gravitational search algorithm. In: 2nd Int. Conf. on Computational Intell. Comm. Sys. and Networks, pp. 7–12 (2010)
5. Hughes, E.J.: Multiple Single Objective Pareto Sampling. In: Congress on Evolutionary Computation (CEC 2003), pp. 2678–2684. IEEE Press (2003)
6. Hughes, E.J.: Many Objective Optimisation: Direct Objective Boundary Identification. In: Rudolph, G., Jansen, T., Lucas, S., Poloni, C., Beume, N. (eds.) PPSN X. LNCS, vol. 5199, pp. 733–742. Springer, Heidelberg (2008)
7. Kauffman, S.A.: The Origins of Order. Oxford University Press (1993)
8. Khan, J.A., Sait, S.M.: Fast fuzzy force-directed simulated evolution metaheuristic for multiobjective VLSI cell placement. The Arabian J. for Sc. and Eng. 32(2B), 264–281 (2007)
9. López-Ibáñez, M., Paquete, L., Stützle, T.: Exploratory analysis of stochastic local search algorithms in biobjective optimization. In: Experimental Methods for the Analysis of Optimization Algorithms, ch. 9, pp. 209–222. Springer (2010)
10. Miettinen, K.: Nonlinear Multiobjective Optimization. Kluwer Academic Publishers (1999)
11. Nobahari, H., Nikusokhan, M., Siarry, P.: Non-dominated sorting gravitational search algorithm. In: Int. Conf. on Swarm Intelligence (2011)
12. Rashedi, E., Nezamabadi-pour, H., Saryazdi, S.: GSA: A gravitational search algorithm. Information Sciences 179(13), 2232–2248 (2009)
13. Verel, S., Liefooghe, A., Jourdan, L., Dhaenens, C.: Analyzing the Effect of Objective Correlation on the Efficient Set of MNK-Landscapes. In: Coello, C.A.C. (ed.) LION 5. LNCS, vol. 6683, pp. 116–130. Springer, Heidelberg (2011)
14. Wagner, T., Beume, N., Naujoks, B.: Pareto-, Aggregation-, and Indicator-Based Methods in Many-Objective Optimization. In: Obayashi, S., Deb, K., Poloni, C., Hiroyasu, T., Murata, T. (eds.) EMO 2007. LNCS, vol. 4403, pp. 742–756. Springer, Heidelberg (2007)
15. Wang, Y., Zeng, J.C.: A constraint multi-objective artificial physics optimization algorithm. In: 2nd Int. Conf. on Computational Intell. and Natural Computing, pp. 107–112 (2010)
16. Wang, Y., Zeng, J.C., Cui, Z.H., He, X.J.: A novel constraint multi-objective artificial physics optimisation algorithm and its convergence. Int. J. Innov. Comput. Appl. 3(2), 61–70 (2011)
17. Zhang, Q., Li, H.: MOEA/D: A Multiobjective Evolutionary Algorithm Based on Decomposition. IEEE Transactions on Evolutionary Computation 11(6), 712–731 (2007)
18. Zitzler, E., Thiele, L., Laumanns, M., Fonseca, C.M., Grunert da Fonseca, V.: Performance assessment of multiobjective optimizers: An analysis and review. IEEE Transactions on Evolutionary Computation 7(2), 117–132 (2003)

Approximation Model Guided Selection for Evolutionary Multiobjective Optimization

Aimin Zhou[1], Qingfu Zhang[2], and Guixu Zhang[1]

[1] Department of Computer Science and Technology, East China Normal University
Shanghai, 200235, China
{amzhou,gxzhang}@cs.ecnu.edu.cn
[2] School of Computing and Electronic Systems, University of Essex
Colchester, CO4 3SQ, UK
qzhang@essex.ac.uk

Abstract. Selection plays a key role in a *multiobjective evolutionary algorithm (MOEA)*. The dominance based selection operators or indicator based ones are widely used in most current MOEAs. This paper studies another kind of selection, in which a model is firstly built to approximate the Pareto front and then guides the selection of promising solutions into the next generation. Based on this idea, we propose two *approximation model guided selection (AMS)* operators in this paper: one uses a zero-order model to approximate the Pareto front, and the other uses a first-order model. The experimental results show that the new AMS operators performs well on some test instances.

1 Introduction

A *multiobjective optimization problem (MOP)* can be formulated as

$$\min F(x) = (f_1(x), \cdots, f_m(x))$$
$$\text{s.t } x \in \Omega$$

where Ω defines the feasible region of the decision space, and $x \in \Omega$ is a decision variable vector. $F : \Omega \to R^m$ consists of m objective functions $f_i(x)$, $i = 1, \cdots, m$, and R^m is the objective space. Since the objectives often conflict with each other, there does not exist a single solution that can optimize all the objectives at the same time. Instead, a set of the best tradeoff candidate solutions among different objectives, which is called *Pareto set (PS)* in the decision space (or *Pareto front (PF)* in the objective space), is of practical interest to a decision maker [1, 2].

Many multiobjective optimization methods aim to find an approximation set which is as diverse as possible and as close to the PF (PS) as possible [2]. Since an *evolutionary algorithm (EA)* works with a population of candidate solutions, it can approximate the PF (PS) of an MOP in a single run. Therefore, EAs are a natural choice for tackling MOPs and recent years have witnessed the rapid development of *multiobjective evolutionary algorithms (MOEA)* [3]. Among different components of an MOEA, the following two are extremely important: (1) reproduction operator, which generates new trial solutions based on current

R.C. Purshouse et al. (Eds.): EMO 2013, LNCS 7811, pp. 398–412, 2013.

population, and (2) selection operator, which selects promising solutions into the next generation. For the former, many popular MOEAs directly use those reproduction operators which were designed for single objective optimization. Recently, some research work has demonstrated that these operators may not always work well for MOPs since, by nature, an MOP is quite different from a single objective optimization problem [4]. Some specific reproduction operators were therefore proposed for dealing with multiobjective optimization [5]. For the latter, most research work on evolutionary multiobjective optimization is on selection operators, and a number of selection operators have been proposed, which shall be introduced shortly in the next section.

Multiobjective evolutionary algorithm based on decomposition (MOEA/D) is a new MOEA framework [6, 7]. It decomposes an MOP into a set of single objective subproblems and optimizes these subproblems simultaneously to approximate the PF (PS). MOEA/D also provides some possibilities for designing and improving reproduction and selection operators. In this paper, we focus on selection and propose an *approximation model guided selection (AMS)* by borrowing the idea of decomposition from MOEA/D. It should be noted that this paper is an extension of our previous work in [8, 9].

The rest of the paper is organized as follows: Section 2 reviews some widely used selection strategies in evolutionary multiobjective optimization. Section 3 presents the proposed selection operators in detail. Section 4 conducts empirical studies of the proposed selection operators. Finally, Section 5 concludes the paper with some suggestions for future work.

2 Related Work on Selection Operators

In single objective optimization, there naturally exists a complete order to compare solutions: for any two feasible solutions x and y, either $f(x) \leq f(y)$ or $f(y) \leq f(x)$. However it is not the case in multiobjective optimization. The Pareto dominance relationship only defines a partial order and not all feasible solutions can compare with each other. Therefore, additional strategies are required in MOEAs to differentiate solutions. Some widely used MOEA selection operators can be classified into the following categories.

2.1 Dominance Based Selection

The dominance based selection operators extend the partial order of Pareto dominance to a complete order by a two-stage strategy. In the first stage, the population is partitioned into several groups by using Pareto dominance. A rank value, x^{rnk}, is assigned to each solution x in this stage. The solutions in the same group are with the same rank value. In the second stage, each solution x is assigned a density value x^{den}, which represents the sparseness of the solution. A complete order, denoted as $\prec_i$, can thus be defined as follows:

$$x \prec_i y, \text{ iff } \begin{array}{l} (x^{rnk} < y^{rnk}), \text{ or} \\ (x^{rnk} = y^{rnk} \text{ and } x^{den} < y^{den}). \end{array}$$

In the rank assignment stage, three strategies are widely used: (1) dominance rank [10, 11], (2) dominance count [12], and (3) dominance strength [13, 14]. In all these strategies, non-dominated solutions in the population will be assigned the lowest rank value.

The density estimation stage is to maintain and encourage population diversity. Some popular methods used in this stage include: (1) niching and fitness sharing [10–12, 15], (2) crowding distance [16], (3) K-nearest neighbor method [14], and (4) grading method [17–20] and its variants such as ε-dominance [21]. The basic idea behind these methods is that solutions in sparse areas are more important than those in dense areas.

2.2   Indicator Based Selection

The indicator based selection operators define a complete order over populations. Let $I(\cdot)$ be a quality indicator which assigns a real value to a population. A complete order, $\prec_p$, is defined as follows:

$$P \prec_p Q \text{ iff } I(P) < I(Q),$$

where P and Q are two populations.

Unlike dominance based selection, indicator based selection considers a population as a whole. This idea was firstly proposed in [22] and it has been proved that maximization of the S-metric [23] over a population is a necessary and sufficient condition for the population to be optimal in a sense. An S-metric guided selection was proposed in [24], and this selection has been widely studied thereafter [25–29].

Although some attempts have been made to reduce the computational overheads [30, 31], a major disadvantage of indicator based selection operators is that calculation of a quality metric is often very time-consuming especially for many-objective problems. .

2.3   Model Guided Selection

Model guided selection operators aim to select solutions into the next generation guided by a model, which approximates the Pareto front. In [32], an estimated PF was first built and then solutions close to the estimated PF were selected. This method only works on problems with convex PFs. Later, it was generalized to generic PFs in [33]. A similar approach called guided hyperplane evolutionary algorithm was proposed in [34]. In [8], a utopian PF guided selection was proposed for bi-objective problems. It uses a line segment to approximate the PF, some evenly distributed points are sampled in the line segment as the target points, and these points, which are close to the target points, are selected into the next generation. Very recently, a reference points guided selection procedure was introduced in [35, 36]. The reference points can be regarded as some points in an estimated PF.

A key issue on designing a model guided selection operator is how to build a model to approximate a PF. It still requires much research effort to develop a simple yet efficient method to model a PF.

3   Approximation Model Guided Selection

It is well known that under mild conditions, the PF (PS) of an m-objective continuous problem forms a piecewise continuous $(m-1)$-dimensional manifold [37]. This regularity property has been used to design reproduction operators for multiobjective optimization in [5]. In this paper, we study how to use this property to design selection operators.

Our basic idea is to build a utopian PF (approximation model) to approximate the true PF, and then use it to guide selection. More specifically, we firstly build a utopian PF based on information extracted from the current population, and then construct a set of single objective functions based on the utopian PF, and finally select solutions according to these single objective function values. We call this selection approach *approximation model guided selection (AMS)*.

Let Q be a set of solutions (e.g., a union of the current population and offspring population in our experiments), the pseudo code of the AMS operator for selecting N solutions to form P (new population in our experiments) is as follows.

Step 0: Set $P = \emptyset$.
Step 1: Build a utopian PF by using information extracted from Q to approximate the true PF.
Step 2: Define N single objective functions $G = \{g^i | i = 1, \cdots, N\}$ based on the utopian PF.
Step 3: Randomly choose $g \in G$, and find:

$$x^* = \arg\min_{x \in Q} g(x),$$

set $Q = Q \backslash \{x^*\}$, $G = G \backslash \{g\}$ and $P = P \cup \{x^*\}$.
Step 4: Repeat *Step 3* until $G = \emptyset$.

It is clear that

- our approach maintains a utopian PF and a set of solutions to approximate the true PF, and
- the convergence and diversity of the population are achieved under the guide of the utopian PF and single objective functions.

The proposed AMS operator is an open framework and there might be different ways to define the utopian PFs and the single objective functions. In the following, we develop two models to approximate the PF: one is a zero-order model and the other is a first-order model.

3.1   Zero-Order Model Guided Selection: AMS0

The simplest model to approximate the PF should be a single point that can be regarded as a zero-order manifold. In this section, we use the ideal point,

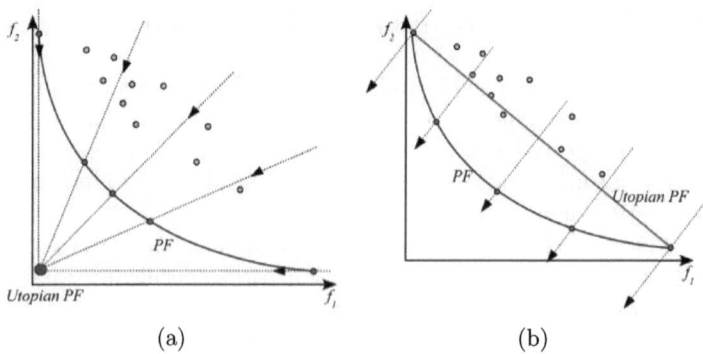

Fig. 1. An illustration to approximate the PF by a model in the case of bi-objective problems: (a) zero-oder model, and (b) first-order model

$z^* = (z_1^*, \cdots, z_m^*)$ where

$$z_i^* = \min_{x \in \Omega} f_i(x), i = 1, \cdots, m,$$

for this purpose. However, the exact ideal point is difficult to obtain in practice. Therefore, we use the best individual objective function values found so far to approximate the ideal point, i.e.,

$$z_i^* = \min_{x \in Q} f_i(x), i = 1, \cdots, m.$$

The approximated ideal point is updated at each generation and hopefully converges to the true ideal point.

We use the Tchebycheff approach to define single objective functions. Let $\{\lambda^i = (\lambda_1^i, \cdots, \lambda_m^i), i = 1, \cdots, N\}$ be a set of predefined weight vectors. The i-th single objective function is

$$g^i(x) = g(x|\lambda^i, z^*) = \max_{1 \le j \le m} \lambda_j^i |f_j(x) - z_j^*|.$$

Fig. 1(a) illustrates this selection operator in the case of bi-objective problems. It is clear that the zero-order model does not consider the shape of the PF. Therefore, the distribution of the final approximation highly depends on the predefined weight vectors, which are hard to be set properly beforehand in some cases.

3.2   First-Order Model Guided Selection: AMS1

A first-order model or linear model can approximate the PF better than a zero-order model. In [8, 9, 38], we introduced a method to construct a utopian PF of an MOP. We generalize this method in this section.

Under the regularity property of continuous MOPs, we can use an $(m-1)$-D simplex S to approximate the PF. Let $NS(Q)$ be the set that contains all the nondominated solutions in population Q, $v^i = (v^i_1, \ldots, v^i_m)$, the i-th vertex of S $(i = 1, \ldots, m)$, is determined as follows:

$$v^i_j = \begin{cases} \min\limits_{x \in NS(Q)} f_j(x) \text{ if } j \neq i \\ \max\limits_{x \in NS(Q)} f_j(x) \text{ if } j = i \end{cases}$$

for $i, j = 1, \cdots, m$.

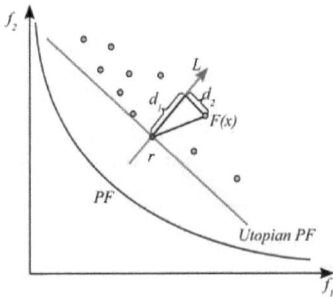

Fig. 2. Illustration of the calculation of the $g^i(x)$ in the case of bi-objective problems

Let $R = \{r^i, i = 1, \cdots, N\}$ be a set of N uniformly distributed points in S. The i-th single objective function is defined as

$$g^i(x) = g(x|r^i) = d_1 + 2d_2,$$

where $d_1 = \frac{|L^T(F(x) - r^i)|}{\|L\|}$, L is a normal direction of S, d_2 is the Euclidean distance from $F(x)$ to the line with direction L and passing through r^i. Fig. 2 illustrates how to calculate $g^i(x)$ in the case of two objectives.

We call the selection based on the above modeling method AMS1. Fig. 1(b) illustrates how it works. It is clear that the first-order model uses more information to approximate the PF than the zero-order model, and therefore the first-order mode is more similar to the PF than the zero-order model.

4   Experimental Results

In this section, the proposed selection operators AMS0 and AMS1 are compared on a popular domination based selection operator on some test instances.

4.1   Experimental Settings

To assess the performance of the proposed two selection operators, AMS0 and AMS1, we compare them with the *nondominated sorting (NDS)* selection operator [16]. For a fair comparison, the three selection operators are used in RM-MEDA [5]. RM-MEDA is based on probability models: it builds a probability model to capture the population distribution in the decision space, and samples new trial solutions from the model thus built; then it selects new generation from the old and new solutions. The process is repeated iteratively until the stopping condition is satisfied. The main procedure of RM-MEDA is as follows.

Step 0: Initialize a population P randomly from the feasible region of the decision space.

Step 1: Build a probability model to capture the structure of the solutions in P in the decision space.

Step 2: Sample a set of new solutions Q, which satisfies $|Q| = |P|$, from the probability model.

Step 3: Repair the infeasible solutions in Q and evaluate the new solutions.

Step 4: Select the new population P for the next generation from $Q \cup P$.

Step 5: Goto **Step 1** if the termination condition is not satisfied.

In **Step 1**, the regularity property [37] of continuous MOPs is considered for model building, which differentiates RM-MEDA from other probability model based MOEAs. In **Step 4**, AMS0, AMS1, or NDS is used to do selection. More details about RM-MEDA are referred to [5].

The problems F_1-F_8 in [5] are used as test instances. The experimental settings are as follows.

- *Number of decision variables:* It is set to be 30 for all the test instances.
- *Number of population size:* The number of population size is set to be 100 and 200 for bi-objective and tri-objective problems respectively.
- *Stopping condition:* All the algorithms stop after 200 generations for F_1, F_2, F_4, F_5, F_6, and F_8, and 1000 generations for F_3 and F_7.
- *Parameter in RM-MEDA Reproduction:* The number of cluster is 5 for all the instances.
- *Number of runs:* Each algorithm is run on each instances for 30 times independently.

The Inverted Generational Distance (IGD) metric [5] is used to assess the algorithm performances in our experimental study. Let P^* be a set of uniformly distributed Pareto optimal points in the PF. Let P be an approximation to the PF. The IGD metric is defined as follows,

$$IGD(P^*, P) = \frac{\sum_{v \in P^*} d(v, P)}{|P^*|}$$

where $d(v, P)$ is a minimum distance between v and any point in P, and $|P^*|$ is the cardinality of P^*. The IGD metric can measure both convergence and diversity although it may not work well in some cases [39]. To have a low IGD value, P must be close to the PF and cannot miss any part of the whole PF. In our experiments, 500 evenly distributed points in PF are generated as the P^* for bi-objective problems and 990 points for tri-objective problems.

The statistical IGD values on the test instances over 30 runs are shown in Table 1. The mean and std. values are provided in this table.

Table 1. Statistical IGD values (mean±std.) on the test instances over 30 runs

	NDS	AMS0	AMS1
F_1	0.0043 ± 0.0001	0.0039 ± 0.0000	$\mathbf{0.0036} \pm \mathbf{0.0000}$
F_2	0.0041 ± 0.0001	$\mathbf{0.0038} \pm \mathbf{0.0000}$	$\mathbf{0.0038} \pm \mathbf{0.0000}$
F_3	0.0057 ± 0.0041	0.0154 ± 0.0102	$\mathbf{0.0032} \pm \mathbf{0.0011}$
F_4	0.0494 ± 0.0015	$\mathbf{0.0490} \pm \mathbf{0.0012}$	0.0830 ± 0.0549
F_5	0.0050 ± 0.0002	0.0042 ± 0.0001	$\mathbf{0.0038} \pm \mathbf{0.0001}$
F_6	0.0122 ± 0.0086	$\mathbf{0.0094} \pm \mathbf{0.0205}$	0.0333 ± 0.0560
F_7	0.1365 ± 0.1703	$\mathbf{0.1011} \pm \mathbf{0.0167}$	0.2044 ± 0.2616
F_8	$\mathbf{0.0657} \pm \mathbf{0.0041}$	0.0769 ± 0.0543	0.0830 ± 0.0405

4.2 Results on Problems with Convex PFs

We firstly consider the bi-objective problems with convex PFs. The final approximations to the PF are plotted in Fig. 3 and the mean IGD values versus generations are shown in Fig. 4.

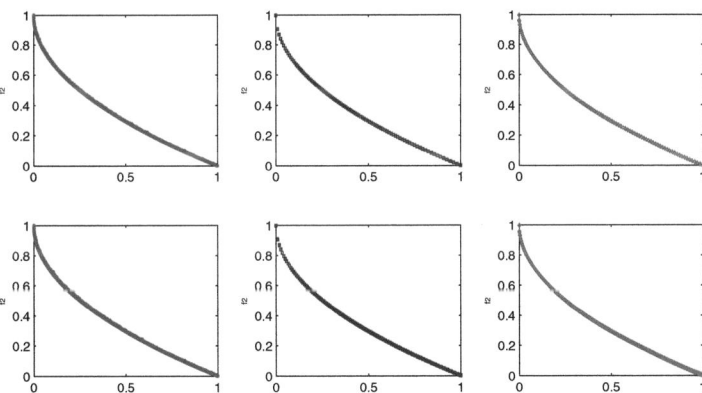

Fig. 3. The final approximations obtained by NDS (left), AMS0(middle), and AMS1(right) on F_1 (1st row) and F_5 (2nd row) respectively

The statistical values in Table 1 show that all the three selection operators work well on both F_1 and F_5, and $AMS1$ performs slightly better than $AMS0$ and NDS. Fig. 4 indicates that by using AMS0, the population converges slightly

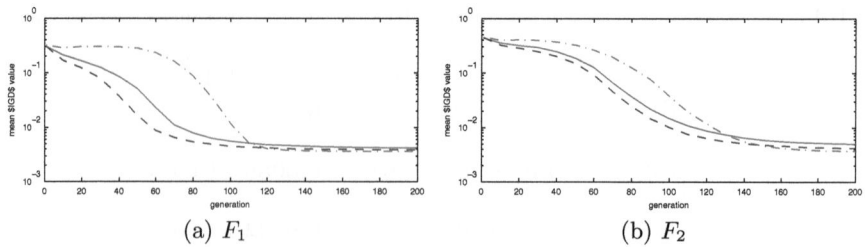

(a) F_1 (b) F_2

Fig. 4. The mean IGD values versus generations on F_1 and F_5. The solid lines are with NDS, the dashed lines are with AMS0, and the dash-dot lines are with AMS1.

faster than those of using NDS and AMS1. The reason might be that the zero-order model is more stable than the first-order model.

4.3   Results on Problems with Concave PFs

Both F_2 and F_6 are bi-objective problems with concave PFs. The final approximations to the PF are plotted in Fig. 5 and the mean IGD values versus generations are shown in Fig. 6.

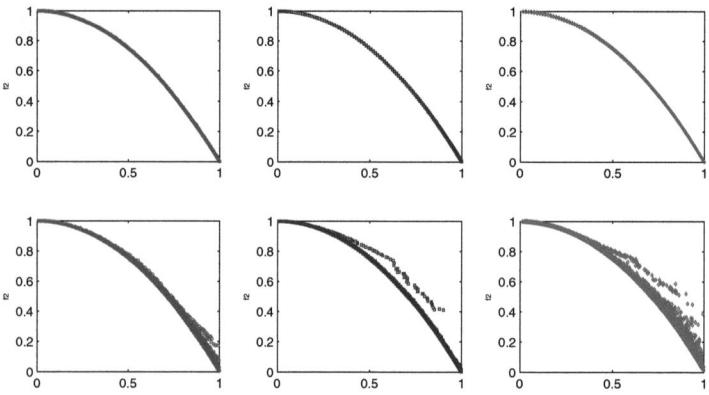

Fig. 5. The final approximations obtained by NDS (left), AMS0(middle), and AMS1(right) on F_2 (1st row) and F_6 (2nd row) respectively

The statistical values in Table 1 show that all the three selection operators work similar on both F_2 and F_6. On F_2, AMS0 and AMS1 have the same performance and on F_6, AMS0 works slightly better. From Fig. 5, we can see that the results on F_6 are not as good as those on F_2. The reason might be that there are nonlinear linkages between the variables on F_6 while variable linkages on F_2 are linear. Therefore, the PS of the F_2 is easier to obtain than that of F_6. Since

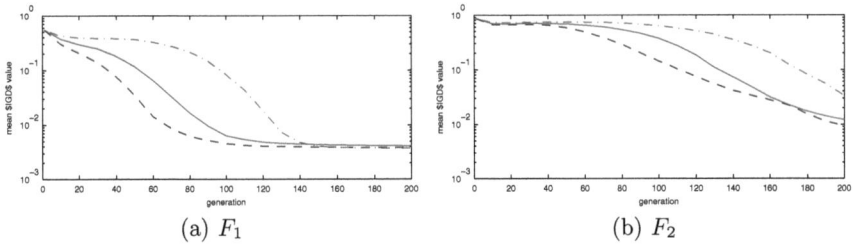

(a) F_1 (b) F_2

Fig. 6. The mean IGD values versus generations on F_2 and F_6. The solid lines are with NDS, the dashed lines are with AMS0, and the dash-dot lines are with AMS1.

the whole PS could not be obtained in some runs on F_6, AMS0 and AMS1 also fail to approximate its PF.

4.4 Results on Tri-objective Problems

Both F_4 and F_8 are tri-objective problems with concave PFs. The final approximations to the PF are plotted in Fig. 7 and the mean IGD values versus generations are shown in Fig. 8.

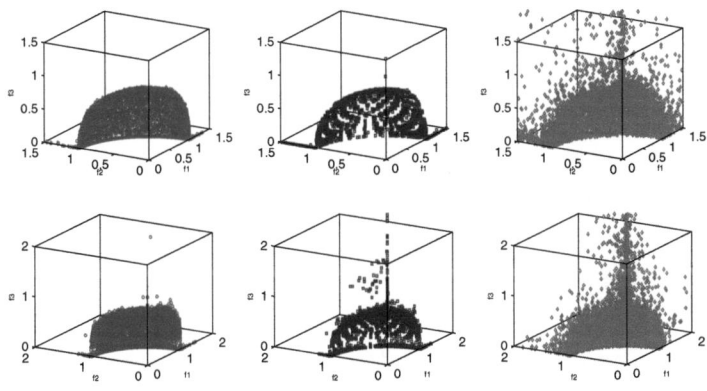

Fig. 7. The final approximations obtained by NDS (left), AMS0(middle), and AMS1(right) on F_4 (1st row) and F_8 (2nd row) respectively

The statistical values in Table 1 show that NDS and AMS0 work similar on both F_4 and F_8 and they perform better than AMS1. The final results in Fig. 7 also show that the solutions obtained by NDS and AMS0 are more close to the PF than those obtained by AMS1. The results indicate that AMS1 is more sensible to the structure of the PF than AMS0. From Fig. 8, we can also observe that AMS1 can not always approximate PF successfully. The reason is that some extreme points, which are used to build the simplex model, mislead

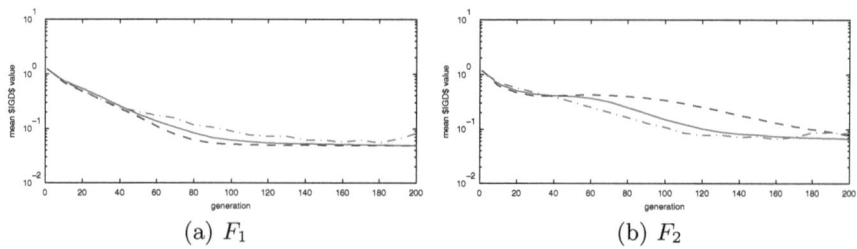

(a) F_1 (b) F_2

Fig. 8. The mean IGD values versus generations on F_4 and F_8. The solid lines are with NDS, the dashed lines are with AMS0, and the dash-dot lines are with AMS1.

the selection. How to improve the performance of AMS1 on high dimensional problems is worth further investigating.

4.5   Results on Problems with Complicated Mappings

Mappings between the PS and the PF on both F_3 and F_7 are complicated, and a uniform distributed PS does not form a uniform distributed PF. The final approximations to the PF are plotted in Fig. 9 and the mean IGD values versus generations are shown in Fig. 10.

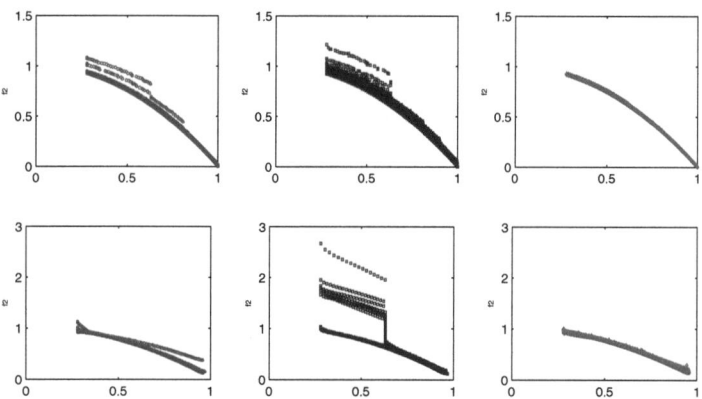

Fig. 9. The final approximations obtained by NDS (left), AMS0(middle), and AMS1(right) on F_3 (1st row) and F_7 (2nd row) respectively

The statistical values in Table 1 show that AMS1 works better than NDS and AMS1 on F_3, while AMS0 has the best performance on F_7. The final results in Fig. 9 show that the solutions obtained by AMS1 are closer to the PF than those obtained by NDS and AMS0. The statical results in Table 1 show that the std. value of AMS1 on F_7 is big. The reason is that in several runs, the algorithm with AMS1 converged to some small parts of the PF. The results suggest that

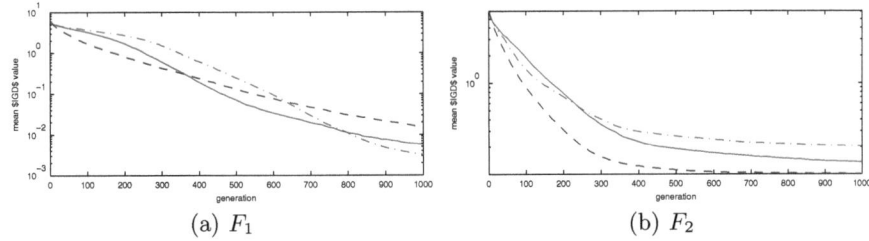

                 (a) F_1 (b) F_2

Fig. 10. The mean IGD values versus generations on F_3 and F_7. The solid lines are with NDS, the dashed lines are with AMS0, and the dash-dot lines are with AMS1.

the final results are highly influenced by the distribution in the decision space. If an algorithm can not produce well distributed solutions in the decision space, the selection operator is not able to approximate the PF well.

5 Conclusion

In this paper, we proposed an approximation model guided selection operator for evolutionary multiobjective optimization. The basic idea is to build a model to approximate the PF and then use the model to guide selection process. We designed two approximation models to implement this idea, one is a zero-order model and the other is a first-order model, by using the selection strategy in MOEA/D framework. The proposed selection operators are tested in the RM-MEDA algorithm on eight test instances with different PF characters.

The experimental results shown that our proposed operators worked slightly better than the non-dominated sorting based selection operator on most of given test instances. Comparing the zero-order and first-order models, we could see that if the first-order model can approximate the PF successfully, it usually works better than the zero-order model because the first-order model offers more information about the PF. However, the experimental results have also shown that our first-order model is more sensitive to the shape of the PF than the zero-order model. Although we mainly focused on selection in this paper, the experimental results indicated again that both the reproduction and selection operators are important for an MOEA.

The work reported in this paper is very preliminary and there are several directions for future work: (1) improving the stability of the first-order model based operator, (2) approximating the PF with higher-order models, (3) testing the approaches on more complicated test instances and many-objective optimization problems.

The Matlab source code of this work can be downloaded from Prof. Zhang's home page: http://dces.essex.ac.uk/staff/qzhang/.

Acknowledgment. This work is supported by the National Natural Science Foundation of China (No.61273313).

References

1. Miettinen, K.: Nonlinear Multiobjective Optimization. Kluwer Academic Publishers (1999)
2. Deb, K.: Multi-Objective Optimization using Evolutionary Algorithms. John Wiley & Sons LTD. (2001)
3. Zhou, A., Qu, B.-Y., Li, H., Zhao, S.-Z., Suganthan, P.N., Zhang, Q.: Multiobjectiveevolutionary algorithms: A survey of the state of the art. Swarm and Evolutionary Computation 1(1), 32–49 (2011)
4. Iorio, A.W., Li, X.: Rotated problems and rotationally invariant crossover in evolutionary multi-objective optimization. International Journal of Computational Intelligence and Applications 7(2), 149–186 (2008)
5. Zhang, Q., Zhou, A., Jin, Y.: RM-MEDA: A regularity model based multiobjective estimation of distribution algorithm. IEEE Transactions on Evolutionary Computation 12(1), 41–63 (2008)
6. Zhang, Q., Li, H.: MOEA/D: A multiobjective evolutionary algorithm based on decomposition. IEEE Transactions on Evolutionary Computation 11(6), 712–731 (2007)
7. Li, H., Zhang, Q.: Multiobjective optimization problems with complicated Pareto sets, MOEA/D and NSGA-II. IEEE Transactions on Evolutionary Computation 13(2), 284–302 (2009)
8. Zhou, A., Zhang, Q., Jin, Y., Sendhoff, B.: Combination of EDA and DE for continuous biobjective optimization. In: IEEE Congress on Evolutionary Computation (CEC 2008), pp. 1447–1454 (2008)
9. Zhou, A.: Estimation of distribution algorithms for continuous multiobjective optimization. Ph.D. dissertation, University of Essex (2009)
10. Goldberg, D.E.: Genetic Algorithm in Search, Optimisation and Machine Learning. Addison-Wesley (1989)
11. Srinivas, N., Deb, K.: Multiobjective optimization using nondominated sorting in genetic algorithms. Evolutionary Computation 2(3), 221–248 (1994)
12. Fonseca, C.M., Fleming, P.J.: Genetic algorithms for multiobjective optimization: Formulation, discussion and generalization. In: Proceedings of the Fifth International Conference Genetic Algorithms, pp. 416–423 (1993)
13. Zitzler, E., Thiele, L.: Multiobjective evolutionary algorithms: A comparative case study and the strength Pareto approach. IEEE Transactions on Evolutionary Computation 3(4), 257–271 (1999)
14. Zitzler, E., Laumanns, M., Thiele, L.: SPEA2: Improving the strength Pareto evolutionary algorithm for multiobjective optimization. In: Evolutionary Methods for Design, Optimisation and Control, pp. 95–100 (2002)
15. Horn, J., Nafpliotis, N., Goldberg, D.E.: A niched Pareto genetic algorithm for multiobjective optimization. In: Proceedings of the First IEEE Conference on Evolutionary Computation, vol. 1, pp. 82–87 (1994)
16. Deb, K., Pratap, A., Agarwal, S., Meyarivan, T.: A fast and elitist multiobjective genetic algorithm: NSGA-II. IEEE Transactions on Evolutionary Computation 6(2), 182–197 (2002)
17. Knowles, J.D., Corne, D.W.: Approximating the nondominated front using the Pareto archived evolution strategy. Evolutionary Computation 8(2), 149–172 (2000)
18. Corne, D.W., Jerram, N.R., Knowles, J.D., Oates, M.J.: PESA-II: Region-based selection in evolutionary multiobjective optimization. In: Proceedings of the Genetic and Evolutionary Computation Conference (GECCO 2001), pp. 283–290 (2001)

19. Yen, G.G., Lu, H.: Dynamic multiobjective evolutionary algorithm: Adaptive cell-based rank and density estimation. IEEE Transactions on Evolutionary Computation 7(3), 253–274 (2003)
20. Coello Coello, C.A., Pulido, G.T., Lechuga, M.S.: Handling multiple objectives with particle swarm optimization. IEEE Transactions on Evolutionary Computation 8(3), 256–279 (2004)
21. Laumanns, M., Thiele, L., Deb, K., Zitzler, E.: Combining convergence and diversity in evolutionary multiobjective optimization. Evolutionary Computation 10(3), 263–282 (2002)
22. Fleischer, M.: The Measure of Pareto Optima Applications to Multi-objective Metaheuristics. In: Fonseca, C.M., Fleming, P.J., Zitzler, E., Deb, K., Thiele, L. (eds.) EMO 2003. LNCS, vol. 2632, pp. 519–533. Springer, Heidelberg (2003)
23. Zitzler, E., Thiele, L.: Multiobjective Optimization Using Evolutionary Algorithms - A Comparative Case Study. In: Eiben, A.E., Bäck, T., Schoenauer, M., Schwefel, H.-P. (eds.) PPSN V. LNCS, vol. 1498, pp. 292–304. Springer, Heidelberg (1998)
24. Huband, S., Hingston, P., White, L., Barone, L.: An evolution strategy with probabilistic mutation for multi-objective optimisation. In: IEEE Congress on Evolutionary Computation (CEC 2003), pp. 2284–2291 (2003)
25. Zitzler, E., Künzli, S.: Indicator-Based Selection in Multiobjective Search. In: Yao, X., Burke, E.K., Lozano, J.A., Smith, J., Merelo-Guervós, J.J., Bullinaria, J.A., Rowe, J.E., Tiňo, P., Kabán, A., Schwefel, H.-P. (eds.) PPSN VIII. LNCS, vol. 3242, pp. 832–842. Springer, Heidelberg (2004)
26. Emmerich, M., Beume, N., Naujoks, B.: An EMO Algorithm Using the Hypervolume Measure as Selection Criterion. In: Coello Coello, C.A., Hernández Aguirre, A., Zitzler, E. (eds.) EMO 2005. LNCS, vol. 3410, pp. 62–76. Springer, Heidelberg (2005)
27. Naujoks, B., Beume, N., Emmerich, M.: Multi-objective optimisation using s-metric selection: Application to three-dimensional solution spaces. In: IEEE Congress on Evolutionary Computation (CEC 2005), vol. 2, pp. 1282–1289 (2005)
28. Basseur, M., Zitzler, E.: Handling uncertainty in indicator-based multiobjective optimization. International Journal of Computational Intelligence Research 2(3), 255–272 (2006)
29. Igel, C., Hansen, N., Roth, S.: Covariance matrix adaptation for multi-objective optimization. Evolutioanry Computation 15(1), 1–28 (2007)
30. Bader, J., Zitzler, E.: HypE: An algorithm for fast hypervolume-based many-objective optimization. Evolutionary Computation 19(1), 45–76 (2011)
31. Guerreiro, A.P., Fonseca, C.M., Emmerich, M.T.M.: A fast dimension-sweep algorithm for the hypervolume indicator in four dimensions. In: Proceedings of the 24th Canadian Conference on Computational Geometry (CCCG 2012), pp. 77–82 (2012)
32. Arakawa, M., Hagiwara, I., Nakayama, H., Yamakawa, H.: Multiobjective optimization using adaptive range genetic algorithms with data envelopment analysis. In: 7th Symposium on Multidisciplinary Analysis and Optimization, pp. 2074–2082. AIAA (1998)
33. Yun, Y., Nakayama, H., Arakawa, M.: Fitness evaluation using generalized data envelopment analysis in MOGA. In: IEEE Congress on Evolutionary Computation (CEC 2004), pp. 464–471 (2004)
34. Rotar, C., Dumitrescu, D., Lung, R.: Guided hyperplane evolutionary algorithm. In: Proceedings of the Conference on Genetic and Evolutionary Computation (GECCO 2007), pp. 884–891 (2007)

35. Deb, K., Jain, H.: An improved NSGA-II procedure for many-objective optimization, part i: Solving problems with box constraints, KanGAL. Tech. Rep. 2012009 (2012)
36. Deb, K., Jain, H.: An improved NSGA-II procedure for many-objective optimization, part ii: Handling constraints and extending to an adaptive approach, KanGAL. Tech. Rep. 2012010 (2012)
37. Hillermeier, C.: Nonlinear Multiobjective Optimization - A Generalized Homotopy Approach. Birkhäuser (2001)
38. Zhou, A., Zhang, Q., Jin, Y.: Approximating the set of Pareto-optimal solutions in both the decision and objective spaces by an estimation of distribution algorithm. IEEE Transactions on Evolutionary Computation 13(5), 1167–1189 (2009)
39. Schütze, O., Esquivel, X., Lara, A., Coello Coello, C.A.: Using the averaged hausdorff distance as a performance measure in evolutionary multiobjective optimization. IEEE Transactions on Evolutionary Computation Evolutionary Computation 16(4), 504–522 (2012)

A Decomposition Based Evolutionary Algorithm for Many Objective Optimization with Systematic Sampling and Adaptive Epsilon Control

Md. Asafuddoula, Tapabrata Ray, and Ruhul Sarker

School of Engineering and Information Technology
University of New South Wales, Canberra, Australia
Md.Asaf@student.adfa.edu.au

Abstract. Decomposition based evolutionary approaches such as MOEA/D and its variants have been quite successful in solving various classes of two and three objective optimization problems. While there have been some attempts to modify the dominance based approaches such as NSGA-II and SPEA2 to deal with many-objective optimization, there are few attempts to extend the capability of decomposition based approaches. The performance of a decomposition based approach is dependent on (a) the mechanism of reference points generation i.e. one which needs to be scalable and computationally efficient (b) the method to simultaneously deal with conflicting requirements of convergence and diversity and finally (c) the means to use the information of neighboring subproblems efficiently. In this paper, we introduce a decomposition based evolutionary algorithm, wherein the reference points are generated via systematic sampling and an adaptive epsilon scheme is used to manage the balance between convergence and diversity. To deal with constraints efficiently, an adaptive epsilon formulation is adopted. The performance of the algorithm is highlighted using standard benchmark problems i.e. DTLZ1 and DTLZ2 for 3, 5, 8, 10 and 15 objectives, the car side impact problem, the water resource management problem and the constrained ten-objective general aviation aircraft (GAA) design problem. The study clearly highlights that the proposed algorithm is better or at par with recent reference direction based approaches.

Keywords: many-objective optimization, generation of reference points, adaptive epsilon comparison, constraint-handling.

1 Introduction

Many objective optimization typically refers to problems with the number of objectives greater than four [1]. There is significant amount of literature discussing the challenges involved in solving them and interested readers may refer to [1] for further details. The commonly used dominance based methods for multi-objective optimization, such as NSGA-II, SPEA2 etc. are known to be inefficient for many-objective optimization as non-dominance does not provide adequate selection pressure to drive the population towards convergence. There has been a number of attempts to modify the underlying selection pressure through the use of substitute distance measures [2][3], average rank

R.C. Purshouse et al. (Eds.): EMO 2013, LNCS 7811, pp. 413–427, 2013.

domination [4], fuzzy dominance [5], ε-dominance [6][7], adaptive ε-ranking [8] etc. without great success. In all the above approaches, while the diversity and the convergence of the population improved during the course of evolution, there is no guarantee that the final non-dominated set spans the entire Pareto surface uniformly.

There are also radically different approaches to deal with many objective optimization, such as attempts to identify the reduced set of objectives [9] or corners of the Pareto front [10] and subsequently solving the problem using these reduced set of objectives. Other attempts include interactive use of decision makers preferences [11], use of reference points [12][13] or solution of the problem as a hypervolume maximization problem [14]. While some progress has been made along these lines, the limiting factors include the inability to obtain solutions close to Pareto set for an accurate identification of redundant objectives, decision making burden associated with preference elicitation and the computational complexity of hypervolume computation.

Decomposition based evolutionary algorithms are yet another class of algorithms originally introduced as MOEA/D [15], wherein the multiobjective optimization problem is decomposed into a series of scalar optimization problems. In a decomposition based approach, one need to generate uniformly distributed reference directions and adopt a method of scalarization. In the context of many objective optimization, the first issue relates to the design of a computationally efficient scheme to generate W uniform reference directions for a M objective optimization problem, where M is typically more than four and W is of the same order as the population size. The second issue relates to scalarization, which essentially assigns the *fittest* individual to each reference direction. The notion of *fittest* is essentially derived using a tradeoff between convergence and diversity measured with respect to any given reference direction. One of the early attempts to generate uniformly distributed reference directions appear in the works of Hughes [13] . The method was not computationally efficient for problems with more than six objectives and often resulted in a large number of reference directions that in turn required a huge population size. More recently, computationally efficient and scalable sampling schemes have been used in the context of many-objective optimization. A systematic sampling [16] scheme has been used in M-NSGA-II [17] while an uniform sampling scheme has been used within MOEA/D [18] to deal with many objective optimization problems.

The second issue related to scalarization has been addressed via two fundamental means i.e. through a systematic association and niche preservation mechanism as in M-NSGA-II [17] or through the use of a penalty function(i.e. an aggregation of the projected distance along a reference direction and the perpendicular distance from a point to a given reference direction) within the framework of MOEA/D. The performance of the penalty function based approach is dependent on the penalty parameter, while the association and the niche preservation process require a careful implementation to address a number of possibilities.

In this paper, we introduce a decomposition based evolutionary algorithm for many-objective optimization. The reference directions are generated using systematic sampling, wherein the points are systematically generated on a hyperplane with unit intercepts in each objective axis. The process of reference point generation is the same as adopted in M-NSGA-II [17]. The fine balance between convergence and diversity along a reference direction is managed using an adaptive epsilon model eliminating the need for the penalty parameter. While M-NSGA-II [17] is a generational model, our proposed algorithm is a steady state form. Furthermore, to deal with constraints, an adaptive epsilon level based scheme is introduced which has been demonstrated to be more effective over *feasibility first* schemes in the context of constrained optimization [19].

The details of the proposed algorithm are presented in Section 2. The performance of the proposed algorithm on benchmark problems (DTLZ1 and DTLZ2 for 3, 5, 8, 10 and 15 objectives) is presented and compared with MOEA/D-PBI and M-NSGA-II in Section 3. In addition to the above set of mathematical benchmarks, the performance of the algorithm is also compared using a number of engineering design problems (car side impact, water resource management and the constrained ten-objective general aviation aircraft (GAA) design). The final section summarizes the contributions and future directions for further improvement.

2 Proposed Algorithm

A many-objective optimization problem can be defined as follows:

$$min. \ [f_1(x), f_2(x), f_3(x), ...f_M(x)], x \in \Omega$$
$$S.t. \qquad g_j(x) \leq 0, j = 1, 2, ...p \qquad (1)$$
$$h_k(x) = 0, k = 1, 2, ...q$$

where $f_1(x), f_2(x), f_3(x), ...f_M(x)$ are the M objective functions, p is the number of inequalities and q is the number of equalities.

The pseudocode of the algorithm is presented below and the subsequent components are discussed in the following subsections.

Algorithm 1. DBEA-Eps

Input: Gen_{max} maximum number of generations, W the number of reference points

---

1: **Generate the reference points and assign their neighborhood**
2: Initialize the population P; $|P|$ = W
3: Evaluate the initial population and compute the ideal point $\bar{z}_j = (f_1^{min}, f_2^{min}, ..., f_M^{min})$ and intercepts a_i's for $i = 1$ to M
4: Scale the individuals of the population
5: **while** $(gen \leq Gen_{max})$ **do**
6:    **for** $i=1$:W **do**
7:        Assign the base parent as P_i
8:        I=Select a mating partner for (P_i)
9:        Create a child via recombination as C_i
10:       Evaluate C_i and compute the distances ($d1$ and $d2$) using all reference directions
11:       Replace the parent P_k with C_i using *single-first encounter*, where k denotes the index of the first parent satisfying the condition of replacement
12:       Update the ideal point ($\bar{z}$), the intercepts and re-scale the population
13:    **end for**
14: **end while**

The algorithm consists of four major components i.e. (a) generation of reference directions and assignment of neighborhood (b) computation of distances along and perpendicular to each reference direction (c) method of recombination using information from neighboring subproblems and finally (d) adaptive epsilon comparison to manage the balance between convergence and diversity.

2.1   Generation of Reference Points and Assignment of Neighborhood

A structured set of reference points (β) is generated spanning a hyperplane with unit intercepts in each objective axis using the algorithm outlined in [16]. The approach generates W points on the hyperplane with a uniform spacing of $\delta = 1/p$ for any number of objectives M. The process of generation of the reference points is illustrated for a 3-objective optimization problem i.e. ($M=3$) and with an assumed spacing of $\delta = 0.2$ i.e ($p = 5$) in Figure 1. The process results in the generation of 21 reference points.

$$W = {}^{(M+p-1)}C_p \tag{2}$$

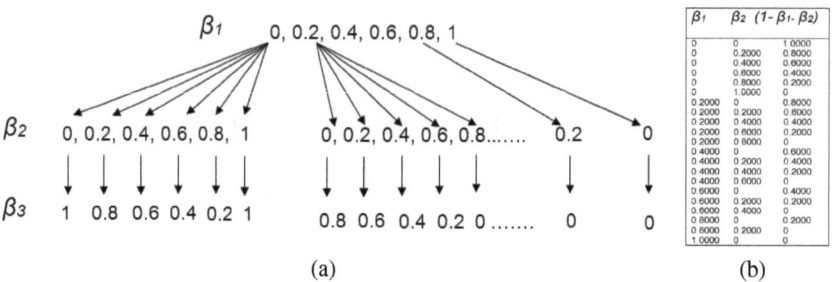

(a)   (b)

Fig. 1. (a) the reference points are generated computing βs recursively (b) the table shows the combination of all βs in each column

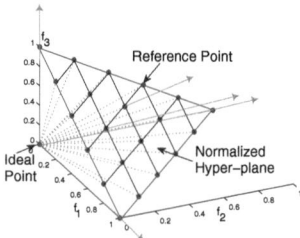

Fig. 2. A set of reference points in a normalized hyper-plane for number of objectives, $M = 3$ and $p = 5$

The distribution of the reference points are presented in Figure 2. The reference directions are formed by constructing a straight line from the origin to each of these reference points. The population size of the algorithm is set to the number of reference points. For every reference point, its neighborhood consists of T closest reference points computed based on a Euclidean distance amongst them. The initial population consists of W individuals generated randomly within the variable bounds. Such solutions are thereafter assigned randomly to a reference direction during the phase of initialization.

2.2 Computation of Distances along and Perpendicular to Each Reference Direction

Since in a generic many objective optimization problem, the objectives may assume negative values or values in varying orders of magnitude, it is important to scale them appropriately. The ideal point of a population is denoted by $\bar{z}_j = (f_1^{min}, f_2^{min}, \ldots, f_M^{min})$ and the extreme point is denoted by $z_j^e = (f_1^{max}, f_2^{max}, \ldots, f_M^{max})$. A hyperplane is created using the solutions that have led to the coordinates of the extreme point. The intercepts of the hyperplane along the objective axes are denoted by $a_1, a_2, \ldots, a_M$. The generic equation of a plane through these points can be represented using the following equation

$$Af_1 + Bf_2 + \ldots + Cf_M = 1 \tag{3}$$

where, $A, B, \ldots, C$ are the unit normal of the plane. The intercepts of the plane with the axis are given by $a_1 = 1/A$, $a_2 = 1/B, \ldots$, and $a_M = 1/C$.

In the event, the number of such solutions are less than M or any of the a_i's are negative, a_i's are set to f_i^{max}. Every solution in the population is subsequently scaled as follows:

$$f_j'(x) = \frac{f_j(x) - \bar{z}_j}{a_j - \bar{z}_j}, \forall j = 1, 2, \ldots M \tag{4}$$

For any given reference direction, the performance of a solution can be judged using two measures d_1 and d_2 as depicted in Equation 8. The first measure d_1 is the Euclidean distance between origin and the foot of the normal drawn from the solution to the reference direction, while the second measure d_2 is the length of the normal. Mathematically, d_1 and d_2 are computed as follows:

$$d1 = w^T f_j'(x) \tag{5}$$

$$d2 = ||f_j'(x) - w^T f_j'(x)w|| \tag{6}$$

where w is a unit vector along any given reference direction. It is clear that a value of $d_2 = 0$ ensures the solutions are perfectly aligned along the required reference direction ensuring perfect diversity, while a smaller value of d_1 indicates superior convergence. These two measures are subsequently used to control diversity and convergence of the algorithm via an adaptive epsilon scheme.

2.3   Mating Partner Selection

The information and similarity of neighboring subproblems are exploited via the process of parter selection. The mating partner for P_i (where i is the index of the current individual in a population) is selected using of the following rules i.e. rule 1: select a parent from the neighborhood with a probability of τ and rule 2: select a random parent from the population with a probability of $(1 - \tau)$.

2.4   Method of Recombination

In the recombination process, two child solutions are generated using simulated binary crossover (SBX) operator [20] and polynomial mutation. The first child is considered as an individual attempting to replace any parent in the population.

2.5   Adaptive Epsilon Comparison to Manage the Balance between Convergence and Diversity

Since every solution is assigned to a reference direction, the average deviation ε_{CD} for the population of solutions is computed using Equation 7, where d_{2i} denotes the d_2 measure of the ith individual in the population.

$$\varepsilon_{CD} = \frac{\sum_{i=1}^{W} d_{2i}}{W} \tag{7}$$

whenever a child solution is created, its d_2 measure is computed along all reference directions and the child solution replaces a single parent based on the following rule Equation 8.

$$(d_1, d_2) <_{\varepsilon_{CD}} (d_1, d_2) \Leftrightarrow \begin{cases} d_1 < d_2, \text{ if } & d_2, d_2 < \varepsilon_{CD} \\ d_1 < d_2, \text{ if } & d_2 = d_2 \\ d_2 < d_2, \text{ otherwise} \end{cases} \tag{8}$$

It is also worth noting that this process is a *single-first encounter* replacement scheme whereby, the child solution can only replace a single parent and the first encountered parent meeting the condition is replaced. Whenever a replacement is successful, a check is performed to identify if there is a need to re-compute the ideal point or the intercepts. The population needs to be re-scaled in the event the ideal point or the intercepts have changed.

The possible epsilon comparison scenarios are presented in Fig. 7 as Case 1, Case 2 and Case 3. Let us assume the parent solution is denoted by ($s1$) and the child solution is denoted by ($s2$)). Case 1: Both the solutions have their d_2 values less than ε_{CD}. One with the smaller d_1 is selected i.e.($s1$). Case 2: Both the d_2 values are more than ε_{CD}. One with the lower d_2 value is selected i.e.($s2$). Case 3: One solution has its d_2 value more than ε_{CD} and the other has its d_2 value less than ε_{CD}. One with the smaller d_2 value is selected i.e. ($s1$).

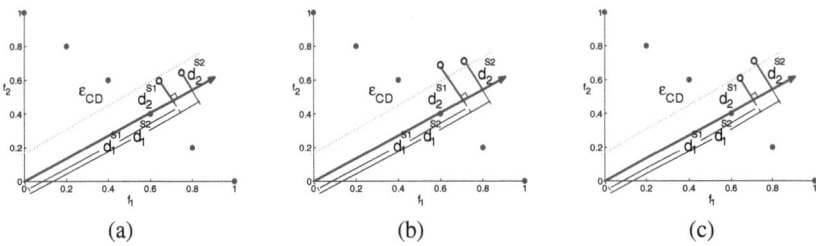

Fig. 3. (a) Case 1 (b) Case 2 (c) Case 3

2.6 Constraint Handling

The constraint handling approach used in this work is based on epsilon level comparison and has been reported earlier in [21]. The feasibility ratio (FR) of a population refers to the ratio of the number of feasible solutions in the population to the number of solutions (W). The allowable violation is calculated as follows:

$$CV = \sum_{i=1}^{p} max(g_i, 0) + \sum_{i=1}^{q} max(|h_i - \varepsilon|, 0) \tag{9}$$

$$CV_{mean} = \frac{1}{W} \sum_{j=1}^{W} (CV_j) \tag{10}$$

$$\text{Allowable violation}(\varepsilon_{CV}) = CV_{mean} * FR \tag{11}$$

An epsilon level comparison using this allowable violation measure is used to compare two solutions. If two solutions have their constraint violation value less than this epsilon level, the solutions are compared based on their objective values i.e. via d_1 and d_2 measures. Such a constraint handling scheme has been demonstrated to be more efficient than *feasibility first* schemes.

3 Experimental Results

In this section, we present the results of proposed decomposition based evolutionary algorithm (DBEA-Eps) and compare its performance with M-NSGA-II and MOEA/D-PBI [22] for DTLZ1 and DTLZ2 problems with 3,5,8,10 and 15 objectives.

The population sizes used in this study are the same as those adopted in [22]. In our proposed algorithm, the probability of crossover is set to 1 and the probability of mutation is set to $p_m = 1/D$, where D is the dimensionality of the problem. The distribution index of crossover is set to $\eta_c = 30$ and the distribution index of mutation is set to $\eta_m = 20$ as in [22]. The probability of selecting parent from its neighborhood (τ) is set to 0.9 and the neighborhood size is set to 20.

To assess the performance, we have selected IGD [23][15] as a performance metric. The IGD metric in our simulation results is calculated by normalizing the approximated set with the theoretical ideal and nadir points for the DTLZ problems.

3.1 Performance on Unconstrained DTLZ Problems

In this comparison, we have reported the best, median and worst IGD results obtained using 20 independent runs for DTLZ1 and DTLZ2. The results are compared against M-NSGA-II and MOEA/D-PBI in Table 3. In Fig 4 and Fig 5, the final Pareto front is shown for three-objective problems of DTLZ1 and DTLZ2.

Table 1. IGD statistics for problems DTLZ1 and DTLZ2 using 20 independent runs

Test Problem	Obj.	MaxGen	Strategy	Best	Median	Worst
DTLZ1	3	400	DBEA-Eps	**8.771e-5**	9.521e-3	5.854e-1
			M-NSGA-II	4.880e-4	1.308e-3	4.880e-3
			MOEA/D-PBI	4.095e-4	**1.495e-3**	**4.743e-3**
DTLZ1	5	600	DBEA-Eps	**1.771e-5**	**2.183e-4**	3.782e-1
			M-NSGA-II	5.116e-4	9.799e-4	1.979e-3
			MOEA/D-PBI	3.179e-4	6.372e-4	**1.635e-3**
DTLZ1	8	750	DBEA-Eps	**4.387e-5**	**3.581e-4**	**1.981e-3**
			M-NSGA-II	2.044e-3	3.979e-3	8.721e-3
			MOEA/D-PBI	3.914e-3	6.106e-3	8.537e-3
DTLZ1	10	1000	DBEA-Eps	**7.691e-4**	**1.504e-3**	**2.700e-3**
			M-NSGA-II	2.215e-3	3.462e-3	6.869e-3
			MOEA/D-PBI	3.872e-3	5.073e-3	6.130e-3
DTLZ1	15	1500	DBEA-Eps	**1.696e-3**	**2.606e-3**	**2.686e-3**
			M-NSGA-II	2.649e-3	5.063e-3	1.123e-2
			MOEA/D-PBI	1.236e-2	1.431e-2	1.692e-2
DTLZ2	3	250	DBEA-Eps	2.040e-2	4.138e-2	6.417e-2
			M-NSGA-II	1.262e-3	1.357e-3	2.114e-3
			MOEA/D-PBI	**5.432e-4**	**6.406e-4**	**8.006e-4**
DTLZ2	5	350	DBEA-Eps	**1.199e-3**	3.024e-3	2.272e-2
			M-NSGA-II	4.254e-3	4.982e-3	5.862e-3
			MOEA/D-PBI	1.219e-3	**1.437e-3**	**1.727e-3**
DTLZ2	8	500	DBEA-Eps	**1.172e-3**	**2.899e-3**	6.915e-3
			M-NSGA-II	1.371e-2	1.571e-2	1.811e-2
			MOEA/D-PBI	3.097e-3	3.763e-3	**5.198e-3**
DTLZ2	10	750	DBEA-Eps	3.656e-3	3.657e-3	3.657e-3
			M-NSGA-II	1.350e-2	1.528e-2	1.697e-2
			MOEA/D-PBI	**2.474e-3**	**2.778e-3**	**3.235e-3**
DTLZ2	15	1000	DBEA-Eps	**5.160e-3**	**5.960e-3**	**5.960e-3**
			M-NSGA-II	1.360e-2	1.726e-3	2.114e-2
			MOEA/D-PBI	5.254e-3	6.005e-3	9.409e-3

One can observe that our algorithm obtained the best IGD values in 8 instances out of 10. In terms of the median performance, our algorithm was the best in 6 instances thereby indicating competitive performance with recently proposed forms.

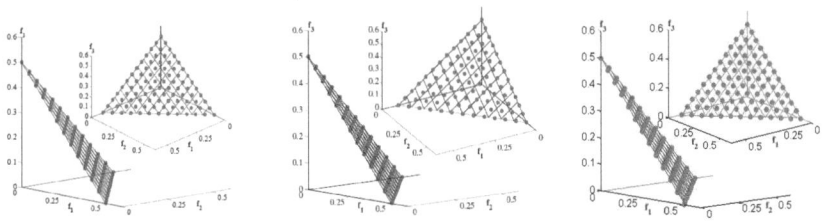

Fig. 4. Obtained solutions by (a) M-NSGA-II (b) MOEA/D-PBI (c) DBEA-Eps for DTLZ1

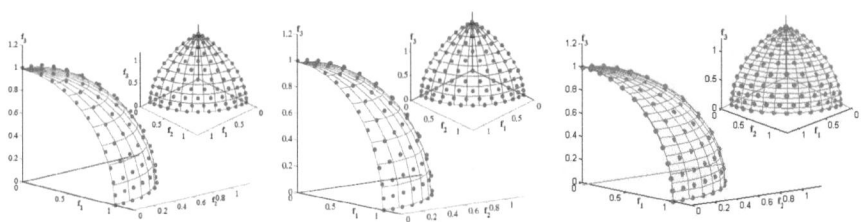

Fig. 5. Obtained solutions by (a) M-NSGA-II (b) MOEA/D-PBI (c) DBEA-Eps for DTLZ2

In order to observe the process of evolution, we computed the average performance of the population i.e. average of the d_1 and d_2 values for the individuals for DTLZ1 (3 objectives). One can observe from Figure 6, that the average d_2 converges to near zero (i.e. near perfect alignment to the reference directions) while the average d_1 measure stabilizes at around 0.5 indicating convergence to the Pareto front.

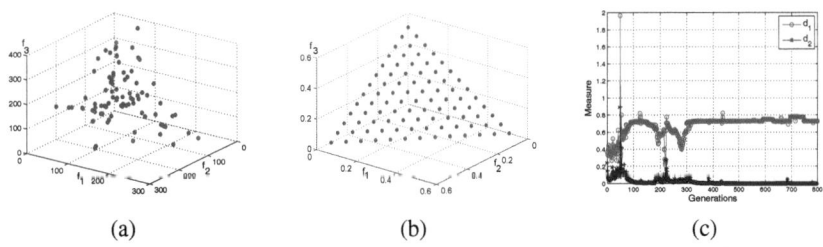

Fig. 6. (a) the initial population of DTLZ1 test problem for number of objectives 3 (b) the final Pareto-front of DTLZ1 test problem for number of objectives 3 (c) the convergence of distance measure over the generations

The association mechanism (i.e. solutions to each reference direction) for a 3-objective DTLZ1 problem is presented in Figure 7. The figure shows the associations in generation 1, 200 and 400 using 15 reference points. One can observe that although initially the association is random, the solutions automatically get associated to the closest reference directions during the course of evolution via the pressure induced by $d2$. This alleviates the need of an extensive niching and association operation as encountered in M-NSGA-II [17].

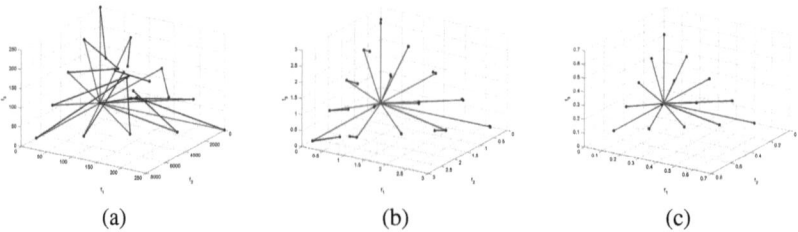

(a)   (b)   (c)

Fig. 7. (a) the initial population of DTLZ1 test problem for number of objectives 3 with 15 reference points (b) at generation 200 (c) at final generation 400

4   Constrained Engineering Design Problems

Since the performance of the proposed algorithm was competitive on unconstrained test problems, we investigated its performance on three constrained engineering design optimization problems i.e. the three-objective car-side-impact problem [24] with ten inequality constraints, five-objective water resource management problem [25] with seven inequality constraints and finally the ten objective general aviation aircraft (GAA) design problem [7] having a single inequality constraint.

4.1   Car Side Impact Problem

The problem aims to minimize the weight of a car, the pubic force experienced by a passenger and the average velocity of the V-Pillar responsible for bearing the impact load subject to the constraints involving limiting values of abdomen load, pubic force, velocity of V-Pillar, rib deflection etc [24].

The problem is solved using DBEA-Eps and MOEA/D-PBI. The algorithms are run for 500 generations and the final non-dominated front is shown in Fig 8.It is important to note that the results of MOEA/D-PBI is derived without scaling which could be a reason among others for poor performance.

4.2   Water Resource Management Problem

This is a five objective problem having seven constraints taken from the literature [25]. The parallel coordinate plot generate using our proposed algorithm (DBEA-Eps) is presented in Fig 9. The best IGD value across 20 runs is $3.29e-2$ and the IGD is computed using the reference set of 2429 solutions [26]. A population of 210 solutions has been used and evolved over 1000 generations.

In Fig 10, a scatter plot-matrix is presented. The results from the DBEA-Eps are shown in the top-right plots vis-a-vis the known reference set of 2429 solutions (shown in bottom-left plots).

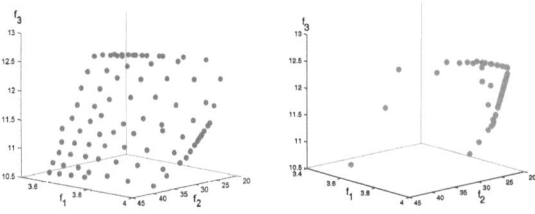

Fig. 8. Solutions obtained using (a) DBEA-Eps (b) MOEA/D-PBI on three-objective car side impact problem

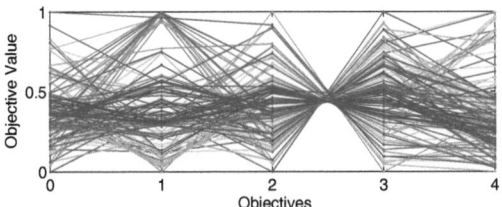

Fig. 9. Solutions obtained using DBEA-Eps on five-objective water problem

4.3   General Aviation Aircraft (GAA) Design Problem

This problem was first introduced by Simpson et al. [27] and has been recently solved using an evolutionary algorithm [7]. The problem involves 9 design variables i.e. cruise speed, aspect ratio, sweep angle, propeller diameter, wing loading, engine activity factor, seat width, tail length/ diameter ratio and taper ratio and the aim is to minimize the takeoff noise, empty weight, direct operating cost, ride roughness, fuel weight, purchase price, product family dissimilarity and maximize the flight range, lift/ drag ratio and cruise speed. Previous studies encountered difficulties in obtaining feasible solutions due to tight constraints [27].

In this example, we have used 100 reference points and the population was allowed to evolve over 5000 generations. A reference set of 412 non-dominated solutions obtained from ε-MOEA and Borg-MOEA is used to compute the IGD metric. The results of the proposed algorithm are compared with four other algorithms i.e. ε-MOEA, Borg-MOEA, MOEA/D and ε-NSGA-II [7]. We have also computed the hypervolume using the ideal point of (i.e.[73.251, 1881.5, 59.114, 1.7977, 359.92, 41879, -2580.2, -16.823, -204.02, 0.26847]) and the extreme point of (i.e.[74.036, 2011.5, 79.993, 2, 483.13, 44590, -2000, -14.408, -189.3, 1.9844]) obtained from the reference set. The performance of the algorithms are compared using the hypervolume in Table 2 and IGD in Table 3. One can observe that the proposed algorithm performs marginally better than others for this problem.

Figure 11 shows the parallel coordinate plot. The figure clearly shows that DBEA-Eps is able to find a widely distributed set of nondominated points for 10-objective GAA design problem.

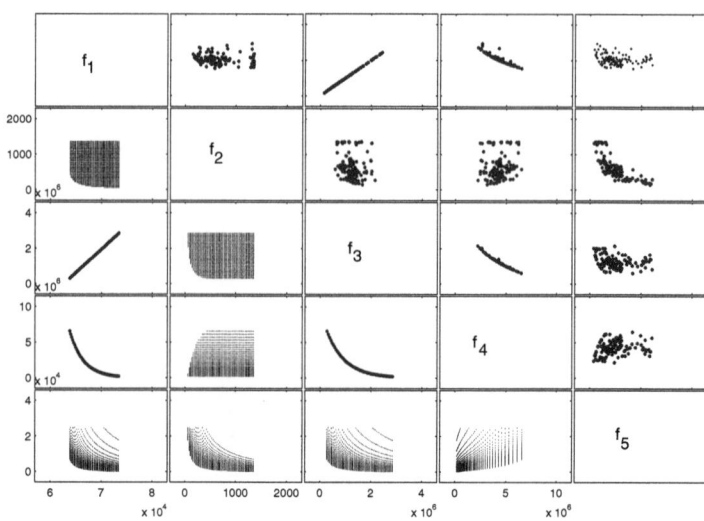

Fig. 10. A scatter plotmatrix showing DBEA-Eps (top-right plots) vis-a-vis the known reference set of 2429 solutions (bottom-left plots)

Table 2. Performance metric value of product family design problem using 50 independent runs

Algorithm	Function Evaluation	Hypervolume			
		Best	Mean	Worst	Std
DBEA-Eps	50,000	0.02899	0.01715	0.00689	0.04561
ε-MOEA		0.02032	0.01032	0.00259	0.04125
Borg-MOEA		0.02245	0.01013	0.00424	0.02327
MOEA/D		0.00092	0.00087	0.00045	0.00145
ε-NSGA-II		0.01636	0.01005	0.00236	0.05232

Table 3. Performance metric value of product family design problem using 50 independent runs

Algorithm	Function Evaluation	IGD			
		Best	Mean	Worst	Std
DBEA-Eps	50,000	0.62070	0.80123	0.82430	0.09210
ε-MOEA		0.98312	0.99123	0.99678	0.10312
Borg-MOEA		0.98211	0.99113	0.99337	0.02321
MOEA/D		0.99117	0.99587	0.99723	0.02145
ε-NSGA-II		0.98571	0.98872	0.99131	0.72123

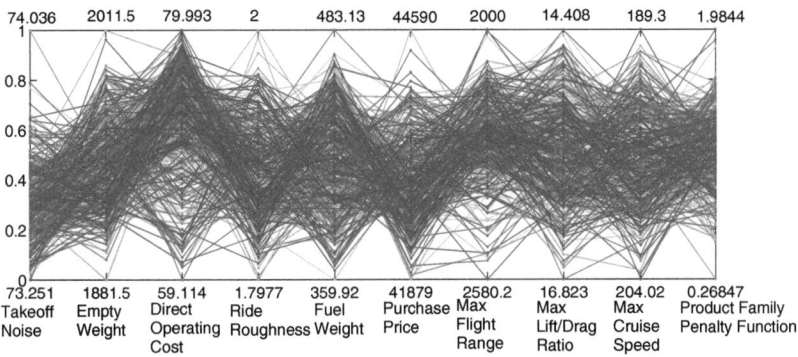

Fig. 11. Parallel coordinate plot of the approximation of Pareto set produced by DBEA-Eps is in a different color trace

5   Conclusion

In this paper, a decomposition based evolutionary algorithm with adaptive epsilon comparison is introduced to solve unconstrained and constrained many objective optimization problems. The approach utilizes reference directions to guide the search, wherein the reference directions are generated using a systematic sampling scheme as introduced by Das and Dennis [16]. The algorithm is designed using a steady state form. In an attempt to alleviate the problems associated with scalarization(commonly encountered in the context of reference direction based methods), the balance between diversity and convergence is maintained using an adaptive epsilon comparison. Such a process also eliminates the need for a detailed association and niching operation as employed in M-NSGA-II. In order to deal with constraints, an epsilon level comparison is used which is known to be more effective than methods employing *feasibility first* principles. The performance of the algorithm is presented using DTLZ1 and DTLZ2 problems with objectives ranging from 3 to 15. Furthermore, three constrained engineering design optimization problems with three to seven constraints (car side impact, water resource management and a general aviation aircraft design problem) have been solved to illustrate the performance of the proposed algorithm. The preliminary results indicate that the proposed algorithm is able to deal with unconstrained and constrained many-objective optimization problems better or at par with existing state of the art algorithms such as M-NSGA-II and MOEA/D-PBI.

Acknowledgement. The second author would like to acknowledge the support of Future Fellowship offered by the Australian Research Council.

References

1. Ishibuchi, H., Tsukamoto, N., Nojima, Y.: Evolutionary many-objective optimization: A short review. In: Proc. IEEE World Congress Computational Intelligence, pp. 2419–2426 (2008)

2. Köppen, M., Yoshida, K.: Substitute Distance Assignments in NSGA-II for Handling Many-objective Optimization Problems. In: Obayashi, S., Deb, K., Poloni, C., Hiroyasu, T., Murata, T. (eds.) EMO 2007. LNCS, vol. 4403, pp. 727–741. Springer, Heidelberg (2007)

3. Singh, H.K., Isaacs, A., Ray, T., Smith, W.: A Study on the Performance of Substitute Distance Based Approaches for Evolutionary Many Objective Optimization. In: Li, X., Kirley, M., Zhang, M., Green, D., Ciesielski, V., Abbass, H.A., Michalewicz, Z., Hendtlass, T., Deb, K., Tan, K.C., Branke, J., Shi, Y. (eds.) SEAL 2008. LNCS, vol. 5361, pp. 401–410. Springer, Heidelberg (2008)

4. Kachroudi, S., Grossard, M.: Average rank domination relation for NSGA-II and SMPSO algorithms for many-objective optimization. In: Proc. Second World Congress Nature and Biologically Inspired Computing (NaBIC), pp. 19–24 (2010)

5. Wang, G., Wu, J.: A new fuzzy dominance GA applied to solve many-objective optimization problem. In: Proc. Second Int. Conf. Innovative Computing, Information and Control (ICICIC 2007), pp. 617–621 (2007)

6. Zou, X., Chen, Y., Liu, M., Kang, L.: A new evolutionary algorithm for solving many-objective optimization problems. IEEE Transactions on Systems, Man, and Cybernetics-Part B 38(5), 1402–1412 (2008)

7. Hadka, D., Reed, P.M., Simpson, T.W.: Diagnostic assessment of the Borg MOEA for many-objective product family design problems. In: IEEE World Congress on Computational Intelligence, Brisbane, Australia, pp. 10–15 (June 2012)

8. Aguirre, H., Tanaka, K.: Adaptive ε-ranking on MNK-landscapes. In: Proc. IEEE Symposium Computational Intelligence in Miulti-Criteria Decision-Making (MCDM 2009), pp. 104–111 (2009)

9. Saxena, D.K., Deb, K.: Dimensionality reduction of objectives and constraints in multi-objective optimization problems: A system design perspective. In: IEEE Congress on Evolutionary Computation (2008)

10. Singh, H.K., Isaacs, A., Ray, T.: A Pareto corner search evolutionary algorithm and dimensionality reduction in many-objective optimization problems. IEEE Transactions on Evolutionary Computation 15(4), 539–556 (2011)

11. Deb, K., Sinha, A., Korhonen, P.J., Wallenius, J.: An interactive evolutionary multi-objective optimization method based on progressively approximated value functions. IEEE Transactions on Evolutionary Computation 14, 723–739 (2010)

12. Wagner, T., Beume, N., Naujoks, B.: Pareto-, Aggregation-, and Indicator-Based Methods in Many-Objective Optimization. In: Obayashi, S., Deb, K., Poloni, C., Hiroyasu, T., Murata, T. (eds.) EMO 2007. LNCS, vol. 4403, pp. 742–756. Springer, Heidelberg (2007)

13. Hughes, E.J.: MSOPS-II: A general-purpose many-objective optimiser. In: Proc. IEEE Congress Evolutionary Computation, pp. 3944–3951 (2007)

14. Bader, J., Zitzler, E.: Hype: An algorithm for fast hypervolume-based many-objective optimization. Evolutionary Computation 19, 45–76 (2011)

15. Zhang, Q., Li, H.: MOEA/D: A multiobjective evolutionary algorithm based on decomposition. IEEE Transactions on Evolutionary Computation 11(6), 712–731 (2007)

16. Das, I., Dennis, J.E.: Normal-bounday intersection: A new method for generating Pareto optimal points in multicriteria optimization problems. SIAM J. Optim. 8(3), 631–657 (1998)

17. Deb, K., Jain, H.: Handling many-objective problems using an improved NSGA-II procedure. In: IEEE World Congress on Computational Intelligence, Brisbane, Australia, pp. 10–15 (June 2012)

18. Tan, Y.Y., Jiao, Y.C., Lib, H., Wang, X.K.: MOEA/D + uniform design: A new version of MOEA/D for optimization problems with many objectives. Computers & Operations Research (January 2012)

19. Takahama, T., Sakai, S.: Constrained optimization by applying the α constrained method to the nonlinear simplex method with mutations. IEEE Transactions on Evolutionary Computation 9(5), 437–451 (2005)
20. Deb, K., Agarwal, R.B.: Simulated binary crossover for continuous search space. Complex Systems 9(2), 115–148 (1995)
21. Asafuddoula, M., Ray, T., Sarker, R., Alam, K.: An adaptive constraint handling approach embedded MOEA/D. In: IEEE World Congress on Computational Intelligence, June 10-15, pp. 1–8 (2012)
22. Deb, K., Jain, H.: An improved NSGA-II procedure for many-objective optimization, Part I: Solving problems with box constraints. KanGAL Report No. 2012009 (June 2012)
23. Lili, Z., Wenhua, Z.: Research on performance measures of multi-objective optimization evolutionary algorithms. In: Proc. 3rd Int. Conf. Intelligent System and Knowledge Engineering (ISKE), vol. 1, pp. 502–507 (2008)
24. Deb, K., Jain, H.: An improved NSGA-II procedure for many-objective optimization, Part II: Handling constraints and extending to an adaptive approach. KanGAL Report No. 2012010 (June 2012)
25. Ray, T., Tai, K., Seow, K.C.: An evolutionary algorithm for multiobjective optimization. Engineering Optimization 33(3), 399–424 (2001)
26. Durillo, J., Nebro, A.: jMetal: a java framework for multi-objective optimization. Advances in Engineering Software 42, 760–771 (2011)
27. Simpson, T.W., Chen, W., Allen, J.K., Mistree, F.: Conceptual design of a family of products through the use of the robust concept exploration method. In: 6th AIAA/USAF/NASA/ISSMO Symposium on Multidisciplinary Analysis and Optimization, vol. 2, pp. 1535–1545 (1996)

Generalized Decomposition

Ioannis Giagkiozis, Robin C. Purshouse, and Peter J. Fleming

Department of Automatic Control and Systems Engineering,
University of Sheffield, Sheffield, UK, S1 3JD
{i.giagkiozis,r.purshouse,p.fleming}@sheffield.ac.uk
http://www.shef.ac.uk/acse

Abstract. Decomposition-based algorithms seem promising for many-objective optimization problems. However, the issue of selecting a set of weighting vectors for more than two objectives is still unresolved and *ad-hoc* methods are predominantly used. In the present work, a novel concept is introduced which we call generalized decomposition. Generalized decomposition enables the analyst to adapt the generated distribution of Pareto optimal points, according to the preferences of the decision maker. Also it is shown that generalized decomposition unifies the three performance objectives in multi-objective optimization algorithms to only one, that of convergence to the Pareto front.

1 Introduction

Decomposition-based methods have been used traditionally in mathematical programming to solve multi-objective problems [1]. These methods use scalarizing functions to decompose a multi-objective problem into several single objective subproblems. These subproblems are defined with the help of weighting vectors. Weighting vectors are k-dimensional vectors with positive components that sum to one, where k is the number of objectives. The *location* on the Pareto front to which each subproblem will tend to converge, depends strongly on the choice of weighting vectors. Therefore the choice of an appropriate set of weighting vectors to *decompose* the multi-objective problem, will determine the distribution of the final Pareto set approximation along the Pareto front. Although a rigorous definition of what is considered a good distribution of Pareto optimal solutions does not exist, there is a consensus about the features that must be present. Firstly, assuming that a decision maker is not involved prior or during the optimisation process, the general tendency is to distribute the Pareto approximation along the entire Pareto front. A second implicit requirement is that Pareto optimal solutions are distributed evenly across the entire front. This emanates from the fact that the preference of the decision maker towards a particular region of the trade off surface is unspecified or unknown. Finally, the distance of the Pareto set approximation must be as close as possible to the true Pareto front. Convergence of the optimization algorithm is measured in terms of that distance.

R.C. Purshouse et al. (Eds.): EMO 2013, LNCS 7811, pp. 428–442, 2013.

In decomposition-based multi-objective algorithms, the first two properties mentioned above are directly controlled by the choice of weighting vectors. So, naturally there have been several suggestions as to their selection. However, the choice of weighting vectors is not independent of the scalarizing method used to decompose the multi-objective problem. Furthermore this problem is non-linear in itself, so in most instances there is no guarantee that a unique solution can be found. In this work we show that for a particular class of scalarizing functions the weighting vectors can be *optimally* calculated, given a clear definition of what is meant by *well* distributed Pareto optimal solutions. Also, knowledge of the Pareto front geometry can greatly increase the accuracy of the generated solutions, as will be explained later.

In contrast with mathematical programming, evolutionary algorithms have tended to use Pareto-based methods to deal with multi-objective problems. However, recent studies [2] strongly indicate that such methods become impractical when solving problems with more than three objectives. The reason for this is that as the dimensionality of the problem is increased, the ability of Pareto-based methods to discriminate between solutions becomes increasingly more difficult. This difficulty stems from the fact that Pareto dominance relations, the basis of Pareto-based methods, induce only a partial ordering. This means that two objective vectors can be either identical, superior or inferior with respect to the other or incomparable. In higher dimensions, the number of incomparable objective vectors increases to such levels that any meaningful selection is simply impossible [2]. An additional difficulty that Pareto-based algorithms are facing for many-objective problems is that the closer the Pareto set approximation is to the Pareto optimal front, the lower the probability that a superior solution will be generated [3]. These problems encountered with Pareto-based algorithms has led some researchers to revisit traditional methods, such as scalarizing functions, to extend algorithms to multi-objective problems. Some examples of promising evolutionary work on decomposition-based methods for multi-objective problems are due to Jaszkiewicz [4], Hughes [5] and Zhang and Li [6].

The remainder of this work proceeds as follows. In Section 2 a problem formulation is given along with some fundamental concepts. In Section 3 decomposition methods are briefly reviewed as well as the methods currently employed in selecting the set of weighting vectors. Generalized decomposition is introduced in Section 4 and in Section 5 the impact that the choice of weighting vectors has on the distribution of points on the Pareto front is studied. In Section 6 a generic reference front geometry is suggested. This work is summarized and concluded in Section 7.

2   Problem Statement and Definitions

A multi-objective problem (MOP) is defined as,

$$\min_{\mathbf{x}} \mathbf{F}(\mathbf{x}) = (f_1(\mathbf{x}), f_2(\mathbf{x}), \ldots, f_k(\mathbf{x}))$$
$$\text{subject to } \mathbf{x} \in \mathbf{S}, \tag{1}$$

where k is the number of scalar objective functions and $\mathbf{x}$ is the decision vector defined in the domain $\mathbf{S} \subseteq \mathbb{R}^n$. The set $\mathbf{S}$ is called the feasible region or decision space and the mapping $\mathbf{F} : \mathbf{S} \to \mathbf{Z}$ where $\mathbf{Z} \subset \mathbb{R}^k$, defines the objective space $\mathbf{Z}$. The optimization algorithm used to solve (1) depends on the properties of the function $\mathbf{F}(\cdot)$. For instance if $\mathbf{F}(\cdot)$ is non-differentiable, gradient-based methods are automatically excluded. In the present work the only assumption is that $\mathbf{F}(\cdot)$ is a continuous function of $\mathbf{x}$.

Pareto-based methods use Pareto-dominance relations [1], to induce a partial ordering in the objective space. These relations, initially introduced by Edgeworth [7] and further expanded by Pareto [8], involve element-wise vector comparison. For example for two vectors $\mathbf{a}, \mathbf{b} \in \mathbb{R}^n$, $\mathbf{a} < \mathbf{b}$ if all the elements in $\mathbf{a}$ are smaller to the corresponding elements in $\mathbf{b}$. This partial ordering, induced by the $\prec$ relation, is denoted as $\mathbf{a} \prec \mathbf{b}$, and, in the context of a minimization problem, this expression is read as: the vector $\mathbf{a}$ dominates $\mathbf{b}$. For a more complete treatment of Pareto-dominance relations the reader is referred to [1].

Definition 1. *Given a set of decision vectors* $\mathbf{A}$ *for which* $\mathbf{F}(\mathbf{A}) \subset \mathbf{Z}$*, the* **non-dominated set**[1] *is defined as* $\mathcal{P} = \{\mathbf{z} : \nexists \tilde{\mathbf{z}} \prec \mathbf{z}, \forall \tilde{\mathbf{z}} \in \mathbf{F}(\mathbf{A})\}$*. If* $\mathbf{A}$ *is the entire feasible region in the decision space,* $\mathbf{S}$*, then the set* $\mathcal{P}$ *is called the* **Pareto optimal set** *or* **Pareto Front** *(PF). A vector* $\mathbf{z} \in \mathbf{Z}$ *is referred to as* **objective vector***.*

Definition 2. *The* **ideal objective vector**, $\mathbf{z}^\star$, *is the vector with elements* $\{\inf(f_1), \ldots, \inf(f_k)\}$*.*

Definition 3. *The* **nadir objective vector**, $\mathbf{z}^{nd}$, *is the vector with elements* $\{\sup(f_1), \ldots, \sup(f_k)\}$*, subject to* f_i *be elements of objective vectors in the Pareto optimal set.*

3   Brief Review of Decomposition Methods

Decomposition methods can be classified according to the interaction with the decision maker. In the present work the focus is on *a posteriori* methods. In this paradigm the aim of the optimiser is to generate a Pareto set approximation that portrays as faithfully as possible the entire Pareto front, prior to introduction of decision maker preferences. *A posteriori* methods do have to make certain assumptions about the decision maker's preferences. One of these assumptions is that the decision maker has no particular preference towards any region of the Pareto front. Following this latter assumption, a reasonable course of action is to produce solutions across the entire front, if possible. Furthermore, this assumption can be used to infer the definition of a *good* distribution of solutions on the front. For instance, some researchers assume that uniformly distributed[2] solutions are preferable [4], while others advocate evenly distributed[3] solutions [6].

---

[1] Or **Pareto Front approximation**.
[2] Distributed according to the uniform distribution.
[3] Even here refers to a distribution of points whose mean distance variance is small.

However a clear resolution of this matter is impossible as it depends on the decision maker. Nevertheless the ability to change this distribution at will can be very helpful.

3.1   Fundamental Methods

As mentioned briefly in Section 1, decomposition methods employ scalarizing functions to divide a multi-objective problem in to a set a single objective subproblems. The premise of this approach is that, upon successful optimization of all the subproblems, a Pareto set will emerge formed by the solutions of these subproblems. There are several scalarizing functions that are available to the analyst [9], however in the present work only the most common methods are discussed.

One of the simplest and perhaps most intuitive scalarizing functions is the weighted sum method [1]:

$$\min_{\mathbf{x}} \mathbf{w}^T \mathbf{F}(\mathbf{x})$$
$$\sum_{i=1}^{k} w_i = 1, \text{ and } w_i \geq 0, \tag{2}$$

where $\mathbf{w} = (w_1, \ldots, w_k)$ is referred to as weighting vector and w_i are the weighting coefficients. The weighting coefficients can be viewed as factors of relative importance of the scalar objective functions in $\mathbf{F}(\cdot)$. The issue with the weighted sum approach is that its ability to produce Pareto optimal solutions depends strongly on the convexity of the Pareto front. For instance locally concave regions of a Pareto front cannot be produced by this method [1]. However the weighted sum method is still employed in practice, and a good reason for that is that (2) preserves differentiability. That is, if the scalar functions $f_i(\cdot)$ are differentiable, then the scalar problem produced by the weighted sum method will also be differentiable.

Another family of scalarizing functions is based on the weighted metrics method [9]:

$$\min_{\mathbf{x}} \left(\sum_{i=1}^{k} w_i |f_i(\mathbf{x}) - z_i^*|^p \right)^{\frac{1}{p}}. \tag{3}$$

Here, as in (2), the weighting coefficients must be $w_i \geq 0$ and $\sum_{i=1}^{k} w_i = 1$, also $p \in [1, \infty)$. However p is usually an integer or equal to ∞. The expression (3) can be read as, minimise the weighted distance of the objective function vector F to a desired point, where the meaning of the term distance depends on the chosen norm. A potential drawback of weighted metrics based scalarizing functions is that the ideal vector, $\mathbf{z}^*$, has to be known *a priori*. However this vector can be estimated adaptively during the process of optimization [6]. For $p = \infty$ the Chebyshev scalarizing function is obtained:

$$\min_{\mathbf{x}} \|\mathbf{w} \circ |\mathbf{F}(\mathbf{x}) - \mathbf{z}^*| \|_\infty. \tag{4}$$

The ∘ operator denotes the Hadamard product which is element-wise multiplication of vectors or matrices of the same size. The key result that makes (4) very interesting is that for every Pareto optimal solution there exists a weighting vector with coefficients $w_i > 0$, for all $i = 1, \ldots, k$ [1]. Meaning that all Pareto optimal solutions can be obtained using (4). This result is quite promising, although in current practice the choice of weighting vectors is made primarily using *ad hoc* methods, see Section 3.2. Therefore direct control of the distribution of solutions on the Pareto front is virtually nonexistent.

The normal boundary intersection method (NBI) introduced by Das [10], presents another formulation of a scalarizing function. The idea in NBI is that by maximizing the distance of a vector normal to the simplex with vertices $\{\mathbf{v}_i : \mathbf{e}_i \circ \mathbf{z}^{nd}\}$, where $\mathbf{e}_i$ is a zero vector with the ith component equal to 1, a solution that is likely to be Pareto optimal is obtained. Using NBI the distribution of solutions on the Pareto front are directly related to the distribution of weighting vectors on the probability simplex[4], thus providing the analyst a clear path in distributing solutions on the Pareto front according to the needs of the decision maker, if these are known. The NBI method is stated as follows:

$$\min_{\mathbf{x}} g_{nbi}(\mathbf{x}; \mathbf{w}, \mathbf{z}^\star) = d$$
$$\text{subject to } \mathbf{z}^\star - d \cdot \mathbf{w} = \mathbf{F}(\mathbf{x}). \tag{5}$$

The equality constraint in the formulation of the NBI method in (5) has to be satisfied in some way; a method proposed by Zhang and Li [6] is the use of a penalty function approach. Therefore, an equivalent formulation of (5) is the following:

$$\min_{\mathbf{x}} g_{nbi}(\mathbf{x}; \mathbf{w}^i, \mathbf{z}^\star) = d_1 + pd_2$$
$$d_1 = \frac{\|(\mathbf{z}^\star - \mathbf{F}(\mathbf{x}))^T \mathbf{w}^i\|_2}{\|\mathbf{w}^i\|_2}, \tag{6}$$
$$d_2 = \|\mathbf{F}(\mathbf{x}) - (\mathbf{z}^\star - d_1 \mathbf{w}^i)\|_2,$$

where p is a tunable parameter which controls the relative importance of convergence, d_1, and position, d_2, in the penalty function. Unfortunately (6) has three significant drawbacks. First, the normal-boundary intersection method does not guarantee that the solutions to the subproblems will be Pareto optimal [10]. Second, NBI has to be solved using a penalty method which introduces one more parameter that has to be tuned for every problem separately. Lastly, it is unclear how this decomposition method can be scaled for problems with many objectives.

3.2   Methods for Generating Weighting Vectors

To solve multi-objective problems using a decomposition method, a set of weighting vectors has to be selected based on the criteria explained in the introduction

---

[4] The simplex with vertices $\mathbf{e}_i$ for all $i = 1, \ldots, k$, is commonly known as the probability simplex.

of Section 3, or perhaps other considerations pertaining to a particular problem. However the real interest is not actually in the weighting vectors but the Pareto optimal solutions that will result by solving the corresponding subproblems generated by the set of weighting vectors. So the question is how to select the weighting vectors in such a way that the desired distribution of Pareto optimal solutions is generated. This question is primarily addressed by two methods.

The first, is to generate a set of weighting vectors that are evenly spaced. This is achieved by discretising every dimension of the objective space so that every weighting coefficient is allowed to assume every value within the set,

$$\left\{ \frac{0}{H}, \frac{1}{H}, \ldots, \frac{H}{H} \right\}, \tag{7}$$

subject to $\sum_{i=1}^{k} w_i = 1$. This approach, first seen in [11, pp. 234], has been adopted by Das [10] for use in NBI where a method to generate weighting vectors for an arbitrary number of objectives is also presented. So for a two objective problem and for $H = 2$ the set of weighting vectors is, $\{(0,1),(0.5,0.5),(0,1)\}$. Although this method seems effective when combined with a normal boundary intersection scalarizing function and perhaps others, its use with the Chebyshev scalarizing function does not produce Pareto optimal solutions that are evenly spaced nor uniformly distributed. This can be seen in [6], and is further explored in Section 5.

The second approach in generating a set of weighting vectors is due to Jaszkiewicz [4]. The idea is to generate a set of weighting vectors that are uniformly distributed on the probability simplex. The assumption is that for a uniformly distributed set of weighting vectors, the corresponding solutions of the associated subproblems, will be uniformly distributed on the Pareto front. To generate a set of weighting vectors according to the suggestions in [4], the following equation can be used as many times as the required size of the weighting vector set,

$$\mathbf{w} = \{w_1, \ldots, w_k\},$$

$$w_i = 1 - \sum_{m=1}^{i-1} w_m - (\mathcal{U}(0,1))^{i-k}, \text{ for all } i = 1, \ldots, k. \tag{8}$$

Here, $\mathcal{U}(0,1)$ is a sample from the uniform distribution in the domain $[0,1]$.

Most other methods either employ the paradigm presented by Das [10], that is, to evenly space the weighting vectors, or by Jaszkiewicz [4] where the weighting vectors are generated at random. What is shown in the present work is that neither approach has the capacity to yield satisfactory results in comparison to the implicit requirements stated above.

4   Generalized Decomposition

As mentioned in Section 3, decomposition methods have two key components: first, the scalarizing function and, second, a set of weighting vectors.

The argument in the present work is that the choice of weighting vectors is very important with respect to the three main objectives of multi-objective optimization. Namely, convergence to the Pareto front, coverage of the entire front and a well distributed Pareto optimal set. All these aspects are directly controlled by the choice of the set of weighting vectors that is used to decompose a multi-objective problem to a set of single objective subproblems. A method, which we refer to as *generalized decomposition*, is presented below that provides an exact solution to the choice of this set of weighting vectors. The version presented here is based on the Chebyshev scalarizing function, due to its guarantee of producing a Pareto optimal solution for every weighting vector [1]. Extension of other scalarizing functions is left for future work. Two other interesting works elaborating on this issue are [12, 13].

4.1 Optimal Selection of the Weighting Vector Set

First, it must be clarified what is meant by *optimal* selection of the weighting vector set. The meaning of the term *optimal* in the present context is that, given a clear mathematical definition of what a well distributed Pareto optimal set is and a way to measure the quality of a candidate set against this definition, then, by using generalized decomposition this quality measure can be maximised. This is subject to some prior information as explained later.

Starting with the Chebyshev scalarizing function as defined in (4) and given a set of weighting vectors, a multi-objective optimization problem can be decomposed to N subproblems as:

$$\min_{\mathbf{x}} g_\infty(\mathbf{x}, \mathbf{w}^s, \mathbf{z}^\star) = \|\mathbf{w}^s \circ |\mathbf{F}(\mathbf{x}) - \mathbf{z}^\star|\,\|_\infty$$

$$\forall s = \{1, \ldots, N\}, \tag{9}$$

$$\text{subject to } \mathbf{x} \in \mathbf{S},$$

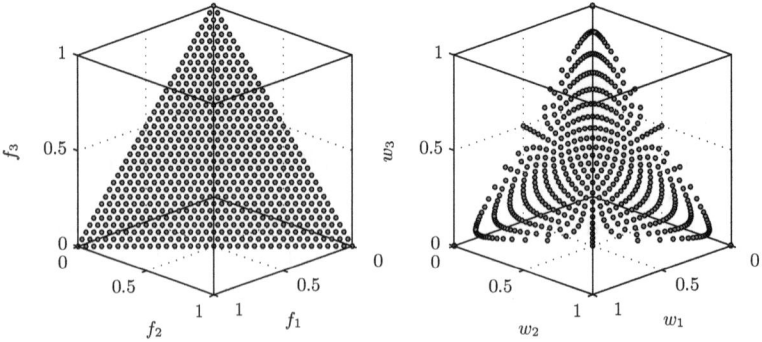

Fig. 1. A reference Pareto front with linear geometry (left) and the corresponding optimal weighting vector set (right)

with $w_i > 0$ for all $i = 1, \ldots, k$ and $\sum_{i=1}^{k} w_i = 1$, for all $\mathbf{w}^s$. For simplicity, let us assume that the ideal vector is $\mathbf{z}^* = (0, \ldots, 0)$ and that the scalar objective functions are normalized in the range $[0, 1]$. This normalization implies that the nadir vector, $\mathbf{z}^{nd}$, is known. Then the question that needs to be answered for a Pareto optimal decision vector $\tilde{\mathbf{x}}$, is whether a weighting vector $\tilde{\mathbf{w}}$ exists, for which the following condition holds,

$$\|\tilde{\mathbf{w}} \circ \mathbf{F}(\tilde{\mathbf{x}})\|_\infty \leq \|\mathbf{w} \circ \mathbf{F}(\tilde{\mathbf{x}})\|_\infty$$
$$\tilde{\mathbf{w}}, \mathbf{w} \in \mathcal{W}, \mathbf{F}(\tilde{\mathbf{x}}) \in \mathcal{P}, \tag{10}$$

where $\mathcal{W}$ is the convex set defined by the vertices $\{\mathbf{e}_i : i = 1, \ldots, k\}$. If such a solution exists, it can be obtained by the solution of the following mathematical program,

$$\min_{\mathbf{w}} \|\mathbf{w} \circ \mathbf{F}(\mathbf{x})\|_\infty,$$

$$\text{subject to } \sum_{i=1}^{k} w_i = 1, \tag{11}$$

$$\text{and } w_i \geq 0, \forall i \in \{1, \ldots, k\}, \mathbf{F}(\mathbf{x}) \geq 0.$$

Notice that in (11), the optimization is with respect to $\mathbf{w}$. In this setting, $\mathbf{F}(\mathbf{x})$ is simply a linear transformation of the vector $\mathbf{w}$, which means that since the weighting vector is convex, then the transformed vector is also part of a convex set. Additionally all norms preserve convexity, hence the problem stated in (11) is convex. Subsequently there is a guarantee that a solution $\tilde{\mathbf{w}}$ exists [14] and hence this solution will satisfy (10). The decomposition method that selects the weighting vector set using (11) we call generalized decomposition.

The assumption in (11) is that there exists a reference Pareto set, $\mathcal{P}_r$, that exhibits the desired properties described in Section 3. Next the mathematical program in (11) is solved for every vector in the set $\mathcal{P}_r$, thus obtaining the optimal weighting vector set. This weighting vector set, in combination with the Chebyshev scalarizing function (9), can then be used with any optimization algorithm, in order that a Pareto front with the desired properties to be obtained. An example of the application of generalized decomposition to a reference Pareto front is seen in *Fig. (1)*. Any convex optimization problem solver can be used for (11), however in the present work CVXGEN [15] is used as it is of several orders of magnitude faster than any other solver. Some alternatives can be found in [16].

4.2   Practical Considerations

The problem that becomes evident with generalized decomposition is that solutions at the extremities of the Pareto front, that is Pareto optimal points for which one of the objective functions is very close to 0, seem to be difficult to obtain. Although this situation is rare in practice, namely it is unusual that one of the scalar objectives in the objective vectors be reduced to zero, nevertheless it is important that the reasons behind this behaviour are understood. The cause

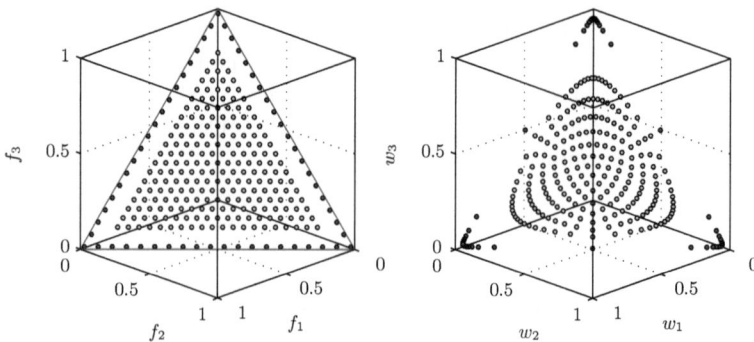

Fig. 2. A reference Pareto front with linear geometry (left) and the corresponding optimal weighting vector set (right)

of this behaviour is linked to the fact that a scalarizing function is used and that the weighting vectors that represent solutions with one 0 component are unavoidably weighting vectors with zero components everywhere except at the location of the 0. For example, in *Fig. (2)*, the left graph illustrates a linear Pareto front. The solutions that lie on the $f_1 f_2$-plane have zero f_3 component, so they are of the form $(a, b, 0)$. Consequently, in this scenario, the mathematical program described in (11), will yield a weighting vector $\mathbf{w} = (0, 0, 1)$ for all solutions of the form $(a, b, 0)$. The same reasoning applies for solutions on the $f_1 f_3$-plane and $f_2 f_3$-plane. This behaviour is also present for problems with many objectives. In practical terms this prevents generalized decomposition from obtaining solutions of this type. This is directly linked to the fact that when using the Chebyshev scalarizing function, although there is a guarantee that every Pareto optimal solution is obtainable via some weighting vector, there is no guarantee that these solutions will not be weakly Pareto optimal [1, pp. 99].

A solution to this is to modify the Chebyshev scalarizing function and, by extension, generalized decomposition. Therefore the problem in (9) can be restated as [1, pp. 101]:

$$\min_{\mathbf{x}} g_\infty(\mathbf{x}, \mathbf{w}^s, \mathbf{z}^\star) = \left\| \mathbf{w}^s \circ \left(|\mathbf{F}(\mathbf{x}) - (\mathbf{z}^\star - \epsilon)| + \rho \sum_{i=1}^{k} |f_i(\mathbf{x}) - (z_i^\star - \epsilon)| \right) \right\|_\infty$$

$$\forall s = \{1, \ldots, N\},$$

$$\text{subject to } \mathbf{x} \in \mathbf{S},$$

$$\tag{12}$$

where ρ and ϵ are sufficiently small scalars. Assuming that the scalar objective functions f_i are normalized in the range $[0, 1]$, the extra term in (12) will be constant for all solutions in the case of a linear Pareto front geometry. For different Pareto front geometries, it will vary within bounds, even if all objectives are normalized. However, if this is applied directly in generalized decomposition it will have a distorting effect on the relative distances of the points of the reference Pareto front. To avoid this, therefore, preserving the desired distribution properties present in the reference front, generalized decomposition is restated as follows:

$$\min_{\mathbf{w}} \|\mathbf{w} \circ (\mathbf{F}(\mathbf{x}) + \rho \cdot C(k))\|_{\infty},$$

$$\text{subject to } \sum_{i=1}^{k} w_i = 1, \tag{13}$$

$$\text{and } w_i \geq 0, \forall i \in \{1, \ldots, k\}, \mathbf{F}(\mathbf{x}) \geq 0.$$

where ρ is a small scalar, as in (12), and $C(k)$ is a linear monotone increasing function of the number of objectives k. Intuitively, the effect of the $\rho \cdot C(k)$ term is that it shifts the reference Pareto front slightly. This, in extension, eradicates solutions that have identically zero components and preserves the relative position of solutions in the reference Pareto front. The penalty for this modification is that all resulting solutions using the weighting vector set produced by (13), will be slightly closer to one another. This effect is directly controlled by ρ and $C(k)$, and can be as small as the machine precision allows for.

An alternative to the modification of generalized decomposition seen in (13), is to simply remove the solutions in the reference Pareto front that have one zero component. This way the original definition, seen in (11), can be used. This can simplify the task, since the extra parameters ρ and $C(k)$ become unnecessary. However this method reduces the number of solutions, N, that are eventually obtained by the optimization, which may be undesirable.

5   The Effect of Weighting Vector Choice in Many Objective Problems

In Section 4 it is stated that the choice of the weighting vector set is very important, and that this set directly controls the distribution of produced Pareto optimal points by a multi-objective optimization algorithm. To test this hypothesis, firstly a definition and a measure of well distributed Pareto optimal solutions is required. A measure that is in common use for evenly distributed points on k-dimensional manifolds is the Riesz kernel, or s-energy [17], defined as:

$$E(\mathbf{Z}; s) = \sum_{1 \leq i \leq j \leq N} \|\mathbf{z}_i - \mathbf{z}_j\|^{-s}, \ s > 0$$

$$\mathbf{z} \in \mathbb{R}^k, \text{ and, } \mathbf{Z} = \{\mathbf{z}_i : i = 1, \ldots, N\}. \tag{14}$$

It has been shown that for a k-dimensional manifold the s-energy is minimized when the distribution of points on that manifold is even, if $s \geq k$ [17]. Therefore, since the Pareto front of a k-objective problem is at most a $(k-1)$-dimensional manifold [6], the s parameter in the s-energy metric used for the following experiment is set to $k - 1$, see Table 1. Generalized decomposition is compared with the methods suggested by [10], and later used by [6] - that is, evenly distributed weighting vectors, and the method suggested by Jaszkiewicz [4], namely the selection of a weighting vector set generated according to (8). The results, shown in *Fig. (3)*, are obtained according to the following procedure:

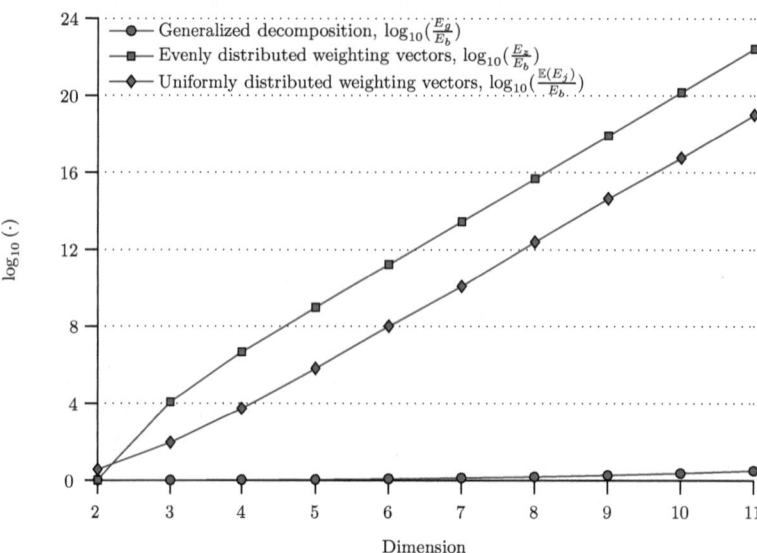

Fig. 3. The $\log_{10}$ energy ratio of Pareto optimal solutions obtained according to the three methods for generating weighting vectors

Table 1. The number of objective vectors, N, for constant H used in the experiment seen in *Fig. (3)*

Obj. #	2	3	4	5	6	7	8	9	10	11
s	1	2	3	4	5	6	7	8	9	10
H	8	8	8	8	8	8	8	8	8	8
N	8	36	120	330	792	1716	3432	6435	11440	19448
$\rho \cdot C(k)$	0.001	0.002	0.003	0.004	0.005	0.006	0.007	0.008	0.009	0.01

Step 1. A Pareto front of linear geometry has been selected for all test instances. This is mainly due the fact that it is straightforward to generate a Pareto front of this geometry with the optimal distribution of solutions, so that these can be used as a reference. The way that this reference front has been generated is identical to the generation of weighting vectors described in Section 3.2. For an example of a Pareto front with this geometry in 3 dimensions, see *Fig. (1)*. This enables a fair comparison with the scheme employed by Zhang and Li [6]. The number of solutions, N, generated in every dimension, is controlled by the H parameter (see Section 3.2). Since this parameter can be seen as the number of subdivisions per dimension, it has been kept constant, see Table 1, for all dimensions. The idea of this setting is to isolate only the effect that the dimensional increase has on the s-energy, and, by extension, the distribution of solutions on the Pareto front.

Step 2. For every dimension, a set of weighting vectors is generated according to the suggestions in [4] and [6]. Since the method suggested by Jaszkiewicz [4] is generating weighting vectors according to the uniform distribution, the s-energy calculated as described in the next step, is averaged over 50 independent weighting vector sets of size N, in each case. The weighting vector set for generalized decomposition is generated according to (13), with $\rho \cdot C(k)$ set as seen in Table 1.

Step 3. A benefit of using a linear Pareto front geometry for test purposes, is that the solutions that will minimize the subproblems defined by the weighting vector set can be directly calculated by solving the following mathematical program:

$$\min_{\mathbf{F(x)}} \| \mathbf{F(x)} \circ \tilde{\mathbf{w}} \|_\infty,$$

$$\text{subject to } \sum_{i=1}^{k} f_i = 1, \tag{15}$$

$$\text{and } f_i \geq 0 \, , \forall \, i \in \{1, \dots, k\}.$$

Note that (15) is a convex problem for the same reasons described in Section 4.1. Therefore, using (15) and a set of weighting vectors, the s-energy can be calculated for the resulting Pareto set. A reference best case energy using the actual Pareto front, E_b, is calculated for every problem instance.

Step 4. The $\log_{10}$ of the ratio of the obtained s-energy (expected energy for [4]) according to every method with the base energy E_b is calculated for all objectives.

From the results shown in *Fig. (3)*, it is apparent that generalized decomposition can follow very closely the desired distribution of solutions in the Pareto set. Additionally, the difference with alternative methods is striking, namely in the range of several orders of magnitude for problems with 3 or more objectives. These results refute the hypothesis that by selecting an evenly distributed set of weighting vectors an evenly distributed Pareto front can be obtained. Furthermore, it is shown that the method proposed by Jaszkiewicz [4], performs consistently better compared to evenly distributed weighting vectors.

The increasing ratio of the s-energy produced by solutions selected using evenly distributed weighting vectors can provide an explanation for the reason that MOEA/D [6] and derivative algorithms seem to perform well in many-objective problems. Namely, for increasing number of objectives MOEA/D-based algorithms find solutions that are more clustered. This means that, relative to the entire Pareto front *area*, such algorithms only focus on a very small part.

6   Reference Pareto Front

A limitation of generalized decomposition seems to be that, since a reference Pareto set is needed to generate the optimal weighting vectors, if that reference is unavailable due to lack of information about the Pareto front geometry, then

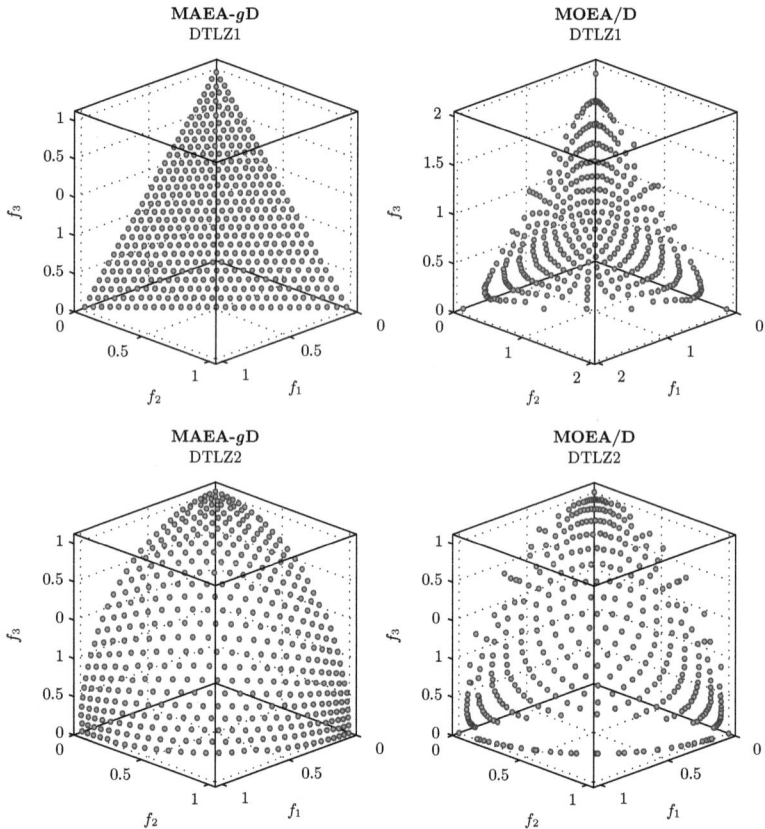

Fig. 4. Attained Pareto optimal front for the DTLZ1 and DTLZ2 3-objective problems by MADE-gD and MOEA/D

the usefulness of the method is restricted. However, this is not entirely true. In the presence of a reference Pareto optimal set, the precision that can be achieved, is exceptional. However, even if a linear Pareto front is used to generate the weighting vectors, very good results can still be obtained regardless of the shape of the true Pareto front. Good in the sense that the distribution of solutions on the Pareto front is close to even. This is because the linear Pareto front geometry seems to be in the middle ground, with respect to the shift in location of the weighting vectors, of concave and convex geometries. Although this hypothesis seems intuitive, it is not clear to what extent it is valid. Further investigation of that matter is left for future research.

To demonstrate that a good distribution of solutions can be obtained using a linear Pareto front geometry as reference for generalized decomposition, we used a revised version of MOEA/D [6] with the neighbourhood distance measured in objective space, instead of the weighting space and weighting vectors generated according to:

– Generate N, evenly distributed points on the probability simplex according to the method described in [10]. Subsequently use generalized decomposition, (11) or (13), to generate the weighting vector set. The evenly distributed points are used as the reference Pareto front.

We refer to this algorithm as a many-objective evolutionary algorithm based on generalized decomposition (MAEA-gD). This is because, as illustrated in Section 5, generalized decomposition scales very well to many objectives. The results are shown in *Fig. (4)*, for the 3-objective instances of the test problems DTLZ1 and DTLZ2 [18].

7   Conclusion

A new concept is proposed, which we refer to as generalized decomposition (gD). Using gD the optimal distribution of solutions across the Pareto front can be achieved, given the geometry of the front is known *a priori*. Furthermore, the obtained results suggest that the method maintains its favourable qualities even for many objectives. It has been shown that gD (see *Fig. (3)*) performs best in comparison with other methods for selecting the weighting vectors. Therefore, selection of weighting vectors in an *ad-hoc* manner can no longer be justified.

Furthermore, gD is not limited to producing evenly distributed solutions. Given a definition and a measure of what is meant by a *well* distributed Pareto set, generalized decomposition can produce optimal results according to that definition. However the assumption that knowledge of the Pareto front geometry is available is to an extent restrictive. For this reason, if the actual Pareto front geometry is unknown, it is suggested that the weighting vector set be produced using a linear Pareto front geometry as the reference front for generalized decomposition. Some future research directions can be the extension of generalized decomposition to other interesting scalarizing functions. Additionally, an adaptive identification of the front geometry seems possible, as it has been observed that the general shape of the front is formed well before all the solutions converge. Such a method would uncouple the dependence of gD on prior knowledge of the Pareto front geometry. Code for generalized decomposition and MAEA-gD is available at: http://ioannis-giagkiozis.staff.shef.ac.uk.

References

1. Miettinen, K.: Nonlinear Multiobjective Optimization, vol. 12. Springer (1999)
2. Ishibuchi, H., Tsukamoto, N., Nojima, Y.: Evolutionary Many-Objective Optimization: A Short Review. In: IEEE Congress on Evolutionary Computation, pp. 2419–2426 (2008)
3. Takahashi, R., Carrano, E., Wanner, E.: On a Stochastic Differential Equation Approach for Multiobjective Optimization up to Pareto-Criticality. In: Takahashi, R.H.C., Deb, K., Wanner, E.F., Greco, S. (eds.) EMO 2011. LNCS, vol. 6576, pp. 61–75. Springer, Heidelberg (2011)

4. Jaszkiewicz, A.: On the Performance of Multiple-Objective Genetic Local Search on the 0/1 Knapsack Problem - A comparative Experiment. IEEE Transactions on Evolutionary Computation 6, 402–412 (2002)
5. Hughes, E.: Multiple Single Objective Pareto Sampling. In: Congress on Evolutionary Computation, vol. 4, pp. 2678–2684. IEEE (2003)
6. Zhang, Q., Li, H.: MOEA/D: A Multiobjective Evolutionary Algorithm Based on Decomposition. IEEE Transactions on Evolutionary Computation 11, 712–731 (2007)
7. Edgeworth, F.: Mathematical Psychics: An Essay on the Application of Mathematics to the Moral Sciences. Number 10. CK Paul (1881)
8. Pareto, V.: Cours D'Économie Politique (1896)
9. Miettinen, K., Mäkelä, M.: On Scalarizing Functions in Multiobjective Optimization. OR Spectrum 24, 193–213 (2002)
10. Das, I., Dennis, J.: Normal-Boundary Intersection: An Alternate Method for Generating Pareto Optimal Points in Multicriteria Optimization Problems. Technical report, DTIC Document (1996)
11. Vira, C., Haimes, Y.: Multiobjective Decision Making: Theory and Methodology, vol. 8. North-Holland (1983)
12. Jiang, S., Cai, Z., Zhang, J., Ong, Y.S.: Multiobjective Optimization by Decomposition with Pareto-Adaptive Weight Vectors. In: International Conference on Natural Computation, vol. 3, pp. 1260–1264 (2011)
13. Gu, F., Liu, H., Tan, K.: A Multiobjective Evolutionary Algorithm Using Dynamic Weight Method. International Journal of innovative Computing, Information and Control 8, 3677–3688 (2012)
14. Boyd, S., Vandenberghe, L.: Convex Optimization. Cambridge University Press (2004)
15. Mattingley, J., Boyd, S.: CVXGEN: A Code Generator for Embedded Convex Optimization. Optimization and Engineering, 1–27 (2012)
16. Grant, M., Boyd, S.: CVX: Matlab Software for Disciplined Convex Programming (2008)
17. Hardin, D., Saff, E.: Discretizing Manifolds via Minimum Energy Points. Notices of the AMS 51, 1186–1194 (2004)
18. Deb, K., Thiele, L., Laumanns, M., Zitzler, E.: Scalable Multi-Objective Optimization Test Problems. In: Congress on Evolutionary Computation, vol. 1, pp. 825–830 (2002)

Evenly Spaced Pareto Front Approximations for Tricriteria Problems Based on Triangulation

Günter Rudolph[1], Heike Trautmann[2],
Soumyadip Sengupta[3], and Oliver Schütze[4]

[1] Fakultät für Informatik, Technische Universität Dortmund, Germany
Guenter.Rudolph@tu-dortmund.de
[2] Fakultät Statistik, Technische Universität Dortmund, Germany
trautmann@statistik.tu-dortmund.de
[3] Jadavpur University, Kolkata, India
senguptajuetce@gmail.com
[4] Departamento de Computación, CINVESTAV-IPN, Mexico D.F., Mexico
schuetze@cs.cinvestav.mx

Abstract. In some technical applications like multiobjective online control an evenly spaced approximation of the Pareto front is desired. Since standard evolutionary multiobjective optimization (EMO) algorithms have not been designed for that kind of approximation we propose an archive-based plug-in method that builds an evenly spaced approximation using averaged Hausdorff measure between archive and reference front. In case of three objectives this reference font is constructed from a triangulated approximation of the Pareto front from a previous experiment. The plug-in can be deployed in online or offline mode for any kind of EMO algorithm.

Keywords: multiobjective optimization, evolutionary multiobjective algorithm, evenly spaced Pareto front approximation, averaged Hausdorff measure, triangulation.

1 Introduction

In multiobjective optimization several (conflicting) objectives have to be optimized simultaneously. Evolutionary multiobjective optimization algorithms (EMOA) have proven to be efficient black-box solvers for these kind of problems. The underlying concept is to generate a finite size approximation of the true Pareto front of the optimization problem which simultaneously maximizes the proximity to the true front (convergence) and ensures a sufficiently high spread of solutions. Several performance indicators (PIs) exist for assessing the quality of a Pareto front approximation [1]. Indicator-based EMOAs internally aim at optimizing a specific PI in the course of the algorithm run. A prominent example is the SMS-EMOA [2] which is based on the dominated hypervolume indicator. Much effort is spent on analyzing the characteristics of solution sets of μ points that minimize specific indicators [3,4] which function as benchmark sets for indicator based EMOA.

R.C. Purshouse et al. (Eds.): EMO 2013, LNCS 7811, pp. 443–458, 2013.

Here, we reverse this approach. We start from the specification of a desired characteristic of the Pareto front approximation for which we would like to select an appropriate PI and design an EMOA which aims at generating Pareto front approximations with the desired characteristics. Specifically, we are interested in generating Pareto front approximations with evenly spaced points which are e.g. required in multiobjective control problems [5]: In case of online control problems it is not possible to fix the control behavior in advance (i.e., offline); rather, it is necessary to gain additional information from the running system before the control behavior can be set. Since the optimization of the control usually cannot be done while the system is running, an offline multiobjective optimization taking into account all possible additional information leads to a reference set of (almost) optimal operating states. These reference states represent the best solutions depending on the additional information gained in online mode. When the additional information changes while the system is running the set of reference states immediately delivers the best possible operating state that is now possible. It may happen that old and new operating points are far apart. This may lead to a big discontinuous change in the operating conditions which in turn may endanger the system's stability. But if there is an evenly spaced approximation of the Pareto set (i.e., the reference set) it is possible to construct a trajectory from old to new operating state with gentle transitions between consecutive intermediate operating points.

For this purpose we make use of the averaged Hausdorff distance Δ_p [6] which measures convergence to and spread of the Pareto front approximation simultaneously. Recently, an algorithmic concept based on optimizing Δ_p was introduced in [7]. The core of this concept is an external archiving strategy that aims for low Δ_p - values using evenly spaced reference fronts sequentially constructed using the current Pareto front approximation of the EMOA. The approach proved to be very successful in generating evenly spaced Pareto front approximations for biobjective problems [7]. In [8], the concept was generalized with special focus on three-objective problems. However, the required sequential generation of reference fronts for the Δ_p - computation in the course of the generations for more than two dimensions is far from straightforward. Therefore, in [8] the archiving strategy is transferred to two dimensions via multi-dimensional scaling. However, although the proposed Δ_p-EMOA was highly successful in generating evenly spaced Pareto front approximations for tricriteria problems and outperformed state-of-the-art EMOA as well as specialized EMOA for this task, it is computationally highly complex.

In this paper, we circumvent the transformation in two-dimensional space and present a sophisticated procedure using specialized triangulation and boundary detection concepts for approximating the 3D-surface of the considered Pareto front approximations. Furthermore, the archive update procedure is extremely accelerated due to a sophisticated strategy for updating Δ_p.

Section 2 gives an overview about multiobjective optimization. The algorithmic framework is then introduced in section 3 followed by a detailed description of the proposed reference front construction in section 4. The results of the conducted experiments are described in section 5 and conclusions together with perspectives for further research are presented in section 6.

2 Multiobjective Optimization

In the following we consider unconstrained multiobjective optimization problems (MOPs) of the form $\min\{f(x) : x \in \mathbb{R}^n\}$ where $f(x) = (f_1(x), \ldots, f_d(x))'$ is a vector-valued mapping with $d \geq 2$ objective functions $f_i : \mathbb{R}^n \to \mathbb{R}$ for $i = 1, \ldots, d$ that are to be minimized simultaneously. The optimality of a MOP is defined by the concept of *dominance* [9].

Definition 1.
Let $u, v \in F \subseteq \mathbb{R}^d$ where F is equipped with the partial order $\preceq$ defined by $u \preceq v \Leftrightarrow \forall i = 1, \ldots d : u_i \leq v_i$. If $u \prec v \Leftrightarrow u \preceq v \wedge u \neq v$ then v is said to be dominated by u. An element u is termed nondominated *relative to $V \subseteq F$ if there is no $v \in V$ that dominates u. The set $\mathsf{ND}(V, \preceq) = \{u \in V \mid \nexists v \in V : v \prec u\}$ is called the* nondominated set *relative to V.*

If $F = f(X)$ is the objective space of some MOP with decision space $X \subseteq \mathbb{R}^n$ and objective function $f(\cdot)$ then the set $F^* = \mathsf{ND}(f(X), \preceq)$ is called the Pareto front *(PF). Elements $x \in X$ with $f(x) \in F^*$ are termed* Pareto-optimal *and the set X^* of all Pareto-optimal points is called the* Pareto set *(PS).*

Moreover, for some $X \subseteq \mathbb{R}^n$ and $f : X \to \mathbb{R}^d$ the set $\mathsf{ND}_f(X, \preceq) = \{x \in X : f(x) \in \mathsf{ND}(f(X), \preceq)\}$ contains those elements from X whose images are nondominated in image space $f(X) = \{f(x) : x \in X\} \subseteq \mathbb{R}^d$.

Since the PS and the PF can typically not be computed analytically, one task in multi-objective optimization is to numerically detect a finite size approximation of $F^* = f(X^*)$. In this work we are particularly interested in a low distance between F and F^* and a sufficiently good spread which is ensured by the averaged Hausdorff distance Δ_p as performance indicator.

Definition 2. *The value $d_H(A, B) := \max(d(A, B), d(B, A))$ is termed the Hausdorff distance between two sets $A, B \subset \mathbb{R}^n$, where $d(B, A) := \sup\{d(u, A) : u \in B\}$ and $d(u, A) := \inf\{\|u - v\| : v \in A\}$ for $u, v \in \mathbb{R}^n$ and a vector norm $\|\cdot\|$.*

The Hausdorff distance is widely used in many fields. It has, however, certain limitations when measuring the distance of the outcome of an EMOA to the PF since outliers generated by EMOAs are punished too strongly by d_H. As a remedy, we follow the suggestion of [6] and use the *averaged* Hausdorff distance.

Definition 3. *The value $\Delta_p(A, B) = \max(GD_p(A, B), IGD_p(A, B))$ with $p > 0$,*

$$GD_p(A, B) = \left(\frac{1}{|A|} \sum_{a \in A} d(a, B)^p \right)^{1/p} \quad and \quad IGD_p(A, B) = \left(\frac{1}{|B|} \sum_{b \in B} d(b, A)^p \right)^{1/p}$$

is termed the averaged Hausdorff distance *between sets A and B.*

The indicator Δ_p can be viewed as a composition of slight variations of the Generational Distance (GD, see [10]) and the Inverted Generational Distance (IGD, see [11]). It is $\Delta_\infty = d_H$, but for finite values of p the indicator Δ_p averages (using the p-vector norm) the distances considered in d_H. Hence, as opposed to d_H, Δ_p does in particular not punish single (or few) outliers in a candidate set.

3   Algorithmic Framework

Suppose that some EMOA has generated an approximation of the Pareto front for some MOP. Typically, this approximation does not yield a finite point set in objective space that is evenly distributed. Therefore, we propose the following *online* approach:

1. Use given approximation of Pareto front to construct a reference front R with evenly spaced elements (see section 4)
2. Run your favorite EMOA with our plug-in using reference front R: as soon as an offspring is generated and evaluated put a copy into the Δ_p-archive updater
3. Print archive after termination of your favorite EMOA

The *offline* version runs as follows:

1. Use given approximation of Pareto front to construct a reference front R with evenly spaced elements
2. Run your favorite EMOA with following add-on:
   - as soon as an offspring is generated and evaluated store a copy in a file
3. After termination of your favorite EMOA:
   - put every offspring from the file into the Δ_p-archive updater sequentially
   - print archive

The idea of the update is as follows: The archive update adds a nondominated solution to the archive if the removal of an archive element leads to a smaller Δ_p distance to the reference set. In this case it removes the archive element leading to maximum improvement. Needless to say, the archive update must be realized in an efficient manner.

This naive approach takes $\Theta(|A| \cdot (|A| \cdot |R| \cdot d))$ time units, whereas the quick update version below (Alg. 2) only needs $\Theta(|A| \cdot |R| \cdot d)$ time units: calculating $d(a, R)$ takes $\Theta(|R| \cdot d)$ whereas calculating $d(r, A)$ takes $\Theta(|A| \cdot d)$ time. Therefore the first loop needs $\Theta(|A| \cdot |R| \cdot d)$ time, the second loop $\Theta(|R| \cdot |A| \cdot d)$ time, and the third loop $\Theta(|A|)$ time. Hence, in total $\Theta(|A| \cdot |R| \cdot d)$ time units are required for an update.

The construction of the reference front can be done in various ways. In case of three objectives a triangulation-based method may be used. The description of this approach is given in the subsequent section.

Algorithm 1. Δ_1-update

Input: archive set A, reference set R, new element x

1: $A = \mathsf{ND}_f(A \cup \{x\}, \preceq)$
2: **if** $|A| > N_R := |R|$ **then**
3:     **for all** $a \in A$ **do**
4:         $h(a) = \Delta_1(A \setminus \{a\}, R)$
5:     **end for**
6:     $A^* = \{a^* \in A : a^* = \mathsf{argmin}\{h(a) : a \in A\}\}$
7:     **if** $|A^*| > 1$ **then**
8:         $a^* = \mathsf{argmin}\{GD_P(A \setminus \{a\}, R) : a \in A^*\}$ {ties broken at random}
9:     **end if**
10:     $A = A \setminus \{a^*\}$
11: **end if**

---

4 Construction of Reference Front

4.1 Parallel Projection of Pareto Front

The Pareto front of problems with three objectives is a surface with dimensionality 2 embedded in $\mathbb{R}^3$. Therefore, triangulations with triangles are possible in principle, but standard triangulation algorithms require the availability of the vertex coordinates in a 2-dimensional coordinate system. The simple projection in the spirit of a multiview orthographic projection simply sets one dimension to zero. The problem with this approach is that points which seem to be close together in 2D can be far apart in 3D. For example, the points $(1, 2, -10^4)$ and $(2, 1, 10^4)$ are close in 2D if they are projected to $(1, 2, 0)$ and $(2, 1, 0)$. If these projected coordinates are used for a triangulation the resulting surface graph in 3D may be quite poor. In sets with dominated solutions it is also possible that a point is hidden by this type of projection: $(1, 2, 1)$ and $(1, 2, 5)$ are identical if they are projected to $(1, 2, 0)$ and $(1, 2, 0)$.

Since Pareto fronts consist of nondominated points we can exploit the property that no point erroneously appears close to another point if we look from such a nondominated point in direction $(1, 1, 1)$ or $(-1, -1, -1)$. Therefore, we apply a parallel projection of the Pareto front approximation in direction $(-1, -1, -1)$ onto the projection plane that is orthogonal to this direction.

The projection plane is given by $v'(x - a) = 0$ where v is the normal and a the support vector. In our case we have $v = (1, 1, 1)'$ for some $a \in \mathbb{R}^3$ (a should be the utopian or ideal point of the Pareto front approximation).

Let $\mathring{x} \in \mathbb{R}^3$ be a nondominated point that is to be projected. The projected point $\tilde{x} \in \mathbb{R}^3$ on the projection plane is the intersection of the line $x = \mathring{x} + \lambda v$ for $\lambda \in \mathbb{R}$ with the projection plane $v'(x - a) = 0$. Insertion leads to the solution

$$\tilde{x} = \mathring{x} - \frac{v'(\mathring{x} - a)}{v'v} v .$$

Although all projected points are on the projection plane, they are still embedded in $\mathbb{R}^3$. Therefore we need a change of basis to express the projected points in the $x_1 x_2$-plane (with $x_3 = 0$ for all points).

Algorithm 2. Quick Δ_1-update

Input: archive set A, reference set R, new element x

1: $A = \mathsf{ND}_f(A \cup \{x\}, \preceq)$
2: **if** $|A| > N_R := |R|$ **then**
3:     $GD_p = IGD_p = 0$
4:     **for all** $a \in A$ **do**
5:        $GD_p(a) = d(a, R)$ // GD_p contribution of archive point a
6:        $GD_p \mathrel{+}= GD_p(a)$ // add GD_p contribution of a
7:        $IGD_p^1(a) = IGD_p^2(a) = 0$ // initialize for later use
8:     **end for**
9:     **for all** $r \in R$ **do**
10:       let $a^* \in A$ such that $d(r, A) = d(r, a^*)$ // closest archive point a^*
11:       $d_1 = d(r, a^*)$ // distance to closest archive point
12:       $d_2 = d(r, A \setminus \{a^*\})$ // distance to 2nd closest archive point
13:       $IGD_p \mathrel{+}= d_1$ // add IGD_p contribution of r
14:       $IGD_p^1(a^*) \mathrel{+}= d_1$ // sum IGD_p contributions with a^* involved
15:       $IGD_p^2(a^*) \mathrel{+}= d_2$ // sum IGD_p contributions if a^* deleted
16:     **end for**
17:     $dp_{\min} = gdp_{\min} = \infty$
18:     **for all** $a \in A$ **do**
19:       $gdp = GD_p - GD_p(a)$ // value of GD_p if a deleted
20:       $igdp = IGD_p - IGD_p^1(a) + IGD_p^2(a)$ // value of IGD_p if a deleted
21:       $dp = \max\left\{\frac{gdp}{|A|-1}, \frac{igdp}{|R|}\right\}$ // Δ_1 if a deleted
22:       **if** $dp < dp_{\min} \vee (dp = dp_{\min} \wedge gdp < gdp_{\min})$ **then**
23:          $dp_{\min} = dp$ // store smallest Δ_1 seen so far
24:          $dp_{\min} = gdp$ // store smallest gdp since last improvement of $dp_{\min}$
25:          $a^* = a$ // save archive point associated with smallest Δ_1 seen so far
26:       **end if**
27:     **end for**
28:     $A = A \setminus \{a^*\}$
29: **end if**

4.2 Coordinate Transformation

Let $\check{x} \in \mathbb{R}^3$ be a point in the standard coordinate system. Let B be a matrix whose columns are the base vectors of the new basis. Then $B^{-1}\check{x}$ expresses the point $\check{x}$ in the new coordinate system.

At first, we must identify the basis of the coordinate system in which the projected points should be transformed. We define that the projection plane becomes the new x_1x_2-plane whereas the normal vector of the projection plane represents the new x_3-axis. Due to this construction the x_3-value in the new coordinate system will be zero for all points.

The basis of the new coordinate system is determined as follows: The projection plane given in normal form $v'(x - a) = 0$ can be expressed in coordinate form $x_1 + x_2 + x_3 = v'a$ or parameter form

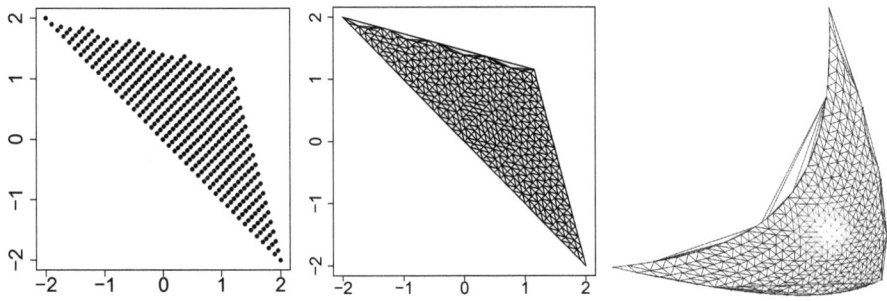

Fig. 1. Example for Viennet: 2D-Projection via parallel projection and coordinate transformation (left), resulting triangulation in 2D (middle) and 3D (right)

$$x = \begin{pmatrix} 0 \\ 0 \\ a_1 + a_2 + a_3 \end{pmatrix} + s \begin{pmatrix} 1 \\ -2 + \sqrt{3} \\ 1 - \sqrt{3} \end{pmatrix} + t \begin{pmatrix} -2 + \sqrt{3} \\ 1 \\ 1 - \sqrt{3} \end{pmatrix} = \begin{pmatrix} 0 \\ 0 \\ 3\bar{a} \end{pmatrix} + s\,b_1 + t\,b_2$$

with parameters $s, t \in \mathbb{R}$. Notice that both directional vectors b_1 and b_2 together with the normal vector $b_3 := v$ represent an orthogonal basis. Division of the base vectors by their lengths leads to an orthonormal basis b_1^0, b_2^0, b_3^0 with $b_i^0 = b_i/\|b_i\|$. Let B denote the matrix whose columns are the orthonormal base vectors above and $p = (0, 0, a_1 + a_2 + a_3)'$. Then the linear transformation $\tilde{x} = B^{-1}(x - p)$ yields points $\tilde{x}$ in the $x_1 x_2$-plane (i.e., the x_3 coordinate is always zero) from x in the projection plane. After these preparations standard triangulation algorithms in 2D can be applied. An exemplary visualization is provided in Fig. 1.

4.3 Triangulations

If we use the original 3D coordinates for the triangulation obtained from the projected and transformed 2D coordinates then it may happen that some edges (actually: triangles) are at "false" positions. The reason for this occurrence stems from the definition of triangulations [12]:

Definition 4. *A collection of simplices $\mathcal{F}$ with vertices in a finite point set A in $\mathbb{R}^d$ is termed a* triangulation *of A if*
a) all faces of simplices of $\mathcal{F}$ are in $\mathcal{F}$;
b) the intersection of any two simplices of $\mathcal{F}$ is a face of both;
c) the union of all these simplices equals the convex hull of A.   □

Since the union of all triangles equals the convex hull of the point set (see Fig. 1), these "false" edges appear if the Pareto front in its 2D projection has concave parts. These false edges must be eliminated next.

4.4 Border Detection of a Triangulation

The aim is to construct a triangulation of a three-dimensional Pareto front (approximation) based on the projection and transformation presented in

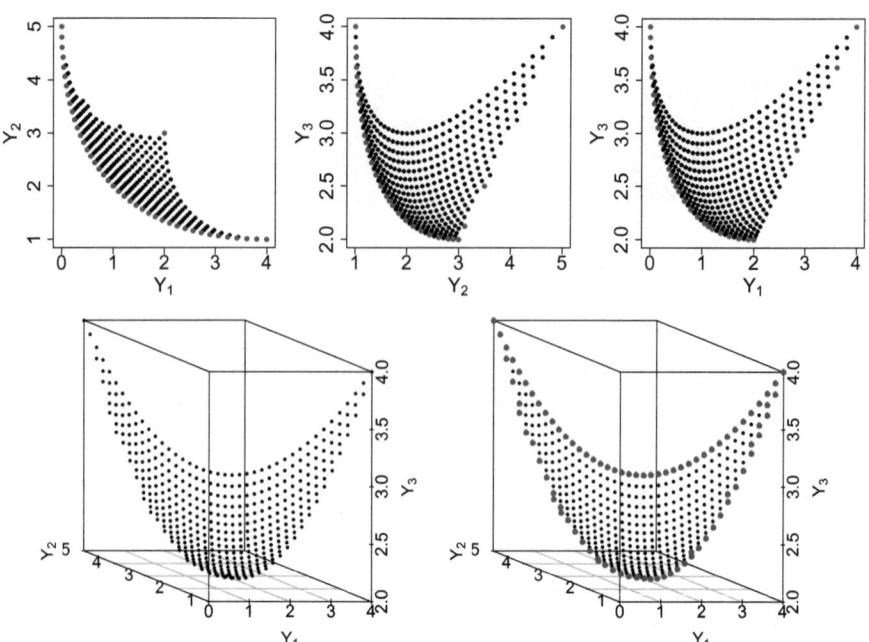

Fig. 2. Example: Detection of border points of a Pareto front approximation for Viennet. Visualized are the three steps of detecting the borders in the three 2D-projections as well as the resulting border of the overall front approximation.

sections 4.1 and 4.2. The challenge in this case is the detection of the border of the front by a procedure which appropriately handles concave regions. Existing methods for detecting so-called "concave hulls" such as the concept of α-hulls [13] heavily depend on the setting of internal parameters representing the desired degree of concavity. As we would like to avoid additional problem-specific parameters we opt for a different approach which exploits the fact that the considered set only contains nondominated points.

The border points are determined individually for the three two-dimensional subsets S^i (i=1,...,3) of the point set by the following procedure: For each edge (u,v) with $u,v \in S^i$ the neighboring vertices $N(u)$ and $N(v)$ are determined. Iff $|N(u) \cap N(v)| = 1$ then both u and v are border vertices.

Depending on the angle of vision and due to the fact that the point set solely consists of nondominated points, there will always be a convex region in this two-dimensional space for which the border will be accurately detected. Thus, by unifying the individual border points an accurate approximation of the border of the three-dimensional point set is obtained (see Fig. 2).

However, the pure knowledge of the border points alone is not sufficient. Additionally, the border edges are required in order to allow for "cutting out" the desired triangulation in the 2D-projection along the border edges. For this purpose the determined border points of the Pareto front in 3D are selected within the 2D-projection. Due to the "circular" resp. boundary structure of the

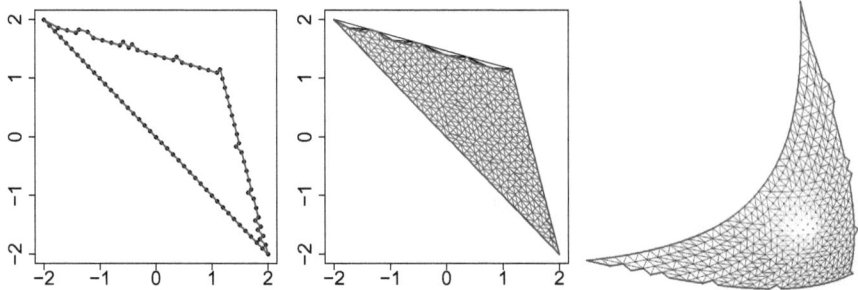

Fig. 3. Detection of border edges of a Pareto front approximation for Viennet. The border points are mapped to 2D-space by parallel projection and coordinate transformation. Left: A TSP-solver determines the path of the border edges (red). Middle / right: The triangulation is clipped to the region within the border edges (red).

resulting point set the detection of the border edges can be transferred to solving a traveling salesman problem (TSP, [14]) on these points. A TSP solver is applied to approximate the optimal tour within the considered set. As it was shown in [15] that for a quite small number of points (i.e. in the range of 25) these kind of instances are easily solvable for the 2-opt heuristic [16], it can be used as a computationally fast replacement for an exact solver such as Concorde [17]. An example is given in the left part of Fig. 3.

4.5 Clipping of False Edges

Let $\mathcal{B}$ be the collection of "true" border edges obtained in the previous step. By construction, these edges form a circuit in the triangulation that bounds a compact set in $\mathbb{R}^2$. All edges in this bordered set are true edges, whereas those outside this set are false edges.

The false edges can be identified as follows: For each border edge $b \in \mathcal{B}$ with coordinates $(x, y) \in \mathbb{R}^2 \times \mathbb{R}^2$ detect if and where the line

$$\frac{x + y}{2} + \lambda \begin{pmatrix} x_2 - y_2 \\ y_1 - x_1 \end{pmatrix} \quad \text{with } \lambda \in \mathbb{R} \tag{1}$$

orthogonal to the line through the border edge (x, y) intersects any other border edge with coordinates $(u, v) \in \mathbb{R}^2 \times \mathbb{R}^2$ on the line

$$u + \gamma (v - u) \quad \text{with } \gamma \in \mathbb{R}. \tag{2}$$

The solution (λ^*, γ^*) of the set of the two linear equations (1) and (2) reveals if there is an intersection ($0 \le \gamma^* \le 1$) and whether the normal vector of the line along edge b points inward ($\lambda^* > 0$) or outward ($\lambda^* < 0$). Next, we check for each border edge $b \in \mathcal{B}$ if there is an intersection along its normal vector in outward direction. If so, we have found false edges and the associated triangles must be removed from the triangulation (Fig. 3).

Fig. 4. Example for Viennet: Points resulting from the subdivision of the triangulation (left), k-means clustering of the points into 100 clusters (middle), reference front generation by computation of cluster centers (right)

4.6   Division of Triangles

Since we aim at the determination of $N_R = |R|$ centroids of disjunct regions C_i with almost equal size, the surface defined by the clipped triangulation must be divided in smaller triangles to assure a small variance in the area sizes of the regions C_i with $i = 1, \ldots, N_R$. For this purpose we recursively split each triangle at the midpoint of its longest side until a targeted area size is reached. The target size is determined as follows considering the consecutive step of clustering the resulting point set: Divide the size of the area defined by the clipped triangulation by the product of the number N_R of desired centroids / cluster and the number of points that should be approximately in each cluster (see Fig. 4).

4.7   Clustering

The subdivision of the triangulation into a huge number of smaller triangles results in a dense approximation of the surface (see Fig. 1 left) which is represented by the nodes of the fine grained triangulation. In the next step the desired number of evenly spaced solutions has to be determined based on this point set. As a heuristic approach we use k-means clustering [18] for this purpose with $k = N_R$. The method aims at minimizing the sum of the within-cluster sums of point-to-cluster-centroid distances over all clusters. The reference front is then formed by the resulting cluster centroids (Fig. 4 right).

5   Experiments

Experiments were conducted for assessing the performance of the proposed algorithmic concept and for comparing the results to our previous approach [8] and competitive algorithms. Both computational efficiency as well as the optimization of Δ_p are focussed. In the following, our proposed algorithm, i.e. the SMS-EMOA as a front-end combined with the new algorithmic framework

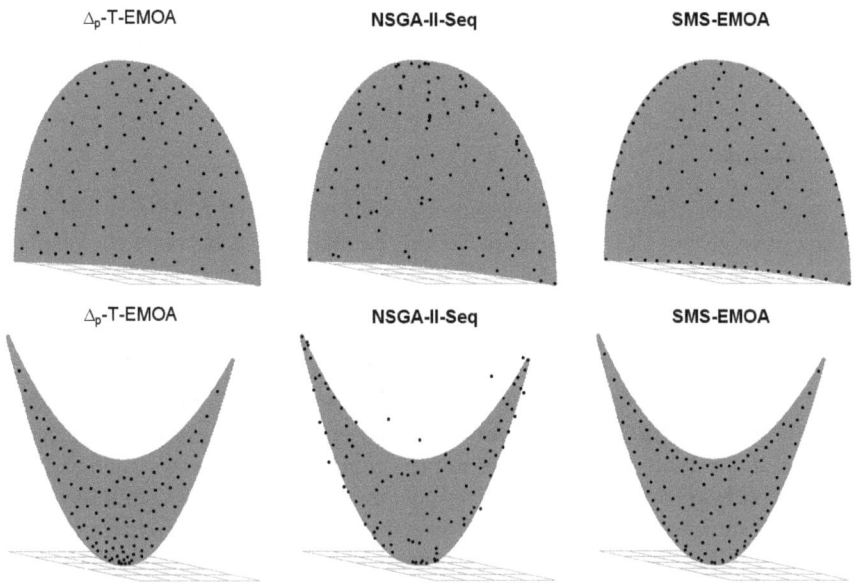

Fig. 5. Best Pareto front approximations of selected algorithms w.r.t. Δ_p for DTLZ2 (top) and Viennet (bottom). The results for DTLZ3 are very similar to DTLZ2.

and construction of the reference front, is denoted as Δ_p-T-EMOA (referring to *T*riangulation) while the previous one [8] is termed Δ_p-M-EMOA (referring to *M*ulti-Dimensional Scaling).

5.1 Setup

The experimental setup coincides with the settings of [8]. Four different test problems with different shapes of the Pareto front and multimodality character-istics are addressed, i.e. DTLZ1 (linear), DTLZ2 (concave), DTLZ3 (concave, multimodal) [19] and Viennet (convex, [20]). The Δ_p-T-EMOA is contrasted to the Δ_p-M-EMOA, the three state-of-the art algorithms SMS-EMOA [2], NSGAII [21] and MOEA/D [22] as well as a NSGAII variant NSGAII-Seq [23] specifically designed for generating evenly distributed Pareto front approximations using se-quential crowding distance updates. Two other respective NSGAII variants were omitted in this setting as they were found to be outperformed by NSGAII-Seq on all considered test functions in [8].

The algorithms were independently run ten times on each test functions for 90000 function evaluations using a population size $\mu = 100$. For Δ_p-T-EMOA, the archive size equals the population size. Therefore, the number of clusters in the k-means clustering (sec. 4.7) are set to 100, and the desired number of points per cluster was chosen as 20 based on preliminary studies. The Δ_p-Indicator is parametrized by $p = 1$ as by this means the influence of outliers is minimized and best algorithm performance results [7]. The parameter settings of the other algorithms can be taken from [8].

5.2   Results

Fig. 6 visualizes the median performance of all considered EMOA in the course of the optimization run. In order to be able to distinguish between the displayed lines, results are only shown for later algorithm stages. The visualization of the complete runs for all algorithms but Δ_p-T-EMOA can be found in [8].

Performance is assessed by calculating the Δ_p values w.r.t. a fixed reference front in the course of the optimization run. Ideal benchmark fronts are composed of the μ-optimal solution set regarding Δ_p to the true PF with μ denoting the population size of the EMOA. However, a procedure to theoretically derive such a set is not available yet, nor is the approximation of such sets straightforward. Moreover, it is yet unclear if such distributions are unique. Thus, in line with [8], we densely sampled the known true Pareto front of the considered test problems using approximately 10000 points. Proposition 7 taken from [6] justifies this approach from a theoretical perspective.

It becomes obvious that for DTLZ2, DTLZ3 and Viennet the Δ_p-T-EMOA even outperforms the Δ_p-M-EMOA which already performs significantly better than all competitors regarding Δ_p. The boxplots in Figure 6 reflecting the distribution of Δ_p values at the final EMOA generation indicate that these results are also statistically significant. Moreover, the proposed algorithmic concept is computationally much more efficient than the Δ_p-M-EMOA. The SMS-EMOA tends to be superior on DTLZ1, i.e. on the linear Pareto front, but the performance differences of Δ_p-T-EMOA, Δ_p-M-EMOA and the SMS-EMOA are not statistically significant based on the Wilcoxon rank sum test. The tendency in fact is not surprising as it is known that the optimal μ-distribution regarding the dominated Hypervolume is equally spaced on linear Pareto fronts. But it has to be kept in mind that a linear Pareto front is a very unlikely special case for practical problems.

Fig. 5 depicts the best Pareto front approximations of the Δ_p-T-EMOA, NSGAII-Seq and the SMS-EMOA. MOEA/D and NSGAII are omitted as they are outperformed by NSGAII-Seq on DTLZ2 and Viennet. The SMS-EMOA is included as it is used as the corresponding front-end of the Δ_p-T-EMOA. While the Δ_p-T-EMOA manages to generate roughly evenly spaced solutions on the Pareto front, the SMS-EMOA focusses on the knee points and accurately approximates the border of the front. We in fact exploit the latter property within the procedure for generating the required reference front for the Δ_p-T-EMOA. The PF approximations of NSGAII-Seq (as well as MOEA/D and NSGAII) do not exhibit a specific structure but rather show gaps in an unstructured manner.

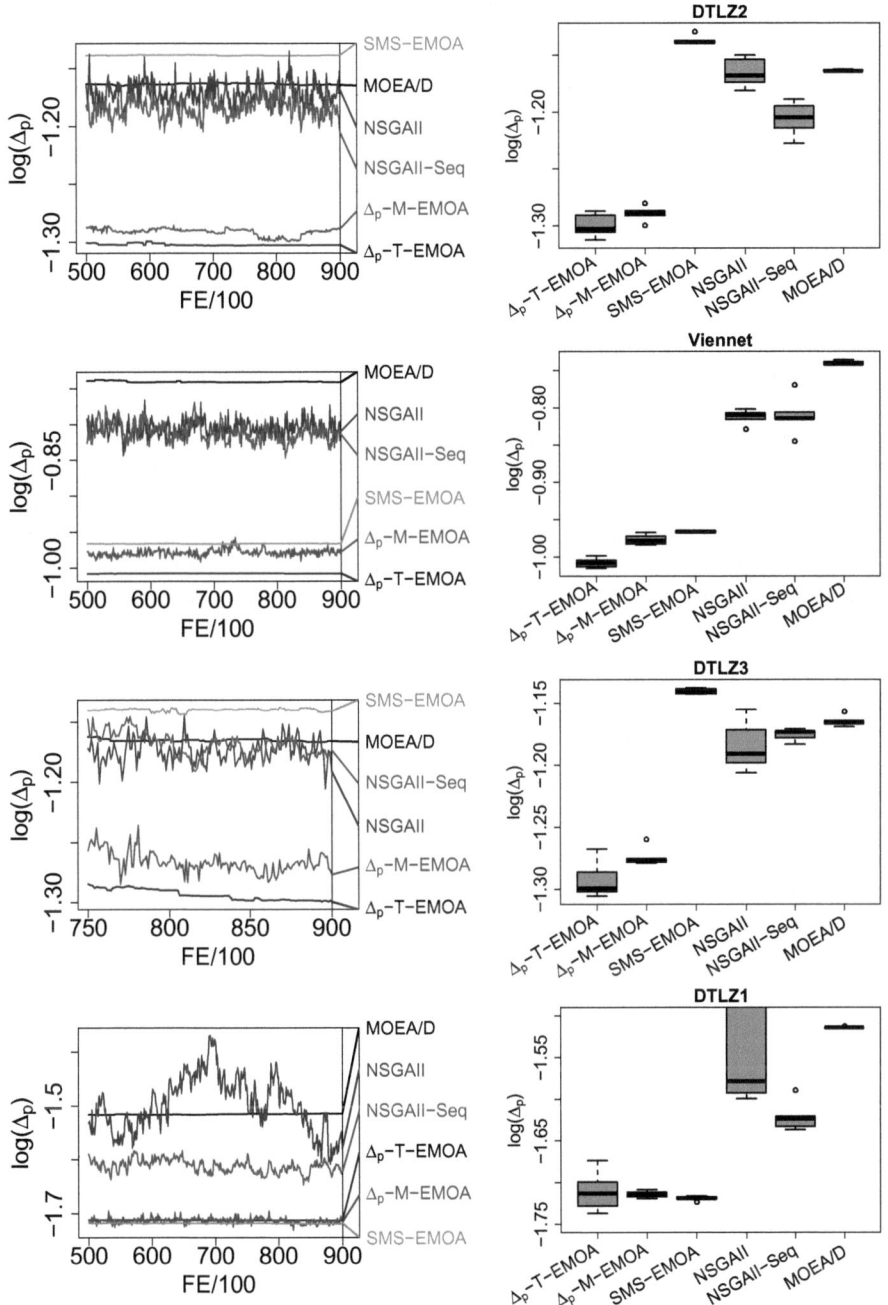

Fig. 6. Median performance results of the considered algorithms w.r.t. Δ_p for DTLZ2 (1st row), Viennet (2nd row), DTLZ3 (3rd row) and DTLZ1 (last row). Boxplots of median Δ_p-values at 90000 FE over all runs for the considered test functions are shown on the right.

6 Conclusions

The introduced algorithmic concept together with the innovative approach for generating the internally required reference front successfully evolved evenly spread Pareto front approximations within the conducted experimental studies. It even outperformed our previous approach which already showed significantly better results regarding Δ_p than competitive EMOA. Moreover, the computational efficiency is extremely increased due to a computationally fast procedure for updating the archive and the conceptual change that the construction of the reference front now is only required once during the EMOA run rather than generation-wise as previously.

However, several issues have to be addressed in future studies. First of all, a theoretical definition of "uniformity" of points on a surface is still missing and is strongly required for theoretical analysis and performance assessment of respective algorithms. An important related aspect is how to appropriately handle the true border of a point set. It could be either desired to place points "uniformly" including the exact border or not while the former case is much more complex to realize. In our approach we did not aim at exactly representing the border of the true Pareto front with the evenly spaced point set but used sophisticated border detection methods for generating the reference front. Based on the definition of uniformity optimal μ-distributions of points on a surface can be generated while the uniqueness of such point sets will have to be investigated. The theoretical derivation of optimal μ-distributions of points w.r.t. Δ_p and the related uniqueness is an open issue as well. Furthermore, the generalization of the procedure to higher dimensions will be investigated. Triangulations are not applicable in these scenarios but have to be replaced by appropriate geometric structures.

Acknowledgments. The third author acknowledges support from the DAAD WISE program. The remaining authors acknowledge support from DAAD project no. 51222288, CONACYT project no. 146776 and 174443 as well as from the DFG, project no. TR 891/5-1. Special thanks to Bernd Bischl for spending much effort on interfacing our C++ code with R.

References

1. Zitzler, E., Knowles, J., Thiele, L.: Quality Assessment of Pareto Set Approximations. In: Branke, J., Deb, K., Miettinen, K., Słowiński, R. (eds.) Multiobjective Optimization. LNCS, vol. 5252, pp. 373–404. Springer, Heidelberg (2008)
2. Beume, N., Naujoks, B., Emmerich, M.: SMS-EMOA: Multiobjective selection based on dominated hypervolume. European Journal of Operational Research 181(3), 1653–1669 (2007)
3. Auger, A., Bader, J., Brockhoff, D., Zitzler, E.: Theory of the Hypervolume Indicator: Optimal μ-Distributions and the Choice of the Reference Point. In: Foundations of Genetic Algorithms (FOGA 2009), pp. 87–102. ACM, New York (2009)

4. Brockhoff, D., Wagner, T., Trautmann, H.: On the properties of the R2 indicator. In: Soule, T., et al. (eds.) Proc. 14th Int'l Genetic and Evolutionary Computation Conference (GECCO 2012), pp. 465–472. ACM (2012)

5. Liu, G., Yang, J., Whidborne, J.: Multiobjective Optimisation and Control. Research Studies Press Ltd., Baldock (UK) (2003)

6. Schütze, O., Esquivel, X., Lara, A., Coello Coello, C.A.: Using the averaged Hausdorff distance as a performance measure in evolutionary multi-objective optimization. IEEE Transactions on Evolutionary Computation 16(4), 504–522 (2012)

7. Gerstl, K., Rudolph, G., Schütze, O., Trautmann, H.: Finding evenly spaced fronts for multiobjective control via averaging Hausdorff-measure. In: Proceedings of 8th International Conference on Electrical Engineering, Computing Science and Automatic Control (CCE), pp. 1–6. IEEE Press (2011)

8. Trautmann, H., Rudolph, G., Dominguez-Medina, C., Schütze, O.: Finding Evenly Spaced Pareto Fronts for Three-Objective Optimization Problems. In: Schütze, O., Coello Coello, C.A., Tantar, A.-A., Tantar, E., Bouvry, P., Del Moral, P., Legrand, P. (eds.) EVOLVE - A Bridge Between Probability. AISC, vol. 175, pp. 89–105. Springer, Heidelberg (2013)

9. Pareto, V.: Manual of Political Economy. The MacMillan Press (1971)

10. Veldhuizen, D.A.V.: Multiobjective Evolutionary Algorithms: Classifications, Analyses, and New Innovations. PhD thesis, Department of Electrical and Computer Engineering. Graduate School of Engineering. Air Force Institute of Technology, Wright-Patterson AFB, Ohio (1999)

11. Coello Coello, C.A., Cruz Cortés, N.: Solving Multiobjective Optimization Problems using an Artificial Immune System. Genetic Programming and Evolvable Machines 6(2), 163–190 (2005)

12. De Loera, J.A., Rambau, J., Santos, F.: Triangulations – Structures for Algorithms and Applications. Springer, Berlin (2010)

13. Pateiro-Lopez, B., Rodriguez-Casal, A.: Generalizing the convex hull of a sample: The R package alphahull. Journal of Statistical Software 34(5), 1–28 (2010)

14. Bertsimas, D., Tsitsiklis, J.: Introduction to Linear Optimization, 1st edn. Athena Scientific (1997)

15. Mersmann, O., Bischl, B., Bossek, J., Trautmann, H., Wagner, M., Neumann, F.: Local Search and the Traveling Salesman Problem: A Feature-Based Characterization of Problem Hardness. In: Hamadi, Y., Schoenauer, M. (eds.) LION 6. LNCS, vol. 7219, pp. 115–129. Springer, Heidelberg (2012)

16. Johnson, D.S., McGeoch, L.A.: The traveling salesman problem: A case study in local optimization. In: Aarts, E.H.L., Lenstra, J.K. (eds.) Local Search in Combinatorial Optimization. Wiley (1997)

17. Applegate, D., Cook, W.J., Dash, S., Rohe, A.: Solution of a min-max vehicle routing problem. INFORMS Journal on Computing 14(2), 132–143 (2002)

18. Jain, A.K., Dubes, R.C.: Algorithms for clustering data. Prentice-Hall, Inc., Upper Saddle River (1988)

19. Deb, K., Thiele, L., Laumanns, M., Zitzler, E.: Scalable multi-objective optimization test problems. In: Fogel, D., et al. (eds.) Proc. Int'l Congress on Evolutionary Computation (CEC 2002), vol. 1, pp. 825–830. IEEE press (2002)

20. Viennet, R., Fontiex, C., Marc, I.: Multicriteria optimization using a genetic algorithm for determining a pareto set. International Journal of Systems Science 27(2), 255–260 (1996)

21. Deb, K., Pratap, A., Agarwal, S., Meyarivan, T.: A fast and elitist multiobjective genetic algorithm: NSGA–II. IEEE Transactions on Evolutionary Computation 6(2), 182–197 (2002)
22. Zhang, Q., Li, H.: MOEA/D: A multiobjective evolutionary algorithm based on decomposition. IEEE Trans. Evolutionary Computation 11(6), 712–731 (2007)
23. Kukkonen, S., Deb, K.: Improved pruning of non-dominated solutions based on crowding distance for bi-objective optimization problems. In: Proc. Congress on Evolutionary Computation (CEC 2006), vol. 1, pp. 1179–1186. IEEE Press, Piscataway (2006)

Relation between Neighborhood Size and MOEA/D Performance on Many-Objective Problems

Hisao Ishibuchi, Naoya Akedo, and Yusuke Nojima

Department of Computer Science and Intelligent Systems, Graduate School of Engineering,
Osaka Prefecture University, 1-1 Gakuen-cho, Naka-ku, Sakai, Osaka 599-8531, Japan
{hisaoi@,naoya.akedo@ci.,nojima@}cs.osakafu-u.ac.jp

Abstract. MOEA/D is a simple but powerful scalarizing function-based EMO algorithm. Its high search ability has been demonstrated for a wide variety of multiobjective problems. MOEA/D can be viewed as a cellular algorithm. Each cell has a different weight vector and a single solution. A certain number of the nearest cells are defined for each cell as its neighbors based on the Euclidean distance between weight vectors. A new solution is generated for each cell from current solutions in its neighboring cells. The generated solution is compared with the current solutions in the neighboring cells for solution replacement. In this paper, we examine the relation between the neighborhood size and the performance of MOEA/D. In order to examine the effect of local mating and local replacement separately, we use a variant of MOEA/D with two different neighborhoods: One is for local mating and the other is for local replacement. The performance of MOEA/D with various combinations of two neighborhoods is examined using the hypervolume in the objective space and a diversity measure in the decision space for many-objective problems. Experimental results show that MOEA/D with a large replacement neighborhood has high search ability in the objective space. However, it is also shown that small replacement and mating neighborhoods are beneficial for diversity maintenance in the decision space. It is also shown that the appropriate specification of two neighborhoods strongly depends on the problem.

Keywords: Evolutionary multiobjective optimization, many-objective problems, MOEA/D, neighborhood size.

1   Introduction

Evolutionary multiobjective optimization (EMO) has been successfully applied to various application fields [4], [5], [26]. Pareto dominance-based algorithms such as NSGA-II [6], SPEA [34] and SPEA2 [32] have frequently been used in the literature. Recently, a scalarizing function-based EMO algorithm called MOEA/D (Multi-Objective Evolutionary Algorithm based on Decomposition [30]) has rapidly increased the popularity due to its simplicity, high search ability, and computational efficiency. In MOEA/D, a multiobjective problem is decomposed into a number of single-objective problems using a scalarizing function with different weight vectors. Each single-objective problem optimizes the scalarizing function with a different

R.C. Purshouse et al. (Eds.): EMO 2013, LNCS 7811, pp. 459–474, 2013.

weight vector. Since fitness evaluation for each individual is based on scalarizing function calculation, it can be efficiently performed even for many-objective problems. High search ability of MOEA/D has been repeatedly reported especially for difficult multiobjective problems in the literature [11], [21], [31].

The main feature of MOEA/D is the decomposition using a scalarizing function with different weight vectors (as its name explicitly shows). Thus the choice of an appropriate scalarizing function is important. Different scalarizing functions work well on different problems. Different scalarizing functions may be effective in different stages of evolution. Adaptive selection of a scalarizing function and the use of multiple scalarizing functions were examined [14], [15]. The population size is also an important parameter since it determines the granularity of weight vectors [13]. Actually the population size is the same as the number of weight vectors in MOEA/D.

Another important feature of MOEA/D is the use of a kind of a neighborhood structure defined by the Euclidean distance between weight vectors. By viewing each weight vector as a point in the weight vector space, MOEA/D can be explained as a cellular algorithm. Each cell has a different weight vector and a single solution. Each cell has a certain number of neighboring cells. A new solution for a cell is generated by choosing a pair of parents from the current solutions in its neighboring cells (i.e., local mating). The generated solution is compared with those solutions in the neighboring cells for solution replacement (i.e., local replacement).

The number of neighboring cells (i.e., neighborhood size) is an important user-definable parameter. However, the importance of its appropriate specification has not been stressed in the literature. This may be because MOEA/D on two-objective and three-objective problems usually has high search ability over a wide range of different specifications of the neighborhood size. In this paper, we demonstrate that its search ability for many-objective problems strongly depends on the neighborhood size. In order to examine the local mating and the local replacement separately, we use a variant of MOEA/D with two neighborhoods as in our former studies on MOEA/D [11], [13], [15]. A pair of parents is selected from a mating neighborhood, and the generated solution is compared with current solutions in a replacement neighborhood. Using such a variant, we examine various combinations of two neighborhoods (e.g., a small mating neighborhood and a large replacement neighborhood). Performance of MOEA/D with two neighborhoods is evaluated with respect to the hypervolume in the objective space and a diversity measure in the decision space.

This paper is organized as follows. First we explain a variant of MOEA/D with two neighborhoods in Section 2. In Section 3, we explain two types of many-objective test problems. One is many-objective knapsack problems, which are used to evaluate the search ability of MOEA/D in the objective space. The other is many-objective distance minimization problems, which are used to visually examine the diversity of solutions in the decision space. Performance measures in the objective space and diversity measures in the decision space are discussed in Section 4. Then we examine the relation between the performance of MOEA/D and the specifications of two neighborhoods through computational experiments in Section 5. Experimental results show that the performance of MOEA/D on many-objective knapsack problems strongly depends on the specifications of two neighborhoods. Different specifications are needed for hypervolume maximization in the objective space and diversity maximization in the decision space. Finally we conclude this paper in Section 6.

2   MOEA/D with Two Neighborhoods

In MOEA/D [30], a multiobjective problem is decomposed into a number of single-objective problems using a scalarizing function with different weight vectors. A set of weight vectors satisfying the following two conditions is used in MOEA/D:

$$\lambda_1 + \lambda_2 + \cdots + \lambda_m = 1 , \tag{1}$$

$$\lambda_i \in \left\{ 0, \frac{1}{H}, \frac{2}{H}, ..., \frac{H}{H} \right\}, \ i = 1, 2, ..., m , \tag{2}$$

where H is a user-definable positive integer. The number of weight vectors can be calculated as $N =_{H+m-1}C_{m-1}$ (i.e., $N = C_{H+m-1}^{m-1}$ [30]). For example, we have 101 weight vectors for a two-objective problem when $H = 100$: $\lambda = (0, 1), (0.01, 0.99), ..., (1, 0)$. In Fig. 1, we show 15 weight vectors for a three-objective problem when $H = 4$.

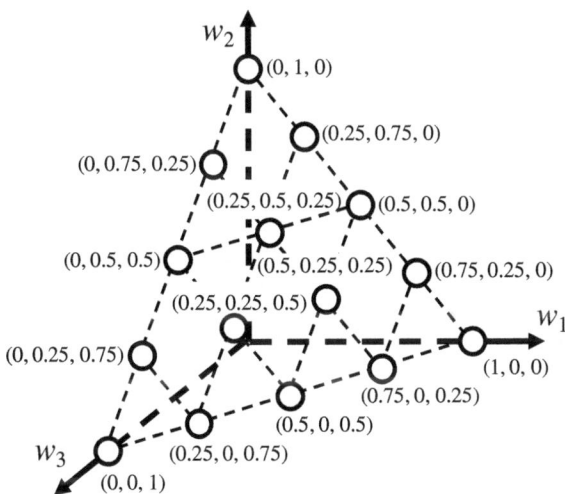

Fig. 1. A set of weight vectors for a three-objective problem ($H = 4$)

As a scalarizing function, we mainly use the weighted Tchebycheff in this paper. Only for many-objective knapsack problems with six and eight objectives, we use the weighted sum since better results were obtained from the weighted sum for those problems in our preliminary computational experiments (e.g., see [15]).

An m-objective maximization problem can be written as

$$\text{Maximize } f(x) = (f_1(x), \ f_2(x), \ ..., \ f_m(x)) , \tag{3}$$

where $f(x)$ is the m-dimensional objective vector, $f_i(x)$ is the ith objective to be maximized, and x is the decision vector.

The weighted sum of the m objectives is written using a weight vector λ as

$$g^{WS}(x \mid \lambda) = \lambda_1 \cdot f_1(x) + \lambda_2 \cdot f_2(x) + \cdots + \lambda_m \cdot f_m(x) . \tag{4}$$

The weighted Tchebycheff is written using a reference point $z^* = (z_1^*, z_2^*, ..., z_m^*)$ and a weighted vector λ as follows:

$$g^{TE}(x \mid \lambda, z^*) = \max_{i=1,2,...,m} \{\lambda_i \cdot \mid z_i^* - f_i(x) \mid\} . \tag{5}$$

In Zhang and Li [30], the reference point z^* was specified for multiobjective knapsack problems (which are multiobjective maximization problems) as

$$z_i^* = 1.1 \cdot \max\{f_i(x) \mid x \in \Omega(t)\}, \; i = 1, 2, ..., m , \tag{6}$$

where $\Omega(t)$ shows the tth population. We use this specification. For multiobjective function minimization problems, the following specification was used in [30]:

$$z_i^* = \min\{f_i(x) \mid x \in \Omega(t)\}, \; i = 1, 2, ..., m . \tag{7}$$

We use this specification for multiobjective distance minimization problems.

As in our former studies [11], [13], [15], we implemented MOEA/D as a cellular algorithm with two neighborhoods. As shown in Fig. 1, a set of weight vectors can be viewed as a grid in the weight vector space where each weight vector corresponds to a cell. Each cell can be viewed as having a weight vector and a single solution.

Let N be the number of weight vectors, which is the same as the number of cells and the population size. We denote N weight vectors as λ^k, $k = 1, 2, ..., N$ where λ^k is a weight vector assigned to the kth cell. In the original MOEA/D [30], each weight vectors has a set of neighbors. Our variant of MOEA/D has two sets of neighbors. That is, each cell has two sets of neighboring cells. One is for local mating and the other is for local replacement of solutions. As in the original MOEA/D [30], the definition of neighbors is based on the distance between weight vectors.

When a solution is to be generated in a cell, a pair of solutions is randomly selected from its mating neighborhood. A new solution is generated by crossover and mutation. The generated solution for the cell is compared with current solutions in its replacement neighborhood. At each cell in the replacement neighborhood, the generated solution is evaluated using the weight vector in that cell. The solution replacement is performed when the generated solution is better than the current one in each cell. It should be noted that the comparison is based on the weight vector at each cell. Thus it is not likely that many current solutions are replaced with a single new solution even when we use a large replacement neighborhood. This is because a new solution is not likely to be evaluated as being better than current solutions at many cells with totally different weight vectors such as (0.2, 0.8), (0.5, 0.5) and (0.8, 0.2).

In the original version of MOEA/D, a parent outside the neighborhood can be probabilistically selected. In our variant, we always choose parents from the mating neighborhood. The upper bound on the number of replaced solutions with a new solution can be specified in the original MOEA/D. We do not use any upper bound on the number of replaced solutions in our variant. The original MOEA/D also has an option of using an archive population to store non-dominated solutions. We do not use any archive population. All of these settings in our variant are to clearly examine the effect of the neighborhood size on the performance of MOEA/D.

It is pointed out in several studies [10], [20], [23] that the recombination of similar parents improves the search ability of EMO algorithms on many-objective problems. This is because the current population of EMO algorithms has a large diversity in the decision space in the case of many objectives [20]. A good solution is not likely to be generated from a pair of totally different parents. MOEA/D has two nice properties as a many-objective optimizer: One is the scalarizing function-based efficient fitness evaluation, and the other is the local mating. It is shown in this paper that an appropriate specification of the mating neighborhood is important in the application of MOEA/D to difficult many-objective problems. The necessity of local replacement is also discussed with respect to the diversity in the decision space.

3   Many-Objective Test Problems

It has been pointed out in the literature that many-objective problems are difficult for Pareto dominance-based EMO algorithms [8], [19], [22]. When EMO algorithms are applied to many-objective problems, almost all solutions in the current population become non-dominated with each other within a small number of generations. This severely weakens the selection pressure of Pareto dominance-based fitness evaluation mechanisms towards the Pareto front. Various approaches have been proposed to increase the selection pressure [16], [17], [23]. EMO algorithms with other fitness evaluation mechanisms such as indicator-based EMO algorithms (e.g., SMS-EMOA [3]) and scalarizing function-based EMO algorithms (e.g., MOEA/D [30]) have been actively studied for many-objective problems. Currently evolutionary many-objective optimization is a hot topic in the EMO community [1], [2], [24], [35].

As test problems, we use two types of many-objective problems. One is knapsack problems, which are difficult many-objective problems for Pareto dominance-based EMO algorithms. The other is distance minimization problems, which are easy many-objective problems. We briefly explain those test problems.

We generated many-objective knapsack problems from the two-objective 500-item knapsack problem of Zitzler and Thiele [34]. This problem is written as follows:

$$\text{Maximize} \quad f_i(\boldsymbol{x}) = \sum_{j=1}^{n} p_{ij} x_j, \quad i = 1, 2, \tag{8}$$

$$\text{subject to} \quad \sum_{j=1}^{n} w_{ij} x_j \le c_i, \quad i = 1, 2, \tag{9}$$

$$x_j = 0 \text{ or } 1, \quad j = 1, 2, ..., n, \tag{10}$$

where n is the number of items (i.e., $n = 500$), x is a binary string, p_{ij} is the profit of item j according to knapsack i, w_{ij} is the weight of item j according to knapsack i, and c_i is the capacity of knapsack i. The value of each profit p_{ij} is a randomly specified integer in $[10, 100]$. This problem is referred to as the 2-500 problem in this paper.

We generated additional objectives $f_i(x)$ for $i = 3, 4, ..., 8$ by randomly specifying the value of the profit p_{ij} as an integer in $[10, 100]$. In this manner, we generated m-objective knapsack problems with up to eight objectives. Each of those test problems is referred to as the m-500 problem in this paper. It should be noted that all of those test problems have the same constraint conditions as the original 2-500 problem. That is, all of our multiobjective knapsack problems have the same feasible solution set. As a result, the Pareto optimal solutions of the original 2-500 problem are also Pareto optimal for the m-500 problems for $m = 3, 4, ..., 8$. This feature is used to visually examine the convergence and the diversity of solutions of many-objective knapsack problems by projecting them onto the two-dimensional objective space with $f_1(x)$ and $f_2(x)$. It has been demonstrated that randomly generated many-objective knapsack problems are difficult for Pareto dominance-based EMO algorithms [11], [23].

We also generated many-objective distance minimization problems with multiple Pareto regions to examine the behavior of EMO algorithms in a two-dimensional decision space [9], [12]. An example of a four-objective problem is shown in Fig. 2. All points in the shaded four squares are Pareto optimal solutions. The ith objective $f_i(\mathbf{x})$ is the minimum distance from a point $\mathbf{x}$ (i.e., solution $\mathbf{x}$) in the two-dimensional decision space to the ith vertexes of multiple polygons:

$$f_i(\mathbf{x}) = \min\{\text{dis}(\mathbf{x}, \mathbf{a}_{i1}), \text{dis}(\mathbf{x}, \mathbf{a}_{i2}), ..., \text{dis}(\mathbf{x}, \mathbf{a}_{ik})\}, \, i = 1, 2, ..., m, \tag{11}$$

where $\text{dis}(\mathbf{x}, \mathbf{a}_{ij})$ is the Euclidean distance between the two points $\mathbf{x}$ and $\mathbf{a}_{ij}$, $\mathbf{a}_{ij}$ shows the ith vertex of the jth polygon, k is the number of polygons (i.e., $j = 1, 2, ..., k$), and m is the number of objectives (i.e., the number of vertexes: $i = 1, 2, ..., m$). In Fig. 2, the four squares have exactly the same shape and the same size. Thus each square is mapped to the same Pareto front in the four-dimensional objective space.

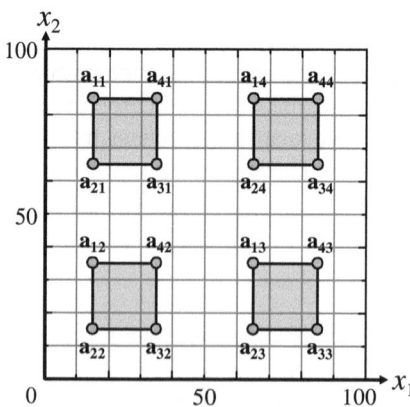

Fig. 2. An example of a four-objective distance minimization problem

In the same manner as in Fig. 2, we generated distance minimization problems with three, five and six objectives as shown in Fig. 3. We also generated four distance minimization problems with a single Pareto optimal region as shown in Fig. 4. In our test problems in Figs. 2-4, all points in each polygon are Pareto optimal solutions.

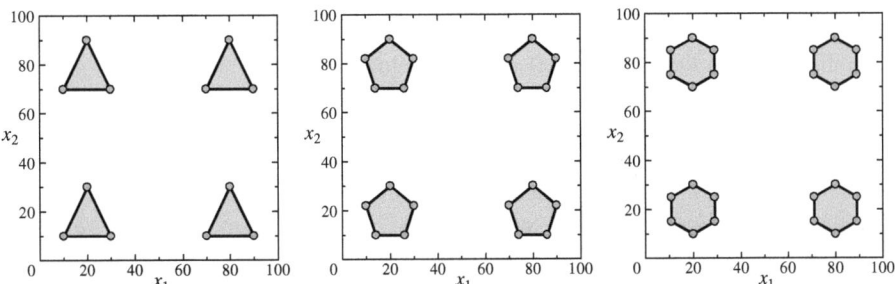

Fig. 3. Distance minimization problems with multiple Pareto optimal regions

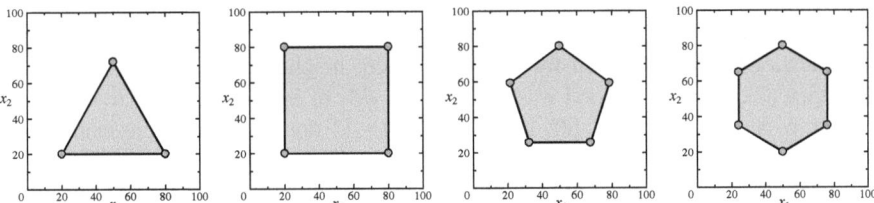

Fig. 4. Distance minimization problems with a single Pareto optimal region

4   Performance Measures and Diversity Measures

A number of performance measures have been proposed to evaluate a set of non-dominated solutions in the objective space [33]. We use the hypervolume measure, which has been frequently used in the literature.

With a few exceptions (e.g., Omni-Optimizer [7]), the decision space diversity has not been used in EMO algorithms. This is because the diversity maintenance in the objective space is very important in EMO algorithms. Recently, the importance of the diversity maintenance in the decision space was stressed in some studies [27]-[29] where the use of the Solow-Polasky diversity measure [25] was suggested. The use of a non-geometric binary crossover was proposed to directly increase the decision space diversity [18]. In this paper, we use the average distance between two solutions in the decision space as a diversity measure since its meaning can be easily understood. The distance between solutions of the knapsack problems (i.e., binary strings) is measured by the Hamming distance while the Euclidean distance is used for solutions of the distance minimization problems (i.e., two-dimensional real number vectors). Whereas we also calculated the Solow-Polasky diversity measure, we only report the average distance between solutions due to the page limitation.

5   Experimental Results

We applied our variant of MOEA/D, NSGA-II and SPEA2 to the 2-500, 4-500, 6-500 and 8-500 knapsack problems using the following parameter specifications:

Coding: Binary string of length 500 (i.e., 500-bit string),
Population size in NSGA-II and SPEA2: 200,
Population size in MOEA/D: 200 (2-500), 220 (4-500), 252 (6-500), 120 (8-500),
Termination condition: 200×2000 solution evaluations,
Parent selection: Random selection from the neighborhood (MOEA/D),
          Binary tournament selection with replacement (NSGA-II and SPEA2),
Crossover: Uniform crossover (Probability: 0.8),
Mutation: Bit-flip mutation (Probability: 1/500),
Number of runs for each test problem: 100 runs.

In MOEA/D, we specified the size of the mating neighborhood as $\alpha\%$ of the population size where $\alpha = 1, 2, 4, 6, 8, 10, 20, 40, 60, 80, 100$. That is, these 11 specifications of α were examined. When $\alpha\%$ of the population size was not an integer, the non-integer value was rounded down. For example, 1% of the population size 220 was handled as 2 by rounding down the calculated value 2.2. The same 11 specifications were also used for the replacement neighborhood. That is, the size of the replacement neighborhood was specified as $\beta\%$ of the population size where $\beta = 1, 2, 4, 6, 8, 10, 20, 40, 60, 80, 100$. All the 11×11 combinations were examined in our computational experiments.

The average value of the hypervolume of the final population over 100 runs of our variant of MOEA/D for each combination is summarized in Fig. 5. The origin of the objective space was used as a reference point in the hypervolume calculation. Good results were not obtained for many-objective knapsack problems by NSGA-II and SPEA2. For example, their results on the 8-500 problem were 1.10×10^{34} (NSGA-II) and 1.03×10^{34} (SPEA2) whereas the best result in Fig. 5 was 1.55×10^{34}. From Fig. 5, we can see that good results were obtained for all the four test problems when the size of the two neighborhoods was specified as follows: 2-10% of the population size for the mating neighborhood and 20-100% for the replacement neighborhood. We can also see from Fig. 5 that the sensitivity of the MOEA/D performance on the neighborhood size increases with the number of objectives. Whereas almost the same results were obtained from a wide range of parameter specifications for the 2-500 problem, very good results were obtained from only a few combinations for the 8-500 problem. These observations suggest that the use of appropriate neighborhoods is important in MOEA/D for many-objective knapsack problems. Fig. 5 also shows that the use of two different neighborhoods improves the performance of MOEA/D.

The average distance between solutions in the final population is summarized in Fig. 6. High average distances between solutions were obtained from a small mating neighborhood for all the six test problems independent of the size of the replacement neighborhood. We can see that there is no clear relation between the hypervolume in Fig. 5 and the decision space diversity in Fig. 6. In Fig. 7, we show the projection of a solution set onto the $f_1(x)$-$f_2(x)$ space obtained by a single run of MOEA/D with a different setting of two neighborhoods. We can see from Fig. 7 that a large diversity in the objective space was obtained from a small mating neighborhood.

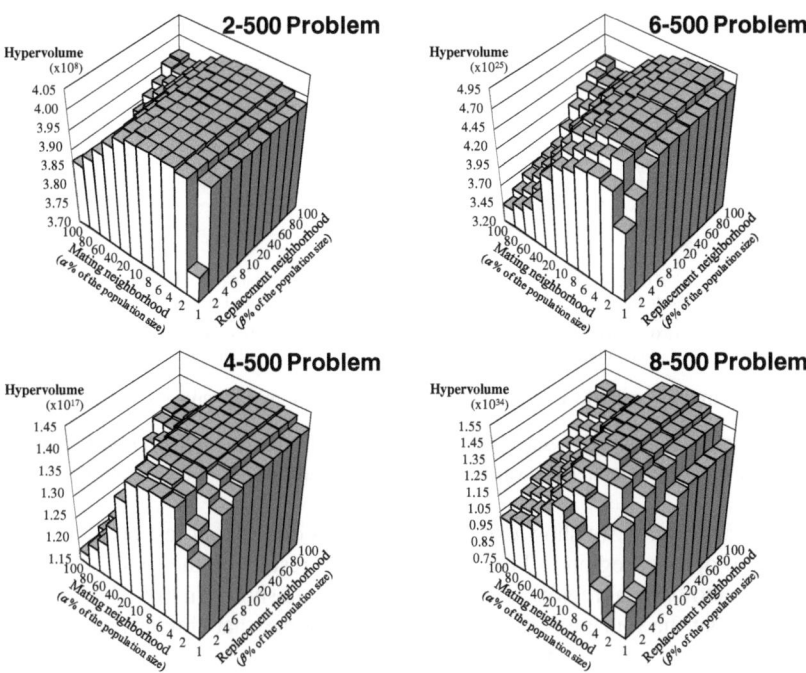

Fig. 5. Experimental results on the knapsack problems (Hypervolume measure)

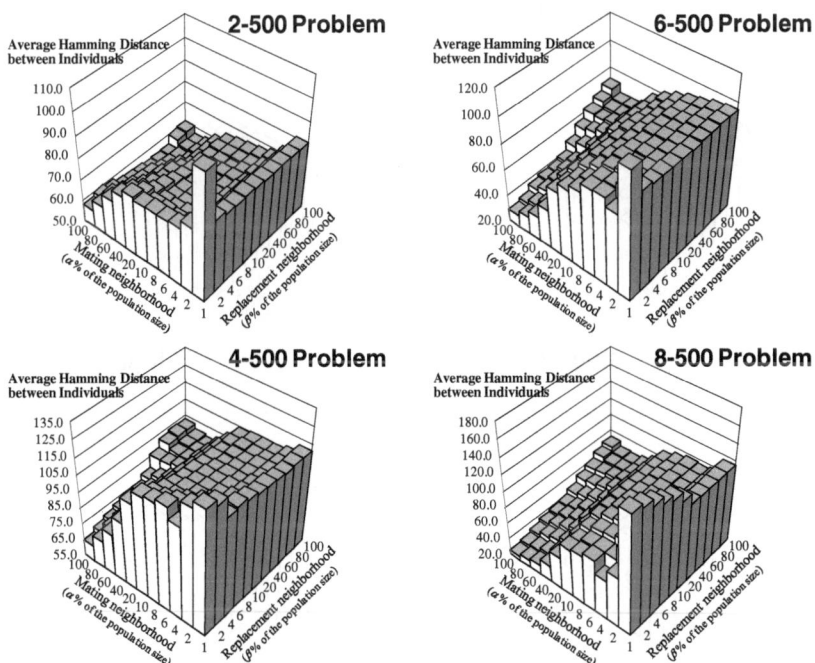

Fig. 6. Experimental results on the knapsack problems (Average Hamming distance)

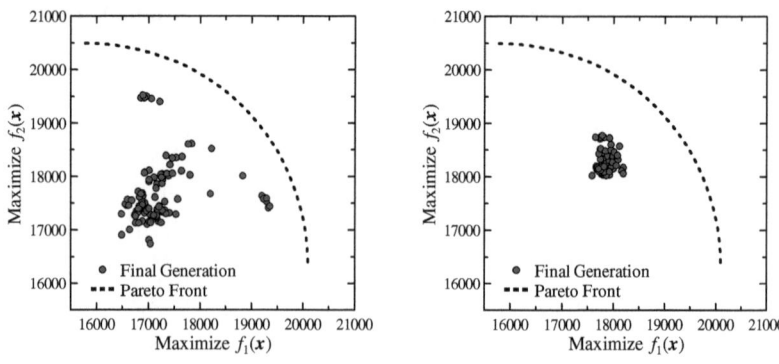

(a) 1% mating and 100% replacement.   (b) 100% mating and 1% replacement.

Fig. 7. Projection of a solution set obtained from a single run of MOEA/D with a different setting on the 8-500 problem. The dashed line shows the Pareto front of the 2-500 problem.

In Fig. 8 and Fig. 9, we show experimental results on the four test problems with a single Pareto optimal region in Fig. 4. Experimental results on the four test problems with four Pareto optimal regions in Fig. 2 and Fig. 3 are shown in Fig. 10 and Fig. 11. We can see that the four plots in each figure show somewhat similar patterns: Good results were almost always obtained from a small replacement neighborhood in Figs. 8-11. These observations are totally different from those for the knapsack problems.

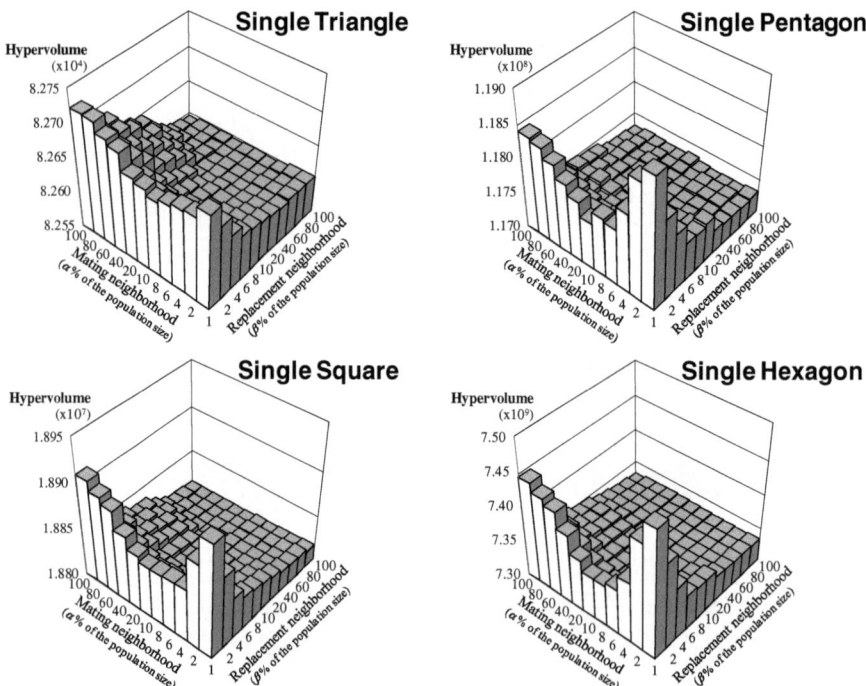

Fig. 8. Results on the minimization problems in Fig. 4 (Hypervolume measure)

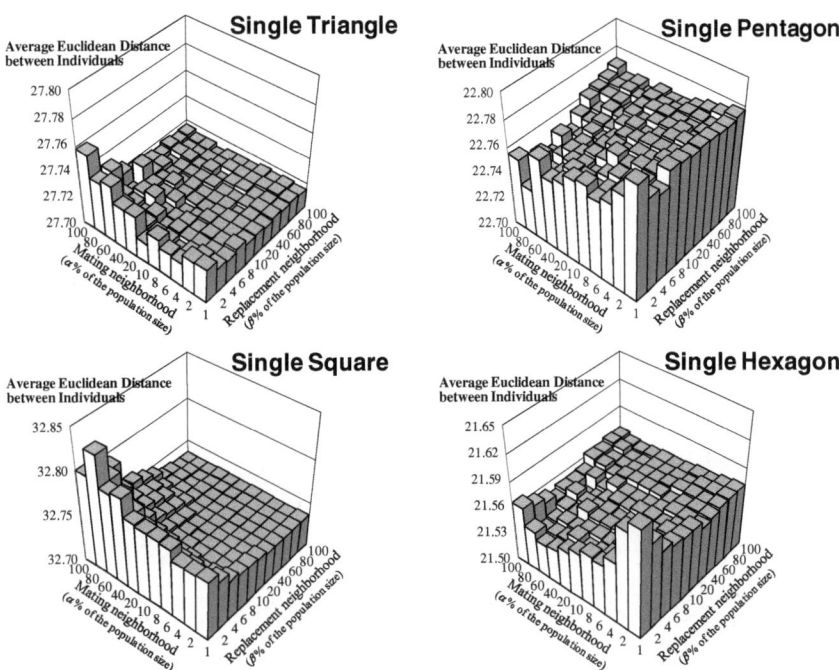

Fig. 9. Results on the minimization problems in Fig. 4 (Average distance)

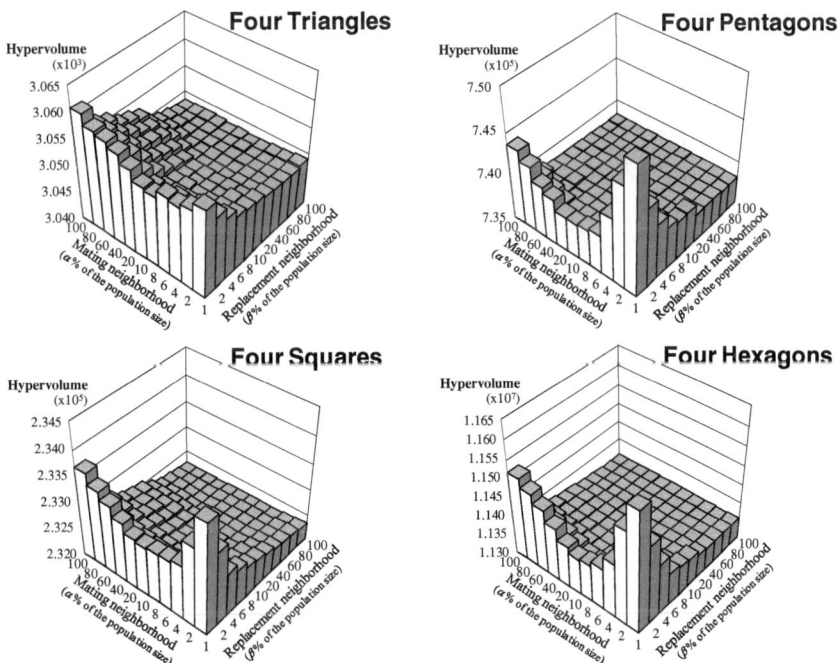

Fig. 10. Results on the minimization problems in Fig. 2 and Fig. 3 (Hypervolume measure)

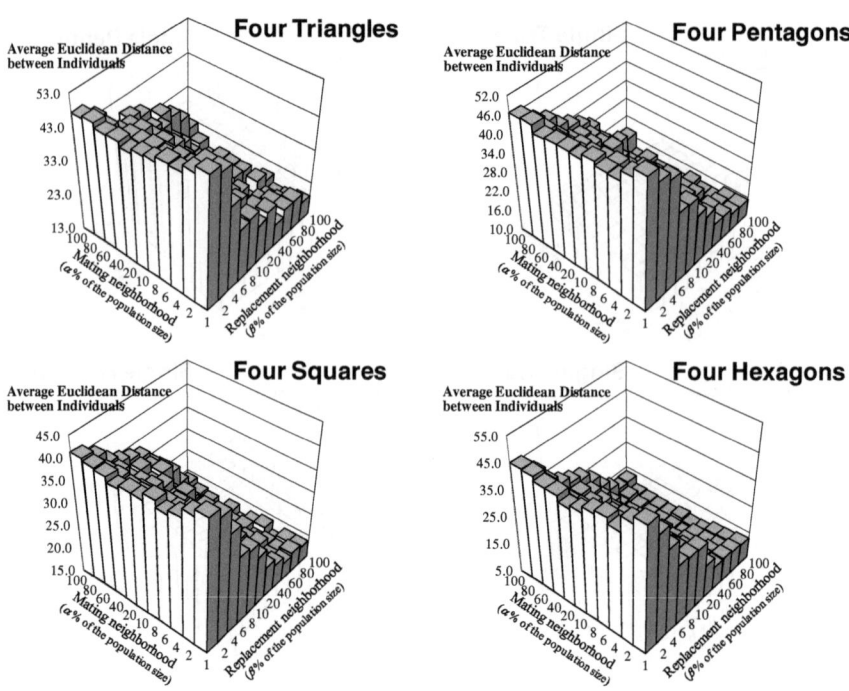

Fig. 11. Results on the minimization problems in Fig. 2 and Fig. 3 (Average distance)

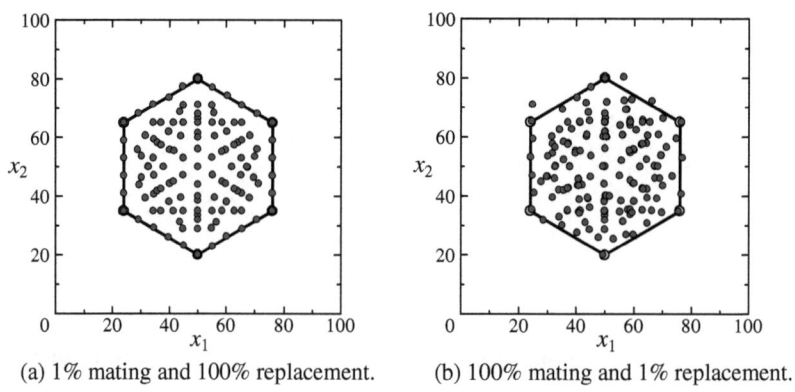

(a) 1% mating and 100% replacement.   (b) 100% mating and 1% replacement.

Fig. 12. A solution set from a single run with a different setting (Single hexagon)

In Fig. 12 and Fig. 13, we show an obtained solution set from a single run of MOEA/D with a different setting. Whereas the result in Fig. 12 (a) looks nice, a higher hypervolume value was obtained from a small replacement neighborhood in Fig. 12 (b) as shown in Fig. 8. Fig. 13 clearly shows that a much larger diversity in the decision space was obtained from a small replacement neighborhood (see Fig. 11).

The average hypervolume values 1.08×10^7 and 1.18×10^7 were obtained for the four-hexagon problem by NSGA-II and SPEA2, respectively, while the best result by MOEA/D was 1.16×10^7 in Fig. 10. These results show that our distance minimization problems are not difficult for Pareto dominance-based EMO algorithms.

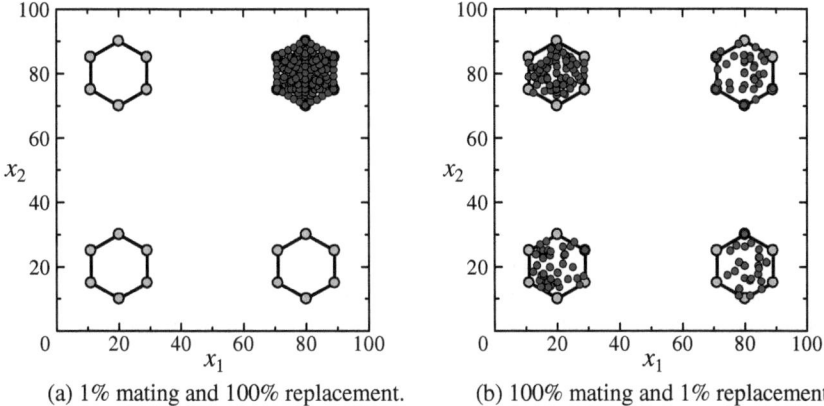

(a) 1% mating and 100% replacement.   (b) 100% mating and 1% replacement.

Fig. 13. A solution set from a single run with a different setting (Four hexagons)

6   Conclusions

In this paper, we explained the relation between the performance of MOEA/D on many-objective problems and the specification of the size of the two neighborhoods (one is for local mating and the other is for local replacement). For many-objective knapsack problems, we obtained the following observations:

(1) Good results with respect to the hypervolume measure were obtained from the following combinations: The mating neighborhood size was 2-10% of the population size and the replacement neighborhood size was 20-100%. The larger the replacement neighborhood was, the better the performance of MOEA/D was.

(2) Good results with respect to the decision space diversity were obtained when the mating neighborhood size was 1% of the population size. The smaller the mating neighborhood was, the larger the decision space diversity was.

Different results were obtained for the distance minimization problems as follows:

(3) Good results with respect to both the hypervolume measure and the decision space diversity were obtained when the replacement neighborhood size was 1%. The smaller the replacement neighborhood was, the better the performance was.

(4) The best results with respect to both the hypervolume measure and the decision space diversity were obtained for the five-objective and six-objective problems when the size of the two neighborhoods was specified as 1% of the population size (whereas the worst results were obtained from this setting for the knapsack problems with respect to the hypervolume measure).

We can see from these observations that an appropriate specification of the two neighborhoods is totally problem-dependent. Moreover, good specifications for the hypervolume maximization are not always good for the decision space diversity maximization. For difficult many-objective problems, high selection pressure toward the Pareto front is needed for efficient search. Thus a large replacement neighborhood is beneficial for MOEA/D. For easy many-objective problems, high selection pressure is not needed. So a large replacement neighborhood is not needed. One potential disadvantage of a large replacement neighborhood is the increase in computation load, which should be further discussed in future studies.

References

1. Adra, S.F., Fleming, P.J.: Diversity Management in Evolutionary Many-Objective Optimization. IEEE Trans. on Evolutionary Computation 15, 183–195 (2011)
2. Bader, J., Zitzler, E.: HypE: An Algorithm for Fast Hypervolume-Based Many-Objective Optimization. Evolutionary Computation 19, 45–76 (2011)
3. Beume, N., Naujoks, B., Emmerich, M.: SMS-EMOA: Multiobjective Selection based on Dominated Hypervolume. European Journal of Operational Research 181, 1653–1669 (2007)
4. Coello, C.A.C., Lamont, G.B.: Applications of Multi-Objective Evolutionary Algorithms. World Scientific, Singapore (2004)
5. Deb, K.: Multi-Objective Optimization Using Evolutionary Algorithms. John Wiley & Sons, Chichester (2001)
6. Deb, K., Pratap, A., Agarwal, S., Meyarivan, T.: A Fast and Elitist Multiobjective Genetic Algorithm: NSGA-II. IEEE Trans. on Evolutionary Computation 6, 182–197 (2002)
7. Deb, K., Tiwari, S.: Omni-Optimizer: A Generic Evolutionary Algorithm for Single and Multi-Objective Optimization. European Journal of Operational Research 185, 1062–1087 (2008)
8. Hughes, E.J.: Evolutionary Many-Objective Optimisation: Many Once or One Many? In: Proc. of 2005 IEEE Congress on Evolutionary Computation, pp. 222–227 (2005)
9. Ishibuchi, H., Akedo, N., Nojima, Y.: A Many-Objective Test Problem for Visually Examining Diversity Maintenance Behavior in a Decision Space. In: Proc. of 2011 Genetic and Evolutionary Computation Conference, pp. 649–656 (2011)
10. Ishibuchi, H., Akedo, N., Nojima, Y.: Recombination of Similar Parents in SMS-EMOA on Many-Objective 0/1 Knapsack Problems. In: Coello, C.A.C., Cutello, V., Deb, K., Forrest, S., Nicosia, G., Pavone, M. (eds.) PPSN XII, Part II. LNCS, vol. 7492, pp. 132–142. Springer, Heidelberg (2012)
11. Ishibuchi, H., Hitotsuyanagi, Y., Ohyanagi, H., Nojima, Y.: Effects of the Existence of Highly Correlated Objectives on the Behavior of MOEA/D. In: Takahashi, R.H.C., Deb, K., Wanner, E.F., Greco, S. (eds.) EMO 2011. LNCS, vol. 6576, pp. 166–181. Springer, Heidelberg (2011)
12. Ishibuchi, H., Hitotsuyanagi, Y., Tsukamoto, N., Nojima, Y.: Many-Objective Test Problems to Visually Examine the Behavior of Multiobjective Evolution in a Decision Space. In: Schaefer, R., Cotta, C., Kołodziej, J., Rudolph, G. (eds.) PPSN XI, Part II. LNCS, vol. 6239, pp. 91–100. Springer, Heidelberg (2010)
13. Ishibuchi, H., Sakane, Y., Tsukamoto, N., Nojima, Y.: Evolutionary Many-Objective Optimization by NSGA-II and MOEA/D with Large Populations. In: Proc. of 2009 IEEE International Conference on Systems, Man, and Cybernetics, pp. 1820–1825 (2009)
14. Ishibuchi, H., Sakane, Y., Tsukamoto, N., Nojima, Y.: Adaptation of Scalarizing Functions in MOEA/D: An Adaptive Scalarizing Function-Based Multiobjective Evolutionary Algorithm. In: Ehrgott, M., Fonseca, C.M., Gandibleux, X., Hao, J.-K., Sevaux, M. (eds.) EMO 2009. LNCS, vol. 5467, pp. 438–452. Springer, Heidelberg (2009)

15. Ishibuchi, H., Sakane, Y., Tsukamoto, N., Nojima, Y.: Simultaneous Use of Different Scalarizing Functions in MOEA/D. In: Proc. of Genetic and Evolutionary Computation Conference, pp. 519–526 (2010)
16. Ishibuchi, H., Tsukamoto, N., Hitotsuyanagi, Y., Nojima, Y.: Effectiveness of Scalability Improvement Attempts on the Performance of NSGA-II for Many-Objective Problems. In: Proc. of 2008 Genetic and Evolutionary Computation Conference, pp. 649–656 (2008)
17. Ishibuchi, H., Tsukamoto, N., Nojima, Y.: Evolutionary Many-Objective Optimization: A Short Review. In: Proc. of 2008 IEEE Congress on Evolutionary Computation, pp. 2424–2431 (2008)
18. Ishibuchi, H., Tsukamoto, N., Nojima, Y.: Diversity Improvement by Non-Geometric Binary Crossover in Evolutionary Multiobjective Optimization. IEEE Trans. on Evolutionary Computation 14, 985–998 (2010)
19. Khare, V., Yao, X., Deb, K.: Performance Scaling of Multi-objective Evolutionary Algorithms. In: Fonseca, C.M., Fleming, P.J., Zitzler, E., Deb, K., Thiele, L. (eds.) EMO 2003. LNCS, vol. 2632, pp. 376–390. Springer, Heidelberg (2003)
20. Kowatari, N., Oyama, A., Aguirre, H., Tanaka, K.: Analysis on Population Size and Neighborhood Recombination on Many-Objective Optimization. In: Coello, C.A.C., Cutello, V., Deb, K., Forrest, S., Nicosia, G., Pavone, M. (eds.) PPSN XII, Part II. LNCS, vol. 7492, pp. 22–31. Springer, Heidelberg (2012)
21. Li, H., Zhang, Q.: Multiobjective Optimization Problems with Complicated Pareto Sets, MOEA/D and NSGA-II. IEEE Trans. on Evolutionary Computation 13, 284–302 (2009)
22. Purshouse, R.C., Fleming, P.J.: On the Evolutionary Optimization of Many Conflicting Objectives. IEEE Trans. on Evolutionary Computation 11, 770–784 (2007)
23. Sato, H., Aguirre, H.E., Tanaka, K.: Local Dominance and Local Recombination in MOEAs on 0/1 Multiobjective Knapsack Problems. European J. of Operational Research 181, 1708–1723 (2007)
24. Schütze, O., Lara, A., Coello, C.A.C.: On the Influence of the Number of Objectives on the Hardness of a Multiobjective Optimization Problem. IEEE Trans. on Evolutionary Computation 15, 444–455 (2011)
25. Solow, A.R., Polasky, S.: Measuring Biological Diversity. Environmental and Ecological Statistics 1, 95–103 (1994)
26. Tan, K.C., Khor, E.F., Lee, T.H.: Multiobjective Evolutionary Algorithms and Applications. Springer, Berlin (2005)
27. Ulrich, T., Bader, J., Thiele, L.: Integrating Decision Space Diversity into Hypervolume-Based Multiobjective Search. In: Proc. of 2010 Genetic and Evolutionary Computation Conference, pp. 455–462 (2010)
28. Ulrich, T., Bader, J., Thiele, L.: Defining and Optimizing Indicator-Based Diversity Measures in Multiobjective Search. In: Schaefer, R., Cotta, C., Kołodziej, J., Rudolph, G. (eds.) PPSN XI, Part I. LNCS, vol. 6238, pp. 707–717. Springer, Heidelberg (2010)
29. Ulrich, T., Thiele, L.: Maximizing Population Diversity in Single-Objective Optimization. In: Proc. of 2011 Genetic and Evolutionary Computation Conference, pp. 641–648 (2011)
30. Zhang, Q., Li, H.: MOEA/D: A Multiobjective Evolutionary Algorithm Based on Decomposition. IEEE Trans. on Evolutionary Computation 11, 712–731 (2007)
31. Zhang, Q., Liu, W., Li, H.: The Performance of a New Version of MOEA/D on CEC09 Unconstrained MOP Test Instances. In: Proc. of 2009 Congress on Evolutionary Computation, pp. 203–208 (2009)
32. Zitzler, E., Laumanns, M., Thiele, L.: SPEA2: Improving the Strength Pareto Evolutionary Algorithm. TIK-Report 103, Computer Engineering and Networks Laboratory (TIK), Department of Electrical Engineering, ETH, Zurich (2001)

33. Zitzler, E., Thiele, L., Laumanns, M., Fonseca, C.M., da Fonseca, V.G.: Performance Assessment of Multiobjective Optimizers: An Analysis and Review. IEEE Trans. on Evolutionary Computation 7, 117–132 (2003)
34. Zitzler, E., Thiele, L.: Multiobjective Evolutionary Algorithms: A Comparative Case Study and the Strength Pareto Approach. IEEE Trans. on Evolutionary Computation 3, 257–271 (1999)
35. Zou, X., Chen, Y., Liu, M., Kang, L.: A New Evolutionary Algorithm for Solving Many-Objective Optimization Problems. IEEE Trans. on SMC - Part B 38, 1402–1412 (2008)

Multiple Criteria Hierarchy Process for the Choquet Integral

Silvia Angilella[1], Salvatore Corrente[1], Salvatore Greco[1], and Roman Słowiński[2]

[1] Department of Economics and Business, University of Catania,
Corso Italia 55, I-95129 Catania, Italy
{angisil,salvatore.corrente,salgreco}@unict.it
[2] Institute of Computing Science, Poznań University of Technology, 60-965 Poznań,
and Systems Research Institute, Polish Academy of Sciences, 01-447 Warsaw, Poland
roman.slowinski@cs.put.poznan.pl

Abstract. Interaction between criteria and hierarchical structure of criteria are nowadays two important issues in Multiple Criteria Decision Analysis (MCDA). Interaction between criteria is often dealt with fuzzy integrals, especially the Choquet integral. To handle the hierarchy of criteria in MCDA, a methodology called Multiple Criteria Hierarchy Process (MCHP) has been recently proposed. It permits consideration of preference relations with respect to a subset of criteria at any level of the hierarchy. In this paper, we propose to apply MCHP to the Choquet integral. In this way, using the Choquet integral and the MCHP, it is possible to compare two alternatives not only globally, but also partially, taking into account a particular subset of criteria and the possible interaction between them.

Keywords: Multiple criteria decision aiding, Choquet integral, Multiple Criteria Hierarchy Process.

1 Introduction

In a multiple criteria decision problem (see [5] for a comprehensive state of the art), an alternative a, belonging to a finite set of m alternatives $A = \{a, b, c, \ldots\}$, is evaluated on the basis of a consistent family of n criteria $G = \{g_1, g_2, \ldots, g_n\}$. In our approach we make the assumption that each criterion $g_i \colon A \to \mathbb{R}$ is an interval scale of measurement. From here on, we will use the terms criterion g_i or criterion i interchangeably ($i = 1, 2, \ldots, n$). Without loss of generality, we assume that all the criteria have to be maximized.

The purpose of Multi-Attribute Utility Theory (MAUT) [12] is to represent the preferences of a Decision Maker (DM) on a set of alternatives A by an overall value function $U \colon \mathbb{R}^n \to \mathbb{R}$ with $U(g_1(a), \ldots, g_n(a)) = U(a)$:

- a is indifferent to $b \iff U(a) = U(b)$,
- a is preferred to $b \iff U(a) > U(b)$.

R.C. Purshouse et al. (Eds.): EMO 2013, LNCS 7811, pp. 475–489, 2013.

The principal aggregation model of value function is the multiple attribute additive utility [12]:

$$U(a) = u_1(g_1(a)) + u_2(g_2(a)) + \ldots + u_n(g_n(a)) \quad \text{for all} \ \ a \in A, \qquad (1)$$

where u_i are non-decreasing marginal value functions for $i = 1, 2, \ldots, n$.

As it is well-known from the literature, the underlying assumption of the preference independence of the multiple attribute additive utility is unrealistic since in real decision problems criteria often interact. In a decision problem one usually distinguishes between positive and negative interaction among criteria, corresponding to synergy and redundancy among criteria, respectively. In particular, two criteria are synergic (redundant) when the comprehensive importance of these two criteria is greater (smaller) than the sum of importances of the two criteria considered separately.

Within Multiple Criteria Decision Analysis (MCDA), the interaction of criteria has been considered in a decision model based upon a non-additive integral, *i.e.* the Choquet integral [3] (see [7,9] for a comprehensive survey on the use of non-additive integrals in MCDA, and [10] for a state-of-the-art survey on Choquet and Sugeno integrals).

A great majority of methods designed for MCDA assume that all evaluation criteria are considered at the same level, however, it is often the case that a practical application is imposing a hierarchical structure of criteria. For example, in economic ranking, alternatives may be evaluated on indicators which aggregate evaluations on several sub-indicators, and these sub-indicators may aggregate another set of sub-indicators, etc. In this case, the marginal value functions may refer to all levels of the hierarchy, representing values of particular scores of the alternatives on indicators, sub-indicators, sub-sub-indicators, etc. Considering hierarchical, instead of flat, structure of criteria, permits decomposition of a complex decision problem into smaller problems involving less criteria. To handle the hierarchy of criteria, the Multiple Criteria Hierarchy Process (MCHP) [4] can be applied. The basic idea of the MCHP relies on consideration of preference relations at each node of the hierarchy tree of criteria. This consideration concerns both the phase of eliciting preference information, and the phase of analyzing a final recommendation by the DM. For example, in a decision problem related to evaluation of students, one can say not only that student a is comprehensively preferred to student b, i.e. $a \succ b$, but also that a is comprehensively preferred to b because a is preferred to b on the subset of subjects (subcriteria) related to Mathematics and Physics, i.e. $a \succ_{Mathematics} b$ and $a \succ_{Physics} b$, even if b is preferred to a on subjects related to Humanities, i.e. $b \succ_{Humanities} a$. Moreover, one can also say that, for example, a is preferred to b on the subset of subjects related to Mathematics because, considering Analysis and Algebra as subjects (sub-criteria) related to Mathematics, a is preferred to b on Analysis, i.e. $a \succ_{Analysis} b$, and this is enough to compensate the fact that b is preferred to a on Algebra, i.e. $b \succ_{Algebra} a$.

In this paper, we apply the MCHP to the Choquet integral. Let us remark that another approach using the Choquet integral on a hierarchy of criteria has been presented in [18] (see also [19]), where the evaluation of an alternative a with respect to a certain criterion $\mathcal{G}_r$ is based on the Choquet integrals of a with respect to all subcriteria of $\mathcal{G}_r$ from the subsequent level. This means that the Choquet integral of a with respect to $\mathcal{G}_r$ is computed as the Choquet integral of other Choquet integrals, one for each subcriterion of $\mathcal{G}_r$ from the subsequent level. For example, let us consider the evaluation of student a with respect to Science and Humanities, with Mathematics and Physics as subcriteria of Science, and Literature and Philosophy as subcriteria of Humanities. In order to compute the comprehensive Choquet integral of a, one has to compute first the Choquet integral of a with respect to Science and the Choquet integral of a with respect to Humanities. Then, the comprehensive Choquet integral of a is obtained as the Choquet integral of the two Choquet integrals previously computed.

In our approach, we do not consider Choquet integrals resulting from aggregation of Choquet integrals representing evaluations at the subsequent level of the hierarchy. Instead of this, we compute the evaluation of an alternative on a certain criterion of the hierarchy as the Choquet integral of the evaluations of the alternative on all elementary criteria descending to the lowest level from that criterion, using the capacity defined on the whole set of elementary criteria only. Coming back to the above example, the comprehensive evaluation of a is calculated as the Choquet integral of the evaluations of a on all considered elementary subjects, i.e. Mathematics, Physics, Literature and Philosophy. The evaluation with respect to Sciences is obtained as the Choquet integral of the evaluations on Mathematics and Physics only, as well as, the evaluation with respect to Humanities is obtained as the Choquet integral of the evaluations on Literature and Philosophy only. In the approach of [18], the evaluations on Humanities and Sciences are also Choquet integrals, but our approach differs in two aspects: we do not need to define two different capacities to compute the two Choquet integrals, one for Science and one for Humanities; we use the two Choquet integrals on Science and Humanities to order students on the basis of Science and Humanities only, and not to aggregate them in order to get the final comprehensive evaluation.

The paper is organized as follows. In Section 2, we present the basic concepts relative to interaction among criteria and to the Choquet integral. In Section 3, we describe the MCHP. In Section 4, we put together the MCHP and the Choquet integral. Section 5 contains a didactic example in which we describe the application of the new methodology, and we compare it with the approach of [18]. Some conclusions and future directions of research are presented in Section 6.

2   The Choquet Integral Preference Model

Let 2^G be the power set of G (i.e. the set of all subsets of G); a fuzzy measure (capacity) on G is defined as a set function $\mu : 2^G \to [0,1]$ satisfying the following properties:

1a) $\mu(\emptyset) = 0$ and $\mu(G) = 1$ (boundary conditions),

2a) $\forall T \subseteq R \subseteq G, \ \mu(T) \leq \mu(R)$ (monotonicity condition).

A fuzzy measure is said to be additive if $\mu(T \cup R) = \mu(T) + \mu(R)$, for any $T, R \subseteq G$ such that $T \cap R = \emptyset$. An additive fuzzy measure is determined uniquely by $\mu(\{1\}), \mu(\{2\}) \dots, \mu(\{n\})$. In fact, in this case, $\forall T \subseteq G, \ \mu(T) = \sum_{i \in T} \mu(\{i\})$.

In the other cases, we have to define a value $\mu(T)$ for every subset T of G, which are as many as $2^{|G|}$. Therefore, we have to calculate the values of $2^{|G|} - 2$ coefficients, since we know that $\mu(\emptyset) = 0$ and $\mu(G) = 1$.

The Möbius representation of the fuzzy measure μ (see [15]) is defined by the function $m : 2^G \to \mathbb{R}$ (see [16]) such that:

$$\mu(R) = \sum_{T \subseteq R} m(T). \tag{2}$$

Let us observe that if R is a singleton, *i.e.* $R = \{i\}$ with $i = 1, \dots, n$, then $\mu(\{i\}) = m(\{i\})$. If R is a couple (non-ordered pair) of criteria, i.e. $R = \{i, j\}$, then $\mu(\{i, j\}) = m(\{i\}) + m(\{j\}) + m(\{i, j\})$.

In general, the Möbius representation $m(R)$ is obtained by $\mu(R)$ in the following way:

$$m(R) = \sum_{T \subseteq R} (-1)^{|R \setminus T|} \mu(T). \tag{3}$$

In terms of Möbius representation (see [2]), properties **1a)** and **2a)** are, respectively, formulated as:

1b) $m(\emptyset) = 0, \ \sum_{T \subseteq G} m(T) = 1,$

2b) $\forall i \in G$ and $\forall R \subseteq G \setminus \{i\}, \ \sum_{T \subseteq R} m(T \cup \{i\}) \geq 0.$

Let us observe that in MCDA, the importance of any criterion $g_i \in G$ should be evaluated considering all its global effects in the decision problem at hand; these effects can be "decomposed" from both theoretical and operational points of view in effects of g_i as single, and in combination with all other criteria. Therefore, a criterion $i \in G$ is important with respect to a fuzzy measure μ not only when it is considered alone, i.e. for the value $\mu(\{i\})$ in itself, but also when it interacts with other criteria from G, i.e. for every value $\mu(T \cup \{i\}), \ T \subseteq G \setminus \{i\}$.

Given $a \in A$ and μ being a fuzzy measure on G, then the *Choquet integral* [3] is defined by:

$$C_\mu(a) = \sum_{i=1}^{n} \left[\left(g_{(i)}(a) \right) - \left(g_{(i-1)}(a) \right) \right] \mu(A_i), \tag{4}$$

where $_{(\cdot)}$ stands for a permutation of the indices of criteria such that $g_{(1)}(a) \leq g_{(2)}(a) \leq \dots \leq g_{(n)}(a)$, with $A_i = \{(i), \dots, (n)\}$, $i = 1, \dots, n$, and $g_{(0)} = 0$.

The Choquet integral can be redefined in terms of the Möbius representation [6], without reordering the criteria, as:

$$C_\mu(a) = \sum_{T \subseteq G} m(T) \min_{i \in T} g_i(a). \qquad (5)$$

One of the main drawbacks of the Choquet integral is the necessity of eliciting and giving an adequate interpretation of $2^{|G|} - 2$ parameters. In order to reduce the number of parameters to be computed and to avoid the difficult description of the interactions among criteria, which is not realistic in many applications, the concept of fuzzy k-additive measure has been considered [8].

A *fuzzy measure* is called *k-additive* if $m(T) = 0$ for $T \subseteq G$ such that $|T| > k$ and there exists at least one $T \subseteq G$, with $|T| = k$, such that $m(T) > 0$. We observe that a 1-additive measure is the common additive fuzzy measure. In many real decision problems, it suffices to consider 2-additive measures. In this case, positive and negative interactions between two criteria are modeled without considering the interaction among any p-tuples (with $p > 2$) of criteria. From the point of view of MCDA, the use of 2-additive measures is justified by observing that the information on the importance of the single criteria and the interactions between two criteria are noteworthy. Moreover, it could be not easy or not straightforward for the DM to provide information on the interactions among three or more criteria during the decision procedure. From a computational point of view, the interest in the 2-additive measures lies in the fact that any decision model needs to evaluate a number $n + \binom{n}{2}$ of parameters (in terms of Möbius representation, a value $m(\{i\})$ for every criterion i and a value $m(\{i,j\})$ for every couple of distinct criteria $\{i,j\}$). With respect to a 2-additive fuzzy measure, the inverse transformation to obtain the fuzzy measure $\mu(R)$ from the Möbius representation is defined as:

$$\mu(R) = \sum_{i \in R} m(\{i\}) + \sum_{\{i,j\} \subseteq R} m(\{i,j\}), \ \forall R \subseteq G. \qquad (6)$$

With regard to 2-additive measures, properties **1b)** and **2b)** have, respectively, the following formulations:

1c) $m(\emptyset) = 0$, $\sum_{i \in G} m(\{i\}) + \sum_{\{i,j\} \subseteq G} m(\{i,j\}) = 1$,

2c) $\begin{cases} m(\{i\}) \geq 0, \ \forall i \in G, \\ m(\{i\}) + \sum_{j \in T} m(\{i,j\}) \geq 0, \ \forall i \in G \text{ and } \forall T \subseteq G \setminus \{i\}, \ T \neq \emptyset. \end{cases}$

In this case, the representation of the Choquet integral of $a \in A$ is given by:

$$C_\mu(a) = \sum_{\{i\} \subseteq G} m(\{i\})(g_i(a)) + \sum_{\{i,j\} \subseteq G} m(\{i,j\}) \min\{g_i(a), g_j(a)\}. \qquad (7)$$

Finally, we recall the definitions of the importance and interaction indices for couples of criteria.

The Shapley value [17] expressing the importance of criterion $i \in G$, is given by:

$$\varphi(\{i\}) = \sum_{T \subseteq G:\, i \notin T} \frac{(|G \setminus T| - 1)!|T|!}{|G|!} \cdot [\mu(T \cup \{i\}) - \mu(T)], \qquad (8)$$

while the *interaction index* [14] expressing the sign and the magnitude of the synergy in a couple of criteria $\{i, j\} \subseteq G$, is given by

$$\varphi(\{i, j\}) = \sum_{T \subseteq G:\, i,j \notin T} \frac{(|G \setminus T| - 2)!|T|!}{(|G| - 1)!} \cdot \tau(T, i, j), \qquad (9)$$

where $\tau(T, i, j) = [\mu(T \cup \{i, j\}) - \mu(T \cup \{i\}) - \mu(T \cup \{j\}) + \mu(T)]$.

In case of 2-additive capacities, the Shapley value and the interaction index can be expressed as follows:

$$\varphi(\{i\}) = m(\{i\}) + \sum_{j \in G \setminus \{i\}} \frac{m(\{i, j\})}{2}, \quad i \in G, \qquad (10)$$

$$\varphi(\{i, j\}) = m(\{i, j\}). \qquad (11)$$

3   Multiple Criteria Hierarchy Process (MCHP)

In MCHP, a set $\mathcal{G}$ of hierarchically ordered criteria is considered, i.e. all criteria are not considered at the same level, but they are distributed over l different levels (see Figure 1). At level 1, there are first level criteria called root criteria. Each root criterion has its own hierarchy tree. The leaves of each hierarchy tree are at the last level l and they are called elementary subcriteria. Thus, in graph theory terms, the whole hierarchy is a forest. We will use the following notation:

- l is the number of levels in the hierarchy of criteria,
- $\mathcal{G}$ is the set of all criteria at all considered levels,
- $\mathcal{I}_{\mathcal{G}}$ is the set of indices of particular criteria representing position of criteria in the hierarchy,
- m is the number of the first level criteria, $G_1, \ldots, G_m$,
- $G_{\mathbf{r}} \in \mathcal{G}$, with $\mathbf{r} = (i_1, \ldots, i_h) \in \mathcal{I}_{\mathcal{G}}$, denotes a subcriterion of the first level criterion G_{i_1} at level h; the first level criteria are denoted by G_{i_1}, $i_1 = 1, \ldots, m$,
- $n(\mathbf{r})$ is the number of subcriteria of $G_{\mathbf{r}}$ in the subsequent level, i.e. the direct subcriteria of $G_{\mathbf{r}}$ are $G_{(\mathbf{r},1)}, \ldots, G_{(\mathbf{r},n(\mathbf{r}))}$,
- $g_{\mathbf{t}} : A \to \mathbb{R}$, with $\mathbf{t} = (i_1, \ldots, i_l) \in \mathcal{I}_{\mathcal{G}}$, denotes an elementary subcriterion of the first level criterion G_{i_1}, i.e. a criterion at level l of the hierarchy tree of G_{i_1},

- EL is the set of indices of all elementary subcriteria:

$$EL = \{\mathbf{t} = (i_1, \dots, i_l) \in \mathcal{I}_\mathcal{G}\} \quad \text{where} \quad \begin{cases} i_1 = 1, \dots, m \\ i_2 = 1, \dots, n(i_1) \\ \cdots \cdots \\ i_l = 1, \dots, n(i_1, \dots, i_{l-1}) \end{cases}$$

- $E(G_\mathbf{r})$ is the set of indices of elementary subcriteria descending from $G_\mathbf{r}$, i.e.

$$E(G_\mathbf{r}) = \{(\mathbf{r}, i_{h+1}, \dots, i_l) \in \mathcal{I}_\mathcal{G}\} \quad \text{where} \quad \begin{cases} i_{h+1} = 1, \dots, n(\mathbf{r}) \\ \cdots \cdots \\ i_l = 1, \dots, n(\mathbf{r}, i_{h+1}, \dots, i_{l-1}) \end{cases}$$

thus, $E(G_\mathbf{r}) \subseteq EL$; in the case $G_\mathbf{r} \in EL$, then $E(G_\mathbf{r}) = G_\mathbf{r}$,

- when $\mathbf{r} = 0$, then by $G_\mathbf{r} = G_0$, we mean the entire set of criteria and not a particular criterion or subcriterion; in this particular case, we have $E(G_0) = EL$,
- given $\mathcal{F} \subseteq \mathcal{G}$, $E(\mathcal{F}) = \cup_{G_\mathbf{r} \in \mathcal{F}} E(G_\mathbf{r})$, that is $E(\mathcal{F})$ is composed by all elementary subcriteria descending from at least one criterion in $\mathcal{F}$,
- given $G_\mathbf{r} \in \mathcal{G}$, $\mathbf{r} \in \mathcal{I}_\mathcal{G} \cap \mathbb{N}^h$ ($G_\mathbf{r}$ is a criterion at level h), $1 \le h < l$, and $k \in \{h+1, \dots, l\}$, we define:

$$\mathcal{G}_\mathbf{r}^k = \left\{ G_{(\mathbf{r}, w)} \in \mathcal{G} : (\mathbf{r}, w) \in \mathcal{I}_\mathcal{G} \cap \mathbb{N}^k \right\}$$

being the set of all subcriteria of criterion $G_\mathbf{r}$ at level k. (For example, in Figure 1, we have that

$$\mathcal{G}_{i_1}^2 = \left\{ G_{(i_1,1)}, G_{(i_1,2)}, G_{(i_1,3)} \right\} \quad \text{and} \quad \mathcal{G}_{(i_1,2)}^3 = \left\{ g_{(i_1,2,1)}, g_{(i_1,2,2)} \right\} \Big)$$

Each alternative $a \in A$ is evaluated directly on the elementary subcriteria only, such that to each alternative $a \in A$ there corresponds a vector of evaluations:

$$\left(g_{\mathbf{t}_1}(a), \dots, g_{\mathbf{t}_n}(a) \right), \quad n = |EL|.$$

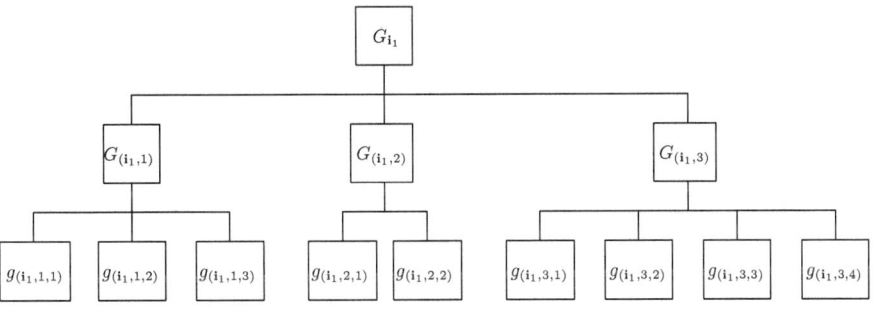

Fig. 1. Hierarchy of criteria for the first level (root) criterion G_{i_1}

Within MCHP, in each node $G_{\mathbf{r}} \in \mathcal{G}$ of the hierarchy tree there exists a preference relation $\succsim_{\mathbf{r}}$ on A, such that for all $a, b \in A$, $a \succsim_{\mathbf{r}} b$ means "a is at least as good as b on subcriterion $G_{\mathbf{r}}$". In the particular case where $G_{\mathbf{r}} = g_{\mathbf{t}}, \mathbf{t} \in EL$, $a \succsim_{\mathbf{t}} b$ holds if $g_{\mathbf{t}}(a) \geq g_{\mathbf{t}}(b)$.

4 Multiple Criteria Hierarchy Process for Choquet Integral Preference Model

In this article, we will aggregate the evaluations of alternative $a \in A$ with respect to the elementary subcriteria $g_{\mathbf{t}}, \mathbf{t} \in EL$, using the Choquet integral as follows.

On the basis of a capacity μ defined on the power set of EL, for all $a, b \in A$, $a \succsim b$ if $C_{\mu}(a) \geq C_{\mu}(b)$, where $C_{\mu}(a)$ and $C_{\mu}(b)$ are the Choquet integrals with respect to μ of the vectors $[g_{\mathbf{t}}(a), \mathbf{t} \in EL]$ and $[g_{\mathbf{t}}(b), \mathbf{t} \in EL]$, respectively.

For all $G_{\mathbf{r}} \in \mathcal{G}, \mathbf{r} \in \mathcal{I}_{\mathcal{G}} \cap \mathbb{N}^h$ ($G_{\mathbf{r}}$ is a criterion at level h), $h = 1, \ldots, l-1$ and for all $k = h+1, \ldots, l$, we can define the following capacity:

$$\mu_{\mathbf{r}}^k : 2^{\mathcal{G}_{\mathbf{r}}^k} \to [0, 1]$$

such that, for all $\mathcal{F} \subseteq \mathcal{G}_{\mathbf{r}}^k$, we have that

$$\mu_{\mathbf{r}}^k(\mathcal{F}) = \frac{\mu(E(\mathcal{F}))}{\mu(E(G_{\mathbf{r}}))} \tag{12}$$

In this way, $\mu_{\mathbf{r}}^k$ is a capacity defined on the power set of $\mathcal{G}_{\mathbf{r}}^k$, that could be computed using the capacity μ defined on the power set of EL.

In the following, we shall write $\mu_{\mathbf{r}}$ instead of $\mu_{\mathbf{r}}^l$.

For all $a, b \in A$, $a \succsim_{\mathbf{r}} b$ if $C_{\mu_{\mathbf{r}}}(a) \geq C_{\mu_{\mathbf{r}}}(b)$, where $C_{\mu_{\mathbf{r}}}(a)$ and $C_{\mu_{\mathbf{r}}}(b)$ are the Choquet integrals with respect to $\mu_{\mathbf{r}}$ of the vectors $[g_{\mathbf{t}}(a), \mathbf{t} \in E(G_{\mathbf{r}})]$ and $[g_{\mathbf{t}}(b), \mathbf{t} \in E(G_{\mathbf{r}})]$, respectively. Observe that for all $a \in A$,

$$C_{\mu_{\mathbf{r}}}(a) = \frac{C_{\mu}(a_{\mathbf{r}})}{\mu(E(G_{\mathbf{r}}))} \tag{13}$$

where $a_{\mathbf{r}}$ is a fictitious alternative having the same evaluations of a on elementary criteria from $E(G_{\mathbf{r}})$ and null evaluation on criteria from outside $E(G_{\mathbf{r}})$, i.e. $g_{\mathbf{s}}(a_{\mathbf{r}}) = g_{\mathbf{s}}(a)$ if $\mathbf{s} \in E(G_{\mathbf{r}})$ and $g_{\mathbf{s}}(a_{\mathbf{r}}) = 0$ if $\mathbf{s} \notin E(G_{\mathbf{r}})$.

The Shapley value expressing the importance of criterion $G_{(\mathbf{r},w)} \in \mathcal{G}_{\mathbf{r}}^k$ being thus a subcriterion of $G_{\mathbf{r}}$ at level k is:

$$\varphi_{\mathbf{r}}^k(G_{(\mathbf{r},w)}) = \sum_{T \subseteq \mathcal{G}_{\mathbf{r}}^k \setminus \{G_{(\mathbf{r},w)}\}} \frac{(|\mathcal{G}_{\mathbf{r}}^k \setminus T| - 1)!|T|!}{|\mathcal{G}_{\mathbf{r}}^k|!} \cdot [\mu_{\mathbf{r}}^k(T \cup \{G_{(\mathbf{r},w)}\}) - \mu_{\mathbf{r}}^k(T)] \tag{14}$$

while the interaction index expressing the sign and the magnitude of the synergy in a couple of criteria $G_{(r,w_1)}, G_{(r,w_2)} \in \mathcal{G}_r^k$ is given by:

$$\varphi_r^k(G_{(r,w_1)}, G_{(r,w_2)}) = \sum_{T \subseteq \mathcal{G}_r^k \setminus \{G_{(r,w_1)}, G_{(r,w_2)}\}} \frac{(|\mathcal{G}_r^k \setminus T| - 2)!|T|!}{(|\mathcal{G}_r^k| - 1)!} \cdot \tau_r^k(T, G_{(r,w_1)}, G_{(r,w_2)})$$

(15)

where

$$\tau_r^k(T, A, B) = \left[\mu_r^k(T \cup \{A, B\}) - \mu_r^k(T \cup \{A\}) - \mu_r^k(T \cup \{B\}) + \mu_r^k(T) \right].$$

In case the capacity μ on $\{g_t, t \in EL\}$ is 2-additive, the Shapley value $\varphi_r^k(G_{(r,w)})$ and the interaction index $\varphi_r^k(G_{(r,w_1)}, G_{(r,w_2)})$, with $G_{(r,w)}, G_{(r,w_1)}, G_{(r,w_2)} \in \mathcal{G}_r^k$, can be expressed as follows:

$$\varphi_r^k(G_{(r,w)}) =$$

$$= \left\{ \sum_{t \in E(G_{(r,w)})} m(g_t) + \sum_{t_1, t_2 \in E(G_{(r,w)})} m(g_{t_1}, g_{t_2}) + \sum_{\substack{t_1 \in E(G_{(r,w)}) \\ t_2 \in E(\mathcal{G}_r^k \setminus \{G_{(r,w)}\})}} \frac{m(g_{t_1}, g_{t_2})}{2} \right\} \cdot \frac{1}{\mu(E(G_r))}$$

(16)

$$\varphi_r^k(G_{(r,w_1)}, G_{(r,w_2)}) = \left\{ \sum_{\substack{t_1 \in E(G_{(r,w_1)}), \\ t_2 \in E(G_{(r,w_2)})}} m(g_{t_1}, g_{t_2}) \right\} \cdot \frac{1}{\mu(E(G_r))}.$$

(17)

Taking into account the expression of the Shapley index in equation (14) and $G_{s_1}, G_{s_2} \in G_{r_1}^k \cap G_{r_2}^k$ (that is G_{s_1} and G_{s_2} are subcriteria of both G_{r_1} and G_{r_2} located at level k), and supposing, without loss of generality, that $r_2 = (r_1, w)$ (that is G_{r_2} is a subcriterion of G_{r_1}), it is worth noting that the following inequalities could be verified:

$$\varphi_{r_1}^k(G_{s_1}) > \varphi_{r_1}^k(G_{s_2}) \quad \text{and} \quad \varphi_{r_2}^k(G_{s_1}) < \varphi_{r_2}^k(G_{s_2}) \quad \text{(or viceversa)}.$$

This means that the importance of the criterion G_{s_1} is greater than the importance of the criterion G_{s_2} if they are considered as subcriteria of G_{r_1}, but the importance of G_{s_2} is greater than importance of G_{s_1} if they are considered as subcriteria of G_{r_2}. In fact, in the computation of $\varphi_{r_1}^k(G_{s_1})$ we take into account not only the interactions between the elementary criteria descending from G_{s_1} but also the interactions between elementary criteria descending from G_{s_1} and elementary criteria descending from G_{r_1}. Because we have supposed that G_{r_2} is a subcriterion of G_{r_1}, and consequently $E(G_{r_2}) \subseteq E(G_{r_1})$, in the computation of $\varphi_{r_1}^k(G_{s_1})$ we take into account more interactions than those considered in the computation of $\varphi_{\mathbf{r}_2}^k(G_{\mathbf{s}_1})$. For example, evaluating students with respect to Science according to their scores on Mathematics and Physics, and with respect to Humanities according to their scores on Literature and Philosophy, one could consider Mathematics more important than Physics whithin Sciences and Literature more important than Philosophy within Humanities. However, taking into consideration that there is a great synergy between Philosophy and Physics, at the comprehensive level, Physics can be considered more important than Mathematics, as well as, at the same level, Philosophy can be considered more important than Literature.

Another interesting situation that can happen with respect to preferences represented by the Choquet integral in case of the hierarchy of criteria is the following. One can have that alternative a is evaluated better than alternative b with respect to all the subcriteria $G_{(\mathbf{r},1)}, \ldots, G_{(\mathbf{r},n(\mathbf{r}))}$ of criterion $G_{\mathbf{r}} \in \mathcal{G}$ from the subsequent level, and, nevertheless b can be evaluated better than a on criterion $G_{\mathbf{r}}$. For example, student a could be evaluated better than b on Science and Humanities but b could be evaluated better than a at the comprehensive level. This is due to the fact that when evaluating a student with respect to Science, we take into account only the interactions among subcriteria of Science as well as in the evaluation of a student with respect to Humanities we take into account only the interactions among subcriteria of Humanities. On the other hand, when evaluating a student comprehensively, we take into account also the interactions among the subcriteria of Science and subcriteria of Humanities. Thus, if there is a strong synergy between one subject from Science (for example, Physics) and another subject from Humanities (for example, Philosophy), and b is better evaluated than a in those subjects, this can result in the overall preference of b over a.

We shall show these situations in the didactic example presented in the next section.

5   A Didactic Example

Let us consider a set of nine students $A = \{a, b, c, d, e, f, g, h, k\}$ evaluated on the basis of two macro-subjects: Science and Humanities. Science has two sub-subjects: Mathematics and Physics, while Humanities has two sub-subjects: Literature and Philosophy. The number of levels considered is two.

Using a formal notation, we have $\mathcal{G} = \{G_1, G_2, G_{(1,1)}, G_{(1,2)}, G_{(2,1)}, G_{(2,2)}\}$, and the elements of $\mathcal{G}$ denote respectively, Science, Humanities, Mathematics, Physics, Literature and Philosophy. The students are evaluated on the basis of the elementary criteria only; such evaluations are shown in Table 1(a).

In the following, we shall consider a 2-additive capacity determined by the Möbius measures in Table 1(b).

Applying the expression (13) of the hierarchical Choquet integral introduced in Section 4, we can compute the evaluation of every student with respect to macro-subjects Science (G_1) and Humanities (G_2), while using the expression (7) of the Choquet integral, we can compute the evaluation of every student with respect to the whole hierarchy of criteria (see Table 2).

Table 1. Matrix evaluation and Möbius measures

(a) Matrix evaluation

	Science		Humanities	
Student	Mathematics	Physics	Literature	Philosophy
a	18	18	12	12
b	16	16	16	16
c	14	14	18	18
d	18	12	16	16
e	15	15	18	14
f	18	14	14	18
g	15	17	18	16
h	10	20	10	20
k	14	14	14	14

(b) Möbius measures

$m(G_{(1,1)})$	0.29
$m(G_{(1,2)})$	0.19
$m(G_{(2,1)})$	0.29
$m(G_{(2,2)})$	0.19
$m(G_{(1,1)}, G_{(1,2)})$	−0.1
$m(G_{(1,1)}, G_{(2,1)})$	0
$m(G_{(1,1)}, G_{(2,2)})$	0
$m(G_{(1,2)}, G_{(2,1)})$	0
$m(G_{(1,2)}, G_{(2,2)})$	0.24
$m(G_{(2,1)}, G_{(2,2)})$	−0.1

For example, looking at the first three rows in Table 2, we get:

- the Choquet integral of a with respect to Science is equal to 18 and it is computed considering the fictitious alternative a_1 having the same evaluations of a on the elementary criteria descending from Science, and null evaluations on all other elementary criteria,
- the Choquet integral of a with respect to Humanities is equal to 12 and it is computed considering the fictitious alternative a_2 having the same evaluations of a on the elementary criteria descending from Humanities, and null evaluations on all other elementary criteria,
- the Choquet integral of a with respect to the whole hierarchy of criteria is equal to 14.28 and it is computed considering the evaluations of a on all elementary criteria.

Hereafter we underline the very interesting inversion of preference regarding alternatives h and k. In fact, looking at Table 2, we can observe that k is better than h with respect to both macro-subjects Science and Humanities $(C_{\mu_1}(k) > C_{\mu_1}(h)$ and $C_{\mu_2}(k) > C_{\mu_2}(h))$ but h is comprehensively better than k $(C_\mu(h) > C_\mu(k))$. The reason of this inversion of preference is explained considering that in the computation of $C_{\mu_1}(\cdot)$ and $C_{\mu_2}(\cdot)$ we take into account only the interaction between elementary criteria descending from Science and Humanities respectively, while in the computation of the comprehensive Choquet integral $C_\mu(\cdot)$ we take into account the possible interactions between all elementary criteria in the hierarchy.

By considering the capacities on the elementary criteria displayed in Table 1(b) and adopting the expression (16) defined in Section 4, we compute the Shapley values of the elementary criteria $G_{(r,i)}$ with respect to their relative parent criterion G_r (see Table 3(a)). Then the overall Shapley values of the elementary criteria (i.e. with respect to G_0) are calculated and shown in Table 3(b). Finally, the Shapley values of subcriteria G_1 (Science) and G_2 (Humanities)

Table 2. Choquet integrals with respect to the macro-subjects Science and Humanities and with respect to the whole hierarchy of criteria

	Science		Humanities			Choquet integrals
	Mathematics	Physics	Literature	Philosophy		
a_1	18	18	0	0	$C_{\mu_1}(a)$	18
a_2	0	0	12	12	$C_{\mu_2}(a)$	12
a	18	18	12	12	$C_{\mu}(a)$	14.28
b_1	16	16	0	0	$C_{\mu_1}(b)$	16
b_2	0	0	16	16	$C_{\mu_2}(b)$	16
b	16	16	16	16	$C_{\mu}(b)$	16
c_1	14	14	0	0	$C_{\mu_1}(c)$	14
c_2	0	0	18	18	$C_{\mu_2}(c)$	18
c	14	14	18	18	$C_{\mu}(c)$	15.52
d_1	18	12	0	0	$C_{\mu_1}(d)$	16.57
d_2	0	0	16	16	$C_{\mu_2}(d)$	16
d	18	12	16	16	$C_{\mu}(d)$	15.26
e_1	15	15	0	0	$C_{\mu_1}(e)$	15
e_2	0	0	18	14	$C_{\mu_2}(e)$	17.05
e	15	15	18	14	$C_{\mu}(e)$	15.54
f_1	18	14	0	0	$C_{\mu_1}(f)$	17.05
f_2	0	0	14	18	$C_{\mu_2}(f)$	16
f	18	14	14	18	$C_{\mu}(f)$	15.92
g_1	15	17	0	0	$C_{\mu_1}(g)$	16
g_2	0	0	18	16	$C_{\mu_2}(g)$	17.52
g	15	17	18	16	$C_{\mu}(g)$	16.58
h_1	10	20	0	0	$C_{\mu_1}(h)$	13.5
h_2	0	0	10	20	$C_{\mu_2}(h)$	13.5
h	10	20	10	20	$C_{\mu}(h)$	15.06
k_1	14	14	0	0	$C_{\mu_1}(k)$	14
k_2	0	0	14	14	$C_{\mu_2}(k)$	14
k	14	14	14	14	$C_{\mu}(k)$	14

and their interaction index (see the expression (17) introduced in Section 4) are computed and displayed in Table 4.

As it has been announced in Section 4, in this example, Mathematics is more important than Physics, when they are considered as subcriteria of Science (see Table 3(a)) and, conversely, Physics is more important than Mathematics when they are considered as subcriteria of the whole set of criteria G_0 (see Table 3(b)).

In order to illustrate the difference between our approach and that of [18], in the following we shall compute the comprehensive evaluations of student g following the approach of [18]. At first, we need to define a capacity for each node of the hierarchy of criteria, which is not an elementary criterion. Because in our didactic example the hierarchy is composed of three nodes being different from the elementary criteria, we need to define three capacities, $\mu_{\{Sci\}}$, $\mu_{\{Hum\}}$, and $\mu_{\{Sci,Hum\}}$ on $\{Math, Phy\}$, $\{Lit, Phi\}$ and $\{Sci, Hum\}$, respectively. Let us suppose that the capacities are defined using the Möbius measures shown in Table 5.

Table 3. Shapley values

(a) Shapley values of every elementary criterion with respect to every macro-subject G_r

	Science		Humanities	
	Mathematics	Physics	Literature	Philosophy
$\varphi_r^k(G_{(r,w)})$	0.63	0.36	0.63	0.36

(b) Shapley values of the elementary criteria

	$\varphi_r^k(G_{(r,w)})$
Mathematics	0.24
Physics	0.26
Literature	0.24
Philosophy	0.26

Table 4. The Shapley values and interaction index of Science (G_1) and Humanities (G_2)

	$\varphi_r^k(G_{(r,w)})$
Science	0.5
Humanities	0.5
	$\varphi_r^k(G_{(r,w_1)}, G_{(r,w_2)})$
Science and Humanities	0.24

Table 5. Möbius measures

Science		Humanities		Science,Humanities	
$m(\{Math\})$	0.7	$m(\{Lit\})$	0.5	$m(\{Sci\})$	0.4
$m(\{Phy\})$	0.5	$m(\{Phi\})$	0.6	$m(\{Hum\})$	0.4
$m(\{Math, Phy\})$	-0.2	$m(\{Lit, Phi\})$	-0.1	$m(\{Sci, Hum\})$	0.2

Computing the Choquet integral of $g = (15, 17, 18, 16)$ with respect to criteria Science and Humanities using the capacities $\mu_{\{Sci\}}$ and $\mu_{\{Hum\}}$, we get:

$$C_{\{Sci\}}(g) = 15m(\{Math\}) + 17m(\{Phy\}) + 15m(\{Math, Phys\}) = 16$$

$$C_{\{Hum\}}(g) = 18m(\{Lit\} + 16m(\{Phi\} + 16m(\{Lit, Phi\} = 17$$

Then, the comprehensive Choquet integral of g is obtained by aggregating the evaluations $(C_{\{Sci\}}(g), C_{\{Hum\}}(g))$ using the capacity $\mu_{\{Sci, Hum\}}$:

$$C_{\{Sci, Hum\}}(g) = C_{\{Sci\}}(g)m(\{Sci\}) + C_{\{Hum\}}(g)m(\{Hum\})+$$

$$+min(C_{\{Sci\}}(g), C_{\{Hum\}}(g))m(\{Sci, Hum\}) = 16.4.$$

The remarkable difference between the method presented in [18] and our approach, is that in the first one, one capacity has to be defined with respect to each node of the hierarchy of criteria being different from the elementary criteria (for example in the didactic example we have defined three different capacities) while in our approach one needs to define only one capacity on the set of all elementary criteria, and the capacities at higher levels are calculated according to formulas given in Section 4.

6   Conclusions

We have proposed the application of the Multiple Criteria Hierarchy Process (MCHP) to a preference model expressed in terms of Choquet integral, in order to deal with interaction among criteria. Application of the MCHP to the Choquet integral permits the handling of importance and interactions of criteria with respect to any subcriterion of the hierarchy. To apply the MCHP to the Choquet integral in real world problems, it is necessary to elicit preference model parameters, which in this case are the non-interactive weights represented by a capacity. The added value of the MCHP is that it permits the DM expressing the preference information related to any criterion of the hierarchy. When MCHP is combined with a disaggregation procedure, the DM can say, for example, that student a is globally preferred to student b, but he can also say that student c is better than student d in Humanities. DM can also say that criterion Science is more important than Humanities, or that the interaction between Physics and Philosophy is greater than the interaction between Mathematics and Literature. Many multicriteria disaggregation procedures have been proposed to infer a capacity from those types of preference information, however, without considering the hierarchy of criteria (see, for example, [13]). Recently, a new multicriteria disaggregation method has been proposed to take into account that, in general, more than one capacity is able to represent the preference expressed by the DM: Non Additive Robust Ordinal Regression (NAROR) [1]. NAROR considers all the capacities that are compatible with the preference information given by the DM, adopting the concepts of possible and necessary preference introduced in [11]. In simple words, a is necessarily or possibly preferred to b, if it is preferred for all compatible capacities or for at least one compatible capacity, respectively. In our opinion, application of NAROR to MCHP for the Choquet integral will permit to take into account interaction among hierarchically structured criteria in a very efficient way, enabling the handling of many complex real world problems.

Acknowledgment. The fourth author wishes to acknowledge financial support from the Polish National Science Centre, grant no. NN519 441939.

References

1. Angilella, S., Greco, S., Matarazzo, B.: Non-additive robust ordinal regression: A multiple criteria decision model based on the Choquet integral. European Journal of Operational Research 201(1), 277–288 (2010)
2. Chateauneuf, A., Jaffray, J.Y.: Some characterizations of lower probabilities and other monotone capacities through the use of Möbius inversion. Mathematical Social Sciences 17, 263–283 (1989)

3. Choquet, G.: Theory of capacities. Ann. Inst. Fourier 5, 131–295 (1953)
4. Corrente, S., Greco, S., Słowiński, R.: Multiple Criteria Hierarchy Process in Robust Ordinal Regression. Decision Support Systems 53(3), 660–674 (2012)
5. Figueira, J., Greco, S., Ehrgott, M. (eds.): Multiple Criteria Decision Analysis: State of the Art Surveys. Springer, Berlin (2005)
6. Gilboa, I., Schmeidler, D.: Additive representations of non-additive measures and the Choquet integral. Ann. Operational Research 52, 43–65 (1994)
7. Grabisch, M.: The application of fuzzy integrals in multicriteria decision making. European Journal of Operational Research 89, 445–456 (1996)
8. Grabisch, M.: k-order additive discrete fuzzy measures and their representation. Fuzzy Sets and Systems 92, 167–189 (1997)
9. Grabisch, M., Labreuche, C.: Fuzzy measures and integrals in MCDA. In: Figueira, J., Greco, S., Ehrgott, M. (eds.) Multiple Criteria Decision Analysis: State of the Art Surveys, pp. 563–604. Springer, Berlin (2005)
10. Grabisch, M., Labreuche, C.: A decade of application of the Choquet and Sugeno integrals in multi-criteria decision aid. Annals of Operations Research 175(1), 247–290 (2010)
11. Greco, S., Mousseau, V., Słowiński, R.: Ordinal regression revisited: multiple criteria ranking using a set of additive value functions. European Journal of Operational Research 191(2), 416–436 (2008)
12. Keeney, R.L., Raiffa, H.: Decisions with multiple objectives: Preferences and value tradeoffs. J. Wiley, New York (1976)
13. Marichal, J.L., Roubens, M.: Determination of weights of interacting criteria from a reference set. European Journal of Operational Research 124(3), 641–650 (2000)
14. Murofushi, T., Soneda, S.: Techniques for reading fuzzy measures (iii): interaction index. In: 9th Fuzzy Systems Symposium, Sapporo, Japan, pp. 693–696 (1993)
15. Rota, G.C.: On the foundations of combinatorial theory I. Theory of Möbius functions. Wahrscheinlichkeitstheorie und Verwandte Gebiete 2, 340–368 (1964)
16. Shafer, G.: A Mathematical Theory of Evidence. Princeton University Press (1976)
17. Shapley, L.S.: A value for n-person games. In: Kuhn, H.W., Tucker, A.W. (eds.) Contributions to the Theory of Games II, pp. 307–317. Princeton University Press, Princeton (1953)
18. Sugeno, M., Fujimoto, K., Murofushi, T.: Hierarchical decomposition theorems for Choquet integral models. In: Proceedings-IEEE International Symposium on Circuits and Systems, vol. 3, pp. 2075–2078 (1995)
19. Tzeng, G.H., Ou Yang, C.T., Lin, Y.P., Chen, C.B.: Hierarchical MADM with fuzzy integral for evaluating enterprise intranet web sites. Information Sciences 169(3-4), 100–126 (2005)

Selection of Inspection Intervals Based on Multi-attribute Utility Theory

Rodrigo J.P. Ferreira, Adiel T. de Almeida, and Cristiano A.V. Cavalcante

Universidade Federal de Pernambuco, Brazil
{rodjpf,almeidaatd,cristianogesm}@gmail.com

Abstract. In the context of complex equipment, inspection has an important role to play in maintaining operations at a suitable performance. In fact, for this kind of system, failures are not immediate. Therefore, the concept of delay time is very useful as a two-step failure process, as a basis for constructing inspection models. Another aspect worth noticing is the fact that, in practice, when the decision-maker decides on the time between inspections, he/she takes different aspects into account in a non-structured way. This happens because the majority of inspection models deal with the problem using an optimization approach, where only one aspect is considered. So, the main contribution of this article is to put forward a multicriteria decision model in order to aid maintenance planning. This model takes the decision maker's preferences into account, as well as, the most important aspects, when considering setting the inspection intervals for periodic condition monitoring, and which are the cost and downtime associated with the inspection policy.

Keywords: Multi-attribute utility theory, maintenance, inspection.

1 Introduction

A multicriteria decision model to aid maintenance planning is investigated in this paper. The novelty of the paper lies in the fact that it considers the delay time concept under a multicriteria approach. It is assumed that the equipment is complex. Therefore, renewal does not take place during maintenance actions. This is because a repair or a replacement of an individual component does not renew the whole equipment.

In fact, having an inspection policy is very useful in order to check the real state of the system. In practice, this kind of policy is translated into several activities to check the state of the system and this might be accomplished with or without instruments. The real contribution of this activity is to identify intermediate states before failure. Despite the real contribution of an inspection, according to Badia (2002), most maintenance policies assume that failures are detected as soon as they occur, and, indeed, for continuous operations this is a realistic assumption. On the other hand, there are some systems where even when failures are visible they are associated with defects that precede them (Cavalcante et al., 2011; Scarf and Cavalcante, 2010). Therefore, any kind of procedure that helps identify such defective items is useful to

R.C. Purshouse et al. (Eds.): EMO 2013, LNCS 7811, pp. 490–499, 2013.

prevent the failure from happening, as well as, to ensure that these items spend as little time as possible in the defective state, since   defects cause losses in the production process and are often not discovered until they take place or, finally, are the cause of failures.

These processes that help to learn more about the state of the items are often called inspection policies. There are many of articles that deal with this problem such as (Valdez-Florez and Feldman, 1989). The standard inspection policy consists of checks on a unit that operates for an infinite span at successive times T_K ($k = 1, 2...$), where $T_0 = 0$. The most common assumption is that any failure is detected at the time of the next check and a replacement is immediately made (Barlow and Proschan, 1965; Nakagawa, 2005). A derivative approach analyzes the delay time. Delay time is the time lapse from when a system defect could first have been noticed until the time when its repair can no longer be delayed because of unacceptable consequences such as a serious catastrophe which might arise due to failure (Christer, 1999). In these models, the focus changes to defects rather failure, since they assume that failures are preceded by defects. These models are very useful for a large number of different situations, since they consider that failures are consequences of two different stochastic mechanisms: the delay in itself and the initial point of identifying defects.

The importance of delay time in maintenance management applications was investigated by Wang (2012). Formally, the delay time concept considers the failure process as a two-stage process, where the first is to do with the emergence of the defect, in which case a defect may be first identified by an inspection, and the second process is about the time lapse from that point until failure, should the defect not be attended to (Wang, 2011) . The time lapse related to the second process, from the initial point of an identifiable defect to failure, is called the delay time of the failure. The delay time concept is shown in Figure 1.

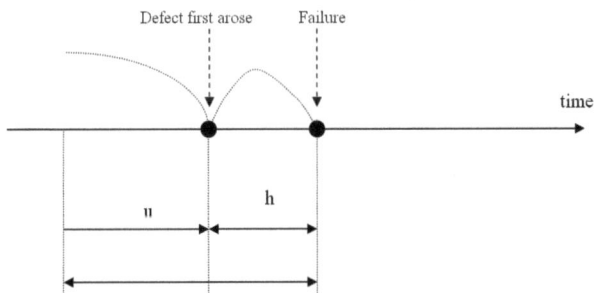

Fig. 1. Delay time concept (Christer, 1982)

The main contribution of the delay time concept arises from the fact that if an inspection is carried out during the delay time of the failure, the defect present might well be detected, and consequently the failure could be avoided. According to Wang (2012) the strong appeal of applying this method can be seen by considering the large number of articles on case studies involving real situations using the delay time concept (Sheils et al (2010);, Berrade, 2012).

However, despite the delay time method pointing to promising horizons in terms of realistic models, another important aspect of the maintenance context needs to be handled in a more effective way; the presence of multiple objectives, in maintenance decisions, in most cases, is still being ignored. There are few studies that deal a maintenance problem under a mutiple objective approach (see, Almeida 2001; Almeida, 2005 and Ferreira 2009).

Therefore, this paper sets out to illustrate the use of a multicriteria model in order to set inspection intervals selection based on multi-attribute utility theory. This model provides decision-makers with a broader view which enables them to explore the different decision dimensions, which thus leads to a better understanding of the conflicts between the criteria, and encourages them to reveal their preferences regarding these criteria.

This paper is organised as follows. A basic introduction and a brief review of the literature are presented in Section 1. The building of a multi-attribute decision model to set inspection intervals is specified in detail in Section 2, and in Section 3, the application of a decision model is presented. Section 4 provides a conclusion of the main results of the paper.

2   Problem Description and Assumptions

For the specific case of an inspection, of a complex system, managers are interested in balancing the costs of the inspection policy with savings arising from improving the performance of the system. In fact, depending on the production process that is supported by a complex system, a very short interruption due to a failure can cause a substantial financial loss. In other contexts, a failure can mean much more than cost. For example, in a service system, a failure can provoke consequences for the users beyond costs. Cavalcante & Almeida (2007) commented on some situations, where cost criteria are not suitable with regard to representing the real objective. They argued that this happens because the policies resulting from aiming only at the minimum cost allow levels of failures that are too high, and therefore unacceptable for these contexts. For example, in the contexts of maintaining medical equipment, the protection of a country by military means and an oil distribution system, a failure could lead to disastrous consequences, and therefore such contexts should not be constrained only by the monetary dimension.

Therefore, the decision about the frequency of inspections should consider all the aspects that are important. In this article, an electric energy system is dealt with where besides costs, availability is very important. In fact, some consequences related to a failure cannot be translated into a monetary scale. Notice that even though consideration of this is complicated, these kinds of consequence, i.e. those beyond cost, should be taken in account, since this aspect, sometimes, is the most relevant in a decision process that seeks to specify at what intervals an inspection should take place. In the electricity sector, the lower the availability, the heavier the fine, but also, in some sense, this aspect indirectly represents all the other dimensions of consequence, since the greater the availability, the smaller the impact of any other consequences associated with the failure. Thus, increasing availability is a way to reduce all the other consequences associated with failure.

Since the model is based on the concept of delay time, the construction of the model should consider the common assumptions defined in the delay time approach. For the particular case of a complex system, this article considered electrical equipment. It is emphasized that a replacement or repair of one component of such equipment is not enough to renew the system. Therefore, the failure process is assumed to be a non-homogeneous Poisson process. The assumptions associated with this process are listed below:

- An inspection takes place every T time units and lasts d_i, where $d_i << T$;
- Inspections are perfect in that any defect present within the system will be identified. Once the defects are identified, the repairs are made immediately after the inspection, the repair for each defects lasts d_d;
- The defects arise at a constant rate per unit of time (λ).
- The delay time (h) is independent of the initial point u;
- The distribution of the delay time is known, and its probability density function $f_h(h)$ is known.

Using these assumptions, the probability of a fault arising as a breakdown is:

$$b(T) = \int_0^T \frac{(T-h)}{T} f_h(h)dh \qquad (1)$$

2.1   The Cost Attribute

The operational cost of applying an inspection policy can be measured. Based on the delay time concept and admitting that the most important assumption that is brought from this fundamental approach is cost, a complex system can be modelled by the following expression:

$$C(T) = \frac{c_i + (c_f)\lambda T b(T) + c_d \lambda T[1-b(T)]}{T + d_i + d_d \lambda T[1-b(T)]} \qquad (2)$$

The assumptions that were assumed are listed below:

- Inspection cost is (c_i)
- Cost of repairs under a failure (c_f);
- Cost of repairing a defect just after the inspection (c_d);

As expected, the cost of inspection is the smallest cost (c_i)<(c_d) < (c_f).

Regarding the third assumption, it was considered that any necessary repair could be planned and made just after the inspection is finished. For a complex system this assumption is quite feasible because, during an inspection, many defects are detected. Having identified these defects, a team could be sent to fix them.

2.2   The Availability Attribute

In terms of availability, as was already mentioned, this criterion is related to the non-monetary aspects associated with the failures. This reflects the ability of the system to perform under the influence of an inspection policy. Availability is about the percentage of time that the system is available. So, in terms of a service system, this availability corresponds to the percentage of time that the service is provided to the client. Therefore, it is easy to see what the percentage of time is that the client is not served. Thus, the lower the availability is, the greater the client's dissatisfaction.

The availability of the complex system can be modelled by the following expression:

$$A(T) = \frac{T - d_f \lambda Tb(T)}{T + d_i + d_D \lambda T [1 - b(T)]} \tag{3}$$

The assumptions made are listed below:

- Inspection time is (d_i)
- The mean time of a repair is (d_f). This time corresponds to repair associated with failures.
- The mean time of repairing a defect just after the inspection is (d_d);

The availability function can obtain identical values for different times between inspections. In short intervals between inspections, high unavailability can occur due to high frequency of time spent on inspections. However for long inspection times, there will be a high proportion of unavailability due to failures. In this situation, the alternatives have the same value in the utility function considering only this criterion. Let $A(t_1) = A(t_2)$, then $U(A(t_1)) = U(A(t_2))$ in their respective unidimensional utility functions.

2.3   The Multi-attribute Utility Theory Model

A multi-attribute utility theory model is proposed considering that attributes of cost and availability are additive independent if and only if the two-attribute utility function is additive. For these criteria, the additive form may be written either as:

$$\max u(C(T), A(T)) = k_c u_c (C(T)) + k_a u_a (A(T)) \tag{4}$$

Once additive independence is observed, the strategy of divide and conquer could be thoroughly be explored when assessing the multi-attribute utility function. Therefore, each one-dimensional utility function for each respective attribute should be elicited. Alternatively, in some cases, a specific analytic function could be used where its shape gives an interesting description of a specific instance of the decision-maker's behaviour for a given attribute. A five steps procedure defined by Ferreira et al. (2009) is indicated to evaluate the consistency of multi-attribute utility function proposed in this paper.

Criteria could not be independent preferentially for a given decision maker. In this case, specific procedures should be applied for deal with this situation, see Keeney & Raiffa (1976). In maintenance management problems, in general, decision makers are familiar to think probabilistically and MAUT can facilitate the use of this type of problem. Other applications of maintenance management problems using multicriteria models are proposed by Brito, et al. (2010), Cavalcante et al. (2010); Alencar & Almeida (2010).

For decision makers who feel difficulty in building one-dimensional utility functions, scale constants and other parameters of multiattribute utility functions other methods can be used, such as UTA methods (Jacquet-Lagreze & Siskos 1982) and Robust Ordinal Regression methods (Greco, Mousseau and Słowiński, 2008).

Selection of inspection intervals can be modelled as group decision making problems when more than one decision makers are involved (Morais & Almeida, 2011; Morais & Almeida, 2006; Morais & Almeida, 2010).

3   Case Study

This paper proposes an inspection model to a safety valve that is responsible for isolating a section. In particular, we considered that the maintenance of pipelines is very complex and depends on high technology tools, since they are underground. Also most failures in pipelines are caused not by the natural degradation, but by external agents, which are not associated with time (Majid et al, 2012). Therefore, in the context of large expansion the operational readiness of protection systems is very important to make sure that, in a leakage situation, the control is manageable and the consequences are acceptable.

We consider a valve as a system comprising the internal parts (trim), body and the pilot, which together are responsible for isolating a given section on demand. Thus, they form a complex system that has to be inspected in order to mitigate a consequence of failure. The problem is that in each time T, in order to see what is happening to this system, by inspection, the valve is tested. The inspection lasts d_i time. If some defects were detected on inspection, the section of gas pipeline is interrupted in order to make a minimal repair on the valve, this repair takes d_d and is done immediately after the inspection. If a failure is observed the valve undergoes a major repair that takes more time (d_f) and is more expensive than the repair that is done when a defect is detected. Notice that for both cases the repair has no effect on the reliability of the valve, since it is complex equipment.

Failures of the operational function of the system are undesirable and incur cost c_f, when a defect is detected before the failure the cost of a minimal repair (c_d) is taken and, when nothing is found, the cost of the inspection is c_i.

We suppose that a defect arises prior to failure, that defects are detectable at inspection, and that any part that is defective on inspection is immediately repaired.

Table 1 presents all the parameters associated with the model.

Table 1. Parameters of the model

λ	0.022 faults per day
$h(mean\ delay\ time)$	Exponential 60 days
d_i	1 hour
d_d	4 hours
d_f	14 hours
c_f	US$ 1,200.00
c_d	US$ 500.00
c_i	US$ 200.00

Based on these parameters from Table 1, the minimum cost is US$ 200,43 for T=52 days and the maximum availability is 0,99298 for T=27 days. Thus, the cost and availability functions for the efficient frontier are plotted in Figure 2 and Figure 3.

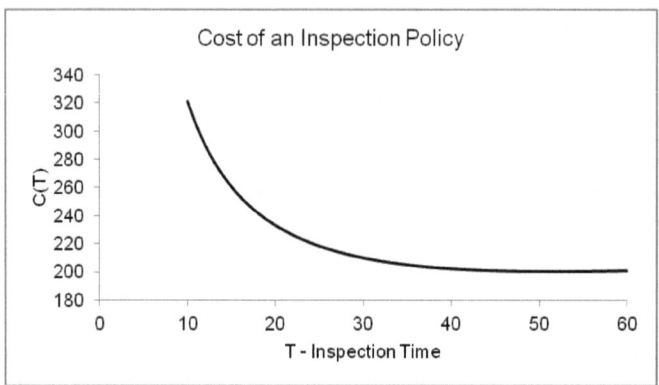

Fig. 2. Behavior of cost attribute as a function of inspection time

Fig. 3. Behavior of availability attribute as a function of inspection time

MAUT was the approach chosen to deal with the uncertainty and the conflicting criteria, as well as to take into account the decision maker's preferences. According to Keeney and Raiffa (1976), there is not an unique series of steps in order to guarantee a properly assessed multiattibute utility function. Thus, for the case of two attributes, the authors propose a five step process for assessing the multiattibute utility function. This process was proposed by the authors in order to provide a better discussion between the authors involved on the decision process. The sequence of steps is as follows: (1) introducing the terminology and idea; (2) identifying relevant independence assumptions (3) assessing conditional utility functions; (4) assessing scaling constants; (5) checking for consistency and reiterating.

As the objective of this process consists of representing the preference of the decision maker by a model, it is important that the decision maker has no doubt about the process, so this first step should not only explain the whole assessment process, but also make sure that the consequence space is completely understood. In our specific case, the set of alternatives comprises any time on the interval $[0, \infty)$. In fact, the alternatives correspond to inspection time t. For each inspection time, t, there is a consequence in terms of cost, $C(t)$, and Availability, $A(t)$. A graphical representation is a relevant instrument to make the decision-maker understand the consequence space.

Once the consequence space is well understood by the decision-maker, the discussion about the his/her preferences should be addressed in order to make it possible to represent these for an utility function. The hypothesis of additive independence was verified. Once the additive independence is observed, the strategy of divide and conquer could really be explored on the assessment of the multi-attribute utility function. In this way, each one-dimensional utility functions for each respective attribute should be elicited. Alternatively, in some cases, a specific analytic function could be used where its shape is interesting to describe a specific preference behaviour of the decision-maker for a given attribute.

The fourth step consists of assessing scaling constants. As the scaling constants form a final expression of the multiattribute utility function, the number and meaning of theses constants is affected by the independence relationships that were observed between the attributes. As we are supposing an additive independence, for our particular case of two attributes we will have two scaling constants. The value of k_c estimated by lotteries is 0.88. This lottery is represented by $u(C^*(T), A^0(T))$ and the value of p where the decision maker becomes indifferent between the lottery with $u(C^*(T), A^*(T))$ and $u(C^0(T), A^0(T))$. The inspection interval with maximum utility is 35 days.

The final step consists of the check of the consistency, so the decision maker has to judge some points of the space of consequence in terms of preference, if the multiattribute utility function is consistent with these judgments the model is working well.

4 Conclusions

This article applied a multi-attribute utility theory to support the planning of an inspection policy. The concept of delay time was used to model the failure process of the valves. In fact, in practice a valve presents a defect before a failure.

For the decision-maker, the model allows a broader view of the problem, in addition to which it weights the alternative that is in accordance with the decision maker's preference, in fact when there is a failure some others non-monetary aspects are involved, such as dissatisfaction. Therefore in order to take into account this aspect the availability was measured for different alternatives of time inspection. In fact, the grater is the unavailability the greater is the dissatisfaction of the decision-maker.

For the specific case of a protection valve a failure of the valve impact on the number of the clients whose gas supplies will be interrupted. Furthermore, in the worst scenario the failure could have a very bad consequence, since the natural gas could cause a explosion in cases where it is not possible to isolate a section with a leakage.

Acknowlegdements. This study has been partially supported by CNPq (Brazilian Research Council).

References

1. Alencar, M.H., de Almeida, A.T.: Assigning priorities to actions in a pipeline transporting hydrogen based on a multicriteria decision model. International Journal of Hydrogen Energy 35, 3610–3619 (2010)
2. Almeida, A.T.: Multicriteria decision making on maintenance: spares and contracts planning. European Journal of Operational Research 129, 235–241 (2001)
3. Almeida, A.T.: Multicriteria modeling of repair contract based on utility and Electre I method with dependability and service quality criteria. Annals of Operations Research 138, 113–126 (2005)
4. Badia, F.G., Berrade, M.D., Campos, C.A.: Optimal inspection and preventive maintenance of units with revealed and unrevealed failures. Reliability Engineering and System Safety 78, 157–163 (2002)
5. Barlow, R.E., Proschan, F.: Mathematical Theory of Reliability. Wiley, New York (1965)
6. Berrade, M.D., Cavalcante, C.A.V., Scarf, P.A.: Maintenance scheduling of a protection system subject to imperfect inspection and replacement. European Journal of Operational Research 218(3), 716–725 (2012)
7. Brito, A.J.M., Almeida Filho, A.T., Almeida, A.T.: Multi-criteria decision model for selecting repair contracts by applying utility theory and variable interdependent parameters. IMA Journal of Management Mathematics 21, 349–361 (2010)
8. Cavalcante, C.A.V., de Almeida, A.T.: A Multicriteria Decision Aiding Model Using PROMETHEE III for Preventive Maintenance Planning Under Uncertain Conditions. Journal of Quality in Maintenance Engineering 13, 385–397 (2007)
9. Cavalcante, C.A.V., Ferreira, R.J.P., de Almeida, A.T.: A preventive maintenance decision model based on multicriteria method PROMETHEE II integrated with Bayesian approach IMA. Journal of Management Mathematics 21, 333–348 (2010)
10. Cavalcante, C.A.V., Scarf, P.A., de Almeida, A.T.: A study of a two-phase inspection policy for a preparedness system with a defective state and heterogeneous lifetime. Reliability Engineering & System Safety 96(6), 627–635 (2011)
11. Christer, A.H.: Developments in delay time analysis for modeling plant maintenance. Journal of Operational research Society 50, 1120–1137 (1999)
12. Christer, A.H.: Modelling Inspection Policies for Building Maintenance. Journal of Operational Research Society 33, 723–732 (1982)

13. Ferreira, R.J.P., Almeida, A.T., Cavalcante, C.A.V.: A multi-criteria decision model to determine inspection intervals of condition monitoring based on delay time analysis. Reliability Engineering & System Safety 94(5), 905–912 (2009)

14. Greco, S., Mousseau, V., Słowińskic, R.: Ordinal regression revisited: Multiple criteria ranking using a set of additive value functions. European Journal of Operational Research 191(2), 416–436 (2008)

15. Jacquet-Lagreze, E., Siskos, J.: Assessing a set of additive utility functions for multicriteria decision-making, the UTA method. European Journal of Operational Research 10(2), 151–164 (1982)

16. Keeney, R.L., Raiffa, H.: Decision With Multiple Objectives: Preferences and Value Trade-offs. Wiley, New York (1976)

17. Majid, Z.A., Mohsin, R., Yaacob, Z., Hassan, Z.: Failure analysis of natural gas pipes. Engineering Failure Analysis 17(4), 818–837 (2010)

18. Morais, D.C., Almeida, A.T.: Water supply system decision-making using multicriteria analysis. Water SA 32(2), 229–235 (2006)

19. Morais, D.C., Almeida, A.T.: Group decision making on water resources based on analysis of individual rankings. Omega (Oxford) 40, 42–52 (2011)

20. Morais, D.C., Almeida, A.T.: Water network rehabilitation: A group decision-making approach. Water SA 36, 487–494 (2010)

21. Nakagawa, T.: Maintenance Theory of Reliability. Springer, London (2005)

22. Scarf, P.A., Cavalcante, C.A.V.: Hybrid block replacement and inspection policies for a multi-component system with heterogeneous component lives. European Journal of Operational Research 206(2), 384–394 (2010)

23. Sheils, E., O'Connor, A., Breysse, D., Schoefs, F., Yotte, S.: Development of a two-stage inspection process for the assessment of deteriorating infrastructure. Reliability Engineering & System Safety 95(3), 182–194 (2010)

24. Silva, V.B.S., Morais, D.C., Almeida, A.T.: A Multicriteria Group Decision Model to Support Watershed Committees in Brazil. Water Resources Management 24, 4075–4091 (2010)

25. Valdez-Flores, C., Feldman, R.M.: A survey of preventive maintenance models for stochastically de- teriorating single-unit systems. Naval Research Logistics 36, 419–446 (1989)

26. Wang, W.: An inspection model based on a three-stage failure process. Reliability Engineering & System Safety 96(7), 838–848 (2011)

27. Wang, W.: An overview of the recent advances in delay-time-based maintenance modelling. Reliability Engineering & System Safety 106, 165–178 (2012)

Minimizing the Compensatory Effect of MCDM Group Decision Additive Aggregation Using the Veto Concept

Suzana de França Dantas Daher and Adiel Teixeira de Almeida

Federal University of Pernambuco, Management Engineering Department
Cx. Postal 7462, Recife, PE, 50.630-970, Brazil
suzanadaher@gmail.com, almeidaatd@gmail.com

Abstract. This paper introduces the concept of a ranking veto based on two new data provided by decision makers (DMs) in an additive model approach: the veto threshold and a value reduction factor (VRF). The use of additive models for aggregating group preferences over available alternatives implies the existence of compensatory effects in the process. In traditional approaches, the final result may assign undesirable or unacceptable alternatives to higher positions in the ranking, thus causing conflicts among DMs. The model proposed softens the compensatory effects on additive models. By way of illustration, a base station allocation problem for a Telecommunication Company is presented as a numerical application. In addition to the model proposed being simple to use, it enables a group to form a collective and/or consensual view on a decision problem.

Keywords: MAUT, additive models, group decision making, ranking veto concept, telecommunication.

1 Introduction

The process of aggregating individual opinions is a crucial problem for any society [1]. According to [2], three major concerns when individual preferences are being put forward must be addressed: first, to decide how information about individual preferences should be collected and represented; secondly, to solve inconsistencies in preferences given by decision makers (DMs); and thirdly, to define how to combine individual preferences to achieve a final group recommendation and/or a consensus. A consensus as unanimity may be a target that might never be reached. This does not mean the absence of a group recommendation.

Group consensus can be achieved by using two types of preference aggregation methods. In the first, an additive ranking rule is the arithmetic mean of the rankings made by all n DMs and the second is based on a multiplicative ranking rule where the product of the rankings made by n DMs is raised to the power $1/n$ [3]. As to trying to achieve a consensus, several studies in the literature analyze different aspects, concepts and models such as conflict resolution [4], fuzzy set theory [5], cost metrics [6,7], ordinal ranking models [8,9,10], imprecise

R.C. Purshouse et al. (Eds.): EMO 2013, LNCS 7811, pp. 500–512, 2013.

information [11], outranking methods [12] and ordinal regression for multiple criteria group decisions [13].

In most situations people are used to making a decision based on intuitive judgment, considering more than one objective in an unstructured way [14]. Group multi-criteria decision making methods and group decision support systems may support decision makers to establishing their criteria, better understanding the decision problem to achieve a final recommendation [15].

Additive value and weighted value models are considered classical models and all variants of these that have been adopted involve the idea of a compensatory decision process, which can also be found in MAUT (Multi-attribute Utility Theory). The compensatory effect means that low scores in one criterion may be compensated by high scores in another. In [16], a group decision model based on a veto ranking concept was proposed to soften the compensatory effect of additive models and to support DMs to accept more consensual alternatives.

This paper extends the study presented in [16] by a modification in the way that a value reduction factor (VRF) is calculated so as to better establish a group recommendation for a final ranking. This paper is organized as follows. Section 2 describes the model proposed and Section 3 presents a numerical application so as to illustrate the model better. The last section contains final remarks, draws some conclusions and makes suggestions for future studies.

2   Model Proposed

One of the major problems of additive models is their compensatory effect where one alternative could be at least desirable for some DMs and have compensation from other DMs. Therefore, a non-balance solution may be proposed. Arrow's Impossibility Theorem describes five properties which are necessary to achieve a social welfare function [17] and later studies had been published discussing that discrete multicriteria problems using additive models for social choices should at least satisfy three of five properties (i.e, unrestricted domain, independence from irrelevant alternatives and monotonicity) [18,19]. Note that, with regard to group decisions, the property of monotonicity may easily be violated [19]. In order to minimize conflicts among DMs, a ranking veto concept is proposed in [16] to support a group of DMs in reaching a final recommendation by using an additive model approach. This model proposes a modification to traditional additive models which aims to lessen their compensatory effect by penalizing conflicting alternatives in their ranking positions and, consequently to promote consensual alternatives to better ranking positions. This model is suitable for ranking and selection problematics.

For a collaborative group decision problem where DMs are willing to give up their most preferred alternatives in favor of more consensual ones, the model described in [16] was proposed and structured in three steps. The first step consists of considering the value system of each DM and therefore, each DM's preference structure over a set of available alternatives to establish individual rankings for the alternatives. To do so, Multiple-Attribute Value Theory (MAVT) was adopted since no uncertainties over the consequences of the alternatives exist. Note that, MAVT is a compensatory technique and, thus, the method does allow one criterion to be compensated for by another (with a better performance). Preference indifference among criteria is considered and, in this case, the additive model as presented in (1) can be adopted:

$$v_{DM}(a) = \sum_{j=1}^{n} k_j v_j(a) \qquad (1)$$

where, a is an alternative, $v_j(a)$ is a value function for each criterion j and k_j represents the scale constant for criterion j. Note that $\sum_{j=1}^{n} k_j = 1$. Based on such data, it is possible to establish the rankings of all alternatives for each DM. The alternative with the best value is placed first in the ranking, and so on. If a traditional additive group model is used, then the next step would be to aggregate all DMs value functions to achieve a final ranking (or selection) using (2).

$$v_{group}(a) = \sum_{i=1}^{n} w_i v_{DM_i}(a) \qquad (2)$$

where a is an alternative, $v_{DM_i}(a)$ is a value function for each DMi and w_i represents its degree of importance for each DMi. Note that $\sum_{i=1}^{n} w_i = 1$. However, due to the compensatory effect, the overall assessment of this model might not represent the opinion of any of the DMs or might promote some alternatives that some DMs consider are unacceptable or undesirable. With a view to minimizing this problem, [16] proposed a modification to (2) by including a new parameter called the *Value Reduction Factor (VRF)* to penalize conflicting alternatives, which are allotted to disagreement zones, and where the global values of these alternatives are higher than the virtual alternative. For a better understanding of these concepts, let us move on to the next step of the model.

This second step is about identifying agreement and disagreement zones among DMs, using a vector space and veto thresholds given by DMs. A vector representation may be useful so that a space of consensus among DMs may be better exploited. Furthermore, for up to three DMs, a graphical resource can be used to visualize DMs' perception of the problem. For instance, Fig.1 illustrates a decision scenario comprising three decision makers (DM1, DM2 and DM3) and four alternatives represented as circles. This form of representation helps DMs to understand how close or far their opinions are from each other's. For the analyst who is supporting a group, this representation can give an initial insight into DMs' perspectives and how to conduct the decision process so as to reach a better global recommendation.

After establishing the vector representation of all alternatives, the analyst should invite all DMs to determine, individually, a veto threshold β_i, $\beta_i \in [0, 1]$, which is the minimum acceptable value for DMi in a decision problem. So each DM must determine a value for which the set of available alternatives will be divided into two subsets, one of which will contain all acceptable alternatives and the other all unacceptable (or undesirable) alternatives. Thus, any alternatives

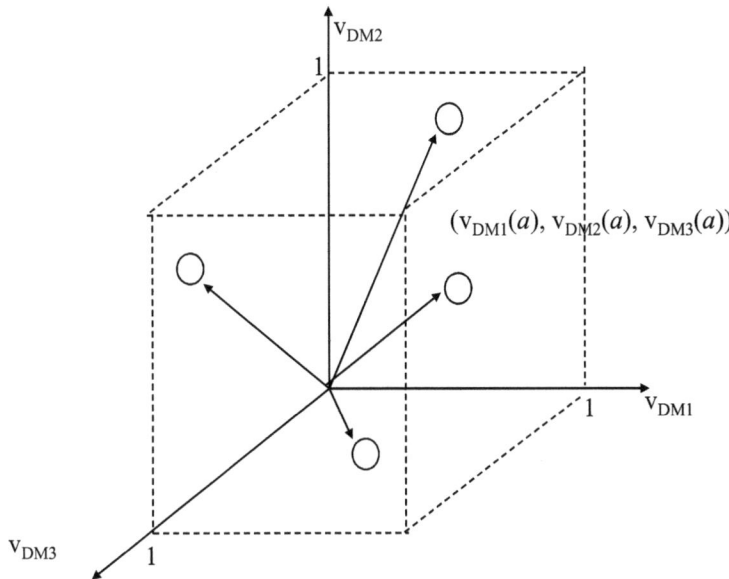

Fig. 1. Vector representation for alternatives

for which the individual value is greater than β_i should be considered as acceptable for DMi and, otherwise, considered as unacceptable or undesirable (and therefore, an alternative of potential conflict). Some DMs may have difficulty in determining a veto threshold, so in such cases, the analyst may support DMs by providing their individual ranking of alternatives and the value they assessed for each alternative. Based on that, the analyst can individually ask them to inform a value which they fill comfortable to considered that any alternative below it must be classify as unacceptable (or undesirable). Moreover, the analyst must conduct a sensitivity analysis for better bear out the decisions of DMs. Note that the veto threshold concept is not an intensity measure of the groups concordance or discordance like the one proposed by outranking methods. Instead, it is additional information provided by DMs about their preferences.

Using the given veto threshold it is possible to identify at least four zones of agreement and disagreement among DMs. Fig. 2 shows these zones both in a 3-dimensional and 2-dimensional spaces (illustrating DMs' point of views, pairwise). There are two agreement zones: a positive one where all DMs are willing to accept the alternatives assigned to this area; and a negative one where alternatives assigned to this area are considered as unacceptable or undesirable for all DMs. The disagreement zones represent regions where all alternatives assigned to these areas are considered as unacceptable or undesirable for at least one DM. The combination of all veto thresholds β_i given by the DMs creates a virtual alternative α.

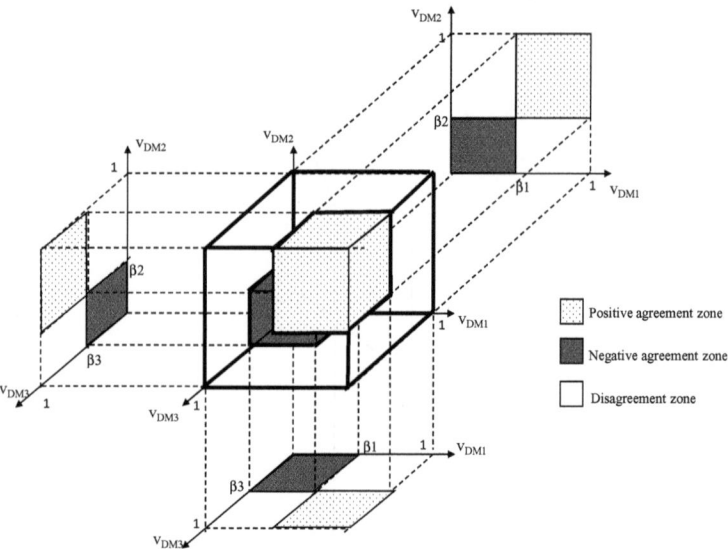

Fig. 2. Graphical representation of agreement and disagreement zones in 2 and 3 dimensions for 3 decision makers

Due to the compensatory effect, alternatives assigned to disagreement zones may have a global value greater than others allotted to positive ones. In order to minimize this problem, a modification to (2) was conducted by introducing a Value Reduction Factor (VRF) as shown in (3).

$$v'_{group}(a) = VRF * v_{group}(a) = VRF * \sum_{i=1}^{n} w_i v_{DM_i}(a) \qquad (3)$$

In [16] the global values of all conflicting alternatives which have a global value greater than the virtual alternative were reduced to the same global value as the virtual alternative. The major problem of adopting that measure was the fact that more than one alternative received the same final value, thus creating ties. This paper proposes another rule for a VRF, where the penalty over the conflicting alternatives will be proportional to their rejection. The procedure to calculate the VRF is presented below:

Procedure to calculate the VRF:
Consider $v_{group}(x)$ as the global value of alternative x calculated using eq. (2). Assume that all DMs have the same importance degree.

Step 1: Identify the alternative y among those totally accepted by all DMs which has the smallest global value. Keep its global value as variable v^*_{group}.

Step 2: Calculate the global value of the virtual alternative. Keep this value as variable $v_{virtual}$.

Step 3: If (*alternative x belongs to a disagreement zone*) and (*its global value is greater than* $v_{virtual}$),
  then, $VRF = (v^*_{group} - v_{virtual}) + [(v_{virtual})/v_{group}(x)]$
  else, $VRF = 1$

This reduction factor assures that all possible conflicting alternatives are going to be penalized by decreasing their global values to one between the last acceptable alternative for all DMs and the virtual alternative. Moreover, these reductions should not introduce any ranking order inversion among those alternatives. This means that, if alternative a was better evaluated (for all DMs) than alternative b, then the model should take this information in account to do not introduce any result distortions. Fig.3 illustrates how VRF acts on the proposed group decision model. Fig. 3(a) indicates no penalty for alternative x (possessed on disagreement zone), since its original global value $v_{group}(x)$ is less than $v_{virtual}$. Fig. 3(b) shows when a penalty for conflicting alternatives should be applied.

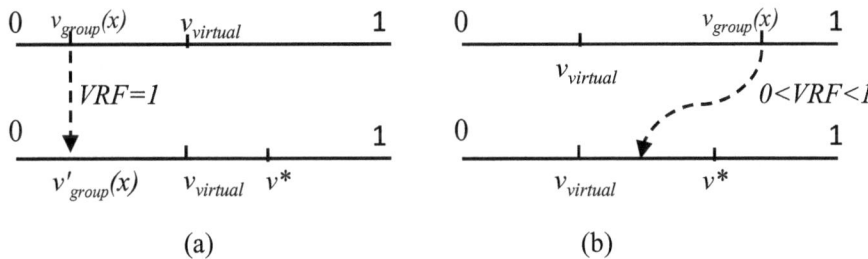

Fig. 3. Value reduction factor

The last step of the model is to create the final ranking by using the VRF. In this phase of the model, the analyst must conduct a sensitivity analysis so as to evaluate better the final group recommendation and also to give DMs more confidence in the outcomes of the model. Fig. 4 summarizes the proposed model. Once DMs become familiar with the model and results are presented, the analyst may need to return to a prior step in the model in order to adjust some data or to revise the preference model given by DMs. Although the model proposed can minimize conflicting situations, it is, unfortunately, not immune to misrepresentation as discussed in [16]. The new rule for VRF put forward in this paper does not (intentionally) promote ties which is much better for a ranking decision problem.

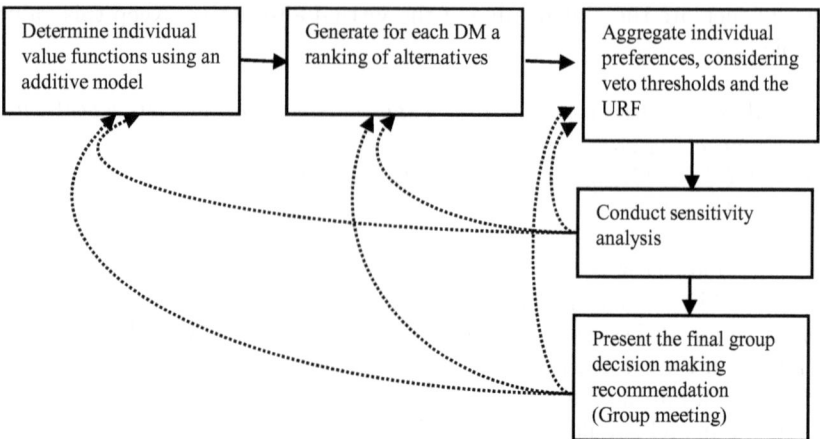

Fig. 4. Group decision model flow

3   Radio Base Station Locations: A Ranking Problem

In order to develop a radio network planning, a telecommunication company must consider how to configure its network resources to assure satisfactory performance for its clients (or technically, the end-users). Three main quality attributes should be taken into account: coverage, traffic capacity and quality of service [20]. Several mathematical and heuristic optimization models and algorithms have been proposed in the literature, which consider selected aspects of network planning [21,22]. Also, there are some factors affecting the choice of optimum locations for base stations in a cell communication system such as non-uniform service areas in complex propagation environments, the distances between base stations (and mutual coverage), signal levels and signal-interference ratio (SIR) [23]. Commonly this kind of decision problem has been dealt with from a mono-user perspective, but the increase in the complexity of systems and of company's strategic goals is encouraging group decision making. Even in a collaborative process where DMs are willing to give up their most preferred alternatives in favor of more consensual ones, divergences may nevertheless still appear among them. This numerical application illustrates how the model proposed in this study can support a group to reach a more consensual final recommendation. An overall cellular network contains a number of different elements as illustrated in Fig. 5. Several Mobile Stations (or mobile phones) are linked to a network infrastructure via a Base Transceiver Station (BTS). Another element is the Base Station Controller (BSC) which manages the routing of calls and decides which base station is the one best suited to accept an end-user call. The BSC is often co-located with a BTS and each BSC can control a small group of BTSs. Other elements are the Mobile Switching Centre (MSC); two location registers: a Home Location Register (HLR) and a Visitor Location Register (VLR);

and the link to the public switched telephone network (PSTN) [24]. The geographical area covered by the telecom service is divided into small cells where the coverage and size of each cell is defined by matters such as the height of the tower (or buildings) where antennas are installed, the irregularities of the relief of the terrain, the power transmission and signal gain of the antennas.

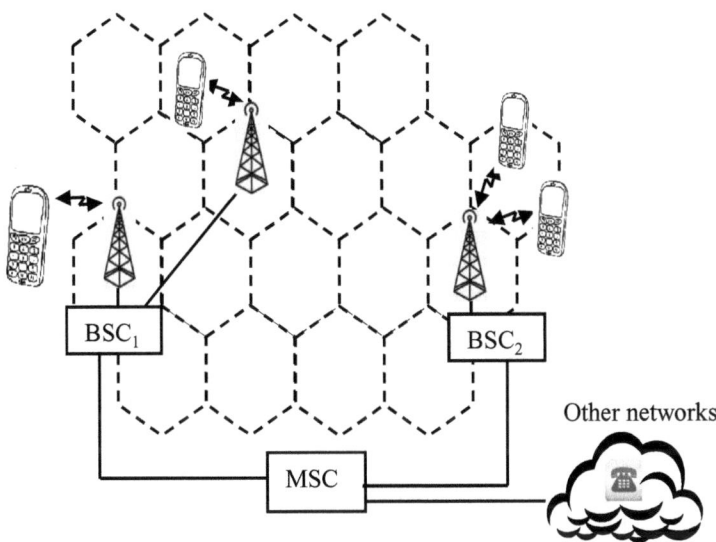

Fig. 5. Cellular network architecture

For the proposal of this study, a selected set of five alternatives (A1, A2, A3, A4, and A5) is already established. Several criteria, such as those mentioned previously, may be considered to define the pre-selected set of alternatives. In this numerical application, although real data are not used, the data used are realistic. The decision problem involves multiple objectives and there are three company senior executives who act as decision makers (DM1, DM2 and DM3) and whose values are taken into account. The final recommendation will be used by the telecommunication company to decide on which base station will be implemented first, second, and so on. An analyst is responsible for conducting the decision process and supporting DMs in identifying their preference structure and establishing their value functions. The company had previously decided to maintain the same equipment suppliers and brands. All locations will receive the same equipment, such as radio, antennas, etc. So the costs of those devices are not considered as a criterion. An attribute range was defined for each criterion and tradeoffs could be evaluated and used as scaling constants. The criteria adopted for the group decision making are:

C1: coverage area (cell size). The range varies from a radius of 1km (in a densely populated area) to 10 km (in a sparsely populated area).

C2: amount of rent/month (in monetary unit) of the area where the items of equipment are installed. The range varies from 2,000.00 up to 8,000.00.

C3: increased capacity of user data (in percentage terms). The range varies from 4 to 55%.

Table 1 evaluates each alternative using the criteria selected.

Table 1. Evaluating the alternatives in accordance with each criterion

Alternatives	C1	C2	C3
A1	4	8,000	35
A2	9	4,500	40
A3	5	2,000	30
A4	3	3,500	55
A5	10	2,000	4

The values of independence between criteria were considered and an additive model was adopted. The analyst conducts the first and second steps of the model proposed which identify the DMs' value functions and he assigns a veto threshold for each DM. The DMs' tradeoffs and the veto thresholds established are shown in Table 2. The combination of all veto thresholds enables the virtual alternative and its group global value to be established. Furthermore, this information enables the VRF to be calculated.

Table 2. Decision makers' tradeoffs

DM	Criteria Tradeoffs			Veto
	C1	C2	C3	threshold
DM1	0.1	0.8	0.1	0.9
DM2	0.25	0.3	0.45	0.57
DM3	0.6	0.3	0.1	0.74

For this numerical application proposal, exponential forms for value functions, $a(1 - b\exp(-cx))$, were obtained using a regression model. The values for each parameter a, b, and c are given in Table 3. The DMs' values for the alternatives are shown in Table 4. The bold numbers in Table 4 indicate which alternatives are considered as undesirable or unacceptable for DMs due to their value being less than the veto threshold set. Table 4 also indicates the group final rankings, whether or not the VRF was considered, i.e, it shows the traditional additive model result and the results from the model proposed. Note that all DMs have the same weight (or degree of importance) for the group decision process.

On using the given veto thresholds, the analyst is able to distribute the alternatives as follows: A2 and A3 are allocated to the positive concordance zone,

Table 3. Parameters used on DM's value functions

Criteria	Parameter	DM1	DM2	DM3
C1	a	0.9	1	1
	b	0.7	0.85	0.4
	c	0.05	0.1	0.023
C2	a	1	1	0.87
	b	0.8	0.77	0.91
	c	0.031	0.062	0.01
C3	a	1	0.43	1
	b	0.8	0.3	0.07
	c	0.04	0.01	0.01

Table 4. DMs' values for alternatives and rankings involved in the group decision problem

Ranking Position	Ranking DM1	Ranking DM2	Ranking DM3	Ranking without VRF	Ranking with VRF
1	A2 0.933678	A2 0.618192	A5 0.763586	A2 0.771017	A2 0.771017
2	A4 0.926911	A5 0.609552	A2 0.761183	A5 0.752251	A3 0.743444
3	A1 0.918692	A3 0.571608	A3 0.741886	A3 0.743444	A5 0.741765
4	A3 0.91684	**A1 0.56015**	**A1 0.737162**	A1 0.738668	A1 0.741673
5	**A5 0.883617**	**A4 0.552584**	**A4 0.732963**	A4 0.737486	A4 0.741665

and alternatives A1, A4 and A5 to the disagreement zones. Although well accepted by two of the DMs, A5 is considered an undesirable alternative by DM1 who ranked it last in his individual ranking. No alternatives are allocated to the negative concordance zone.

Moreover, the analyst must conduct a sensitivity analysis to evaluate the robustness of the model or how much the veto threshold could vary without modifications to the final group recommendation. For this to happen, the value of the value thresholds informed by DMs may vary in range by 10%. Basically, the results are still the same if all DMs' thresholds change simultaneously and in the same direction. On analyzing individual variations (i.e. modifying only one DM's veto threshold and keeping the others the same), it was found that in a scenario with a 10% variation, the final group ranking remains the same as that of the traditional additive model (whether or not the VRF is considered) if only DM1's or DM3's veto threshold varies. Unfortunately, it was found that the model proposed is not immune to misrepresentations, and DMs can manipulate the veto threshold they set in order to benefit their most preferred alternatives. It is expected that the analyst can address and overcome this kind of situation.

4   Final Remarks

In this paper, the concept of a ranking veto based on a veto threshold and that of a value reduction factor (VRF) were introduced in order to propose a group decision model for a collaborative decision problem, based on traditional additive models. The additive function aggregates individual preferences into a global one. For the purposes of this study all DMs have the same weight and no power relations are considered. The individual preferences structure is elicited using MAVT, thus defining a value function for each DM. By considering the VRF as part of the group final value function, the model softens the compensatory effect of additive models and also enables DMs to form a collective view on the decision problem. As discussed in [16], this model differs from previous veto models found in the literature and its approach fits better to MCDM additive models when integrated with an additive group decision aggregation model. Furthermore, spaces of consensus based on agreement and disagreement zones are established using the veto threshold established.

Although important for the decision process, defining the set of alternatives lies outside the scope of the model. To do so, depending on the nature of the problem, a multi-objective optimization model can be adopted. This kind of model could be especially interesting if no alternatives appear in the positive concordance zone which means that the set of alternatives could be incomplete and needs to be reviewed. In such cases, multiple objectives should be put forward as a number of objective functions which are to be minimized or maximized, subject to a number of constraints which must be satisfied. As discussed in [25], Evolutionary Multi-objective Optimization (EMO) and MCDM may work together to enhance both areas. One possibility is to use EMO to generate a set of non-dominated solutions and then a MCDM approach can be used to select the alternative most preferred. Another possibility could be to embed the principles of MCDM within an EMO through a value (or utility for MAUT approach) measure.

As to future studies, the authors plan to extend the concept of concordance zones and discordance zones by creating other spaces of consensus. Therefore, other methodologies such as robust ordinal regression may be useful. Moreover, the authors of this article wish to work on how this model acts when different degrees of importance between DMs are considered.

Acknowledgment. This study is part of a research project funded by the Brazilian Research Council (CNPq).

References

1. Montero, J.: The impact of fuzziness in social choice paradoxes. Soft Computing - A Fusion of Foundations, Methodologies and Applications 12(2), 177–182 (2008)
2. Wang, Y.M., Yang, J.B., Xu, D.L.: A preference aggregation method through the estimation of utility intervals. Computers & Operations Research 32(8), 2027–2049 (2005)

3. Bui, T.X.: Co-oP: A group decision support system for cooperative multiple criteria group decision making. LNCS, vol. 290. Springer, Berlin (1987)
4. Li, K., Hipel, K., Kilgour, M., Fang, L.: Preference uncertainty in the graph model for conflict resolution. IEEE Transactions on Systems, Man and Cybernetics, Part A: Systems and Humans 34(4), 507–520 (2004)
5. Parreiras, R., Ekel, P., Morais, D.C.: Fuzzy set based consensus scheme for multi-criteria group decision making applied to strategic planning. Group Decision and Negotiation 21(2), 153–184 (2012)
6. Ben-Arieh, D., Easton, T., Evans, B.: Minimum cost consensus with quadratic cost functions. IEEE Transactions on Systems, Man, Cybernetics, Part A: Systems and Humans 39(1), 210–217 (2008)
7. Ben-Arieh, D., Easton, T.: Multi-criteria group consensus under linear cost opinion elasticity. Decision Support Systems 43, 713–721 (2007)
8. Cook, W.D., Kress, M., Seiford, L.M.: A general framework for distance-based consensus in ordinal ranking models. European Journal of Operational Research 96, 392–397 (1996)
9. Morais, D.C., Almeida, A.T.: Group decision making on water resources based on analysis of individual rankings. Omega (Oxford) 40, 42-52 (2011)
10. Morais, D.C., Almeida, A.T.: Water network rehabilitation: A group decision-making approach. Water S.A (online) 36, 487–493 (2010)
11. Brito, A.J.M., Almeida Filho, A.T., Almeida, A.T.: Multi-criteria Decision Model for Selecting Repair Contracts by applying Utility Theory and Variable Interdependent Parameters. IMA Journal of Management Mathematics 21, 349–361 (2010) (Print)
12. Silva, V.B.S., Morais, D.C., Almeida, A.T.: A Multicriteria Group Decision Model to Support Watershed Committees in Brazil. Water Resource Management 24, 4075–4091 (2010)
13. Greco, S., Kadzinski, M., Mousseau, V., Slowinski, R.: Robust ordinal regression for multiple criteria group decision: UTAGMS-GROUP and UTADISGMS-GROUP. Decision Support Systems 52, 549–561 (2012)
14. Morais, D.C., Almeida, A.T.: Water supply system decision making using multi-criteria analysis. Water S.A 32(2), 229–235 (2006)
15. Daher, S.F.D., Almeida, A.T.: Recent Patents Using Group Decision Support Systems: A Short Review. Recent Patents on Computer Science 3, 81–90 (2010)
16. Daher, S.F.D., Almeida, A.T.: The Use of Ranking Veto Concept to Mitigate the Compensatory Effects of Additive Aggregation in Group Decisions on a Water Utility Automation Investment. Group Decision and Negotiation 21(2), 185–204 (2012)
17. Arrow, K.: A difficulty in the concept of social welfare. Journal of Political Economic 58(4), 328–346 (1950)
18. Arrow, K.J., Raynaud, H.: Social choice and multicriterion decision making. MIT Press, Cambridge (1986)
19. Munda, G.: Social multi-criteria evaluation for a sustainable economy. Springer, Berlin (2008)
20. Khalek, A.A., Al-Kanj, L., Dawy, Z., Turkiyyah, G.: Site placement and site selection algorithms for UMTS radio planning with quality constraints. In: IEEE 17th International Conference on Telecommunications (ICT), pp. 375–381 (2010)
21. Chamberland, S.: An efficient heuristic for the expansion problem of cellular wireless networks. Computer & Operations Research 31, 1769–1791 (2004)

22. Amaldi, E., Capone, A.: Planning UMTS Base Station Location: Optimization Models with Power Control and Algorithms. IEEE Transactions on Wireless Communications 2(5), 939–952 (2003)
23. Anderson, H.R., McGeehan, J.P.: Optimizing microcell base station locations using simulated annealing techniques. In: 44th IEEE Vehicular Technology Conference, vol. 2, pp. 858–862. IEEE Press (1994)
24. Schiller, J.H.: Mobile Communications, 2nd edn. Addison-Wesley (2003)
25. Wallenius, J., Dyer, J.S., Fishburn, P.C., Steuer, R.E., Deb, K.: Multiple Criteria Decision Making, Multiattribute Utility Theory: Recent Accomplishments and What Lies Ahead. Management Science 54(7), 1336–1349 (2008)

A Dimensionally-Aware Genetic Programming Architecture for Automated Innovization

Sunith Bandaru and Kalyanmoy Deb

Indian Institute of Technology Kanpur, Kanpur, UP 208016, India
{sunithb,deb}@iitk.ac.in
http://www.iitk.ac.in/kangal

Abstract. Automated innovization is an unsupervised machine learning technique for extracting useful design knowledge from Pareto-optimal solutions in the form of mathematical relationships of a certain structure. These relationships are known as design principles. Past studies have shown the applicability of automated innovization on a number of engineering design optimization problems using a multiplicative form for the design principles. In this paper, we generalize the structure of the obtained principles using a tree-based genetic programming framework. While the underlying innovization algorithm remains the same, evolving multiple trees, each representing a different design principle, is a challenging task. We also propose a method for introducing dimensionality information in the search process to produce design principles that are not just empirical in nature, but also meaningful to the user. The procedure is illustrated for three engineering design problems.

Keywords: genetic programming, dimensional awareness, automated innovization, multi-objective optimization, design principles.

1 Introduction

In recent years there has been a growing interest in the field of post-optimality analysis. In a single objective scenario, this usually concerns the optimality, sensitivity and robustness studies on the obtained solution. Multi-objective optimization on the other hand, poses an additional challenge in that there are a multitude of possible solutions (when the objectives are conflicting) which are all said to be Pareto-optimal. The data-mining of Pareto-optimal solutions has received particular attention as it can reveal certain characteristic features exclusive to these solutions. In practical problem solving, the knowledge of these features can give the designer a better understanding of the problem structure. Most studies in this direction rely on visual means of identifying the features. A summary of these studies can be found in [1].

Deb and Srinivasan [5] describe the concept of *innovization* by defining the special features as commonalities among the Pareto-optimal solutions. These commonalities (or invariants) are given mathematical forms, called *design principles*, by performing regression between variables and/or objectives using appropriate functions. However, regression can only be performed when a correlation

R.C. Purshouse et al. (Eds.): EMO 2013, LNCS 7811, pp. 513–527, 2013.

is observed between the regressed entities. Innovization, in its original form, required users to identify this correlation visually through two and three dimensional plots. This *manual innovization* task is therefore limited to features of Pareto-optimal solutions present in humanly perceivable dimensions.

Automated innovization [2] is an unsupervised machine learning technique that can identify correlations in any multi-dimensional space formed by variables, objectives, etc. specified by the user and subsequently performs a selective regression on the correlated part of the Pareto-optimal dataset to obtain the design principle $\psi(\mathbf{x})$. The procedure was later extended [1] so that design principles hidden in all possible Euclidean spaces formed by the variables and objectives (and any other user-defined functions) can be obtained simultaneously without any human interaction. The regression assumes the following mathematical structure for the design principle,

$$\psi(\mathbf{x}) = \prod_{j=1}^{N} \phi_j(\mathbf{x})^{a_j b_j}, \tag{1}$$

where ϕ_j's are N basis functions (variables, objectives functions, constraints etc.) specified by the user which can have Boolean exponents a_j and real-valued exponents b_j. It has been argued that since many natural, physical, biological and man-made processes are governed by formulae with the same structure (power laws [10]), most correlations are expected to be mathematically captured by it. By definition, $\psi(\mathbf{x})$ is a design principle if it is invariant, i.e. $\prod_{j=1}^{N} \phi_j(\mathbf{x})^{a_j b_j} = c$ is true for a majority of the Pareto-optimal solutions, for some constant c. However, due to the approximate nature of Pareto-optimal datasets, the equality relation may not hold strictly and hence the extent of commonality of a design principle $\psi(\mathbf{x})$ is obtained by clustering the set of c-values. The minimization of equal-weighted sum of (i) number of clusters ($\mathcal{C}$), and (ii) percentage coefficient of variance ($c_v = \sigma/\mu$) within these clusters, has been proposed in [2] to obtain design principles. An optimization problem with this objective function,

$$\text{Minimize} \quad \mathcal{C} + \sum_{k=1}^{\mathcal{C}} c_v^{(k)} \times 100\% \quad \text{where } c_v^{(k)} = \frac{\sigma_c}{\mu_c} \quad \forall\, c \in k\text{-th cluster}, \tag{2}$$

is formulated with a_j's represented by an N-bit binary variable string and b_j's as N real variables. The algorithmic calculation of the objective function requires the use of a derivative-free optimization method like genetic algorithms (GA). The population based approach of GA also enables obtaining multiple design principles simultaneously using a niching strategy [1]. Given enough GA generations, the final population will contain all possible design principles that fit the form in Eq. (1).

The complete details of the above algorithm and a pseudocode can be found in [1]. In this paper, we generalize the mathematical structure of the design principles in Eq. (1) using parse tree representation. The overall automated innovization problem is solved using a system that integrates a GA with a genetic programming algorithm for handling parse trees.

2   Genetic Programming for Automated Innovization

At an abstract level, genetic programming (GP) is a *weak* search algorithm for automatically generating computer programs to perform specified tasks [9]. Weak search methods do not require the user to know or specify the form or structure of the solution in advance [8]. Most GP implementations employ an evolutionary algorithm as the main search engine. However, simulated annealing, hill climbing approaches and estimation of distribution algorithms (EDAs) have also been used in literature [12]. Like other evolutionary computation techniques, a typical GP starts with a population of randomly created individuals, which in this case are programs. The fitness for each individual is determined by running the program. High fitness individuals are selected to form the mating pool, on which primary genetic operations, namely crossover and mutation, are applied to create a new population of programs. The process is repeated until some stopping criterion (like maximum number of generations) is met.

Most GP systems evolve programs in a domain-specific language specified by primitives called *functions* and *terminals*. The terminal set ($\mathcal{T}$) may consist of the program's external inputs, ephemeral random constants, and nullary (zero-argument) functions/operators where as the function set ($\mathcal{F}$) may contain operators (arithmetic, Boolean, conditional, etc.), mathematical functions and constructs (loops, for example) that are defined in the language being used. Computer programs are traditionally represented in the memory as parse trees made up of such primitives. Other common ways of expressing programs include linear and graph-based representations.

The most common application of GP has been to the process of induction of mathematical models based on observations. This process is known by the names model induction, system identification and symbolic regression depending on the purpose. The power of GP algorithms to evolve models in a symbolic form without assuming the functional form of the underlying relationship can also be applied to automated innovization. In this paper, we generalize the mathematical structure of the design principles in Eq. (1) by representing $\psi(\mathbf{x})$ using parse trees composed of the N basis functions and real-valued ephemeral constants as terminals, i.e $\mathcal{T} = \{\phi_1, \phi_2, \ldots, \phi_N, \mathcal{R}\}$ and a user-specified function set $\mathcal{F}$. Fig. 1 shows two examples of parse trees and their corresponding $\psi(\mathbf{x})$ expressions obtained by inorder depth-first tree traversal. By starting with a population of such trees and using the objective function in (2), it is possible to evolve design principles of a generic form using a GP system. In the following sections we discuss each step of the (GA + SmallGP [11]) system used in this work. Some of these steps are standard and hence described only briefly, while others which have been modified to suit the requirements of automated innovization are explained in more detail.

2.1   Initialization

Two initialization methods are very common in GP, the `Full` method and the `Grow` method [12]. The `Full` method always generates trees in which all leaves (end nodes) are at the same user-specified $MAXDEPTH$ value. This is achieved by randomly selecting nodes only from the $\mathcal{F}$ set until the depth limit is reached, at which nodes are selected from the $\mathcal{T}$ set. On the other hand, the Grow method creates trees of varied sizes and shapes by randomly selecting nodes from the full primitive set $(\mathcal{F}+\mathcal{T})$ for all nodes until the depth limit is reached, at which nodes are chosen from $\mathcal{T}$ as in the case of Full method. In Fig. 1, the tree on the left could have been the result of either the Full or the Grow method, but the one on the right can only be created by the latter. SmallGP uses a mix of Full and Grow methods. When selecting from the full primitive set the probability of choosing from the terminal set is,

$$p(\mathcal{T}) = \frac{|\mathcal{T}|}{|\mathcal{F}| + |\mathcal{T}|} \times SF,$$

where $|.|$ denotes the set size and SF is the scaling factor which scales initialization between Full (when $SF = 0$) and Grow (when $SF = 1$). It is to be noted that the ephemeral constants only contribute one virtual terminal symbol to the $\mathcal{T}$ set so that $|\mathcal{T}| = N + 1$. The initialization also takes into account the maximum program (tree) length $MAXLEN$ which can also be specified by the user.

2.2   Fitness and Constraint Evaluation

This step involves the use of a grid-based clustering algorithm for evaluating the fitness function given in (2) for the trees created above. Each tree is decoded to obtain the design principle $\psi(\mathbf{x})$ that it represents. The resulting mathematical expression is evaluated for all m trade-off solutions provided as input to the GP algorithm to obtain the corresponding c-values. Grid-based clustering [2] involves sorting these c-values into a set $\mathbf{C}$ and dividing their range into d equal divisions (a parameter of the clustering routine). Elements in $\mathbf{C}$ which belong to divisions with less than $\lfloor m/d \rfloor$ c-values are categorized as *unclustered*. Adjacent divisions with more than $\lfloor m/d \rfloor$ c-values are merged to form clusters. Thus, the number of clusters $\mathcal{C}$ and the number of unclustered points $\mathcal{U}$ can be obtained and used in (2) to calculate the fitness for any given tree. Instead of asking the user to choose the parameter value for d, it is evolved alongside the GP trees using a GA. Therefore, each population member of the proposed system consists of a GP tree variable for $\psi(\mathbf{x})$ and an integer variable for d, which is also initialized (in the range $[1, m]$) in the previous step. This is the reason for integrating GA with SmallGP in this paper.

It has been suggested in [2] that for obtaining the most accurate design principles, the constraint

$$\mathcal{U} = 0, \tag{3}$$

should be imposed during clustering. This forces unclustered c-values to form one-element clusters by increasing the value of d, which in turn causes $\mathcal{C}$ and c_v within clusters to increase. The optimization of the weighted objective compensates for this by producing more accurate design principles [1,2].

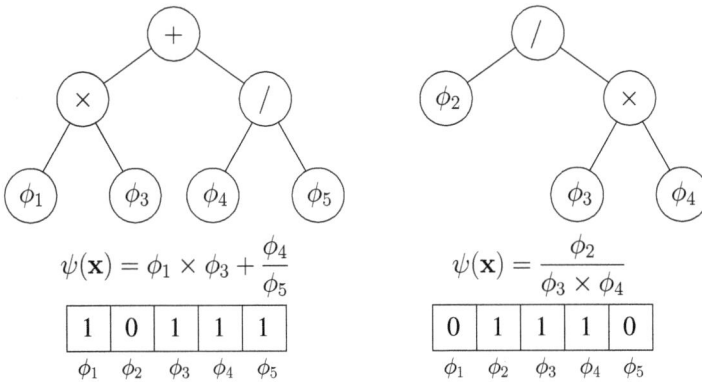

Fig. 1. Two examples of parse trees obtained from the `Full` (left) and `Grow` (right) methods for $MAXDEPTH = 2$. The corresponding $\psi(\mathbf{x})$ expressions and binary strings are also shown.

A measure for the degree of commonality of $\psi(\mathbf{x})$ is also calculated in this step. This measure is called the significance S of the design principle and is defined as,

$$S = \frac{(m - \mathcal{U}')}{m} \times 100\%. \tag{4}$$

Here $\mathcal{U}'$ is the total number of elements in $\mathbf{C}$ which belong to divisions with less than $(\lfloor m/d \rfloor + \epsilon)$ c-values. By choosing a small integer value for ϵ, c-values which barely formed clusters due to imposition of (3) can be identified [2].

2.3   Niched-Tournament Selection

In order to maintain multiple design principles in the population, [1] proposes the use of niched-tournament selection for creating the mating pool. We adopt a similar niching technique in the present paper. A binary string of length N is associated with each GP tree. If the l-th basis function is present in a tree then the l-th bit of the corresponding binary string is assigned a value of 1, else it takes a value of 0. Tournaments are only allowed between trees which have exactly the same binary string. This allows different *species* of design principles to co-exist in the population, while still promoting the better individual (fitness wise) when the trees have exactly the same basis functions. Fig. 1 shows such binary strings for the two trees. Since they are different, both individuals are equally competent irrespective of their fitness values and tournament selection is not performed between them.

2.4   Crossover and Mutation Operators

SmallGP uses a size-safe subtree crossover [12] to recombine trees in the mating pool. It ensures that the created children do not exceed the maximum tree length

$MAXLEN$. This is accomplished by first determining the number of nodes by which the smaller parent tree can be extended. Then, a random subtree satisfying this requirement is cut from the larger parent tree. Similarly, a random subtree (whose maximum size is determined by taking the new size of the larger tree into account) is cut from the smaller tree. The two subtrees are exchanged at the cut locations to produce the offspring trees. Implementation details can be found in the manual [11].

Point mutation [12] is the simplest form of mutation in GP where a random node is selected and the primitive at that node is replaced with a different random primitive of the same kind (function or terminal) and arity to maintain the closure property [9]. Like the bit-flip mutation in GAs, point mutation is applied on a per-node basis, thus allowing multiple nodes to be mutated independently.

The GA variable d is recombined using the discrete version of simulated binary crossover (SBX) and mutated using the discrete version of polynomial mutation both of which are suggested in [7].

3   Dimensional Awareness

GP systems are known to produce exceptionally good models in symbolic regression applications, given that an appropriate set of primitives is provided. This has inspired the use of GP for scientific knowledge discovery from datasets obtained through physical processes, experiments, phenomena, etc. Computer-aided scientific knowledge discovery differs from standard symbolic regression in that the obtained model, in addition to fitting the data well, is also expected to be novel, interesting, plausible and understandable [14]. The key to achieving this is to incorporate the semantic content that is encapsulated in the data into the search process.

The foremost application of automated innovization has been for engineering problems [2,3]. It may be beneficial in these cases to extract design principles which are not just empirical in nature but also meaningful to the designer. In GP this is usually achieved by constraining the tree structures [12]. For example, if a model is known to be periodic *a priori*, then the search may be constrained to models that take the form $a \times \sin(b \times t)$ through strong typing or grammar-based constraints When no such domain-specific information is available, one can still generate meaningful and syntactically correct tree structures by taking into account the most basic requirement for relationships governing all physical systems, namely *dimensional consistency* or *commensurability*, which states: (i) Only commensurable quantities (quantities with the same dimensions) may be compared, equated, added, or subtracted and (ii) One may take ratios of incommensurable quantities (quantities with different dimensions), and multiply or divide them. Previous work on dimensionally-aware genetic programming proposed a *weakly typed* or *implicit casting* approach [8] where dimensionality is not enforced, but promoted through an additional objective. We incorporate a similar strategy in the proposed (GA + SmallGP) system using constraints for penalising dimensional inconsistency.

Table 1. Transformed terminal operations for calculating the exponent E from the exponents e_i and e_j of two terminals. Note that Z can be any value greater than E_{max}.

Operation	Transformed Operation
$T_i + T_j$	$E = e_i$, if $e_i = e_j$ $E = Z$, otherwise
$T_i - T_j$	$E = e_i$, if $e_i = e_j$ $E = Z$, otherwise
$T_i \times T_j$	$E = Z$, if $\max(|e_i|, |e_j|) > E_{max}$ $E = e_i + e_j$, otherwise
$\dfrac{T_i}{T_j}$	$E = Z$, if $\max(|e_i|, |e_j|) > E_{max}$ $E = e_i - e_j$, otherwise
$T_i^{T_j}$	$E = e_i T_j$, if $|e_i| \leq E_{max}$ & $e_j = 0$ $E = Z$, otherwise

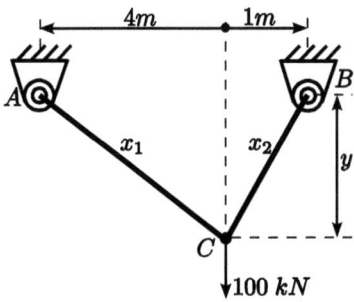

Fig. 2. Two-bar truss configuration showing the cross-sectional areas x_1 and x_2 and the vertical length y. A load of $F = 100$ kN is applied.

Datasets obtained from multi-objective optimization of engineering systems consist of Pareto-optimal values of variables, constraints and objectives, for all of which the dimensions are known *a priori*. Each basis function $\phi(\mathbf{x})$ provided as input to the proposed system can therefore also carry its dimensionality information. For a given tree the dimensional consistency of $\psi(\mathbf{x})$ is checked using the following procedure. Noting that the largest possible absolute value of the exponent for any fundamental dimension (mass, length, time, etc.) among all known physical quantities is four, it can be established that in a tree of depth $MAXDEPTH$ the maximum absolute value that a dimension exponent can have in the corresponding $\psi(\mathbf{x})$ is $E_{max} = 4 \times 2^{MAXDEPTH}$. For two terminals T_i and T_j having dimension exponents e_i and e_j, undergoing various operations, the resulting exponent E is calculated using the transformed terminal operations shown in Table 1. Note that when adding or subtracting incommensurable quantities, E is assigned an arbitrary value Z greater than E_{max}. This indicates to subsequent operations that the tree being evaluated (or $\psi(\mathbf{x})$) is already dimensionally inconsistent. The absolute value of the exponent E obtained after completely evaluating the tree is constrained to be at or below E_{max} for all basic dimensions, thus imposing dimensional consistency.

The optimization problem for (GA + SmallGP) based automated innovization can now be formulated as,

$$\text{Minimize} \atop \{\psi(\mathbf{x}), d\} \quad TS\left(\mathcal{C} + \sum_{k=1}^{\mathcal{C}} c_v^{(k)} \times 100\%\right) \quad \text{where } c_v^{(k)} = \frac{\sigma_c}{\mu_c} \quad \forall \, c \in k\text{-th cluster,}$$

$$\text{and} \quad c = \psi(\mathbf{x}) \quad \forall \, m$$

$$\text{Subject to } \left\{ 1 \leq d \leq m; \, \mathcal{U} = 0; S \geq S_{reqd}; |E| \leq E_{max} \; \forall \text{ basic dimensions} \right\}$$

$$(5)$$

The tree-size TS (number of nodes in the tree) is multiplied to the weighted objective function in (2) in order to promote smaller trees. The minimum significance required by the user is specified by S_{reqd}. Theoretically, a given S_{reqd} value should include all design principles that can be obtained with any higher S_{reqd} value. However, this is not true in practice owing to limited population size and generations. Therefore, in general, a high value of S_{reqd} should be used. If design principles are not obtained then it may be lowered. In addition, elitism is introduced by performing niched-tournament selection on the combined parent-child population. Both $MAXDEPTH$ and $MAXLEN$ values should be set based on the size of the primitive set. Higher values tend to form longer trees and hence more redundant design principles than values closer to $|\mathcal{F}| + |\mathcal{T}|$.

4   Results

We now illustrate the working of the proposed algorithm on three engineering design problems. In all cases the design optimization problem is solved using NSGA-II [4] with the following parameters: Population size $= 500$ (truss and welded-beam), 1000 (metal-cutting); Number of generations $= 500$; SBX for real variables with $p_c = 0.9$ and $\eta_c = 10$; Polynomial mutation for real variables with $p_m = 0.05$ and $\eta_m = 50$.

4.1   Two-Bar Truss Design

The two-bar truss design problem involves three variables as shown in Fig. 2. The bi-objective formulation is,

$$
\begin{aligned}
\text{Minimize} \quad & f_1(\mathbf{x}) = \text{Volume } (V) = x_1\sqrt{16 + y^2} + x_2\sqrt{1 + y^2}, \\
\text{Minimize} \quad & f_2(\mathbf{x}) = \text{Max. Stress } (S) = \max(\sigma_{AC}, \sigma_{BC}), \\
\text{Subject to} \quad & \left\{ S \le 10^5 \text{ kPa}; 0 \le x_1, x_2 \le 0.01 \text{ m}^2 \text{ and } 1 \le y \le 3 \text{ m} \right\}
\end{aligned}
\tag{6}
$$

NSGA-II gives $m = 500$ non-dominated solutions at the end 500 generations. For automated innovization we choose the function set $\mathcal{F} = \{+, -, \times, \%, ^{\wedge}\}$ where $\%$ represents protected division and $^{\wedge}$ represents the power function. The objectives and the variables are chosen as the basis functions, i.e, $\Phi = \{\phi_1, \phi_2, \phi_3, \phi_4, \phi_5, \} = \{V, S, x_1, x_2, y\}$. and so the terminal set is $\mathcal{T} = \{V, S, x_1, x_2, y, \mathcal{R}\}$. The following parameters are used in the proposed (GA + SmallGP) algorithm to solve the optimization problem in (5):

1. Population size $= 1000$,
2. Number of generations $= 100$,
3. Discrete SBX for variable d with $p_c = 0.9$ and $\eta_c = 10$,
4. Discrete polynomial mutation for variable d with $p_m = 0.05$ and $\eta_m = 50$,
5. SmallGP crossover probability $= 0.9$, mutation probability (per node) $= 0.2$,
6. Maximum program depth $(MAXDEPTH) = 10$,
7. Maximum program length $(MAXLEN) = 10$,

Table 2. Design principles obtained using (GA + SmallGP) algorithm for truss design problem

Notation	Design Principle (DP) $\psi(\mathbf{x}) = constant$	Significance	Basic Dimensions		
			Mass	Length	Time
DP1	$y = constant$	86.60%	0.0	1.0	0.0
DP2	$S \times V = constant$	87.00%	1.0	2.0	-2.0
DP3	$S \times x_1 = constant$	85.00%	1.0	1.0	-2.0
DP4	$S \times V \times y = constant$	87.00%	1.0	3.0	-2.0
DP5	$(V \times y)/x_2 = constant$	86.20%	0.0	2.0	0.0
DP6	$(V \times y)/x_1 = constant$	88.20%	0.0	2.0	0.0
DP7	$V/x_1 = constant$	86.40%	0.0	1.0	0.0
DP8	$V/(S \times x_1 \times x_2) = constant$	87.20%	-1.0	0.0	2.0
DP9	$V^2/(x_1 \times x_2) = constant$	87.40%	0.0	2.0	0.0
DP10	$y/(S \times x_1) = constant$	88.00%	-1.0	0.0	2.0
DP11	$x_2/x_1 = constant$	83.80%	0.0	0.0	0.0
DP12	$(S \times V \times x_2 \times y)/x_1 = constant$	88.00%	1.0	3.0	-2.0
DP13	$V/x_2 = constant$	86.80%	0.0	1.0	0.0
DP14	$(S \times V^2 \times y)/x_1 = constant$	87.20%	1.0	4.0	-2.0
DP15	$(x_2 \times y)/x_1 = constant$	86.40%	0.0	1.0	0.0
DP16	$x_2/(S \times x_1^2) = constant$	86.40%	-1.0	-1.0	2.0
DP17	$V^2/(x_1 \times x_2 \times y) = constant$	91.40%	0.0	1.0	0.0
DP18	$(S \times V^2)/x_2 = constant$	87.20%	1.0	3.0	-2.0
DP19	$S \times x_2 \times y = constant$	87.00%	1.0	2.0	-2.0
DP20	$(x_2 \times y)/(S \times x_1^2) = constant$	86.80%	-1.0	0.0	2.0

8. Ephemeral constants, $\mathcal{R} = \{-10.0, -9.5, -9.0, \ldots, 9.0, 9.5, 10.0\}$,
9. Threshold significance $S_{reqd} = 80\%$, Clustering constant $\epsilon = 3$.

Table 2 shows the obtained design principles, their significance values and the exponents of their basic dimensions. A total of 26 principles were obtained, which were symbolically simplified in MATLAB and only the unique ones are presented here.

The truss design problem can be mathematically solved using the identical resource allocation strategy in order to verify the obtained design principles. Increasing the cross-sectional area of one member reduces the stress induced in it and so the second objective takes the other member into account at some point. But since both the objectives are equally important, this cannot be allowed. A balance can be obtained only when the stresses in both the members are equal.

$$S = \sigma_{AC} = \sigma_{BC} \Rightarrow S = \frac{100}{5} \frac{\sqrt{16 + y^2}}{yx_1} = \frac{4 \times 100}{5} \frac{\sqrt{1 + y^2}}{yx_2}. \tag{7}$$

Following a similar argument for the volumes we get,

$$V = 2 \times x_1 \sqrt{16 + y^2} = 2 \times x_2 \sqrt{1 + y^2}. \tag{8}$$

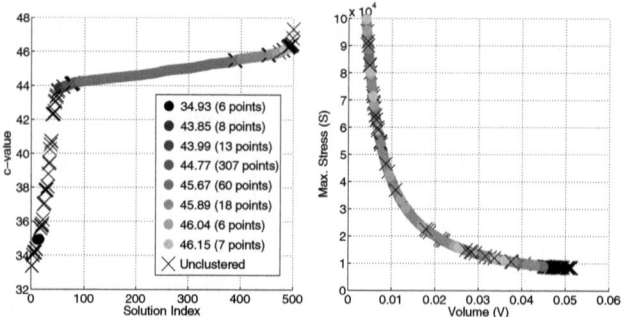

Fig. 3. Cluster plot (left) and the mapping of clusters in the objective space (right) for DP3. 425 out of 500 (85.00%) c-values obtained from $\psi(\mathbf{x}) = S \times x_1$ form eight clusters shown in shades of gray. The largest cluster has 307 points and an average c-value of 44.77.

Solving (7) and (8) gives the following relationships, all of which must be true for Pareto-optimality,

$$y = 2, \quad x_2 = 2x_1, \quad V = 4\sqrt{5}x_1 = 2\sqrt{5}x_2 \quad Sx_1 = 20\sqrt{5}, \quad Sx_2 = 40\sqrt{5}. \quad (9)$$

All design principles obtained by our approach conform to the above relationships. On the other hand, a dimensionally *unaware* GP produced relationships such as,

$$0.5 - (x_2 - S)^y = constant, S \times (V - x_2) = constant, (x_1 \times S) + x_2 = constant, \quad (10)$$

which, although numerically satisfy the requirements of a design principle, are of no practical value to the designer.

The next question to investigate is whether the 20 design principles can be reduced to the few shown in (9). To answer this, we first need to look at the c-value cluster plots of all design principles. For illustration let us consider DP3 and DP16. In each case, $\psi(\mathbf{x})$ is evaluated for all $m = 500$ trade-off solutions. The resulting c-values are sorted and plotted as shown in Figs. 3 and 4.

The right side plot in each figure shows that both design principles are applicable on (approximately) the same part of the trade-off front, indicating that they can be combined. Indeed reducing DP16 with DP3 results in $x_2/x_1 = constant$ which in itself is another design principle (DP11). By considering the largest clusters of DP16 and DP3, the approximate value of the constant in DP11 is found to be $0.0444 \times 44.77 = 1.99 \approx 2$, which agrees with the second relationship in (9). In fact, all design principles in Table 2 form clusters in the same part of the trade-off front and hence they can be combined in any way to eliminate the redundant ones.

The tree structures of DP12 and DP16 are shown in Figs. 5 and 6 for illustration.

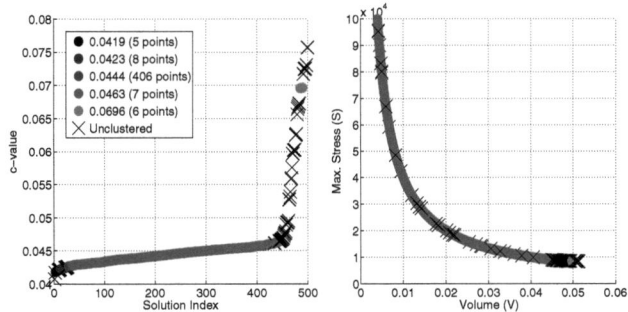

Fig. 4. Cluster plot (left) and the mapping of clusters in the objective space (right) for DP16. 432 out of 500 (86.40%) c-values obtained from $\psi(\mathbf{x}) = x_2/(S \times x_1^2)$ form five clusters shown in shades of gray. The largest cluster has 406 points and an average c-value of 0.0444.

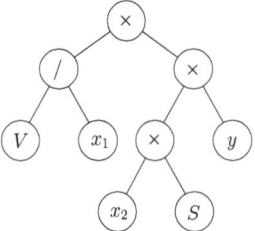

Fig. 5. GP tree for DP12 of truss design problem. Tree depth $= 3$ and size (or length) $= 9$.

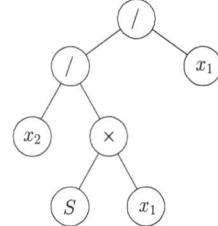

Fig. 6. GP tree for DP16 of truss design problem. Tree depth $= 3$ and size (or length) $= 7$.

4.2   Welded Beam Design

As our next example, we consider the bi-objective welded-beam design problem. It involves the minimization of welding cost C and end deflection D of a welded cantilever beam carrying an end load of 6000 lb. The design variables (in inches) are: beam thickness b, beam width t, length of the weld l and weld thickness h. The allowable bending stress (σ), shear stress (τ) and buckling force (P_c) are limited by constraints. The problem formulation [5] is:

$$
\begin{aligned}
\text{Minimize} \quad & f_1(\boldsymbol{x}) = C = 1.10471h^2l + 0.04811tb(14.0 + l), \\
\text{Minimize} \quad & f_2(\boldsymbol{x}) = D = \frac{2.1952}{t^3b}, \\
\text{Subject to} \quad & \left\{
\begin{array}{l}
\tau(\boldsymbol{x}) \leq 13,600 \text{ psi}; \sigma(\boldsymbol{x}) \leq 30,000 \text{ psi}; b \geq h; P_c(\boldsymbol{x}) \geq 6,000 \text{ lb} \\
0.125 \leq h, b \leq 5.0 \text{ in.}; 0.1 \leq l, t \leq 10.0 \text{ in.}
\end{array}
\right\}
\end{aligned}
$$

(11)

Table 3. Design principles obtained using (GA + SmallGP) algorithm for welded-beam problem

Notation	Design Principle (DP) $\psi(\mathbf{x}) = constant$	Significance	Basic Dimensions			
			Mass	Length	Time	Cost
DP1	$(D + t) = constant$	95.20%	0.0	1.0	0.0	0.0
DP2	$t = constant$	95.60%	0.0	1.0	0.0	0.0
DP3	$D \times b = constant$	95.00%	0.0	2.0	0.0	0.0
DP4	$D \times b \times t = constant$	95.60%	0.0	3.0	0.0	0.0
DP5	$\sigma \times b = constant$	94.80%	1.0	0.0	-2.0	0.0
DP6	$\sigma \times b \times t = constant$	95.60%	1.0	1.0	-2.0	0.0
DP7	$D/\sigma = constant$	95.60%	-1.0	2.0	2.0	0.0
DP8	$D/(\sigma \times t) = constant$	95.60%	-1.0	1.0	2.0	0.0

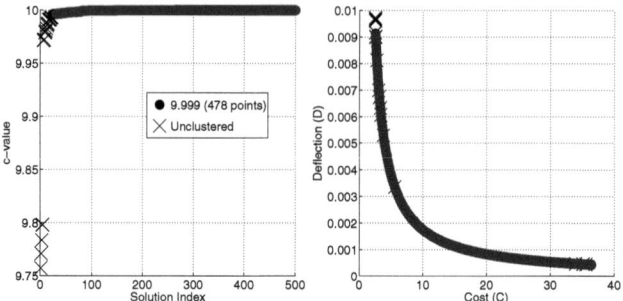

Fig. 7. Cluster plot (left) and the mapping of clusters in the objective space (right) for DP2. 478 out of 500 (95.60%) c-values obtained from $\psi(\mathbf{x}) = t$ form a single clusters with an average c-value of 9.999.

$$\text{where } \tau(\mathbf{x}) = \sqrt{(\tau')^2 + (\tau'')^2 + (l\tau'\tau'')/\sqrt{0.25(l^2 + (h + t)^2)}},$$

$$\tau' = \frac{6,000}{\sqrt{2}hl}, \quad \tau'' = \frac{6,000(14+0.5l)\sqrt{0.25(l^2+(h+t)^2)}}{2[0.707hl(l^2/12+0.25(h+t)^2)]},$$

$$\sigma(\mathbf{x}) = \frac{504,000}{t^2b},$$

$$P_c(\mathbf{x}) = 64,746.022(1 - 0.0282346t)tb^3.$$

The problem is solved using NSGA-II to obtain $m = 500$ trade-off solutions. For automated innovization, we choose the same parameters as before, except the population size which is increased to 3000. The function set for GP remains the same. The terminal set is chosen as $\mathcal{T} = \{C, D, b, t, l, h, \sigma, \tau, P_c, \mathcal{R}\}$. An extra 'Cost' dimension is added to the basic dimensions' set for providing the dimensional information of C to the algorithm. Table 3 shows the obtained design principles for $S_{reqd} = 90\%$.

All relationships are found to be applicable over approximately the same part of the trade-off front and therefore some of them are redundant. The independent design principles DP2, DP3 and DP5 have been obtained previously [1,5].

DP1 requires special attention because it exposes a limitation of the current approach. The deflection D is of the order of 10^{-4} whereas the cluster plot of DP2

Table 4. Design principles obtained using (GA + SmallGP) algorithm for metal-cutting problem

Notation	Design Principle (DP) $\psi(\mathbf{x}) = constant$	Significance	Mass	Length	Time	Life
DP1	$v/(f^2 \times \xi) = constant$	72.70%	0.0	-1.0	-1.0	-1.0
DP2	$(a \times v)/f = constant$	74.60%	0.0	1.0	-1.0	0.0
DP3	$v/(f^2 \times T_p \times \xi) = constant$	73.40%	0.0	-1.0	-2.0	-1.0
DP4	$f = constant$	72.90%	0.0	1.0	0.0	0.0
DP5	$a/(f \times T_p) = constant$	72.90%	0.0	0.0	-1.0	0.0
DP6	$(a^{5.5} \times f \times \xi)/T_p = constant$	77.50%	0.0	6.5	-1.0	1.0
DP7	$(a \times T_p \times v)/f = constant$	74.20%	0.0	1.0	0.0	0.0
DP8	$a^{5.5} \times T_p \times \xi = constant$	82.60%	0.0	5.5	1.0	1.0
DP9	$a \times T_p \times v = constant$	74.10%	0.0	2.0	0.0	0.0
DP10	$(a^2 \times T_p \times \xi)/v = constant$	74.40%	0.0	1.0	2.0	1.0
DP11	$(a^2 \times \xi)/v = constant$	76.00%	0.0	1.0	1.0	1.0
DP12	$a^{5.5} \times f \times \xi = constant$	76.80%	0.0	6.5	0.0	1.0

in Fig. 7 reveals that the value of t is clustered around 10 in. for most solutions. This leads to an ambiguity where $D + t$ numerically satisfies the requirement of a design principle. The current approach does not handle these situations. A possible remedy for future analysis could be to normalize each basis function (ϕ) with its order of magnitude.

4.3   Metal-Cutting Process Optimization

Next, we consider the metal-cutting process optimization problem described in [13]. A steel bar is to be machined using a carbide tool of nose radius $r_n = 0.8$ mm on a lathe with $P^{max} = 10$ kW rated motor to remove 219912 mm^3 of material. A maximum cutting force of $F_c^{max} = 5000$ N is allowed. The motor has a transmission efficiency $\eta = 75\%$. The total operation time (T_p) and the used tool life (ξ) are to be minimized by optimizing the cutting speed (v), the feed rate (f) and the depth of cut (a) while maintaining a surface roughness of $R^{max} = 50\mu m$. The problem is formulated as,

$$\text{Minimize} \quad f_1(\mathbf{x}) = T_p(\mathbf{x}) = 0.15 + 219912 \left(\frac{1 + \frac{0.20}{T(\mathbf{x})}}{MRR(\mathbf{x})} \right) + 0.05 \quad \text{min}$$

$$\text{Minimize} \quad f_2(\mathbf{x}) = \xi(\mathbf{x}) = \frac{219912}{MRR(\mathbf{x})T(\mathbf{x})} \times 100\%$$

$$\text{Subject to} \quad \left\{ \begin{array}{l} P(\mathbf{x}) \leq \eta P^{max}; F_c(\mathbf{x}) \leq F_c^{max}; R(\mathbf{x}) \leq R^{max} \\ 250 \leq v \leq 400 \text{ m/min}; 0.15 \leq f \leq 0.55 \text{ mm/rev}; 0.5 \leq a \leq 6 \text{ mm} \end{array} \right\}$$

$$(12)$$

where $T(\mathbf{x}) = \frac{5.48 \times 10^9}{v^{3.46} f^{0.696} a^{0.460}}$, $MRR(\mathbf{x}) = 1000vfa$

$$P(\mathbf{x}) = \frac{vF_c(\mathbf{x})}{60000}, \quad F_c(\mathbf{x}) = \frac{6.56 \times 10^3 f^{0.917} a^{1.10}}{v^{0.286}}, \quad R(\mathbf{x}) = \frac{125 f^2}{r_n}.$$

NSGA-II results in 1000 trade-off solutions. (GA + SmallGP) is used with the same parameters as for truss design problem for $S_{reqd} = 70\%$ to obtain the design

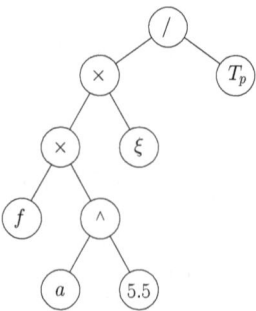

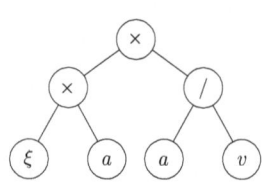

Fig. 8. GP tree for DP6 of metal-cutting problem. Tree depth = 4 and size (or length) = 9.

Fig. 9. GP tree for DP11 of metal-cutting problem. Tree depth = 2 and size (or length) = 7.

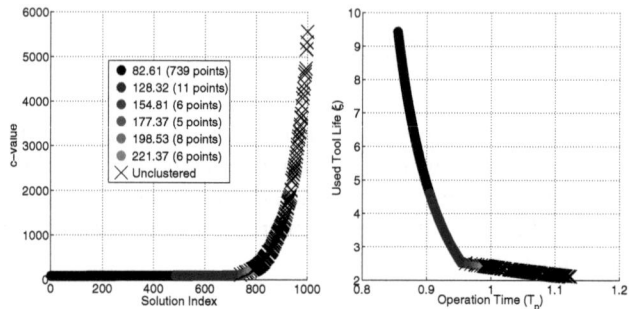

Fig. 10. Cluster plot (left) and the mapping of clusters in the objective space (right) for DP6. 775 out of 1000 (77.50%) c-values obtained from $\psi(\mathbf{x}) = (a^{5.5} \times f \times \xi)/T_p$ form six clusters shown in shades of gray. The largest cluster has 739 points and an average c-value of 82.61.

principles shown in Table 4. A new dimension 'Life' is introduced to denote used tool life which is expressed as a percentage of total tool life.

Empirical (and more accurate) forms of DP4 and DP9 have previously been reported in [6]. Here, we sacrifice the accuracy of the design principle in favour of ease of interpretability for the designer, by making use of a dimensionally-aware GP.

The tree structures of two of the design principles, DP6 and DP11, are shown in Figs. 8 and 9. Fig. 10 shows the cluster plot for the former.

5   Conclusions

This paper introduced a generalization to the automated innovization framework proposed previously by the authors. A tree-based representation is used to evolve generic design principles for extracting knowledge from multi-objective trade-off datasets. The underlying algorithm for automated innovization remains the same, but introduction of the parse tree representation required the use of a genetic programming system. We integrated the SmallGP system with a standard genetic algorithm for evolving the design principles. In order to obtain only physically meaningful design principles, we made the (GA + SmallGP) system dimensionally aware by penalising operations performed between incommensurable quantities. The dimensional consistency is checked at each step of tree evaluation for all fundamental dimensions. The proposed algorithm was tested on three engineering design problems. While syntactically correct principles were obtained in both cases, a limitation concerning different magnitudes of basis functions was identified.

References

1. Bandaru, S., Deb, K.: Automated Innovization for Simultaneous Discovery of Multiple Rules in Bi-objective Problems. In: Takahashi, R.H.C., Deb, K., Wanner, E.F., Greco, S. (eds.) EMO 2011. LNCS, vol. 6576, pp. 1–15. Springer, Heidelberg (2011)
2. Bandaru, S., Deb, K.: Towards automating the discovery of certain innovative design principles through a clustering-based optimization technique. Engineering Optimization 43(9), 911–941 (2011)
3. Deb, K., Bandaru, S., Greiner, D., Gaspar-Cunha, A., Tutum, C.: An integrated approach to automated innovization for discovering useful design principles: Three engineering case studies. Tech. rep., Indian Institute of Technology Kanpur (2012)
4. Deb, K., Pratap, A., Agarwal, S., Meyarivan, T.: A fast and elitist multi-objective genetic algorithm: NSGA-II. IEEE Transactions on Evolutionary Computation 6(2), 182–197 (2002)
5. Deb, K., Srinivasan, A.: Innovization: Innovating design principles through optimization. In: GECCO 2006 - Proceedings of the 8th Annual Conference on Genetic and Evolutionary Computation, pp. 1629–1636. ACM, New York (2006)
6. Deb, K., Datta, R.: Hybrid evolutionary multi-objective optimization and analysis of machining operations. Engineering Optimization 44(6), 685–706 (2012)
7. Deb, K., Goyal, M.: A combined genetic adaptive search (GeneAS) for engineering design. Computer Science and Informatics 26, 30–45 (1996)
8. Keijzer, M., Babovic, V.: Dimensionally aware genetic programming. In: Proceedings of the Genetic and Evolutionary Computation Conference, vol. 2, pp. 1069–1076 (1999)
9. Koza, J.: Genetic Programming, On the Programming of Computers by Means of Natural Selection. A Bradford Book. MIT Press, Cambridge (1992)
10. Newman, M.: Power laws, pareto distributions and zipf's law. Contemporary Physics 46(5), 323–351 (2005)
11. Planatscher, H.: SmallGP - A lean genetic programming system for symbolic regression (2004), http://planatscher.net/smallgp.html
12. Poli, R., Langdon, W., McPhee, N., Koza, J.: Genetic programming: An introductory tutorial and a survey of techniques and applications. University of Essex, UK. Tech. Rep. CES-475 (2007)
13. Quiza Sardiñas, R., Rivas Santana, M., Alfonso Brindis, E.: Genetic algorithm-based multi-objective optimization of cutting parameters in turning processes. Engineering Applications of Artificial Intelligence 19(2), 127–133 (2006)
14. Valdés-Pérez, R.: Principles of human computer collaboration for knowledge discovery in science. Artificial Intelligence 107(2), 335–346 (1999)

A New Multiobjective Genetic Programming for Extraction of Design Information from Non-dominated Solutions

Tomoaki Tatsukawa, Taku Nonomura, Akira Oyama, and Kozo Fujii

Institute of Space and Astronautical Science, Japan Aerospace Exploration Agency,
3-1-1 Yoshinodai, Sagamihara, Kanagawa 252-5210, Japan
{tatsukawa,nonomura,oyama,fujii}@flab.isas.jaxa.jp

Abstract. We propose a new type of multi-objective genetic programming (MOGP) for multi-objective design exploration (MODE). The characteristic of the new MOGP is the simultaneous symbolic regression to multiple objective functions using correlation coefficients. This methodology is applied to non-dominated solutions of the multi-objective design optimization problem to extract information between objective functions and design parameters. The result of MOGP is symbolic equations that are highly correlated to each objective function through a single GP run. These equations are also highly correlated to several objective functions. The results indicate that the proposed MOGP is capable of finding new design parameters more closely related to the objective functions than the original design parameters. The proposed MOGP is applied to the test problem and the practical design problem to evaluate the capability.

Keywords: Multi-Objective Genetic Programming, Multi-Objective Design Explolation, CFD, NSGA-II.

1 Introduction

Multi-objective design exploration (MODE)[1] is proposed as an approach to extract design information from multi-objective design optimization (MOO) problems. In MODE, a multi-objective evolutionary algorithm (MOEA) is used to efficiently find a set of optimal solutions, known as Pareto-optimal solutions or non-dominated solutions. Then, various data analysis techniques are used to extract the design information from the non-dominated solutions[2–5]. Design information includes the trade-off information between objective functions, relationship between objective functions and design parameters, and constraint conditions among design parameters. It is relatively easy to obtain reasonable non-dominated solutions using MOEAs when the number of objective functions and design parameters remains small. However, it is difficult to retrieve the design information from non-dominated solutions when nonlinear relations between objective functions and design parameters exist.

Genetic programming (GP) is an evolutionary algorithm that is capable of automatically revealing the relationship between parameters as expressions in

R.C. Purshouse et al. (Eds.): EMO 2013, LNCS 7811, pp. 528–542, 2013.

symbolic form without prior knowledge of the problem[6]. The unique feature of GP is that it finds not only the linear relationship between parameters, but also the nonlinear relationship automatically.

In MODE, two-objective GP(TOGP) that handle the residual and the number of nodes can be used as a symbolic regression technique[7]. An input data set of TOGP is a colloection of an objective function and design parameters of non-dominated solutions. Two-objective GP is capable of extracting the relationship between objective functions and design parameters of non-dominated solutions as symbolic equations. However, if TOGP is used, it should be applied separately to each objective function. In such a case, symbolic equations that fit to one of the objective functions of MOO problems are obtained in a single GP run. To find symbolic equations for all objective functions, TOGP has to be executed as many times as the number of objective functions, and when the residual measure is used, TOGP has to optimize not only the terms but also the coefficients and constants to improve accuracy. Thus, it takes a long time for GP to produce symbolic equations for all objective functions of MOO problems.

The objectives of this study are to present a new type of multi-objective genetic programming(MOGP) to extract the relationship between objective functions and design parameters from non-dominated solutions and to evaluate the capability of the new MOGP. One advantage of the proposed MOGP is shorter computational time than TOGP. Another advantage is that the proposed MOGP enables us to simultaneously solve the symbolic regression problems for all objective functions using correlation coefficients and find symbolic equations that are highly correlated to the multiple objective functions. In this study, the test problem and the multiobjective aerodynamic design optimization problem of a bi-conical shape reusable launch vehicle (RLV) are considered to evaluate the proposed MOGP.

2   Two-Objective Genetic Programming

In [7], two objective functions are considered to produce symbolic equations. One of the objective functions represents the accuracy of an evolved symbolic equation, e.g., the sum of absolute error or the mean square error. This objective function characterizes how well an evolved equation matches the given data set. Another objective function is the complexity of an evolved equation, e.g., the order of nonlinearity[8], the depth of the syntax tree(Fig. 1).

Figure 2 shows the flow chart of MODE using TOGP prepesed in [7]. First, multi-objective evolutionary algorithms are used to find as many non-dominated solutions as possible. Second, two-objective GP is used as the data analysis method where it is separately applied to each objective function. Input data set for TOGP is the objective function and design parameters of the non-dominated solutions. The objective functions of TOGP are as follows.

Objective function 1: the minimization of the mean absolute error between one objective function of the MOO problem (the measure of accuracy).

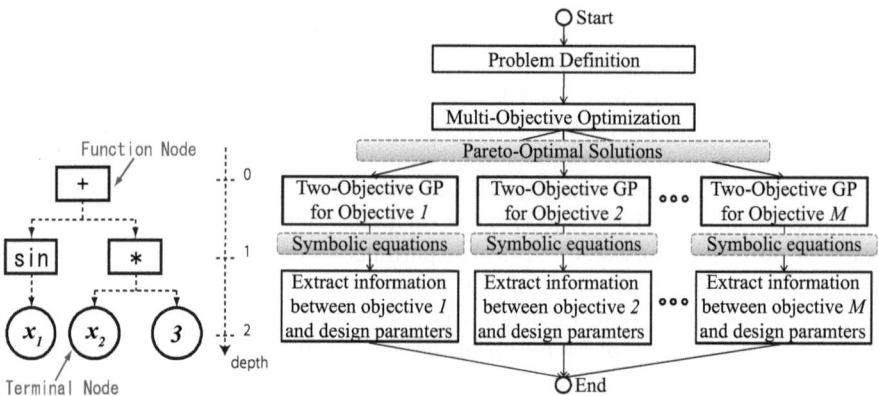

Fig. 1. GP syntax tree representation $sin(x_1) + 3x_2$

Fig. 2. Flow chart of MODE using TOGP

Objective function 2: the minimization of the number of nodes in the syntax tree (the measure of complexity).

After one TOGP run, a set of non-dominated "equations" that fits the ith objective function of the MOO problem is obtained. One of the non-dominated equations is as follows.

$$f_i = \hat{f}_i(\boldsymbol{x}) \tag{1}$$

Where f_i is the ith objective function of the MOO problem, $\hat{f}_i$ is one of the approximated symbolic equation of the ith objective function for optimization, $\boldsymbol{x} \in R^n$ is the design parameter vector, and n is the number of design parameters. When the number of objective functions of the MOO problems is M, after M TOGP runs, a set of approximated symbolic equations for all objective functions is obtained. Finally, the information such as the relation between all objective functions and design variables is obtained from symbolic equations. However, TOGP has to be executed as many times as the number of objective functions. Thus, it takes a long time to obtain symbolic equations for all the objective functions.

3   New Multi-Objective Genetic Programming

Here, we propose a new type of multi-objective GP to extract the design information from non-dominated solutions more efficiently than the two-objective GP. The new MOGP is capable of simultaneously finding the approximated symbolic equations to multiple objective functions. There are two differences between the proposed MOGP and two-objective GP used in [7].

The first difference is the measure of accuracy. The proposed MOGP does not use a residual characteristic but the squared correlation coefficient of fitting one

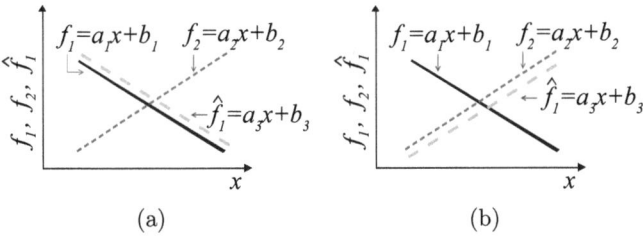

Fig. 3. Plot of two objective functions and one candidate equation of TOGP

of the multiple objective functions of MOO problems. In the proposed MOGP, it is important to use the squared correlation coefficient as the measure of accuracy. When the residual measure is used, it is necessary to optimize not only the terms but also the coefficients and constants to improve accuracy, and it is difficult to simultaneously obtain a small mean square error for several objective functions. However when the squared correlation coefficient is used, it is not necessary to optimize the coefficients and constants. Moreover, if common terms between objective functions are found, it is possible to improve correlation for both objective functions.

For example, two objective functions and one candidate equation are considered as shown in Fig. 3. In this figure, $f_1(x)$ and $f_2(x)$ are objective functions, $\hat{f}_1(x)$ is the candidate equation, and x is the design parameter. The objective functions($f_1(x)$ and $f_2(x)$) have different coefficients(a_1, a_2) and different constants(b_1, b_2) respectively. From Fig. 3(a), when the mean square error between $\hat{f}_1(x)$ and $f_1(x)$ is small, the mean square error between $\hat{f}_1(x)$ and $f_2(x)$ is large. From Fig. 3(b), when the mean square error between $\hat{f}_1(x)$ and $f_2(x)$ is small, the mean square error between $\hat{f}_1(x)$ and $f_1(x)$ is large. It is difficult to simultaneously obtain a small mean square error for both objective functions.

In contrast, when squared correlation coefficients are used, it is not necessary to optimize the all coefficients and constants. If x that is the common term between the two objective functions is set, the squared correlation coefficient between $\hat{f}_1(x)$ and $f_1(x)$ and the squared correlation coefficient between $\hat{f}_1(x)$ and $f_2(x)$ are the same. Moreover, if common terms between objective functions are found, it is possible to simultaneously obtain high correlation for the objective functions. It is expected that the computational time to obtain the symbolic equations for all the objective functions will decrease. Therefore, the correlation coefficient is used as the measure of accuracy in the proposed MOGP.

Second, the number of objective functions of MOGP is different. In the proposed MOGP, the maximization of the squared correlation coefficient between all the objective functions of the MOO problem and the candidate symbolic equation is considered. Objective functions characterizes how well an evolved equation matches all the objective functions of the given data set. The other objective function minimizes the number of nodes in the syntax tree and characterizes the level of simplicity of the expression of an evolved equation. For example, when the number of objective functions in the MOO problem is M, the total number of objective functions in the proposed MOGP is $M + 1$. The objective functions of the proposed MOGP are as follows.

Objective function 1: the maximization of the squared correlation between the first objective function of the MOO problem and the candidate symbolic equation (the measure of accuracy).

Objective function 2: the maximization of the squared correlation between the second objective function of the MOO problem and the candidate symbolic equation (the measure of accuracy).

...

Objective function M: the maximization of the squared correlation between the Mth objective function of the MOO problem and the candidate symbolic equation (the measure of accuracy).

Objective function M+1: the minimization of the number of nodes in the syntax tree (the measure of complexity).

The squared correlation coefficient is

$$Cor^2(f_i, \hat{f}) = \left(\frac{\sum_{j=1}^{N}(f_{i,j} - \bar{f}_i)(\hat{f}_j - \bar{\hat{f}})}{\sqrt{(f_{i,j} - \bar{f}_i)^2}\sqrt{(\hat{f}_j - \bar{\hat{f}})^2}} \right)^2 \to max, \quad i = 1, 2, ..., M \quad (2)$$

$$\bar{f}_i = \frac{\sum_{j=1}^{N} f_{i,j}}{N}, \quad \bar{\hat{f}} = \frac{\sum_{j=1}^{N} \hat{f}_j}{N} \quad (3)$$

where N is the number of the given data sets, $Cor^2(f_i, \hat{f})$ is the ith objective function of the proposed MOGP, f_i is the ith objective function of the MOO problem, $\bar{f}_i$ is the average value of the ith objective function of the MOO problem, $\hat{f}$ is the candidate equation that MOGP generates, and $\bar{\hat{f}}$ is the average value of the candidate symbolic equation. The value of $Cor^2(f_i, \hat{f})$ is between $+1$ and 0. Figure 4 is the psedo code of MOGP. In Fig. 4, the squared correlation coefficients and the number of nodes are evaluated in calc_objs() function. Other functions are same as conventional GA.

```
void gp_evolve() {
  create_random_pop(); // Create initial random populations
  for (gen=0; gen < gen_max; gen++) {
    eval_objs();        // Evaluate the objective functions
    calc_pareto();      // Calculate the pareto-ranking
    select_parent();    // Select parent population
    genetic_operation(); // Create new populations by crossover and mutation
  }
}
```

Fig. 4. Pseudo code for the evaluation of MOGP

As a result of MOGP, a set of non-dominated equations is obtained in a single GP run. Some non-dominated equations are highly correlated to one of

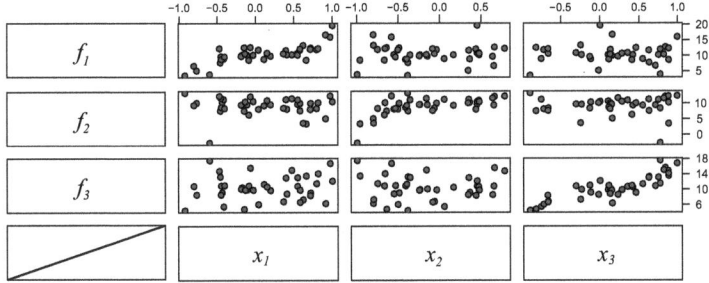

Fig. 5. Data set of test functions

the objective functions of the MOO problem. In addition, other non-dominated equations are highly correlated to some of the objective functions of the MOO problem at the same time.

4   Test Problem

4.1   Problem Definition

The test problem considered here has three objective functions and three design parameters. We suppose the non-dominated solutions satisfy the following relations.

$$f_1(x_1, x_2, x_3) = 10x_1^3 + 5x_2x_3 + 10 \qquad (4)$$

$$f_2(x_1, x_2, x_3) = 10x_2^3 + 5x_3x_1 + 10 \qquad (5)$$

$$f_3(x_1, x_2, x_3) = 10x_3^3 + 5x_1x_2 + 10 \qquad (6)$$

Sample points that imitate non-dominated solutions are randomly created by substituting values in $[-1, 1]$ to variable x_1, x_2 and x_3. The number of data set is 40. Figure 5 shows the scatter plots between the objective functions and design parameters. From Fig. 5, it is difficult to find the relations between each objective function and the design parameters.

Two type of the two-objective GP(TOGP MAE, TOGP SCC) and the proposed multi-objective GP(MOGP-SCC) are applied to the test problem. In the TOGP-MAE, the minimization of the mean absolute error (MAE) is used as one of the objective functions for measuring accuracy. In the TOGP-SCC, the maximization of the squared correlation coefficient (SCC) is used as one of the objective functions for measuring accuracy.

4.2   Approach

In this study, the symbolic equation of GP is expressed as a syntax tree, which is composed of terminal and function nodes. The terminal nodes are the design parameters(x_1,x_2, and x_3) and constants. The function nodes are arithmetic operators ($+$, $*$, $/$, and $-$). The GP procedure is based on the non-dominated sorting genetic algorithm-II (NSGA-II)[9]. Non-dominated sort ranking is used as the ranking method. Crowded tournament selection is applied to select pairs. Subtree crossover[10] and subtree mutation[11] are applied to the selected pairs to create the population. The size of the population is 1000, and the number of generation is 1000. The crossover rate is 0.75, and the mutation rate is 0.2. Two-objective GP(TOGP-MAE, TOGP-SCC) have to be separately applied to each objective function. The number of each trial is 15. Therefore, the total number of trials in TOGP is 45, whereas the proposed multi-objective GP(MOGP-SCC) is simultaneously applied to all objective functions. Therefore, the total number of trials in MOGP is 15.

4.3   Results and Discussions

Figure 6 shows the history of the most accurate individual of each trial and the average of that of the trials. The horizontal axis is generation and the vertical axis is the measure of accuracy. When the mean absolute error is used, the equations that have the residuals of zero are the most precise ones, and when the squared correlation coefficient is used, the equations that have the correlation coefficient of one are the most precise ones. Figures 6 (a), (b), and (c) show the results of separately applied TOGP-MAE to each objective function. From Figs. 6 (a), (b), and (c), no equation has been found that has the residuals of zero. The non-dominated equations that have the smallest residuals among all trials are shown as follows,

$$\hat{f}_1(x_1, x_2, x_3) = 10x_1^3 + 4x_2x_3 + x_1 + x_2 - 0.3x_1x_3 - x_1x_3^2 + 10 \qquad (7)$$

$$\hat{f}_2(x_1, x_2, x_3) = 10x_2^3 + 5x_1x_3 - 0.2x_3 + x_2x_3 + 10 \qquad (8)$$

$$\hat{f}_3(x_1, x_2, x_3) = 11.1x_3^3 + 4x_1x_2 - x_3 + 10 \qquad (9)$$

These equations include not only correct terms such as x_1^3 and x_2x_3 but also additional terms such as x_1 and $x_1x_3^2$. It is difficult to recognize which terms are truly included in the original objective functions. Figures 6 (d), (e), and (f) show the results of separately applied TOGP-SCC to each objective function. From Figs. 6 (d), (e), and (f), there are some trials in which the squared correlation coefficient goes to one. The equations that have the squared correlation coefficient is 1 have the correct terms and the correct ratio of coefficients between each term. The non-dominated equations that have the smallest residuals among all trials are shown as follows.

$$\hat{f}_1(x_1, x_2, x_3) = x_1^3 + 0.5x_2x_3 \qquad (10)$$

$$\hat{f}_2(x_1, x_2, x_3) = 2.85x_2^3 + 1.428x_1x_3 \qquad (11)$$

$$\hat{f}_3(x_1, x_2, x_3) = 6.66x_3^3 + 3.33x_1x_2 + 3.33 \qquad (12)$$

These equations only have correct terms, and the ratios of the coefficients between each term(e.g., $\frac{x_1^3}{x_2x_3} = 2$) coincide with those of the original objective functions. However, to obtain the optimal equations for all the objective functions, it is necessary to conduct at least three TOGP runs. Figures 6 (g),(h), and (i) show the results of applying MOGP-SCC to all the objective function at one time. From Figs. 6 (g),(h), and (i), there are some trials that in which a squared correlation coefficient goes to one.

$$\hat{f}_1(x_1, x_2, x_3) = x_1^3 + 0.5x_2x_3 \tag{13}$$

$$\hat{f}_2(x_1, x_2, x_3) = 2x_2^3 + x_1x_3 \tag{14}$$

$$\hat{f}_3(x_1, x_2, x_3) = 2x_3^3 + x_1x_2 \tag{15}$$

From Fig. 6, we see that the convergence velocity of MOGP-SCC is almost the same as that of TOGP-SCC. Figure 7 shows the computational time of GP normalized by MOGP-SCC time. Where, the time of MOGP-SCC is 1. From Fig. 7, the computational time of the proposed MOGP is shorter than that of TOGP. Table 1 shows the minimum, average and standard deviation of the accuracy measurement in the last generation. From Table 1, we see that the search performance of MOGP-SCC is almost the same as that of TOGP-SCC.

Table 1. Minimum, average and standard deviation of the accuracy measurment in the last generation. In TOGP-MAE, the accuracy measurment is the mean absolute error. In TOGP-SCC and MOGP-SCC, the accuracy measurment is the squared correlation coefficient.

(a) f_1

	TOGP-MAE	TOGP-SCC	MOGP-SCC
Min.	0.39	1	1
Avg.	0.65	1	0.999
Std.	0.21	0	0.00022

(b) f_2

	TOGP-MAE	TOGP-SCC	MOGP-SCC
Min.	0.28	1	1
Avg.	0.69	0.99	0.98
Std.	0.18	0.017	0.022

(c) f_3

	TOGP-MAE	TOGP-SCC	MOGP-SCC
Min.	0.28	1	1
Avg.	0.80	0.99	0.98
Std.	0.20	0.0005	0.03

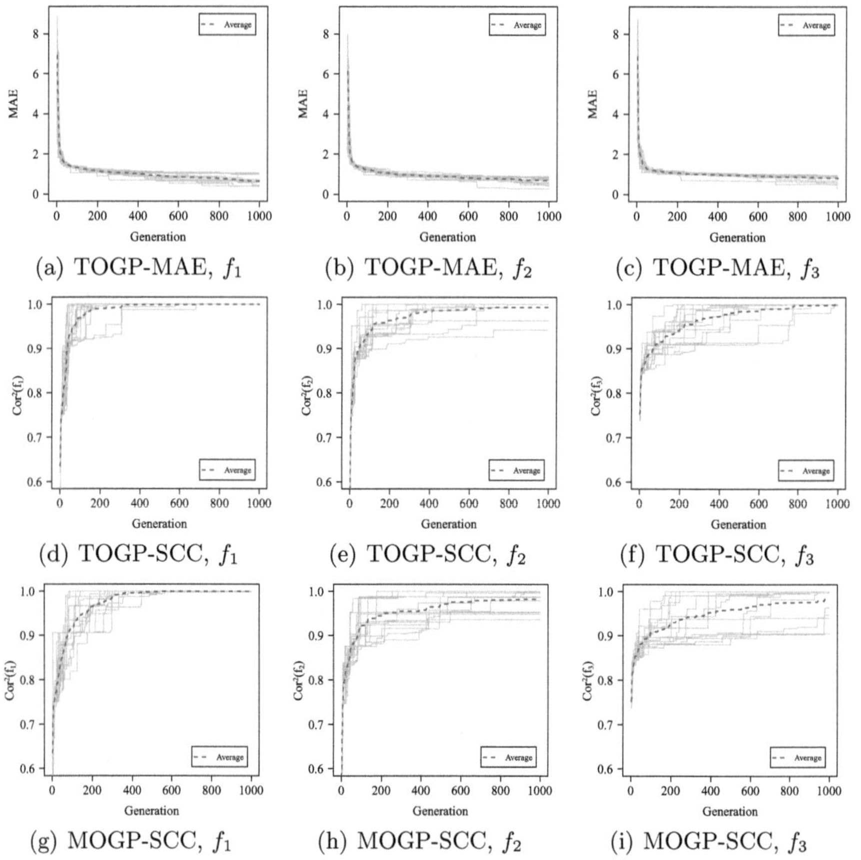

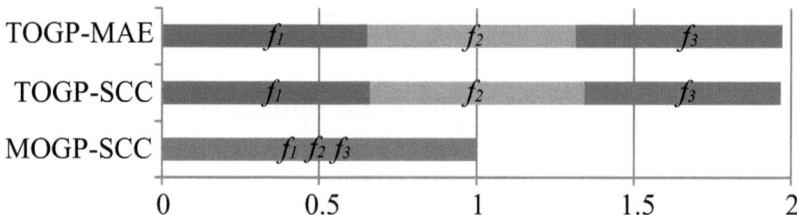

(a) TOGP-MAE, f_1 (b) TOGP-MAE, f_2 (c) TOGP-MAE, f_3

(d) TOGP-SCC, f_1 (e) TOGP-SCC, f_2 (f) TOGP-SCC, f_3

(g) MOGP-SCC, f_1 (h) MOGP-SCC, f_2 (i) MOGP-SCC, f_3

Fig. 6. Accuracy versus generation of TOGP-MAE, TOGP-SCC, and MOGP-SCC

Fig. 7. Comparison of the computational time of MOGP-SCC, TOGP-MAE, and TOGP-SCC

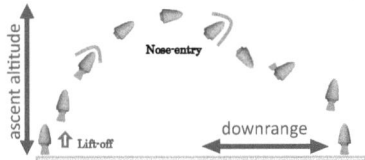

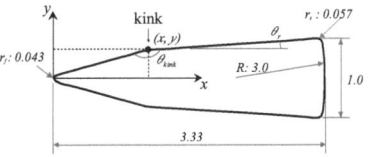

Fig. 8. Flight sequence of RVT   **Fig. 9.** Definition of the body geometry

5   Practical Design Problem

Here, the non-dominated solutions of the multi-objective aerodynamic design optimization problem of a bi-conical shape reusable launch vehicle (RLV)[12, 13](Fig. 8) is analyzed with MOGP.

5.1   Problem Definition

The following four objective functions are considered in the multi-objective aerodynamic design optimization. (1) The aerodynamic drag during the ascending flight that greatly affects the ascent altitude. (2) and (3) The maximum lift-to-drag ratios (L/D) during the return phase that greatly affects the downrange. There are two flight regions to consider for maximum L/D - the subsonic flight region with Mach number of 0.8 and the supersonic flight region with Mach number of 2.0. Because the aerodynamic characteristics of the subsonic and supersonic flight regions differ significantly, both flight regions should be considered for the maximum lift-to-drag ratio. (4) The maximization of the body volume that is also necessary to enlarge the on-board capability for payloads, fuel, and equipment.

Objective function 1: the minimization of the drag at Mach number of 2.0 and zero angle of attack conditions.

Objective function 2: the maximization of the maximum L/D at Mach number of 0.8(subsonic condition).

Objective function 3: the maximization of the maximum L/D at Mach number of 2.0(supersonic condition).

Objective function 4: the maximization of the body volume.

The shape of vehicle body is axisymmetric and has one kink, the so-called biconical configuration. Figure 9 shows the body geometry. The lengths in the figure are non-dimensionalized by the base diameter. The design parameter is the position of the kink, which is represented in a two-dimensional Cartesian space, as shown in Fig. 9. The body geometry is defined by the position of the kink, namely x and y. The exploration range of x is between 0.33 and 3.0 and that of y is between 0.15 and 0.5.

5.2   Approach

In [13], 276 non-dominated solutions were obtained. Here, the proposed MOGP is applied to the non-dominated solutions to obtain many non-dominated equations that are highly correlated to the objective functions of the MOO problem. The objective functions for MOGP are summarized as follows.

Objective function 1: the maximization of the squared correlation between zero-lift drag at supersonic conditions.
Objective function 2: the maximization of the squared correlation between maximum L/D ratios at subsonic conditions.
Objective function 3: the maximization of the squared correlation between maximum L/D ratios at supersonic conditions.
Objective function 4: the maximization of the squared correlation between body volumes.
Objective function 5: the minimization of the number of nodes

The terminal nodes are the design parameters(x and y) and constants. The function nodes are arithmetic operators (+, *, /, and -). The GP procedure, crossover and mutation are the same as the test problem. The population size is 1500, and the generation size is 1500. The crossover probability is 0.8, and the mutation probability is 0.2. Arithmetic operators (+ , - , * , and /) are considered as function nodes. The kink position and constants defined in [−1, 1] are considered as the terminal nodes. The constraint condition is that the number of nodes is greater than 1. Crowded tournament selection is applied to select pairs. Subtree crossover and subtree mutation are applied to the selected pairs to create the population. The number of trials is 15.

5.3   Non-dominated Solutions

Figure 10 shows the results of the multi-objective optimization. Red squares represent the non-dominated solutions. Grey circles represent the dominated solutions. From Fig. 10, we extract the relationship between the objective functions and design parameters. However, it is not easy to efficiently extract the relation between each objective function and the design parameters.

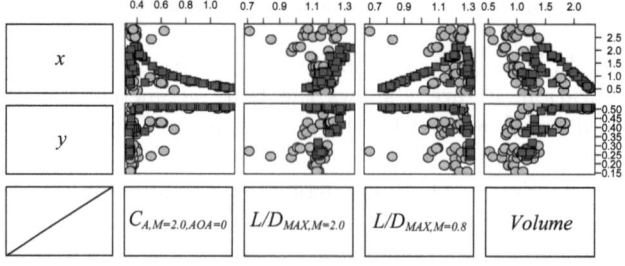

Fig. 10. Dominated(grey) and non-dominated(red) solutions-scatter plots

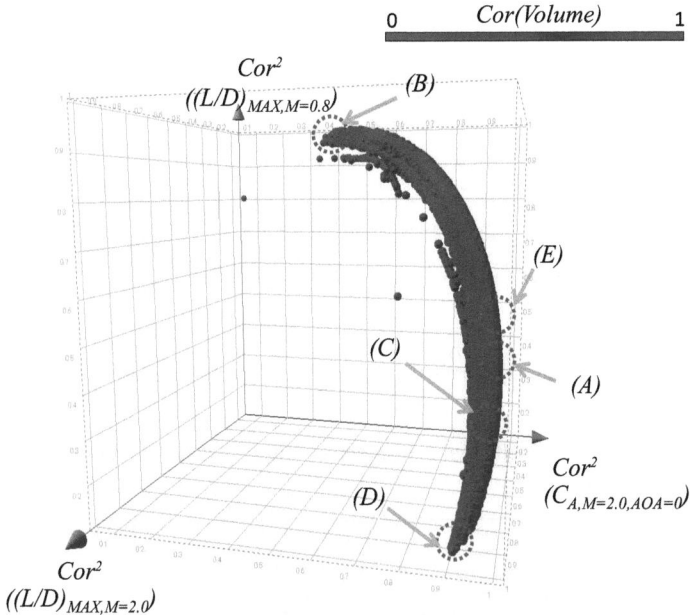

Fig. 11. Three-dimensional scatter plot-All non-dominated equations that include the equations of the highest squared correlation to each objective function of the MOO problem ((A)−(D)), and the equation that has the maximum sum of all squared correlation coefficient(E)

5.4   Results and Discussions

As a result of the proposed MOGP, more than 10000 non-dominated equations are obtained per trial. Figure 11 shows the non-dominated equations of all trials. The x, y, and z axes are the squared correlation between zero-lift drag at supersonic condition and the non-dominated equations, the squared correlation between maximum L/D at subsonic condition and the non-dominated equations, and the squared correlation between maximum L/D at supersonic condition and the non-dominated equations, respectively. The plots are colored according to the squared correlation between the body volume and the non-dominated equations. Each plot corresponds to an evolved symbolic equation. The symbolic equations from (A) (B) (C) and (D) in Fig. 11 have the highest squared correlations to each objective function of the MOO problem. The symbolic equation (E) has the maximum sum value of all squared correlations.

Equation (A) that has the highest squared correlation coefficient with the zero-lift drag at supersonic condition is expressed as

$$\hat{f}_A(x,y) = -4.1y + 3.1xy + 2.75x - 2.6x^2$$
$$-2y^2 - x^2y + x^3 - 1.65 \tag{16}$$

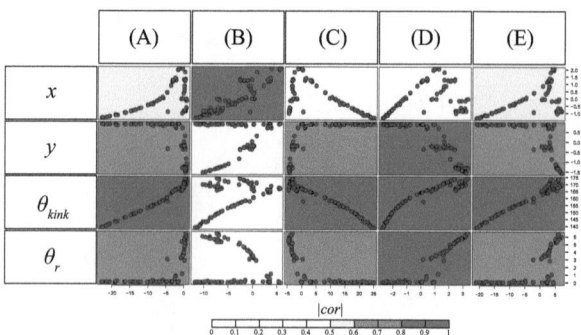

Fig. 12. Scatter plots of $x, y, \theta_{kink}, and \theta_r$ of the non-dominated equations

Table 2. Correlation coefficients between each objective function and the non-dominated equations

	A	B	C	D	E
$C_{A,M=2.0,AOA=0}$	-1.0	-0.44	0.98	0.89	-0.99
$(L/D)_{MAX,M=0.8}$	0.46	0.97	-0.36	-0.078	0.44
$(L/D)_{MAX,M=2.0}$	0.98	0.35	-1.0	-0.94	0.98
$BodyVolume$	-0.91	-0.11	0.96	1.0	-0.93

Equation (B) that has the highest squared correlation coefficient with the maximum lift-to-drag ratios at subsonic condition is

$$\hat{f}_B(x,y) = 10.3xy - 8.8y^2 - 0.48x + 0.28 + \frac{10xy^3}{1+x+y} \tag{17}$$

Equation (C) that has the highest squared correlation coefficient with the maximum lift-to-drag ratios at supersonic condition is

$$\hat{f}_C(x,y) = 11y + 6.8y^2 - 6.8xy - 5.4x + 2x + 3.3 \tag{18}$$

Equation (D) that has the squared correlation coefficient with the body volume is

$$\hat{f}_D(x,y) = -2y + 0.2y^2 + 0.1xy + x + 0.3 \tag{19}$$

Equation (E) that has the maximum sum value of all squared correlations is

$$\hat{f}_E(x,y) = -5.33y + 4.3xy + 4.3x - 1.28x^2 + xy^2$$
$$-0.8x^2y - 0.33y^2 \tag{20}$$

From Table. 2, the correlation coefficient between the zero-lift drag at supersonic condition and Eq. 16 is -1.0. Furthermore Equation 16 is highly correlated to not only the lift-to-drag ratio at supersonic condition but also the body volume. This illustrates that Eq. 16 has high squared correlation coefficients between multiple objective functions at the same time, which demonstrates the importance of this equation. From Table. 2, the correlation coefficient between the lift-to-drag ratios at subsonic condition and Eq. 17 is approximately 1.0. However, Eq. 17 exhibits low correlation between other objective functions. From Table. 2, Equation 18, Eq. 19 and Equations 20 are highly correlated to the zero-lift drag at supersonic condition, the lift-to-drag ratio at supersonic condition and the body volume. This also shows the importance of these equations to this problem.

In the practical design problem, the good design parameters are highly correlated to the objective functions. However, it is not always clear which parameter set is good as design parameters. When we have to select design parameters, the results of the proposed MOGP can be used to point out the more suitable parameters. In this study, kink angle(θ_{kin}) and aft body angle(θ_r) are the candidate design parameters. Figure 12 shows the relation between each non-dominated equation and the design parameters that include θ_{kin} and θ_r. From Fig. 12, θ_{kin} is highly correlated to the non-dominated equtations. A Taylor expansion of θ_{kin} is

$$\theta_{kin} = 180 - \frac{18}{\pi}arctan\left(\frac{y}{x}\right) + \frac{180}{\pi}arctan\left(\frac{0.5 - y}{3.33 - x}\right)$$
$$= (188.53 - 16.8y - 0.74y^2 + O[y]^3)$$
$$+ \left(\frac{180}{\pi y} + 2.52 - 4.83y - 0.64y^2 + O[y]^3\right) x$$
$$+ \left(-\frac{60}{\pi y^3} + O[y]^2\right) + O[x]^4 +$$
$$(-1)^{Floor\left(\frac{\pi + 2Arg[X] - Arg[y^2]}{2\pi}\right)} \left(-90 + O[y]^3\right) \tag{21}$$

Some terms in Eqs. 16, 18, 19, and 20 are included in Eq. 21. These results imply that θ_{kin} has good potential to be a design parameter. θ_{kin} has the linear relationship among three objective functions. Thus when we choose θ_{kin} as the design parameter, the value of objective functions can be easily estimated by θ_{kin}.

6   Conclusions

In this study, a new type of MOGP is proposed, which is capable of finding the symbolic equations to multiple objective functions through one GP run. In the proposed MOGP, the correlation coefficient is used as the measure of accuracy and the maximization of the correlation coefficient of each objective function of the MOO problem is considered simultaneously. One advantage of the proposed MOGP is shorter computational time than TOGP. Another advantage is that the proposed MOGP enables us to simultaneously solve the symbolic regression problems for all objective functions using correlation coefficients and find symbolic equations that are highly correlated to the multiple objective functions.

We compared the proposed MOGP and traditional two-objective GP in the test problem. The convergence speed and search performance of the proposed

MOGP is almost the same as that of two-objective GP. However, the total computational time of the proposed MOGP is shorter than that of TOGP.

In addition, we applied the proposed MOGP to the practical aerodynamic design optimization problem of a biconical shape reusable launch vehicle. The results of MOGP are symbolic equations that are highly correlated to each objective function. These equations are also highly correlated to the kink angle of the body geometry. Although it is difficult to directly translate the MOGP results to the kink angle of the body geometry, this implies that MOGP is capable of finding composite new design parameters from the original design parameters.

References

1. Obayashi, S., Jeong, S., Chiba, K.: Multi-objective design exploration for aerodynamic configurations. In: AIAA-2005-4666 (2005)
2. Chiba, K., Jeong, S., Obayashi, S., Yamamoto, K.: Knowledge discovery in aerodynamic design space for flyback-booster wing using data mining. In: AIAA-2006-7992 (2006)
3. Sugimura, K., Obayashi, S., Jeong, S.: Multi-objective design exploration of a centrifugal impeller accompanied with a vaned diffuser. In: 5th Joint ASME/JSME Fluid Engineering Conference (2007)
4. Oyama, A., Okabe, Y., Shimoyama, K., Fujii, K.: Aerodynamic multiobjective design exploration of a flapping airfoil using a navier-stokes solver. Journal of Aerospace Computing, Information, and Communication 6(3), 256–270 (2009)
5. Oyama, A., Nonomura, T., Fujii, K.: Data mining of pareto-optimal transonic airfoil shapes using proper orthogonal decomposition. Journal of Aircraft 47, 1756–1762 (2010)
6. Schmidt, M., Lipson, H.: Distilling free-form natural laws from experimental data. Science 324(81), 81–85 (2009)
7. Tatsukawa, T., Oyama, A., Fujii, K.: Extraction of design information from pareto-optimal solutions using genetic programming: A first report. In: International Workshop on Future of CFD and Aerospace Sciences (2012)
8. Vladislavleva, E., Smits, G., Den Hertog, D.: Order of nonlinearity as a complexity measure for models generated by symbolic regression via pareto genetic programming. IEEE Transactions on Evolutionary Computation 13(2), 333–349 (2009)
9. Deb, K., Pratap, A., Agarwal, S., Meyarivan, T.: A fast and elitist multiobjective genetic algorithm: NSGA-II. IEEE Transactions on Evolutionary Computation 6(2), 182–197 (2002)
10. Angeline, P.J.: Subtree crossover: Building block engine or macromutation? Genetic Programming 97, 9–17 (1997)
11. Koza, J.R.: Genetic Programming: On the Programming of Computers by Means of Natural Selection. MIT Press (1992)
12. Nonaka, S., Ogawa, H., Inatani, Y.: Aerodynamic design considerations in vertical landing rocket vehicle. In: AIAA-2001-1898 (2001)
13. Tatsukawa, T., Nonomura, T., Oyama, A., Fujii, K.: Aerodynamic design exploration for reusable launch vehicle using genetic algorithm with navier stokes solver. In: 28th International Symposium on Space Technology and Science (ISTS) (2011) ISTS 2011-e-10

Evidence Accumulation in Multiobjective Data Clustering

Julia Handl[1] and Joshua Knowles[2]

[1] Manchester Business School, University of Manchester, UK
julia.handl@mbs.ac.uk
[2] School of Computer Science, University of Manchester, UK
j.knowles@manchester.ac.uk

Abstract. Multiobjective approaches to data clustering return sets of solutions that correspond to trade-offs between different clustering objectives. Here, an established ensemble technique (evidence-accumulation) is applied to the identification of shared features within the set of clustering solutions returned by the multiobjective clustering method MOCK. We show that this approach can be employed to achieve a four-fold reduction in the number of candidate solutions, whilst maintaining the accuracy of MOCK's best clustering solutions. We also find that the resulting knowledge provides a novel design basis for the visual exploration and comparison of different clustering solutions. There are clear parallels with recent work on 'innovization', where it was suggested that the design-space analysis of the solution sets returned by multiobjective optimization may provide deep insight into the core design principles of good solutions.

1 Introduction

Data clustering is the problem of identifying groups (clusters) of similar data items within collections of unlabelled data. One of the key challenges in this respect is the mathematical description of a good cluster, which may then be used to define an actual clustering objective. Existing objective functions for data clustering typically make fairly strong assumptions about the properties of a good cluster, and therefore lack robustness towards data that are violating those assumptions. Recently, multiobjective approaches to data clustering have been introduced with the aim of optimizing not one but several clustering objectives simultaneously. It has been argued that this use of several objectives facilitates a more natural (and robust) definition of the clustering problem, and recent work has shown that the sets of optimal trade-off solutions generated by multiobjective clustering techniques do indeed contain solutions that improve upon the quality of the solutions obtained by optimizing a single clustering objective only [6, 11].

As the objectives used in multiobjective clustering are typically conflicting, even a single run of a multiobjective clustering method (for a given number of clusters) will return a set of different trade-off solutions. Some multiobjective clustering methods, such as Multiobjective Clustering with Automatic k-Determination (MOCK, [10, 11]) additionally generate solutions across a range of different numbers of clusters and, therefore, return solution sets that cover both a range of different numbers of clusters and different trade-offs between the clustering objectives. In practical applications, a user of a

R.C. Purshouse et al. (Eds.): EMO 2013, LNCS 7811, pp. 543–557, 2013.

multiobjective clustering technique will select one (or a few) preferred solution from the final set of optimal trade-off solutions. Evidently, there is a strong need to support the user during this process of *model selection*, and dedicated approaches to this end have been proposed in the literature. For the multiobjective clustering method MOCK, we previously devised an automated technique of model selection [10] that selects a single most promising solution from the set of trade-off solutions. The technique is based on an analysis of the location of solutions in objective space relative to a background of unstructured 'control data' [see 17]. When applied to the multiobjective clustering technique MOCK, this approach has been shown to outperform more traditional techniques of model selection such as the Silhouette Width [15].

A potential criticism of MOCK's standard model selection approach [10] is the following: The analysis is based entirely in objective space and does not fully utilize the information captured by the approximation set as a whole. Recent research in the field of evolutionary multiobjective optimization has shown the potential value of identifying features in design space that are overrepresented within the approximation set returned by an EMO algorithm [1, 2]. This raises the question of whether further improvements in the accuracy, presentation and selection of multiobjective clustering solutions may be feasible by integrating the information provided from the entire set of optimal trade-off solutions.

Although not applied to the multiobjective clustering algorithm MOCK before, the idea of integrating sets of solutions is not novel to the field of clustering and has been adressed in the form of cluster ensemble techniques [7, 8, 16]. Cluster ensemble techniques typically operate on sets of cluster assignments that are returned by a range of clustering methods and attempt to integrate these labels into a single 'consensus clustering'. In this context, the technique of evidence accumulation has been shown to be particularly effective [7], and this is the method we will adopt in our work. Specifically, we aim to investigate whether evidence accumulation provides a suitable means of integrating the set of trade-off solutions returned by multiobjective data clustering, whether this leads to an improvement in solution accuracy, and whether this enables us to obtain a better understanding of the relationships between solutions and the features shared by different optimal trade-off solutions.

In the following (Section 2), we briefly review a number of key concepts related to this work. Section 3 describes the experimental setup, including details of the algorithms and the data sets employed. Section 4 reports our results and discusses the key findings from our experiments. Finally, Section 5 concludes.

2   Background

In this section, we first discuss the principles of model selection for data clustering. We then provide some background on multiobjective clustering, and consider the use of model selection in multiobjective clustering. Finally, ensemble techniques for data-clustering are reviewed.

2.1   Model Selection

Model selection, i.e., the identification of the most suitable solution or algorithm parameter, is a fundamental problem in data clustering. When a single, deterministic clustering technique is used (and all available partitionings are obtained for the same set of input features), the problem reduces to that of identifying the number of clusters k in a data set. More generally, however, the problem of model selection will also include choices between different possible partitionings with the same number of clusters, such as different solutions returned (for the same k) by a non-deterministic method such as k-means, or the results returned (for the same k) by different algorithms.

Model selection in clustering has been addressed using a variety of different techniques [see 9, 12, for reviews]. One of the most common approaches to model selection is the evaluation of all clustering solutions using a specialized internal validation index and the subsequent selection of the top scoring solutions. These indices of cluster validation typically assess the balance between some measure of intra-cluster and inter-cluster variation, and prominent examples include the Silhouette Width [15], the Dunn index [3] and the DB-Index [9]. Alternative approaches to model selection consider the stability of the partitionings under re-sampling [14] or the relative quality of a partitioning compared to a partitioning obtained on unstructured data [17].

2.2 Multiobjective Clustering with Automatic K-determination

The multiobjective clustering method MOCK [10] is based on the evolutionary multiobjective algorithm PESA-II [5] and has been designed for the optimization of two different clustering criteria. The first of these, *overall deviation*, measures the compactness of clusters, whereas the second objective, *connectivity*, considers whether adjacent data items are placed in the same clusters. See [10] for formal definitions.

A single run of the multiobjective clustering method MOCK returns a set of solutions that correspond to different trade-offs between these two objectives. One of MOCK's parameters is an upper limit on the required number of clusters (typically, $k = 25$ is used), but apart from this, the number of clusters is kept open. Many of the solutions returned by MOCK therefore correspond to different numbers of clusters, in addition to providing different trade-offs between the clustering objectives.

2.3 Model Selection in Multiobjective Clustering

As multiobjective approaches to data-clustering typically return a set of possible clustering solutions, some previous work on these methods considered automatic ways of selecting a single preferred clustering solution.

In MOCK, an integrated method of model selection is used [11] which works, briefly, as follows: Given a data set of interest, MOCK is first run to determine an initial set of optimal trade-off solutions. MOCK then produces several sets of 'control data', which are unstructured data sets that are generated randomly within the bounds of the original data set. MOCK determines a set of optimal trade-off solutions for each of these sets of control data. After a normalization of the objective values, the distances between the initial solutions and the solutions on the control data can be compared in objective space. The initial solution that is furthest away from the control points is selected as the best solution. The approach is described in more detail in [11].

In the context of multiobjective fuzzy clustering, a different approach to model selection has been described by Maulik et al. [13]. For the multiobjective data clustering method MOGA (which returns possible partitionings for a single, fixed number of clusters), the authors (ibid.) developed an approach that utilizes an analysis in decision space: they use a re-labelling strategy to maximize the overlap between all of MOGA's output partitionings, and to identify those data points that are consistently assigned to the same cluster (and also have a significant degree of membership with that cluster). The cluster labels of those points are then used as the class labels in the training of a support vector machine, which is applied to the prediction of cluster membership for all remaining data points. Using this approach, the method was shown to achieve an improvement in terms of the Silhouette Width of the final clustering solution, though no external validation of the clustering results was performed.

2.4   Ensemble Techniques

Methods designed for the combination of the output of different clustering techniques are often referred to as ensemble methods. Similar to bagging and boosting in supervised classification [4], clustering ensembles are designed to improve the performance of clustering techniques by combining the results from several different runs, parameterizations or types of algorithms. Ensemble techniques typically operate on sets of cluster assignments (the outputs from clustering algorithms) only and do not consider the original input data. One of the best-known groups of ensemble techniques are the methods introduced by Strehl and Ghosh [16], which use the idea of hypergraphs to collect information from various partitionings; they then apply graph partitioning methods to obtain a final consensus clustering.

A relatively recent development in ensemble clustering is the technique of *evidence accumulation*, introduced by Fred and Jain [7]. Similarly, to Strehl and Ghosh's approaches [16], the method starts with the cluster assignments returned by all algorithms, but the algorithm then proceeds to count co-associations between all data items. This information is used to construct a new dissimilarity matrix, which can then be partitioned using a standard hierarchical clustering approach. The dendrogram returned by the hierarchical algorithm can be cut to obtain a pre-specified number of clusters. The resulting partitioning provides a new consensus clustering, and this approach has been shown to outperform ensembles based on graph partitioning.

For our purpose, which is the aggregation of the solutions returned by multiobjective clustering, the method of evidence accumulation is appealing, as (i) it appears to be one of the best ensemble techniques currently available; (ii) it can be used to combine partitionings with different numbers of clusters; and (iii) it provides an output with a straightforward and intuitive interpretation: the height of a branch directly reflects information about the minimum strength of co-association between data items within that branch.

3   Method

We experimentally explore the use of evidence accumulation for the aggregation of solutions in multiobjective clustering. First, we assess the quality of the final solutions returned from evidence accumulation on MOCK's solution sets, and compare the quality of these solutions to those obtained using alternative approaches. We then discuss the potential of evidence accumulation to help in the visualization of clustering solutions and to reduce the problem of model selection in multiobjective clustering.

3.1 Sets of Clustering Solutions

In addition to the solution sets returned by MOCK, we generate alternative sets of solutions using a range of established clustering techniques. This is done in order to compare the performance of evidence accumulation for inputs derived from a range of different methods. Overall, five different sets of clustering solutions are used:

- MOCK (**M**): This set contains the solutions returned by MOCK for $k \in [1, 25]$. For the data sets considered, the output set of MOCK typically contains between 80 to 120 solutions (also see Figure 3 in the Results section). MOCK is run using standard parameter settings as described in [11].
- k-means (**K**): This set contains the solutions returned from the standard R implementation for k-means for $k \in [1, 25]$ (i.e., the set contains 25 solutions in total).
- Average-link (**A**): This set contains the solutions returned from the standard R implementations of average-link hierarchical clustering for $k \in [1, 25]$ (i.e., the set contains 25 solutions in total).
- Single-link (**S**): This set contains the solutions returned from the standard R implementations of average-link hierarchical clustering for $k \in [1, 25]$ (i.e., the set contains 25 solutions in total).
- Combined (**C**): This set combines the solutions sets of k-means, average-link and single-link (above). Overall, this set therefore contains 75 solutions.

3.2 Evidence Accumulation

The next step of the experiments is to process some of the above sets as follows: Each of the sets is, individually, used as the input to Fred and Jain [7]'s method of evidence accumulation. We then generate a new set of output solutions by applying the appropriate cuts to the dendrogram and generating partitionings for $k \in [1, 25]$.

As single-link and average-link are hierarchical (and deterministic) methods, the application of evidence accumulation to their output alone does not lead to any new clustering solutions. Consequently, sets of inputs based on their individual outputs only are not used in these experiments. Evidence accumulation therefore generates three new sets of solutions only, which are denominated as MOCK with Evidence Accumulation (**MEvAcc**), k-means with Evidence Accumulation (**KEvAcc**) and Combined with Evidence Accumulation (**CEvAcc**), and contain 25 solutions each.

Evidence accumulation is implemented as described by Fred and Jain [7]. Given a set of input clustering solutions for a data set containing N items (e.g. from a single run of MOCK), the $N \times N$ co-association matrix is constructed as

$$C(i, j) = \frac{m_{ij}}{M},$$

where M gives the number of clustering solutions contained in the set, and m_{ij} indicates the number of times (within those M partitions) that data items i and j have been assigned to the same cluster. A new dissimilarity matrix is then obtained as $D(i,j) = 1 - C(i,j)$, and two different hierarchical clustering methods (single-link and average-link agglomerative clustering) are used to construct the consensus partitions of the data. In line with Fred and Jain [7], the results for single-link agglomerative clustering are consistently worse than the results for average-link agglomerative clustering, so results for this are not shown in the experimental section.

3.3   Solution Selection Methods

Using the sets of solutions generated in the previous stages, we further investigate whether evidence accumulation may present a suitable approach for model selection in multiobjective clustering. For this purpose, we compare a number of alternative techniques of model selection. The first of these is MOCK's established approach [11], which identifies a single partitioning based on distances (in objective space) to random control data.

As a second option, we explore the use of the solution sets returned by evidence accumulation: The output from evidence accumulation is, initially, a set of solutions that contains a single solution for each possible number of clusters (here, $k \in [1, 25]$). As a result, the spacing of solutions along the Pareto front is more even than the spacing in the fronts returned directly from multiobjective clustering (which usually contain several solutions for each value of k). Knee detection based on the local shape of the Pareto front may therefore become more feasible, and we test this by calculating the angles between triplets of adjacent clustering solutions, and selecting the 'middle' solution with the smallest angles as the final solution.

Finally, as a third option, we consider the fact that evidence accumulation uses a hierarchical clustering algorithm to partition the co-association matrix, and that its output is, therefore, best represented using a dendrogram. In previous work, Fred and Jain [7] suggest that branch length within this dendrogram can be used for model selection: they propose to identify the cut that eliminates the longest branch in the dendrogram and select the associated partitioning as the best solution. We explore the potential of this approach for the dendrograms returned from evidence accumulation on MOCK's clustering solutions.

3.4   Data Sets

The techniques discussed above are compared using a test suite of data sets that contain multiple Gaussian clusters in various dimensions. These data sets are generated using the cluster generator described in [11] and available online. The parameterization of the generator is shown in Table 1. Data sets are generated in three and ten dimensions and contain four, six or eight Gaussian clusters. Ten different instances are generated for each combination of dimension and cluster number, resulting in a total of 60 different instances. Individual instances are denoted as Dd-Cc-no$I$, where D indicates the dimensionality of the data, C indicates the number of clusters and I is the index of the

Table 1. Parameters of the synthetic data generator, where N_k gives the number of points in the kth cluster, μ_{kd} defines the mean of the kth cluster in the dth dimension and σ_{kd} defines the variance of the kth Gaussian cluster in the dth dimension. The parameters of individual clusters are generated randomly within the bounds shown below.

Min N_k	Max N_k	Max μ_{kd}	Min μ_{kd}	Min σ_{kd}	Max σ_{kd}
10	100	10	-10	0	$20\sqrt{D}$

instance. All experimental results reported are obtained over 21 independent runs per algorithm per instance, and the Euclidean distance function is used in all experiments.

3.5 Comparison Metrics

A range of techniques are used to evaluate the quality of the solution sets and individual clustering solutions. First, a visualization of the sets of clustering solutions in bi-objective space is used to understand the actual effect of evidence accumulation. As we are dealing with sets of clustering solutions in bi-objective space, some of these results are summarized in the form of attainment fronts. Results are obtained over 21 runs for each data set, so the first and eleventh attainment front are employed to indicate top and median performance.

Furthermore, the agreement of the partitionings with the known cluster memberships is determined using an an external validation technique. The Adjusted Rand Index is used for this purpose, as it provides an established way of comparing partitionings with different numbers of clusters [12]. It returns values within the range $[0, 1]$, where a value of 1 indicates a perfect agreement with the known cluster memberships. During the evaluation of results, the Adjusted Rand Index is utilized in two different ways. For the comparison of solution sets, we are interested in evaluating the algorithms performance at generating high-quality solutions. Hence, the comparison focuses on the best clustering solution found within each solution set (i.e., the solution that scores highest with respect to the Adjusted Rand Index is identified directly). When comparing techniques for model selection, evaluation is based on the Adjusted Rand Index of the final (single) solution selected.

Finally, we also consider the sizes of the solution sets returned by the different techniques.

4 Results

Figure 1 shows the evaluation of the solution sets for a three-dimensional data set with eight clusters. This visualization in bi-objective space (using MOCK's clustering objectives) reveals an interesting phenomenon regarding the effect of evidence accumulation: For the solution sets generated by k-means or the combination of algorithms, evidence accumulation generates results that dominate the original solutions with respect to MOCK's clustering objectives. Unlike the original input solutions, the solutions resulting from evidence accumulation tend to be mutually non-dominated. This

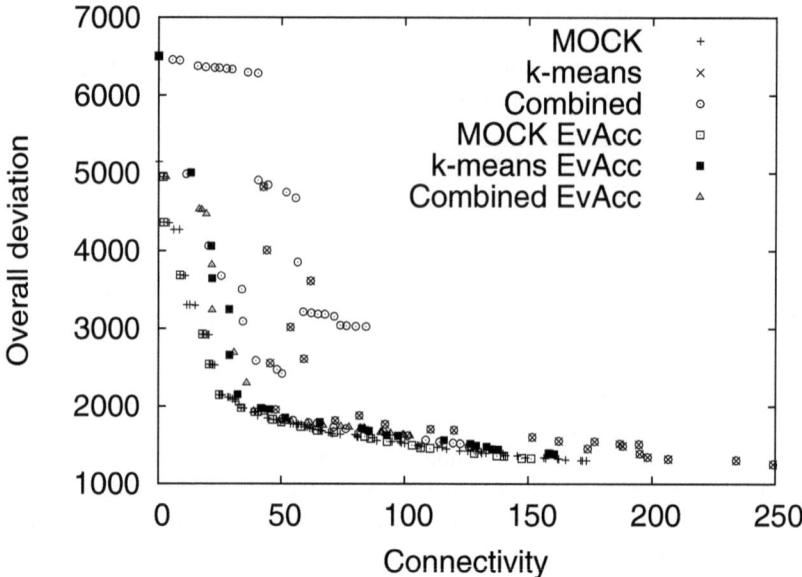

Fig. 1. Results for instance 3d-8c-no0. Sets of clustering solutions obtained by a single run of MOCK, MOCK with evidence accumulation (MOCK EvAcc), k-means (k-means), k-means with evidence accumulation (k-means EvAcc), the ensemble of three traditional algorithms (Combined), and the ensemble of three traditional algorithms with evidence accumulation (Combined EvAcc).

is surprising, as the objective of connectivity is not directly optimized by any of these algorithms. The results suggest that the technique of evidence accumulation produces solutions that implicitly optimize this measure. Interestingly, the same effect is not seen when evidence accumulation is applied to MOCK's solutions: Evidence accumulation does not generally produce solutions that dominate those contained in MOCK's original approximation front. This may be because MOCK's solutions are already close to optimal with respect to both objectives.

To provide a better idea of the stochastic variation in these results, Figure 2 shows the first and eleventh attainment fronts for all six algorithms on the same data set. It can be seen that there is no substantial difference in terms of the attainment of MOCK's solutions before and after evidence accumulation. On the other hand, it is clear that both sets of results dominate the solution sets returned by alternative techniques.

Next, we consider the size of the solution sets and the quality of the best solutions in terms of the known cluster memberships. Summary results over all 60 instances are show in Figures 3 and 4, in the form of boxplots. Consistent with the observations in objective space and the results in [7], the application of evidence accumulation results in improved solutions (compared to the original input solutions) for the use with k-means

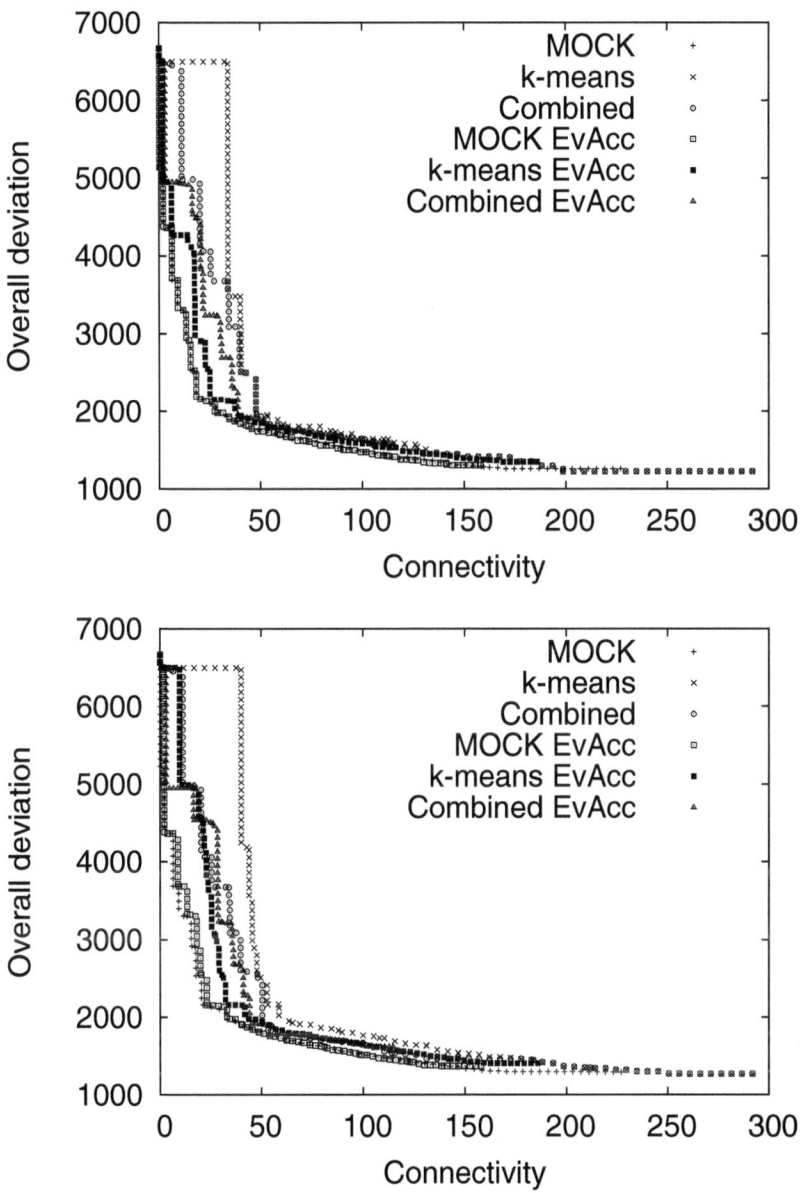

Fig. 2. Attainment fronts on instance 3d-8c-no0 for MOCK, MOCK with evidence accumulation (MOCK EvAcc), k-means (k-means), k-means with evidence accumulation (k-means EvAcc), the ensemble of three traditional algorithms (Combined), and the ensemble of three traditional algorithms with evidence accumulation (Combined EvAcc). (Top) First (best) attainment front. (Bottom) Eleventh (median) attainment front.

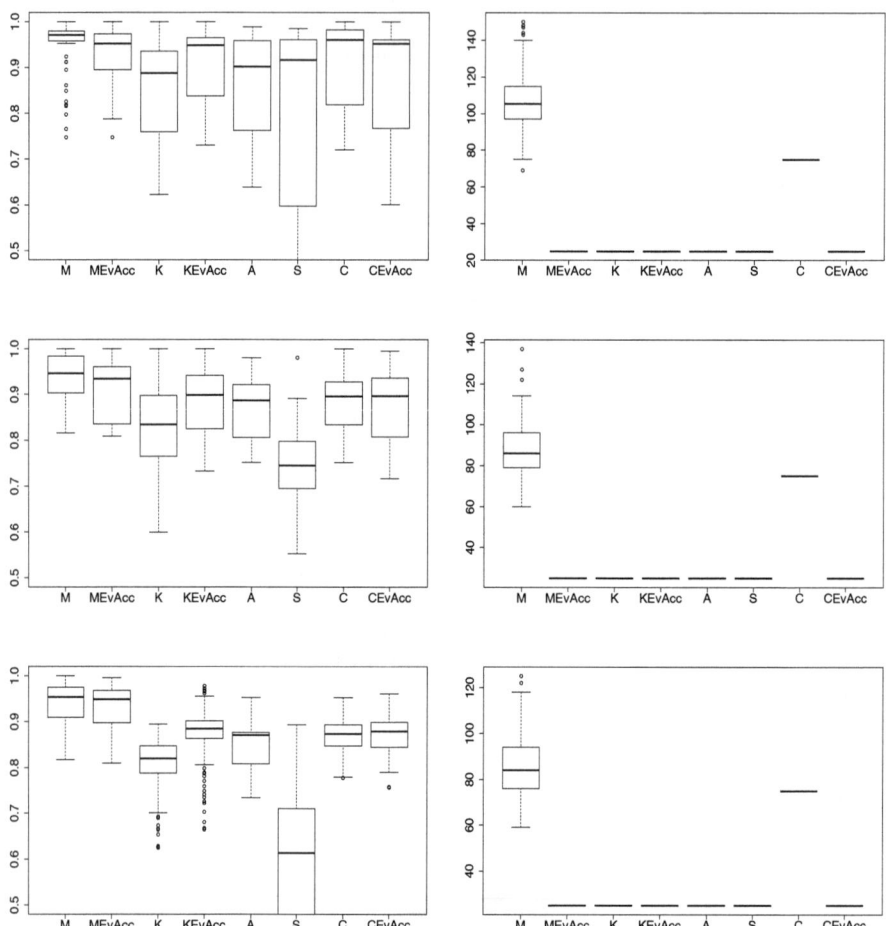

Fig. 3. Results for 21 runs each across ten different instances with three dimensions and (top) four clusters; (centre) six clusters; and (bottom) eight clusters. (Left) Adjusted Rand Index of the best solution in the final set of clustering solutions for each algorithm; (right) Number of solutions in the final set of clustering solutions. for each algorithm.

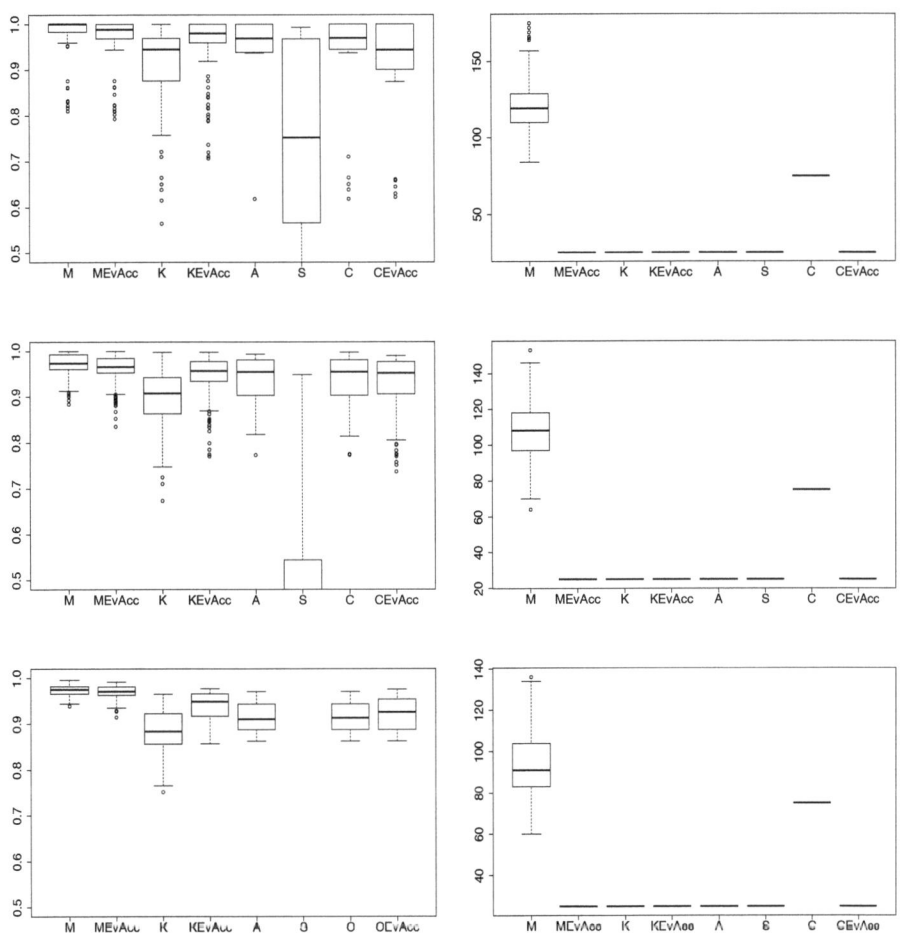

Fig. 4. Results for 21 runs each across ten different instances with ten dimensions and (top) four clusters; (centre) six clusters; and (bottom) eight clusters. (Left) Adjusted Rand Index of the best solution in the final set of clustering solutions for each algorithm; (right) Number of solutions in the final set of clustering solutions. for each algorithm.

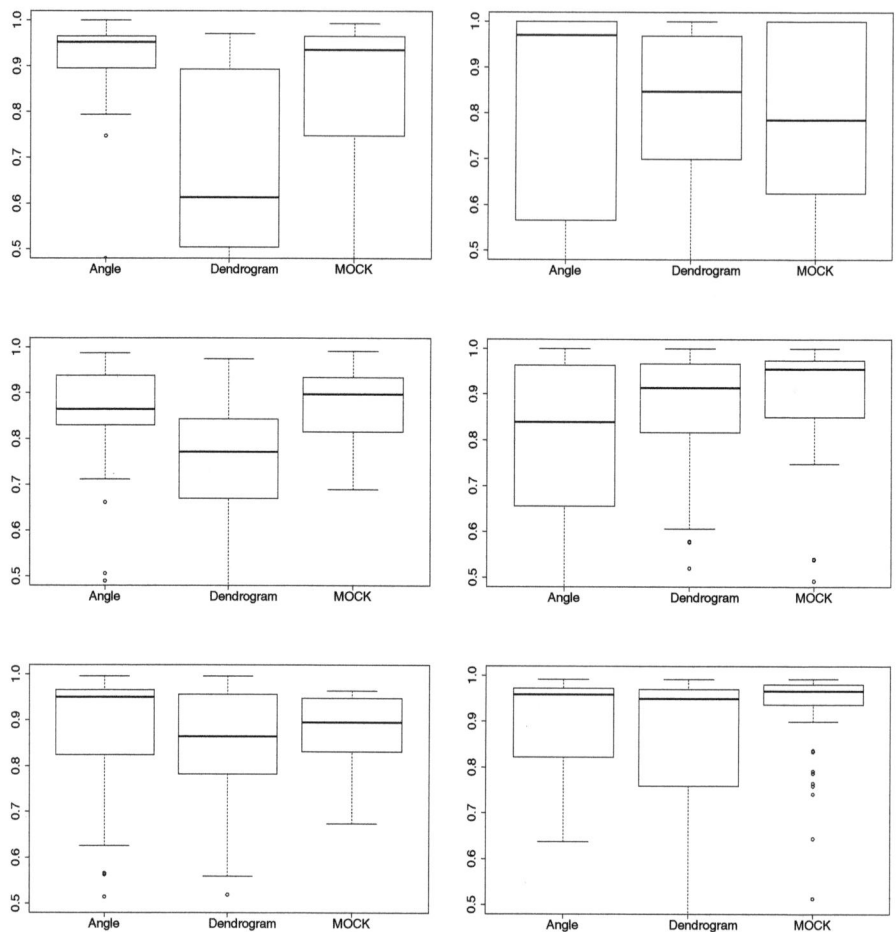

Fig. 5. Results for 21 runs each across ten different instances with (left) three dimensions; (right) ten dimensions. (Top) four clusters; (centre) six clusters; (bottom) eight clusters. Adjusted Rand Index of the solution selected by different methods of solution selection.

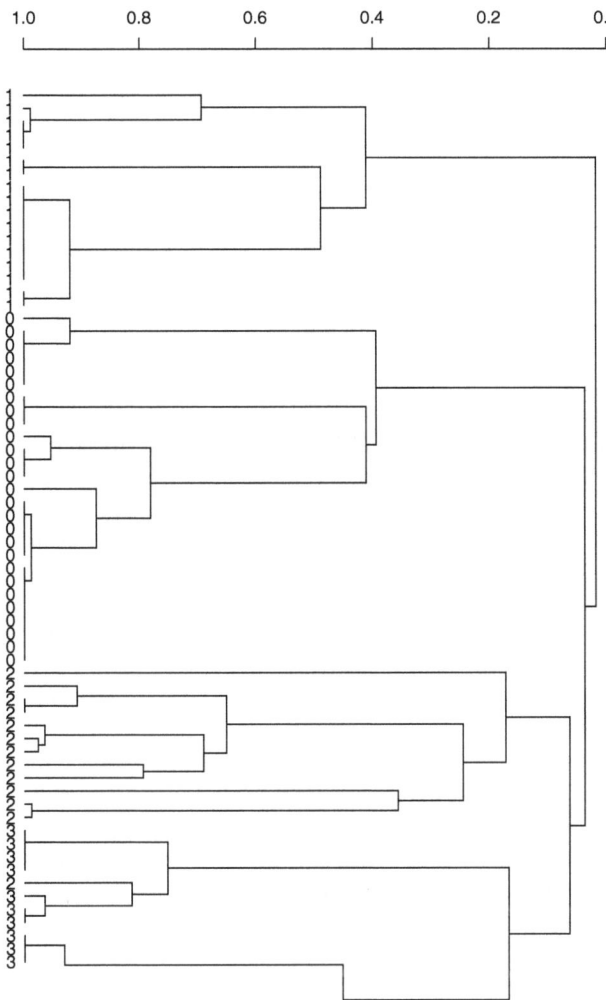

Fig. 6. Visualization of one of MOCK's clustering solutions for a sub-sampled instance 3d-4c-no0. The dendrogram structure is obtained based on evidence accumulation of all of MOCK's trade-off solutions, wheras the numbers displayed at the leaf nodes reflect the assignments made by a single, selected clustering solution. Using this visualization, a user can easily identify discrepancies between this particular solution and the 'majority opinion': here, the dendrogram is almost entirely consistent with the labelling provided by the selected solution (it can be seen that a cut of the dendrogram for $k = 4$ would result in an almost identical clustering solution), indicating that the particular solution is in strong agreement with the majority of solutions in MOCK's complete set of trade-off solutions. There is one discrepancy in the fourth cluster (note the single label of "2" within a series of "3"s), which highlights a data point that has been misclassified by the clustering solution selected. The visualization also helps in identifying data items for which there is particularly low or high uncertainty in the cluster assignment, e.g. the length of the branches in the dendrogram indicates that, overall, there is higher consensus in the assignments to clusters 0 and 1, relative to assignments to clusters 2 and 3. This provides additional information about the level of definition of individual cluster structures in the underlying data.

solutions. For the output of MOCK and the ensemble of algorithms, we see no such effect in terms of the accuracy of the best clustering solutions. For MOCK, this result is consistent with our observations in objective space: It seems that evidence accumulation is not able to improve upon the solutions returned by multiobjective clustering, which may be due to the strong performance of MOCK on these data sets.

We next investigate the size of the solution sets in Figures 3 and 4. From these data, it is evident that the application of evidence accumulation results in a significant (about four-fold) reduction in the size of MOCK's solution sets, which is an important advantage. The results also show that this reduction comes at no significant expense in terms of solution quality: in terms of the Adjusted Rand Index, the best solutions returned by both MOCK and MOCK EvAcc are usually comparable and reliably outperform the best solutions returned by the six alternatives considered.

We are further interested whether evidence accumulation will allow for more effective means of model selection, and Figure 5 shows the related comparisons. The performance of the three model selection techniques is mixed. While, overall, MOCK's original strategy shows the most consistent performance, the angle and the dendrogram-based technique show very good performances for some of the data sets. The angle and dendrogram-based techniques are conceptually different and exploit different types of information, which leads us to hope that, in future work, higher robustness may be achieved through the integration of both approaches. Compared to MOCK's established selection strategy, an important advantage of both of these approaches is reduced computational expense, as they do not rely on the costly generation and clustering of control data.

Finally, we consider how the information derived from evidence accumulation may be used to support a user in the exploration of the solution sets returned by a multiobjective clustering algorithm. Evidence accumulation captures valuable information about the frequency of co-assignment of different items, which is displayed in the resulting dendrogram. We suggest to use this dendrogram for the visualization of individual clustering solutions. In Figure 6, this concept is illustrated for MOCK's output on a four-cluster data set.

5   Conclusion

Evidence accumulation is a state-of-the-art ensemble technique that has been shown to provide an effective way of combining and improving the results of traditional clustering techniques. This manuscript investigates evidence accumulation as a means to support the post-processing of the clustering solutions returned by the multiobjective clustering method MOCK. On the data sets considered, we find that evidence accumulation does not improve the accuracy of MOCK's clustering solutions, but that it achieves a substantial reduction in the number of trade-off solutions to be considered (with no loss of accuracy). We further demonstrate how the knowledge generated by evidence accumulation may be used in the selection, visualization and analysis of the solutions returned by multiobjective clustering. Future work may look into the integration of evidence accumulation into MOCK's search, as well as the development of more robust approaches to solution selection.

References

[1] Bandaru, S., Deb, K.: Automated discovery of vital knowledge from Pareto-optimal solutions: First results from engineering design. In: 2010 IEEE Congress on Evolutionary Computation (CEC), pp. 1–8. IEEE (2010)

[2] Bandaru, S., Deb, K.: Automated Innovization for Simultaneous Discovery of Multiple Rules in Bi-objective Problems. In: Takahashi, R.H.C., Deb, K., Wanner, E.F., Greco, S. (eds.) EMO 2011. LNCS, vol. 6576, pp. 1–15. Springer, Heidelberg (2011)

[3] Bezdek, J.C., Pal, N.R.: Cluster validation with generalized Dunn's indices. In: Proceedings of the Second New Zealand International Two-Stream Conference on Artificial Neural Networks and Expert Systems, pp. 190–193. IEEE (1995)

[4] Brown, G., Wyatt, J., Harris, R., Yao, X.: Diversity creation methods: a survey and categorisation. Information Fusion 6(1), 5–20 (2005)

[5] Corne, D.W., Jerram, N.R., Knowles, J.D., Oates, M.J.: PESA-II: Region-based selection in evolutionary multiobjective optimization. In: Proceedings of the Genetic and Evolutionary Computation Conference, pp. 283–290 (2001)

[6] Delattre, M., Hansen, P.: Bicriterion cluster analysis. IEEE Transactions on Pattern Analysis and Machine Intelligence 2(4), 277–291 (1980)

[7] Fred, A.L.N., Jain, A.K.: Combining multiple clusterings using evidence accumulation. IEEE Transactions on Pattern Analysis and Machine Intelligence 27(6), 835–850 (2005)

[8] Ghaemi, R., Sulaiman, M.N., Ibrahim, H., Mustapha, N.: A survey: clustering ensembles techniques. In: Proceedings of Computer, Electrical, and Systems Science, and Engineering (CESSE), vol. 38, pp. 644–653 (2009)

[9] Halkidi, M., Batistakis, Y., Vazirgiannis, M.: On clustering validation techniques. Journal of Intelligent Information Systems 17(2), 107–145 (2001)

[10] Handl, J., Knowles, J.: Exploiting the Trade-off — The Benefits of Multiple Objectives in Data Clustering. In: Coello Coello, C.A., Hernández Aguirre, A., Zitzler, E. (eds.) EMO 2005. LNCS, vol. 3410, pp. 547–560. Springer, Heidelberg (2005)

[11] Handl, J., Knowles, J.: An evolutionary approach to multiobjective clustering. IEEE Transactions on Evolutionary Computation 11(1), 56–76 (2007)

[12] Handl, J., Knowles, J., Kell, D.B.: Computational cluster validation for post-genomic data analysis. Bioinformatics 21(15), 3201–3212 (2005)

[13] Maulik, U., Mukhopadhyay, A., Bandyopadhyay, S.: Combining Pareto-optimal clusters using supervised learning for identifying co-expressed genes. BMC Bioinformatics 10(1), 27 (2009)

[14] Roth, V., Lange, T., Braun, M., Buhmann, J.M.: A resampling approach to cluster validation. In: COMPSTAT, pp. 123–128 (2002)

[15] Rousseeuw, P.J.: Silhouettes: a graphical aid to the interpretation and validation of cluster analysis. Journal of Computational and Applied Mathematics 20, 53–65 (1987)

[16] Strehl, A., Ghosh, J.: Cluster ensembles—a knowledge reuse framework for combining multiple partitions. The Journal of Machine Learning Research 3, 583–617 (2003)

[17] Tibshirani, R., Walther, G., Hastie, T.: Estimating the number of clusters in a data set via the gap statistic. Journal of the Royal Statistical Society: Series B (Statistical Methodology) 63(2), 411–423 (2001)

Visualising High-Dimensional Pareto Relationships in Two-Dimensional Scatterplots

Jonathan Fieldsend and Richard Everson

University of Exeter, Exeter, UK
{J.E.Fieldsend,R.M.Everson}@exeter.ex.ac.uk

Abstract. In this paper two novel methods for projecting high dimensional data into two dimensions for visualisation are introduced, which aim to limit the loss of *dominance* and *Pareto shell* relationships between solutions to multi-objective optimisation problems. It has already been shown that, in general, it is impossible to completely preserve the dominance relationship when mapping from a higher to a lower dimension – however, approaches that attempt this projection with minimal loss of dominance information are useful for a number of reasons. (1) They may represent the data to the user of a multi-objective optimisation problem in an intuitive fashion, (2) they may help provide insights into the relationships between solutions which are not immediately apparent through other visualisation methods, and (3) they may offer a useful visual medium for *interactive* optimisation. We are concerned here with examining (1) and (2), and developing relatively rapid methods to achieve visualisations, rather than generating an entirely new search/optimisation problem which has to be solved to achieve the visualisation– which may prove infeasible in an interactive environment for real time use. Results are presented on randomly generated data, and the search population of an optimiser as it progresses. Structural insights into the evolution of a set-based optimiser that can be derived from this visualisation are also discussed.

Keywords: Dimension reduction, Pareto optimality, data visualisation.

1 Introduction

The visualisation of a set of solutions maintained by modern evolutionary multi-objective optimisation (EMO) algorithms is of interest to researchers wishing to track the behaviour of algorithms, decision makers who use the output of EMO algorithms, and those wishing to develop *interactive* multi-objective optimisers. Most EMO practitioners are comfortable with visualising a set of solutions with 2 or 3 objective dimensions as a scatter plot of points, and can rapidly determine the non-dominated subset (and those associated with dominated shells [5]) from this. Visualisation of sets with more objectives is often more difficult to interpret via a single scatter plot, and a range of other approaches has been used to visualise these populations in the multi-objective optimisation literature (e.g.

R.C. Purshouse et al. (Eds.): EMO 2013, LNCS 7811, pp. 558–572, 2013.

parallel coordinate plots [7,15,12] heatmaps [22,25], directed graphs [24], Chernoff faces [1], and self-organising maps [21,11]). Dominance relations and shells are not always apparent in these visualisations however (or are only presented between adjacent shells). We are concerned with visualising more than just the estimate of the Pareto front that comes out of most modern EMO algorithms, but more broadly any general set of points (e.g. a search population), from which a visualisation thereof can inform us of the *structure* of the set. Such visualisations can give us extra information relating to the Pareto front estimation, and convey to the problem holder visually *how* an optimisation is progressing.

Here we are concerned with producing a visualisation in the plane, which may be relatively rapidly computed, and is interpretable quickly by both experienced practitioners in EMO, and by problem owners who may not be as familiar with the interpretation of the methods mentioned above. We focus on a single scatter plot of points representing solutions (unlike pairwise coordinate plots [4], which uses $D(D-1)$ separate scatter plots). We shall shortly provide a brief discussion of some existing examples of these, and introduce our two new approaches, but before this we will more formally define Pareto dominance, which is crucial to most modern EMO algorithms, and our visualisation approaches.

2   Pareto Dominance

Pareto dominance is used extensively within the search processes of most modern multi-objective optimisation algorithms [4], and, even if not used explicitly in the search process (if aggregation techniques are used for fitness assignment for instance), it is still used to define the properties of the final output set from the optimisers. EMO algorithms are concerned with exploring a *decision space* for design solutions, where an evaluation of a particular design results in an associated point in *objective space*. If we consider (without loss of generality) that all objectives are to be minimised, an objective vector $\mathbf{y}$ of D objectives $(y_1, \ldots, y_D)$ is said to dominate another $\mathbf{y}'$, written $\mathbf{y} \prec \mathbf{y}'$, iff:

$$(y_i \leq y_i', \forall i) \wedge (\exists i, y_i < y_i'). \tag{1}$$

Succinctly, the best set of solutions to a multi-objective problem (the Pareto set) are the maximal set for which it is impossible (given the problem constraints) to improve any single objective (or group of objectives) of a set member by varying its parameters without having to decrease its performance on one or more other objectives. The image of this set in objective space is known as the Pareto front, $\mathcal{F}$. Given any objective vector set $Y = \{\mathbf{y}_i\}_{i=1}^N$, the non-dominated subset of Y is determined as $\mathcal{S}_0 = \{\mathbf{y} \in Y | \nexists \mathbf{z} \in Y, \mathbf{z} \prec \mathbf{y}\}$. This can be taken one step further (as for instance in the popular NSGA-II algorithm [5]), where not only is a dominance relationship put on members of Y (i.e. where any two members are mutually non-dominating, $(\mathbf{y}' \nprec \mathbf{y}) \wedge (\mathbf{y} \nprec \mathbf{y}')$, or one dominates the other), but also every member of Y is assigned to a *Pareto shell*. Here members of $\mathcal{S}_0$ are said to be in the zeroth Pareto shell (an estimate of the Pareto front, $\hat{\mathcal{F}}$). Subsequent shells are defined iteratively in the same manner, subject to the previous shell being removed from Y until the empty set $\emptyset$ is obtained. That is

$$S_j = \{\mathbf{y} \in Y'_j | \not\exists \mathbf{z} \in Y'_j, \mathbf{z} \prec \mathbf{y}\} \tag{2}$$

where $Y'_j = Y \setminus \bigcup_{k=0}^{j-1} S_k$, and $Y'_j = \emptyset$ for $j \geq k^*$ with some $k^* \in \{1, 2, \ldots\}$. Note that under (1) and (2) it is possible for two members of Y to be mutually non-dominating, but for one to be in a *better* shell than the other.

3   Approaches for Visualising Multi-dimensional Solution Sets Via Scatter Plots

If we wish to project an objective vector $\mathbf{y} \in \mathbb{R}^D$ into $\mathbb{R}^2$ to enable visualisation as a point in a plane we must utilise a dimension reduction technique of some form, and, unless there are redundant or perfectly correlated objectives, some information loss is inevitable.

One of the most popular linear dimension reduction techniques is principal component analysis (PCA, [16]), which identifies the directions of objective space that capture the maximum amount of variance in the solutions. Neuroscale [20,19] has also been used for multi-objective visualisation [11,8] – but unlike PCA it provides a non-linear mapping. However, although popular across many application domains, both Neuroscale and PCA are oblivious to whether solutions dominate each other, or are mutually non-dominating in multi-objective populations, or what their Pareto shell is. We recently defined a new distance measure, the *dominance distance*, that captures the similarity of the dominance relations of solutions, and we have used this to project mutually non-dominating sets using multi-dimensional scaling [23,26] to points on the plane [25]. In the same work we also investigated the use of Radviz [13,14] for this mapping. However even with these representations is is not geometrically apparent which solutions are in which shell or which dominate others.

In [18] a visualisation is presented which does map the S_0 solutions in a multi-dimensional objective space to a mutually non-dominating shell in $\mathbb{R}^2$, with all other mapped solutions being dominated by members of the planar representation of S_0 (although subsequent shells are not explicitly represented). We will discuss the method described in [18] further in Sect. 5, as it is conceptually the close to the methods we propose.

4   Desired Properties When Visualising Shells in the Plane

Given a set $Y^D = \{\mathbf{y}_i\}_{i=1}^N \subset \mathbb{R}^D$, we wish to find a mapping to $Y^2 = \{\mathbf{u}_i\}_{i=1}^N \subset \mathbb{R}^2$ such that if $\mathbf{y}_i \prec \mathbf{y}_j$, then $\mathbf{u}_i \prec \mathbf{u}_j$, and if $\mathbf{y}_i \not\prec \mathbf{y}_j$, then $\mathbf{u}_i \not\prec \mathbf{u}_j$. In general a mapping $\mathbf{u} = \mathbf{g}(\mathbf{y})$ with this property does not exist (the reader is directed toward the proof provided in [18] for further details). Instead here we shall concern ourselves with a mapping with two properties, one of which we *can* guarantee, and the second of which we seek a good approximation to, namely:

1. Ensure that the mapping preserves Pareto shells. That is, if we denote by $\mathcal{S}_j^D$ the jth Pareto shell in an ambient space of D dimensions, then $\mathbf{u} \in \mathcal{S}_j^2$ (where $\mathbf{u} = \mathbf{g}(\mathbf{y})$). The superscript on $\mathcal{S}_j$ denotes the dimensionality of the space which it inhabits.
2. Minimise dominance misinformation. We describe three ways to quantify dominance misinformation in Sects. 5, 6 and 7.

Computational methods for quickly determining shells are well-known (see, e.g. [5]) – and are embedded in many EMO algorithms [4]. Furthermore, ensuring that shell members are maintained via a projection into a lower dimension is actually fairly trivial: a very simple approach would be to distribute each shell as illustrated in Fig. 1. Here there are three shells projected from $\mathbb{R}^D$, $D > 2$, with the number of members in each shell being $|\mathcal{S}_0^D| = 4$, $|\mathcal{S}_1^D| = 6$ and $|\mathcal{S}_2^D| = 3$. When projecting these into $\mathbb{R}^2$ each shell member is projected to a point in the positive quadrant, which lies on the circumference of the circle with radius equal to its shell rank plus one. As long as the mapping is such that the minimum values of the objectives in both dimensions of $\mathcal{S}_j^2$ are greater or equal to the minima in $\mathcal{S}_{j-1}^2$, then this will have the effect that every member of $\mathcal{S}_j^2$ is dominated by at least one member of $\mathcal{S}_{j-1}^2$, and the members of each $\mathcal{S}_j^2$ are mutually non-dominating as required.

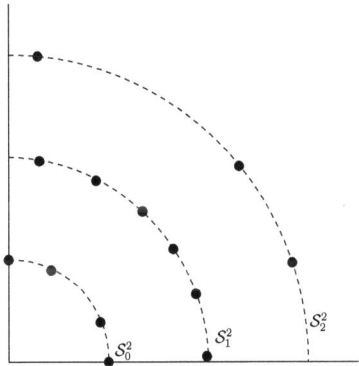

Fig. 1. Simple projection preserving shells but not necessarily dominance relations

This mapping provides the first property mentioned above, but still leads to the issue of *where* to place the $\mathbf{u}_i$ to minimise whichever dominance misinformation objectives may be defined. It is the definition of this property, and methods to incorporate it within a planar visualisation we shall now discuss. The first new approach we consider uses proximity to *domination rays* to convey dominance. The second we introduce uses a direct geometric transference of the dominance relation. First however we will describe the visualisation of Köppen and Yoshida [18].

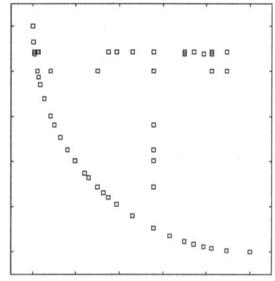 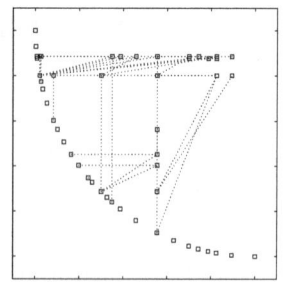 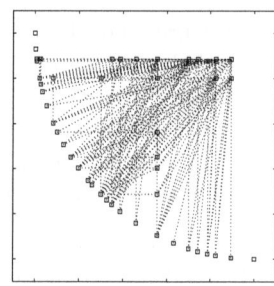

Fig. 2. Visualisation using the approach of [18] of 100 randomly generated points in 4 dimensions. *Left*: Dominance links not shown. *Middle*: Dotted lines show dominance relations between the members of adjacent shells. *Right*: Dotted lines show dominance relations between all members of Y.

5   Visualisation of Köppen and Yoshida

In [18] the non-dominated set from Y^D was mapped to the positive quadrant, lying on the circumference of a circle whose centre point is the origin, and the objectives were maximised. In keeping with the rest of this paper, where objectives are to be minimised, we have 'flipped' the representation from the original and project instead to the negative quadrant. Once the non-dominated subset of Y^D is mapped, for every *dominated* point $\mathbf{y}_i \in Y^D$, the subset of $\mathcal{S}_0^D$ which dominates it is determined, and the worst objective values in the mapping of this set are used to fix the position of $\mathbf{u}_i$ in two dimensions. The exact order of the solutions mapped to $\mathcal{S}_0^2$ was treated as a permutation problem for a multi-objective evolutionary optimiser in [18]. The location of projected solutions on the curve of $\mathcal{S}_0^2$ was determined such that the separation between points was proportional to the distances of immediate $\mathcal{S}_0^2$ neighbours in the original $\mathbb{R}^D$ space. Let $\boldsymbol{\pi}$ be a permutation of the integers $1, \ldots, |\mathcal{S}_0^D|$ describing the order in which the solutions are arranged along $\mathcal{S}_0^2$, so that $\mathbf{u}_{\pi_1}$ is placed on the extreme left, with $\mathbf{u}_{\pi_2}$ next, and so on. Then the two objectives that Köppen and Yoshida seek to minimise in the selection of an optimal permutation are

$$\sum_{k=1}^{|\mathcal{S}_0^D|-1} d(\mathbf{y}_{\pi_k}, \mathbf{y}_{\pi_{k+1}}) \tag{3}$$

where $d(\mathbf{x}, \mathbf{z})$ is the Euclidean distance between $\mathbf{y}_{\pi_k}$ and $\mathbf{y}_{\pi_{k+1}}$, and, denoting by $\mathbf{v}_l \in Y^D$ members of Y^D which are dominated by members of $\mathcal{S}_0^D$,

$$|\{k| \exists 1 \leq i < k < j \leq |\mathcal{S}_0^D|, \mathbf{v}_l \text{ with } (\mathbf{y}_{\pi_i} \prec \mathbf{v}_l) \wedge (\mathbf{y}_{\pi_j} \prec \mathbf{v}_l) \wedge (\mathbf{y}_{\pi_k} \not\prec \mathbf{v}_l)\}| \tag{4}$$

As such, for each element $\mathbf{y}_{\pi_k}, 1 < k < |\mathcal{S}_0^D|$, (4) checks if *any* two elements lower and higher in the permuted order both dominate a subset of Y^D which the kth ordered solution does not. The minimisation of (3) and (4) is approximated using a real-valued sorting encoding in the NSGA-II algorithm [5]. However, how the final visualisation permutation is selected from the set of trade-off permutations is not described.

To illustrate this visualisation, we draw 100 points from an isotropic four-dimensional Gaussian distribution, and then map them down to $\mathbb{R}^2$. We first optimise the permutation in the same fashion as [18], using the NSGA-II algorithm, with a population size 100, for 500 generations. We then select the solution on the returned $\hat{\mathcal{F}}$ which minimises (4) as the permutation to use in the final visualisation. We chose this permutation as this objective is essentially a form of dominance misinformation, which is one of the key properties we are concerned with. The resultant visualisation is presented in Fig. 2.

6   Representing Dominance in $\mathbb{R}^D$ by *closeness* in $\mathbb{R}^2$

Once we have determined the shell membership of solutions in the original space, the problem is where to place these solutions on their projection to equivalent shells in the lower dimensional space. The first set of novel transformations we present are based upon converting the dominance relation in a higher dimension to a distance relationship in the two dimensional mapping. That is, we attempt to place dominated solutions *close* to those solutions which dominate them, whilst maintaining correct shells. Here we represent the distance to dominating individuals in a different fashion to [18], which does not require the running of a multi-objective optimiser to generate the mapping. Each shell is mapped to a distinct shell (as illustrated in Fig. 1). We then place the solutions, as close as possible to the solutions which dominate them. One way of conceiving of this is that each solution is placed on the curve corresponding to their shell and connected via a spring to all those points which dominate it. These springs act to pull together points which are dominated by the same solutions.

This approach is illustrated in Fig. 3. As in [18] the problem arises as to how to distribute the solutions in $\mathcal{S}_0^2$, however, instead of casting this as a problem to tackle with an evolutionary optimiser, we instead order the solutions using spectral seriation. For a set of $K = |\mathcal{S}_0^D|$ solutions we require a $K \times K$ similarity matrix A describing the similarity between any pair of solutions of this set. Given A, to place similar solutions together, we seek a permutation $\boldsymbol{\pi}$ over the solutions in $\mathbf{S}_0^D$ that minimises:

$$\gamma(\boldsymbol{\pi}) = \sum_{j=1}^{K} \sum_{k=1}^{K} A_{kj} (\pi_k - \pi_j)^2. \tag{5}$$

$\gamma(\boldsymbol{\pi})$ is minimised when similar solutions are placed close to each other, and dissimilar solutions far apart. In general, this is NP-hard because the permutation is discrete [2]. Instead, [2] suggests finding an approximation obtained by relaxing the permutation $\boldsymbol{\pi}$ to a continuous variable $\mathbf{w}$ and minimising:

$$h(\mathbf{w}) = \sum_{j=1}^{K} \sum_{k=1}^{K} A_{kj} (w_k - w_j)^2 \tag{6}$$

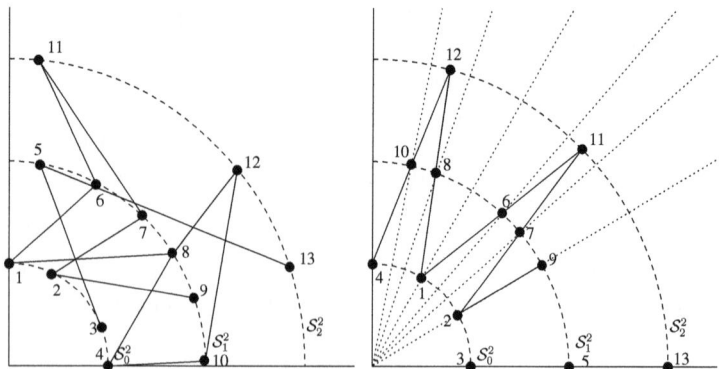

Fig. 3. *Left*: Illustration of the initialisation of a distance-based visualisation, dominance relationships between points on adjacent shells are shown via solid connecting lines. *Right*: Minimal distance rays are plotted projected from the origin through members of $Y \setminus \mathcal{S}_2^2$, which indicate where on each shell a solution must be placed to be the minimal distance away from the dominating point.

with respect to $\mathbf{w}$. This relaxed objective is subject to two constraints. Firstly, to ensure that adding a constant to all w_n does not change the order of the individuals the constraint $\sum_n w_n = 0$ is imposed. Also, in order to avoid the trivial solution in which all $w_n = 0$, we require $\sum_n w_n^2 = 1$. The solution to the constrained problem can be found with linear algebra via the graph Laplacian [10,3] (further details on how to do this efficiently can be found in [17]). The similarity measure we choose to use here is the *dominance similarity*, which we have used previously for MDS visualisations of multi-objective sets [25,9].

The dominance similarity between two solutions $\mathbf{y}_j$ and $\mathbf{y}_k$, relative to a third solution $\mathbf{y}_p$, is defined as being proportional to the number of objectives on which $\mathbf{y}_j$ and $\mathbf{y}_k$ have the same relation (greater than, less than, or equal) to $\mathbf{y}_p$. That is:

$$
S(\mathbf{y}_k, \mathbf{y}_j; \mathbf{y}_p) = \frac{1}{D} \sum_{d=1}^{D} \Big[I((y_{pd} < y_{kd}) \wedge (y_{pd} < y_{jd}))
$$
$$
+ I((y_{pd} = y_{kd}) \wedge (y_{pd} = y_{jd}))
$$
$$
+ I((y_{pd} > y_{kd}) \wedge (y_{pd} > y_{jd})) \Big] \tag{7}
$$

where $I(q)$ is the indicator function that returns a value of 1 when the proposition q is true and 0 otherwise.

The dominance similarity across the set $Y = \{\mathbf{y}_i\}_{i=1}^N$ is obtained by averaging $S(\mathbf{y}_k, \mathbf{y}_j; \mathbf{y}_p)$ across all the elements of the set:

$$
A_{kj} = \frac{1}{N-2} \sum_{\substack{p=1 \\ p \notin \{k,j\}}}^{N} S(\mathbf{y}_k, \mathbf{y}_j; \mathbf{y}_p). \tag{8}
$$

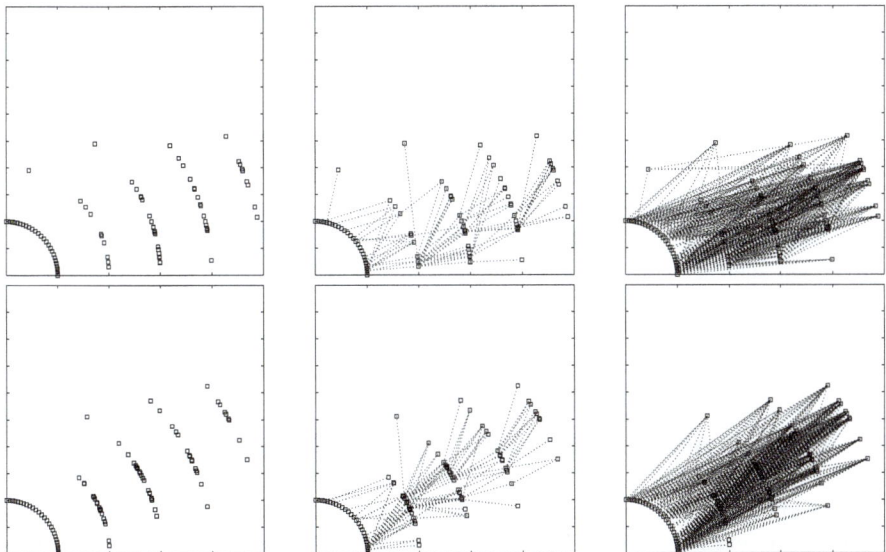

Fig. 4. Visualisation using closeness approach of 100 points randomly generated in 4 dimensions. *Top*: Initial pass. *Bottom*: After refinement iterations. *Left*: No domination links shown. *Middle*: Dotted lines show dominance relations between members of adjacent shells. *Right*: Dotted lines show dominance relations between all members of Y^D.

Utilising (8) to calculate A for just the $\mathcal{S}_0^D$ members of Y^D (but averaging across their similarity to all members of Y^D), gives us an order on the elements of $\mathcal{S}_0^D$ with minimisation of (6), which we transfer to $\mathcal{S}_0^2$. We space the $\mathcal{S}_0^2$ solutions on the curve proportional to their Euclidean distance in $\mathcal{S}_0^D$ (as in [18]).

The distance between shells in the mapping is arbitrary, so we use the angle of the ray passing through a mapped point and the origin to determine the placement of dominated solutions. Specifically, the location of a $\mathbf{u}_i$ is initially placed on the ray through the origin whose angle is the average of the angles of the rays associated with the mapped points which dominate it. As the position of $\mathcal{S}_0^2$ is determined using spectral seriation (as detailed above), the rays defining $\mathcal{S}_1^2$, can be rapidly computed, which, along with $\mathcal{S}_0^2$ can then be used to fix $\mathcal{S}_2^2$, and so on. A schematic of this is in shown the right-hand panel of Fig. 3, and an empirical example is provided in the top panels of Fig. 4 (using the same data as Fig. 2). However, as only the *dominating* points are considered for determining the angle of the ray on which a solution resides, if two solutions are dominated by exactly the same subset of Y^D, then they will lie at the same point – even if the subsets that they both *dominate* are not the same.

In order to resolve the issue of mapping points to the same location when their dominance relationships with Y^D as a whole are not identical, an iterative procedure is used to adjust the locations of $\mathcal{S}_i^2$ points (where $i > 1$), such that the mean of the angles in $\mathbb{R}^2$ of those points which are *dominated* in $\mathbb{R}^D$, as well as those which dominate in $\mathbb{R}^D$, are used to set the location angles of $\mathcal{S}_i^2$ members. Each shell is evaluated in turn until all the shells have been processed ($\mathcal{S}_0^2$ remaining unchanged). This is repeated until the positions no longer vary. Empirically the number of complete passes before stabilisation is reached has proved small – in the example shown here for instance the location changes were negligible (10^{-3}) within six passes. The bottom panels of Fig. 4 shows the result of this iterative location smoothing – note how a number of individuals in $\mathcal{S}_1^2$ which dominate many elements of $\mathcal{S}_2^2$ have been pulled to a more central region of the $\mathcal{S}_1^2$ shell by this process. On the other hand, the refinement process has left the shells in the same general region as the single pass algorithm, so the single pass seems to give a reasonable approximation (on this instance) to the final refined visualisation.

7 Representing Dominance in $\mathbb{R}^D$ by *Dominance* in $\mathbb{R}^2$

The second new approach we consider here attempts to *directly* translate the dominance relationships in the higher dimensional space into the two dimensions in a way that is conceptually more akin to [18]. Again, the ordering of solutions mapped to $\mathcal{S}_0^2$ is determined via spectral seriation using dominance similarity, but instead of placing individuals on dominated shells using angles to dominating and dominated solutions, we attempt to minimise the divergence between the dominance relations implied by the lower dimensional visualisation and the true dominance relations in the original space. That is, if an individual $\mathbf{u} = \mathbf{g}(\mathbf{y})$ has the relationship $\mathbf{y}' \prec \mathbf{y}$, then as far as possible we would like $\mathbf{u}' \prec \mathbf{u}$ to hold (and vice versa). To this end we propose a deterministic iterative procedure which attempts to arrange the solutions in each $\mathcal{S}_j^2$ to accomplish this.

When deciding on the placement of the $\mathcal{S}_1^2$ individuals, the members of $\mathcal{S}_0^2$ effectively delimit a number of regions on the feasible curve for $\mathcal{S}_1^2$. Any point in one of these regions has an equivalent dominance relation with $\mathcal{S}_0^2$; that is, any point in a particular curve segment r_k is dominated by the same subset of $\mathcal{S}_0^2$. This is illustrated in the left panel of Fig. 5 – the members of $\mathcal{S}_0^2$ partition $\mathcal{S}_1^2$ into $2|\mathcal{S}_0^2| - 1$ segments into which members of $\mathcal{S}_1^2$ can be placed. In selecting which region to map a solution $\mathbf{y} \in \mathcal{S}_1^2$ to, a natural approach would be to find the one which yields the smallest dominance error. If we denote by $\mathbf{r}_i$ any point in the ith region, and by R_1 the set of these points (one point for each region) for the oneth shell, then we can define a dominance error as having two parts:

$$e_1(\mathbf{r}_i, \mathbf{y}, \mathcal{S}_0^D) = |\{\mathbf{y}' \in \mathcal{S}_0^D | \mathbf{y}' \prec \mathbf{y} \wedge \mathbf{g}(\mathbf{y}') \nprec \mathbf{r}_i\}|, \qquad (9)$$

the number of members of $\mathcal{S}_0^D$ which dominate $\mathbf{y}$ but fail to dominate $\mathbf{r}_i$ in their $\mathcal{S}_0^2$ projection and

$$e_2(\mathbf{r}_i, \mathbf{y}, \mathcal{S}_0^D) = |\{\mathbf{y}' \in \mathcal{S}_0^D | \mathbf{y}' \nprec \mathbf{y} \wedge \mathbf{g}(\mathbf{y}') \prec \mathbf{r}_i\}|, \qquad (10)$$

the number of members of $\mathbf{S}_0^D$ which do not dominate $\mathbf{y}$ but incorrectly dominate $\mathbf{r}_i$ in their $\mathbf{S}_0^2$ projection. Empirically we find that simply summing these two

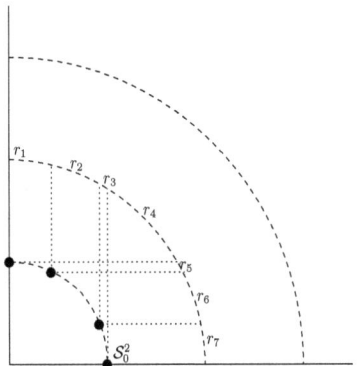 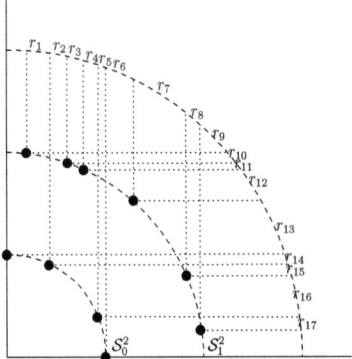

Fig. 5. *Left*: Illustration of potential intervals for placement of $\mathcal{S}_1^2$ individuals, once the order of $\mathcal{S}_0^2$ has been determined. *Right*: Illustration of potential intervals for placement of $\mathcal{S}_2^2$ individuals, once the order of $\mathcal{S}_1^2$ has been determined. Note there are no intervals on the extremes, as the interval ranges must be dominated by at least one member of the previous shell.

penalty terms to generate a combined error (to minimise) does not lead to a satisfying projection. This is because (10) tends to outweigh (9) with the result that all the solutions in dominated shells tend to be pushed close to the axes. Therefore, we find the subset of R_j which minimises e_1, and then choose the element of this subset with the lowest e_2. By geometry, we can see that $e_1 = 0$ can be achieved for *any* dominated element Y^D projected onto $\mathcal{S}_i^2$, as long as the shell radius for $\mathcal{S}_i^2$ is $\sqrt{2}$ times the shell radius for $\mathcal{S}_{i-1}^2$, or greater. This will mean that there is a region on the $\mathcal{S}_i^2$ curve which is dominated by *all* elements of $\mathcal{S}_{i-1}^2$, therefore we choose the shell radii accordingly to guarantee this.

It is also possible (and indeed inevitable if $2|\mathcal{S}_0^D| - 1 < |\mathcal{S}_1^D|)$[1] for some solutions in $\mathcal{S}_1^2$ to be placed in the *same* region. We would not however wish to place them on exactly the same point, as they may not dominate the same subset of Y^D. If more than one solution is placed in a region, then they are spaced evenly across the curve segment that region defines, otherwise it is placed in the centre of the segment. After the $\mathcal{S}_1^2$ shell is assigned, subsequent shells are assigned in order in a similar way to that described for $\mathcal{S}_1^2$; that is in (9) and (10) $\mathcal{S}_0^D$ is replaced by $\bigcup_{k=0}^{j-1} \mathcal{S}_k^D$ (where the jth shell is being assigned).

This still leaves the problem of how to order multiple solutions mapped to the same region. This is, however, another permutation problem, and as such we simply construct the dominance similarity matrix for solutions in this region, and order them according to the order suggested by spectral seriation.

---

[1] Generally, in the jth shell ($j > 0$), there are *at most* $1 + \sum_{i=0}^{j-1} 2(|\mathcal{S}_i^2| - 1)$ regions where a shell member may be placed, this growth is illustrated in the right panel of Fig. 5.

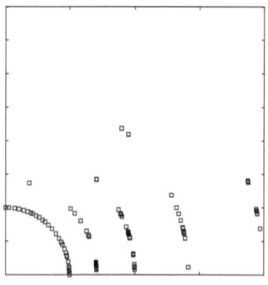 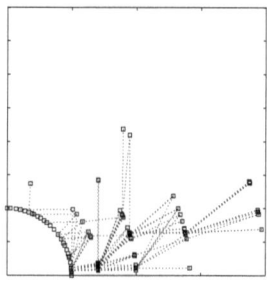 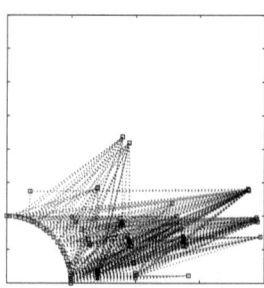

Fig. 6. 100 randomly generated points in 4 dimensions as in previous illustration, visualisation using dominance approach but with modified minimisation function. *Left*: Dominance links not shown. *Middle*: Dotted lines show dominance relations between members of adjacent shells. *Right*: Dotted lines show dominance relations between all members of Y^D.

Table 1. Property comparison of the three scatter plot visualisation methods. $\mathbf{y}_i$ and $\mathbf{y}_j$ are original objective vectors drawn from Y^D, and $\mathbf{u}_i$ and $\mathbf{u}_j$ are their corresponding projections using the methods examined into $\mathbb{R}^2$.

	Köppen & Yoshida	Distance-based	Dominance-based
(I) If $\mathbf{y}_i \in \mathcal{S}_k^D$ then $\mathbf{u}_i \in \mathcal{S}_k^2$	✗	✓	✓
(II) If $\mathbf{y}_i \prec \mathbf{y}_j$ then $\mathbf{u}_i \prec \mathbf{u}_j$	✗†	✗	✓
(III) If $\mathbf{y}_i \not\prec \mathbf{y}_j$ then $\mathbf{u}_i \not\prec \mathbf{u}_j$	✗	✗	✗

† If solutions in $\mathcal{S}_0^2$ can be arranged so that (4) is equal to zero, then (II) holds for any pair of points which are not mapped to the same location in $\mathbb{R}^2$. If (4) is not equal to zero then (II) cannot be guaranteed to hold anywhere in the mapping of [18].

The visualisation approach, using our running example, leads to the projection shown in Fig. 6. All linked points in this visualisation can be seen to dominate in a geometric sense.

8   Visualisation Comparisons

The dominance and shell properties of the three visualisations we have illustrated here (that of [18], and our two new visualisations) are presented in Table 1. Assuming a permutation of $\mathcal{S}_0^2$ can be found such that (4) is equal to zero, then the method of [18] guarantees property (II) through the placement of the dominated solutions in Y^2 using the worst values of the mapped dominating subset of Y_0^D. In practice an ordering which obtains (4) equal to zero is rare however, and it still allows points to be placed on the same location when one dominates the other. Our distance/angle-based visualisation guarantees (I), however as it reinterprets geometric dominance into angles it does not attempt to provide (II) or (III). Our dominance-based visualisation guarantees both (I) and (II), and tries to minimise (III) (subject to (II)), by minimising (10) and the corresponding

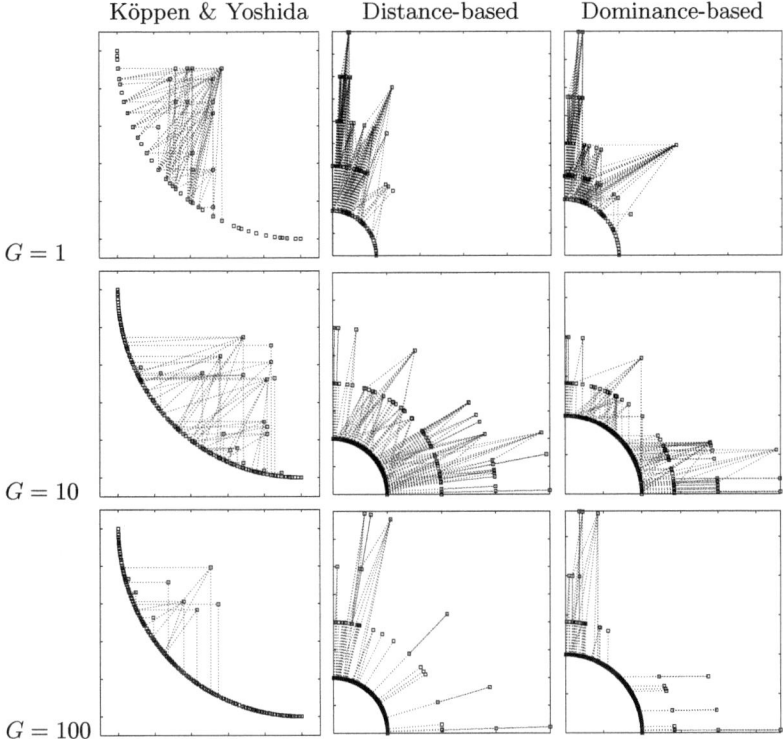

Fig. 7. Visualisation of SPEA2 4-objective problem search populations. G indicates the generation. Dotted lines show dominance relations between *all* members of Y^D.

objective functions for later shells. Note, other mappings to the plane previously used in the EMO field (e.g. PCA, MDS, Neuroscale, RadViz) do not guarantee *any* of the properties listed in the Table 1.

We now provide a further brief comparison of the two methods we have introduced here, along with the method of [18], using the run time population of an EMO algorithm. We visualise the combined archive and search population of the popular SPEA2 algorithm [27] as it progresses through the optimisation of a 4-objective optimisation problem (the DTLZ2 test problem [6]). The algorithm is run with an archive size of 100 and a population size of 100, and we visualise the combined population of 200 solutions after 1, 10 and 100 generations in Fig. 7. A number of structural properties are immediately apparent from the runtime results presented in Fig. 7. The two visualisations introduced here clearly show the number of shells, and the proportion of points on these can be reasonably gauged. It is interesting to note that the method for placing dominated individuals from [18] visually loses many dominated individuals in the population entirely because it maps them onto the same location as a solution which dominates them. On the other hand, [18] does push the elements of $\mathcal{S}_0^2$ (the projection of the estimated Pareto front, $\hat{\mathcal{F}}$) which do not dominate *any* other set members to the two extremes of the shell, so it is clear which non-dominated members are structurally unsupported. This is not so immediately determined from the other two visualisations, however it can be coarsely judged by looking at the number of $\mathcal{S}_0^2$ members which do not have lines attached to them.

All visualisation approaches show that the number of dominated points in the search population is decreasing as the search progresses, indicating that the search population is spreading out and advancing slowly (rather than making big jumps forward – which would lead to a larger proportion of Y being dominated).

The distance- and dominance-based visualisations could be modified to use more of the plotting space, by making the dominance errors (used in fixing point locations) concerned with only relationships between *adjacent* shells – however this would reduce the structural inference possible from the plots. For example, it can be seen in the dominance-based visualisation that at $G = 1$, the members of shell 4 (and all bar one member of shell 3) are exclusively dominated by only a small number of members in shell 0, as the members of these shells are gathered to the top left of the plot, and property (II) means that only members of shell 0 below and to the left of them can dominate them. This kind of structural information is not readily apparent in the method of [18], and completely lacking from approaches which attempt to visualise $\hat{\mathcal{F}}$ alone.

9   Discussion

We have introduced two related novel visualisations of multi-dimensional sets of points, which endeavour to preserve Pareto shell and dominance information. As with all point mappings which reduce the dimensionality of the data, there is inevitably some loss in information, and assessing the quality of the visualisations presented is by its nature subjective. However, we believe they are a useful contribution to the group of methods in the field; because they exhibit some useful properties (listed in Table 1), and have an advantage over some other approaches in their speed of computation. Of the two, we have a slight preference for the dominance-based approach. It guarantees two useful relationships in Y^D are preserved in Y^2 and endeavours to translate the geometric properties practitioners are already familiar with. That said, as long as the user is comfortable inferring dominance by angle similarity (and/or links), then the angular/distance-based approach is generally quicker to compute (as all the various candidate $\mathbf{r}$ of the dominance method do not need to be computed and compared). The method of Köppen and Yoshida has the advantage of being compact, however shell information is lost and it can become more cluttered than the other two. It is also expensive to generate, as (4) is not quick to compute and the ordering of solutions in $\mathcal{S}_0^2$ requires the use of an evolutionary optimiser. We note however that spectral seriation could be used to obtain a permutation for $\mathcal{S}_0^2$ here also.

It is possible that 'better' point locations may be found given the fitting objectives of our two methods using evolutionary optimisation approaches, however this would likely undermine their speed benefits if used to visualise search populations *during* a multi-objective optimisation. There are however further avenues of research that may prove useful. There may be useful information that can be conveyed in the magnitude of the shell radii (as used in [18] to convey the range and magnitude of $\mathcal{S}_0^D$). The shape of the shells being mapped to is also arbitrary; by allowing a greater freedom in location we may be able to convey more information, and it may also improve the false positive rate if some solutions can be *closer* to their dominating shell and therefore be erroneously dominated by fewer solutions.

We also look forward to examining the use of these visualisation approaches in *interactive* optimisation, for example using the structural information presented to select population members for further examination and/or variation.

Acknowledgements and Resources. The authors would like to thank the anonymous reviewers for their very useful and constructive comments. MATLAB code for the work presented here can be obtained from http://emps.exeter.ac.uk/staff/jefields/.

References

1. Agrawal, G., Lewis, K., Chugh, K., Huang, C.-H., Parashar, S., Bloebaum, C.L.: Intuitive Visualization of Pareto Frontier for Multi-Objective Optimization in n-Dimensional Space. In: Proceedings of 10th AIAA/ISSMO Multidisciplinary Analysis and Optimization Conference (2004)
2. Atkins, J.E., Boman, E.G., Hendrikson, B.: A Spectral Algorithm for Seriation and the Consecutive Ones Problem. SIAM Journal on Computing 28(1), 297–310 (1998)
3. Chung, F.R.K.: Spectral Graph Theory. American Mathematical Society (1997)
4. Deb, K.: Multi-Objective Optimization using Evolutionary Algorithms. Wiley-Interscience Series in Systems and Optimization. John Wiley & Sons, Chichester (2001)
5. Deb, K., Agrawal, S., Pratab, A., Meyarivan, T.: A Fast Elitist Non-Dominated Sorting Genetic Algorithm for Multi-Objective Optimization: NSGA-II. KanGAL report 200001, Indian Institute of Technology, Kanpur, India (2000)
6. Deb, K., Thiele, L., Laumanns, M., Zitzler, E.: Scalable Multi–Objective Optimization Test Problems. In: Congress on Evolutionary Computation (CEC 2002), vol. 1, pp. 825–830 (2002)
7. D'Ocagane, M.: Coordonnées parallles et axiales: Méthode de transformation géométrique et procédé nouveau de calcul graphique déduits de la considération des coordonnées parallèlles. Gauthier-Villars, reprinted by Kessinger Publishing (1885)
8. Everson, R.M., Fieldsend, J.E.: Multi-class ROC Analysis from a Multi-objective Optimisation Perspective. Pattern Recognition Letters 27, 531–556 (2006)
9. Everson, R.M., Walker, D.J., Fieldsend, J.E.: League Tables: Construction and Visualisation from Multiple Key Performance Indicators. Technical report, The University of Exeter (2012)
10. Fiedler, M.: Algebraic Connectivity of Graphs. Czechoslovak Mathematical Journal 23(98), 298–305 (1973)

11. Fieldsend, J.E., Everson, R.M.: Visualisation of Multi-class ROC Surfaces. In: Proceedings of the ICML 2005 Workshop on ROC Analysis in Machine Learning, pp. 49–56 (2005)
12. Fonseca, C.M., Fleming, P.J.: Genetic Algorithms for Multiobjective Optimization: Formulation, Discussion and Generalization. In: Proceedings of the Fifth International Conference on Genetic Algorithms, pp. 416–423. Morgan Kauffman (1993)
13. Hoffman, P., Grinstein, G., Marx, K., Grosse, I., Stanley, E.: DNA Visual and Analytic Data Mining. In: Proceedings of the 9th Conference on Visualization, VIS 1997, pp. 437–441. IEEE Computer Society Press, Los Alamitos (1997)
14. Hoffman, P.E.: Table Visualisation: a Formal Model and its Applications. PhD thesis, University of Massachuesetts Lowell (1999)
15. Inselberg, A.: N-dimensional Coordinates. Picture Data Description & Management, IEEE PAMI, 136 (1980)
16. Jolliffe, I.T.: Principal Component Analysis. Springer (2002)
17. Kaveh, A., Rahimi, H.A.: Bondarabady. Finite Element Mesh Decomposition Using Complementary Laplacian Matrix. Communications in Numerical Methods in Engineering 16(379-389) (2000)
18. Köppen, M., Yoshida, K.: Visualization of Pareto-sets in evolutionary multi-objective optimization. In: Proceedings of the 7th International Conference on Hybrid Intelligent Systems, pp. 156–161. IEEE Computer Society, Washington, DC (2007)
19. Lowe, D., Tipping, M.E.: Feed-Forward Neural Networks and Topographic Mappings for Exploratory Data Analysis. Neural Computing and Applications 4(2), 83–95 (1996)
20. Lowe, D., Tipping, M.E.: Neuroscale: Novel Topographic Feature Extraction Using RBF Networks. In: Advances in Neural Information Processing Systems 9, NIPS 1996, pp. 543–549 (1996)
21. Obayashi, S.: Pareto Solutions of Multipoint Design of Supersonic Wings using Evolutionary Algorithms. In: Adaptive Computing in Design and Manufacture, pp. 3–15. Springer (2002)
22. Pryke, A., Mostaghim, S., Nazemi, A.: Heatmap Visualization of Population Based Multi Objective Algorithms. In: Obayashi, S., Deb, K., Poloni, C., Hiroyasu, T., Murata, T. (eds.) EMO 2007. LNCS, vol. 4403, pp. 361–375. Springer, Heidelberg (2007)
23. Sammon, J.W.: A Nonlinear Mapping for Data Structure Analysis. IEEE Transactions on Computers 18(5), 401–409 (1969)
24. Walker, D.J., Everson, R.M., Fieldsend, J.E.: Visualisation and Ordering of Many-objective Populations. In: IEEE Congress on Evolutionary Computation, pp. 3664–3671 (July 2010)
25. Walker, D.J., Everson, R.M., Fieldsend, J.E.: Visualising Mutually Non-dominating Solution Sets in Many-objective Optimisation. IEEE Transactions on Evolutionary Compuation (2012) (to appear), http://cis.ieee.org/
26. Webb, A.R.: Statistical Pattern Recognition, 2nd edn. John Wiley & Sons (2002)
27. Zitzler, E., Laumanns, M., Thiele, L.: SPEA2: Improving the Strength Pareto Evolutionary Algorithm. Technical Report TIK-Report 103, Swiss Federal Institute of Technology Zurich (ETH) (May 2001)

A Comparative Study of Multi-objective Evolutionary Trace Transform Methods for Robust Feature Extraction[*]

Wissam A. Albukhanajer[1], Yaochu Jin[1], Johann A. Briffa[1], and Godfried Williams[2]

[1] University of Surrey, Faculty of Engineering & Physical Sciences,
Department of Computing,
Guildford, Surrey, GU2 7XH, United Kingdom
{w.albukhanajer,yaochu.jin,j.briffa}@surrey.ac.uk
http://www.surrey.ac.uk/computing
[2] Intellas UK Ltd.
Level 37, One Canada Square,
London, E14 5AA, UK
g.williams@intellas.co.uk
http://www.intellas.biz

Abstract. Recently, Evolutionary Trace Transform (ETT) has been developed to extract efficient features (called triple features) for invariant image identification using multi-objective evolutionary algorithms. This paper compares two methods of Evolutionary Trace Transform (method I and II) evolved through similar objectives by minimizing the within-class variance (S_w) and maximizing the between-class variance (S_b) of image features. However, each solution on the Pareto front of method I represents one triple features (i.e. 1D) to be combined with another solution to construct 2D feature space, whereas each solution on the Pareto front of method II represents a complete pair of triple features (i.e. 2D). Experimental results show that both methods are able to produce stable and consistent features. Moreover, method II has denser solutions distributed in the convex region of the Pareto front than in method I. Nevertheless, method II takes longer time to evolve than method I. Although the Trace transforms are evolved offline on one set of low resolution (64×64) images, they can be applied to extract features from various standard 256×256 images.

Keywords: Evolutionary algorithms, multi-objective optimization, Pareto optimality, Trace transform, image identification, invariant feature extraction.

1 Introduction

Identification of digital images is challenging as pictures of the same object will look very different taken from different angles, distances and lighting conditions.

[*] This work is supported by EPSRC and Intellas UK Ltd.

R.C. Purshouse et al. (Eds.): EMO 2013, LNCS 7811, pp. 573–586, 2013.
© Springer-Verlag Berlin Heidelberg 2013

Further, images acquired by cheap consumer cameras are usually noisy and differ by different camera specifications [1]. Therefore, a robust image identification requires extracting image features independent of the way the objects are presented in the image.

Correspondingly, extracted features should be insensitive to variations in geometric transformations such as rotation, scale and translation (RST). Additionally, features derived from different samples of the same image class should be similar. Conversely, features derived from samples of different image classes (see Fig. 1) should considerably differ from each other.

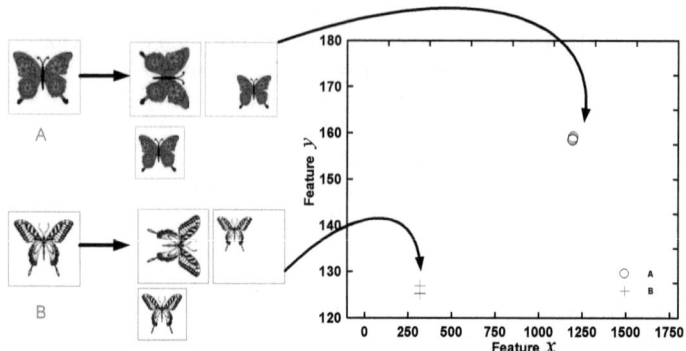

Fig. 1. Mapping of two image classes into a 2D invariant feature space

Trace transform (TT) [2] calculates functionals on image pixels along straight lines projected in different orientations. For example, Fig. 2 depicts an image and its Trace transform produced by using the first Trace functional in Table 1 (the integral of a function). Refer to Fig. 2, the transform matrix in Fig. 2c is obtained by tracing the image (Fig. 2b) with straight lines characterized by a length ρ and an angle θ, and calculating a functional called "Trace" T over parameter t (see Fig. 2a). Therefore, the transform matrix is a 2D image of $\rho \times \theta$ pixels (Fig. 2c). Different transforms can be obtained by using different Trace functionals.

With the help of a second functional called "Diametric" D evaluated along columns of Trace matrix (i.e. along parameter ρ), a string of numbers is created

Table 1. List of some Trace functional

No.	Functional	Description
1	$\int f(t)dt$	Radon transform
2	$\int \left\| f(t)' \right\| dt$	Integral of Gradient
3	$\left(\int \|f(t)\|^p \, dt \right)^q$	p-Norm, $p = 0.5, q = 1/p$
4	$max - min(\|f(x)\|)$	Maximum-minimum of the function

(a) The Trace parameters [2]   (b) Fish image.   (c) Its Trace Transform.

Fig. 2. The Trace Transform

with a length of θ. Finally, a third functional called "Circus" C is applied to the final string of numbers (over parameter θ) to produce a scalar value "real number". This number is termed a "Triple" feature denoted by $\prod$ and it can be used to form a unique identifier for the image [3].

To characterize an image, features derived by Trace transform are not necessarily transparent to the human perception such as brightness or color. In fact, thousand of features can be constructed using different Trace transform functionals by using different combinations of these functionals. Therefore, one may consider Trace transform as a mathematical tool to represent brain sub-conscious which we can not usually identify [4].

Trace Transform has been successfully applied to many image processing tasks such as image database retrieval [2], texture classification [4] and characters recognition [5]. Evolutionary Trace Transform (ETT) [6] has recently been developed to construct efficient Trace transform triple features to represent an image. The main idea is to find optimal combinations of the functionals together with the number of projections in Trace transform to achieve fast and robust feature extraction. It has been shown that evolutionary Trace transform is more robust and efficient than the traditional Trace transform [6].

This paper compares two methods of Evolutionary Trace Transform, method I and II, developed using multi-objective evolutionary optimization of Trace transform to produce candidate features of digital images. The remainder of the paper is organized as follows. Section 2 presents a brief overview of the ETT. Both methods are given as two different evolutionary methods to evolve the Trace transform. Section 3 depicts the experimental results for performance evaluations. Finally, a conclusion is given in Section 4.

2   Evolutionary Trace Transform

A variety of Trace functionals can be employed in Trace transform to extract features that may represent an image. However, robustness and computational speed are two important factors for efficient image analysis. Therefore, it is crucial to design an algorithm to construct efficient features.

An attempt to use evolutionary algorithms in Trace transform was reported by Liu and Wang [7] for face recognition. The authors have introduced a hybrid Trace features from multiple rotation-based Trace functionals and a traditional Genetic Algorithm (GA) to optimize a scalar variable associated with each trace feature. In [8] a reinforcement learning algorithm was applied to the weighted Trace transform (WTT) to find the optimal threshold in the WTT space to minimize the within-class variance only. Recently, in [6], an Evolutionary Trace Transform (ETT) is developed for invariant feature extraction. It has been shown that ETT outperforms the traditional TT in extracting robust triple features from images.

ETT employs a Pareto optimization method, the NSGA-II to search for optimal functional combinations that trade off between minimizing within-class variance and maximizing between-class variance of triple features represent an image. However, the extracted features are one-dimensional. If two-dimensional features are to be extracted from images, then two pairs of functionals need to be constructed. An alternative is to optimize Trace functionals by directly extracting two-dimensional features. Therefore, in the following, we will discuss these two approaches in greater details.

2.1   Method I

In this method, each solution in the Pareto-front represents a single triple feature. Then, two solutions from the final Pareto-front are randomly selected (denoted by $\prod_x^{(I)}$ and $\prod_y^{(I)}$) to form a 2D feature space. In the following, the main components of method I are presented.

- *Chromosome*: Each chromosome in method I encodes 4 integer parameters for each triple feature, namely, Trace $T1$, Diametric $D1$, Circus $C1$ and $\theta1$.

- *Population*: The population size is initialized randomly with constraints on the design variables. For example, there are 14 Trace functionals, then $T1$ change from 0 to 13.

- *Fitness*: The fitness function is characterized by two objectives which are set for minimization in the evolutionary algorithm. The two objectives are defined in (1):

$$f_1 = S_w^I \tag{1a}$$
$$f_2 = 1/(S_b^I + \epsilon) \tag{1b}$$

where ϵ is a small quantity to avoid division by zero. S_w^I and S_b^I are the within-class variance and between-class variance defined in (2):

$$S_w^I = \sum_{k=1}^{K} \sum_{j=1}^{N_k} (x_{jk} - \mu_k^x)^2 \tag{2a}$$

$$S_b^I = \sum_{k=1}^{K} (\mu_k^x - \mu^x)^2 \tag{2b}$$

where

$$\mu_k^x = \frac{1}{N_k} \sum_{j=1}^{N_k} x_{jk}, \ \mu^x = \frac{1}{K} \sum_{k=1}^{K} \mu_k^x$$

and K: number of classes, N_k: number of samples in class k, μ_k^x: mean of class k of x triple features, x_{jk}: the jth sample of class k of x triple features, and μ^x: mean of all classes of x triple features.

2.2   Method II

Pareto-optimal solutions in method II are a set of optimal triple features that describe an image by extracting a pair of triple features (denoted by $\prod_x^{(II)}$ and $\prod_y^{(II)}$) instead of a single triple feature. That is, each Pareto-optimal solution represents a complete 2D feature space. In the following, the main components of method II are presented.

- *Chromosome*: Each chromosome in method II encodes 8 integer parameters instead of 4 only in method I, namely, trace $T1, T2$, diametric $D1, D2$, circus $C1, C2$ and $\theta1, \theta2$ for each triple feature.

- *Population*: The population size is fixed and similar to method I and initialized randomly with constraints on all design variables. For example, there are 14 Trace functionals, then $T1$ and $T2$ change from 0 to 13.

- *Fitness*: The fitness function in (3) is similar to method I where f_1 and f_2 are minimized in the evolutionary algorithm. However, the two objectives S_w^{II} and S_b^{II} are determined based on a pair of triple features defined in (4).

$$f_1 = S_w^{II} \tag{3a}$$

$$f_2 = 1/(S_b^{II} + \epsilon) \tag{3b}$$

where ϵ is same as in (1).

$$S_w^{II} = \sum_{k=1}^{K} \sum_{j=1}^{N_k} (x_{jk} - \mu_k^x)^2 + (y_{jk} - \mu_k^y)^2 \tag{4a}$$

$$S_b^{II} = \sum_{k=1}^{K} (\mu_k^x - \mu^x)^2 + (\mu_k^y - \mu^y)^2 \tag{4b}$$

where

$$\mu_k^x = \frac{1}{N_k} \sum_{j=1}^{N_k} x_{jk}, \mu_k^y = \frac{1}{N_k} \sum_{j=1}^{N_k} y_{jk}, \ \mu^x = \frac{1}{K} \sum_{k=1}^{K} \mu_k^x, \ \mu^y = \frac{1}{K} \sum_{k=1}^{K} \mu_k^y$$

and K: number of classes, N_k: number of samples in class k, μ_k^x: mean of class k of x triple features, μ_k^y: mean of class k of y triple features, x_{jk}: the jth sample of class k of x triple features, y_{jk}: the jth sample of class k of y triple features, μ^x: mean of all classes of x triple features and μ^y: mean of all classes of y triple features.

The following operations are similar for both methods:

− *Selection*: Selection operation is performed twice in the evolutionary loop. The first selection is performed to select parents for mating. It has been shown [9] that the *tournament selection* has better or equivalent convergence and computational time complexity compared to any other selection operators that available in the literature and, therefore, it is adopted in this work. In a tournament selection, two solutions are chosen from the population and a tournament is played, a good solution wins and placed in the mating pool. This operation repeats and two other solutions are played. The better solution wins and placed to fill the mating pool. Each solution will participate twice in the tournament and better solutions will win twice, and therefore two copies may exist in the population to replace the bad solutions. The second selection occur after mating to produce new population for the next generation. The elitist NSGA-II based selection [9] is adopted which consists of four steps. First, parents and offsprings are merged in one population. Second, a non-dominated sorting (Pareto-front assignment) is performed. Each non-dominated solution is assigned a Pareto-front rank number 1 (first non-dominated front), then the next non-dominated solutions in the population are identified and assigned Pareto-front rank number 2 (second non-dominated front). By repeating this procedure, a set of r Pareto-fronts are generated. The third step involves sorting all solutions in an ascending order according to the assigned Pareto-front rank number, the solutions that have the same Pareto rank number are sorted in an decreasing order according the the crowding distance, and a solution with larger (better) crowding distance survives. The reader is referred to [9] for details about *Crowding Distance* calculation. Finally, the top individuals that fit the population size are selected and passed to the next generation.
The selection operation is performed on the *combined* population to generate new parents for the next generation. This preserves the good parents to survive to the next generations.

− *Recombination*: Two parents are selected for mating by exchanging (*crossing over*) a portion of information between parents in the mating pool. The *crossover* performed in variable ways depending on the position of the allels to be exchanged, i.e. at single point (*one-point crossover*), two points (*two-points crossover*) or at an allel level (*uniform crossover*). In this work we adopt *uniform crossover*. Crossover operation occurs during the evolution at *crossover probability* P_c.

- *Mutation*: The next operation during evolution is Mutation operation which is performed to prevent the population from falling into a local optimum. The mutation is performed by inverting the gene value in an individual at *mutation probability* P_m. Some types of mutation operation includes Flip bit, Uniform and Gaussian mutations. In this work we adopt the *uniform mutation*, the value of the gene is changed between predefined upper and lower limits.

  It is not necessary that good solutions will be created through crossover and mutation. However, only better solution will survive through the selection operator [9]. At the end of the evolution, the final non-dominated solutions are analyzed and used as feature extraction on unseen images.

3   Experiments on Method I and II

For robust image identification, triple features of an image should be very close to triple features of the distorted version of the same image. On the other hand, triple features of two different images should differ as much as possible.

In the experiment, a set of trace, diametric and circus functionals are used, which consists of 14 trace functionals (T), six diametric functionals (D) and six circus functionals (C). Some of these functionals are listed in Table 1. Methods I and II are run individually to search for the best combinations of the Trace functionals for 200 generations. During the evolutionary stage, a set of five image classes are used with a low resolution of dimensions 64×64. Each class contains four images: original image and three distorted versions: rotated, scaled and translated (20 images in total). The original five images are displayed in the first row in Fig. 3. The population size and number of generations in method I and II are set to 150. There is no classifier training involved in this work. Table 2 depicts the parameters used in method I and II.

It should be mentioned that several independent runs of the two algorithms are performed and almost the same Pareto-fronts are achieved and the hypervolumes from different runs are almost the same. However, the final solutions may be reached at earlier generations but it continue as set to 200 generations. This conclude the randomness, yet, guided search of the evolutionary algorithms.

The experiments are performed on the same machine with Intel® Core™2Duo 3.1GHz processor with 3GB RAM using Microsoft Visual C++ compiler. The optimization time using method II took about 62 hours for 200 generations which is about a double the time in method I, which took about 29 hours. This is expected due to the double length of chromosomes in method II that requires running the Trace algorithm twice for each solution in the population. It is worth mentioning that this is an offline optimization of Trace transform aimed at finding out the good combinations of Trace transform functionals that might produce better triple features for image analysis in the online Trace algorithm which is itself takes a few seconds to process an image. The Pareto fronts at the final generation for both methods are depicted in Fig. 4. Undoubtedly, less solutions from method I are distributed at the convex area compared to method II which shows greater density in its Pareto front.

After 200 generations, there are nine solutions in total in method I. Each solution on the Pareto front of method I represents one triple feature to be combined with another solution to form a 2D feature space, whereas 19 solutions in the Pareto front of method II, each solution is equivalent to a pair of triple features which can form a 2D feature space directly. In method I, we construct up to 36 different pairs from the nine solutions to be evaluated using (4) and compared with solutions from method II. At this level, Fig. 5a shows the two equivalent fronts from method I (36 solutions) and method II. Obviously, both fronts are approximately identical, this can be concluded from the hypervolume indicator in Fig. 5b, where as little as $\simeq 0.11\%$ hypervolume increase in method II is observed.

In the following, we investigate solutions on Pareto front of method I and II for the both objectives i.e. the within-class variance S_w and the between-class variance S_b^{-1}. First, Fig. 6a depicts S_w for solutions from both methods. A minimum (better) value can be seen from method II, whereas a maximum value can be identified from method I. Second, a maximum value of S_b^{-1} can be identified from method II as shown in Fig. 6b. Generally, solutions have minimum value in one objective are not necessary have a minimum in the second objective. Additionally, one may also calculate the ratio of the two objectives as S_w/S_b and are shown in Fig. 7. From the figure, a minimum value can be found in method I, whereas a greater maximum of this ratio can be found from solutions in method II. Keeping in mind, the 36 solutions in method I are thoroughly calculated by all possible combinations $\binom{9}{2} = \frac{9!}{2(9-2)!}$ from the original nine solutions from the final Pareto front. This may be an easy task as few (nine) solutions were found in the final front of method I. However, it would have been a hard task if there were more solutions in the final front. Consequently, one may choose any two preferred solutions from the final front of method I to form a pair of triple features for image analysis, whereas solutions from method II can be used directly.

Table 2. Parameter Set-up for method I and II

Parameter	Value
Population size N_p	150
Mutation probability	0.125
Crossover probability	0.9
Number of generations	200
ϵ	10^{-5}

Next, we test the two methods on different images from fish database deformed by random rotation, scale and translation. Figure 3 depicts 20 original images (20 classes) used in the experiments. At this stage, each image has a standard dimension 256×256 and is subject to rotation, scaling and translation (distorted versions are omitted from the figure). Therefore, a total of 80 images are used (i.e. 20 classes, 4 images in each class).

Recall that the images in rows 2-4 of Fig. 3 (and their distorted versions) were not used in the evolutionary stage. Only the five images displayed in the first row of the figure (and their distorted versions) with a low resolution of dimensions 64×64 were used in the evolutionary stage of each method.

An example of features constructed from one solution picked up from each method I and II is shown in Fig. 8a and 8b respectively. We scaled these features to the interval [0,1] by dividing features by a constant number. Assuredly, both figures show stable features and there is no overlap between any different classes. Moreover, each class shows compact features for different deformations of images belong to the same class.

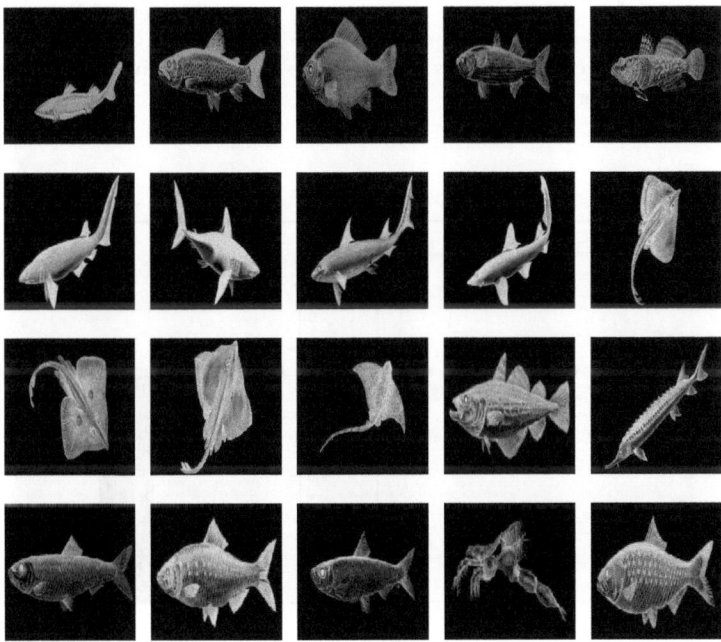

Fig. 3. Fish database [2]. Each image subjected to a random RTS deformation to form 80 image in total.

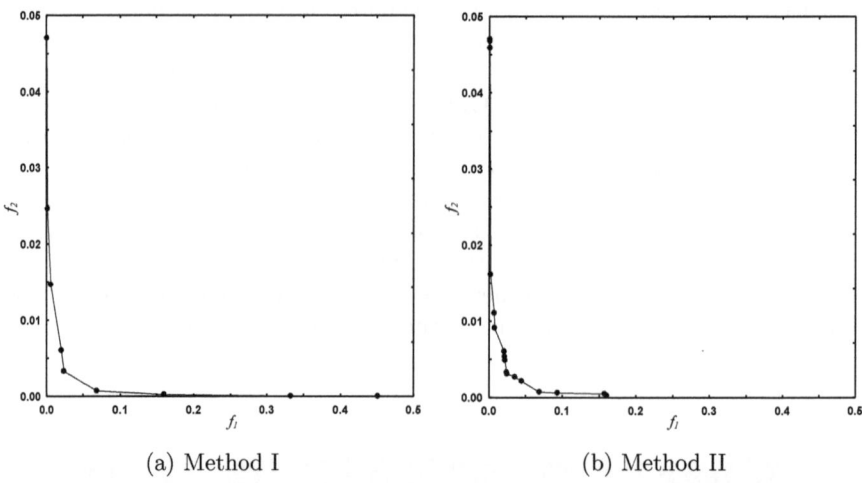

(a) Method I   (b) Method II

Fig. 4. Non-dominated solutions as Pareto fronts in the objective space for method I and II after 200 generations

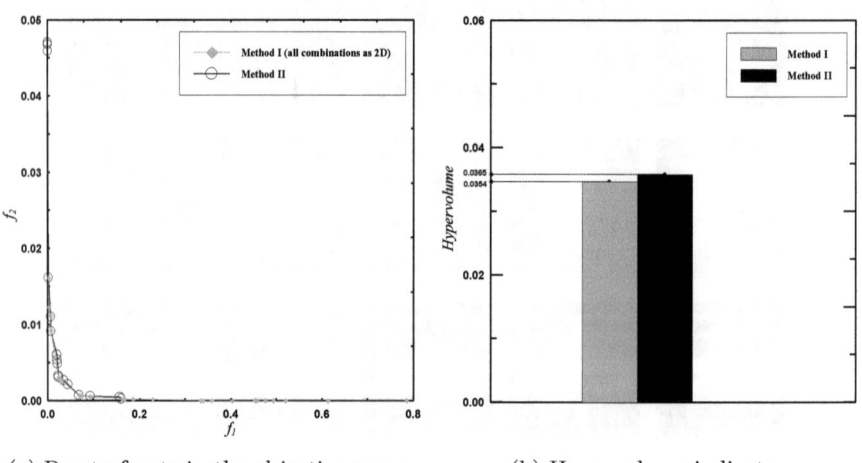

(a) Pareto fronts in the objective space.   (b) Hypervolume indicator.

Fig. 5. Non-dominated solutions from method I(combined as 2D) and method II

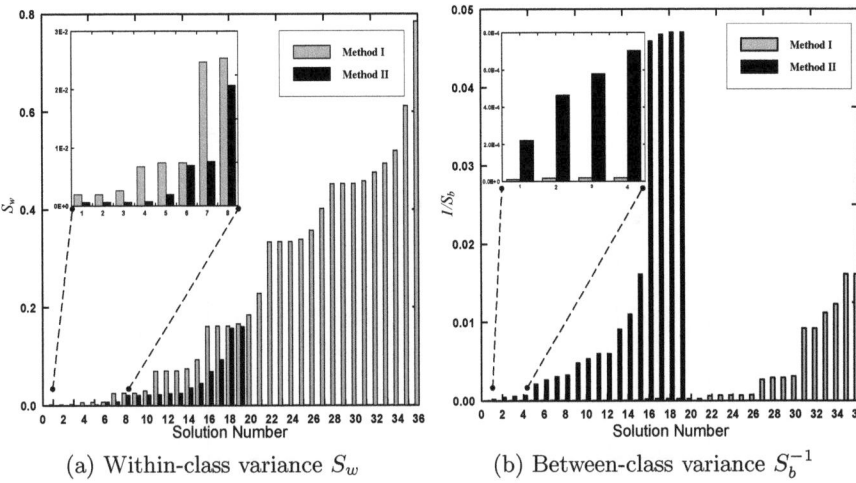

(a) Within-class variance S_w (b) Between-class variance S_b^{-1}

Fig. 6. The S_w and S_b^{-1} for solutions from method I and II

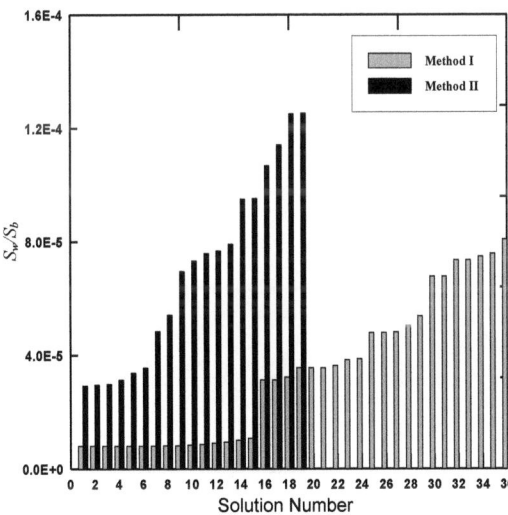

Fig. 7. The ratio of S_w/S_b for method I and II

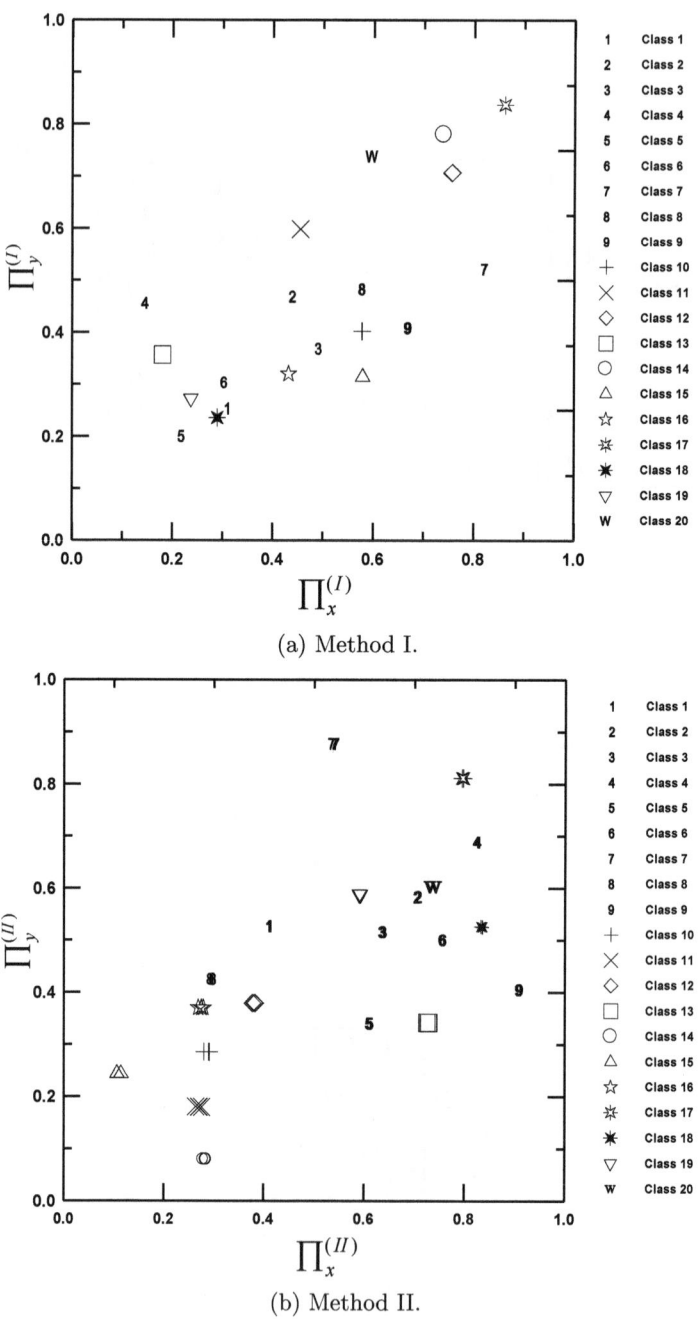

(a) Method I.

(b) Method II.

Fig. 8. Feature space using method I(combined as 2D) and method II

4   Conclusion

Two methods of evolving the Trace transform (method I and II) for robust image feature extraction are compared. In method I, one combination of functionals are optimized to extract 1D features. If 2D features are needed, two sets of functionals can be selected from the Pareto solutions to construct two 1D features. In method II, the Trace transform is optimized to directly extract 2D features. Therefore, each Pareto-optimal solution represents a pair of combinations of functionals, which can be applied to extract 2D features. A multi-objective evolutionary algorithm, NSGA-II, is employed to optimize the functionals in Trace transform. Method I and II uses a small number of low resolution (64×64) images for optimizing the functionals. Nevertheless, the optimized functionals have shown to work effectively to extract features from images of a high resolution. This indicates that functionals optimized offline using an evolutionary algorithm are able to extract features for robust identification of unseen images. The two methods are shown comparable results in terms of performance while method I is faster to evolve than method II. A set of 80 images form the fish database have been used to verify the effectiveness of method I and II. In the future, we plan to test the methods on various larger databases, and classifiers ensembles will be built based on the features extracted by ETT for image identification.

Acknowledgements. The authors would like to thank the anonymous reviewers for their insightful comments and suggestions that have significantly improved the quality of this paper.

References

1. Ruf, B., Kokiopoulou, E., Detyniecki, M.: Mobile Museum Guide Based on Fast SIFT Recognition. In: Detyniecki, M., Leiner, U., Nürnberger, A. (eds.) AMR 2008. LNCS, vol. 5811, pp. 170–183. Springer, Heidelberg (2010)
2. Kadyrov, A., Petrou, M.: The trace transform and its applications. IEEE Transactions on Pattern Analysis and Machine Intelligence 23(8), 811–828 (2001)
3. Petrou, M., Kadyrov, A.: Affine invariant features from the trace transform. IEEE Transactions on Pattern Analysis and Machine Intelligence 26(1), 30–44 (2004)
4. Kadyrov, M.P.A., Talepbour, A.: Texture Classification With Thousands of Features. In: 13th British Machine Vision Conference, BMVC, September 2-5, pp. 656–665 (2002)
5. Nasrudin, M.F., Petrou, M., Kotoulas, L.: Jawi Character Recognition Using the Trace Transform. In: Seventh International Conference on Computer Graphics, Imaging and Visualization, CGIV, pp. 151–156 (2010)

6. Albukhanajer, W.A., Jin, Y., Briffa, J.A., Williams, G.: Evolutionary Multi-Objective Optimization of Trace Transform for Invariant Feature Extraction. In: 2012 IEEE Congress on Evolutionary Computation, CEC, Brisbane, Australia, June 10-15, pp. 401–408 (2012)
7. Liu, N., Wang, H.: Extraction of hybrid trace features with evolutionary computation for face recognition. In: IEEE Congress on Evolutionary Computation, CEC 2007, pp. 2493–2500 (2007)
8. Srisuk, S., Fooprateepsiri, R., Petrou, M., Waraklang, S., Sunat, K.: A general framework for image retrieval using reinforcement learning. In: The Image and Vision Computing 2003, November 26-28, pp. 36–41 (2003)
9. Deb, K.: Multi-Objective Optimization using Evolutionary Algorithms, 1st edn. John Wiley & Sons. Ltd., England (2002)

Performance of Specific vs. Generic Feature Sets in Polyphonic Music Instrument Recognition

Igor Vatolkin[1], Anil Nagathil[2], Wolfgang Theimer[3], and Rainer Martin[2]

[1] Department of Computer Science, TU Dortmund
`igor.vatolkin@tu-dortmund.de`
[2] Institute of Communication Acoustics, Ruhr-Universität Bochum
`{anil.nagathil,rainer.martin}@rub.de`
[3] Research In Motion, Bochum
`wolfgang.theimer@ieee.org`

Abstract. Instrument identification in polyphonic audio recordings is a complex task which is beneficial for many music information retrieval applications. Due to the strong spectro-temporal differences between the sounds of existing instruments, different instrument-related features are required for building individual classification models. In our work we apply a multi-objective evolutionary feature selection paradigm to a large feature set minimizing both the classification error and the size of the used feature set. We compare two different feature selection methods. On the one hand we aim at building specific tradeoff feature sets which work best for the identification of a particular instrument. On the other hand we strive to design a generic feature set which on average performs comparably for all instrument classification tasks. The experiments show that the selected generic feature set approaches the performance of the selected instrument-specific feature sets, while a feature set specifically optimized for identifying a particular instrument yields degraded classification results if it is applied to other instruments.

Keywords: Polyphonic instrument recognition, multi-objective feature selection.

1 Introduction

Instrument recognition in audio recordings is a basic task in music information retrieval (MIR) as it further enables many different applications. For instance, if it is known which instruments are frequently played in songs of certain genres and substyles, the role of instrumentation in those music categories can be described by corresponding classification models. Furthermore, the structuring and organization of large song collections can be solved in an automatic way and instrument-related features may provide interpretable and small feature sets for supervised learning. Also, the transcription from audio to score or musical source separation require the exact knowledge of the audio sources.

However, instrument identification remains one of the most challenging tasks in MIR. On the one hand, a lot of work on the automatic identification of instruments for single playing notes has been succesfully carried out in the last

R.C. Purshouse et al. (Eds.): EMO 2013, LNCS 7811, pp. 587–599, 2013.

decade and earlier (please refer to [9] for a list of publications). If on the other hand several sources are playing at the same time, different harmonic and noise components (the impact of the latter for the identification of single instrument samples is investigated in [16]) may interfere with each other at the same frequencies making it more difficult to extract the amplitude contribution of partials which belong to different instruments. Also, the temporal evolution of frequency distributions varies for different sources. Moreover, the models built for classifying singular instrument samples are not suitable any more for the identification of instruments in polyphonic recordings.

A common approach to deal with instrument recognition is to design the features for categorizing specific instruments. To name just a few important recent works on instrument identification in polyphonic audio, the explicit learning from both attack and steady-state intervals in polyphonic recordings is discussed in [22]. Characterization of envelope dynamics by Gaussian models was investigated in [6]. In [14] the robustness of audio features was measured by their dependancy on the overlapping sound of different simultaneously played instruments. In our previous study [21] we have successfully applied multi-objective feature selection for instrument recognition in audio intervals (two tones playing at the same time) and chords (three and four tones) based on a large up-to-date feature set. Here the classification error and the subset size were minimized at the same time.

However, it can be argued that it is also reasonable to search for generic features which are suitable for training classification models for different instruments since the existing number of instruments and digital effects is nearly unlimited. In this work we therefore extended the study from [21] with the target to explicitly compare the performance of specific (best features for identification of a concrete instrument) and generic (best compromise features for different instruments) feature sets. As testbed we concentrated on instrument recognition from chords. It is a good compromise between the easier recognition of instruments in monophonic recordings and the rather complex problem of instrument detection in recordings with a larger number of playing instruments such as orchestra pieces. Another advantage of our data set is that the ground truth for training the classification model is completely available providing exact learning in contrast to songs where the exact times of instrument onsets and offsets are not available (unless the exact score representation is present). In the following sections, we describe the complete feature set, then introduce the multi-objective feature selection approach and conclude with the discussion of the experimental study on the recognition of four instrument groups in chords.

2   Audio Features

For each music tone a so-called attack-decay-sustain-release envelope describes the progress of energy, timbre and frequency properties [17]: *Attack* phase is characterized by increasing energy and instrument-dependent non-harmonic components (e.g. a violin bow strike). *Decay* corresponds to the succeeding energy decrease and is followed by longer stable *sustain* phase where the harmonics (fundamental frequency which corresponds to perceived pitch and its whole number multiple frequncies, or *overtones*) play the most important role. The last *release* phase conforms to final energy decrease as the sound quiets down.

The features which are used throughout this study and which are fed into the multi-objective feature selection stage can be grouped into three categories.

The first feature group consists of mel-frequency cepstral coefficients (MFCCs) and linear prediction coefficients (LPCs) extracted from a larger frame of 1.3s by segmentation into several 4096-sample blocks as introduced in [8]. The blocks should describe the different properties of the starting sound event and the 1.3s frames are positioned at the beginnings of attack phases previously estimated by MIR Toolbox [15]. Also the complete spectral envelope amplitudes are saved for the large frame. The overall number of these features is 353.

The second feature group consists of 265 commonly used and mostly short-time based signal descriptors extracted from the time, phase, spectral and cepstral domains of a music signal which are jointly computed within the AMUSE (Advanced Music Explorer) framework [19]. They can be divided into timbre and energy features, harmony features as well as correlation features. Timbre and energy features comprise e.g. the time domain zero-crossing rate, root mean square, low energy, normalized energy of harmonic components, tristimulus, various spectral statistics such as the spectral centroid, irregularity, bandwidth, skewness, kurtosis, crest factor, flatness measure, extent, flux and brightness, the spectral slope and features extracted from spectral sub-bands which are aggregated according to equivalent rectangular bandwidths. Harmony characteristics encompass the fundamental frequency, inharmonicity, the key and its clarity, different variants of chroma features (frequencies mapped to 12 pitch classes of an octave), amplitudes of strongest spectral peaks, the tonal centroid or the harmonic change detection function. Correlation features measure the relative periodicity and the sum of correlated components. For all these features we estimate the 'onset' frame with the highest energy of chords and save the features from the middle of the attack interval, the onset frame and the middle of the release interval separately, so that the final number of feature dimension for classification is 795.

The third feature group is based on cepstral coefficients derived from a third octave constant-Q spectral representation which were proposed for music instrument identification [5]. In contrast to the commonly used discrete Fourier transform the constant-Q transform [4] matches the frequencies of musical notes and yields a spectral estimate with improved spectral resolution for music analysis while keeping the number of sub-bands comparatively low. Therefore, it can be argued that cepstral coefficients obtained from a constant-Q spectrum are better suited to describe timbral characteristics of music instruments. Further, in order to account for temporal variations of musical timbre we propose to compute a sliding modulation spectrum of the constant-Q cepstral coefficients whose magnitude part is averaged over time. A low-rank approximation of the temporally averaged modulation spectrum is then computed using a singular

value decomposition, only retaining the first and second order singular values
and vectors which provide 102 features.

3 Multi-objective Feature Selection Framework

Feature selection (FS) is a meaningful process for classification in general and becomes essential if a very high number of available numerical data characteristics
exists but it is not clear which of them are relevant for the concrete categorization task. For general discussion of FS and method classification please refer to
[12]. As we could verify in a previous study [3], even a classifier with integrated
feature pruning techniques such as decision tree C4.5 [18] may be overwhelmed
by a large amount of features and produce more misclassifications when using a
complete feature set. All of classifiers applied in [21] (C4.5, random forest, naive
Bayes and support vector machine with linear kernel) performed significantly
worse using a complete feature set without FS.

A formal definition of FS is given as [12]

$$\theta^* = \arg\max_{\theta} \left[I\left(Y; \Phi(X, \theta)\right) \right], X \in \mathbb{R}^d, \tag{1}$$

where X is the complete feature set, θ are the indices of selected features, Y the
classification target, $\Phi(X, \theta)$ the selected feature set and I the relevance function
for the reduced feature set, e.g. correct recognition rate or accuracy.

In [20] we discussed several groups of optimization criteria (metrics) which
make sense for MIR classification tasks. Since the optimization of one criterion
often leads to decreased performance with regard to others, the idea behind
the multi-objective optimization (MOO) is to optimize O selected metrics at
the same time and search for the best compromise solutions. In that case the
definition of FS is extended to:

$$\theta^* = \arg\max_{\theta} \left[I_1\left(Y; \Phi(X, \theta)\right), ..., I_O\left(Y; \Phi(X, \theta)\right) \right],$$
$$X \in \mathbb{R}^d. \tag{2}$$

For the instrument recognition study we selected two criteria to be minimized.
The first one is the mean squared error (MSE), which is defined in general as
follows:

$$E^2 = \frac{1}{L} \sum_{i=1}^{L} \left(\hat{s}_i - s_i\right)^2, \text{ where} \tag{3}$$

L is the number of chords, $s_i \in [0; 1]$ is the labeled and $\hat{s}_i \in [0; 1]$ is the predicted
instrument relationship. Since we build exactly one classification instance per
chord and identify binary if a certain instrument in a chord is existing, in that
case $s_i, \hat{s}_i \in \{0; 1\}$.

The second is the percentage of selected features from the full set

$$f_r = \frac{|\Phi(X, \theta)|}{|X|} \tag{4}$$

where $|\cdot|$ denotes the size of a feature set.

Whereas the reason for MSE minimization is rather obvious, small f_r values provide also several advantages. Smaller feature sets lead to less storage demands, which is crucial for mobile devices with limited hardware resources. Furthermore, smaller feature sets mean faster model training and classification. Also they may help to avoid overfitting when the models are constructed from large feature sets and consequently are over-optimized for a concrete data set.

Since both criteria are taken into account, some of the solutions are not directly comparable - consider e.g. a very small feature set with higher MSE and another large feature set with a smaller classification error. A dominance relation is defined for two solutions $\mathbf{x}' := \Phi(X, \theta'), \mathbf{x}'' := \Phi(X, \theta'')$ as follows [7] ($\mathbf{x}''$ dominates $\mathbf{x}'$):

$$\mathbf{x}' \prec \mathbf{x}'' \text{ if } \quad \forall j \in \{1, ..., O\} : I_j(\mathbf{x}') \leq I_j(\mathbf{x}'')$$
$$\text{and} \quad \exists k \in \{1, ..., O\} : I_k(\mathbf{x}') < I_k(\mathbf{x}'') \tag{5}$$

The ultimate target of multi-objective FS is to find an optimal *Pareto front* $\mathcal{P}_f$ of trade-off solutions, which are not dominated by any other solution:

$$\mathbf{x} \in \mathcal{P}_f \text{ if } \nexists \mathbf{x}' : \mathbf{x} \prec \mathbf{x}' \tag{6}$$

In this work we apply the s-metric selection evolutionary multi-objective algorithm (SMS-EMOA) [1] for minimizing both metrics Equ. (3) and Equ. (4). Here the solutions are rated by areas (for two-objective scenarios) or hypercubes (for three and more objectives) which are dominated by them as defined in Equ. (5). The complete solution front can be evaluated by the *dominated hypervolume* [23]:

$$\mathcal{S}(\mathbf{x}_1, ..., \mathbf{x}_N) = \bigcup_i vol(\mathbf{x}_i). \tag{7}$$

Here $vol(\mathbf{x}_i)$ corresponds to the hypercube between the solution $\mathbf{x}_i$ and the reference point which should be set to the worst possible solution responding to all metrics (in our case [1;1] corresponding to the maximal feature set and the maximal classification error). For the optimization of FS we use a binary vector $\mathbf{v}$ of dimensionality $|X|$ as a representation of the current feature subset: $\mathbf{v}_i = 1$ means, that the feature i is selected for classification and $\mathbf{v}_i = 0$ means, that the feature is not used. The rough sketch of SMS-EMOA (please refer to [1] for detailed explanation) is: at first a *population* of N solutions, or *individuals* is created by random initialization of $\mathbf{v}$. In each optimization step a new offspring solution is created by mutation of a randomly selected parent individual flipping several bits on or off. Here we use an asymmetric mutation from [13], so that the flip probability $p_m(i)$ favors the reduction of feature set flipping more bits off than on:

$$p_m(i) = \frac{\gamma}{|X|}|\mathbf{v}_i - p_{01}|, \qquad (8)$$

where the global probability of switching 0 to 1 $p_{01} = 0.01$ and the step size $\gamma = 32$ are set with regard to previous experiments [3,21]. Then the new solution is evaluated and integrated into the population. From the extended population with the size $N + 1$, the solution with the smallest $vol(\mathbf{x}_i)$ is sorted out and the process continues with the next offspring generation. We do not apply a recombination operator since three different crossovers used in [21] did not lead to any significant improvement of performance.

It is worth to mention the validation of FS algorithms. In [10] it is discussed in particular for feature selection in MIR, that it is very important to evaluate the results using an independent test set avoiding the overfitting danger. Overfitting means here, that some features, which are indeed noisy or irrelevant for a classification task are recognized as relevant for a concrete set of data instances (here chords): With larger feature numbers the probability of such situation increases. Therefore we distinguish between three independent data sets:

- The *training* set is used for building classification models from the audio features.
- The *optimization* set is used for optimizing the feature selection by SMS-EMOA. The models created from the training set are evaluated for classification instances (chords) from the optimization set, and the both criterions are minimized for this set during the optimization.
- The *holdout* set is used for the validation of the optimized feature sets and is neither involved in the model training nor the feature selection: The optimization criterions are estimated for the models created from the training set and optimized for the optimization set.

For an overview of different data mining evaluation techniques in general see e.g. [2].

4   Experimental Study

The chord database with three or four simultaneously playing tones has been originally created for our previous work [21] and enables currently four instrument identification tasks: piano, guitar (acoustic and electric), wind (flute and trumpet) and strings (violin, viola and cello). The original tones were taken from the McGill[1], RWC [11] and Iowa[2] sample databases. 2000 chords were used for training and optimization based on a 10-fold cross-validation principle: at each fold the classification models were trained on 9/10 of the labeled chord feature

---

[1] http://www.music.mcgill.ca/resources/mums/html
[2] http://theremin.music.uiowa.edu

set and evaluated on the remaining 1/10 during the optimization process. The averaged MSE Equ. (3) across 10 folds was the 1st optimization criterion and f_r Equ. (4) the 2nd. Another 1000 chords were used for holdout validation.

Behind the four above mentioned classification tasks identifying the 'specific' features best matched for the identification of a concrete instrument, we run experiments with the target to optimize the mean error across several differ-ent classification tasks in search for 'generic' features suitable for classifying the different instrument groups. In that case the formal definition of the 1st opti-mization criterion should be extended to:

$$\widehat{E^2} = \frac{1}{C} \sum_{k=1}^{C} \left(\frac{1}{F} \sum_{j=1}^{F} \left(\frac{1}{L} \sum_{i=1}^{L} (\hat{s}_i - s_i)^2 \right) \right). \tag{9}$$

Here E^2 is at first averaged across $F = 10$ folds of subsequent partitioning into training and optimization sets. At the next level, it is averaged also across $C = 4$ different classification tasks.

Classification instances were created by extracting 1250 features described in Sect. 2. A random forest classifier was used, since it provided the best com-promise between speed and quality in [20,21]. The reasonable number of FS optimization steps by SMS-EMOA was limited to 2000 and the population size was set to 30 individuals.

5   Discussion of Results

5.1   Optimization Performance

Figure 1 shows the increase of the mean dominated hypervolume size during the optimization process. At first, it can be clearly seen from the left subfigure, that the tradeoff solutions become better with respect to both criteria during the optimization. This process does not saturate at 2000 evaluations. The bottom figure depicts the hypervolume increase also for the holdout set which was nei-ther involved in the classification model building nor in the evolutionary feature selection. Despite of a slightly worse performance on the holdout set compared to the optimization set, it can be assumed that the classification models are gener-alizable and perform well on an independent holdout set. Since the small increase in the dominated hypervolume size can be stated also for the last evaluations, the situation of over-optimization is avoided: in that case the performance on the holdout set would decrease whilst the performance on the optimization set would increase further.

Another observation is the different increase in mean hypervolume size for the different problems. The attainment of the non-dominated solutions with the higher dominated hypervolume occurs better for strings (dash-dotted line) and piano (dashed line) recognition. The optimization experiments in which the mean MSE across all classification tasks were optimized are marked by circles and show a performance which - as expected - is in between the performances of the instrument-specific feature sets.

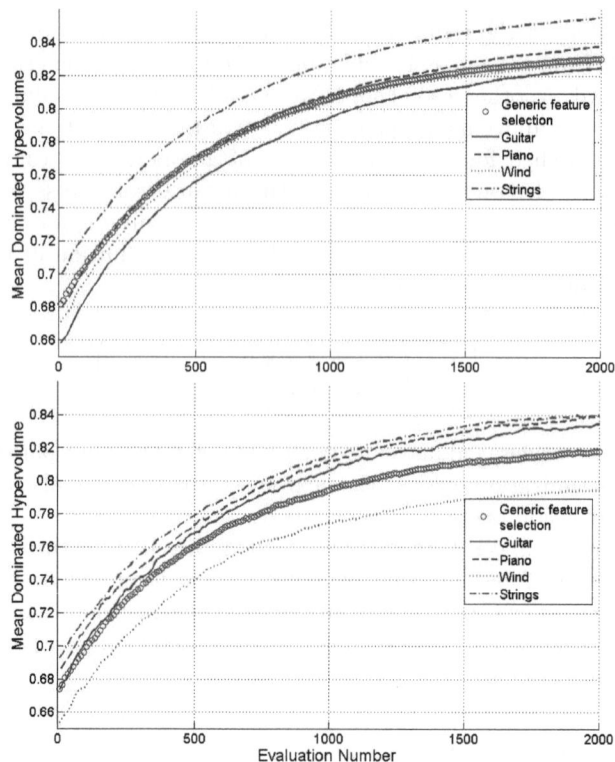

Fig. 1. Optimization set (top subfigure) and holdout set (bottom subfigure) dominated hypervolumes for all classification tasks, averaged across 20 statistical runs

5.2   Analysis of Tradeoff Solutions

For Fig. 2, we have plotted at first the best non-dominated solutions for each of the four classification tasks (specific features for a current task) after 20 statistical repetitions and connected them by a thin line. E.g. in the upper subfigure (guitar identification) the best solution responding to MSE has $E^2 = 0.142$ and requires 0.028% of the total features amount; the smallest feature set has a feature rate of $f_r = 0.004$ (which corresponds to only 5 features) but $E^2 = 0.358$ is in that case very high. Even if such solutions may be hardly interesting for decision maker, many other solutions can be taken into account: for example the 4th left solution ($f_r = 0.008$, $E^2 = 0.15$) is only slightly worse responding to E^2 than the MSE-best solution but requires almost four times less features. Such solutions can be interesting especially for classification on devices with limited resources (less features need to be stored; classification is done faster), and the models built from smaller feature sets may have reduced danger of being overfitted. Comparing the smallest classification errors for each task it can be

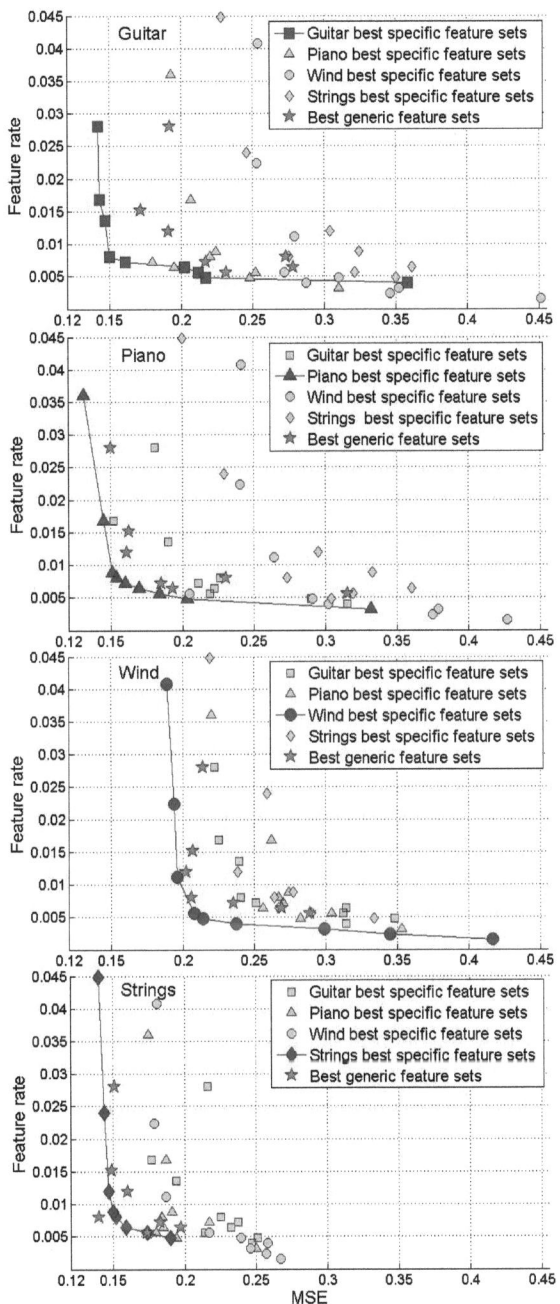

Fig. 2. Non-dominated solutions from 20 statistical multi-objective optimization runs

seen, that wind instrument identification is the most complex task, whereas the recognition of piano and strings are more simple.

In the next step we took the non-dominated solutions obtained for a specific instrument, i.e. a set of optimally selected instrument-specific features, and applied these for classifying other instrument groups for which these features are not optimal. They are marked by non-connected squares, circles, diamonds and triangles. One can clearly see, that feature subsets optimized for recognizing a specific instrument are not suited well for the classification of other instruments. In other words, instrument-specific FS makes sense, because each categorization task requires its own features.

The last step was to integrate the generic non-dominated solutions where $\widehat{E^2}$ was used as optimization criterion (pentagrams). We can observe that the performance of the generic feature set clearly approaches the non-dominated instrument-specific solutions. Therefore, an important conclusion is that the generic feature sets are well suited in many instrument classification tasks: for strings one feature set (bottom subfigure, pentagram at $[0.14; 0.008]$) seems to perform even significantly better than strings-specific feature sets due to both criterions - the best MSE solution of specific features $[0.14; 0.04]$ has the same E^2 but requires 5 times more features. The possible explanation that a generic solution is here better than the own specific feature set is that the figures plot the performance on the holdout chord set which was not involved into optimization process. For piano and strings the generic features are also rather close to the original specific feature sets; only for the guitar task the own specialist features are significantly better than generic features.

Concerning the almost unlimited number of existing real and virtual instruments, we argue that the search of generic features for the identification of different instrument groups is indeed reasonable. The advantages are that the optimization is not required for each instrument class and only a small feature set must be stored. On the other side it is clear, that the adjusted FS optimization for a concrete instrument identification will produce the smallest errors in almost all cases. More studies are also required for more precise statements: we cannot expect a good performance of our generic features for e.g. drums identification, which differ too much from the instruments involved in our experiments. But we can recommend to search for such generic features within some group of more or less similar instruments: e.g. a 'string generic feature set' can be convenient for classification of several instruments of this class which may sound slightly different.

Figure 3 lists the features from the non-dominated specific and generic feature sets (selected features are marked by dashes). For each task (separated by horizontal lines), the upper solution (row of dashes) corresponds to the smallest E^2 and largest f_r. The following solutions (from up to down) require less and less features but produce on the other side larger errors. Some of the features seem to be very important for the corresponding task - e.g. the 1st envelope amplitude for guitar (222) is in 8 of 9 non-dominated sets, and 1.3s-frame MFCC 2 (207) is in 6 of 9 wind instrument tradeoff sets. However, many features not only from

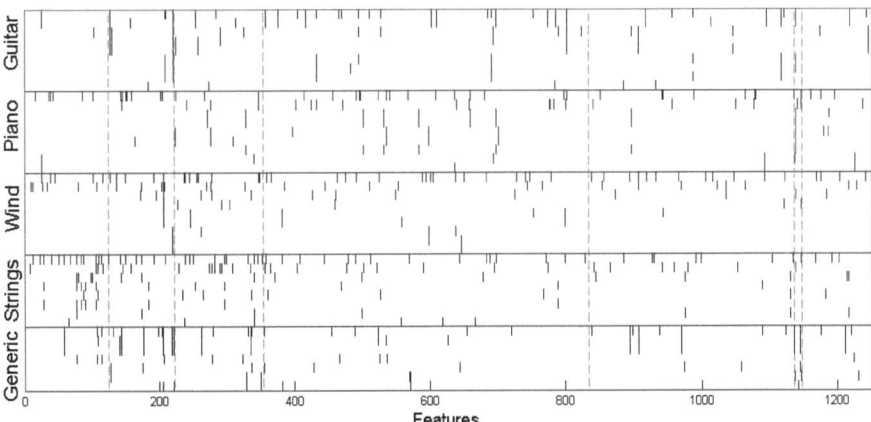

Fig. 3. Feature sets from the non-dominated fronts. The groups are separated by dashed vertical lines from left to right: Blocked LPCs, blocked and overall MFCCs, envelope amplitudes, timbre and energy features, harmony features, correlation features and constant-Q coefficients.

the first MSE-best solution (with the highest feature number and smallest error) belong to other tradeoff solutions. Therefore it could not be recommended just to save only the features from these first solutions for each classification task if the larger region of the non-dominated front should be considered.

6   Conclusions and Outlook

In general, three strategies can be applied for feature design in supervised classification of music instruments. One can develop category-specific features based on extensive knowledge of the physical characteristics of a certain instrument. This approach was beyond our scope, but may indeed help to create the best suitable classification models. Since it requires a lot of human effort for each new instrument to categorize, we concentrated rather on automatic feature selection methods starting from a large feature number in search for instrument-specific features, which are well suited for the current task. The last approach yields a generic feature set which omits the FS optimization rerun for each task and which can be used for the identification of several different instruments. As the results have shown, the last two methods provide reasonable classification quality for polyphonic mixtures. Further, the generic features are relatively close in their performance to the best specific features. However one must keep in mind that the performance of such generic sets may suffer with the increasing number of different instruments - for that case a possible solution is to detect a limited number of 'representing' classification tasks.

The optimization of both criteria - classification error and feature subset size - enabled the search for several tradeoff solutions from which one or more can be selected due to concrete application situation and user preferences. One can then select the proper balance between as small error as possible against small feature sets, faster classification and less danger of overfitting the models towards a certain data set.

A part of our current research is to integrate the instrument classification models as high-level features for prediction of music genres and styles. In future works, we are going to extend the number of instruments, instrument groups or certain playing styles (e.g. open or fretted strings) searching for different specific and generic features. Another important direction is to improve the models for more robust classification by increasing the number of simultaneously playing sources or by conducting experiments with strongly varying loudness of instruments.

Acknowledgments. This work was partly supported by Klaus Tschira Foundation.

References

1. Beume, N., Naujoks, B., Emmerich, M.: SMS-EMOA: Multiobjective Selection Based on Dominated Hypervolume. European Journal of Operational Research 181(3), 1653–1669 (2007)
2. Bischl, B., Mersmann, O., Trautmann, H., Weihs, C.: Resampling Methods for Meta-Model Validation with Recommendations for Evolutionary Computation. Evolutionary Computation 20(2), 249–275 (2012)
3. Bischl, B., Vatolkin, I., Preuss, M.: Selecting Small Audio Feature Sets in Music Classification by Means of Asymmetric Mutation. In: Schaefer, R., Cotta, C., Kołodziej, J., Rudolph, G. (eds.) PPSN XI. LNCS, vol. 6238, pp. 314–323. Springer, Heidelberg (2010)
4. Brown, J.C.: Calculation of a constant Q spectral transform. Journal of the Acoustical Society of America 89(1), 425–434 (1991)
5. Brown, J.C.: Computer identification of musical instruments using pattern recognition with cepstral coefficients as features. Journal of the Acoustical Society of America 105(3), 1933–1941 (1999)
6. Burred, J.J., Röbel, A., Sikora, T.: Dynamic spectral envelope modeling for timbre analysis of musical instrument sounds. IEEE Transactions on Audio, Speech, and Language Processing 18(3), 663–674 (2010)
7. Coello Coello, C.A., Lamont, G.L., van Veldhuizen, D.A.: Evolutionary Algorithms for Solving Multi-Objective Problems. Springer, Heidelberg (2007)
8. Eichhoff, M., Weihs, C.: Musical Instrument Recognition by High-Level Features. In: Proceedings of the 34th Annual Conference of the German Classification Society, pp. 373–381. Springer (2012)
9. Eronen, A.: Signal Processing Methods for Audio Classification and Music Content Analysis. PhD thesis, Tampere University of Technology, Departmrnt of Signal Processing (2009)
10. Fiebrink, R., Fujinaga, I.: Feature Selection Pitfalls and Music Classification. In: Proceedings of the 7th International Conference on Music Information Retrieval, ISMIR, pp. 340–341 (2006)

11. Goto, M., Hashiguchi, H., Nishimura, T., Oka, R.: RWC Music Database: Music Genre Database and Musical Instrument Sound Database. In: Proceedings of the 4th International Conference on Music Information Retrieval, ISMIR, pp. 229–230 (2003)
12. Guyon, I., Gunn, S., Nikravesh, M., Zadeh, L.: Feature Extraction, Foundations and Applications. Springer (2006)
13. Jelasity, M., Preuß, M., Eiben, A.E.: Operator Learning for a Problem Class in a Distributed Peer-to-Peer Environment. In: Guervós, J.J.M., Adamidis, P.A., Beyer, H.-G., Fernández-Villacañas, J.-L., Schwefel, H.-P. (eds.) PPSN VII. LNCS, vol. 2439, pp. 172–183. Springer, Heidelberg (2002)
14. Kitahara, T., Goto, M., Komatani, K., Ogata, T., Okuno, H.G.: Instrument identification in polyphonic music: Feature weighting to minimize influence of sound overlaps. EURASIP Journal on Advances in Signal Processing (2007)
15. Lartillot, O., Toiviainen, P.: MIR in Matlab (ii): a Toolbox for Musical Feature Extraction from Audio. In: Proceedings of the 8th International Conference on Music Information Retrieval, ISMIR, pp. 127–130 (2007)
16. Livshin, A., Rodet, X.: The significance of the non-harmonic "noise" versus the harmonic series for musical instrument recognition. In: Proceedings of the 7th International Conference on Music Information Retrieval, ISMIR, pp. 95–100 (2006)
17. Park, T.H.: Introduction to Digital Signal Processing: Computer Musically Speaking. World Scientific (2010)
18. Quinlan, J.R.: C4.5: Programs for Machine Learning. Morgan Kaufmann (1993)
19. Vatolkin, I., Theimer, W., Botteck, M.: AMUSE (Advanced MUSic Explorer) - A Multitool Framework for Music Data Analysis. In: Proceedings of the 11th International Society for Music Information Retrieval Conference, ISMIR, pp. 33–38 (2010)
20. Vatolkin, I., Preuß, M., Rudolph, G.: Multi-Objective Feature Selection in Music Genre and Style Recognition Tasks. In: Proceedings of the Genetic and Evolutionary Computation Conference, GECCO, pp. 411–418. ACM Press (2011)
21. Vatolkin, I., Preuß, M., Rudolph, G., Eichhoff, M., Weihs, C.: Multi-objective evolutionary feature selection for instrument recognition in polyphonic audio mixtures. Soft Computing - A Fusion of Foundations, Methodologies and Applications 16(12), 2027–2047 (2012)
22. Wu, J., Vincent, E., Raczyński, S.A., Nishimoto, T., Ono, N., Sagayama, S.: Polyphonic Pitch Estimation and Instrument Identification by Joint Modeling of Sustained and Attack Sounds. IEEE Journal of Selected Topics in Signal Processing 5(6), 1124–1132 (2011)
23. Zitzler, E., Thiele, L.: Multiobjective Optimization Using Evolutionary Algorithms - A Comparative Case Study. In: Eiben, A.E., Bäck, T., Schoenauer, M., Schwefel, H.-P. (eds.) PPSN V. LNCS, vol. 1498, pp. 292–301. Springer, Heidelberg (1998)

Multi-objectivization of the Tool Selection Problem on a Budget of Evaluations

Alexander W. Churchill, Phil Husbands, and Andrew Philippides

University of Sussex, Brighton, United Kingdom
{a.churchill,philh,andrewop}@sussex.ac.uk

Abstract. Tool selection for roughing components is a complex problem. Attempts to automate the process are further complicated by computationally expensive evaluations. In previous work we assessed the performance of several single-objective metaheuristic algorithms on the tool selection problem in rough machining and found them to successfully return optimal solutions using a low number of evaluations, on simple components. However, experimenting on a more complex component proved less effective. Here we show how search success can be improved by multi-objectivizing the problem through constraint relaxation. Operating under strict evaluation budgets, a multiobjective algorithm (NSGA-II) is shown to perform better than single-objective techniques. Further improvements are gained by the use of guided search. A novel method for guidance, "Guided Elitism", is introduced and compared to the Reference Point method. In addition, we also present a modified version of NSGA-II that promotes more diversity and better performance with small population sizes.

Keywords: NSGA-II, Guided Elitism, Preferential Search, Micro GA, Tool Selection, Evolutionary Multiobjective Optimization, Evolutionary Multicriterion Optimization, Roughing, CAM.

1   Introduction

The tool selection problem in Computer Aided Manufacturing (CAM) involves finding the best tool or sequence of tools that will manufacture a component whilst satisfying a number of objectives. Common objectives include manufacturing time, surface finish and total cost. The majority of previous attempts to solve this problem use single objective search methods (e.g. [1,3,8,14,15,18]). In these cases, objectives are weighted and combined into a single aggregate function, or punishment factors are added to ensure that solutions with objective values falling outside certain predefined regions are filtered out. In previous work [4,5], we used a single objective approach that was shown to work well on a relatively simple component. However, in more recent experiments we found that it was difficult for the algorithms to find optimum solutions for a part with a more complicated structure.

Here, we investigate multi-objectivization through constraint relaxation as a way of escaping local optima and increasing search success. A feature of the tool selection problem is that solutions are evaluated using a computationally expensive simulator.

R.C. Purshouse et al. (Eds.): EMO 2013, LNCS 7811, pp. 600–614, 2013.

While this provides solutions that are much more industrially relevant than other approaches, the number of fitness function evaluations used becomes an issue of particular importance. Recognizing this challenge, all of the algorithms tested below have their performance analyzed when the number of fitness evaluations are restricted to 150, 250, 350 and 500.

One of the features of multiobjective search is to provide the decision maker with a set of solutions. However, in our case we are still looking to find a single solution, the one that is globally optimal in the single objective approach. Algorithms that implement preference guidance are evaluated to see if this improves their ability to find this specific solution. NSGA-II [6] is used as the base level multiobjective search technique and preference guidance is applied using the reference point extension [7] and an alternative novel method termed "Guided Elitism". A correction to the diversity protection function in NSGA-II, which under certain conditions is shown to prevent diversity in small populations, is also presented and compared to the other techniques.

2   The Tool Selection Problem in Rough Machining

A milling machine is used to cut solid materials into a desired shape. The first stage of this process is roughing (or rough machining) which attempts to quickly cut a solid material until it closely resembles a predefined shape.

This paper concentrates on the tool selection problem for roughing. We attempt to find the sequence of tools that will cut a solid material so that it resembles a specified component to within a predefined surface tolerance, in the shortest amount of time.

Many researchers have tackled the tool selection problem, using a variety of approaches [1,3,8,14,15,18]. These have mainly restricted machining to use only flat end mill cutters, which means that solutions may not be truly relevant to industry. Many also do not take tool paths and their related cutting speeds into account (e.g. [14,15]), an aspect that leads to the complex search landscape present in real world applications of the problem. Existing models, such as those found in [1,14], have also made assumptions that can mean that the optimal solution cannot be found when using multiple cutters types [4].

A single objective metaheuristic approach was introduced by the authors in [5]. The problems described above were avoided by evaluating solutions using an industrially relevant simulator and applying fewer restrictions on tool sequences. In this work (as described in section 3 below) we found a component that creates a highly epistatic and difficult to traverse search space, which meant that the metaheuristic algorithms used regularly got stuck at suboptimal solutions.

The main motivation for the multiobjective approach proposed in this paper is to improve the search success for this difficult component. There has not been a great deal of research into applying multiobjective Evolutionary Algorithms to the tool selection problem. A notable exception is found in [13], which uses a combination of a micro GA and SPEA2 to calculate a Pareto front formed of three objectives – time, surface tolerance, and surface uniformity. The goal of the search in [13] is to find the set of single tools, with associated cutting parameters, that are Pareto optimal with respect to the three objectives. In this paper, the problem is expanded to allow a sequence of tools but machining parameters are not optimized.

3   Multi-objectivization

Multi-objectivization comprises of transforming a single objective problem into one with two or more objective functions. This is achieved in three main ways. The first technique uses helper objectives, for example in [16]. A second and very common method is decomposition [12]. The final method, and the one used in this paper, is constraint relaxation. Multi-objectivization has been shown to improve search success by creating a more efficient exploration of the search landscape [12].

Multi-objectivization was used in this work for two reasons, to allow for a better exploration of a highly epistatic and rugged search landscape and to remove the problem of dealing with constraints. A component was found that created a search space that prevented the algorithms in [4,5] from reliably finding the optimum solution. **Fig. 1** shows the fitness of the final solution returned by a Steepest Ascent Hill Climbing algorithm [17], when using each unique feasible solution as a different starting position for the component used in [4] and the component used here. It is clear that the second part has a tougher search space, with many more local optima and far fewer routes to the optimum. The motivation for the work in this paper was to investigate whether a multi-objectivization approach to the tool selection problem could achieve more successful search on this difficult part.

The single objective used by the algorithms in the authors' previous work [4], is:

$$f(x) = total\ machining\ time\ +\ c \tag{1}$$

where c is a punishment factor determined by the maximum distance of excess stock, d. If $d < 1mm$, then $c = 0$. If $d \geq 1mm \leq 1.5mm$ then $c = k$. Otherwise if $d > 1.5mm$ then $c = 2k$. k is a user defined constant. In [4], k was set equal to the time taken by the smallest single tool in the tool library. Punishment factors are used to keep unfeasible solutions within the population as they may have properties that help in finding good feasible solutions. However, they can severely distort the search landscape. A small change in genotype can have a large change in phenotype and neighbors in parameter space are distant in objective space. Another issue is raised by how to choose a good value for k. It becomes another parameter that must be chosen in advance, and one that is difficult to set without prior knowledge of the search space. To get around these issues, $f(x)$ was separated into two objective functions:

$$f_1(x) = total\ machining\ time \tag{2}$$

$$f_2(x) = maximum\ distance\ of\ excess\ stock \tag{3}$$

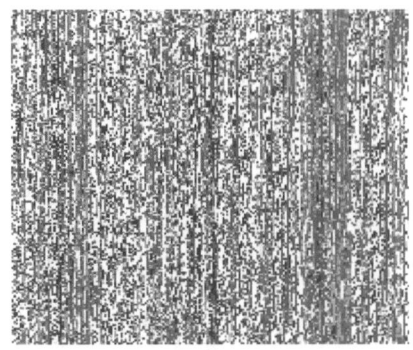

Fig. 1. Pseudo-color plots (created using the *imagesc()* function in Matlab) showing the fitness of the best solution found using Steepest Ascent Hill Climbing from every starting position, for the component used in [4,5] (left) and the component used here (right). The ordering of solutions was based on the index of the run and so structure in the images should be ignored. Black shows the optimum solution was found, and lighter shades show solutions with worse fitness. The left plot shows that the majority of solutions found are optimal. The right plot shows many different solutions are found, with few being optimal, suggesting that the search space for this component is more complex.

4   Experimental Setup

4.1   Test Case

The task in this experiment was to find a sequence of 5 or fewer tools that, in the shortest amount of time, could machine a component so that the maximum vertical distance of material (excess stock) that deviates from the desired final shape in any one place is less than 1mm. Tools were chosen from a library of 18, consisting of 6 end mill, 6 ball nosed and 6 toroidal type cutters. The component used is shown in **Fig. 2(a)**. The interaction between variable length sequences of different tool types, with different cutting speeds makes this a difficult combinatorial problem.

4.2   Evaluations

Tool sequences were evaluated using Vero Software's Machining Strategist CAM software, which gave a full tool path for each tool and a 3d representation of the final surface of the machined part. Evaluations could take between 10 seconds and 10 minutes on an Intel Core i7 @ 2.7ghz depending on the length of the tool sequences and the size of the tool paths. Caching (only evaluating individual solutions once in an independent search run) considerably speeds this up but it is clear that the evaluation function is expensive. Consequently, an important part of the experiment was testing how algorithms worked with a budget of evaluations. Their performance was tested under forced limits of 150, 250, 350 and 500 evaluations.

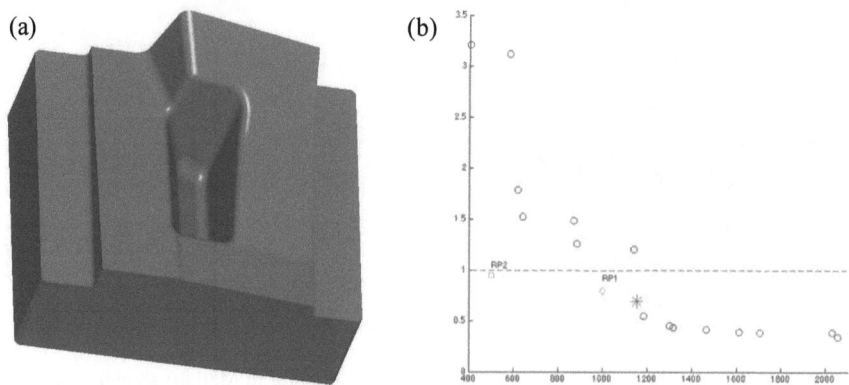

Fig. 2. (a) The part used for the experiments and (b) The Pareto front for this part. The x-axis shows total manufacturing time in minutes and the y-axis shows excess stock in mm. The solution that is optimal in the single objective formulation of the problem, and considered to be the optimum is marked with a *. The dotted line separates solutions under 1mm. The two reference points used in the R-NSGA-II algorithm are marked with a diamond (RP1) and a square (RP2).

4.3   Population Sizes, Diversity Protection and Preferential Search

The algorithms were tested with different population sizes (and in the case of Random Restart Stochastic Hill Climbing, different allowed numbers of evaluations within a restart), to see how they performed under the different evaluation limits. An intuition was that smaller population sizes would converge quicker, which could make them more successful under the lower limits. However, the algorithms tested can only return as many solutions as their population size. **Fig. 2(b)** shows the Pareto front for the part used in the experiments. It contains 16 solutions. As we have no constraints on the search, there are solutions on the Pareto front that we are not interested in. To overcome this problem preferential strategies were used to guide the search towards the region that we are interested in, close to the * in **Fig. 2(b)**. Another method used to improve search is an amendment to the crowding distance assignment in NSGA-II, which should create more diversity when used with smaller population sizes.

4.4   Algorithms

Three main algorithms were tested on the task described above. These are the single objective Genetic Algorithm (GA) [9], Random Restart Stochastic Hill Climbing (RRSHC) [10], and Non Dominated Sorting Genetic Algorithm II (NSGA-II) [6]. Two variations of NSGA-II were tested that included preferences in the search. These are the established Reference Point NSGA-II (RP) [7] and a novel method that will be referred to as the Guided Elitism NSGA-II (GE). In addition, the crowding distance assignment in NSGA-II has been modified to prevent certain duplicate solutions being given an infinitely high diversity measure. Algorithms that use this will be denoted with a *, e.g. NSGA-II*. All of the algorithms used the same methods for representation (a string holding 5 elements, each indicating either a specific tool or a blank, ordered from left to right), gradual mutation and recombination (this is not present in RRSHC). The operators and representation scheme are described in full in [4,5].

Single Objective Genetic Algorithm. A standard implementation of a generational GA was tested. The GA used mutation and recombination, as well as Roulette Wheel Selection and elitism [9]. A more complete description of the algorithm used here is given in [4,5]. Many permutations were tested for the mutation and crossover rates, over 100 trials and they were set to 0.4 and 0.7 respectively. The search was stopped when the evaluation limit was reached.

Random Restart Stochastic Hill Climbing. RRSHC is a single objective search algorithm that executes a Stochastic Hill Climbing (HC) run, restarting from multiple starting positions when certain stopping conditions are met [10]. A more complete description of the algorithm used in this paper can be found in [4,5]. A new HC run was started when the number of evaluations in a run reached a certain limit, with many different values being tested in the experiments. The search ended when the total evaluations reached a predefined limit.

Non Dominated Sorting Genetic Algorithm II. NSGA-II is an elitist multiobjective algorithm [6]. The algorithm creates an initial population in the same way as a GA but ranks each member using Pareto dominance [11] and crowding distance, which is a measure of diversity. At each generation, a new population is created using binary tournament selection. The new population is added to the old population, and then re-ranked and sorted. The top members are kept and used in the next generation. Mutation and recombination are used as in the GA. Many different permutations were tested for the mutation and crossover rates over 100 runs each, finding the best values to be 0.6 and 0.05 respectively. Pseudo-code for NSGA-II can be written as:

Algorithm 1. Pseudo-Code for NSGA-II

```
0:      create a random population, p
1:      while evaluations < budget, do
2:              while p' < pop_size, do
3:                      parents = binary_tournament_selection(p)
4:                      create new offspring through mutation and recombination
5:                      add offspring to p'
6:              end while
7:              p = p + p'
8:              assign a Pareto ranking to each member of the population
9:              assign a crowding distance value to each member of the population
10:             sort the population using Pareto ranking and then crowding distance
11:             while new_population < pop_size, do
12:                     remove top ranked members of p and add to new_population
13:             end while
14:             p = new_population
15:     end while
```

Reference Point NSGA-II. R-NSGA-II is a method used for providing preferences in multiobjective search. The decision maker (DM) sets a point in objective space and solutions should be returned that are close to this region, therefore giving the DM a set of solutions more relevant to their search goal. The algorithm modifies the crowding distance assignment in NSGA-II, so that solutions are ranked based on how far away they are from their closest reference point, rather than their distance from each other. In the original NSGA-II, for each Pareto front, pseudo-code for crowding distance assignment can be written as:

Algorithm 2. Pseudo-Code for the crowding distance assignment in NSGA-II

0:	initialize the crowding distance, cd, to 0 for each solution
1:	**for** each objective, **do**
2:	normalize the objective
3:	sort the solutions in the Pareto front according to their value for this
4:	set cd for the first and last solutions to infinity
5:	**for** solutions from solutions[start + 1] to solutions[end − 1], **do**
6:	subtract the objective value to the left of the solution from the
7:	**end for**
8:	**end for**

The crowding distance assignment in RP changes this to:

Algorithm 3. Pseudo-Code for the crowding distance assignment in RP

0:	initialize the crowding distance, cd, to 0 for each solution
1:	**for** each solution, **do**
2:	set the solution's crowding distance to the normalized Euclidian distance
3:	**end for**

In this way, solutions with the same Pareto ranking are treated more favorably if they are closer to the DM defined reference point. The choice of reference point is important, and in this work two different reference points are compared, referred to in the text and figures as RP1 and RP2. It should be noted that the reference point used here should not be confused with the reference point used to calculate hypervolumes in other multiobjective optimization algorithms.

Modification to NSGA-II's Crowding Distance Assignment. In the Crowding Distance Assignment method seen above, for each objective the highest and lowest solutions have a crowding distance score (CD) set to infinity. For these solutions, the crowding distance will remain infinite no matter what their other objective values are. In the original algorithm there is no explicit sorting between duplicate solutions, which we consider to be solutions with identical objective values. This can lead to a case where non-dominated duplicate solutions that score best in one objective and worst in another can both have their crowding distance set to infinity, meaning that are guaranteed to remain in the population. This is seen in an example run of NSGA-II used with a population of 6 in **Fig. 3**. Solutions 0 and 1 are duplicates, as are

Id	$f_1(x)$	$f_2(x)$
0	1951.4	0.4
1	1951.4	0.4
2	641.7	1.5
3	1395.1	1.3
4	403.7	3.2
5	403.7	3.2

Unsorted Population

Id	$f_1(x)$
4	403.7
5	403.7
2	641.7
3	1395.1
0	1951.4
1	1951.4

Sorted by $f_1(x)$

Id	$f_2(x)$
0	0.4
1	0.4
3	1.3
2	1.5
4	3.2
5	3.2

Sorted by $f_2(x)$

Fig. 3. Tables showing in grey shading which members of the population, marked by their id, are set to infinity when using NSGA-II's crowding distance assignment

4 and 5. This gives 4 unique solutions out of the six members of the population. As no explicit method is specified for sorting, duplicate solutions can be given an arbitrary ordering amongst themselves. To correct this, we propose an additional procedure that ensures that only one of the duplicate solutions is given an infinite crowding distance. For example, in **fig. 3** where solutions 0, 1, 4 and 5 would be given infinite crowding distance normally, our procedure assigns this value only to solutions 0 and 5.

Guided Elitism Applied to NSGA-II. The Reference Point method applies DM preferences in the selection mechanism, both in binary tournaments and in sorting new populations. Pareto dominance takes precedence over the DM's preferences. One problem with RP is that an extra parameter needs to be set – the reference point. Without a full understanding of the search space this can be difficult to set. Also, changing points can significantly affect search success, from the DM' perspective.

In this work, we have made a single objective problem multiobjective by treating our constraint, the amount of excess material, as its own objective. By relaxing constraints it means that the NSGA-II may return some undesirable but Pareto optimal solutions, e.g. those with large amounts of excess stock but short machining times. The evaluation function in the single objective version, described in section 3 above, puts constraints on these undesirable solutions but made it difficult to traverse the search space. As an alternative to the Reference Point method, a novel strategy was devised to guide search using the original evaluation function, combining the advantages of the single and multiobjective techniques. This is introduced at the elitism stage. At the point in NSGA-II where it creates a new population by merging the previous generation with a population of new children, in our version we sort the population according to the single objective evaluation function. The best y members are removed from this temporary population and added into the population of the next generation. Solutions are then added to this new population using the normal NSGA-II methods. The value of y is set using:

$$y = round(\frac{population\ size}{10})\qquad(4)$$

We refer to this method as "Guided Elitism". Pseudo-code for this stage in NSGA-II can be expressed as:

Algorithm 4. Pseudo-Code for Guided Elitism applied to NSGA-II
0:      append the newly generated child population to the old population, p
1:      initialize new_population as an empty list
2:      sort p using the single objective evaluation function
3:      **while** size(new_population) < y, **do**
4:              remove top ranked members from p and add to new_population
5:      Pareto rank the modified population, p
6:      assign crowding distances to the modified population, p
7:      re-sort p according to Pareto rank and crowding distance
8:      **while** new_population < pop_size, **do**
9:              remove top ranked members of p and add to new_population
10:    **end while**

4.5   Performance Assessment

The algorithms were tested on the task described above in section 4.1. Due to the stochastic nature of these search techniques, 1000 independent executions were performed, for each population size (or restart evaluation limit) and function evaluation budget, on a cached version of the search space. RRSHC was tested with 16 different restart evaluation limits, 10 – 160 evaluations in an independent run, in increments of 10. The other algorithms were tested with 16 different population sizes from 5 – 15 in increments of 1 and additionally 20, 25, 30 and 40. These population sizes were chosen based on the tight evaluations budgets. The algorithms were judged on how many times they were able to find the solution, which is marked with a * in **Fig. 2(b)**. It should be noted that while in normal multiobjective optimization the goal is to find a set that finds a good approximation of the Pareto optimal front, here we are applying multiobjective techniques to find the solution that is optimal under the single objective formula, which we refer to as the optimum throughout the rest of the paper.

5   Results

5.1   Single Objective Algorithms

The boxplot in **Fig. 4** shows the number of times that the optimum solution was found, from each of the 16 configurations that the algorithms were tested with. Looking firstly at the single objective algorithms, it is clear that they compare favorably at the 150 evaluations limit (which we will notate as 150-*EL*) but are greatly outperformed at higher evaluation levels. In terms of the scores from the best configurations, seen in the bar chart in **Fig. 5**, and the median score from all the configurations, the GA is more successful than RRSHC until 350-*EL* where RRSHC narrowly wins. At 500-*EL*, a large gap emerges between the two algorithms. In the best configurations, RRSHC finds the optimum 540 times compared to 413 by the GA. There is only a small improvement in the GA's search success when moving from 350-*EL* to 500-*EL*, which implies that the algorithm has had enough time to converge but struggles with this search space.

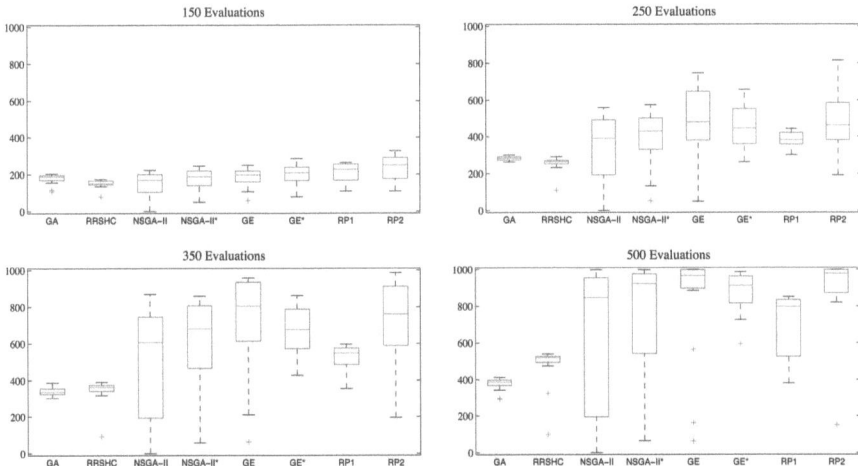

Fig. 4. Box plots showing the median radius of catchment areas for all 16 configurations of the 8 algorithms (box: 25th and 75th percentile; central line: median; whiskers: extent of the data; crosses: outliers, defined as points that are beyond the quartiles by more than 1.5 times the inter-quartile range). The y-axis displays the number of times that the optimum solution was found out of 1000 runs for each algorithm.

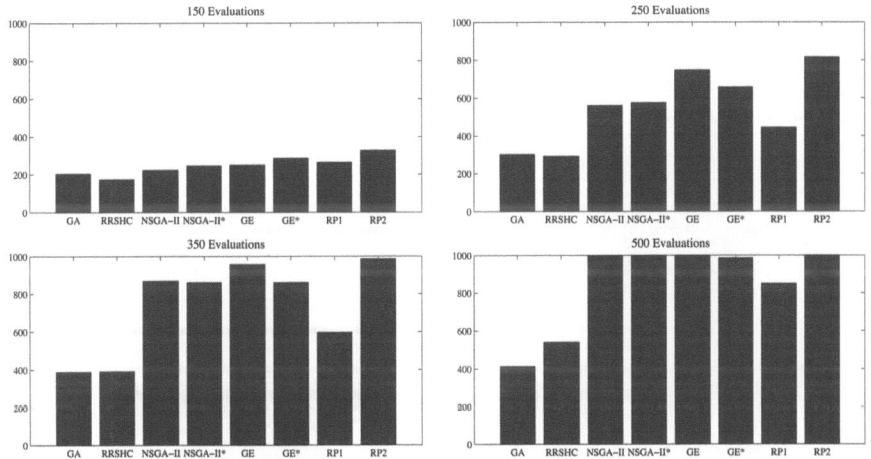

Fig. 5. Bar charts showing on the y-axis the number of optimum solutions found by the best configuration of the algorithm labeled on the x-axis

From the pseudo-color plots in **Fig. 6**, we can see that both of the single objective algorithms have a lot less variability in terms of the different configurations compared to their multiobjective compatriots. Within each evaluation band there is much less variation in shade. This means that there is a lot less reliance on choosing optimal parameters, which can be difficult when there is no great advance knowledge of the search space.

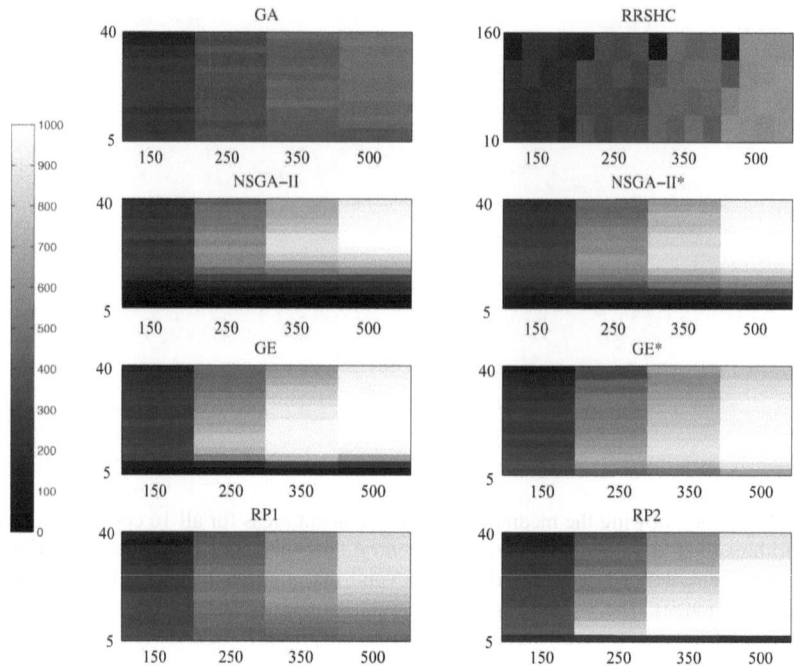

Fig. 6. Pseudo-color plots showing for each algorithm, the number of optimum solutions found for each population size (or number of evaluations in a run for RRSHC). Population size is on the y-axis and the evaluations limit is on the x-axis. Black shows low search success, while white shows high success.

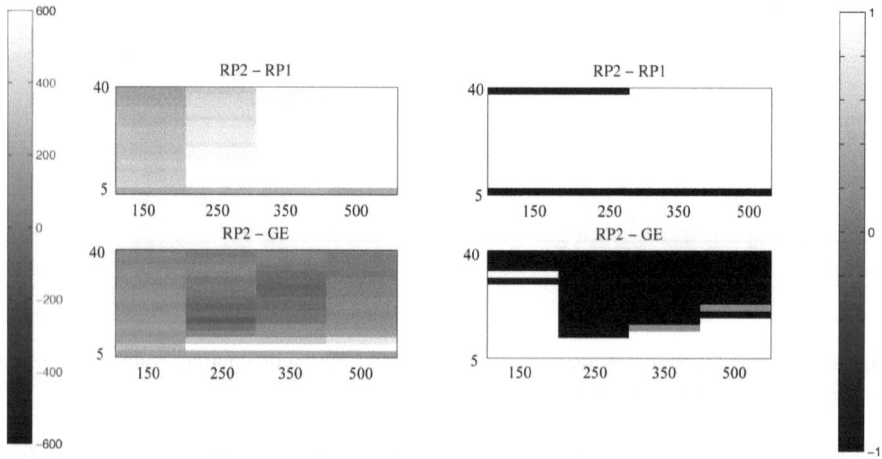

Fig. 7. Pseudo-color plots showing on the left, the number of optimum solutions found for each population size by RP2 subtracted by RP1 (top) and GE (bottom). Population size is on the y-axis and the evaluations limit is on the x-axis. White shows that RP2 performs much better, while black represents a much worse performance. On the right, the plots show for each population size if RP2 is better (white), worse (black), or the same (grey) as RP1 (top) or GE (bottom).

5.2   Multiobjective Algorithms

Moving on to the multiobjective algorithms, we will look at NSGA-II first. Concentrating on the best configurations, the algorithm achieves a good rate of optimal solutions compared to the single objective techniques. Whilst comparable to the GA at 150-*EL*, success increases greatly at higher evaluations, achieving 56%, 87% and 100% at 250-*EL*, 350-*EL* and 500-*EL* respectively. However, as with the other multiobjective algorithms, there is great variability between the success rates of different population sizes. As can be seen in the dark patches in the heat map in **Fig. 6**, NSGA-II performs particularly badly with small populations and cannot find the optimum a single time with a population of 5, regardless of the number of evaluations used. This is likely to be due to the crowding distance problem described above. The modified version, NSGA-II*, improves on the worst scores of NSGA-II and achieves much better median and upper quartile scores, with comparable best scores, at all evaluation levels. Looking at **Fig. 6**, NSGA-II* has lighter regions in the smaller population sizes, and less variability, which is also reflected in the boxplot.

Incorporating "Guided Elitism" (GE) into the NSGA-II has created significant improvements. Median and 3rd quartile scores are higher than NSGA-II and NSGA-II* for all evaluation limits and the best configuration can reach the optimum almost 97% of the time with 350 evaluations compared to 87% for NSGA-II and 39% for the GA. The color plot and boxplot shows that there is a lot of variability created by the different population sizes but less so than with NSGA-II. Similarly GE also performs badly with very small populations. Applying the sorting modification in GE* improves the performance on the smallest populations, with much lighter bands on the color plot. GE* performs better at 150-*EL*, which is likely to be due to working better with these populations. However, unlike with NSGA-II*, GE* is much less successful than GE at 150-*EL*, 250-*EL* and 500-*EL* in terms of best, median and 3rd quartile scores.

Two reference points were tested for the R-NSGA-II, which are labeled in the figures as RP1 and RP2. The optimum solution takes 1153 minutes, with 0.7mm of excess stock. RP1 uses a reference point very close to this with 1000 minutes and 0.8mm of excess stock. RP2 uses an unreachable point, 500 minutes and 0.95mm excess stock. Our intuition was that RP1 would outperform RP2 because it is closer to the optimum solution. However, the results went against this. RP2 performs far better than RP1 in terms of best, median and 3rd quartile scores. The best configuration of RP2 beats GE for every evaluation limit. This includes very high scores of 82% and 99% at 250-*EL* and 350-*EL*. The variability between population sizes is similar to GE and GE* and much improved on NSGA-II*. RP2 deals well with smaller population sizes, although it does not perform well with a population size of 5.

RP1 performs worse than all the other multiobjective algorithms, apart from at the lowest evaluation limit, where it is beaten only by RP2. This is especially the case at 250-*El* and 350-*EL*. The only advantage to RP1 seems to be its consistently good performance with smaller populations. This is seen in the color plot and also in its worst scores, which are often better than the other multiobjective algorithms. In **Fig. 7**, two color plots are shown, where the number of optimums found by RP1 are subtracted from RP2. If we look at the top right of **Fig. 7**, light areas show populations where RP2 outperforms RP1. This occurs almost everywhere apart from with a population of 5, where RP2 performs badly and with a population of 40 at 150 and 250-*EL*.

A similar analysis is shown in the bottom half of **Fig. 7**, where GE's scores are subtracted from RP2. In the bottom right of **Fig. 7**, white patches signify RP2 performing better, black patches where GE dominates and grey patches where they are equal. It appears that GE outperforms RP2 in more places. RP2 performs better than GE with the smallest population sizes, which is likely to be due to the sorting problem in NSGA-II. RP uses a different method for crowding distance, tied to the reference point, so it does not suffer from this problem. The bottom left of **Fig. 7** shows how much better the algorithm performs. There are few extremely dark or light areas, which means that generally the algorithms do not greatly outperform each other.

6   Discussion and Future Work

The results in section 5 show that multi-objectivization is very successful in improving search success rates on the difficult search space used in this work. This is particularly the case for higher evaluation limits and success is still rather small at the 150 evaluations level, with the best algorithms finding the optimum solution around 30% of the time. With a limit of 350 evaluations, the best results showed a success rate above 90% and at 500 evaluations all of the multiobjective algorithms could find the optimum solution at least 99% of the time.

NSGA-II search performed admirably on this problem but the modifications introduced in this paper, "Guided Elitism" and the sorting correction achieved an even better success rate. The modified algorithm NSGA-II* could be applied to any existing problems that NSGA-II has been used on, so it would be interesting to evaluate it on standard benchmarks. This is likely to make more of a difference when using small evaluation budgets. The color plots in **Fig. 6** show for this problem, smaller population sizes are more effective when using tight evaluation limits.

GE performed very well on this problem and was considerably more successful than NSGA-II. This shows that, in this problem instance, it is a very good method for providing search preferences. It is likely that improvements gained over NSGA-II are created by increasing the survival rate of solutions that may only weakly dominate or be indifferent to others but contain useful properties. More analysis of the algorithm is needed to see how it performs in a more general setting.

The difference in success between the two reference points in RP1 and RP2 is a cause for concern. The worse method uses a reference point that is very close to the optimum but the algorithm seems to converge on suboptimal regions of search space. However, RP2 is arguably the most successful search algorithm, so the reference point method is an important tool for this type of problem. We would like to further explore the effect that different reference points have on search success and discover what causes the surprising difference in performance between the two used here, which we believe is likely to be due to RP1 guiding search towards regions of the search space which contain difficult routes to the optimum solution.

For this problem, it could be argued that "Guided Elitism" is a better strategy to use than the reference point method because it has fewer parameters and could be more practical when there is not enough time available to tune parameters. It seems to be particularly well suited for multi-objectivization problems, where a single objective evaluation function already exists. A future area of research we would like to explore is testing "Guided Elitism" on other problem domains.

Population size had a large effect on all of the multiobjective algorithms, which is very noticeable in the pseudo-color plots in **Fig. 6**. More work is needed on setting guidelines or creating a function to recommend population sizes to use with a given evaluation limit.

7   Conclusions

To the best of the authors' knowledge this paper presents the first use of multiobjective algorithms to optimize the selection and sequence of multiple tools in rough machining. The results show that this approach is successful. The difference between single and multi objective search becomes particularly pronounced with a limit of 250 evaluations and above. A modification to the crowding distance assignment method in NSGA-II is also shown to improve performance, particularly when using small populations. This could be useful to micro-GA researchers.

Using preferences accelerates convergence time. This is likely to be useful in multi-objectivization cases where we are looking for a specific single solution or a set of solutions in a small, well-defined region of the search space. This could also be helpful for dealing with expensive evaluation functions by using small population sizes. The success of the reference point method has been shown to be dependent on choosing a good value for the reference point, which could make it unreliable. More research is needed to analyze this further. The novel "Guided Elitism" method is very successful, and with one fewer parameter setting could be seen as more reliable and stable than the reference point method, in this problem domain. We would like to investigate the success of this method with different problems to see if the trend continues.

Acknowledgements. Special thanks to Steve Youngs who provided invaluable help in setting up simulations and also to Gerry O'Driscoll, John Cockerill and Adrian Thompson. AWC is supported by an EPSRC studentship in partnership with Vero Software.

References

1. Ahmad, Z., Rahmani, K., D'Souza, R.M.: Applications of genetic algorithms in process planning: tool sequence selection for 2.5-axis pocket machining. Journal of Intelligent Manufacturing 21(4), 461–470 (2008)

2. Angantyr, A., Andersson, J., Aidanpaa, J.: Constrained Optimization based on a Multiobjective Evolutionary Algorithms. In: Proceedings of the Congress on Evolutionary Computation 2003 (CEC 2003), pp. 1560–1567 (2003)

3. Chen, Z.C., Fu, Q.: An optimal approach to multiple tool selection and their numerical control path generation for aggressive rough machining of pockets with free-form boundaries. Computer Aided Design 43(6), 651–663 (2011)

4. Churchill, A., Husbands, P., Philippides, A.: Metaheuristic Approaches to Tool Selection Optimisation. In: Proceedings of the Fourteenth International Conference on Genetic and Evolutionary Computation Conference (GECCO 2012), pp. 1079–1086 (2012)

5. Churchill, A., Husbands, P., Philippides, A.: Metaheuristic Approaches to Tool Selection Optimisation in Rough Machining. The International Journal of Advanced Manufacturing Technology (2012) (under review)

6. Deb, K., Pratap, A., Agrawal, S., Meyarivan, T.: A Fast and Elitist Multiobjective Genetic Algorithm: NSGA-II. IEEE Transactions on Evolutionary Computation 6(2), 182–197 (2002)

7. Deb, K., Sundar, J.: Reference Point Based Multi-Objective Optimization Using Evolutionary Algorithms. In: GECCO 2006: Proceedings of the 8th Annual Conference on Genetic and Evolutionary Computation, pp. 635–642 (2006)

8. D'Souza, R.M., Sequin, C., Wright, P.K.: Automated tool sequence selection for 3-axis machining of free-form pockets. Computer-Aided Design 36(7), 595–605 (2004)

9. Goldberg, D.E.: Genetic Algorithms in Search, Optimization, and Machine Learning. Addison-Wesley, Reading (1989)

10. Hoos, H.H.: On the run-time behaviour of stochastic local search algorithms for SAT. In: Proceedings of the Sixteenth National Conference on Artificial Intelligence and the Eleventh Innovative Applications of Artificial Intelligence Conference, pp. 661–666 (1999)

11. Horn, J., Nafpliotis, N., Goldberg, D.E.: A Niched Pareto Genetic Algorithm for Multiobjective Optimization. In: Conference on Evolutionary Computation (CEC), pp. 82–87. IEEE Service Center, Piscataway (1994)

12. Knowles, J.D., Watson, R.A., Corne, D.W.: Reducing local optima in single-objective problems by multi-objectivization. In: Proceedings of the First International Conference on Evolutionary Multi-Criterion Optimization, pp. 269–283 (2001)

13. Krimpenis, A., Vosniakos, G.-C.: Rough milling optimisation for parts with sculptured surfaces using genetic algorithms in a Stackelberg game. Journal of Intelligent Manufacturing 20(4), 447–461 (2008)

14. Lim, T., Corney, J., Ritchie, J.M., Clark, D.E.R.: Optimizing tool selection. International Journal of Production Research 39(6), 1239–1256 (2001)

15. Lin, A.C., Gian, R.: A Multiple-Tool Approach to Rough Machining of Sculptured Surfaces. The International Journal of Advanced Manufacturing Technology 15(6), 387–398 (1999)

16. Lochtefeld, D.F., Ciarallo, F.W.: Multiobjectivization via Helper-Objectives With the Tunable Objectives Problem. IEEE Transactions on Evolutionary Computation 16(3), 373–390 (2012)

17. Russell, S.J., Norvig, P.: Artificial Intelligence: A Modern Approach, 2nd edn. Pearson Education (2003)

18. Spanoudakis, P., Tsourveloudis, N., Nikolos, I.: Optimal Selection of Tools for Rough Machining of Sculptured Surfaces. In: Proceedings of the International MultiConference of Engineers and Computer Scientists, vol. II, pp. 1697–1702 (2008)

Applying Bi-level Multi-Objective Evolutionary Algorithms for Optimizing Composites Manufacturing Processes

Abhishek Gupta[1,3], Piaras Kelly[1,3], Matthias Ehrgott[1], and Simon Bickerton[2,3]

[1] Department of Engineering Science,
The University of Auckland, New Zealand
agup839@aucklanduni.ac.nz
[2] Department of Mechanical Engineering,
The University of Auckland, New Zealand
[3] Centre for Advanced Composite Materials,
The University of Auckland, New Zealand

Abstract. Resin Transfer Molding (RTM) and Compression RTM (CRTM) are popular methods for high volume production of superior quality composite parts. However, the design parameters of these methods must be carefully chosen in order to reduce cycle time, capital layout and running costs, while maximizing final part quality. These objectives are principally governed by the filling and curing phases of the manufacturing cycle, which are strongly coupled in the case of completely non-isothermal processing. Independently optimizing either phase often leads to conditions that adversely affect the progress of the other. In light of this fact, this work models the complete manufacturing cycle as a static Stackelberg game with two virtual decision makers (DMs) monitoring the filling and curing phases, respectively. The model is implemented through a Bi-level Multi-objective Genetic Algorithm (BMOGA), which is integrated with an Artificial Neural Network (ANN) for rapid function evaluations. The obtained results are thus efficient with respect to the objectives of both DMs and provide the manufacturer with a diverse set of solutions to choose from.

Keywords: Bi-level Multi-objective Optimization, Composite Manufacturing.

1 Introduction

Resin Transfer Molding (RTM) and Compression RTM (CRTM) together form a sub-category of composites manufacturing methods, commonly known as rigid-tool Liquid Composite Molding (LCM) processes. These processes can be described over four phases: preform (fibrous reinforcement) preparation, partial mold closure (or dry fiber compaction), mold filling and finally resin cure. For RTM, the mold filling phase consists of injecting a liquid thermosetting resin into a completely closed mold containing the fibrous reinforcement. In case of CRTM, the mold filling phase is subdivided into parts: resin injection and complete mold closure (or wet fiber compaction). The filling and curing phases are the focus of the present study as they determine the efficiency of the manufacturing cycle and the final part quality.

R.C. Purshouse et al. (Eds.): EMO 2013, LNCS 7811, pp. 615–627, 2013.

A successful process design depends on an appropriate combination of a large number of process variables which are primarily distributed over the filling and curing phases. For an entirely non-isothermal process these phases are strongly coupled to one another through the resin cure and temperature distributions reached within the part at the end of filling. A majority of the published papers dealing with comprehensive optimization of LCM processes concentrate solely on the curing phase by assuming minimal resin polymerization during filling and a uniform temperature profile [1]. Such simplifications are rendered invalid in generic situations involving preheated molds and/or resin systems, which lead to significant cure and temperature variations within the part. Another common feature of most optimization algorithms suggested in the literature is that they convert the multi-objective optimization problem into a single objective one by using a weighted sum of the normalized objective values [1; 2]. Deciding on such weights implies some prior knowledge of the behavior of the objective functions over the design space, which is often not available to the manufacturer.

The design of the filling phase of the process governs the fill time, the magnitude of clamping forces (force requirements determine the capacity and cost of manufacturing equipment) [3] and the existence of dry spots due to partial fiber impregnation [4]. On the other hand, the design of the curing phase affects the total time required for the resin to cure satisfactorily and the generation of residual stresses within the part due to large temperature gradients [1; 2]. Although the objectives of both phases seem largely disparate, they are fundamentally linked to each other in the case of fully non-isothermal processing. Optimization of the filling phase, without considering the cure phase, can lead to thermal conditions which adversely affect the progress of the cure phase. Moreover, the generally weak interactions between *the majority* of the process variables of one phase and objectives of the other, preclude a classical, centralized optimization approach. This difficulty is overcome by designing a decentralized, game-theoretic framework to model the problem. Two virtual decision makers (DMs) are assumed to monitor the critical phases and interact as the players of a static Stackelberg game. The DM controlling the cure phase (called the upper level problem) is assumed to be the leader, while the one controlling the fill phase (called the lower level problem) is assumed to be the follower. The design variables are distributed among the DMs such that final solutions are not skewed in favor of the objectives of either phase.

In order to implement the game-theoretic framework a Bi-level Multi-objective Genetic Algorithm (BMOGA), based on the popular elitist Non-dominated Sorting Genetic Algorithm (NSGA-II) [5], is developed. Taking a multi-objective approach leads to the final result being produced as a set of Pareto efficient solutions [6]. Prior weights need not be assigned to the objective functions, thereby making any *a priori* knowledge of the problem unnecessary. However, it is noted that a GA typically requires a few thousand function evaluations to find good solutions. Using a mold-filling simulation code for the function evaluations of the lower level problem would render the approach infeasible in terms of solution time. This obstacle is overcome by implementing an Artificial Neural Network (ANN) as a surrogate to the original simulation code.

This paper is organized as follows. Section 2 introduces the differential equations governing the resin flow and heat transfer phenomena, along with various empirical models, including the fiber compaction model. Section 3 contains brief descriptions of the simulation algorithm and the surrogate model. A mathematical description of the bi-level multi-objective optimization problem is presented in Section 4. It is also shown that under the given formulation the composite manufacturing process fits perfectly into this framework. Section 5 describes the BMOGA for solving the bi-level optimization problem, which is then applied to a test case in Section 6. The last section concludes the work presented.

2   Governing PDEs and Empirical Models

2.1   Equation of Fluid Flow

Darcy's law is commonly used to model viscous flow through porous media. When combined with the equation for mass conservation it takes the following form for a process with temporally varying thickness:

$$\nabla \cdot \left(h \frac{K}{\mu} \nabla p \right) - \frac{\partial h}{\partial t} = 0 . \tag{1}$$

Here, the thickness h, resin pressure p and the permeability K are in general coupled. The viscosity μ is likely to vary spatially with the temperature and degree of resin conversion. This response can be captured by the following widely used rheological model:

$$\mu = A_\mu e^{E\mu/R\,T_{abs}} \left(\frac{\alpha_g}{\alpha_g - \alpha} \right)^{a+b\alpha} , \tag{2}$$

where α_g is the resin gel conversion, E_μ is the activation energy, R is the universal gas constant, T_{abs} is the absolute resin temperature, α is the instantaneous resin conversion. A_μ, a and b are determined experimentally.

2.2   Energy Equation

Assuming that the resin and fiber phases have the same local temperatures, the volume averaged energy equation is written as:

$$\rho\, C_p \frac{\partial T}{\partial t} + \rho_r C_{pr}(\boldsymbol{u} \cdot \nabla T) = \nabla \cdot (k \nabla T) + \varphi \dot{H} . \tag{3}$$

The material properties ρ, C_p and k represent, respectively, the average density, specific heat capacity and thermal conductivity of the resin-fibre system. T is the local temperature, $\boldsymbol{u}$ is the volume averaged (or Darcy) velocity, φ is the local reinforcement porosity and $\dot{H}$ is a source term representing the thermal energy generated by the resin as it cures.

2.3   Species Equation

The volume averaged species equation can be expressed as follows:

$$\varphi \frac{\partial \alpha}{\partial t} + u.\nabla \alpha = \varphi R_\alpha \; . \tag{4}$$

To solve Eq. 4 the rate of resin polymerization (R_α) must be accurately modeled. The following general model is widely used to describe the polymerization reaction:

$$R_\alpha = \left(A_1 . e^{(-E_1/R\,T_{abs})} + A_2 . e^{(-E_2/R\,T_{abs})}.\alpha^{m_1}\right).(1 - \alpha)^{m_2} \; , \tag{5}$$

where A_1 and A_2 are constants; E_1 and E_2 are activation energies; m_1 and m_2 are catalytic constants.

2.4   Fiber Compaction Model

An accurate prediction of the mold clamping force requires a fiber compaction model which mimics the response of the fibrous reinforcement to applied load. Although the fibrous preform generally behaves in a viscoelastic manner, a mixed-elastic model [7] is simple and works well. The mixed-elastic model describes the behavior of a single material using four nonlinear stress-volume fraction curves –"Dynamic & Dry", "Dynamic & Wet", "Static & Dry" and "Static & Wet". A five-term polynomial (Eq. 6) provides excellent fit to the experimental data for a wide range of fiber volume fractions and may subsequently be implemented for the force simulations.

$$\sigma\left(V_f\right) = a'V_f^4 + b'V_f^3 + c'V_f^2 + d'V_f + e' \; , \tag{6}$$

where σ is the fiber compaction stress, V_f is the fiber volume fraction. a', b', c', d' and e' are experimentally determined constants for the particular material in use. The values of these constants for a glass-fiber chopped strand mat (CSM) are given in Table 1.

Table 1. Mixed-elastic compaction model parameters [7]

	Dynamic dry	Dynamic wet	Static dry	Static wet
a'	4.68e+7	2.96e+7	2.97e+7	4.55e+7
b'	-5.14e+7	-2.65e+7	-3.60e+7	-5.80e+7
c'	2.27e+7	9.50e+7	1.72e+7	2.84e+7
d'	-4.62e+6	-1.54e+6	-3.68e+6	-6.17e+6
e'	3.58e+5	9.03e+4	2.943e+5	4.95e+5

3   Simulation and Surrogate Modeling

The filling phase of the process is modeled using a hybrid Finite Element/Finite Difference methodology. However, since the simulation code is too expensive to be considered for every lower level function evaluation, a trained ANN is employed as a computationally efficient surrogate for rapid function evaluations. During meta-modeling and optimization, it is considered desirable to have an automated framework which requires minimum interactions with the manufacturer. In order to achieve this, the Cascade-Correlation Learning Architecture Neural Network [13] is used as it builds its own neural topology according to the requirements of the problem (a more detailed discussion on the construction, training and verification of the network is available elsewhere [3; 14]).

The simulation of the cure phase is inexpensive and is integrated with the BMOGA. To predict the evolution of the thermal field, Eq. 3 and Eq. 4 (without the convection terms) are solved using an unconditionally stable Finite Difference scheme. Ordinarily, the initial condition for the cure phase, i.e. the temperature and cure distribution, is obtained through the fill simulation. However, since the BMOGA does not use the actual fill simulation for the lower level function evaluations, the temperature and cure distributions at the end of filling are also predicted through the ANN.

4   Bi-level Multi-objective Optimization

The mathematical description of a bi-level multi-objective minimization problem is as follows [8; 9]:

$$Minimize \; F(x_u, x_l),$$

$$subject \; to \; x_l \in argmin \; \{f(x_l) \mid g(x_l) \geq 0, h(x_l) = 0\},$$

$$G(x_u, x_l) \geq 0, H(x_u, x_l) = 0, \tag{7}$$

$$x_u = (x_1, \ldots, x_r); \; x_l = (x_{r+1}, \ldots, x_n),$$

$$x_i^{(L)} \leq x_i \leq x_i^{(U)}, i = 1, \ldots, n.$$

In the description, F and f are the objective function vectors of the upper and lower level problem. G, H and g, h are, respectively, the upper and lower level constraints. x_u and x_l are the upper and lower level design vectors that together form x ($x = (x_u, x_l)$), which is the n dimensional design vector of the overall problem. It is important to note that the lower level problem is optimized with respect to x_l only, while x_u acts as a fixed parameter. Therefore x_l can be considered to be a function of x_u.

This formulation of the problem corresponds to an optimistic approach [8] wherein it is assumed that the DM of the lower level problem chooses among all Pareto efficient solutions, that which is best suited for the upper level. It can be argued that in general this approach is overly optimistic. Several points on the Pareto front of the lower level problem may be totally undesirable to the lower level DM. For example, in case of the composites manufacturing problem, minimization of the clamping forces and the resin fill time are considered as the lower level objectives. It is possible to have very low fill times at the cost of unrealistically large clamping forces, or vice versa, but such a configuration is unlikely to be chosen by the manufacturer. Therefore, it would be incorrect to include such solutions as feasible solutions for the upper level problem, even if they are Pareto optimal. As an alternative, we propose a more subdued-optimistic approach by introducing a slight modification to the mathematical description:

$$x_l \in \{argmin \{f(x_l) \mid g(x_l) \geq 0, h(x_l) = 0\} \cap \{x_l : f(x_l) \in RI\}\}, \qquad (8)$$

where *RI* represents a region of interest in the objective space of the lower level DM. Therefore only those Pareto efficient solutions which lie within the region of interest are considered feasible for the upper level problem. Although it is difficult to make an *a priori* prediction of the region of interest, in [10] it was noted that *"from practical experience... the user or designer usually picks a point on the middle of the surface...where the Pareto surface bulges out the most"*. Mathematically it was defined to be a point on the Pareto front, below and farthest from the convex hull defined by the individual function minima. This point, also designated as the *"knee"*, is assumed to be the center of the region of interest, the extent of which can be manually altered. It should be noted that selecting the *knee* as a unique solution of the lower level problem is also a viable option; however it may be too restrictive for the upper level DM. Selecting a set of equally desirable solutions within a preferred region lends greater robustness to the final Pareto set.

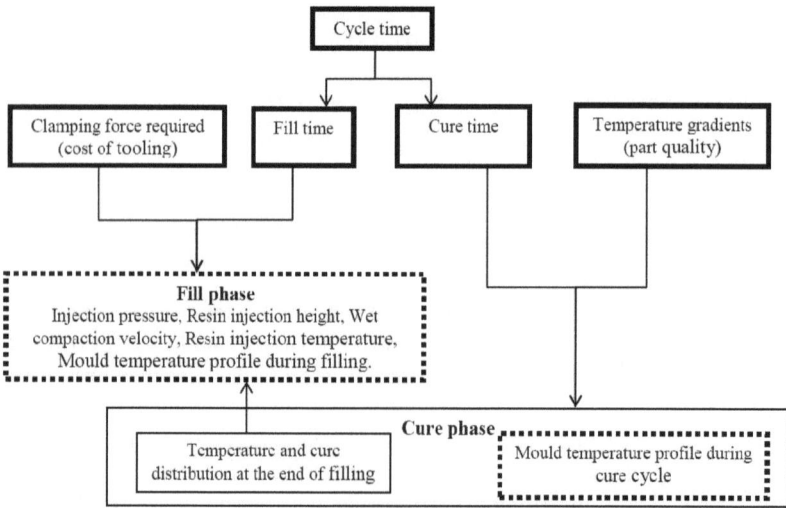

Fig. 1. Manufacturing problem summary. Bold solid blocks contain objective functions; Dotted blocks contain design variables; Arrow heads indicate dependency.

A summary of the manufacturing problem structure is depicted in Fig. 1. Clearly, it is possible for the filling phase to be optimized without taking the cure phase into account (as discussed in the introduction, this may lead to skewed results). Vice versa is not always true as the progress of the cure phase depends on the temperature and cure distributions at the end of filling.

One possible simplistic approach to optimize the complete problem would be to include all the objectives, i.e. the maximum clamping force required, the total cycle time and the maximum temperature gradient, into a single (centralized) tri-objective minimization algorithm. However, it should be observed that the objectives of the fill phase are completely independent of the design variables of the cure phase. Similarly, a majority of the design variables of the fill phase have little, if any, effect on the objectives of the cure phase. It is only the thermal conditions of the injected resin and the preheated mold that interlink the two phases. For such a scenario, adopting a decentralized decision making methodology is more compatible with the requirements of the problem at hand. Moreover, the different objectives of either phase lead to a situation which is best addressed by game theory.

In order to design such a decentralized system which maintains the link between the phases it is important to redistribute the design variables among the DMs such that the objectives of the fill phase may be affected by the choices of the upper level DM. To this end, it is considered that the mold temperature profile for the entire process, including the filling phase, is controlled by the upper level DM. In other words, the mold temperature profile may be viewed as x_u while all other design variables constitute x_l. The objectives of the lower level DM are to minimize the fill time and the clamping force requirements, whereas the upper level DM attempts to minimize the cycle time (fill time + cure time) and reduce residual stresses by minimizing the through-thickness temperature gradients within the part during processing.

5   The BMOGA

Fig. 2 shows a diagrammatic summary of the steps involved in the proposed BMOGA. It is constructed by nesting one NSGA-II simulation within another. The outer simulation corresponds to the cure phase (upper level problem) whereas the inner simulation corresponds to the filling phase (lower level problem). For the sake of brevity, usual concepts such as non-dominated sorting, crowding distances, selection, cross-over, mutation etc. are not discussed; details can be found in [5].

The algorithm starts with the generation of a population of mold temperature profiles at the upper level. During every function evaluation at the upper level, the mold temperature profile is sent to the lower level NSGA-II simulation. The mold temperature during filling is extracted from the profile and it acts as a constant parameter according to which the lower level design variables are optimized. Once convergence onto the lower level Pareto front is achieved, the knee is used as a reference point to migrate the Pareto efficient solutions such that they satisfy Eq. 8 (algorithms that allow such migration have been suggested in [11; 12]). These solutions are returned to the upper level. Therefore, a single mold temperature profile leads to a large subpopulation of solutions.

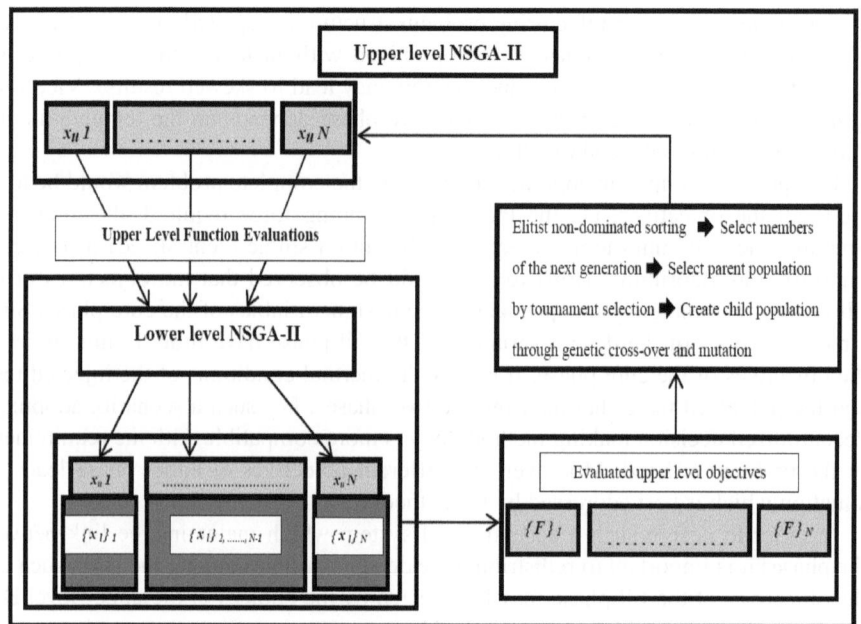

Fig. 2. The proposed BMOGA procedure

Cure simulations are carried out on each member of every subpopulation, based on the initial conditions of temperature and cure prescribed by the lower level design variables, to obtain the upper level objective functions. Further, all subpopulations at the upper level are concatenated and the usual NSGA-II procedure is performed to assign each member of every subpopulation a non-dominated rank and crowding distance, based on the upper level objectives only. A particular mold temperature profile is recognized by the lowest (best) non-dominated rank among members of its subpopulation and the highest crowding distance among solutions with the lowest rank (all solutions in the subpopulation with higher ranks are eliminated). Selection operations at the upper level are based on these values of recognition. Cross-over and mutation operations are performed on selected temperature profiles to generate new populations of solutions which are evaluated by recalling the lower level NSGA-II simulation. This procedure is repeated for a prescribed number of generations or till convergence to the Pareto front is achieved.

Although a nested approach may be deemed computationally more intensive, as compared to an intertwined approach (suggested in [9]), it is found to be necessary when lower level preference information, in its present form, is incorporated into the algorithm.

6   A Test Case

The proposed framework is now applied to a test case for the manufacture of an axisymmetric part of moderate thickness by the CRTM process. The chopped strand mat

(CSM), the compaction model for which is detailed in Table 1, is taken to be the pre-form material. The permeability of this material is expressed by the following empirical exponential relation:

$$K = 3.07 \times 10^{-8} e^{-12.97 V_f} \, (m^2),\tag{9}$$

where V_f is the fibre volume fraction. Thermal properties of the CSM and the reactive epoxy resin system considered can be found in [3; 13]. The diameter of the part is 1 m with a central injection hole of 2 cm. The final part thickness is 1.5 cm and the final fibre volume fraction is approximately 35%.

Prior to building the ANN a preliminary study is conducted over the design space in order to determine feasible bounds of the design variables within which the resin is unlikely to gel prematurely. Details of these bounds and other constraints considered in this study are given in Table 2. For simplicity only a single temperature ramp is considered during the cure phase. For complex resin systems, a more intricate mold temperature profile, with multiple ramps and dwelling periods, is often applied. The phase specific enlargement of the design variable space that this would entail, further justifies the decentralized decision making approach being used.

While running the NSGA-II simulation the following parameter setting is used: crossover probability of 90%, mutation probability of 10%, distribution indices for the real-coded genetic operators are chosen to be 15 and 5 for the crossover and mutation operators, respectively. The lower level solutions are evolved using a population size of 40 individuals, initially over 60 generations to obtain a rough approximation of the complete Pareto front, and then for 30 further generations to focus the search within the region of interest. For the upper level problem a population size of 30 individuals is considered, which is evolved over 75 generations.

Table 2. Design variable bounds and constraints

Lower level design variables (X_l)	
Injection pressure (P_{inj})	[200, 500] kPa
Injection height (H_{inj})	[1.6, 2] cm
Wet compaction velocity (V_{wet})	[0.5, 2.5] mm/min
Resin temperature	[293, 348] K
Upper level design variables (X_u)	
Mold temperature during filling	[293, 333] K
Mold heat rate	[2, 30] K/min
Final mold temperature	[393, 433] K
Maximum exothermic temperature allowed	500 K

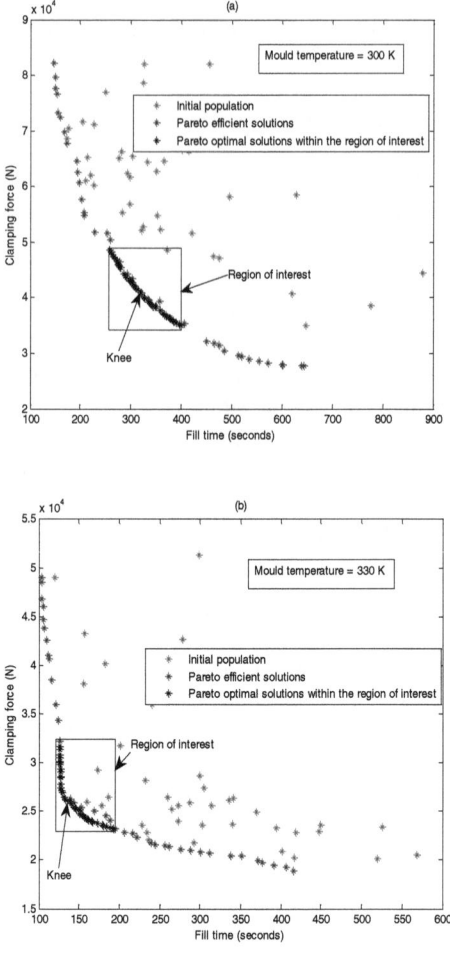

Fig. 3. Results of lower level NSGA-II performed for two different mold temperatures

Figs. 3a and 3b show the results obtained from the lower level NSGA-II by considering the mold temperature to be a constant parameter. The figure also depicts the evolution of the solutions over three steps: the initial population; solutions converged onto the Pareto front and finally, Pareto efficient solutions migrated within the region of interest around the *knee*.

In this study the cure simulations at the upper level are continued until a minimum of 90% cure is reached. Fig. 4 shows the Pareto efficient solution set obtained after the complete bi-level optimization procedure. The figure also depicts the solutions of the initial upper level population. The initial population is seen to form several clusters, each corresponding to a lower level NSGA-II simulation call. All of these solutions are far away from the obtained Pareto front, thus highlighting the importance of the optimization study.

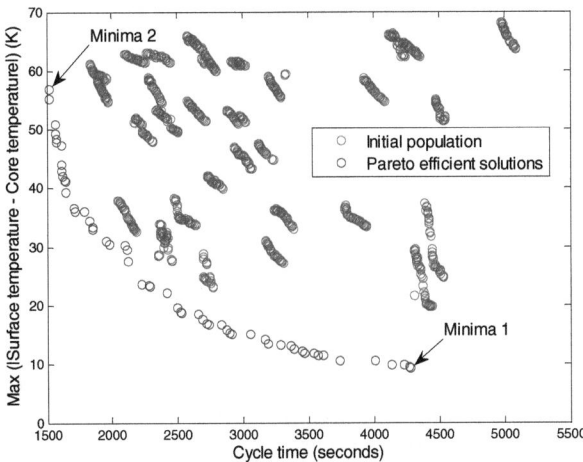

Fig. 4. Results after complete BMOGA simulation

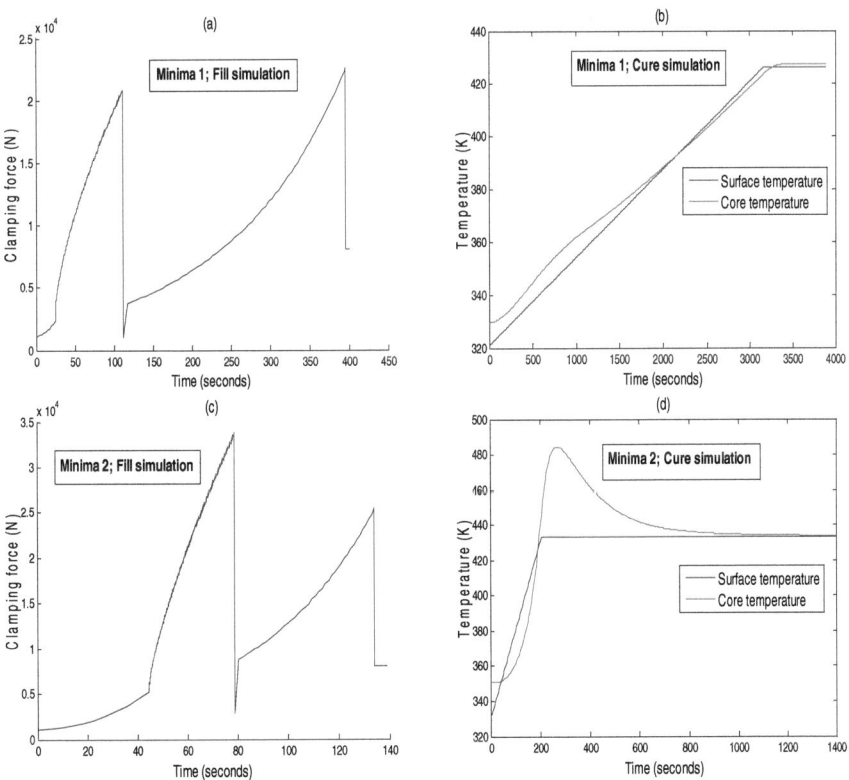

Fig. 5. Force (a & c) and temperature (b & d) evolution curves of the solutions representing the individual function minima in Fig. 4

Figs. 5a-5d depict the force and the mid-radius temperature evolution curves for the solutions representing the individual function minimum of the upper level problem. In Fig. 5b although the resin curing time is high, the temperatures at the surface and core of the part evolve uniformly, which minimizes residual stresses within the part. On the other hand, in Fig. 5d, although the resin rapidly reaches the desired cure level, large temperature gradients are induced which may cause high residual stresses within the part.

7   Conclusions

In this study a multi-objective evolutionary game-theoretic framework is proposed for the optimization of non-isothermal CRTM processes. Modeling the problem in the form of a Stackelberg game prevents the final solutions from being skewed in favor of the objectives of either the filling or curing phase. These final solutions may be regarded as sub-game perfect Nash Equilibria, but not in the traditional sense as a unique objective, and consequently a unique course of action/strategy, cannot be defined. Crucially, presenting the final result as a diverse set of Pareto efficient solutions allows the manufacturer to choose the most preferred solution *a posteriori*, without requiring any prior knowledge of the problem.

References

1. Ruiz, E., Trochu, F.: Multi-criteria thermal optimization in liquid composite molding to reduce processing stresses and cycle time. Composite Part A: Applied Science and Manufacturing 37(6), 913–924 (2006)
2. Yu, H.W., Young, W.-B.: Optimal design of process parameters for resin transfer molding. Journal of Composite Materials 31(11), 1113–1140 (1997)
3. Gupta, A., Kelly, P.A., Bickerton, S., Walbran, W.A: Simulation the effect of temperature elevation on clamping force requirements during rigid-tool liquid composite moulding processes. Composites Part A: Applied Science and Manufacturing 43(12), 2221–2229 (2012)
4. Ruiz, E., Achim, V., Soukane, S., Trochu, F., Bréard, J.: Optimization of injection flow rate to minimize micro/macro-voids formation in resin transfer molded composites. Composite Science and Technology 66(3-4), 475–486 (2006)
5. Deb, K., Pratap, A., Agarwal, S., Meyarivan, T.: A fast and elitist multiobjective genetic algorithm: NSGA-II. IEEE Transaction on Evolutionary Computation 6(2), 182–196 (2002)
6. Deb, K.: Multi-Objective Optimization Using Evolutionary Algorithms. Wiley, Chichester (2001)
7. Walbran, W.A., Verleye, B., Bickerton, S., Kelly, P.A.: Prediction and experimental verification of normal stress distributions on mould tools during Liquid Composite Moulding. Composites Part A: Applied Science and Manufacturing 43(1), 138–149 (2012)
8. Eichfelder, G.: Multiobjective bilevel optimization. Mathematical Programming 123(2), 419–449 (2010)

9. Deb, K., Sinha, A.: An Efficient and Accurate Solution Methodology for Bilevel Multi-Objective Programming Problems Using a Hybrid Evolutionary-Local-Search Algorithm. Evolutionary Computation 18(3), 403–449 (2010)
10. Das, I.: On characterizing the "knee" of the Pareto curve based on Normal-Boundary Intersection. Structural and Multidisciplinary Optimization 18(2-3), 107–115 (1999)
11. Bechikh, S., Said, L.B., Ghédira, K.: Searching for knee regions of the Pareto front using mobile reference points. Soft Computing – A fusion of foundations, methodologies and applications 15(9), 1807–1823 (2011)
12. Deb, K., Sundar, J., Uday, N., Chaudhuri, S.: Reference point based multi-objective optimization using evolutionary algorithms. International Journal of Computational Intelligence Research 2(6), 273–286 (2006)
13. Fahlman, S.E., Lebiere, C.: The Cascade-Correlation Learning Architecture. Carnegie Mellon University, Repository (1990)
14. Gupta, A., Kelly, P.A., Ehrgott, M., Bickerton, S.: A Surrogate mode based evolutionary game-theoretic approach for optimizing non-isothermal compression RTM processes. Composites Science and Technology (under Review)

Multi-criteria Optimization for Parameter Estimation of Physical Models in Combustion Engine Calibration

Susanne Zaglauer and Michael Deflorian

BMW Group,
Petuelring 130, 80788 Munich, Germany
susanne.zaglauer@bmw.de,
michael.deflorian@bmw.de

Abstract. In the automotive industry the calibration of the engine control unit is getting more and more complex because of many various boundary conditions, like the demand on CO_2 and fuel reduction. One important calibration problem is the parameter estimation for the model of the intake system of the combustion engine. This system is modeled by a physically motivated system, which can be parameterized by black-box models, like neural nets and characteristic diagrams, whose parameters must be set by an intelligent optimizer. Further, two contradictory aims must be considered and the engineer expects at the end of the optimization a pareto-front, where he can choose the best settings for the application from the pareto-optimal parameter estimations. To solve this multi-criteria optimization task an evolutionary algorithm is used, which is a combination of a genetic algorithm and an evolutionary strategy. This evolutionary algorithm is like all other stochastic searching methods leaned on the naturally biological evolution. It combines the well-known covariance matrix adaption for the mutation of the individuals with the S-metric selection for the multi-criteria fitness assignment of the individuals. It also improves these combination by the use of many subpopulations, which work parallel on various clusters, and by the use of an intelligent *DoE*-strategy for the initialization of the start individuals. With these improvements the developed evolutionary algorithm can easily fit the model of the intake system to test bed measurements and can provide the user a pareto-optimal set of parameters, on which he can choose on his own that ones, which he find most plausible.

Keywords: Evolutionary algorithms, Multi-criteria optimization, Adaptive algorithm, intake system, Automobile industry, Engine modeling, Global optimization, Simulation, Engine control.

1 Introduction

The automobile manufacturers have to meet the requirements of customers (consumption, dynamic performance) and the legislative body (consumption, emissions). Power train development makes a crucial contribution to reach these

R.C. Purshouse et al. (Eds.): EMO 2013, LNCS 7811, pp. 628–640, 2013.

partly opposed goals. For a given power train it is important to find optimal settings for the engine to reach high efficiency and low emissions. Thus, the calibration process from data acquisition at the test bed to control unit parameterization is crucial for the engine performance.

In the last years the complexity of the combustion engine has grown continually and the development cycles became shorter. The increased flexibility implicates a more challenging and time consuming calibration process and thus new calibration and optimization techniques are required to cope with the grown complexity in reasonable time.

One possible approach to reduce the measurement and calibration effort is to use physical models. A physical model adapted to the measurements can be used to calculate the optimal settings for given control tasks. The challenge is the estimation of the model parameters. Usually it is not straight forward to say which parameters are the best ones and an experienced engineer has to choose the best settings from different parameter sets.

The example discussed in this contribution is the intake system of the combustion engine. The fuel pedal position and the ancillary units determine the required torque to realize the requested performance. Thereby, the torque provided by the engine depends directly on the injected amount of fuel. Modern combustion engines use an air-fuel ratio equal one at most operating points. This means, exactly enough air is provided to completely burn all of the fuel. Therefore, the air provided is proportional to the torque and thus a very important factor that in series-production vehicles cannot be measured. To estimate the available air in the cylinders a physically motivated model of the intake system is used that must be fitted to test bed measurements. Beside the model quality the plausibility of the parameters is essential. Some of the parameters are stored in characteristic diagrams and it is known that the characteristic diagrams have to be smooth to be feasible. On the other hand, smooth characteristic diagrams usually result in inferior models and therefore a tradeoff between model quality and maps smoothness has to be found. The goal is to provide a pareto-front for the engineer such that the best settings can be chosen from the pareto-optimal parameter estimations.

The rest of the paper is organized as follows. First, the model of the intake system is described and the optimization problem is formulated. Then the optimization algorithm used to solve the optimization problem is presented. Next the optimization results are discussed and finally the main results are summarized.

2   Optimization Problem

As mentioned in the introduction, the goal of the optimization is the identification of an accurate model of the air flow through the intake system. Since a direct measurement of the fresh air mass in series-production vehicles is not possible the only way is the use of an accurate model. The physically motivated model of the intake system described in the next section is parameterized by neural nets and characteristic diagrams whose parameters have to be adapted to the measurement data. Since the characteristic maps can be physically interpreted, the model quality is evaluated by the fitting quality of the model and the feasibility of the parameters. Only a model with plausible parameters can be expected to provide accurate estimations of the fresh air mass in any driving situation. Therefore, the optimization process has to provide an accurate model and also alternative solutions so that the calibration engineers are able to choose the best compromise between model fitting and most plausible parameters.

Next, the model is described roughly and then the criteria for plausible parameters are formulated.

2.1   Physically Motivated Model of the Intake System

The model of the intake system in the control unit is a physically motivated observer model that calculates the fresh air in the cylinders using values of the calibration unit and values that can be measured with justifiable effort in any series-production vehicle. These are the valve timing of the intake and exhaust valves, the valve lift of the intake valve, the exhaust gas back pressure, the manifold pressure and the engine speed. The model plays a central role in the control unit calibration because the output of the model is used for example to calculate the injected fuel mass, the start of injection and the ignition timing.

Basically, the model of the intake system describes the relation between the manifold pressure and the relative filling of the cylinders under consideration of the mentioned values. The relative filling is the fraction of the mass of the maximal cylinder volume filled with air under standard condition and the current fresh air mass. The characteristic curve of the relative filling can be explained by superposition of two effects. First, the mechanical work required to pump fresh air in and out of the cylinder is almost proportional to the manifold pressure if the other quantities remain constant and no exhaust-gas turbo charging is taking place. This results in the relation displayed on the left side of figure 1.

The second effect is caused by the residual gases since it is not possible to pump out all the exhaust gas after the combustion without turbo charging. The exhaust gas back pressure and residuals due to the clearance volume inhibits a complete filling of the cylinder with fresh air. These effects are also called reflow. However, the turbo charging can also cause the contrary effect. If the intake and exhaust valve are open simultaneously and the pressure in the intake manifold is higher than in the exhaust manifold, the exhaust gas is scavenged out completely and the fresh air mass in the cylinder is higher than the maximal cylinder volume under standard condition. The offset due to the back pressure and clearance volume and the nonlinearity because of scavenging are the reasons for the characteristic nonlinear curve as displayed on the right side of figure 1.

The resulting characteristic curve (figure 2), which originates from the superpostion of the pump and the overflow/reflow model, can be determined by four parameters. The first parameter is the intake manifold pressure where the curve intersects the axis of abscissa (theoretical value). The second is the slope of the linear part. The other both parameter are the scavenging rate as well as the curvature. The shape of the curve and therefore the parameters depend mainly on

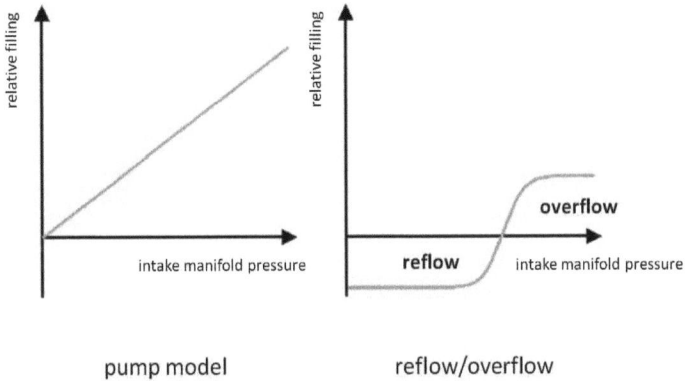

Fig. 1. Superposition of the two effects of the characteristic curve. Left: The linear pump model. Right: The nonlinear overflow and reflow model.

the engine speed, the valve timing of the intake and exhaust valves, the intake valve lift and the exhaust gas back pressure. The dependence of the four parameters of the curve on the described quantities is stored in two characteristic diagrams and two neural networks. To fit the intake model to the measurement data the optimization algorithm has to adapt the characteristic diagrams and the parameters of the neural networks.

The resulting model has the following form:

$$rf = \frac{\theta_1(u)}{p_n - \theta_2(u)} \left(\frac{p_i - \theta_2(u)}{p_n} + \theta_3 \left(\frac{v_1}{v_2} + \theta_4(v) \right) \right), \quad u \in \mathcal{R}^5, \quad v \in \mathcal{R}^2,$$

where rf is the relative filling, p_i and p_n are the intake manifold and ambient pressure, θ are the parameters of the curve and u, v the quantities on which θ depend. Thereby, $\theta_1(u)$, $\theta_2(u)$ are neural networks and $\theta_3(v), \theta_4(v)$ are characteristic diagrams. A more detailed description of the model can be found in [5].

The values of the characteristic diagrams can be interpreted and have to meet the criteria described in the next paragraph to be plausible.

2.2 Criteria for Physical Plausible Model Parameters

After the description of the model it is important to define the optimization criteria, which have to be optimized. The first criterion for the optimization algorithm is the model error D in respect of the model output z_M and test bed measurements z_{tr}, which are also called training data. The model error D can be defined as the mean absolutely relative error:

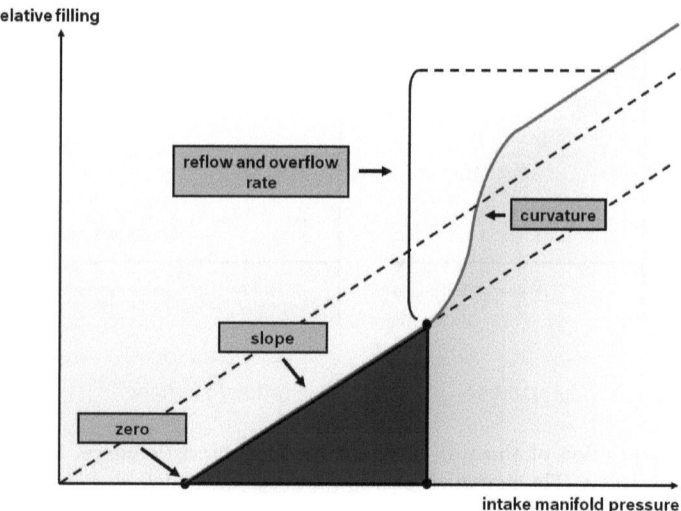

Fig. 2. Characteristic curve of the relation intake manifold pressure and relative filling

$$D(\theta_1, \theta_2, \theta_3, \theta_4) = \frac{1}{N} \sum_{i=1}^{N} \left| \frac{z_{tr}^{(i)} - z_M^{(i)}}{z_{tr}^{(i)}} \right|, \qquad (1)$$

where N is the number of measurements. Beside the model error we have a second criterion to be optimized. It is known that the characteristic diagrams have to be smooth to be feasible, so the smoothness is the second optimization criterion. As reference value for the smoothness we take the energy of the characteristic diagrams.

Definition: For two adjacent matrix dots x_i, x_j and a characteristic map F the connection energy $x_i - x_j$ in the characteristic diagram is given by

$$E_{(i,j)}(F) = \mathbf{1}_{|y^{(i)} - y^{(j)}| > c}$$

where the connection energy is 1, if the y-distance is greater than a constant c and otherwise 0. Without restrictions we set $c = \frac{1}{2}$.

Definition: The energy of the characteristic map F is then given by the sum of all connection energies:

$$E(F) = \sum_{i < j \, adjacent} E_{(i,j)}(F).$$

This definition has a natural interpretation in practice: Is the distance between two adjacent matrix dots greater than a certain limit value, so the corresponding regulating variable cannot be adapted in real time, if a change of the operating point take place with a given speed. Summarizing we have the final optimization problem:

Optimization Problem

$$\min_{\theta_1,\theta_2,\theta_3,\theta_4} \quad D(\theta_1,\theta_2,\theta_3,\theta_4), E(\theta_3,\theta_4) \tag{2}$$

3 Optimization Algorithm

To solve this optimization problem an evolutionary algorithm is used. Evolutionary algorithms are characterized by a multiple of changeable parameters. This includes the choice of a suitable representation of the problem, the various evolutionary operators within the parameters, the right population size and also the weighting function itself. The parameters allow on the one hand a high adaptability of the algorithm to the given problem but on the other hand they can make the algorithm very sensitive about changed properties of the optimization problem. In the following the structure and the design and the various components of the used algorithm for the multi-criteria parameter optimization of the model of the intake system are presented.

3.1 Structure and Design of the Algorithm

As mentioned before the configuration of the parameters is not easy and the parameters cannot usually be understood as individual and independent controllers but they build a connected network. The modification of any parameter has an essential effect on other parameters. Good parameter settings are different for each problem and cannot be transferred to other algorithms with other evolutionary operators. For the given problem the used algorithm is divided into the components initialization, fitness assignment, selection, recombination, mutation, reinsertion and termination (see figure 3), in which a few of the components are taken from literature and a few are developed by ourselves, for example the initialisation. Also this combination of components does not exist in literature and many other additional features are implemented in the algorithm, like a treatment with various constraints.

3.2 Initialisation of the Individuals

At the beginning of the algorithm the individuals of the start population must be calculated. In many other evolutionary algorithms this is done on the basis of a random initialization in the input space. But we have asked ourselves how the measuring points, the initial individuals, should be distributed *efficiently* in the experimental space. The answer to this question is provided by the Design of Experiments (*DoE*). The goal is to identify the connections between target and

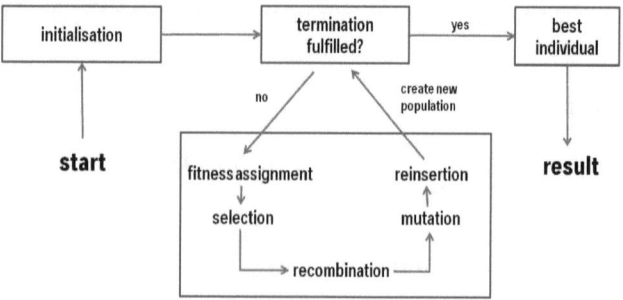

Fig. 3. Structure and design of the evolutionary algorithm

influence factors systematically with as few experiments as possible. This means to achieve a maximum of information of the system under investigation with every measurement. The kind of *DoE* determines the distribution of the initial individuals in the experimental space. In this contribution we explain the Latin-Hypercube and the S-optimal Design, because a combination of both have the best performance in practice (see [6]). They also do not require any derivatives from the technical design problem, which we do not have, and they do not care about the used problem structure and try to cover the input space as equal as possible (space filling designs).

The Latin Hypercube Design. The Latin Hypercube design is a statistical sampling method and in this scheme only one sampling point is in every column and row of a grid. For this purpose one sampling position is placed randomly in every cell along the grid diagonal. After this the rows of the grid are changed in that way, that a chosen criteria, like the maximizing of the minimal distance of the points, is fulfilled:

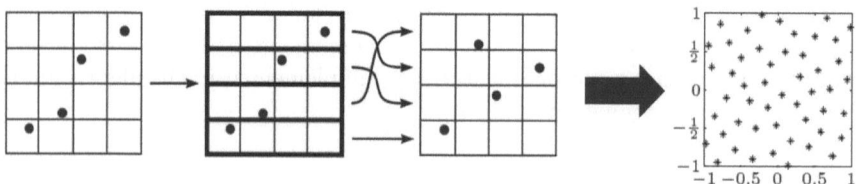

Fig. 4. Representation of the maxmin Latin-Hypercube-Distribution for the choice of the design points

The Latin Hypercube design makes sure that the ensemble of random numbers represents the real oscillations and the points are distributed well in the input space.

S-Optimal Design. The S-optimal design is a distance-based criterion, which is based on the distance $d(x, \mathcal{A})$ from a point x in the p-dimensional Euclidean space $\mathbb{R}^p$ to a set $\mathcal{A} \subset \mathbb{R}^p$. This distance is defined as follows:

$$d(x, \mathcal{A}) = \min_{y \in \mathcal{A}} \|x - y\|,$$

where $\|x - y\|$ is the usual p-dimensional Euclidean distance:

$$\|x - y\| = \sqrt{(x_1 - y_1)^2 + ... + (x_p - y_p)^2}.$$

S-optimality seeks to maximize the harmonic mean distance from each design point to all the other points in the design:

$$\max_{y \in \mathcal{C}} \frac{N_\mathcal{D}}{\sum_{y \in \mathcal{D}} \frac{1}{d(y, \mathcal{D} - y)}},$$

where $\mathcal{C}$ is the set of candidate points and $\mathcal{D}$ is the set of design points. For an S-optimal design, the distances $d(y, \mathcal{D} - y)$ are large, so the points are as spread out as possible. Since the S-optimality criterion depends only on the distances between design points, it is usually easier to compute and optimize than other distance-based criteria, like the U-optimal one, which depends on the distances between all pairs of candidate points.

For the parameter estimation of the model of the intake system a combination between the Latin-Hypercube and the S-optimal design is used. Therefore, the rows of the Latin-Hypercube grid are changed in that way, that the harmonic mean distance from each design point to all the other points in the design is maximized.

3.3 Fitness Assignment

After the initialization the individuals must be evaluated on the basis of a defined fitness function. Therefore the fitness assignment defines how many offsprings every individual produces. Generally the distribution of the fitness values should guarantee, that less good individuals produce no exorbitant number of offsprings and with many good individuals a significant distinction of the fitness values of the good individuals still take place. A robust solution, which is also implemented in the used algorithm, is the nonlinear ranking [1] and in use of multi-criteria problems the S-metric or hypervolume measure [3].

3.4 Selection and Recombination

The direct selection of the individuals is implemented in the next step. At this step, the selection, the individuals, who serve as parents for the next generation, are chosen according to their fitness. The selection probability of an individual is calculated by its fitness normalized by the whole fitness of the selection pool (stochastic universal sampling [2]).

After the selection follows the recombination, in which the information of the parents is combined and the offsprings are created of the parents. For the parameter optimization a discrete recombination is used. For every variable position i it must be chosen from which parent ($par1$ or $par2$) the variable value (var) of the offspring (off) is used:

$$var_i^{off} = var_i^{par1} \cdot a_i + var_i^{par2} \cdot (1 - a_i) \quad i \in (1, 1, ..., N),$$

$a_i \in \{0, 1\}$ with the same probability   and a_i for every i new calculated.

3.5 Mutation

After the recombination the mutation of the offsprings is performed. Therefore, the variables of the offsprings are changed by little disturbances (mutation step). In the used algorithm a mutation of real variables with adaption of the step sizes is realized, which is known as covariance matrix adaption [4]. This mutation method learns the direction and the step size by adaption. To store these step sizes and directions additional variables are attached to the individuals. In addition, this mutation operator works with a huge population, in which only the best individuals produce offsprings. Nearly all parents are replaced by the offsprings, which build the new population. This allows a good evaluation because all offsprings are in the population at least for a short time.

To solve the given multi-criteria optimization problem described in section 2 the covariance matrix adaption is improved through the initialization of a few subpopulations, which find various parts of the pareto-front. At the end of the optimization we have therefore many parts of the pareto-front and all together are a good approximation of it.

3.6 Reinsertion and Termination

After the offsprings are produced and evaluated, they are included into the population. This is done by the reinsertion. This step is particularly important if the offsprings not easily replace the parents. It is decided by the reinsertion which offsprings are reinserted in the population and which individuals of the population are replaced ("'die"'). A selection of the offsprings must only be done if not all produced offsprings are reinserted in the population. The parameter therefore is the reinsertion rate. How many offsprings are produced depends on the parameter generation gap. Both parameters decide how many offsprings are reinserted. Is the reinsertion rate $\geq$ the generation gap, then all offsprings are reinserted in the population. If the reinsertion rate is $<$ the generation gap, then only a part of the offsprings are reinserted.

Then the algorithm continues as long as a fixed termination criterion is fulfilled. The maximum number of generations (or objective value calculations) is the most popular termination criterion in the use of evolutionary algorithms. Its advantage is the good manageability and the guaranteed termination of the optimization.

3.7   Parallelization of the Algorithm

A huge advantage of evolutionary algorithms is the time saving through the great performance for parallelization. One possibility of parallelization is the coarse-grained parallelization, at which more than one subpopulations work parallel on different clusters. The management of the population, the selection, the recombination and the mutation take place for every subpopulation on a different slave processor. On the master processor the migration operator serves to exchange individuals between the subpopulations. This kind of parallelization is simple to implement and is worth if the objective value calculation is very time intensive. It also gives a better performance in terms of diversity and processing time than other parallelization methods like the farming model. In picture 5 the parallelization of various clusters and processor is shown by a multi agent system (MAS). Thereby, this is a system, which consists of many similar or different acting units, which solve the underlying problem collectively.

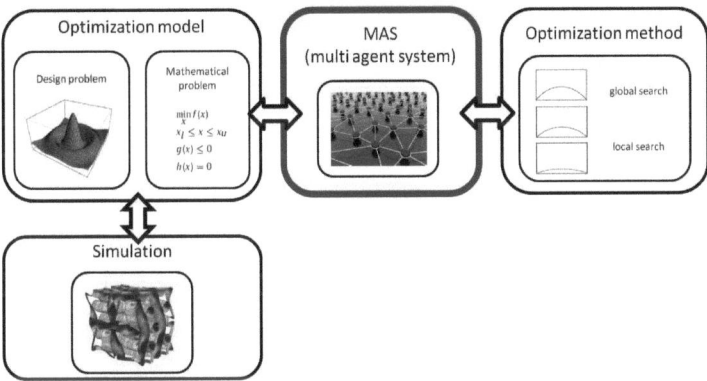

Fig. 5. Schematic Representation of the MAS (multi agent system)

4   Results

In the following the introduced evolutionary algorithm is evaluated by fitting the model for the intake system of the combustion engine to the measurement data. Therefore, the parameters of the neural nets and characteristic diagrams are optimized with the goal of smooth characteristic diagrams and a low model error (see equation (2)). To initialize the model with good parameters, the characteristic maps of a predecessor engine are used. Thus, at first only the parameters of the neural nets are optimized (one-criteria optimization) and then only the characteristic diagrams are fitted (multi-criteria optimization). Finally, the whole model of the intake system with both the neural nets and the characteristic diagrams are optimized (multi-criteria optimization). This three step optimization has proven to be advantageous. Due to this three different optimizations the paramter settings of the optimizer change for each optimization.

In the first optimization the selection pressure is 1, the generation gap is 5 and a population with 5 individuals is used. The optimization is terminated after 2000 generations. In the second and third optimization the selection pressure is 17, the generation gap is 0.9 and 2 subpopulations with 25 individuals each are used. Every subpopulation is calculated parallel on one cluster and has the same components described in the last chapter and after 50 generations they compete with each other and exchange the best individuals. Finally the optimization is terminated after 1000 generations.

After these three optimization runs the user have the choice of many graphical plots. One of these shows the finally pareto front.

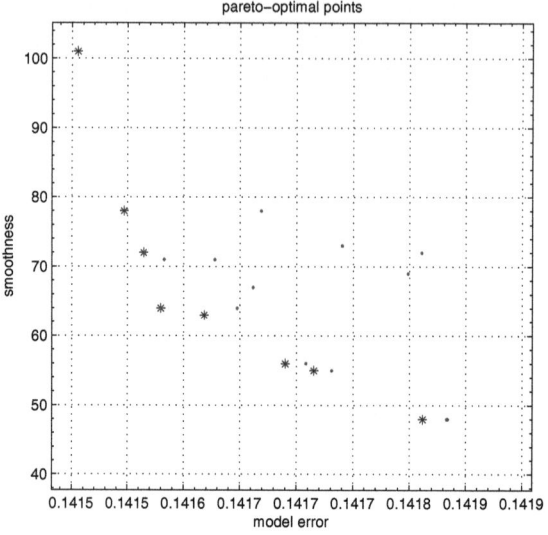

Fig. 6. Pareto-optimal points regarding model error and smoothness

On figure 6 the pareto optimal points of the last optimization, the optimization of all parameters, are presented. The "'stars"' show the pareto optimal points and the "'points" show the dominated individuals of the best individuals per subpopulations. On the x-axis the model error is represented and on the y-axis the energy of the characteristic diagrams is shown. There a low value indicates a high smoothness and vice versa. As you can see, with the subpopulation strategy we achieve a very good diversity of the solution and a good representation of the pareto front. All together we get 16 different pareto optimal points on the pareto front, where now the application engineer can choose the best point for him and his application task.

In figure 7 the true-predicted compares the output of the intake model with the measurement data for the absolute filling. A perfect model would match noise free measurements exactly and all dots would be located on the main diagonal.

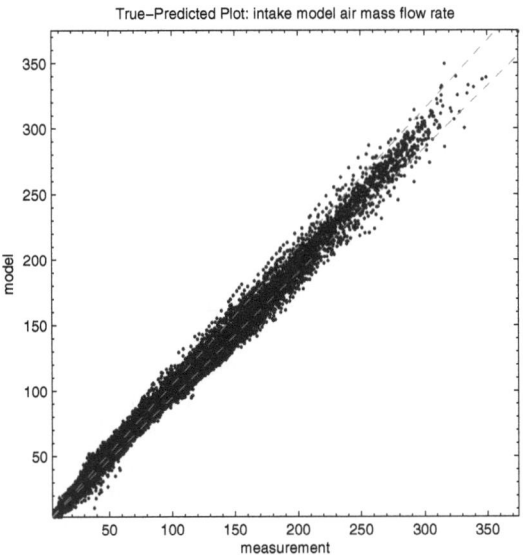

Fig. 7. Representation of the true-predicted plots

In this example the measurement noise is correlated with the measurement and therefore higher measurement errors can be expected if the absolute filling is higher. Thus, the model quality is measured relatively to the measurement values (see also equation (1)). The dashed lines show the 10% tolerance

$$\left| \frac{z_{tr}^{(i)} - z_M^{(i)}}{z_{tr}^{(i)}} \right| < 0.1.$$

This tolerance level specified by the engineers is the basis for the evaluation of the model quality. Within a large region the dots are located on the main diagonal. The relative error for 75% of the data is less than 18% and for 50% less than 9%. This is a quite good result since the measurement data is quite noisy and also outliers are present (in the range 55-65 some measurements are positively biased). For low absolute fillings the measurement noise is higher than the required model quality of 10%. For example, absolute fillings of 5 would require a measurement accuracy of 0.5. This measurement accuracy cannot be reached in practice. As a result of the model error shown in figure 6 is quite high (around 14%). The model accuracy for low relative filling compared to higher filling seems to be worse, but the relative error is clearly higher in the presence of measurement noise for minor filling values.

The good representation of the pareto front and the satisfactory model quality show the performance of the presented optimization algorithm. The presented optimization strategy facilitates the work of the calibration engenieer because he gets a set of pareto-optimal model. Based on his experience he can choose the best model for his application from the pareto-optimal set and is not longer forced to find a good weighting between model error and characteristic diagram smoothness.

5   Conclusion

In this contribution the model of an intake system of a combustion engine is discussed. This model is fitted to test bed measurements whereby the tradeoff between model quality and characteristic diagram smoothness has to be found. This multi-criteria optimization problem is solved by an evolutionary algorithm. The structure and the design of this algorithm are presented and all new components are explained. It combines the well-known covariance matrix adaption for the mutation of the individuals with the S-metric selection for the multi-criteria fitness assignment of the individuals. It also improves these combination by the use of many subpopulations, which work parallel on various clusters, and by the use of an intelligent *DoE*-strategy for the initialization of the start population. The results are very promising and show the good performance of the algoritm. The intake model can be fitted adequately to the test bed measurements under consideration of the smoothness of the characterisitc diagrams. A comparison with other evolutionary algorithms, like MOEA or NSGA-II, is not necessary up to now because the developed algorithm can find robustly the optima of the given problems in practice and the optimizer provides as result a pareto-front for the engineer such that the best settings can be chosen from the pareto-optimal parameter estimations.

References

[1] Bagot, B., Pohlheim, H.: Complementary selection and variation for an efficient multiobjective optimization of complex systems. In: GECCO 2004 - Proceedings of the Genetic and Evolutionary Computation Conference (2004)

[2] Baker, J.: Reducing Bias and Inefficiency in the Selection Algorithm. In: Proceedings of the Second International Conference on Genetic Algorithms and their Application, pp. 14–21. Lawrence Erlabaum Associates, Hillsdale (1987)

[3] Beume, N., Naujoks, B., Emmerich, M.: SMS-EMOA: Multiobjective selection based on dominated hypervolume. European Journal of Operational Research (2006)

[4] Christian Igel, N.H., Roth, S.: Covariance matrix adaptation for multi-objective optimization. Evolutionary Computation 15, 1–28 (2007)

[5] Roithmeier, C.: Virtuelle Applikation von Motorsteuerungsfunktionen am Beispiel der Lasterfassunfsfunktion und der Fahrdynamikfunktionen. PhD thesis, Institut für Kolbenmaschjinen, Karlsruher Institut für Technologie (2011)

[6] Zaglauer, S.: The Evolutionary Algorithm SAMOA with Use of Design of Experiments. In: GECCO 2012, Genetic and Evolutionary Computation Conference (2012)

A Real-World Application of a Many-Objective Optimisation Complexity Reduction Process

Robert J. Lygoe[1], Mark Cary[1], and Peter J. Fleming[2]

[1] Powertrain Calibration & Development, Ford Motor Company, 15/2A-F13-C,
Dunton Technical Centre, Laindon, Basildon, Essex. SS15 6EE, U.K.
{blygoe,mcary}@ford.com
[2] Automatic Control & Systems Engineering, The University of Sheffield,
Mappin Street, Sheffield. S1 3JD, U.K.
p.fleming@sheffield.ac.uk

Abstract. In this paper a real-world automotive engine calibration problem has been distilled into a ten-objective many-objective optimisation problem. The objectives include dynamic measures of combustion quality as well as sensitivity quantities related to a control system actuator, which exhibits significant variation. To address the computational demands of such a high-dimensional problem, use was made of parallel computing. The objective reduction process consisted of four stages and progressively reduced objective dimensionality where evidence of local objective harmony existed. It involved the advice of the calibration engineer at various stages on objective priorities and on whether to discard clusters containing solutions of no apparent interest. This process culminated in two sub-problems, one of three and one of four conflicting objectives. From the corresponding Pareto-optimal populations (POPs), visualisation together with objective priorities was used to identify preferred solutions. A comparison of the resulting POP, preferred solution and an independently generated, manually tuned calibration was made for each of the two sub-problems. In general, the preferred solution outperformed the independent calibration.

Keywords: Many-objective, optimization, decision-making, high-dimensional, objective reduction, engine calibration.

1 Introduction

With ever more stringent emission standards being imposed upon passenger vehicles, calibration approaches that minimise emissions during start and immediately after start or 'run-up' are becomingly increasingly important to vehicle manufacturers [1]. Minimisation of hydrocarbon (HC) emissions during start-up, prior to the exhaust catalyst achieving a sufficiently high working temperature, is of particular significance [2].

Due to the many degrees of freedom associated with typical Gasoline Direct Injection (GDI) controls, significant time and effort is needed to develop an engine start and run-up characterised by: an instantaneous first fire, followed by a stable and

R.C. Purshouse et al. (Eds.): EMO 2013, LNCS 7811, pp. 641–655, 2013.
© Springer-Verlag Berlin Heidelberg 2013

reliable engine run-up, while simultaneously minimising the delivered fuel quantity in an attempt to lower HC emissions. Apart from geometric design variables, such as the spark plug and injector location, it is essential to optimise calibration parameters such as injection pressure, injection and ignition timing.

Ideally, at any temperature, the start profile should remain repeatable, regardless of the background variation or noise, e.g. fuel type, engine age or variations in fuel pressure. Hence, from a user-perspective, the objective is to discover a robust calibration, i.e. one which shows relatively low, or ideally no, variation to this background noise. Such effects can be formulated as noise sensitivities and can conflict with their performance counterparts. This justifies the application of multi-objective optimisation approaches.

Furthermore, the inclusion of robustness measures in the optimisation process significantly increases the number of objectives. To reduce the dimensionality in this problem to a manageable level, engineering knowledge is applied *a priori* to limit the focus to robustness measures associated with the dominant noise variable, which is considered to be fuel pressure. Nevertheless, even with this simplification a constrained ten-objective optimisation problem results.

This paper extends the application of the complexity reduction strategy, proposed in Lygoe et al. [3], to a high-dimensional real-world study, comprising ten objectives and one constraint. Three important enhancements are also introduced. The principle behind the proposed process together with a literature review of relevant approaches is provided in Section 2. Section 3 describes the complexity reduction process with enhancements outlined in Section 4. The optimisation problem formulation and results are presented in Section 5 followed by conclusions and next steps in Section 6.

2   Background

The proposed objective reduction process is based on the principle that for many-objective problems, there may exist local objective harmony or redundancy [4]. In other words, for the Pareto-optimal solutions within the Decision Maker's (DM's) region of interest, there may be objectives which are sufficiently positively correlated.

To reveal any local dependency between objectives, it is necessary to partition or cluster the Pareto-optimal front into groups of like-solutions. A suitable clustering algorithm should be able to efficiently generate the correct number and location of clusters in high-dimensional objective space. The k*-Means algorithm [5] is such an algorithm as evidenced by simulation testing carried out in Lygoe [6].

Subsequently, any objective dependency needs to be identified and quantified to justify potential objective dimension reduction. Surveys of dimension reduction methods exist, *e.g.* Fodor [7], and are categorised into linear and non-linear methods. With regard to linear dimension reduction approaches, Factor Analysis [8] assumes that there is some random error in the data being analysed when there is no such component in the mathematical models, which are evaluated to generate solutions. By contrast, Principal Components Analysis (PCA) does not make such an assumption, but is cited [9] as not being suitable for non-linear data such as that typical of

Pareto-optimal fronts. Higher order methods which allow for non-Gaussian data, can be computationally expensive or may rely on other methods [10]. With non-linear methods, the DM may need to specify additional information such as the non-linear transform required or distributional assumption. In addition, there are known problems with i) Multi-Dimensional Scaling [11] not being able to project onto lower dimensions, with ii) Self-Organising Maps [12], which have issues with subjectivity involved in hierarchical clustering, convergence and interpretation, and with iii) Vector Quantisation, where the DM must specify target dimension a priori and no consideration is given to objective harmony and conflict [13].

PCA may still be useful in identifying local harmony for objective reduction however, if the Pareto-optimal front is first partitioned into groups of like-solutions. This reason, together with its widespread usage and computational efficiency, justified using PCA to help identify any objective redundancy.

3   Process

To place the process enhancements in context, it is necessary to recall the Multi-Objective Optimisation Decision-Making (MOODM) process first published and fully detailed in Lygoe [6], which comprises a number of steps:

1. Generate the Pareto-optimal population (POP).
2. Cluster the POP using the k*-Means algorithm [5]. Verify the number and location of clusters using verification rules.
3. For each cluster, apply Principal Components Analysis (PCA) to identify local harmony and conflict for potential objective reduction. Use heuristic rules including objective priorities, if specified, to retain only the dominant/preferred conflicting objectives.
4. If no objective reduction is achieved, the process terminates.
5. If objective reduction is achieved, then continue optimisation within each cluster with reduced objectives, but constrained by the cluster boundary in an attempt to preserve objective correlations.
6. Go to Step 2   and continue until no further objective reduction is achieved.

A number of observations from the previous application of the MOODM process to the diesel calibration problem [3] have relevance to higher dimensional optimisations:

• This six-objective optimisation involved only one stage of objective reduction. It is possible that for problems with a larger number of objectives, the number of stages increases also. In such a scenario, the application of the clustering verification and objective reduction rules will become lengthy. A more compact form for these rules, which lends itself to being automated, would be useful.
• Higher-dimensional problems may require larger populations to provide effective search. Larger populations in more objectives may generate more clusters. Both place significant demands on computational efficiency. Parallel computing is one approach to address this requirement.

- As the number of objectives increases so does the number of Principal Components (PCs). A PCA on a larger number of objectives may reveal a finer gradation in the percentage of variation represented by the PCs. In other words, it may be possible that the threshold for selecting PCs could be varied slightly from the suggested 95% to retain a different number of PCs and potentially, a different degree of objective reduction.

4   Enhancements

The complexity reduction strategy is extended with several significant enhancements to support higher-dimensional multi-objective problems. These are:

- **The introduction of sensitivity objectives.** These have been added to the problem formulation so that the optimiser simultaneously searches for solutions, which are optimal for performance and which minimise the sensitivity to background noise.
- **Variations on thresholds for reducing the number of objectives.** Varying the threshold used for selecting Principal Components may affect the number of objectives retained using the objective reduction rules. This can provide flexibility in the dimension reduction process.
- **The use of parallel computing methods.** The computational demands on the process under investigation are now sufficiently high to justify a parallel computing approach. A parallel MOEA has been developed to evaluate large populations distributed across a cluster of processors. Batch processing in parallel has also been utilised to accelerate the clustering task.

However, fundamentally the concept of local harmony is exploited to allow various degrees of complexity reduction in several local domains of the POP. The resultant sequence of optimisations, clustering and objective reduction processes enables the decision maker, working in conjunction with an experienced calibration engineer, to propose potential solutions. These results, developed systematically using the methods described, are shown to out-perform the existing calibration developed using empirical approaches.

5   Cold Start Engine Calibration Optimisation

5.1   Implementation of Process Enhancements

More information is now provided on the implementation of the previously described process enhancements with full details available in Lygoe [6].

As with many engineering problems, in engine calibration studies it is desirable to achieve a solution that is not only optimal in some sense, but also robust to variation. For the purposes of this study, a robust, optimal calibration is defined as some optimal trade-off of competing engine responses, for which the solution is relatively insensitive to noise, i.e. piece-to-piece variation in control system sensors and actuators, external environmental factors and customer duty [14]. In practice, it may not be possible to simultaneously achieve optimal performance and low sensitivity to noise and therefore some compromise may be necessary. The primary aim for this problem is to include sensitivity and engine response objectives in the optimisation problem formulation so that both can be simultaneously searched for Pareto-optimal solutions. Requirements to achieve this aim comprise a computationally efficient, easy-to-implement approach, which can be integrated into the optimisation and is sufficiently general for engine calibration optimisation problems. The Direct Derivatives method (e.g. finite difference methods [15]) most closely matches these requirements and is the selected sensitivity analysis approach for this problem. A review of sensitivity analysis approaches together with a definition of the sensitivity functions used is provided in Lygoe [6].

As previously described, the main purpose of using PCA is to reduce the dimensionality of the problem by replacing the objectives by a smaller number of PCs which account for most of the variation. Several rules have been developed to identify this smaller number of PCs including *ad hoc* rules of thumb, rules based on hypothesis tests and statistically-based rules [16]. For this study, a cumulative percentage of variation threshold is used where the number of PCs retained is the smallest number of PCs whose cumulative percentage of variation exceeds this threshold. Varying this threshold can change the number of objectives retained and so it may be possible to reduce this threshold by a relatively small amount (whilst still accounting for most of the variation) to achieve greater objective reduction.

Historically, parallel or distributed computing has been an important initiative in solving time-consuming real-world optimisation problems. The proportion of non-dominated solutions in the Pareto-optimal front becomes large as the number of objectives is increased [17], and the selection pressure correspondingly reduces. In addition, in order to generate a diverse Pareto-optimal front, a large population is required, which can be computationally expensive with serial Multi-Objective Evolutionary Algorithms (MOEAs), but may be much less time-consuming with parallel MOEAs (pMOEAs). There are three broad paradigms for parallelisation: the Master-Slave model, the Island model and the Diffusion model [18]. Due to the fact that a compute cluster was available and the execution speed of the objective functions used is very fast (*i.e.* msecs), the most suitable approach that could be applied to this research is the island-based pMOEA. As a result, a parallelized version of NSGAII was developed. A validation test was carried out on the six-objective problem in Lygoe [3], resulting in good agreement between the POPs from the serial and the parallel NSGAII and a significant speed-up was achieved (~x80) reducing execution time from approximately 21h to 15min [6]. This speed-up was achieved through the parallelisation (20 processors used) and the resulting efficiency of cache speed-up arising from the parallel configuration.

5.2   Problem Formulation

The objective functions used in the cold start case study were based on empirical engine models. These were developed from experimental data taken from a 2-litre in-line four cylinder turbocharged direct injection gasoline passenger car engine. The data comprised a series of cold start tests each from an initial engine coolant temperature of 20 deg. C. After each start, the engine was fully warmed-up to a stabilised temperature to burn off any residual hydrocarbon emissions and fuel in the oil [19]. The engine was then switched off and chilled back down to the initial coolant temperature in preparation for the next start. Fifty-seven starts were conducted as part of a designed experiment with ten validation tests.

The engine test facility used was a dynamic dynamometer encapsulated test cell, a photograph of which is shown in Fig. 1. Such facilities can provide very efficient, cost effective and realistic testing on a rig as opposed to building expensive prototype vehicles, which require specialised vehicle-based test facilities or testing in remote cold climate locations.

The optimisation was formulated as a ten-objective, single constraint problem as follows:

Minimise the objectives listed in Table 1.

Table 1. Objective description

Label	Description	Units
Obj1	Combustion variation metric for cycles 2-5	Bar
Obj2	Combustion variation metric for cycles 6-12	Bar
Obj3	Negative run-up combustion intensity for cycles 2-5	Bar
Obj4	Negative run-up combustion intensity for cycles 6-12	Bar
Obj5	Fuel quantity	Unitless
Obj6	Maximum engine speed flare after start	RPM
Obj7	Absolute value of sensitivity of combustion variation metric for cycles 2-5 to Fuel Pressure	Bar/MPa
Obj8	Absolute value of sensitivity of combustion variation metric for cycles 6-12 to Fuel Pressure	Bar/MPa
Obj9	Absolute value of sensitivity of run-up combustion intensity for cycles 2-5 to Fuel Pressure	Bar/MPa
Obj10	Absolute value of sensitivity of run-up combustion intensity for cycles 6-12 to Fuel Pressure	Bar/MPa

These are subject to a constraint on the mild extrapolation of valid domain or boundary of the models. This model boundary is an envelope wrapped around the boundary of the data used to build the models and allows, in this case, mild extrapolation [20]. The constraint is defined as model boundary ≤ 0.15.

All models had the following inputs, all of which were used as decision variables:

- AIR - inducted air mass flow (kg/h) as controlled by the engine throttle.
- DEC - exponential decay (unitless) in injected fuel quantity.
- SPK2 - crankshaft angle timing (degrees before piston top dead centre) of ignition.
- F - injected fuel quantity, expressed as a factor (unitless).
- FP - fuel pressure (MPa). Limited control on this control system actuator is available during cold start operation.
- EOI - crankshaft angle timing (degrees before piston top dead centre) of end of fuel injection. This was fixed at a value of 75 degrees BTDC from a previous optimisation.

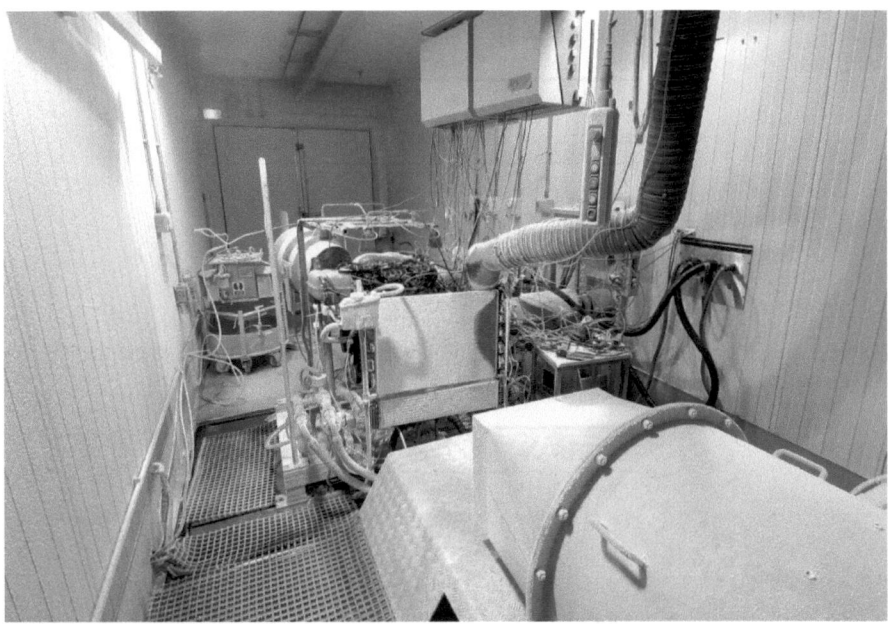

Fig. 1. Encapsulated dynamic dynamometer cell 109 in the West Wing Laboratory at the Ford Dunton Technical Centre, Essex, U.K.

The pMOEA previously described was used with the following options and parameters specified:

- Number of generations: 50000.
- Population size: 20000. (Both of these first two parameters were chosen to provide a reasonable computational effort and a reasonably large initial population is used to allow cluster-based sub-sampling consistent with the cluster verification rules.)
- External archive used and updated every 1 generation.
- Number of migrants: 2% of island population.
- Migration frequency: every 1 generation.
- Selection operator: tournament of size 2.
- Crossover operator: SBX [21] with probability: 0.7 and distribution index: 20.
- Mutation operator: polynomial [22] with probability: 0.17 and distribution index: 20 [23].
- Minimum and maximum range limits on the decision variables are listed in Table 2. These were not explicitly included in the problem formulation. Instead, a so-called *boundary constraint model* (referred to as model_bdry) was incorporated, which represents a convex hull envelope around the data. Beyond this model_bdry, models built from the data are extrapolated.
- Initially, all the objectives were minimised subject to the constraint, which corresponds to the *constrained minimisation* formulation in the Progressive Preference Articulation method of Fonseca and Fleming [24] (PPA$_{FF}$). The resulting initial goals and priorities are shown in Table 3.

Table 2. Decision variable ranges

Decision Variable	Units	Ranges	
		Minimum	Maximum
AIR		25	45
DEC		0	0.104
SPK2	°BTDC	-10	10
F		1.49	3.49
FP	MPa	2.02	3.5

Table 3. Initial goals and priorities for the ten-objective cold start problem, where the last row represents the constraint

Objective	Goal	Priority
Obj1	$-\infty$	1
Obj2	$-\infty$	1
Obj3	0	1
Obj4	0	1
Obj5	0	1
Obj6	0	1
Obj7	0	1
Obj8	0	1
Obj9	0	1
Obj10	0	1
model_bdry	0.15	2

5.3   Results

Results from the optimisation, comprising three stages of successive objective reductions process, are presented here. The analysis involved is summarised in a flowchart representation. For clarity, only the results from the 1st stage of objection reduction after the initial optimisation are described and depicted in the flowchart in Fig. 4. Full details are provided in Lygoe [6]. The resulting POP from the optimisation generated 18,552 solutions, which were robustly clustered, suitable sub-sample clusters determined and then PCA applied to potentially reduce the number of objectives within each cluster. Mathematical notations are used in the clustering and objective reduction blocks, with an example of the former provided.

Clustering and Verification

- **Establish Reference Clusters.** The first step was to cluster the reference POP generated from the initial optimisation. The results can be summarised using the notation, $V_1^{10}(18552, lr, cs, 5000, 0.1) = 4$. That is, the reference clustering analysis in ten objectives, of the reference POP of 18,552, from various learning rates, *lr*, and initial number of clusters, *cs*, maximum iterations of 5000 and a convergence tolerance of 0.1 generated a *reference solution* of four converged clusters.

- **Establish sub-sampled POP size.** Subsequently, the reference solution was randomly sub-sampled per cluster to generate smaller POPs of 10000, 5000, 2000 and 1000. Clustering was run on all of the sub-sampled POPs to test for agreement with the reference solution clusters. As a result, it was decided that the POP of 10,000 was the smallest sub-sampled POP that provided acceptable agreement with the reference POP. This is denoted by:

  - $V_2^1(10000, lr, cs, 5000, 0.1) = [4, 4, 4]$ - a clustering analysis on a sub-sampled POP of 10,000 with various *lr* and *cs* resulted in three alternative solutions of four clusters. A run with the best convergence was selected from those in the first solution, which was the most frequently occurring solution within the batch of clustering runs.
  - $\{_2D_i \subset {}_1D_i\}^4_{i=1} \wedge \Omega_2(0.04) \leq 0.05 \wedge \Phi_2(0.06) \leq 0.1$ – with respect to the corresponding reference clusters ($_1D_i$), the selected sub-sampled 10,000 POP clusters ($_2D_i$) are a subset AND have cluster centres (Ω) in close agreement AND have cluster correlation matrices (Φ) in close agreement.

- **Check cluster bounds.** The only engineering limit that was specified by the Cold Start calibration engineer, was applied to Objective 6 (Peak Flare Speed), where only solutions in the 1300 to 1500rpm range were of interest. The Peak Flare Speed data within Cluster 1 violated this limit and so was discarded, while the other clusters satisfied this limit and were retained.

PCA and Potential Objective Reduction

The objective reduction rules were applied to each cluster to identify any opportunity for potential objective reduction. In the process of discussing and selecting clusters to be retained, the Cold Start calibration engineer provided revised objective priorities and these were taken into account, where necessary, when applying the objective reduction rules. Due to space constraints, a detailed explanation of how these rules are applied cannot be included (see [6] for a full description). Nevertheless, as an example, six objectives were retained (denoted by the set I) in cluster 2: 1, 3, 6, 7, 9 and 10 and therefore, the remainder discarded (denoted by the set E), i.e. objectives: 2, 4, 5 and 8.

The Effect of a Reduced Percentage of Variation on Objective Reduction

Consequently, it was decided to consider two further such scenarios where PCs are retained that account for approximately 90% and 86% of the variation. The results from the objective reduction process for varying thresholds of cumulative percentage of variation are collated in Table 4. It can be seen that while some objective reduction was achieved using a threshold of 95%, using 86% gives significantly more reduction. As a result, it was decided to proceed with retaining the objectives per cluster corresponding to the 86% threshold in subsequent optimisations. This stage culminated in three Clusters (2, 3 and 4) with six, seven and five objectives being retained, respectively.

Table 4. 1st Stage retained objectives from objective reduction process for varying thresholds of cumulative percentage of total variation

Threshold for cumulative % of total variation	No. of objectives retained		
	Cluster 2	Cluster 3	Cluster 4
95	7	10	7
90	6	8	7
86	6	7	5

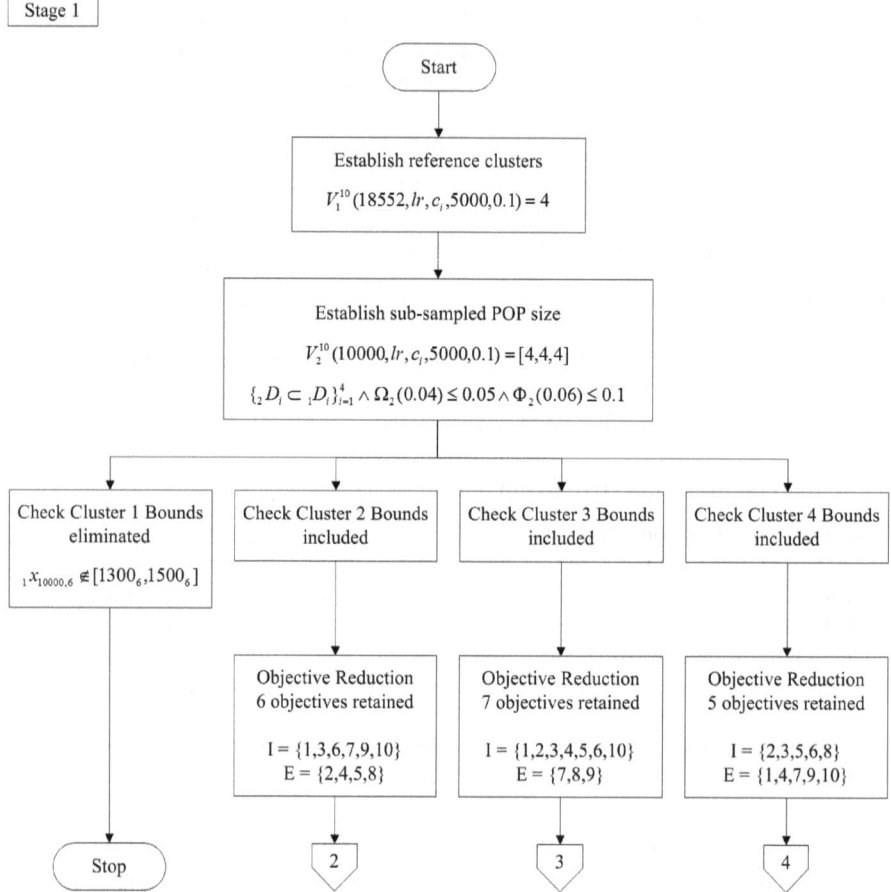

Fig. 2. Flowchart of the results from objective reduction - 1st stage: clusters 2, 3 and 4 are retained leading to 6, 7 and 5 objectives being retained, respectively. This flowchart starts after the first optimisation has been completed.

5.4   Conclusions from the Objective Reduction Process

The results from the final objective reduction (from the 3rd Stage) are displayed in Table 5, which shows the number of objectives retained in each cluster at each stage.

Table 5. Number of objectives retained at each stage of the objective reduction process

Objective reduction Stage	No. of objectives retained			
	Cluster 2	Cluster 3	Cluster 4	
1st	6	7	5	
			Cluster 4_1	Cluster 4_2
2nd	4	5	4	4
3rd	4	4	4	3

A Parallel Coordinates plot of the POPs per cluster generated and resulting from the final objective reductions is shown in Figure 3. This plot was reviewed with the Cold Start calibration engineer and the following conclusions were arrived at:

- In Cluster 3:
  - Obj5 was in the range 1.5-1.6. At these low levels of Fuel quantity, the engine cold start performance was erratic when lower quality fuels (available in some markets) were tested.
  - While Obj2 was relatively high, (combustion intensity for cycles 6-12 was strong), this was at the expense of Obj1, which was comparatively low (weak combustion intensity for cycles 2-5).
  - When the sensitivity objective, Obj10 is plotted next to the objective to which it relates (Obj2), the resulting sensitivity values for Obj10 are in some cases almost as large for those for Obj2, indicating that these solutions show high sensitivity.
- The resulting POPs in Clusters 4_1 and 4_2 display somewhat similar parallel coordinates profiles, which is to be expected given that they have the same parent cluster. Nevertheless, it can be seen that Cluster 4_1 (red) performs worse in the sensitivity objectives than Cluster 4_2 (gold). In this case, more sensitivity means the start performance is less robust to variations in Fuel Pressure, which is not tightly controlled. In a mass-production environment, this variation is likely to increase and may lead to poor customer satisfaction with start performance and potentially, warranty cost.

Consequently, it was decided, in consultation with the calibration engineer, to discard Clusters 3 (green) and 4_1 (red) and to select preferred solutions from the retained Clusters 2 (blue) and 4_2 (gold).

In Cluster 2, of the retained objectives, Obj6 (Peak_Flare_Speed) was the highest priority. A preferred solution (no. 271) that was relatively insensitive as measured by Obj7 and Obj10 and also with a relatively low value of Obj5 (Fuel quantity - a surrogate measure for HC emissions) was selected.

In Cluster 4_2, of the retained objectives, the highest priority objective was Obj5 (Fuel quantity), then Obj6, then Obj8 (sensitivity of cycles 6-12 combustion variation to fuel pressure). A preferred solution (no. 355) that was relatively insensitive as measured by Obj8 was selected.

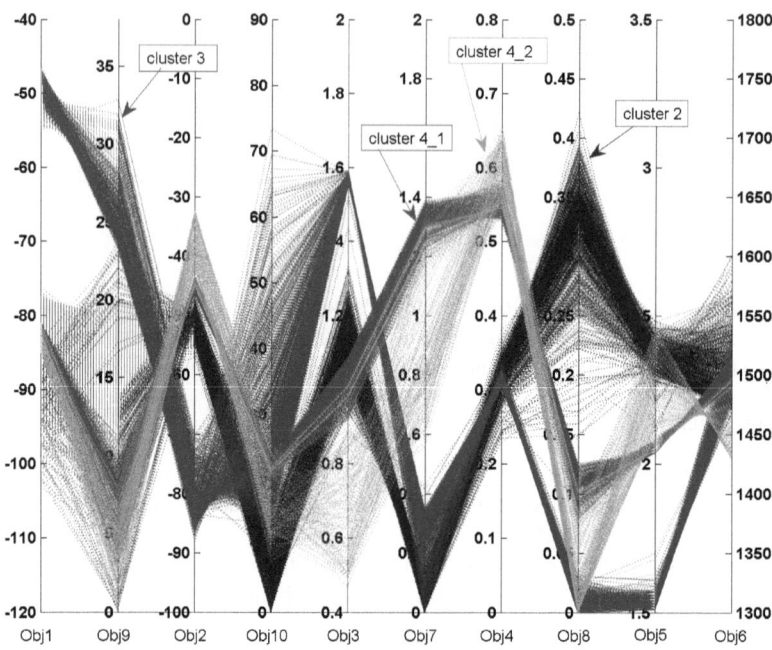

Fig. 3. Parallel Coordinates plot of the final POPs resulting from objective reduction in each cluster. The objectives have been re-ordered so that each sensitivity objective is adjacent to the objective to which it relates[1].

Of further interest was a comparison of these solutions against their respective parent cluster of solutions and against a recent calibration generated by the Cold Start calibration engineer using a manual, iterative tuning process. Parallel Coordinates plots of this data are shown for Cluster 2 and 4_2 in Figs. 6 and 7, respectively. For each figure, the final POP (labelled as Cluster 2 and Cluster 4_2 data) resulting from objective reduction, the selected solution and an independently, manually generated calibration are overlaid. For both clusters, it can be seen that the calibration is inferior with respect to the POP and the selected solutions. The exception is in Cluster 2, where the calibration is slightly better (smaller) than selected solution 271 with respect to Obj1.

---

[1] Colour version of plot available in [6]: Fig. 6.11, p.165.

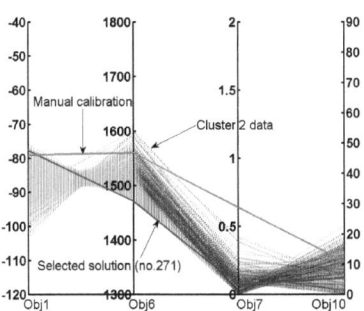 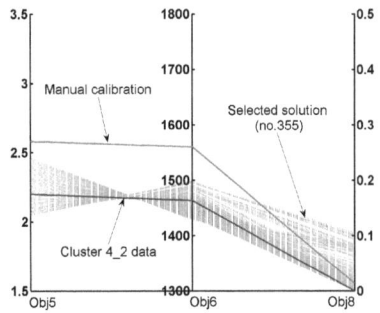

Fig. 6. Cluster 2, selected solution no. 271   **Fig. 7.** Cluster 4_2, selected solution no. 355

In summary, the selected solution is a significant improvement compared to the calibration in respect of:

- In Cluster 2: Obj6 (Peak_Flare_Speed) and Obj7 (combustion variation sensitivity for cycles 2-5). The former is considered to be very important for customer satisfaction with the starting process and the latter indicates much improved start robustness and quality.
- In Cluster 4_2: Obj5 (Fuel quantity) and Obj6 (Peak_Flare_Speed). Fuel quantity is very important with regard to ever-increasing customer expectations of good fuel economy. Also, as fuel quantity has been used as a surrogate for legislated HC emissions, reducing fuel reduces HC. So, in this case, significantly reduced HC emissions is considered especially important as the vast majority of HC emissions are produced before the exhaust after-treatment system (e.g. catalytic convertor) has reached operating temperature, i.e. at, and after, engine start.

6   Conclusions and Future Work

Based on previous work [6,3], the MOODM process has been extended and applied to a real-world automotive engine problem of increased complexity, i.e. that of a ten-objective engine cold start optimisation. The process enhancements comprised defining and embedding sensitivity objectives into the optimisation to yield a robust calibration; application of parallel computing to make the process efficient and the use of varying PC thresholds to explore the potential to achieve greater objective reduction.

In general, the preferred solutions that resulted from this MOODM process compared favourably to those generated from a manual tuning calibration approach.

Although significant objective reduction and favourable optimisation solutions have been realised with this complexity reduction process, there are a number of enhancements which can be made. Improvements to the existing process include:

- Revisit the objective reduction process where there is evidence of independence between objectives. In such scenarios, an increased number of lower dimension optimisations may result.

- Explore alternative migration settings and schemes for the island-based pMOEA implementation.
- While the k*-Means clustering algorithm automatically determines the number of output clusters, its parameters may require a lot of calibration, i.e. lots of runs, to generate converged clusters. Other clustering approaches should be evaluated.
- The two main stages of the search (evolution of an initial population for clustering and further evolution of populations within clusters) have involved many thousands of objective function evaluations. Further research is required to determine whether this significant computational investment is justified or whether fewer objective function evaluations would suffice.

More general many-objective optimisation research opportunities include:

- Implement mathematical notation for the Clustering verification and Objective reduction Rules in software. This will make a high dimensional multi-stage objective reduction process more efficient, less error-prone and potentially fully automated including documentation of results at each stage. Different degrees of objective reduction and different objective priority orders could easily be explored, which may be subject to available computational resources.
- This software could be implemented as a collection of routines underlying a Graphical User Interface in for example, a MATLAB© Toolbox. This could be designed to easily allow alternative optimisation, clustering or dimension reduction algorithms to be 'plugged-in' as well as being able to interface to a parallel computing facility.
- The 'toolbox' software concept could be extended into a more general 'Many-Objective Optimisation' Toolbox, which not only supported the proposed dimension reduction process, but also provided for other processes, based on alternative methods to be implemented. Furthermore, the Toolbox could provide a convenient and easy-to-use capability to compare the results of various alternative approaches to optimising the same problem.

Acknowledgments. The first two authors are grateful to Ford Motor Company Limited for their resources and support to research, develop and apply this MOODM process.

References

1. Wiemer, S., Kubach, H., Spicher, U.: Investigations on the start-up process of a disi engine. Powertrain and Fluid Systems Conference and Exhibition (2007); SAE Technical Paper Series, SAE International, Rosemont, Illinois. SAE paper no.:2007-01-4012.
2. Bielaczyc, P., Merkisz, J.: Exhaust emission from passenger cars during engine cold start and warm-up. In: SAE International Congress & Exposition. SAE International, Detroit (1997)
3. Lygoe, R.J., Cary, M., Fleming, P.J.: A Many-Objective Optimisation Decision-Making Process Applied to Automotive Diesel Engine Calibration. In: Deb, K., Bhattacharya, A., Chakraborti, N., Chakroborty, P., Das, S., Dutta, J., Gupta, S.K., Jain, A., Aggarwal, V., Branke, J., Louis, S.J., Tan, K.C. (eds.) SEAL 2010. LNCS, vol. 6457, pp. 638–646. Springer, Heidelberg (2010)

4. Purshouse, R.C.: On the Evolutionary Optimisation of Many Objectives. PhD thesis. University of Sheffield, Sheffield, UK (2003)
5. Cheung, Y.M.: k*-Means: A New Generalized k-Means Clustering Algorithm. Pattern Recognition Letters 24(15), 2883–2893 (2003)
6. Lygoe, R.J.: Complexity reduction in high-dimensional multi-objective optimisation. Ph.D. thesis. University of Sheffield, Sheffield, U.K (2010),
   http://delta.cs.cinvestav.mx/~ccoello/EMOO/
   thesis-lygoe.pdf.gz
7. Fodor, I.K.: A survey of dimension reduction techniques, Technical report, Center for Applied Scientific Computing, Lawrence Livermore National Laboratory (2002)
8. Kendall, M.: Multivariate Analysis. Charles Griffin & Co. (1975)
9. Saxena, D.K., Deb, K.: Non-linear Dimensionality Reduction Procedures for Certain Large-Dimensional Multi-objective Optimization Problems: Employing Correntropy and a Novel Maximum Variance Unfolding. In: Obayashi, S., Deb, K., Poloni, C., Hiroyasu, T., Murata, T. (eds.) EMO 2007. LNCS, vol. 4403, pp. 772–787. Springer, Heidelberg (2007)
10. Hyvärinen, A.: Survey on Independant Component Analysis. Neural Computing Surveys 2, 94–128 (1999)
11. Morrison, A., Ross, G., Chalmers, M.: Fast Multidimensional Scaling through Sampling, Springs and Interpolation. Information Visualization 2(1), 68–77 (2003)
12. Kohonen, T.: Self-Organizing Maps. Springer, Berlin (1995)
13. Kambhatla, N., Leen, T.K.: Dimension reduction by local principal component analysis. Neural Computation 9(7), 1493–1516 (1997)
14. Davis, T.P., Grove, D.M.: Engineering Quality and Experimental Design. Longman Scientific and Technical (1992)
15. Delinchant, B., Wurtz, F., Atienza, E.: Reducing sensitivity analysis time-cost of compound model. IEEE Transactions on Magnetics 40(2) (2004)
16. Jolliffe, I.T.: Principal Component Analysis, 2nd edn. Springer, New York (2002)
17. Deb, K., Zope, P., Jain, A.: Distributed Computing of Pareto-Optimal Solutions with Evolutionary Algorithms. In: Fonseca, C.M., Fleming, P.J., Zitzler, E., Deb, K., Thiele, L. (eds.) EMO 2003. LNCS, vol. 2632, pp. 534–549. Springer, Heidelberg (2003)
18. Coello, C.A.C., Lamont, G.B., Veldhuizen, D.A.V.: Evolutionary Algorithms for Solving Multi-Objective Problems, 2nd edn. Springer (2007)
19. Heywood, J.B.: Internal Combustion Engine Fundamentals. McGraw-Hill Series in Mechanical Engineering. McGraw-Hill, Singapore (1988)
20. MathWorks: Model-Based Calibration Toolbox™: Model Browser User's Guide. The MathWorks, Inc. (2008a)
21. Deb, K., Agrawal, R.B.: Simulated Binary Crossover for Continuous Search Space. Complex Systems 9(2), 115–148 (1995)
22. Deb, K., Goyal, M.: A Combined Genetic Adaptive Search (GeneAS) for Engineering Design. Computer Science and Informatics 26(4), 30–45 (1996)
23. Khare, V., Yao, X., Deb, K.: Performance Scaling of Multi-objective Evolutionary Algorithms. In: Fonseca, C.M., Fleming, P.J., Zitzler, E., Deb, K., Thiele, L. (eds.) EMO 2003. LNCS, vol. 2632, pp. 376–390. Springer, Heidelberg (2003)
24. Fonseca, C.M., Fleming, P.J.: Multiobjective Optimization and Multiple Constraint Handling with Evolutionary Algorithms — Part I: A Unified Formulation. IEEE Transactions on Systems, Man, and Cybernetics, Part A: Systems and Humans 28(1), 26–37 (1998a)

Knowledge Extraction for Structural Design of Regional Jet Horizontal Tail Using Multi-Objective Design Exploration (MODE)

Hiroyuki Morino[1] and Shigeru Obayashi[2]

[1] Mitsubishi Aircraft Corporation, 455-8555 Nagoya, Japan
hiroyuki_morino@mitsubishiaircraft.com
[2] Institute of Fluid Science, Tohoku University, 980-8577 Sendai, Japan
obayashi@ifs.tohoku.ac.jp

Abstract. An multi-objective optimization (MOO) and knowledge extraction were conducted for a structural design of a regional jet horizontal tail using Multi-Objective Design Exploration (MODE). MODE reveals the structure of the design space from the trade-off information and visualizes it as a panorama for a Decision Maker. The present form of MODE consists of Kriging model, Adaptive Range Multi-Objective Genetic Algorithm and Self-Organizing Map (SOM). Combination between the stringer-pitch and the rib-pitch was optimized based on detailed buckling evaluation and MSC/NASTRAN static analysis using realistic aircraft structural model. The resulting Kriging model provided several solutions with improvements, both in the structural weight and the number of structural components, compared with the baseline design. Furthermore, SOM divided the design space into clusters with specific design features. The acquired design knowledge from the present application has been utilized for the horizontal tail design of Mitsubishi Regional Jet (MRJ).

1 Introduction

Recently, an multidisciplinary design optimization (MDO) has become one of the essential tools for aircraft design. A typical MDO problem involves multiple competing objectives. While single objective problems may have a unique optimal solution, multi-objective problems (MOPs) have a set of compromising solutions, largely known as the trade-off surface, Pareto-optimal solutions or non-dominated solutions. These solutions reveal trade-off information among different objectives. They are optimal in the sense that no other solutions in the search space are superior to them when all objectives are taken into consideration. A designer will be able to choose a final design with further considerations. Multi-objective optimization obtains only non-dominated solutions. However, it is essential for designers to have information regarding the design space, such as the relations between design variables and objective functions. The design information directly helps the designer determine the next geometry.

R.C. Purshouse et al. (Eds.): EMO 2013, LNCS 7811, pp. 656–668, 2013.

An MDO system denoted multi-objective design exploration (MODE) was proposed in [1] and is illustrated in Fig. 1. The MODE is not intended to provide an optimal solution. The MODE reveals the structure of the design space from trade-off information and visualizes it as a panorama for a decision maker. The form of MODE in [1] consists of the Kriging model [2–4], Adaptive Range Multi Objective Genetic Algorithms (ARMOGAs), analysis of variance and a self-organizing map (SOM) [5]. An ARMOGA is one of Evolutionary algorithms (EAs) [6] suitable for finding many Pareto-optimal solutions. To alleviate the computational burden of EAs, the Kriging model has been introduced as a surrogate model [7]. An SOM divides the design space into clusters. Each cluster represents a set of designs containing specific design features. A designer may find an interesting cluster with good design features. Such design features are composed of a combination of design variables. If a particular combination of design variables is identified as a sufficient condition belonging to a cluster of interest, it can be considered as a design rule. Obayashi and Sasaki investigated design tradeoffs for two multi-objective aerodynamic design problems of supersonic transport by using visualization and cluster analysis of the non-dominated solutions based on the SOMs [8]. They successfully revealed correlation of the cluster of design variables with aerodynamic objective functions and their relative importance. However, actual aircraft design requires not only aerodynamic but also structural evaluation since pure aerodynamic optimization provides wings with a low-thickness-to-chord ratio and a high aspect ratio, suffering lack of strength and undesirable aeroelastic phenomena from the low bending and torsional stiffness.

In 2003, Mitsubishi Heavy Industries, Ltd. (MHI) started an R&D project to develop an environmentally friendly high performance small jet aircraft. The purpose of this project was to build a prototype aircraft using advanced technologies, such as low-drag wing design, and lightweight composite structures, which were necessary for the reduction of environmental burdens. In March 2008, MHI decided to bring this conceptual aircraft into commercial use. This commercial jet aircraft, named the Mitsubishi Regional Jet (MRJ, Fig. 2), has a capacity of about 70-90 passengers. This project focused on environmental issues, such as reduction of exhaust emissions and noise. Moreover, in order to bring the jet to market, lower-cost development methods using computer-aided design were also employed in this project. Under this project, Tohoku University participated as a collaborator and applied the MODE approach to the wing design [9, 10] and the engine-wing integration [11]. In these applications, not only aerodynamic performance such as aerodynamic drag under cruising conditions but also structural weight were optimized with constraints of strength and flutter requirements, and the useful knowledge regarding aerodynamic and structural wing design was successfully extracted. However, structural design knowledge obtained in these applications was not so practical as to be directly utilized in the actual aircraft structural design because the wing structural models used in these applications were simplified, where the skin and the stringer were integrated and treated as an equivalent plate element. Moreover, in these applications, strength evaluation was also simplified, where compressive and tensile stress on the skin-stringer equivalent plate, spar plate and rib plate were roughly evaluated without considering detailed buckling modes that were important for actual aircraft structural design [12, 13].

In this paper, the MODE approach based on detailed buckling evaluation using realistic aircraft structural model is presented and applied to a structural design of regional jet horizontal tail in order to obtain knowledge that can be directly utilized in the actual aircraft structural design. In the present study, combination between the stringer-pitch and the rib-pitch is optimized for two objective functions, along with interactive criteria for various buckling modes. This multi-criterion optimization requires the capability of finding global optimal solutions and it will provide a good application field for Evolutionary Multi-Criterion Optimization (EMO).

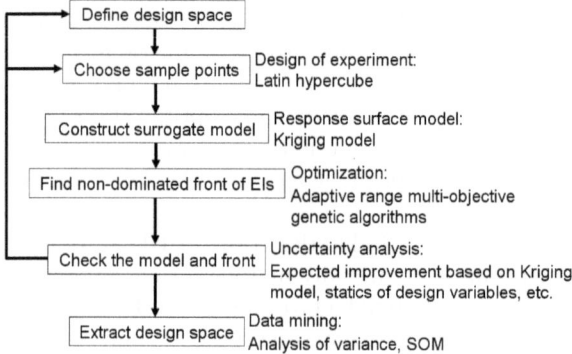

Fig. 1. Flowchart of MODE with component algorithms

Fig. 2. Mitsubishi Regional Jet (MRJ)

2   Multi-Objective Structural Design Exploration for the Regional Jet Horizontal Tail

The MODE approach shown in Fig. 1 was applied to a structural design optimization of a regional jet horizontal tail. It should be noted that the optimized tail is not the exact MRJ tail; rather, the acquired design knowledge from the present application has been utilized for the MRJ tail design. The main steps of the present application are as follows:

1. Define a design space (design parameters, objective functions and constrains, etc.).

2. Choose initial sample points using Latin hypercube method.

3. Calculate objective functions for each sample point based on a strength evaluation and then construct the Kriging model. The correlation function used in the current Kriging model [2] is the Gaussian function. From the theoretical point of view, the correlation function in Kriging model should be determined according to the variogram of sample points. However, the variogram analysis requires tremendous number of sample points. Thus, in real-world application, the Gaussian function is widely used as correlation function of Kriging model without the variogram analysis [14]. The strength evaluation used in the present application is described in detail in later section.

4. Find the non-dominated front of expected improvements using an ARMOGA developed by Sasaki et. al [15]. ARMOGA can find non-dominated solutions efficiently because of the focused search in design space, while maintaining diversity. The population size is set to sixteen in one generation and the population is re-initialized at every five generations for the range adaptation. The total evolutionary computation of 200 generations is carried out in the present application.

5. Check the Kriging model and the front, and then choose additional samples if necessary to improve the Kriging model accuracy.

6. Extract the design knowledge from the design space based on the Kriging model using a SOM.

2.1   Definition of Optimization Problem

The present application is a structural optimization for a regional jet horizontal tail made of composite materials. Use of composite materials is gaining great importance in recent commercial aircraft to reduce airframe weight and to lower the maintenance costs. As for MRJ, composite materials are used for the empennage and the control surfaces, and the weight fraction of composite structure is about 12% of the aircraft. Combination between stringer-pitch and rib-pitch is a key issue in aircraft structural design in order to realize not only weight reduction but also decrease of number of structural components for manufacturing cost saving. To find the best combination between them, the structural optimization has been conducted based on realistic strength evaluation with considering various buckling modes. The following design objectives are considered here.

– Objective functions

  Minimize

  - Structural weight of the horizontal tail
  - Number of the main structural components (= Number of stringers + Number of ribs)

– Design variables (see Fig. 3)
  - Stringer-pitch = 1 variable
  - Rib-pitch = 1 variable

  2 variables in total

– Assumptions for structural sizing
  - Material properties
    * 1 ply thickness = 0.188 mm
    * Density = 1.6×10^{-6} kgf/mm^3
    * Poisson's ratio = 0.34
    * $E_L = 14,131$ kgf/mm^2, $E_T = 773$ kgf/mm^2, $G_{LT} = 471$ kgf/mm^2 where E_L , E_T and G_{LT} are longitudinal, transverse elastic modulus and longitudinal shear modulus, respectively.

  - Laminate configurations
    * Stacking sequence:
      $[0°/\pm 45°/90°] = [X\%/Y\%/Z\%]$ = specified value
    * Stiffness ratio:
      $EA_{skin}/(EA_{skin} + EA_{stringer})$ = specified value
      where EA_{skin} and $EA_{stringer}$ are cross-sectional stiffness of skin and stringer, respectively.

    * Stringer configuration: (see Fig. 4)
      T-type, $t_b = t_w$, W_b = specified value, H_{in} = specified value

  - Sizing criteria
    * Material allowable strain: ϵ_c = specified value
    * Assumed buckling modes:
      Euler and skin buckling, stringer crippling, stiffened panel buckling, spar web shear buckling

– Constraints
  - Stringer thickness = $t_w = t_b >$ specified value
  - Strain margin > specified value

It should be noted that the present optimization was simplified by assuming that the aerodynamic shape and design loads were previously given and not changed in the optimization process. The design loads were estimated based on gust and maneuver load computations and the SMT (Shear, Moment, Torque) loads introduced into the tail-box structure were obtained. Fig. 5 and Fig. 6 present the load definition and the resulting design load distributions, respectively. As presented in Fig. 6, seven load conditions were applied to the present application.

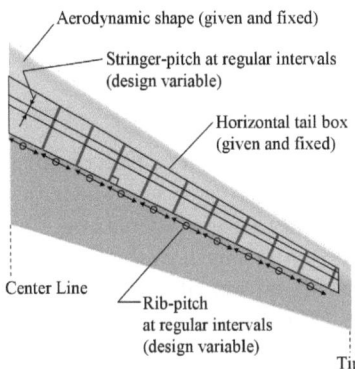

Fig. 3. Design variables for the tail optimization

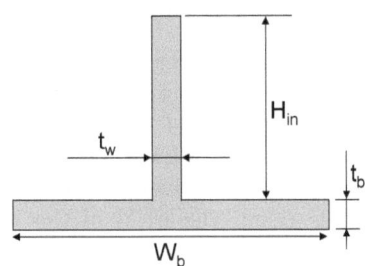

Fig. 4. Cross-sectional definition of T-type stringer

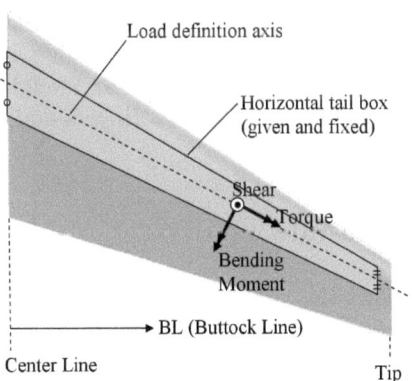

Fig. 5. Design load definition

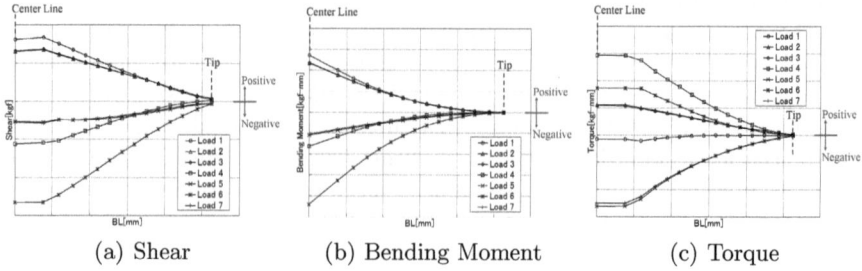

(a) Shear   (b) Bending Moment   (c) Torque

Fig. 6. Design load distributions

2.2   Strength Evaluation

In the present strength evaluation, main buckling modes such as Euler and skin buckling, stringer crippling, stiffened panel buckling and spar web shear buckling, are considered based on a static analysis using detailed structural Finite Element Model (FEM). The flowchart of the present strength evaluation is presented in Fig. 7. Given the design parameters (stringer and rib pitches) for each sample point, the FEM of the horizontal tail (Fig. 8) is generated automatically by a model generator and then automated structural sizing is conducted to realize minimum weight with constraints imposed on strain, buckling and minimum thickness using both MSC/NASTRAN [16] static analysis (Fig. 9) and a structural sizing code. The strain and buckling strength for each component are evaluated from the internal stress computed by the static analysis together with sizing criteria for spar web shear buckling [17] and the other buckling modes [12, 13]. In the sizing process, thickness of each structural component such as skin, stringer, rib and spar is increased or decreased, depending on whether each strength margin is smaller or larger than required value. Iterating the sizing process together with updating the FEM, thickness and weight of the each structural component are converged. The numbers of stringers and ribs are also evaluated in the automated FEM generation.

3   Optimization Results

During the optimization, the update of the Kriging model was performed twice. A total of 36 sample points were used. Fig. 10(a) and Fig. 10(b) present the resulting response surface model of the weight and the expected improvement of the weight, respectively. Expected improvement of the weight in Fig. 10(b) showed maximum value around minimum weight region in Fig. 10(a) indicating that the Kriging model was well constructed using the sample points. Fig. 11 shows the objective functions of the baseline design and those of additional

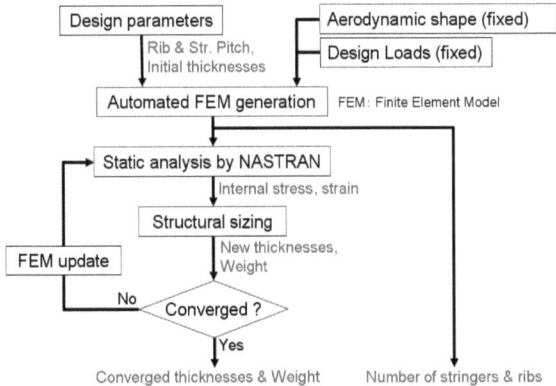

Fig. 7. Flowchart of the present strength evaluation

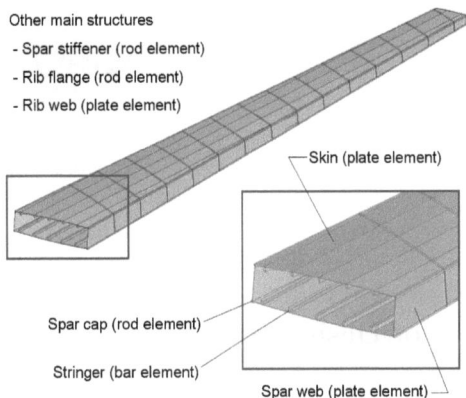

Fig. 8. FEM of the regional jet horizontal tail

sample points at every Kriging model update. As the update progressed, sample points moved toward the optimum direction indicating that the additional sample points for update were selected successfully. Several solutions with improvements in both objective function values compared with the baseline design were obtained. One of the solutions was improved by 5.5 kg in weight and by 10 in number of the structural components compared with those of the baseline design.

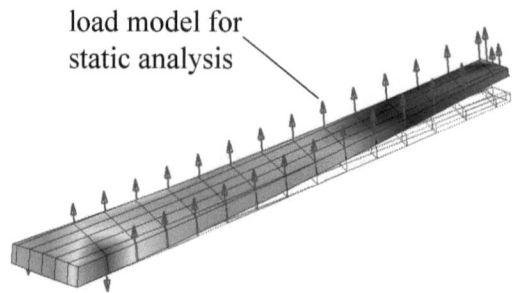

Fig. 9. NASTRAN static analysis for structural sizing

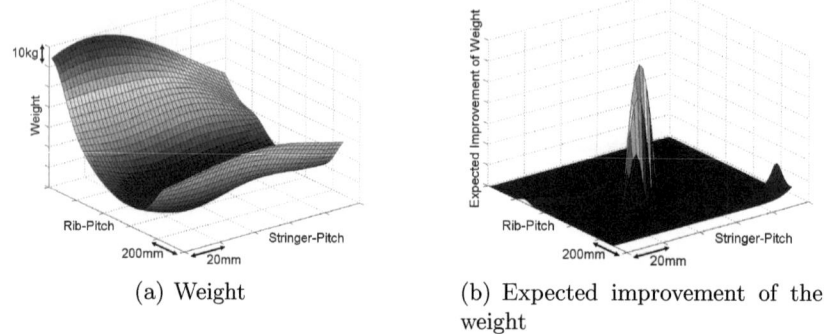

(a) Weight

(b) Expected improvement of the weight

Fig. 10. Response surface model for design variables

4   Visualization of Design Space

In order to visualize the design space, SOMs proposed by [5] were employed. The following SOMs were generated by Viscovery SOMine (http://www.viscovery.net/somine. accessed July 16, 2012). Once a user specifies the size of the map, this software automatically initializes the map based on the first two principle axes. The aspect ratio of the map is also determined according to the ratio of the corresponding principle components. The size of the map is usually 2000 neurons, which provides a reasonable resolution within a reasonable computational time.

Solutions uniformly sampled from the design space were projected onto the two-dimensional SOM. In the present application, the SOMs were calculated on the Kriging model. Fig. 12 shows the resulting SOM with 8 clusters considering not only total weight (one of the objective functions) but also inboard and outboard weight of the horizontal tail. It should be noted that in the present study, focusing on extracting design knowledge regarding weight reduction, the second objective function (number of the structural components) was

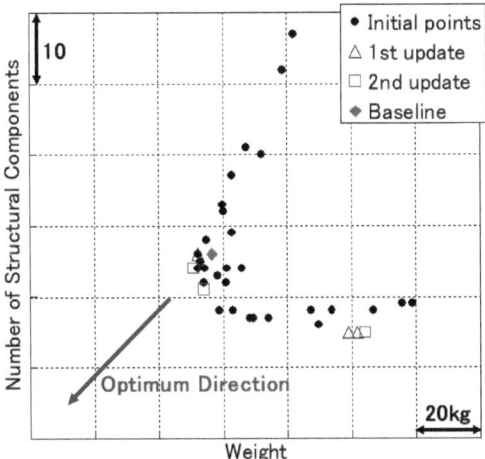

Fig. 11. Comparison of objective function among the baseline and sample points through Kriging updates

not considered for the SOM generation. Fig. 13 shows the same SOM colored by the total, the inboard and the outboard weight. These color figures show that the SOM indicated in Fig. 12 can be grouped as follows:

– The region between the center and the right area corresponds to the designs containing light total weight and light inboard weight.
– The left region corresponds to designs with light outboard weight.

Fig. 14 shows the same SOM colored by the stringer-pitch and the rib-pitch. In Fig. 14(a) colored by the stringer-pitch, the yellow-green-colored pitch values can be found in the middle area across the left and the right side and they cover the whole areas of the light total, inboard and outboard weight. This indicates that an optimum stringer-pitch can be selected around this yellow-green-colored-values to minimize the inboard and the outboard weight simultaneously. Furthermore, in Fig. 14(b) colored by the rib-pitch, smaller and larger rib-pitch values can be found at the right and the left side, respectively. These areas correspond to the light inboard and the light outboard weight, respectively. This signifies that smaller and larger rib-pitch are effective to reduce the inboard and the outboard weight, respectively. This design knowledge suggests that larger outboard rib-pitch compared to the inboard rib-pitch could lead to additional weight reduction from the present design candidates whose ribs are placed at regular intervals.

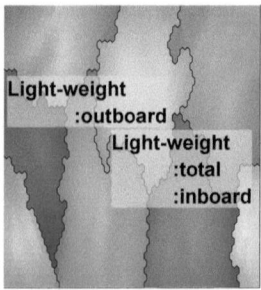

Fig. 12. Self-organizing map based on the total, the inboard and the outboard weight of the horizontal tail uniformly sampled from the design space

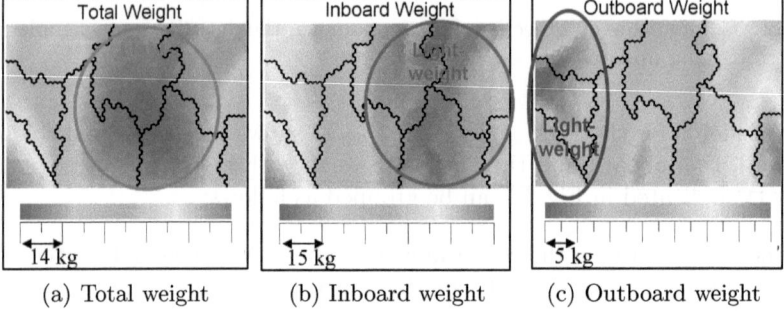

(a) Total weight   (b) Inboard weight   (c) Outboard weight

Fig. 13. Self-organizing map based on the total, the inboard and the outboard weight of the horizontal tail colored by each weight

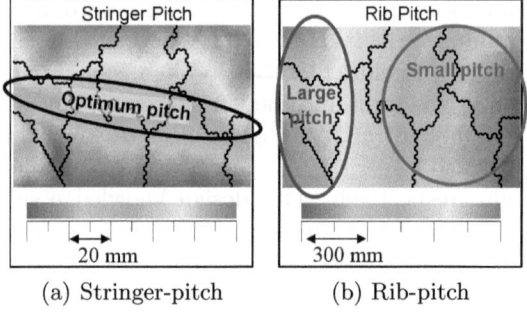

(a) Stringer-pitch   (b) Rib-pitch

Fig. 14. Self-organizing map based on the total, the inboard and the outboard weight of the horizontal tail colored by stringer-pitch and rib-pitch

5   Conclusions

We discussed the application of MODE to the structural design optimization for the regional jet horizontal tail. The resulting Kriging model provided several solutions with improvements, both in the structural weight and the number of structural components, compared with the baseline design. Furthermore, visual data mining for the design space was performed using SOM. SOM divided the design space into clusters with specific design features. As for the stringer-pitch, the particular pitch size has been found effective to reduce both the inboard and the outboard weight. On the other hand, the smaller and the larger rib-pitch has been effective to reduce the inboard and the outboard weight, respectively.

The acquired design knowledge from the present application had been utilized in the MRJ horizontal tail design.

References

1. Obayashi, S., Jeong, S., Chiba, K.: Multi-objective design exploration for aerodynamic configurations. In: AIAA-2005-4666, 35th AIAA Fluid Dynamics Conference and Exhibit, Tronto, Ontario (2005)
2. Jeong, S., Obayashi, S.: Efficient global optimization (ego) for multi-objective problem and data mining. In: IEEE Congress on Evolutionary Computation, Edinburgh, Scotland, pp. 2138–2145 (2005)
3. Jones, D.R., Schonlau, M., Welch, W.J.: Efficient global optimization of expensive black-box functions. Journal of Global Optimization 13, 455–492 (1998)
4. Keane, A.J.: Wing optimization using design of experiment, response surface, and data fusion methods. Journal of Aircraft 40, 741–750 (2003)
5. Kohonen, T.: Self-Organizing Maps. Springer (1995)
6. Deb, K.: Multi-Objective Optimization Using Evolutionary Algorithms. John Wiley & Sons (2001)
7. Queipo, N.V., Haftka, R.T., Shyy, W., Goel, T., Vaidyanathan, R., Tucker, P.K.: Surrogate-based analysis and optimization. Progress in Aerospace Sciences 41, 1–28 (2005)
8. Obayashi, S., Sasaki, D.: Visualization and Data Mining of Pareto Solutions Using Self-Organizing Map. In: Fonseca, C.M., Fleming, P.J., Zitzler, E., Deb, K., Thiele, L. (eds.) EMO 2003. LNCS, vol. 2632, pp. 796–809. Springer, Heidelberg (2003)
9. Chiba, K., Oyama, A., Obayashi, S., Nakahashi, K., Morino, H.: Multidisciplinary design optimization and data mining for transonic regional-jet wing. Journal of Aircraft 44, 1100–1112 (2007)
10. Kumano, T., Jeong, S., Obayashi, S., Ito, Y., Hatanaka, K., Morino, H.: Multidisciplinary design optimization of wing shape for a small jet aircraft using kriging model. In: AIAA-2006-0932, 44th AIAA Aerospace Sciences Meeting, Rene, Nevada (2006)
11. Kumano, T., Jeong, S., Obayashi, S., Ito, Y., Hatanaka, K., Morino, H.: Multidisciplinary design optimization of wing shape with nacelle and pylon. In: European Conference on Computational Fluid Dynamics (ECCOMAS CFD 2006), Egmond aan Zee, The Netherlands (2006)
12. Niu, M.C.Y.: Airframe Stress Analysis & Sizing, 2nd edn. Adaso Adastra Engineering Center (1998)

13. Niu, M.C.Y.: Airframe Structural Design: Practical Design Information and Data on Aircraft Structures, 2nd edn. Adaso Adastra Engineering Center (1999)
14. Isaaks, E.H., Srivastava, R.M.: An Introduction to Applied Geostatistics. Oxford University Press, USA (1990)
15. Sasaki, D., Obayashi, S.: Efficient search for tradeoffs by adaptive range multi-objective genetic algorithms. J. Aerospace Comput. Inform. Commun. 2, 44–64 (2005)
16. Lee, J.M.: MSC/NASTRAN Version 69+ Linear Static Analysis User's Guide. The MacNeal-Schwendler Corporation (1997)
17. ESDU80023: Buckling of Rectangular Specially Orthotropic Plates. Engineering Sciences Data Unit

Application of the MOAA to Satellite Constellation Refueling Optimization

Valerio Lattarulo[1], Jin Zhang[2], and Geoffrey T. Parks[1]

[1] University of Cambridge
Department of Engineering
Engineering Design Centre
Cambridge, Trumpington Street, CB2 1PZ, UK
{vl261,gtp10}@cam.ac.uk
[2] National University of Defense Technology
Changsha, People's Republic of China, 410073
zhangjinxy@yahoo.com.cn

Abstract. This paper presents a satellite constellation refueling optimization problem. The design variables, composed of both serial integers and real numbers, are the refueling sequence, service time and orbital transfer time, while the objectives are the mean mission completion time and propellant consumed by orbital maneuvers. The problem is solved by a mixed-integer version of the MOAA, a recently introduced multi-objective variant of the Alliance Algorithm. This approach is compared, using the epsilon and hypervolume indicators, with a hybrid-encoding genetic algorithm (GA) composed of NSGA-II and an integer-coded GA for classical traveling salesman problems (TSP). The results show that the MOAA is able to outperform the hybrid approach based on NSGA-II by finding a better Pareto front which provides more useful information to the decision-maker.

Keywords: Multi-objective optimization, Evolutionary Algorithms, MOAA, NSGA-II, Satellite Constellation Refueling.

1 Introduction

Satellite constellation technologies are now widely used in satellite communication, Earth observation and satellite navigation, while on-orbit refueling is an important direction of development for the aerospace industry due to its considerable economic value [1]. It can be economically advantageous to refuel multiple satellites in a constellation with one service spacecraft. Constellation service missions have been studied as combinatorial optimization problems by [2–5]. These studies sought to minimize only one objective (the total propellant consumption). However, a constellation refueling mission is by its nature a mixed-integer nonlinear programming (MINLP) problem because the refueling sequence of satellites is combinatorial while the time of flight of each refueling activity is a continuous variable. Designers usually want to complete the refueling mission as soon as possible while consuming as little propellant as possible.

R.C. Purshouse et al. (Eds.): EMO 2013, LNCS 7811, pp. 669–684, 2013.

Two conflicting design objectives, the total time of flight and total propellant consumption, need to be minimized simultaneously. In consequence, a multi-objective MINLP technology should be employed. Metaheuristic approaches are particularly good for the resolution of complex real-world problems and many multi-objective evolutionary algorithms have been created for the purpose. In books such as [6] and [7] some of the most widely used multi-objective evolutionary algorithms, such as NSGA-II [8] and SPEA 2 [9], are introduced.

The Alliance Algorithm (AA) is a recently developed single-objective optimization algorithm that has been applied successfully to different problems [10–13]. These promising results motivated the development of a multi-objective variant [14], the performance of which has been compared with NSGA-II [8] and SPEA 2 [9]. That study revealed a certain complementarity because the three approaches offered superior performance for different classes of problems.

In this paper a mixed-integer version of MOAA is presented with hybrid components that enable it to outperfom another hybrid method based on NSGA-II. The use of hybrid approaches, which combine different methodologies that are complementary in their strengths and weaknesses, has improved the performance of many metaheuristics, as shown in [15] and [16] where metaheuristic approaches are combined with local search in order to improve convergence.

The rest of the paper is structured as follows: Section 2 introduces the satellite constellation refueling mission; Section 3 describes the optimization model; Section 4 presents the mixed-integer MOAA; Section 5 provides details of the specific problem studied and introduces the indicators and statistical test used for performance comparison; Section 6 reports the MOAA's performance, a comparison with the approach based on NSGA-II and discusses the results; Section 7 concludes the paper and suggests possible future work.

2   Satellite Constellation Refueling Mission

A typical refueling mission for a satellite constellation is illustrated in Fig. 1. Several satellites belonging to a constellation run on a near-circular orbit, their propellant has been exhausted, and some of them have drifted from their nominal constellation positions. These satellites are referred to as target satellites and need propellant refueling. Let $p \in [1, 2, \cdots, Q]$ be the serial numbers of the target satellites, where Q is their total number. A chaser, i.e. the service spacecraft, also runs on this near-circular orbit and is required to visit and refuel target satellites one by one. Each refueling mission consists of three steps. First, the chaser maneuvers to rendezvous with one target satellite. Second, the chaser serves and resupplies propellant to this target. Third, the chaser moves on and maneuvers to rendezvous with the next target, while the resupplied target maneuvers to return to a desired constellation reconfiguration position. These steps are repeated until each target has been visited. After these operations, the normal configuration and capability of the constellation can be restored. Let $s \in [1, 2, \cdots, Q]$ be the serial numbers of reconfiguration positions.

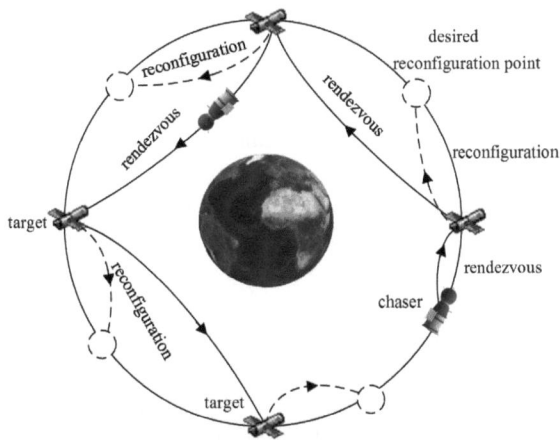

Fig. 1. A refueling and reconfiguration mission for a satellite constellation

The initial and final times of the qth rendezvous operation are given by equations (1) and (2), where t_0 is the initial time of the entire task, dur_q and ser_q are the rendezvous orbital transfer time and service time respectively, and $k \in [1, 2, \cdots, q-1]$ denotes the serial numbers of the refueling missions already completed.

$$t_{q0} = t_0 + \sum_{k=1}^{q-1}(dur_k + ser_k) \tag{1}$$

$$t_{qf} = t_{q0} + dur_q \tag{2}$$

The initial and end times of the qth reconfiguration operation are shown in equations (3) and (4), where dur'_q is the reconfiguration orbital transfer time.

$$t'_{q0} = t_{qf} + ser_q \tag{3}$$

$$t'_{qf} = t'_{q0} + dur'_q \tag{4}$$

2.1   Orbital Transfer

The state of a spacecraft can be expressed as follows:

$$E = (a, u, \xi, \eta, i_{inc}, \Omega)^T \tag{5}$$

Here: a is the semi-major axis; i_{inc} is the orbital inclination; Ω is the right ascension of ascending node (RAAN); u is the argument of latitude; $\xi = e\cos(\omega)$ and $\eta = e\sin(\omega)$ are the modified orbital elements suitable for describing near-circular orbits, where e is the eccentricity and ω is the argument of perigee. Let $E_C(t_0)$ be the initial state of the chaser, $E_{Tp}(t_0)$ be the initial state of the target numbered p, and $E_{Rs}(t_0)$ be the initial state of the reconfiguration position numbered s. In the qth refueling mission, let p_q and s_q be the target serial number and the corresponding reconfiguration position serial number respectively, and then the target's initial and end states for each rendezvous operation are functions of its initial state and the time:

$$E_{Tp_q}(t_{q0}) = G[E_{Tp_q}(t_0), t_{q0} - t_0] \tag{6}$$

$$E_{Tp_q}(t_{qf}) = G[E_{Tp_q}(t_0), t_{qf} - t_0] \tag{7}$$

In equations (6) and (7): G is the function propagating orbital states by solving Kepler's equation [17]. The chaser's initial state for the first rendezvous operation is $E_C(t_0)$ and that for the qth ($q > 1$) rendezvous operation is equal to the target's state during the $(q-1)$th refueling mission at the chaser's departure time:

$$E_C(t_{q0}) = \begin{cases} E_C(t_0) & (q = 0) \\ G[E_{Tp_{q-1}}(t_0), t_{q0} - t_0] & (q > 0) \end{cases} \tag{8}$$

The initial and end states of the reconfiguration position are given by equations (9) and (10):

$$E_{Rs_q}(t'_{q0}) = G[E_{Rs_q}(t_0), t'_{q0} - t_0] \tag{9}$$

$$E_{Rs_q}(t'_{qf}) = G[E_{Rs_q}(t_0), t'_{qf} - t_0] \tag{10}$$

The target's initial state for the qth reconfiguration operation is:

$$E_{Tp_q}(t'_{q0}) = G[E_{Tp_q}(t_0), t'_{q0} - t_0] \tag{11}$$

When the initial and end states of an orbital transfer are specified, a linear dynamics model is used to calculate maneuver impulses. The state variable used to express orbital differences between the chaser and a target is shown in equation (12), where the subscript r denotes the reference orbit, Δa is the difference in semi-major axis, $\Delta\theta$ is the difference in argument of latitude, Δi_{inc} is the difference in orbital inclination, $\Delta\Omega$ is the difference in RAAN, and $\Delta\xi$ and $\Delta\eta$ give the differences in the eccentricity vector.

$$X = (\Delta a/a_r, \Delta\theta, \Delta\xi, \Delta\eta, \Delta i_{inc}, \Delta\Omega)^T \tag{12}$$

Using first-order approximations, the orbital transfer of the qth rendezvous operation can be expressed as described in [18] and [19]:

$$\Phi(\Delta t_{q0})X_0 + \sum_{j=1}^{2} \Phi_v(\Delta t_{qj}, u_{qj})\Delta v_{qj} = X(t_{qf}) = 0_{1\times 6} \tag{13}$$

$$\Phi(\Delta t_{q0}) = \begin{bmatrix} 1 & 0\,0\,0\,0\,0 \\ -\frac{3}{2}n_r\Delta t_{q0} & 1\,0\,0\,0\,0 \\ 0 & 0\,1\,0\,0\,0 \\ 0 & 0\,0\,1\,0\,0 \\ 0 & 0\,0\,0\,1\,0 \\ 0 & 0\,0\,0\,0\,1 \end{bmatrix} \tag{14}$$

$$\Phi_v(\Delta t_{qj}, u_{qj}) = \begin{bmatrix} 0 & 2 & 0 \\ 0 & -3n_r\Delta t_{qj} & 0 \\ \sin(u_{qj}) & 2\cos(u_{qj}) & 0 \\ -\cos(u_{qj}) & 2\sin(u_{qj}) & 0 \\ 0 & 0 & \cos(u_{qj}) \\ 0 & 0 & \sin(u_{qj})/\sin(i_r) \end{bmatrix} \tag{15}$$

Here: the subscript '0' denotes the initial state; $\Delta t_{q0} = t_{qf} - t_{q0} = dur_q$ is the orbital transfer time; $t_{q1} = 0$ and $t_{q2} = dur_q$ are the two burn times; $\Delta t_{qj} = t_{qf} - t_{qj}$ $(j = 1, 2)$; u_{qj} and Δv_{qj} $(j = 1, 2)$ are the argument of latitude and the impulse of the jth maneuver respectively; $n_r = \sqrt{\mu/a_r^3}$ is the mean angular motion rate, where μ is the geocentric gravitation constant. Equation (13) is a linear dynamics equation-set, with six unknown impulse components for six equations in total, and its solution can be obtained using Gaussian elimination. Let t'_{qj} and $\Delta v'_{qj}$ $(j = 1, 2)$ be the burn time and impulse of the jth reconfiguration transfer maneuver. The reconfiguration impulses are calculated using the dynamics model given above.

2.2 Propellant Consumption

Let m_{p0} be the initial mass of the target numbered p, Δm_{p0} be the propellant required for each target's daily operation, and m_{cqf} be the chaser's mass after the qth refueling mission. Using the Tsiolkovsky rocket equation [17], the target's mass at the initial time of the qth reconfiguration operation is calculated based on its dry mass, daily operation propellant and reconfiguration maneuver impulses:

$$m_{p_q}(t'_{q0}) = (m_{p_q0} + \Delta m_{p0})\exp\left(\frac{\sum_{j=1}^{2}\|\Delta v'_{qj}\|}{g_0 I_{sp}}\right) \tag{16}$$

Here: g_0 is the sea-level standard acceleration of gravity and I_{sp} is the specific impulse of thrusters. Therefore, the propellant resupplied from the chaser to this target is expressed by:

$$\Delta m_{p_q} = m_{p_q}(t'_{q0}) - m_{p_q0} \tag{17}$$

The chaser's mass after the $(q-1)$th refueling mission is equal to its initial mass in the qth mission, and can be calculated based on the maneuver impulses of the qth rendezvous operation as:

$$m_{c(q-1)f} = m_{cq0} = (m_{cqf} + \Delta m_{p_q})\exp\left(\frac{\sum_{j=1}^{2}\|\Delta v'_{qj}\|}{g_0 I_{sp}}\right) \tag{18}$$

In consequence, the chaser's initial mass for the entire task is given by:

$$m_{c0} = (m_{c1f} + \Delta m_{p_1}) \exp\left(\frac{\sum_{j=1}^{2} \|\Delta v'_{1j}\|}{g_0 I_{sp}}\right)$$
(19)

3   Optimization Model

There are two types of design variable. A solution Y is made up of two groups of serial integers Y_1 and Y_2 and a set of real numbers Y_3 that consists of rendezvous orbital transfer times, service times and reconfiguration orbital transfer times:

$$Y = (Y_1, Y_2, Y_3)$$
(20)

where

$$
\begin{aligned}
Y_1 &= (p_1, p_2, \cdots, p_Q) \\
Y_2 &= (s_1, s_2, \cdots, s_Q) \\
Y_3 &= (dur_1, dur_2, \cdots, dur_Q; ser_1, ser_2, \cdots, ser_Q; dur'_1, dur'_2, \cdots, dur'_Q)
\end{aligned}
$$
(21)

The sequence of the elements of Y_1 represents a refueling order. The search space of Y_1 is therefore discrete and its elements must be manipulated in combination. These characteristics also apply to Y_2. The first objective is to minimize the mean mission completion time (equation (22)); the second is to minimize the propellant consumed by orbital maneuvers (equation (23)):

$$\min \ f_1 = t'_{Qf}/Q$$
(22)

$$\min \ f_2 = m_{c0} - m_{cQf} - Q\Delta m_{p0}$$
(23)

In equation (23): m_{cQf} is the chaser's mass after the last refueling mission and also denotes the chaser's dry mass in this study.

4   Multi-Objective Alliance Algorithm

The MOAA is a metaheuristic optimization algorithm based on the metaphorical idea of a number of tribes struggling to conquer an environment that offers resources that enable them to survive. The tribes are characterized by two features: the skills and resources necessary for survival. Tribes try to improve skills by forming alliances, which are also characterized by the skills and resources needed, but these now depend on the tribes within an alliance. The two main search elements of the algorithm are the formation of alliances and the creation of new tribes. One AA cycle ends when the strongest possible alliances of existing tribes have been created. The algorithm then begins a new cycle starting with new tribes whose creation is influenced by the previous strongest alliances.

4.1 The Main Entities

Two main entities play important roles in the MOAA: the tribes and the alliances. A tribe t is a tuple (x_t, s_t, r_t, a_t) composed of:

- a point in the solution space x_t;
- a set of skills $s_t = [s_{t,1}, s_{t,2}, \ldots, s_{t,N_S}]$ dependent on the values of the N_S objective functions $S = [S_1, S_2, \ldots, S_{N_S}]$ evaluated at x_t:

$$s_{t,i} = S_i(x_t) \quad \forall \ i = 1, 2, \ldots, N_S \tag{24}$$

- a set of resource demands $r_t = [r_{t,1}, r_{t,2}, \ldots, r_{t,N_R}]$ dependent on the values of the N_R constraint functions:

$$r_{t,i} = R_i(x_t) \quad \forall \ i = 1, 2, \ldots, N_R \tag{25}$$

- an alliance vector a_t containing the IDs of the tribes allied to tribe t. Initially an alliance is composed of just one tribe, thus $a_t(1) = ID_t$.

An alliance is a mutually disjoint partition of tribes. The tribes within an alliance perform actions as a unique entity. Each alliance a forms a new point x_a in the solution space defined by the tribes in the alliance. The sets of skills s_a and resource demands r_a of the alliance consist of the objective and constraint functions S and R evaluated at x_a.

4.2 Algorithm Steps

The procedure followed by the MOAA can be divided into several steps which can be performed differently according to the problem at hand and user preference. For reasons of space, only the differences between this mixed-integer version of the MOAA and the previous version (described fully in [14]) are detailed.

Solution Generation. In the MOAA's first cycle the continuous variables of the tribes (solutions) are chosen randomly (with a uniform distribution), as shown in equation (26); the discrete variables are composed of two groups of random serial integers (Y_1 and Y_2), as shown in equation (27):

$$x_{t,i} = U(L_i, H_i) \quad \forall \ i = 1, 2, \ldots, 3Q \tag{26}$$

$$x_{t,Y_1} = (p_1, p_2, \cdots, p_Q) \quad x_{t,Y_2} = (s_1, s_2, \cdots, s_Q) \tag{27}$$

Here: $x_{t,i}$ is the ith continuous variable of tribe t, and L_i and H_i are respectively the lower and upper bounds on this variable; x_{t,Y_1} and x_{t,Y_2} are respectively the random sequences for Y_1 and Y_2; p and s are vectors where each element has a unique number from 1 to Q.

In subsequent MOAA cycles, new continuous solutions are sampled from a normal distribution with defined mean and standard deviation σ described in equation (28), and new discrete solutions modify the sequences of the non-dominated solutions found, as shown in equation (29):

$$
\begin{aligned}
var_{pos}(i) &= U_d(1, 3Q) & &\forall\, i \in [1, 2, \ldots, N_{var}] \\
\text{If } \exists\ j &\in var_{pos}\ \ j = 1, 2, \ldots, 3Q: & x_{t,j} &= N(b_{r,j}, \sigma) \\
\text{else}: & & x_{t,j} &= b_{r,j}
\end{aligned}
\tag{28}
$$

$$
\begin{aligned}
n_{change,i} &= U_d(1, Q) \quad n_{pos,i} = U_d(1, Q - n_{change}) \quad \forall\, i = 1, 2 \\
x_{t,Y_i} &= b_{r,Y_i} \quad x_{t,(n_{pos,i},n_{pos,i}+n_{change,i})} = inv(x_{t,(n_{pos,i},n_{pos,i}+n_{change,i})})
\end{aligned}
\tag{29}
$$

In equation (28): N_{var} is the number of possible changes to a tribe's variables; var_{pos} is a vector of the positions of the variables to be modified; $U_d(1, 3Q)$ is the discrete uniform distribution between 1 and $3Q$; r is a random integer between 1 and N_P, the number of Pareto-optimal (PO) points found; $b_{r,j}$ is the normalized jth variable of the rth PO solution found.

In equation (29): i is the number of the sequence (Y_1 or Y_2); $n_{change,i}$ is the number of discrete variables in sequence i that need to be inverted; $n_{pos,i}$ is the start position for the inversion of sequence i; b_{r,Y_i} is the sequence Y_i of the rth PO point found; $inv(start, end)$ is the function that inverts the discrete variables (i.e. $[1324] \rightarrow [4231]$). This cycle is repeated until all N tribes are generated.

Formation of an Alliance. An alliance/tribe (A/T) is chosen randomly and given the chance to forge an alliance by being given a token. Meanwhile all the other A/Ts wait their turn. The A/T t with the token chooses another tribe to become an ally, thus forming a new alliance (a movement in solution space).

The alliance formed is composed of variables from the tribes within the alliance: given an alliance composed of N_a tribes, every continuous variable has $1/N_a$ probability to be equal to the corresponding variable of any tribe within the alliance:

$$
\begin{aligned}
c &= U_d(1, N_a) \\
x_{a,i} &= x_{c,i}
\end{aligned}
\tag{30}
$$

Here: c is the index of the chosen tribe within the alliance; $U_d(1, N_a)$ is the discrete uniform distribution between 1 and N_a; $x_{a,i}$ is the value of the ith component of the alliance; $x_{c,i}$ is the value of the ith component of the chosen tribe. This is repeated until all continuous components of the alliance are defined.

The case of the discrete variables is different: the components of the sequences Y_1 and Y_2 of tribes within the alliance are summed and then ranked so the component with the smallest sum is 1 and that with the largest Q. Ties in ranking are broken randomly.

$$
\begin{aligned}
x_{tot,Y_i} &= \sum_{t=1}^{N_a} x_{t,Y_i} \quad \forall\, i = 1, 2 \\
x_{a,Y_i} &= rank(x_{tot,Y_i})
\end{aligned}
\tag{31}
$$

Here: i is the number of the sequence (Y_1 or Y_2); x_{t,Y_i} is the ith sequence of tribe t within the alliance; x_{tot,Y_i} is the summation of ith sequences; $rank(x)$ is the function that ranks components of a given sequence; x_{a,Y_i} is the ith sequence of the alliance.

The new alliance will only be confirmed if the skills s_a of x_a are all individually better than (or at least as good as) the skills s_{t_1} of the solution representing the A/T with the token x_{t_1} and the skills s_{t_2} of the tribe chosen to become an ally, i.e. if solution x_a *dominates* x_{t_1} and x_{t_2}.

The resource function $R(x)$ plays no role here because the problem tackled in this particular study is unconstrained.

Alliance and Data Structure Update. There are two possible outcomes from the previous actions: the chosen tribe joins the entity with the token forming a new alliance, or the tribe does not join and the new alliance is not confirmed. Next there is an update of the data structures necessary for the low level system to function, such as the necessity to provide a unique ID to created alliances. The cycle termination conditions are also checked. The cycle finishes when each A/T has tried to form a new alliance with every other tribe and remains unchanged (because there is no advantage in changing). If this condition is not met, the token is given to another A/T.

Selection of the Strongest Alliances and Termination. At the end of the interactions between tribes, many alliances will have been formed but only the strongest A/Ts will conquer the environment. Therefore the A/Ts selected are the non-dominated points in objective space. These correspond to the best solutions to the problem found thus far. They can be used as the input to another MOAA cycle or, if the algorithm has ended, they represent the final results.

Generally, there is a limit n to the number of best solutions saved in the archive of PO solutions, but for this problem no limit has been set, allowing as many solutions as possible to be saved in order to provide a clear picture of the Pareto front. The MOAA is terminated when a specified limit on the number of solution evaluations is reached. The output of the algorithm is then the best solutions and the Pareto front found.

Support Functions. This version of the algorithm is supported by several functions that hybridize it. These functions generate new solutions and the archive of PO solutions is updated appropriately. The functions used here are:

- *Differential Search*: Two PO solutions are randomly selected. The integer part is created randomly (as in the first cycle of Solution Generation). The continuous part is set equal to the absolute difference between the continuous parts of the two selected solutions divided by 2; component values below lower bounds assume the values of the bounds.
- *Local Search*: One PO solution is randomly selected. The components of two randomly selected positions in sequence Y_1 are swapped. This process is repeated for Y_2. The continuous part of the solution is not modified.

- *Sequence Search*: One PO solution is randomly selected. The integer parts of this solution (Y_1 and Y_2) are replaced in turn by the integer parts of all the other PO solutions. The integer parts giving the best f_2 result are then substituted into all the other PO solutions.
- *Single Variation Search*: One PO solution is randomly selected. The integer part remains unchanged, while one component of the continuous part is changed with a uniform distribution.
- *Gap Filling Search*: The PO solutions are ordered according to their f_1 values. The pair of consecutive solutions that are furthest apart in objective space is identified. A number of new solutions are generated. These inherit the integer parts (Y_1 and Y_2) of the first of the pair, while the components of the continuous part are generated randomly in the range spanned by the corresponding components of the pair. Taking these new solutions into account, this process is repeated a number of times. This function exploits a characteristic of the problem under consideration, whereby solutions that lie between others in solution space also lie between them in objective space.

These functions introduce new solutions that are used by the MOAA for the creation of new tribes and alliances: Differential Search helps to create new types of solutions, especially on the edges of objective space; Local Search helps to improve the characteristics of one particular solution by introducing small changes in the discrete solution space; Sequence Search helps to find new sequences that could benefit all the population in order to improve overall convergence; Single Variation Search helps to improve the characteristics of one particular solution by changing one continuous variable; Gap Filling Search helps to fill in the gaps between solutions creating a well-spaced Pareto front.

5   The Problem and Performance Indicators

In this section details of the specific problem studied and of the indicators and statistical test used to compare the performance of the two approaches are given.

5.1   Problem Configuration

The initial time of the entire task in Gregorian universal coordinated time (UTCG) format is 1 June 2015 00:00:00.00. The target spacecraft run on an orbit similar to the operational orbit of the Globalstar constellation [13]: $Q = 8$, $a_r = 7792.137$ km and $i_r = 52°$. The initial arguments of latitude of the target spacecraft and reconfiguration positions are provided in Table 1. The chaser's initial argument of latitude and RAAN are 20° and 100° respectively. The initial RAANs of the target spacecraft are all equal to 100°. The dry mass of the chaser $m_{cQf} = 500$ kg, the initial mass of each target spacecraft $m_{p0} = 350$ kg, the propellant used for each target's daily operation $\Delta m_{p0} = 100$ kg, and the capacity of each target's propellant tank $\Delta m_{max} = 200$ kg. The thruster parameter $g_0 I_{sp} = 3000$ m/s, and the geocentric gravitation constant

Table 1. Initial target spacecraft and reconfiguration position arguments of latitude

Serial number	Target spacecraft	Reconfiguration position
1	10°	0°
2	55°	45°
3	100°	90°
4	145°	135°
5	190°	180°
6	235°	225°
7	280°	270°
8	325°	315°

$\mu = 3.986004418 \times 10^{14}$ m^3/(kgs^2). The reference orbital period $T_0 = 6845.353$ s, and the bounds on rendezvous orbital transfer time, service time and reconfiguration orbital transfer are $dur_q \in [10T_0, 600T_0]$, $ser_q \in [10T_0, 600T_0]$ and $dur'_q \in [10T_0, 600T_0]$ respectively.

5.2 Indicators and Statistical Test

The performance measures chosen to evaluate the algorithms are the *epsilon indicator* [20] and *hypervolume indicator* [21] taken from the PISA package [22]. Given a reference set of points (ideally the true Pareto front, if available) the epsilon indicator measures the minimum amount ϵ necessary to translate all the points of the found Pareto front to weakly dominate the reference set. The hypervolume indicator [21] measures the hypervolume of the space dominated by the found Pareto front and compares it to the hypervolume of the space dominated by a reference set (again, ideally the true Pareto front). This indicator needs a reference point which is dominated by all the found points in order to bound the hypervolume.

The statistical test chosen for result evaluation is the Mann-Whitney test, provided in the PISA package [22]. This is a non-parametric rank-based test that can be used to compare two independent sets of sampled data. It outputs p-values that estimate the probability of rejecting the null hypothesis of the study question when that hypothesis is true. Here the p-values can be interpreted as the probability that the performance of one algorithm is superior with statistical significance to that of the other.

6 The Tests

In this section the performance of the mixed-integer version of MOAA is compared with that of the hybrid algorithm based on NSGA-II. The latter is a combination of an integer-coded GA for classical TSP [23] and the real-coded version of NSGA-II [8]. The design variable vector $Y = (Y_1, Y_2, Y_3)$ is used directly as the chromosome of an individual: the arithmetical crossover and non-uniform mutation operators are applied to Y_3 [8], while the ordered crossover (OX) and three-point displacement mutation operators are both applied to Y_1 and Y_2 [23]. An elitist strategy is employed alongside tournament selection during the algorithm's selection phase.

The two algorithms were tested with limits of 1000 and 2000 function evaluations. Every test was repeated 20 times. The reference set used for the comparison was composed of the PO solutions found by both algorithms. The algorithm parameters used are shown in Table 2. The same MOAA parameter values were used in both cases. For the hybrid NSGA-II the number of generations was changed. In MOAA Gap Filling Search 5 solutions were generated for 20 gaps.

Table 2. Algorithms parameters

Mixed-integer MOAA		Hybrid NSGA-II	
Parameter	Value	Parameter	Value
Number of tribes	8	Population size	40
Standard deviation used for the creation of tribes	0.2	Number of generations	25 (1) 50 (2)
Maximum number of variations in new tribes	3	Tournament selection scale	3
Solutions created by Differential Search	20	Crossover probability of real variables	0.8
Solutions created by Local Search	40	Crossover probability of integer variables	0.7
Solutions created by Single Variation Search	100	Mutation probability of real variables	0.4
Solutions created by Gap Filling Search	100	Mutation probability of integer variables	0.3

Table 3 shows the means and standard deviations of the epsilon and hypervolume indicators. It is evident that the mixed-integer MOAA has outperformed the hybrid NSGA-II with respect to both indicators, for both of which smaller values indicate better performance.

Table 3. Performance comparison between the two algorithms

	Mixed-integer MOAA				Hybrid NSGA-II			
	Epsilon		Hypervolume		Epsilon		Hypervolume	
Evaluations	Mean	Std	Mean	Std	Mean	Std	Mean	Std
1000	0.0460	0.0171	0.0238	0.0111	0.0959	0.0227	0.1490	0.0357
2000	0.0266	0.0091	0.0114	0.0093	0.0384	0.0103	0.0249	0.0147

The 20 values of the epsilon and hypervolume indicators after 1000 function evaluations (top two graphs) and 2000 function evaluations (bottom two graphs) sorted from best to worst are shown in Fig. 2. In all the graphs there is a clear

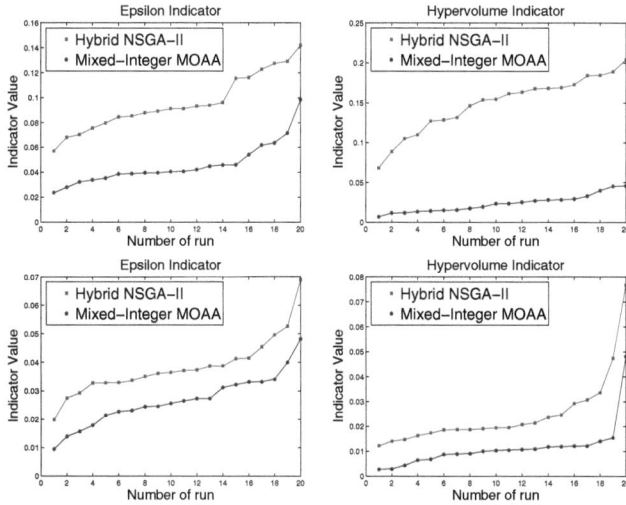

Fig. 2. Comparison of the two indicators for both algorithms on both tests

Table 4. Mann-Whitney test between the two algorithms

	$p(A > B)$	$p(A > B)$
Evaluations	epsilon	hypervolume
1000	0.99	0.99
2000	0.99	0.99

gap between the MOAA and hybrid NSGA-II lines; this is particularly apparent
in the hypervolume graph for 1000 evaluations.

Table 4 shows the probability that the mixed-integer MOAA (A) performs
better than the hybrid NSGA-II (B) applying the Mann-Whitney test to the results obtained for the epsilon and hypervolume indicators by the two approaches.
This test confirms that the MOAA is consistently better (probability very near
to 1) in both tests for both indicators.

Fig. 3 shows the results (PO solutions found in individual runs) combined for
the 20 runs of the two algorithms after 1000 (on the left) and 2000 (on the right)
function evaluations. In both cases many MOAA solutions dominate the hybrid
NSGA-II solutions, showing better convergence. This characteristic is confirmed
in Fig. 4 where it is shown that the non-dominated solutions found by the MOAA
dominate those found by the hybrid NSGA-II, providing the decision-maker with
higher quality solutions to choose from.

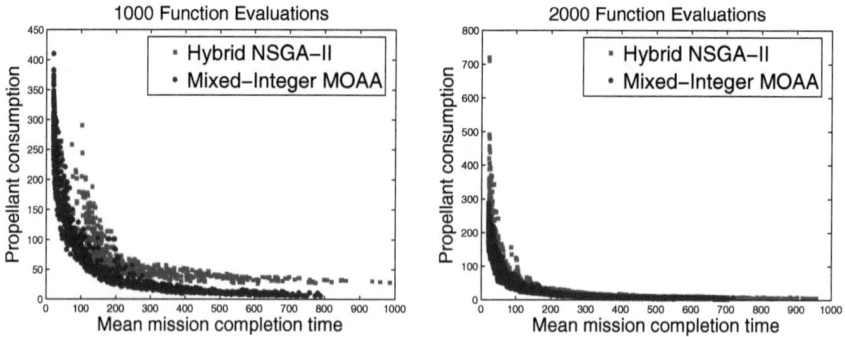

Fig. 3. Comparison of the algorithm results from 20 runs in both tests

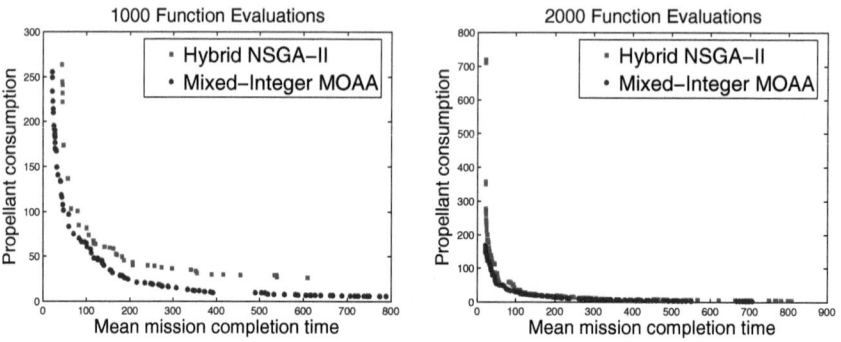

Fig. 4. Comparison of the non-dominated solutions found in both tests

7   Conclusions and Future Work

The paper has presented a satellite constellation refueling optimization problem. The problem was solved by a mixed-integer version of the MOAA, a recently introduced multi-objective variant of the Alliance Algorithm. This performance is compared with that of a hybrid-encoding GA, which combines NSGA-II and an integer-coded GA for classical TSP. The results show that the MOAA is able to outperform the hybrid approach based on NSGA-II over runs of 1000 and 2000 function evaluations' duration. Although the number of function evalutions was limited, MOAA was able to identify a useful set of solutions, making the algorithm a sensible choice for the type of problem under consideration.

In future work the performance of the algorithm will be tested with an increased allowed number of function evaluations, on more advanced forms of satellite constellation refueling optimization problem with more practical dynamics models and considering engineering constraints, and the properties of the refueling orders in PO solutions will be analyzed. Moreover, other types of support functions will be tested in order to seek further improvements in overall performance.

References

1. Saleh, J.H., Lamassoure, E.S., Hastings, D.E., Newman, D.J.: Flexibility and the value of on-orbit servicing: New customer-centric perspective. Journal of Spacecraft and Rockets 40(2), 279–291 (2003)
2. Alfriend, K.T., Lee, D.J., Creamer, N.G.: Optimal servicing of geosynchronous satellites. Journal of Guidance, Control, and Dynamics 29(1), 203–206 (2006)
3. Shen, H., Tsiotras, P.: Peer-to-peer refueling for circular satellite constellations. Journal of Guidance, Control and Dynamics 28(6), 1220–1230 (2005)
4. Dutta, A., Tsiotras, P.: Egalitarian peer-to-peer satellite refueling strategy. Journal of Spacecraft and Rockets 45(3), 608–618 (2008)
5. Dutta, A., Tsiotras, P.: Network flow formulation for cooperative peer-to-peer refueling strategies. Journal of Guidance, Control and Dynamics 33(5), 1539–1549 (2010)
6. Coello Coello, C.A., Lamont, G.B., Veldhuizen, D.A.V.: Evolutionary Algorithms for Solving Multi-Objective Problems (Genetic and Evolutionary Computation). Springer, Secaucus (2006)
7. Deb, K.: Multi-Objective Optimization Using Evolutionary Algorithms, 1st edn. Wiley, Chichester (2001)
8. Deb, K., Pratap, A., Agarwal, S., Meyarivan, T.: A fast and elitist multiobjective genetic algorithm: NSGA-II. IEEE Transactions on Evolutionary Computation 6, 182–197 (2002)
9. Zitzler, E., Laumanns, M., Thiele, L.: SPEA2: Improving the strength pareto evolutionary algorithm for multiobjective optimization. In: Evolutionary Methods for Design Optimization and Control with Applications to Industrial Problems, pp. 95–100 (2001)
10. Lattarulo, V.: Application of an innovative optimization algorithm for the management of energy resources. BSc thesis, University of Salerno (2009)
11. Calderaro, V., Galdi, V., Lattarulo, V., Siano, P.: A new algorithm for steady state load-shedding strategy. In: 12th International Conference on Optimization of Electrical and Electronic Equipment (OPTIM), pp. 48–53 (2010)
12. Lattarulo, V.: Optimization of biped robot behaviors by 'alliance algorithm'. Master's thesis, University of Hertfordshire (2011)
13. Lattarulo, V., van Dijk, S.G.: Application of the "Alliance Algorithm" to Energy Constrained Gait Optimization. In: Röfer, T., Mayer, N.M., Savage, J., Saranlı, U. (eds.) RoboCup 2011. LNCS, vol. 7416, pp. 472–483. Springer, Heidelberg (2012)
14. Lattarulo, V., Parks, G.T.: A preliminary study of a new multi-objective optimization algorithm. In: International Conference on Evolutionary Computation (CEC) (2012)
15. Deb, K., Goel, T.: A Hybrid Multi-objective Evolutionary Approach to Engineering Shape Design. In: Zitzler, E., Deb, K., Thiele, L., Coello Coello, C.A., Corne, D.W. (eds.) EMO 2001. LNCS, vol. 1993, pp. 385–399. Springer, Heidelberg (2001)
16. Yu, G., Chai, T., Luo, X.: Multiobjective production planning optimization using hybrid evolutionary algorithms for mineral processing. IEEE Transactions on Evolutionary Computation 15, 487–514 (2011)

17. Vallado, D.A.: Fundamentals of Astrodynamics and Applications, 2nd edn. Microscosm Press, Torrance (2001)
18. Labourdette, P., Baranov, A.A.: Strategies for on-orbit rendezvous circling Mars. Advances in the Astronautical Sciences 109, 1351–1368 (2002)
19. Zhang, J., Luo, Y.Z., Tang, G.J.: Hybrid planning for LEO long-duration multi-spacecraft rendezvous mission. Science China Technological Sciences 55(1), 233–243 (2012)
20. Zitzler, E., Thiele, L., Laumanns, M., Fonseca, C.M., Grunert da Fonseca, V.: Performance assessment of multiobjective optimizers: An analysis and review. IEEE Transactions on Evolutionary Computation 7(2), 117–132 (2003)
21. Knowles, J., Thiele, L., Zitzler, E.: A Tutorial on the Performance Assessment of Stochastic Multiobjective Optimizers. TIK Report 214, Computer Engineering and Networks Laboratory (TIK), ETH Zurich (February 2006)
22. Bleuler, S., Laumanns, M., Thiele, L., Zitzler, E.: PISA – A Platform and Programming Language Independent Interface for Search Algorithms. In: Fonseca, C.M., Fleming, P.J., Zitzler, E., Deb, K., Thiele, L. (eds.) EMO 2003. LNCS, vol. 2632, pp. 494–508. Springer, Heidelberg (2003)
23. Chatterjee, S., Carrera, C., Lynch, L.A.: Genetic algorithms and traveling salesman problems. European Journal of Operational Research 93(3), 490–510 (1996)

Parametric Design Optimisation
for Planetary Landing Systems

David Riley[1], Dave Northey[1], Rodrigo Haya Ramos[2],
Mariano Sánchez Nogales[2], Davide Bonetti[2], and Dave Dungate[1]

[1] Tessella Ltd., Elopak House, Rutherford Close, Stevenage, SG1 2EF, UK
david.riley@tessella.com
[2] Elecnor Deimos, Ronda de Poniente 19, 28760 Tres Cantos, Madrid, Spain

Abstract. The problem of design optimisation for planetary landing systems, which must address multiple criteria and constraints covering, for example, mass, trajectory, risk, and the cost of technology developments, is introduced. The approach used to solve this problem in the context of the real world issues faced and the decisions made is presented, and we demonstrate a tool which has been used in this way to explore the parametric space of possible landing systems for a number of planned and possible European missions to Mars, for use in early phase sizing studies.

Keywords: Entry descent and landing systems, parametric optimization.

1 Introduction

Planetary science depends on *in situ* measurements. In the case of Mars, bringing a probe safely to rest on the surface provides a number of challenges, including a significant gravity, and an atmosphere which is thin but not negligible [1]. The entry, descent and landing system (EDLS) is responsible for meeting these challenges, as the probe goes from hypersonic speeds at the top of the atmosphere to landing on the surface without damaging the fragile scientific equipment it carries. The design of this subsystem is essentially a multi-criteria constrained optimization task.

1.1 Entry, Descent and Landing Systems Design

A number of conflicting driving requirements must be considered in the design of the EDLS:

- **Minimise mass:** At heart, the objective is to minimise the total mass of the EDLS: given that the mass which can be delivered to Mars is limited, every extra kilogram of EDLS mass means an equivalent reduction in the scientific payload.
- **Constrain height loss:** The EDLS must slow the probe down sufficiently for safe landing in the available height between entering the atmosphere and reaching the ground.

R.C. Purshouse et al. (Eds.): EMO 2013, LNCS 7811, pp. 685–695, 2013.

- **Constrain volume:** Stowage constraints are also significant: the system must be able to fit on the launcher.
- **Reliability/complexity:** There are potentially a number of single points of failure, and so the overall system design must consider the risk of mission loss.
- **Cost**: This is also an issue, both that required in developing a particular technology under consideration to bring it to a sufficient Technology Readiness Level for use on a planned mission, and manufacturing costs.
- **Schedule:** For any given mission, the timescales required to develop needed technologies, and the schedule risks this brings, must also be considered.
- **Landing Site Accuracy:** For planetary science, another requirement is: how close to a selected site of scientific interest can the lander be placed?

Different EDLS configurations can be considered. Atmospheric entry is always effected using a heat shield, which aerodynamically slows the probe from interplanetary speeds and dissipates the heat generated. In most cases, the entry phase is followed by deployment of a parachute once the probe has reached a sufficiently low speed, slowing it even further. Different parachute systems have been used, which may include one or multiple parachute stages, and use parachutes of different types. Retro-rockets may be used at this point to reduce the speed even further. Finally, touch-down on the surface may be effected using airbags, crushable structures, or landing legs.

Each of the different components carries out a different task within the overall EDLS. For example a heatshield is suitable for use at hypersonic speeds, but to reduce the speed adequately to transition to airbags, as drag is proportional to frontal area, would require heatshield far larger that could realistically be launched; a parachute can provide the required drag area for much less mass, and can be packed to take up only a relatively small volume.

The overall EDLS design needs to be self-consistent. The final velocity that is achieved by one component needs to be compatible with the conditions at which the next component can be safely triggered. Likewise the sections of the trajectory for each stage of the descent must join up consistently, ensuring that there is enough altitude to reach the final velocity and that the atmosphere density at each stage is consistent with that assumed for sizing.

The mass of the EDLS can be comparable to the mass of the landed payload, and so the mass of EDLS components being carried must be allowed for when sizing other components. In some cases, the self-mass term can also be significant: a parachute must support both the payload and its own mass.

1.2   Previous Mission Experience

Different past missions have chosen different EDLS configurations. The first successful NASA landers, the Viking missions of the 1970's, used a heatshield followed by a small parachute, and then used rockets to bring them to a soft landing on landing legs [1]. More recent NASA missions (Mars Pathfinder [2], Mars Exploration Rovers

(MER, the *Spirit* and *Opportunity* rovers) [3]) have used a heatshield followed by a single stage parachute and retro-rockets, but used airbags for landing.[1] The unsuccessful European Beagle 2 lander, on the other hand, used a two-stage parachute system and airbags without an intervening rocket stage [5]. NASA's most recent mission, Mars Science Laboratory (MSL, the *Curiosity* rover), used a single-stage parachute followed by an innovative "sky-crane" system which lowered the lander to the ground from a rocket-propelled carrier [1].

The different choices of EDLS have been driven by several different factors. The optimum system is heavily dependent on the lander mass: the best system to use for a small lander is different to that for a large lander. For example, scaling the airbag system successfully used for MER for the considerably heavier MSL was determined to be impossible: the airbag mass grew much more rapidly than the lander mass. Improvements in knowledge about Mars and in technology have also led to evolution in descent and landing systems. The Viking mission added greatly to our knowledge of the Martian atmosphere, reducing the margins required for the landing system. Over the same period, airbag technology improved significantly, mainly driven by automotive applications.

1.3   Planned Missions

We have thus shown that the problem of designing and entry, descent and landing system for a planetary mission is driven by multiple requirements, and that the optimal solution for one mission may look quite different to that for another. Within this paper we consider particularly the missions to Mars that the European Space Agency (ESA) is considering over the next decade:

- ExoMars: the planned European / Russian mission which aims to search for signatures of past or present life on Mars, which will deliver a 600 kg surface platform in October 2016 [6].[2]
- Mars Network Science Mission: a mission to place three small (~ 150 kg) landers onto the Martian surface, to make simultaneous measurement from multiple locations for seismology, geodesy and meteorology studies, which is currently undergoing early-phase concept studies [7].

---

[1] All Mars missions which have used airbags have used the non-vented type, commonly referred to as "bouncy ball airbags", which entirely surround the lander, and where the lander will bounce multiple times before coming to rest. Airbags which are designed to release gas on impact and so only have a single impact ("vented airbags") have been studied, but not flown; such designs have more in common with the airbag systems used in aircraft and automobiles, and have the advantage of only needing to cover one face of the lander [4].

[2] ExoMars consists of several spacecraft to be sent to Mars on two separate launches, in 2016 and 2018. The entry, descent and landing system for the 2016 mission is the responsibility of ESA. The entry, descent and landing system for the 2018 mission is joint the responsibility of the Russian Federal Space Agency, Roscosmos, and ESA. The 2018 mission is not considered in detail here.

- Mars Sample Return, a long-term objective of the science community to return macroscopic samples to Earth, which would use one or more larger landers. A number of studies of various technological developments which will be needed for sample return are ongoing [8].

Each of these future missions has quite different requirements in terms of the mass to be landed, the landing site accuracy required, tolerance of risk and development cost, and so on, demonstrating the advantages of a tool which can be used for trade-offs in support of early-phase sizing studies, identifying the regions of parameter space where each should focus their efforts.

1.4   EDLS Design Problem

Overall, the EDLS design question can be split into two parts. Firstly, sizing calculations are used to find the masses for all components, canopy sizes, airbag designs, etc., for a self-consistent design. These calculations can look at different overall configurations (e.g. single-stage vs two-stage parachute, and with or without retrorockets) to find the optimal design within each configuration, for example varying the size of the two parachutes and the amount of retro fuel to find the system which minimises the overall EDLS mass and stowed volume. The overall optimum design is then found by comparing different configurations. Secondly, the proposed design must be tested through trajectory simulation. The ideal design process would have tight integration of the two parts, using trade-off studies and optimisation to rapidly converge on best design, and then feeding the results of simulation into improved design, to find the best overall choice of design parameters.

Such design trade offs are needed in many engineering problems. Initial trade-off studies, using simplified models, can be used to get a rapid understanding of the complete parameter space. For this purpose, the models need the right balance between detail and complexity: they must capture the essential aspects of the real-world problem, while being simple enough to allow rapid evolution of design. The choice of parameters to trade off is critical, allowing identification of the most promising regions of parameter space, providing a starting point for more detailed investigation.

The decision maker in this case is typically an experienced EDLS engineer, needing to provide supporting justification to the engineering team responsible for the overall mission. The design process required is therefore one which keeps the decision maker very much at the centre, allowing a wide range of designs to be explored and traded off, rather than focusing on the detailed numerical optimisation.

2   Parametric Sizing Tool

The Parametric EDLS Sizing and Design Optimisation tool (PESDO) provides a framework for the design of Entry, Descent and Landing Systems (EDLS). PESDO can be used to perform parametric sizing and end-to-end trade-off studies of entry, descent and landing systems, to estimate component sizes and masses, and to consider trade-offs, such as between mass and landing site accuracy, and finding the optimal balance between the parachute and retro rocket systems. Such trade-offs must be made in the choice of an optimal system.

PESDO is intended for use in relatively early-phase system studies, to explore a wide range of potential mission options. In such studies, PESDO can be used to compare different possible EDLS configurations (e.g. comparing different types of parachutes or different strategies for powered descent) and broadly identify preferred regions within the parameter space of all possible EDLS designs.

PESDO is based on an internal tool developed at Tessella over a number of years to support EDLS design work on missions including Beagle 2 and ExoMars, and for studies such as the Robust Entry, Descent and Landing Guidance and Control Techniques study.[3] Its original form was based on Excel and incorporated engineering estimates, sizing approaches and input data gathered from considerable previous work and expertise in the field (both in and out of the space industry). The tool was upgraded during 2006 to a MATLAB-based form, built on Tessella's generic simulator structure. This has enormously expanded the capabilities of the tool, allowing very large parameter spaces of design points to be investigated easily. The underlying structure brings complete traceability and repeatability and greatly reduces the risk of errors in setting up the calculations. Where possible, the calculations have been validated against independently determined design points, such as past missions.

Having recognised that this capability would prove valuable in studies for a wide range of future missions, ESA asked Tessella to turn PESDO from an internal tool into a deliverable piece of software suitable for use internally by ESA. Tessella have also investigated ESA's general requirements for EDLS design and optimisation capabilities, in view of expected and possible future missions. In recent years, the tool has been used on preliminary studies being run by ESA in preparation for the Network Science and Sample Return missions. During the evolution of the tool, new capabilities have been added to PESDO based on feedback from decision makers regarding the requirements of different missions.

The central concept of the tool is the "design point", a mapping from some specified inputs (such as the chosen EDLS sequence, masses of non-EDLS elements, required triggering values) to the outputs (masses and sizes of the EDLS elements, landing accuracy achieved). By examining suitable ranges and grids of the input values, the sensitivity of the design to these variations can be established, allowing trade-offs to be examined and optimal designs to be identified.

PESDO is not limited to considering obvious parameters such as mass and stowed volume but can also include such aspects as development cost and risk. The sensor requirements for the different designs can also be studied.

The EDLS components available in PESDO include single- and two-stage parachute systems using a number of parachute types, retro-rocket systems including solid- and liquid-fuelled rockets, and landing system, including vented and non-vented

---

[3] For technical background on the EDLS problem and an overview of the role that Tessella's parametric analyses played in ExoMars, the reader is directed to Ref. [9].

airbags.[4] By selecting an appropriate set of components, end-to-end sizing calculations for many different EDLS configurations can be performed.

A simple trajectory model is used to determine a figure of merit for landing site accuracy or height loss performance. The trajectory calculation can also check that the system reaches near-terminal velocity under the parachute, in order to evaluate whether duration of the parachute phase was long enough (i.e. that deployment was high enough). Different triggers for parachute deployment (e.g. timer based, acceleration based or range based) can be considered, and their impact on the dispersion assessed. The calculation can also look at the effects of atmospheric variability, sensor errors and wind unpredictability on landing site accuracy as well as height loss performance. These calculations are not intended to be used for evaluation of the actual system performance (which requires more accurate trajectory simulations), but simply to give a measure of comparison between the different design points. This provides qualitative understanding of the effects of the input parameters on the trajectory performance in a consistent way, a key input for the error budgeting, and indicates invalid areas of the parameter space.

PESDO is designed to be a flexible tool, able to cope with the needs of different users to consider different missions and to perform studies for several phases within a particular mission. It is extensible: the user can define and use new components, allowing the use of PESDO on a wide range of future missions involving as-yet undetermined EDLS components. Existing tools can be easily interfaced to the PESDO framework.

2.1   Example PESDO Visualisations

Some typical PESDO outputs are shown in the following figures. These are the types of outputs which are used by the decision maker to understand the effect that changing the input design parameters has on the outputs (mass, size, etc.) as part of the overall design process as controlled by the decision maker.

Figure 1 shows how the tool can be used to find the mass of the different EDLS components as two key input parameters are varied: the acceleration being provided by the first stage (drogue) parachute when the second (main) parachute is triggered (a measure of drogue size), and the ratio between the ballistic coefficients of the main parachute and the heatshield (a measure of main parachute size). Because the initial position and velocity are the same in all cases, and the three components must between them reduce the velocity to zero just before landing, adjusting these two parameters allows the user to perform a three-way trade-off between the two parachutes and the retro system. As can be seen from the figure, the different components change mass at different rates: using a bigger parachute means that less fuel is required for the retro system, and vice versa. Hence the total descent and landing system mass varies, making it possible to select control parameters to produce the lightest system.

---

[4] The PESDO tool excludes the entry phase, and does not at present consider heatshield sizing, as the entry phase is quite decoupled from the descent phase, and the solution is largely the same for all missions.

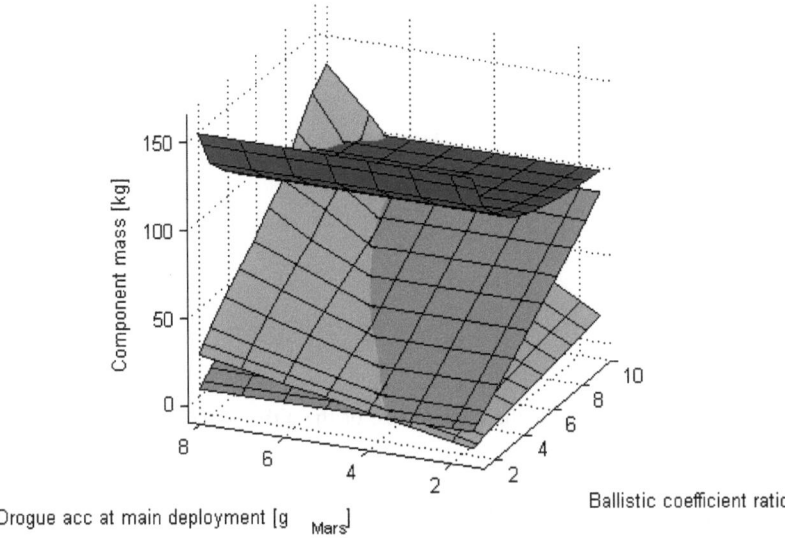

Fig. 1. EDLS component masses as a function of ballistic coefficient ratio and drogue deceleration at main parachute deployment: drogue parachute mass (cyan), main parachute mass (green), retro system mass (red)

It should be noted that PESDO is a tool for early-stage design work. As such, it tends to be used to explore a wide range of the available parameter space, rather than to "optimise" the design parameters to a high level of precision. Beyond the level of accuracy shown here, the differences between the final built system and the sizing estimate will become more significant than the differences between nearby design points, particularly as any engineering design evolves over the course of a lengthy project.

In order to get meaningful sizing estimates of the different EDLS elements, the tool needs to have engineering models of an appropriate level of fidelity. Figure 2 demonstrates the non-vented airbag model; the sizing estimates are based on scaling from the design used by NASA's successful Mars Expedition Rovers, which used six inflated spheres on each face of an approximately tetrahedral lander. The airbag model selects the number of lobes on each face and the size of the lobes, based on the minimum stroke length (minimum distance that the deceleration must take place over in order to keep the loads on the lander below the permitted value, which in this case means a deceleration of less than 40 g) and the size of the lander. The different layers of material (bladder layers, abrasion layers and restraint layers) are treated separately, as is the inflation system needed to produce the gas to inflate the airbag system.

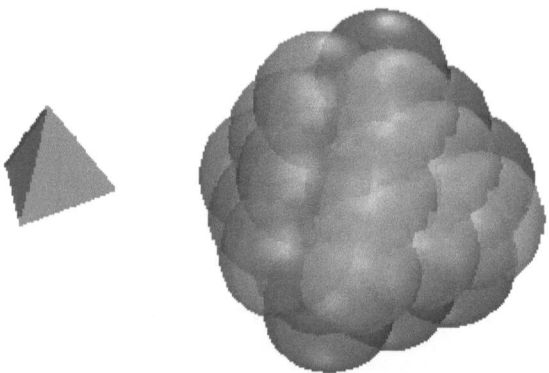

Fig. 2. Airbag design. To show the relative size of the airbags, the lander without airbags is shown on the left.

The tool considers a degree of *local* optimisation within individual EDLS components. In the case of the airbags, this means selecting the optimal number of lobes per side which gives the lightest overall system. In the MER case shown above, there are 6 lobes per side, arranged in billiard rack formation. It would be possible instead to come up with a design with 3 or 10 lobes instead. In this case, a system using 10 spherical lobes is not possible: the spheres are all smaller, reducing the stroke length below the minimum. Three large lobes would be possible, but leads to a heavier airbag system. Having only a small number of options to choose from, this local optimisation is carried out simply by running the sizing calculation for each option and picking the one which gives the lowest mass. We refer to this question as a local optimisation because the number of lobes in an airbag system is decoupled from all of the other EDLS components, unlike the trade-off between parachute and retro mass discussed previously where the choice of control parameter has direct consequences across the entire descent sequence.

The outputs shown in Figure 1 are smoothly varying functions of the inputs, which is relatively common for the EDLS sizing calculations performed in PESDO. However, Figure 2 demonstrates an example of a system which will have step changes in the results: adding an extra row of lobes will lead to a discontinuity in the system properties. Other discrete quantities include the number of layers of material used in the airbags, or the number of lines used in a parachute. Similarly, it is usually necessary to make certain components out of commercially-available materials such as parachute fabrics and cables. There are only a limited range of these available, and so, for example, as the shock load at deployment is increased, we reach a point when we need to make the parachute using a stronger, and hence heavier, fabric, thus step changes occur in the parachute mass.

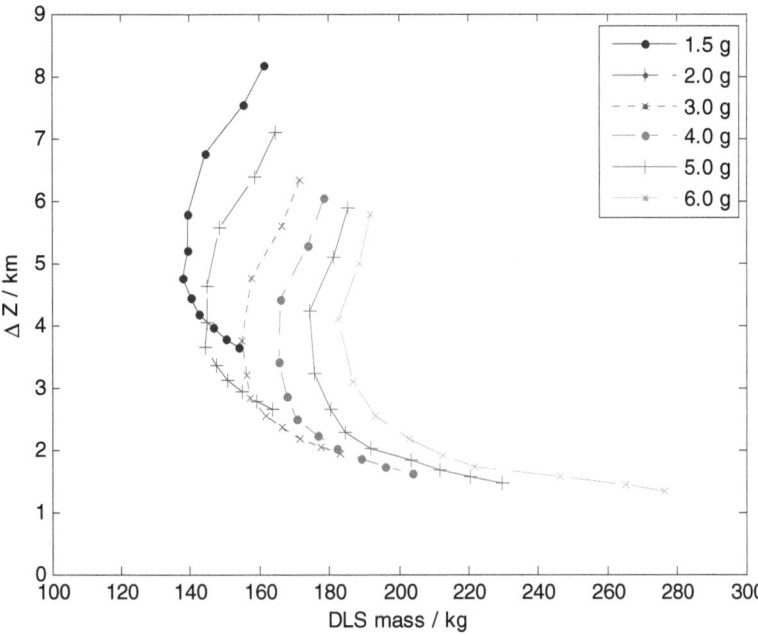

Fig. 3. Trade off between total descent system mass and the height loss performance. The plot shows all of the points in parameter space requested by the decision maker, including both dominated and non-dominated results. The lines between points are drawn to guide the eye.

Figure 3 shows how the tool can be used to study the trade off between descent and landing system mass and the height loss performance that can be achieved, as the sizes of the first and second parachute stages (the same parameters as in Figure 1, corresponding to different lines and different points on the same line respectively) are varied. Systems with bigger main parachutes (on the right) are heavier; however, they also tend to provide more deceleration throughout and hence less height loss. Generally a heavier system can stop in less altitude, permitting its use at higher altitude landing sites, although it can be seen that the curves "turn around" at the left side: making the second-stage parachute smaller leads to more mass being required for the retro system, which beyond a certain point leads to a heavier system overall.

The ideal system would be as close to the bottom left corner as possible. A clear Pareto frontier can be seen in the plot: to get any particular height loss performance requires a minimum EDLS mass; other systems require higher EDLS mass to achieve the same height loss performance. For the present purpose, it is found to be sufficient to produce a plot such as Figure 3 so that the Pareto frontiers can be identified by eye; the plot is useful in that it aids the decision maker directly, rather attempting to provide an "automatically" optimised decision.

The types of investigation discussed above allow the user to compare different design points within a single configuration (a two-stage parachute system followed by retro-rockets and airbags in the example shown). PESDO also allows the user to try out different configurations, to see which one can give the best overall design.

In section 1.1 we listed a number of driving requirements which the tool must address. Of these, the outputs shown above have looked mainly at mass and height loss. Packed volume and landing site accuracy are included similarly. As the tool is currently used, issues such as cost, schedule and reliability are are treated more qualitatively, as matters for the decision maker to consider when choosing between different overall designs, rather than being quantified within the tool.

2.2   Optimisation

We have demonstrated above how PESDO can be used to find "optimal" solutions which consider both local optimisation (within a single EDLS component, i.e. able to be decoupled from the wider problem) and the global picture.

As noted previously, the results of the sizing calculations tend to be relatively smoothly varying functions of the inputs, though sometimes with (a relatively small number of) step discontinuities. Numerical optimisation on such a function is generally not difficult. However, the problem that we face here is that it would be difficult to define a single numerical measurement for optimality, taking into account the different drivers: mass, stowage volume, performance, cost, development required, schedule risk, etc. Although it might be possible to define such a measurement for a specific mission, the balance required between the driving factors varies between missions, and hence a different single measurement would be appropriate for each mission. A tool such as PESDO can provide valuable support to expert users in the process, enabling them to understand and justify the design choices made, rather than attempting to provide the user with a single "answer". As such there is a balance to be found between what the tool can do and what the user needs to do.

3   Conclusions

We have demonstrated that the problem of systems engineering for planetary entry, descent and landing is driven by multiple, sometimes conflicting, criteria. We have demonstrated an approach, and a supporting software tool, which is being successfully used on a number of planned ESA missions to Mars. We have also discussed the limitations to this approach, both engineering (the limits to the accuracy with which it is sensible to model the system) and in terms of the desire of decision makers to have a tool which they can use to investigate and understand trade-offs, rather than to provide a definitive (but actually quite debatable) answer.

Acknowledgements. We thank our colleagues on Entry, Descent and Landing projects for many valuable discussions, in particular Stefano Portigliotti and Fabio Calantropio (Thales Alenia Space Italy), and Leila Lorenzoni, Kelly Geelen, Eric Bornschlegl, Rémi Drai and Christian Philippe (European Space Agency).

References

1. Braun, R.D., Manning, R.M.: Mars Exploration Entry and Landing Challenges. Journal of Spacecraft and Rockets 440 (2), 310–323 (2007)
2. Spencer, D.A., Braun, R.D.: Mars Pathfinder atmospheric entry - Trajectory design and dispersion analysis. Journal of Spacecraft and Rockets 33(5) (1996)
3. Roncoli, R.B., Ludwinski, J.M.: Mission Design Overview for the Mars Exploration Rover Mission. In: AIAA/AAS Astrodynamics Specialist Conference and Exhibit, AIAA-2002-4823. AIAA, Washington, DC (2002)
4. Northey, D., Morgan, C.: Improved Inflatable Landing Systems for Low Cost Planetary Landers. In: 5th IAA International Conference on Low-Cost Planetary Missions. Acta Astronautica (2005)
5. Fallon, E.J. II., Sinclair, R.: Design and Development of the Main Parachute for the Beagle 2 Mars Lander. In: AIAA (2003)
6. Capuano, M., et al.: ExoMars Mission 2016 EDL Technology Demonstrator Mod-ule for landing on Mars surface IAC-2010-A3.3A (2010)
7. European Space Agency CDF Study Report: Assessment of Mars Network Science Mission (MNSM) (2011)
8. European Space Agency, "Mars Sample Return", http://www.esa.int/esaMI/Aurora/SEM1PM808BE_0.html
9. Northey, D., et al.: ExoMars Phase B1: Descent and Landing System Sizing and Sensitivities. In: 7th International ESA Conference on Guidance, Navigation & Control Systems (2008)

Single and Multi-objective *in Silico* Evolution of Tunable Genetic Oscillators

Spencer Angus Thomas and Yaochu Jin

Department of Computing,
University of Surrey
Guildford, Surrey, GU2 7XH, UK
{s.thomas,yaochu.jin}@surrey.ac.uk

Abstract. We compare the ability of single and multi-objective evolutionary algorithms to evolve tunable self-sustained genetic oscillators. Our research is focused on the influence of objective setup on the success rate of evolving self-sustained oscillations and the tunability of the evolved oscillators. We compare temporal and frequency domain fitness functions for single and multi-objective evolution of the parameters in a three-gene genetic regulatory network. We observe that multiobjectivization can hinder convergence when decomposing a period specific based single objective setup in to a multi-objective setup that includes a frequency specific objective. We also find that the objective decomposition from a frequency specified single objective setup to a multi-objective setup, which also specifies period, enable the synthesis of oscillatory dynamics. However this does not help to enhance tunability. We reveal that the use of a helper function in the frequency domain improves the tunability of the oscillators, compared to a time domain based single objective, even if no desired frequency is specified.

Keywords: Gene regulatory networks, *in silico* evolution, sustained oscillation, evolutionary algorithms, multiobjectivization.

1   Introduction

An important area of computational science is systems biology, and over recent years there have been many contributions to the field of biology from computer scientists and mathematicians. Many biological systems lack a global theoretical basis and one way to improve our understanding is to analyse the dynamics of the system *in silico*, i.e. in a computational environment. The evolutionary synthesis, or production, of these dynamics can be tested in computational simulations to investigating biological hypothesis that may be subject to experimental, theoretical and timescale limitations.

Nature is full of complex biological systems and those that are of interest to biologists and computer scientists often consist of large numbers of interacting genes. These complex networks can be broken down into smaller subnetworks

R.C. Purshouse et al. (Eds.): EMO 2013, LNCS 7811, pp. 696–709, 2013.
© Springer-Verlag Berlin Heidelberg 2013

often containing repeating patterns that appear more often than they would in a random network [1, 2]. These repeating patterns are known as network motifs and are believed to be the building blocks of complex biological networks, which are modular in structure [1, 2]. Motifs that produce self-sustained oscillations are particularly important in biological systems [3, 4] and are involved in circadian rhythms [5] and the active transport of hydrogen ions [6]. The ability to accurately tune the period of a genetic oscillator is vital in biological modelling due to the range of oscillator periods observed. Biological oscillatory time scales range from seconds for neuronal and cardiac rhythms [5], to minutes for mitosis cell cycles [7], to hours for the sleep/wake cycle [8] and circadian rhythms [6], to weeks for the ovarian cycle [5] and to years for predator-prey population cycles [9]. Further examples and details about biological oscillations can be found in [5, 10, 11].

To produce oscillations, biochemical systems require negative (repressive) regulatory circuits [4], which also improves robustness to environmental perturbations [12], a necessity of many biological systems [1]. Moreover negative autoregulation (NAR), where a gene will repress its own protein production, results in a rapid response to an input signal which is important for biological systems and are common in biology as it can help reduce noise [1]. Positive feedback loops have been demonstrated to enhance frequency tunability in biological oscillators, with little cost to amplitude, biological systems therefore often contain both positive and negative regulatory circuits [13].

Gene regulatory networks (GRNs) can be modelled through gene-protein interactions, where genes produce proteins, which interact with genes and affect protein production. The dynamics and structure of GRNs is important in the understanding of natural evolution [3] and the analysis of GRN motifs is a growing area of importance in systems biology [1, 14]. A common way to model gene regulatory dynamics is to use differential equations [15]. To generate typical regulatory dynamics *in silico*, evolutionary algorithms (EAs) have widely been used to evolve the parameters and structure of GRNs [3, 16, 17, 18, 19]. However, it is noted that the evolution of oscillatory dynamics is non-trivial and many different objective functions have been suggested to facilitate the evolution of oscillation [3, 16].

In this work, we investigate the use of techniques such as multi-objective optimisation and multiobjectivization to improve the success rate and tunability in evolving sustained genetic oscillators, which have been used to accelerate convergence speed and obtain the global optimum [20, 21, 22].

The differential equations describing the dynamics of the genetic networks studied in this work are introduced in Section 2. Various single and multi-objective fitness setups are proposed in Section 3. The description of the single- and multi-objective EAs adopted in this work is given in Section 4, followed by the experimental results and discussions in Section 5. Section 6 concludes the paper and suggests a few topics for future work.

2   Gene Regulatory Networks

Here we model the GRN, shown in Fig. 1, using the interaction of the genes through their protein production. The GRN contains a negative feedback loop, as this can generate sustained oscillations for interacting genes, and a positive feedback loop, which has been demonstrated to aid evolvability, robustness and tunability of the oscillator when used with a negative loop [13]. All genes in the network also use negative auto-regulation (NAR), where the protein of a gene represses its own production [3]. The use of NAR can decrease response time to an input signal [1], which is an essential function in biological systems. This functionality is important in many prokaryotic transcription networks, such as *E. coli*, which uses NAR in up to 56% of it's expressed transcription factors [23].

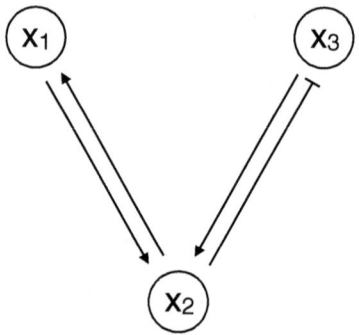

Fig. 1. A consistently regulated motif consisting of three genes with x_2 as the target gene. Here arrows and lines with bar ends represent activating and repressor interactions between the genes respectively. All the genes here also have negative auto-regulations (not shown).

Here we model the protein interactions between the genes using the following differential equations,

$$\dot{x}_1 = a_{12}H_{12}(x_2) - a_{11}x_1 , \tag{1}$$

$$\dot{x}_2 = a_{23}L\Big(H_{23}(x_3), H_{21}(x_1)\Big) - a_{22}x_2 , \tag{2}$$

$$\dot{x}_3 = a_{32}H_{32}(x_2) - a_{33}x_3 . \tag{3}$$

Where $\dot{x}_i$ is the time derivative of x_i, x_2 is the target gene, a_{ij} are regulatory parameters, $H_{ij}(x_j)$ is the Hill function and $L\big(H_{23}(x_3), H_{21}(x_1)\big)$ represents the logic function which combines the interaction of two regulatory genes to the target gene. The Hill function can represent either an activating, $H_{ij}^a(x_j)$, or repressive gene interaction, $H_{ij}^r(x_j)$, as shown in (4).

$$H_{ij}^a(x_j) = \frac{\beta x_j^n}{\theta_i^n + x_j^n}; \quad H_{ij}^r(x_j) = \frac{\beta}{1 + (x_j/\theta_i)^n} , \tag{4}$$

where i and j represent the gene interaction pair, x_i and x_j, n is the Hill coefficient, β is the production rate and θ_i is the threshold for gene x_i. The interactions from x_1 and x_3 to x_2 are combined using summation logic, $L(x, y) = \frac{1}{2}(x + y)$.

3 Single Objective versus Multi-objective Approach

3.1 Single Objective

Time Domain. It is possible to produce self-sustained oscillations in (1), (2) and (3) by reducing the error between the dynamics of only the target gene, described by (2), and a desired oscillatory state of a specified period. As an oscillation can be described using a simple sine wave, we define the desired state of the target gene as

$$x_{tg}^d(t) = \sin\left(\frac{2\pi t}{T}\right) , \tag{5}$$

where T is the period of the oscillator. Using this desired state we define the mean specified time domain fitness, f_{t_s}, for R number of runs of each individual as

$$f_{t_s} = \frac{1}{R} \sum_{r=1}^{R} \sum_{t=0}^{N} \left(x_{tg}^i(r, t) - x_{tg}^d(t)\right)^2 . \tag{6}$$

Here r is the run number, N is the number of time steps and $x_{tg}^i(r, t)$ is the state of the target gene for the *ith* generation of run r at time t. This mean squared error (MSE) method provides a simple, tunable fitness function in the time domain.

Frequency Domain. One difficulty with the MSE method in the time domain is that an individual solution that produces a sustained oscillation that is out of phase with the desired state, given in (5), will lead to a high MSE. This is a consequence of the simple fitness function defined in (6). This high MSE may lead to these individuals being removed from the population, possibly in favour of solutions with a lower MSE value that do not produce oscillations. In order to avoid this potential loss of desired solutions, a fitness function based on the frequency of the oscillation is required. We perform a Fourier Transform on (5) to determine the frequency of the oscillation using the FFTW3 algorithm, as it has been demonstrated to be very efficient for discrete Fourier Transforms [24]. The maximum value in the frequency spectrum denotes the main frequency component of the wave, which is equal to the integral of the curve for a pure sine wave. For the GRN dynamics there may be some chaotic behaviour initially in the temporal plane so the distribution of frequencies may not be a delta-like function as in the case for a pure wave. Therefore in order to produce a sustained oscillation we first define the desired frequency for the target gene,

$$\omega_{tg}^d = MAX \left\{ \hat{F} \left[\sin \left(\frac{2\pi t}{T} \right) \right] \right\} , \tag{7}$$

where MAX denotes the peak in the frequency domain, $\hat{F}[\sin(2\pi t/T)]$ is the Fourier Transform of the desired state of the target gene from (5). To determine the fitness for solution i, we apply a Fourier Transform to the target gene and calculated the MSE as in (6),

$$f_{\omega_s} = \frac{1}{R} \sum_{r=1}^{R} \sum_{t=0}^{N} \left(\hat{F} \left[x_{tg}^i(r, t) \right] - \omega_{tg}^d \right)^2 , \tag{8}$$

where $x_{tg}^i(r, t)$ is the state of the target gene. This is a slightly more complex objective setup than for the time domain, however, due to the efficiency of FFTW3, there is no noticeable effect in computational performance.

3.2   Multi-objective

Time and Frequency Domains. Here we use the MSE fitness definition for both the time and frequency domains in a multi-objective setup. Here we use two objectives, f_1 and f_2, which are equivalent to (6) and (8) respectively. These objectives have the same goal due to the inverse relationship between frequency and period. This setup can therefore be considered a decomposition of the single objectives described in Section 3.1, in to a multi-objective setup. This decomposition is referred to as multiobjectivization and has been shown to aid convergence for certain problems [25, 26].

Time Domain and Non-Specific Fourier Transform. We also investigate a multi-objective setup which uses a Fourier Transform, but not for a specified frequency. In this method we use the MSE fitness function for the time domain as defined in (6) and a frequency domain fitness different from (8). Here we use the Fourier Transform not to specify a desired frequency, but simply to produce an oscillation. To produce an oscillation of a non-specific frequency we use the following fitness function with no need for a desired state

$$f_{\omega_u} = \frac{1}{R} \sum_{r=1}^{R} \frac{1}{MAX \{ \hat{g}(r, \omega) \}} \int \hat{g}(r, \omega) \ d\omega , \tag{9}$$

where $\hat{g}(r, \omega)$ is the Fourier Transform of the target gene dynamics for the rth set of random initial conditions,

$$\hat{g}\left(r,\omega\right) = \hat{F}\left[x_{tg}^{i}\left(r,t\right)\right] \ . \tag{10}$$

The oscillator is tuned using the time domain fitness function, whereas the Fourier Transform is to improve the oscillator success rates and remove the non-oscillatory solutions. In this setup the two objectives are given by (6) and (9) for f_1 and f_2 respectively. Again this is an example of multiobjectivization as the addition of the frequency objective, although it does not specify a frequency, is a helper function that could aid the algorithm in converging to oscillatory solutions. This could help avoid solutions with a low fitness in the time domain objective, such as an equilibrium that corresponds to the point of highest gradient for the oscillations. As with objective decomposition, multiobjectivization through adding helper functions has been shown to aid convergence [25, 27] and may also provide more non-dominated solutions with no extra cost to functional evaluation [22].

4   The Evolutionary Algorithms

To ensure fairness of the comparison of the single and multi-objective fitness setups, we use the same crossover and mutation operators in the single and multi-objective EAs. In addition, both algorithms adopt an elitism strategy. The main difference is that in the single objective optimisation, a deterministic elitism similar to the plus strategy in evolution strategies is adopted, whereas in the multi-objective case, elitist non-dominated sorting operations are used, i.e. NSGA-II [21]. NSGA-II has been shown to be successful at solving a wide range of optimisation problems. In the following, we will present the details of the two EAs.

4.1   Genetic Variations

Simulated Binary Crossover. The standard recombination operator used in binary genetic algorithms (GAs) is the crossover operator, in which segments are taken from the string of values of the parents to form the offspring. For real-coded GAs, however, a simulated binary crossover (SBX) operation can be used. For parent solutions $x_i^{(1,t)}$ and $x_i^{(2,t)}$ to produce offspring solutions $x_i^{(1,t+1)}$ and $x_i^{(2,t+1)}$ first a random number, u_i, between 0 and 1 is chosen and used to determine

$$\beta_{qi} = \begin{cases} \left(2u_i\right)^{\frac{1}{n+1}} & \text{if } u_i \leq 0.5 \\ \left(\frac{1}{2(1-u_i)}\right)^{\frac{1}{n+1}} & \text{otherwise} \ . \end{cases} \tag{11}$$

The offspring solutions are then calculated as follows:

$$x_i^{(1,t+1)} = 0.5\left[\left(1+\beta_{qi}\right)x_i^{(1,t)} + \left(1-\beta_{qi}\right)x_i^{(2,t)}\right] \ , \tag{12}$$

$$x_i^{(2,t+1)} = 0.5 \left[(1 - \beta_{qi}) \, x_i^{(1,t)} + (1 + \beta_{qi}) \, x_i^{(2,t)} \right] . \tag{13}$$

In all simulations we set the recombination probability to 0.9.

Polynomial Mutation. For binary coded algorithms, which are encoded by a fixed length string, discrete mutation operations are used [28]. These operations use a mutation probability to determine if the value of the parameter is flipped. However for real-coded GAs, a polynomial mutation operator is used. For these operations the mutation probability is based on the number of dimensions, Δ, in the problem $p_m = \Delta^{-1}$. The distribution of a spread factor is defined as,

$$P(\beta_{mi}) = 0.5 \, (n + 1) \, (1 - |\beta_{mi}|)^n , \tag{14}$$

where β_{mi} is given by,

$$\beta_{mi} = \begin{cases} (2u_i)^{\frac{1}{n+1}} - 1 & \text{if } u_i \leq 0.5 \\ 1 - (2 \, (1 - u_i))^{\frac{1}{n+1}} & \text{otherwise} . \end{cases} \tag{15}$$

If a mutation in the individual occurs, the parameter value is given as

$$x' = x + (\alpha - \delta) \, \beta_{mi} , \tag{16}$$

where α and δ are the upper and lower bounds for the mutation values respectively. Further details of both the SBX and polynomial mutation operations can be found in [29].

4.2   Single-Objective Selection

For the single objective (SO) setup we use an elitist strategy known as $(\mu + \lambda)$, where after each generation μ parents and λ offspring solutions are combined and ranked in terms of their fitness. The fittest λ solutions of this combination are selected as the parent population for the next generation. We adopt this strategy for both the objective problems given in (6) and (8) and use a population size of 100 for both parent and offspring solutions. This strategy ensures that good solutions are not discarded after each generation, and has been demonstrated to aid convergence in many optimisation problems [17, 21, 30]. It has been widely reported that elitist strategies can lead to premature convergence at local, rather than global, optima for some optimisation problems. However, here we are interested in the production of self-sustained oscillations rather than obtaining a global optimal solution and therefore all solutions that produce oscillatory dynamics are considered successful.

4.3   Multi-objective Selection

The two multi-objective (MO) setups described in Section 3.2 are solved using the elitist non-dominated operations in NSGA-II. Here, after fitness evaluation, parent and offspring solutions are combined and sorted into non-dominated fronts, where front 1 is comprised of non-dominated solutions, front 2 is comprised of solutions that are only dominated by the solutions on front 1, etc. Next the crowding distance operation is applied to all solutions, which is the average distance between a solution and its nearest neighbours on the same non-dominated front. A new population of size μ is then filled from the non-dominated fronts starting with the solutions on front 1, then from successive fronts if there are spaces in the new population. If there are more solutions on a front than spaces in the population the most diverse solutions, i.e. those with the largest crowding distance, are selected. Once this new population is full, two randomly selected solutions are compared in a tournament selection, with a low front wining the tournament. In the case that the solutions are from the same front, the solution with the higher crowding distance is selected as the better solution to promote the diversity of the population. The resulting solutions from the tournament selection form the mating pool for the next generation and the crossover and mutation operations described in Section 4.1 are applied to produce the next generation of offspring solutions. Further details on this algorithms and the operations used can be found in [21].

5   Result and Analysis

5.1   Success Rates: Untuned Oscillators

Single Objective. All objective setups are simulated 50 times for different random number seeds to investigate the success rates of the method at producing self-sustained oscillations. We run all simulations for 100 generations. Oscillations, of varying periods, are observed in 31 runs for the time domain fitness setup described in Section 3.1. For the frequency domain setup, also described in Section 3.1, no oscillations of any period were observed. Although the single objective time domain setup can potentially lead to the loss of oscillatory solutions due to phase shifts, it is still successful at producing oscillations. The frequency domain method is not only unable to avoid the potential problem of phase shift in the time domain, it is unable to lead to oscillatory dynamics at all, sustained or damped.

Multi-objective. For the multi-objective setups, the combination of the time and frequency domain single objectives described in Section 3.2, lead to 24 observed oscillations out of the 50 test cases. The decrease in observed oscillations may be due to the frequency domain objective, which is unable to produce oscillations unaided. To further test the effect of multiple objectives in oscillation production success rates, we also used the setup of the single objective time domain fitness and an untuned Fourier Transform fitness as described in Section 3.2. Here we observed 22 oscillations of the 50 simulations, showing a slight decrease compared to the other multi-objective setup.

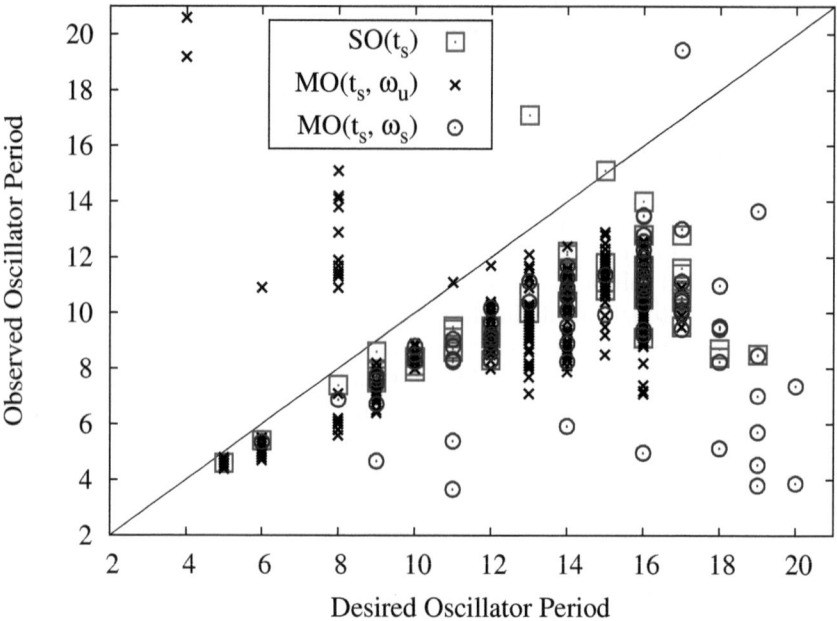

Fig. 2. Tunability of the oscillator for single objective (SO) and multi-objective (MO) setups. The objective setup is indicated by t_s, w_s and w_u for (6), (8) and (9) respectively. Here Included is a $y = x$ line for clarity.

5.2   Success Rates: Tuned Oscillators

We also investigated the tunability of the oscillator period using the four setups detailed in Section 3. Here we simulate oscillator periods varying from 1 to 20 for 50 test cases. The oscillators are tuned according to (6) and (8), however the observed oscillations produced appear to vary in period for each test case. The periods of the successful oscillations are measured and compared to the desired oscillator period shown in Fig. 2. Here we compare the observed oscillator periods for all setups against the desired periods. The single objective frequency domain method is not included as there were no observed successful oscillations. The multi-objective (MO) setups provide more solutions than the single objective (SO) setup because the MO simulations contain an archive of solutions which cannot be compared directly and so all oscillatory solutions are plotted. For a given MO simulation there may be several solutions in the archive that produce oscillations, but not necessarily all. This leads to a range of observed periods for each simulation at each desired period. For MO simulations with more than one oscillatory solution, the observed periods are averaged to give the datum point in Fig. 3. The error bars are the maximum and minimum periods observed in that simulation and represent the spread of that values. For the MO simulations that only have one successful solution a single point only is included.

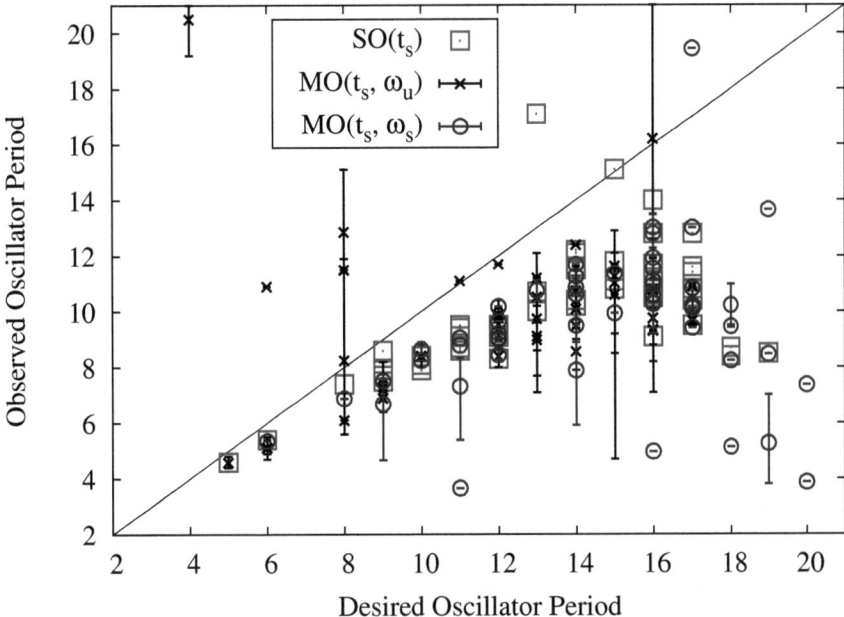

Fig. 3. Averages for multi-objective (MO) results. The MO points represent the average oscillator periods and the error bars represent the maximum and minimum values for each simulation; see text for details.

These results indicate that neither the single or multiple objective setups are tunable for the small three gene configuration, Fig. 1, used here. The SO setup in the time domain appears to be tunable for low period oscillations $T \leq 9$. At higher oscillator periods however, this setup becomes less tunable and begins to diverge from the desired period. The MO methods both demonstrate a large spread of periods for successful runs. Though the average values appear to follow a similar pattern to the single objective setup, with the exception of a few outliers, this does not indicate a tunable oscillator. Average values are used here as there is no direct way to compare archive solutions from the MO setup and so we cannot say if a solution is better than another without further analysis. Even when including all MO archive solutions only two solutions fall on the line indicating the observed period is the desired period in Fig. 2. These two solutions are at $T = 11$ and $T = 12$ and are the only oscillatory solutions from the archive for their respective runs. This is represented by the fact that they do not have bars on the values in Fig. 3. It is however worth noting that the SO setup was unable to produce an oscillation at the period.

5.3 Discussion

No objective setup investigated in this work is able to reliably produce tunable oscillations. One notable observation here was the inability for a simple frequency based objective function to produce an oscillation when an equivalent time domain objective is successful. This may be a result of the desired state used as the Fourier Transform of a pure sine wave will produce a single peak in the frequency domain. However the dynamics of a randomly initialised GRN may initially be completely different and thus the EA is not able to find a parameter set that is able to produce a pure sine wave and therefore a single peak in the frequency plane. When considering these aspects, and that the fitness only determines the location of the peak with no width constraints, this objective is more complex than the time domain objective. This is in contrast to (9), where the width constraint comes from the minimisation of the MSE and therefore the area of the curve. Thus, for the specified frequency objective, very broad peaks that have low maximum values are considered good solutions if they peak at the correct frequency regardless of the rest of the frequency profile. Adding other objectives such as minimising the width of the peak could also be used. However incorporating this into an objective that also specifies the location of the peak, and thus the frequency, may be nontrivial. Further investigation in to the frequency domain objective setup is needed in order to tune the oscillator.

The addition of a second objective based on an unspecified frequency, (9), leads to a spread in the observed oscillator period. This is a consequence of the second objective not specifying an oscillator period. Solutions that are oscillatory, but not of the required period, will be retained if the value of (9) is low despite the value of the other objective, (6). This gives an indication as to why there is a large spread of observed oscillations and is likely to be a consequence of the selection operations in NSGA-II. The crowding distance operation used in the non-dominant sorting part of NSGA-II favours more diverse solutions to provide a wide-spread Pareto front. This, however, will ultimately lead to solutions with a low value for the unspecified frequency objective, (9), even at the cost of a more optimal time domain objective. Thus in the Pareto optimal solutions there may be many different oscillator periods representing the optimal solutions for the unspecified frequency based objective. The addition of this objective has however aided the convergence to tuned oscillators. It is able to produce solutions that are close to the desired period for a larger range than the SO setup using the same time domain objective. The improvement can be seen for periods $T = 5, 6, 10, 11, 12$ and 14, with only a slight decrease in desired period for $T = 8$ and 9 compared with the SO setup (see Fig. 2). However there are potentially more solutions for this setup due to the Pareto front, thus more analysis is required for this method than for a SO method.

In comparison with the above setups, the MO method, which specifies the period and frequency, should help remove the spread of observed periods. However, this is not the case, and this setup performs worse with increasing oscillator period. This cause of failure is likely to be the same as the SO specific frequency setup. However it is interesting that the addition of another objective, that specifies the period, enables this set to produce oscillations, though does not improve tunability. This demonstrates that the decomposition of a SO problem into a MO one can aid convergence for the case of the SO frequency domain objective compared with the MO time and specified frequency domain setups. Here the

SO frequency domain was unable to produce any oscillations but the addition of a time domain objective enabled oscillatory dynamics to be produced from the GRN.

6   Conclusions and Future Work

Both the decomposition of a single objective problem into a multiple objective problem and the use of a helper function have effects on the convergence of the problem investigated here. For the case of objective decomposition, we find that the addition of a specified frequency based method to a time based setup does not aid convergence, and performs worse in some cases. However for the case of adding a time based objective to a frequency based setup, objective decomposition enables the evolution of oscillations though does not aid tunability. We also observe that the use of an unspecified frequency based helper objective to a time based objective shows improvements in tunability at some periods compared with the single objective setup. This indicates that the effect of multiobjectivization on performance is dependent on the problem and a general statement on the effects cannot be made, which was also observed in [22, 25]. The results here also indicate that the effects of multiobjectivization may also depend on the objective setup and fitness domain used in the optimisation problem.

Objective setup in the *in silico* evolution of oscillatory dynamics of gene regulatory networks requires much more investigation, particularly the single objective frequency domain objective. Further investigation into objective setup could lead to a tunable oscillator using a small gene regulatory networks and evolutionary algorithms. Network size and loop combination have also been shown to be important in tuning oscillators [13], therefore a wider range of gene regulatory network structures, such as gene number, logic functions, auto-regulation and feedback setups should also be investigated.

Acknowledgment. This research is funded by an EPSRC DTC studentship.

References

[1] Alon, U.: Network motifs: theory and experimental approaches. Nat. Rev. Genet. 8, 450–461 (2007)
[2] Milo, R., Shen-Orr, S., Itzkovitz, S., Kashtan, N., Chklovskii, D., Alon, U.: Network motifs: Simple building blocks of complex networks. Science 298(5594), 824–827 (2002), http://www.sciencemag.org/content/298/5594/824.abstract
[3] Jin, Y., Sendhoff, B.: Evolving in silico bistable and oscillatory dynamics for gene regulatory network motifs. In: IEEE World Congress on Computational Intelligence Evolutionary Computation, CEC 2008, pp. 386–391 (June 2008)
[4] Gonze, D.: Coupling oscillations and switches in genetic networks. Biosystems 99(1), 60–69 (2010), http://www.sciencedirect.com/science/article/pii/S030326470900149X
[5] Goldbeter, A.: Biochemical Oscillations and Cellular Rhythms, vol. 1. Cambridge University Press (April 1997), http://dx.doi.org/10.1017/CBO9780511608193

[6] Chay, T.R.: A model for biological oscillations. Proceedings of the National Academy of Sciences of the United States of America 78, 2204–2207 (1981), http://ukpmc.ac.uk/abstract/MED/6264468

[7] Francis, M.R., Fertig, E.J.: Quantifying the dynamics of coupled networks of switches and oscillators. PLoS ONE 7(1), e29497 (2012), http://dx.doi.org/10.1371%2Fjournal.pone.0029497

[8] Rempe, M., Best, J., Terman, D.: English A mathematical model of the sleep/wake cycle. Journal of Mathematical Biology 60, 615–644 (2010), http://dx.doi.org/10.1007/s00285-009-0276-5

[9] Ennes, R.H., McGuire, G.C.: Nonlinear Physics with Mathematica for Scientists and Engineers. Birkhäuser, Boston (2001)

[10] Fall, C.P., Marland, E.S., Wagner, J.M., Tyson, J.J.: Compuatational Cell Biology. In: Antman, S.S., Marsden, J.E., Sirovich, L., Wiggins, S. (eds.), vol. 20. Springer (July 2002)

[11] Berridge, M., Rapp, P.: A comparative survey of the function, mechanism and control of cellular oscillators. The Journal of Experimental Biology 81, 217–279 (1979), http://europepmc.org/abstract/MED/390080

[12] Jin, Y., Meng, Y., Sendhoff, B.: Influence of regulation logic on the easiness of evolving sustained oscillation for gene regulatory networks. In: IEEE Symposium on Artificial Life, ALife 2009, pp. 61–68 (April 2009)

[13] Tsai, T.Y.-C., Choi, Y.S., Ma, W., Pomerening, J.R., Tang, C., Ferrell, J.E.: Robust, tunable biological oscillations from interlinked positive and negative feedback loops. Science 321(5885), 126–129 (2008), http://www.sciencemag.org/content/321/5885/126.abstract

[14] Alon, U.: An Introduction to Systems Biology: Design Principles of Biological Circuits. Chapman & Hall/CRC (2006)

[15] Ito, S., Izumi, N., Hagihara, S., Yonezaki, N.: Qualitative analysis of gene regulatory networks by satisfiability checking of linear temporal logic. In: Proceedings of the 2010 IEEE International Conference on Bioinformatics and Bioengineering, ser. BIBE 2010, pp. 232–237. IEEE Computer Society, Washington, DC (2010), http://dx.doi.org/10.1109/BIBE.2010.45

[16] Chu, D.: Evolving genetic regulatory networks for systems biology. In: IEEE Congress on Evolutionary Computation, CEC 2007, pp. 875–882 (September 2007)

[17] Jin, Y., Meng, Y.: Emergence of robust regulatory motifs from in silico evolution of sustained oscillation. Biosystems 103(1), 38–44 (2011), http://www.sciencedirect.com/science/article/pii/S0303264710001693

[18] Sirbu, A., Ruskin, H.J., Crane, M.: Comparison of evolutionary algorithms in gene regulatory network model inference. BMC Bioinformatics 11, 59 (2010)

[19] Thomas, S.A., Jin, Y.: Combining genetic oscillators and switches using evolutionary algorithms. In: 2012 IEEE Symposium on Computational Intelligence in Bioinformatics and Computational Biology (CIBCB), pp. 28–34 (May 2012)

[20] Handl, J., Kell, D.B., Knowles, J.: Multiobjective optimization in bioinformatics and computational biology. IEEE/ACM Trans. Comput. Biol. Bioinformatics 4(2), 279–292 (2007), http://dx.doi.org/10.1109/TCBB.2007.070203

[21] Deb, K., Pratap, A., Agarwal, S., Meyarivan, T.: A fast and elitist multiobjective genetic algorithm: Nsga-ii. IEEE Transactions on Evolutionary Computation 6(2), 182–197 (2002)

[22] Handl, J., Lovell, S.C., Knowles, J.D.: Investigations into the Effect of Multiobjectivization in Protein Structure Prediction. In: Rudolph, G., Jansen, T., Lucas, S., Poloni, C., Beume, N. (eds.) PPSN X. LNCS, vol. 5199, pp. 702–711. Springer, Heidelberg (2008), http://dx.doi.org/10.1007/978-3-540-87700-4_70

[23] Silva-Rocha, R., de Lorenzo, V.: Noise and robustness in prokaryotic regulatory networks. Annual Review of Microbiology 64(1), 257–275 (2010), http://www.annualreviews.org/doi/abs/10.1146/annurev.micro.091208.073229

[24] Frigo, M., Johnson, S.G.: The design and implementation of FFTW3. Proceedings of the IEEE 93(2), 216–231 (2005); Special issue on "Program Generation, Optimization, and Platform Adaptation

[25] Brockhoff, D., Friedrich, T., Hebbinghaus, N., Klein, C., Neumann, F., Zitzler, E.: Do additional objectives make a problem harder?". In: Proceedings of the 9th Annual Conference on Genetic and Evolutionary Computation, ser. GECCO 2007, pp. 765–772. ACM, New York (2007), http://doi.acm.org/10.1145/1276958.1277114

[26] Knowles, J.D., Watson, R.A., Corne, D.W.: Reducing Local Optima in Single-Objective Problems by Multi-objectivization. In: Zitzler, E., Deb, K., Thiele, L., Coello Coello, C.A., Corne, D.W. (eds.) EMO 2001. LNCS, vol. 1993, pp. 269–283. Springer, Heidelberg (2001), http://dl.acm.org/citation.cfm?id=647889.736521

[27] Jensen, M.T.: Helper-objectives: Using multi-objective evolutionary algorithms for single-objective optimisation. Journal of Mathematical Modelling and Algorithms 3, 323–347 (2004), http://dx.doi.org/10.1023/B:JMMA.0000049378.57591.c6, doi:10.1023/B:JMMA.0000049378.57591.c6

[28] Goldberg, D.E.: Genetic Algorithms in Search, Optimization and Machine Learning, 1st edn. Addison-Wesley Longman Publishing Co., Inc., Boston (1989)

[29] Deb, K., Agrawal, R.B.: Simulated binary crossover for continuous search space. Complex Systems 9, 115–148 (1995)

[30] Mendoza, M.R., Bazzan, A.L.C.: Evolving random boolean networks with genetic algorithms for regulatory networks reconstruction. In: Proceedings of the 13th Annual Conference on Genetic and Evolutionary Computation, ser. GECCO 2011, pp. 291–298. ACM, New York (2011), http://doi.acm.org/10.1145/2001576.2001617

An Application of a Multicriteria Approach to Compare Economic Sectors: The Case of Sinaloa, Mexico

Juan Carlos Leyva López, Diego Alonso Gastélum Chavira, and Margarita Urías Ruiz

Universidad de Occidente
Blvd. Lola Beltrán y Blvd. Rolando Arjona
Culiacán, Sinaloa, México
{juan.leyva,diego.gastelum}@udo.mx,
maggie_1777@uaim.edu.mx

Abstract. In this paper, a multicriteria approach for ranking the performance of the economic sectors of the Sinaloa economy is proposed, and the most attractive sectors are identified. To achieve this goal, the software SADAGE was used for solving ranking problems, which require one to rank a set of alternatives - given evaluations in terms of several criteria - in decreasing order of preference. The approach uses the ELECTRE III method to construct a valued outranking relation and then a multiobjective evolutionary algorithm (MOEA) to exploit the relation to obtain a recommendation. The retail and manufacturing sectors were ranked first in all the rankings; the utilities sector was ranked second in all the rankings; the mining sector and the management of companies and enterprises sector were ranked lowest. The results of this application can be useful for investors, business leaders, and policy-makers. This study also contributes to an important, yet relatively new, body of application-based literature that investigates multicriteria approaches to decision making that use fuzzy theory and evolutionary multi-objective optimization methods.

Keywords: Multicriteria Decision Analysis, Economic Sectors, Ranking Problem, ELECTRE III, Multiobjective Evolutionary Algorithms.

1 Introduction

One of the most important requirements for planning the economic development of developing countries is to be able to promote different economic sectors appropriately to contribute most effectively toward solving social, economic and other related problems. (Sudaryanto, 2000).

The Mexican economy, like economies in other parts of the world, must address the new realities, challenges, and opportunities presented by the globalization of business activities.

Firms, industries and entire sectors operating within the Mexican economy have experienced varying degrees of success in coping with the competitive global economic environment. Therefore, investors and policy-makers must assess economic performance in a relatively new context.

R.C. Purshouse et al. (Eds.): EMO 2013, LNCS 7811, pp. 710–725, 2013.

The relative performance of sectors within a given economy can be assessed using different types of traditional methods. It is important to select a method that is systematic, practical and proven. Such an evaluation method should be multicriteria in nature because of the multidimensional nature of economic and business performance.

This paper addresses the application of a Multicriteria Decision Aiding (MCDA) method to the evaluation of the relative performance of sectors in the Sinaloa economy. The multicriteria analysis is intended to offers stakeholders (in particular, policy-makers, investors and business leaders) a structured approach for the evaluation of the relative performance of economic sectors. To do this, specific economic indicators are considered and thus the assessment of the relative performance of economic sectors is made more transparent. The decision analysis performed by this approach aids the decision-making process because it permits a relatively large problem to be broken down into a set of less complex situations (Autran et al., 2011).

In recent years, multicriteria-based methods have been employed to assess the performance of economic sectors and have yielded decision-making implications (e.g., Augusto et al., 2005; Balezentis et al., 2012; Sudaryanto, 2000). However, such applications are still limited in number and scope. This relatively small number of applications is interesting because multicriteria methods can be adapted to the economic and social sciences (Treadwell, 1995).

This study utilizes a multicriteria approach to construct an aggregation model of preferences and then a multiobjective evolutionary algorithm to exploit the model to rank the performance of economic sectors of the Sinaloa economy. While such an application has practical implications, the method has not yet been sufficiently developed. This study also contributes to an important, yet relatively new, body of application-based literature that concerns a multicriteria, and multiobjective evolutionary approach to decision-making.

This paper is organized as follows: the second section presents a brief description of the relevant literature concerning the performance of economic sectors. The third section describes a study and focuses on the procedure and method used. The fourth section describes a sensitivity analysis of the final result. The fifth section presents results and a brief discussion. The final section presents concluding comments.

2   Literature Review

Most social, economic, biological and environmental systems are complex in nature; therefore, measuring their performance is a multifaceted and difficult task (Augusto et al., 2005). Thus, economic sectors are not easy to compare. In practice, several approaches can be used to measure the performance of economic systems. These approaches include multiple criteria optimization (Steuer, 1986), multiple attribute decision theory (Keeney and Raiffa, 1993) and multicriteria decision aiding (Roy, 1996). The ELECTRE methods (Roy, 1996) are a group of well-known decision aiding methods. In recent years, a vast number of applications with ELECTRE methods to performance ranking problems were developed (Karagiannidis and Moussiopoulos, 1997; Rogers and Bruen, 1998; Salminen et al., 1998; Teng and Tzeng, 1994;

Martel et al., 1988; Beccali et al., 1998; Georgopoulou et al., 1997; Siskos and Hubert, 1983; Blondeau et al., 2002; Colson, 2000; Augusto et al., 2005; Leyva, 2005).

Fuzzy set theory (Zadeh, 1965) is also significant in the social sciences and humanities because it can treat ambiguities, uncertainties, and vagueness that cannot be treated by methods that use crisp values. Balezentis et al. (2012) presented an integrated assessment of Lithuanian economic sectors based on financial ratios and fuzzy Multicriteria Decision Making (MCDM) methods. Three fuzzy MCDM methods were applied in this study: VIKOR (Kaya and Kahraman 2011), TOPSIS (Yu and Hu, 2010), and ARAS (Turskis and Zavadskas, 2010).

Sudaryanto (2000) described the application of a fuzzy multi-attribute decision-making model for the empirical identification of the key sectors of the Indonesian economy. Diaz et al (2006) presented a fuzzy clustering approach to identify the key sectors of the Spanish economy. Furthermore, Misiūnas (2010) analyzed the performance of Lithuanian economic sectors using financial analysis. As demonstrated in previous studies (Xidonas and Psarras 2009; Xidonas et al. 2009, 2010), the application of multicriteria decision making methods significantly improves the robustness of financial analysis and business decisions. Balezentis et al (2012) proposed a method of inter-sectoral comparison based on financial indicator analysis that uses multicriteria decision aiding methods.

Finally, evolutionary algorithms are beginning to be used in the outranking approach to address large-scale problems and to mitigate the complexity of some computations in the outranking methods; the complexity is primarily due to the nonlinearity of the formulas used in these methods (Figueira et al., 2010).

3   The Study

3.1   Research Framework

A decision-aiding method is only relevant for decision processes that involve decision makers. In this paper, we will focus our attention on the set of activities (steps) occurring within such a setting. Tsoukias (2007) called such a set of activities a "decision aiding process". The ultimate objective of this process is to arrive at a consensus between the decision maker and the analyst. The decision maker has domain knowledge concerning the decision process. In contrast, the analyst has methodological, domain-independent knowledge. Given the decision maker's domain knowledge and the analyst's methodological knowledge, the analyst must interpret the decision maker's concerns and knowledge so that he or she can improve his or her perceived position compared with the reference decision process. Such an interpretation ought to be "consensual" (Tsoukias, 2007).

The multicriteria approach utilized in this study combines the logic of outranking models (the ELECTRE III procedure (Roy, 1996)) with multiobjective evolutionary algorithms (MOEA) (Leyva and Aguilera, 2005), aided by the SADAGE Software (Leyva et al., 2008), to solve the ranking problem.

Configuration of the Decision Aid Process. In a systematic decision aid process, there is a continuous flow of activities between the different phases, but at any phase, there may be a return to a previous phase (this is referred to as feedback). The general scheme of the ELECTRE III–MOEA method is schematically represented in Figure 1. A decision aiding process is not a linear process where the stages follow one another. Instead, it should be noted that the procedure is iterative rather than simply sequential. If the decision maker is unsatisfied with the result at any stage, he or she may return to any step and redo it.

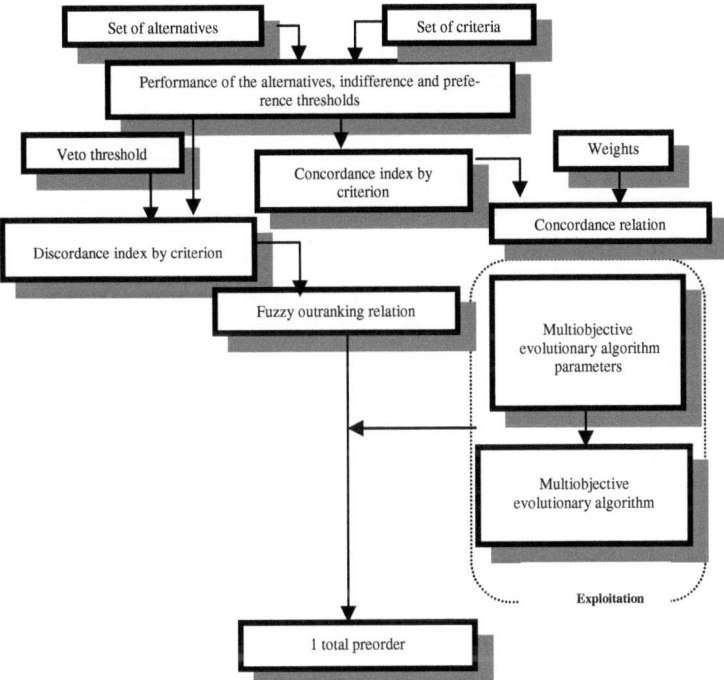

Fig. 1. General scheme of the ELECTRE III–Multiobjective Evolutionary Algorithms

3.2   Data Source

The data used in this study were obtained from a database supplied by The National Institute of Statistics and Geography (Instituto Nacional de Estadística y Geografía, INEGI, http://www.inegi.gob.mx), which performs the economic census in Mexico. The data are part of the 2009 Economic Census.

The objective of the census is to obtain updated and reliable basic statistical data about establishments that manufacture goods, trade merchandise and render services to generate various detailed geographic, sectoral, and thematic economic indicators for Mexico. The classification used for the census is the North American Industry Classification System (NAICS) 2007.

Table 1 presents the dominant economic sectors in Sinaloa, Mexico.

Table 1. Dominant economic sectors in Sinaloa, Mexico

Sector code (alternative)	Economic sector
A_1	21 (212) Mining (except Oil and Gas)
A_2	22 Utilities (Electricity, Water and Gas Distribution to Final Customer)
A_3	23 Construction
A_4	31-33 Manufacturing
A_5	42 Wholesale trade
A_6	44-45 Retail trade
A_7	48-49 Transportation and Warehousing
A_8	51 Information
A_9	52 Finance and Insurance
A_{10}	53 Real Estate and Rental and Leasing
A_{11}	54 Professional, Scientific and Technical services
A_{12}	55 Management of Companies and Enterprises
A_{13}	56 Administrative and Support and Waste Management and Remediation Services
A_{14}	61 Educational services
A_{15}	62 Health Care and Social Assistance
A_{16}	71 Arts, Entertainment, and Recreation
A_{17}	72 Accommodation and Food Services
A_{18}	81 Other Services (except Public Administration)

3.3   Criteria

According to Bouyssou (1990) the criteria family should be legible (containing sufficiently small number of criteria), operational, exhaustive (containing all points of view), monotonic and non-redundant (each criterion should be counted only once). These rules provide a coherent family of criteria. The criteria family used to rank the economic sectors are primarily economic. The criteria used in this study are reported in Table 2. These criteria are designed to capture the multidimensional nature of the performance of the studied sectors. These criteria include the following, where the last six are expressed in millions of Mexican pesos and all of them are defined with increasing preference direction in Table 2:

- Number of employees
- Remunerations
- Total gross production
- Intermediate consumption
- Gross fixed capital formation
- Gross value added
- Total fixed assets

Table 2. Values of the criteria for each economic sector

Sector code (alter-native)	Economic sector	Number of em-ployees	Remu-nera-tions	Total gross pro-duc-tion	Inter-me-diate con-sump-tion	Gross fixed capital forma tion	Gross value add-ed	Total fixed as-sets
A_1	21 (212) Mining (except Oil and Gas)	1192	64	400	156	83	243	373
A_2	22 Utilities (Electricity, Water and Gas Distribution to Final Customer)	6257	1235	15607	7137	904	8469	3962 4
A_3	23 Construction	22440	1172	11150	6910	188	4240	2384
A_4	31-33 Manufacturing	58804	2729	35553	24376	824	11176	1447 8
A_5	42 Wholesale trade	32044	1933	13103	4915	386	8187	5779
A_6	44-45 Retail trade	130186	3031	17728	8625	2344	9103	1958 8
A_7	48-49 Transportation and Warehousing	22529	976	6708	3396	275	3312	5832
A_8	51 Information	5869	914	7897	4668	407	3229	4618
A_9	52 Finance and Insurance	3906	3471	1329	938	17	390	238
A_{10}	53 Real Estate and Rental and Leasing	6331	198	2473	970	36	1502	1575
A_{11}	54 Professional, Scientific and Technical Services	9710	527	1797	574	42	1223	656
A_{12}	55 Management of Companies and Enterprises	931	0	433	257	3	176	29
A_{13}	56 Administrative and Support and Waste Management and Remediation Services	17789	956	2424	810	288	1613	1423
A_{14}	61 Educational Services	11941	835	2102	498	45	1604	999
A_{15}	62 Health Care and Social Assistance	14461	304	1378	605	32	772	1122
A_{16}	71 Arts, Entertainment, and Recreation	6286	128	1199	558	47	640	1017
A_{17}	72 Accommodation and Food Services	43916	1082	6348	3889	235	2458	4406
A_{18}	81 Other Services (except Public Adminis-tration)	32533	663	2890	1422	92	1467	2987

Table 2 reports the values of the criteria for sector. The results in Table 2 unders-core the differences that exist among the studied sectors based on the different meas-ures used.

3.4   Procedure and Methodology

Multiple factors motivated the selection of the ELECTRE III method for the assess-ment of the performance of the economic sectors of Sinaloa, Mexico.

First, Leyva and Aguilera (2005) presented a MOEA to exploit a valued outranking relation, but it is interesting to demonstrate the functionality of the combination of ELECTRE III and MOEA to a real-world application. This method was systematized using the SADAGE software (Leyva et al., 2008), which was used to analyze the problem addressed in this study.

Second, there exist a set of discrete alternatives and a set of economic dimensions that can be easily converted into a set of criteria. Additionally, the problem type ad-dressed in this study can be modeled as a ranking problem. Based on the literature, the

ELECTRE family of methods is considered appropriate for addressing a problem type such as the one addressed in this study (see Roy, 1996). This is especially true for the ELECTRE III method.

Third, ELECTRE was originally developed by Roy to incorporate the fuzzy (imprecise and uncertain) nature of decision-making by using thresholds of indifference and preference. This feature is appropriate for solving this problem.

Fourth, the decision maker is required to assign numerical values to the inter-criteria parameters associated with the different criteria (Roy, 2006).

Fifth, another feature of ELECTRE that distinguishes it from many multicriteria solution methods is that it is fundamentally non-compensatory. This means that good scores on some criteria cannot compensate for a very bad score on a different criterion.

Finally, another feature is that ELECTRE models allow incomparability. Incomparability, which should not be confused with indifference, occurs between some alternatives a and b when there is no clear evidence in favor of some type of preference or indifference.

Two important concepts that underline the ELECTRE approach, thresholds and outranking will now be discussed. Assume that there exist defined criteria $g_j, j=1,2,...r$ and a set of alternatives A. Traditional preference modeling assumes that the following two relations hold for the two alternatives $a,b \in A$:

$$aPb \quad (a \text{ is preferred to } b) \quad \Leftrightarrow \quad g(a) > g(b)$$
$$aIb \quad (a \text{ is indifferent to } b) \quad \Leftrightarrow \quad g(a) = g(b)$$

In contrast, the ELECTRE methods introduce the concept of an indifference threshold, q; then, the preference relations are redefined as follows:

$$aPb \quad (a \text{ is preferred to } b) \quad \Leftrightarrow \quad g(a) > g(b) + q$$
$$aIb \quad (a \text{ is indifferent to } b) \quad \Leftrightarrow \quad |g(a) - g(b)| \leq q$$

Whereas the introduction of this threshold partially accounts for how a decision maker actually feels when making real comparisons, a problem remains. Namely, there is a point at which a decision maker changes from indifference to strict preference. Conceptually, it is justified to introduce a buffer zone between indifference and strict preference that corresponds to a decision maker hesitating between preference and indifference. This zone of hesitation is referred to as weak preference; it is also a binary relation like P and I above and is modeled by introducing a preference threshold, p. Thus, we have a double threshold model with an additional binary relation Q that measures weak preference:

$$aPb \quad (a \text{ is strongly preferred to } b) \quad \Leftrightarrow \quad g(a) - g(b) > p$$
$$aQb \quad (a \text{ is weakly preferred to } b) \quad \Leftrightarrow \quad q < g(a) - g(b) \leq p$$
$$aIb \quad (a \text{ is indifferent to } b; \text{ and } b \text{ to } a) \quad \Leftrightarrow \quad |g(a) - g(b)| \leq q$$

The choice of thresholds intimately affects whether a particular binary relation holds. Although the choice of appropriate thresholds is not easy, in most realistic decision-making situations, there are good reasons for choosing non-zero values for p and q.

Note that we have only considered the simple case where thresholds p and q are constants instead of functions of the values of the criteria; the latter is the case of variable thresholds. While the simplification of using constant thresholds aids the utilization of the ELECTRE method, it may be worth using variable thresholds in cases where criteria with larger values lead to larger indifference and preference thresholds. In this study, a government official acted as the decision maker and the authors of this paper acted as the analyst. Table 3 reports the indifference and preference thresholds for the criteria used in this study. The veto threshold was not considered.

Using thresholds, the ELECTRE method seeks to build an outranking relation S. aSb means that according to the global model of decision-maker preferences, there are good reasons to believe that "a is at least as good as b" or "a is not worse than b". Each pair of alternatives a and b is then tested to check whether the assertion aSb is valid. This yields one of the following four situations: aSb and $not(bSa)$; $not(aSb)$ and bSa; aSb and bSa; $not(aSb)$ and $not(bSa)$.

The third situation corresponds to indifference, whereas the fourth corresponds to incomparability.

Table 3. Indifference (q) and preference (p) threshold values

	Criterion (g_j)	Indifference (q_j)	Preference (p_j)
g_1	Number of employees	6000	14000
g_2	Remunerations	250	400
g_3	Total gross production	200	500
g_4	Intermediate consumption	300	600
g_5	Gross fixed capital formation	200	400
g_6	Gross value added	250	500
g_7	Total fixed assets	1100	2100

Using thresholds, the ELECTRE method seeks to build an outranking relation S. aSb means that according to the global model of decision-maker preferences, there are good reasons to believe that "a is at least as good as b" or "a is not worse than b".

The thresholds and weights represent the subjective input provided by the decision maker. Weights used in the non-compensatory ELECTRE model are significantly different from weights used in compensatory decision modeling approaches. Weights in ELECTRE are "coefficients of importance" and, as Vincke (1992) notes, they can be considered votes for each of the criterion "candidates." Roger et al. (2000) reviewed existing weighting schemes for ELECTRE and provided a useful discussion of the weighting concept in ELECTRE. Care also must be taken in determining threshold values, which must relate specifically to each criterion and reflect the preferences of a decision maker. Procedures for choosing appropriate threshold values were addressed by Roger and Bruen (1998). The decision maker was assisted in defining the 7 criteria weights, which are shown in Table 4. Personal Construct Theory (PCT), as suggested by Rogers et al. (2000), was used for the weight definition.

Table 4. Criteria weights

	g_1	g_2	g_3	g_4	g_5	g_6	g_7	RtG	$RtG+1$	Final Weight
g_1	-----	O	X	O	X	O	O	2	3	1.07
g_2	X	-----	X	O	X	X	O	4	5	1.79
g_3	O	O	------	O	X	O	O	1	2	0.71
g_4	X	X	X	------	X	X	O	5	6	2.14
g_5	O	O	O	O	------	O	O	0	1	0.36
g_6	X	O	X	O	X	------	O	3	4	1.43
g_7	X	X	X	X	X	X	------	6	7	2.50
							Total	21	28	10.00

Notes

1. $RtG \leftarrow RtG+1$ to account for criterion 5.
2. For every cell ij, {X, E, O} signifies that criterion g_i is {more, equal, less} important than criterion g_j.
3. The weight for every criterion g_i is obtained by dividing RtG_i+1 by the total.

The input data used in the calculations are the values presented in Table 2 (the performances of the alternatives). All compared alternatives and criteria have been used in the calculation. Information about the preferences of the decision maker – namely, the values of the indifference and preference thresholds for each criterion and the values of the relative importance of the criteria – are presented in Table 3 and Table 4. The values of the relative importance of the criteria indicate that the total fixed assets (g_7) and the intermediate consumption (g_4) criteria are most important to the decision maker.

The computation has been performed on the input data (Table 2) and on the information about the preferences of the decision maker (Table 3 and Table 4) using the ELECTRE III method. According to the additional information noted above, we applied ELECTRE III to construct a valued outranking relation, which has been omitted for the lack of space.

This concludes the construction of the outranking model. The next step in the outranking approach is to exploit the model and produce a ranking of alternatives from the valued outranking relation. Our approach for exploitation is to use a multiobjective evolutionary algorithm-based heuristic method, which is explained in the work by Leyva and Aguilera (2005).

The valued outranking relation was processed using the MOEA to derive the final ranking and systematized using the SADAGE software. The MOEA used the following parameters: the number of generations was set to 10,000; the population size was set to 40; the crossover probability was 0.85; and the mutation probability was 0.35. The restricted Pareto front, $PF_{known}^{restricted}$, that was determined and the associated final set of solutions returned by the MOEA at termination, $P_{known}^{restricted}$, are presented in Table 5. u , f , and λ are the objective functions of the MOEA.

Table 5. Restricted Pareto front determined and the associated individuals of the solution space

Ran king	$\tilde{p}_1$	$\tilde{p}_2$	$\tilde{p}_3$	$\tilde{p}_4$	$\tilde{p}_5$	$\tilde{p}_6$	$\tilde{p}_7$	$\tilde{p}_8$	$\tilde{p}_9$	$\tilde{p}_{10}$	$\tilde{p}_{11}$	$\tilde{p}_{12}$	$\tilde{p}_{13}$	$\tilde{p}_{14}$	$\tilde{p}_{15}$	$\tilde{p}_{16}$
1	A_6	A_4	A_4	A_6	A_6	A_6	A_2	A_4	A_2	A_6	A_6	A_4	A_6	A_6	A_4	A_{17}
2	A_4	A_6	A_6	A_4	A_5	A_4	A_6	A_2	A_4	A_4	A_4	A_5	A_5	A_8	A_6	A_6
3	A_2	A_2	A_2	A_2	A_4	A_5	A_4	A_6	A_6	A_2	A_2	A_3	A_8	A_7	A_2	A_{13}
4	A_5	A_5	A_5	A_5	A_2	A_3	A_5	A_5	A_5	A_5	A_{17}	A_6	A_4	A_4	A_8	A_4
5	A_3	A_3	A_3	A_3	A_3	A_2	A_3	A_3	A_7	A_3	A_5	A_2	A_2	A_2	A_5	A_{14}
6	A_8	A_8	A_8	A_8	A_8	A_7	A_8	A_8	A_3	A_8	A_3	A_8	A_3	A_{18}	A_1	A_{11}
7	A_7	A_7	A_7	A_7	A_7	A_8	A_7	A_7	A_8	A_7	A_8	A_7	A_7	A_5	A_3	A_2
8	A_{17}	A_{17}	A_{17}	A_{17}	A_{17}	A_{17}	A_{17}	A_{17}	A_{17}	A_{17}	A_7	A_{17}	A_{17}	A_3	A_7	A_{10}
9	A_{18}	A_{18}	A_{18}	A_{18}	A_{18}	A_{18}	A_{18}	A_{13}	A_{18}	A_{13}	A_{18}	A_{18}	A_{18}	A_{17}	A_{17}	A_{15}
10	A_{13}	A_{13}	A_{13}	A_{13}	A_{13}	A_{13}	A_{13}	A_{18}	A_{13}	A_{18}	A_{13}	A_{13}	A_{11}	A_{13}	A_{18}	A_5
11	A_{14}	A_{14}	A_{14}	A_{14}	A_{14}	A_{14}	A_{11}	A_{14}	A_{14}	A_{15}	A_{14}	A_{14}	A_{13}	A_{14}	A_{13}	A_3
12	A_{11}	A_{11}	A_{11}	A_{11}	A_{11}	A_{11}	A_{14}	A_{11}	A_{11}	A_{14}	A_{11}	A_{11}	A_{14}	A_{11}	A_{14}	A_8
13	A_{10}	A_{10}	A_{10}	A_{10}	A_{10}	A_{10}	A_{10}	A_{10}	A_{10}	A_{11}	A_{10}	A_{10}	A_{10}	A_{10}	A_{11}	A_7
14	A_{15}	A_{15}	A_{15}	A_{15}	A_{15}	A_{15}	A_{15}	A_{15}	A_{15}	A_{10}	A_{15}	A_{15}	A_{15}	A_{15}	A_{10}	A_{18}
15	A_9	A_9	A_9	A_9	A_9	A_9	A_9	A_9	A_9	A_9	A_9	A_9	A_9	A_9	A_{15}	A_9
16	A_{16}	A_{16}	A_{16}	A_{16}	A_{16}	A_{16}	A_{16}	A_{16}	A_{16}	A_{16}	A_{16}	A_{16}	A_{16}	A_{16}	A_9	A_{16}
17	A_{12}	A_{12}	A_{12}	A_{12}	A_{12}	A_{12}	A_{12}	A_{12}	A_{12}	A_{12}	A_{12}	A_{12}	A_{12}	A_{12}	A_{16}	A_{12}
18	A_1	A_1	A_1	A_1	A_1	A_1	A_1	A_1	A_1	A_1	A_1	A_1	A_1	A_1	A_{12}	A_1
u	0	0	0	1	2	2	2	3	3	3	4	4	6	11	12	36
f	0	1	2	3	0	1	2	0	1	3	2	3	3	1	1	3
λ	0.56960	0.63880	0.71870	0.73780	0.57000	0.63960	0.71880	0.57000	0.63990	0.73880	0.72000	0.73960	0.73990	0.64000	0.64000	0.74000
fit-ness	37.00851	37.00851	37.00851	37.00851	37.00851	37.00851	37.00851	37.00851	37.00851	37.00851	37.00851	37.00851	34.30998	18.71453	17.15499	5.718329

Table 6 shows the number $T(i, j)$, $(1 \le i, j \le m)$, of times (i.e., the position frequencies) that an alternative was found at a certain place in the ranking of the individual $\tilde{p}_i$ associated with the members of the final restricted Pareto front. Based on Table 6, we found a compromise solution using the following procedure: because the ranking of the alternatives is of significant importance, the number of times that an alternative is found at a certain place in the ranking is weighted according to the importance of the alternatives to be ranked. Then, we calculate the weighted sum $\sum_{i=1}^{m} w_i T(i, j)$, $j=1,2,$ ..., m. Finally, we obtain a succession in decreasing order of preference generated in this manner and a recommendation for the decision maker.

Table 6. The number of times that an alternative was found at a certain place in the ranking

Weight w_i	Rank	A_1	A_2	A_3	A_4	A_5	A_6	A_7	A_8	A_9	A_{10}	A_{11}	A_{12}	A_{13}	A_{14}	A_{15}	A_{16}	A_{17}	A_{18}
18	1	0	2	0	5	0	8	0	0	0	0	0	0	0	0	0	0	1	0
17	2	0	1	0	6	3	5	0	1	0	0	0	0	0	0	0	0	0	0
16	3	0	7	1	2	1	2	1	1	0	0	0	0	1	0	0	0	0	0
15	4	0	1	1	3	8	1	0	1	0	0	0	0	0	0	0	0	1	0
14	5	0	4	8	0	2	0	1	0	0	0	0	0	0	1	0	0	0	0
13	6	1	0	3	0	0	0	1	9	0	0	1	0	0	0	0	0	0	1
12	7	0	1	1	0	1	0	0	3	0	0	0	0	0	0	0	0	0	0
11	8	0	0	1	0	0	0	2	0	0	1	0	0	0	0	0	0	12	0
10	9	0	0	0	0	0	0	0	0	0	0	0	0	2	0	1	0	2	11
9	10	0	0	0	0	1	0	0	0	0	0	1	0	11	0	0	0	0	3
8	11	0	0	1	0	0	0	0	0	0	0	1	0	2	11	1	0	0	0
7	12	0	0	0	0	0	0	0	1	0	0	11	0	0	4	0	0	0	0
6	13	0	0	0	0	0	0	1	0	0	13	2	0	0	0	0	0	0	0
5	14	0	0	0	0	0	0	0	0	0	0	2	0	0	0	13	0	0	1
4	15	0	0	0	0	0	0	0	0	15	0	0	0	0	0	1	0	0	0
3	16	0	0	0	0	0	0	0	0	1	0	0	0	0	0	0	15	0	0
2	17	0	0	0	0	0	0	0	0	0	0	0	15	0	0	0	1	0	0
1	18	15	0	0	0	0	0	0	0	0	0	0	1	0	0	0	0	0	0
$\sum_{i=1}^{m} w_i T(i, j)$		28	248	213	269	236	276	191	208	63	99	119	31	151	130	87	47	185	155

Minimum λ:

0.5696

Table 6 suggests the following final ranking:

$$A_6 \succ A_4 \succ A_2 \succ A_5 \succ A_3 \succ A_8 \succ A_7 \succ A_{17} \succ A_{18} \succ A_{13} \succ A_{14} \succ A_{11}$$
$$\succ A_{10} \succ A_{15} \succ A_9 \succ A_{16} \succ A_{12} \succ A_1 \tag{1}$$

The above multicriteria method was performed 50 times using the SADAGE software with the same data (performance matrix, inter-criteria parameters and MOEA parameters) to produce 50 rankings. Then, using the same procedure as in the above paragraph, we calculated the number $T(i, j)$, $(1 \le i, j \le m)$, of times (i.e., the position frequencies) that an alternative was found at a certain place in the 50 rankings, which are shown in Table 7.

Table 7. The number of times that an alternative was found at a certain place in the 50 rankings

Weight w_i	Rank	A_1	A_2	A_3	A_4	A_5	A_6	A_7	A_8	A_9	A_{10}	A_{11}	A_{12}	A_{13}	A_{14}	A_{15}	A_{16}	A_{17}	A_{18}
18	1	0	0	0	24	0	26	0	0	0	0	0	0	0	0	0	0	0	0
17	2	0	0	0	26	0	24	0	0	0	0	0	0	0	0	0	0	0	0
16	3	0	0	44	0	0	6	0	0	0	0	0	0	0	0	0	0	0	0
15	4	0	6	1	0	43	0	0	0	0	0	0	0	0	0	0	0	0	0
14	5	0	0	46	0	1	0	0	2	0	0	0	0	0	0	0	0	1	0
13	6	0	0	2	0	0	0	7	37	0	0	0	0	0	0	0	0	4	0
12	7	0	0	1	0	0	0	27	7	0	0	0	0	0	0	0	0	15	0
11	8	0	0	0	0	0	0	15	4	0	0	0	0	0	0	0	0	27	4
10	9	0	0	0	0	0	0	1	0	0	1	0	0	7	2	0	0	3	36
9	10	0	0	0	0	0	0	0	0	0	12	7	0	11	10	1	0	0	9
8	11	0	0	0	0	0	0	0	0	4	7	9	0	19	8	2	0	0	1
7	12	0	0	0	0	0	0	0	0	6	16	7	0	8	11	2	0	0	0
6	13	0	0	0	0	0	0	0	0	6	5	11	0	3	11	12	2	0	0
5	14	0	0	0	0	0	0	0	0	10	4	8	2	2	7	11	6	0	0
4	15	0	0	0	0	0	0	0	0	11	2	4	2	0	1	10	12	0	0
3	16	4	0	0	0	0	0	0	0	9	3	3	10	0	8	13	0	0	0
2	17	16	0	0	0	0	0	0	0	4	0	1	19	0	0	3	7	0	0
1	18	22	0	0	0	0	0	0	0	0	0	0	17	0	0	1	10	0	0
$\sum_{i=1}^{m} w_i T(i,j)$		98	794	697	874	755	876	590	637	239	353	317	103	405	356	237	153	573	493

Minimum λ:

0.5016

Table 7 suggests the following final ranking:

$$\{A_6, A_4\} \succ A_2 \succ A_5 \succ A_3 \succ A_8 \succ A_7 \succ A_{17} \succ A_{18} \succ A_{13} \succ \{A_{14}, A_{10}\} \succ A_{11} \succ \{A_9, A_{15}\} \succ A_{16} \succ \{A_{12}, A_1\} \quad (2)$$

4   Sensitivity Analysis of the Final Result

In most cases, arriving at the final ordering accepted by the decision maker does not conclude the decision aiding process. The analyst can additionally propose performing a sensitivity analysis. Examples of employing a sensitivity analysis have also been presented elsewhere (Briggs et al., 1990; Goicoechea et al., 1982; Rios Insua and French, 1991, Leyva 2005).

A sensitivity analysis is used to characterize the influence of changing the values of parameters, which consist of information about the decision maker's preferences (the various methods use different parameters to reflect the decision maker's preferences), on the final result. Sensitivity analysis is useful for interpreting results that have been achieved by modifying the values of the appropriate parameters reflecting the decision maker's preferences and in estimating the influence of the modifications on the final result. The decision maker supplies a range of values that he considers still consistent with his preferences.

Using this input, the range of sensitivity analysis is defined. The analysis considers the following types of changes in the parameters:

- changes in the values of the relative importance (w) of a single criterion,
- simultaneous changes in the values of the relative importance (w) of multiple criteria,
- changes of the values for threshold functions, which include the thresholds of indifference (q) and preference (p), for a single criterion, and
- simultaneous changes of the values for the thresholds of indifference (q) and preference (p) for multiple criteria.

The results of the sensitivity analysis performed, which depend on the allowed range of values for the selected parameters that describe the decision maker's preferences, are not presented in this paper for lack of space.

Changing the values of the relative importance of a criterion, w, had the least influence on the final order of alternatives. Of the 17 cases in which changes were introduced, in the majority of the cases, the final result typically preserved the final ranking selected by the decision maker (but the alternatives were not always in the same rank). For the ranges of changes in the values of parameters suggested by the decision maker, the sensitivity of the final result (the ranking) was insignificant.

The final ranking, shown in (2), was still achieved when the values of relative importance (w) were changed for both a single criterion and for multiple criteria simultaneously. Basing on the sensitivity analysis, we conclude the following: the decision maker can accept a different final ranking when the influence of the parameter changes on the final result can be justified and when the result changes only slightly compared with the final ranking accepted by the decision maker before the sensitivity analysis was performed. Performing a sensitivity analysis ends the decision aiding process.

5   Results and Discussion

Table 7 presents a summary of the results of this study. Based on these results and the proposed final ranking given in (2), we find the following:

- the retail trade (A_6) and manufacturing (A_4) sectors were consistently ranked first;
- the utilities sector (A_2) was ranked second;
- the wholesale trade sector (A_5) was consistently ranked third;
- the mining sector (A_1) and the management of companies and enterprises sector (A_{12}) were consistently ranked at the bottom of the ranking;
- the art, entertainment, and recreation sector (A_{16}) was consistently ranked in one of the lowest positions, just above the A_1 and A_{12} sectors; and
- the remaining sectors were consistently ranked in the middle.

Based on these results, the retail trade and manufacturing sectors are the most attractive for potential investors. The weak performance of the mining sector may be attributed to the lack of technological innovations and infrastructure investment. However, in the last 5 years, there has been an important revival of this sector, which is primarily due to direct foreign investment. In contrast, the weak performance of the management of companies and enterprises sector may be attributed to the centralized economic activity in some Mexican states. Thus, private and public policy initiatives aimed at improving the performance of these subsectors are needed.

The art, entertainment, and recreation sector ranks low in terms of its attractiveness to investors because of the lack of infrastructure investment and the violent crime and public insecurity in Sinaloa in the last 10 years. Business innovations and policy-making linked with the federal government of Mexico are needed to stop the deterioration of this sector. In the middle of the ranking, we find a large set of economic sectors. These sectors present stable investment opportunities.

6   Concluding Comments

The aim of this study was to offer a novel procedure for integrated assessment and comparison of Sinaloa economic sectors using a Multicriteria Decision Aiding Approach. The proposed procedure for multicriteria comparison of economic sectors uses the ELECTRE III method to construct a valued outranking relation and then a multiobjective evolutionary algorithm (MOEA) to exploit it to obtain a ranking of the economic sectors in decreasing order of performance. The results suggested that the best-performing sector is the retail sector. Furthermore, enterprises operating in the sectors of manufacturing industries, wholesale trade, utilities, and construction work more efficiently than an average Sinaloa enterprise. In contrast, the mining sector; the arts, entertainment and recreation sector; and the management of companies and enterprises sector were ranked below the average alternative.

The multicriteria method utilized in this study to rank the Sinaloa economic sectors is both practical and adequate. The proposed multicriteria assessment framework can provide a rationale for interested stakeholders, including government institutions and policy-makers; investors, financial institutions, and businessmen; employees and trade unions; and clients and suppliers related to certain sectors.

The application presented in this study underscores the applicability of multiobjective evolutionary algorithms to real-life business problems in a multicriteria decisional context. Thus, this study contributes to a growing body of application-based knowledge, which was until very recently the exclusive domain of engineering and the natural sciences.

References

1. Augusto, M., Figueira, J., Lisboa, J.: An Application of a multi-criteria approach to assessing the performance of Portugal´s economic sectors. European Business Review 17, 113–132 (2005)

2. Autran Monteiro Gomes, L.F., Gomes Correa, M., Duncan Rangel, L.A.: Sustainability in Mining: An Application of Promethee II. In: Trzaskalik, T., Wachowicz, T. (eds.) Multiple Criteria Decision Making´10-11. University of Economics in Katowice, Poland (2011)
3. Baležentis, A., Baležentis, T., Misiunas, A.: An integrated assessment of Lithuanian economic sectors based on financial ratios and fuzzy MCDM methods. Technological and Economic Development of Economy 18, 34–53 (2012)
4. Beccali, M., Cellura, M., Ardente, D.: Decision making in energy planning: the ELECTRE multi-criteria analysis approach compared with a FUZZY-SETS methodology. Energy Conversion and Management 39, 1869–1881 (1998)
5. Blondeau, P., Spérandio, M., Allard, F.: Multi-criteria analysis of ventilation in summer period. Building and Environment 37, 165–176 (2002)
6. Bouyssou, D.: Building criteria: A prerequisite for MCDA. In: Bana e Costa, C.A. (ed.) Readings in Multiple Criteria Decision Aid, pp. 58–80. Springer, Berlin (1990)
7. Briggs, T., Kunsch, P.L., Mareschal, B.: Nuclear waste management: An application of the multicriteria PROMETHEE methods. European Journal of Operational Research 44, 1–10 (1990)
8. Colson, G.: The OR's prize winner and the software ARGOS: how a multijudge and multicriteria ranking GDSS helps a jury to attribute a scientific award. Computers & Operations Research 27, 741–755 (2000)
9. Diaz, B., Moniche, L., Morillas, A.: A fuzzy clustering approach to the key sectors of the Spanish economy. Economic Systems Research 18, 299–318 (2006)
10. Figueira, J., Greco, S., Roy, B., Slowinski, R.: ELECTRE Methods: Main Features and Recent Developments. Cahier Du LAMSADE 298, Dauphine Université Paris (2010)
11. Georgopoulou, E., Lalas, D., Papagiannakis, L.: Multi-criteria decision aid approach for energy-planning problems: the case of renewable energy option. European Journal of Operational Research 103, 38–54 (1997)
12. Goicoechea, A., Hansen, D.A., Duckstein, L.: Multiobjective Decision Analysis with Engineering and Business Applications. J. Wiley, New York (1982)
13. Karagiannidis, A., Moussiopoulos, N.: Application of ELECTRE III for the integrated management of municipal solid wastes in the Greater Athens Area. European Journal of Operational Research 97, 439–449 (1997)
14. Kaya, T., Kahraman, C.: Fuzzy multiple criteria forestry decision making based on an integrated VIKOR and AHP approach. Expert Systems with Applications 38, 7326–7333 (2011)
15. Keeney, R., Raiffa, H.: Decisions with Multiple Objectives: Preference and Value Tradeoffs. Cambridge University Press, Cambridge (1993)
16. Leyva López, J.C., Dautt, L., Aguilera Contreras, M.A.: A Multicriteria Decision Support System with an Evolutionary Algorithm for Deriving Final Ranking from a Fuzzy Outranking Relation. Operations Research: An International Journal 8, 47–62 (2008)
17. Leyva López, J.C.: Multicriteria Decision Aid Application to a Student Selection Problem. Pesquisa Operacional 25, 45–68 (2005)
18. Leyva-Lopez, J.C., Aguilera-Contreras, M.A.: A Multiobjective Evolutionary Algorithm for Deriving Final Ranking from a Fuzzy Outranking Relation. In: Coello Coello, C.A., Hernández Aguirre, A., Zitzler, E. (eds.) EMO 2005. LNCS, vol. 3410, pp. 235–249. Springer, Heidelberg (2005)
19. Martel, J., Khoury, N., Bergeron, M.: An application of a multi-criteria approach to portfolio comparisons. Journal of the Operational Research Society 39, 617–628 (1988)
20. Misiūnas, A.: Financial ratios of the country's enterprises in the face of economic growth and decline. Ekonomika 89, 32–48 (2010)

21. Rios Insua, D., French, S.: A framework for sensitivity analysis in discrete multiobjective decision-making. European Journal of Operational Research 54, 176–190 (1991)
22. Roger, M., Bruen, M., Maystre, L.: Electre and Decision Support. Kluwer, Academic Publishers (2000)
23. Rogers, M., Bruen, M.: Choosing realistic values of indifference, preference and veto thresholds for use with environmental criteria within ELECTRE. European Journal of Operational Research 107, 542–551 (1998)
24. Roy, B.: Multi-criteria Methodology for Decision Aiding. Kluwer Academic Publishers, Dordrecht (1996)
25. Salminen, P., Hokkanen, J., Lahdelma, R.: Comparing multi-criteria methods in the context of environmental problems. European Journal of Operational Research 104, 485–496 (1998)
26. Siskos, Y., Hubert, P.: Multi-criteria analysis of the impacts of energy alternatives: a survey and a new comparative approach. European Journal of Operational Research 13, 278–299 (1983)
27. Steuer, R.: Multiple Criteria Optimization: Theory, Computation, and Applications. John Wiley & Sons, New York (1986)
28. Sudaryanto: Empirical Identification of Key Sectors in Indonesian Economy: A Multi Attributes Decision Making Approach. In: Proceeding of the Indonesian Student Scientific Meeting (ISSM 2000), Paris, pp. 98–102 (2000)
29. Teng, J., Tzeng, G.: Multi-criteria evaluation for strategies of improving and controlling air quality in the super city: a case study of Taipei City. Journal of Environmental Management 40, 213–229 (1994)
30. Treadwell, W.A.: Fuzzy set theory movement in the social sciences. Public Administration Review 55, 91–98 (1995)
31. Tsoukias, A.: On the concept of decision aiding process. Annals of Operations Research 154, 3–27 (2007)
32. Turskis, Z., Zavadskas, E.K.: A new fuzzy additive ratio assessment method (ARAS-F). Case study: the analysis of fuzzy multiple criteria in order to select the Logistic Center location. Transport 25, 423–432 (2010)
33. Vincke, P.: Multicriteria Decision Aid. Wiley, Chichester (1992)
34. Xidonas, P., Psarras, J.: Equity portfolio management within the MCDM frame: a literature review. International Journal of Banking, Accounting and Finance 1, 285–309 (2009)
35. Xidonas, P., Ergazakis, E., Ergazakis, K., Metaxiotis, K., Askounis, D., Mavrotas, G., Psarras, J.: On the selection of equity securities: an expert systems methodology and an application on the Athens Stock Exchange. Expert Systems with Applications 36, 11966–11980 (2009)
36. Xidonas, P., Mavrotas, G., Psarras, J.: A multiple criteria decision making approach for the selection of stocks. Journal of the Operational Research Society 61, 1273–1287 (2010)
37. Yu, V.F., Hu, K.J.: An integrated fuzzy multi-criteria approach for the performance evaluation of multiple manufacturing plants. Computers and Industrial Engineering 58, 269–277 (2010)
38. Zadeh, L.A.: Fuzzy sets. Information and Control 8, 338–353 (1967)

Multi-objective Optimisation for Social Cost Benefit Analysis: An Allegory

Robin C. Purshouse[1] and John McAlister[2]

[1] Department of Automatic Control and Systems Engineering,
University of Sheffield, Sheffield, S1 3JD, UK
r.purshouse@sheffield.ac.uk
www.sheffield.ac.uk/acse
[2] Strategy and Decision Science Practice, PA Consulting Group,
123 Buckingham Palace Road, London, SW1W 9SR, UK
john.mcalister@paconsulting.com
www.paconsulting.com

Abstract. Social cost benefit analysis often involves consideration of non-monetary outcomes. Multi-objective optimisation is an appropriate method for handling problems of this type, but many decision-makers have a strong mistrust of the approach. Reflections by the authors on real experiences supporting decision-makers suggest that the key barriers to using multi-objective methods for social cost benefit analysis include: (i) the inadequacy of current social systems models for measuring the end benefits provided by a candidate solution; (ii) the lack of appropriate societal preference estimates for resolving the inherent trade-offs between objectives; and (iii) the lack of practical examples, case studies and guidance which demonstrate that the approach works well.

Keywords: multi-objective optimisation, decision support systems.

1 Introduction

Social cost benefit analysis is concerned with appraising the effects on society of potential government investments or policies. This type of analysis is the orthodoxy for decision-making in many Western economies, including the United Kingdom [1]. The ultimate aim is to estimate, for each investment or policy option of interest, the *net benefit* of that option to society in cash terms, taking account of the value and timing of all the outcomes (costs and benefits) that arise [2]. However decision-makers are often faced with situations in which some aspects of option quality cannot readily be converted to monetary terms. The concept of multi-objective optimisation offers an appropriate means for dealing with these non-monetary outcomes [3,4].

Despite the advantages that multi-objective optimisation offers, in terms of transparency and auditability, it is still often regarded as an avant-garde alternative to traditional approaches. It is more often the case that decision-makers will prefer to engage in a deliberative discussion based on rhetoric, or weigh such aspects in an internal manner (cognitively speaking), or simply ignore these outcomes altogether (thereby implicitly assuming they have no value).

R.C. Purshouse et al. (Eds.): EMO 2013, LNCS 7811, pp. 726–740, 2013.
© Springer-Verlag Berlin Heidelberg 2013

This paper seeks to explore the possible reasons why multi-objective optimisation is not regarded as the standard choice for social cost benefit analysis. The exploration is based on reflections by the authors on our attempts to use multi-objective optimisation in helping decision-makers to resolve the real problems facing them. We describe our experiences through a fictionalised example:

IMAGINE A TRIBE living on a small, forested, island. Some of the tribespeople are *gatherers* - they cultivate small market gardens in the forest and harvest the crops using scythes. Other tribespeople are *hunters* - they roam the forest looking for animals to catch and kill using spears. The tribe can survive by eating either the crops or the meat from the animals. A disadvantage of the crops is that each year there is a small risk that the crops will fail, leaving the gatherers with insufficient produce to feed the whole tribe. A disadvantage of the meat is that it cannot be eaten directly, but must be given to the tribal *chefs* who produce edible food by either cooking or curing the raw meat delivered by the hunters. Sometimes the animals caught by the hunters turn out to be inedible.

Up until now, both the heads for the scythes used by the gatherers and the heads for the spears used by the hunters have been produced by a forge on the island, smelting copper ore mined from one of the hills. The demand for new scythe heads and spearheads is high, due to population growth and damage to old tools, but the supply of copper ore to the forge is running out. The tribal *elders* predict that within five years there will be no functioning scythes or spears on the island, with dire consequences for the tribespeople.

Recently, the tribe has discovered a source of iron ore in another large hill on the island. This iron ore could be used to smelt new scythe heads and spearheads, but to do this would require an upgrade to the island's forge. It would be relatively straightforward to upgrade the forge to smelt iron scythe heads - the tribe elders believe that the gods would require a reasonably small number of blood sacrifices to give their blessings to this forge. However, additional forge complexity would be needed to smelt spearheads, and the elders believe that substantially more blood sacrifices would be needed in this case. To operate the new forge without the required sacrifices would be heresy and, as such, is inconceivable.

The elders are now faced with a decision. They can carry on using the existing copper forge - this would avoid the need for any blood sacrifices but would mean that the tribe faces starvation in five years' time. Alternatively, the elders can sanction a new iron forge to smelt scythe heads for the gatherers, either with or without the capability to smelt spearheads for the hunters. This is a difficult decision for the elders, with potentially major repercussions for the wellbeing of the tribe. The tribe only has a limited number of virgins available for sacrifice in any year and the elders need to be sure that the blood sacrifices spent on the forge could not be better used on other areas of tribal life where the gods must also be appeased.

The remainder of the paper takes our story into those parts of the multi-objective optimisation process that are, in our experience, crucial to the success or failure of the enterprise. In Section 2, we consider the overall governance arrangements for multi-objective optimisation, highlighting the roles of decision-makers and stakeholders. In Section 3, we look at the process of identifying a set of objectives against which the performance of the various solution options is to be appraised, whilst in Section 4 we look at the challenge of measuring performance against those objectives. In Section 5 we consider the thorny issue of preferences. In Section 6 we conclude.

2   On Governance

Problems requiring solutions and decisions tend not to exist in splendid isolation from the rest of the world; rather they are situated in complex organisational and social contexts that need to be accounted for in the optimisation process. In the UK, the *Office of Government Commerce* (OGC) imposes a formal process on solution development and decision-making for major public investment decisions. Decision-makers and stakeholders are required to develop compelling and robust business cases for change. A business case develops in an incremental fashion, with the OGC imposing a set of formal assessments (known as *gateways*) which the business case must successfully pass through before a decision is finally approved. Gateway processes can be found in many organisations in both the public and private sectors across Western economies.

Business cases in the UK follow a *five case model* [5], with the five dimensions being:

strategic case explains why solving the problem is essential in supporting the strategic objectives of the sponsoring organisation (in this case, the objectives for society in the UK, as expressed through the goals of the Government).

economic case estimates the overall impact of each solution option, in terms of costs and benefits. Ideally the impact should be expressed as a scalar quantity in monetary terms (in current prices): a *net present value* (NPV).

commercial case describes the different options for how solutions will be procured.

financial case estimates the affordability of preferred solutions, in terms of the impact on the organisation's financial accounts.

management case describes how the solution will be implemented, how risks and issues will be handled, and what evaluation processes will be enacted to measure the actual costs and benefits of the recommended solution.

The cost benefit analysis lies at the heart of the economic case. However costs also form the basis for the financial case (through the translation of theoretical opportunity costs into practical budgetary implications), with benefits forming the basis for the rhetoric of the strategic case. Whilst the ultimate decision is made at the level of the five cases, it is likely that some devolved decision-making, and expressions of preference, will be made during the design and economic assessment of the solution options.

The wider context around the cost benefit analysis (and, by implication, any supporting multi-objective optimisation) means that engagement with stakeholders is important. These include the designers of the solution options, the individuals or organisations who are expected to deliver the benefits, those whose budgets will be impacted by the solution's costs, colleagues involved in the other dimensions of the business case, and the assessors whom the business case must satisfy. This engagement should be ongoing and used to help steer the analysis. For example, it is wise to check that the assessors are comfortable with any intended use of multi-objective methods.

> THE ELDERS DECIDED that they needed more information before making a decision, and instructed the tribal *thinkers* to appraise the costs and benefits of each option. Some of the hunters and gatherers were irritated when they learned of the elders' actions – surely the thinkers knew nothing about either hunting or gathering? Nevertheless, when the thinkers asked for representatives from each group of hunters and gatherers to join a working party to appraise the options, no group wanted to be left out.

3   On Objectives

> FROM THEIR INITIAL conversations with the elders, the thinkers knew that the strategic objective for the forge problem was the maintenance of a healthy tribe. They also knew that there were essentially two functional requirements for the forge: to smelt scythe heads and to smelt spearheads. But how did the availability of scythe heads and spearheads actually go on to support a healthy tribe?

The objectives for a problem are those aspects of solution quality against which all the candidate solutions will be judged. Developing a coherent set of objectives for a government investment or policy problem can be challenging, because the effects of such interventions are played out in the social world, which is inherently complex [6]. We adopt, at least in spirit, the approach of Hammond et al. [7], which is to steadily progress from intermediate outcomes to end outcomes by successively asking "why?" until the question no longer has an answer. In practice, the end outcomes will typically be some subset of an organisation's strategic objectives. An effective communication tool, especially when working with stakeholders to understand the problem environment, is to visually map out the flow of cause-and-effect, from the functional requirements of a solution (often known as *enablers*), through the intermediate chain of benefits, to the strategic objectives. These visual representations are known as *benefit dependency maps* or, simply, *benefits maps* [8].

Note that the strategic objectives will not necessarily be the objectives used in the multi-objective optimisation. Strategic objectives can be difficult to appraise or evaluate in practice, and so more tangible intermediate benefits close (in a causal sense) to the strategic objectives may be selected instead. There is often a tendency, particularly when working with solution designers, to define the objectives in a region of the benefits map close to the enablers. This is natural, since this is the part of the problem that is most well understood and easily quantifiable, but these objectives offer no guarantee of being good proxies for the actual value that a solution offers to an organisation or society.

Human factors are also important considerations when constructing a set of objectives. Human decision-makers have limited cognitive abilities in processing information and therefore large numbers of simultaneous objectives are to be avoided where possible (although the golden rule of having no more than seven categories may not be as robust as once thought [9]). Also, particularly where causal pathways are tortuous, stakeholders may have ownership only of intermediate benefits close to the enabler end of the benefits map, but be expecting to see these benefits explicitly represented as objectives.

THE THINKERS VISITED one of the forest gardens and mapped out with the gatherers how a forge would help to keep the tribe fed. Then the thinkers went through the same process with the hunters. A consolidated benefits map, shown in Fig. 1, was constructed.

With the map finalised, the thinkers convened the working party in a clearing in a forest to sit down and agree on the objectives for the forge problem. The thinkers arrived at the clearing with a proposal: there should be two objectives: (1) the cost of the forge; (2) based on the benefits map, the health of the tribe (denoted feed tribe). The hunters were not impressed: it was their job to catch and kill animals, but it was down to the chefs to prepare the food that fed the tribe. The chefs always expressed satisfaction with the quality of the hunters' catch, but what the chefs actually did with the animals was up to them. Given that the thinkers didn't appear to have invited any chefs to join the working party, argued the hunters, it was essential that the capacity of the hunters to catch and kill animals be included as an objective. The thinkers had not expected this opposition to their proposal and were placed on the back foot. They knew that the methods they were about to use would be sensitive to double-counting of objectives and so wanted to resist the hunters' demands, but they were also worried about their own ability to quantify the link between the catch of the hunters and the food produced by the chefs. Reluctantly, the thinkers conceded that the capacity to catch and kill animals – denoted kill animals – be included as an objective. This decision, in turn, upset the gatherers. Initially happy with just the feed tribe objective, since they could see exactly how the harvest fed the tribe, the gatherers now demanded parity with the hunters: that their labour – harvest crops – be included as an objective. Faced with the otherwise unappetising prospect of telling the elders that the objectives could not be agreed, the thinkers conceded to the gatherers' demands as well.

4   On Models

Optimisation methods tend to rely on mathematical models that enable the performance of each option against each objective to be estimated quantitatively. In our experience, decision-makers often prefer to simply commission a model, implemented as a user-friendly software tool, that they can experiment with in order to find satisficing solutions to their problem, rather than also commissioning the extra work required to perform a formal optimisation of the solution. Without a clear demonstration of the benefits of optimality (or, in reality, an approximation to it), satisficing is – by definition – likely to be seen as good enough.

Mathematical models for the appraisal of government investment and policy options can be challenging to build when working under limited resource constraints, due to the complex nature of the social systems they are seeking to represent. Even if resource were available to synthesise all the available primary evidence, the gaps in that evidence base tend to produce high levels of modelling uncertainty when attempting to link the enablers all the way through to the strategic objectives. Modelling of intermediate benefits may be a more realistic prospect, but assessors will need to be convinced that these benefits are reasonable proxies for estimating solution value. An alternative approach is to use expert opinion to score the solutions against the objectives. The lack of transparency is a key concern here, particularly where pilot or prototype evaluations are not possible. Such legitimacy issues may undermine the whole analysis. A further disadvantage is that the burden on the experts increases with the number of competing options, although it may be possible to mitigate the burden through the use of meta-modelling techniques to estimate scores for intermediate solutions.

A notable modelling issue in UK Government decisions is that estimates of costs must be explicitly increased (and also estimates of benefits correspondingly reduced) to account for the demonstrated tendency of project appraisals to be overly optimistic in their assessments [10]. This phenomenon is known as *optimism bias*. The adjustments are based on historical data of business case evaluations, and their magnitude may be reduced (but not eliminated entirely) through demonstrable good practice in estimation and implementation, and through successively more detailed iterations of the business case.

EARLY IN THEIR conversations with the hunters and the gatherers, the thinkers realised that building a mathematical model of the relationship between forge requirements and tribe health was going to be difficult. The actual locations and number of forest gardens was unknown. The hunters refused to reveal anything about their activities for fear the thinkers would come crashing in and scaring the animals. The supply networks through which the garden produce and meats reached the hungry tribespeople had never been formally recorded. And the thinkers hadn't even begun to consider the role of the chefs. Knowing that the elders needed information quickly, the thinkers decided to ask each hunter

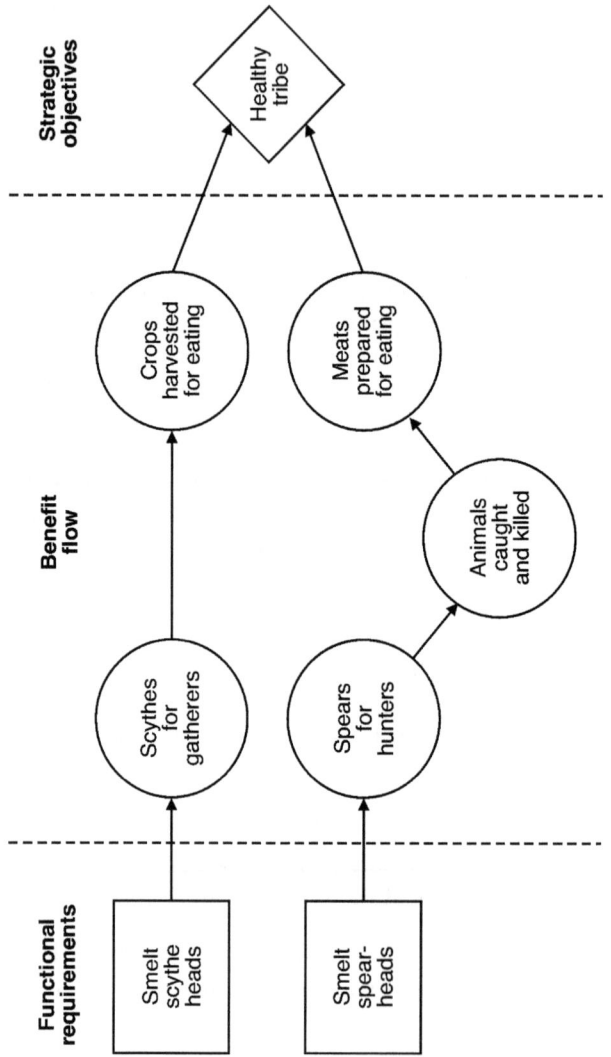

Fig. 1. Forge benefits map

and gatherer group for their expert opinion on how well each option for the new forge would satisfy the three benefit-related objectives. They also consulted the tribal *shaman* for estimates of the number of blood sacrifices that would be required to separately bless the scythe and spear functionality, and increased these estimates to account for the suspected optimism of the shaman.

Next, the thinkers went to the beach and collected five shells. Then, for each option – no forge, scythe-only forge, and scythe-and-spear forge – the thinkers asked the groups to indicate with shells how well the option would support the objectives compared to the current copper forge: 0 shells = no support, 1 = very limited support, 2 = limited support, 3 = slightly limited support, 4 = same support, 5 = better support. At this stage it became clear that both the hunters and gatherers had other means of delivering their benefits than using scythes and spears. The gatherers could collect berries using their bare hands but this would need an increase in the number of berry pickers. The hunters could catch fish in the reefs surrounding the island but this would also mean more rafts would need to be built. The shaman was able to estimate the cost of blessing these alternative activities. The mitigations provided two further options: no forge (mitigated) and scythe-only forge (mitigated). The thinkers took an average of the opinions across the various groups of hunters and gatherers to produce the benefit scores shown in Table 1.

Table 1. Benefit scores for forge options

Option	kill animals	harvest crops	feed tribe
no forge	0	0	0
no forge (mitigated)	1	1	1
scythe-only forge	0	4	2
scythe-only forge (mitigated)	1	4	3
scythe-and-spear forge	4	4	4

5   On Preferences

It is typically quite rare in a multi-objective problem that there will be a single logically 'optimal' solution offering equal or superior performance across every objective, when compared to every other possible option. Rather, there will exist a *trade-off surface* of 'Pareto-optimal' solutions for which improvement in one or more objectives cannot be made without performance sacrifices in other objectives. In these circumstances, to identify a single solution to recommend, we need to understand the relative worth of a particular level of performance across one or more objectives compared to some other level of performance across one or more objectives. These value judgments are known as decision-maker *preferences*.

The inescapably subjective nature of preferences, by contrast to the ostensibly objective nature of the objectives, options, and the models that estimate performance of options against objectives, serves to make the elicitation and use of preferences arguably the most challenging aspect of multi-objective optimisation.

Having noted that preferences are difficult to avoid, the next question is whose are the preferences that are to be elicited and used. In social cost benefit analysis, a working definition given by the UK's *Department for Communities and Local Government* (DCLG) is that the preferences should be "the **informed** preferences of people as a whole, to the extent that these preferences can be measured and averaged" [11]. Such preferences may be elicited directly from a representative sample of the population or may be reflected through the views of experts or officials. Interesting examples of both the former [12] and latter [13] approaches can be seen in the area of healthcare resource allocation.

In our experience, preferences are likely to be incorporated into the decision process in a staged and hierarchical manner, reflective of the governance arrangements. Specifically, preferences for differing levels of performance between the benefit-related objectives may be delegated by senior decision-makers (in the higher echelons of government) to experts in the field. The aim is to reduce the dimensionality of these objectives to a scalar overall benefit score, where the underpinning vector of performance remains available for inspection if required. The senior decision-makers are then in a position to examine the critical trade-off between cost and overall benefit.

> AT SUNRISE THE elders found the thinkers sat in a small circle on the beach, apparently messing around with some animal bones and a heap of pebbles. Due to emerging strategic considerations elsewhere in the life of the tribe, the elders explained to the thinkers, a decision on the forge would be needed earlier than first anticipated – could the thinkers find out from the hunters and gatherers the relative importance of the `kill animals`, `harvest crops` and `feed tribe` objectives before sunset? The thinkers felt under-prepared: from their earlier experiences with the hunters and gatherers, they knew that obtaining a single expression of preference might be difficult. Nevertheless, the elders' wisdom was not to be challenged; dragging the skull of a shark out from under a pile of driftwood, the thinkers hastily made their final preparations.

The most conventional expression of preference is to use a non-negative weight, w_i, to describe the relative importance of each objective, i. The weights are then incorporated into a functional form that provides the overall value of an option; the most common function is the *weighted sum* of individual performance across each objective, where the weights are normalised to sum to unity (less formally, the overall value can be seen as a measure of average performance). The use of the weighted sum is theoretically troublesome because it cannot find solutions in concave regions of trade-off surfaces and also, if the problem is non-convex, offers no guarantee of finding solutions even in the convex regions [3]. The method is also prone to double-counting biases, but it has a simplicity that is easy to communicate to stakeholders and, during procurement exercises, the potential suppliers of solutions.

Weight-based methods involve the forced cohesion of non-commensurate objectives, thus requiring the objectives to be normalised. If the normalisation is done without preferences (which is usually the case since the bounds of what is good and bad are not known a priori with certainty) then the importance of an objective cannot exist independently from the range of performance exhibited for that objective. For this reason, weight-based approaches tend to elicit and apply preferences *after* the performance of each option is known. *Swing weights* are used in which the decision-maker is firstly asked to specify the objective with the most important observed variation between worst performance and best performance (combined across all options); and secondly to weight the importance of the variation seen in other objectives relative to this reference objective [14]. This approach is prone to the perception that the weights are being manipulated by participants to fix the results of the analysis to a pre-determined, favoured, solution. For this reason, in multi-objective optimisation approaches where providers must compete to offer solutions, it is often a regulatory requirement to determine and publish the weights in advance. However, without a priori knowledge of the range of solution performance, there is a risk of using weights that are an incorrect expression of decision-maker preferences, potentially leading to the selection of undesirable solutions. Publishing the weights in advance can also lead to potential gaming by suppliers.

A major discomfort that decision-makers and stakeholders tend to have is the difficulty to state preferences precisely. This leads to a sense of arbitrariness about the method and undermining of confidence in the results. Sensitivity analysis on the preferences can help reduce such anxieties – by showing how much preferences would have to change before the ordering of options by overall benefit would change [14].

IN ORDER TO elicit the swing weights for the three benefit-related objectives, the thinkers asked the elders to invite five hunters and five gatherers to a special meeting in a clearing in the forest, close to the entrance to a cave. Inside the cave, the thinkers placed three skulls: those of a wolf, a boar, and a shark to represent, respectively, the `kill animals`, `harvest crops` and `feed tribe` objectives. Each hunter and gatherer was then given a polished stone, where the stones were selected to be as indistinguishable from each other as possible.

Next, the thinkers revealed to the ten tribespeople the consolidated benefit scores for the five forge options, as shown in Table 1. The tribespeople were then asked to consider: (i) the nadir outcome of no support to killing animals, or harvesting crops, or feeding the tribe; and (ii) the ideal outcome of full support for these three benefits. The thinkers then asked the tribespeople to imagine a situation in which they could improve just one objective from its worst position to its best position. Each tribesperson was invited in turn to enter the cave and place his or her stone inside the skull that symbolised the most important objective to improve.

The hunters and gatherers looked at each other in disbelief. Yes, they were experts in hunting and gathering, but surely the judgment of which objective was the most important was a matter for the whole tribe? Shouldn't it be for the elders to make this decision? Then again, it was the elders who had invited them to this crazy meeting. The hunters and gatherers decided to humour the thinkers, at least for the time-being, and prepared themselves to enter the cave.

The hunters knew that the feed tribe objective was ultimately the most important but were worried that the elders might opt for a scythe-only forge and so decided to put all their stones into the wolf skull rather than the shark skull. Meanwhile the gatherers, whilst sympathetic to the extra resilience that spears would bring, were worried that if the benefits of a scythe-only forge were seen as tiny compared to a scythe-and-spear forge then the elders, balking at the number of blood sacrifices required

Table 2. Social cost benefit analysis results

Option	Blood sacrifices	Overall benefit score
no forge	0	$\frac{1}{3} \times 0 + \frac{1}{3} \times 0 + \frac{1}{3} \times 0 = 0$
no forge (mitigated)	2	$\frac{1}{3} \times 1 + \frac{1}{3} \times 1 + \frac{1}{3} \times 1 = 1$
scythe-only forge	5	$\frac{1}{3} \times 0 + \frac{1}{3} \times 4 + \frac{1}{3} \times 2 = 2$
scythe-only forge (mitigated)	6	$\frac{1}{3} \times 1 + \frac{1}{3} \times 4 + \frac{1}{3} \times 3 = \frac{8}{3}$
scythe-and-spear forge	20	$\frac{1}{3} \times 4 + \frac{1}{3} \times 4 + \frac{1}{3} \times 4 = 4$

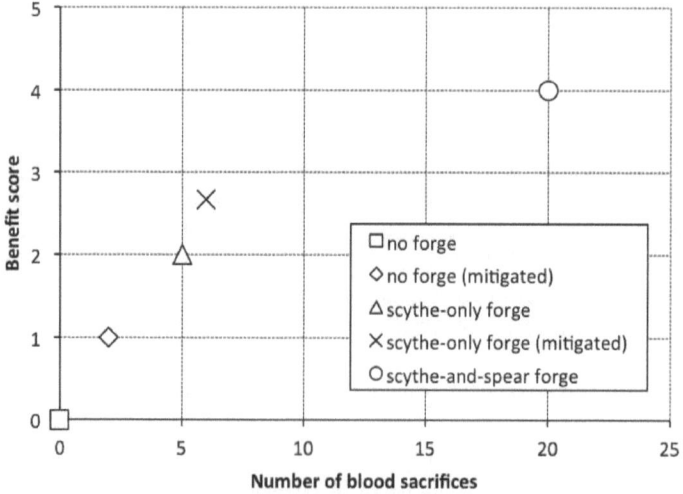

Fig. 2. Forge options cost-benefit scatterplot

for the latter, would choose not to build any new forge at all. So the gatherers placed all their stones into the boar skull.

When the voting was complete, the thinkers retrieved the three skulls from the cave and counted out the results in front of the ten tribespeople: five stones for the `kill animals` objective and five stones for the `harvest crops` objective. The thinkers looked nervous. They asked if any of the hunters and gatherers would like to say how they had voted and explain their reasons why. The tribespeople looked at their feet. Then one of the gatherers spoke up: why didn't the gatherers have more stones than the hunters, given that, on the island, they outnumbered them by a factor of ten to one? Without blinking, the thinkers replied that the stones were equally balanced to reflect the equal expertise of the two roles. The tribespeople were then asked to consider the initial results and subsequent discussion, and to vote once more for the most important objective to improve. The tribespeople entered the cave and placed their stones once more. The thinkers emerged again with the results: five stones in the wolf skull and five stones in the boar skull, precisely as before. The thinkers shrugged their shoulders and declared that both objectives had been assessed as equally the most important.

Next, the thinkers gave to each of the ten tribespeople ten tiny polished pebbles. The tribespeople were asked to think about how important it was to improve the remaining `feed tribes` objective from its worst position to its best position, compared to the two most important objectives already identified. Each hunter and gatherer was then invited to enter the cave as before, and to place into the shark skull as many tiny pebbles as they felt were proportional to the comparative importance of the `feed tribes` objective, where 0 pebbles = no importance and 10 pebbles = equal importance. The hunters and gatherers again complained that it was not their place to judge the importance of feeding the tribe, relative to other matters. However, even a tribal youngling could easily grasp that feeding the tribe was an important outcome which (since the hunters and gatherers had already each secured the importance of the intermediary benefits that they brought) should be associated with a lot of pebbles. Each tribesperson entered the cave and placed all ten pebbles inside the shark skull. The thinkers staggered out of the cave carrying the laden shark skull. After a few minutes, they were able to announce that the skull contained all one hundred possible pebbles and so the `feed tribes` objective had equal importance to the `kill animals` and `harvest crops` objectives.

With the preferences for the three benefit-related objectives finally elicited as $w_1 = w_2 = w_3 = \frac{1}{3}$, the thinkers were able to compute the overall benefit score for each option. The thinkers now also revealed to the tribespeople what the estimated number of blood sacrifices was in each case. The results are shown in Table 2. The thinkers laid out two twigs perpendicularly to each other on the forest floor to indicate scales of benefit and sacrifice, and then used the leaves of different plants to indicate the bi-objective outcomes for each option. This scatterplot is shown in Fig. 2.

Tired and unhappy, the hunters and gatherers studied the pattern of twigs and leaves on the ground. The thinkers were allegedly quite clever, but did they really think the elders would be using this information to make any kind of decision about the forge? Shaking their heads, the hunters and gatherers returned to their homes.

6   Conclusion

THE THINKERS INVITED the elders to the cave entrance to present the findings of the analysis. The elders studied the collection of twigs and leaves on the forest floor. They understood the benefits scores and were satisfied with the process by which they had arisen – the hunters and gatherers did a good job of serving the tribe's needs and their estimates and preferences could be respected. The elders had heard rumblings of complaint about the thinkers – particularly about the meeting at the cave – but everything here appeared to be in order. Well, that was life. The elders felt that it was important to invest in the four-shell `scythe-and-spear forge` that maintained existing capability, since the tribe should be protected wherever possible against starvation, and so the required number of blood sacrifices would just have to be managed.

The elders were less comfortable with the estimates presented by the thinkers for the blood sacrifices. If they were to sanction the `scythe-and-spear forge` option then it would be very important to develop afford-able proposals, particularly since budgets for blood sacrifices in other areas of tribe life would need to be cut back. The shaman had a noto-rious track record in under-estimating the number of sacrifices required to appease the gods. Whilst the thinkers claimed to have accounted for the over-optimism of the shaman, the elders were not convinced. They asked the thinkers to return to the shaman and work with him again to obtain improved sacrifice estimates. When this was done, declared the elders, a plan for sacrifices could be put in place and the new iron forge could at last be sanctioned.

In this paper, we have reflected on our practical experiences of supporting real-world decision-making. The discussions in the preceding sections, and particularly the events arising in the fictionalised account, all point towards the critical role that human factors play in multi-objective optimisation. Most government decision-makers, at least in the UK, are schooled and skilled in rhetoric; these people are wary of formal analytical methods. "Over-analysis" is a common criticism of government business cases, whilst the term "over-writing" would be seen as an oxymoron. There is a cultural barrier that leads to a basic mistrust of the findings of multi-objective optimisation.

This cultural barrier can lead to reluctance on the part of analysts to use multi-objective methods, even when to do so would clearly be appropriate. Typically, analysts are rewarded for business case progression; there is an incentive to pragmatically choose methods that are likely to see business cases pass successfully over the hurdles of the governance process (such as the OGC gateways), even when these methods are known to be poor.

The OGC process does tend to force large programmes into grand solutions requiring complex assessment under great uncertainty, which is discouraging to quantitative analysis. It may be wiser to take time to consider any activities that could resolve some of the uncertainty. This type of analysis is sometimes seen in medical decision-making, where formal *value of information* methods are used to identify useful activities, at least in cases where the uncertainty can be parameterised [15]. Such activities could be performed prior to the decision, or progressively built into the solution itself via evaluation [16].

To overcome the lack of confidence that many decision-makers have in multi-objective methods, analysts need to demonstrate that the methods are robust to repeated application. These demonstrations must be accessible to non-experts, and should include practical examples, case studies and guidance. Arguably, the main messages that a decision-maker would receive from browsing a copy of the current manual [11] is that the methods are complex and can only be understood through, or applied to, the most trivial of examples: the selection of a toaster.

Whilst human factors are important considerations, they are not the only key barriers to the uptake of multi-objective optimisation methods for social cost benefit analysis. It can be difficult to measure the performance of a candidate solution in terms of benefits rather than functional requirements. The causal chain by which a solution supports strategic objectives is often not well understood. There are a lack of available methods for the mathematical modelling of social systems, where the issue of causality remains a major area of debate [17]. In the story, to model the relationship between the requirements for the forge and the survival of the tribe would have required a major research project in its own right, in terms of both methodological development and application.

A further key barrier is the lack of societal preference estimates for use in trading-off performance levels between objectives. Experts are often very reluctant to express preferences between benefits, since their expertise tends to lie in the functional requirements and the preferences are fundamentally societal in nature. An alternative to using experts is to sample a broader cross-section of society, separately to the optimisation process, and build a model of preferences that can, in principle, be applied to multiple decisions. This type of approach has been explored perhaps the most in the area of health economics, although the research is still in its infancy (see, for example, [12]). A remaining issue is that part of the advantage of multi-objective optimisation is that it permits progressive preference articulation - that is, it permits the philosophical assumption that preferences are actually some function of available performance.

In summary, multi-objective optimisation has real potential for supporting government decision-makers in making informed choices for investment and policy decisions. The choices are informed to the extent that a solution should be Pareto-optimal and the performance against the multiple objectives should be reflective of societal preferences. Whilst there has been much work on the development of methods for multi-objective optimisation, application to *real* problems with *real* decision-makers is somewhat limited, at least to the extent that these applications are documented in the literature. We have sought to identify and discuss some of the reasons for this, and look forward with optimism to future practical implementations informed by developments in the academy.

References

1. HM Treasury, The Green Book. The Stationary Office (2003)
2. Winter, H.: Trade-offs: An introduction to economic reasoning. The University of Chicago Press (2005)
3. Miettinen, K.M.: Nonlinear Multiobjective Optimization. Kluwer International Series in Operations Research & Management Science (1998)
4. Deb, K.: Multi-objective Optimization using Evolutionary Algorithms. John Wiley & Sons (2001)
5. Flanagan, J., Nicholls, P.: Public Sector Business Cases using the Five Case Model: A toolkit. HM Treasury (2007)
6. Byrne, D.S.: Complexity Theory and the Social Sciences: An introduction. Routledge (1998)
7. Hammond, J.S., Keeney, R.L., Raiffa, H.: Smart Choices: A practical guide to making better life decisions. Harvard Business School Press (1999)
8. Bradley, G.: Benefit Realisation Management: A practical guide to achieving benefits through change, 2nd edn. Gower Publishing Limited (2010)
9. Baddeley, A.: The magical number seven: Still magic after all these years? Psychological Review 101(2), 353–356 (1994)
10. HM Treasury, Supplementary Green Book Guidance - Optimism Bias. HM Treasury (2003)
11. Department for Communities and Local Government, Multi-criteria Analysis: A manual. Communities and Local Government Publications (2009)
12. Watson, V., Carnon, A., Ryan, M., Cox, D.: Involving the public in priority setting: A case study using discrete choice experiments. Journal of Public Health 34(2), 253–260 (2012)
13. Matrix Insight, Prioritising Investments in Preventative Health," tech. rep., Health England (2009)
14. Goodwin, P., Wright, G.: Decison Analysis for Management Judgment, 3rd edn. John Wiley & Sons (2004)
15. Yokota, F., Thompson, K.M.: Value of information literature analysis: A review of applications in health risk management. Medical Decision Making 24, 287–298 (2004)
16. HM Treasury, The Magenta Book: Guidance for evaluation. HM Treasury (2011)
17. Sawyer, R.K.: Social Emergence: Societies as complex systems. Cambridge University Press (2005)

Biobjective Optimisation
of Preliminary Aircraft Trajectories

Christos Tsotskas, Timoleon Kipouros, and Anthony Mark Savill

Cranfield University, School of Engineering,
Department of Power and Propulsion,
MK430AL, Bedfordshire, UK
{c.tsotskas,t.kipouros,mark.savill}@cranfield.ac.uk

Abstract. The protection of the environment against pollutants produced by aviation is of great concern in the 21st century. Among the multiplicity of proposed solutions, modifying flight profiles for existing aircraft is a promising approach. The aim is to deliver and understand the trade-off between environmental impact and operating costs. This work will illustrate the optimisation process of aircraft trajectories by minimising fuel consumption and flight time for the climb phase of an aircraft that belongs to A320 category. To achieve this purpose a new variant of a multi-objective Tabu Search optimiser was evolved and integrated within a computational framework, called GATAC, that simulates flight profiles based on altitude and speed.

Keywords: trajectory optimisation, multi-disciplinary design optimisation, optimisation framework, Tabu Search.

1 Introduction

The protection of the environment is of great concern in the 21st century. The continuous operation of aircraft severely affects the climate, environment and humans, especially in the vicinity of the airports. Numerous studies confirm that the produced emissions and noise are an important threat. Currently, air-transport is the world's fastest growing sector. Aviation is responsible for 2% of man-made $CO2$ and aircraft production increases approximately by 5% per year. Due to continuous and increasing air-traffic the emissions will reach 3% by 2050. These pollutants have also great impact on health. In order to be environmentally and human friendly, the energy trends in aviation industry are related to reduction in noise footprint, improvement of fuel consumption and $CO2$ efficiency[1].

There are three possible options in order to reduce pollution [2] : a) decrease the overall number of operations, b) change aircraft type, c) alter aircraft trajectories. The first option is not feasible, as discussed in [1, 3]. Therefore, a combination of the last two options seems a viable approach with respect to the initial problem. However, changing the type of the aircraft is a difficult task and will significantly affect the aircraft operations as a whole. So, this turns out just

R.C. Purshouse et al. (Eds.): EMO 2013, LNCS 7811, pp. 741–755, 2013.

to be an alternative solution in the long-run. One can find several studies carried out that discuss the feasibility of setting new or modifying operational rules, regulations and procedures that decrease the impact of aircraft to the environment and climate, for instance [4]. Hence, the optimisation of trajectories could be readily performed as a more promising short-term target. This is a straightforward technique that could considerably reduce the effect of aircraft operation on the environment without the need to change[1] any of the participating entities of aviation industry (aircraft, airports, human resources, other equipment, etc.). This research activity focuses on this direction by employing the most modern computational methods and tools in this multi-disciplinary field.

Aviation problems are governed by multiple disciplines [5, 6]; overhead, various types of costs, multiple performance metrics, further desired properties, etc. Definitely, aircraft-trajectory optimisation belongs to this category and Air Traffic Management (ATM) is an important issue. Several studies focus on flight efficiency factors such as noise and fuel consumption, advanced trajectory technologies such as prediction and management, and other air-transport economics. Ultimately, all these technologies aim at mitigating environmental impact by relieving over-crowed airspace and airports[1, 7].

Besides academia and research centres, industrial sectors show high interest in trajectory optimisation. In 2001, European Union started CLEAN SKY [8], which is a Joint Technology Initiative (JTI), with focus on the sustainability of the increasing air-traffic demand while developing breakthrough technologies so as to drastically reduce the environmental impact by the aviation industry. This work is carried out under Systems for Green Operations (SGO), one of the six Integrated Technology Demonstrators (ITDs) that compose CLEANSKY, which attempts to improve the contribution of the aircraft systems and reduce their impact to the environment.

Following the purposes of SGO that coincide with the research goals of Advisory Council for Aviation Research in Europe (ACARE), the target is to minimise CO2 and NOx emissions by determining more efficient trajectories for all the flight phases. This is achieved by employing a computational framework named Greener Aircraft Trajectories under ATM Constraints (GATAC)[9]. It is developed in collaboration between Cranfield University and University of Malta. This work contributes to the improvement of the framework by developing a new optimisation module, which will be available in the future versions of GATAC and is expected to bring significant benefits to aircraft trajectory optimisation. More specifically, aircraft flight trajectories are optimised considering that the aircraft/engine configuration is already designed and in operation.

The purpose of this work is to investigate and assess the performance of a new variant of Multi-Objective Tabu Search (MOTS), called (MOTS2), in the field of trajectory optimisation and present a new methodology to tackle this type of problems. The test case includes the optimisation of the trajectory of a single commercial passenger aircraft of the A320 family under the climb phase, which is the most demanding and important in terms of energy requirements

---

[1] In some cases minor changes might occur.

and pollutants production. The performance of the newly developed optimiser will be compared against the existing one, which belongs to Genetic Algorithms (GAs). The tools and methods employed are intended to be further improved and integrated into an aircraft so as to carry out trajectory optimisation in real time, while in flight. The ultimate goal is to help in shaping the future of aviation.

2   Aircraft Trajectory Simulation

A trajectory is defined by the area navigation (RNAV) method, which is based on coordinates. Aircraft trajectory denotes the RNAV route an aircraft flies on while passing through specified geographical locations, which are called waypoints. It is assumed that the aircraft passes through a waypoint at a certain speed without deviating. The trajectory breaks down into smaller parts/periods - also known as phases of flight. The most usual phases are take-off, landing, climb (or ascent), cruise, descent. The first two are the shortest parts of the flight and heavily depend on the current environmental conditions, ATM constraints (imposed by local authority) and pilot's judgement[7]. So, there is no point in optimising these parts. The remaining phases could be automated and relative to take-off and landing are significantly larger. Therefore, they attract higher interest since it is less intuitive for the pilot to take into consideration all of the parameters and operate the aircraft in the most optimal way in terms of fuel consumption and flight time. Since climb is the most fuel consuming flight phase, this work will focus on that part of the flight.

The formulation of the trajectory, type and number of waypoints involved, affects the complexity of the optimisation process. In reality, these trajectories are 3D paths of curvilinear shape. In this work, a simplified trajectory approach has been employed: The considered trajectories are in 2D, hence vertical trajectories, and the range - distance flown on a given amount of fuel - is split in small straight-line-segments. In fact, the third dimension is mainly used for turning and will be considered in future studies in the next phase of CLEANSKY project. Each segment is defined between two waypoints. The target trajectory is formed by connecting these segments in a very specific order so that the total energy at the boundaries of two adjacent segments is the same. Moreover, the following segment depends on the end of the previous one, a principle from control theory.

Considering that the waypoints and respective speed values have been set for a single phase, the performance indices that characterise the flight are resolved by the Aircraft Performance Model(APM) and the Engine Performance Model(EPM). These models are coupled and applied on every segment and the corresponding indices are aggregated for the whole phase. Based on the aircraft performance characteristics (size, lift, etc.), the APM used in this study (provided by University of Malta [10]) calculates the required thrust throughout the target segment and the respective flight time. Then, the EPM (provided by Cranfield University [11]) is invoked to calculate the fuel consumption of the engine over the same flight period. This method is iteratively repeated for every segment. It is important to mention that at the end of the simulation of a single segment, the APM calculates the exiting flight path angle and the EPM computes the mass of the consumed fuel. These values will be used as input for the simulation of the following segment for the entering flight path angle and the new (reduced) total aircraft mass, respectively. Therefore, a single phase of the flight has been simulated. This procedure (Fig. 1a), which is automatically handled by GATAC, will be called evaluation of the trajectory, or simply evaluation. It will be repeated several times under different altitude and speed values in order to obtain the optimum behaviour.

The number of segments to which the trajectory breaks down is an important factor, which is related to the complexity of the case, as it increases the dimensionality. The trajectory simulation consists of two types of parameters; control and state parameters. Initially, the state variables (such as aircraft weight, range number of segments) are pre-defined by the user and the control variables (altitude and airspeed) are handled systematically by an external algorithm (the optimiser).

In this work, 4-segment trajectories will be simulated, where each segment is defined between two waypoints. Although five waypoints compose the trajectory, the beginning and end waypoint and their speed values remain constant. Therefore, the 4-segment trajectory is parametrized by setting 6 variables; 3 for the altitude (ATL1, ALT2, ALT3) and 3 for the speed (SPD1, SPD2, SPD3). This set of 6 variables will be the design vector. The range of each segment is constant and equals 40000m. During each segment the speed remains constant at the originally set value. The simulated aircraft belongs to A320 category, which carries two typical two spool, high bypass turbofan air-engines with separate exhaust, model CFM56-5B4, and weights of 70 tonnes.

When all the aforementioned variables are set, the simulation generates a vertical trajectory and calculates two performance indices; the weight of the consumed fuel and the respective endurance. These are the considered objectives for this work. However, fuel efficiency is related to endurance, and vice-versa. It seems obvious that saving up fuel and shortening the endurance - maximum length of flight time - are conflicting objectives. The targets are to minimise fuel consumption and endurance. Hence, this is a multi-objective optimisation problem, which will be resolved by employing native multi-objective optimisers.

3   Multi-objective Aircraft Trajectory Optimisation

When calculating optimal flight profiles, complex optimisation techniques have to be used. The optimisation of aircraft trajectories is a constrained, non-linear, multi-disciplinary and multi-objective problem. The parameters are dynamic, deterministic and real-valued. In addition, it involves principles of optimal control theory. The integration of trajectory simulations along with optimisation algorithms under GATAC has been presented in [12–14]. Currently, the framework provides a variety of trajectory simulation models, but only one optimiser. This work expands the portfolio of optimisation algorithms available to the user by adding a new optimiser. This is common practice and more flexibility is provided to the end-user to choose between the two optimisers for different cases, because no optimiser is equally good for every possible scenario [15].

In the majority of published literature, flight paths are optimised by transforming the original problem into an optimal control problem, such as [16]. Then, the new problem is resolved by employing standard techniques from optimal control theory. This method is partially chosen because it is easy to access the formulae that describe the problem which then turns out to be one of numerical analysis. However, the necessary information is not always available, mainly due to the complexity of the simulation and the number of participating factors (individuals and codes). Here, the authors follow a different approach, which is more flexible and easily extensible. For a given trajectory, the aforementioned APM and EPM are coupled together and deliver the output metrics. Then, MOTS2 collects and handles this pair of input and output for the optimisation phase. This is therefore a modular approach of three individual modules, which are managed by GATAC. Each part operates independently of the other part and can be manipulated separately, as described above.

Following the description of the trajectory simulation in Sect. 2, which represents the objective function evaluation of the design vector in terms of optimisation, the case specification is presented in Table 1. The aircraft is subject to a number of constraints regarding its structural (e.g. maximum travel speed) and operational (e.g. maximum angle of attack) limitations and ATM restrictions (e.g. flight within certain altitude margin). All these constraints affect the range of components of the design vector. Furthermore, for climb, a continuous ascending altitude must be used. The lower and upper bounds for both altitude and speed delimit the design space, wherein the optimiser should locate the best designs based on the objective values. In addition, hard constraints are imposed by the APM and EPM whenever the design vector produces irregular trajectories.

Regarding the trajectory optimisation, the combination of the parameters ALT1, ALT2, ALT3, SPD1, SPD2 and SPD3 defines the design. Each component of the design varies within a continuous range of real numbers, which denotes the design space, $\mathbb{R}^6$. In an analogous way, the objectives fuel consumption (FUEL) and endurance (TIME) belong to a different space, namely objective space, $\mathbb{R}^2$. Any single point of the design space maps to a point of the objective space. The aim of the optimiser is to try different combinations of these variables on the given simulation model and detect which areas express the best performance, defined by the objectives. Following a number of successful iterations through the optimisation phase, the best discovered Pareto Front is presented to the designer to choose the final design. This is known as the decision phase. The time required in order to establish the variables-to-objectives mapping is the evaluation time of the given variables via the simulation. This is the most critical part of the optimisation process, as it affects the overall execution time of the whole optimisation and can take several days. In fact, the overall execution time can be expressed as the summation of multiples of the execution time required for a single evaluation and the overhead of the optimiser, which is practically negligible. In this case, each design evaluation can take up to 2 minutes, which is prohibitive for real time optimisation.

The communication time among the participating modules is an important factor, too. The interface between the optimisers and the trajectory simulation is handled by GATAC, as depicted in Fig. 1a. Part of evaluation time is spent in exchanging files among the modules, which will be improved by employing direct communication methods. The (black box) communication is achieved by using special dictionaries of extensible mark-up language (XML) and via directly exchanging files. Using sampling techniques and parallelism can significantly speed-up the optimisation process.

In this work, two native multi-objective optimisers will be used on the same test case, as discussed in Sect. 2, and their performance will be analysed and assessed. Handling all of the objective functions at the same time, without using any other kind of transformation is of paramount importance, since this can deliver an unbiased trade-off[17]. The first optimiser, called Non-dominated Searching Genetic Algorithm - Multi-Objective (NSGAMO)[13], is a variant of the Non-Dominated Sorting Genetic Algorithm (NSGA-II)[18]. The second optimiser is based on Multi-Objective Tabu Search (MOTS)[19, 20] and will be described below. Both optimisers can operate in constrained and unconstrained problems and they will run for 20000 objective function evaluations. The configuration settings (Table 2) were chosen based on authors' experience so as to explore the design spaces sufficiently and generate feasible trajectories.

Table 1. 4 segments climb case specification

node	Range	Min Altitude	Max Altitude	Min Speed	Max Speed
0	0 m	1500 m	1500 m	130 m/s CAS	130 m/s CAS
1	40000 m	3000 m	5500 m	130 m/s CAS	190 m/s CAS
2	80000 m	3500 m	6000 m	130 m/s CAS	190 m/s CAS
3	120000 m	5500 m	6000 m	140 m/s CAS	190 m/s CAS
4	160000 m	6000 m	6000 m	0.8 Mach	0.8 Mach

3.1   Multi-Objective Tabu-Search 2

Tabu search methods can be classified as stochastic search optimisers. The original and multi-objective versions were presented in [20] and [19], respectively. A new variant, namely MOTS2, has been evolved by the lead author, based on the original MOTS scheme developed and deployed to a range of aerodynamic optimisations [21–24]. In general, the design space is explored in a stochastic way, while recently visited points (stored in a Tabu memory) are avoided so as to guarantee more exploitation of the unknown design space. In fact, the local search scheme (Hooke and Jeeves [25]), which is particularly efficient for continuous parameters, is combined with stochastic elements. Different hierarchical memories (Short-term, Medium-term, Long-term and Intensification[2]) are used to assist

---

[2] This memory was defined in [19].

critical decisions during the optimisation process. It also keeps track of certain statistics during the process, which direct the search according to the discovered landscape of the design space. In addition, the optimiser employs a mechanism for local (Intensification Move) and global (Diversification Move) search. The statistics detect design points around the current search point, within relatively short distance, whereas the search mechanisms attempt to discover good design points in the entire design space. The outline of MOTS2 is depicted in Fig. 1b. In addition, MOTS2 includes the improvements (local search enhancements for Diversification Move) discussed in [21] and, given any parallel framework such as GATAC, it can operate in parallel mode saving elapsed time.

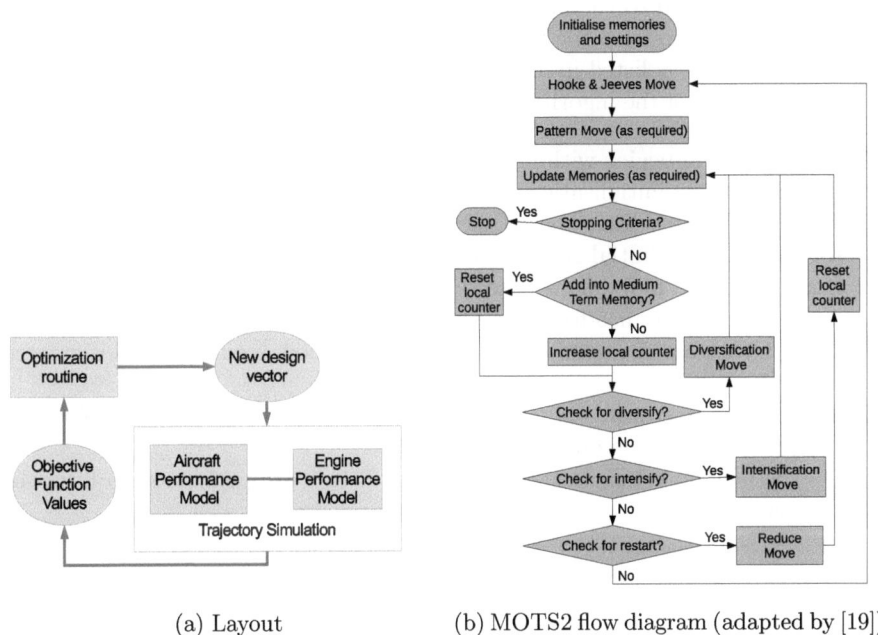

(a) Layout   (b) MOTS2 flow diagram (adapted by [19])

Fig. 1. Optimisation Process

The search is guided by the current base point and the aforementioned memory containers. Around the base point, adjacent candidate design points are investigated and evaluated. Then, the corresponding objective values are sorted according to domination criteria of multi-objective optimisation [26] into the appropriate containers for the current iteration and the next base point is resolved. The previous base point and all the recently generated points are inserted into the appropriate memory containers. Aggregated information (e.g. multiplicity of certain objectives) will be used in future steps, when certain conditions (e.g. exceed a predefined threshold) are met. The optimisation process keeps repeating until stopping criteria (such as number of evaluations) are met. During every iteration a fraction of the flow of the algorithm is executed every time, and the rest runs when certain conditions are met. During the execution of MOTS2, several statistics are gathered to monitor the optimisation progress and explain the functionality of the optimiser.

3.2   The GA Optimiser in GATAC

The development of NSGAMO was discussed in [13] and it performance was demonstrated in [12–14]. It has been considered to be the state-of-the-art GA for multi objective optimisation, appropriate for trajectory optimisation. The method is designed for lower computational complexity during Non-dominated sorting using the concept of Elitism which improves the ability to retain good solutions. Compared to the original NSGA-II, NSGAMO uses a different selection process so as to form the mating pool and an entirely different sequence of genetic operators, as listed in Table 2.

The main flow of the algorithm consists of the following stages:

1. The algorithm begins with an initial population of N individuals and multiplies it with an initialisation ratio for the 1st generation.
2. This population is sorted based on the principle of Constraint Non-dominated sort to form the initial generation P_t.
3. If after sorting, individuals exceed N, then N individuals are selected based on Crowding Distance from the final Front.
4. Using Constrained Crowded Tournament Selection, a mating pool is created from P_t.
5. The genetic operators (Mutation and Crossover) are then used to form an Offspring Population Q_t.
6. On the merged set $R_t = (Q_t + P_t)$, steps 2 to 3 are performed to form the next generation.

4   Optimal Aircraft Trajectories - Results and Discussion

Both optimisers ran for 20000 evaluations under their individual settings listed in Table 2 and the final Pareto Front is shown in Fig. 2. Among the discovered designs one from each optimiser is selected approximately from the middle of the plane and the corresponding trajectory, also known as the compromise design, is visualised. The designs of the Pareto Front revealed by NSGAMO spread almost uniformly over the Objective Space, whereas the ones discovered by MOTS2 lie on two narrower regions but their population is larger. So, depending on the requirements of the application, it is impossible to chose one optimiser from the other. In general, NSGAMO delivered a large range of designs, which informs the user about the performance of the aircraft in many different scenarios. Contrary, the designs revealed by MOTS2 are denser and very close to each other. So, there is information about near-similar performance on different settings of altitude and speed.

Table 2. Configuration settings for trajectory optimisation

(a) MOTS2

(b) NSGAMO

(a) MOTS2	(b) NSGAMO
diversify after 15 iterations	% for creep mutate with decay 0.01
intensify after 10 iterations	% for dynamic vector mutate 1.01
reduce after 30 iterations	% covered dynamic vector mutate 0.75
search step {0.3,0.3,0.3,0.1,0.1,0.1}	% covered for Vector mutate 1
search step retained factor 0.5	% covered for element mutate 0.6
6random samples	Convergence Fitness Tolerance 1.0E-6
6 variables	Inflationary scheme population limit 3
2 objectives	Inflationary scheme starting point 1.6
4 regions in Long Term Memory	Initialization Factor 30
Tabu Memory size 15	Maximum Generations 150
	Population Size 100
	Element mutation Probability 0.05
	Creep mutate Probability 0.1
	SBX Distribution Coefficient 1.0
	Selection Pressure 2.0

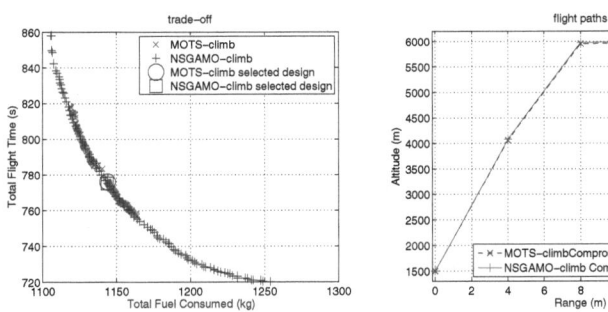

Fig. 2. Comparing the trade-offs and the compromise designs

Despite the difference in the shape of Pareto Front, the trajectory of the compromise design generated by MOTS2 is slightly lower in altitude and quicker than NSGAMO's but the difference in objective values is negligible, as shown in Fig. 3. Hence, the compromise design behaves equally well in both cases. It is obvious, that the margin between the extreme designs for NSGAMO is larger. This is expected, since the variation of the variables is wider and the extrema are quite distant.

Two methods that will assist in assessing the performance between the optimisers are the Principal Component Analysis (PCA)[27] and the hypervolume indicator [28]. The first is applied only to the design vectors of the Pareto Front for each optimiser. This method can detect which components of the design vector are the most energetic. This is achieved by calculating which parameter

Table 3. Statistics

	Principal Components Analysis						Hypervolume Indicator
	ALT1	ALT2	ALT3	SPD1	SPD2	SPD3	(reference : {1253.8, 858.1})
NSGAMO	0.9316	0.0658	0.0024	0.0001	0.0000	0.0000	15109.666
MOTS2	0.8958	0.1011	0.0028	0.0003	0.0001	0.0000	12579.178

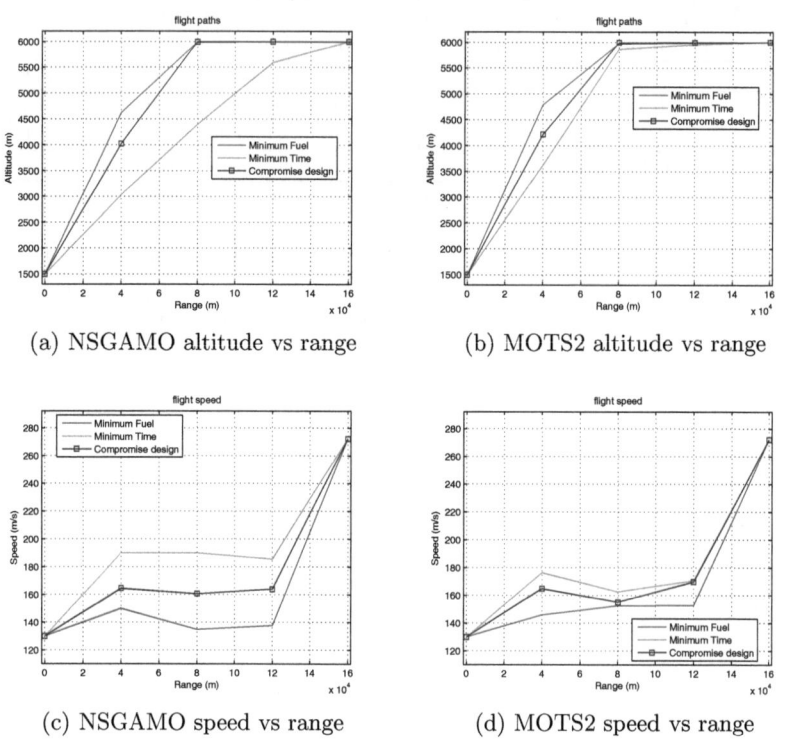

(a) NSGAMO altitude vs range   (b) MOTS2 altitude vs range

(c) NSGAMO speed vs range   (d) MOTS2 speed vs range

Fig. 3. Generated trajectories

has the highest variance. Consequently, these parameters are mainly responsible for the current instance of the trade-off and their importance is quantified. The left part of Table 3 presents the variance of the score of PCA for each parameter over the total variance. So, ALT1 contains by far the highest percentage of variance for both cases and it is considered the most significant parameter. Hence, the optimisation process should mainly focus the search based on this parameter. This has a double implication: either the search step could be very fine for the specific parameter, or the optimisation could be performed again with fewer variables, which would suppress the dimensionality and speed-up the process. Because the number of iterations is very large, setting ALT1 as the most

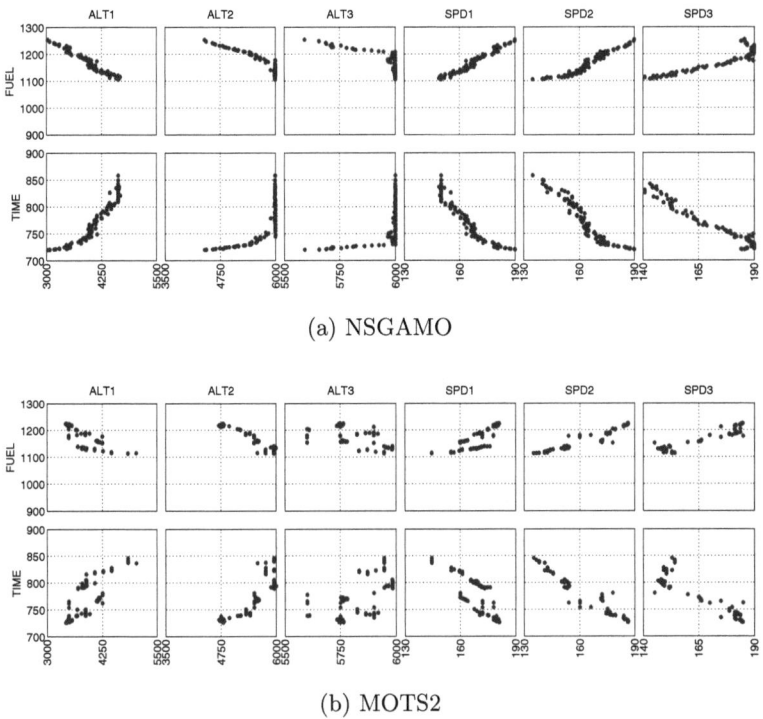

(a) NSGAMO

(b) MOTS2

Fig. 4. Design variables to objective functions/metrics relationships

important variable seems to be a safe choice, but this is not guaranteed to be true until the whole search space is explored. The hypervolume indicator (right part of Table 3) is related to the objective space and is used to quantitatively compare the performance of the trade-off. More specifically, the richness and the span of the Pareto Front are combined in one metric; the higher the value is, the better the trade-off. The reference point used is the combination of the worst objectives from both Pareto Fronts. According to hypervolume indicator, NSGAMO achieved a higher value against MOTS2. However, by definition, the indicator favours the trade-off that spreads over the design space. Therefore the user is informed about the importance of the discovered variables and the overall performance of the revealed Pareto Front.

The pairwise relationship between each variable against each objective for each optimiser is depicted in Fig. 4. The situation for NSGAMO is straightforward. Whenever the altitude increases the fuel consumption is reduced and the flight time is longer and vice versa, not in linear way. The same statement is true for speed, but with smoother response. It is noteworthy that, since ALT1 was identified as the most significant variable, for less than 1200 kg of consumed fuel the second and third altitude parameter are almost constant. Similarly, if flight

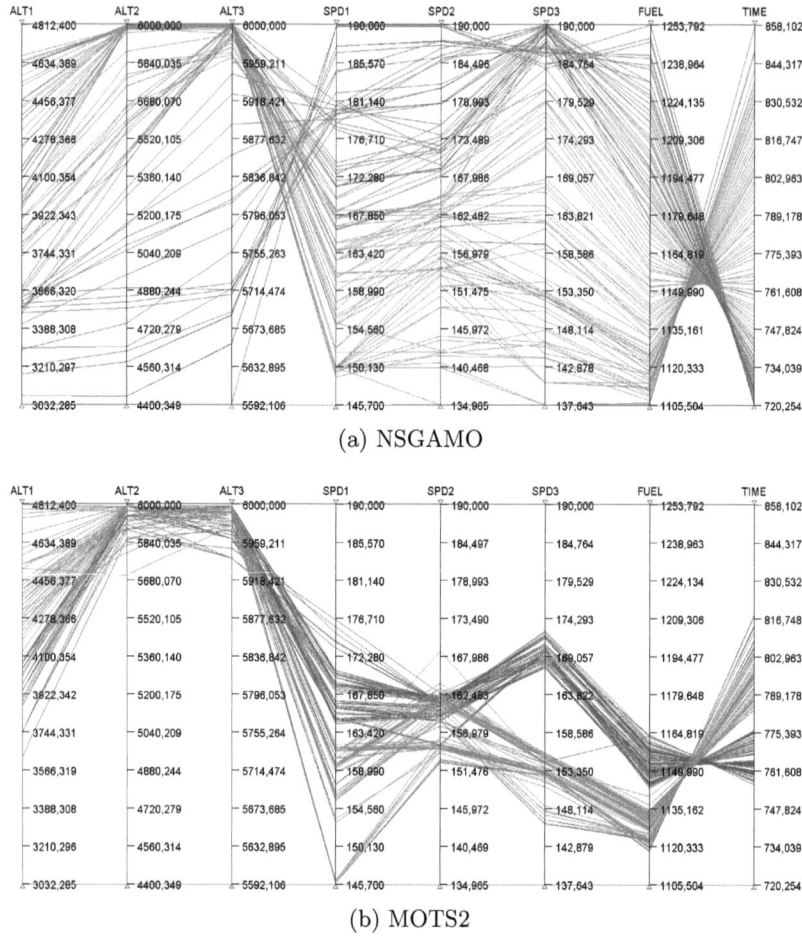

(a) NSGAMO

(b) MOTS2

Fig. 5. Parallel coordinates projection

time is longer than 750s the same components remain constant. Interpreting the results from MOTS2 is less intuitive. Although the Pareto Front is denser, the design variables to objective functions relationship presents two separate zones of performance. Again, as ALT1 increases fuel consumption drops and flight time increases. Then as speed increases more fuel is consumed and the flight is shorter. However, trends appear smoother. This is partly true because the trade-off is not as wide as with NSGAMO. Hence, the reduced distance between the extrema explains this behaviour. In both cases, the response of speed parameters against FUEL seems to follow a linear pattern.

The revealed Pareto Fronts are further investigated by using the Parallel Co-ordinates Projection[29], which is illustrated in Fig. 5. It is confirmed again that NSGAMO's trade-off is wider and equally spaced, whereas MOTS2 discovered two rich and different zones of performance. The domain is clustered based on the fuel efficiency high and low consumption. Moreover the user is informed about which regions each optimiser explored.

Parallel Coordinates can interpret the behaviour of the optimal profiles. Initially, for NSGAMO, the population (which is 36% of the whole set) of trajectories that belongs to the upper-half of FUEL corresponds to a very thin range for TIME. This means that increasing the fuel consumption does not equally improve the flight time. Besides the ALT3 axis, the majority of designs do not mix and thus the flight performance could be easily separated. NSGAMO discovered designs at a larger fraction of the axes, which also justifies the span of the Pareto Front.

Two distinct zones of performance were recognised by MOTS2 with complicated interactions between the variables. In terms of the range of objectives, Tabu-Search is more balanced. The local search scheme also affects the (mixed) shape of the Parallel Coordinates plot, where the concentration of coordinates is more intense from region to region of the axes.

In general, the ratio of high to low fuel efficiency for NSGAMO is $64/36$ whereas the same ratio for MOTS2 is $112/217$. Although the designs seem scrambled, the classification of FUEL can be directly resolved from ALT1. In addition, both cases revealed that fuel-efficient (lower half FUEL axis) trajectories present very small variation in ALT2 and ALT3. Also, a similar pattern between ALT1 and ALT2 axis is presented.

5   Conclusions

This work demonstrated that the newly developed MOTS2 is a competitive alternative against NSGAMO when applied on aircraft trajectory optimisation problems, with very distinct results. The performance of these optimisers was compared in a 4 segments climb of commercial passenger aircraft of the A320 family and different zones of performance were highlighted, which necessitate the need for further exploration. Although the revealed Pareto Front is not as wide as NSGAMO's, the discovered designs seem to be more focused within a short range. Compromise and extreme designs were discovered and discussed. It was found that the first variable (ALT1) is by far the most significant parameter for 4-segments climb trajectory and severely affects the performance of the flight. The findings are in agreement with the flight physics and provide additional insight between the control parameters and the objective values.

Future work will involve more segments and more flight phases, so as to optimise more realistic trajectories. This leads to the development of optimisation algorithms that can manage large number of design parameters. Since the ultimate goal is to carry out optimisation in real-time, further developments that eliminate the evaluation cost are required. This can be achieved either by developing/adapting new algorithms or by employing alternative computer architectures.

Acknowledgments. The research project has received funding from the European Union's Seventh Framework Program (FP7/2007-2013) for the Clean Sky Joint Technology Initiative under grant agreement N0 CJSU-GAMSGO-2008-001. The lead author would like to thank Bodossaki Foundation. The authors would like to thank David Zammit-Mangion, Vishal Sethi, William Camilleri, Kenneth Chircop, Hugo Pervier, Matthew Sammut and Matthew Xuereb for their contribution in providing APM, EPM, NSGAMO and GATAC.

References

1. EUROCONTROL. PERFORMANCE REVIEW REPORT An assessment of Air Traffic Management in Europe during the Calendar Year 2011. Technical Report PRR2011, EUROCONTROL (2012)
2. Clarke, J.P.: The role of advanced air traffic management in reducing the impact of aircraft noise and enabling aviation growth. Journal of Air Transport Management 9(3), 161–165 (2003)
3. ACARE. The Addendum to the Strategic Research Agenda (2008)
4. Portillo, Y.: Pre-Tactical Trajectory Compatibility Determination to Reduce Air Traffic Controller's Tactical Workload, MSc Thesis, Cranfield University, UK (2012)
5. Henderson, R.P., Martins, J.R.R.A., Perez, R.E.: Aircraft conceptual design for optimal environmental performance. The Aeronautical Journal 116(1175) (2012) (in press)
6. Kroo, I.: Innovations in Aeronautics - AIAA Dryden Lecture. In: 42nd AIAA Aerospace Sciences Meeting, Reno, Nevada, January 5-8, pp. 1–11. AIAA (2004)
7. Cook, A.: European Air Traffic Management: Principles, Practice and Research, Ashgate, Aldershot (2007) ISBN 978-0754672951
8. CLEANSKY. Innovating together, flying greener, http://cleansky.eu/ (accessed November 2012)
9. Chircop, K., Xuereb, M., Zammit-Mangion, D., Cachia, E.: A generic framework for multi-parameter optimization of flight trajectories. In: 27th Congress of the International Council of the Aeronautical Sciences, September 19-24 (2010)
10. Camilleri, W., Chircop, K., Zammit-Mangion, D., Sabatini, R., Sethi, V.: Design and Validation of a Detailed Aircraft Performance Model for Trajectory Optimization. In: AIAA Modeling and Simulation Technologies Conference, Minneapolis, Minnesota, August 13-16, pp. 2012–4566. AIAA (2012)
11. Palmer, J.R.: The TURBOMATCH Scheme for Gas-Turbine Performance Calculations, User's Guide, Cranfield University, UK (1999)
12. Sammut, M., Zammit-Mangion, D., Sabatini, R.: Optimization of fuel consumption in climb trajectories using genetic algorithm techniques. In: AIAA Guidance, Navigation, and Control Conference, Minneapolis, Minnesota, August 13-16, pp. 2012–4829. AIAA (2012)
13. Pervier, H., Nalianda, D., Espi, R., Sethi, V., Pilidis, P., Zammit-Mangion, D., Rogero, J.M., Entz, R.: Application of genetic algorithm for preliminary trajectory optimization. SAE International Journal of Aerospace 4(2), 973–987 (2011)
14. Chircop, K., Camilleri, W., Sethi, V., Zammit-Mangion, D.: Multi-Objective Optimisation of a constrained 2000 km Trajectory using Genetic Algorithms. In: 3rd CEAS Air & Space Conference, October 24-28 (2011)

15. Wolpert, D.H., Macready, W.G.: No free lunch theorems for optimization. IEEE Transactions on Evolutionary Computation 1(1), 67–82 (1997)
16. McEnteggart, Q., Whidborne, J.: A Multiobjective Trajectory Optimisation Method for Planning Environmentally Efficient Trajectories. In: 2012 UKACC International Conference on Control (CONTROL), September 3-5, pp. 128–135 (2012)
17. Deb, K.: Multiobjective optimization using evolutionary algorithms. Wiley, Chichester (2001) ISBN 9780471873396
18. Deb, K., Agrawal, S., Pratap, A., Meyarivan, T.: A Fast Elitist Non-Dominated Sorting Genetic Algorithm for Multi-Objective Optimization: NSGA-II. IEEE Transactions on Evolutionary Computation 6(2), 182–197 (2002)
19. Jaeggi, D.M., Parks, G.T., Kipouros, T., Clarkson, P.J.: The development of a multi-objective tabu search algorithm for continuous optimisation problems. European Journal of Operational Research 185(3), 1192–1212 (2008)
20. Glover, F., Laguna, M.: Tabu search, 3rd printing edn. Kluwer Academic Publishers (1999) ISBN 9780792381877
21. Kipouros, T., Jaeggi, D.M., Dawes, W.N., Parks, G.T., Savill, A.M., Clarkson, P.J.: Biobjective Design Optimization for Axial Compressors Using Tabu Search. AIAA Journal 46(3), 701–711 (2008)
22. Kipouros, T., Jaeggi, D.M., Dawes, W.N., Parks, G.T., Savill, A.M., Clarkson, P.J.: Insight into High-quality Aerodynamic Design Spaces through Multi-objective Optimization. Computer Modeling in Engineering and Sciences (CMES) 37(1), 1–44 (2008)
23. Ghisu, T., Jarrett, J.P., Parks, G.T.: Robust Design Optimization of Airfoils with Respect to Ice Accretion. Journal of Aircraft 48(1), 287–304 (2011)
24. Ghisu, T., Parks, G.T., Jarrett, J.P., Clarkson, P.J.: Robust Design Optimization of Gas Turbine Compression Systems. Journal of Propulsion and Power 27(2), 282–295 (2011)
25. Hooke, R., Jeeves, T.A.: Direct Search Solution of Numerical and Statistical Problems. Journal of the ACM (JACM) 8(2), 212–229 (1961)
26. Geilen, M., Basten, T., Theelen, B., Otten, R.: An algebra of pareto points. Fundamenta Informaticae 78(1), 35–74 (2007)
27. Jolliffe, I.T.: Principal Component Analysis, 2nd edn. Springer Series in Statistics. Springer (2010)
28. Auger, A., Bader, J., Brockhoff, D., Zitzler, E.: Theory of the Hypervolume Indicator: Optimal μ-Distributions and the Choice of the Reference Point. In: Foundations of Genetic Algorithms (FOGA 2009), pp. 87–102. ACM, New York (2009) ISBN 978-1-60558-414-0
29. Inselberg, A.: Parallel Coordinates: Visual Multidimensional Geometry and Its Applications. Springer, Dordrecht (2009) ISBN 9780387686288

A Case Study on Multi-Criteria Optimization of an Event Detection Software under Limited Budgets

Martin Zaefferer[1], Thomas Bartz-Beielstein[1], Boris Naujoks[1], Tobias Wagner[2], and Michael Emmerich[3]

[1] Faculty for Computer and Engineering Sciences
Cologne University of Applied Sciences, 51643 Gummersbach, Germany
firstname.lastname@fh-koeln.de
[2] Institute of Machining Technology (ISF)
TU Dortmund University, 44227 Dortmund, Germany
wagner@isf.de
[3] Leiden Institute for Advanced Computer Science
Leiden University, The Netherlands
emmerich@liacs.nl

Abstract. Several methods were developed to solve cost-extensive multi-criteria optimization problems by reducing the number of function evaluations by means of surrogate optimization. In this study, we apply different multi-criteria surrogate optimization methods to improve (tune) an event-detection software for water-quality monitoring. For tuning two important parameters of this software, four state-of-the-art methods are compared: S-Metric-Selection Efficient Global Optimization (SMS-EGO), S-Metric-Expected Improvement for Efficient Global Optimization SExI-EGO, Euclidean Distance based Expected Improvement Euclid-EI (here referred to as MEI-SPOT due to its implementation in the Sequential Parameter Optimization Toolbox SPOT) and a multi-criteria approach based on SPO (MSPOT).

Analyzing the performance of the different methods provides insight into the working-mechanisms of cutting-edge multi-criteria solvers. As one of the approaches, namely MSPOT, does not consider the prediction variance of the surrogate model, it is of interest whether this can lead to premature convergence on the practical tuning problem. Furthermore, all four approaches will be compared to a simple SMS-EMOA to validate that the use of surrogate models is justified on this problem.

1 Introduction

The time required for a process feedback can play a crucial role in many fields of industrial optimization. Complex and expensive real-world processes or time consuming simulations lead to large evaluation times. This restricts optimization processes to only a very limited number of such evaluations. Moreover, almost all industrial optimization tasks feature more than one quality criterion.

R.C. Purshouse et al. (Eds.): EMO 2013, LNCS 7811, pp. 756–770, 2013.

Techniques from multi-criteria decision making, evolutionary multi-criteria optimization (EMO) in particular, were developed during the last decade to solve such tasks. The necessity to combine EMO techniques and optimization methods such as EGO [13] or SPO [1], which require a very small number of function evaluations only, should be self-evident. The application of such methods to real-world problems in industrial optimization provides a reasonable way to assess their feasibility. In contrast to artificial test functions, it allows for an assessment of the practical relevance for these kinds of problems.

In this paper we focus on four different tuning methods which are applied to tune an anomaly detection software for water quality management. This problem is usually handled by receiver operator characteristic (ROC) analysis. Due to specific limitations of the software concerned, this can not be applied in the classical way. Rather, the ROC curve should be approximated by Multi-Criteria Optimization (MCO) methods. That means, the ROC curve can be interpreted as a Pareto front. Interpreting ROC curves from the multi-criteria optimization perspective is an established approach in computational intelligence, see, e.g., [17].

In Sec. 2, we will summarize the former work performed in relevant research fields. The specific problem is presented in Sec. 3. The tuning algorithms (based on different SPO and EGO implementations) are described in Sec. 4. Section 5 describes the experimental setup, whereas the analysis is presented in in Sec. 6. Finally, Sec. 7 gives a summary of findings and an outlook on future work.

2 Former Research

Surrogate modeling is not a new topic in optimization. Jin [12] provides a comprehensive overview of single-objective optimization with surrogate models. While methods like EGO or SPO for single criteria optimization are well established, the application of surrogate modeling procedures for multiple objectives is more recent.

2.1 Surrogate Modeling in Multi-Criteria Optimization

In MCO, several approaches employ surrogate modeling. One example is the well established ParEGO by Knowles [15]. An overview of surrogate modeling in MCO is given by Knowles and Nakayama [16]. To balance exploration and exploitation in case of a limited budget, several methods employ infill criteria based on expected improvement (EI). Two things are required for defining such a criterion: the definition of the improvement and an algorithm to compute its expectation [22]. Since negative improvements are not possible, dominated solutions should yield an improvement of zero. As large variances potentially result in large improvements and large deteriorations are not penalized, these criteria also focus on the exploration of uncovered areas of the search space. It is of interest to see if this kind of additional exploration is desirable for the problem at hand. In particular, it remains to be seen whether there is already sufficient exploration done due to the initial design or due to the requirement of covering a whole set of Pareto optimal points.

2.2 ROC Analysis

This work deals with tuning the event detection software CANARY [10,19][1] which tries to detect anomalies in water quality data. The core algorithm in CANARY compares the difference between a predicted value and the most recently measured value to a user defined threshold. If the threshold is exceeded, an alarm is triggered. ROC provides means to select a threshold of a classifier based on trade-off between its True Positive Rate (TPR) and False Positive Rate (FPR). In the case of CANARY, TPR is the hit rate which is based on the number of correctly recognized events. FPR on the other hand is the false alarm rate. False alarms occur whenever the algorithm detects an event when actually none exists.

The ROC curve shows the trade-off between TPR and FPR. Usually, it is drawn based on the threshold value of the classifier. This means, depending on the chosen threshold value one receives different pairs of TPR/FPR values which can be connected to a curve. To evaluate the performance of a classifier, the Area Under Curve (AUC) can be used. The worst possible classifier will have an AUC of 0.5, since all pairs of TPR and FPR will be on the straight line between the two extreme points of the curve. This performance would be equal to random guessing. The best possible classifier will have an AUC of 1, which means there is a configuration where no false alarms occur, all events are identified (cf. Fig. 1).

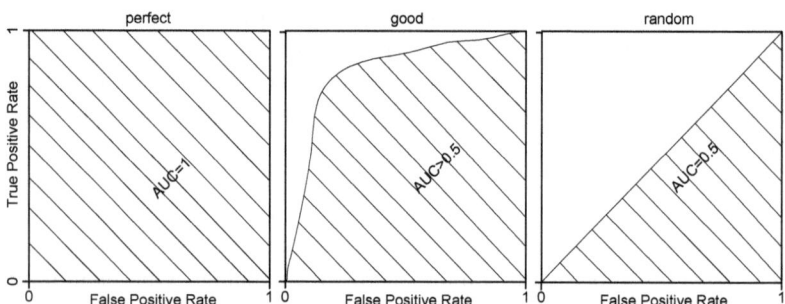

Fig. 1. ROC curves for classifiers of different quality. Leftmost is the case of perfect classification, the rightmost is random guessing.

In the case of CANARY, this form of measuring the performance can not be used, since the threshold value used in CANARY cannot be chosen independently. Therefore, each different setting of the threshold has to be considered as

---

[1] For documentation, manuals and source code of CANARY see:
https://software.sandia.gov/trac/canary

a new classifier. The ROC curve can then be used to compare performance of the different classifiers. Consequently, the threshold value is one of the parameters to be optimized.

2.3   MCO in ROC Analysis

The ROC curve can be interpreted as a Pareto front, although it would classically only represent the Pareto front of an MCO problem with one dimensional decision space (i.e. the decision threshold being the only decision variable). However, it is reasonable to apply MCO methods for other cases, for instance when different classifiers are to be compared, or the threshold is not independent of the classification process. This is the case in the problem described in this paper.

Applying MCO for ROC analysis is not a new topic. Kupinski and Anastasio [17] considered performances of the solutions returned from a multi-criteria objective genetic optimization as series of optimal (sensitivity, specificity) pairs, which can be thought of as operating points on a ROC curve. Fieldsend and Everson [7] used MCO to construct the ROC curve for a vector of parameters (including the threshold) of a binary classifier, thus analyzing several different classifiers. In a second study they discuss the application of MCO for the ROC analysis of a multi class problem [5]. A survey of MCO in ROC analysis can be found in the work of Everson and Fieldsend [6].

3   Problem Description

The problem to be solved in this paper is the tuning of a software designed for anomaly detection in water quality management: CANARY. It was developed to detect anomalies (or events) in water quality time series data. It implements several different algorithms for time series prediction, pattern matching, and outlier detection. The main concept is to employ a time series algorithm to predict the next time step, and afterwards to distinguish whether the real value deteriorates from the predicted value sufficiently to declare it an outlier or anomaly.

We will tune the two relevant parameters window size and threshold value. The window size defines how many values are used for the prediction, while the threshold value defines how much deviation between measured and predicted value are sufficient to declare an outlier. Both parameters have previously been tuned in different ways. Firstly, they have been tuned by a step-by-step procedure [19] which unfortunately does not consider interactions between parameters. Secondly, another study [23] tuned them with model based optimization, considering interactions, but only used a single criteria approach, which basically combined the objectives False Alarm Rate and Hit Rate to a weighted sum.

Usually, as described by Murray et al. [19], a classical ROC analysis would be performed. The AUC would be used as a single quality criterion. This approach is not perfectly viable in this case, as the threshold value is not independent of the prediction process. Therefore, it is a more reasonable approach to add the threshold to the list of tuned parameters and apply multi-criteria optimization.

For this reason, we will mainly use MCO-terminology in the following (e.g., Pareto front instead of ROC curve).

4   Algorithm Description

Four different tuning algorithms are in the focus of this study. Due to the similarity to the AUC, the hypervolume is applied as a criterion in all but one of these approaches. Two of them are based on R-code (SPOT package), two are MATLAB implementations (SMS-EGO and SExI-EGO). All four share the following basic workflow:

1. Evaluate an initial design of n points on the target problem (CANARY)
2. Build models (here: Kriging) for each objective
3. Use models to determine the next design point to be evaluated, based on a certain infill criterion
4. Evaluate design point and update non dominated set
5. Iterate 2-4

The four tuning algorithms differ in the type of the invoked infill criterion. Three algorithms use different multi-criteria EI concepts. The fourth is a straightforward approach that, instead of aggregating the objective values from the models, tries to optimize these separately with common MCO methods.

4.1   MEI-SPOT

This multi-criteria expected improvement approach is the only approach that does not use hypervolume as a criterion. The implementation is based on MATLAB code of Forrester et al. [9]. MEI-SPOT is based on the integration over the non-dominated area and an Euclidean distance to the next point on the front. While Forrester et al. use a dominating variant (e.g. improvement considers only points that dominate existing Pareto-optimal solutions), the implementation used here uses an augmenting variant (i.e. improvement is also reported when a point is added to the front, without dominating an existing Pareto-optimal solution). The different formulations for this distinction are detailed by Keane [14]. This approach is time consuming due to the integration. It can also have issues with the scaling of different objectives, since it is based on the Euclidean distance.

4.2   SExI-EGO

The S-Metric Expected Improvement [3] computes the expected increment in hypervolume for a point, given a non-dominated set. Its exact computation is described in [4]. It is differentiable, rewards high variances [4], and is continuous over the whole search domain. A disadvantage is the high effort of its exact computation, in particular when more than two objectives are considered.

4.3 SMS-EGO

SMS-EGO, as suggested by Ponweiser et al. [20], employs a hypervolume based infill criterion as well. Thereby, a potential solution is computed using the lower confidence bound $\hat{y}_{pot} = \hat{y} - \alpha\hat{s}$, where $\hat{y}$ is the mean value predicted by the Kriging model, $\hat{s}$ is the variance, α is a gain factor for the variance. This approach may also explore unvisited regions of the design space, but without requiring the tedious integration of the previous approaches. It thus scales better with increasing objective dimension.

If the resulting $\hat{y}_{pot}$ is ϵ-dominated or dominated, SMS-EGO will assign a penalty value. If it is non-dominated, the hypervolume contribution will be used. This approach avoids plateaus of the criterion, but integrates non differentiable parts. For more details see Ponweiser et al. [20] and Wagner et al. [22].

4.4 MSPOT

MSPOT is a multi-criteria approach based on the Sequential Parameter Optimization Toolbox SPOT (cf. Zaefferer et al. [24]). It does not employ any form of expected improvement, or other forms of using the variance for exploration. The surrogate models of the different objectives are exploited by using a multi-criteria optimization algorithm (for instance: SMS-EMOA or NSGA-II).

This will yield a population of promising points. One or more points of these are chosen for evaluations on the real target function. This selection is based on non-dominated sorting and the individual hypervolume contribution. As the original approach [24] could lead to clustering of solutions in the objective space, the available points have to be considered when calculating the hypervolume contributions. For this purpose, the known points are reevaluated on the surrogate model.

In contrast to the other approaches in the study, this one does not promote exploration as much, since the variance measure computed by the Kriging model will not be used. On the other hand, the approach is not limited to surrogate modeling methods that yield a variance for each candidate. Of course, the variance can easily be added to MSPOT, as well as be removed from SMS-EGO ($\alpha = 0$) or the integration-based algorithms ($\hat{s} = 0$).

The optimization process of MSPOT is not a completely new idea. Especially, two similar approaches suggested previously have to be mentioned. Firstly, Voutchkov and Keane [21] employed NSGA-II to generate promising solutions in a quite similar optimization loop. In contrast to MSPOT, they used Euclidean distance to ensure evenly spaced points on the front. Instead of considering distance to known points, they suggest a larger number points in each loop, which also ensures a wider spread on the final front. The second similar approach is presented by Jeong and Obayashi [11]. While they also optimize the objectives separately, they employ the single objective EI criterion for each objective, thus optimizing a vector of EI values.

4.5   SMS-EMOA

In addition to the four approaches above, a simple SMS-EMOA will be considered (cf. Beume et al. [2]). The results from this optimizer are used as a baseline for the comparison. In general, surrogate optimization methods are expected to outperform a non-surrogate SMS-EMOA, particularly on small budgets.

5   Experimental Setup

The following research questions are to be treated for the CANARY problem in this study.

1. Can multi-criteria methods produce a front of parameter settings that help an operator to choose parameters for the CANARY event detection software?
2. Which kind of tuner is recommendable?
3. What aspects of a tuner affect its performance?
4. Is the use of surrogate models advantageous?
5. Can previous findings about the tuners be confirmed?
6. How are Pareto optimal solutions spread in the design space?

To answer these questions, several experiments were conducted. Their setup is described in the following.

5.1   Time Series Data

Two different sets of raw data are used. The first set is used to train CANARY (i.e. to tune the parameters), the second is used for validation of the resulting settings on unseen data. Additionally, from each of those sets, 3 different instances are generated, where each contains simulated (i.e. superimposed) events to be detected by CANARY. The data sets considered are available within the CANARY software package.

Training Data. The data recorded over a first month at a specific measurement station is used as training data. Four different sensor values are used (pH-Value, Conductivity, Total Organic Carbon, Chlorine). The time interval between measurements is five minutes. This results in about 9 000 time steps for each of the four sensors.

As the data-set contains no events known beforehand (which is a typical problem for any available real-world data), events have to be simulated and incorporated in the time series. Therefore, 3 data sets are created from the raw data, each containing superimposed square waves (with smoothed transition) of different event strengths: 0.5, 1, and 1.5. These strengths indicate the amplitude of the events, and are multiplied to the standard deviation of the original signal. Figure 2 presents raw data and data with events for two sensor value as an example.

As can be seen from the left part of Fig. 2, the raw data (i.e. without events) is rather strongly affected by background changes. In general, these background

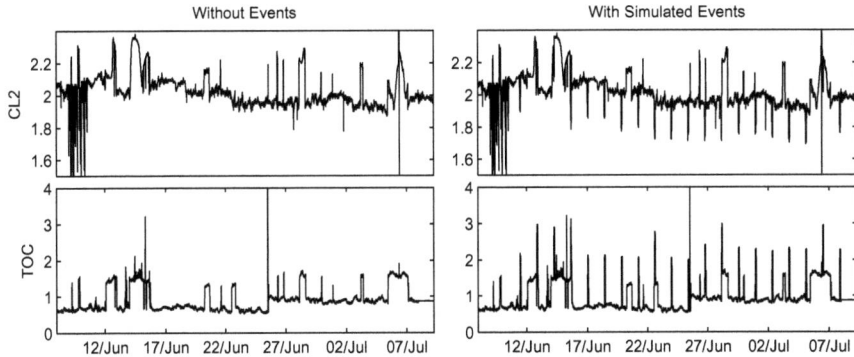

Fig. 2. Example of time series data as used in the experiments. Raw data without events (left) and with superimposed events of strength 1.5(right). CL2 is Chlorine, TOC the Total Organic Carbone.

changes are irregularly distributed over time and always switch back and forth for each of the signals. Obviously, such changes make event detection extremely difficult.

Validation Data. The validation data is similar to the training data, as it is the second month of data from the same measurement station. As could be expected, it provides a very similar background behavior with some sudden jumps. These jumps, however, are more numerous than in the training data, which is expected to lead to higher false alarm rates on the validation data.

5.2 Optimization Problem Configuration

As mentioned earlier, three different data sets are considered, each with a different event strength. Additionally, CANARY is tuned in 3 different configurations, where each configuration uses a different time series prediction algorithm. These are: Time Series Increment TSI, Linear Prediction Correction Filter LPCF and Multi-Variate Nearest Neighbor MVNN. For more details on these algorithms, which are implemented in CANARY, see the corresponding documentation [19] and the manual [10]. Therefore, $3 \times 3 = 9$ instances are to be optimized. The optimization problem is multi-objective, where both decision and solution space are two dimensional: The window size and the threshold are tuned, to yield a minimal FPR and a maximal TPR value. Since all tuning methods in this study do minimization, TPR is negated. The problem is not noisy, as the algorithms employed in the event detection software are deterministic.

There are two nice features of the problem, which avoid issues of algorithm configuration.

1. The choice of the reference point. With this problem the worst case is known: Zero for TPR and one for FPR. To avoid extreme points overlapping with

the reference point, the latter was chosen to be [0.1,1.1] since TPR ranges from -1 to 0 due to the negation.

2. Scaling of objective space is not an issue here, as both objectives have the same range. They only differ in that way, that TPR is maximized and FPR minimized. Scaling has not to be considered in the algorithms.

5.3   Tuning Methods Configuration

All algorithms are configured to use approximately the same settings, i.e.:
- initial design size: 21
- number of points added in each step: 1
- number of maximal evaluations of the target function: 80
- surrogate model: DACE-Kriging [18]

The optimization method to find the best point (with or without expected improvement) differs. Some criteria aggregate the different objectives into a single-objective infill criterion, which is then optimized. Therefore, SExI-EGO, SMS-EGO and MEI-SPOT invoke a local optimization method, restarting in several partitions of the design space. In contrast to this, MSPOT uses SMS-EMOA to optimize the surrogate models of the objectives without aggregating their information.

While both the R and the MATLAB implementations use DACE-Kriging [18] there are small differences in the implementations. This includes differences between the inbuilt local optimization methods (e.g. simplex, gradient based) used during model building and optimization.

The SMS-EMOA employed as a baseline comparison is configured to also use a starting population of 21 points. All other settings are left at defaults.

6   Analysis

The results of the experiments are depicted in Fig. 3. It shows the resulting hypervolume of each tuner for each problem instance. The hypervolume values are recalculated with respect to the reference point [0, 1] to have the ranges comparable to the AUC values. Plots with the original reference point used during tuning look alike and do not show major differences. As can be seen, there are no significant differences between the performance of SMS-EGO, MSPOT, and SExI-EGO. In comparison, MEI-SPOT and SMS-EMOA perform worse. For the SMS-EMOA, this was expected and can be blamed to not invoking a surrogate model. The Euclidean EI criterion employed in MEI-SPOT, on the other hand, was already reported to be less viable due a non-monotonicity with the dominance relation [22].

It can be observed that the event strength has an improving influence on the detection performance of the Pareto optimal solutions. This is expected, as stronger events should be easier to identify. The same can be observed for the algorithm MVNN, which provides best overall detection results. Both observations are in line with earlier reported behavior in the work on tuning CANARY single objectively [23]. An optimal performance would be leading to a hypervolume of exactly one. Realistically, this is not obtainable. The gap between the best front's hypervolume and the theoretical optimum is largely due to the fact that the FPR rate is strongly affected by the sudden jumps in the time series data. Besides this, the results are in a similar range of TPR and FPR values as found in the earlier mentioned single-objective tuning of CANARY.

The results discussed above have been received on the training data set only. To validate our findings, we invoked the validation data set as well. All points of the final Pareto front were re-evaluated on the validation data set. It can be observed that performance differences become smaller while variances increase. The former particularly holds for incorporating MVNN in CANARY. In general, it can be noticed that the received hypervolume decreases on the validation data. This can mainly be blamed on a slightly different background behavior of the validation data set, as described in Sec. 5.1. While there is a strong performance drop for nearly all instances, the results for MVNN and an event strength of 1.5 only decrease slightly. A similar behavior was observed for single-objective tuning of CANARY in an earlier work by Zaefferer [23]. Still, the validation data shows the same relations as the training data, considering the performance of different tuners. In the briefness of this paper we focus on the training data results, since differences can mainly be blamed on different background behavior in the two time frames (e.g. more sudden jumps in the second month, compared to the first month).

Table 1 provides the average number of points on a single Pareto front. Note, that 30 to 50 percent of the points on a front are actually dominated, if being re-evaluated on the validation data.

Table 1. Average number of points on a Pareto front. The second line shows how many of those points remain, after being reevaluated on validation data.

	MEI-SPOT	MSPOT	SMS-EGO	SExI-EGO	SMS-EMOA
Training Data	38.70	29.27	23.11	22.92	21.00
Validation Data	22.91	18.13	14.54	14.27	12.54

As mentioned above, MSPOT, SMS-EGO, and SExI-EGO do not show significant differences in their results. One main distinction between these approaches is that MSPOT does not make use of the variance produced by the DACE model, and therefore lacks exploration. It might therefore be the case that MSPOT performs as good as the other methods, because the initial design already provides enough exploration of the design space so that it is sufficient to spent all sequential evaluations on purely exploiting the surrogate models prediction.

To test this, two additional experiments are performed. Firstly, the MSPOT experiment is repeated with a much smaller initial design of just 5 points (labeled MSPOTSMALL), to validate whether a smaller initial design can deteriorate results. Secondly, the SMS-EGO experiment was repeated disregarding any variance information. To this end, the gain α is set to zero. Therefore, instead of

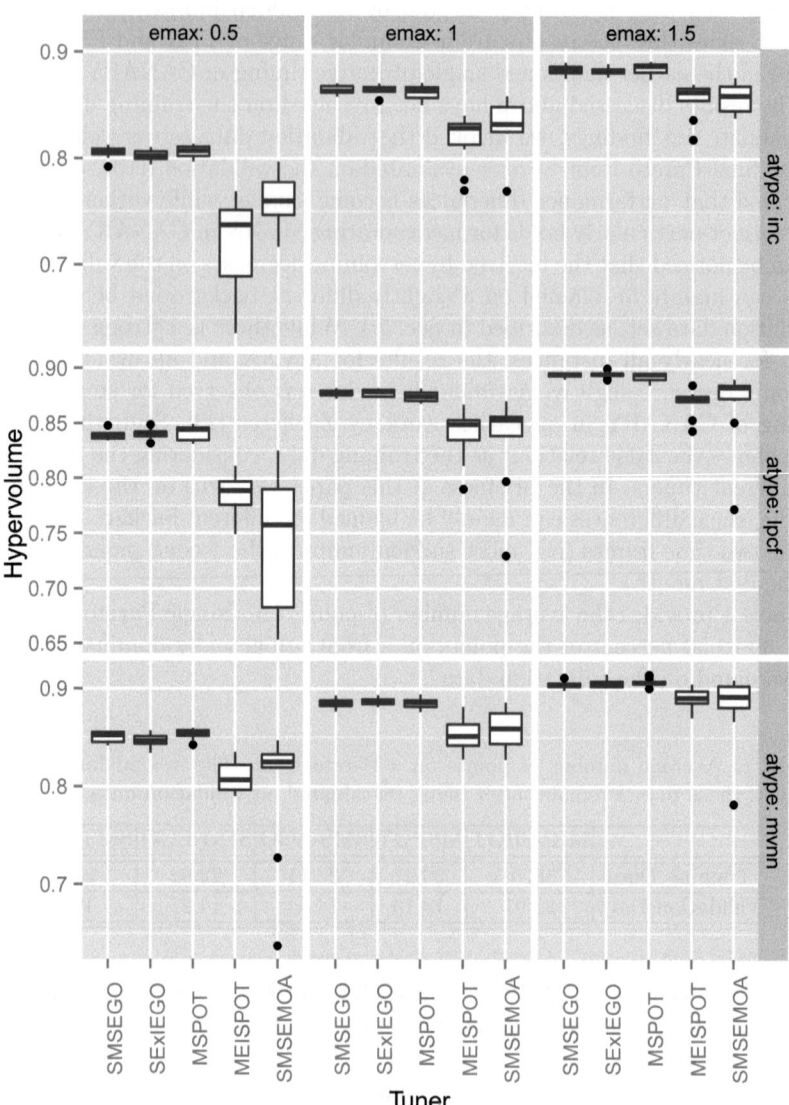

Fig. 3. Boxplot of first results on training data. The hypervolume is computed with a reference point of zero for TPR and one for FPR. Larger hypervolumes are better. emax is the event strength, atype is the algorithm type used in CANARY.

the lower confidence bound $\hat{y}_{pot} = \hat{y} - \alpha\hat{s}$ the potential solution will be $\hat{y}_{pot} = \hat{y}$. This approach will be labeled as SMS-EGOg0. The resulting hypervolumes on training data are depicted in Fig. 4. The smaller initial design in fact decreases performance of MSPOT, however, the margin is quite small. Furthermore, a comparable performance of SMS-EGO with or without taking the variance in consideration is observed.

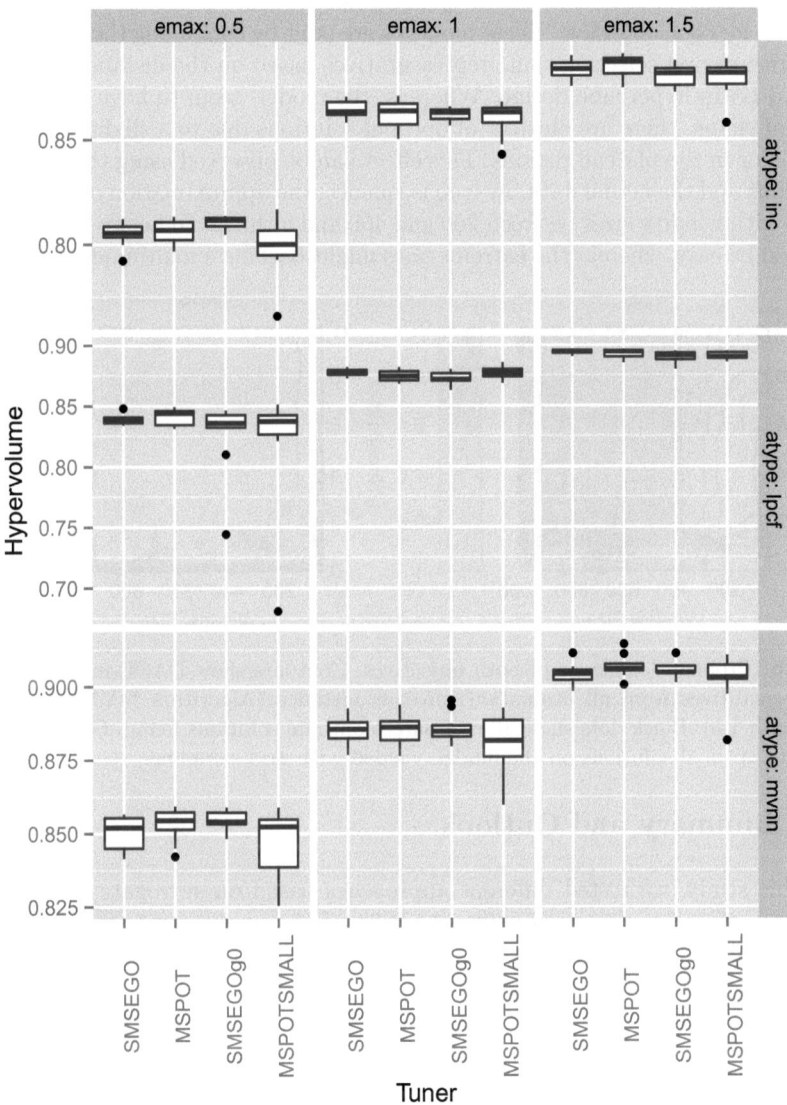

Fig. 4. Boxplot of additional results on training data, comparing results of follow-up experiments (i.e. MSPOTSMALL and SMSEGOg0). The hypervolume is computed with a reference point of zero for TPR and one for FPR. Larger hypervolumes are better. emax is the event strength, atype is the algorithm type used in CANARY.

In some instances SMS-EGO even performs better without incorporating the variance information. There seems to be no strong need for the additional exploration in this case. Such observations are normally expected for unimodal problems, while more exploration should be profitable on multi modal problems. It might further be considered that additional exploration is already inherent in the selection process as not one single optimum, but a set of points is demanded.

To visualize the problem landscape, Fig. 5 shows contour plots of reference DACE-models for each objective. These models were built by combining the designs of all algorithms and selecting some representatives based on the distance to an optimized Latin hypercube design. Whereas, the models seem to have a rather unimodal shape, there are clusters of optimal solutions due to a slightly oscillating behavior in the plateau regions. This effect can be observed using the model predictions and the actual data. As a consequence, the approximation of the knee region with window sizes between 200 and 400 and a threshold between 1.0 and 1.5 should be easy, whereas the extreme ones might become a multimodal problem.

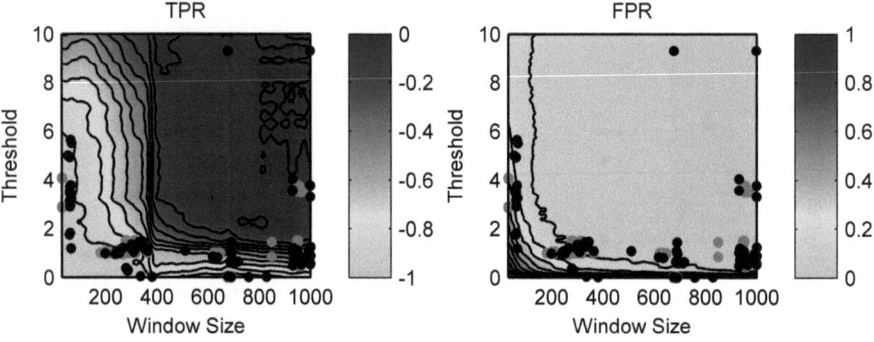

Fig. 5. Problem landscape of both objectives. Contours show DACE-model based on representatives from all evaluations on this instance (Algorithm MVNN and event strength 1.5). Black dots show all real pareto optimal solutions found. Grey dots show Pareto optimal solutions on the model, yielded with grid sampling.

7   Summary and Outlook

In this study, we tested different approaches based on surrogate optimization to tune an event detection software. Most of the analysis was focused on the results of training data, since the results on validation data mainly provided similar results as on the training data. The surrogate optimization approaches are mostly able to outperform a baseline SMS-EMOA. The MEI-SPOT approach proved to be the exception from this observation, which confirms earlier findings by Wagner et al. [22]. This approach of calculating the expected improvement for multiple criteria seems to be unfavorable.

There was no decisive difference between the other tested approaches, regardless whether variance was used in the approach (SMS-EGO and SExI-EGO) or not (MSPOT and SMS-EGO with zero gain). Plots of the model structures seem to indicate an almost unimodal fitness landscape for both objectives. This indicates that the additional exploration by variance might not be needed here, since the fitness landscape is easy to approximate without additional exploration of the design space. Since it was neither a disadvantage, it might be interesting to test the lower confidence bound in MSPOT for future experiments.

This study showed that the problem of tuning CANARY can reasonably be solved by multi-criteria methods. The produced results yield reasonable FPR and TPR values, which are comparable to previous results achieved by single-objective optimization. Here, however, the approximation of a Pareto front offers more flexibility for the operator in charge.

It has to be noted that only points on the convex hull of the ROC curve can be considered to be optimal in some sense. Any point below that hull might be considered to be improvable [8]. Future work should investigate if concavities in the ROC curve can be repaired for the application described here.

The concentration on certain regions of a Pareto front might be a topic for future research as well. An operator might be more interested in the knee region of the Pareto front, and less on extreme values, which might cause intolerable numbers of false alarms. Focusing on a subset of the Pareto front might also reduce the required budget.

Acknowledgments. This work has been kindly supported by the Federal Ministry of Education and Research (BMBF) under the grants MCIOP (FKZ 17N0311) and CIMO (FKZ 17002X11). In addition, the paper is based on investigations of project D5 "Synthesis and multi-objective model-based optimization of process chains for manufacturing parts with functionally graded properties" as part of the collaborative research center SFB/TR TRR 30, kindly supported by the Deutsche Forschungsgemeinschaft (DFG).

References

1. Bartz-Beielstein, T., Parsopoulos, K.E., Vrahatis, M.N.: Design and analysis of optimization algorithms using computational statistics. Applied Numerical Analysis and Computational Mathematics (ANACM) 1(2), 413–433 (2004)
2. Beume, N., Naujoks, B., Emmerich, M.: SMS-EMOA: Multiobjective selection based on dominated hypervolume. European Journal of Operational Research 181(3), 1653–1669 (2007)
3. Emmerich, M.: Single- and Multi-objective Evolutionary Design Optimization: Assisted by Gaussian Random Field Metamodels. PhD thesis, Universität Dortmund, Germany (2005)
4. Emmerich, M., Deutz, A., Klinkenberg, J.: Hypervolume-based expected improvement: Monotonicity properties and exact computation. In: 2011 IEEE Congress on Evolutionary Computation (CEC), pp. 2147–2154. IEEE (2011)
5. Everson, R.M., Fieldsend, J.E.: Multi-class ROC analysis from a multi-objective optimisation perspective. Pattern Recognition Letters 27(8), 918–927 (2006)
6. Everson, R.M., Fieldsend, J.E.: Multi-objective Optimisation for Receiver Operating Characteristic Analysis. In: Jin, Y. (ed.) Multi-Objective Machine Learning. SCI, vol. 16, pp. 533–556. Springer, Heidelberg (2006)
7. Fieldsend, J.E., Everson, R.M.: ROC Optimisation of Safety Related Systems. In: Hernández-Orallo, J., Ferri, C., Lachiche, N., Flach, P.A. (eds.) ROCAI, pp. 37–44 (2004)

8. Flach, P.A., Wu, S.: Repairing concavities in ROC curves. In: Proceedings of the 19th International Joint Conference on Artificial Intelligence, IJCAI 2005, pp. 702–707. Morgan Kaufmann Publishers Inc., San Francisco (2005)

9. Forrester, A., Sobester, A., Keane, A.: Engineering Design via Surrogate Modelling. Wiley (2008)

10. Hart, D.B., Klise, K.A., Vugrin, E.D., McKenna, S.A., Wilson, M.P.: Canary user's manual and software upgrades. Technical Report EPA/600/R-08/040A, U.S. Environmental Protection Agency, Washington, DC (2009)

11. Jeong, S., Obayashi, S.: Efficient global optimization (EGO) for multi-objective problem and data mining. In: Corne, D., et al. (eds.) IEEE Congress on Evolutionary Computation, pp. 2138–2145. IEEE (2005)

12. Jin, Y.: A comprehensive survey of fitness approximation in evolutionary computation. Soft Computing 9(1), 3–12 (2005)

13. Jones, D., Schonlau, M., Welch, W.: Efficient global optimization of expensive black-box functions. Journal of Global Optimization 13, 455–492 (1998)

14. Keane, A.: Statistical improvement criteria for use in multiobjective design optimisation. AIAA Journal 44(4), 879–891 (2006)

15. Knowles, J.: Parego: A hybrid algorithm with on-line landscape approximation for expensive multiobjective optimization problems. IEEE Transactions on Evolutionary Computation 10(1), 50–66 (2006)

16. Knowles, J.D., Nakayama, H.: Meta-Modeling in Multiobjective Optimization. In: Branke, J., Deb, K., Miettinen, K., Słowiński, R. (eds.) Multiobjective Optimization. LNCS, vol. 5252, pp. 245–284. Springer, Heidelberg (2008)

17. Kupinski, M.A., Anastasio, M.A.: Multiobjective genetic optimization of diagnostic classifiers with implications for generating receiver operating characteristic curves. IEEE Transactions on Medical Imaging 18, 675–685 (1999)

18. Lophaven, S., Nielsen, H., Søndergaard, J.: DACE—A Matlab Kriging Toolbox. Technical Report IMM-REP-2002-12, Informatics and Mathematical Modelling, Technical University of Denmark, Copenhagen, Denmark (2002)

19. Murray, R., Haxton, T., McKenna, S.A., Hart, D.B., Klise, K., Koch, M., Vugrin, E.D., Martin, S., Wilson, M., Cruz, V., Cutler, L.: Water quality event detection systems for drinking water contamination warning systems—development, testing, and application of CANARY. Technical Report EPA/600/R-10/036, National Homeland Security Research Center (May 2010)

20. Ponweiser, W., Wagner, T., Biermann, D., Vincze, M.: Multiobjective Optimization on a Limited Budget of Evaluations Using Model-Assisted $\mathcal{S}$-Metric Selection. In: Rudolph, G., Jansen, T., Lucas, S., Poloni, C., Beume, N. (eds.) PPSN X. LNCS, vol. 5199, pp. 784–794. Springer, Heidelberg (2008)

21. Voutchkov, I., Keane, A.: Multiobjective optimization using surrogates. In: Adaptive Computing in Design and Manufacture ACDM, pp. 167–175 (2006)

22. Wagner, T., Emmerich, M., Deutz, A., Ponweiser, W.: On Expected-Improvement Criteria for Model-based Multi-objective Optimization. In: Schaefer, R., Cotta, C., Kołodziej, J., Rudolph, G. (eds.) PPSN XI. LNCS, vol. 6238, pp. 718–727. Springer, Heidelberg (2010)

23. Zaefferer, M.: Optimization and empirical analysis of an event detection software for water quality monitoring. Master's thesis, Cologne University of Applied Sciences (May 2012)

24. Zaefferer, M., Bartz-Beielstein, T., Friese, M., Naujoks, B., Flasch, O.: Multi-criteria optimization for hard problems under limited budgets. In: Soule, T., et al. (eds.) GECCO 2012 Proceedings, Philadelphia, Pennsylvania, USA, pp. 1451–1452. ACM (2012)

Multi-Objective Evolutionary Algorithm with Node-Depth Encoding and Strength Pareto for Service Restoration in Large-Scale Distribution Systems

Marcilyanne Moreira Gois[1], Danilo Sipoli Sanches[2], Jean Martins[3],
João Bosco A. London Junior[1], and Alexandre Cláudio Botazzo Delbem[3]

[1] São Carlos Engineering School of University of São Paulo, São Carlos, SP, Brazil
{mmgois,jbalj}@usp.br
[2] Federal Technological University of Paraná, Cornélio Procópio, Brazil
danilosanches@utfpr.edu.br
[3] Institute of Mathematics and Computer Science, University of São Paulo, São
Carlos, SP, Brazil
{jean,acbd}@icmc.usp.br

Abstract. The network reconfiguration for service restoration in distribution systems is a combinatorial complex optimization problem that usually involves multiple non-linear constraints and objectives functions. For large networks, no exact algorithm has found adequate restoration plans in real-time, on the other hand, Multi-objective Evolutionary Algorithms (MOEA) using the Node-depth encoding (MEAN) is able to efficiently generate adequate restorations plans for relatively large distribution systems. An MOEA for the restoration problem should provide restoration plans that satisfy the constraints and reduce the number of switching operations in situations of one fault. For diversity of real-world networks, those goals are met by improving the capacity of the MEAN to explore both the search and objective spaces. This paper proposes a new method called MEA2N with Strength Pareto table (MEA2N-STR) properly designed to restore a feeder fault in networks with significant different bus sizes: $3\,860$ and $15\,440$. The metrics R_2, R_3, Hypervolume and ϵ-indicators were used to measure the quality of the obtained fronts.

1 Introduction

There are many Multi-objective optimization problems (MOP) in real world such as: vehicle routing [1], phylogenetic reconstruction [2] and service restoration in distribution systems [3]. MOP are characterized by the presence of multiple objective functions to be optimized simultaneously, since such objectives can be conflicting and there is no single optimal solution that satisfies all objectives equally [4].

In order to find feasible solutions for MOP, Multi-objective Evolutionary Algorithms (MOEAs), such as Nondominated Sorting Genetic Algorithm II (NSGA-II)

R.C. Purshouse et al. (Eds.): EMO 2013, LNCS 7811, pp. 771–786, 2013.

[5] and Strength-Pareto Evolutionaty Algorithm 2 (SPEA2) [6] were proposed in the literature. These algorithms search for an approximated Pareto-optimal set and both are based on elitism, i.e. the best solutions in the population are preserved to the next generation. Despite the similarities, these techniques differ in the way that they implement the elitism and in the strategy used to select the best solutions according to multiple objective functions. The NSGA II based on Non-dominated Sorting (NS), usually, fails in combinatorial optimization problems with many objectives [7, 8].

To improve the performance of MOEAs for large-scale network design problems, new dynamic data structures (encodings) that exclusively generate feasible solutions have been investigated [9]. Those encodings allow a suitable exploration of the search space, increasing the quality of solutions provided by MOEAs. Among the encondigs from the literature, the Node-depth encoding (NDE) [10] better scales, enabling its use for optimization methods applied to large networks. In this sense, some MOEAS using NDE have been investigated: MoEA with Node-depth encoding (MEAN) [3], NSGA-II with NDE (NSGAN) [11] and MEAN with Non-dominated subpopulation table (MEA2N) [12]. The MEAN and MEA2N uses subpopulation table [12] store the best found solutions according to distinct evaluation criteria. The MEAN and MEA2N have a common feature, they search for solutions that simultaneously optimize each objective separately and one or more aggregation function(s) combining objectives. The MEA2N includes features of NSGA-II in the MEAN by adding a subpopulation tables that stores non-dominated solutions. NSGAN and MEA2N have shown to be able to solve combinatorial problems with two or more objectives.

Those approaches have been evaluated for network design problem called DS reconfiguration. The Network reconfiguration of a DS is a combinatorial optimization problem, which consists in the process of opening and closing some switches to modify the topology of the network that represents the distribution system. The network reconfiguration for Service Restoration (SR) is classified as an NP-Hard problem, which it is i) highly combinatorial, due to the large number of switching elements; ii) nonlinear, since the equations governing the electrical system are in general nonlinear; iii) non-differentiable, because a switch status change may result in crisp variations of values in objectives and constraints; iv) constrained, due to the electrical and operational restrictions; v) multi-objective, considering that the plan should maximize the number of restored costumers and minimize the number of switching operations and, when not conflicting with the two previous objectives, Ohmic losses are also considered. Thus, the design of an optimal network configuration for SR require the investigation of several switching status vectors.

To obtain a Pareto-optimal front with better convergence and also preserving the diversity, this paper proposes a new method called MEA2N with subpopulation table related to solution STRength (MEA2N-STR), which extends the strategy of the subpopulations tables of MEA2N and incorporates a table of non-dominated solutions based on SPEA2[1] to improve the capacity of investigating the objective space. Besides obtaining adequate restoration plans for large DS, MEA2N-STR also finds plans for small or relatively large networks with similar quality.

This paper is structured as follows: Section 2 describes the Service Restoration Problem Formulation; Section 3 explains the NDE; Section 4 presents the main concepts of Multi-objective Evolutionary Algorithms (MOEAs); Section 5 shows the test problems and experimental results and, finally, Section 6 presents the conclusions.

2   Service Restoration Problem Formulation

The Network reconfiguration for the Service Restoration Problem is the process of opening and closing of some switches to modify the topology of a distribution network modeled by a forest. Fig. 1 (a) illustrates an example of SR in a DS with three feeders that are represented by nodes 1, 2 and 3. Each feeder supplies a subset of consumer load points (sectors) represented by other nodes . The sectors are interconnected by edges that indicate the switches (feeder lines). The switches can be Normally Closed (NC) (solid lines) and Normally Opened (NO) (dotted lines). Each tree of the forest corresponds to a feeder with its sectors and Normally Closed switches.

Assuming that a fault occurred in sector 10 (Fig. 1 (a)), all the switches connected to sector 10 (switches 10-11, 10-7 and 10-9) must be opened in order to isolate the sector in fault, thus, Sectors 11, 9 and 28 are in an out-of-service area. One way to restore energy for those sectors is by closing the switches 24-28 and 28-11 (Fig. 1 (b)).

The SR problem emerges after the faulted areas has been identified and isolated. Its solutions is the minimal number of switching operations that results in a configuration with minimal number of out-of-service loads, without violating the DS operational and radialily constraints. The minimization of the number of switching operations is important since the time required by the restoration process depends on the number of switching operations. The SR problem can be formalized as follows:

$$Min. \; \phi(G), \gamma(G) \; and \; \psi(G, G^0)$$
$$s.a.$$
$$Ax = b$$
$$X(G) \leq 1$$
$$B(G) \leq 1 \tag{1}$$
$$V(G) \leq 1$$
$$G \; is \; a \; forest,$$

---

[1] Individuals in the table based on SPEA2 correspond to the best found solutions according to the strength value, i.e. the strength of dominance of an individual in relation to other individuals.

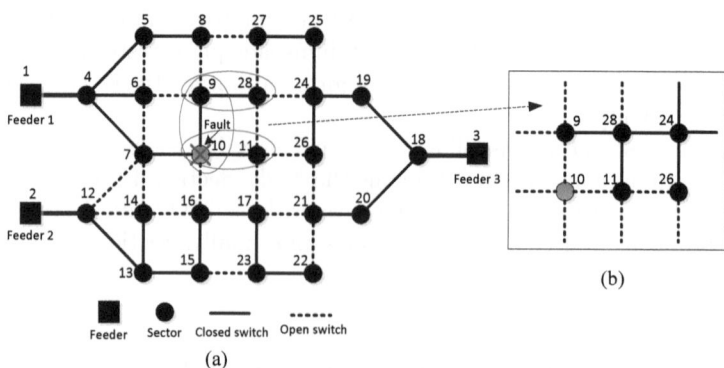

Fig. 1. DS modeled by a graph, a fault is simulated in the sector 10 (a) and then is restored energy for Sectors 11, 9 and 28 by closing the switches 24-28 and 28-11 (b)

where G is a spanning forest of the graph representing a system configuration [13]; $\phi(G)$ is the number of consumers that are out-of-service in a configuration G (considering only the reconnectable system); $\psi(G, G^0)$ is the number of switching operations to reach a given configuration G from the configuration just after the isolation of the fault G^0; $\gamma(G)$ are the power losses of configuration G; A is the incidence matrix of G [14]; x is a vector of line current flow; b is a vector containing the load complex currents (constant) at buses with $b_i \leq 0$ or the injected complex currents at the buses with $b_i > 0$ (substation); $X(G)$ is called network loading of configuration G, that is, $X(G)$ is the highest ratio $x_j/\overline{x}_j$, where $\overline{x}_j$ is the upper bound of current magnitude for each line current magnitude x_j on line j; $B(G)$ is called substation loading of configuration G, that is, $B(G)$ is the highest ratio $b_s/\overline{b}_s$, where $\overline{b}_s$ is the maximum current injection magnitude provided by a substation (s means a bus in a substation); $V(G)$ is called the maximal relative voltage drop of configuration G, that is, $V(G)$ is the highest value of $|v_s - v_k|/\delta$, where v_s is the node voltage magnitude at a substation bus s in pu and v_k the node voltage magnitude at network bus k obtained from a SLFA for DSs, and δ is the maximum acceptable voltage drop. Formulation of Eq. (1) can be synthesized by considering:

i) Penalties for violated constraints $X(G)$, $B(G)$ and $V(G)$;

ii) The use of the NDE [3], i.e. an abstract data type [15] for graphs that can efficiently manipulate a network configuration (spanning forest) and guarantee that the performed modifications always produce a new configuration G that is also a spanning forest (a feasible configuration);

iii) The nodes are arranged in the Terminal-Substation Order (TSO) for each produced configuration G in order to solve $Ax = b$ using an efficient SLFA for DSs. The NDE stores nodes in the TSO;

iv) $\phi(G) = 0$. The NDE always generates forests that correspond to networks without out-of-service consumers in the reconnectable system.

Eq. (1) can be rewritten as follows:

$$Min. \; \psi(G, G^0), \; \gamma(G) \; and$$
$$\omega_x X(G) + \omega_b B(G) + \omega_v V(G)$$

$s.a.$ \hfill (2)

$$Load \; flow \; calculated \; using \; the \; NDE,$$
$$G \; is \; a \; forest \; generated \; by \; the \; NDE,$$

where ω_x, ω_b and ω_v are weights balancing among the network operational constraints. In this paper, these weights are set as follows:

$$\omega_x = \begin{cases} 1, & \text{if, } X(G) > 1 \\ 0, & \text{otherwise;} \end{cases}$$

$$\omega_b = \begin{cases} 1, & \text{if, } B(G) > 1 \\ 0, & \text{otherwise;} \end{cases}$$

$$\omega_v = \begin{cases} 1, & \text{if, } V(G) > 1 \\ 0, & \text{otherwise.} \end{cases}$$

3   Node-Depth Encoding

A graph can be seen as a mathematical formalism for system modeling, where each element of the system is called node (in general graphically drawn as a dot or a circle) and the potential relationships between nodes are called edges (represented by a straight line connecting the nodes). The whole set of nodes and edges involved in a system composes the graph of the system. Formally, a graph G is a pair of two sets, V containing the nodes and E containing the edges of G. A DS can be modeled by a kind of graph called tree[2], which the nodes represent sectors[3] and the edges represent the sectionalizing and tie-switches. For large systems, only subsystems must be connected.

These systems can be represented by an dynamic data structure called Node-Depth Encoding (NDE). Basically, the NDE consists of a linear list containing a pair of values (n_x, d_x), where (n_x) are the tree nodes and (d_x) their depths[4]. The order in which the pairs are disposed in the array can be obtained by a Depth Search Algorithm [15]. Thus, the NDE of a tree is an array of nodes (with the corresponding depths) in the order the Depth Search Algorithm visits each node when traversing the tree. This processing can be executed off-line.

---

[2] A tree structure is a connected and acyclic subgraph of a graph and we can access all nodes from any other node and there is no cycles, i.e. there exists only one route to reach each node.

[3] A sector corresponds to a group of buses and lines without sectionalizing and tie-switches.

[4] The depth of a node is the number of nodes in the path from it to the root (a reference node) of tree, where path is a sequence of nodes with each adjacent pair connected by an edge.

The proposed forest representation is composed of the union of the encodings of all trees that compose the forest. Therefore, the forest data structure can be easily implemented using an array of pointers, where each pointer indicates the NDE of a tree. Fig. 2(a) presents a graph and highlights one spanning forest, which has three trees (T_1, T_2 and T_3). Straight lines represent edges of a tree and dotted lines are edges of the graph not used by any tree. Nodes 1, 2 and 3 are the root nodes of trees T_1, T_2 and T_3 respectively. Fig. 2(b) illustrates the NDE corresponding to each spanning tree.

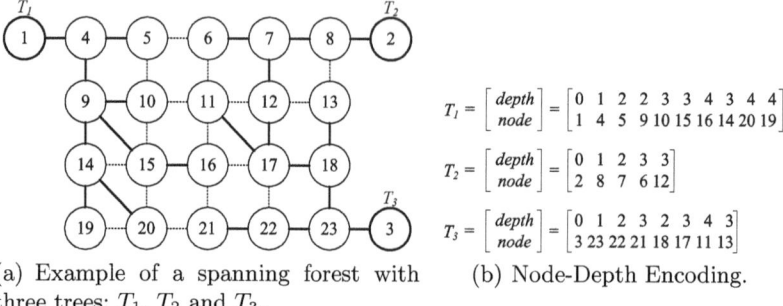

(a) Example of a spanning forest with three trees: T_1, T_2 and T_3.

(b) Node-Depth Encoding.

Fig. 2. NDE arrays for three trees of the spanning forests

From the NDE, two operators were developed to efficiently manipulate a forest producing a new one: the Preserve Ancestor Operator (PAO) and Change Ancestor Operator (CAO). Each operator performs modifications on an NDE that are equivalent to prune and graft a forest generating a new forest. The CAO produces more complex modifications than PAO in a forest, as described in [3]. Both operators are computationally efficient, requiring $O(\sqrt{n})$ average time to construct a new NDE [16].

4   Multi-Objective Evolutionary Algorithms

Multi-Objective Evolutionary Algorithms (MOEAs) are stochastic search metaheuristics based on principles of Evolutionary Theory [17] and strategies used for multi-objective optimization. MOEAs can deal with multiple objective functions, which may be multimodal, non-continuous or non-differentiable. Similarly to an usual evolutionary algorithm [18], MOEAs search for the optimal solution of a problem by manipulating multi-sets (populations) of candidate solutions (individuals) [18].

Usually the first population is randomly generated. Then, a subset of the population is selected to apply operators that modify individuals of the subset producing a new population. The MOEA literature [17] has shown that an adequately designed algorithm can find the optimal or a near-optimum solutions for hard optimization problems.

The fundamental feature of a MOEA is the strategy used to select promising solutions when considering multiple objective functions. One of the most successfull strategies is the Non-dominated Sorting (NS) used by the NS Genetic algorithm II (NSGA-II) proposed by [5]. In NSGA II, for each solution a NS is used to obtain a Pareto ranking. The NS consists in dividing a set of M solutions into several fronts $(\mathcal{F}_1, \mathcal{F}_2, ..., \mathcal{F}_k$, where k is the number of fronts) according to the dominance ranks of each solution. NSGA II has a fitness assignment based on the estimation density of a solution. The density estimation of NSGA II is performed by a truncation operator based on crowding distance which for each individual of the population computes values relative to distance between two points on either side of this point along each of the objectives [5].

Another MOEA used is the Strength-Pareto Evolutionaty Algorithm 2 (SPEA2) proposed by [6]. It is an improved version of SPEA [19]. In SPEA2, each solution is associated to a strength value that defines the strength of dominance of an individual in relation to other individuals. The individuals with higher strength values are preserved and maintained in the population. SPEA2 also uses the nearest neighbor as a density estimation technique [6].

The NSGA-II and other approaches based on NS in general fail for combinatorial optimization problems with many objectives [7, 8]. Recently methods, as MOAE/D [8] and MEAN [20, 3, 12] have shown to be able to solve such problems. In fact, both methods have a common feature, i.e. they search for solutions that simultaneously optimize each objective separately, and one or more aggregation function(s) combining objectives. This paper investigates advances from the MEAN, since it has been successfully employed as search technique in approaches for reconfiguration of DS [21, 3].

4.1 Multi-Objective EA with NDE and Strength Pareto

The proposed method, called Multi-objective EA with NDE and Strength Pareto (MEA2N-STR), combines the main characteristics of MEA2N and SPEA2.

MEA2N-STR explores the objective space using the concept of subpopulation tables, i.e. each subpopulation stores the best solutions found according to an objective or an aggregation function of objectives. The MEA2N-STR possesses subpopulation tables which store different levels of non-dominated solutions and incorporates a strength Pareto table based on SPEA2 to provide more diversity among solutions.

Reproduction operators used to generate new individuals (as for example, PAO and CAO, Section 3) are applied to solutions that are selected as follows: first a subpopulation S_i is randomly chosen, then, an individual from S_i is also randomly taken. A new generated individual (I_{new}) is included in the subpopulation table S_i if this table is not full (since it subpopulation has a fixed size) or if I_{new} is better (according to the objective or criterion associated with S_i) than the worst solution in S_i, then replacing it.

S_i is related to non-dominance and filled according to Pareto fronts. Solutions from $\mathcal{F}_1$ (non-dominated solutions considering the whole population) are stored in table S_{F_1}. S_{F_2} maintains the non-dominated solutions in $\mathcal{F}_2$; while S_{F_3}

stores solutions from $\mathcal{F}_3$. Finally, the strength Pareto table are filled according to number of the solutions that each individual dominates. It is considered the best individual who dominates most solutions. If the size of the strength Pareto table exceeds a predefined limit, the worst individual is deleted.

Table 1 summarizes the main features of NSGA-II, SPEA2, MEAN, NSGAN, MEA2N and MEA2N-STR.

Table 1. Main features among MEAN, NSGAN, MEA2N and MEA2N-STR

MOEA	External set	Genetic operator
MEAN	Uses subpopulation tables	PAO and CAO (NDE)
NSGAN	Truncation uses crownding distance	PAO and CAO (NDE)
MEA2N	Uses subpopulation tables with 3 tables of nondominated solutions	PAO and CAO (NDE)
MEA2N-STR	Uses subpopulation tables similar to MEA2N, but it is included a Strenght Pareto table based on SPEA2	PAO and CAO (NDE)

4.2 Performance Assessment

The performance between MOEAs is usually assessed by the quality of the approximated Pareto fronts found by the algorithms. In general, three characteristics are taken into account to evaluate an approximated Pareto front: 1) proximity to the Pareto-optimal front, 2) diversity of solutions along the front and 3) uniformity of solutions along the front. These three criteria guide the search to a high-quality and diversified set of solutions which enable the choice of the most appropriate solution in a posterior decision-making process [2].

To quantify these three characteristics in a set of non-dominated solutions, various measures have been developed, as example, Error Ratio [22], Generational Distance [22], the R_2 and R_3 [23] Hypervolume (HV) [24] and ϵ-indicator [24]. In this paper, R_2, R_3, HV and ϵ indicators are used to assess the performance of the proposed algorithm, each of them is based on different preference information, then by using them all we provide a range of comparisons intead of just one point-of-view.

Considering two approximation sets (A and B) and some utility function U aggregating k objective functions, $u : \mathbb{R}^k \to \mathbb{R}$, the R_2 measure calculates the expected $E(.)$ difference value in the utility of an approximation A in relation to another one B, considering B as an estimative of the true Pareto front P. Formally, R_2 evaluates the proximity of A in relation to P [2], as follows.

$$R_2(A, B, U, p) = E(u(\lambda, A)) - E(u(\lambda B)),$$
$$R_3(A, B, U, p) = (E(u(\lambda, A)) - E(u(\lambda B)))/E(u\lambda B),$$

The utility $u(\lambda, A)$ of the approximation set A, on the scalarizing vector λ is the minimum distance from A to the reference point B according all scalarizing vectors λ (uniformly distributed across the objective space). The R_3 measure is very similar, however, instead of using the minimum distance the ratio is used. In both cases lower values of R_k measure the mean distance of the attainment sufraces A and B from a user-defined reference point p.

Another relevant quality indicator is the HV [25, 26], which uses the covered volume dominated by an approximated Pareto Front A as a measure of quality of such front. The calculus of the covered volume requires a reference point, which usually consists of an anti-optimal point or "worst values" point in the objective space [27]. For each generated decision vector vec_i a hypercube vol_i is constructed in relation to the reference point, after this, the hypercubes of all decision vectors are joined. Higher values of hypervolume are expected to mean a larger scattering of solutions and a better convergence to the true Pareto front. The HV is given by Eq. (3) [2].

$$Hypervolume = \sum_i vol_i, \quad vec_i \in PF_{known} \qquad (3)$$

Another quality indicator used in this paper is the unary ϵ-indicator. Basically, it quantifies a value ϵ by which we can multiply each objective value of an approximated front A, such that the resulting front is still weakly-dominated by another approximated front B. Consequently, the ϵ-indicator measures how much an approximation set is worse than another with respect to all objects.

5   Test Problems and Results

In order to analyze how the methods MEA2N, MEAN, NSGAN and MEA2N-STR performs for SR problem, the real DS Sao Carlos city (called System 1) was used to compose other DS with size of four times the original DS (called System 2). System 2 is composed of four Systems 1 interconnected by 49 NO new additional switches (the data of the two DSs are available in [28]).

These DSs have the following general characteristics:

System 1 (S1): 3860 buses, 532 sectors, 632 switches (509 NC and 123 NO switches), three substations, and 23 feeders;

System 2 (S2): 15 440 buses, 2128 sectors, 2577 switches (2036 NC and 541 NO switches), 12 substations, and 92 feeders.

A fault in the largest feeder of Systems 1 and 2 interrupts the service for the whole feeder. Note that a single fault in System 2 increases the complexity of searching the best solution due to the size of the system.

The experiments carried out using both DSs evaluates the methods according to: a) the relative performance of those MOEAs concerning R_2, R_3, HV and ϵ-indicators; b) the performance of them for SR problem.

Table 2 synthesizes the results of four approaches for the SR point of view. They were run 50 times, then the averages and each objective and constraint were calculated. Observe that MEA2N-STR obtained a better average for lower switching operations than NSGAN, MEAN and MEA2N. Note that such objective is the most important one for SR if all constraints have been satisfied.

Table 2. Simulation Results - Single Fault in Systems 1 and 2

		MEAN		NSGAN		MEA2N		MEA2N STR	
		Avg[1]	Dev.[2]	Avg	Dev.	Avg	Dev.	Avg	Dev.
S1	Power Losses	299.13	7.38	361.79	38.96	356.71	33.65	370.54	36.68
	Voltage Ratio(%)	3.25	0.01	4.21	0.83	3.89	0.84	4.15	0.86
	Network Loading (%)	77.79	3.23	86.57	8.18	82.24	5.71	80.52	5.82
	Transformer Loading (%)	55.15	7.38	52.77	3.02	52.93	2.20	53.24	2.21
	Switching Operations	24	2.27	16	12.21	11	2.73	**9**	2.11
	Running Time	14.38	1.32	5.11	0.23	9.17	0.13	14.68	1.27
S2	Power Losses	1014.24	27.84	1165.22	28.76	1170.92	40.50	1170.02	44.11
	Voltage Ratio(%)	5.76	2.43	4.11	0.99	4.18	0.87	3.86	0.72
	Network Loading (%)	92.45	8.15	92.11	8.01	88.71	9.17	86.70	9.31
	Transformer Loading (%)	70.00	27.84	55.16	2.38	61.41	9.95	44.82	2.24
	Switching Operations	87	15.88	50	32.79	28	16.26	**24**	14.41
	Running Time	21.67	2.42	6.52	0.58	9.46	0.54	16.13	0.71

[1] Average.
[2] Standard Deviation.

Moreover, analyses of the results according to metrics used to compare MOEAs show that the MEA2N-STR outperforms MEAN, NSGAN and MEA2N for both test problems (System 1 and 2) in terms of approximating the Pareto optimal set while preserving a diverse, evenly-distributed set of nondominated solutions. Figs. 3(a) and 3(b) indicates that MEA2N-STR is able to evolve individuals near to the Reference Front (which is composed using solutions of all found fronts obtained from 50 trials with each method)when compared with the approaches MEAN, NSGAN and MEA2N.

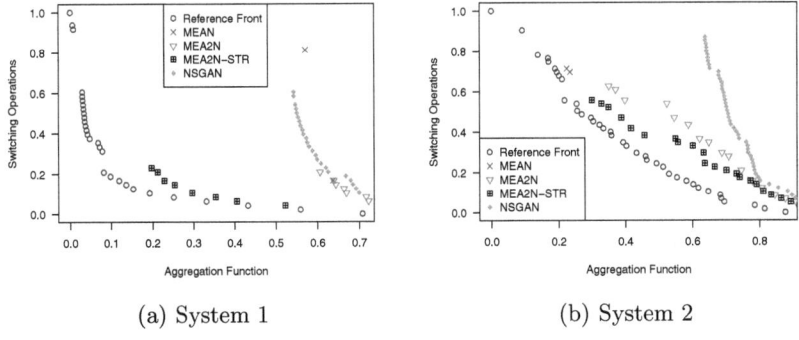

(a) System 1   (b) System 2

Fig. 3. Pareto fronts obtained from Systems 1 and 2

Table 3. Rank of each method in 50 trials for System 1

	MEAN	NSGAN	MEA2N	MEA2N-STR
Hypervolume	-94	-24	0	118
ϵ-indicator	-74	-6	-16	96
R_2	-102	-26	12	116
R_3	-102	-26	12	116

Table 4. Rank of each method in 50 trials for System 4

	MEAN	NSGAN	MEA2N	MEA2N-STR
Hypervolume	-112	-86	54	144
ϵ-indicator	-102	-86	56	132
R_2	-140	-60	60	140
R_3	-138	-62	60	140

The distribution of the performance metrics R_2, R_3, HV and ϵ-indicators for System 1 are shown in Figs 4(a), 4(b), 4(c) and, 4(d), respectively. The MEA2N-STR can find in average a front that is diverse and uniformly distributed for System 1 when compared with other approaches.

Moreover, Figs 5(a), 5(b), 5(c) and, 5(d) corroborate such performance for System 2.

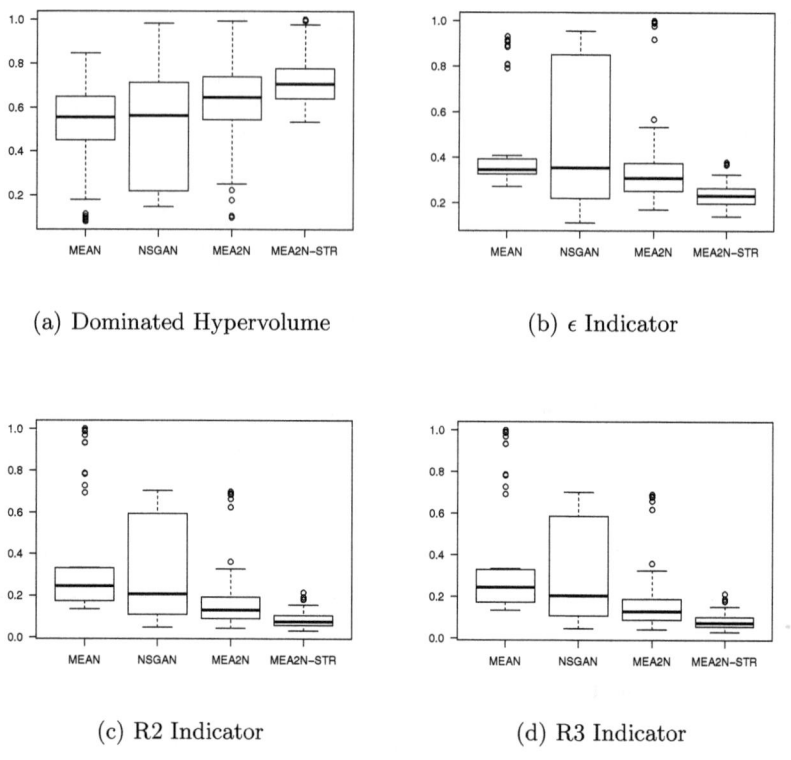

(a) Dominated Hypervolume

(b) ϵ Indicator

(c) R2 Indicator

(d) R3 Indicator

Fig. 4. Box plots for metrics R_2, R_3, HV and ϵ-indicators obtained using System 1

Finally, the method performances are summarized in Table 3, which shows the dominance ranking among the four evaluated methods, which describes the number of experiments in which each algorithm has been found the best Pareto approximation. Tables 3 and (4) shows the ranks (the number of wins subtracted by the number of losses in 50 trials) using System 1 and (System 2) [29]. Results from both tables clearly indicate that MEA2N-STR superates other methods in the performed test.

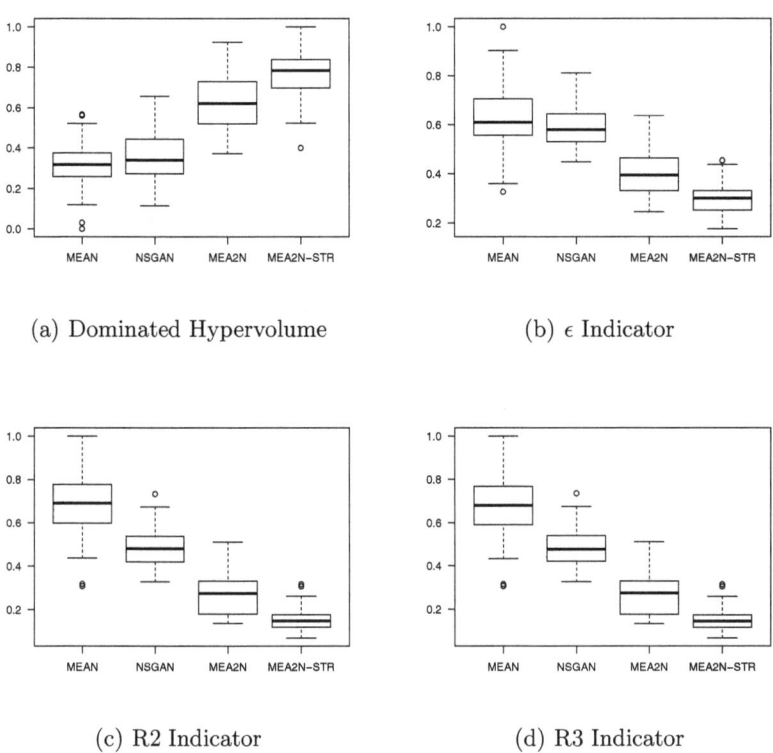

(a) Dominated Hypervolume (b) ϵ Indicator

(c) R2 Indicator (d) R3 Indicator

Fig. 5. Box plots for the metrics R_2, R_3, HV and ϵ-indicators obtained in System 2

6 Conclusions

This paper presented a new MOEA using NDE to solve the SR problems in large-scale DSs (i.e., DSs with thousands of buses and switches).

The proposed approach, called MEA2N-STR, combines the main characteristics of MEAN, NSGA2 and SPEA2. Similarly MEAN, MEAN-STR is based on the idea of subpopulation tables. However, it has additional subpopulation tables to store non-dominated solutions called non-dominated subpopulation tables (similar MEA2N) and incorporates a table of non-dominated solutions based on SPEA2, where each individual is associated to a strength value, which means the strength of dominance of an individual in relation to other individuals. These tables ensure diversity among the solutions significantly improving the performance of MOEAs for SR problem.

In the experiments four approaches, named NSGAN, MEAN, MEA2N and MEA2N-STR were applied to four test DSs. The results show that they enabled SR in large-scale DSs and solutions were found where: energy was restored to the entire out-of-service area, the operational constraints were satisfied, and a reduced number of switching operations was obtained. Moreover, from the relatively low running time required to restoration plans for the Systems 1 and 2, we can conclude that those approaches can generate appropriately SR plans for large-scale DSs.

A statistical analysis performed using 50 experiments for each approach shows MEA2N-STR performs better than NSGAN, MEAN and MEA2N for SR problem, since MEA2N-STR has obtained the best average results for lower switching operations, converging to the true Pareto optimal solution set while preserving the diversity of solutions.

To measure the quality of obtained solutions of MOEAs using NDE, the metrics R_2, R_3, HV and ϵ-indicators were used. According to the simulation results, MEA2N-STR showed a better performance in terms of R_2, R_3, HV and ϵ-indicators in relation to other approaches analyzed.

Finally, this study forms a good basis for combining promising aspects of different algorithms into a new approach that shows good performance on all DSs used.

Acknowledgments. The authors would like to acknowledge CAPES, CNPq and FAPESP for the financial support given to this research.

References

[1] Toth, P., Vigo, D.: The vehicle routing problem. Society for Industrial and Applied Mathematics, Philadelphia, PA, USA (2001)

[2] Coelho, G., Von Zuben, F., da Silva, A.: A Multiobjective Approach to Phylogenetic Trees: Selecting the Most Promising Solutions from the Pareto Front. In: Seventh International Conference on Intelligent Systems Design and Applications, ISDA 2007, pp. 837–842 (October 2007)

[3] Santos, A., Delbem, A., London, J., Bretas, N.: Node-Depth Encoding and Multiobjective Evolutionary Algorithm Applied to Large-Scale Distribution System Reconfiguration. IEEE Transactions on Power Systems 25(3), 1254–1265 (2010)

[4] Martins, J.P., Soares, A.H.M., Vargas, D.V., Delbem, A.C.B.: Multi-objective Phylogenetic Algorithm: Solving Multi-objective Decomposable Deceptive Problems. In: Takahashi, R.H.C., Deb, K., Wanner, E.F., Greco, S. (eds.) EMO 2011. LNCS, vol. 6576, pp. 285–297. Springer, Heidelberg (2011)

[5] Deb, K., Pratap, A., Agarwal, S., Meyarivan, T.: A fast and elitist multiobjective genetic algorithm: NSGA-II. IEEE Transactions on Evolutionary Computation 6(2), 182–197 (2002)

[6] Zitzler, E., Laumanns, M., Thiele, L.: SPEA2: Improving the Strength Pareto Evolutionary Algorithm. Technical report (2001)

[7] Deb, K., Sundar, J.: Reference point based multi-objective optimization using evolutionary algorithms. In: Proceedings of the 8th Annual Conference on Genetic and Evolutionary Computation, GECCO 2006, pp. 635–642. ACM, New York (2006)

[8] Zhang, Q., Li, H.: MOEA/D: A Multiobjective Evolutionary Algorithm Based on Decomposition. IEEE Transactions on Evolutionary Computation 11(6), 712–731 (2007)

[9] Rothlauf, F.: Representations for Genetic and Evolutionary Algorithms. Springer (2006)

[10] de Lima, T.W., Rothlauf, F., Delbem, A.C.: The node-depth encoding: analysis and application to the bounded-diameter minimum spanning tree problem. In: Proceedings of the 10th Annual Conference on Genetic and Evolutionary Computation, GECCO 2008, pp. 969–976. ACM, New York (2008)

[11] Mansour, M., Santos, A., London, J., Delbem, A., Bretas, N.: Node-depth Encoding and Evolutionary Algorithms applied to service restoration in distribution systems. In: Power and Energy Society General Meeting, pp. 1–8. IEEE (2010)

[12] Sanches, D., Mansour, M., London, J., Delbem, A., Santos, A.: Integrating relevant aspects of moeas to solve loss reduction problem in large-scale Distribution Systems. In: PowerTech, 2011 IEEE Trondheim, pp. 1–6 (June 2011)

[13] Diestel, R.: Graph Theory. Third edn. Graduate Texts in Mathematics, vol. 173. Springer, Heidelberg (2005)

[14] Ahuja, R.K., Magnanti, T.L., Orlin, J.B.: Network Flows: Theory, Algorithms,and Applications. Printce Hall, Englewood Cliffs (1993)

[15] Cormen, T.H., Leiserson, C.E., Rivest, R.L., Stein, C.: Introduction to Algorithms, 2nd edn. MIT Press (2001)

[16] Delbem, A.C.B., De Lima, T., Telles, G.P.: Efficient Forest Data Structure for Evolutionary Algorithms Applied to Network Design. IEEE Transactions on Evolutionary Computation PP(99), 1 (2012)

[17] Deb, K.: Multi-objective optimization using evolutionary altorithms. Wiley, New York (2001)

[18] De Jong, K.: Evolutionary computation: a unified approach, pp. 2245–2258. ACM, New York (2008)

[19] Zitzler, E., Thiele, L.: Multiobjective evolutionary algorithms: a comparative case study and the strength Pareto approach. IEEE Transactions on Evolutionary Computation 3(4), 257–271 (1999)

[20] Santos, A., Delbem, A., Bretas, N.: A Multiobjective Evolutionary Algorithm with Node-Depth Encoding for Energy Restoration. In: Fourth International Conference on Natural Computation, ICNC 2008, vol. 6, pp. 417–422 (October 2008)

[21] Delbem, A., de Carvalho, A., Bretas, N.: Main chain representation for evolutionary algorithms applied to distribution system reconfiguration. IEEE Transactions on Power Systems 20(1), 425–436 (2005)

[22] Van Veldhuizen, D.A.: Multiobjective Evolutionary Algorithms: Classifications, Analyses, and New Innovations. PhD thesis, Wright-Patterson AFB, OH (1999)

[23] Hansen, M., Jaszkiewicz, A.: Evaluating the quality of approximations to the non-dominated set. Technical report, Poznan University of Technology (March 1998)

[24] Zitzler, E., Thiele, L., Laumanns, M., Fonseca, C., da Fonseca, V.: Performance assessment of multiobjective optimizers: an analysis and review. IEEE Transactions on Evolutionary Computation 7(2), 117–132 (2003)

[25] Zitzler, E., Brockhoff, D., Thiele, L.: The Hypervolume Indicator Revisited: On the Design of Pareto-compliant Indicators Via Weighted Integration. In: Obayashi, S., Deb, K., Poloni, C., Hiroyasu, T., Murata, T. (eds.) EMO 2007. LNCS, vol. 4403, pp. 862–876. Springer, Heidelberg (2007)

[26] While, L., Hingston, P., Barone, L., Huband, S.: A faster algorithm for calculating hypervolume. IEEE Transactions on Evolutionary Computation 10(1), 29–38 (2006)
[27] While, L., Bradstreet, L., Barone, L., Hingston, P.: Heuristics for optimizing the calculation of hypervolume for multi-objective optimization problems. In: The 2005 IEEE Congress on Evolutionary Computation, vol. 3, pp. 2225–2232 (September 2005)
[28] Source Project (2009), http://lcr.icmc.usp.br/colab/browser/Projetos/MEAN
[29] Kuncheva, L.I., Rodríguez, J.J.: An Experimental Study on Rotation Forest Ensembles. In: Haindl, M., Kittler, J., Roli, F. (eds.) MCS 2007. LNCS, vol. 4472, pp. 459–468. Springer, Heidelberg (2007)

A DSS Based on Multiple Criteria Decision Making for Maintenance Planning in an Electrical Power Distributor

Adiel Almeida-Filho, Rodrigo J.P. Ferreira, and Adiel Almeida

Universidade Federal de Pernambuco, Management Engineering Department, Brazil
{adieltaf,rodjpf,almeidaatd}@gmail.com

Abstract. This paper presents a real world application that lead to the construction of a decision support system built to support maintenance planning in an electrical power distribution company. The outcome of the DSS is to determine the order of priority in which potential failure repair orders should be conducted. This company-specific problem has been structured and is similar to that of most electrical power distribution companies, since this company has a complex distribution network with different types of customers which leads it to considering different criteria such as income losses, the probability of service interruptions, national regulation criteria, and so forth. There is a specific budget for potential failure repair orders, while functional failure repairs are not modeled at this managerial level since once the service should not be interrupted and involves long term planning. During periodical inspections the entire distribution network is inspected and a database with all this data is maintained. Thereafter, the potential failures in the distribution network are converted into maintenance orders which typically amount to more than 4 times the annual sum allocated for potential failure repairs budget. After having structured the problem, a DSS was built to support maintenance planning that will result in an annual maintenance plan based on the company´s strategic criteria and national regulations.

Keywords: Maintenance Planning, MCDM, Promethee.

1 Introduction

Maintenance is a fertile field which provides several research modeling opportunities for practical problems. Usually uncertainty is inherent in the modeling due to factors such as failure probabilities and reliability nature. In addition, Multiple Criteria Decision Making (MCDM) is also been considered once there are several factors besides costs to be evaluated in real applications, such as availability, quality, dependability and other issues that may affect the company's image or the service perception.

In this paper is presented a real problem and the modeling to deal with this situation, resulting on a Decision Support System (DSS) built to aid managers in this particular problem of maintenance planning in an electrical power distributor.

R.C. Purshouse et al. (Eds.): EMO 2013, LNCS 7811, pp. 787–795, 2013.

This paper is structured in five sections, starting with an introduction and followed by the problem context description. In the third part there is a discussion on the MCDM approach adopted with a brief description of the Promethee II method, followed by the presentation of the DSS, finally final remarks are presented with future work directions with Evolutionary Multiple Objective to be incorporated in the model.

2   Problem Context

One of the main characteristics of this real problem is its size. The electrical power distributor network comprehends 128,412.5 km supplying almost 200 cities, which means about 98,546.70 km² with almost 3.1 million customers consuming 12,266,246 MWh per year.

Along this power distribution network there is large number of components (voltage transformers, isolators,…) subjected to weather conditions and the increasing age of components. The maintenance policy adopted follows a schedule for some preventive maintenance services and an inspection schedule to assess components health state, identifying potential failures.

The particular maintenance policy adopted follows some concepts described in Moubray (1994), considering three typical states related to a failure mode: the normal state, a defect state and a failure state. Depending on the outcome of an item failure mode on the system functioning, the failure mode is classified in a potential failure or a functional failure.

A potential failure is an observable condition that will anticipate a functional failure, in other words, if no preventive action is taken a functional failure will occur. The potential failure is a defect that changes the equipment operation characteristics but is not enough to interrupt the system operability but may reduce its efficiency.

Moubray (1994) points that a functional failure is the inability of an item to perform a specific function within desirable operational limits, thus leads to a disruption on the system operability.

Considering this maintenance culture there is a backlog of preventive maintenance services of about 25 thousand orders to repair potential failure built based on the company inspection calendar and recorded in the company's database for its enterprise resource planning (ERP) system. The inspection calendar covers the entire power distribution network in a period of 10 years, which is based in periodical inspection activities that take place at one, two, five and ten year intervals.

After each inspection the state of equipment is updated on the maintenance module of the company's ERP. Functional failures are not recorded from these inspections because it causes service disruptions and a corrective maintenance service is performed to restore the system. Due to the service characteristics there is no budget limit to restore the system. Although potential failure does not cause an immediate service disruption, this is just a matter of time in the while to the defect state turn into a functional failure.

In order to assure the quality of service there is a budget for preventive maintenance orders of potential failures. However the budget is not enough to perform the entire maintenance orders for the set of potential failures. Thus is necessary to prioritize the items which have more impact on the perceived quality of service, revenue and other operational performance indexes, in other words, is necessary to avoid disruption to the service and its consequences to strategic and operational objectives.

The Brazilian electric power market is regulated by Aneel, a Brazilian Government agency responsible for the regulation of electrical power generation, transportation and distribution. Aneel is responsible to set operational and service performance levels for companies involved with electrical power generation, transportation and distribution, regulating also the electric power market negotiation and tariffs. In its attributions, Aneel may levy fines according with regulatory rules and has also the power to set electrical power tariff based on the quality of the service provided. Thus, improving the service level and Aneel performance indexes reflects directly on the company's income (Aneel, 2012).

The main performance indexes considered by Aneel to measure quality of service are the DEC and FEC. DEC is related to the duration of service disruptions whenever these occur and FEC considers the frequency of disruption to the service (Aneel, 2012).

In this particular situation, repairs are usually carried out under an outsourcing contract and for a matter of the contract structure geographical issues are not considered for the repairs cost, which means that there is no difference in the costs of two repairs if it takes place in the same neighborhood or in different cities. Since there is no interaction between maintenance orders, for this particular problem there is not mandatory to model it as a portfolio problem. The backlog of maintenance orders is evaluated under multiple criteria that represent operational and strategic objectives, establishing a complete ranking of this potential failure maintenance orders set.

In the following sections, the MCDA structure used to evaluate alternatives priority is presented followed by DSS characteristics.

3 MCDM Approach

MCDA methods are tools to combine preferences over multiple criteria or multiple objectives. There is a set of methods available in the literature and different methods based on the same paradigm concepts such as the ELECTRE methods and PROMETHEE methods (Roy, 1996; Brans and Mareschal, 2002).

Applications using MCDA are frequent in several areas such as maintenance (Almeida and Souza, 1993; Almeida and Bohoris, 1995; Almeida, 2001; Almeida 2005; Cavalcante et. al, 2010), risk evaluation (French et. al, 2005; Brito and Almeida, 2009; Alencar and Almeida, 2010; Brito et. al, 2010), outsourcing and logistics

(Almeida, 2005; Brito et. al, 2010), project management (Mota and Almeida, 2011; Mota et. al, 2009; Alencar and Almeida, 2010), water resources management (Morais and Almeida, 2006; Morais and Almeida, 2010; Silva et. al, 2010; Morais and Almeida, 2011) and many others.

According to Campos et al (2011) since the aim of organizations is to become or remain competitive in a global society, there is a need to consider many aspects in the decision process, which justifies the development of models and applications which highlights decision makers' preferences over several objectives.

For this specific DSS, the Promethee II method was used. This is one of the methods of the Promethee family which have been evolving since 1982 (Brans and Mareschal, 1984; Brans and Mareschal, 2002). The choice of this method is justified as it can provide a complete ranking order that considers a wide range of value functions that may be available to the decision maker by means of a DSS. One important factor in choosing this MCDA method is related to the simplicity with which it elicits parameters. This is important as it consolidates the decision maker's readiness to use the DSS. Another characteristic is the calculation process. Given that there are about 25 thousand alternatives and that this number may grow, an MCDM method needs to be able to give a response within an appropriate interval of time so decision makers may build scenarios and conjectures and use sensitivity analysis.

The Promethee family consists mainly of Promethee I (which provides a partial pre-order), Promethee II (which provides a complete pre-order), Promethee III (which extends the notion of indifference and provides an interval order), Promethee IV (which provides a complete pre-order and is an extension of Promethee II for a continuous case), Promethee V (an extension of Promethee I and II for portfolio problems given a set of constraints), Promethee VI (an extension of Promethee I and II considering partial information in its parameters) and Promethee Gaia (an extension with visual and interactive procedures using Promethee elements) (Brans and Mareschal, 1984; Brans and Mareschal, 2002).

Recently some consistency issues have been identified in Promethee V (portfolio problematic), some of which were reported on by Vetschera and Almeida (2012). Vetschera and Almeida (2012) also proposed a c-optimal portfolio concept to improve this method. Almeida and Vetschera (2012) presented an analysis on the scale transformations used in Promethee V by evaluating how it influences the results obtained by the Promethee V procedure.

In the literature Promethee II is often questioned regarding rank reversal new alternatives are added to the sets of alternatives due to the pairwise comparison process, although Mareschal et al (2008) presented conditions when this situation may occur, which is restricted to very limited situations.

The Promethee II method allows the decision maker to choose between six different value functions, namely, defining each criterion as the usual criterion, a u-shape criterion, a v-shape criterion, a level criterion, a v-shape with an indifference criterion or a Gaussian criterion (Brans and Mareschal, 1984; Brans and Mareschal, 2002). Promethee II is based on pairwise comparisons and on aggregated preference indices and outranking flows.

$$\pi(a,b) = \frac{1}{W} \cdot \sum_{j=1}^{k} w_j P_j(a,b) \tag{1}$$

Equation (1) represents the preference indices, and express to what degree a is preferred to b over all the criteria, $W = \sum_{j=1}^{k} w_j$, and w_j represents the weight of criterion j, $w_j \geq 0$. Equation (2) represents the net outranking flow, which consists of the difference between the positive and the negative flow of an alternative a. Based on the net outranking flow, a complete pre-order is provided that ranks all alternatives.

$$\phi(a) = \frac{1}{n-1} \left[\sum_{\substack{b=1 \\ b \neq a}}^{n} \pi(a,b) - \sum_{\substack{b=1 \\ b \neq a}}^{n} \pi(b,a) \right] \tag{2}$$

4   Decision Support System

A DSS is an information system (IS) used to support decision makers at any organizational or strategic level for semi-structured or non-structured problems (Davis and Olson, 1985; Shim et al, 2002). The main reasons that justify building a DSS are: the complexity of the decision process; the interactive relation between the DSS and the decision maker, and the alternative; to provide simplicity through a friendly interface; to process data in a decision model and build information, scenarios and solutions for a semi-structured or non-structured problem.

According to Sprague and Watson (1989) and Bidgoli (1989), the architecture of a DSS usually consists of a data base to support the system, a model base to provide analytics and dialogs to support interaction between the user and the system.

The data base is responsible for the physical storage of consistent data which are of significant value to an organization. In this case, the DSS data base comprises data provided from the organization's ERP maintenance module and deals with equipment, maintenance services, resources required for these maintenance services, service costs, performance of managerial objectives, decision model parameters which represent the decision maker's preferences and the decision environment including information about the decision maker.

The base of the DSS model consists of a multiple criteria decision model based on the Promethee II method, including a module to build scenarios and sensitivity analysis. Dialog is an important component and is based on the simplicity of the principles and a friendly interface. It is drawn up so that it works in an integrated way with the ERP maintenance modules using spreadsheet reports to import and export data as far as the existing systems are concerned. The flexibility in the DSS dialog allows the decision maker improves the interactions of the decision process.

The main output from the DSS is preventive maintenance orders sequenced according to the objectives and take the decision maker's preferences into account. On importing data, the DSS is supplied with all the data on observed potential failures such as location, equipment, local network consumption, revenue losses including from disruptions to the service, regulatory fines, emergency and healthcare services affected by the specific failure, an expert's prior estimation for the mean time to functional failure and other factors.

The main sets of criteria identified for the decision model are: Degree of Damage (to installation and people, verbal scale), Average Affected Consumption, Electric Charge, % of Regional Network Electric Charge (considering the network branch), Special Clients Affected (subjected to regulatory special rules), Healthcare Services, Slack on DEC (difference between branch DEC and Aneel target for DEC), Slack on FEC (difference between branch FEC and Aneel target for FEC), Political Consequences of a Failure.

In addition to the MCDA methodology which considers more than one criterion simultaneously , all concepts have been adjusted to the company's maintenance culture. This is related to Moubray's reliability-centered maintenance critical levels which uses verbal scales to determine the level of degradation of the equipment. Figure 1 shows the DSS main screen for the decision model.

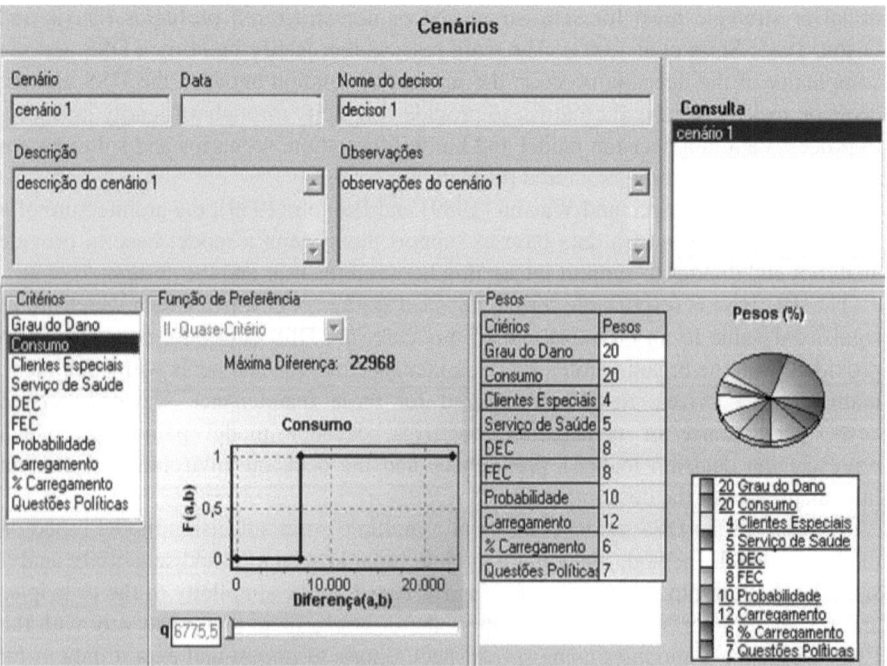

Fig. 1. DSS main screen for decision model

After inputting data into the fields required for the decision model and performing the Promethee II method to obtain the sequencing of the priority given to preventive maintenance orders, the decision maker can generate reports regarding preventive maintenance covered by the budget constraint and analytics over sensitivity analysis.

Next, the scenarios are evaluated using the DSS support graphs to compare the effectiveness of each action while considering costs and the managerial objectives (DEC and FEC). Figure 2 presents one of these graphs that considers the costs to perform the preventive maintenance order over potential failures and its losses (such as, to revenue and in fines) as a consequence of the functional failure.

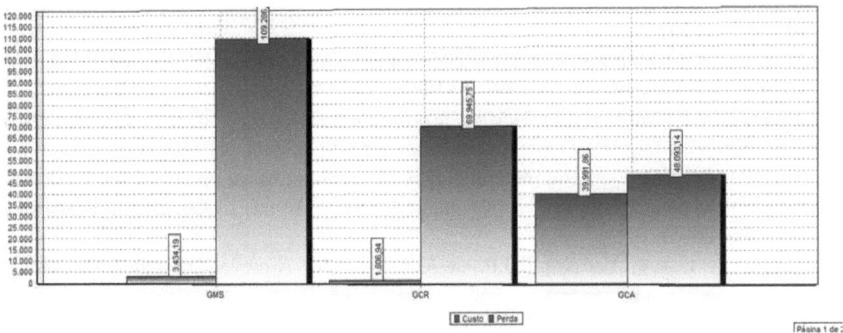

Fig. 2. Preventive maintenance order over the effect of potential failures

It is interesting to observe that some prioritized maintenance actions may not prove to be financially effective. However, thy do prevent losses in other dimensions, such as service quality which is monitored by the regulatory agency (Aneel) or any special clients affected, for example.

5   Conclusions

This paper presented a DSS to support maintenance planning in an electrical power distribution company. There are several studies in the literature that present models to establish preventive maintenance intervals, inspection intervals and optimal maintenance policies, mostly with regard to a single piece of equipment or a single subsystem. Nevertheless, drawing up a schedule for the maintenance department and programming activities are hard tasks and depend on some strategic and operational objectives being in place. In this specific case, there were thousands of activities to be scheduled and prioritized in a complex distribution network.

The decision maker regarding this specific problem in the power distribution company felt satisfied within the decision process, however other issues may be considered in future, such as the interaction amongst service orders location, which may be addressed with an evolutionary multiobjective approach.

Acknowledgments. This work has been partially supported by CNPq (the Brazilian Bureau that Funds Research).

References

1. Moubray, J.: Reliability-centered maintenance. Butterworth Heinemann, Oxford (1994)
2. Aneel, Qualidade do serviço (2012),
   http://www.aneel.gov.br/area.cfm?idArea=79&idPerfil=2
3. Roy, B.: Multicriteria Methodology for Decision Aiding. Kluwer Academic Publishers (1996)
4. Brans, J.P., Mareschal, B.: Promethee-gaia, une methodologie d´aide à la décision em présence de critères multiples. Bruxelles, Éditions Ellipses (2002)
5. Almeida, A.T., Souza, F.M.C.: Decision theory in maintenance strategy for a two-unit redundant standby system. IEEE Transactions on Reliability, IEEE Reliability Society 42(3), 401–407 (1993)
6. Almeida, A.T., Bohoris, G.A.: Decision theory in maintenance. Journal of Quality in Maintenance Engineering 1(1), 39–45 (1995)
7. Almeida, A.T.: Multicriteria decision making on maintenance: spares and contracts planning. European Journal of Operational Research 129(2), 235–241 (2001)
8. Almeida, A.T.: Multicriteria modeling of repair contract based on utility and Electre I method with dependability and service quality criteria. Annals of Operations Research 138, 113–126 (2005)
9. Cavalcante, C.A.V., Ferreira, R.J.P., de Almeida, A.T.: A preventive maintenance decision model based on multicriteria method PROMETHEE II integrated with Bayesian approach. IMA Journal of Management Mathematics 21, 333–348 (2010)
10. French, S., Bedford, T., Atherton, E.: Supporting ALARP decision making by cost benefit analysis and multi-attribute utility theory. Journal of Risk Research 8(3), 207–223 (2005)
11. Brito, A., Almeida, A.T.: Multi-attribute risk assessment for risk ranking of natural gas pipelines. Reliability Engineering & Systems Safety 94, 187–198 (2009)
12. Alencar, M.H., Almeida, A.T.: Assigning priorities to actions in a pipeline transporting hydrogen based on a multicriteria decision model. International Journal of Hydrogen Energy 35, 3610–3619 (2010)
13. Brito, A.J.M., Almeida, A.T., Miranda, C.M.G.: A multi-criteria model for risk sorting of natural gas pipelines based on ELECTRE TRI integrating Utility Theory. European Journal of Operational Research 200, 812–821 (2010)
14. Brito, A.J., de Almeida-Filho, A.T., Almeida, A.T.: Multi-criteria decision model for selecting repair contracts by applying utility theory and variable interdependent parameters. IMA Journal of Management Mathematics 21(4), 349–361 (2010)
15. Mota, C.M.M., Almeida, A.T.: A multicriteria decision model for assigning priority classes to activities in project management. Annals of Operations Research, Online (2011)
16. Mota, C.M.M., Almeida, A.T., Alencar, L.H.: A multiple criteria decision model for assigning priorities to activities in project management. International Journal of Project Management 1, 1–10 (2009)
17. Alencar, L.H., de Almeida, A.T.: A model for selecting project team members using multicriteria group decision making. Pesquisa Operacional 30, 221–236 (2010)
18. Morais, D.C., Almeida, A.T.: Water supply system decision making using multicriteria analysis. Water S.A. 32(2), 229–235 (2006)
19. Morais, D.C., Almeida, A.T.: Water network rehabilitation: a group decision-making approach. Water S.A. 36, 487–493 (2010)
20. Silva, V.B.S., Morais, D.C., Almeida, A.T.: A multicriteria group decision model to support watershed committees in Brazil. Water Resources Management 24, 4075–4091 (2010)

21. Morais, D.C., Almeida, Teixeira, A.: Group decision making on water resources based on analysis of individual rankings. Omega 40, 42–52 (2011)
22. Campos, A.C.S.M., Daher, S.F.D., Almeida, A.T.: New patents on business process management information systems and decision support. Recent Patents on Computer Science 4, 91–97 (2011)
23. Brans, J.P., Mareschal, B.: PROMETHEE: a new family of outranking methods in multi-criteria analysis. Operational Research 84, 408–421 (1984)
24. Vetschera, R., Almeida, A.T.: A PROMETHEE-based approach to portfolio selection problems. Computers & Operations Research 39, 1010–1020 (2012)
25. Almeida, A.T., Vetschera, R.: A note on scale transformations in the PROMETHEE V method. European Journal of Operational Research 219, 198–200 (2012)
26. Mareschal, B., De Smet, Y., Nemery, P.: Rank reversal in the PROMETHEE II method: some new results. IEEE Proceedings, 959–963 (2008)
27. Davis, C.B., Olson, M.H.: Management information systems: conceptual foundations, structure and development. McGraw-Hill (1985)
28. Shim, J.P., Warkentin, M., Courtney, J.F., Power, D.J.: Past, present, and future of decision support technology. Decision Support Systems 33, 111–126 (2002)
29. Sprague Jr, R.H., Watson, H.J.: Decision support systems – putting theory into practice. Prentice-Hall, Inc. (1989)
30. Bidgoli, H.: Decision support systems – principles and practice. West Publishing Company (1989)

Multi-objective Optimization under Uncertain Objectives: Application to Engineering Design Problem

Céline Villa, Eric Lozinguez, and Raphaël Labayrade

Université de Lyon, Lyon, F-69003, France
Ecole Nationale des Travaux Publics de l'Etat,
Département Génie Civil Bâtiment 3, rue Maurice Audin,
Vaulx-en-Velin, F-69120, France

Abstract. In the process of multi-objective optimization of real-world systems, uncertainties have to be taken into account. We focus on a particular type of uncertainties, related to uncertain objective functions. In the literature, such uncertainties are considered as noise that should be eliminated to ensure convergence of the optimization process to the most accurate solutions. In this paper, we adopt a different point of view and propose a new framework to handle uncertain objective functions in a Pareto-based multi-objective optimization process: we consider that uncertain objective functions are not only biasing errors due to the optimization, but also contain useful information on the impact of uncertainties on the system to optimize. From the Probability Density Function (PDF) of random variables modeling uncertainties of objective functions, we determine the "Uncertain Pareto Front", defined as a "tradeoff probability function" in objective space and a "solution probability function" in decision space. Then, from the "Uncertain Pareto Front", we show how the reliable solutions, *i.e.* the most probable solutions, can be identified. We propose a Monte Carlo process to approximate the "Uncertain Pareto Front". The proposed process is illustrated through a case study of a famous engineering problem: the welded beam design problem aimed at identifying solutions featuring at the same time low cost and low deflection with respect to an uncertain Young's modulus.

Keywords: Multi-objective optimization - Objective function uncertainty - Reliable solutions - Engineering design problem.

1 Introduction

To solve Multi-objective Optimization Problems (MOP), the Pareto optimality approach is usually employed in order to identify a set of Pareto optimal solutions that make it possible to obtain the best tradeoffs between different objectives. Such a set is defined according to the Pareto dominance notion: a solution S, defined by specific decision variable values, dominates another solution S' if and only if S is better than S' for all objectives and if there exists at least

R.C. Purshouse et al. (Eds.): EMO 2013, LNCS 7811, pp. 796–810, 2013.

one objective for which S is strictly better than S'. The set of non-dominated solutions is called Pareto optimal. The image of the Pareto optimal solution set in objective space is known as the Pareto Front and delimits the feasible and unfeasible tradeoffs [1].

Real-world case studies are usually affected by different kinds of uncertainties that should be taken into account in the optimization process. Jin and Branke [2] present four categories of uncertainties in Evolutionary Computation: noise on evaluations of objective functions, uncertainties on decision variables and environmental parameters, approximation errors on objective functions in use of meta-model and dynamic environments. In engineering, errors and uncertainties appear during designing, manufacturing and operating processes. The devoted optimization technics are aimed at studying the expected performances of solutions that can actually be obtained in real-world under these uncertainties [3]. So, in the related literature, uncertainties are taken into account in the optimization process in such a way that optimal solutions are identified in the uncertain environment. In MOP, such uncertainties are described by perturbations on decision variables or on environmental parameters [4, 5]. In this context, the target is to develop methods that determine the most robust solutions, *i.e.* the less sensitive solutions to these perturbations. Robust solutions present low variations of their performances in objective space while undergoing variations in decision space or in environmental parameter values. In previous work, two approaches have been proposed. The first one consists in determining a robust front, by optimizing effective functions [3, 4, 6], or by optimizing initial objective functions under constraints on variations of performances [4, 7, 8]. In the second approach, the goal is to identify the best tradeoffs between optimal performances and robustness. Thus, additional objective functions related to robustness are introduced in the MOP [9–11]. Robustness measures rely mostly on neighbourhood performances average [4, 9] or variance [10]. However, in the related literature, no method has been proposed to deal with uncertainties affecting solution performances, inherent to the solution and for which the exact value is unknown in the real-world. Uncertainties directly related to objective functions are handled in the literature about optimization in noisy environment. The noise usually appears when the objective function evaluation device is not deterministic (*e.g.* measurement errors, stochastic simulations, uncertainties on model parameters, use of neural network, etc.). So different performances can be assigned to the same solution during the optimization run. In this case, contradictory evaluations can appear during the run producing convergence troubles. Thus, some methods introduced in previous work are aimed at reducing noise effects in order to converge to the most accurate solutions. Random sampling (using Monte Carlo approach) is employed to estimate the correct value of performances of each solution [2, 12–14]. However, these approaches implementing *Explicit Averaging* are time-consuming ; strategies are still being investigated with the aim to avoid sampling of every individual of every generation in evolutionary algorithm. Instead of random sampling, *Implicit Averaging* is aimed at increasing the population size, assuming that there are individuals close to each other at each generation, and thus at the end of the run the best individuals are evaluated many times [2, 15–17]. Modification of the selection process is also investigated [12, 18–21], by means of a modified Pareto-dominance that allows to take into account the PDF of additional noise on candidate values for fitness evaluation and selection in Evolutionary Computation [19, 21]. In all these methods, uncertainties on objective functions are handled as biasing errors to be removed before determining the most accurate set of optimal solutions. Moreover, there is a loss of information due to the fact that the influence of uncertainties is somewhat aggregated at a stage of the algorithm.

In this paper, we adopt a different point of view and consider that some uncertainties affecting objective functions cannot be reduced to biasing errors in the results of the optimization process, but are inherently part of the system to optimize. This positioning leads to the proposal of a new framework aimed at preserving all the information along with presenting impacts of uncertain objective functions on optimization results in an all-comprehensive way.

For that purpose, in section 2 we first introduce a new framework and its implementation method, that we will call the "Uncertain Pareto Front" defined as a "tradeoff probability function" in objective space and a "solution probability function" in decision space. The first one allows to determine the probability for each tradeoff in objective space to be one of the best tradeoffs according to uncertain objective functions. The second one, the "solution probability function", is defined in decision space and provides the probability for each possible solution to be Pareto optimal according to uncertain objective functions. Then, the method is illustrated through an engineering problem introduced in section 3 and processed in section 4. Finally, the method relevance is discussed in section 5.

2   New Formalisation and Implementation Method

2.1   "Uncertain Pareto Front"

The uncertain objective function can be defined as a random variable X distributed according to its PDF. The use of a function X_i as an objective function to optimize in the optimization algorithm results in one Pareto Front F_i and one set of Pareto optimal solutions SF_i, each defined by a vector of decision variable values, related to the function X_i. Thus, the obtained Pareto Front and the set of Pareto optimal solutions can be respectively defined as random variable F and SF as illustrated in Figure 1. The "Uncertain Pareto Front" corresponds to the set of all possible Pareto Fronts when uncertain objective functions vary according to their PDFs. The explanation above is presented in the case of one uncertain objective function but it can easily be generalized: for more than one uncertain objective function, the PDF of X is determined by the product of the PDF of each uncertain objective function. Table 1 includes all the notations used in Equations 1-12.

Table 1. Notations used in Equations 1-12

	Objective space	Decision space
Continuous space	T: tradeoff (a vector function)	S: solution (a vector)
Discretized space	px: discrete element	sx: discrete element
One draw	F_i : Pareto Front	SF_i Set of Pareto optimal solutions corresponding to a given F_i
n draws	F: "Uncertain Pareto Front"	SF: Uncertain Optimal Set of solutions
Probability	P_t: "Tradeoff probability function"	P_d: "Solution probability function"
	n: the number of Monte Carlo draws *i.e.* each draw corresponds to one possible objective function according to its PDF	

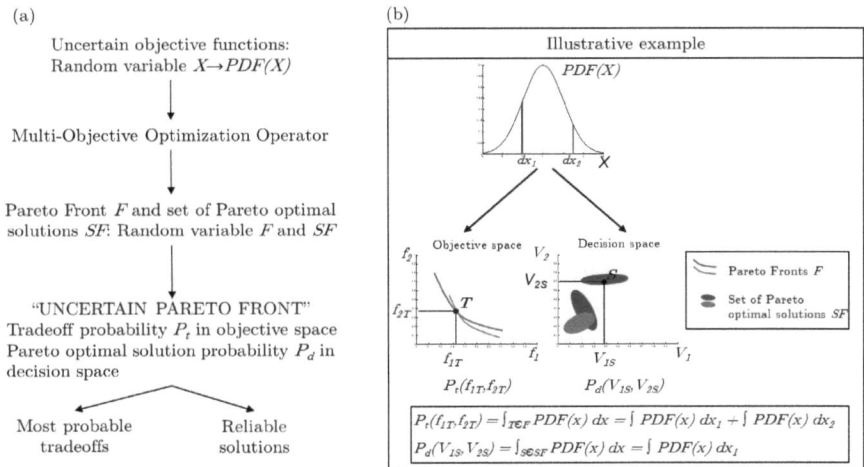

Fig. 1. "Uncertain Pareto Front".(a) Proposed approach (b) Illustrative example.

"Tradeoff Probability Function"

In objective space, the probability P_t for each tradeoff T to be one of the best tradeoffs, *i.e.* belonging to possible Pareto Fronts defined by a random variable F, is expressed as:

$$P_t(T) = \int_{X/T \in F} PDF(X) dX \qquad (1)$$

Equation 1 is the theoretical formulation of the "tradeoff probability function". It means that the probability for a tradeoff T to be one of the best tradeoffs is deduced from the probability of the realizations of the random variable X modeling the uncertain objective functions, that results in the Pareto Fronts which include the tradeoff T. An illustrative example is given in Figure 1 with the point T.

In order to approximate the "Uncertain Pareto Front", we introduce a method based on the integration of Monte Carlo method [22] along with the optimization process. The Monte Carlo method allows to estimate the occurrence of specific Pareto Fronts in the "Uncertain Pareto Front". To that purpose, a discretisation of the objective space is employed (*e.g.* in two dimensions, the space is divided in pixels). For each Monte Carlo draw, the counter of discrete elements (*e.g.* pixels in two dimensions) crossed by Pareto Front F_i is increased by one. The goal is to determine for each discrete element px of the objective space the occurrence of the proposition "Best tradeoffs are included in the discrete element px" (or "Pareto Front F_i crosses the discrete element px"). Thus, the "tradeoff probability function" can be approximated: probability P_t of obtaining Pareto Fronts in the discrete element px is expressed as follows:

$$P_t(px) = \frac{\sum_{i}^{n} \delta_{F_i,px}}{n} \quad \text{with} \quad \delta_{F_i,px} = \begin{cases} 1 & \text{if } F_i \in px \\ 0 & \text{if otherwise} \end{cases} \tag{2}$$

Equation 2 can be seen as the discrete form of Equation 1.

"Solution Probability Function"

In the decision space, the probability P_d for each solution S, defined by a vector of decision variable values, to belong to the set of Pareto optimal solutions SF can be determined according to Equation 3.

$$P_d(S) = \int_{X/S \in SF} PDF(X)dX \tag{3}$$

Equation 3 is the theoretical formulation of "solution probability function". An illustrative example is given in Figure 1 with S.

A Monte Carlo approach can be employed to approximate the "solution probability function". The decision space is discretised and probability P_d for each discrete element sx to belong to the set of Pareto optimal solutions SF_i is expressed as follows:

$$P_d(sx) = \frac{\sum_{i}^{n} \delta_{SF_i,sx}}{n} \quad \text{with} \quad \delta_{SF_i,sx} = \begin{cases} 1 & \text{if } SF_i \in sx \\ 0 & \text{if otherwise} \end{cases} \tag{4}$$

Equation 4 can be seen as the discrete form of Equation 3.

2.2 Approaches to Extract Information from "Uncertain Pareto Front"

Much more information can be derived from the "Uncertain Pareto Front".

Information Extracted from the "Solution Probability Function" in Decision Space: Reliable Solutions

In presence of uncertain objective functions, every solution in the decision space can achieve different performances in objective space. Depending on its performances, which are dependent on each probable uncertain objective function, a

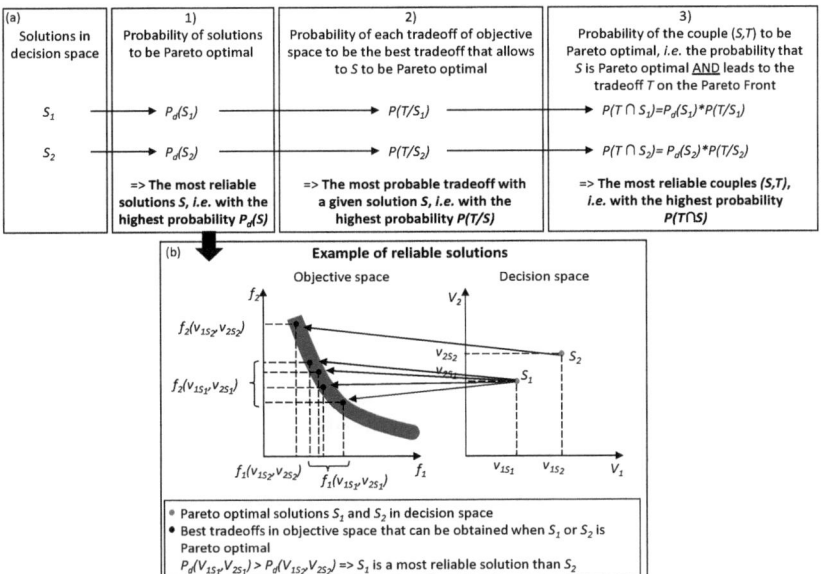

Fig. 2. (a) Approaches to obtain information through the "solution probability function" (b) Illustrative example of reliable solutions

given solution may or may not allow to obtain a best tradeoff between the objective functions. Reliable solutions can be defined as solutions that are Pareto optimal for many different realisations of the random variable X (*i.e.* for different values of uncertain objective functions). The "Solution probability function" provides the probability for each solution $P_d(S)$ to be Pareto optimal in decision space. An example of this approach is illustrated in Figure 2(b), where S_1 is more reliable than S_2 because it belongs to the set of Pareto optimal solutions for more different realisations of random variable X.

As presented in Figure 2(a), a decision maker could go through the following steps:

1. First, identify the most reliable solutions in decision space from $P_d(S)$ described in Equation 3 and approximated in Equation 4;
2. Then, obtain in objective space the probability $P(T/S)$ (approximated by $P(px/sx)$) for each tradeoff T to be a best tradeoff achieved with S as a Pareto optimal solution as follows:

$$P(T/S) = \frac{\int_{X/S \in SF \wedge T \in F} PDF(X)dX}{P_d(S)} \quad (5)$$

$$P(px/sx) = \frac{\sum_{i}^{n}(\delta_{SF_i,sx} * \delta_{F_i,px})}{\sum_{i} \delta_{SF_i,sx}} \quad (6)$$

3. Finally, obtain the probability of the couple (S,T), *i.e.* the probability $P(S \cap T)$ (approximated by $P(sx \cap px)$)for S to be Pareto optimal <u>AND</u> to allow to obtain tradeoff T as follows:

$$P(S \cap T) = P_d(S) * P(T/S) \quad (7) \qquad P(sx \cap px) = \frac{\sum_{i}^{n} (\delta_{SF_i,sx} * \delta_{F_i,px})}{n} \quad (8)$$

Consequently, the decision maker will know for each solution the probability of being Pareto optimal $P_d(S)$, the probability of obtaining the optimal performances $P(T/S)$, and the probability of obtaining a given optimal solution with a given tradeoff $P(S \cap T)$.

Information Extracted from the "Tradeoff Probability Function" in Objective Space: Most Probable Tradeoffs

From the "tradeoff probability function $P_t(T)$" formalized in Equation 1, the tradeoffs that present the highest probability of being one of the best tradeoffs can be found in objective space and will be called "most probable tradeoffs" in the remainder of the paper.

As presented in Figure 3(a)&(b), in practice, if a decision maker selects one tradeoff in objective space, the user will be able to:

1. First, directly assess the probability $P_t(T)$ for this tradeoff T to be one of the best tradeoffs under uncertainties from Equation 2;

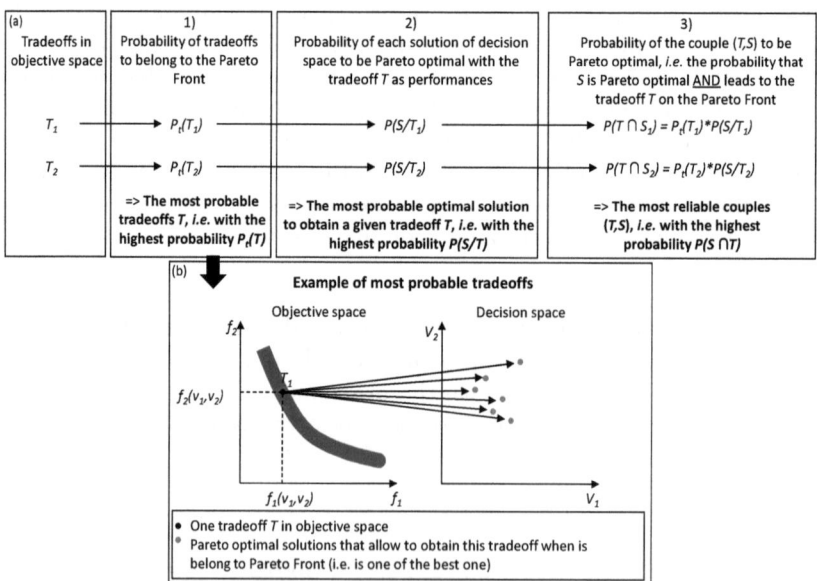

Fig. 3. (a) Theoretical approaches to obtain information through the "tradeoff probability function" (b) Illustrative example of most probable tradeoffs

2. Then, identify Pareto optimal solutions in decision space that allow to obtain this specific tradeoff T. The probability $P(S/T)$ (approximated by $P(sx/px)$) for each Pareto optimal solution S to obtain the given tradeoff T (i.e. f(S)=T) and to be Pareto optimal with these performances can be derived as follows:

$$P(S/T) = \frac{\int_{X/T \in F \, \wedge \, S \in SF} PDF(X)dX}{P_t(T)} \quad (9) \qquad P(sx/px) = \frac{\sum_i^n (\delta_{F_i,px} * \delta_{SF_i,sx})}{\sum_i^n \delta_{F_i,px}} \quad (10)$$

3. Finally, compute the probability of obtaining the couple (T,S), *i.e.* the probability $P(T \cap S)$ (approximated by $P(px \cap sx)$) of obtaining the tradeoff T with the given Pareto optimal solution S, expressed as follows:

$$P(T \cap S) = \begin{cases} P_t(T) * P(S/T) \\ \int_{X/T \in F \, \wedge \, S \in SF} PDF(X)dX \\ P(S \cap T) \end{cases} \quad (11)$$

$$P(px \cap sx) = \frac{\sum_i^n (\delta_{F_i,px} * \delta_{SF_i,sx})}{n} \quad (12)$$

The couple (T,S) with the highest probability $P(T \cap S)$ is the most reliable.

3   Application of the Method

The proposed method is implemented on an engineering problem and the purpose is to identify optimal and reliable solutions.

3.1   Case Study

The method is applied to the welded beam design problem illustrated in Figure 4, which has already been studied in multi-objective optimization in uncertain environment [4, 23, 24]. The goal is to design a welded beam that minimize the cost and the deflection. In reality, the Young's modulus E is inherently uncertain, thus the deflection (objective function f_2 defined in System 13) is also uncertain. In this study, E is implemented as a normal distribution (*i.e.* E $\sim Normal(\mu_E, \sigma_E^2)$ illustred in Figure 5).

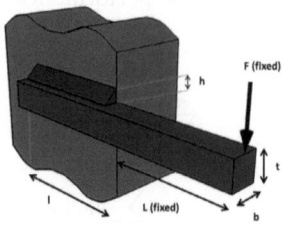

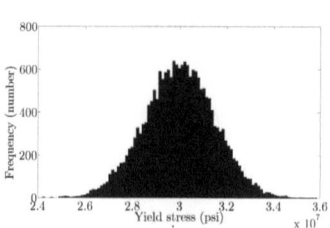

Fig. 4. The welded beam design problem   **Fig. 5.** Distribution of values of E with $\mu_E = 30.10^6$ psi and $\sigma_E = 3.10^6$ psi (20.000 draws)

Decision Variables

In what follows, the four decision variables are : h the thickness of the weld, l the length of the weld, t the width of the beam and b the thickness of the beam.

Constraints

Four nonlinear constraints related to normal stress, shear stress, buckling limitations and geometry are used (see System 14).

Environmental Parameters

The beam carries a fixed load F (F=6000 lb) and the overhang portion of the beam (noted L) has a fixed length of 14 inches (see Figure 4).

The lower bound of h and b (resp. l,t) is 0.125 inch (resp. 0.1 inch) and the upper bound is 5 inches (resp. 10 inches).

Objective Functions. The problem is a two-objective optimization problem :

- f_1 is the fabrication cost in \$ (to minimize);
- f_2 is the uncertain end deflection of the structure in inches (to minimize).

f_1 and f_2 are defined in System 13.

Table 2. Settings of NSGA-II

Population size	Number of generations	Crossover rate	Mutation rate
200	200	0.9	0.25

$$\begin{cases} f_1(x) = 1.104h^2l + 0.048tb(14 + l) \\ f_2(x) = \dfrac{2.1952}{t^3 b} = \dfrac{4FL^3}{Et^3 b} \end{cases} \quad (13)$$

$$\begin{cases} g_1(x) = 13600 - \tau(x) \geq 0 \\ g_2(x) = 30000 - \sigma(x) \geq 0 \\ g_3(x) = P_c(x) - 6000 \geq 0 \\ g_4(x) = b - h \geq 0 \\ with \\ \tau(x) = \sqrt{(\tau')^2 + (\tau'')^2 + \dfrac{l\tau'\tau''}{\sqrt{0.25(l^2 + (h + t)^2}}} \\ where \\ \tau' = \dfrac{6000}{\sqrt{2}hl} \\ \tau'' = \dfrac{6000(14 + 0.5l)\sqrt{0.25(l^2 + (h + t)^2)}}{2(0.707hl(\dfrac{l^2}{12} + 0.25(h + t)^2))} \\ \sigma(x) = \dfrac{504000}{t^2 b} \\ P_c(x) = 64746.022(1 - 0.0282346t)tb^3 \end{cases}$$

$$(14)$$

3.2   Evolutionary Multi-objective Optimization Algorithm

The genetic algorithm NSGA-II (Non-dominated Sorting Genetic Algorithm- II) was used. It was developed by Deb and al. [25]. NSGA-II Matlab code available online was employed with the real-coded GA. Each gene corresponds to one decision variable. The settings of NSGA-II algorithm are summarized in Table 2.

4   Results of Multi-objective Optimization with Uncertain Objective Function (f_2)

To go further, the objective space and the decision space are discretized. The size of discrete elements is indicated in Table 3.

Figure 6 shows the "Uncertain Pareto front" and Figure 7 represents the set of solutions obtained with 20.000 draws ($\mu_E = 30.10^6$ psi, $\sigma_E = 4.5.10^6$ psi). Note that the decision space is a four-dimensional space which is represented by two two-dimensional spaces in Figure 7.

The decision maker can go throuh the steps presented in section 2.2.

Table 3. Pixel size used for "Uncertain Pareto Front" approximation

Objective space		Decision space (inch)
$f_1(\$)$	$f_2(inch)$	h l t b
0.5	0.0001	0.1

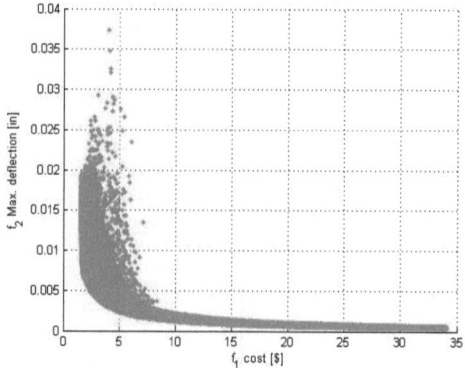

Fig. 6. "Approximation of Uncertain Pareto front" (20.000 draws $\mu_E = 30.10^6$ psi, $\sigma_E = 4.5.10^6$ psi)

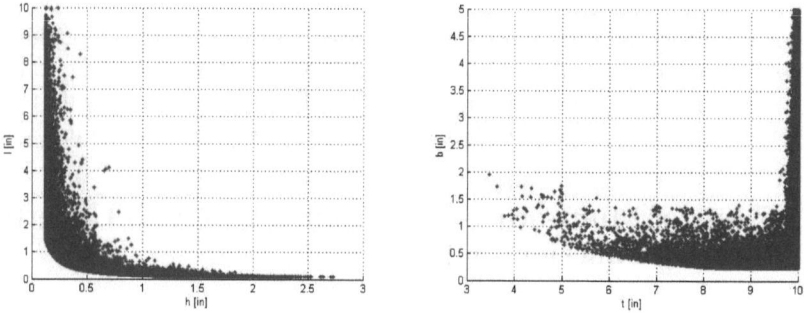

Fig. 7. Projections of Set of optimal solutions (20.000 draws $\mu_E = 30.10^6$ psi, $\sigma_E = 4.5.10^6$ psi)

Information Extracted from the "Tradeoff Probability Function" in Objective Space: Most Probable Tradeoffs.

1. In objective space, the most probable tradeoff given by $P_t(px)$ for each pixel can be determined. Figure 8 shows the "tradeoff probability function", the third axis representing the probability $P_t(px)$.

   The pixel noted T_1 [69,5] (i.e. a cost between 34\$ and 34,5\$ and a deflexion between 0.0004 and 0.0005 inch) is the most probable tradeoff with $P_t(T_1)$ = 0.85635.

2. In decision space, we can find the solution S to maximize $P(S/T_1)$. The discrete element noted S_1 [13,2,100,50] (i.e. h $\in$ [1.2;1.3] inch, l $\in$ [0.1;0.2] inch, t $\in$ [9.9;10] inch, b $\in$ [4.9;5] inch) is the most probable solution that is bound to the tradeoff T_1 with $P(S_1/T_1)$ =0.9990.
3. Finally, we obtain the probability of the most probable couple $(T_1 \cap S_1) = P_t(T_1)*P(S_1/T_1) = 0.8555$.

Information Extracted from the "Solution Probability Function" in Decision Space: Reliable Solutions

1. In decision space, the most reliable solution can be determined by $P_d(sx)$. The discrete element noted S_2 [13,2,100,50] is the most reliable solution with $P_d(S_2)$ =1.
2. In objective space, we can find the best tradeoff T that maximizes $P(T/S_2)$. The pixel noted T_2 [69,5] is the most probable tradeoff that allows S_2 to be a Pareto optimal and $P(T_2/S_2)$ =0.8554.
3. Finally, we obtain the probability of the most probable couple $(S_2 \cap T_2) = P_d(S_2)*P(T_2/S_2)=0.8554$.

This example is a specific case because the analysis results with $T_1=T_2$ and $S_1=S_2$. The analysis presented above has been made with the most probable tradeoff and with the most probable solution but it can be conducted for any tradeoff and any solution.

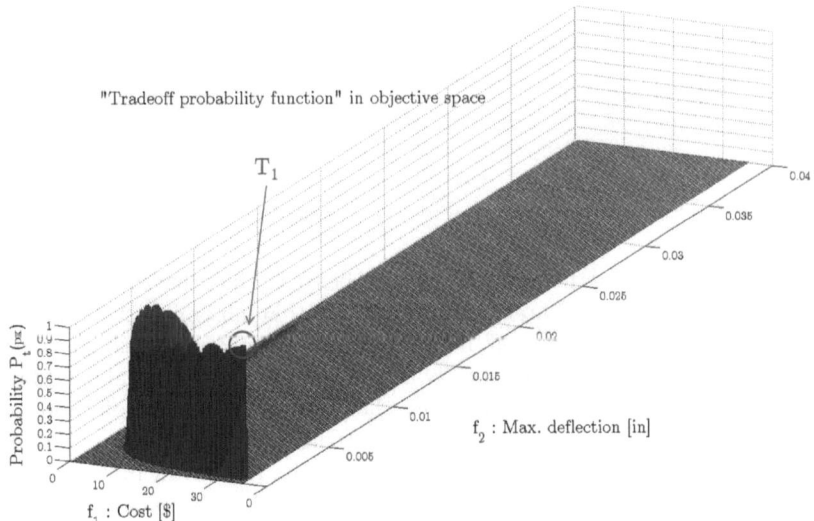

Fig. 8. $P_t(px)$: most reliable solutions with the "Uncertain Pareto front"

In conclusion, the decision maker learned that :

1. the most probable tradeoff is T_1 and $P_t(T_1)$ (85,63%);
2. the most reliable solution is S_1 and $P_d(S_1)$ (100%);
3. with S_1 as Pareto optimal solution, T_1 has a probablity to be the best tradeoff equal to $P(T_1/S_1) = 85.54\%$;
4. with T_1 as tradeoff, S_1 has a probablity to be Pareto optimal solution equal to $P(S_1/T_1) = 99.90\%$;
5. the probability to obtain T_1 (Pareto optimal front) and S_1 (Pareto optimal solution) is equal to $P(S_1 \cap T_1) = 85.55\%$.

Pratical Advantages and Additional Results
The decision maker can set the cost to a certain value and search for the most probable tradeoff :

with f_1 fixed at 20 \$, tradeoff T_3 is the most probable with $P_t(T_3) = 62.54$ % in pixel [40,8] (cf. Figure 9). In decision space, the discrete element noted S_3 [13,2,100,29] is the most probable solution that is bound to tradeoff T_3 with $P(S_3/T_3) = 35.83$ %. The probability of the couple $(T_3 \cap S_3) = 22.41$ %.

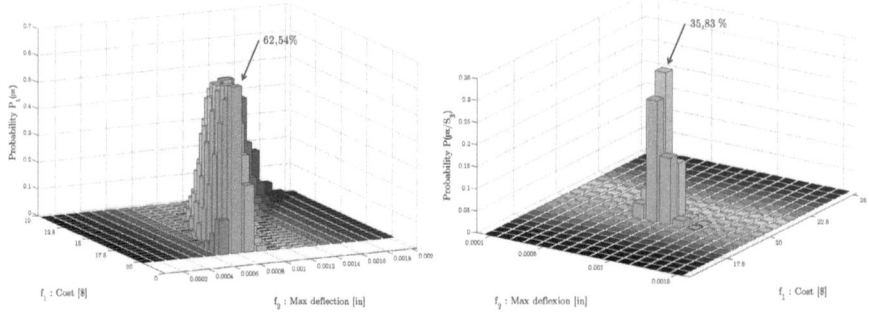

Fig. 9. Section of $P_t(px)$ ($10\$ \leq f_1 \leq 20\$$)   **Fig. 10.** Section of $P(px/S_3)$

A major advantage of the proposed method lies in the fact that for a given trade-off, we can not only get the Pareto optimal solution that maximizes statistically the chances of obtaining this tradeoff, but also obtain other Pareto optimal solutions that achieve this tradeoff. Reciprocally, for a chosen Pareto optimal solution, we know the most probable tradeoff and also the other probable tradeoffs and therefore the worst possible performance. Figure 10 illustrate that with S_3, the maximum deflexion is 0.0011 inch (99.99% to be lower than 0.0011 inch, 99.89 % to be lower than 0.0010 inch, 97.63 % to be lower than 0.0009 inch, 69.34 % to be lower than 0.0008 inch and 9.15 % to be lower than 0.0007 inch).

5   Discussion and Conclusion

A new framework, based on the "Uncertain Pareto front", has been introduced in this paper to handle multi-objective optimization under uncertain objective functions. It provides all-comprehensive information about the influence of uncertainty on the system to optimize, including range values of all possible tradeoffs and all possible Pareto optimal solutions. Moreover, it gives the probability of achieving each tradeoff and indicates the most reliable solutions to the decision maker.

A Monte Carlo based implementation was integrated in the optimization process in order to approximate the "Uncertain Pareto front" from random draws. The proposed method can be applied to real-world systems that are inherently uncertain. This framework could be useful with objective functions natively uncertain in real-world: objective functions estimated from data collected from a panel of persons, prices, experimental measures, environmental parameters, etc. Therefore, this method was applied to the study of the welded beam to identify the most probable couples of tradeoffs and solutions. This point is crucial for the decison maker. Last but not the least, the user may know exactly the probability of his choice and the most reliable solutions.

Also, this method is generic and any evolutionary algorithm can be used (for example here NSGA-II was employed). Even though the number of draws requested by this method is large, solutions are already investigated to accelerate the computation time. Approximately 20 hours are required to compute 10 000 draws using non-optimised Matlab code and one processor (3 GHz) with a popultation of 100 and 100 generations (this duration is proportional to the number of generation and to the square of the population, the evolution of the error in Monte Carlo process is in $1/\sqrt{n}$ according to the number of draws n). C code may be about 50 times faster than Matlab code and Monte Carlo method is highly parallelisable. GPU implementation could also be considered to further improve the applicability of the method. That may lead to computing time of few minutes. However, more efficient numeric schemes will be investigated to reduce computation time.

References

1. Coello Coello, C.A., Lamont, G.B., Van Veldhuizen, D.A.: Evolutionary algorithms for solving multi-objective problems (2007)
2. Jin, Y., Branke, J.: Evolutionary optimization in uncertain environments - a survey. IEEE Trans. on Evol. Comput., 303–317 (2005)
3. Tsutsui, S., Ghosh, A.: Genetic algorithms with a robust solution searching scheme. IEEE Trans. on Evol. Comput., 201–208 (1997)
4. Deb, K., Gupta, H.: Introducing robustness in multi-objective optimization. Evol.Comput., 463–494 (2006)
5. Goh, C.K., Tan, K.C., Cheong, C.Y., Ong, Y.S.: An investigation on noise-induced features in robust evolutionary multi-objective optimization. Expert Syst. with Appl., 5960–5980 (2010)
6. Lim, D., Ong, Y.S., Lee, B.S.: Inverse multi-objective robust evolutionary design optimization in the presence of uncertainty. In: Proceedings of the 2005 Workshops on Genetic and Evolutionary Computation, pp. 55–62 (2005)
7. Barrico, C., Antunes, C.H.: A New Approach to Robustness Analysis in Multi-Objective Optimization. In: 7th International Conference on Multi-Objective Programming and Goal Programming (2006)

8. Gunawan, S.: Parameter Sensitivity Measures for Single Objective, Multi-Objective, and Feasibility Robust Design Optimization (2004)
9. Gaspar-Cunha, A., Covas, J.A.: Robustness in multi-objective optimization using evolutionary algorithms. Comput. Optim. and Appl., 75–96 (2008)
10. Jin, Y., Sendhoff, B.: Trade-Off between Performance and Robustness: An Evolutionary Multiobjective Approach. In: Fonseca, C.M., Fleming, P.J., Zitzler, E., Deb, K., Thiele, L. (eds.) EMO 2003. LNCS, vol. 2632, pp. 237–251. Springer, Heidelberg (2003)
11. Ray, T.: Constrained robust optimal design using a multiobjective evolutionary algorithm. In: Proceedings of the 2002 Congress on Evol. Comput., pp. 419–424 (2002)
12. Babbar, M., Lakshmikantha, A., Goldberg, D.E.: A Modified NSGA-II to Solve Noisy Multiobjective Problems. In: Cantú-Paz, E., Foster, J.A., Deb, K., Davis, L., Roy, R., O'Reilly, U.-M., Beyer, H.-G., Kendall, G., Wilson, S.W., Harman, M., Wegener, J., Dasgupta, D., Potter, M.A., Schultz, A., Dowsland, K.A., Jonoska, N., Miller, J., Standish, R.K. (eds.) GECCO 2003. LNCS, vol. 2723, pp. 21–27. Springer, Heidelberg (2003)
13. Tan, K.C., Goh, C.K.: Handling uncertainties in evolutionary multi-objective optimization. In: Proceedings of the 2008 IEEE World Conference on Computational Intelligence: Research Frontiers, pp. 262–292 (2008)
14. Wu, J., Zheng, C., Chien, C.C., Zheng, L.: A comparative study of Monte Carlo simple genetic algorithm and noisy genetic algorithm for cost-effective sampling network design under uncertainty. Adv. in Water Resour., 899–911
15. Fitzpatrick, J.M., Grefenstette, J.J.: Genetic algorithms in noisy environments. Mach. Learn., 101–120 (1988)
16. Goldberg, D.E., Deb, K., Clark, J.H.: Genetic algorithms, noise, and the sizing of populations. Complex Syst. Champaign, 333 (1992)
17. Miller, B.L., Goldberg, D.E.: Genetic algorithms, selection schemes, and the varying effects of noise. Evol. Compu., 113–131 (1996)
18. Branke, J., Schmidt, C., Schmec, H.: Efficient fitness estimation in noisy environments. In: Proceedings of Genetic and Evolutionary Computation (2001)
19. Hughes, E.J.: Evolutionary Multi-objective Ranking with Uncertainty and Noise. In: Zitzler, E., Deb, K., Thiele, L., Coello Coello, C.A., Corne, D.W. (eds.) EMO 2001. LNCS, vol. 1993, pp. 329–343. Springer, Heidelberg (2001)
20. Roy, R., Azene, Y.T., Farrugia, D., Onisa, C., Mehnen, J.: Evolutionary multi-objective design optimisation with real life uncertainty and constraints. CIRP Ann. Manuf. Technol., 169–172 (2009)
21. Teich, J.: Pareto-Front Exploration with Uncertain Objectives. In: Zitzler, E., Deb, K., Thiele, L., Coello Coello, C.A., Corne, D.W. (eds.) EMO 2001. LNCS, vol. 1993, pp. 314–328. Springer, Heidelberg (2001)
22. Metropolis, N., Ulam, S.: The monte carlo method. Journal of the American Statistical Association, 335–341 (1949)
23. Herfani, T., Utyuzhnikov, S.V.: Control of robust design in multiobjective optimization under uncertainties. Struc. Multidisc. Optim (2011)
24. Chaudhuri, S., Deb, K.: An interactive evolutionary multi-objective optimization and decision making procedure. Applied Soft Computing, 496–511 (2010)
25. Deb, K., Pratap, A., Agarwal, S., Meyarivan, T.: A fast and elitist multiobjective genetic algorithm: NSGA-II. IEEE Trans. on Evol. Comput., 182–197 (2002)
26. Ong, Y.S., Nair, P.B., Lum, K.Y.: Max-min surrogate-assisted evolutionary algorithm for robust design. IEEE Trans. on Evol. Comput., 392–404 (2006)

Decision-Maker Preference Modeling in Interactive Multiobjective Optimization

Luciana R. Pedro and Ricardo H.C. Takahashi

Universidade Federal de Minas Gerais,
Belo Horizonte, Minas Gerais, Brasil
lu.ufmg@gmail.com, taka@mat.ufmg.br

Abstract. This work presents a methodology for modeling the information concerning preferences which is acquired from a Decision-Maker (DM), in the course of one run of an interactive evolutionary multiobjective optimization algorithm. Specifically, the Interactive Territory Defining Evolutionary Algorithm (iTDEA) is considered here. The preference model is encoded as a Neural Network (the NN-DM) which is trained using ordinal information only, as provided by the queries to the DM. With the NN-DM model, the preference information becomes available, after the first run of the interactive evolutionary multiobjective optimization algorithm, for being used in other decision processes. The proposed methodology can be useful in those situations in which a recurrent decision process must be performed, associated to several runs of a multiobjective optimization algorithm over the same problem with different parameters in each run, assuming that the utility function is not dependent on the changing parameters. The main point raised here is: the information obtained from the DM should not be discarded, leading to a new complete interaction with the DM each time a new run of a problem of the same class is required.

Keywords: progressive preference articulation, preference model, neural networks, interactive multiobjective optimization.

1 Introduction

It is well-known that the process of optimizing two or more conflicting objectives usually leads to a set of solutions, the Pareto-optimal solutions, which cannot be ordered by the simple comparison of their objective function values. These incomparable solutions, also called the non-dominated solutions, are delivered by the multiobjective optimization algorithms. The first canonical versions of algorithms for evolutionary multiobjective optimization were intended to deliver a detailed uniform sampling of the Pareto front [1,2,3]. Once this sampling was available, it was assumed that a Decision-Maker (DM) would compare those solutions, indicating the preferred one as the final solution of the problem.

In recent years, a new approach started to receive a growing attention. Due to the high cardinality of a detailed sampling of the entire Pareto front in some

R.C. Purshouse et al. (Eds.): EMO 2013, LNCS 7811, pp. 811–824, 2013.

problems, with particular emphasis on the cases with more than three objectives, some works have proposed procedures that concentrate the sampling in some regions of the Pareto front, based on information which is obtained from interaction between the optimization algorithm and the DM [4,5].

Among the algorithms which consider the DM interaction with the optimization process, the work by Karahan and Köksalan [5] receives a special mention here. That work proposed the Interactive Territory Defining Evolutionary Algorithm (iTDEA), a preference-based multi-objective evolutionary algorithm which identifies the preferred region interacting with the DM on pre-determined generations. In each interaction with the DM, a new best individual is chosen and a new preferred region is stipulated, with a smaller territory for each individual in that region. Individuals falling in that region are assigned smaller territories than those located elsewhere, making the sampling density of the preferred regions higher.

It should be noticed that the information extracted from the DM by the iTDEA is useful only within the scope of the optimization process in which such information is obtained. Whenever the same (or a similar) problem needs to be solved, the DM has to answer the queries about the same region again. However, it should be noticed that, very often, a multiobjective optimization problem might be solved for slightly different conditions, which makes the Pareto-front to become different from one run to the other, with the DM's preferences kept unchanged. For instance, a product may be produced in different instances with different constraints in the resources availability, or with different parameters in some objective functions.

The work by Pedro and Takahashi [6] proposed the construction of a model for the DM's preferences considering the utility function level sets, the NN-DM. The preference information extracted from the DM involves ordinal description only, and is structured using a partial ranking procedure. An artificial neural network is constructed to approximate the DM's preferences in a specific domain, approximating the level sets of the underlying utility function. The proposed procedure was stated with the aim of helping in situations in which recurrent decisions are to be performed, with the same DM considering different sets of alternatives.

This paper presents the results of the hybridization of the iTDEA with an enhanced version of NN-DM. Using the same amount of preference information required by the iTDEA, the NN-DM is able to construct a model for the DM's preferences, so that no more queries are required from the DM related to that specific region of the objective space. This model can now solve similar decision-making problems that comes from optimization problem leading to Pareto-optimal fronts in the same region of the space. Once this preference model is adjusted, it can be used inside the optimization process to guide the search without demanding more information from the DM.

This paper is organized as follow. Section 2 presents some introductory discussion about decision-making theory and multi-objective optimization. Section 3 presents the Interactive Territory Defining Evolutionary Algorithm (iTDEA), according to [5]. At each interaction with the DM, the iTDEA requires the indication of a best solution to guide the optimization process. Considering the same DM's availability, Section 4 introduces the NN-DM, which constructs a model for the DM's preferences that can be used during the optimization process and which can be re-used in similar optimization problems without the need of further interaction with the DM. Section 5 studies the complexity of iTDEA and NN-DM, concerning the number of DM calls. Section 6 provides the solutions for illustrative multi-objective optimization problems using the iTDEA and exhibits the final model for the DM's preferences in the considered cases. Section 7 brings forward a discussion about the results and some future works.

2   The Multi-objective Optimization Problem and the Decision-Maker

A multi-objective optimization problem can be written as

$$\min f(X) = (f_1(X), f_2(X), \ldots, f_m(X))$$

subject to

$$g_i(X) \leq 0, \qquad i = 1, 2, \ldots, k$$
$$h_i(X) = 0, \qquad i = 1, 2, \ldots, r$$

where the f_i are the objective functions, the g_i are the inequality constraints, the h_i are the equality constraints and $X = (x_1, x_2, \ldots, x_N)$ is the vector of decision variables. The minimization is performed with regard to the partial ordering established by the $\leq$ operator, which it is defined, for vectors, as resulting *true* when the inequality is true for each vector component. The minimal elements of this partial ordering are the solutions for this problem – those solutions are called Pareto-optimal solutions, or non-dominated solutions.

A multicriteria decision making problem considers multiple criteria in decision-making situations and involves the following basic elements: a set $\mathcal{A}$ of alternatives, each one with its attributes; a DM, with its associated preferences; and a decision procedure which formulates queries to de DM, obtaining preference information. Usually, it is not possible to assume that the DM would be able to inform the cardinal value of the preference on any alternative; instead, the DM is usually able to furnish only ordinal information, stating that A is better than B or that B is better than A, or yet that the DM is indifferent to those alternatives. Also, the DM is usually able to perform comparisons about a set with some few alternatives only, being unable to process large sets properly.

It is assumed that there is a utility function $\mathcal{U}$ which encodes the preference relations among all alternative pairs, such that if $\mathcal{U}(A) > \mathcal{U}(B)$ then A is better than B, under the DM's viewpoint. The best alternative $x^* \in \mathcal{A}$ is the one that maximizes the function $\mathcal{U}$ in the set $\mathcal{A}$.

In the context of the multi-objective optimization problems, the Pareto-optimal set is not ordered with regard to the objective functions. Therefore, the choice of one alternative within this set becomes a decision problem, in which the most preferred solution should be chosen by a DM.

3   The Interactive Territory Defining Evolutionary Algorithm

The Territory Defining Evolutionary Algorithm (TDEA) [7] was proposed by Karahan and Köksalan. The TDEA is a steady-state elitist evolutionary algorithm to approximate the Pareto-optimal front in multi-objective optimization problems based on a territory around each individual. Introducing the DM's preferences *a priori* in TDEA, the preference-based TDEA (prTDEA) [7] is an algorithm that obtains a detailed approximation of the desired regions within the entire Pareto-optimal front. Improving this idea, the authors proposed the Interactive Territory Defining Evolutionary Algorithm (iTDEA) [5], an algorithm that interacts with the DM during the course of optimization at predetermined generations, finding the best recent solution and guiding the search toward the neighborhood of that solution. The next paragraphs present a brief explanation about these three algorithms. For further information about these methods, including a detailed overview of the algorithms, check the reference [4].

The Territory Defining Evolutionary Algorithm (TDEA) is an algorithm which maintains two populations: a regular population, which has a fixed size, and an archive population, which has flexible size and contains the non-dominated individuals copied from the regular population. In each generation, a single offspring is created and tested using the dominance for the acceptance in the regular population. If the offspring is accepted in the regular population, the individuals in the archive population dominated by the offspring are removed from the archive. If the offspring is dominated by one individual in the archive population, it is rejected, otherwise a territory is defined around the individual closest to the offspring. The offspring is accepted in the archive population only if it does not violate this territory.

Let $y = (f_1, f_2, \ldots, f_m)$ be an individual in the archive population. The territory of the individual y is defined as the region within a distance τ of y in each objective among the regions that neither dominate nor are dominated by y. Mathematically, the territory of y contains all points in V defined by

$$V = \{y' : |f_j - f_j'| < \tau, \quad \text{for } j = 1, 2, \ldots, m \ \land \\ y \text{ and } y' \text{ do not dominate each other}\} \tag{1}$$

where f_j and f_j' are the j-th objective values of y and y', respectively, and τ determines the territory size.

This territory defining property is responsible for the archive population diversity, since each individual in the archive population controls a territory and disallows other individuals in its territory. The idea of favorable weights is employed to identify the location of an individual. The favorable weights of an individual are a set of weights that minimize its weighted Tchebycheff distance from the ideal point.

In TDEA, the parameter τ defines the territory size, which bounds the maximum number of individuals in the archive population. By changing the territory size parameter τ, the authors introduce a version of TDEA, the preference-based TDEA (prTDEA).

The prTDEA possesses a mechanism to incorporate the DM's preference and to modify the territory size of an individual depending on its location on the Pareto-optimal front. Before the optimization, the algorithm requires the DM to specify her/his preferred region R_P, defined by a set of Tchebycheff weight ranges, and sets the remaining space as R_U. Therefore, two values for the parameter τ are stipulated, respectively: τ_P and τ_U. The usage of a small τ_P maintains more individuals from the preferred region in the archive population, while individuals located elsewhere have the eventual neighbors eliminated by a larger τ_U. The prTDEA still requires a change in the acceptance procedure for the archive population: the τ value is now determined by the region that contains the offspring. An illustration of different territory sizes is given in Figure 1.

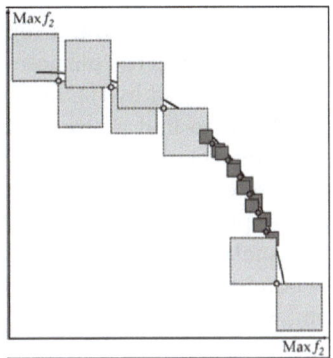

Fig. 1. Different territory sizes. (Figure of [7])

Improving the prTDEA, the authors developed the Interactive Territory Defining Evolutionary Algorithm (iTDEA), an interactive approach that converges to the preferred solutions by progressively obtaining preference information from the DM.

The iTDEA identifies the preferred region interacting with the DM at predetermined generations. Interaction stages $h = 1, 2, \ldots, H$ are scheduled at the generations $G_1, G_2, \ldots, G_H$, respectively. The starting territory size is τ_0 and the final is τ_H, with the intermediate τ values calculated by an exponential decrease. A filtering procedure that utilizes a modified dominance scheme similar to ϵ-dominance is employed for selecting individuals to be presented to the DM. For m objectives, the number of solutions in each interaction stage is $P = 2m$, except in the first and in the last stage, in which $P = 4m$.

At interaction stage h, the DM chooses the preferred individual among the filtered sample of individuals found so far by the algorithm. The preferred individual determines the preferred weight region R^h, which is defined by a set of Tchebycheff weight ranges and it has a specific τ value, τ_h. Individuals falling in these regions are assigned smaller territories than those located elsewhere, so that the density of the preferred regions is higher. For the acceptance in the archive population, the algorithm determines all R^h regions to which the offspring belongs and selects the last created region k, which has the smallest τ_k.

The DM's preferences are simulated using Tchebycheff, linear and quadratic underlying utility functions. The algorithm is tested in three problems with two and three objectives and the number of interactions with the DM is 4 or 6 times. The runs are made with and without filtering and tests are performed with the incorporation of a Gaussian noise in the utility function calculations. The iTDEA converges to the final preferred region of the DM interactively in all selected test problems.

4   The Decision-Maker Model

Pedro and Takahashi [6] introduced the NN-DM, a representation for the DM's preferences using neural networks that has the purpose of reproducing the ordering of alternatives that would be delivered by the utility function $\mathcal{U}$ in a specific domain. The goal of the NN-DM is to find a representation $\hat{\mathcal{U}}$ that models the DM's preferences using ordinal information obtained from queries presented to the DM. The $\hat{\mathcal{U}}$ function must preserve the ordinal relationship between any two points, which is equivalent to stipulate that the level sets of $\mathcal{U}$ and $\hat{\mathcal{U}}$ must be the same. As only ordinal information can be obtained from the DM about $\mathcal{U}$, a partial ranking procedure is employed in order to establish the problem of finding $\hat{\mathcal{U}}$ as a regression problem, which is settled using an artificial neural network. The role of such a function is to replace the DM in new instances of the same multi-objective optimization problem, avoiding the formulation of new queries.

The proposed methodology consists of three main steps:

Step 1: Choose the domain $\mathcal{A}$ for approximation.
Step 2: Build a partial ranking, assigning a scalar value to each alternative and finding a partial sorting for the alternatives.
Step 3: Construct an artificial neural network $\hat{\mathcal{U}}$ which interpolates the results and approximates the DM utility function $\mathcal{U}$.

A more detailed description of those steps is presented now:

Step 1: As the decision problem is related to the choice of a final solution of a multiobjective optimization problem, the domain for $\hat{\mathcal{U}}$ is induced from the domain of the Pareto-optimal available estimative during the optimization process. The domain is defined as the box constructed considering the minimum and maximum values of the Pareto-optimal alternatives in each problem dimension. In this domain, a simulated decision-making problem is built, in which the alternatives are randomly located considering an uniform distribution. The queries to the DM are presented over these simulated alternatives. The number of random alternatives is related to the quality of the approximation $\hat{\mathcal{U}}$: the larger this number, the better the approximation, but, in this case, many queries will be asked to the DM. An estimate of this trade-off is presented in section 6.

Step 2: The partial ranking is a technique used to find a partial sorting for the alternatives, assigning a scalar value to each alternative. Considering a set $\mathcal{A}$ with n alternatives, this process is performed through the following steps:

- Choose randomly $p = \log n$ alternatives[1] from the set $\mathcal{A}$; these alternatives are called *pivots*.
- Sort the pivots in ascending order of the DM's preferences, using ordinal information obtained from yes/no queries. A rank is assigned to each pivot, corresponding to its position in this sorted list.
- For each one of the $n - p$ remaining alternatives, assign a rank that is the same one of the pivot immediately better than the alternative, in the DM's preference. If the current alternative is better than the rank p pivot, it receives rank $p+1$, and p is increased. Each remaining alternative is compared with the middle pivot and, based on the result, compared with the middle pivot of the higher or lower sub-partition. This process continues until a rank is assigned.

This procedure creates a partition of the set $\mathcal{A}$ in at least p disjunct subsets. As the number of pivots is less than the number of alternatives, many alternatives should have the same ranking, providing a partial sorting. The ranking-based classification offers a quantitative (cardinal) way to compare the alternatives, a kind of information which is not provided directly by the DM. In any case, an alternative which is assigned a level $i + 1$ is necessarily better than an alternative with a level i, although two alternatives with the same level i may be not equivalent under the utility function $\mathcal{U}$.

Step 3: In this paper, the regression technique used is the radial basis function (RBF) networks. A radial basis function (RBF) is a real-valued function whose value depends only on the distance from the origin, so that $\phi(\mathbf{x}) = \phi(\|\mathbf{x}\|)$; or alternatively on the distance from some other point x_i, called a center, so that $\phi(\mathbf{x}, x_i) = \phi(\|\mathbf{x} - x_i\|)$. Sums of radial basis functions are typically used to approximate given functions. This approximation process can also be interpreted as a kind of artificial neural network. Radial basis functions networks are typically used to build up function approximations of the form

$$y(\mathbf{x}) = \sum_{i=1}^{N} w_i \cdot \phi(\|\mathbf{x} - \mathbf{x}_i\|),$$

where the approximating function $y(x)$ is represented as a sum of N radial basis functions, each one associated with a different center x_i, and weighted by an appropriate coefficient w_i. The weights w_i can be estimated using linear least squares, since the approximating function is linear in those weights.

It can be shown that any continuous function on a compact interval can be interpolated with arbitrary accuracy by a sum of this form, if a sufficiently large

---

[1] The $\log x$ is used as the same of $\log_2 x$.

number of radial basis functions is used. In this paper, a commonly used type of radial basis function is employed, a Gaussian given by

$$\phi(r) = e^{-r^2}, \text{ where } r = \|\mathbf{x} - \mathbf{x}_i\|.$$

For training the RBF network $\hat{\mathcal{U}}$ which approximates the utility function $\mathcal{U}$, the alternatives within the domain are used as inputs and the ranking level of each alternative, as outputs. It is not necessary to model $\mathcal{U}$ exactly, because the partial ranking keeps the partial sorting of the alternatives, providing a resulting function whose level sets are similar to the ones of $\mathcal{U}$. Figure 2 presents an example of an underlying utility function and the model found by the NN-DM.

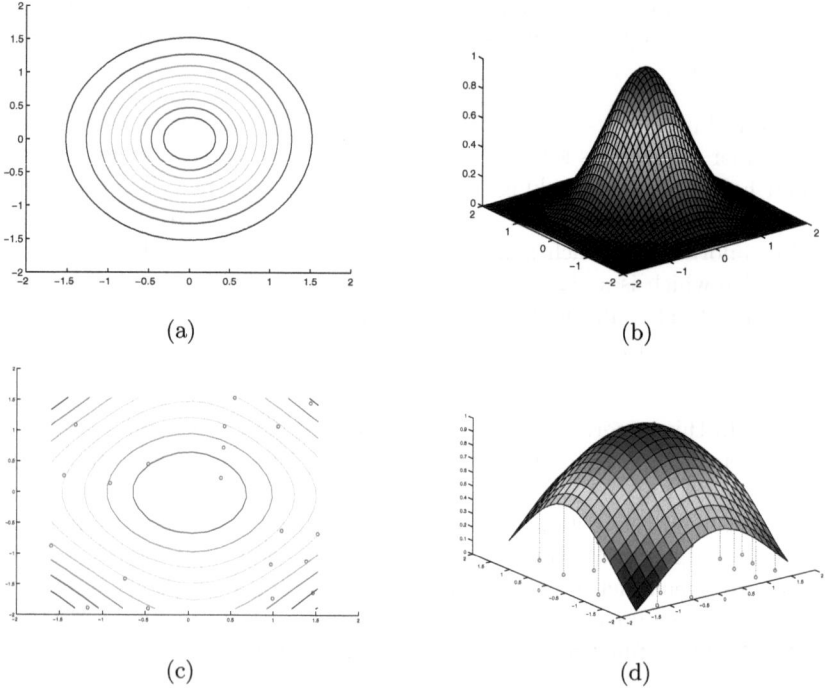

(a)

(b)

(c)

(d)

Fig. 2. Underlying utility function: (a) level sets (b) surface. Resulting estimated utility function: (c) level sets (d) surface.

4.1 Algorithm

The following algorithm presents the pseudocode for the DM model construction.

```
program NN-DM
{Construction of the DM model}

1.  Find the domain
2.  Create n random alternatives
3.  Select the pivots
4.  Sort the pivots in ascending order
5.  Assign a rank to each pivot
6.  Assign a rank to each n - p remaining alternatives
7.  Train the artificial neural network
8.  Measure the efficiency

end
```

5   Number of DM Calls

This section discusses the number of calls of the DM, while interacting with iTDEA and NN-DM methodologies. The DM is considered able to provide only ordinal information about the alternatives. For comparing the methodologies, an estimate of the average number of queries presented to the DM is calculated considering both models.

5.1   DM Calls in NN-DM

In the generation of NN-DM model, the interaction with the DM is necessary in two occasions: the pivot total sorting and the position selection of each remaining alternative. From now on, consider n as the total number of alternatives and $p = \log n$ the number of pivots for the partial sorting.

Pivot Total Sorting. The number of queries the DM has to answer is equal to the number of comparisons that a sorting algorithm must execute. A method as *Quicksort* is known to perform, on average, $p \cdot \log p$ comparisons between the alternatives to sort them, therefore this value represents a good estimate for the number of queries presented to the DM for the total sorting of the pivots.

Position Selection of Each Remaining Alternative. For selecting the position of each remaining alternative, the alternative must be compared with the pivot alternatives. Using a tournament procedure, on average $\log p$ queries are required for each alternative. So, as there are $n - p$ remaining alternatives, then $(n - p) \cdot \log p$ queries are made, on average, during this procedure.

Therefore, the average total of queries to the DM is given by

$$
\begin{aligned}
A(n) &= p \cdot \log p + (n - p) \cdot \log p \\
&= p \cdot \log p + n \cdot \log p - p \cdot \log p \\
&= n \cdot \log p \\
&= n \cdot \log(\log n).
\end{aligned}
$$

Now another advantage of this methodology is pointed out. By considering the alternatives in a domain instead of the Pareto-optimal solutions only, the dominance can be used to replace the DM in some decisions, making the process cost-effective. Considering the answers provided by the dominance, only the information about non-dominated solutions is required from the DM, reducing the total number of queries. The results presented in Section 6 are in agreement with the value $n \cdot \log(\log n)$ as an upper bound for the number of queries presented to the DM.

5.2   The iTDEA

In iTDEA, the DM is required to interact with the optimization process at predetermined generations. The chosen number of interactions with the DM is 4 or 6 times and those interactions are used to identify the preferred regions.

For m objectives, the number of solutions presented to the DM in each interaction stage is $P = 2m$, except in the first and in the last stages, in which $P = 4m$. The DM is required to find the best solution among those P filtered solutions. Considering that only ordinal information is available with binary comparisons, for each set of n elements, at least $n - 1$ queries are made to the DM [8]. Thus, a lower bound for the number of queries presented to the DM is $10m - 4$ in 4 interactions and $14m - 6$ in 6 interactions. Those estimates were used, in this work, as a reference for the number of queries that may be used to construct the model for the DM's preference.

6   Computational Experiments

Some computational experiments in problems with 2 and 3 objectives are reported in this section. The parameters employed in the algorithms are displayed in Table 1.

Table 1. Test parameters

	2D	3D
Ideal vector, f^*	$(0,0)$	$(0,0,0)$
Number of interactions H	4	4
Population size	200	200
τ_0	0.1	0.1
τ_H	0.001	0.001
Number of iterations T	10 000	10 000
Number of replications T	50	50
Number of training points T	12	18
Estimate number of queries	20	44

In all cases, the DM utility function is simulated, considering the following function:

$$\mathcal{U}(p) = \exp(-\mathbf{p} \cdot A \cdot \mathbf{p}^t) \tag{2}$$

6.1   Bi-objective Optimization

As a first example, the following bi-objective optimization problem with two decision variables is considered:

$$\mathbf{p} = \{p_1, p_2\}, \mathbf{f} = (f_1, f_2), \tag{3}$$

$$f_i(\mathbf{p}) = (\mathbf{p} - \mu_i) \cdot M \cdot (\mathbf{p} - \mu_i)^t, i = 1, 2 \tag{4}$$

$$M = \begin{bmatrix} 1 & 0 \\ 0 & 1 \end{bmatrix} \qquad \begin{matrix} \mu_1 = [1\ 0] \\ \mu_2 = [0\ 1] \end{matrix}$$

The utility function $\mathcal{U}$ is instantiated with:

$$A_{10} = \begin{bmatrix} 1 & 0 \\ 0 & 0 \end{bmatrix} A_{11} = \begin{bmatrix} 1 & 0 \\ 0 & 1 \end{bmatrix} \tag{5}$$

The resulting samplings of the Pareto-optimal fronts are presented in figure 3, both for the Territory algorithm and for the NN-DM algorithm. This figure does not present any relevant difference between the results of the two algorithms. The NN-DM network has a Kendal-Tau Distance[2] of 0.1 in relation to the ideal utility function $\mathcal{U}$, with error of $\pm 1\%$, with a number of calls of the DM which is similar to the one performed by the iTDEA algorithm.

6.2   Three-Objective Optimization

An optimization problem with three objectives and three variables is also considered:

$$\mathbf{p} = \{p_1, p_2, p_3\}, \mathbf{f} = (f_1, f_2, f_3), \tag{6}$$

$$f_i(\mathbf{p}) = (\mathbf{p} - \mu_i) \cdot M \cdot (\mathbf{p} - \mu_i)^t, i = 1, 2, 3 \tag{7}$$

with:

$$M = \begin{bmatrix} 1 & 0 & 0 \\ 0 & 1 & 0 \\ 0 & 0 & 1 \end{bmatrix} \qquad \begin{matrix} \mu_1 = [1\ 0\ 0] \\ \mu_2 = [0\ 1\ 0] \\ \mu_3 = [0\ 0\ 1] \end{matrix}$$

The utility function $\mathcal{U}$ is instantiated with:

$$A_{100} = \begin{bmatrix} 1 & 0 & 0 \\ 0 & 0 & 0 \\ 0 & 0 & 0 \end{bmatrix} A_{111} = \begin{bmatrix} 1 & 0 & 0 \\ 0 & 1 & 0 \\ 0 & 0 & 1 \end{bmatrix} \tag{8}$$

---

[2] The Kendal-Tau Distance is a measure of preference inversions, counted over a set of alternatives.

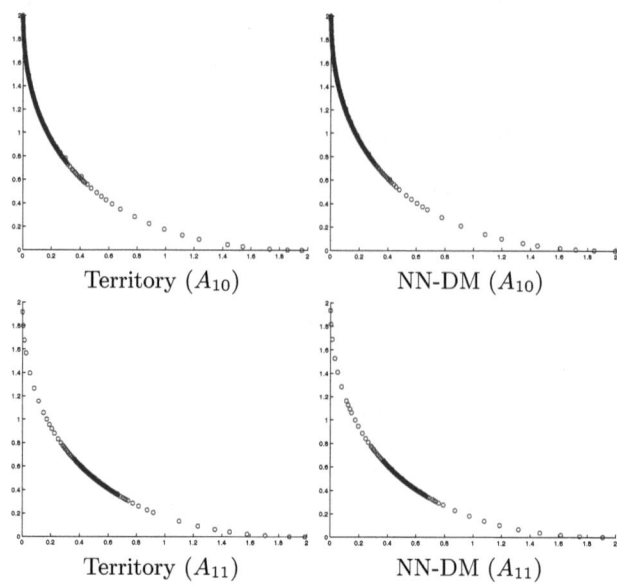

Fig. 3. Resulting samples of the Pareto-optimal front from the Territory algorithm and from the NN-DM algorithm

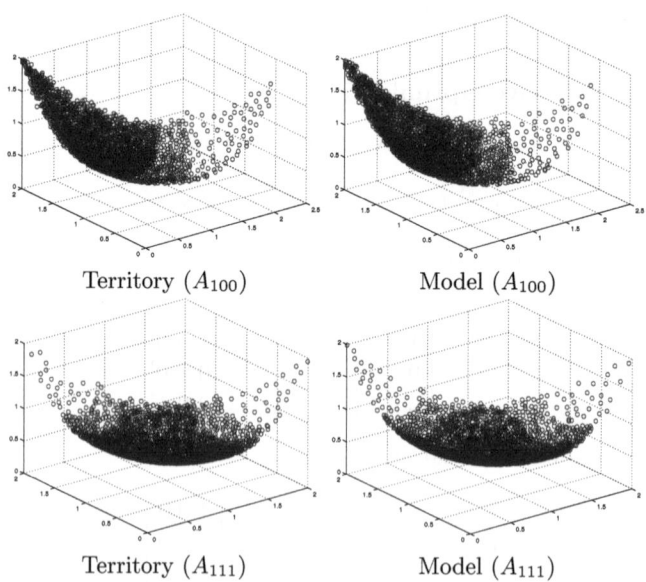

Fig. 4. Results

6.3   Number of Queries

The figure 5 presents a comparison between the total number of queries presented to the DM, the number of queries solved by the dominance, and the expected number of queries, calculated from the complexity analysis.

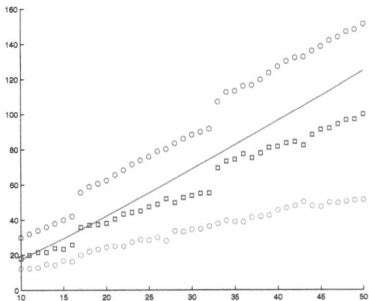

Fig. 5. Number of queries for solving the A_{11} instance. Blue: number of queries presented to the DM. Green: number of queries solved by dominance. Red: total number of queries (blue + green). Magenta line: expected number of queries.

7   Conclusion

This work presented a methodology that allows to get the information concerning preferences which is acquired from a Decision-Maker (DM), in the course of one run of an interactive evolutionary multiobjective optimization algorithm, such that this information becomes available, after that specific run, for being used in other decision processes. The preference information is stored in a Neural Network (the NN-DM), which is trained using ordinal information only, as provided by the queries to the DM. The NN-DM is intended to approximate the level sets of an underlying utility function associated to the DM.

The proposed methodology can be useful in those situations in which a recurrent decision process must be performed, associated to several runs of a multiobjective optimization algorithm over the same problem with different parameters in each run, assuming that the utility function is not dependent on the changing parameters. Some examples of such a situation should be mentioned: (i) the choice of the operation point of an electric power system, under different load constraints (intra-day or intra-week periods); (ii) the manufacturing of a compound which may be composed of different raw materials, under different relative prices of such materials; (iii) the choice of routes, in any routing problem, under different situations of the costs associated to the problem arrows, or under situations of unavailability of some arrows; etc. The main point raised here is: the information obtained from the DM should not be discarded, leading to a new complete interaction each time a new run of this class of problems is required. It is also straightforward to notice that it is possible to perform either a refinement of the NN-DM model or its validation, through some interaction with the DM in new runs of the algorithm.

The authors are currently investigating the possibility of usage of the NN-DM structure for a stronger interaction with interactive EMO algorithms, as an attempt to enhance the convergence properties of such algorithms.

References

1. Fonseca, C.M., Fleming, P.J.: An overview of evolutionary algorithms in multiobjective optimization. Evolutionary Computation 7(3), 205–230 (1995)
2. Zitzler, E., Thiele, L.: Multiobjective evolutionary algorithms: A comparative case study and the strenght pareto approach. IEEE Trans. on Evolutionary Computation 3(4), 257–271 (1999)
3. Deb, K., Pratap, A., Agarwal, S., Meyarivan, T.: A fast and elitist multiobjective genetic algorithm: NSGA-II. IEEE Transactions on Evolutionary Computation 6, 182–197 (2002)
4. Karahan, I.: Preference-based flexible multiobjective evolutionary algorithms. Master's thesis, Dept. Ind. Eng., Middle East Technical University, Ankara, Turkey (2008)
5. Köksalan, M., Karahan, I.: An interactive territory defining evolutionary algorithm: itdea. IEEE Trans. Evolutionary Computation 14(5), 702–722 (2010)
6. Pedro, L.R., Takahashi, R.H.C.: Modeling Decision-Maker Preferences through Utility Function Level Sets. In: Takahashi, R.H.C., Deb, K., Wanner, E.F., Greco, S. (eds.) EMO 2011. LNCS, vol. 6576, pp. 550–563. Springer, Heidelberg (2011)
7. Karahan, I., Köksalan, M.: A territory defining multiobjective evolutionary algorithms and preference incorporation. IEEE Trans. on Evolutionary Computation 14(4), 636–664 (2010)
8. Knuth, D.E.: The art of computer programming, 3rd edn., vol. 2. Addison-Wesley Longman Publishing Co., Inc., Boston (1997), http://portal.acm.org/citation.cfm?id=270146

Automatically Improving
the Anytime Behaviour
of Multiobjective Evolutionary Algorithms

Andreea Radulescu[1], Manuel López-Ibáñez[2], and Thomas Stützle[2]

[1] LINA, UMR CNRS 6241, Université de Nantes, Nantes, France
`andreea.radulescu@etu.univ-nantes.fr`
[2] IRIDIA, CoDE, Université Libre de Bruxelles (ULB), Brussels, Belgium
{`manuel.lopez-ibanez,stuetzle`}`@ulb.ac.be`

Abstract. An algorithm that returns as low-cost solutions as possible at any moment of its execution is said to have a good anytime behaviour. The problem of optimising anytime behaviour can be modelled as a bi-objective non-dominated front, where the goal is to minimise both time and cost. Using a unary quality measure such as the hypervolume indicator, the analysis of the anytime behaviour can be converted into a single-objective problem. In this manner, available automatic configuration tools can be applied to improve the anytime behaviour of an algorithm. If we want to optimise the anytime behaviour of multi-objective algorithms, we may apply again unary quality measures to obtain a scalar value for measuring the obtained approximation to the Pareto front. Thus, for multi-objective algorithms, the anytime behaviour may be described in terms of the curve of the hypervolume over time, and the quality of this bi-objective tradeoff curve be evaluated according to its hypervolume. Using this approach, we can automatically improve the anytime behaviour of multi-objective evolutionary algorithms (MOEAs). In this article, we first introduce this approach and then experimentally study the improvements obtained considering three MOEAs, namely, IBEA, NSGA-II and SPEA2.

1 Introduction

In many real world problems, the quality of solutions is evaluated according to multiple objective functions. The goal of algorithmic approaches to their solution typically is to provide an as good as possible approximation to the unknown Pareto front of tradeoff solutions. Among the most successful such algorithmic approaches are multi-objective evolutionary algorithms (MOEAs).

In practical settings, the user will have a limited time to run an MOEA, and the available amount of time is not always known in advance when deciding for the MOEA's parameter settings. Thus, a goal in the design of MOEAs and other multi-objective optimizers is to find algorithm parameter settings that allow finding the best possible Pareto front approximations for any stopping criterion. More in general, algorithms that provide as good solutions as possible

R.C. Purshouse et al. (Eds.): EMO 2013, LNCS 7811, pp. 825–840, 2013.

independent of a specific termination criterion are referred to as having good anytime behaviour [18].

In this paper, our goal is to examine and, in particular, to improve the anytime behaviour of MOEAs by automatic algorithm configuration techniques [2,6,9,11, 13]. To this aim, we need to model the anytime behaviour of MOEAs. For single objective algorithms, the anytime behaviour may be modelled as a bi-objective non-dominated front, where both solutions quality and computation time must be considered [15, 16]. By using a unary quality measure of the obtained non-dominated front, the analysis of the anytime behaviour is converted into a single-objective problem [16]. Here, we follow [16] and use the hypervolume indicator for this task [22].

In addition, we need to determine the quality of the set of solutions returned by the MOEA. For this task, again the hypervolume indicator can be used and each pair of (time, hypervolume) represents an improvement of the best solutions found at a particular time since the algorithm start. The objective of optimising the anytime behaviour reduces then to find MOEA configurations that produce the best possible set of sequences of points (time, hypervolume). Hence, by making this double usage of the hypervolume, we can apply standard algorithm configuration techniques for the automatic configuration of MOEAs.

We analyse the impact of automatic configuration on the anytime behaviour of MOEAs. In particular, we selected three classical MOEAs: the indicator-based evolutionary algorithm (IBEA) [19], the nondominated sorting genetic algorithm (NSGA-II) [4], and the strength Pareto evolutionary algorithm (SPEA2) [20]. These algorithms are among the best-known MOEAs and they have been thoroughly studied in the literature.

This paper is organized as follows. Section 2 introduces basic notions of multiobjective optimization and MOEAs. In Section 3, we describe the method used for tuning anytime behaviour of MOEAs. The experimental setup and the results are described in Sections 4 and 5, respectively. We conclude in Section 6.

2   Multi-objective optimization

A multiobjective optimization problem (MOP) can be formulated as

$$\text{minimize} \quad \boldsymbol{f}(x) = (f_1(x), \ldots, f_m(x))^T \quad \text{subject to} \quad x \in \Omega$$

where Ω is the search space, $\boldsymbol{f} \colon \Omega \to \mathbb{R}^m$ consists of m real-valued objective functions and $\mathbb{R}^m$ is called objective space. A continuous MOP is an MOP where each of the D variables is a continuous variable with $x_i \in \mathbb{R} \; \forall x_i, i = 1, \ldots, D$ and possible constraints restrict the set of feasible solutions.

Typically, the objectives of an MOP are conflicting and there is no solution $x \in \Omega$ that minimizes all objectives simultaneously. The solutions representing the best compromise between the objectives can be defined in terms of Pareto optimality. Let u, v be two vectors in $\mathbb{R}^m$; u is said to dominate v if and only if $u_i \leq v_i$ for every $i \in 1, \ldots, m$ and $u_j < v_j$ for at least one index $j \in 1, \ldots, m$. This definition applies without loss of generality to minimization problems. A point $x^* \in \Omega$ is Pareto optimal if there is no point $x \in \Omega$ such that $\boldsymbol{f}(x)$ dominates $\boldsymbol{f}(x^*)$. A Pareto set is the set of all Pareto optimal points and the Pareto front is the set of the objective vectors of all Pareto optimal points.

In order to measure the quality of a Pareto front, we can use the hypervolume indicator [22]. The unary hypervolume indicator measures the quality of a set P of n non-dominated objective vectors produced in a run of a multiobjective optimizer. For a minimization problem involving m objectives, this indicator measures the region that is simultaneously dominated by P and bounded above by a reference point $r \in \mathbb{R}^m$ such that $r > (\max_p p_1, \ldots, \max_p p_m)$, where $p = (p_1, \ldots, p_m) \in P \subset \mathbb{R}^m$.

3 Anytime Optimization

An anytime algorithm returns as high quality solutions as possible at any moment of its execution [18]. One characteristic of anytime algorithms is that, independently of the termination criterion, the best solution found so far is steadily improved, eventually finding the optimal. This implies that anytime algorithms should keep exploring the search space and avoid getting trapped in local optima. Moreover, good solutions should also be discovered as early as possible. This implies that the algorithm should converge to good solutions as fast as possible.

Normally, there is a trade-off between the quality of the solution and the runtime of the algorithm. There are two classical views when analyzing this trade-off. One view defines a number of termination criteria and analyzes the quality achieved by the algorithm at each termination criterion. In this quality-versus-time view, the anytime behaviour is often analyzied as a plot of the (average) solution quality over time, also called SQT curve (e.g., Fig. 1). A different view defines a number of target quality values and analyzes the time required by the algorithm to reach each target. In this time-versus-quality view, algorithms are often analyzed in terms of their runtime distribution [10, Chapter 4].

We consider here a third view that does not favor time over quality or viceversa, but models the anytime behaviour as a bi-objective problem [16] where the first objective, the solution quality, has to be maximized, while the second, the time, has to be minimized. If we consider the points (*quality, time*) that describe at which time the quality of the best solution has been improved by the algorithm, then the set of points that describe a run of an algorithm is, by definition, a nondominated set of solutions. Moreover, we can definitely compare the anytime behaviour of two algorithms by comparing their respective nondominated sets of (*quality, time*) points.[1] If the nondominated set of one algorithm dominates (in the Pareto-optimality sense) the nondominated set of another algorithm, we say that the anytime behaviour of the former is better than the

---

[1] In our approach, we do not give more importance to either *time* or *quality*, and, hence, the order in which they are plotted is irrelevant. However, we visualize later the results in terms of SQT curves, and, hence, the order (*quality, time*) is more natural for such purpose.

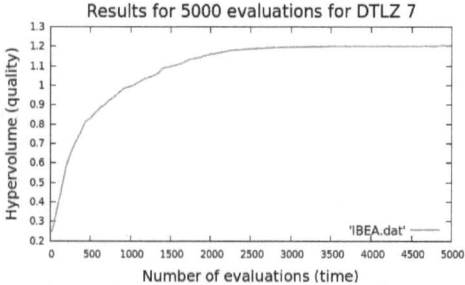

Fig. 1. An example of SQT curve

anytime behaviour of the latter. More importantly, this model allows us to apply the same unary quality measures used in multi-objective optimization to evaluate the anytime behaviour of algorithms. In particular, we have identified the hypervolume measure [22] as being the most suitable for this purpose [16].

In the case of single-objective optimisers, the output of an algorithm is a single solution, and its quality is a unique scalar value. In the proposal described above, their anytime behaviour is evaluated as a bi-objective problem with two objectives (*quality, time*). In the case of multi-objective optimisers such as MOEAs, the output of the algorithm is a nondominated set of points, and, hence, we need an additional step that assigns a unique scalar quality value to each nondominated set by means of a unary quality measure. Once each nondominated set is assigned a scalar quality value, we can proceed as described above and evaluate the anytime behaviour of MOEAs as the hypervolume of the nondominated set of points (*quality, time*) that describes a run of an algorithm.

In summary, our proposal for assessing the anytime behaviour of a run of a MOEA consists of two main steps:

- *Compute the quality of the best nondominated set found by a run of an algorithm at each moment of its execution.* This entails recording every improvement of the best nondominated set. Since this may turn out to be computationally expensive for long runs, a good approximation is to record improvements of the best nondominated set only at specific time intervals. Then, we may compute the quality of these nondominated sets by means of any unary quality measure. For simplicity, we have chosen here the hypervolume measure. In this way, we obtain a nondominated set of (*quality, time*) points that describe the anytime behaviour of the run of the algorithm.
- *Compute the quality of the anytime behaviour curve.* At the end of the run of an algorithm, its anytime behaviour is evaluated by computing the hypervolume measure of the nondominated set of (*quality, time*) points obtained in the previous step. In the case of a fixed frequency of time steps, the hypervolume computation in this step can be simplified as the sum of the qualities over each time step. However, the use of the hypervolume is more general

and it allows the introduction of preference information [16], although we do not examine this possibility in this paper.

Although one may use a different quality measure in the first step than the hypervolume measure used in the second step, we have chosen to use the hypervolume in both steps for simplicity. In this case, the above steps are equivalent to extending the original multi-objective problem with an additional objective (*time*) and replacing the two steps above by simply computing the hypervolume of the extended problem. We prefer the two-step approach described above because it clearly separates between the computation of the quality of a solution (set) to the multi-objective problem at hand, and the evaluation of the anytime behaviour of the algorithm.

A practical application of the above proposal is the automatic configuration of the anytime behaviour of MOEAs. In all parametrized algorithms, such as MOEAs, the search behaviour is heavily influenced by their parameter setting. The goal of automatic algorithm configuration is to determine the settings of both numerical and categorical parameters before the algorithm is actually deployed in order to have an algorithm that is as high performing as possible. Automatic algorithm configuration is crucial in the design phase of parametrized algorithms. It is also relevant in practical applications when known algorithms are applied to specific classes of problems, in order to find the parameter settings that optimise performance for such problems.

Automatic algorithm configuration consists of two main phases:

- *tuning phase*: the algorithm is tuned on a representative set of problem instances;
- *production (or testing) phase*: a chosen algorithm configuration is used to solve unseen problem instances.

In recent years, a number of automatic configuration methods have been developed and recent overviews are available in the literature [2,6,8,9]. The method proposed above to automatically improve the anytime behaviour of MOEAs is mostly independent of the automatic configuration method used. In this paper, we use as automatic configuration method the implementation of I/F-Race [2] provided by the **irace** software package [13]. We combine this method with a publicly available implementation of the hypervolume measure [7] to tune the anytime behaviour of MOEAs.

4   Experimental Setup

In this section, we first present the three analysed MOEAs, explain the main parameters, and introduce the benchmark problems used in the experiments.

NSGA-II [4] uses nondominated sorting and a density estimator to rank the generated solutions and to construct a fixed-size elite population.

SPEA2 [20] keeps the best solutions in an fixed-size elite archive. After each generation, the archive is either truncated with an operator based on k-th nearest neighbour or completed with dominated solutions from the current population.

Table 1. Parameter space for tuning of IBEA, NSGA-II and SPEA2

Parameter	Role	Type	Range	Default
p_c	probability of mating two solutions	real	(0.0,1.0)	1.0
p_m^{ext}	probability of mutating a solution	real	(0.0,1.0)	1.0
p_m^{int}	probability of mutating a variable in a solution	real	(0.0,1.0)	0.0833
N	number of solutions	integer	[10, 1000]	20
DI_c	distance between children and their parents	integer	[0, 100]	15
DI_m	distribution of the mutated values	integer	[0, 100]	20
l	scaling factor (IBEA)	real	(0.0, 1.0)	0.05
$\overline{N}$	archive size (SPEA2)	integer	[10, 1000]	100
k	k-th nearest neighbour (SPEA2)	integer	[1, 50]	10

IBEA [19] uses a binary quality indicator, in particular, the binary additive ϵ-indicator, in order to assign a fitness value to each solution and to keep a fixed-size elite population.

In this paper, we use the implementation of these three algorithms available in *ParadisEO* [12], a software framework dedicated to the flexible design of meta-heuristics. For the experiments in this paper, all algorithms use the simulated binary crossover (SBX) operator [3] and polynomial mutation.

MOEAs Parameters. Table 1 summarises the parameters of the three MOEAs tested (IBEA, NSGAII and SPEA2), their default values and the range considered for tuning. The default values are the ones used in the ParadisEO framework [12], and mostly correspond to the values suggested in the literature [4,5,19,20,21]. There are six common parameters: the population size (N); the probability of crossover (p_c); the probability of external mutation (p_m^{ext}), which determines whether a solution will be mutated; the probability of internal mutation (p_m^{int}), which determines which variables of a solution will be mutated; the crossover distribution index (DI_c), which determines the amount of exploration outside the parents, and the mutation distribution index (DI_m), which determines the distance between the original and the mutated value of a variable. Besides these common parameters, IBEA has a parameter l called the fitness scaling factor, which is used for computing the fitness values and it depends on the indicator used in the algorithm. SPEA2 has two additional parameters: the archive size $\overline{N}$, and the k-th nearest neighbour, which affects the density estimation operator.

Benchmark Problem Instances. As benchmark instances, we consider real-valued functions from two well-known benchmark sets: ZDT [21] and DTLZ [5]. The original ZDT set contains six bi-objective functions, but ZDT5 was not included in our setup. From the DTLZ set we used the seven functions with three objectives. Note that these benchmark sets are scalable to any number of decision variables (D) and they pose different difficulties to multi-objective algorithms such as non-convex fronts or discontinuous fronts.

Monotonicity of Hypervolume in MOEAs. MOEAs store the best non-dominated set found in an elite population that acts as an archive of solutions. Ideally, the quality of this archive should monotonically increase over time. However, MOEAs often limit the size of this archive. Most archiving algorithms, e.g. the one used by SPEA2 and NSGA-II, are not monotonic with respect to dominance [14], and even if the archiving algorithm is monotone with respect to dominance, such as the one of IBEA, the hypervolume of the archive does not need to be strictly monotonic (see below).

Figure 2 shows SQT curves of the three MOEAs considered here, in terms of hypervolume development over the number of function evaluations. In the left column, we plot the SQT curve corresponding to the elite population, whereas the right column corresponds to the SQT curve of an external, unbounded archive that stores all the dominated solutions found so far within a single run. The plots show that an unbounded archive results in a monotonic increase of the hypervolume over time. Moreover, the quality of the unbounded archive is significantly better than the quality of the elite population. Therefore, in the following, we always make use of an external unbounded archive.

Tuning Setup. We use Iterated F-race [2], as implemented by the `irace` software [13], to tune the parameters of the MOEAs. Each tuned parameter configuration is obtained by running `irace` with a budget of 2 000 runs of the MOEA being tuned. Each MOEA run is stopped after $100D$ function evaluations, where D is the number of variables of the problem.

In order to increase the effectiveness of the tuning, each run of `irace` was repeated 10 times with different random seed, and the ten resulting MOEA configurations were compared using F-race [2] in order to select the best one.

Each MOEA is tuned separately on each set of benchmark instances, namely, one run of `irace` uses DTLZ instances and another ZDT instances. Due to the diversity of the benchmark sets, the training instances are setup in a special way. We split each benchmark set into training instances used for tuning and testing instances used for comparing the configurations obtained after tuning. For tuning, the input of `irace` is a stream of instances that is structured in blocks, each block containing one function of each type (5 functions for ZDT and 7 functions for DTLZ) and random $D \in [10, 100]$. We setup `irace` in such a way that configurations are run at least on two blocks of instances before discarding any configuration, and we only discard configurations after evaluating each surviving configuration on a whole block of instances. For testing, we select each function with $D \in \{20, 35, 50, 65, 80, 95\}$ (these values for D are excluded from the training set).

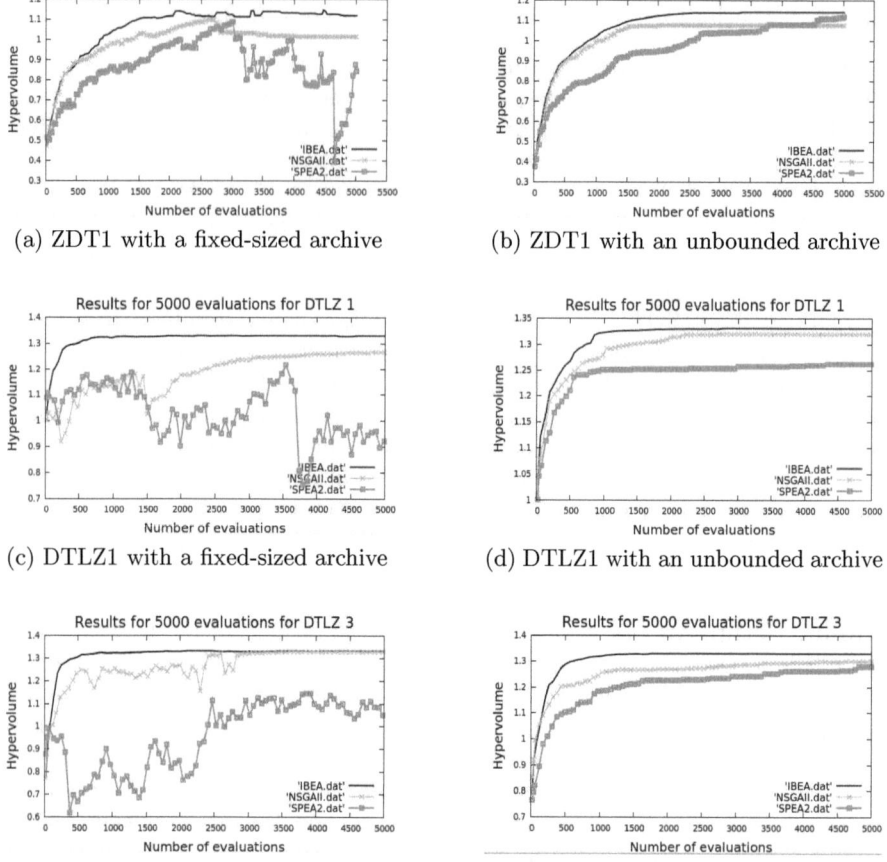

Fig. 2. Monotonicity preservation of the hypervolume through an unbounded external archive

5   Experimental Results

Tuning for Anytime Behaviour. Figure 3 illustrates the difference between the results obtained with the default values and with the best configuration tuned for the anytime behaviour. The values plotted represent the hypervolume value of the SQT curves for each test problem (that is, for each function in each of $D \in \{20, 35, 50, 65, 80, 95\}$). Each point represents the mean quality value obtained after 15 runs of a particular configuration. A larger hypervolume value indicates a better anytime behaviour. The plots clearly show that, as expected, the hypervolume of the SQT curves obtained by the MOEAs improve significantly after tuning.

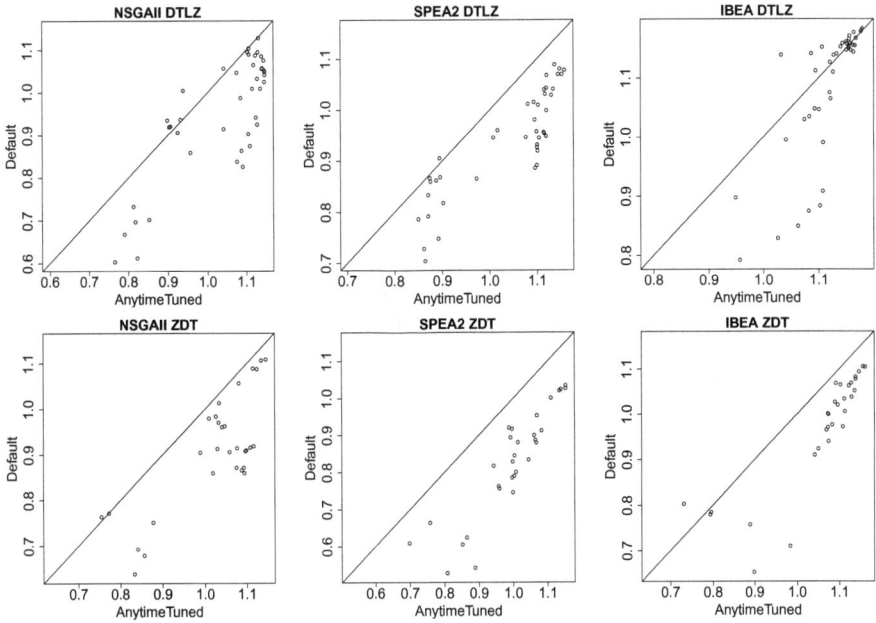

Fig. 3. Anytime behaviour quality for the default configuration versus configurations tuned for anytime behaviour. Each point represents an "unseen" instance from the ZDT set or the DTLZ set and gives the mean hypervolume of the SQT curves (measured across 15 runs) for the default configuration (y-axis) and the anytime tuned configuration (x-axis). A point below the diagonal indicates better results by the anytime tuned configuration.

In order to assess whether improving the hypervolume of the SQT curve results in a visible improvement of the anytime behaviour of the MOEAs, we plot in Fig. 4 the mean SQT curve over 15 runs of each MOEA on individual benchmark instances. Moreover, we plot for each curve the 95% confidence interval around the mean as a grey shadow to give an idea of the variation over multiple runs. In all cases, the anytime behaviour of the MOEAs visibly improves after tuning over default settings. The improvement is very strong for NSGA-II and SPEA2, which shows that their default parameters are far from ideal.

Next, we examine whether the final quality of the non-dominated set obtained at the maximum termination criterion ($100 \cdot D$) is improved or not by tuning for anytime behaviour. Figure 5 compares the hypervolume of this final non-dominated set when generated by the default configuration versus the one generated by the configuration tuned for anytime behaviour. In the case of NSGA-II and SPEA2, the final quality is improved in most cases. In the case of IBEA, the final quality is clearly improved in a few cases, but the differences are often rather small.

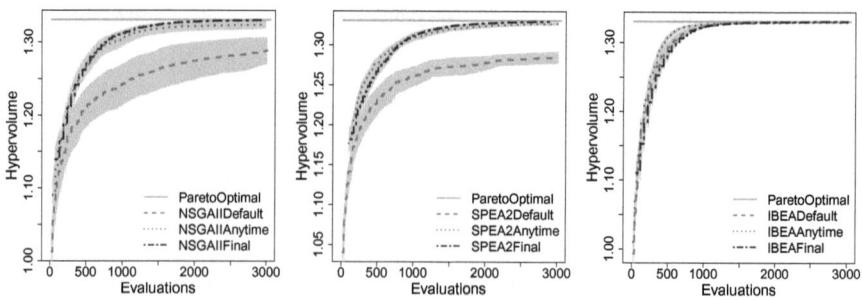

Fig. 4. Variation of the quality of the Pareto front obtained for the three different configurations: the default parameter set, the best parameter set tuned for the anytime behaviour and the best parameter set tuned for the quality of the final Pareto front. The instance DTLZ1 with 30 variables was executed 15 times for each MOEAs.

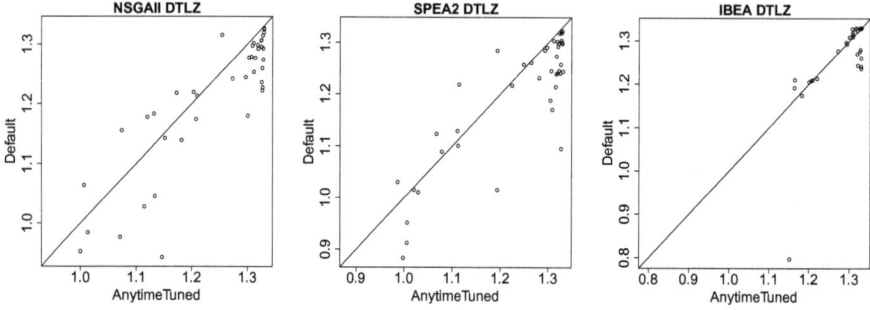

Fig. 5. Final Pareto front quality for the default configuration versus configurations tuned for anytime behaviour. Each point represents an "unseen" instance from the DTLZ set and gives the mean hypervolume of the final Pareto front approximation (measured across 15 runs) for the default configuration (y-axis) and the anytime tuned configuration (x-axis). A point below the diagonal indicates better results by the anytime tuned configuration.

Tuning for Final Quality. We also evaluate the MOEAs tuned for anytime behaviour relative to how much improvement could be reached by tuning for final quality, the latter being a more traditional approach. When tuning for final quality, the tuning procedure ignores the SQT curve and only takes into account the quality of the non-dominated set obtained after $100 \cdot D$ function evaluations. We measure the quality of this final non-dominated set according to the hypervolume. Otherwise, we follow the same tuning setup as described above. After tuning, we obtain a parameter configuration for each MOEA and each benchmark set and run the configurations 15 times on each test instance.

Figure 6 compares the configurations tuned for anytime behaviour and the configurations tuned for final quality when evaluated with respect to their anytime behavior for the three MOEAs and the two benchmark function sets. (The

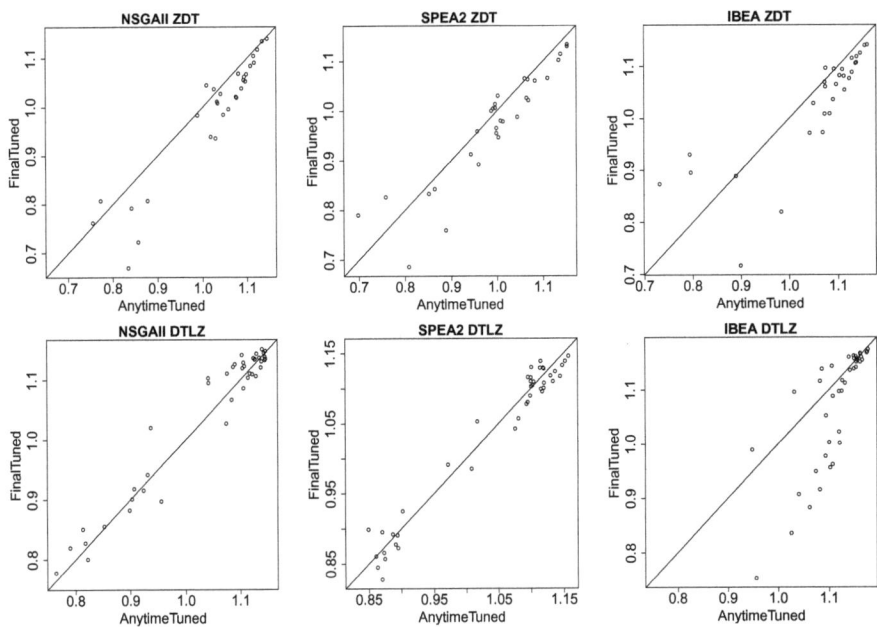

Fig. 6. Difference for **anytime behaviour quality** obtained for the best parameter set tuned for anytime behaviour and the best parameter set tuned for quality of the final Pareto front. Each point represents an "unseen" instance from the ZDT and the DTLZ benchmark sets.

configurations tuned for anytime behaviour and final quality turned out to be far better than the default configurations and therefore the plots comparing to the default configurations are omitted here.) As the plots show, the configurations tuned for anytime behaviour obtain generally better results for the anytime behaviour than those tuned for the final quality.

Figure 7 shows that, in terms of final quality, the differences between the configurations tuned for the anytime behaviour and the configurations tuned for the final quality are only slightly in favor of the latter with the exception of few outliers in the cases of NSGA-II on the DTLZ set and IBEA and SPEA2 on the ZDT sets. Thus, the improved anytime behaviour does not necessarily incur a strong loss with respect to the final quality reached.

By comparing the three MOEAs on the same function (here as example DTLZ1, $D = 30$), we can observe that IBEA is typically the best performing algorithm for the default settings, and the configurations tuned for anytime behaviour or final quality (Fig. 8). However, while for the default settings the advantage of IBEA is often substantial, after tuning NSGAII and SPEA2 are typically strongly improved and can become competitive.

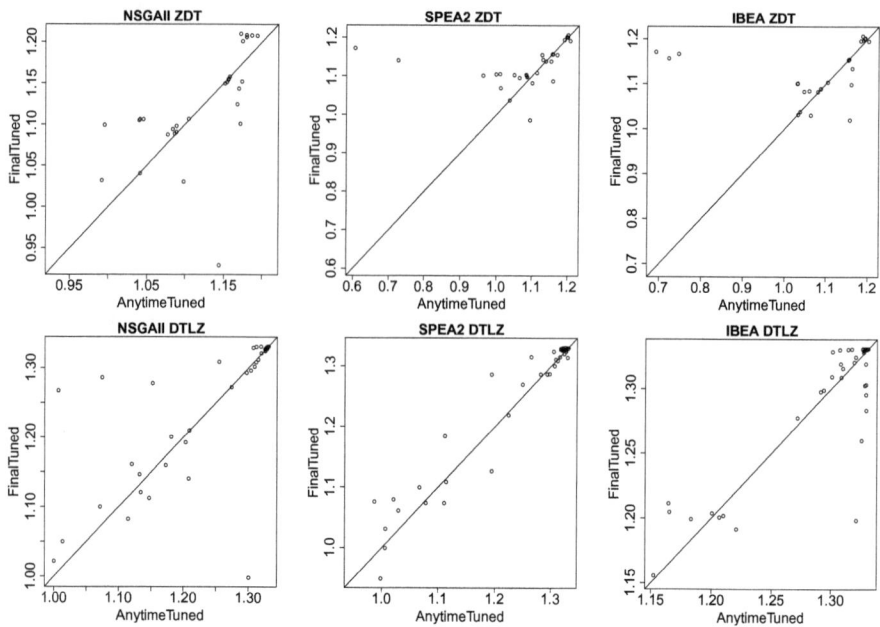

Fig. 7. Difference for the quality of the **final Pareto front** obtained the best parameter set tuned for anytime behaviour and the best parameter set tuned for quality of the final Pareto front. Each point represents an "unseen" instance from the ZDT and the DTLZ benchmark sets.

The differences in the behaviour of the algorithms before and after tuning may be explained by the different parameter settings. Table 2 shows the default configuration, and the parameter configurations found when tuning for anytime behaviour and when tuning for final quality. The most notable differences with respect to the default parameters is the larger population size for the

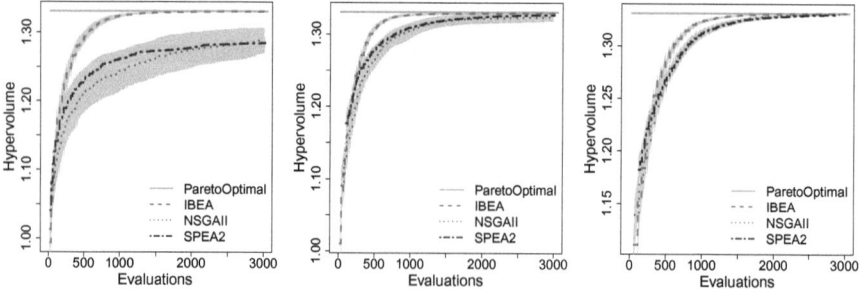

Fig. 8. Variation of the quality of the Pareto front approximation obtained by the three MOEAs with default parameter sets (left plot), parameter sets tuned for anytime behaviour (middle plot) and parameter sets tuned for final quality (right plot).

Table 2. Best values of the tuned parameters for IBEA, NSGAII and SPEA2. Fo each algorithm and benchmark set the tuned values for optimising the anytime behavior (any) or the final quality (final) are given together with the default values.

Para-meter	default values	IBEA DTLZ any	IBEA DTLZ final	IBEA ZDT any	IBEA ZDT final	NSGAII DTLZ any	NSGAII DTLZ final	NSGAII ZDT any	NSGAII ZDT final	SPEA2 DTLZ any	SPEA2 DTLZ final	SPEA2 ZDT any	SPEA2 ZDT final
p_c	1.0	0.824	0.844	0.935	0.915	0.928	0.755	0.398	0.843	0.172	0.190	0.823	0.790
p_m^{ext}	1.0	0.468	0.371	0.799	0.796	0.131	0.066	0.290	0.921	0.027	0.099	0.823	0.743
p_m^{int}	0.083	0.848	0.756	0.915	0.779	0.578	0.894	0.807	0.966	0.099	0.130	0.876	0.857
N	20	23	60	32	60	39	72	46	79	101	123	49	71
DI_c	15	81	71	96	69	94	99	65	31	30	87	69	37
DI_m	20	18	69	0	1	15	11	1	0	20	36	0	0
l	0.05	0.136	0.065	0.317	0.148	-	-	-	-	-	-	-	-
$\overline{N}$	100	-	-	-	-	-	-	-	-	184	482	209	340
k	10	-	-	-	-	-	-	-	-	11	16	20	41

configurations optimising anytime behaviour and final quality. Interestingly, when optimising performance for final quality, the population sizes are in all cases larger than when optimising for anytime behaviour. This suggests to us that an incremental population approach could probably help to improve the anytime behaviour of these MOEAs. The effect of other parameters is more difficult to interpret, but the lack of a clear trend suggests that simple time-varying parameter adaptation strategies may not be successful. A more profound analysis, for example, by tuning the parameters for various termination criteria, could help to reveal some exploration versus exploitation trade-offs. A more direct approach would be to implement a large number of parameter variation strategies and find the best strategy by means of the technique demonstrated in this paper, that is, by automatic configuration with respect to anytime behaviour.

6   Conclusions

In this article, we have shown that the quality of the anytime behaviour of MOEAs can be improved significantly by using the combination of Iterated F-race [2,13] and the hypervolume quality measure. For this task, the hypervolume measure is used in two places. First, to measure the quality of the Pareto-front approximations generated by the MOEAs; second, to measure the quality of the anytime behaviour as defined by the trade-off curve of the solution quality over time. We have applied the resulting methodology to three different MOEAs: IBEA, NSGA-II and SPEA2.

We have combined the above approach with an automatic configuration method (irace) in order to automatically improve the anytime behaviour of the MOEAs. However, the proposed method is not restricted to irace, and other automatic configuration methods could be used for this purpose. Future work should investigate whether some configuration methods are more suited for this task than others.

The experimental results presented in this paper showed that a considerable improvement of the anytime behaviour could be obtained for all three algorithms.

Additionally, tuning for the anytime behaviour improves also (with respect to anytime behaviour) over a version of the three algorithms that is tuned for optimising the quality of the final Pareto front approximation. This can be seen as a confirmation that the optimisation goal of improving anytime behaviour actually leads to algorithms configurations that are more robust to different termination criteria without sacrificing much of the solution quality that may be obtained by the algorithm. Most importantly, these improvements are obtained by an automatic method, which is saving substantial human effort.

In initial research efforts, we have considered the automatic tuning and configuration of the anytime behaviour of single objective optimisation algorithms [15, 16]. This here is the first attempt to apply the automatic configuration of the anytime behaviour to multi-objective algorithms, especially MOEAs. There are several directions that can be taken to extend this work. A first one is to extend the analysis of the impact of automatic tuning to others MOEAs, such as SMS-EMOA [1] or MOEA/D [17]. The quality of the anytime behaviour can be tested also with others benchmark problems, with more complicated objective functions. A second direction is to consider more parameters, including also categorical parameters such as the choice of the cross-over and the mutation operators. In this paper, we only consider static parameters as done in the original MOEAs. A promising approach when dealing with anytime optimization is to consider parameter variation strategies, and automatically configure the parameters of such strategies w.r.t. anytime behaviour [16]. Moreover, we have limited ourselves to parameters independent of problem instance features. Nevertheless, it would be straightforward to define parameters as functions of instance features and tune the parameters that define such functions. A more challenging question is how to detect automatically which features should influence which parameters. We are convinced that further work in the directions indicated here, will lead to the design of multi-objective algorithms that are more robust with respect to the choice of specific termination criteria and, thus, will improve the practice of multi-objective evolutionary algorithms.

Acknowledgements. This work was supported by the META-X project, an *Action de Recherche Concertée* funded by the Scientific Research Directorate of the French Community of Belgium. Manuel López-Ibáñez and Thomas Stützle acknowledge support from the Belgian F.R.S.-FNRS, of which they are a postdoctoral researcher and a research associate, respectively. Andreea Radulescu acknowledges support through the Laboratoire d'Informatique de Nantes-Atlantique (LINA) and Nantes university. The authors also acknowledge support from the FRFC project *"Méthodes de recherche hybrides pour la résolution de problèmes complexes"*.

References

1. Beume, N., Naujoks, B., Emmerich, M.: SMS-EMOA: Multiobjective selection based on dominated hypervolume. European Journal of Operational Research 181(3), 1653–1669 (2007)

2. Birattari, M., Yuan, Z., Balaprakash, P., Stützle, T.: F-race and iterated F-race: An overview. In: Bartz-Beielstein, T., et al. (eds.) Experimental Methods for the Analysis of Optimization Algorithms, pp. 311–336. Springer, Berlin (2010)
3. Deb, K., Agrawal, R.B.: Simulated binary crossover for continuous search spaces. Complex Systems 9(2), 115–148 (1995)
4. Deb, K., Pratap, A., Agarwal, S., Meyarivan, T.: A fast and elitist multi-objective genetic algorithm: NSGA-II. IEEE Trans. Evol. Comput. 6(2), 181–197 (2002)
5. Deb, K., Thiele, L., Laumanns, M., Zitzler, E.: Scalable test problems for evolutionary multi-objective optimization. TR 112, Computer Engineering and Networks Laboratory, Swiss Federal Institute of Technology, Zürich, Switzerland (2001)
6. Eiben, A.E., Smit, S.K.: Parameter tuning for configuring and analyzing evolutionary algorithms. Swarm and Evolutionary Computation 1(1), 19–31 (2011)
7. Fonseca, C.M., Paquete, L., López-Ibáñez, M.: An improved dimension-sweep algorithm for the hypervolume indicator. In: Congress on Evolutionary Computation (CEC 2006), pp. 1157–1163. IEEE Press, Piscataway (2006)
8. Hoos, H.H.: Automated algorithm configuration and parameter tuning. In: Hamadi, Y., Monfroy, E., Saubion, F. (eds.) Autonomous Search, pp. 37–71. Springer, Berlin (2012)
9. Hoos, H.H.: Programming by optimization. Communications of the ACM 55(2), 70–80 (2012)
10. Hoos, H.H., Stützle, T.: Stochastic Local Search—Foundations and Applications. Morgan Kaufmann Publishers, San Francisco (2005)
11. Hutter, F., Hoos, H.H., Leyton-Brown, K., Stützle, T.: ParamILS: an automatic algorithm configuration framework. Journal of Artificial Intelligence Research 36, 267–306 (2009)
12. Liefooghe, A., Jourdan, L., Talbi, E.G.: A software framework based on a conceptual unified model for evolutionary multiobjective optimization: ParadisEO-MOEO. European Journal of Operational Research 209(2), 104–112 (2011)
13. López-Ibáñez, M., Dubois-Lacoste, J., Stützle, T., Birattari, M.: The irace package, iterated race for automatic algorithm configuration. Tech. Rep. TR/IRIDIA/2011-004, IRIDIA, Université Libre de Bruxelles, Belgium (2011)
14. López-Ibáñez, M., Knowles, J., Laumanns, M.: On Sequential Online Archiving of Objective Vectors. In: Takahashi, R.H.C., Deb, K., Wanner, E.F., Greco, S. (eds.) EMO 2011. LNCS, vol. 6576, pp. 46–60. Springer, Heidelberg (2011)
15. López-Ibáñez, M., Liao, T., Stützle, T.: On the Anytime Behavior of IPOP-CMA-ES. In: Coello, C.A.C., Cutello, V., Deb, K., Forrest, S., Nicosia, G., Pavone, M. (eds.) PPSN 2012, Part I. LNCS, vol. 7491, pp. 357–366. Springer, Heidelberg (2012)
16. López-Ibáñez, M., Stützle, T.: Automatically improving the anytime behaviour of optimisation algorithms. Tech. Rep. TR/IRIDIA/2012-012, IRIDIA, Université Libre de Bruxelles, Belgium (2012)
17. Zhang, Q., Li, H.: MOEA/D: A multiobjective evolutionary algorithm based on decomposition. IEEE Trans. Evol. Comput. 11(6), 712–731 (2007)
18. Zilberstein, S.: Using anytime algorithms in intelligent systems. AI Magazine 17(3), 73–83 (1996)

19. Zitzler, E., Künzli, S.: Indicator-Based Selection in Multiobjective Search. In: Yao, X., Burke, E.K., Lozano, J.A., Smith, J., Merelo-Guervós, J.J., Bullinaria, J.A., Rowe, J.E., Tiňo, P., Kabán, A., Schwefel, H.-P. (eds.) PPSN 2004. LNCS, vol. 3242, pp. 832–842. Springer, Heidelberg (2004)
20. Zitzler, E., Laumanns, M., Thiele, L.: SPEA2: Improving the strength Pareto evolutionary algorithm for multiobjective optimization. In: Giannakoglou, K., et al. (eds.) Evolutionary Methods for Design, Optimisation and Control, pp. 95–100. CIMNE, Barcelona (2002)
21. Zitzler, E., Thiele, L., Deb, K.: Comparison of multiobjective evolutionary algorithms: Empirical results. Evolutionary Computation 8(2), 173–195 (2000)
22. Zitzler, E., Thiele, L., Laumanns, M., Fonseca, C.M., Grunert da Fonseca, V.: Performance assessment of multiobjective optimizers: an analysis and review. IEEE Trans. Evol. Comput. 7(2), 117–132 (2003)

Author Index

MIX
Papier aus verantwortungsvollen Quellen
Paper from responsible sources
FSC® C105338

If you have any concerns about our products,
you can contact us on
ProductSafety@springernature.com

In case Publisher is established outside the EU,
the EU authorized representative is:
Springer Nature Customer Service Center GmbH
Europaplatz 3, 69115 Heidelberg, Germany

Printed by Libri Plureos GmbH
in Hamburg, Germany